The Reunification of Science and Philosophy

Organised Project Research Notes Volume 0

Copyright © Michael Pitman 2020
ISBN 978-1-8380618-0-7

A catalogue record for this book is available from the British Library.

The right of Michael Pitman to be identified as the Author of this Work has been asserted by him in accordance with the Copyright, Designs and Patents Act 1988.

All Rights Reserved. Apart from any use expressly permitted under UK copyright law no part of this publication may be reproduced, stored in an alternative retrieval system to the one purchased or transmitted in any form or by any means electronic, mechanical, photocopying, recording or otherwise without the prior permission in writing of the Author.

Published by Merops Press
websites: www.scienceandphilosophy.co.uk
www.cosmicconnections.co.uk
www.michaelpitman.co.uk

Acknowledgements: Suzanne, my wife, Françoise, my daughter, Dave Gant, Rick Pulford.

TOTAL CONTENTS

Summary:

Part 0: Setting the Scene

Book 0   A Systematic Start
- Polar Perspectives
- Natural Dialectic's ABC
- Information
- Information

Book 1   The Stage before a Play
- Energy (Physics)

Part 1: Playtime

Book 2   A Player's Mind
- Psychology: Consciousness
- Psychology: Subconsciousness

Book 3   A Player's Body
- Biology

Book 4   The Whole Cast
- Community

Contents

In keeping with the nature of this book's framework contents are, except for the continuity of chapter numbers and illustrations, assigned binary (0, 1) digits.

Volume 0: **Setting the Scene**
Book 0: **A Systematic Start**
Part 0: **Natural Dialectic's ABC**

Science and the Soul: Reception 11

0 Poles Together *39*
   The Light in Your Eye. 39
   Experience. 39
   Perhaps It's All in Mind. 39
   World-view. 40
   Natural Dialectic. 42
   Opposite Directions of Mind. 47
   Two Sorts of Implement. 53
   Archetypal Greeks. 57
   Two Pillars: a Dialogue of Faith. 60
      The Objective Pillar 67
      The Subjective Pillar 73
      Pure Objectivity. 77
   Scientific Delusions. 81
   Religious Delusions. 88
   Models. 92
   The Cosmological Axis. 106
   Polar Perspectives: *Top-Down/ Bottom-Up.* 114
   Comparative Analysis. 121
   Antagonism. 122
   A Complementary Course. 124

1. Natural Dialectic's ABC 130
   First Principles 130
      Nothing. 132
      Something. 134
   Two Existential Principles. 136
      Causality. 137
      Polarity 147
   Information. 151
   Energy. 152
   Informed Energy. 153

2. Cosmic Fundamentals, Stacks and Ziggurats 161
   Ranges. 161
   Hierarchy. 163

Event Vectors. 169
Three in One. 172
    Primary and Secondary Dialectical Stacks 173
    The Tri-logical Form of Natural Dialectic. 175
Cosmic Fundamentals - Three Qualities 176
The Essential or Central (*Sat*) Quality.. 185
The Existential Vectors (*raj* and *tam*). 187
Hierarchical, Triplex Construction of the Cosmic Pyramid. 194
Sliding Up and Down a Slope. 199
Subtendence. 206
Transcendence... 216

3.   *Truth, Appearance and Reality* *221*
Paradox.. 221
Two Kinds of Paradox 227
Switches, Twists or Inversions. 229
    Primary Inversion. 229
    Scale Switch. 234
    Perspective Switch..238
Semantic Switch.. 240
Truth, Appearance and Reality. 242
Two Value Systems. 254
Rights and Wrongs.. 258
Is There an Absolute Morality? 263
From Science to Conscience.. 265
    Conscience in Principle 267
    Conscience in Practice 270

4.   *Cause and Coincidence* *272*
Coincidence or No Coincidence? 273
Integrity, Coordination and Coherence. 276
An Act of Creation. 277
Stingers. 287
How Can Science Sensibly Cope? 289

Volume 0:   **Setting the Scene**
Book 0:   **A Systematic Start**
Part 1:   **Information**

5.   *Information is Immaterial*.. *292*
The God Delusion?.. 293
Intelligent Design... 302
Authorisation.. 314
Information, Messages, Arrangement. 319
Information is Immaterial. 326
Information is Hierarchical. 327
Orderly Creation. 336
A Cyclical Order of Information. 343
Alignment with Truth. 351
(*Sat*) Potential or Transcendent Information. 356

  Top Teleology. 359
 6. *Design Dissected*..365
  Computation. 366
  The Calculations of Mind. 367
  Phased Intent. 371
   Purposely Down to Earth. 373
   Getting Your Way - Pragmatics. 375
   What Do You Mean? 376
   Information's Infrastructure - Code. 380
   The Lowest, Physical Level. 386
  Active on Passive. 388
  Music. 390
  Machines. 392
  Mind Machines. 399
  Universal Authorship. 409

Volume 0: *Setting the Scene*
Book 1: *(Energetic) Stage Before a Play*
Part 0: *Physics*
 7. *Towards a Theory of Physic and Metaphysic*................418
  At Least Two Sides to a Single Story. 420
  Towards a Theory of Physic and Metaphysic? 423
  Can Mathematics Help Us? 428
  Any More Questions? 434
  What Answers? 436
  Are There Any Other Kinds of Answer? 440
  The Principles of a Unified Theory of Matter. 446
   The Essential Pair 454
   The Infinite 455
   The Order of the Infinite 463
   The Finite 466
   The Order of the Finite 471
  Order Before a Physical Start. 484
  Tugs and Wobbles. 488
  Order Below a Physical Start. 490
  Bulk Reality. 500
 8. *Lord Deliberate or Lady Luck?*.......................................506
  Accidentally or On Purpose? 506
  An Age of Unreason. 507
  The Order of Invariance. 508
  Precondition/ Potential. 510
  Is a Finely Tuned Universe Coincidental? 516
  Is the Invariance a Fix? 517
  Symmetry. 519
  Valuable Constancy. 521
  Emergence from the World's Nest. 524
  Metaphysical Evasion. 530

Getting to Grips with Lady Luck. 535
    Are You Certain? 538
    Punting on the Cosmic Stream 545
    Up Horseshoe Creek 547
    Shots in the Dark 553
    An Unholy Apparition 561
    The Chance-Killer 563
Puppets, Mummery and Drama. 569
No Information from Confusion. 572
Systematic Clarity. 581
A Flight from Science? 587
A Culture of Doubt. 590
The Matrix. 593
Is The Match Friendly? 595
Atheism's Last Refuge. 605

9. Holy Grails 615
Unity and Unification. 615
Holy Grails. 628
High Wires. 632
Snip, Snip. 639
Flow. 641
Split Flow. 645
The Spring of Inter-polar Motion. 655
Boxing the Infinite. 660

10. Nothing 668
The Nature of Nothing. 668
The Order of Nothing. 680
There's More to Nothing. 686
Nihilisms. 689
Pay Attention. 690
Is Nothing Important? 691
How Does Nothing Physical Work? 694
Space for Space. 698
    Old Vacuums 698
    New Vacuums 708
    Dialectical Vacuum 714
Gazing into Space 715
Is Space a Waste? 724

11. Time 731
Time. 731
Species of Time. 737
    Super-time. 740
    Psycho-time. 741
    Archetypal Time. 744
    Bio-time. 745
    Material Times. 748
The Geometry of Time 754

>   AC or DC? 755
>   DC-time. 757
>   AC-time. 761
> Day-Age Controversy. 763
> Grades, Principles and Times. 765

12. *Physical Energy* *771*

> Pure, Polar Energy 771
> Matter's Holy Ghost 774
> *tam/ raj   Sat* 774
> How Absolutely Holy is the Ghost? 782
> Grit in the Ghost 786
> Alpha Points. 791
> The Labours of an Empty Womb. 800
> *PAM*'s Miracle 805
> Twinkle, Twinkle 821
> (*Raj*) Levity 831
> (*Sat*) Equilibrium 832
> (*Tam*) Gravity 835
> It Takes Three to Tango 838
> Points Omega 842

Glossary *851*
Index *877*
Bibliography *895*

Illustrations

0.0	Abbreviated Contents Box	16
0.1	Directions of Focus: Linear Model	48
0.2	Directions of Focus: Circular Model.	49
0.3	Intensity of Focus and its Polar Mode	50
0.4	Models of Energy	95
0.5	Pyramid, Ziggurat and Concentric Spheres.	98
0.6	Materialistic Ziggurat	102
0.7	Cosmological Bearings (i)	107
0.8	Cosmological Bearings (ii).	109
0.9	Cosmological Bearings (iii).	110
0.10	Cones, Loops and Lines of Focus and Dispersal	111
0.11	Cosmic Rings Showing Three Grades of Matter.	115
0.12	Types and Limits of Cosmic Expertise.	116
0.13	*Top-down* and *Bottom-up* Perspectives.	121
0.14	Cut-off or Involvement?	127
1.1	Duality within Unity (i).	136
1.2	Duality within Unity (ii).	138
1.3	Three Tiers of Mount Universe.	154
1.4	Cosmic Divisions and Subdivisions.	156
2.1	Idealised Models, Old and New: Cosmos and Atom.	167
2.2	Event Vectors.	169
2.3	Cosmic Vectors.	179
2.4	Cosmic Fundamentals, Vectors and Balances.	181
2.5	Pivoted Existence.	182
2.6	Vector Tendencies.	188
2.7	Psychological Pivot: Informative Swing	189
2.8	Psychological Pivot: Comprehension/Mood Swing	190
2.9	Physical Pivot.	192
2.10	Cosmic Fundamentals and Their Ziggurat	195
2.11	Regular Subdivisions of the Ziggurat.	196
2.12	Upper Subdivisions of the Ziggurat.	197
2.13	Lower Subdivisions of the Ziggurat.	198
2.14	Trapped in Rings.	199
2.15	Subtendence and Transcendence	206
2.16	Plenitudes and Voids.	207
3.1	Inversion.	229
3.2	Changing Gear.	235
3.3	Truth in the Balance.	244
4.1	Components of Science and Consciousness.	278
4.2	Involvement of Mind with Matter.	282
5.1	Accidental Variations	299
5.2	Hierarchical Information.	328
5.3	The Hierarchical Act of Creation.	336
5.4	The Act of Creation in More Detail.	339
5.5	*Top-down* Dialectic and the Scientific Method	341
5.6	Dialectical Homeostasis: Vibration	343

5.7	Biological Information Lop.	345
6.1	The Order of an Act of Creation	371
6.2	Dialectical Perspective Rephrased	410
7.1	The Diamond Capstone.	427
7.2	Crystallisation of Principles	447
7.3	The Infinite	456
7.4	The Order of the Infinite	463
7.5	Dialectical Perspective; Ziggurat and Rings	476
7.6	Menus of Polarity	480
7.7	Psychological Poles	483
7.8	Apparent Infinity: Archetypal Linkage.	485
7.9	The Physical Pole	496
8.1	The Dialectical Order of Emergence	511
8.2	Archetypal Nesting of Physical Principle	526
8.3	Projection: Egg's Development from Nest	528
9.1	The Fall of Unity	620
10.1	Pregnant Voids	671
10.2	Vectored Voids	675
10.3	Source and Sink are Zero	681
10.4	Out of Nothing	708
11.1	Species of Time	737
11.2	Physical Grades of Time	766
11.3	Physical Grades and Eras	767
12.1	Primary and Secondary First Causes	774
12.2	The Symmetrical Geometry of Light	780
12.3	Where Alpha is Omega	792
12.4	Alpha Answers	792
12.5	*PAM*'s Miraculous Projection	805
12.6	A Physical Menu?	832
12.7	Omega Loops	842

Volume 0:   Setting the Scene

Book 0:   A Systematic Start

Part 0:   Natural Dialectic's ABC

Science and the Soul: Reception

I woke up and found myself here. Ordinary? How extraordinary! What on star-enveloped earth is all this? Who in heaven's name am I? Where did I come from, where am I going and what's the point? These child-like questions have excited and still excite the whole of philosophy and science. I look for a pattern in which I can understand my part and in the light of whose logic I can derive a self-consistent set of answers.

The basis of this book's construction is very simple. You are conscious. You know full well, subjectively, what consciousness is - but is it proven physical? Your body is without doubt physical and you accept a cosmos made of matter. Is your consciousness not metaphysical and, like your body, also part of something universal? **I have simply added immaterial, as a second fundamental cosmic ingredient, to material. Or, conversely, I have added material to immaterial.** The results are astounding. But, you have to check, is such dualism preposterous? Or true? Are you ready for a shift of paradigm? Then welcome to Reception at the project 'Reunification of Science and Philosophy'. This book is the finally organised source for the project. It contains research notes arranged, indexed and with glossary in preparation for reduction to the first book of the series, Science and the Soul. From this, staged abbreviations including, hopefully, clarifications lead to 'The Reunification of Science and Philosophy', a summary whose text is about one-twentieth the size of these notes. It contains ideas (some later rejected) and possibly editorial issues (such as typos), factual errors, conceptual errors (science changes fast) and repetition - each as any preparation for an essay might. But it also means that the direction of travel can be reversed and an interested student branch back up to source to find more details about omissions and curtailments that the abbreviated versions make.

Enough said! Two propositions are central to this title and, as far as I can see, the cosmos it describes.

There exists a material universe and, which is central to each of us, a psychological appreciation of it. **Thus, the first proposition is that realistic comprehension of the world includes *two* primary components - immaterial and material or, as obvious to everyone, mind and matter.**

Is there really any difference? Isn't a material brain the same as, or at least the generator of, your mind? Aren't you your body? It is made of cells, cells are made of chemicals, chemicals of atoms and atoms aren't alive. If atoms, molecules and cells aren't then your body isn't. It might be a marvellous machine but it is not alive. **So who are you? Are you alive or dead? We'll see.**

And if the universe indeed involves an immaterial as well as a material aspect how might the whole thing fit? How might such composition work?

What, moreover, was before the world began? What is the nature of such nothingness whose logic or its lack substantiates creation, chemicals and bodies? Existence as a whole, the science in us feels, is 'logical'. **The second proposition is that existence as a whole *is* 'logical'.** By 'logical' is not only meant the mathematical quality of formal logic or the non-symbolic, informal logic of language with its associated fallacies of presumption, irrelevance and ambiguity - although these enter in. Nor is the word entirely involved with inductive and deductive practices, though these will help provide this volume with its spine. Inductive logic works from observation to a possible conclusion; whereas if the premises of an apology are shown to be definitely true and conclusion follows by necessity then the kind of argument is called deductive. As we'll discover (Chapter 0: Polar Perspectives) two major forms of counter-current running up and down the body of the book's spine are expressed as *bottom-up* (materialistic) and *top-down* (holistic) vectors of discussion. One rises, according to this view, from creation's periphery towards its centre and the other, conversely, falls towards the cosmic edge. But Natural Dialectic also uses the term with respect to *Logos* or First Cosmic Cause - a power as wordless as a harmony. This form of *Logic* is the ultimate standard that 'opposes' any notion of matter/ energy as the primary cause of this world.

Is cosmic genesis devoid of rationale? If there are abstract, immaterial laws of mathematics and of logic how did these appear; and, from atomic maelstrom, how did mind arise that realised them? Does such abstraction just reflect the physical behaviour of things or are patterns of behaviour drawn by prior metaphysic? In other words, did the natural order of existence spring by accident or not? Certainly, the rationale of this book's 'polar logic' is, as opposed to any species of academic sophistry, simple and natural. Indeed, its polarity conforms to an easy structure that involves union, division and motion between complementary opposites (e.g. 'light and dark', 'conscious/ non-conscious' or psychology/ physic). Into this structure, Natural Dialectic, can be translated every philosophy and fact. Could such Dialectic's to-fro swing really reflect the cyclical, oscillatory nature of the world?

All these questions that are not immediately answered! Of course, they may provoke a line of thought but also, in due course, as the nested order of this book unfolds, they will be answered. Indeed, this Foyer will post directions leading straight to 'areas' where answers to main issues, perennial and topical, are dealt with. The idea is that you can simply click and link to matters you're most interested in. So, in short, questions may be fired rhetorically but answered in their proper place according to the logic Natural Dialectic generates.

Logic is, of whatever kind, a form of thought. How physical is that? Might you propose that mind is nothing physical; and that psychological appreciation is, as opposed to the material bodies of the world, an immaterial (or non-material) thing? Are not information, intelligence and subjective meaning, though they can be reflected by specific arrangements of matter, in fact metaphysical factors? The sense of the physical world is not of but lies outside it. **I repeat the primary assertion - mind and matter are two separate elements. But, you respond, materialism's primary axiom is that there is only one.** Precisely so. That is non-materialism's simple null hypothesis. But let us at the outset be completely clear. **These assertions are both philosophical;**

neither is a scientific one. Materialism is a philosophical and not a scientific posture.

OK, OK, hands up! I understand naturalism's aghast response. Science assumes materialistic answers; it works with a methodology of repeated testing; testing can't occur where there is no material to test. I therefore wholly understand and sympathise with fears an open door to immaterialism might excite. Would not all kinds of bogeyman, fundamentalists of any sect and creed, swarm upon our ordered class? Might they not chaotically upset the finest, most objective use of reason mankind has devised?

Such fear misunderstands, as it will be seen, what 'immaterial' means. Firstly, immaterial factors such as human logic or incorrect ways it can distort the interpretation of facts cannot ever change the mode of nature's operations or, when tests pin precisely, scientific truths. In this way naturalism's door remains inexorably closed against a would-be interloper hawking any claim about material events that cannot be materially engaged. However, science tests contemporary operations and, when it comes to history and, especially, origins, the means to test does not exist. Educated guesses are the way. Science would prefer a default naturalistic framework into which to cast an explanation but regarding origins, as we shall see, this may not entirely fit the bill. *An immaterial element of information may indeed be called upon to draw best inference from evidence.* The nature of this element is what will be explored.

Can everything be detected by physical equipment? It may, secondly, be the case that immaterial elements naturally, universally exist. If so, as no-one throws out baby with bath water, they need orderly and non-religiously motivated inclusion. **The purpose of this tightly ordered book is to provide a non-sectarian, irreligious framework for a whole consideration of creation.**

Thirdly, everyone is human. Who's not driven by agenda, wish for recognition and remuneration? Communities reflect the thinking of the day and, although ideally in pursuit of perfect truth about the universe, the community of scientific workers is the same. What's the first line of the scientific creed? Is it truth at all conceptual cost or only in a naturalistic mind-set that's allowed? Not all statements that are issued, even by the famous, stick to nitty-gritty work in field and lab alone. *For example, scientism's adoption of materialism and its consequent, reductive claims that naturalistic science is the only logical arbiter of reality and that life is a molecular configuration are each metaphysical, philosophical statements.* Nor, as Albert Einstein, Richard Feynman and host of other scientific luminaries concur, can those most vital entities, ethical values, ever be reduced to scientific formulae. Can even life's centralities of consciousness and reasonable mind? And if they can't is intellectual chaos what ensues?

If, in spite of these points, your assertion of materialism's right then I am wrong and the conclusions of this book are faulty. However, you cannot assert I'm wrong just because, metaphysically and thus unscientifically, you have decided that I am - that way leads the circle of tautology. And such void circle is the only means you have to wind assertion with. *There is, for example, no scientific proof at all that consciousness is made of subatomic particles, a force of physics or of chemicals.* Even a neurologist can't say (though some have faith and live in hope) that your experience is wholly a molecular construction made of nerves. **If, therefore, I'm right you're wrong. The table's turned. Materialism promulgates but half the whole truth. Indeed, the gauntlet's thrown down! Granted my prime axiom show me where I've got it wrong - or else accept this book's elaboration of such possibility. Expansion of such consequence dramatically affects us all.**

One can't, by definition, prove or disprove any immaterial factor using scientific methods; nor can one infer its presence from materialism's mind-set. If you dislike the 'immaterial idea' the most that you, un-academically, can do is close discussion down. You might proclaim (some do) a kind of intellectual bull forbidding its consideration. The recruitment of some high profile names as witness in support of such an edict by no degree diminishes its non-scientific nature. No doubt that science employs naturalistic methodology but only prior philosophical decision supports the interpretation of all phenomena - including origins of cosmos, mind and forms of life - according to 'exclusive materialism'. Only such decision excludes and denies the Dialectic's primary, inclusive axiom. If, on the other hand, the duality of material and immaterial is accepted then there flows the logic of Natural Dialectic. Such polar logic may be paradoxical where 'mono-polar' unity's involved (Chapter 2: The Primary Dialectic) but, from what is called a *top-down* point of view, it still turns out impeccably. What remains, therefore, but to identify the natural character of any immaterial element?

If I am right and you are partially wrong I therefore ask who lives in some degree of structural fantasy, in some locality of philosophical illusion. The framework in which thoughts are set affects their quality, conclusions and the thinker's subsequent behaviour. Whose is the far-off mirage, whose clear definition closer to full sanity?

Thus if materialism is your axiomatic creed, don't be too quick to rubbish any argument whereby, according to such measure of philosophy, the facts are not interpreted your way; or simply since you don't or won't believe it.

In this respect it's a relief to note that, though experimental science has traditionally studied the behaviours of energy and matter alone, a reluctant and inertial tide seems, perhaps, on the turn. Physicist Paul Davies and Niels Hendrik Gregersen are the editors of a recent symposium (Information and the Nature of Reality - From Physics to Metaphysics) which inverts the currently fashionable explanatory scheme so that information becomes regarded as 'the foundation on which physical reality is constructed'. The work is highly intellectual and the contributors (from the fields of physics, biology, philosophy and theology) are not necessarily in agreement with each other. However, its thrust marks a cogent case for the re-inclusion, henceforth, of immaterial information into our scientific, or even super-scientific, world-view. This inclusion is confirmed by the fine-tuning paradox (that finds the strength of forces and masses of particles 'just right' for life to exist), considerations from quantum physics and multiple biological arguments (not least from minimal functionality, molecular biology and genetics). Not only is information included in the scientific paradigm but is, in our post-mechanistic age, elevated to a prime conceptual position. Can we, it is asked, even in principle reduce the language of biology and psychology to that of chemistry and physics? And when authors can apply such phrases as 'the ultimate ontological reality is indeed information' and 'primordial consciousness that is ontologically prior to all physical realities' we know the ground has moved: and it has moved in just the seismic direction that Natural Dialectic has more comprehensively developed. There now exists the scaffolding, perhaps more, round which a more inclusive science for the future can be built.

Base-line needs underlining. ***If the axiom that mind and matter are two different kinds of element is true the logic of this book in its entirety is unassailable***. **It is drawn, over the chapters, into a self-consistent, polar model of creation; and, paradoxically transcending such polarity, the presence of a causal singularity.**

So, what's *your* sense of our base reality? In no man's land, when bullets fly between Darwinian naturalism and Biblical creationism, is anybody safe? The fact is Natural Dialectic recommends both parties modify their view. Both are going to have to give a bit - perhaps more than either's hard-line missionaries would like.

No doubt, today's fount of scholarship flows overwhelmingly with naturalistic songs. With the refrain that 'all is physical and physically explicable' the discipline of natural science has transformed itself into, some preach, sole arbiter of reason and enlightenment. It has become a vehicle that seeks the whole truth through exclusively materialistic explanations of not only nature's operation but also primal generation. It is, therefore, the view of secularism and especially its most apt protestant, neuroscience, that mind *is* brain. Consciousness, the heart of what we know as life, springs from material patterns and no psychological entity called soul exists. **Such 'scientism' (see the Glossary) is based on the premise of a biological theory of evolution.** Some swear this theory, atheism's answer with respect to where you and all other forms of life once started out, is a scientifically self-evident and proven fact. In case of any questions life throws up shouldn't evolutionary mind-set work from Darwin's Primary Presumption? The Primary Corollary of Materialism's Axiom is, simply, that 'man is the product of natural selection and random mutations'. The IQ of this pair is zero and, therefore, if you agree do you not choose a scientific faith in 'No Intelligence'? At least, on pain of exile to an academic gulag, you hadn't better disagree!

Truth in our time! Evolution is the core idea round which amoral scientific atheism's spun. Its tang pervades most scholarship and thence, amplified uncritically, education systems, journalism and the media. It is contemporary reason's story of creation, a relentlessly repeated cult of origin that in the name of science seeks to block out other points of view. Thus evolutionary scientism has become a sacred cow and challenging that sort of cow may, history records, transform it into bull-like rage. **The red rag that provokes emotional response is, as regards their omnipresent coding and subsequently integrated functional parts, that life forms might have been conceptually developed and therefore, according to a Theory of Intelligence, be actually (and not 'as if') designed.** Red rag, rank heresy, dismissal from the company - but is the general theory of naturalistic evolution wholly right? Does it explain or just -

using science as a cudgel to club metaphysic into philosophical submission - explain away the origins of man and beast? **To check this out we shall interrogate four revolutionary ideas - the generation of a cell from chance agglomeration of some chemicals, natural selection, genetic mutation and, last but far from least, embryonic development (or evo-devo) - by which the current form of Darwin's evolutionary theory is sparked and propped.** And yet, to focus properly, we need to venture far more broadly in our scheme.

Abbreviated Contents Box

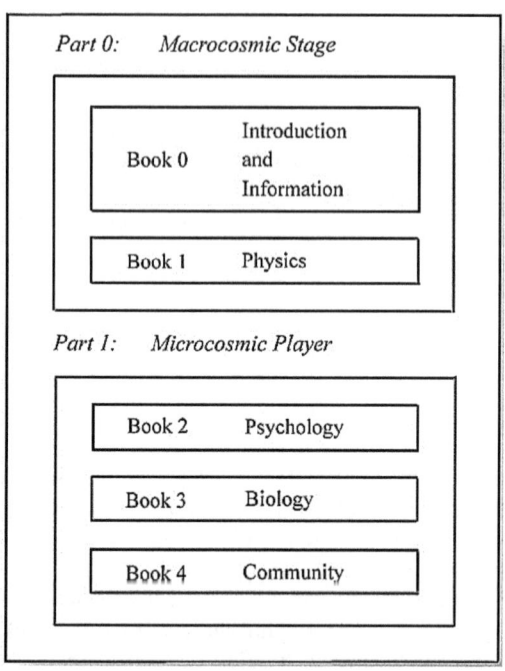

Here, briefly, is the way we'll do it. Look at the Abbreviated Contents Box. Films, musicals and theatre are just three of many vehicles that artists use to project ideas. In this theatrical case Part 0 sets the scene. It outlines the nature of our own polarity and then the polar operation of this book's dynamic vehicle, Natural Dialectic. Ideas precede expression; plays precede performance; sets do not exist without their soul, their genius, their prior playwright. Therefore the narrative proceeds to qualify the immaterial half of nature's existential dipole, information. What informs the universal action? Who creates the cosmic play? Could you infer design behind the show or not? Hence discussion turns to how, in principle, the stage is set for things to roll - things that will include its players, that is, animation that includes our own! This section, 'Order of the Macrocosmic Stage', belongs to physics. Silence! But things are about to liven up. 'Action!' nods the director. 'Action' means 'Playtime'. The band strikes up, curtain rises, shadows fade and Part 1 springs to life. Enter role, plot and each microcosmic player's mind - the *mise-en-scène* of individual 'Character'. What is the fundamental nature of that element of information, mind? Is there a cosmic logic to psychology? Next the focus of attention switches from character to looks and behaviour; it turns from psychological to physical, from mind to player's body - the 'Dialectic of Bio-logical Form'. But a player does not only examine his or her own person. Introspective soliloquy is tedious drama so that, finally, we turn to incorporate all the characters in 'Logic of the Cast', that is, the nature of Community.

This, in a nutshell, is the order of play. You can expand it using 'Total Contents' (at the start) and then by Chapters (28) and sections of each.

This Gate is a foyer from which you can choose to visit places of special personal interest. Or, if you like, the Foyer is a main routine from which the various subroutines are linked. The idea now is to define the issues so that you can hyperlink to sections of your choice.

Chapter 0 introduces the notion of polarity, dialogue, Natural Dialectic and, arising from the dualistic way that we experience life, two major world-views. These are visited as pillars of human faith. Each is subjected to interrogation; their various delusions, scientific and religious, are in turn laid bare. Next, since the Greeks (along with later Christianity) laid foundations for the western mode of thought - including, though you may not be aware of it, your own - the archetypal role of Greek philosophy is very briefly browsed. You need hang concepts and ideas on hooks. These hooks often take the shape of models. Three basic models that define a three-tier cosmos are revealed. They will henceforth be employed throughout the book. Next, who do you think you are and where? This question addresses what your place is in the universe. Finally we turn to describe the two perspectives of dialectical dialogue. These anti-parallel vectors, cast as characters elaborating basic mind-sets, are called *top-down* and *bottom-up*. You might think that they express antagonistic mind-sets and, to some extent, that's true; but lastly it is noted that, in fact, they must compose a complementary pair. Which, however, of these opposites is ultimately true, which false?

Chapter 1 draws up the Dialectic's ABC. First principles are now discussed. Did cosmos spring from nothing? Quantum vacuum isn't nothing. Stable or unstable, it is something. Hence let's initiate consideration of one vital factor in a binary pair - 0. Nothing's nature, we shall see, is at the heart of cosmic mystery. Why, indeed, does anything exist? We move from nothing to the unit of our basic pair - 1. Something. Things/ events exist. Existence is the second principal/ principle of Natural Dialectic. Existence, it is argued, fundamentally embodies in its unity duality. How this works in practice is elaborated through the book. Of course, events have starts and consequences. Cause and effect; this pair chains the universe together. If there was nothing first then what first cause made anything? The principle of causality is introduced; and then polarity. Thence, we are led to ask, what are the agents that, combined, express creation? They are material energy and immaterial information. A brief introduction to the nature of information will expanded in Chapters 5 and 6. The case of energy takes all Book 1. Mind and matter, energy and information - how can these pairs (or this pair) ever be combined? A section called 'Informed Energy' outlines the situation. The way the couple works together consumes the polar dialectic of books following this first one, that is, takes up Books 1, 2, 3 and 4.

Chapter 2 sets Natural Dialectic's 'grammar' out. It shows how stacks are built, vectors included and polarity, by means of fields and hierarchy, energised. Three cosmic fundamentals, basic to the way that nature and the dialectic work, are introduced. So central to the formulation are they that they head each and every stack. What, therefore, do the qualities (*sat, raj, tam*) represent? Next the infrastructure is applied to generation of a cosmic pyramid, that is, the ordered universe. Steps of this pyramid, which might be termed a ziggurat, are one by one explained. Of course, the world's dynamic and its energy is radiated in the form of oscillations, cycles, 'sway' or waves. Such waves vibrate through fields of influence; and fields extend between their poles of source and sink. Thus, finally, Natural Dialectic treats the universal poles it names transcendence and subtendence.

Chapter 3. The columns of the Dialectic's stacks are like a backbone. A muscular philosophy is up and running yet, when you consider physical and metaphysical together paradox appears. What is the nature of such paradox and how is it resolved? What are its essential and then existential forms? As part of its embodiment of paradox the cosmos also seems, from pole to pole, to invert; it twists, once, turning upside down from top to toe. How does such inversion show? Does it affect a reading of the graduated steps or, if you like, the conscio-material scale of Natural Dialectic's stacks? Having settled the effect of twist and turn debate investigates the character of truth, appearance and reality. This leads to ask if polar logic accurately reflects, as many thought and think, the nature of a universe that's understandable. Its order, uniformity of law and mathematical predictability are what keeps science on the hook but what about morality? What do 'right' and 'wrong' mean? Are they values that are fixed? What is conscience and is there any absolute morality abroad? This chapter starts to treat these questions; there's more to read in Chapter 26.

Chapter 4 takes stock. It summarises what has gone before. It asks whether cosmos came about by chance. Three answers to this question are addressed. Two of these answers (not the hybrid one) will be elaborated from now on. A high degree of order, both in physical constraints and biological configurations, indicates the possibility of a non-accidental world. In this theatre of action players and their stage could be interpreted as evidence of authorship. How, from your own experience, do acts of creation work? If *top-down* perspective indicates creation on a cosmic scale might that reflect the way you work things out? Or might your mind reflect the cosmic way? However, such adoption raises ticklish issues that, to make sense of its perspective, need to be resolved. These issues, so-called 'stingers', are like nettles that the Dialectic's metaphysic has to grasp. Six are listed. If, however, reasonable response ensues it automatically challenges a current faith, materialism, and its basic prop, the theory of evolution. Thus such challenge generates the full range of emotion. How could science and, especially, scientific atheism rationally cope?

The groundwork's done. It's time to fly into more detailed, scientific kinds of sky. Firstly, and not just because its age is on us, we take off into information's zone. What computers would tell you, if they could, is that without it you cannot control or order anything. Its piece is therefore prior to all the others and its immaterial sky's in part alive.

Chapter 5 apprises you of information. Thus it presses metaphysic harder. For Natural Dialectic information is potential for a course of action; potential is, effectively, a field whence certain latent possibilities can actually, each on its own, occur; without prior information nothing orderly is understood or done. Thus, to kick off metaphysics, we investigate The God Delusion. Straightway we turn to its anathema, the illusion - for Darwinians - of purposeful design. How cogent is the argument for natural, intelligent design? How 'science-hating' is it, if at all? This section's flint ignites a spark which, by Book 3, will have flared into discursive fire. What, in any case, does 'information' mean? To inform is to give meaning, send a message or instruct. Authors instruct dramatic narrative; they create their works. What instructs our social norms? What might inform the norms of physics, that is, natural law? We turn to study information as an immaterial entity. Its order is construed as hierarchical; it leads to programmed and, where mind's involved, intended creativity. It's also construed as cyclical; it radiates and runs in vibratory fields and feedback loops. Information is the basis of biology; its messages are myriad; bodies are, vibrating in dynamic equilibrium called health, in every cell exchanging signals that control the action 'rightly'. Psychology is based on feedback loops as well. If cause precedes effect then prior, higher truth secretes itself as cause. What is the nature of First Cause? Can we consciously align or even merge with such a Truth? If it precedes existence then it stands beyond the mind; if, however, it is intellectually incomprehensible then a philosopher can't get his mental hands on it. *Tant pis*, it seems, for academe. What, therefore, *can* grasp hold of transcendent or potential information? What is the nature of Top Teleology?

Chapter 6 inspects the way creation works; also the way computing both reflects and simulates it. After differentiating between non-purposive and purposive complexity, upward and downward information loops and the order of creation's act it turns to sliding an important scale. This scale is in two major parts - active and passive modes of information. The former is mental; it includes intent, design and the transmission of design by way of message coded to convey what's meant. The second part, passive information, involves material expression of active's psychological constructions. Code is spoken, written, electronically or otherwise conveyed from messenger to a recipient. Such information may be stored on disc, in book or memory. Whichever way, code is information's infrastructure; it is the mode of meaningful exchange, the medium of message, signal and communication through non-conscious nature. Has such 'automatic' nature any meaning? Are its objects and reactions saying anything to us? Of course, as far as an observer is concerned they obviously transmit such information as his instruments can sense and mind make sense of. Minds, by nature, try to understand what's going on. Musicians, artists, everyone relates what's psychologically going on by their behaviours and through work. The constructions are reflections of their minds; they act as media by which you can interpret an artistic, technological or less elevated purpose. Could natural data be the product of cosmic intent? Is it possible non-conscious phenomena could speak a language or reflect a mind whose principles we crave to understand? Science blossomed by precisely following this line. Is, on the other hand, this world the product of material accidents - a meaningless, oblivious extent? Active on passive, informed energy is now explored. Purpose is reflected in creations. Discussion turns to music, machines and mind machines, computers that reflect the psychological intentions of mankind. From this reflection there emerges certain possibility that, logically, cosmos and its forms of life might be interpreted as made. Thus physicality would constitute expression of an inward gradient of creativity; it would represent, theatrically, the purposes of mind. Active on passive - the result is that we're living here. Finally, therefore, an explicit inference of design is drawn. Two major thrusts well indicate a cosmic scheme of anti-chance. The physic-chemical absorbs Book 1; biological appreciation of the issue animates Books 3 and 4.

Things began with a projection from 'beyond'. If this 'beyond' was metaphysical (which, if there wasn't any physical, it was) we ask where immaterial and material meet. What transaction could occur between? A theory of physic *and* metaphysic needs to be outlined. In this The Diamond Capstone illustrates where cosmic pyramid begins. Five facets dominate preordination. These are mind, potential, unity, infinity and void. How on earth could these translate to physics? They sparkle in rotation turned through Chapters 7 (Infinity), 8 (Potential), 9 (Unity), 10 and 11 (The Nature of Nothing) and Book 2 (Psychology). Chapter 12 deals with the 'pack' that chemistry and physics shuffle round – energy; it coruscates with how things may have started, how astronomical phenomena appear and how the world might end.

Chapter 7 enters physics' realm; energy sustains the cosmic stage. The history of physics is, in founding part, research into the nature of illumination. Microscopes probe minuscule configurations. Telescopes access the canopy of stars and, since light's their only messenger, it's all we've got to know them by. In all that's matter physics seems to serve us very well so what's the need to 'shoehorn' metaphysic in - unless, of course, what is unknown but could be suspected counts as well? What's seen, we're told, amounts to less than 5% of what there is, so even physically, it's claimed we've 95% to learn about; and if any aspect of creation is *not* non-consciously physical we need to count this fraction too. Could another kind of light be shed and psychological illumination play a part in how the stage is set and drama run? Whence, for example, do the stalwarts of our explanations, immaterial laws of logic and mathematics, stem? Of the questions men have asked what answers have been formulated; what is standard theory today; and, at this ephemeral moment, what anomalies appear, what areas are incomplete and, what, if any, are objections to the standard version of events? The current account is grouped into a composition that includes theories of relativity, classical and quantum mechanics, quantum electrodynamics, particle physics and so on. Outstanding issues, to be elaborated later, concern big bang and its competitors, Higgs fields, *ZPE* (zero point energy), dark matter and dark energy and, unless you guess it's inexplicably by chance, the origin of nature's very finely-tuned 'correctness' when it comes to constancies. Natural dialectical suggestions towards a combined theory of physic and metaphysic are proposed. The unification process involves, after introduction of a global, polar pair, facets of The Capstone. First in line The Infinite is placed beneath a microscope. Has this entity an order? It's argued that a hierarchical expression is unfurled, in file, 'down' creation; a cosmic menu of division can be pictured with respect to finite sorts of part. Indeed, The Finite and its order follow; the whole range of cosmos, mind and matter, is made clear. What is immaterial is, of course, physically formless and, therefore, by definition infinite; it is suggested that potential matter, in the form of causal archetype, is 'physical infinity'. This brings us to a primary consideration of what, if anything, sourced energy's big bang. What high ghost predisposes cosmos, what immaterial precedent cooked matter up? What kind of tugs and wobbles might have generated what we now know must have been a most extraordinary and highly ordered start? The last drop is into a phase we call normality, the stage of bulk reality. This is the aggregate of basic particles and forces we, even with our naked eyes, observe. We thus fall to an appraisal of the world of sense in which we live.

Chapter 8 considers precondition, that is, the potential for a world created out of energy. First to ask is how its order was established - by accident or purposely. Is Lady Luck or Lord Deliberate originator of the universe and, therefore, is the place irrational or not? Is it hierarchical or not? What is the order of potentiality's emergence into what we spy with scopes and eyes? Does so-called symmetry play more than an aesthetic part and are there any constant constants such that we could nail the universe by immaterial numbers or (say, 0, 1 or other) a Single, Basic Number? We have to press as far as intellect can penetrate towards understanding how objects flew the cosmic nest. Can metaphysic be, although evaded, quite avoided? To answer this we need to sniff materialism's evanescent but essential perfume; we need to get to proper grips with Lady Luck. How does she relate to certainty? Horseshoe Creek, Shots in the Dark and Chance's Last Saloon are visited. Surely no Unholy Apparition startles us? And, if the haunting Lady's to be exorcised, what is the nature of Chance-Killer? Where, moreover, in this reflex drama, struts the player? Its stage is set with interlocking frame of elemental force and piece but, even greater riddle, how does life fit in? Is the actor puppet, robot or free spirit? Puppets are oblivious without a sense of order or experience. They can't even be confused. Whence, therefore, springs confusion in our minds and what's its salve? Is not, where mindless matter is oblivious, information salve of curiosity? And only information leads to systematic clarity. If the actions of creation are discerned by logic, math and reason in the form of natural law, whence accidentally arose (and perhaps evolved) such sublime coherence? You might think, in resolute denial of Deliberate's metaphysic and steadfast betrothal to Luck's hazy, lazy explanation for the world of physics, science in the larger sense had been betrayed. No doubt, healthy science rides a sea of doubt and probabilities but blur and doubt are not the same as faithlessness. Science has (see Chapter 0: Scientific Delusions) its own delusions. Could such a culture bring itself to doubt its own foundations? In this case what would a dose of doubt about materialism do? Could mind be the matrix of all matter; could information be potential's leading light? For sure, the cosmic generator's knobs are finely tuned, interlocking and extremely friendly set for course of life. What sort of settings are they that some accidental 'bang' refined? How might atheism's plot shake off the burden of their seeming proof of more-than-matter and thereby escape? What could seem plausible and feasible enough? An invention called a multi-verse was born in mind. Might evolution of this concept spawn the ultimate evasion of Deliberate's metaphysic or is such rampant speculation only cosmic fiction well beyond the rational pale? Physical phenomena are a projection from non-physicality. Is such everlasting and all-powerful nothingness luck's realm or order's latent field? For sure, polar camps debate the world's potentiality.

Chapter 9. Unity and unification are cherished goals of science and philosophy. One law to govern all, one law from which all others automatically flow - can such ultimate, defining source be known? When known what form might it take? What is the nature of enlightenment? A look is taken at First Cause, 'the fall of unity' and three main aspects of physical unification. Holy grails involve, for physics, a teasing out of suspected simplicity that lies, in practice as well as principle, behind our star-embroidered universe. Several kinds of unified theory are inspected – not least strings. String theory is a mathematical spectacular that seeks to combine general relativity with quantum theory, that is, all things great with small. Indeed, quantum theories are, in effect, a recipe to connect classical and quantum worlds; they fail insofar as their equations tend towards infinities which must then be 'normalised' and because the quantisation of gravity (a theory of quantum gravity) has not yet been achieved. However, the formulation includes invisible minutiae, extra dimensions, *p*-branes, gravitons and more exotica - and as regards the standard theory of particle physics works well.

Next, the principles of continuity, concordance and liberation from constraint are treated with respect to psychology as well as physics. Would such principles expressed in practice not be unifying entities, most unified in fields of influence, most unified of all in latency? Fields are a medium of communication and convey the patterns of relationship. Between the poles of opposites a range of tenor runs; the special case, converse of continuity, is discontinuity. Separation, lock-out and isolation by degrees from wholeness into clear individuality - these principles describe how things stand out. We turn to examine three fundamental mechanisms that split space apart - Planck units, electric charge and an Exclusion Principle. How, in this case, do the motions that make things obey the principles of Natural Dialectic? This is explained by using a simple analogy with the action of a spring. Things run down; loss of energy or information seems to box the infinite; they cause the lock-up whereby unity becomes divided into myriad dualities; they cause the bonding that creates all cosmic aggregates and bring us down to earth.

Chapter 10 elaborates on one of Natural Dialectic's two first principles - nothing. First Causal Impetus would seem to be a cosmic starting-point and all creation an effect. A starting-point is primal poise. An origin. If there *was* cosmic start and nothing gone before what is the nature of its Prior Inaction? What is the nature of Void Absolute? Voids are pregnant with potentiality; they are unmoving fields of latency - or are they impotent with total emptiness? Is void a source or sink - or both? Vacuous space, in Natural Dialectic's terms, is the physical expression of an absolute - but is it really void or, subliminally, full of relativity (say, *ZPE*)? In a three-tiered cosmos tiers above are 'nothing' to the tiers below; a thought, for example, isn't anything material. Thus, it seems, there more to nothing than immediately meets the eye. Don't confuse it, though, with nihilism or annihilation. Natural Dialectic is, for sure, nihilism's absolute converse; and annihilation breaks up things to nothing but how, technically, can you do that? Always dust or energy is left. Thus pay attention to The Void. It may well be the Essence of Reality - at least, the mystics say so. 'Nothing' is important since, in Pure Consciousness, we find it at the source of mind. It is 'being' that substantiates the actions of psychology. If Void is the source of mind what, logically, is source of matter? It would be nothing physical but, perhaps, a metaphysical preordination - an archetypal memory in universal mind. How, we ask, might such projection work? We turn to how mankind has thought of space. Firstly, 'old' vacuums - emptiness as Indians, Greeks and others used to think of it; then 'new' vacuums as they are conceived by scientific physics. How does Natural Dialectic rate material void? What kind of models might we use to lend it form? How does mass weigh in? Finally, we ask if space is wasted. They are not a part of main-stream science but various ideas for 'mining' space for energy are mentioned. So far they've not been proved correct but, the instant that one is, a whole new field of physics and technology will open up. And endless source of power will beckon. Watch this space!

Chapter 11 involves a second absolute, energetically impotent yet as hard to grasp as space. Time. Time is nothing you can use your organs on yet something, closely, you perceive. Not physical but psychological, it isn't nerve but mind clocks time. Intangible, implacable, what is time but space for motion and your turn of mind? What is its nature? Are there species of it? What are super-time, psycho-time, archetypal time and bio-time? What's the history of clocks and how we measure standard time? If time has any kind of geometry is it best considered linear or cyclical? *DC* (time's straight progression) and *AC* (cycles) are examined; so also is the long and short of how, in different times, we've viewed geology. We go further, noting eras of our universe have different sorts of time; and these same sorts relate to archetypal, quantum and bulk grades of things. Mass-less light can travel fastest of them all but do immaterial archetypes, if omnipresent and omnipotent, make any move at all? And does bulk matter, representing the exhausted epoch of creation, really slow time down?

Chapter 12 inspects the third of three material absolutes - pure polar energy that is conserved, transformed and, variously expressed, constitutes the substance of events and things. Energy is physics' basis; it is matter's 'holy ghost'. All vibrates but whence does power to vibrate come from? What sustains an oscillating world? Of all substance the most pure is light; but after light how absolutely holy is the ghost? What about the 'grit' in space - electronic charge, neutrinos, maybe 'strings' and particles of various complexion that combine to form the dust of earth, the stage on which we life-forms play? If energy's conserved then what about its alpha point - the cosmic genesis? The universe and understanding of it are both projects starting with initial conditions. What are these? What are the prior assumptions; what transcendence has projected energy that is not conscious and makes all material things? Big-bang? Was there an eternal steady-state? Are there any faults with such scenarios? How did the cosmic missile, issued from a non-existent gun, obtain initial, fine-tuned definition of coherent parts that has not changed since shot? Alpha answers are perused. 'The Labours of an Empty Womb' and '*PAM*'s Miracle' are sections that devote discussion to the mystery. Problems and devices that might rescue comprehension (such as inflation, cold dark matter and dark energy) are considered one by one. Hypotheses might thus obtain a universe that's clearly fact but, even then, how does expanding energy devolve into the galaxies, stars, comets, moons and planets that we see? *Ad Hoc*'s a masseur who can oil the wheels of plausibility; he can conjure answers to the puzzlement and massage reason into bodies, heavenly bodies, which astronomers observe. Are his reasons clear or conflicting? Does each conundrum fizz with different speculations? What exactly is the star-eyed state of play? Next, physical expressions of three cosmic fundamentals (*sat, raj, tam*) are worked out; these show as equilibrium, levity and gravity respectively. Thus we see how Natural Dialectic fits an explanation for our world. Finally we drive to how it all might end - point omega. There are, we'll see, several options for apocalypse. Fading whimpers vie with bouncy yo-yo comebacks, dissolution or collapse. Take your pick. As often is the case, the one you choose depends upon interpretation and perspective, that is, on point of view.

Do you have anything particular in mind? Book 2 deals with psychology.

Chapter 13: 'Psyche' means 'soul'. Is immaterial soul what matter-based psychology is now about? Perhaps mind is only nerves and soul a nervous figment of imagination. It's a myth; or else there's neurological delusion that it is. What does neuroscience say? Has an atomistic mind-set laid the ghost of consciousness and exorcised its metaphysical accessory, the soul? It is suggested that, in fact, brain mediates and not originates. It mediates between mind and the body it controls. It exchanges coded information in between this pair of cosmic zones. No doubt, therefore, electrochemistry reflects our thoughts and, on the other hand, sensations as they're passed up to the mind; but perhaps brain neither does nor ever did enjoy a subjective, seamless, unified experience. Mapping every last electron or ionic pattern in grey matter would, as if you'd registered the signals on communication wires as actual thoughts, have missed the point. It's argued that you'll never find a thought or memory, that is, a mind in brain. Materialism's lens, through which molecular activity of nerves is seen as generator of 'illusion', is a myopic, topsy-turvy one. If so, what about your consciousness? What is the nature of the 'thing'? From soil to soul, did not life's consciousness evolve so that the thrust of mystic practice, 'vaporising' mind and reaching towards some cosmic essence, is the view awry? What, therefore, is that nonsense by the name of yoga all about? If you're still unsure we open up the brain-box and expose the contents in a dialectic style. A *top-down* view of its construction makes a lot of sense. Doubting Thomas, even laureate or double doctorate, should by now at least be open-minded. Thus, while metaphysics is a standpoint that's materialistically (and scientifically?) debarred, there follow rational steps to understand psychology and mental links with body by developing a Unified Theory of Science and Psychology. Firstly, macrocosm (universe) is related to the microcosm (organism in the form of man) by use of ziggurat, cones and concentric rings that signify the sheaths of soul, sheaths layered out through mind to a periphery of body (that includes non-conscious molecules of nervous system and its brain). Next, how does essential psychology, that's ordered round three verbs (to be, to know, to do) relate to how mind works; and, in the swing of it, what are the five main states of consciousness that summarise the oscillating balance of a mind? We turn from state to the ingredients of experience and its connection, through world-suit we were born in, with surroundings near and far. Where western lacks them Sanskrit terms from yoga's oriental sense are used. Yoga aims to reach transcendence. What light is to physics, you could say, transcendence is to psychology. It is the measure absolute of consciousness and thus criterion by which to gauge the relativities of thought. Finally, as tribute to the engineering we're endowed with to inform and be informed about the world, let's build a brain. Systems analyst precedes the programmer; in this case code will generate a form. The actual architecture, let alone development of its complexity from just one cell, commands an admiration that extends to awe. Could this greatest wonder of the world, most complex object in the universe, have been haphazardly cooked up by chance? Is it rational to even think the thought?

Chapter 14 lays out Natural Dialectic's hierarchical perspective on psychology. Five states of consciousness (including deep sleep's special case of lack of it) have been identified. This Chapter deals with states 1 to 3. You may remember (Chapter 5: Top Teleology) a condition - first in dialectical psychology - called transcendence or, as mystics know it, communion, *samadhi*, super-consciousness, *nirvana* and so on. This state, mind's 'Alpha Moment', is explained. It is no exaggeration to declare that civilisation is constructed from and around such supra-religious tryst with the eternal moment. Human faith, hope and ideals are derived from the materially meaningless experience of transcendence. Second down in line comes normal, 'centaur' consciousness with its dual prospects, 'up' and 'down'. Such normality exhibits higher tendencies towards ideals, an intermediate condition using reason to perform the intellectual tasks and practical necessities of worldly life and, lower, egotistical and negative propensities. We each experience each such phase, sometimes within an hour or less; but general character is an amalgam usually involving bias towards particular expression. Are you, for example, generally optimistic, pessimistic, enthusiastic or depressed? Next, we note that mind acts as a balance. It seeks equilibrium (called equanimity) in the forms of poise, peace and contentment; but in pursuance and relief of such satisfaction choice of his desires will help or hinder anyone's success. Desires are weights that force and sway your mental pans about; purposes, if you don't choose them carefully, can soon knock you off a balanced perch. In order to achieve its end - optimally painless, stimulated yet contented living - mind operates as an exchange. It exchanges immaterial information, it communicates in code or, *via* code, by muscular manipulation. In fact, life is replete with information. What is experience if not communication by exchange of information? Furthermore, each organism incarnates sheer information in the way it's coded, built and operates. There are two major loops that cycle information to make sense of our continuous experience and, thereby, the world that it's connected to. The upper, metaphysical loop deals in ideas, ideals and all forms of contemplative research. The lower, physical loop extends from brain to lower nervous and hormonal systems that sustain the body's balance in the face of tugs and pushes from inside mind and outside milieu. Reflex systems help to keep the body toned and fit. What, after all, is health but balance of the body's operations such that you scarcely notice them? Of course, science deals in quantity and measurement but how do you measure quality of mind? What units are assigned to moral character? What kind of mind is sanest and, knowing which, how might one obtain its quality? How, if at all, does such a quest relate to quality of information and intelligence (a function of attentive focus and ability to spot the principle behind a practice or the rules behind behaviour? Are there grades regarding quality and, if so, which is highest grade - the best? What is it most worthwhile to understand and transform into life's priority? Finally, we turn to the ascent of man. Is this a psychological or evolutionary affair? To understand psychology one needs to grasp its origin and, in this task, *top-down* and *bottom-up* stretch poles apart. *Bottom-up*, minds have developed over eons from scum of slime on early ponds; *top-down*, creation was developed by projection of the light of immaterial purity, that is, from pristine, undiluted consciousness. In this case an ascent involves mind's evolution back to its original tone; otherwise ascent involves haphazard 'progress' leading who knows where - but for certain most uncertainly to humans; and, most certainly, not beyond the world's end, that is, any creature's time of death!

Chapter 15 leads down to the darkness of a Hades called sub-consciousness. Materialism sees subconscious action as a product of an organism's nervous circuitry; its ionic programs stimulate sorts of metabolism and muscular responses that a body automatically needs. *Top-down*, the program's also crucial but differently viewed. Subconscious mind is understood to operate as a psychosomatic interface between an aspect of the immaterial mind and its body that is serviced and controlled. What aspect of mind? As fluid gas de-energised drops into solid phase what does mind de-energised become? Asleep? Heavy with ignorance or dark of mood? Conscious mind is fluid but sub-consciousness is, going nowhere, stuck. Is it possible that mental image may be fixed? Can present-day or instant's moment become filed or filmed - to be recalled as history? It is; fixation is called memory. Memory is the natural way that mind files images; and these remain subconscious until they are retrieved. Who knows how the undercurrents of a reef of memories affect the tides of mind but, certainly, when one remembered 'pops' to surface, it can influence body by hormonal, nervous or a muscular reflex. Personal memories are, in the black box of unconscious mind, one thing; memories of a certain type of organism, instinct, are another. It is argued that 'sub-consciousness' is a descriptive generality that covers, as a part of the subjective non-state known as unconsciousness, all the files of any living type. Such memory is seen as a thought-object or thought-video; as a record of the type's 'statistics' all kinds of organism, even individual cells, involve, like their material *DNA*, such 'holographic' record of their 'way to be'. In other words, it's logically argued, there's a link between the lower, fixed or 'solid' aspect of a mind and body's nerves and/ or other chemical configurations; this link specifies the way an organism will develop and behave. Firstly, however, the subjective nature of sub-consciousness is opened up. A light is shone into the states of dream and deep sleep. What use are these oscillations? How did they evolve? Slumber's black box includes complex, switching chemistry; what's the way that random dreams are woken up? From subjective to objective, next comes 'frozen time'. Memory is frozen time. Thought-objects lack life of their own but are like files stored 'off-screen'. How without a history could anything go forward; how without such reference would anything know what to do? Thus such files are necessary, not just for you, but for every organism and, much larger, the entire operation of the universe. How do your remembrance and retrieval systems work? Are brains and other chemical 'antennae' like transmitter/ receiver sets? As well as electronic is the process radio? With light involved? And, as already mooted in Book 1, if immaterial mind's an element could universal mind, like universal body called the cosmos, not exist? Universal mind would, in its subconscious phase, consist of files of memories, that is, of patterns; these patterns would inform material behaviours and thus, since information acts as the potential for an action, the Dialectic calls this phase 'potential matter'. You might add a synonym - archetypal memory in universal mind. Such archetypes would act as instinct for material bodies both inanimate and animate; they would specify both form and energy's relationships in simple (physical) and complex (biological) configurations. Is the phrase 'law of nature' rising into consciousness? Psychosomatic linkage of this kind is, dialectically, a logical component of creation. Since, somewhat akin to gravity, the influence of universal mind is omnipresent so the sway effects, according to its 'wavelength', any individual particles or form of life. Such immaterial construction is 'outside' material science but has mankind never picked the concept up? From what 'shape' of memory are mathematics, logic and other universally effective abstracts drawn? They're not physical devices but remarkably describe the way the natural world is run. Whence arose psychosomatic healing schemes or yoga using *ch'i* or *prana* as specific, patterned forms of energy? Indeed, could psychosomasis employ the intermediates of quantum vacuum and electro-dynamic process in the way that your own mind-body linkage works? This requires, of course, a fresh perspective that materialism's loathe to look along. Natural Dialectic takes the baton up. Wireless man and synchromesh (conscious to subconscious gearing; and then subconscious to physical) are driven through a logical progression. Architecture of sub-consciousness in both conscious and unconscious organisms is suggested - every cell must, on this reckoning, reflect its species' template, that is, archetype. Cells don't have nervous sensibility but are subservient to a 'mnemone' governing their instinct, chemical dynamics and, with the conceptual framework of development, their shape. While Book 3, building on Book 2, elaborates the physical conclusion (which includes a coded 'aerial' called *DNA*) we now turn, after briefly peeping at the personal mnemone, to the typical.

Chapter 16 inspects archetype (or universal memory or typical mnemone) in terms of man. What does your image look like? Is it as you biologically appear or, as storage of an image on a hard-drive, not at all the same as the image that's obtained on screen? Your brain, as well as other parts, would be programmed on this metaphysical component; body would reflect conceptual origin. Such notion is, of course, anathema to scientific atheism, evolutionism and such philosophical band-wagons; however, Natural Dialectic rates the search for truth above dogmatic obfuscation and thus presses on. Thus we elaborate mnemonic architecture that was introduced in Chapter 15. What is 'signal translation'; how does resonance play a part? Can instincts really be explained by genes, that is, by a chemical prescription that's composed of phosphates, carbohydrates and nitrogen all in a ring? Why do complex banks of switches target, in a most accurate, determined way, developmental stages and their product, adult form? What is, dialectically, a morphogene and how might this relate to morphogenesis? Of course, interlinkage is essential and thus, once more, we turn to examine further aspects of psychosomasis - mind-body interaction - and how this interaction works. Electro-dynamics, as in the works of Bose and Burr, plays an invisible but crucial part. Each living cell is surrounded and controlled by its electric fields; healthy voltage is maintained across all living membranes and various instruments (such as extra-cellular matrix) provide a medium for body-wide bioelectrical coherence. Could extrapolation 'downwards', mind to matter, lead to the conclusion that life's integrated information networks, the precision of cell operations and *DNA* hard-coding are no accident.

Chapter 17 relates the logic of embodiment as it applies to you - an incarnation of supremely functional logic. This would make a nucleic acid database, like any book, c-drive or physical CD, a symbolic expression of metaphysical ideas. Materialism cannot cope with concept of this kind; it is not dialectically structured. Core principles are next discussed; the cosmic fundamentals are unfurled, from top to bottom, in all organisms (as Chapter 19 will explain) but especially clearly defined in the human form. For this reason human body, by its clear reflection of the conscio-material gradient, is said to be 'a microcosm that reflects the macrocosm'. Is this true? Is the universe in man? No doubt, as a whole, bodies work harmoniously; their coherent logic *is* their health. Material science isn't much enamoured of the pin prick acupuncture's system of philosophy evokes. It's metaphysical in practice thus it shouldn't work; but the Chinese NHS will tell you doctors, after centuries of practice that it works. It even works on animals who don't know what it's all about! What confusion! 'Chase the nerves' and, neurologically, try and bring the explanation down to earth - but western reflex misses the conceptual point. Evolution says it cannot or, at least, it shouldn't be. How can western science and the sharp prick of a silver-plated pin be reconciled? Man's morphogene, the shape-constructing aspect of an archetype, is called 'caduceus'. The way that vibrant morphogene might call up from its various subroutines a body with coherent systems, organs and cellular parts well illustrates the myth of primary, all-powerful genes. Archetype is a most rational idea; it is derived conceptually and not, like Darwinian evolution's thesis, from irrational chance. Nevertheless it will possibly turn out to be the most controversial and therefore ferociously attacked holistic element in Natural Dialectic's paradigm. This is because an archetypal memory, like any state of immateriality, spells materialism's death. The latter world-view will obviously attempt to spike such mortal enemy!

Next, we discuss the logic of development. Further than conceptual morphogene the linkage of causal bioelectrical patterns with genetic, physiological and anatomical effects will further be confirmed and pursued in Chapter 25. Nor now we study the logical shape of man and its relationship with an ancient, symbolic double helix, badge of the medical profession, a caduceus. Its expression well reflects those cosmic fundamentals (Chapter 2) that underwrite the whole of Natural Dialectic. What are its informative and informed, energetic domains? How might one illustrate the wireless framework of your 'sheath' called memory-man? Historically the double helix that encompassed glands expressed the path of archetypal energy called *ch'i* or *prana*. How, employing as a model the Vitruvian man, might this insight be updated? Pyramidal, information and universal images of man serve to press home the point. Informative and informed domains express the pattern that, by means of Natural Dialectic, it is easy to articulate. Part of its complementary, polar nature includes balance and we see that, biologically, symmetry and homeostasis are central to life's schemes; they variously control the orderly, oscillatory way that bodies work. A simple example of a coded metabolic pathway, one involved in hormonal homeostasis with its feedback mechanisms, is related - the allied controls of calcium levels and of metabolic rates. How did metabolic pathways, organs, systems or developed bodies evolve by happenstance? How, evolving, were half-way-there parts able to survive? A body's health is, from a dialectical perspective, due to its conceptual integration in a plan that doctors everywhere work hard, when it is compromised, to reinvigorate. It was called '*vis medicatrix*'; for Natural Dialectic read 'conceptual template' or 'an archetype'. Materialism doesn't like this rational idea; instead of good design degrading due to entropy it prefers that no design, against the stream, can by little bits irrationally specify life's vast complexity. It replaces rational information of life's engineering with chance in the form of mindless reconfigurations of genetic text. It then rounds on anybody who might disagree with such blatant lack of logicality and stabs them with the epithet 'irrational'! Oh, dear; what can you do when archetypal memory is not a concept making any sense? In the next chapter we shall turn from life to, briefly, death; but after that (in Chapters 19 to 26) resurrection that is biological will further illustrate why, when it comes to specified complexity of bio-mechanisms, reason trounces chance. If life expresses rationale, if bio-logic implies concept then Lady Luck and time's great gamble crumble back to their home ground- they bite their oblivious progenitor, the dust.

Chapter 18 deals in death. As cycles roll what's it to be - life afterwards or not? Of course, if you believe that mind is made of nerves then only molecules exist; dust to dust means only that. If, however, brain's a medium modulating earthbound consciousness what happens when the pot cracks and its fluid immaterial flies free? An incarnate cycle is considered; what about excarnate ones as well? If 'out-swing' swept you into physical configuration where might 'in-swing' lead? Are you ready for its D-day; could disembodiment, with you intact, exist? Bio-logic of embodiment is Book 3's line but is your disembodiment a reasonable corollary? Natural Dialectic ventures towards a Unified Theory of Life and Death. Death might seem a closed box, last horizon of our human ignorance. What, however, is the nature of its black hole? Will you peer from outside or, crossing its event horizon, experience within? Is this possible? What happens as you fall inside? Can *OBE*s and *NDE*s explain? We now engage, post-mortem, old yet fresh psychology. How will this spooky visitation work? Have mystics, such as Buddha, anything to add? Next, down to earth, we fall upon anathema. Theodora didn't like the notion of reincarnation and so it ceased to be recycled round the Christian faith. Can one woman from Byzantium pronounce forever on the case of life and death? Let's see if metempsychosis, Greek for reincarnation, makes any sense and, on the way, discuss various brands of immortality. High-level death, however, is what every faith is aiming for. What is the nature of this super-state such that you come to rest, beyond existence, in essential light? If a dier doesn't hit the bell, what then? We close by looking to the ante-natal element of full psychology. Where will out-swing from the centre hurl? According to what order will gravity grip soul and pull it towards embodiment? Of course, once in a body we can turn to biological Book 3.

Let's get physical. Let's run through this Book's exercise.

Chapter 19 starts upon the study of embodiment. The principles of a unified theory, wherein immaterial and material elements are (as in you) conjoined, are bio-logically proposed. Information, homeostasis and the three main principles that drive construction of all bodies are defined. Indeed, these principles drive *all* events - of which a body's life is just a very complex, long and special one. Firstly, is not material biology a precipitate of immaterial information? Next, bodies need to generate sufficient energy to hold dynamic equilibrium in place. Lastly, they exhaust waste products or, built from a single cell, fresh bodies that ensure survival past their own. In its event biology is hierarchical and cyclical and we discover how. Analogy is made of cell with a machine, a mind machine (computer), factory, government and industry. The element of concept, archetype, is thoroughly discussed. Finally, the dialectical construction of a body is laid out in terms of those three cosmic fundamentals first described in Chapter 2.

Chapter 20 asks whether naturalistic theory applied to origins of life makes any rational sense. Could evolution that is merely chemical be true? If not then Darwin's theory could turn out a less than half-truth that is, in the whole view, wrong. At minimum what does a cell require in order to respire, reproduce and otherwise metabolise? The least, we find, is a tall order and thirty cumulative reasons are adduced that aggregate to the impossibility of its completion just by chemistry and physics. Formless, colourless, inedible - the red herring swimming evolution's downstream tide is time. Time by itself can't write linguistically coded plans but shelve that minor problem for a moment. What about conditions on an early earth? Fluid factors include atmosphere, volcanic action and the sea. What constitutional problems could there be? Miller, Urey and their 'hand of God' (i.e. experimental strategy) had seemed to score a goal but, after hesitation, analysts cried foul. Therefore, could evaporated soup or fool's gold whistle up life's kicking-off spot? If not hot-water vents, clays, cool springs, salt pools or what not? Cells are bio-factories; they survive by working cybernetic chains of chemical reactions mediated by chained chemicals called polymers. Each type of chain is critically precise and, according to instructions, links with all the rest. The whole opera is orchestrated by a score. A score's a program of instruction to achieve harmonious and purposive effect. It's argued that unplanned production of a biochemical outside a cell is thus irrelevant to life-from-matter (an atheist's suggestion called abiogenesis). How, anyway, could carbohydrates, lipid membranes, correctly bonded nuclear and amino acids and other cytological enchantments simply 'pop together' at the same time in a micrometric and yet agitated volume carelessly? At that magic but imaginary instant how did correlated code for future metabolic manufacture just appear with specific proteins synthesised and all correctly folded ready to clock on? We check the qualities of *DNA*. Its density of data storage, elegance of structure, subtlety of operation and the flexibility with which its 'books of life' are written are each awesomely revealed. Code rules; yet code's symbolic and a symbol always stands for what it *means*. *DNA means* proteins, protein bio-logic *means* the body its designed to help compose. How can atoms set up such linguistic order; how do molecules perchance make functional machines? How, at root, does any chemical anticipate and thus develop coded metabolic pathways that are crucial operators to sustain life's chemistry? For example, *DNA* can't replicate itself. There's no such thing as '*SRM*' (self-replicating molecule) except, arguably, in the highly artificial case of a complex ester that dissolves in chloroform and is wholly irrelevant to life. A complex operating system is required to perform nucleic acid tasks - direction of protein synthesis and replication for cell reproduction. Minimal functionality decrees no cell exists without its various sub-machineries intact. What machine can work half-built, inaccurately lathed or only one or more (up to perhaps 90) percent complete? If *DNA* cannot perform the trick why not try phosphates, sugars and the necessary bases to compose a single line of *RNA* instead. No more success? The fact is, anyway, raw energy destroys but life needs its fuel refined. Whence sprang its *ATP* refineries? The only species of complexity to defy entropy is in fact informative. It is conceptual. Thus evolution struggles, back-to-front, to pump complexity increasing in its specificity up delusion's hill. Such topsy-turvy logic is examined. Noise, monkeys and imagination can't, it seems, generate life's catalytic chemistry.

Chapter 21 transcends chemical components to reach the level of a cell and organelles. How were metabolic pathways 'lifted' clear of primal slime? How did energy metabolism break from muddled, salty pools? At every step an atheist is forced into contorted explanations of how concept is replaced by vagaries of chance. Continual vagaries of this-or-that hypothesis abound while obvious interpretation that includes a mental source of information is stubbornly/ philosophically ignored. Natural nanotechnology is, in this vein, explored. Does it display, in various constructions smaller and more powerful than man has yet conceived, irreducible complexity or functionality that won't, below a minimum configuration, work acceptably. Innovation is the problem. Variations on existent factors easily occur but creating new ones where they weren't before is a bar beyond the leap of accident - but life in all its forms is full of bespoke, coded mechanisms that must, fresh or ancient in design, perform. Can't science biosynthesise new life? Biotechnology and biosynthesis, employing vast and pooled intelligence, can vary or rebuild a variation on the theme of what wild nature has supplied. How much more intelligence will need to be invested until an innovation actually occurs; and then it will have been constructed for a purpose - one belonging to the men who made it. Perhaps, since nothing normal seems to make life up, life issued as 'extremophiles' or even flew in out of space! Surely a bacterium can 'just occur spontaneously' and thence evolve out of its most successful rut? Wishful thinking coupled with a promissory faith is insufficient evidence of truth, rescuing devices simply instruments to keep belief alive and the atheistic vessel left, by storm of reason, all at sea. In hypothesising life from matter perhaps materialism's axiom is incorrect. *Believe it if you want to but there's no necessity.* Yet its adoption is, officially, uncompromising unto flight by many reasonable scientists from science. You have just begun to look and see.

Chapter 22 moves 'up' from molecules to living bodies, fossil bodies, natural selection and family trees. Do the iconic illustrations of Darwinian evolution (finch beaks, peppered moths, ammonite successions and so on) correctly make their point or not? If not, is there any other explanation for homology, convergence or developmental embryology? Are evolutionary tales of brilliant yet accidental innovation, transformation and addition fact or fiction? It all depends, according to your faith, upon interpretation of the facts. The Law of Biogenesis fought off assault but The Law of Heredity, whereby all organisms without known exception reproduce according to their type, is also under siege. Is limitlessly plastic speciation really the natural case? There exists, as well as the evolutionary lens, a *top-down* paradigm. Though some would lock it in the dark it needs illumination to be clearly seen. The first step is to promote to first place in the order of consideration information. Immaterial information constitutes the rational basis of biology, that is, of bio-logic. The wind and rain don't dream up schemes - not even if a bolt of lightning strikes. Material oblivion cannot anticipate a need - indeed, it doesn't have one. Therefore, we ask, can evolutionary flounder actually create a thing; are natural selection, gene mutation and development (all facts) really indicators that the living world has risen, randomly and meaninglessly, from dust and water? Is your being really only muddled but exotic mud? Could non-conscious atoms that aren't even dead have issued from their deadness life? First natural selection is grilled. What is it, what does it explain? Does it explain an origin of species - if in this blur you're clear exactly what a species is? Variation happens all the time but what are limits, if they're any, to plasticity? We visit Crufts, The Chelsea Flower Show and Galapagos. Who was Edward Blyth; what was it turned him on his head? Do peppered moths exhibit evolution fluttering towards the future; and what's been happening to the tree that Darwin doodled into academic life? If this tree illustrates the way we're all related to bacteria (or some imaginary such-like) how might homology be thought of - a product of descent or of design? *E. coli*, dogs, mice, fruit flies and malarial parasites don't seem to morph enough to squeeze any macro-evolution from their tubes. Could Chinese boxes serve as well as trees and, without transformation, use tree-less nests in order to relate the different types of life? Could you apply such package to the fossil world? We ask if there is any way such evidence might vary from the evolutionary frame and, if so, what might your fresh paradigm predict? We take a flier on the back of birds including *Archyopteryx*; we check mosaic patterns in a lungfish, platypus and plants. Can you swallow down that acrid mouthful, irreducible complexity? Must a mechanism thus composed serve purpose? How can it become informed by mindless fits and starts with no idea of what they're doing let alone what they'll do next? To point, how can specified bio-complexity develop due to random changes in its how-to-build-and-keep-me text? Now look in the mirror. Are your eyes to see? How can purposive complexities, coordinated with the other fractions of a body, gradually, as if from blindness into clear daylight, emerge from long-time fogs of error into fully-working view? Eyes of distinctly different optical construction are placed on the slide of our conceptual microscope. In summary, when surveying this whole chapter what is it we see? Is evolutionary atheism reasonable or could, as we progress, irrational become a better word?

Chapter 23, since natural selection alone most definitely won't work, deals with its partner, random generator of all novelty, mutation. Are the couple, yoked together, strong enough to plough earth's field and yield the harvest of all life? First we take a look at language of the genes. Language isn't something you'd expect from wind and rain; you'd not expect that natural forces published volumes that transformed their information into material reality! How are such volumes printed out in organisms they describe? What happens when a 'magic manual' is infected by a mutant misprint so it can't perform its trick? Or is 'misprint' the word that should be used? There's nothing wrong with natural forces so how, lacking all objective, can they ever make mistakes? After all, without its load of purpose, meaning and its ordered information, what's gene but chemical - *DNA* that's nothing more to life than ink and paper are to books? Could, however, innovation and mutation ever be combined and thus drive evolution gradually, totally haphazardly not 'forward' but 'any which way'? What *is*, if it exists, a 'beneficial mutation'? How might such an item be defined? You'd better make it clear since all of secular academy depends, for its verbose existence, on this evanescent gleam of hope. Can '*BM*s' rise to the occasion? They must do so to a high degree. As with energy there's entropy of information too. How might this affect an evolutionary state of play? Indeed, no mutations have been caught creating any enzyme or new metabolic pathway, biochemical, organelle, cell type, tissue, organ, system or a body-plan! Yet these are solely what a 'rational man' supposes generated his own being! After a discussion of *BM*s and entropy the catch-phrase 'evolution seen in action' is discussed. Reference is made to *HIV*, *E.coli*, fruit flies and malarial parasites. Do claimants mean, as usually they do, not innovation but some variation on existent features that, in certain well-defined conditions, confers advantage as regards survival? Such variation can't be cited as the source of macro-evolution unless by imaginary extrapolation of the facts. How far, therefore, can you stretch mutation until death, not transformation into a fresh kind of organism, snaps genetic elasticity? Next we discuss what was called 'junk' but now, more circumspectly if not wisely, designated as non-protein-coding *DNA*. Of course, in the course of many reprints down the years printings will by entropy occur; but high fidelity is still retained and Natural Dialectic would suggest such *DNA* may well comprise part of a genetic operating system crucial to expression of the programs that have always underwritten living. Indeed, a library of *DNA* is complex in the way it's written out. It may be, for example, that adaptations don't, for the most part, need mutation; the necessary information is already, as anticipatory buffer, coded in. Layers of code, super-codes, epigenetic codes lend flexibility to life's undying, underlying certainty. The more we learn the more we learn that life is not by chance. Thus the normal sort of adaptation isn't then construed as due to accident but foresight! It's clear more drastic action is required. Tinkering with natural selection and mutation cannot innovate. Thus so-called 'natural engineering is invoked. This bolt-on includes mobile genetic elements, regulatory modules and other lately discovered co-factors in the bio-operating system's kit of messengers; or, it's suggested, you might stumble on correct configurations that could generate novel function using gene duplication, high-level developmental gene mutation, genome transformation and so forth. Can random changes, this time 'chunky', really service evolution?

Chapter 24 revives discussion of inversion or (see Book 0) reflective asymmetry. In creation, for example, there is dialectical inversion from consciousness to base non-consciousness; and such centrifugal devolution is matched by its anti-parallel, centripetal vector of evolution. *Top-down* and *bottom-up* perspectives also turn each other upside down - they invert each other's way of seeing things. Each dialectical stack describes the way creation 'oscillates' between antagonistic vectors. It might be said nucleic acid anti-parallels reflect this two-way flow; plaited strands reflect the geometry of information as a double helix. The extent of 'twist' includes our nervous system. We recall (from Chapter 13: To Build a Brain) the complementary nature of our information-centre's hemispheres and thence proceed to map the asymmetrically reflective architecture of its sensory and motor nervous systems of control. Moving on, we notice just such 'mirror' in the way that sex is split;

and how inversions of nucleic acid code occur, in protein synthesis, from informer (*DNA*) through medium (*RNA*) to the informed outcome (protein). Is it simply accident that helical motif is so widespread? It appears as a dynamic structure at the heart of cosmic creativity. Could such consistent geometry, combined with structures that incorporate the golden mean, reflect a higher rationality? No doubt that death-defying reproduction's crucial for survival through the universal tide of entropy. One temporary life-bubble seeds the next; one evanescent cycle hands its baton to wheel, with another runner, round the track again. How did asexual reproduction start? In what way, precisely as it works, could the components of its critical machinery have obliviously 'self-constructed' so death was overcome first time. Remember, one mistake, you die! How, therefore, did this miracle of 'resurrection' make itself appear? The problem's multiply compounded when it comes to sex; we examine how severe the situation actually is. Such severity makes rational nonsense of the claim it happened randomly, that is, irrationally. How, indeed, could sex evolve? Yet if polarity is archetypal then such architecture might be deeply natural to re-creation's plans. Could modern biological research be in the process of revealing sex as a reflection of a whole that's split to complementary, polar parts? Can *yin* and *yang* take on more than conceptual meaning? No doubt, science gets embroiled in detail but maybe there's a larger, general plan. Sex, of course, is often thought of in the same sigh as is love. Quick, erotic, stable or devotional - at last we ask the character of love, true love.

Chapter 25 leaves sex and passes to the course of reconstruction. What is the origin of growth? How did development piecemeal evolve? No doubt, a process that anticipates a target product is not easy to explain in terms of mindless, random push-and-pull. Nor do molecules have any wishes; DNA, no more than ink and paper, has a penchant to survive. Elaboration of a most efficient, systematic code that's coupled with amazing chemical machinery that has in turn (on pain of naturally selected death) to work immediately crushes mindless explanations as if ants by metal swats! How *do* plans planlessly hove into view? How does teleological persuasion by instructive code press purposive configurations, of which life abounds, each into its allotted place? We check the nature of the dialectic's bio-logic; we see this information is conceptually expressed. Some have claimed that an evolutionary process, macro-evolution, must occur because development depends on banks of switches hierarchically generating system-code. Since natural selection and mutation won't, as previously conceived, produce the large-scale transformations Darwin's needs an addition has been made. Evolution by development will, it is claimed, easily perform the trick. Mutations to high-level switches must produce a wealth of variant detail further down development's cascade. Can evo-devo cut the mustard where all else has failed? We check exactly what's proposed and, set against a dialectical interpretation of the facts, how much sense it makes. Eggs are top potential; out of a single, tiny egg came everything that's physically you. Did an egg precede its adult; an adult come before its egg; or did both arrive together? How were initial conditions for development established accidentally? How did a most rational, complex process simply spill from chance? No, natural selection isn't chance - agreed; nor is artificial selection - entirely different because it has a breeder's wit behind deliberate variation on a given type. But natural selection cannot change a thing; the driver of an evolutionary process must be mutation, that is, *random* change to *DNA*. Chance must create fine order. From the all-powerful, preordained and preordaining order of an egg we move to bio-logical increase in size and in complexity. Does phylogeny explain this matter; could entirely plastic, accidental variation over many generations yield a route from molecules to men? Bio-logic's specified by code; creed aside, does evolution offer a superior report of how life's 'operational intelligence' was, as in the case of all machines, injected at the start? Or, lacking in conceptual information due to matter having none, does it render a less reasonable, inferior account? The weakness of a case for accidental aggregate is pressed into acute an episode by metamorphosis. Of course, hardly plausible accounts have been concocted how, overnight, a caterpillar might evolve, by first dissolving almost all its tissues in a chrysalis, into a perfect butterfly. Surely dotty butterflies are not the symbol of Darwinian nemesis and, at the same time, fluttering flags of life's innate, original intelligence? Preserve us! Pickle me! Yet, unfortunately, the world will not - once come of age an adult form 'runs down'. Old age and, at some hour, death overtakes each one. Why has evolution worked amiss? Immortality might seem a great solution for survival but the process of progress hasn't seemed to twig it yet. If, therefore, the cosmos spins more like a gradually slowing carousel than heads up any great progressive scheme where does it leave you, a mutant ape? Enter now a hall of smoke and mirrors. We take a fairly detailed look at problems plaguing primate paleontology, stories that the discipline evolves and what the fossils are interpreted to show. Could it be that, in the last analysis, man is man and ape is ape - and always has been? What form, therefore, after this discussion does a third and final flight from science take? Has Darwin had his day? Finally, since sitting on the fence might seem inclusive of both sides, some persons seek to accommodate both theories of Intelligence and No Intelligence. They seek to run with hare and hounds and thereby unify, in terms of origins, both pillars of contrary faith we met at the beginning (Chapter 0). Is this embrace a possibility? The futility of such accommodation's exercise is categorically laid clear.

Chapter 26 extends debate from individual bodies to embrace community. A Unified Theory, covering both essential and existential kinds of union, is outlined. No matter where you dwell in space or time the question of association is the same - with what do you most advantageously associate? The reckoning involves not year or distance but the conscio-material gradient with dynamic vectors 'in' (↑) and 'out' (↓). If such loop includes the concept of 'return' then which direction does the word imply, how does one set out on the way and, if the cycle is a cosmic one, what represents the point towards which space-travellers such as you and I might profitably steer? Next we turn to 'wider body' than an individual one - environment with living and non-living parts. We examine the ecology of mother earth. Does life upon the marbled surface of our ball appreciate how fortunate it is? Through atmospheric gases, waters and solids of land masses we list hi-fi, hi-felicity of multiple conditions that are required in stable concert for a single seed to sprout or any organism grow or move and inch. Rare earth! The list is long and yet, though woven into strong, elastic and coherent tapestry, now stretched and torn by careless human industry. A red light winks and silent siren howls; yet was there, is there any perfect world? What would it resemble if there was? Men complain when struck by natural tragedies. How fair is that when nature's a 'machine' without a mind? If deity inclined to rearrange the torrent of reflex events to satisfy each imprecation or each clashing wish what chaos there would have to be! We check out the belief that 'God's creation' shouldn't suffer catastrophic pain or inflict, on innocents, 'unjust' illness or 'accursed' accident. What, after all, is a Malthusian creation but continual struggle for survival, hard and mean, against fierce nature's hostile odds? In this struggle only strong, fit, fertile or sheer lucky organisms seem to win; on pain of death a savage world is driven forward. Good God, how could you be? How, on the non-religious hand, does socio-biology assess materialism's 'scientific' view of life? Selfish genes and so-called memes are pulled apart to see what's in them. After dissection who can put them back again? In fact, it seems the laws of logic, reason and morality, not being made of atoms, light or gravity, can't properly explain the way that callous selfishness - whose reasons Darwinism justifies - is not the character most humans love the best. It seems, in fact, that selfish lusts are at the root of violence and passions we identify as sin. Sin and evil are not fashionable atheistic words; what's immoral if our genes are full of crime? It's not me, guv; you should nick my nucleotides! Yet there are actions everyone calls evil. What is the nature of this anti-quality? How do fascism, racism, communism and other species of gang warfare fit into its theme? Most thinking atheists are moral men; they do not advocate that humans go to war. Is, however, this morality consistent with their core, materialist (and, therefore, evolutionary) beliefs or do they think inconsistently? Immorality and evil are, for victims and criminals alike, facts of life. How can a benevolent creator even once permit a scene of worse-than-bestiality? Such 'benevolence' should be rejected while we make our own rules up. Thus atheism makes up on-the-rebound rationality. Wooden puppets, certainly, can only dance as puppet-masters pull the strings; but what relationship is there with puppet-child? Love is not control. You give conscious children, not least adult children, choice with guidance. In your absence they must make free choices anyway. Thereby, in freedom, lies the risk. In free will lies the risk that, out of ignorance or accident, pain-bearing choices will be made. Mistakes. All the while we forge our fates and destinies by choice. What, therefore, is the nature of free will? How does the paradox between free will and determinism dialectically play out? It's hard to see but we can take a focused overview. The issue must be soluble and so we seek sharp resolution. This solution leads us to propose a Theory of Religion, Politics and Law. This triplex (as, for example, English history shows) is closely bound in regulating its community. Upon what principles are social contracts based and with what actions are they signed and sealed? How men deliver principle in practice is, of course, paramount; and it depends, above all other, on what a group believes is true about itself, its purposes and its relationship with natural law. Natural law includes, of course, its own sort of maker. What is a man's relationship with this to be? It's down to origins; to sort of origin. Thus you can believe your origin's oblivion or that you issue, at base, from an immaterial element of consciousness. If it's the latter then your source would be alive. You can adopt peripheral or nuclear religion as you choose but which is *right*? Check out faith's atom. On which vector will you ride? And, if you slide down, what is (↓) fall; what does the drop towards negativity entail? The stacks of Natural Dialectic clearly illustrate the way. The fittest winner clearly (↑) rises in reverse. This positive solution is simply, systematically applied to categories of politics and law. Such broad brush amounts to infrastructure into which their verbose detail perhaps best fits. From government, however, to self-government; the next move is, with reasonable goals in mind, to forge a pact with wayward currents of the mind - indeed, to control them. Such control leads to a focus towards achieving fuller sanity. In what does sanity consist? What is a metaphysical relationship? How finally, we ask, can men attain Communion, the goal of Real Philosophy?

Chapter 27 summarises all that's gone before. Where does the data nature, friends and family provide actually lead? Is it, meaninglessly, round a hopelessly material carousel or is there more to nature than meets mortal eye? Bottom line, top conclusion will reveal the way that Natural Dialectic indicates is right. A crunch applies, upon inspection and as demonstrated, to atheism's gross fragility; the top conclusion follows up the Dialectic's right-hand path. Once round the universe is quite enough. Thus here I rest my case and bid farewell.

This in brief is what the chapters say and, hyperlinked, how you can reach the subjects of your interest in a click.

It's clear by now that one of Science and the Soul's primary objectives, having scythed through tangled myth and an undergrowth of cant and dogma, is to make clear an impartial yet inclusive middle way - to include both metaphysical pot and physical kettle. *It is not that Natural Dialectic adds to the exploding volume of scientific data.* **Its intention is rather to generate a neutral framework within which such data can, from opposing materialistic and holistic viewpoints, be assigned then reasonably assessed. Further, it is to generate a philosophical routine, a 'dynamic' within which physic and metaphysic are accommodated and may be reconciled.**

This fresh perspective doesn't deal in cultural war but in clearly defined reconciliation. Of course, religious and anti-religious 'fundamentalists' think, by definition, in a rigid way. They *know* by appeal to their authorities that their interpretation of the world is wholly right. Their truth is absolute and outsiders range from relatively to ridiculously wrong. Therefore they either will not want to hear or only hear in order to destroy. Is this not basic lack of academic grace? Therefore, although nature itself may include extremes called 'absolutes', fundamentalists of neither camp are likely to appreciate the politic but neutral, natural balance of this book. At the same time its central ground is no blurred compromise. Dialectical resolution is as sharp as it is distinctive.

The compass of human knowledge, now so wide and detailed that a single person cannot know all scientific or other facts, has been sectioned. Instead of a polymath familiar with the whole picture, specialists now marshal the details of each sector. In this respect the point of Natural Dialectic is to simplify. Its polar scale introduces the idea of fundamental vectors that propel existence, an existence swinging round the pivot of an essence. Such motions help to show that while the facts are myriad the principals are few enough for a diligent, intelligent layman to grasp; and while this book may appear complex its main game is, like life's, very simple - to clearly grasp core principles and break through to a *top-down*, comprehensive vision that is (see Chapter 0: Polar Perspectives) easy. Thus the Dialectic provides a preliminary sketch, a philosophical principle and a framework into which data may be fed. At whatever level you study life and the world, it aims to help promote rather than expertly obfuscate an understanding. **The basic ideas behind good science and philosophy are simple.** They are, even if their ramifications are complex, simple enough for an intelligent person to grasp. **This is certainly also the case, in principle, with this book.** Science and the Soul's second objective thus involves, in the wisdom of Albert Einstein's words, "*... seeking the simplest possible scheme of thought that will bind together the observed facts*". To work from the complexity of this sensible world and develop simplicity of principle within a tightly woven, highly ordered structure has not been easy. *For a narrative to unfold according to the lines of nature has meant the issue became, time and again, one of order, order and more correctly nested order.* Without such order no archetypal consequence, however mundane or revolutionary, is inexorably derived. The structure of Natural Dialectic unwinds (or you might say 'devolves') a hierarchical, *top-down* perspective. *Grasp the principles and details will 'self-organise' around them. Despite so many trees the whole wood is still visible.* The book as a whole is divided into volumes and nested 'levels' that reflect the fundamentally binary, complementary construction of both contrived and natural creations.

Does this sound difficult? From a seed develops a plant. From potential (or 'capability' as the landscape gardener Capability Brown would have said) issues expression, action or behaviour. From Essential Nothing issues something called existence which includes all changeful things. *Natural Dialectic deals in polar opposites.* To reflect such fundamental duality the two volumes are labelled 0 and 1; its books begin with 0; and the thread is introduced in Chapter 0. The narrative erupts *ex nihilo*. Its world is switched on by a pulse from zero, nothing, nix. Thenceforth, whether or not binary Dialectic is *the* perfect format to describe the nature of things, there unfolds a framework of principle for, consequently, practical discussion. <u>*I hope that its essential simplicity, developed gradually through the first five chapters, will afford the reader a clear, easy and commanding perspective from which to address the subsequent, more complex issues of science and philosophy.*</u>

So how, in summary, do things stack up? Could a dialectical system accurately reflect not only universal but personal reality? Might its understanding spring surprises? *It's worth checking. Take the plunge.* By the time you surface and have swum to the other side you will have seen things in an old but at the same time new, obvious and yet unexpected way. If, meanwhile, travel gets tough and you do not fully understand a point at first reading, keep going; and if you have not 'done science' or meet an unfamiliar detail don't be put off by a few words. *Consult the Glossary, a textbook or the internet but, again and above all, simply press on. Grasp the principles and details will 'self-organise' around them.* You can even skip along the string of stepping-stones, skate across a surface made of text *italicised* and **bold**. A learning curve's a journey. This book is its map. Facts, of course, don't change but how you see them can. Press on and in less than an athletic month you may engage a healthy, fresh world-view. I hope, because it's entertaining, you enjoy the exercise!

0 *Poles Together*

The Light in Your Eye.

Rays of light are passing into each of your eyes. They travel through a lens, across the vitreous humour and strike the retina. From here impulses are transmitted via the optic nerve to your brain.

With the right instruments this activity can be traced. If you were correctly attached you could note the effects of your sensation on a screen. So far the process can be monitored scientifically. It can be quantified in terms of mathematics whose descriptions underlie physics and chemistry. Or it can be pictured as a biological phenomenon. In short, the process can be viewed from 'outside' as 'not-me'. Call it *objective*.

Experience.

At this moment something extraordinary happens. Sensation flips beyond the grasp of science, of 'outside' or 'not-me'. It becomes me. Your sensation becomes you. You are conscious of your surroundings. You enjoy a light, lucid, colourful and accurate perception of this book. 'Conscious' means 'knowing-as-a-whole'. The experience, which integrates sensory and motor functions, thought, emotion and memory into a conscious stream, is the most important thing in your life. It is what you are. It unifies and therefore simplifies a complex, bit-made world. But is the experience you have at this moment simply electronic charge in action or an interactive pattern of neurons? That is, for some, a belief. It is the 'outside' or objective interpretation of events. There is, however, no proof that consciousness is essentially and only physical, that brain equates with mind. The view simply underpins a faith, materialism, to which you may or may not subscribe. *It is not, unless you assume from the start there is nothing but matter and equate science with such exclusive materialism, a scientific view.* <u>Be clear, there is faith whose authority derives from the philosophers of scientific materialism but, although materialism (or naturalism as it is often called) is the official doctrine of contemporary scientism, there is neither proof nor knowledge that it is more than partially true</u>.

Be that as it may your experience at this moment, so lucid and clear, is real. Who is the knower, the seer that observes? Consciousness, in which you indissolubly and indivisibly live, is central. It *is* you. You experience existence from 'within' looking 'outwards'. Science can, from within, neither examine nor create it; it can't even be objectively observed. Call experience *subjective*.

objective   *subjective*

Perhaps It's All in Mind.

outside   *inside*

Moreover, look around. Since rays of light have entered eyes where else except inside the head can pictures of the world be found? Outside objects (which include your body and its brain) are brought together in perception by sensation.

That is not to say these objects don't have individual identities; or that different nerves do not submit their gathered information for collation in your brain. It is to say that you experience a unity of many different objects and events; and this experience (including thoughts and memories and dreams) is registered within your skull. Does such unified experience extend beyond the brain? Because it seems the world inside your head is clearly outside, where it really is, as well. Is mind outside too? If it is not, why don't you simply watch a cinematic 'movie' screened inside the darkness of your head? And if it is, then mind is not the same as brain. Perhaps, therefore, the world's not all in brain. How, either way, does such projection work; surely it is not the gift of atoms, molecules or electronic charge - those non-conscious ingredients that constitute a nerve?

Now concentrate upon a thought. Close your eyes and contemplate on, say, a candle and its flame. You are, somehow, inside your head but open up your eyes again. You're not. Nervous process mediates between externals that you're looking at and you; but you don't sense them in the same way as the conjured candle flame. The world, including body that I'm clothed in, is 'out there'; how, though, from my 'cockpit' do I sense things far and near? And if I'm psychologically outside as well what is the range of 'psycho-space' or 'field of mind'? Look up. Surely it cannot include the sky? Again, at night, is the starry firmament inside your skull? It doesn't feel that way. It doesn't feel like just an image that is represented by a pattern in atomic matter - nerves. And yet, if you subtract your senses, only darkness is inside your thoughtful head.

What, therefore, is the constitution of this 'mind-field' that envelopes all I know? What is the nature of perceptive consciousness; and the nature of my contemplation? What, in essence, is the nature of the 'ground' of mind? In Book 2 we return to these essential, unsolved mysteries; but, before that, we can build a framework within which to better understand. For now we simply note the paradox of 'mind-in-matter' or 'everything-made-one-by-me'. Such is experience. Knowing, which is of the living not the lifeless, is never actually objective; it cannot be materially observed. Do we therefore live in mind or brain; or have experience using both?

Whichever way, we certainly inhabit a subjective part. We live in mind with brain attached; nor is there evidence (although there's dogma) that these two are indeed the same. You are inside looking out; the world is known inside out; in this sense everything's subjective. *All experience, which is the only way that anything is known, is subjective.* **This subjective aspect is an abstract - 'nothing' in material terms: but without such immaterial 'nothingness' there's nothing! Paradoxically, as far as we're concerned, it's everything!**

World-view.

A world-view is a network of basic beliefs about reality in the light of which one's observations are interpreted. Everyone has their own 'logic'; no evidence 'speaks for itself' but is interpreted according to its finder's view. Presupposition, prejudice and bias are embedded into every mode of thought. We all wear goggles when we look at things. The question is - what is the clearest optic; what glass least distorts reality; what is the angle of best vision? This depends, in turn, on what foundation a world-view is ultimately built.

Nobody argues physical creation is not fact but whence, at root, do its phenomena derive? Is it eternal matter, an unconscious 'deity'; the atheistic final truth lies in unconsciousness. Is it, on the other hand, an eternal, conscious deity; theism's absolute resides in consciousness. Which one will, on full investigation, make more reasonable sense?

unconscious   *conscious*

Is to be awake a lightness and asleep a darkness? If so the oblivion of non-consciousness is darkness. In this sense, as opposed to an illumined, pure and eternal heart of life, naturalistic atheism's faith in matter might be called the heart of darkness. Of course, if eternal, heartless matter can somehow create a conscious stew (called mind) then hearts of either kind are rendered obsolete!

It is, therefore, imperative to ask, 'Is consciousness unconscious?' Is mind matter? Is it sprung from an atomic complex called a brain? Or is such contradiction fundamental, such division - mind from matter - inescapable? **Arbitrary guesses and hypotheses abound but, at very least, it follows that a scientific world-view that does not profoundly and completely come to terms with the nature of conscious mind can have no serious pretension of wholeness.** Perhaps our present position with respect to physics, chemistry, biology and psychology is a staging-post and we need a paradigm-shift, the turning of a corner, shaking of the foundations, a reversal of polarity from the way, through the last couple of centuries, modern thinking has been drifting. Such a shift might not necessarily conform to our prescription. For example, we study the neurobiological processes by which information in light rays is presented, via the brain, to our minds but it may never prove possible to explain vision or its correlate, mind, completely objectively.

Equally, it may be possible to describe a fundamentally important property hidden unobserved (or even observed but ignored) in the ordinary behaviours of our world. Take, for example, the nature of matter, time, space and the laws that govern them. Sir Isaac Newton treated omnipresent, omnipotent gravity as a force; for Albert Einstein it became something different - space-time whose curvature embraces all particles and forces. Einstein did not diminish Newton. While the latter remains reliable, the former added a rewarding new perspective. Max Planck's 'quantisation' that stipples all of matter into tiny parts added another. *We will need to supersede all three with a theory that includes consciousness.* Moreover, just as some of the aforementioned theories include counter-intuitive elements (such as the speed of light being constant for all observers), so we may need to supersede our intellectual notion that normal human consciousness is the only or even the supreme variety. In each case, having found an appropriate new perspective we may have gained without losing anything.

We've already made significant advance. Materialism claims, as noted in The Gate, everything including consciousness is physical. This is a belief; its promissory faith and hope are that belief, with scientific proof, will one day be converted into nervous fact. If, on the other hand, you believe mind is in fact an immaterial element then what about the nature of its cosmic concentrate? <u>*The central feature of this book is no more recondite, eccentric or arcane than*</u>

anything you're well aware of; it is no more exotic, no less central and essential than the 'substance' of your being. Are you not most intimately acquainted with the element in which you live - consciousness?

Thus the primary axiom of materialism is that everything is made of matter. Is this right? The primary axiom of Natural Dialectic responds that mind and matter are two separate elements. And, just as the non-conscious, objective element of energy might be identified in pure form as radiant light, so the subjective basis of mind, consciousness, may also be discovered in its purity. In this respect there springs a faith that life is physical, a body part of universal body but also metaphysical a part of Universal Life. Aspiration's therefore to obtain subjective knowledge of such Elemental Purity.

Thus we've established that both parties set out on an equal footing; each of the two main pillars of world faith we're going to explore is founded on belief; and each philosophical belief involves a lens of creed with which to interpret the workings of the world. Whose lens is least distorting?

Natural Dialectic.

Dialectic or dialectical method is a way of resolving actual or apparent contradiction by way of dialogue, back-and-forth interaction or to-fro cycles of communication. Dialectic, dealing in polarity, is historically and geographically central to many strands of philosophy. In the west perhaps Zeno and Heraclitus (for whom the unity of things was characterised by intrinsic pairs of opposites) seeded the Socratic and Platonic search for truth. The seeker's method was by use of reason and of logic in a dialogue. Islamic scholars used (and use) Platonic dialogue; and in medieval, Christian universities part of the curriculum (the *trivium*) was dialectical logic.

objective	*subjective*
material action/ deed	*immaterial consideration*
matter	*mind*
pragmatism	*idealism*

There followed Georg Hegel whose dialectic involved thesis, antithesis and synthesis which, transcending the opposing pair, might be represented as a 'spiral' resolution by fresh information rather than an endless, to-fro 'cycle' of argument. For Hegel history was seen in terms of dialectic; so also by Karl Marx who, as opposed to Hegel's idealism, explained the development of human history in terms of class struggle, aggressive revolution and thus domination of 'the other side' by superior strength. In this respect Marxist dialectical materialism, beloved of Lenin and Stalin, was central to the thrust of soviet ideology. It treated opposition to its chosen 'positive' as an excluded 'negative' with which it was at war and wanted to eliminate. This hard, aggressive kind of dialectic, taking materialism and atheism as its 'positive' thesis, thereby demonizes metaphysic as its enemy. It puts matter prior to mind and the complementary consideration of opposites inside out; it flips the positive Hegelian absolute of spirit on its head. Reversing the Hegelian order (spirit → mind → matter) of all religions it likewise logically negates them - including oriental Hindu and Buddhist practices that deal specifically in education by dialogue, debate and dialectic.

'Science and the Soul' is cast in Natural Dialectic. **Its dynamic format oscillates, with respect to mind and matter, between two poles; and it continually transcends the flux between them at a point of balance.** Thus it finds polar existence intrinsic in its essence, that is, relativity within its absolution. If cosmos *is* rooted in polarity then you will have to judge for yourself how closely Natural Dialectic comes to frame its truths; and how closely or otherwise it aligns with other ways that to-fro, polar interactions have historically been expressed. This volume certainly revises, modernizes and swings science centre-stage. Its Dialectic is, as opposed to Marxist negativity, positive in that it treats opposites as two sides of the same coin, only *apparently* contradictory. Including them as *complementary* it incorporates both sides of an equation and expresses a conclusion that, in its unifying/ holistic wholeness, transcends each. As such (we'll understand from Chapter 3) it is essentially paradoxical. Since this style of dialectic well expresses the nature of *three* cosmic fundamentals (Chapter 2) we may come to understand the structure of our cosmos as essentially paradoxical as well. In this way Natural Dialectic well reflects relationships between all kinds of complementary pair; and, beyond them, links each couple with its pivotal control, that is, its government. One in two or three, or three in one, or one in two might indicate prime paradox embedded in the way things work and in the nature of the way they are. **Indeed, the Dialectic is a theory of principle that, setting up its framework of polarity, makes possible a universal and yet self-consistent description of the natural order.**

As an aid to understanding we shall learn to build 'dialectical stacks' which embrace flows of traffic in between all kinds of polar opposite. **At this point please check the entry 'dialectical stack' in the Glossary. A consistent arrangement of opposites (or polarities) forms the basis not only, it is argued, of discussion but of nature itself.** Its perspective is already glimpsed in the previous mention of three pairs of opposites - subjective/ objective, inside/ outside and mind/ matter. Such stacks do not necessarily list synonyms or make equations; their perusal is intended to promote connections; and consideration of connections tends to help unify/ collate/ organise one's working comprehension of any matter in hand. *The next few chapters will develop, as the backbone or spinal operation of this book, a description of dialectical construction and the way it works.*

The energy of light or sound is *physical*; and our *psychological* perception, interwoven with memory, reason and emotion, is subjective; thus two themes interweave this book. The first deals with whatever occurs independent of my perception. It includes everything that is not my subjective experience, which latter is the second theme. Physical body and its information centre, psychological mind, are well met. *They meet at a point studied by neurological science.* What is called science today is limited by human definition (but not necessarily by reality) to the objective, physically measurable side of things alone. For Science and the Soul such limitation is unacceptable. Physical is near but metaphysical still nearer. You know, I am sure, your thoughts, memories and consciousness. *While these are connected by an intricate nervous construction to their physical environment they are not themselves proven material - as will be seen, far from it.* But both body *and* its informant, mind, are natural. **Mind is the natural metaphysic, neither unnatural nor super-natural.**

Hidden in plain sight the revolutionary element that opens up holistic paradigm is immaterial consciousness. Material *and* immaterial exist. No philosophy is complete without consideration of *both* elements. Neither scientific atheism nor metaphysical theism is, in isolation, enough. Each has its perspective. Both demand inclusion even if, in the final analysis, only one of them proves fundamental. *Indeed, the Primary Axiom of Natural Dialectic states creation is a compound of the pair; this PAND (*ref. Glossary*) sees the compound as a hierarchy, a dynamic conscio-material gradient (fig.0.8, 0.11 and 0.14).* Set within the basic construct of polarity the notion of a **conscio-material gradient** is, although revolutionary, simplicity itself. Rainbows reflect a coloured gradient of energy. By analogy the conscio-material gradient, composed of a graduated, elemental, complementary mix of energy and information, is 'a rainbow spectrum'; the level of informative activity decreases in proportion as, on a sliding scale, material condensations multiply. Such falling slope is called existence and, as we'll show (in Chapter 17) is well reflected in the human form. **Thus, in this view, mind and matter are the mix with which to grade the whole creation; and an exchange of Natural Dialectic, in between your immaterial and material parts, dynamically informs a microcosmic parallel - yourself.** We'll address how such a vector translates into science and philosophy.

Are there other central principles round which the details of this work are magnetised? From spectrum to a pair of scales; **you can also see the cosmos as a pair of scales.** In this picture ups and downs swing round a pivot by degrees. Two pans of a kitchen balance oscillate about their starting-point. Opposing vectors swing around a point of balance. The motion of creation weighs upon a fulcrum of potential. Although a thorough expression of nature in terms of its balance takes the whole book, the basic idea behind Natural Dialectic is extremely simple. *It is to weigh the cosmos in a way, at base, as simple as the scale you have in mind.*

A third idea, as basic as the law of equilibrium round which the cosmos spins, follows hard on the heels of the first. You can think of intelligence, information, energy or bulk materials in terms of concentration gradients. Think, for example, of an energetic concentrate, a star; and then the radiant diffusion of its light. Or, if you like, think of electrical or chemical diffusion down a slope from concentration, source or origin. Prior to action any single thing is, almost egg-like, in 'potential equilibrium'; such stillness, readiness or poise is full of certain possibilities. There follows (if some fluid circumstance permits) a spread of chemicals, a reaction or the flow of current pole to pole. The event rolls to its close; exhaustion is an opposite, inertial kind of equilibrium. Could you, from such 'flat death', regain original charge? You would have to pump with effort up 'against the flow' but you'd 're-charge your battery', re-concentrate lost energy and thus complete the cycle. **In this view potential, whether in the form of energy or information, is a source.** It is replete with charge that is by action discharged into impotence, its sink.

Strong concentrations spread towards weaker ones; chemicals diffuse into inertial equilibrium. In the case of conscio-material slope where do you start? Which - consciousness or matter - sources which; which embodies the potential for creation? Or, put another way, whose concentrate streams from the universal mountain-top? Does a strong concentration of awareness drop towards sleep's oblivious sink? Or *vice versa*? Does concentrated energy 'diffuse' into exhausted matter? Or *vice versa*? And on the gradient as a whole does body follow mind or make it; is non-conscious, will-less matter dropped from consciousness or *vice versa*? From what kind of apex, mind or matter, does creation fall? From atoms to the galaxies 'hard science' uses rainbows in the form of spectra to establish how the world is made. What, as Chapters 7-12 investigate, might be the order of a conscio-material rainbow out of which a three-tiered universe is built?

Apart from principal ideas we ask whom Natural Dialectic's discourse is between. Protagonist and antagonist, we'll learn, are *top-down* and *bottom-up* perspectives (Chapter 0: Polar Perspectives). It is through the eyes of holist and materialist we'll check the best interpretation of our universal facts and faith. What you hold as true is your belief. What is best belief when all we know has been considered? Love of such wisdom is philosophy! Natural Dialectic is a compass in the fields of fact and meaning; it will help us orienteer.

In this the Dialectic's exposition steers clear and free of all religion - whether that religion's secular or not. *Within the external differences of vehicle, behind the prismatic refractions of each faith we need to seek and find a single centre, a heart, a perennial truth.* Such supra-religious singularity, this trans-religious goal, is Natural Dialectic's aim. *If this ,truth is single, various descriptions will sparkle like the facets of a diamond. Natural Dialectic will reverberate with elements from different descriptions of the way it is.*

One cannot thread simpler than a single axiom (mind and matter are separate elements). From this are woven answers old and insights new. The Dialectic is unequivocally, hierarchically structured in a thorough but very easy, binary way; thus it amounts to a new yet also ancient 'digital' philosophy as well as scientific paradigm. For this reason its two volumes are presented as the binary digits, bits 0 and 1. Zero and something. Things are things but nothing - what is that? Is cosmos an infinite series of causation or did it have a start? If so, what precedes all things? What causal, starting impetus makes all creation its effect? If a source is 'higher' than its issue, then the source called Causal Impetus presides 'above' all subsequent effects. Indeed, its 'Apex' would initiate the Dialectic's conscio-material gradient, that is, the existential stream. If existence includes everything what sourced First Cause? Or was the Cause of causes Uncaused? What sort of essential absence seems to have projected what we call real things; what kind of relative reality are they suspended in? What is 'The Nature of Original Nothing'?

Does, you may well ask, a Grand Theory of Phenomena have to include None-of-Them, that is, make room for Void? Or must, on the other hand, a Theory of Nothingness envelope anything at all? The list of students involved in attempts to define void, space and nothingness is long and eminent. It includes all oriental philosophy, religion and mystic practice (Vedanta, Buddhism, Taoism and so on), the Babylonians, the Greeks (such as Thales, Plato, Aristotle, Pythagoras, Stoics and atomists such as Democritus and Leucippus), Lucretius, neo-Platonists (such as Plotinus and Porphyry) and a host of medieval philosophers. At the heart of Semitic tradition (Judaism, Christianity and Islam) also rests the Notion and Essential Nature, beyond existing objects and events, of Nothing - or should it be The Infinite One?

Is such ado about all things or nothing? Wits and writers, not least William Shakespeare himself, spun pun, paradox and nothing as you like it into practically any clever thing - until Galileo (who trusted experiment above verbal *legerdemain*) decided it was time to find a physical vacuum and translate nothing into something seriously real. There followed Torricelli, Pascal, Spinoza, Leibniz, Newton, Planck and Einstein - whose leading probes have each furthered an understanding of the shadowy creation of something from, apparently, nothing. Now science, having swept aside nothing (in the formlessness of 'ether'), has refilled its vacuum with quantum fields, *ZPE*, space granulations and so forth. Rich a seam. Is there any more to mine? Has *nothing*, a key paradoxical space under the spotlight of modern physics, any more to give away? You might certainly wonder how to best grasp nothing and, in grasping firmly, make the most of it. *Nothing is a key principle of Natural Dialectic and there is, I can assure you, more to its nature than meets the eye!*

What, therefore, was our universe projected from? How did life on earth (or anywhere) begin and thence develop? Theories vie for popularity but popular approval isn't a criterion of truth. In its day 'phlogiston theory' was all the rage. Ptolemy's (not Biblical) astronomy was geocentric; Copernicus, Galileo, Kepler, Halley, even Isaac Newton thought the sun was at the centre of the universe. 'Ether' used to be the stuff of space. A famous theory of inheritance by Jean-Baptiste Lamarck has also hit the bin. Millions have believed what isn't true. In the present millions more devote their lives to the embellishment of theories also perhaps just partly true. Big bang will cross our sights. Are blind forces really capable of fabricating forms of life? Did it self-organise from chemicals and thence evolve in a Darwinian sort of way?

We'll explore these issues later in a more extensive way. For now let's simply bear in mind the basic digital nature of our world. Fundamental polarity is a concept new neither to philosophy (e.g. in the complementary opposition of *yin* and *yang*) nor science (e.g. + and − electrical charge). Of tried and tested pedigree, its dynamic reflects the range and interplay of attractive and repulsive tensions that exist between opposites. Such dialectic is not abstract or detached. Gottfried von Leibniz , who devised calculus and the binary system of mathematics that today underwrites information technology, might agree. Binary machinery's of great import. From this seed notion Alan Turing built the Bombe and Tommy Flowers Colossus. Thus was ushered in our 'information age'; all electronic, digital computing grew from these. And nature always had an age of digital polarity. Chemistry embodies interactive charge exchange; so, we may yet find (Chapters 19-25), does coded, computational biology; and (Chapters 5, 13 and 14) psychology's information systems - input, functional decision-making process, output - well reflect a dialectical polarity. *And if, more generally, the seed idea of binary description is found to fundamentally and accurately reflect the structure of creation, it may come to represent a necessary alteration to the underpinnings of our scientific viewpoint of the nature of reality.* It is from natural philosophy that, as a materialistic subset, modern science has evolved. What is a PhD. but a Doctor of Philosophy? Including both physical and metaphysical, Natural Dialectic is a philosophical grid into which we can simply sort both kinds of data. It might also offer the basis for a paradigm shift in our thinking. *And it might equally represent a restoration of, in its fullness, natural order.*

The claim is science fits within the Dialectic's oscillatory paradigm; the challenge is, if this is so, to work out how.

Opposite Directions of Mind.

The light in your eye is transformed to life. We started by identifying the kind of continuum through which physical data informs, via the nervous system, living experience. This experience is based on awareness and, therefore, levels of attention. Can one map attention, what are its dimensions, what are the parameters of consciousness? Is there any easy way to 'get the gist'? In fact, philosophers fish by using kinds of metaphor to hook what is invisible; and a psychologist will try and model how the immaterial - your own experience - works. 2-d graphical (e.g. Tart, Varela), 3-d boxes (e.g. Hobson's brain-mind space) and a pictorial host of other metaphors are used. And two 'cosmic metaphors' of Natural Dialectic, ones that dominate the illustrations of this book, will be introduced shortly.

There follow, however, three elementary gauges. *These, it needs be emphasised, are introductions; they are starter models; they serve as simple, early essays which a practised or professional intellect may find, as yet, imprecise.* If you're sophisticated then repair to the next section; if you're not they'll preface ideas with a gist that's going to be elaborated - especially in Chapter 13.

Gauges let a driver know what's going on. There follow four broad, initial but tentative suggestions how to measure that elusive entity, that unifier of duality, subjective consciousness. They are crude types of 'conscio-meter'. The first 'line gauge' illustrates the character of inward and outward directions Gauges let a driver know what's going on. There follow four broad, initial but tentative suggestions how to measure that elusive entity, that unifier of duality, subjective consciousness. They are crude types of 'conscio-meter'. The first 'line gauge' illustrates the character of inward and outward directions of a focus of attention. Its needle points to the condition of your consciousness at any given instant. Such condition is composed of changing permutations of sensual, emotional, intellectual or, in deep contemplation, supra-rational experience. In *fig.* 0.1 the needle indicates that the subject might be in the 'middling' process of working out how to do something.

external/ outward attention	*internal/ inward attention*
centrifugal	*centripetal*
objects/ objective parts	*subject/ subjective consciousness*
specifically focused attention	*globalised attention*
disparate world/ body	*unifying mind*
physical/ material	*metaphysical/ immaterial*
involuntary/ reflex/ perforce	*voluntary/ by choice*
physical space	*'psycho-space'*
material objects	*mental objects/ forms of thought*
chemistry of sensation	*feeling/ thought*
instinctive perception/ response	*learning/ understanding*

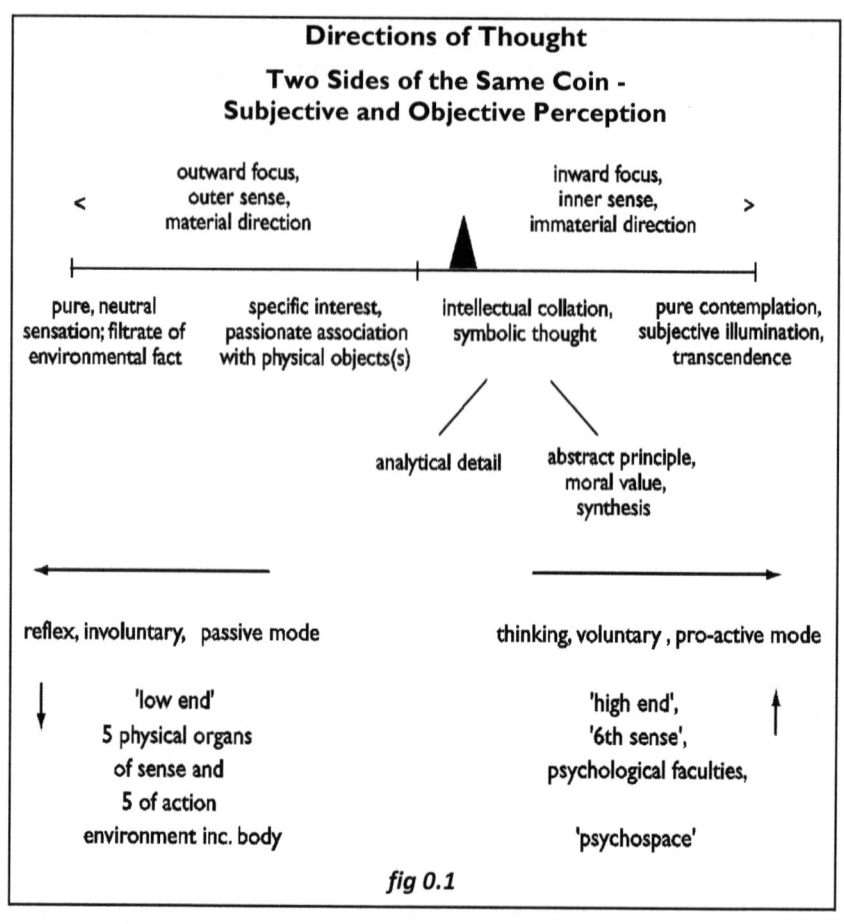

fig 0.1

You could take the horizontal line and switch it vertical. The material base of this gauge would register reflex sensation and the top transcendent absorption. Such vertical measure represents degrees of subjectivity; it illustrates the hierarchical order of what might be called a conscio-material gradient.

Since different levels of mental operation (e.g. sensation and calculation) may occur simultaneously on this gradient the single needle would register the current average position of the mind.

Check *fig*. 0.2. Engine drivers are used to circular dials and you can, in the cab of your psychological engine, register the gradient on such a consciometer.

Left-hand arcs register focus that is channelled outward. The sub-rational arc represents 'passive' aspects of mind - instinctive automation as opposed to the right-hand, 'active' psycho-space of thought (or cogitation). It includes conscious but instinctive responses. Such 'animal' program is concentrated on external distractions; such 'exterior events' include, centrally, one's own body's changing circumstance. In this case *fig*. 0.2's arrow registers some emotional reflex; it indicates what might be fear or lust. Further round the reflex arc includes both unconscious, subliminal perception (things registered of which we are unaware) and sensation 'uncontaminated' by the general context of memory, preconception, thought, emotion or particular point of sensory interest. This is the condition of raw, pre-filtered information as gathered by sense organs; its motor correlate is released, post-processing and equally reflexively, at the 'muscular' end of behaviour by the organs of action. Thus start and finish of this information cycle might be labelled *pure sensation* or *pure action*; this pair are as physical as psychological can get.

On the other hand, circling to the right, intellect is ranged against its lack. The indicator registers manipulation of symbol, calculation and executive planning to achieve goals. This is the domain of problem-solving, business and mathematics. Rising to the right it engages creativity, the consideration of general principles and maybe moral qualities that govern any object studied, that is, which is being thought about. Such understanding reason should, according to the Dialectic, at best reflect universal patterns. If man was in fact created as a universal image then you might expect the deepest patterns of his mind would, by natural resonance obtained in contemplation, intuitively reflect them.

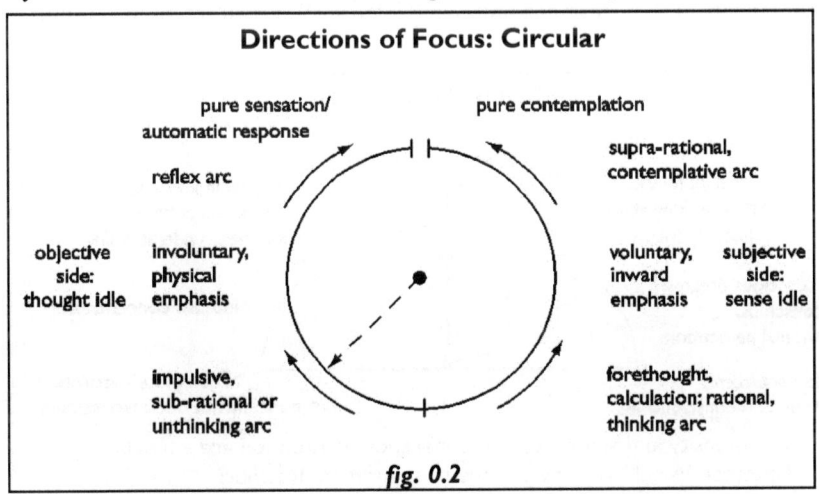

fig. 0.2

On the other hand, circling to the right, intellect is ranged against its lack. The indicator registers manipulation of symbol, calculation and executive planning to achieve goals. This is the domain of problem-solving, business and mathematics. Rising to the right it engages creativity, the consideration of general principles and maybe moral qualities that govern any object studied, that is, which is being thought about. Such understanding reason should, according to the Dialectic, at best reflect universal patterns. If man was in fact created as a universal image then you might expect the deepest patterns of his mind would, by natural resonance obtained in contemplation, intuitively reflect them.

The power of reason may be shallow, distracted and weak or, as an increasingly laser-like focus is brought to bear at a single point, become concentrated and profound. *The deeper an interest the sharper my attention, the more light it sheds and the more I come to understand.* I ascend the contemplative arc and enjoy its 'bliss' when, for example, I am totally absorbed in a book, film, game, music or piece of work.

Now subtract external aids to concentration and voluntarily ascend into the 'zone' of wholly internalised absorption in your own thoughts. Deleting even thought, transcend the mind. This sort of internal coherence, this one-pointed, single-mindedness will have precluded interference by the relatively imperfect conditioning processes called reason, emotion and fantasy. From things the object of absorption has swung to an extreme self-absorption. *Attention is focused only on attention.* You have thrown off all clothes and coverings; naked of thought you rise in contemplative detachment from your 'worldly business' and into an absorption void of thoughts and the sensations representing things. This absorption amounts, beyond thinking, to a direct perception of the root nature of mind (and, therefore, mind-in-principle); it is knowledge undistorted by the lenses of individual impressions, memories and thoughts. It might be labelled pure contemplation or total subjectivity. How is such experience unnatural? How is what it knows not real and, if real, the most important thing in cosmic life? If there is subjectivity then what precisely is the nature of Pure Subjectivity? The nature of contemplative, subjective non-thing is, paradoxically, a subject of encompassing interest. We shall return to scrutinise its sort of 'void'.

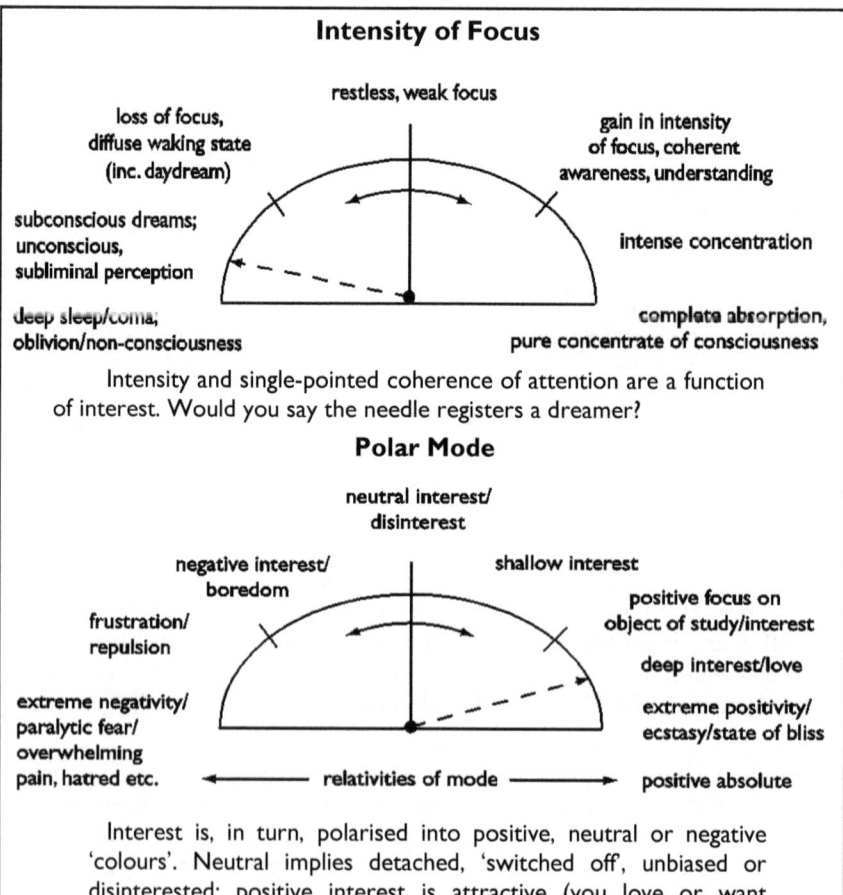

Intensity and single-pointed coherence of attention are a function of interest. Would you say the needle registers a dreamer?

Interest is, in turn, polarised into positive, neutral or negative 'colours'. Neutral implies detached, 'switched off', unbiased or disinterested; positive interest is attractive (you love or want something); negative is repulsive (you dislike, fear or don't want something). Intensity and polarity of mode of focus might be said to

> change like facial expression, chameleon-like colours of emotion or the lights of interest-blown weather across the sky of mind. This particular indicator shows a subject in the heat of creativity, intense attraction or engrossment in some enjoyed activity. Does this mean swings to right are, far from an illusion, very healthy? What does that make transcendence and enlightenment? Should evolutionary psychology awaken from the nerves-make-thoughts, matter-makes-mind mind-set and countenance such central aspects of our mental being? Book 2 checks the case.
>
> You will, incidentally, by now have noticed these 'psycho-meters' have nothing to do with brain scan images that map the nervous correlates of experience. It may be materialism's illusion that brain generates subjective consciousness. For its part Natural Dialectic treats the brain as a modulator and not a generator of consciousness. Being the physical component of mind-body mediation this modulator and its neural activity would not register the subjective dimension of experience but, simply, reflect the objective, somatic side of psychosomatic effects.
>
> **fig. 0.3 (see also 13.5)**

As opposed to direction and object there is also *intensity* of focus.

'Doing nothing in particular' is a popular pastime. 'Thinking nothing in particular' is, unless you mean sleep, more difficult. It usually describes an unfocused drift, daydreaming or a stream of uncontrolled thought - a leaf in the wind but a mind that is far from empty. The opposite of such diffuse, 'passive' lack of focus is gathering of attention - concentration.

Sharp concentration may accrue from shock, repulsion or fear; then it is reflex, hard and negative. Or it may accrue from attraction and desires; such interest flows coherent, voluntary, warm and positive. These are modes of focus. They are attentive qualities that 'colour' the chameleon of mind.

negative	positive
no	*yes*
inverted love/ hate	*love*
fear	*confidence*
frustration	*relief*
repulsion	*attraction*

A positive focus includes poise. This kind of neutrality is 'ready-and-waiting', detached but not 'running out' in any particular direction. Active focus pursues a course of interest. Its enthusiasms (e.g. hobbies, sports and other kinds of love affair) provide an externalised but limited form of absorption. Such interests constitute the secular, social base of a range of contemplative depths that can be 'sharpened up' into complete, meditative absorption.

In other words, absorption's current is concentrated. It runs fast, coherent and clear. We seek and seek to prolong the experience. Indeed, a mystic asks the eminently rational question, "What is the most 'switched-on', positive frame of mind? How, in your experience, have you 'flown' with greatest warmth, intensity and willingness? How might this flight be reinforced? Is there a way to maximise this overdrive, prolong the flight or even generate a brilliant permanence of love?" If the question is of interest what's his answer? Is its framework scientism's? You might like to hear.

As opposed to 'switched on' is 'switched off'. 'Negative awakening', sleep, is the inverted, polar opposite of Transcendent Absorption. Its equilibrium is inertial, without poise, impotent and its lack of focus, its dispersal, is in deep sleep complete. You also lose your mind to some extent in negative kinds of neutrality or detachment such as boredom, disaffection or sluggish, slow-flowing disinterest. Negative focus, on the other hand, tends to 'freeze'. It ranges from anger through ill will to obsessive hatred and to paralysing fear. The current of negative concentration freezes and is frozen; its thought is incoherent or static; its life-force darkens, sinks or stops. While positive thought excites, negative 'gets you down'. It imposes, it depresses. Health-wise, we strive to avoid or terminate the *'rigor mortis'* of those kinds of focus negatively forced on us.

outward	*inward*
diffusive/ centrifugal	*centripetal/ concentrative*
'catching the world'	*'catching its operating principle'*
reflex/ involuntary	*deliberate/ voluntary*
sensation	*reason*

Direction and qualities of awareness (that is, its intensity and mode) combine with grades that range between polar opposites. From moment to moment, lifelong, the needles flicker and fluctuate to and fro gauging our involvements. These are, no doubt, simplistic models. It's possible to sketch more complex, 3-D models of a prototype psychology; and in Chapter 13 we'll explore how Natural Dialectic's kind of cubic psycho-space might be to some extent reflected by a brain-in-skull, that is, by polarised connections in our hemispheric calculator. However, detail's not the order of this moment so let's usefully rephrase the vectored situation.

Attention is life's 'ground of being'. It is mind's light. Is your radiance mostly scattered centrifugally, down and out into the drama of sensation? Is it like a torch that flits across cave walls, a restless beam that, never stopping, never clearly sees? In fact, even animals don't rove continuously. Objects and activities of interest fix their gaze. And it's the same with us. We even crave acute engagement with a fine sensation or an intellectual pursuit. You can use a magnifying glass to focus ordinary light into a spot that soon bursts into flame. To engage a focus of attention that is laser-like upon an object that we're interested in is just the same - except it is the fire of knowledge bursts alight.

By habit and material circumstance man's attention is drawn strongly towards his earthly circumstance; out and in, by action and sensation, oscillates our centrifugal tendency. Weak, on the other hand, the inward, centripetal bearing; education strengthens contemplative thrust that, normally, is focused outward on the plans and principles of mundane circumstance. Intellect and reason chew forever worldly fat. Only, in detachment beatific contemplators seek to strengthen inward, centripetal concentration such that centrifugal pull's eliminated and the mind detached completely from its bodily environment. In short, we use and gather information in two ways; we oscillate each side in various degrees. In some the bias drops towards animal condition; in others rises into intellectual learning and such natural principles as guide the work of scientist or, centripetally complete, the life of saint.

In other words, the needle of an average human consciousness hovers half along the dials; it roughly flits around the front-central volume of a psycho-box. Its direction oscillates between emotional response and weakly concentrated reasoning. It darts between reflex and egocentric calculation.

Which segment of the dials, which zone of psycho-space do *you* accentuate? The area with needles of intensity and polar mode pushed right; with vector in mind's volume arrowed towards transparency? Universities and schools have all devolved from an extreme monastic focus towards such luminous, transparent origin. Learning's vector is an arrow towards its Nothingness; right action is a vector leading from its Truth. *Lux et Veritas*. Coherence of attention is the educational attribute that's prized above all others. A first-class mind is judged by its ability to focus inwardly, concentrate, grasp principle and problem-solve. It is this that elevates a (wo)man. At the same time this ability to 'lose yourself' is in itself a joy or, if 'ecstasy' means 'standing beyond your normal, mundane self', a literal ecstasy. It is a condition which men, however misguided their methods and behaviours, crave to achieve.

Therefore again and most importantly, which perspective do you voluntarily amplify, where do you aim as much as possible to be? In darkness or, clouds cleared and sunshine out, in light? What seed of perception might a wise man choose to water and so yield most lovely flowers and richest fruit?

Two Sorts of Implement.

outward/ downward focus	*inward/ upward focus*
action/ sensation	*contemplation*
stepwise logic	*idea/ grasp of principle*
analysis/ breakdown	*grasp of whole*
intellect	*intuition*
rationale	*insight*
protocol	*creativity*
left hemispheric 'nuance'	*right hemispheric 'nuance'*

The two natural, complementary directions of human focus have been developed into two apparently contradictory ways of looking at the world - subjective-and-objective holism and objective materialism. Both enjoy the same faculties of mind - intellect and contemplation - but the emphases of these spheres of influence are opposite. It will be argued, in Chapter 13: To Build a Brain, that their complementary, counter-balancing nature is well expressed by two hemispheres of the brain interacting through a panel (the *corpus callosum*) set between.

Psychological attachment to material objects by means of sense, passion and intellect is 'the way of the world', physical science and technology. Thus, for objective materialism, great emphasis is placed on sense or technologically-obtained data (science moves behind the vanguard of its progress, instrumentation). From experience to hypothesis and experiment, the scientific method is well documented. Its stimulus is an idea or observation that provokes a question. From here events follow (see especially *fig.* 5.5, also 5.3, 5.4 and 6.1) the order of creation. They follow the common-sense gradient of any rational venture - exploration, business, government or your own aspirations great and small. What if? A hypothetical explanation (in physics often one that takes the form of a mathematical relationship) is suggested. A mental framework for the possible development of the idea or answer to the question is sketched. It follows the pattern of causality. A cause (called an independent variable) is suspected. All else being equal what is the effect of this cause on a second factor called a dependent variable? A physical test must be devised. After this test and an analysis of the results a conclusion is reached. The original hypothesis is either accepted or rejected. If repeated experimentation confirms the original result, such hypothesis is elevated to the status of theory and, if no exceptions are found after extensive research, predictive law. In science the material sense is powerfully refined.

Coupled with sensation is one's intellect. Intellect involves the symbolic (psychological) manipulation of data. As eye and hand are excellent instruments for their own purposes, so intellect is an instrument of mind. Its function is to 'get a grip' on events, marshal and manipulate details, reason and thereby reduce them to patterns and first principles. The process of intellect is linear, from cause to effect. Its analysis and manipulations, both verbal and mathematical, represent both comprehension and empowerment. If I can predict, might I not control and exploit the object of my attention? Intellectual reason, allied with natural and technological instruments of sensation, provides objective science with its purposes and implements.

<u>Objective science</u>, *with the use of reason, focuses out through the senses and their technological extensions.*

Conversely, <u>subjective science</u> *involves detachment from the external whirl of events and steadfast, meditative attachment to the internal centre.* Called the 'contemplative way' or 'science of the soul' such practice is, with its core metaphysical values, supported in formal ways by all world faiths. *Its technology is naturally prefabricated - yourself; its focus is inward towards one's own 'third eye'; and its understandings can, although numberless and wordless, be reasonably phrased by intellect.*

Thus both objective and subjective science are systematic but (*fig.* 0.5) each is directed towards an opposite pole of cosmos. One's object is to define 'diverse, peripheral phenomena' (the world of matter) and the other's to unite with a Single, Central Subject. Both involve mentors (skilled 'adepts' qualified by successful practice), employ the process of experiment and log self-consistent, universally agreed results. Both work in a laboratory; you go in and close the door. In the objective case a lab is stocked with kit and chemicals; in the subjective case the inexpensive crucible is body and, from experiment to experience, its instrument is mind cleansed by the practice of an upright life. *Mystic technique, taught and encouraged by its own professors, is meditation.* Is conscious attention not the heart of what you call your life? Various techniques of contemplation each have the same end - a centripetal channelling of attention towards its own focal point and, at the same time, origin. Insight not outsight is its strength; its concern is not with things *per se* but with principles, quality of mind and the improvement of its experience.

While this book is no more a manual or guide to contemplative techniques than it is a teacher of science, it is useful to definitely locate the instrument of meditation and identify it as focus of attention. Subjective regimes - improving mental fitness by some form of exercise - involve a difference of emphasis but yield fine, 'concrete' results. Its focus of attention is reversed; it's turned against the centrifugal tide that's hurling all creation down; as it approaches source so it recedes from objects of this world and, eventually, draws beyond even the most refined objects of thought. The inward focus of subjective science is its vehicle of elevation. Based on reason it eventually transcends all existential logic. *Just as exercise strengthens muscles and reason improves with practice, so the quality of psychological balance improves with meditation.*

Of course, contemplation precedes effective action but an inward meditative focus, collected to itself for its own sake, is from the view of physical activity and material design nonsense. It is an inactive emptiness, an unworldly waste of time. Furthermore its techniques are associated with the word 'mysticism'. For some people this implies a vague, irrational and impractical philosophy imbued with a farrago of superstitious mumbo-jumbo - just the shambles objective science rightly eliminates. In fact the word is variously derived from Greek verbs 'to shut the eyes and ears' (muo, μύω), 'to murmur with closed lips' (muzdo, μύζω) or 'to initiate' (mueo, μυέω). An initiate intones a given, sacred and secret mantra whose voluntary repetition is able to systematically cut the heads off mind's Hydra, that is, its ever-rising thoughts and desires. It is a practice that permits one-pointed concentration in preparation for mystic experience. The word mysticism is connected with 'mystery' and what is 'concealed' from mundane sensations; and because its concentration is not on outer, objective but inner, subjective phenomena it might be termed 'the science of the soul'. Initiates of the historical Eleusinian mysteries included most Athenian freemen. Socrates was one of the latter and probably one of the former. The Delphic injunction "Know thyself" - effectively an order to answer the subjective question "Who am I?" - is also a Socratic one. There are clear hints of so-called mysticism, not least its key concepts of a metaphysical creation and 'dying while living', in Greek philosophy. For example, an in-the-know priest at Delphi, Plutarch, writes of the Great Mysteries that 'at the moment of death the soul experiences the same impressions and passes through the same process as is experienced by those who are initiated'. And Plato (*Phaedo* 67C) quotes his teacher in discussion, just before he drank hemlock, about death. Socrates says "And purification...consists in separating the soul as much as possible from the body, and accustoming it to withdraw from all contact with the body, and concentrate itself by itself". There can hardly be a clearer hint of subjective science, although without description of technique, than this. Does precise description of such Greek technique exist?

Any 'farrago of mumbo-jumbo', which subjective scientists are as keen to sweep clear as their objective counterparts, includes the jumble of memories, sensation, opinion, desires, superstition, preconceived dogma and relative ignorance - in a word, '*ego*'. Murmuring with closed lips betrays a psychological technique used by contemplatives to help achieve this 'cleanliness'. It is, let us repeat, a method called *repetition*. It amounts, in effect, to a focal device, a switch diverting the diffuse and rambling stream of '*ego*' to a single, coherent, non-egocentric track. From Orphic, Pythagorean or Eleusinian roots, it was probably what gave mysticism its western name. *There is absolutely nothing mysterious about it.* Alexander the Great's gymno-sophist yogis (see below) would have called it '*mantra*'. All faiths employ prayer beads. Is this all there is to it?

It will become plain that the subjective, mystic course, far from being mysterious, whimsical or weak-minded, is rational, logical, clear, purposeful and highly focused. Its practitioner's aim is to steadily increase his/ her understanding of who, what and where we are; and why we are here. *An experient's aim might also be to conquer death but it is, above all, to obtain Communion with the Infinite One.* With such profound and estimable goals subjective science plays, as far as human thinking goes, the premier league.

Physical anaesthesia is one thing. Imagine you are suspended in a sense-deprivation tank. You are blindfold and your ears are plugged. You cannot taste, smell or touch a thing. You might as well be afloat in outer space. You are detached from the physical universe and, senseless, alone with inner space - your mind. Your senselessness is the primary condition, the kindergarten stage of subjective research, whose first step is to switch off contact with the physical world. The next job will be to switch off the thoughts that continually rise like waves in water, like flames in fire. This is more difficult. *Metaphysical anaesthesia* is called stilling the mind. It is the elimination of emotional or intellectual turbulence that continually distorts any clear reflection of truth. Repetition is part of the preparatory procedure helping to clear the mind of unwanted intrusions. From this point essentially similar practices of meditation are employed by different schools to advance and deepen the process of focus. As interference fades and concentration proportionately intensifies new kinds of lesson are learnt. These lessons are, as the Buddha reiterated, not so much the result of scholarship as experience.

A rationalist may, through lack of practical exercise, label meditation an inconsequential, emotional or irrational (in)activity. Reason, however, works from cause to effect. It is linear in process. Is science itself not therefore based on the fallacy that the concepts of linear thought are a genuine description of nature's interrelationships? Isn't it delusion to believe that intellect can fully comprehend the multi-layered network of synchronous events? Or that its own brainy thinking is the result of non-conscious matter, that is, of evolved non-sense? No doubt it is capable of partial truths; and it can stitch together partial truths to try and build a model of the whole. But can analytical science reveal the final truth if its analysis cannot even tell us the true nature of the mind performing it; if it pretends (or even persuades itself) that the psychological dimension is physical; if, in other words, it almost entirely ignores the subjective coordinate?

A rationalist may, unlike Socrates, forget through ignorance or inexperience to look above and beyond reason. He may, as Plato noted, turn his back to light. Indeed, a person schooled and rewarded for intellectual prowess may strongly resist its 'abandonment'. *To abandon reason might appear irrational until he or she is reminded of the lift to supra-rational state.* Supra-rational perception is non-linear. Like an idea or inspiration, its seed is whole, immediate and complete. It is, by contrast, the rational outworking of an intuition that, as

Einstein for one realised, takes time. It is in this outworking that the subjectivist, like the objectivist, employs intellectual reason, pictures, words and numbers. In fact, in our particular form of terrestrial body intellect is a predominant frame of mind. Humans are more or less intellectual animals. You might, noting this ambivalence, call us centaur-(wo)men. Part man, part horse; half low animal, half high divine; the centaur is a myth whose seed, we'll see, is fact.

Archetypal Greeks.

Perennial philosophy may well have passed, by way of Asia Minor and the eastern Mediterranean, from the orient. Records are sparse but, as far as western academy is concerned, its traditions began in Greece. From there the major bodies of work that have been preserved, through Roman, Christian and Islamic scholarship, are those of Plato (a disciple of Socrates) and Aristotle (a disciple of Plato). These three along with Parmenides, Sophocles, Euripides, Cicero and the emperors Augustus and Marcus Aurelius were among initiated luminaries of the mysteries at Eleusis. These rites lasted for two thousand years until the neo-Platonists and Plotinus in the 5th century ad. Teaching included the Hindu-style doctrines of divine return, immortality and reincarnation/ transmigration (Greek: metempsychosis) of the soul. Such initiates applied intellect to their intuitions and interests in a way that, because it set a course for western and thence scientific thought, identifies them as co-founders with Christianity of our own particular culture. With child-like and at the same time profound simplicity the Greeks wanted to know who they were, where they came from, how wild nature works and how to maximise the life it gives.

For example, what makes things the shape they are? The world is full of shapes yet morphogenesis, that is, the origin and change of forms, still challenges the new physics. Form includes both internal and external shapes of any body. It is neither a vector nor a scalar quantity, nor is it conserved. Both static and vibratory forms interested the Greeks. Fascination with the former took the shape of geometry, of the latter harmony. Plato, for example, took great interest in the mystic school of Pythagoras (a philosopher, mathematician and musician). One of Pythagoras' teachings was that shapes arise from variations and combinations of a couple of transcendent geometrical units derived from the mobile extension of an infinitesimal dot - when it returns to itself (a circle) and when it never does (a straight line). The shape of the world was explicable in terms of numbers associated with the development of these units into spheres, triangles, cubes etc. At root, pure numbers run the show. Pythagoras, more or less thinly disguised, still works in Departments of Mathematics and Physics.

Mathematics does not 'underpin the fabric of the universe' but, like harmonic rules of music, it seems able to *describe* the fundamental operations whence creation works. Thus what causes cosmos is the same as causes, as an aspect of it, wranglers' rules. Divinity would not communicate in any human language or in visual specificities and, therefore, by what medium could nature's causal principles conveyed? By vibration, resonance and thereby music metaphysical - by natural, inner harmony called *Logos* or 'creative stream'? And does this stream continually flow, like air through instrumental stops, through archetypal instruments of form? This couple, organ of creation and creative stream, might comprise the blueprint whence, at root, the music of a dancing world derives. *From actuality to abstract, nobody finds equations in the street or graven on a mountainside. Such abstractions, physics' language, only live in mind.* Thus, as mathematician Michael Heller has proposed, the laws of mathematics might be one way to describe high-level thoughts of God as they are lodged in universal mind. **For Natural Dialectic physics' immaterial equations constitute the information locked in how things have to play; they describe what will be termed 'archetypal memories'.**

Plato (*Phaedo* 97A ff.) shows keen interest in Anaxagoras' concept of universal mind Duality within Unity. Existence is binary but contained within Singular Essence. Essence is nothing existential and yet is the cause of existence. It is 'beyond' yet creates the existential system. From an absolute viewpoint (called Enlightenment) all is quintessentially one.

Dependent on the Essential, Independent Monopole is its issue, the second pole of the Essential/ existential dipole, existence. The existential pole of the primary dipole is itself, as the *yin-yang* formulation illustrates, polar.

We say, therefore, that from the Monopole of Potential issues the dipole of motion, actualisation or changing existence. Change involves relativity and, from the viewpoint of duality (or polarity) things appear separate, different and locked into a process of transformation. Such transformation we call 'nature'. Mind is nature's metaphysic; experience, thought and memory its psychological components. Matter, on the other hand, is nature's physic. 'Eternal factors', such as specks of mass and polar charge called a proton (or three quarks) and an electron, form our outer world's foundation. Let us start to express unity and duality in separate columns.

as a prime mover of things physical; and keen disappointment in the way that, like a modern materialist, the Clazomene proceeds to dilute its causal power away. He also thought shapes physical derive from eternal blueprints, archetypes, *ideas* or 'ideal entities'. These are not abstractions; just as memories really populate your mind so they are real inhabitants of universal mind. The world of physics and its bodies is seen (Chapters 8, 16 and 19) as a reflection by material, contextual variation of the way that such objective metaphysic guides; we call it natural law. Perhaps, as Heller might agree, mathematics is one way you can describe its archetypal operation; and perhaps 'impure approximations' of our physical reality are more or less distortions that derive from 'meta-world perfections'. **Is cosmos not a wobbly kind of variation round the axis of its subtle, unseen, archetypal themes; and our circumstance therefore a case of 'as above, below'?** Later we shall take a closer look. Premier philosopher Plato also, using dialectical discourse, laid a rational groundwork for the higher 'mystic' search after such absolutes as Goodness, Beauty, Truth. To lay down such a trans-mathematical foundation was, he believed, the highest use of the instrument of reason - the preparation for its own supra-rational abandonment. *From archetypal ideas derive the shapes of matter but from Archetype the ideal shapes of mind.* Thus he mapped the inner journey; he oriented soul towards Transcendence, that is, Ideal Life.

Later neo-Platonists certainly saw Socrates as a mystic. St. Augustine, formerly a neo-Platonist, did not so much disavow polar dialectic and philosophy as transcend it through the vehicle of Christian faith. Plato and

Aristotle (one of whose pupils was Alexander the Great) were translated and taught in the great Islamic universities. In Spanish Cordoba, for example, Muslim, Jew and Christian peacefully cohabited. Ibn Arabi, Averroes, Maimonides and Alfonso X absorbed Greek philosophy, with its mystical connections, into their curricula. Complementing contemporary art, architecture, music and science it formed a key strand in the establishment of their 'renaissance with God'. From Islamic climax blossomed out the Christian cultural development we call 'renaissance'. The same key thread was woven into the tapestry of medieval European culture. At the schools of Paris, Oxford and Cambridge (where dons still don the same black gown as an *imam* or *mullah*) Greek thought complemented Christian theology. Modern scientism is forgetful. Our cultural roots are monasteries. Just as Plato is still exercised at Qom (the Shi'ite university town in Iran), so medieval monks debated Aristotle for a *tripos* (three-legged stool) degree. What, though, of spirituality? '*Dominus Illuminatio Mea*' (God is my light) remains, despite scepticism, the motto and therefore at least the nominal inspiration of Oxford University. Christianity and natural philosophy combined. A Doctor of Philosophy came to represent the highest level of academic prowess: and from such philosophy developed modern science and, this time without God, the orthodox scientific/ secular world-view.

Such view, emphasizing its roots in Greek mathematics and philosophy, observes that primitive, polytheistic cultures (including that of the common Greek) deify the forces of nature. It is the objective of science to remove this false veil of divinity and reveal the automatic, godless workings of our world. Gods of gaping ignorance and worship of non-conscious, material events are, of course, entirely separate from consideration of another, older style. The monotheistic worship of a single, non-material non-object - an Intelligent, Creative Being - is promoted by Sikhs, Muslims, Christians, Jews, Buddhists (as the Ground State of *Nirvana*) and Hindus (*Atman-Brahman*). What is *Atman* but eternal self, world soul or essential being? Like a banner, cast across the entrance to the temple of Apollo at the centre of the classical religious world, Delphi, was that injunction 'Know Thy Self'. What Self is that; who should you know? Is it Psyche (see Book 2)? Secular interpretation, by classicists or humanists who grasp at complexities of the Olympian pantheon as if a stick with which to beat all 'superstitious religion', misses the perennial, simple, mystic point of view.

Asia Minor and the Greek west undoubtedly knew the east. Siddharta Gautama lived in the same (5th) century as Socrates and the rise of Delphi. Acquaintance with the Persians was more than passing. Archaeological remains of a Graeco-Buddhist culture that flourished in the wake of Alexander's marches survive in the Punjab; Gandharan coins bearing the heads of Greek satraps are regularly found; and in the Swat valley there flourish the offspring of Macedonian pips (olive trees) and seed (blue-eyed, red-haired individuals). Aristotle's pupil also enjoyed the company of oriental 'gymno-sophists' or 'naked contemplators'. This seems to be a reference to Hindu holy men, *saddhus* who lived then as now. Strabo (Geography Book 15) relates how the Greeks were surprised and fascinated by the similarity of 'gymno-sophist' teachings to Plato's. Oriental mystic practice thus resonated with the same Platonic and Aristotelian tradition that was eventually to underwrite western science!

The Magi (from which our word 'magic' is derived) were Zoroastrians whose creed expresses fundamental duality or polarity in nature. A wise triplet followed the star to Bethlehem. And did not the Silk Road run from Buddhist China to Antioch, Tyre and Constantinople? The influence of the east travelled, usually bearing gifts not war, westward. But never say the traffic was not to-fro. The founder of influential neo-Platonism, an Egyptian called Plotinus, joined a military campaign eastwards to learn more about the ancient philosophy of wise men and the naked contemplatives known as yogis. He, perhaps the closest of all Greek philosophers to Natural Dialectic, ended his days teaching in Rome. Note, for later reference, his 'One' (Plato's 'Good') as the ground of existence; which latter, in both metaphysical and physical aspects, may be regarded as a series of concentric rings expanding from a Central Axis. Each ring stands in a relation of dependence to the one nested within it; and matter, the limiting case of creation's outer boundary, is represented by the peripheral circle. Matter, a far-flung circumference, is the antithesis of Universal Centre-point.

Either way, Science and the Soul is inclusive. Natural Dialectic embraces both east and west. However, the mystical angle in original Greek philosophy has by now been so revolved and objectified by an ethos of 'scientific rationalism' that its mind-set has (like that of Christian cosmology) been culturally sidelined. Indeed, it is notable that materialism spotlights, far out of proportion to the whole panoply of philosophers, the influence of a handful of minor but apparently like-thinking players (such as Democritus, Leucippus and Epicurus). Nor were Greeks (except a few fragmentary and/or debatably-interpreted pieces from, say, Empedocles) stomping the *stoa* on behalf of random-based notions such as Darwinian-style evolution. The reverse - chance as a cause was generally derided. Socrates is portrayed by Xenophon (Memoirs of Socrates: Book 1 Part 4) as a clear proponent of intelligent design. Later Cicero (Nature of the Gods: Book 2, 93) expounds a similar anti-chance, teleological view. Order was the order of their cosmic day. Plato, Aristotle and the classical 'top table' are still studied but materialism's *zeitgeist*, understandably, condemns and sweeps their import to the bin of missed-the-point analysis. Is that any wonder if your perspective's diametrically opposed? What relevance has holistic whole if, popularly overweening, you deny there's any but material part? More ancient than Greek, the mother tongue of the yogic 'gymno-sophists' was Sanskrit; from this a raft of Indo-European languages derive. Their philosophy shared common points with western thought - from which developed the naturalistic offshoot of science. Later a few yogic terms will help clarify the dynamic of Natural Dialectic.

To summarise the last three sections, there are two directions that your focus of attention takes; and continually, between their poles, various degrees of concentration fluctuate. *From these directions are derived a fundamental couple, two main implements of knowing; and, in turn, from these two anti-parallel mind-sets, pillars that support all human faith.* **Materialism is a mind-set; and holism too. What are these elaborations, why do they oppose each other, which is quintessentially true?**
Two Pillars: a Dialogue of Faith.

Would you judge the Greeks were predominantly philosophical (like Socrates) or scientific (as with Aristotle)? Had they bias? *Scientia* is a Latin word meaning 'knowledge'. Science *is* knowledge. In our time, however, the word has been reduced to describe a particular (objective) kind of knowledge and its experimental method of discovery. This biased appropriation, this exclusive definition of information applies to what is physical alone. Neither the Greeks nor our own, closer forefathers made such division, a divisiveness whose materialism presently sires misunderstanding and ignorance of metaphysics. **This book, Science and the Soul, is a dialogue *between* the two pillars of faith; but it does not end up, on the fence, *in between* them.**

Why only two? What fundamentals does the Dialectic choose as 'anti-parallels'? What is each world-view's First Cause? Is it an immaterial or material form of 'deity', a purely conscious or non-conscious one - in which latter case creation would be due to purpose or an accident? Which does the evidence fit best? Or, between the contradictory pair, is there accommodating ground? You may interpret evidence with atheistic or theistic eyes but what about a 'neutral' or unbiased lens?

If the latter, 'neutral' lens were not correct why bother with its blur? If it were then neither of the contradictory pair would be. So what's neutrality's foundation? How could creation's fundamental unity fall *in between* the obvious pair - unless its 'deity' were polar and, at once conscious and unconscious, its First Cause was some form of duality. But such duality seems inconsistent, self-refuting and thereby irrational. Reason notwithstanding, it might still be true. According to theistic faith both immaterial and material parts of universe derive from a holistic Cause - but Chapter 22: Theories of Accommodation will reveal the weakness of theistic evolution's game. And some 'holistic' atheists accept, as well as a material, an immaterial element exists as well. Abstractions such as laws of logic, mathematics, information, innovation and, perhaps, even purpose are allowed. However, do innovation, information generation and intentional behaviour reflect reflex, atomic ways? Is such abstraction as mind deals in seen in mindless chemistry? How could immateriality 'emerge' from matter? In short, are critical abstractions, such as standards of reasoning, dependent on things of the automatic, material universe or does it depend on them - with metaphysic in control?

'Of course it's not,' an atheist cries. Mutations and selection naturally evolved the conscious mind and rationality. Abstraction's just a complex twitch of brain; and, since each brain differs from all others and in its life-time even changes in its own conditions, cerebral sets of logic shouldn't ever be the same! Indeed, why should eternal abstracts, like the laws of logic, perfectly align with how the world's created and its 'reasons' run? And how, if not from eternal matter then from an abrupt, initial accident, might cosmic regularity and uniformity appear intact? While cosmos ever changes why should mathematics and the laws of nature always stay the same? Why shouldn't each evolve? And, finally, how does the 'holistic' atheist justify his arbitrary inclusions of some kinds of immateriality yet, just as arbitrary, deny those such as higher states of consciousness, universal mind, archetype or a transcendent nature of divinity? Capricious pick 'n' mix is, when it comes to cosmic shopping, plain irrational. 'Neutrality' from either camp does not stack up. More straight-forward is to invert, right-side up, and thus conclude that physical reflects the psychological, material mirrors immaterial and thus first, fundamental cause is not unconscious but eternally aware. That would, though, leave metaphysic in control.

Thus, returning to a choice of two, are the natural outlooks of a scientist and saint mutually exclusive; are they incompatible or is there any bridge between them? Could some dynamic conscio-material logic, accurately reflecting all reality, combine the pair?

Both camps unveil what is already here. Physical and metaphysical, both seek substance under superficial show. They seek the principles on which mundane details and circumstantial data are based. Although their directions of focus differ, both seek revelation. *Both want the truth. Both seek it with concentrated rigour.*

One leads a quest for inner light. Who, though, would pay to test what can't be physically tested? You can't have a professional saint but science is a real professional possibility! The outer quest casts light in terms of measurement and understanding of non-conscious affairs. Thus the first, critical step towards a naturalistic world-view is taken when a scientist seeks to satisfy his curiosity. Can an idea be translated into experimental test for truth? If not, chuck it out. Adopt objective methodology. Don work-goggles which exclude any non-materialistic possibility.

However a clear, initial distinction needs to be drawn between physical science and materialism. **It cannot be overemphasised that materialism is a special case of metaphysical philosophy - one that denies the metaphysical.**

There are, moreover, overlapping brands of materialism. *Naturalism* (Natural Dialectic's *PAM* with *PCM* - see Glossary) involves the notion that life is the product of accidentally generated laws of nature acting over time. This view, labelled *bottom-up*, will be dealt with throughout Science and the Soul.

Science is performed empirically. Its methodological naturalism may, for the sake of experiment, pretend that no Elemental Intelligence exists but cannot, at the same time, justify the treatment of pretence as truth. *Empiricism*, however, is the notion that only what is physically observable and thereby testable is acceptable as fact. What is not material is not a fact; it is imagination. Of course, eyes see and sight's a sense. It's therefore common sense that what we cannot see (or sense otherwise by body or by instruments) does not exist. With what, however, do you sense thought, qualities, morality or feeling? How does an empiricist account for the laws of logic, mathematics or his own rationality? None of these belongs to the observable, physical world. You can't find them anywhere. Indeed, how does an empiricist know that all knowledge is gained through information since such an observation is clearly metaphysical? It is not something he has ever seen! Indeed, how does he know, by his own standards, that he knows anything? How can *anyone* know anything if knowledge cannot be observed? In this way the world is unknowable and thus reduced to senselessness - which is clearly nonsense. And by this criterion the laws of logic can't exist and thus neither materialism, atheism nor any other -ism could ever be tested. If, moreover, an empiricist should by any other method 'test' or 'prove' his view, this 'proof' were self-refuting. Even if it were true he could never know it (how could he empirically observe its truth?) and thus never hold it proven! Empiricism cannot self-attest. It cannot stand the check of logic. As a doctrine it abides confused.

Another form of materialism, *relativism*, holds that no absolutes are possible. No ultimate presupposition, no first cause or cosmic potential is possible.

However, to assert that 'everything is relative' is in itself a 'proposition absolute'. You may claim, 'It is absolutely true that final truth is not absolute'; but if such relativism were absolutely true then it were, by its own admission, false. If it were true it would be false! Such doctrine self-refutes. Why, moreover, just because you say so, can't final truth be absolute? For example, you would have to know it all to prove that God did not exist; but isn't 'God' the label that we give to such omniscience? You'd have become God just to prove that He does not exist! Such proof now self-refutes. Worse still, if you are 'God' then how does He make way? If you are right then you, a separate entity, could not exist as well! Relativism, whether it applies to matter or to morals, is an arbitrary, arrogant and, in self-refutation, an irrational creed.

Thus, if it is insisted that science be identified with materialism the best that can be said of any interpretation of objects or events is that, *within the limited context of science*, it is reasonable. Not necessarily complete. Less than satisfactory, of course, is the insistence that all forces and objects are material; and the subsequent 'logical' but prejudicial identification of study taking any other line as 'pseudo-science'. Indeed, Sigmund Freud and others were soldiers in an uphill battle to establish the study of mind (psychology) as a member of the 'hard' club of objective sciences such as physics and chemistry. The price was a reduction of perspective to the statistical observation of behavioural patterns and an emphasis on animal motivations, evolutionary explanations and neurology. In the reduced, wholly objective paradigm where 'science = materialism' subjective science is not 'science'. All knowledge is, however, essentially the gift of consciousness and therefore subjective; and such subjectivity is, in the context of faith, based on one of the two pillars. Can materialism arrogate the whole truth? Is heterodox holism delusory, fraudulent or deceptive? Are its methods (including the experiment of meditation) always misdirected, its discoveries invalid or its wisdom void of truth? Philosophy means 'love of wisdom'; science ('knowledge') is the name of its material fraction. Why should physic, ousting metaphysic, make a cuckoo's nest of truth? Indeed, though science sports materialism's badge many professional scientists have been and are today holistic when it comes to personal philosophy. Why ever not?

Material science is, like the phenomena it studies, in an incessant state of change but may eventually deliver a complete description, in changeless mathematical and genetic principle, of physical operations. Such description may well present decisive evidence for best interpretation according to one of the two basic pillars of faith. Science *per se* is not, however, a good arbiter in the court of origins; it cannot judge fundamental truth. **Indeed, subjective science sees 'scientific materialism' as a subset of the whole truth, not wrong but only partly right**. It conceives of the physical world as a superficial exterior, an outer shell or lifeless crust. It applies techniques to examine the interior, metaphysical motivator - mind. It attempts to cool perception of sensation, overcome the welling of desires, cut reason's calculations and thereby calm down the waves of thought. Such dissolution of an intellectual kind of reasoning is not the low-grade or inconsequential version of irrationality that all philosophers despise. Irrational sub-reason wallows in dysfunctional logic, emotional turbulence or, in the case of thoughtless reflex, instinct. Refined super-reason, on the other hand, is a goal not so much to be explained (i.e.

confined in reason's various walls) as to be experienced. Transcendence seeks to understand what principles command the outward show but its peak experience, enlightenment, is, as brilliant Buddha underneath his bo-tree clarified, the ultimate form of supervision.

Be transcendence as it may, the two directions of focus have, at the extremes of materialistic logic and holistic super-reason, given rise to our ancient philosophical divide. So let's rephrase. There exist, *on the one-hand*, materialists whose tough-minded, no-nonsense rationalism seeks to fully understand the nature of objects. Such worldlings seek to *raise*, as it were, the inanimate universe of objects into the realm of reason, reason which understands what they are, how they work and (within its own mechanistic terms) where they came from. Because every cause is conceived of as material, a mechanistic universe is considered *causally closed*. There is no metaphysical input. Even prior void is of material kind. Matter is its own cause. And from 'cosmic autogenesis' it tends to naturally organise itself into more and more complex 'systems'. In this view mind is an effect of neurochemistry and mental states have no reality apart from matter. Indeed, mind is just an 'extra layer' of such matter when configured in a complex manner called a brain. The cause of life (if mind is where you live) is, therefore, material. Mind (the part that knows and, knowing, we call 'live') is no more than a secondary, dependent effect called, in the jargon, an epiphenomenon. *No less than Nobel prize-winners have brazenly reduced your 'real' identity to nervous molecules!*

Perhaps, on the other hand, they're less than half right. Can neuroscience or genetics ever tell you how to live? Are hormones all you feel and fluctuating brain behaviours other than *reflections* of your thoughts? Surely 'secondary' mind is primary to your experience of everything? <u>Indeed, the idea that psychological phenomena are explicable in terms</u> of chemistry <u>alone is purely speculative</u>. It may be inferred, as may the reverse, but neither inference is physically proven nor, in materialistic terms alone, provable. The possibility of a *causally open universe* (and therefore a metaphysical origin) exists. Supra-rationalists seek clarity and understanding by eliminating their own interference in the process of understanding. When, therefore, they explain what has been revealed they bring *down* their knowledge of principles (as opposed to bringing *up* objective data) into the realm of reason. The fact is, supra-rationalists say that ultimate truth lies beyond reason and physical science.

The split between materialism and holism, with its momentous conclusions, can be identified as early as Plato and Aristotle. The former tended towards idealism and his pupil, called by some westerners 'the first scientist', towards an analysis of nature. Yet neither man rejected metaphysics; both believed a common logic, recognisable by the intellect, runs through all aspects of god-given data. An extreme form of philosophical schism was neatly encapsulated in a curt rejoinder between two English philosophers. Said Bishop Berkeley (as he passed the rationalist Thomas Hobbes in the street one day), "No matter." Hobbes' quick riposte was, "Never mind!"

Same problem. A theologian can no more scientifically prove the existence of big G than an atheistic anti-theologian falsify it. Both pillars of faith are essentially metaphysical. So how, with respect to origins, can you know for certain which is true? Is final, absolute truth a chimera? Is Ultimate Science (or

Complete Knowledge) knowable in human terms or not? In 'objective' investigation matter predominates and the observer is theoretically but not, as informed experimenter, really excluded. In 'subjective' or psychological investigation consciousness predominates and matter matters less. *In one case the experimental laboratory is the universe and in the other it is the mind that comprehends the universe.* Both ways employ strict methodology and take count only of repeatable results. In each case these results generate self-consistent, logical and predictive world-views. *They may reasonably be called physical and metaphysical science.* Can their truths unite?

Physical science deals in tangible facts; metaphysical, dealing with the immaterial, has no recourse but to resort to metaphor for explanation. Man to man imaginations differ. For such intellectual reason this book studiously avoids the 'theological'; it eschews the various mental constructs known generically by the word of 'God'. You will find this word excised. G-word apart, however, both types of science seek as far as possible to eliminate the noisy, opinionated *ego* that would always strut centre-stage. Sceptical materialism and the rigour of physical science, for example, are indispensable filters that strain unverifiable speculation; they are blades that prune exaggeration, extirpate superstition and eliminate, as far as possible, the weeds of cultism, occultism, charlatan, crank and fraudster. Reason can flourish unchecked, free.

Can materialistic rationalism weed out any other kind? We have seen that, with its science, it cannot. Is there, after what is physical, nothing left? What, then, is nothing physical but metaphysical? Is not exclusion of an immaterial metaphysic therefore unreasonable, arbitrary and thus irrational? *For example, the whole logic of Natural Dialectic devolves from the simple premise of duality - two fundamental components of existence - derived from a single prior essence.* Metaphysical consciousness interacts with non-conscious energy; and the nature of singular essence is our very point. Is it physical or metaphysical? Much depends upon decisive truth.

Both metaphysical and physical science emphasise practice and trust only in the results of experimentation. The former seeks, within the myriad transformations of mind, pure consciousness; and the latter, within the myriad transformations of matter, the nature of pure, mind-free energy. Both parties push to the limits. In both cases many observations take place in realms inaccessible to the normal senses. While they equally seek open-mindedness, both also claim truths that sound fantastic and extreme to an egotistical, sense-based, mid-scale inhabitant of cosmos. As physicist Fritjof Capra said, "Neither is comprehended in the other, nor can either of them be reduced to the other; but both of them are necessary, supplementing one another for a fuller understanding of the world".

outward	*inward*
objective	*subjective*
sensation	*contemplation*
atheistic tendency	*theistic tendency*

Yet these two perspectives derive from nothing more than a difference of focus. One has focused on objects and the other, as far as objects are concerned, has vectored in on nothing. One eschews sensation, the other embraces and enhances it. Perhaps, therefore, neither alone relates the truth, both in principle and detail, about everything. Perhaps, like sensation and contemplation, as complementary truths they make up a whole. Better view them, like *yin* and *yang* as complementary, interactive parts of this whole universe. In other words, perhaps the ancient divide - between reductionist materialism and holistic metaphysics - is illusory.

Except as regards one aspect - *origins*. Only one fundamental can be absolutely true; the other may be relatively so. Only one ultimate can provide the most coherent set of answers to our interrogation of the world. A materialistically objective perspective brings information into the field of reason (or processing) from 'outside' or 'below'. It denies, which the holist avers, that there exists a metaphysical 'inside' or 'above'. Thus, for a reductionist, things originate by 'chance' within the fields of time and natural law; their origin is, as with laws of logic or the natural laws themselves, material and therefore a non-hierarchical, evolutionary scenario is forced. It becomes inevitable. For a holist the reverse is true. The difference is critical. Pressed to their extremes the modes, objective and subjective forms of creativity, have evolved into two ancient pillars of faith, two perspectives each of which lays claim to be the whole truth and nothing but the truth. Each fervently believes its own perspective's right. As we'll see, each stakes claim to a complete explanation of the cosmos - physical science with a *TOE* (Theory of Everything - but only anything material) and metaphysical science with a *TOP* (Theory of Potential - that includes both immaterial and material elements). Yet from *TOP* to *TOE*, from head to foot, their perspectives are polarised. Inverted. One looks outward, the other inward. They enjoy the same world at the same time but, especially regarding the historical issues of origin and moral issues of fundamental goal, see life from opposing sides. They can carry on in working harmony but their underlying views are irreconcilable. *Nor, in this tension, can both be wholly right.* <u>Which is the ultimate criterion against which your own truth is measured</u>?

materialism	*holism*

In short, there are ultimately only two ways to view the universe. **Either, materialistically, everything is composed of non-conscious matter/ energy or, non-materialistically, everything is not composed of such non-consciousness; there exists a conscious, immaterial element. <u>The latter is non-materialism's single, simple axiom (of which materialism's is the simple null hypothesis)</u>. Accordingly, creation either originates without reason (atheism) or planned reasonably (holism/ theism).** Its origin and construction, perhaps through a hierarchy of stages, is either conscious or not. The information, rule or regulation that shapes the universe and its creatures derives from an intelligent *or* a material source. From such division arise the two pillars of faith; we call them *materialism* and, because spirituality is a word that evokes such emotional and conceptual differences, *holism*. From each girding arises one of the two contrary citadels of philosophy, psychology and their derivatives - law, politics and religion.

exclusive	*inclusive*
empirical/ inductive	*logical/ deductive*
bottom-up	*top-down*

experimental	*axiomatic*
physic	*metaphysic*
physics	*mathematics/ logic*
science	*philosophy*

The trusting adoption of one of these opposing principals either consciously or unconsciously colours, in a way Natural Dialectic will develop, all subsequent thought. While the implements, methods and primary assumptions of the two pillars of faith are different, so are their conclusions; and thus their fundamental purposes. One seeks to develop models and exploit the world; the other to realise, experience and unite with the highest Natural Truth.

The sacred and profane? Since the basic perspective of an individual or a community produces its behaviours the issue of a correct choice of faith is critical. Are there ways in which the two pillars can compromise, complement or both be true? Paradoxically, could both be simultaneously right? Or only one? If so, which is right? Which wrong? **It matters since to choose the wrong way is, as all detectives know, to err in your interpretation of the clues all down the line and reach, at last, a false conclusion.**

The Objective Pillar

Philosophy means 'love of wisdom'. Of course, there is a philosophy of contemporary science but its focus is on the arrangement or order of physical phenomena. Its premise is physical, not metaphysical. Indeed its first, major reduction is to enter an exclusively circular mode of reasoning. Having *a priori* excluded from consideration any metaphysical or spiritual element of existence, it may then systematically deny such possibility! A tautology! It is a closed loop that, however you pass off the presence of a mind, does not transcend inanimate matter. *But to decide something is true does not mean it is.* In open court any evidence is examined from the perspectives of both prosecution and defence. Is not **materialism**, having debarred one half of the equation, having reduced its vision to the use of a one-eyed lens, a single perspective on anything it examines, like a 'kangaroo' court? **It is indeed one-eyed philosophy.** This is because, ideally, science is 'objective' in its quest. *It examines objects that are 'not-self'; it interrogates materials 'outside-me', that is, outside my mind.* Indeed, is this one-eyed philosophy correct? Is conceptual removal of your mind from all equations and experiments for real? Is it more than a pretension?

Mind feints withdrawal from its own conceptual world yet that same world is painted, by the senses, in that mind. You are absented from such picture but, at once, *are* it; and the canvas, sense and thought, aren't physical. Nervous process cannot generate a sense of colour, sound, taste, intent or value. Where are nerves that understand, say, 'yellow'; atoms, even billions of them, never feel or understand a thing. Yet, purpose-built, all the kit that registers oblivious events needs sense to register the measurements and thought to order them. What, according to 'the understand-ability of nature', do they mean?

Which picture of the world is, therefore, real - yours or lifeless objectivity beyond your comprehension? Mind is the only all-world stage but, as its immaterial self, appears nowhere; and, if conceptually denied as well, no wonder that with personality removed materialistic methodology evinces lack of purpose, meaning and all feeling. Such detachment is a figment left at the laboratory door. *Indeed, such shortcoming, along with already-noted reservations concerning scientific empiricism, needs be clearly borne in mind.*

Science as a concept does itself involve such immaterial categories as honesty, integrity and truth. Yet another kind of background frame of reference, *scientism*, is a creed as one-eyed as they come. Science generally treats 'as if' there's nothing but material; then, however, some get carried to believing it. They're surprised you dare suggest that this belief is not the only 'rational' one. There is, such scientism claims, no more for sure; everything is shoe-horned to material fit; and what we don't yet know or have the instrument to sense will one day also be explained this way. *Scientific atheism is a promissory faith, a chit that's served in hope that cosmos and its life-forms will be only physically explained.* **It is a creed dependent on appeal to three main forms of plausibility - that, behind an opaque veil of eons, huge complexity or vast improbability, anything can happen.** *Impossibility can be converted into certain actuality.* This faith's patron saint, a sagacious elder with a beard and seat of honor on the scientific pantheon, is called Charles Darwin. Who was it, after all, suggested life evolved from matter (that is, non-life) and thus could not have sprung from elemental life itself? Such pillar of non-consciousness is set, as explanatory theories tend to shift, in shifting sand. It is, as we shall come to clearly see, no more than faith hung out and strung along material gaps and a great intellectual guess - the guess that no distinct and immaterial element exists, that information isn't an intrinsic, independent part of nature as a whole and subjectivity is only nerves! Is this the objective message life and cosmos really sends? What, again, is *meaning*?

If 'natural' is equated with what's physical can there be super-nature or sub-natural forms? Is matter conscious or not? Is conscious *mind* a natural thing; or could what you are thinking with be metaphysical and therefore, by scientism's definition, unnatural, sub-natural or super-natural? The laws of physics don't allow for immateriality. If, therefore, scientism discounts metaphysic as a fantasy you must be, in thinking now, nothing more than mental fantasy. Your experience counts, objectively, as physically unreal. Is your mind real or not? Are your thoughts atomic, electronic or ionic; or illusions conjured by electromagnetism? Who or what, objectively, are you? And if there exists, relative to current level of awareness, subliminal awareness (called sub-consciousness) then why not also higher, super-consciousness? If mind were real, natural and metaphysical then 'naturalistic rationalism' could have spun misunderstanding simply by the use of verbal prejudice.

Doesn't one tier of existence (matter) seem simpler and more obviously sensible than complex ones including tiers of metaphysic or vague concepts such as 'spirit', 'soul' or 'psyche'? Such materialistic paradigm includes a notion of the universe as a mechanical or field-based system made of elementary particles, the human as a complex, soft machine, life as struggle for survival and a topping faith in progress through technology. Its adherent's chosen to believe 'objective reason' is the only winner on the road to truth. Is the winner therefore naturalism or empiricism? Is it relativism, scientism or exactly what? Whatever, the ideal is pure detachment, revelation nature's pristine logic and the faith's objective pillar a religion (or, if you prefer, a non-religious religion) called materialism. Reductionism, humanism and scientific atheism are the sort of sects that hug together under this umbrella. *For many intellectuals such exclusive, saturated materialism is an accepted, unexamined and habitual way of thinking and, thereby, framework of 'objective' interpretation and explanation; yet in the last analysis, as we shall see, all the weight of such presumption presses on the crutch of hope and faith in guesswork. <u>Why is such a crutch less brittle than an atheist claims the theist's is</u>?*

What is it? What is it made of? How does it work? Although child-like in their simplicity these questions drive physical science. The honourable, analytical intention is to lay ghosts of superstition, evaporate the fogs of rumour, flatten tall tales and replace metaphysical considerations with tangible and measurable truths. Therefore classical science, as a first premise, believes only what its instruments can register. It operates in an involuntary, purposeless 'clock-and-ruler' world. Objectivity is in this sense a function of sensation but sensation that excludes from its account the voluntary, purposeful nature of an experimental set-up or its system's personnel. Having observed and measured aspects of a phenomenon in such 'detached' and 'non-subjective' way mind's aim is to reduce it to an admirably objective set of mathematical equations, hopefully a 'neat' or 'elegant' one or, in the more complicated case of biology, at least concordance with a reason and not chaotic inexplicability. Fact stripped of all fiction rules this admirable sort of day. An experimental hypothesis may, after sufficient confirmation, be termed a theory. If such principles can be used to accurately predict outcomes or make machines the similar words regularity, norm or law are used. That is, without known exception, something seems to be the case. Thus one discovers how things function but not necessarily, it must be cautioned, how either the material universe or any living organism came to be. First causes bear 'with caution, danger' signs.

What about the mind that churns all this objective thinking, a subjective factor based on consciousness? Easy! Everything is actually determined by constraints of 'natural law'. It is just electrochemistry of brain that 'generates' a mental person. ***PAM*'s the gambit.** At first strike and in accordance with The Primary Axiom of Materialism reductionism treats its consciousness as a non-conscious material - perhaps something like an image on a television screen or a state of brain. Forget about subjective context that surrounds electrons making patterns on the screen; forget about the television's information content - makers, broadcasters and watchers. Just as a TV program could be framed as only charge-in-motion that depends for its existence on a set so thought is framed as consciousness-in-motion and, prior to thought, there needs a brain, a body. Mind depends on body. Matter is the primary 'stuff'; it comes before. There is, in this view, no creative agency. **Such is central dogma. From hydrogen to human, from molecules to man, from prime simplicity materials evolve complexity; last of all and only from complexity called brain is mind, perhaps like a gas from chemical reaction, noiselessly evolved. Indeed, the theory of evolution is the central, essential corollary (see *PCM*) of such materialism**. Mind is a complex pattern etched in matter; thought is an extrusion that's dependent on a brain. No doubt, when cutting to the chase of intricate 'design', the plainness of a one-tier, mindless universe has no alternative but chance, governed by describable but inexplicably created statutes of necessity, as its 'creator'.

The reason for such laws is chance. Thus, by this chancy form of reasoning, the first and fundamental 'cause' of anything, including the apparition of reason itself, is non-reason - even mind, whose nature is to order, rose from unpredictability. Chance is a lack of cause, at least a lack perceived; it is not so much a reason as a missing link. You cannot, therefore, press-gang 'therefore' or 'because' into a reason for the origin of such a strange illogicality. Is this story, based on primordial emptiness of reason, void or valid? Is such objective 'reason' why you came to be here orderly and working right?

It takes madness to deny the obvious. Did Stephenson design 'The Rocket' or Edison a light bulb? The obvious presence of creators with subjective, scientific minds is one thing but a Theory of No Intelligence specifically denies intelligence in natural designs. Therefore the 'anti-church' of scientism typically resorts to ridicule *ad hominem* or similar 'political' attack whenever an appeal is made to an intelligence inferred from 'laws', that is, the infrastructure and designs you find in nature. A Theory of Intelligence strikes at its heart of thought - so what do you expect? If senior sophisticates disdain it as 'naïve' their juniors patronise in suit. Indeed, *PAM* will not allow such unintelligence, such heresy; the law of naturalistic laws is 'anything except such Theory'. Order, information and a natural plan amount to teleology. Such rationale is totally proscribed by those for whom materialism's is the only kind of reason; and excoriation reaches fever pitch when debate involves the obviously rational codes and mechanisms that inhabit every cell in all biology. Darwin is, of course, materialism's answer and relief. You can simply not have understood the power of natural selection and the *PCM*. Your imaginations therefore are not 'scientific'. Take a dunce cap and repent!

mass	*energy*
outer appearance	*inward reality*
effect	*cause*
objects	*driving forces*
superficial overlay	*substrate*
mass-dominated level	*energy-dominated level*
particle	*wave*
classical	*quantum*

How did materialistic rationale develop? Scientists such as Bacon and Galileo decided 'to place knowledge beyond belief' by practical experimentation. Within this frame at first only solids, liquids and gases were 'real'. Then atoms, particles and fields (see Glossary) hove into conceptual view. As materialism was modified new perspectives were added. Perhaps the major principles of equivalence are Albert Einstein's equation of matter with energy and gravity with acceleration. Without a doubt classical determinism has made a profound, positive departure into a 'dynamic', modern era. Hard, mechanistic reductionism has mellowed until conventional theory now seeks to explain everything in terms of the quantum mechanical properties of excited fields and, by chaos theory, 'inevitable self-organisation' without a clear view how embodied life's ingenious and coded mechanisms ever 'self-assembled' in the first place.

isolation	*interaction*
division/ analysis	*union/ synthesis*

serial/ broken up   *immediate/ integrated*
detachment   *relationship*

Such advance has led, we've seen and will more clearly come to see, to a greater awareness that material objectivity may be a form of illusion. An observer affects his observation. His method of observation, its consequent results and his interpretation of these are all affected by his preconceptions. Might you even glimpse an element of man-centred 'empathy' a-creep?

If the oriental outlook is synthetic then the west's is analytic. Scientific analysis breaks objects down to their component parts and, still further, fundamental constituents. Its powerful focus isolates a factor from the complex real world and, by experimental artifice, observes cause wreak effect; then, recognising patterns, the seeker after truth abstracts collected data into abbreviated formulae. These formulae can powerfully transform a myriad of factual details into principle. They describe those constant themes behind the multitude of changes we observe, the ones we call the 'laws of nature'. The process all depends on automation, that is, the involuntary character of energy.

In the hierarchy of reductionism biology is seen as based on chemistry and chemistry on physics and its laws. Indeed, you can reduce all things. Isn't each and every body basically composed of forces, particles and atoms? Aren't bodies biological just 'bags of complex molecules'? Aren't you twopence worth of common chemicals? In this view nothing is more than the sum of its parts and, in fact, these parts are the whole substance of any event, the fabric of every object, the very basis of material being. Reductionism understands a whole by an analysis of parts. *A whole, even a whole machine or body informed by purpose and created as a unit of many parts integrated in a complex way, counts for no more than the sum of those parts.* This view has become prevalent in the discipline of particle physics; and in the disciplines of biochemistry, genetics and molecular biology to the point of eclipsing the morphology, instinct and purpose of a whole organism. "Come on down", cajole some Nobel prize-winners, "to the fundamental levels where we work. Can't you see life's game is chemical? Aren't chemicals alone its lowest commonality and, not least, the substance of its information pack, the *DNA*?" Genes are thus, if there's no mind without a brain, promoted to 'top dogs', creators and the bosses of communicating bio-matter. What other species of intelligence is, materially speaking, possible? Life is essentially its genome, an array of molecules that 'self-assemble' to evolve what coded sense there is by 'accident within necessity' (necessity is a constraint whose other name is the deciphered laws of physics).

Does such ritual reductionism actually explain the properties and capabilities of different levels of complexity - especially when it comes to systems, purpose and technology? Does the *arrangement* of your molecules not count? Is *operation* no more than a sum of parts? Whence (Chapter 8) does even the arrangement of those parts derive? For Natural Dialectic rules of games are a conceptual factor, no less fact for that. Cosmic agents (forces, particles and atoms) play creation's game; and each type of life-form is construed a complex statute - law incorporating integrated clauses that biologists call homeostatic norms. In this case if there's any more to life or to machine than molecules then, sir, yours is a '*reductio ad absurdum*'. It is naïve and, especially when you factor in the mind, as false as it's absurd.

Such a pattern of 'objective' thought does, though, pervade a scientific mind-lock and obscure the possibility (indeed, the fact) that things are, in percept and concept, more than sums of separate, simple parts. This species of objective philosophy, which lacks a view of information as essentially intelligent and, therefore, an intelligent view of information, is also contrary to dialectical holism. The re-discovery is that the 'whole' of a system may be, in fact, greater than the sum of parts available for a material analysis; that there exists a coherent, self-regulating interconnectedness of all things; and that humans, therefore, comprise but a single strand in the psychology of life on earth. Moreover the roots of hard, divided substances and their predictable motions have been evaporated into 'quantum ontology', that is, a concept of nature built from events *and* informative signals that create tendencies for their occurrence. Mind affects matter; not only in respect of conscious choice but of universal mind, whose archetypes might be described as 'a holistic informative structure representing tendencies for physically real, localised events to occur'. This, the quantum state of cosmos, sums to archetypal possibilities. **Such archetypes the Dialectic calls 'potential matter'.** Such a psycho-physical perspective allows that conscious choice may, both in individual and universal cases, set up initial conditions for action, that is, for creations of definite kinds but indefinite appearances.

Such 'quantum epistemology' satisfies the intuition that choice *does* enter into at least our individual worlds. Not only is the mind-matter dichotomy relieved; classical determinism is reduced (nature 'chooses' in a given context from fixed pieces and their set of possibilities) and a sensible, immaterial place for a field of consciousness (whose aggregates are thoughts and forces are intentions) is secured. Without such an informative approach to reality science, lacking any cogent explanation of how mind and matter interact, remains radically unsatisfactory. Its objectivity alone is insufficient, incomplete.

For example, confronted with a machine many materialists will deny there is mind in it. Exhaustive tests can't find a single thought, intention, choice or even moral element. Do you recall the case of television sets in their subjective context? Are not such sets much more than just the sum of off-the-shelf, screwed-and-glued-together parts? Does a machine, in working to fulfil its maker's purpose, not transcend what it is made of? Does, for example, a space shuttle not transcend its pieces? Does the meaning of a letter not transcend ink, paper, single letters of the alphabet, individual words and even sentences? None of these is anything without its purpose and, therefore, a meaning. Machines all work with coherent and cooperative parts - to serve a purpose. Mind, great chooser, cuts the set of possibilities to its desire. If not the cosmos could at least a human body fit this reasonable way of making sense of what is certainly its purposive and specified complexity? **For holism, therefore, an extra, immaterial part of any whole is information. This, on which the rest depends, is the purposeful design, development and arrangement of its parts, each contingent on the other's operation; and, beyond even bodies, is not the experience of life itself the root of meaning in a scripted universe?**

Lacking account of this immaterial, objectively inaccessible ingredient reductionism is only applicable to the oblivion of physical forces, objects and events. In the eyes of holism it is, therefore, a 'flat-earth', one-dimensional or one-tiered perspective on the world.

A battle rages, we shall see, between Theories of Intelligence and No Intelligence as chosen explanations of our origins. How comes it these opposing theories are both held by sane men of the highest scientific calibre and academic credentials? Could the facts be left to speak?

The Subjective Pillar

Is there no more to life than meets the eye? Experience, emotion, purpose, knowledge, dreams - that's immaterial life. *The subjective part is seen as 'self'.* How well does objective science deal with subjectivity?

energy	*information*
matter	*immaterial mind*
objective/ outward side	*subjective/ inward side*
physical	*metaphysical*
non-conscious dimension	*conscious dimension*

The subjective pillar of faith is called 'holistic'. As well as the physical it includes the metaphysical or psychological element. In this way materialism is a subset of its totality, a part of the whole truth. Where matter's underlying factor is identified as *energy* that of mind (and therefore metaphysic) is known as *information. Only with the reconciliation and coordination of these two complementary factors can a whole picture of yourself and existence begin the struggle to emerge. It can be elaborated in terms of these basic, polar coordinates alone. Creation is a dipole.* **The concentration gradient or, if you like, the field of influence of this dipole has already been identified. It is a conscio-material gradient whose effect is called existence.**

A thought is not, as we've begun to see, physical; nor is an emotion or sensation that you feel. What precisely is a 'physical experience'? A complicated but material twitch of nerves? *What if an immaterial element were consciousness?* **This might not be a fashionable assertion but since when was fashion arbiter of truth?** Such cosmic element is not a quantity but, at the root of knowledge and all information, a fundamental quality; it is not objective but subjective; information's centre is not physical but metaphysical. The ground of all experience is conscious self; subjective self is at the heart not biological periphery of life but science, on its current reading, rules its metaphysic out! Surely science and especially psychology would not, materialistically, delete the immaterial soul of life's dimension?

In other words, the informative, non-physical element of subjectivity is an *inner* dimension. Nowhere in the outward, non-conscious dimension of the material universe will either individual or universal mind be found. Imagine mind 'superposed' on the material dimension; its *influence* is 'entangled' everywhere but at same time is secreted nowhere to be seen; it is together yet apart. The influence is more obvious in forms of animation than of automation. You might view such complex forms as valves where-through, to various extent, effects of mind are leaked. And where a leak from the subjective reservoir is strong enough its local stream of consciousness can light the dark vale of oblivion; a filtrate of life's essence livens up the lifeless stage; a drop of soul takes shape, a 'particle of God' acts out its role. In short, psychological is hidden 'within' yet 'beyond' the physical dimension. The source and truth of mind does not exist in physicality.

To a materialist holism's inclusion of a metaphysical dimension is an unfashionable, unwanted, retrograde and 'anti-scientific' step. But it seems to others, in the way that Einstein broadened Newton's work, a progressive, more enlightened and inclusive version of the whole truth - version whose voluntary, psychological part cannot, however, be dealt with mathematically. *A choice of materialistic or holistic framework is one of preference; but only one of them reflects the whole truth.*

Who am I? Childlike in simplicity, this question drives subjective science. The question leads directly on to others. 'Am I a spiritual being having an earthly experience or a complex form of earth somehow knowing it?' 'What, if anything, does soul mean?' and 'Whence did I originally derive?' If I once answer these then I might answer this: '**Why am I here; what is the meaning or the purpose of my life?**' These questions, although critical, are not - except for one - ones material science pretends to address. Matter is non-conscious and materialism, having *a priori* rejected an Information Centre, can only ascribe existence to chance, automated randomness and therefore a lack of purpose. Scientism's answer, therefore, adds down to a hopeless, dismal negative. An internal void. Materialism owns the world's material coordinate but for a graph you need at least two axes. *On its half-graph the information coordinate is missing.*

chance	*design*
disorder	*order*
accident	*purpose*

Objectively material; subjective immateriality. Whether or not I incorporate other tiers of existence (see *figs.* 0.5 and 1.3: A Three-tiered Universe) I am at least and undeniably 'embodied awareness'. And, as a pilot's in a plane or captain's on a ship, human experience is impossible without its vehicle, human body with a brain. Your current sensation and motion depend on it. Mind and body work as one. **If, however, your information factor (mind) is intimately linked but as an entity distinct from your energetic factor (body) then your present combination may not last intact; and if, in their disintegration, mind is not brain then subjective experience may occur *independently*.**

For its objective part my body (Chapters 17 and 19-25) is a biological machine. Machines are designed with a purpose in mind. Materialism's simple, one-tier universe has only reflex matter in it. *To invoke design to explain such a machine is, therefore, materialism's great heresy. It is an 'objective sin'.* 'Counter-intuition', on the other hand, is an ebullient, fashionable noun to use; it's been invented to persuade you how to see that purpose (including your own purposes) or purposeful machinery might happen accidentally. How precisely in some detail can you rationally imagine that? Just stand on your head. An upside down perspective makes irrational ideas seem plausible. For example, Darwin plausibly designed designer-less biology. *How do you conceive that subjectivity began?* Is information a material factor that evolved from a 'primordial slime-ball' or an immaterial one devolved from, say, Pure Consciousness?

outward form	inner principle
outward focus	inward focus
detachment from metaphysical	detachment from physical
physical science	metaphysical science

Maybe the question 'Who am I?' has, depending on my identification, more than one answer. Do I identify with body, mind or Pure Awareness; with only one or two of these; or are they all one substance, matter? **What is the nature of pure consciousness? In what does subjectivity consist?** These are questions that, to partial and differing extent, all faiths try to explain.

I was in a hard place on The Rock. I was in Alcatraz State Penitentiary, San Francisco. As I toured the 'holes' for solitary confinement in the Treatment Block I was listening to an audio. Prisoners reminisced. How did you cope with the dark, steel-floored isolation into which you were, with only a blanket, slammed? Suddenly an ex-inmate came on line. He had discovered if he looked 'inside' he found a faint light in the darkness of his head. If he focused hard (and it was not easy) the light grew until he emerged into a 'garden' where you could do pretty well anything as if for real. Was this a trip to a disembodied, 'astral' phase of mind, a psychotic episode or simply the operation of his neurons uninhibited by sense input? Was his subjectivity a waking dream and, if so, how real is such wakefulness? What, imprisoned in a body, is anybody's wakefulness but a sense-restricted dream? Then how can one wake up? Not wake within the dream but wake from it? Is your dream's end oblivion or Wakefulness?

For what it's worth the prisoner's report throws our normal immersion in body and its sensations of a vibrant, external world into sharp relief. Who believes that such immersion is the sole reality, the only frame of mind? The prisoner was deprived of sensation by force; and indefinite, untrained, involuntary isolation harms the human soul - or mind, at least. But might you *voluntarily* transcend his Alcatraz experience? Do you remember that self-deprivation tank? It left you senseless, physically anaesthetised and, in inner space, with thoughts alone. Could you press further? Is there detachment purer than sensational; purer than an intellectual form of objectivity such as mathematics; at its purest on detachment from the body when the latter's drag is broken at the time of death? Turmoil of the mind must, states the mystic's code of practice, go. Even mind must melt to find the peace. Since the aggregate of mind and matter add up to existence, he seeks detachment from the maelstrom called existence. What, as he prolongs and deepens his detachment, is there left - a dark hole of oblivion, blackest death, material absolution? Or does he find a well-spring and a pivot round which every changeful thing is swung? Is the fulcrum of existence Essence? Does he realise a Being that's Supreme, whose other name is Pure Awareness?

He finds, such is The Word passed down, complete detachment from existence in Essential Being. Since Essence is Pure Subjectivity he finds, as was previously surmised, the purest Objectivity in Subjectivity - and at the same time, paradoxically, most positive involvement from The Point of Balance. Is this sort of poise what Lao Tzu called 'masterly inactivity' or 'inactivity in action'? What are the other words we give to positive, enlightened interest? Enthusiasm? Wisdom? Love?

Whether I am body, mind, Pure Consciousness or all three rolled in one, the first question 'Who am I?' leads to a second. 'What should I do? How should I act to best lead life, one maximising good and happy influence?' This runs directly to a third. 'What do I mean by good? What is bad? What is right or wrong?' Some people never ask themselves such questions. Others think that they can do without them. Still others ask but never think the matter through. Such persons live in broad daylight but, to their extent of ignorance, in mental shadow-land. They do so because such matters underwrite the ethical philosophies whence flow religion politics and law. Society in which they live is based on them. *Rather than the valuable instruments science makes to promote our physical well-being, such invisible business informs the hub of human community.* Whether individualised or institutionalised, the answers to subjective questions are central to human happiness. Perhaps, from the vantage point of Subjectivity, the answers are completely obvious but most of us are not sufficiently detached. And, anyway, the business of a body politic involves external, social as well as internal, individual order. It engages physical survival, financial practicality and therefore, often, moral compromise. Some of the unenlightened think, of course, they know it all. They, from various angles, always know what's best. This is the arrogance of politics. Humility instead prefers a guide and guidance. Does best practice stem direct from Central Subjectivity? Is this why prime ministerial majesty, to try and legislate for what is best, used to call 'wise men' to counsel and enshrine a saint's subjective message at the heart of constitution as the guiding principle of government? From such priority, it's sure, Britain, Europe and America have each derived their histories; for caliph, maharajah and imperial orient such honorific principle applied as well.

Is it true, therefore, 'maturity' means only scepticism? Is growing old not growing wise but cynical? As opposed to a political leader responsible for his community's well-being, is it true that a hands-on scientist wants to know everything, try out anything but take ethical responsibility for nothing? To experiment technologically without a moral, social guarantee? Atheist Lewis Wolpert exclaimed, "I am adamant that reliable science has no ethical content". Ethics has no place in the descriptions of true science. This detachment from morality occurs because the scientific form of objectivity is incomplete. Its impersonality engages the calculation of things but disengages, intellectually, from subjectivity. It can therefore, only engage material existence; it cannot find, in Subjectivity, true Objectivity. For all its strengths science *per se* does not pretend, nor should one rely on it, to address the central questions of humanity. <u>Scientists like Newton, Maxwell and Robert Boyle (founder of The Royal Society) saw no difficulty in this. Their impetus was to reveal how its Creator made the physic of the cosmos tick.</u> Even religiosity's great critic, Thomas Huxley, was not atheistic but agnostic. What has changed? Natural science worries matters outside mind; it looks outward to the world. Subjective teachings, whether occidental or oriental, address the inmost nature of the seer; and at this hub of selves they seek to find the central self. They seek subjective purity. The immaterial core of their subjective pillar is adamant, is indestructible and is Essential Absolution.

Pure Objectivity.

What's pure objectivity? Does its reality consist of objects or detachment from them; does it involve mathematical appreciation or is it just the state that matter's in - existence as an object in non-conscious oblivion; is objectivity a physical or psychological affair?

Maybe there's wholly abstract and, in this sense, pure objectivity. 'To me,' wrote Roger Penrose in Shadows of the Mind, 'the [Platonic] world of perfect forms is primary... its existence being almost a logical necessity.' He conceded that 'perhaps a Platonic reality should be assigned to other abstracts, not just mathematical ones.' How far is this from archetypes (or archetypal memories) in universal mind? Physical cosmos, in both animate and inanimate respects, is seen as a reflection of such archetypes (or ideals), a reflection wrought by context into myriad different expressions throughout space and time.

David Bohm, the physicist, believes that individual things are all, in their 'deep nature', connected to some fundamental 'archetype'. A hologram might help you understand this style of thought. Holograms are 3-d images recorded with the help of lasers. Unlike a normal photographic image each part of one contains the image held by the whole. If the causal aspect of the universe, wherein 'the whole is in each part and each part in the whole', is holographic then you might in fact see it as a glorious, highly detailed apparition. Each part of this phantom, including you, would be a projection, an extension connected to its causal archetype. Indeed, for Natural Dialectic nature is a nested hierarchy leading back, in various forms, to the centre of all implication, an Essential Egg. Such archetypal objectivity approaches, in its distance from a world that's tangible, the mystical.

You may, in addition and in the company of psychologist Karl Pribram, view your nervous system as a personal hologram, an image-storing device and a lens so constructed that it only subtracts a human perspective from the totality of an environment. This is our 'consensus of reality' but, it would seem, just a fraction of the possibilities. Is Pribram right? If you can't take brain's goggles off then how detached is that? Such a view is partly what a Buddhist means by '*maya*' or illusion. Buddhist practice would transcend man's optic and see nature 'as it really is', that is, at its deepest layer, at its source. **His pure objectivity is thus 'outside'/ 'not-of'/ 'beyond' existence; or, better phrased, such an experience is 'sited' at the creation's core looking from the 'inside' out.** Is this what faiths see as Communion, *Samadhi* or The Heart of Knowledge? From such detached but superhuman point you may realise that the phenomena of mind and matter are but lesser truths, projections of an Inner, Central Truth. Such case transcends both mind and matter and, therefore, the combination of this pair - existence. Its view is from beyond creation. <u>Essential Self</u> surveys from Central Apex. In this sense creation is an object; mind and matter are dependents on the Independent Truth. Less than Truth is relative illusion; how objective can you be (see Chapter 3: Truth, Appearance and Reality) about what isn't ultimately Real? This Reality is therefore, by transcendence, in a state of Pure Objectivity; but since its purity is also conscious it is, at the same time, Pure Subjectivity! The nature of this Central Paradox will be explored in Chapters 5, 6 and 13.

No-one likes to feel deluded. One might, therefore, prefer less radical, more prosaic a grasp of what an object really is - even if it's only partly true. After all, does not a scientist need some event or object to detach from and thus analyse 'objectively'? *Detachment* is a philosophical and scientific holy grail. From such objective vantage point each of these disciplines seeks, like mysticism, to efface subjective interference and obtain without distortion, natural and universal truths.

Human vision isn't that of Pure Objectivity. Thus, in its lesser state, what do we think that precious quality, detachment, is detached from? Is it from unproved imaginations, pain or the various, lashing passions of desire; ort an escape from heavy existential weather or acute self-consciousness - from self and subjectivity? In this case sleep, repulsion and unconsciousness are forms of negative detachment. Nothing drags you into body like a pain; the transparency of health is an expression of detachment too. So, from the tribulations of one's daily life, are positive engrossment, interest and enthusiasm. In this case subjective, psychological detachment from the 'real', hard world is a common process men dip more or less profoundly in and out of all the time. The question resolves to one of depth and quality. Ecstasy or 'out-of-selfishness' depends on quality of attention *and* its object. In what do you prefer to lose yourself? Indeed, isn't loss of ego objectivity's main trick? This trick is the annihilation of your personal self. How does a scientist lose himself? How does a mystic do the same?

An ideal scientific loss of self would gain you perfect, superhuman objectivity. The way's to try and isolate experimenter from experiment; and to use 'numerical philosophy', that is, self-consistent mathematical abstraction. Such abstraction is a universal language, elements of which are used by science to describe relationships that theoretically could or actually do predictably recur in time and space. It ties down quantities of shape and changes in the form of energy. Theorems, formulae and abstract wrangling render everlasting truths; nor are numbers swayed emotively.

Is abstract objectivity restricted to the facts? Is, for example, geometry derived from just bond angles, crystals and the architecture of atoms and their compounds? Does objectivity demand, especially where the exotic speculations of cosmology and quantum physics are concerned, that a mathematical construction should apply to measureable reality alone? Can it range as widely as extrapolations of sophisticated science fiction? Subjectivity is not entirely squeezed.

An exciting triumph of physical science has been to precisely elaborate the tightly knit, interlocking way in which each working part of cosmos is defined in terms of the others. There exists no absolute point of reference except, perhaps, the invariant constants of nature whose origin is unknown. One transformation of energy leads to another until, in a circular and self-consistent way, the whole business has been described in its own terms - almost. The conditions that approach micro-infinity on the one hand, complexity in the middle and macro-infinity on the other have yet to be reconciled. No doubt, one form or other of scientific *TOE* (Theory of Everything) will be discovered; and, while its structure and ramifications may or may not be unexpected, they will include previous mathematical reliabilities. *However such a TOE will only, as if maths and matter made up everything, apply to physicalia; but such claim to omniscience would, metaphysically, mislead.*

Surely, though, despite its limitations mathematics is the purest form of objectivity? Who can argue with its ageless propositions; who can wrangle out of its equations or can slip the nature of its logic? Mind discovers and invents it yet, remarkably, it also seems attuned to nature in a 'pre-established harmony'. And thus, although he may not claim mind made it, a theorist studies cosmos in his head. Science, having captured observations in a web of numbers, wants to accurately and timelessly predict from its equations and include the past within its 'law'. Words lead to imprecision, doubt and ambiguity; only maths is rigorous enough. *No physicist feels an idea has become a theory until it is cast in the framework of maths and then, with difficulty in the case of cosmology, tested.* Because maths involves equations where technicalities and formulae can render it indecipherable except to the specialist, the non-specialist is understandably marginalised in the same sort of way as a congregation round a Latin-speaking priest. A relatively ignorant flock must, as regards the physical universe, follow where a scientific shepherd speculates; for which intellectual shepherd only a materialistic explanation is, on principle, allowed. Therefore because both parties breathe, think and have somehow arrived on earth the driving principle round which a shepherd's speculation must revolve is the so-called 'fact' of evolution. His universe revolves round Darwin's great idea. So all permitted lines of reasoning converge upon this much-loved principle, this preference. Is preference part of pure objectivity? Or does bias shatter it?

It is sometimes claimed that 'natural mathematics underpins the fabric of the universe'. Immaterial abstracts, numbers, run the show. However, while numbers may describe behaviour they do not *create* it. Maths is like a fire without its flame, an empty bowl that begs of energy for sustenance. It is like description of a chamber's music rather than the sound itself. Indeed, its own 'reality' is no real check and, given assumptions within a self-consistent framework, mathematical abstractions as fantastic yet interesting as fiction can be spun. Of these a restricted subset seems to represent, as far as we can see, the patterns of automata. These reflections, now well wrangled, are not 'laws'. Mathematics cannot even flick a ping-pong ball across a net. It does nothing, governs nothing but describes the principles of mindless shape and physical behaviour. In this conceptual, metaphysical, reflective element resides its awe-inspiring power. Is there any truth maths might not catch? You might love numbers; they won't love you back. In any scene they're nowhere to be found; nature only prints their absence in a mind. So can maths ever circumscribe all that there is to know?

If something caused existence then the latter's an effect. *Science is the study of effects. Its equations describe neither fundamental cause, emergent properties such as colour and texture or, most importantly, the activity of what it perceives as simply another 'emergent property' - voluntary, affective mind.* The more voluntary the less predictable and more inaccessible to mathematics a situation becomes. Neither ideas, purpose, creativity, discovery nor learning are amenable to numerical equation; nor ethical equations; nor, overall, is that basic, subjective reality - consciousness. How can you count, even statistically, a meaningful experience? Yet consciousness is *the* central part, the essence of humanity or any other knowing kind. It is, in a way to be explained, the undeniable core of life. It is the causative nature of mind that, outside the scope of mathematical science, a mystic addresses. Is this address, an unscientifically subjective one, therefore 'unreal', 'incorrect' or 'deluded'? Whose monopoly is truth?

We can go further. You can count on it. The rigour of mathematics is relevant when dealing with involuntary, physical automata. For objective purposes its brilliant computations underwrite a comprehensive description of the non-psychological world and generate dazzling technologies. Knowledge of things has greatly improved. So (but not without potentially devastating backlashes such as climate change, overpopulation or ecological devastation) has the biological circumstance of most humans. On the other hand, science cannot claim to have achieved or ever be able to achieve more than a provisional insight into any profound aspect of our paradoxical existence. *This is especially the case where mind, is concerned.* <u>Mind is subjective. For subjective purposes, a mathematical description of the voluntary operations and calculations of consciousness (volition, imagination, emotions) is, if not impossible, irrelevant.</u>

Some scientists over-enthusiastically confuse their explanations and models with the reality of nature itself. They genuinely feel that the 'pure objectivity' a *TOE* would provide will be 'the greatest intellectual triumph of all time'. *They forget that it will deal only with involuntary, non-conscious entities. Not everything that is real can be quantified let alone qualified by mathematical abstraction.* **Such researchers labour, therefore, under the delusional faith that science, underpinned by mathematics, will one day be able explain everything. If information, mind and consciousness are metaphysical what we call science (physical science) never will.**

<u>*Nor is Natural Dialectic a mathematical but a conceptual tool*</u>. Its self-consistent logic may lead to fresh insights but it is not so much an instrument of number as of paradox. Moreover it is inclusive. It includes both elements of any paradoxical duality that contribute to the structure of our world. It includes the central, metaphysical realm of mind - mind *and* matter, voluntary *and* involuntary, metaphysical *and* physical aspects of existence. Indeed its emphasis is so clearly on the primacy of mind over matter (as is our own circumstance) that, rather than mind being dubbed 'a ghost in the machine', matter might be dubbed 'a ghostly machine'. Natural Dialectic, in a manner contemptible only to a philosophical sector called scientific materialism, complements and unites both worlds. **Indeed, the moments that have most profoundly shaped human history have certainly been, as opposed to scientifically objective, entirely subjective in nature - the inspirational ideas of scientists and the enlightenment of Christ, the Buddha and other communicators of Truth.**

Thus we cycle round the universe again. To be objective is to be 'detached'. To be detached you 'stand apart'. No ties or dependencies. What, except what's finite, stands different from Infinity? Source is independent of its issue; what apart from Essence, is untied to all existence? Natural Dialectic would identify Essential Independence, on which existence hangs and all the worlds depend, as Pure Subjectivity. Thus might there rest, in this impartial detachment from creation, Pure Objectivity as well? A concentrate of Subjectivity is, paradoxically, Pure Objectivity! The single species of Complete Detachment. In brief, if I've subtracted all objective focus in one-pointed meditation, I might logically obtain detachment from the world that in its total subjectivity is found to be the only way to fully understand things. *Enlightenment is centred at the Hub, an Axis motionless amid the movement. From The Essential Top-down Vista, at the Apex of Mount Universe, I can alone survey existence, that is, all creation.*

By such reckoning it is not the detailed, comprehensive collection of data but profound insight that yields a fuller truth. The more one-pointed the mind, the higher the principle perceived. There is, at the top of cosmic menu, an overarching Super-truth. For that vista climb to Essence. To recap, if you observe existence (mind and matter) from its Centre is not that pure objectivity? **You'd find, at Point of Purest Subjectivity, the Purest Objectivity. How's that for paradoxical enlightenment!**

Scientific Delusions.

Each of the two pillars of faith incorporates, as well as truths, its own denominational delusions. Science, like law, politics or teaching, is a profession. It may have one exhilarating difference, that in principle it is dedicated to the discovery of fresh material truths, but it is prone like any human undertaking to professional deformation as regards its mind-set, peer-group pressures, lapse from ideal and, thinking or unthinkingly, to individual and collective delusion; and, with delusion, imposition of taboo. However, while a few excited militants stridently proclaim their 'scientific atheism' a majority of persons simply, coolly and without much thought accept, by a kind of intellectual osmosis, philosophical materialism, its central prop (the theory of evolution) and its world-view. Don't think twice, it's alright; but the problem is it's not. Some may believe that science is too down-to-earth for mind (an exchange of information) and other hard-to-pin-down non-materials to plague their cogitations. *Yet such materialism is actually as well as philosophically unjustified; and modern science has constructed a whole version of reality from the limited constituents of its carelessness.* The nonchalance amounts, if not to delusion, to a currently common form of sleep-walking.

Maxwell saw the light; Newton did not need a wake-up call; nor did Faraday, Kelvin, Boyle and a host of other scientific luminaries fall half asleep. Many still do not. A commonly touted 'scientific' delusion is that religious faith is blind. Its beliefs and trusts lack evidence. *Does science, as we'll see, provide no evidence for an informed creation - both as regards its physics and biology?* **Indeed, the discipline is wholly based upon a metaphysical assumption, faithfully assumed, that the universe is rationally comprehensible.** 'The most incomprehensible thing about the universe is that it is comprehensible' hinted faithful Einstein. Further, in so far as we are not omniscient we are *all*, fundamentally, creatures of faith. Of course, material science works on the naturalistic assumption that everything you can test, measure or observe with senses and technology can be explained without resort to other than the behaviours and thus laws of physic. Such naturalistic methodology is healthy when it involves the discussion, experimental examination and mathematical description - however complex - of physical operations, that is, of all reflex physical phenomena. It is the shield against incursion of the metaphysical, an obligatory trammel of ideas that keeps the mind-set of materialistic science on its right lines.

However, some practitioners extrapolate. A fraction of the scientific community, men and women well respected for their scholarship, have also taken up philosophy. *For a start their rules of play invoke Clause 1 - that 'natural is equal to material'.* Lifeless energy's behind all things, mindlessness rules where no facts metaphysical/ supernatural/ immaterial exist. To shield, thereby, is added rapier attack. *If there is mystery only science solves it. You only come to truth assuming its material way.* Its single rationale applies not only to the operation but (outside experimental science proper in forms of historical extrapolation) to the origins of cosmos and our lives within it. That rules out metaphysic for a start! It neatly though maybe incorrectly rules out any immaterial factor! The first sleight-of-hand that's passed is, simply, that there's no such factor. By now, therefore, you might have guessed. **The *leading* science delusion is, in the naturalistic heads of those who suffer it, that there is only unconscious energy/ matter. This alone makes up reality.**

As you will. Now you see, depending on your own perspective, that the phrase 'blind faith' can cut both ways. No-one can physic metaphysic, see it with the body's eyes or treat it in a lab. A holist cannot prove or disprove immateriality; nor, in the naturalistic way, can a materialist. *Naturally enough, there's no material evidence at all that immateriality is an illusion nor, therefore, that the delusional 'matter-only' half-truth is, in spite of scientism's protestation, fully true.*

Yet from this primary error spring, perforce, the rest. **The *second* great delusion is that mind is, somehow, brain.** 'Unconscious matter makes up consciousness.' *Mind-is-meat's the final, crucial sealant that entombs holism and therewith, of course, the basic premise of this book - root immateriality*! In this view consciousness must somehow rise, like atoms dreaming, out of nervous chemistry; it must not be different in kind but only in degree from, say, quantum electronic states of chance. It is non-conscious matter, in the form of brain, that generates what we call consciousness and sometimes, at the mind's extreme, a God delusion. 'You, that is, your experience, are nothing but a bunch of nerves that work by mindless chemistry'; this is scientism's grand, deflationary idea. But atoms and their molecules are not alive. How, therefore, can multiplying lifeless atoms make them live as mind, memory and aware *experience*? How, moreover, if you claim awareness is illusion doesn't such awareness thus deny its own reality? Who, since we're all illusions, dares to disagree? So consciousness is an illusion suffering an illusion! Is this mad kind of nonsense what materialism tips you to - so that emerges grand delusion into which it straightway, po-faced pitches any seeker after truth?! Such search is, like its reasoning, the true illusion! However, it makes perfect sense to neuroscience or (*for which there's not an ounce of evidence - nobody's seen a thought or, come to think of it, a memory*) to any mind-set thinking its own thoughts are somehow made of matter. Think of barking, trees and gum.

Of course, brain-is-mind cuts out any form of immortality. Brain decomposition is the grave of post-mortality; hell and heaven, though they might inhabit life, in death are definitely dead. If, on the other hand, you fail to solve the brain-is-mind equation then after-life may not be a delusion after all.

For a moment let's hark back to memory. Aren't memories also nervous traces in the brain? Yet despite billions of dollars' worth of intensive research over more than a hundred years (well spent or wasted if mind isn't physical) no trace of actual memory has ever been discovered. *There's not a shred of evidence that any memory's located in some nerves.* Indeed, if a memory were nervous traces you would have to have another trace for its retrieval; then another for retrieval of retrieval and so on *ad infinitum*. If logic from regression's not enough then how, amid dynamic changes that we know continually occur in brains, is permanence of long-term trace of memory sometimes with crystal clarity maintained? *Materialism, when it comes to information, wishes, purpose - any feature of a mind, is not a good idea. Its explanation for the seat of memory is illusory.*

Now you see it, now you don't. Only objective no subjective actuality exists. Of course, brain links you with body-world but look out to the stars. Are you reduced to just inside your skull; is the universe screened in the darkness there? *Isn't such extreme and literal closed-mindedness, for which there's not the slightest evidence, itself a grand delusion*? If psychic phenomena (like perceptions, thoughts, emotions and imaginations) are illusions certainly the universe itself should, locked in brain, be a delusion too. What, in that case, is logic, reason or an artist's poetry? 'Concept' is a case of cells connected and your subjectivity, like morals and aesthetics, only electronic relativity! This is because psychological thought, having been 'logically proven' chemical, is thenceforth snared in the net of physical science. In such manner a materialist, at the confident moment he believes metaphysical mind disproved and a chemical 'non-soul' proved, has actually reached his wit's end!

For example, a physiological replica of external objects in the brain is not the case; nor replica of bodily sensations or spatial sensations. Nerves simply don't make shapes of sights or sounds or smells! Nor do cerebral correlations with a body's parts bear the slightest resemblance in shape to the body that they're sensing in (see *fig*. 24.4). Therefore what is it, if not ionic flickering in brain (where everything is registered), that we imagine we can see in mind and thought-wise reason with? Why, indeed, should mind need to form a psychical image from dissimilar physiological changes if these nervous motions - which can be scanned by instruments - are sufficient by themselves to determine the course of activity or thought?

"It is, a reductionist asserts, an axiom of empirical science, self-evident when impartially considered, that any physical change must be causally determined by some antecedent physical change - whether or not you are yourself able to predict it. The changes in the human brain, including those that are accompanied by consciousness, are essentially physical changes. All conscious processes must *therefore* be due to antecedent physical changes. There is no room for any such concept as mind or any such purely mental process as free choice or free will. Conscious processes must *therefore* either be introspective concomitants of the corresponding brain processes or else (which seems more probable) they must themselves be generated by, and *therefore* really consist of, the accompanying cerebral processes."

How, though, can a set of physical changes, physically caused, possibly 'correspond' to such conscious experience as seeing that the abovementioned 'axiom is self-evident' or to conscious logical transition as implied by the word 'therefore'? If a whole sequence of logical steps, mathematical algorithms or reasonable statements were indeed merely the effect of a causal chain of physical processes, all blindly and mechanically or electrochemically determined, it would follow that the speaker could not think what he wanted or help saying what he did. His statements, as reasoned arguments, should therefore carry no weight. *Indeed, why should we believe a word he says?* **In this respect it is common knowledge that, on his own terms, the serious atheist is talking gobbledegook.** Physicist Sir John Polkinghorne rams the point home. 'Thought is replaced by electro-chemical neural events. Two such events cannot confront each other in rational discourse… The very assertions of the reductionist himself are nothing but blips in the neural network of his brain. The world of rational discourse dissolves into the absurd chatter of firing synapses.' If thought's like that why trust its propagator's rationality? Not least Crick, Watson, Harris, Hitchings, Dawkins, Dennett and others of an atheistic crew - especially since the theory of evolution's only, at its biochemical roots, electrical activity; surely atoms that they argue have evolved into a brain are not as logical as that! How do molecules make language? What is a neuron-generated theory worth? *An atheist unwittingly insists that his own automated arguments aren't logically worth a fig!*

Moreover, what of his experience? Or yours as you survey the world? Is life a vast illusion made of molecules? *Experience is your central truth; if subjective mind is not made up of the four forces, particles and atoms it is immaterial.* Is it, therefore, unnatural, metaphysical or super-natural? **Of course, if you believe that *somehow*, endlessly arguably, thought and intention might be made of chemicals then that's your faith; no matter if you want to frame it as a 'scientific' faith, it is your binding, blinding faith.** The word 'religion' *means* a binding faith. Maybe, however, in this case materialism's stark equation is absurd. It can't be proved and yet its own Clause 1 scythes immaterial subjectivity! **If this Clause 1, the basic axiom of *PAM*, is wrong, I can repeat that this whole book falls logically into the order of correctness.** Natural is not only equal to material. Information's immaterial. Your thoughts *are* 'super-material' and natural mind is metaphysical!

And yet, as all psychiatrists including saints are well aware, disentanglement from a delusion that the patient thinks is true is not an easy task. Some, for example, argue that material science is the only way to understand the world. Subjective science is a species simply not allowed. You dissect a thing and not a thought – unless your cogitation is, computer-like, an electronic trace. In which case please explain how hydrocephalic persons think. Hydrocephalus is a condition wherein up to 95% of grey matter is replaced by cerebrospinal fluid. Such special water, for one student who had practically no brain, still gained him a first-class degree in maths! Surgeons have cut nerves but who has cut a thought up? If thought is not material then can reductionism get you everywhere? Especially delusory is the notion that its analytic methods can supply an answer that's complete; or that knowing parts means understanding wholes. For example, if a TV was dismantled in a lab and then these parts ground up for further, chemical analysis would this research complete your comprehension of TVs? Or tell you anything about the broadcasts that you bought it for? Likewise, in the Department of Biochemistry, if you were reduced to molecules would that define the 'basic you'? **A *third* delusion, called reductionism, thinks it might.**

Scientific methodology demands objective answers, detachment of its '*magisterium*' commands authority and, free of ideology or dogma, its church will in times to come dispense a total, god-like knowledge of all cosmic facts. 'Knowledge is power', claimed Francis Bacon. Already, perhaps anticipating such omniscient omnipotence, professional institutes of science and of state are seamlessly combined. Advice issued from a scientific body is stamped with, supposedly, the hallmark of impartiality; experimental evidence is what has won and wins respect. **But a *fourth* delusion is that material science might, alone and in the future, yield omniscience.**

A *fifth*, **that we've already met, is vested in delusion of its total objectivity**. Of course, striving and pretension co-exist but is mind disembodied? Does hypothesis pull clear of expectation? Is subjectivity squeezed wholly out of science? In other words, does an 'observer', though hypothesizing, devising, performing, measuring and then interpreting what he observes, detach completely from all interest, enthusiasm or attachment to results? Impersonal voice (not 'I did' but 'it was done') adds tone of *gravitas*, lends suggestion weight of fact and neutralises personality but is the aura of detachment more than a professional face? Isn't ideal, super-human aspiration thwarted by its human execution? Do experimenters not affect results - not only in the quantum field but especially when it comes to quest involving aspects of psychology? Do 'right' results and expectations play no part - even if researchers sometimes cherry-pick their best results and throw the rest away? Do editors of journals suffer no delusion, prejudice or preference in the papers that they choose? Does the same apply, when consensus and a certain paradigm may rule the temporary day, to peers who review what a career has been invested in? Are double checks and rigorous self-regulation what, every time, applies? The answer's no; and with that negativity there exits total objectivity. Objectivity's diluted further when, through the media, 'cutting-edge' discoveries are filtered to a scientifically-challenged public. Is sufficient scrutiny applied, are inaccuracies and bias never waved uncritically through? For different reasons scientists, like journalists, aren't always thoroughly objective chaps. We note, in brief, that 'pure objectivity' is just materialistic superstition science never will achieve.

Egocentric vision is as natural to humans as the geocentric notion of earth's place; we strive for universal objectivity but with questionable success. Some, certainly, idealise progressive, liberated science and technology as leading towards a brave new world but social, economic and ideological pressures act as de-objectifying forces in the world of science as in any other industry. Detachment beckons but attachments hold us back. Assumptions, paradigms and prejudice ensure that what does not 'fit in' is dismissed or else explained away. The plain fact is that, especially where observations rest on theory's mind-set, there is no lack of preconception with respect to questions, answers and interpretations. **Lack of bias, when it comes to forms of life, mind and their origins, is especially a myth. The mind-set of such prejudicial mythology applies with force to origins and consciousness and biological constructions.** An immaterial, subjective element is consensually, consistently ignored. **Naturally creative information has no place in scientific nature; purpose is taboo so that, as yet, formulations such as Natural Dialectic have not been permitted shape.**

Do you, however, think eliminating the subjective is objective after all? Or think materialism's primary equation is simply verbal fun and intellectual games? It is promulgated, in the name of science, just as seriously and tirelessly as any other faith and is, in fact, none other than the heartless heart of a religion. This species of religion infers a 'principle of continuity' in space and time that can, without inventive intervention by intelligence, eventually deliver minds in humans who can then dream up how they came, perchance, to be. **Yet is this naturalism able fully to describe not operations but the *origins* of life and cosmos?** *From physics you could possibly infer a cosmic origin that's metaphysical but from a cell's biology, as we shall see, it's easy and the case is very strong.* To deny, as official science does, such origin might warp the evidence into a thousand tales - some as fantastic as they are exotic - of denial; it might constitute a blockage, an inhibiting infection that, fighting tooth and nail, will never recognise another possibility. Why, after all, should mindless unpredictability be able to produce man's logic or the universe's comprehensibility? Is such logic, born of matter's maelstrom, absolute or just an unanticipated, unintended incident of change? And, if reason's reason is irrationally by chance constructed, why should I commit to its conclusions? Such contradictory, naturalistic mode of packaging the facts seems radically diseased. Yet, as in madness, thinkers cannot penetrate their own delusion, a delusion that is 'scientific' and wears evolution as the jewel in its imperious crown.

No doubt, science should be open-minded. Yet, it seems, no-one finds it harder 'to think outside the box' than evolutionists with 'evolutionary solutions'. No doubt there are, as we'll see, elements of truth in Darwinism so the theory can't be called delusory. **Indeed, if you're a naturalist or wish to offer matter-only plausibility, then life's Darwinian development is rooted as the mainstay of your faith.** Such religious belief in the sufficiency of natural cause must pay continual homage to evolutionary 'inventiveness' and, reflexively, reject a common-sense appraisal of biological programs and designs; these, theory insistently insists, involve no psychological/ purposive component - since matter isn't purposive! Some even swear the theory of progressive evolution is a *fact* and, it's tub-thumped, only madness disagrees with *facts* - but, as we'll see, you'd be unwise to nail your reputation to that mast. **A *sixth* delusion is, therefore, that specific, complex, systematic information (such as codes convey) can ever physically perchance 'self-organise'.** The most important trick is, thereby, to persuade yourself and thence your audience that bodies biological might seem 'designed' but really aren't. *Their coded forms, you claim, have just 'self-organised'.* Since materialism's essence, Clause 1, is at stake so-called 'intelligent design' is diagnosed as a contrary, unboxed and unhinged delusion. Darwinians continually, forcefully on an industrial scale repeat this diagnosis till, by dint of repetition, one might be drummed into believing that it's true. Religious scientism, atheism and such congregation press the case. Thus an alliance party of materialists defames the holist camp. It was not always so but now evangelism of unholy 'anti-church' promotes its godless creed; you might evangelise the idol of this cult who is, by rite and right, Charles R. Darwin. **Without his 'bible' secularity could never reasonably exist.** Thus the media's humanistic message, from nature programs unto science magazines, exclusively promotes materialism's evolutionary grain of 'rationality', that is, 'design' of bodies isn't what it seems. It isn't what the word's intent implies.

How else, except by natural law and creativity called chance, could we be here? Thus, twisted round, 'design' becomes a mindless freak of time, a vast improbability that's landed up the certainty (if not illusion) that is you. Having banned the thought of immaterial mind this mind-set press-gangs animistic language ('code', 'function', 'program' and so on) to smuggle immaterial information back into its explanations. Yet if, and only if, an immaterial element of information exists in nature we would see the tables turned. As this book at length and in some detail demonstrates the notion of design might be a valid one. Armed with such rational plausibility in whose mind would delusion and irrational illusion seem to be?

Is nature actually purposeless? If not then how is any (if there's any) immaterial, conceptual side expressed? **This book confronts materialism's *seventh* main delusion and, in discussion that is normally taboo, evokes the greatest heresy of all. This is to question whether evolution theory, in such confusion as the Glossary defines, is in its biological department true.** This, today, is where the battle-lines of faith, materialistic or holistic, are most sharply drawn.

Is it not instructive? In August 1845 the oldest continuously published magazine in the United States - Scientific American - was started by Rufus Porter. The intention of its founder-editor, a prolific inventor and staunch Christian creationist, was to promote an understanding of God's creation in the same way as the founders of renaissance science as a whole. In November 1869 Sir Norman Lockyer first published a journal called 'Nature'. This editor was, by contrast, more interested in pagan history than its 'overlay', Christianity; and the journal was conceived and raised to serve polemic ends. Many of its early articles were written by Lockyer's fellows in the Royal Society, including members of the X Club (see also Chapter 23) that included staunch anti-Christian evolutionists such as Thomas Huxley, Joseph Hooker and Herbert Spencer. Isn't it the same today? No doubt that on materialism's terms Porter was 'irrational'; did he not credit supernatural nonsense with priority? Rationalism's faith is secular; naturalism only looks at nature in a 'scientific', that is, a materialistic way.

You're in denial if you think the issue isn't one of faith. 'Objective reason' has become faith's godless hope. Evolutionary science, whose conscientious devotees adhere exclusively to theory that's experimentally unproven and as this book well demonstrates may be delusory, has metamorphosed to religion. Is *PAM*'s catechism a contortion that compresses and distorts reality? Don't this religion's half-truths systematically flip the logic of the whole truth on its head? Throughout this book we'll take a look and see.

What are the facts? You can't squeeze much from nothing, can you? Perhaps you can. Perhaps, as you claim, nothing physical produced the physical, that is, the whole non-conscious universe. One's eyebrows rise. But wait! There's still one crucial bubble to 'emerge'. Sometime, in a nifty but momentous moment chock-a-block with serendipity, life emerged from matter - so that in time its atoms brewed up brows to raise and 'minds' to wryly realise the mindlessness of everything. Both 'factual emergences' originate from the material presumption 'metaphysical is nonsense'. **You might conflate them as an overall, *eighth* wonder - or delusion - of the universe.**

Naturalistic and thus scientific script includes no more than partial truth. In other words, if there is immateriality what is the nature of its 'nothingness'? What might 'metaphysic' actually mean? Nature preceded man and man preceded his philosophy, religion and, lately, scientific methodology. No doubt that for some the third and latter category of labour swallows up the other two. Its 'methodology of knowledge', specifically excluding other than phenomenal reality, renders them redundant. You objectify by looking from outside but you can't see a mind from there. This is fine for matter but for mind is not. Immaterial experience is understood by empathy but not by microscopes or spectroscopic scan. Undaunted, 'scientific method' still subordinates all metaphysics to the status of 'imagining', 'illusion' and 'physical unreality'; by which it means twitching of a nervous kind. This is, for example, how professional 'detachment' by neuroscience, psychology and sociology is philosophically forced to force religion into chemistry of limited component and location in grey matter of a madman's throbbing brain. Yet isn't such conclusion an illusion? Could it entirely miss the real point? **And thus, by sociology, invent another grandiose delusion, this time a religious delusion, from an evolutionary point of view?**

But, in a broad, holistic sense, the first two of the three pursuits remain, despite the cuckoo's protestations and attempts to heave them from man's intellectual nest, alive. They include an immaterial, subjective element and their goal, more or less successfully achieved, is the promotion of clear thinking in a truthful framework, that is, one that coincides with metaphysical *and* physical aspects of reality. In short, wholesome philosophy is not constrained by its materialistic scion and, thereby, the various science delusions. Its formal yet subjective arm, religion, is developed specifically to include and preserve man's sense of inner truth, that is, of a Central Metaphysical Heart. Its metaphors, first created to 'think the unthinkable', must then be literalised in order to create external observances. Such rituals are intended to help recreate an *experience* of understanding the true order of things. Many are derived from universal gestures of thanks, praise, reverence, obedience and so on. In his 'Book of Ceremonies' the Byzantine emperor Constantine VII Porphyrogenitus used his style of reason to define ceremony as 'the outward form of inward harmony'. How sane or deluded was he?

Religious Delusions.

For materialistic faith the *first* and greatest of religious delusions is, in the heads of those that suffer it, that a Live Creator (or an administrative hierarchy of Transformers) made the world; and that, having sourced existence, this Creator constitutes the Ultimate Reality. From this primary error spring, perforce, the rest. Of course, from an upside-down perspective atheism would reverse theism's case. Thus, it is claimed, we aren't big G's creation but, simply, He is ours. **No metaphysic, no creator; thus religions with creators simply wallow in a *second*, consequent delusion.** Divinity is an idea alone; mind makes up God, matter makes up mind and brain, like everything, evolved. He is, like you, reduced to a few chemicals. So what evolved the world from nothing? Did chance create life's very subtle, complex chemistry and, in its finest hour, eventually strike up such spiritual delusion as arises in that finest accident, a human brain? The issue is not one, for either camp, of fact. Each, atheist or theist, interprets in a circle from his *PAM* or *PAND*.

Atheism's a religion. An atheist claims restriction to one-tiered materialism *ought* to satisfy all men's belief. **Note straightaway how self-contradictory and inconsistent that such reasoning is**. His world is moving matter; perhaps four forces and twelve particles compose the lot. Thus there exists no 'ought' or 'should' except the laws of physics. Does gravel have morality? *Thus, if he claims 'religion is a force for evil' or 'it's wrong to teach there's God' he does so not according to his atheistic tenets but in stark, clear inconsistency.* Why, by that token, *shouldn't* any person strive to grasp the self-consistency of a three-tiered, theistic creed that, this time, self-consistently includes, as Natural Law, morality and immateriality? Who, theist or atheist, is in delusion?

Maybe, however, theistic delusions do not stem from fundamental error or the immaterial reality of subjectivity but from formalisms, rituals and cultural styles; and from deep egotism that, insisting 'my (or our) way is only the right one', insists on conflict too. An outward form dries hard around the moment of its energetic making. The inward images and tales that frame subjective truth grow dull with repetition. Habit dries, ritual rigidifies, convention mummifies. They choke love and understanding off. Original excitement fades, inspiration cools and revelation dies into formality. Now only schism, only fresh experience can raise an intellectual ghost.

The symbol of science is number. Based on empirical, objective research, material science tries to reach its truths by, as far as possible, the exclusion of any subjective, metaphysical element. Its formal explanations are rigorously mathematical, more flexibly statistical or at least hard factual. They are developed specifically, in the interest of eternal truths, to filter mumbo-jumbo out and set in stone the principles of scientific thought.

Science isn't poetry but surely you have feelings? How do you define your inner self by number? How can you explain - except by using metaphor or model - emotion, concept, moral or ideal; how express a state of mind or your experience of life? *The symbolic tools of the humanities are picture and language.* In the pictorial case the question of religious *delusion* really boils down to your own imagination(s) of God. These may be positive, negative or neutral but if such Essence transcends mind, as all spiritual advice (not least the third commandment) warns it does, ideas or models cannot capture its reality. At best they'll reach approximation of the truth.

Metaphor is also loose and wordy symbolism imprecise - especially when, as mystics or a poet may, one tries explaining love or meaning that transcends objective fact. Is, for example, love a red, red rose or G a patriarchal astronaut who hovers up above or perhaps outside the gravity of spatial sky? In describing our centrality, emotions, how can simile not blur reality? Talking of religious truth you have to 'break the inner symmetry of silence' and deliver 'graven images' to which ideas the faithful cling but over whose interpretations squabble both conceptually and, often, physically. You have to use conceptual metaphor but aren't these same idolatrous imaginations what the faithless denigrate? If higher truth transcends the intellect then talking, just like babbling, won't achieve Essential Experience - if it exists. In Chapter 5 we'll treat a special form of God Delusion or of Truth, depending on your point of view, in more detail. This is the idea of a Designer God, an omniscient, omnipotent repository of first and final knowledge, know-how and authority. For now let's turn back down to earth.

Metaphysic is described by metaphor. The immaterial is thus granted form but how enduring or reliable is that? Once milked by repetition of the juice of novelty, an image withers to a husk, a shell, a shadow of its nutrient inspiration. Symbols, literally interpreted or simply misinterpreted, soon host useless theological polemic. Holistic creed can quickly set, by repetition, into rites and dogma that will glue you down. The fire of transcendence fades; enlightenment is pulverised to intellectual dust - priestly renderings of 'what the teacher really meant'. **Thus is ghosted, in dogmatic argument, the *third* and ritualistic species of delusion.**

Naturalists can appear as blinkered in their own religion, scientism, as a priestly zealot. Nor is such faithful fixation aloof from partisan interpretation and a central dogma - *PAM*. The partiality decrees a study of the physical alone; equation of all nature with materials; and sometimes an interpretation of religious practice as, at best, an evolutionary redundancy that knowledge of the natural sciences can heal us from. Nor do moral values enter into measurements of such formality - except, maybe, to think a moral absolute unthinkable and in the feast of reason that ensues take up arms for free men's relative morality. In fact, in order to establish a comprehensive, exclusively scientific world-view the secular mind even sets out to embrace various social, ethical and political schools of theory accurately termed 'pseudo-scientific'. In this respect a scientific atheist's polemic tags along. No doubt, with all this rationality, fanatics, religious or secular, spring like anxiety. Fundamentalists, if God is a viral object of imagination that can drive you mad, are an irrationality that, when they're spitting violence, how can humanism rationally cure? Yet fundamentalism is, in fact, of two basic kinds - holism and materialism. A zealot can attach to either one; then in its name pervert the social peace. Does not a mission secular in style also strive to propagate, educate and see its cult and culture win? It may project its creed informally but its 'rational' delusion is, by dint of driving faith, religious just the same.

Entropy (see Glossary) is closely coupled with materialisation. It describes the inexorable process of exhaustion that inhabits all events. Not only energy but also information is subject to such loss of motion, 'downward vector' or, in terms of life, the slide towards oblivion and its rigidity. *Formalisation might therefore be construed as an inevitable process - entropy of information, increase in noise-to-signal ratio or, like Chinese whispers, loss of original meaning from a message*. The heat of inspiration cools. Struck while the iron is hot an idea's flux cools into 'solid' creed and 'set' convention. Principle diverges into multiplicity of explanation and, eventually, frozen dogma. Emphasis now falls upon the exterior rites and rituals rather than the practice of its inner truth. This is entropy of idea and ideal. **In short, the perspective of either of the two pillars of faith tends, when presented in public, canonical form, to ossify and degenerate into cast-iron dogma, unquestioned and unquestioning, of one sort or the other.**

In principle the application of religious formality, whether theistic or not, always seeks to encourage the angel and discourage the beast in centaur-man; the charge of organised religion is to promote good and diminish evil. But in the hands of insensitive, ignorant, unscrupulous or power-hungry individuals it easily degrades into shades of divisive politics, polemic, 'thought-police', an agency of fear and, for the unwilling subject of constraint, a hatred of creed or manifesto-based repression and punishment. Man's long sentence shouts truth's travesty. *Iron-hard religious delusions hurt.* History is replete with social swings involving rebellion and revolution that, in liberation from one such harsh corruption, turns in a while (perhaps even by the way of intermediate, liberal regime) into the next.

Not only social climates change. Sects come and go but even 'monolithic' fashions change. Intellectual cycles wheel around. Whether or not they convey eternal truths the vehicles of philosophy are constantly renewed. **Religions, in whose curtained palanquin Subjective Knowledge of Communion is borne, are casings, just as body cages mind.** They too grow old. Subject to entropy of information even they are vulnerable to misinterpretation, superstition and incomprehension. They seize up. Therefore, what constitutes the fount of youth? What fire melts iron, what sun melts ice and energy rekindles faith? What anti-entropy arises like a phoenix from the ashes?

In casting out motes of tradition, overturning planks of mammon and clearing the weeds of dogmatic bigotry that inevitably clog The Garden of Establishment mystics (see Glossary: mysticism) have always been iconoclastic. **They subvert and refute that *fourth*, oft-bellicose delusion of any particular organised religion to be 'the only way' or 'sole repository of truth'. Religions are nurtured, cultural phenomena. Spirituality is natural and innate.** Such supra-religious anti-bigotry is truly catholic in the literal ('universal') and holistic sense of the word. Transcending the ways of normal human mind they leave its egotism, images of God and group mentality. These transcendent revolutionaries cleanse the perennial philosophy. They have consequently often been reviled and persecuted by the priests and politicians of their time. Evaporating man's delusions with the fire of trans-religious truth they fall foul of 'the establishment's authority. They challenge its corruption and decay with blossoms freshly sprung from nature's own informal, upturned roots. These roots reach, as the tree of life, from above. In time, however, freshness withers. Inspiration atrophies, ritual and formal canon once again invade the social psycho-space. Just what schism or escape from psycho-structures is there for a man or his society? Wasn't each contemporary consensus once resisted? Isn't the 'accepted wisdom' (such as a theory dominant in science) often in its youth decreed eccentric or heretical? Even reformation that arises, phoenix-like, from intellectual ashes only generates, in time, another pile of ritual ash.

With religion, as all else, irregularities are plainer and negativities most obviously crystallised upon its outer crust. Who takes this superficial crust for prize is one who values pearls according to their shell. He estimates an outer truth. Open and yet closed, his eyes incessantly assess exteriors but miss the vital, central point. The point is wisdom's sight - whose pearls are gathered by a single, inner eye.

What, therefore, is the nature of such pearl? What is the fire that melts traditional stabilities or, on the other hand, rigidities? Science tries to write objective principles on changeless tablets made of stone but what Subjective, Vital Information do the saints refresh? Mystics take the 'upward vector'. They concentrate their life, their focus of attention. They ascend and scale creation's peak. A mystic has achieved a pinnacle that fame and fortune's spotlight cannot find. (S)he is more literally 'atop the world' than any climber on the summit of Mount Everest. The summit is eternal life; it is the singularity from which duality of form proceeds. How can formlessness be other than informal? It is, beyond imagination, pure informality. In the Subjectivity of Sanctified Experience formality dissolves. You will always find a mystic at the inmost sanctum of your sacred heart.

Is learning wrong or education bad because two children in the playground fight? But egotism in both camps is a delusion that wields war. *Do you blame pure science for a wish to dominate, exploit or go to war? Or blame religion for violence and perversion in the hearts of fallen men?* If ignorant or vengeful persons battle using the discoveries of science are scientists not as sad as mystics, whose fine truths are also twisted and subverted by power-hungry politicians, mob rule and criminals? Trust in science is politicised as well. Is not humanism's socialistic aspiration - secular society - based on faith that science answers everything? Even the official mind of science, as devoted to its cause as any other faith, is not entirely light and liberal. Its establishments reflect the attitudes, social and economic pressures, prejudice, hopes, fears and even fads of its participants. *The downside of objective science mirrors that of subjective science; and scientism shows the same characteristics as any other system of belief. It is simply secular religion.*

In summary, for non-believers the religious delusion is that metaphysic has existence. For believers various delusions stem from encrustation of ideals, misunderstandings, misinterpretations, fanatical conformities and their hypocrisies. Yet, even in a dusty state of *rigor mortis*, positive ideals outshine their lack. Behaviours in both camps, sacred and profane, range from idealistically 'pure' through missionary to evidence of 'gross impurities' like bias, dogma, counterfeit integrity, downright trickery and war. On the other hand more positive, thoughtful adherents will, within the strictness of their discipline, consider both sides of a well-rehearsed and self-consistent argument. Fair play is, for the open-minded, a flexible formality! It helps to cure both humanistic and religious strains of error!

Models.

Scientific metaphor! Metaphors and models are conceptual hooks. What's invisible or immaterial is hung on images (or, if you like, imaginations). These can mislead, approximate or usher towards a clearer model and a closer truth. In short, mind needs models like a handle so that it can grasp abstraction. Can you handle metaphor? Think of the sun. Call a farmer, nuclear physicist, biologist, musician and a child to join you and describe it each one using media most suited to his mind-set. Although 'imaging' the same object, their descriptions will vary considerably in depth, quality and type; and, since words are less precise than numbers, verbal pictures suffer imprecision more. Think of an atom. From plum puddings through mini-solar systems to electronic pathways images have changed. Have they reached perfection yet? Suns, atoms, an emotion or a virtue are no less *actually* true even if described by several different models; nor, if you don't like a model, can you necessarily deny the subject that it represents.

The piquant problem here is that different societies in other centuries have adopted different model/ metaphors in order to explain both their material and immaterial truths. No doubt, scientific accuracy closes towards perfection when it comes to current physicality; and, if you presume omniscience and don't allow for any unknown factor, it can extrapolate. When, though, it comes to immateriality (mind, information, feelings and so on) models wobble; metaphysical statistics aren't in vogue. Yet, as we'll see, metaphysic fundamentally affects both microcosmic human and the cosmos as a whole. Thus you grasp a model or a metaphor but have to grasp behind it too; you need to distinguish inner truths of physics, metaphysics, philosophy and faith as partial views of one reality.

Thus how shallow seems complaint that metaphorical expressions of immaterial religion vary whilst the model rules of rational, material science don't; or claims that immaterial is irrational or, at most, obtains a lower class of truth. The roots of mind and matter may involve invariant principles but science knows the world expresses massive variation on its basic themes; and so, according to their differing cultural perspectives, do the faiths of men. Idealistic Plato sought, as scientists and mystics do, to divine the principles by which the world is made and operates. These include metaphysical, psychological principles while more Aristotelian, down-to-earth 'realism' seeks physical principles that underlie the origin and operation of bodies. But whether you try to *explain* metaphysical universals (say, archetypes or other principals), personals (say, inspiration or sensation) or physical universals/ impersonals (any material object or event) you use intellect. You bring intuition down or sense data up into that arena. In each case you need concepts, pictures or (we come round to their centrality again) metaphors and models on which to hang the explanations you afford. What fresh models might well serve the polar yet dynamic Natural Dialectic well?

Think of yourself as a reflection of the cosmos - a microcosm of the macrocosm. Your body is demonstrably a fragment of the universal body. Perhaps the greatest discontinuity in nature is between animate and inanimate bodies; nevertheless both are reducible to an arrangement of atoms and chemical reactions. Yet aren't you more than this? **Aren't you cosmos made conscious?** What models might best illustrate the way the *microcosm* and its correlate, the *macrocosm*, work?

Materialism, although it allows three dimensions of space and time's past, present and future, is in a dialectical sense one-dimensional - because it just admits material energies. When a biological form dies it is a case of 'dust to dust'. That change is, for a materialist, the end of it. The information centre, whether mental or genetic, is simply a shape of matter. You can picture atoms as if mini solar systems, model expansive big bangs and develop evolutionary scenarios then clothe this comprehensive infrastructure, layer on non-conscious layer, with the details of the world around.

Holism, on the other hand, includes both subjective and objective conditions. These are universal information (mind) as well as energy (or matter): and, prior to each of these, their Essence - the Supra-mental Infinite (*figs.* 0.4, 0.7 and 1.3). *Its dimensions are, therefore, three. The subjective aspect of life (consciousness or experience) may, moreover, relate to matter without being material.* By this token it might be argued that your mind is as small, impotent yet complex a fragment of universal mind as your body is of the physical universe.

The previous pair, materialism and holism, are basic models covering the universe and life. Most, however, concern specific issues. In this respect it needs be remembered you can only push an analogy, a model or a symbol so far. Indeed, while it makes that useful peg on which to hang a way of understanding anything, is any concept ever absolutely right? It presents a relative or partial picture of the whole truth. This is no less true of each and every separate object whose aspects, as science well knows, you can peel like the layers of an onion. There may be more to things than meets the eye.

Indeed, the 'additional value' that can be squeezed from incomplete models constitutes a sober part of both scientific advance and the intellectual fashion industry. Academic fortunes wax and wane and, as battles are waged, so pictures that compose a mind-set are revised repeatedly.

Incompleteness means that a model is, whether mathematical or otherwise, only as good as its primary axioms, its assumptions. For example, the primary axiom of scientific materialism is that nature is wholly explicable in terms of material objects and forces alone. Any model is designed to satisfy various purposes and, generally, its assumptions will serve those purposes. It may promote an intuitive grasp of a statistically complex situation or provide a framework within which new questions can be reasonably asked and consequent research undertaken; or it may support a particular theory. In this latter, 'political' respect there may arise a need to rephrase issues, obfuscate or even 'change the goalposts' in order to preserve the apparent validity of a theory or primary assumption. Re-working a model occurs especially in the case of origins - the way materialisation of things inanimate and animate first came about. What forces and mechanisms were involved? You might construct, for example, an inflationary model (see Chapter 12); and having pictured how things possibly began then follow up by modelling life's inauguration (see Chapter 20). How can you hypothesise the extensive accumulation of biological information against a tide of entropy that breaks it down to 'noise'? Is it, for example, feasible to claim that 'beneficial mutations' - the scientific thread on which material philosophy entirely hangs - can create novel, coherent genomes? Such benefits of doubt might let the evolutionary model live to fight another day but is the idea actually accurate? You can mathematically model populations, gene pools and genetic drifts but, though these factors play a part in variation, can they cause novel types of organism to appear? Using nothing else but time can you extrapolate from chemicals through cells to the profuse complexity of life on earth? You assume mutation is enough to turn the reproductive trick. Does this assumption pass the acid test; do its consequences match the way that nature really works? We'll take a closer look in Chapter 23.

Other dangers of conceptual understanding are that models may ossify, seem to contradict each other or cause dissension and in this way sometimes distract from the original truth they were seeking to approach. This is especially true of cosmological models of origins - for example, so-called big bang theory. A kind of intellectual 'rigor mortis' seems to gradually pervade each fresh wave of insights, concepts and the terminology that's coined to tie down universal truths. This sort of encrusted exclusivity with consequent self-limitation and dogmatic squabbling is common in departments of religion, science and philosophy.

Light, Bell or Pool

Rings, 'vortical' cone and expanding sphere are three correlated, energetic models of the macrocosm. Another important expression of energy levels or wavebands is a spectrum (which is the Latin word for 'ghost'). A visible spectrum, for example, is derived from 'white' translucency and drops from rainbow violet through red lights to out of sight. Or, in the image of a stone dropped in a pond, concentric rings radiate from a high-energy, central source to lower, peripheral frequencies. Arguably, the natural cosmos drops from Infinite Centre through a conscio-material spectrum of mind to matter; or, in the image of waves, it spreads from a Source to its periphery. Is this all too simple?

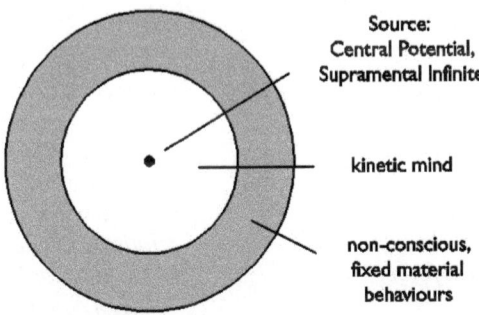

Concentric rings: radiant energy

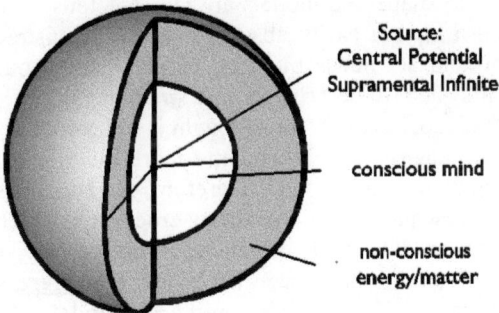

In terms of 3-dimensional volume the rings are seen as concentric spheres

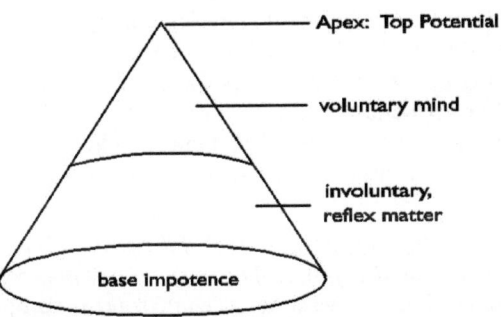

> Take the source as a peak. Leaving the most peripheral ring at base draw this central source up to an apex; the concentric rings now describe a **cone**
>
> *fig. 0.4*

finite	*Infinite*
existence	*Essence*
created/ creation	*Uncreated*
inert/ kinetic expressions	*Potential*
concentric rings	*Radiant Source/ Centre*
spectrum of colour	*White/ All-in-One*
steps down/ up	*Apex*
consequence	*Source/ Cause*
further from centre	*towards/ nearer centre*
down/ below	*up/ above*
more passive/ inert	*more active/ kinetic*
outer rings	*inner rings*
informed creation	*informing creativity*
relatively exhausted	*relatively dynamic*
shaped form	*energetic shaping*
black sink	*range of colour*

What choice is there? For intellectual comprehension there is, it seems, none. Thoughts need shape and models are the skeletons on which they are fleshed. In school a top set easily absorbs a subject's abstract thought and principles; the bottom set needs pictures, parables and models (like plum pudding or a solar system to describe atomic structure) that relate to ordinary life. Indeed, the less capable of abstract thought is a class of intellect, the more down-to-earth the images it needs and the more literally it interprets any metaphor. What encompassing models, therefore, can holism use? How can it relate a microcosm (such as you or I) to the macrocosm in which we take part? Three preliminary models that might help define our own dynamic information centre - mind - have been described. **But there are also three main, simple images or, if you prefer, universal models of creation used by Natural Dialectic.** In macrocosmic terms Chapters 5-12 are based around them: in microcosmic terms Chapters 13-26. The first two, when conjoined, yield a transcendent third.

Images of radiant energy familiar to physics and mystics alike include **The Light, The Bell and The Pool**. In each case concentric rings radiate from a central point of origin. A source generates waves of light, sound or water in a still medium. After a kinetic phase vibration fades away. What is the source of cosmic radiance, from what sound or light or other primal energy is all existence made? In what form (or forms) appears the limit of such influence, the cosmic sink?

In such respect the *first, energetic* model of the cosmos is one of concentric rings (or, three-dimensionally, spheres) losing power as they recede from their *Source*. This model is 'light'. It describes a central source, a projection of power (rather than a once-for-all creation) and a gradient of vibrant energy - just as light dims with distance or the ripples on a pond fade.

It is sometimes possible to understand the same thing in different ways. A classic example is light. Physics tells us that light behaves as a wave or a particle depending on the way you look at it.

Radiation is also a useful analogy for understanding information or mind. In this case mind is wireless or 'radio'; and normal human consciousness is, like the waveband of visible light, just a fraction of the whole spectrum. The range runs from 'high energy' (high consciousness) through to low. If you liken Pure Consciousness to transparent (or white) then mind is spectral, matter black. Awareness is entirely absent in the darkness of unknowing, which oblivion is the very nature of our universe. Non-conscious matter is entirely passive; it can only be informed. Awareness actively informs and is informed. The greater such awareness the more comprehensive is the quality of its information. Information is, dialectically, combined with material energy in a range of proportions, a sliding scale, a conscio-material spectrum; or, in the sense that spectrum marks a gradient, the conscio-material gradient. Existence is an expression of this scale, spectrum or gradient. Gradients, motions between poles, diffusions and so on are scientific second nature; but energy's half-story simply cannot recognise and therefore lacks account of immaterial subjectivity.

The sound of bell or gong fades into silence. Dynamic rings pulse from a central store of power but radiation, even in a vacuum, fades away. If your metaphor derives from heat, then fluid waves diffuse according to the nature of their concentrated cause; or else are locked by cooling into crystalline precipitate. *<u>Potential energy flows through kinetic down to an exhausted level</u>*. Preparation passes through action to completion; things run out from source to sink. Is such dynamic not the heart of physics? Might not a cosmic hierarchy drop from Apical Infinity through finite, existential levels of the mind to an unconscious periphery?

Cosmic hierarchy 'precipitated out' from Great Infinity? Is existential being made from an essential non-being or is it formulated from Supremacy of Being, a Single, Basic One? This paradoxical notion is basic to Natural Dialectic and so before turning to the next model of a pyramid it is time to lay a common schoolboy ghost - the argument from 'from 'infinite regression'. If chains of causes always have a cause then what came First; which Independent Candidate - Intelligence or Mindless Matter - qualifies the best to weave from primordial simplicity such purposive complexity as you yourself embody? If you decide the universe was made by mind, don't minds inhabit only brains? The world is complex, precise and most improbable; you might be tempted to appeal to engineering. In which case what created Super-Brainy Engineer's Complexity? Who or what created God? Was it Super-G then Super-Duper-G *ad infinitum*? How foolish, the polemic runs, can you become? Even if you seek reduction in a modest Theory of Everything (a *TOE*) whence emerged your all-encapsulating Formula? Was it from a Super-Formula or, carved in stone then blazed across the sky, Super-Duper-Formulae *ad infinitum*?

Tetrahedral Pyramid: Material Solidity

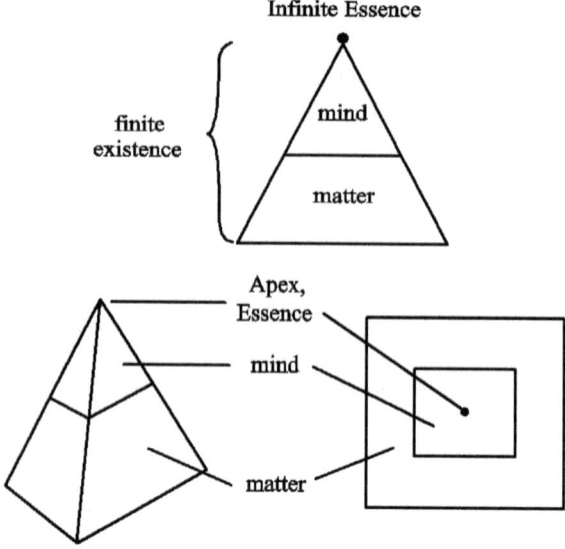

Energy radiating from a source is gradually lost. The concentric waves emanating from a pebble dropped into a pool die out. You saw (fig. 0.4) that cosmic rings can, dropping from their central instigation, be represented as a cone. Conversely described, the cone is a picture of concentric rings with their centre raised smoothly to an apex. **A cone 'squared' is a tetrahedral pyramid and the top-down view from apex is a 'Chinese box'.** The two models are, therefore, essentially the same - except that rings express flux and straight lines fixity. A straight line is a special case of curve - its absence since this curve's 'circle' has an infinite diameter. One illustrates dynamic motion and the other immobility. Whereas curves are associated with vibration, wave and energy, straight lines symbolise the spent phase of energy - locked-up solidity. What about a dot? **At apex or at centre a dot stands for potential, pre-condition or a possibility.** It represents a start, the point of nothing from which lines of shapes are drawn.

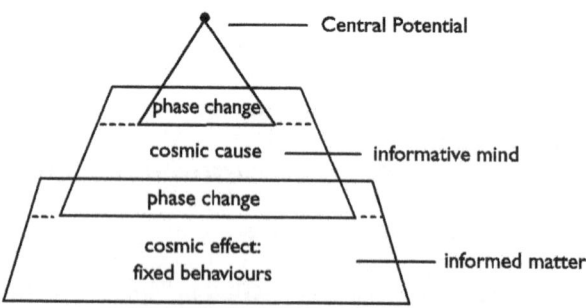

Top-down, viewed from above, the hierarchical grades of pyramid show as the angular equivalent of concentric spheres - **nested squares or 'Chinese boxes'** (see, for example, the Contents Box and *figs*. 8.2, 13.2 or 21.3). Straight lines are a special case of curve, a flat one. Indeed, a 3-dimensional square is a cube and, as opposed to concentric spheres, the 'dead' aspect of solidity could be expressed as nested cubes. Such a nest, where inmost shows as highest, is a ziggurat.

The square cone, a pyramid, describes 'static' hierarchy. A very useful alternative representation is the stepped pyramid, also called the abovementioned **ziggurat**. In this case each step of a **ziggurat** stands clearly for a phase or stage; and the apex of its capstone, a point that points beyond the finite grades below, implies infinity. This capstone, axis or central source of **concentric ring structure is, dialectically, the same as the peak of what we are about to meet as** Mount Universe. Simple geometry. Is it a simple cosmos they describe?

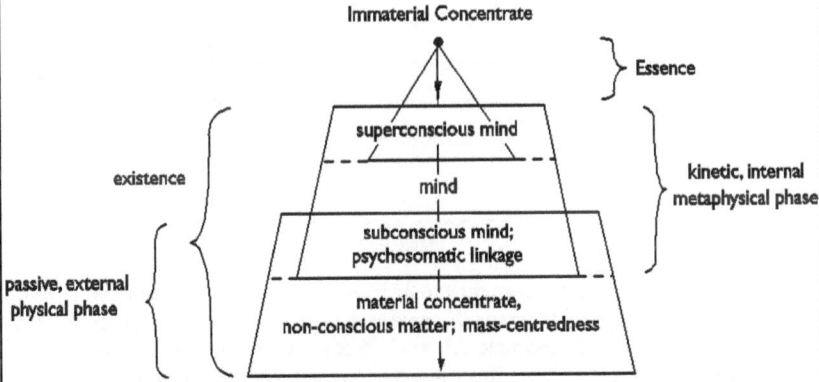

Hierarchy is a fundamental aspect of creation. *If you disagree with that, look no further than your own construction.* It is (see later in Book 2) built hierarchically. In this ziggurat the vertical 'vector' represents a consciomaterial gradient of creativity, that is, descent into materialisation (see also figs. 0.8 and 0.14). Levels are expressed both in terms of energetic continuity of spectrum and 'flat' discontinuity of phase, grade or step. In other words the boundaries between levels are, spectrally, unnoticeably smooth or, phase-wise, noticeably compressed and discrete. Therefore treat levels either way - as different-level perturbations of a single, seamless field or distinct and individual packets, steps or states.

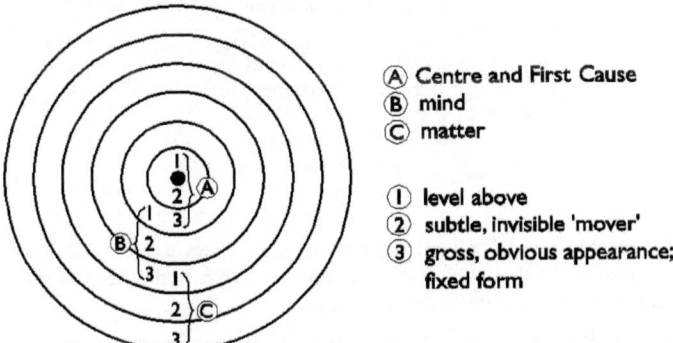

You know rainbows. Their scientific name is '**spectrum**'. A ziggurat is the frozen equivalent of a spectrum. On this phased range of energy each step is seen as a level that intergrades with the next. Each intermediary intergrades with 'phase-change' at its head and base. *It integrates such that the lowest grade of one level is the highest grade of the next.* In cosmic terms the ziggurat divides as follows:

Concentric rings illustrate overlap in a continuous, energetic, spectral form.

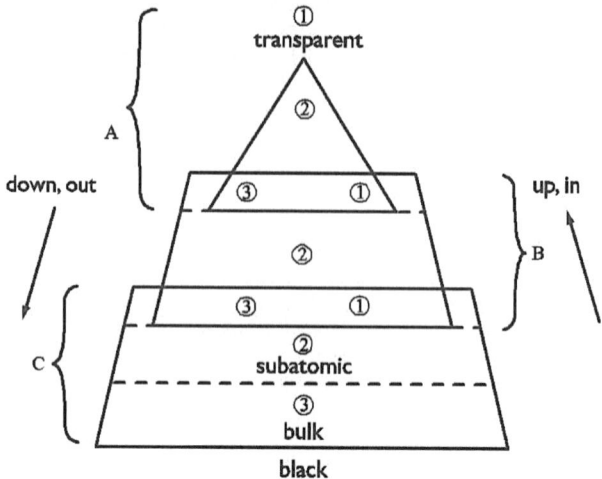

On the other hand, **ziggurat** boundaries are defined by overlapping steps wherein the lowest sector of one discrete step *is* the highest sector of the one below. For example A3 = B1. If concentric rings or spheres (*fig.* 0.4) were imagined as circular or spherical rainbows then the colours of a ziggurat would run from unseen, transparent apex through violet to deep red and, at base, a rim of black.

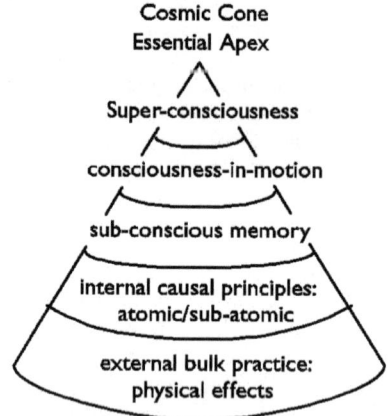

The **cosmic cone**. You see how from Apex things pan out. You see how cosmos is illuminated from a source. What if this source, like nothing else, were numberlessly infinite?

The cosmic cone is a dynamic form of **ziggurat** but you know reality is rougher than smooth lines or regularities of geometry. From now use but also read behind the ciphers. Grasp what the models represent. **Behold, Mount Universe!**

fig. 0.5

Yet an atheist is subject, in the anti-wisdom of his imagery, to the same charge. If you beg the question by assuming your Essential Cause is energetic, then what made energy (or matter)? Was it super-matter then a super-duper-species - pre-bang 'stuff-on-stuff' *ad infinitum*? What came before, *ex nihilo*? What, on the other hand, if *nihil* (nothingness) is non-existent and energy eternal? How then, conceptually, might everlasting, uncreated matter differ from a natural godhead save for its entire non-consciousness? The nature of such metaphysic vexes all cosmology. *The current way an atheist argues from infinite regression is simply to propose, of course, an endlessly creative multiverse (see Chapter 8)*. Such a figment of imagination takes the place of God! The substitute is matter.

If God's improbable so also is the universe; but it exists so why not God? Why, anyway, should an unlikely cosmos not demand improbable a God? You perform, by mind, continual improbabilities. Physics *per se* could not, from natural law, predict invention of conceptually fresh machines nor pre-determine actions rising from desires; nor can it completely understand biology. Aren't your creative thoughts improbable or are you just robotically, purposelessly predictable? If, therefore, the mind of God has specified improbable an actual world why think its rules are random? Is everything improbable by chance or due to information born of purposeful design? You decide but, having once decided, are you naturally right?

Is, at this rate, immaterial God or godless matter the eternal, uncreated one? In fact, an atheist has it worse because the Dialectic's Apex is an actual, numberless Infinity. You can't recede beyond Essential Infinity, beyond the Full Potential that, until it moves, is nothing existential. It is a real Non
-Entity whose motion is The Cause substantiating everything. No doubt that, in creation, it's causality that rules supreme but why should that which sources every motion be created? How could you create a Real Infinity? Why shouldn't such First Cause of every other cause be Uncreated? Nor can Boundless Essence be composed, like some eruption out of physical vacuity, of energy. The character of energy (and thus existence) is mobility. Mobility makes relativity and relativity is not the same as absolution. Nor is perpetual motion or - because by basic law all motions run out of steam - some absolute and un-sourced energy a possibility that modern physics rationally entertains. So there you have it - motion absolute (what drives it?): or substance absolute that simply *is* - Potential Information, Pure Consciousness, the Dialectic's *top-down* choice of Uncreated Boundlessness. If you assert that Uncreated was created then it may be argued (Chapter 1: Causality) you suffer from defective logic, misconception or 'created-god delusion'. This might be difficult to intellectually grasp but mind's forever knotted in its own complexity! What simpler as First Cause than Side-less Singularity, The (N)One; or, if you like, than Nothing that's potential for all other, finite ones?

From this point the dialectical view will emerge increasingly clearly. To clarify its Apex/ Basis let's repeat the case. Prior to finite forms and events was their Infinite Source. Regression cannot fall back further than Infinity - not a numerical, scientific sort but one of boundless 'substance'. What is that? What is the nature of the starting-point beyond which no recession is possible? Could this unassailable buffer, Precedent No-thing, be made of endless pre-material energy? But energy, however pure, involves changeful motion and, therefore, finite forms, time, space and relativity. What involves relativity and relatives cannot be called absolute. If not matter why not immaterial mind? But mind also changes and events exist in time; it cannot be the timeless, placeless essence whence existence flows. What, however, is the central and unchanging base of mind? **Is it possible that Essence is a Concentrate of Consciousness; and that from this unbroken field of possibilities, from this Omni-potential, ideas first arose?**

You want a cosmic organising principle? Preposterously, try mind. Mind before matter! Is there, as this book explores, any explanation how this works?

"What arrant nonsense", atheistic naturalists cry! Science knows, of course, that mind is brain and brain evolved. Mind succeeds matter. You cannot have consciousness before its substance, matter. You might squeeze purpose out of mind but how, since cosmos and especially its life forms seem complex and organised, might 'organised complexity' have been designed by other than a complex mind? So how in heaven's name could G's intelligence evolve before his cosmic panorama did? Where's the necessary brain? How, by the Theory of No Intelligence, can a creator follow his creation? *Ergo* 'God' is nonsense, Darwin sense. Natural Dialectic, on the other hand, has turned the whole thing right side up. It indicates complexity derives from an initial simplicity. Space yields content. From primal unity of silence a few simple notes may be developed, as any composition of a melody can show, into orderly and complex symphony. If an atheist has ill-conceived the act and current of creation so be it. If he misconceives the vibrant roots from which events of cosmos all harmoniously derive whose fault is that? A conscio-material Dialectic indicates, however, the intrinsic nature of The Silence is the Substance of Pure Consciousness. From silence sound; still air vibrates; consciousness-in-motion is the nature of the mind - a vibrant organising principle.

Whilst a vibratory world involves the notion of dynamic cycles and concentric rings its 'ups and downs' are also represented by a concrete illustration of stability, a model of solidity - the pyramid.

From fig. 0.5 the <u>second image</u> to represent the gradient of creation is <u>Mount Universe</u> or, in the conceptual geometry of a cone or pyramid, The Cosmic Pyramid. This model is 'heavy'. It is an inertial, solid picture of existence. Its grades are not smooth; its phases change in 'jumps'; welcome, then, unto the Dialectic's tetrahedral ziggurat.

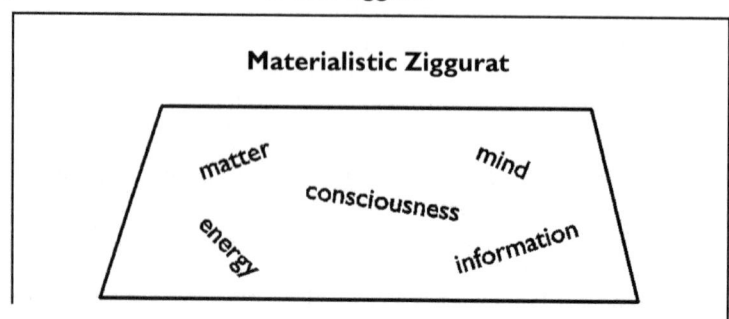

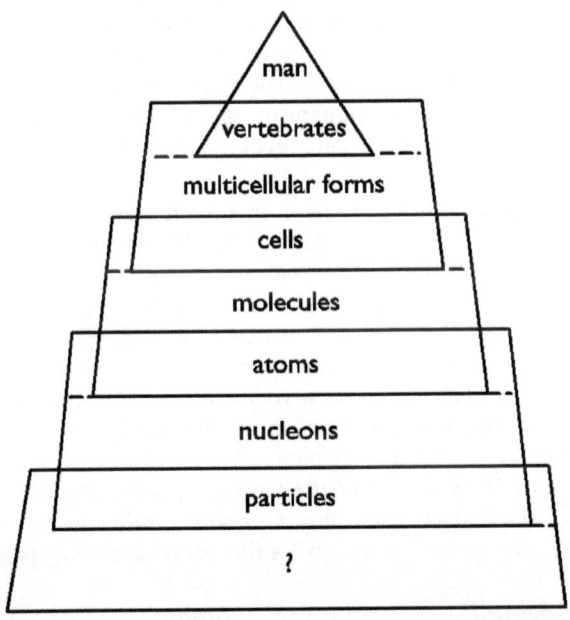

There exists no perception of hierarchy because everything, including mind, is made of material energy. Materialism, therefore, just includes a single, flat domain into which the world is 'squeezed'. It lacks holistic hierarchy (figs. 0.4 and 0.5) but includes a staged crescendo of complexity that, replacing concept, is called natural evolution. Some might write a 'progressive ziggurat of evolution' up like this:

As the section 'Polar Perspectives' will soon clarify, the evolutionary vector 'particles-to-man' builds '*bottom-up*' from a material base. This sort of ziggurat has been dubbed, by a leading proponent of evolution, Mount Improbable. It is. Very. However, what technology's impossible by chance mind's purpose easily makes probable. The brighter mind, the greater is the probability. Could there possibly devolve, as its '*top-down*' antithesis, a conceptual vector dropping 'mind-to-matter'?

fig. 0.6

If you look from above on rings drawn at different levels round the cone, you see concentric rings; and if you look at the squares of a pyramid you obtain 'Chinese boxes'. Such boxes represent systems nested like Russian dolls. They can represent classification running from general to particulars (as with biological classification) or power structures with a king at their heart. And, in the tiers of a pyramid, you obtain a clear picture of hierarchy.

Movement 'up' or 'down' any power structure or concentration slope is 'vectored'. Shifts in mood, quality of information or grade of energy can climb or slide but, spatially, go nowhere. Prepositions are used to describe such 'changes of gear' in a non-spatial way. There are, for Natural Dialectic as for a scale-balance, two fundamental cosmic vectors, a two-way flux or sway. They sway around a pivot.

As games swing round the axis of their rules what, we shall ask, might nature's fulcrum be? How does an ordered world proceed?

At this point it is worth interjecting a one-tier, materialistic as opposed to holistic ziggurat into the equation. There is no difference, materially speaking, between materialism and holism, because the immaterial part does not exist! Mind is therefore material; creation is as flat as the oblivion of things. And if the only differences are those of shape and force whence do mind and purposing come into play? Brain evolved from matter, thoughts are functions of the brain and purpose obviously derives from special matter. It evolved so that, from a scientific point of view, holistic ziggurats are a misleading prompt, mistaken models and an evolutionary profanity. They are medieval where 'material alone' is modern; and 'evolution' constitutes materialism's hierarchy of creation.

Stick with 'medieval' for a moment. In holistic circumstance let the apex of a cosmic cone (*fig.* 0.5 vi) stand for Infinity. Below its tip divide the cone into upper and lower sections. As a gas is gradually solidified imagine these sections represent the two main stages on a creation gradient or two main divisions of a spectrum through which, by degrees, the Infinite is materialised. Not energy but subjectivity is lost; consciousness is thickened, life gradually 'lignified'; matter is a very colourful but lifeless shadow of the Day. The scale slides from prior potential through kinetic to a solid, static base or, in other words, diffuses through a hierarchy of existence from mind to matter. *Centre is its apex and the base periphery. Down and out from centre are, for the hierarchical images of rings and cone, the same thing. So, conversely, are up and in.* To give a flavour of antagonistic vectors you might write:

>   *outward*   *inward*
>   *downward/ descent*   *upward/ ascent*
>   *passive*   *active*
>   *gravity*   *levity*
>   *body*   *mind*
>   *gross*   *subtle*
>   *heavy*   *light*
>   *mass*   *energy*

'Down' therefore implies materialisation. Its vector gravitates towards darkness, tension, confinement or blockage of flow. Levitation, on the other hand, reverses this polarity. It ascends towards increasing lightness, subtlety, freedom and unification in principle. To elicit a flavour of psychological complements (or anti-parallel inversions or antagonists) write:

>   *down/ low*   *up/ high*
>   *disaffected/ bored*   *keen*
>   *sad*   *happy*
>   *depressed*   *exhilarated*
>   *inner deadness*   *liveliness*
>   *separation/ isolation* *connection/ togetherness*
>   *loathing*   *loving*

A pyramid exists. The celestial point of infinity into which it disappears we call Essence. *The distinction between essence and existence is fundamental. Existential duality springs from essential unity. Three in one. Tri-unity.* Natural Dialectic will subject both principals to scrutiny but for the moment just hold in mind a three-tier structure both of man and, which his structure reflects, the universe. Body, mind and soul.

Now imagine you are on the pyramid. From the top, looking down, you can see that everything is connected. As you descend the connections become less and less obvious. *Descent is a separator; ascent a connector.*

Human mind pivots at its 'cosmological axis' between inner psychological and outer physical worlds. Below that point you turn objectively outward and downward to the material levels of separate things. Above it you turn within; subjective ascent is a connector. The heights of Mount Universe therefore represent an inward, invisible gradient. The subjective rather than objective coordinate of existence predominates. As you climb inward and upward from external details to inward principle you see more and more connections. And at the Top, the Inmost Centre will not all be unified? Will all not be seen as aspects of The One? One Truth? Everyone, holist or materialist, is in the same predicament. **All seek the truth and there ascend many paths towards it.**

Mount Universe appears, like any other mountain, different according to your position on it. Its gradient of creation disperses, *top-down*, from the Peak through a metaphysical (mental) region to the peripheral, physical detail of foothill terrain. Science, on the other hand, starts from detail at the base and works the slope from *bottom up*. Out of myriad difference and detail it rises to discern underlying patterns, principles and connectivity. For example, it has simplified the basic 'kit' from which every object in the universe is composed to about a hundred elements; reduced these to three 'normal' kinds of particle - the proton, electron and neutron; then further simplified - just quarks and electrons seem, at root, to make up everything. But what makes these? The higher it ascends the more it collects, collates and interlinks. Its holy grail is a Theory of Everything (*TOE*) that explains the world in terms of a single principal from which all others are derived. Will this *TOE* have reached the cosmic climax?

If mind is real yet subjective how can objectivity describe it? How can physical measure metaphysical? The materialistic presumption is that no metaphysical 'ghost in its machine' exists. The diagram of concentric rings indicates that such a view, angled from the outer circle and including only outer circles, is literally eccentric. And the fact remains that experience *per se* cannot be reduced to mathematical construction. In this way, although it may seem imminent that a unified theory will close the final gaps in our physical knowledge, the intangible, subjective element of consciousness may yet blow matters wide open; it may drive a coach and horses through materialism's yard; it may, reversing its polarity, depolarise our thinking. Meanwhile science may never, unless it lifts its own philosophical constraints, reach above the lowest tier of the pyramid. Its climax is the top of bottom grade. There may be climax but there is no Climax there.

Natural Dialectic chooses for its models sacred geometries. Concentric spheres are radiated from a source; you think of struck gong, bell or eternal flame. Nested steps rise to the apex of a ziggurat; holy mountains stand for hierarchy. Mythical Meru might exemplify Mount Universe but other really rocky symbols - Kailash, Fuji, Sinai, Olympus and *Jabal-al-Nur* (the hill in whose cave of Hira Mohammed received the angel Gabriel's first Koranic revelation) - also rise up round the world. *For complex thinkers these two models, radiant rings and cosmic 'cone', may seem a laughably simple way to picture and thence detail the processes of cosmos - until you probe a little deeper.*

In summary, two iconic models covering Natural Dialectic as a whole are radiant Bell/ Pool and static Ziggurat. Soon, after an important spot of orienteering, it will be time to introduce a *third*,

The Cosmological Axis.

Who do you think you are and where? What is the direction of your short-lived pass through time and space? A traveller wanting to arrive need first locate exactly where he is and where he thinks he's going. Every explorer needs to *orientate*. Only then he can plot with military precision the bearings of his destination and begin the journey. Isn't life a journey? But a disoriented traveller simply wanders aimlessly before he dies. It is therefore important to identify constant points of reference, compass bearings that can unfailingly let you know where you stand.

We saw that for a materialist, whose focus is predominantly outward towards the sensible, physical world, it appears that a human is an animal composed of chemical parts. It is a child of chance, a minuscule creature cradled in the vast, dark, cold and sometimes violent abyss of space. Its life is simply the outcome of molecular arrangements and its consciousness, where manifest, derives from one of these (a brain). Because life is a physical body which is, in turn, simply a unique, very complex but temporary atomic configuration there can be no life before or after death. The 'purpose' of such bodies seems to be to survive in their present form as long and pleasantly as possible. Whatever best frustrates frustration's bitter rain and gains this end is 'good'. This, an axis of survival, is the bearing of a body's compass.

This is certainly part but is it your whole truth? The holistic view is different. It includes an axis of subjective mind. With respect to mind the previous and next two simple, nested diagrams are of interest. Three points on these simple models can act as a map to obtain cosmological bearings.

First is the psychosomatic (mind-matter) interface. This, labelled *PSI*, exists at the base of mind's least lively part - subconsciousness. Waggle the little finger of your right hand. This specific act identifies a matter of cosmic relevance. You have registered a higher level of order on matter. Somehow information has caused matter to move *on purpose*. Touch something to reverse the process; now an arrangement of matter has registered on your consciousness. You used, down and out, a motor system; and, up and in, a sensory. In each case, it seems, information has crossed a psychosomatic border; the exchange involves a localised, psycho-physical axis of events. Natural Dialectic would (*figs.* 0.8, 0.11 and 1.3; also Book 2: Chapters 15-17) suggest this frontier zone 'entangles' with the sort of physics we call 'quantum'.

Cosmological Bearings (i)

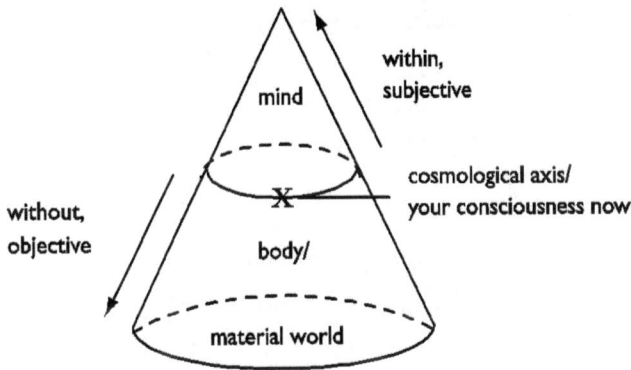

Could this cone stand for you and me? Your visible body is a special case composed of elements of universal body. Is your invisible mind a special, compound case of universal mind?

Central Pole;
Essential Cosmological Axis;
Cosmic 'Pole Star'

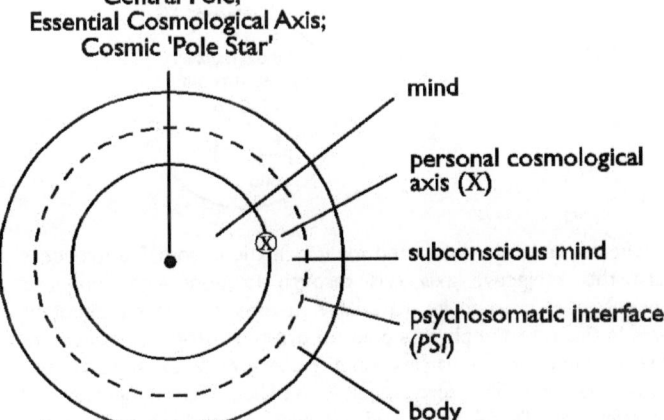

This is the cone viewed from above. Could such a 'micro-cosmic onion' reflect a 'macro-cosmic' one?

You could even represent the Dialectic as a 'cosmo-space', a 3-dimensional graph with conscio- and energetic axes set in space-time. An event of any kind moves from its point of origin. **What value do we put on origin?** Zero? Nothing? Do things start with nothing? Things or events derive from prior potential; this is their transcendent cause; form possibility they are reduced to single actuality.

First Cause rests 'on high'. Mark the Essential Origin as Infinite Capacity. Its pre-creative immanence is out of view above the scheme; it is the Origin of the whole three-dimensional, existential graph. Such Transcendent Zero is Most Positive. It is creation's culmination and, at once, its *top-down* starting-point - Potential Information or, as was noted, Pure Consciousness. Locate this (N)One atop the y-axis, the informative coordinate.

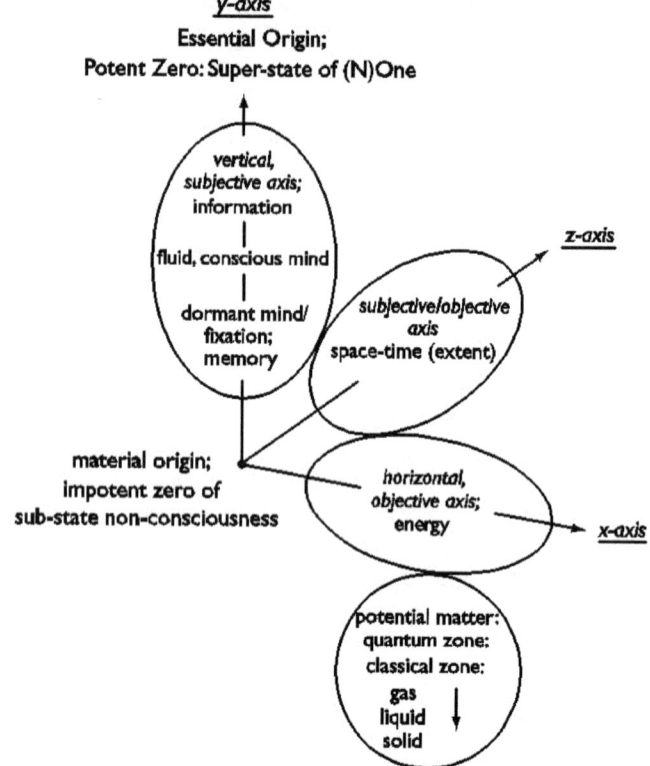

Let the vertical y-axis stand for information; from Transcendent Zero this subjective axis falls through lowering conscious and subconscious states to its material, non-conscious base - impotent zero. In this case the physic's point of origin is marked as 'no extent of space and time', no manifest kinetic/ energetic factor and a zero at the base of mind. This zero is, below the filing place of memory and archetype, mind's sub-state void. Oblivion. From here extends the whole objective/ material coordinate.

Let the z-axis represent extent in terms of psychological and physical space and time.

Finally, let the horizontal, x-axis stand for non-conscious energy. From origin (potential) arises cause; cause creates effect. Where once was nothing (or no action) things occur. In cosmic terms Transcendent First Cause drops through its effects. A void gives rise to subtle (quantum) and thence gross, aggregate effects; States of high energy drop into low solidity; this is the effect of entropy.

Including mind and matter, this diagram graphically illustrates the different levels on which change - the fundamental nature of existence - operates. Note that within a scaffold of **conscio-material coordinates** material events lack all subjective contour; the mindless, passive transformations chemistry and physics map are 3-dimensional but 'flat'.

fig. 0.7 (see also figs. 0.10, 1.3 and 10.3)

If such an interface exists between mind and matter, your consciousness (that is, you) is near but not at the border. Closer than breathing, this crossing-point's subtle nature needs to be explored especially if, by extension, it exists on a cosmic scale. Its study may reveal a real connection between physics and subjective psychology - the point at which mind and matter meet. *A spectrum or continuum that, crossing the psychosomatic border, includes awareness might form the basis of a Grand Unified Theory with some serious claim to completeness.*

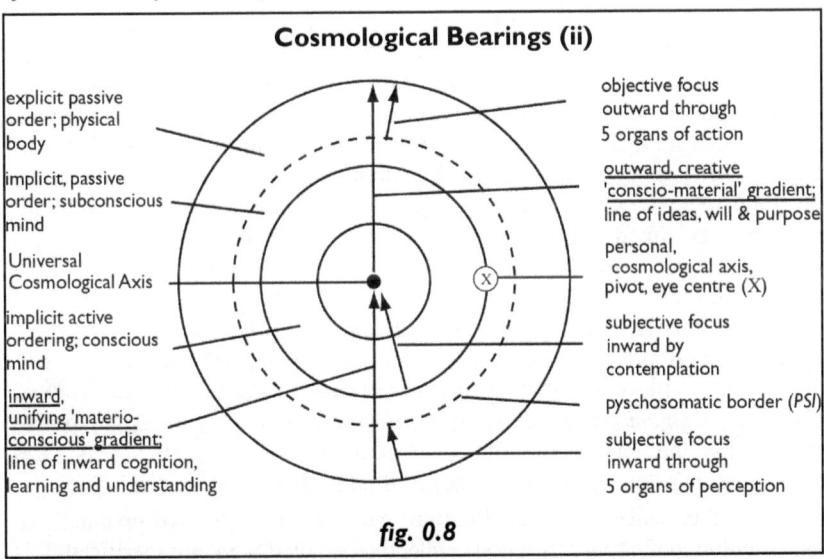

fig. 0.8

X marks the spot. Touch your forehead at a central point above the nose, just above and between the eyes. Just behind this, marked X on the diagram, is where you think. *This location, identified as the **second** point of interest, is called the eye-centre.* Though not physical this centre is your vantage point, a cockpit of consciousness sometimes also called the 'third' or 'single' eye. This is the non-sense eye. It is information's point of concentration, the hub of embodied life, a metaphysical pivot round which the scales of human judgment swing. All campaigns, from trivial to grand, are led from this centre of command and control. *It is not the central reality of the whole universe but it is certainly yours.* Just as a geocentric view of cosmos used to dominate men's idea of the universe, so an egocentric vision dominates our lives. This axis of mind, local and dynamic, is involved in changeful relativity. **On the conscio-material gradient of Mount Universe, it also marks the interface between conscious and unconscious, metaphysical and physical, immaterial and material dimensions.** It marks, from a holist's point of view, your point of exchange between inner and outer worlds, your pivotal link between the furthest reach of physical space and creation's Inmost Core. *This 'third eye', where coordinates of mind and matter meet, is your constant point of reference, pole star or personal cosmological axis.*

Look at X, the light of human life, another way. Let the form of cosmos (illustrated in *figs*. 0.8-0.10) be expressed by you. 'Outside' (which includes your body) find the energetic world of mindless things; brain and subconscious mind control your interaction with this world except for what you call your 'conscious self'. No more than our sun is central to the universe is this 'self' central either. Its 'cosmological axis' is a relative position; it represents the changing truth of where and who you physically and psychologically are; it is your temporal experience. Move 'inward' towards the central super-state or Universal Cosmological Axis. Could you climb up to the Central Apex of creation; is it possible that you could realise The Absolute? At least, from Natural Dialectic's point of view, you are at once at all locations on the cosmic dial. You *could* reach it but the centre is for now beyond your scope; the place you presently reside is lower down at X.

From X, therefore, concentric rings of your environment radiate. Outward rings spread into a world of bodies - the physical; but inwardly, the metaphysical direction, rings evolve towards the Centre of their Concentration. In this sense a Hindu or Buddhist takes the third, internal eye as a sacred, inner axis around which human thought and, therefore, life revolves. This fundamental axis of your symmetry is sometimes outwardly pinpointed on the forehead not by an X but by a painted red spot (*tilak* or *bindu*). A Muslim *sufi* calls it 'Ka'bah's arch' and Christ exclaimed 'If thine eye be single...' This is your thinking point. It is the single eye of knowledge set, inwardly and symbolically, at the apex of a triangle that includes the two outer, physical seers. Unity bestrides duality. To *concentrate* means, literally, to centre thought. The Greek word '*temenos*' means a 'temple precinct'. From it probably derives the Latin word '*templum*' which means a place set apart for divine inspection (augury) or a consecrated holy place. *Contempl*ation has come to mean observation, careful thought or focus inside the temple of your own skull at the eye-centre. This (*sat*) centralising, *middling* process of concentration is also called *medi*tation.

In a further sense, therefore, the third eye becomes a 'sacred precinct', an altar whereon individual restrictions collectively called '*ego*' are sacrificed. It is the starting-point for inward, contemplative travel. From toes to nose - when attention is withdrawn from the body to this point the outside world disappears. From sole to soul - this state, absorption, is the precondition for mystic ascent. His absorption is the opposite of dispersal; dispersal is 'down' from X through the senses, absorption comes from withdrawal 'up' into X. His ascent is one from relativity to Absolution, that is, to the Natural Centre of the Universe. Thus, as was suggested, according to *figs.* 0.7- 0.9 X marks a point of balance, your personal pivot between inner and outer worlds. *Such a cosmological axis can become, if you wish, the launch-pad towards a **third** point of interest - The Universal Cosmological Axis.*

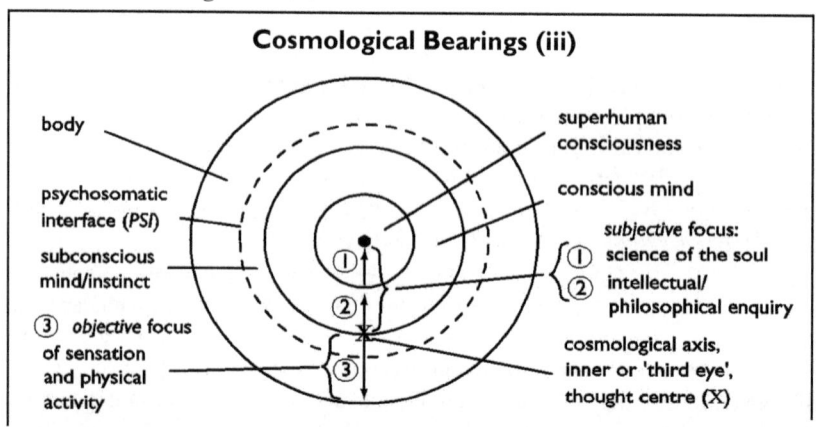

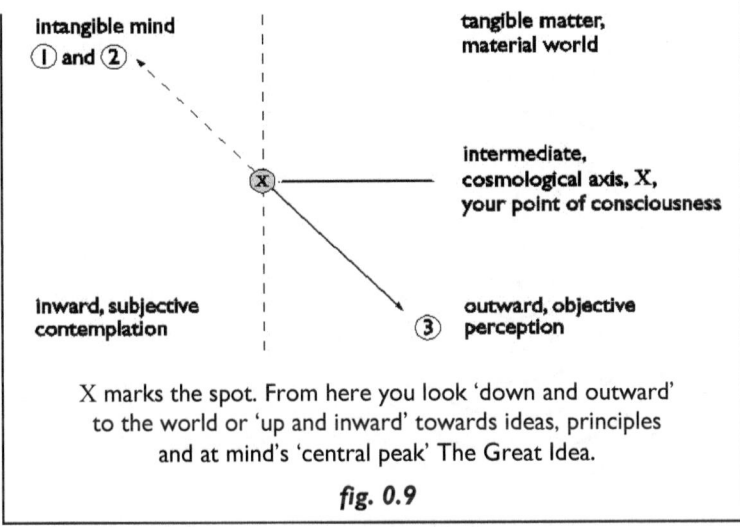

X marks the spot. From here you look 'down and outward' to the world or 'up and inward' towards ideas, principles and at mind's 'central peak' The Great Idea.

fig. 0.9

This Pole Star, Absolute Point of Reference or Pivotal Truth is within your own more outward cosmological axis. This Axis is *within* the ceaseless hive of mind. Mystic practice is thus, literally, to concentrate on centralising. The program is a gathering up and targeting; it is withdrawal of the mind towards its Inward Centre rather than dispersal in material circumstance of sense; it is to 'set the dials' from lesser, individualised aXis, 'make towards the Heart' or, in terms of *fig.* 0.8, no less than climb towards the Source of three-tiered universe; if Axis is the Origin then such ascent amounts to a return. In Christian terms a human body is a 'temple of the living God' wherein 'if your eye is single your whole body will be full of light'. Sub-human does not understand the human way. Do humans understand their superhuman possibility? Such luminous condition is, even if unrealised, entirely natural. It is neither rare nor exclusive. And, because it is incorporated in the potential of every person, you also have this Essential Pole within yourself. Indeed at Heart you *are*, unbeknown, this Invisible Super-state. If anything is unavoidable it is the Heart of nature, the all-encompassing Essence of existence!

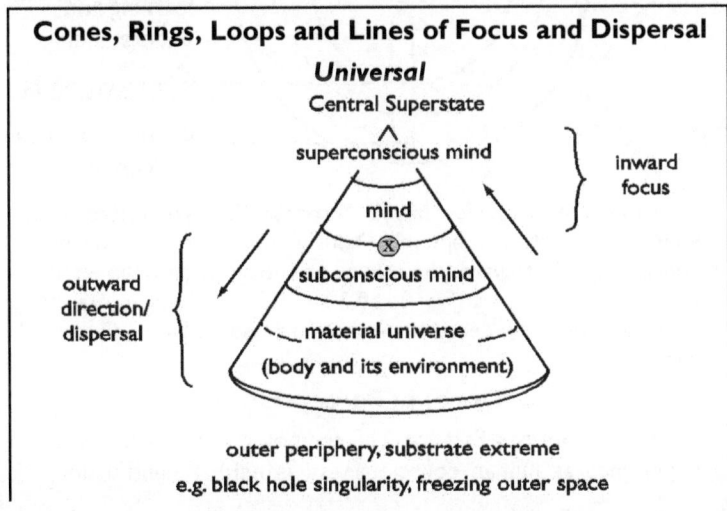

This universal cone is looked at from the side. You are identified as the microcosm of its macrocosm. Now look at it from above.

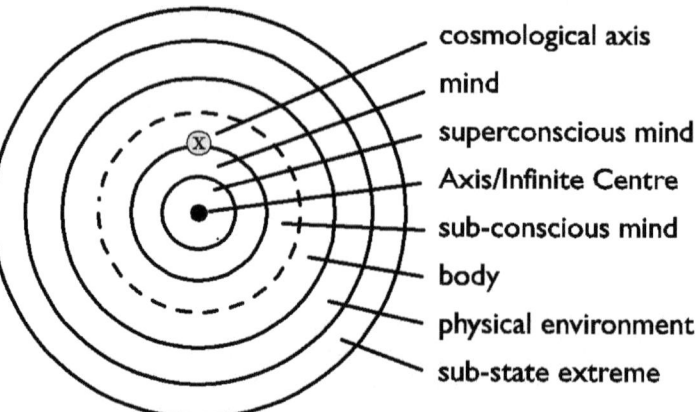

- cosmological axis
- mind
- superconscious mind
- Axis/Infinite Centre
- sub-conscious mind
- body
- physical environment
- sub-state extreme

It radiates, *top-down*, from centre to periphery. Infinite Potential is the Centre, Pivot or the Axis round which cosmos 'orbits'. It is the Source or Sun on which existence depends. Body, local environment, earth and universe are, being all material, of the same peripheral zone. They compose the circumference, that is, outer limit that encompasses creation. Chapters 5 - 17 explain how the perspective of a **conscio-material gradient** works.

Personal

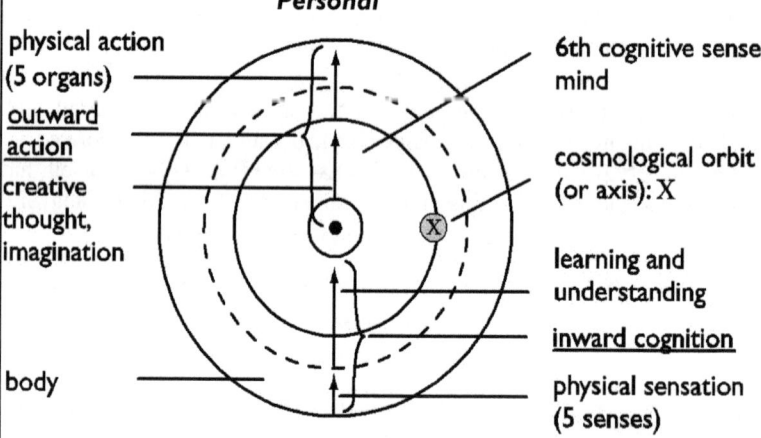

physical action (5 organs)
outward action
creative thought, imagination
body

6th cognitive sense: mind
cosmological orbit (or axis): X
learning and understanding
inward cognition
physical sensation (5 senses)

This is a *top-down* view of you. **X marks the eye-centre**; it is your 'door of perception'. You may open it inwardly, contemplatively, towards the centre; or outwardly onto the world. Normal consciousness (*figs.* 0.1, 0.2 and 0.3) oscillates across the threshold with a tendency skewed towards sense-based dispersal into the world.

Doors of Perception

But X is not this Essence. The cosmological axis which we experience as human consciousness, is neither central nor peripheral. In universal terms it is neither at the base nor peak of Mount Universe but, like the waveband of visible light in a full electromagnetic spectrum, somewhere in the middle. X marks the spot. *From this position you can look down and out (through the senses) to the lower slopes of physical matter; or up and in (through contemplation) to the higher peaks.*

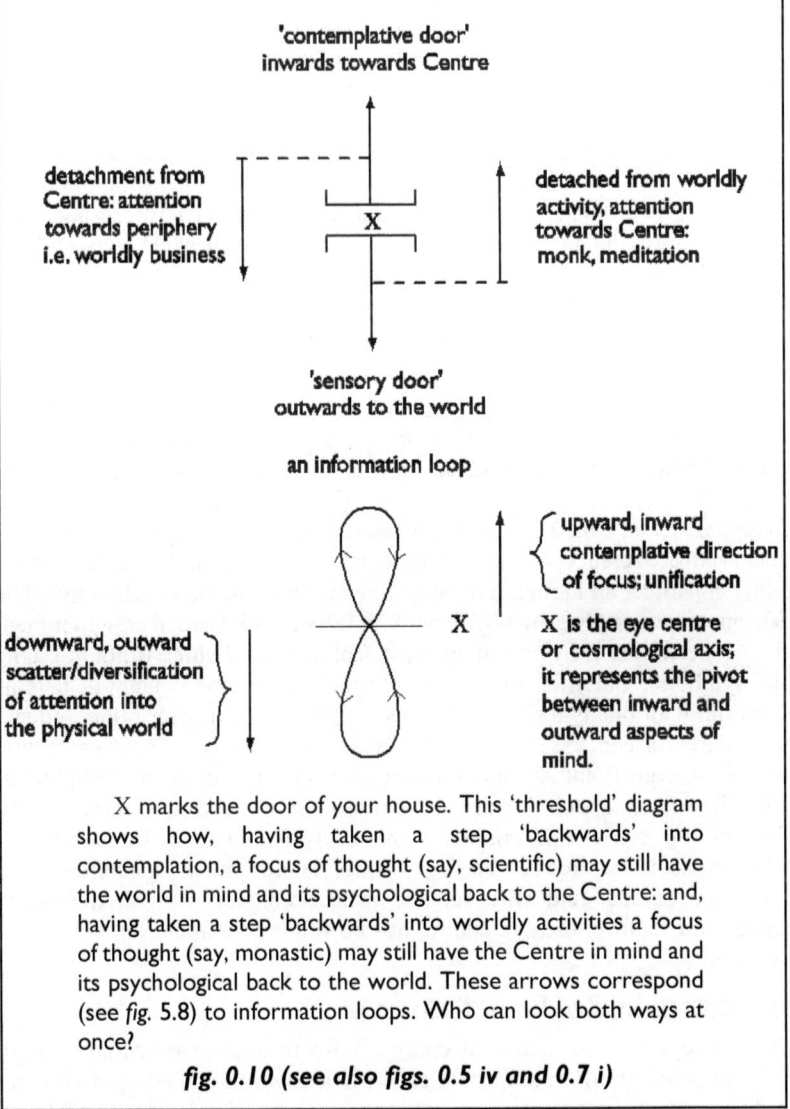

X marks the door of your house. This 'threshold' diagram shows how, having taken a step 'backwards' into contemplation, a focus of thought (say, scientific) may still have the world in mind and its psychological back to the Centre: and, having taken a step 'backwards' into worldly activities a focus of thought (say, monastic) may still have the Centre in mind and its psychological back to the world. These arrows correspond (see *fig. 5.8*) to information loops. Who can look both ways at once?

fig. 0.10 (see also figs. 0.5 iv and 0.7 i)

These two directions of your focus (which constitute your life) are alternatively represented in a second diagram.

Existence, in this scheme of things, is a dependency. A pendent. A suspension. Prior to existence there is only Independent Essence. Check *figs*. 0.5 iii and 0.8. Prior to creation there was Nothing. Below Essential Nothing hangs the world. If this is Nihilism, let it be! You could exclaim that round the

Hub of Nothing turns the wheel of worlds. At that Universal Axis, says the mystic, let me be! Or, yet again, the Hub's a Seed from which, inverted, springs the cosmic tree. If this tree is all around me, every cell connected up, then let me know the power and greatness of its grain!

The first direction ranges outwards from your sensational body through the variegated environment of your habitat, planet and, more usually thought of in astronomical terms, the starry heavens. The exchange of information is incessant. You affect and are affected by your terrestrial home. This objective focus is the way of sensation, science and technology. It is interaction with the world at large.

The second, contemplation, takes the principle of focus and exercises it. Its inner gym is where you practise laser-like, coherent focus of attention. Education plays its part. Study concentrates the mind. The more intense a focus of interest the faster time flies; a voluntary, unbroken coherence of concentration is at the root of happiness; and the specific purpose of the meditative science of the soul is to perfect concentration. Its laboratory is not down the road. It is incorporated at the heart of the system, at the cosmological axis, your eye-centre.

Thus the first step in 'middling' meditation is to achieve a highly coherent or concentrated focus of attention at the eye-centre. From this point the goal, following an inward transport or ascent, is to reach the top of Mount Universe and thereby achieve identity with the Centre. *This identity is called Enlightenment - in which a completely top-down perspective is obtained.* It is not materialistic science; teacher, method, practice and completion-in-identity together constitute an immaterial study you can call 'the top psychology'. The path is one of voluntary growing up or, with full potential called enlightenment achieved, becoming fully grown up. Such Communion (which is not the same as the symbolic, Christian sacrament, Holy Communion) is natural. It is as clear as sunlit air of any trace of religious or philosophical description. Indeed, it is the pure, unseen, essential air of everyone's subjective life. In short, two 'doors' lead from Point X. One leads out through the senses to the material world; the other opens inwards towards the Subjective Centre. Every individual both senses and thinks but this analysis explains why the mystic, whose emphasis is 'above' his eye-centre, is not much concerned with materialistic details; and why science, whose concern is with the world 'outside' and 'below' the eye-centre, has no time for immaterial, subjective realities.

Polar Perspectives: *Top-Down/ Bottom-Up.*

Source to sink: the course of energy flows from prior potential through activity to completion. Pole to pole such hierarchy's modelled by a ziggurat (***top-down* from apex**) or by concentric rings (***central outwards* to periphery**). You can take these metaphors for cosmos or (*figs.* 19.1 and 19.2) for microcosmic bodies made of cells but at this point we turn from energy to information. We contrast two perspectives, polar angles from which people tend to view the world.

range of expertise   Top

imperfect	*Perfect*
incomplete grasp	*Truth/ Principle*
student/ teacher	*Illumined One*
degrees of understanding	*Knowledge*
learner/ less knowledge	*more knowledge/ teacher*
lower/ more peripheral	*higher/ more central*
doubtful/ ignorant/ in the dark	*more certain*
working towards light	*relatively illumined*
bottom-up	*top-down*

Is there Truth Absolute (see Chapter 3)? Aren't psychological and physical perceptions all a case of relativity, of partiality in scope and certainty? There may be an Expert Teacher; other teachers' knowledge is, relatively, less than all-embracing. At any rate, in any field, top-down is an expert's view. Those who 'know the lot' look down on those who don't - but not at all in snooty ways. For example, you speak your native language fluently and are, as such, an expert who can immediately and accurately express anything you want; but you no more treat someone of lesser fluency arrogantly than a teacher his pupil. On the other hand, when you hear a foreign language it seems like smog at first. It's 'Chinese'. You don't understand a thing but gradually, bottom-up, you grasp grammar and its meanings so the mist dissolves. Soon you're standing, expert now, in the clear air of fluency.

Of course, the higher the level of your understanding, the more competent become your expertise and your ability to express things accurately. It is the same in the case of creation. Experts create according to the level of their expertise. The nearer you are to 'knowing the lot', the more effective your artistic, technological or ministerial work.

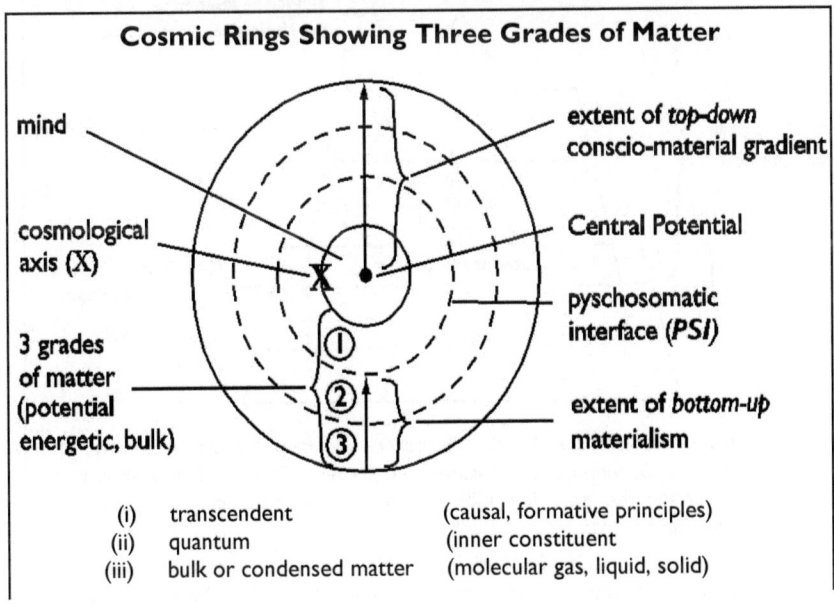

(i)	transcendent	(causal, formative principles)
(ii)	quantum	(inner constituent
(iii)	bulk or condensed matter	(molecular gas, liquid, solid)

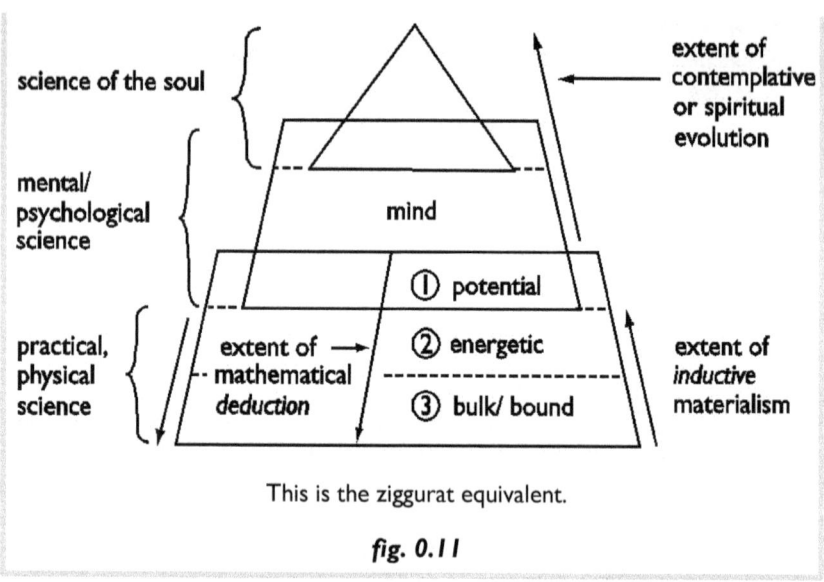

This is the ziggurat equivalent.

fig. 0.11

The fact is that, while a Buddha might obtain the summit of existence, most humans are relatively ignorant. They dwell at Point X, well down the slopes. Point X and all 'above it' is subjective; below extends the objective, physical world. As this applies to your own microcosmic construction, so it applies in universal terms.

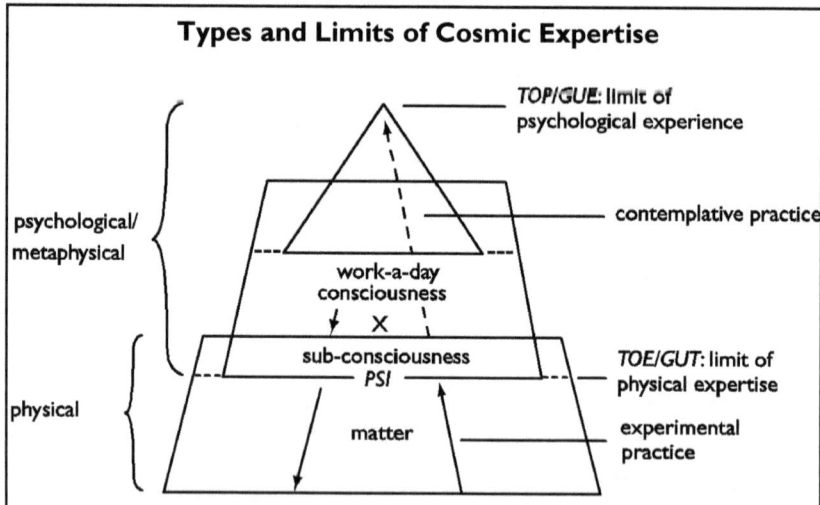

Turn on, tune in, arise! In this diagram two kinds of learning are displayed. Arrow rises, *bottom-up*, from student to expert. The 'hard' line indicates physical study; the 'soft' line indicates inward, metaphysical technique and practical experiment.

Note that the limit of objective, scientific expertise ends at the source of matter (*PSI*), that is, where the appearance of material phenomena can first be recorded. This is generally assumed absolute at the 'grainy' level of Planck dimensions. Even mind and consciousness are assumed to be, somehow, physical entities (fig. 0.6 i). There is, in this view, nothing metaphysical or 'supernatural'.

On the other hand, arrows drop, *top-down,* from expert to student. The expert, knowing his subject, creates and/or teaches accordingly. This perspective includes subjective mind, knowledge and the nature of consciousness. From this perspective objective, sense-based observations involve looking darkly through a nervous glass. Sensation sees the superficial nature of the world and science views material cosmos in this way. And the bracket "psychological' is only supernatural in such objective, naturalistic terms; in fact, mind is natural, immaterial and the start of its inward, meta-scientific observations begin at a point of conscious experience, the aforementioned cosmological axis, Point X.

Expert knows, student learns. 'Experts' vary in degree of understanding but whosoever understands metaphysical *GUE* or physical *GUT* is, in his realm, the ideal, omniscient expert. Does he yet exist? Students, not least physics students (which includes professionals), work towards grasping the most profound, underlying truth or truths pertaining to matter. Who, on the upper hand, is expert, who is student when it comes to *TOP/GUE,* that is, cosmos in its subtle, immaterial part?

Natural Dialectic would place informative potential prior to energetic expression. It thus places mind prior to or 'above' non-conscious, purposeless matter. There is however, given this prospect, no pejorative implication that one of these two kinds of practice/ expertise is higher or lower, better or worse than the other. *Top-down* creates with competence; *bottom-up* learns how things work. Both perspectives involve knowledge; one seeks, the other's found. There exist many levels of comprehension but for each vector the source-and-target, alpha-and-omega or be-all-and-end-all is a *GUE* or *GUT*. This way you can understand the maximum efficiency by which the smallest clutch of principles (or one) may regulate the largest range of things (perhaps all). While there is clear indication that their subjects/ objects of interest, direction of focus, goals and therefore ultimate truths are different, there is no indication that both do not, in their own domain, draw entirely true conclusions.

The banal message of *fig.* 0.12 might be 'seek out experts; obtain a teacher with the highest understanding of his subject'. His grasp will involve its central axioms, the ones that let you work out how the object of your interest works.

Finally, there exists a band of natural opacity. Between the upper limits of experimental science and Point X is sandwiched a dialectical band of existence called sub-consciousness. This band (X to *PSI*) is difficult to research; but its logical, archetypal nature, derived from its position in the pyramid, is dealt with in Chapters 7-12 and, especially, Chapters 15 and 16.

fig. 0.12

Humans observe a vast astronomical universe and its myriad aspects through senses and their technological extensions (microscopes, oscilloscopes, telescopes etc.). We gather information and then, thinking about it, contextualise and incorporate it into a personal, cultural or religiously held 'encyclopaedia' of knowledge. In this way an inquisitive child learns by observation, consideration, question, experiment and instruction. It learns a language, how to read and write, carry out tasks and relate to other lives. It learns its place in society and the world. Child, adult, student or expert, it is the same thing. Inquisitive explorers strike out from imperfect frontiers into fresh and rough terrain; soon they colonise, incorporating their advances in the human fold. Trial and error, suck and see. Yes or no. There is no guide and the method is, therefore, entirely empirical. Observations are condensed to principles from which a qualified bushwhacker can predict. The ancient nature he discovers and exploits is only new to him. *This bushwhacker method is the student's.* **A student's *bottom-up* approach is at the heart of scientific practicality; it is also why, in mystic practice, student 'sits at teacher's feet' means, simply, 'attends with devotion looking upward' in that class.**

Therefore, with childlike but careful curiosity, seekers after either physical or metaphysical knowledge track from fragmented events to pattern, from apparent chance to reason and from practice to principle. Having observed some physical object or behaviour one asks a question, hypothesises an answer, tests this possibility and comes to a conclusion. Was my informed guess true or false, did it pass or fail the test? By a gradual accretion of well-tested information it is hoped to progress until the mists of ignorance and doubt finally, perhaps in a *GUT* or *TOE*, evaporate. **At least, we trust our expert tells us what for sure is true; or lesser experts qualify their insecurity with honesty.**

left-hand hemispheric tendency	*right-hand hemispheric tendency*
detail	*global view*

In Chapter 13: To Build a Brain we'll look at hemispheric tendencies of brain reflecting an innate polarity of mind. These complementary and yet potentially antagonistic tendencies affect the way that we attend to life and thus experience its show. Here let's simply note that right-hand bias tends towards inclusiveness, global grasp and a deductive frame of principle-that-leads-to-practice.

reductionist	*anti-reductionist/ holist*
part-by-part	*as a whole*
experimental	*contemplative*
exclusive	*inclusive*
physical only	*physical/ metaphysical*

Science, born of left-hand bias that concentrates outwards towards creation's physical periphery, is experimental and inductive. Its practice leads to a discovery of principles. We call this mode of thinking *bottom-up*. Thus, starting at the solid base of Cosmic Pyramid, its tendency eschews the metaphysical and even queries immaterial existence. Thus, in seeking after truth, scientific testing will not lift you further than a physical phenomenon. Renaissance science confined itself to measurements of undeniably obvious, sensible aspects of the non-metaphysical world. It penetrated inwards to the subtleties of atoms, atomic principles and, always testing, the underlying nature of events. It is now, locked against a 'glass ceiling' of its own exclusively materialistic making, reaching for the top ledge of a *TOE*. What will this *TOE*-hold show? You might have grasped a vantage point encapsulating principle that governs all the works below but which took precedent, principle or works? Did things or rules that order them come first? Or did this couple rise together from the mists of chaos, from the 'time' before time's start?

Which experiment, however, has conclusively displayed the absence of a cosmic but unseen, immaterial element of information; or demonstrated

physical empiricism is the only way to know for sure? For sure it's not as, in Two Pillars, we've already seen. Thus the issue's down to faith. Natural science is convinced both rules and game emerge the same; thus 'immutable perfection' sprang from a 'big bang' of sorts - unless, of course, you're saying cosmic rules 'evolve' the way that universe, according to the laws of entropy, 'evolves'. Either way, our cosmic grammar just 'appeared'. *By contrast Natural Dialectic, as a framework, slots in rules before the game.* In Book 1 we'll follow evidence to check out which interpretation it supports the best.

If, though, Mount Universe comprises two grades, metaphysical and physical, then science cannot 'know the lot'. The physical level below Point *PSI* represents the limit of science. Only hypotheses that you are able to subject to 'clock-and-ruler' types of test are scientific. What is immaterial (which may well include the nature of information, meaning, consciousness and mind itself) is not, by this definition, scientific. Either everything is accountable in a material, scientific book or else science, in its contemporary terms, cannot address certain issues. In other words, up to Point-of-*PSI* a scientist might come to 'know the lot'. It is simply that such 'completeness' is incomplete. *If immaterial, subjective truth exists then material science must, contrary to claims for TOE, forever prosper in a realm of partial truth.*

What does 'know the lot' mean? Does it mean cognisance of every, single detail? Or is a thorough grasp of principals and principles enough to qualify? Is to grasp the potential, capacity or capability of something all you really need to know? **Laws describe potential**. They relate what can and cannot happen. Possibilities. If, therefore, you grasped the first principle from which all laws derived might you be said to 'know the lot'? A *TOE* is, in this sense, a theory of potential; but its principles and derived sub-principles can never tell exactly how a local game will play. Nor, above all, can it calculate the operations of a conscious mind (such as purpose, reasoning and desire). Psychology is not in physics' disposition; nor is it crystallised by chemistry.

For Natural Dialectic, on the other hand, a calculator is more important than the calculated, a creator than creation. From this perspective subjective mind precedes objective matter and The First Principle is conscious. Therefore its theory of everything includes subjective potential. Just as such a theoretician has a different starting-point ('Who am I?' rather than 'What is it?'), so he has a different closure', a different sort of perfect comprehension. While a scientific *TOE* includes a *GUT* (Grand Unified Theory which unites the electro-weak and strong nuclear forces but still aches to include gravity), Natural Dialectic's *TOP* (Theory of Potential) includes a Subjective Principal. How, therefore, could its experient be said to 'know the lot'? In this case (*fig* 0.12) through a *GUE* (Grand Unified Experience), a fully *top-down* vision from the peak of Mount Universe, that is, from the radiant source of mind.

From this point, therefore, we complement objective *GUT* with subjective *GUE* and, comprehending cosmos top-to-toe, subjective *TOP* with objective *TOE*. *These are the top-down and bottom-up versions of perfect knowledge.*

In short, one learns, *bottom-up*, by experiment and experience; and at the same time is guided, *top-down*, by the certainties of a higher authority, a teacher. **One might call these information vectors (shown in *figs*. 0.12 and 0.13) anti-**

parallel. They compose the anti-parallels of knowledge. Instruction, questions, tests, discovery and disclosure, facts and rules and principles - this is the way of education. You learn, up to a point, and from that particular level you can turn, look down, create and teach. However, it is equally clear that expertise is relative until you reach an absolute peak. Imperfect experts and teachers will disseminate from different heights (or, if you insist, depths) of understanding.

Is there any speciality that education should prize most? Which of, say, oncology, accountancy, philately or neuroscience is the most important? Are all equally important or is an intrinsic grade of worth (as well as the depth of a practitioner's expertise) applicable? How can you judge intrinsic worth? Which discipline digs nature deepest and, therefore, takes priority? Of cosmology, quantum physics, cooking or commercial quarrying which, on what basis, supersedes the others? You might even claim that the study of subjective life, psychology, was most important. Man neither made nature nor fully understands it. Nor, certainly, can understanding be complete without full knowledge of the nature of the very basis of understanding itself - consciousness. What, therefore, about the metaphysical study of this substrate, ground-state or, in Dialectic terms, informative overlay? *If it transpires that the cosmos ('ordered system') represents the realisation of interior principles and the creation gradient runs from mind to matter, then its source of order will be found at source of mind.* In this case what more natural than subjective, holistic psychology - the goal of whose practice is to discover the fundamental nature of mind and, consequently, its projection of matter. What more important than discovering the source and origin of what you're naturally part of?

In short, because my knowledge is imperfect, I accept the authority of qualified teachers. A professor says, "Trust me, I know from experience. Believe me; practice what I say then you won't have to believe me - you'll come to know for yourself". In this respect there is no difference between professors of objective and subjective science; but their direction of focus and therefore perception of truth differs. *In the holist case, a professor has not just studied the artist's canvas. He has identified with the artist's mind to the point of essential communion.* ***If the canvas is creation then, in top-down descent from the peak, he will gravitate in line with its ideas and principles.*** His student has faith that his description of descent and, equally, the method of any possible ascent is a reliable and accurate reflection of the truth. From such professors we can take a cosmic lead. This is because, unlike the scientific professor, they have taken the whole and not simply the non-conscious, physical part of creation into account. It is clear from all such accounts that their priority and emphasis weigh on the information-rich, mental side of life rather than any exclusive interest in the inanimate and reflex, passive quality of material energy and its bodies. Both sides of things are, however, real and full of interest. Indeed, it will be argued that, *top-down*, metaphysical will meet physical science as the latter, with its field theories, approaches the 'roots' of matter at the base of universal mind. This base has already been identified as the psychosomatic, sub-conscious border, Point *PSI*. Its spot might be the summit and the source of physical phenomena but no Bodhisattva waits in meditation at its ledge; he lives, like the scientist himself, much higher up. From a *bottom-up* perspective, however, having reached the zenith of material exploration science has no option but ascent into the ancient halls of metaphysic and its transcendent explanations for the way that matter is projected and the universe is made. In fact, we'll see that such ascension could be judged to have begun.

Comparative Analysis.

Holist and materialist are, in spite of their different perspectives and goals, both honest and keen seekers after truth. We know the problem. A holist commends scientific discovery and accepts, as far as it is proven, scientific fact; he also accepts, however, an immaterial, informative domain from which physical shapes and behaviours are eventually derived. A materialist, on the other hand, accepts no grade of creation other than material.

For this reason, as *fig*. 0.13 explains, the Natural Dialectic of Polarity consistently contrasts anti-parallel vectors of information; it contrasts *top-down* and *bottom-up* perspectives in a specific way. Indeed, so different are the 'world-views' derived from the 'opposing' perspectives that from now on, whenever the counterpoint of this contrasting Dialectic is expressed, each party is habitually italicised.

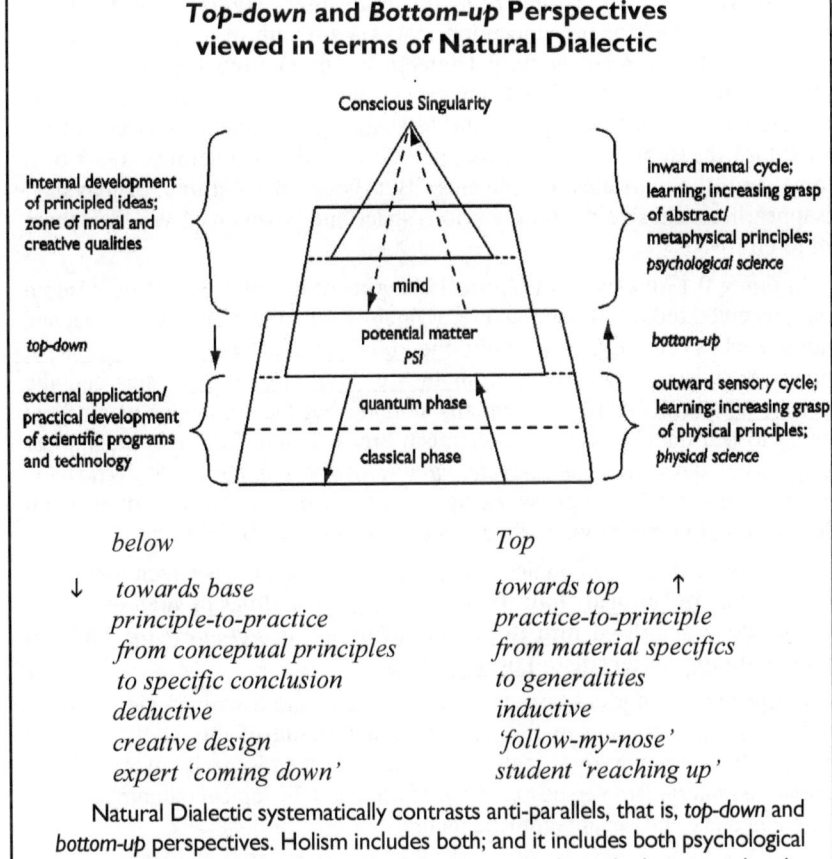

Top-down and *Bottom-up* Perspectives viewed in terms of Natural Dialectic

below   *Top*

↓ *towards base*   *towards top* ↑
principle-to-practice   *practice-to-principle*
from conceptual principles   *from material specifics*
to specific conclusion   *to generalities*
deductive   *inductive*
creative design   *'follow-my-nose'*
expert 'coming down'   *student 'reaching up'*

Natural Dialectic systematically contrasts anti-parallels, that is, *top-down* and *bottom-up* perspectives. Holism includes both; and it includes both psychological and physical phases of each vector. Materialism includes only the material cycle; and so, since it can observe no immaterial factor, does science.

> At the top of Mount Universe there is, for Natural Dialectic, a Conscious Singularity. At base non-conscious phases of creation are expressed. Science works within materialism's cycle; it wishes, thus far in its development, no more than to entirely understand all physical phenomena.
>
> Of course, a seeker after inner as opposed to outer truth works bottom-up as well. This is learning's way. **However, for working purposes this book a decision is made; top-down is identified with deductive holism (left-side) and bottom-up with inductive materialism (bottom-right).**
>
> In Chapter 13 it will be elaborated how these 'top-down' and 'bottom-up' pillars of faith are derived from modes of looking at the physical world as prefaced by complementary *left* and *right*-hand hemispheres of the cerebral cortex.
>
> **fig. 0.13**

Top-down implies you've got the information that you need; it is the expert's view. Thus it is construed as the perspective of an Expert Engineer, positioned at the very Start of cosmic creation and thence understanding all consequent principles and practices.

***Bottom-up*, on the other hand, is taken as the empirical method of a humble student who, from child-like ignorance, starts from knowing nothing**. Such lack of preconception marks a strength of scientific method. Such method works (in Natural Dialectic's sense) from the periphery of nature; it mines information from gross, obvious factors first. From the tangible world of solids, liquids and, less easy, gases science moves inward to explore the subtle nature of matter - forces, fields, chemical reactions, atoms and the particles of physics. But when the 'grains of physics' disappear it reaches void, non-conscious space and 'virtuality'. Where is there left to go?

In *figure* 0.13 the cycle of physical experiment is noted. Scientific wisdom that's received today - in lab, lecture, written word and broadcast - is couched within such cycle. <u>Experimentally the mind-set's *bottom-up*; in terms of universals it is *top-down* only from the 'ceiling' quantum physics and the vacuum represent</u>. There is essentially nothing but the behaviour of physical energy to learn about. Nor does scientism brook dissent from this incomplete perspective. *We call its mini-cycle, almost overwhelming in its exclusivity, scientific materialism.* **For working purposes this book identifies such philosophical perspective, through-a-glass-darkly, as '*bottom-up*'.**

Socratic dialogue embodies an exchange of views between opposing personalities. <u>**In Natural Dialectic's case the personalities of such exchange become depersonalised into the theme of opposing *top-down* (or holistic) and *bottom-up* (materialistic) perspectives.**</u>

Perspective changes as consciousness changes; and his world-view colours a person's basic attitude to life. An important strand of this book therefore amounts to a comparative analysis of the two 'hemispheric' world-views, *bottom-up* materialism versus *top-down* holism, and the logical outcome of each complementary and yet antagonistic tenure.

<u>Antagonism.</u>

Despite agreement on the manner matter works the heat of opposition, friction, burns. The two views rub each other wrong way up. **Each systematically inverts the other's concepts and perspective when it comes to how the cosmos is constructed and how *physicalia* began.**

For one, which works *bottom-up* from detail to principle, matter is real and consciousness as insubstantial as a 'ghost'; mind is an imagined vapour in the material machine. The order of its evolution, from matter to mind, is astronomical followed by neo-Darwinian. And, in accordance with a true but only half-true doctrine 'molecules make man', the bright fraternity of biochemists and molecular biologists is used to rationalise discoveries in a way as obdurate as any kind of *bottom-upper* when it comes to what it studies - life. Isn't any living body, inspected darkly through materialism's thinking-glass, just 'spontaneous ingenuity' composed of most unlikely molecules? Isn't life, from such a reasonable angle, obviously chemical assemblies, geared reactions, mindless cells and systems 'built' from them? Tap on the bio-microscopic door. Is stony silence the material response; or, as a neurologist might claim, does the machinery of nerve cells generate conceptual answers worthy of a Nobel Prize? <u>*What on earth is an atomic frame of mind? Is reason simply electronic shunting and, subjectively, experience electromagnetism in a complicated form*</u>? Indeed, do e-m brainwaves generate such reason (in the form of scientific cerebration) as displaces childish faith? The sharper edge of science overpowers belief; a bench-test marks the line of truth; an atheologist pursues big G's redundancy.

On the other hand a holist, *top-down*, places mind in front of matter. In this way human projects all evolve. Could one also reckon natural outcomes are projections of an insubstantial, background mind? Not human but a universal, macrocosmic mind. An evolved rationalist deprecates this view as an anthropomorphic, that is, 'unobjective', human-centred projection of human being onto the cosmic whole. Is it? Perhaps *top-down* reality is the reverse. Perhaps (Chapter 17) it includes a very rational projection of the cosmos into an image, a microcosmic representation, quintessential summary or living symbol of the universe - man. In this case man were also the measure, not just the measurer, of all things. **A holist's holy grail is therefore the source of plan, principle and purpose.** His peak is, beyond mind, the Essence of existence. A whirling dervish calls it *fenafillah* - the science or knowledge of God. It is what Buddhists identify as Enlightenment, Hindus *samadhi* and Christians Communion with the Father. Descriptions differ of the same; but intellect can never know it. The grail is actually its seeker's Origin.

For example, Christ's reality was in spirit and not body. For the Buddhist existence (*samsara*) is, in its aspects both of mind and body, insubstantial, fleeting and illusory. It is Projector not projection that a Hindu holds as real. For a materialist, on the other hand, physical and not metaphysical is real. *Nirvana*, the point of enlightenment, is nothing; it is self-deceptive nonsense. Creation is not grand illusion but Creator is the great delusion.

appearance	*reality*
away from truth	*towards truth*
illusion	*truth*

In short, there is scarcely any conflict when it comes to the question of obvious facts and mundane activities; but conflict is sharp when it comes to the larger picture, one including origins. Origins are critical because, just as things start from them, so does basic perspective. In this case there seems to exist an incompatibility, a clear, extreme and irresolvable polarity of belief. What *is* the truth when each calls real what the other calls, at best, no more than partial truth and, at worst, an unreality, delusion or a lie?

It could be claimed that scientific descriptions of the universe are correct but lacking some fundamental property or dimension and, as such, are fundamentally incomplete. Appearance of their truths as absolutely basic is akin to an illusion. You see something and presume to recognise its character; but, from superficial recognition to profound, there exist different levels of understanding. To grasp anything fully one needs first grasp its origin and, thereby, its plan. Its basic concept. This applies as well to teacups as to lips, tongue, china clay or cosmos as a whole. *Metaphysics gives plan to physics*. If there is a subjective, conscious dimension to the world (and surely you are living proof?) we should include it. Indeed, consciousness is really all you have. It is the essence of your life, the centre of your frame. But could it be included at the centre of a natural, 'unmade' universal frame? Of course, there's no more mind in matter than in a machine; but there's mind behind machines so what's behind the cosmic automation? And what's the way its blueprint is expressed?

dark   *light*
ice   *fire*

Plato said it and his image falls upon the Dialectic. Suppose that cosmos is a chiaroscuro, a shadow play. If I face into a cave with a fire behind me can I see the cause of shadows flickering on its inmost walls? What if I turned? I'd see the cause of changefulness and understand that *my direction of attention had been wrong*. Nothing's changed today. For several centuries western thought has swivelled. It has gradually reversed attention. Lecture pews now face the back; the lectern's at the rear. Who'd, turning round, negate the teaching? 'Turn back round! We nail repeat offenders to an intellectual cross!' Materialism concentrates upon objective matters to, practically, exclusion of subjective mind itself. It has concentrated on the observed almost to the exclusion of the observer, on a sink of shadow-play instead of source of light. Its grasp of physical principles and consequent technological exploitation of the outer layer of creation now seems to far outmuscle any antique, oriental way. Modernism thereby flits with shadows and refuses Plato's fire. It watches film but icily, dismissively denies illumination. Is the source of cosmic information one of any interest?

A Complementary Course.

Remember, though, that opposites can also make a complementary pair. A range of shades can run between their absolutes. A dialogue can run between the parties in debate. As well as bias dialectic can bring balance to the table; such equilibration is, at root, Natural Dialectic's way.

Such Dialectic naturally identifies the complement to a material coordinate. <u>It identifies a missing, immaterial component and includes, in the following</u>

chapters, a description of the Information Coordinate. Indeed, it links the two coordinates in a 'graphic combination' as the axes of an abovementioned 'conscio-material gradient' or 'conscio-material spectrum of existence'.

Science follows instruments of measurement; but the subjective trial needs no technology. All the apparatus for its practice, to develop extreme inward focus, was present in the first man. And although the preponderance of its interest tips the subjective pan it also, in the scale of things, recognises physical phenomena. We saw that Aristotle, Islamic science and the monks that founded our universities, from whose curricula natural philosophy and science developed, certainly recognised τα φύσικα, physics or the physical world. It may therefore be that the principles of modern science are compatible with the essence of the wisdom of the east which simply adds to them - adds a rewarding, essential perspective, a component more ignored than unsuspected, more unobserved than unobservable. Closer than breath, it would be foolish to exclude it.

Is it scientific to deny a claim but then refuse experiment? It would be foolish, not to say ironic, if a 'scientific' atheist were, without sufficient research or sincere, long-term commitment to experimentation, to dismiss holistic, subjective conclusions. **It were as if a scientist refused to accept the results of an experiment and, at the same time, unscientifically refused to whole-heartedly carry out the same. Paradoxically such a person would have denied an application of the very scientific methods he himself espouses!** *Is he not hoisted highly on his own petard?*

intellect	*intuition*
manipulation	*sympathy*
cleverness	*wisdom*

'I want to believe in science, not in non-science. I prefer scientific atheism to any metaphysical assertion'. Some shout it from the roof-tops. This is fine as long as the speaker clearly understands that his is not a scientific but simply a philosophical, even anti-metaphysical position. It is a belief, a fundamental or first-order profession of faith that infuses every aspect of his thought, interpretations of his observations and therefore whole life.

imbalance	*balance*
discord	*harmony*

Scientific materialism has bred scepticism and we now seek to establish heaven on earth by means of material progress. We worship at the shrines of commerce, technology and a '24/7' life-style based on power and exploitation. We worship intellectual scholarship, extol competition and drive a single deal - with wealth. Men toil for economic growth and, by pressurising nature to the limit, squeeze out an ever-higher material standard of living. They chase the shadows of these interests all day long but often forget that luxury, although relieving stress, also creates it. Mammon is neither essential for happiness nor guarantees wisdom; in fact, natural calamity could, easily as power cuts, strike the pride of modern cities out. There is perhaps, between the weights of cleverness and wisdom, need for balance.

pyramid	*concentric rings/ spheres*
discontinuity	*continuity*

separation/ disconnection	relationship/ connection
solidity/ fixity	radiance/ flow
flatness	vibration
straight-line	cycle
object	energetic field
outer appearance	inner order
confusion	synergy
exhaustion	drive
cacophony	harmony

If the connections made in this 'Musical Stack' are indistinct, they will soon reverberate. Of course, the opposite of harmony is noise. If, however, you consider both objective and subjective kinds of integration you can complement them:

linguistic abstraction	whole context
numerical analysis	vibratory synthesis
symbolic manipulation	music

Mathematics is an abstract, archetypal language expressing everything numerically; music is such language moving energetically. One dances intellectually, the other actually.

Do you remember (*figs.* 0.4 and 0.5) why energetic, concentric spheres and steps of ziggurats are used to model our Mount Universe? Rings or spheres that ripple outward are combined with movement up and down a scale. The combination makes a vibratory scale. There springs to mind the seamless continuum of a spectrum combined with individual notes, chords and colours. Sound and light fit this particular bill. Both carry information. Have you, for example, heard music that touched you, that seemed to strike your soul? Such harmony to which, as it struck, you gave yourself completely? Such resonance. Such emotional awakening. What about the composer, whose inspiration now is yours? **We can coordinate the two other models and, including an informant composer, introduce the third, transcendent model of creation - greatest of the three - musical vibration.**

Orderly arousal, growth, emergence into matter's day. Musicians, engineers, businessmen - all creators work in the same order - from principle to practice. The inspiration of an idea gives way to the intention to express it. Such expression involves careful, detailed planning before the project is ready for physical production. Purpose, coordination and coherence are all psychological; subjective precedes objective, animate orders inanimate phase. **A good example of well-informed, variable energy-in-motion is a cinematic movie but, because it includes fundamental harmony, the Master Analogy of Natural Dialectic for creation and the cosmos is Music**.

Music is coherent; its harmonic energy's informed. Wordless, it touches with meaning; and word that's musically voiced is, every singer knows, a song. Could such happy order be analogised with what, at root, projects the cosmos? Could it actually, a subtle stream, be the cause? This would be a song of songs, a melody of logic and of love. If such an operatic operation made things work you might expect to find, at heart, a joyous universe. This is indeed what mystics say. Are they the world's fools or its wise men?

At the start of this chapter we followed the path of optical vision through a continuum from the outside, objective world to our inner, subjective one. From the time of Descartes (who thought only humans possessed his definition of 'life' or 'soul') post-renaissance science has increasingly split these worlds apart. By now it has come to polarise them, black and white. Non-existence of the immaterial! Metaphysical redundancy! It tends to deny, on the basis of an inability to physically handle it, the metaphysical. A modern rationalist scrubs all except materiality. His scientific method is, of course, a powerful route to knowledge physical. It delves the mechanisms of non-conscious universe but (Chapter 0: Opposite Directions of Mind) is this outward mode of reference the only one? Is there no inner world, no half that's been forgotten by the business of bodies? Or is faith's fool entirely irrational - as unreasoning, for example, as holistic reason of the Natural Dialectic way?

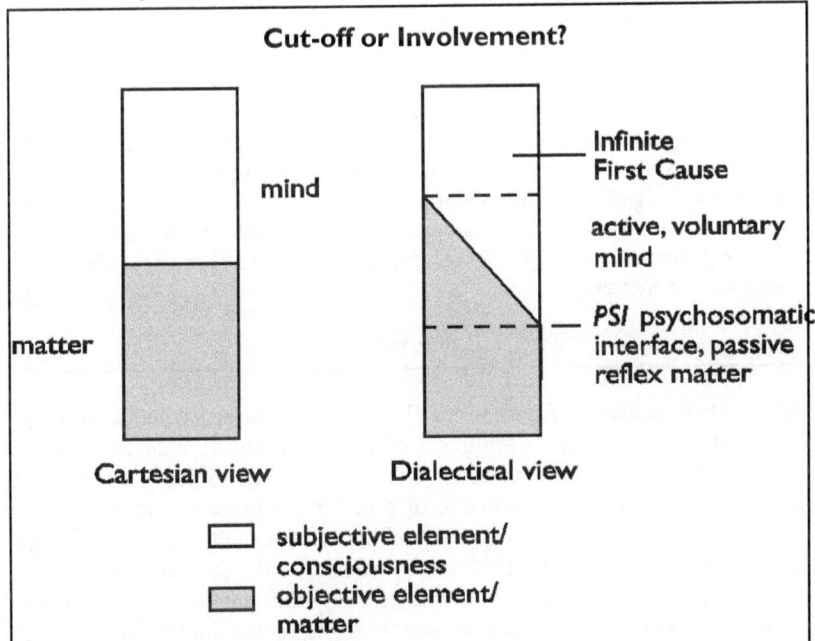

Is the dualistic Cartesian view that mind and matter are distinct, parallel entities? In fact, Descartes claimed, in a monistic way, that his mind and body were composed of a single whole. Natural Dialecti certainly treats mind and matter as distinct but interactive component of the universe. Immaterial consciousness can exist apart from bod (matter) and non-conscious body apart from consciousness; but, *top down*, the source of both physical and metaphysical reality is primordi consciousness. In dialectical terms mind becomes involved wit increasingly dense forms of energy; alternatively, physical devolves from subtle or metaphysical energy. But it is not the other way round; mind and consciousness are not 'emergent', *bottom-up*, from matter. Thus, at cosmic root, active mind informs passive matter.

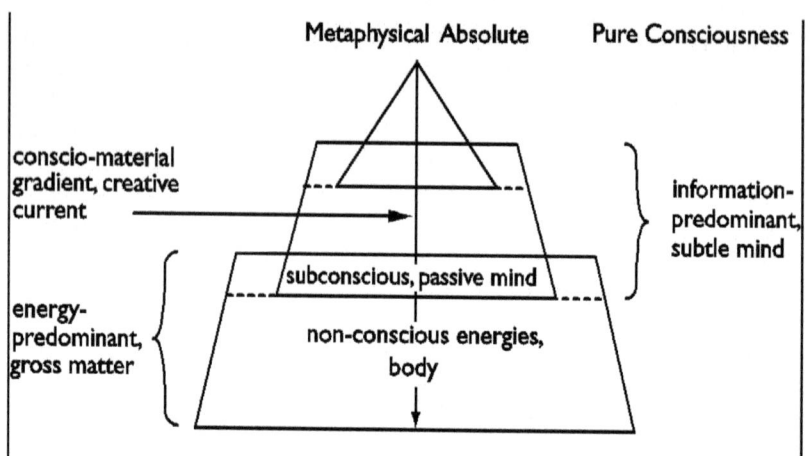

A commonly used, alternative metaphor is one that sees creative energy in terms of wave, current or flow. Surge from the Infinite Centre, called *Logos*, cascades through 'sedimentary layers' of creation. These precipitates range from subtle (regions of mind) to coarse (regions of unawareness and inflexible, reflex behaviours). All is flow and, in a way apparently confirmed by physics, objects are simply a congealed or bonded form of flow. Where objects are concerned the very lowest, coarsest 'layer' of cascade is frozen solid. There naturally, you might agree, the cosmos has to stop. World's end is solid matter.

fig. 0.14

Materialism prefers, of course, to find an explanation (whether right or wrong) that utterly compacts the ghost of immateriality. It squeezes flat the mind's dimension and has therefore come to classify 'objective' real and 'subjectivity', because it is incapable of proof that's physical, unreal. Never mind that the subjective *is* our point of knowledge and reality. Although neuroscience cannot distinguish in between 'hard, factual' and 'wild, imaginary' modes of thinking it has labelled every thought (and especially that of immaterial 'God') an artefact of neural wiring or the subtle shuttle of an orderly but senseless chemical refrain! Has not cosmos therefore forged its self-awareness from oblivion? Aren't genes made up of *DNA* the root of knowledge? If they alone make brains (and therefore 'emergent' consciousness) then many reckon so. Such stark, relentlessly objective emphasis has frog-marched western thought into lopsidedness. It has evolved 'reductionist religion' whose cutting undertow has dragged academy off-balance. The bias has infected philosophy, psychology and biology so that a sophisticated malaise pervades the universities and, from their educational effect, the media. Thought precedes action and unhealthy thinking, applied with the increasingly sophisticated instruments of technology, is wreaking several kinds of havoc across the globe. Does belief in the single, hopeless oblivion of matter not leave space for moral vacuum? And, because you always need ideals to fill such vacuum up, then rationalities have sparked, by flint of various political excuse, war and fanatical oppression vaster than religion's used to be. Technology supplies such exploits with an ever more controlling and destructive weaponry. A paradigm that, without addressing the root potential for evil in men's hearts, is able to create such outcomes cannot be the whole answer.

matter   *mind*
objective   *subjective*

For Natural Dialectic it is definitely not. In this view mind is at least as real as matter. Subjective is as real as objective and, although metaphysical and therefore immune to a scientific type of proof, is its **complement**. Each is the other's complement. Both individually (when you sense anything or waggle your little finger) and universally (in the nature of creation), interaction between the two involves a two-way range of transmissions. If metaphysical and physical *do* meet in such an easy way as your finger-waggle showed, how could such connection ever work? And, born as it is of the proposition that awareness and matter are two fundamentally separate entities, what are the implications of such an observation? This whole book is dedicated to an answer.

Finally, therefore, the Dialectic is a work of reconciliation between physics *and* metaphysics, occidental *and* oriental, scientist *and* saint. It is not a question, for anyone, of 'either/ or' but 'and'. This book is keen to identify an overall infrastructure within whose apparent logic nature works; to investigate the construction of both subjective and objective aspects of the world; and to shed light on the character of such psychosomatic interface as may exist between them. Integration and cooperation between sciences, in an unrestricted sense of the word, is the only intelligent way forward. In other words, it is time that subjective and objective, in theory and in practice, were seamlessly coordinated and thereby the balance of the whole truth given better chance.

1. Natural Dialectic's ABC

Albert Einstein summarised the goal of Natural Dialectic when he said, "We are seeking the simplest possible scheme of thought that will bind together the observed facts".

If you relish intellectual complexity then maybe dialectical simplicity will not appeal to you. For whom, however, truth is not a function of complexity let's sketch the principles, framework or chassis of this philosophical vehicle. The process will be like learning a language. The picture will emerge gradually from a grammar whose basic structure is polar. *Its simple outlines will act, in an unfolding narrative, as the framework for more complex physical, biological and psychological arguments.* **The next two chapters sketch the Dialectic's ABC.**

First Principles.

Any engineer will tell you that there is no avoiding first principles. However abstract or impractical they may appear, they constitute the basis for whatever system follows. *This chapter is devoted to the principles of Natural Dialectic and the next to the grammar that expresses these 'profundities'.* If its structural design is symmetrical, strong and works then details should fall into place; if it accurately reflects nature, including mind, it should be functional; and, although its logic is not typically mathematical, it should be self-consistent. It should 'work'. Such 'functional logic' should provoke patterns of thought, questions and, within its framework, solutions.

thing   *nothing*

The Dialectic's fundamental approach is binary. It is polar. Switch on, switch off; switched on, switched off. Before things were 'switched on' there was nothing - they were 'switched off'. The question is whether nothing (0), from whose potential derives anything and everything (1), is potent or impotent. How could impotent absence make a thing? What, in that case, is potent nothingness?

Most science and philosophy agree things had a start. Before things there was nothing. *Space, time and things make physical existence up. But existence, in the image of Mount Universe, includes mind.* You assert there's only matter; I retort, as a first principle, that both material and immaterial factors interweave our world. In other words, the Cosmic Mountain has aspect physical *and* metaphysical.

material	*immaterial*
physical	*metaphysical*
matter	*mind*

Of what is your own subjectivity composed? What kind of thing is mind? If mind is not a physical thing it is a physical nothing. In other words, mind is nothing or 'hidden' with respect to physical phenomena. But it exists. It is a metaphysical thing. It is the very basis of your own existence and you use it to arrange materials around you how you want them.

In one matter all matter is identical - its non-consciousness. Material essence is non-conscious. Is, on the other hand, immaterial essence therefore conscious? What is mind's essence? Is it consciousness? If mind is immaterial then could the essence of pure immateriality be consciousness?

1   *0*
existence   *essence*

What's this about pure no-thingness? Non-existence, called essence, is nothing relative to mind or matter. It is nothing with respect to existence; in its purity nothing existential moves; it is, devoid of form or action, purely immaterial. But is such Purely Immaterial Essence, the dialectical precursor of existence, empty or omnipotent? A single thing's a unit (1); is Essential Nothing (0) actually the source of every single 'happening' or not? Is it or is it not The Central Peak of Universe? What is the real nature of such existential void?

Is this all a slip of sophistry? Essentially empty paradox? Arcane quibbles adding up to little more than nonsense? For saturated materialism it may be but for drier, immaterial wit far from it. Just consider how our universe began. Does not what starts have cause for starting? If so cosmos had a cause. Yet premier cosmologist Stephen Hawking has suggested that space, time and things stemmed from 'absolutely nothing'. Causelessly. Can nothing in the sense of absence be a cause of anything? Could there exist an emptier explanation, more vacuous a conclusion for such all-embracing fullness of a start? *The fact is, though, that empty nothing is a negative that never makes a thing; what can nothing ever make but nothing?* Such void is void of sense, such creative absence nonsense! And if what happens without cause happens so by chance what, by itself, is chance? Chance implies unknown cause; but chance itself (except in common, quasi-scientific myth) is neither substitute nor understudy for a cause. *The two are not the same. Randomness has never organised itself; neither randomness nor solely physical behaviour ever entered into purposive complexity.* How, therefore, did you evolve to reading this?

Does Hawking therefore really claim, fantastically unrealistically, that absolutely nothing nowhere for no reason rippled; the spontaneous ripple swelled and from this inflation everything exploded - not least, given time galore, you, me and all else that lives? Or was it two most orderly inflations did the trick (see Chapters 7 and 8: Is a Finely-tuned Universe Coincidental?)? How 'natural' does it sound to hear that in a fraction of a second secondary inflation blasted off the first, blew cosmos up a zillion times and filled itself with all the matter and the energy there is? Or is 'inflation' as unnatural a moment as they come? No matter; any fabrication beats considering as fact a universal mind. Mind is the natural order-maker but, in spite of this, do you believe that order issues from the chaos of explosion or that, borne on causal void's expansion, cosmic order's simply an effect? What nonsense and, like nothing, getting nowhere fast. It sounds less plausible than classical cosmogony but 'big bang scenario' is actually the latest, cutting-edge, scientific attempt to clarify the mystery of the origin of the universe. Common law is a presumption that extends across creation's fabric. However, what is both fantastic and historical but lacking proof and unobserved we call mythological; so isn't a prime mover physical as mythological as one that's metaphysical? It may turn out that from the view atop Mount Universe material cosmogony appears truncated and, excluding immaterial factors, incomplete. Either way, we'll be prodding a good deal more at nothing in particular. Its nature underlies the whole discussion of this narrative. Let's take it, framed within the parameters of Natural Dialectic, as a starting-point.

Nothing.

Nothing is, in personal terms, the absence of a thing I seek; or it's a form of ignorance - what I don't know or haven't yet discovered doesn't mean a thing to me. My present state of knowledge does not mean (as science keeps on showing) what I conceive as 'nothing' actually is. In fact, is non-existence an imagination and creation really nowhere empty? If so, our mental and material voids are no more than appearances. They are relative. If not, however, is there absence absolute - nothing existential out of which existence might have sprung? And, if so, what might be the character of Universal Void?

Such Void (in Sanskrit *sunya*), the starting-point for all creation, implies no action, influence or thought. No form, either physical or psychological. Not even vacuum; no extension, neither space nor time. No change. Therefore did formless nothingness precede creation's forms of influence and patterns of behaviour? Or were things always here?

Of course, if the universe is tiered - a possibility to which Natural Dialectic is committed - something on one tier (say, immaterial mind) may be nothing, that is, may not exist, upon another (e.g. matter).

Do you still think nothing's nothing? Do you still think nothing of it? Think again. **What are time and space but, respectively, nothing physical and, by definition, nothing. Immaterial Nothing seems to be the very fabric of our physicality!** We've two whole chapters (10 and 11) dedicated to this ghostly, fundamental pair or, as Einstein well observed, this space-time union. The basis of our modern, relativistic physics surely can't be nonsense; but did the phantom couple actually precede all 'real' forms of physics? For now let's simply, briefly make a starting-point of start-point's absence. Let us take a stab at nothingness and see if anything is hit.

thing	*Nothing*
with form	*Formless/ Structureless*
finite	*Infinite*

Several issues have already risen. **Firstly, if Nothing (which by definition's formless, boundless and thus infinite) trumps things then let us write its column with a capital.**

existence	*Essence*
1/ on/ action	*0/ Off/ Peace*
expression	*Potential*
action-to-completion	*Poise/ Readiness*
result	*Cause*
order	*Preordination*

Secondly, which follows on, there exists polarity; and its primary duality (*fig.* 1.2) will be identified in Chapter 2 as the Main Dialectic between Essence and existence. It is to highlight this most basic of all polarities that from now on essential principles or characteristics will, it is re-emphasised, be prefixed with a capital letter; existential ones will not. We therefore write that Essential Nothing is all-powerful, super-existent and, as it moves, kinetically substantiates existence.

Thirdly, you can think of nothing in two crucially different ways. *Potential* and *exhaustion*. Pre-active and post-active absences of change. *Fig. 2.5 iii might help to explain.*

In a *negative* way it describes a state(lessness) that occurs when everything has been subtracted; or when nothing has been added to nothing in the first place. Zero. The sub-state of exhaustion, abstraction, absence or impotent void. Inert and empty space. Nothing left comes after.

In a *plus way* nothing comes before. It describes the state of pre-condition that anticipates an actual behaviour and without which the behaviour, pattern or event could not occur. Such poised readiness is the source of possibilities prior to the realisation of any particular one. For example, potential energy precedes kinetic energy; it precedes any particular creations of that energy and, of course, its exhausted effects. In a similar way the potential for a mission is its plan - the information. Command and control executes orderly strikes according to such immaterial material, such non-physical intelligence. Could potential information precede its own kinetic motion (motion we call mind)? What is the prerequisite of thought; what is the nature of potential thought? And could immaterial mind precede material arrangement? Such 'nothingness' is prior, potent and pre-active; it is action's 'super-state'. **This 'form' of nothing plays a primary role in Natural Dialectic. Its immanent latency is called Potential or, as Absolute Potential, Essence.**

Fourthly, *a principal comes first in line; a principle describes a source of order or a guiding force.* What has form exists; it has a start. If neither mental nor material form exists there's nothing - their non-being. Yet prior Nothing, whence they must have formed, is Essential Being. Call it metaphysical necessity. What first and final paradox - non-being is The Being! **Fundamental Nothing is therefore the Principal and the Essential Principle of Natural Dialectic.**

Formless, being-without-predicate, this Essence simply is. It is one-without-a-second. Formless, it has neither beginning nor end. It is, therefore, by definition infinite. *In fact it is at once zero, one and infinity.* If everything materialised from this then Nothing must embody every form in latent formlessness. *Essence, which precedes existence, must be pure, undeveloped potential.* As the Chinese sage Lao Tzu noted, "Non-existence is called the antecedent of heaven and earth; subsequent existence is the mother of all things". Existence can therefore be seen as a dependency of Essence, as a restriction of Infinity or an expression of Absolute Potential. *Infinity is, by definition, unconditioned.* It is the only absolute but at the same time non-existent frame of reference. In other words, in the last analysis what is conditioned is measured against the unconditioned.

It therefore remains to note, **fifthly**, that Main Dialectic is perfused with paradox; and that a basic thrust of this book is to make space for space and explore the Nature of Nothing.

Hawking's 'void', whence he imagines everything has come, is negative. Be positive. If there was always Something Prior what is its character? The 'plus way' indicates potential predates action, leading even on a cosmic scale. If, on the other hand, something (or all things) took shape from Nothing what's the nature of Informative First Cause? Could, in this case, Something and Nothing be the same? Surely (while created something's something else) they're not both the Uncreated, Quintessential (N)One?

No joke! For a taste of what's to come glance forward to, for example, *figs.* 2.3 and 13.4. These triangles express, in another way, the Taoist divisions of *figs.* 1.1, 1.2 and 3.1. The Central Singularity is Essence (Lat. *esse* 'to be', *essentia* 'being'); the Apex of Mount Universe *is* Supreme Being.

finite	*Infinite*
limited/ conditioned	*Unlimited/Unconditioned*
created/ creation	*Uncreated*
relative	*Absolute*
every other unit/ thing	*(N)One*
peripheral	*Central*
lesser being	*Supreme Being*
existence	*Essence*

Lastly, not as proof but prelude in consideration of the nature of this (N)One think of a dot. Aren't dots nearer nothing than a circular periphery or line? Inmost essence is, as atoms demonstrate, less visible than outward show and yet supports appearances. Mind is less tangible than even subtlest quantum matter; could it support non-conscious, physical behaviours? Not your own mind, of course; but if universal mind supports the cosmos how does such atlantean labour effortlessly work? And what, in turn, substantiates its action? Surely not, as light a film or fuel machine, pure consciousness? Surely you can't call refined potential - yet again - the (N)One?

For Natural Dialectic only Essence *is*; changeless, it lacks any predicate. Existence is restriction of Essential Nothing; the nature of restriction is just motion; existence is Essence-in-motion, that is, Essence *becoming* something else. Thus, strictly, nothing existential is; all, in flux, *becomes*. Such becoming is creation; it is Potential being worked out; all different kinetic and exhausted states, all objects and events, psychological and physical, are expressions of Essential Potential. Existence (Lat. *ex-sisto* 'come forth' or 'stand out') is, in this view, neither more nor less than the dynamic, changeful 'realisation' of Essential Reality. Finite, transient 'beings' (call them 'lesser' or restricted essences) compose its parts. And in this whirlpool of events the only constants are those abstract principles by which continual flux occurs; the only fixity is archetypal regulation or 'a transcendent framework' within which the cosmic game is played. Seen this way the whole creation boils down to the interplay of just three cosmic fundamentals as explained in Chapter 2. Its serial logic is a story of relationships between such complementary opposites as Nothing and something, Unity/ polarity or Essence and existence. This latter couple constitute (as Chapter 7: The Essential Pair elaborates) the Dialectic's basic pair of poles.

Something.

Big question. Why's there something and, it seems, not nothing everywhere? Why does anything exist?

Existence is the second basic principal/ principle of Natural Dialectic. Of the Essential Pair it is the 'other', non-essential pole.

An actual infinity of forms, events or things cannot exist. There'd always be more, infinitely more. You'd never reach the end. Yes, a potential infinity of numbers can be mathematically proposed but such abstract collections are a fiction. They don't prove material endlessness. You might, instead, proclaim existence made of lesser, relative infinities (say, 'indefinities') of space, time, number and whatever you define as 'individual thing'. Such conditioned 'indefinity' numerically reflects the Infinite. Finite beings, that compose the universe, must each have had a start. What is not composed of finite parts? Why should universe be infinite? Why should it exist before its parts? Therefore, cosmos must have had a start.

Wait, you cry. An example of specific endlessness is progress round a circle; *spheres radiant and vibrant cycles are ways the infinite can enter finity.* However they begin, they endlessly repeat or grow - until their energy of motion wanes, power is cut or shape withdrawn. These shapes of existence are, to repeat, reduced infinities, *'conditioned indefinities'*, localities. However, essential characteristics such as 'unity' and 'centrality' are reflected throughout the restricted, local circumstances and individual events that sum to existence. They are, in this sense, 'lesser essences', finite appearances of Real Infinity. **Could creation's *finities* be shadows clouding Sky; or, through the prism we call mind, projections of an Inmost Light?**

In my present consciousness I neither know nor understand Essential Void. It may, like the vacuum of space, be neither empty nor powerless. As atoms or X-rays are nothing to the senses so Essence is nothing to our minds. In this way Essence, the origin and centre of existence, is itself non-existent. Pre-existent. It is no thing. Yet, paradoxically, it is the seed of every single thing.

Where Essence is Potential existence is development of this Potential. Such expression, something from nothing, is forever becoming. Its nature is motion whose dynamic is derived from 'hidden' information and energy. 'Hidden' potential pre-exists whatever derives from it. It is, with respect to that thing, nothing; it is, with respect to physic, its superior metaphysic. The nature of informative and energetic precedents is worth a close examination. For a start we say that Essence is potential existence and, within existence, mind is potential matter. As Essence is an existential nothing, so mind is a physical nothing. Mind is a subjective absence of matter. Metaphysic is a positive or pregnant void. Such superior void informs what follows.

The void of space is starlit; but Void is not lit by any concentrated point of energy because its Concentrated Source is Infinite. Innumerable shadows of existence cloak Bright Essence. Together they compose kaleidoscopic cinema. Also called *maya* (see the Glossary), the cosmic film's degrees of shade represent degrees of what, set against the Truth, is relative illusion; its colours show, against Transparency, as spectrum; and its objects, dotted over time and space, as discontinuous localities. The pitch-black, wholly non-subjective grade is called material oblivion.

Check *figs*. 0.13 and 1.4. Existence is, according to the Dialectic's Primary Axiom (*PAND*: see Glossary), a compound of mind and matter. It is, therefore, the umbrella-word for all psychological and physical constructions. It includes every formulation, realisation or thing. It is, paradoxically, both within and without Essence. For materialism such Essence makes nonsense but existence makes sense. Supreme Being is an absurdity but what are 'lesser' beings - objects and events - if they are sensible? They have 'ordinary' being without superiority, inferiority or any sense of hierarchy. For a holist, on the other hand, Essence and existence both make sense but only existence is sensible. Essence is paradoxical. It is beyond the senses but at once the heart of mind and body. It is the Unmoving, Uncreated Axis round and yet within which existence orbits; it is the Potential whence actuality appears. In your case everything you know appears in consciousness. How, though, did and does the universe appear; what is its fundamental field?

In brief, Absolute Potential is pure, motionless and unbroken. It is at peace. Existence 'stands out' from this nothingness; it springs from the division of its unity. Motion breaks from unity into duality; such motion implies relativity both with respect to things in motion and, which is not, Essence. Motion implies energy. Existence is Pure Essence moved to impurity by shapes and relativities of activated energy. These myriad forms show as perceptions, forces, objects and events. Corollaries of motive power are space, time and change. We observe interactions and transformations within an extension called the space-time continuum but have not identified the underlying source of power: we live in the thoughts, dreams and other experiences of mind but do not properly understand how they are known. We have not got to the bottom of it. We do not really understand the cause of things.

Two Existential Principles.

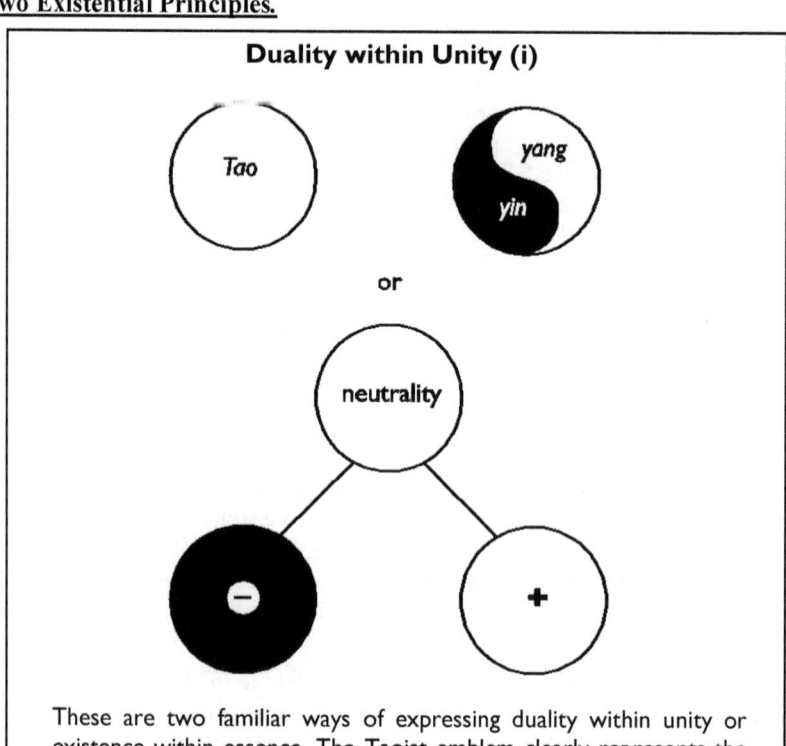

Duality within Unity (i)

These are two familiar ways of expressing duality within unity or existence within essence. The Taoist emblem clearly represents the principle of polarity within complementary unity. The mystic quest, Taoist or otherwise, is to transcend the duality of existence.

fig. 1.1 (see also figs. 1.2 and 3.1)

yin/ yang	*Tao*
existence	*Essence*
becoming	*Being*
non-unity/ duality	*Unity*
dipoles	*Monopole*
relativity	*Absolution*
expression	*Potential*
opposites	*Neutrality*
motion	*Peace*
conditions	*Precondition*
yin	*yang*
negative pole	*positive pole*
passive	*active*
receptor	*donor/ doer*
sink	*source*
foil	*stimulus*
static/ inertial part	*kinetic part*

Unity + duality makes trinity. Tao, yin and yang are three. But one, the Tao, transcends duality. We have already met Essence (Latin: Being) and learned to denote all essential characteristics (e.g. unity, balance, neutrality) with a capital letter. We place these in what is called a Primary Stack.

Each group of columns like this is called a 'stack'. So next, when the dualistic nature of the left-hand column is split, we note the polar aspects in a lower, secondary stack. You can also place duality within unity inside two binary stacks. Notice that 'dialectical stacks' (whose character is developed in the next Chapter and the Glossary) are formed by sets of pairs of polar characteristics. In this example of two stacks the one above sets relatives against Absolutes; in the one below each pair of opposites implies a range, spectrum or covalency of relative values between them. For example, 'Peace' and 'motion' are poles; in motion's duality, itself polar, there exists a range of possibilities between positive and negative extremes (for example, the shades between fast through slow to static). In short, the lower stack expresses existential relativities.

Potential is realised by changes in energy which, in a broad sense of the word, we can call motion; motion causes effects; cause and effect are differences between which transformation occurs; they are 'source and sink' or 'poles of change'. The basic existential principles are Causality (the driver) and Polarity (opposites between which degrees of change occur). These two principles order, at every level, every aspect of existence.

Causality.

You could also represent Duality-in-Unity (1.1) like this (1.2).

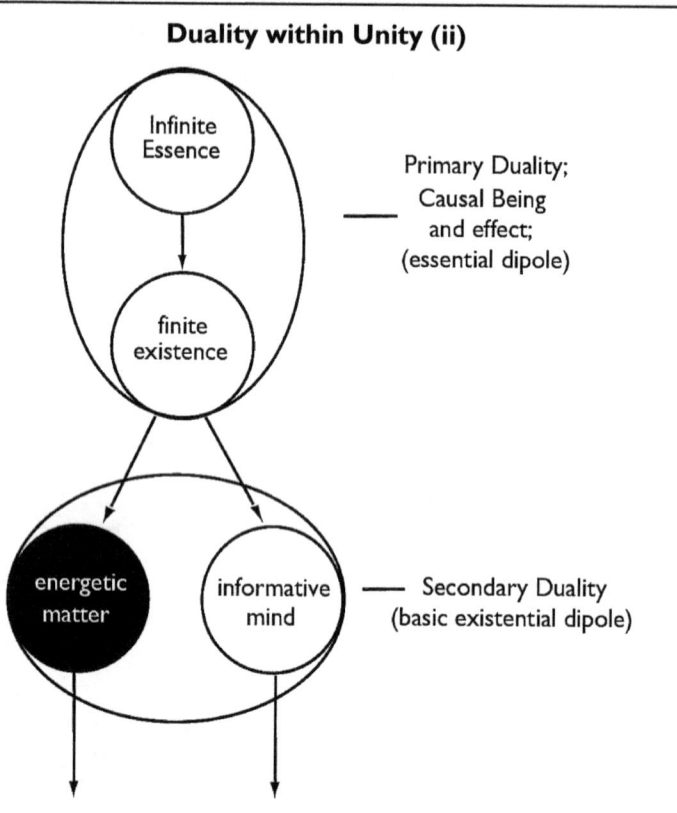

all subsequent polar subdivisions
(see Chapter 2 for Primary and Secondary Dialectic, also
this diagram's relationship with stacks and ziggurats)

The philosophical term '**monism**' means that you stipulate a single, fundamental and universal quality or entity. 'Dualism' means that you stipulate two.

Saturated materialists are 'monists' whose single element is energy/ matter.

On the other hand Natural Dialectic proposes '**Essential Monism**' *with* '**existential dualism**'. The single, infinite 'substance' is identified as The Concentrate of Pure Consciousness. Motion of this essence gives rise to the dualism of existence; the basis of this dualism is information/ energy or mind/ matter. The couple is hierarchical in that the former generates the latter; immaterial generates material; informant mind generates (in a way that Natural Dialectic describes) informed patterns of energy. Finally, just as there exist universal matter and specific biological forms, so there exist universal mind and specific embodied minds.

fig. 1.2

And as you read this correlated stack, bear both the images in mind.

existential duality	*Essential Unity*
conscio-material gradient	*Pure Immateriality*
objective/ subjective mix	*Complete Subjectivity*
subsequent effects/ causes	*First Cause*
effect	*cause*
physical bodies/ events	*mind*
particular definition	*shades/ range of mix*
result/ product	*excitation*
counteraction	*action*

Existence is composed of finite events. Whatever begins to exist, asserted the influential *sufi* Algazel, has a cause; 'something which begins has a sufficient cause' is also the modern principle of causality. This principle is constantly verified and never falsified. The physical universe began to exist and therefore has a cause. What is caused is not eternal. It is finite. Its effect becomes a further cause. Thus all existence is a changeful network made of causes and effects; creation is an action and reaction zone.

Where nothing is an absence, nothing comes from nothing. Did you think space was nothing? Wrong. Vacuum did not come before its cosmos; nor, since particles arise from fluctuations in it, is it nothing. Why, therefore, should cosmos as a whole derive from nothing? If it didn't, did the universe create itself? Then it existed prior to itself. Cause caused itself. Such is the kind of incoherent, 'boot-strap' logic some display. Perhaps, you claim, there was no cause of physicality! But, if there were, such cause could not be physical. It must transcend the physical. Its physical non-being must be immaterial; its being must be metaphysical. This Uncaused Being is not absent. It is self-sufficient, potent and with presence; its causal level of reality is 'higher' than non-conscious, physical phenomena; and its timeless time is Now.

Let Gottfried von Leibniz help. Whatever exists, he states, has an explanation for its existence and, if the universe has such an explanation, it is God. The universe indeed exists so has an explanation; that explanation must therefore be God. Oh dear! Surely this is deep as a delusion drops! What, Leibniz, is the cause of God? 'God is uncaused, a metaphysical necessity. God is Nature that transcends the nature of materials,' comes L's reply. Objection! The universe is all there is; it is exempt from principles of cause. These are arbitrary assumptions. If, however, God does not exist there rests no explanation for the universe - except its own eternal self-sufficiency. 'How, if cosmos had a start, can matter be an uncaused and eternal, physical necessity?' Easy. I will generously invent a multiverse. Put that in your hookah, Algazel; stuff it, Leibniz, in your clay and smoke it!!

These matters are the substance of elaboration that Chapters 7 to 12 involve. They include the scientific instance of fine-tuning and appearance of design. Are these due to physical necessity, chance or a mind-behind? Here promissory atheistic faith abounds. As we'll see, it's not the number on a chit that counts in cosmic lottery; it's whether it's a single sort - the rare, life-permitting sort. The chance against such universe is so remote we can reasonably ignore it. We are here, yes, I agree! But how? Do M-theory's fantastic worlds, forever changing constants and initial quantities, exist in 11-dimensional fact? Won't *TOE*s one day supply the answer to my waking up this morning?

In summary, it's as simple as it's crystal clear. **What starts to exist is always caused. Material cosmos, known as nature to the natural sciences, started to exist and therefore has a cause**. This cause, preceding physicality, is physical non-being. It is nothing physical but causes cosmological effects. What came (or comes) before the latter's matter, space and time and must itself be time-less, space-less, immaterial. **Preceding its secondary causes and effects the primary, first cause must be physically uncaused; it is uncreated in a naturalistic sense, thus super-natural and metaphysical. In this respect the term preferred by Natural Dialectic to 'big bang' is '*transcendent projection*'**. Shot from its archer's super-natural bow this projection undergoes, according to time's arrow, change and decay. Why should what precedes the arrow ever suffer physical demise? Better try and understand the nature of an uncreated immanence from whose causation galaxies and men have sprung.

Can there be infinite, non-existential being - physical non-being but being metaphysical? Numerical infinities are made of bounded units and, at least as you start to count, have starts; but they never have an end. You might call them 'false' or 'relative' infinities. This is not to say that Formless Nothing, which is Actual Infinity, has any cause. What is intrinsically boundless hasn't start or end. Why, therefore, should Infinite Essence need 'sufficient cause'? What is uncreated, as Algazel and Leibniz noted, transcends the natural existence of causality. No boundaries *means* infinity; no natural thing *means* nothing. Is physical infinity and nothingness First Cause of every universal thing? Are such abstractions actually the power above this world? Materialism really doesn't like ideas of First Cause any more than cosmic hierarchy down the line, and in Causal Presence atheism has no place, except that figment of imagination called a multiverse, in which to hide. It therefore strives, in pseudo-scientific ways we'll see, to rationalise 'eternal, uncreated matter'. **But motion is potential that's expressed; cosmic action is Uncaused Potential worked.**

Rephrased, the primal motion of Essence is First Cause; such First Expression is the Start of starts and purest form creation issues in. All existence, psychological and physical, is the Start's effect. Although, within the whirligig of space and time, we choose to call some actions 'causes' and others their 'effects' in fact they're all, traced back to origin, effects of Cause. **First Cause is linked with everything that follows its beginning; and it forms the link between the pair - Essence Absolute and relative existence - that comprise the basic, prime polarity of Natural Dialectic.**

Enmeshed in cycles of reaction we forget the Fresh and Primal Spring. Yet does a spring, once sprung, dry up and cease its issue from above? Is his coronation all a king's reign is? Does a boss, upon establishing the power structure of his business, straightaway retire? Electric through a grid is not supported by a one-off cycle of its dynamo. Why should 'primordial first principal' have only had a moment in the past? You might expect that timeless cause remained behind its timed effect and, if it 'stopped', that consequent effect would also disappear. If cosmic motion is perpetual why should the universal party's life and soul, its central part, be absent now? It should be, as substance and support, in front of us as well. In this view existence is the great effect. First Cause seeds the development of things in time; as such its location as the Source and Apex heads the existential ziggurat, that is, it tops the conscio-material gradient. First Motion links the Infinite with finite powers; it cuts The Whole One down to sizes. It is thereby the central and supportive spring of 'lower' and dependent worlds.

OK. Now let's briefly introduce a mystery that later through the book we shall increasingly explore. **No doubt, energetic causes push effects; they bump you from behind; their arrow, physic's arrow, runs from past to present.** And things suffer (though you'd hardly think it as regards a proton or electron) from increasing weariness called entropy. They run out of steam.

Try this initial, dialectical description of such energetic phenomena:

practice	*Principle*
'right' behaviour	*Law*
outworking	*Plan/ Program/ Theme*
shaped behaviour	*Attractor/ Archetype*
its expression	*Information*
development	*Causal Seed*
outer rings	*Innermost Core*
2° effect	*1° effect*
matter-in-practice	*matter-in-principle*
outward passivity	*inward action*
macroscopic	*microscopic*
gross precipitate	*subtle governor*
classical effects	*quantum/ atomic effects*

This stack appears to say that inward, quantum energies are what 'inform' the gross world's fixed and fluid shapes. Subtle causes gross but what is meant by Attractor, Inmost Principle or Causal Seed? This is another matter that this book explores. For now let's understand that energetic or material causes (call them 'passive information' if you wish) are not the only kind.

Who 'causes', for example, a computer? Or should we say, what sets in motion its recurrent operations? Who inputs 'causes' in the form of programs or sets up machinery then lets it work its purpose out? First Cause is the Inventor. What follows are the programmed physical effects. These might be serial but also, dialectically as *fig* 0.7 iii shows, hierarchical. Creation obviously precedes the automatic operation that accords with a machine's capacity. Thus lesser causes (called by Thomas Aquinas 'delegated causes') administer the order of a natural bureaucracy; they constitute a cascade of causality wherein the lower minions may be unaware of what has set the manner of their motions up. What less aware than physical phenomena?

Could you not say, however, cause had been implanted into operations digital? Could you say that in the way that a computer program unwinds subroutines *top-down* from its masterful or main routine so the creation's mode is ramifying 'downward' through subtle quantum into secondary, sensible effect? Yet how on earth can an attractor, law or program be implanted in the natural world? It's materialistic heresy to think that physical phenomena are, were or ever will be psychologically informed. We assume that, without proof either way, natural law arose by chance. According to the core of what our science studies and discovers, the masterful regime of natural law, there continually link up and then diversify chains of action and reaction, networks of cause and effect. If you reversed the flow you could trace them back to their source. You could pull the strands together and see how it worked.

What, though, about a cause that is conceptually implanted. What about *informative causation*? **This is goal-oriented; and goals are in the future pulling you their way. They pull you from the future; they are metaphysical attractors, guides that govern your behaviour as they lead you through the world.** Not material but immaterial, such leadership is not by force of gravity, electric charge or magnetism; information's metaphysical not physical; it's psychological and, although at this point you will want to know exactly how, be patient - as the book's elaborated we shall come to see. **Mind is negentropic and thus metaphysic's arrow flies, from future back to present, anti-parallel to physic's.** Thus you are guided, present to the future, by plans realising goals.

This stack details informative, metaphysical phenomena:

degrees of ignorance	*Knowledge*
relative/ variegated awareness	*Awareness*
shades/ forms of mind	*Consciousness*
lesser purposes	*Purpose*
creations	*Causal Seed of Mind*
shaped behaviour	*Attractor/ Archetype*
expression	*Potential*
2° effect	*1° effect*
dimness	*brilliance*
inertia	*aspiration*
fixity	*fluid/ dynamic mind*
forgetfulness	*learning*
mindless automation	*reason*
sleep	*normal waking*
unconsciousness	*awareness*

Check *figs.* 1.3, 1.4 and 2.6. Never say, even though it breaks a paradigm, that nature has no purpose. What's past is materialistic ideology; next, as we'll see, Natural Dialectic is iconoclastic but most reasonably. Purpose is what pulls you forward and, even if they're fixed, we'll see how creation's vibrant archetypes, its morphological attractors, pull things forward the way they have to go. **Push and pull cooperate. Energetic and informative causation, running anti-parallel, are the way the world proceeds.** You still ask, however, how this can apply to physicality.

In future we shall not much use the word 'attractor'. **The Dialectic's word is 'archetype'.** Attractors/ archetypes, the source of natural order, purpose and coherence are of two kinds although both found in mind.

The conscious attractor, causal or potential mind, is identified as

Archetype; this Primary Archetype is Logical; it is Single, First Cause or (see Chapter 5: Top Teleology) *Logos*.

The lower, unconscious attractors, causal or potential matter, are identified as archetypal memories or multiple, fixed forms in universal mind. In the human case these secondary, passive records are reflected by fixed patterns in your own subconscious mind; they are memories known, in part, as instinct. The way such morphological attractants, that is, archetypes are linked to polar, cosmic infrastructure is an issue we'll elaborate. Gradually we'll pin the crucial, causal notion down.

individual	*general*
somewhere	*nowhere/ everywhere*
local	*non-local*
factual	*conceptual*

An archetype is general not specific; it is immaterial not material, conceptual not factual. It's mysterious yet obvious, as hard to grasp yet easy as mathematics, logic or a way of nature - nowhere, everywhere and always. Thus its 'shadow' is more real than transient shadows flitting at its bidding, shadows we call physical reality. Indeed, various behavioural patterns, archetypes, substantiate all form - including yours. If this idea is old perhaps quantum physics indicates it's new as well. For now, let's simply note that archetypal memory is something that (especially in Books 2 and 3) we'll learn much more about.

Informative causation and, in forms of life, attractant archetypes are where natural teleology obviously resides. Just as genetic code is not 'as if' a code but *is* one, so natural design is obvious and needs no counter-intuitive 'as if' spraining of the intellect to comprehend it is not virtual, imaginary or metaphorical but real. It is, except to naturalism's philosophical agenda, easily intuitive. Moreover, you might see an archetype as unexpressed potential; it is the 'shape' of latency that lies behind a field of action; you could even say it's the conceptual field itself. On fields the actors play. Electrons, protons, forces and their various agents; and you live in a field not only matter (body) but of immaterial mind. Archetypal fields are omnipresent possibility whose actual, expressed behaviours will, by energetic push and informative, attractant pull, be individually realised in contexts that occur. In this whole business, what is certain, what is not? We need to briefly introduce a causative antinomy - at least an apparent one.

existence	*Essence*
expression	*Potential*
change	*Certainty*
practices	*Principle*
2° effect	*1° effect*
fixed	*fluid*
worked out	*working out*
become	*becoming*
certain	*uncertain*
determinism	*indeterminism*
automatic	*creative*
none	*possibilities*

Mind and matter will emerge, like information paired with energy, as the base pairs of this Dialectic. They constitute polarity within Essential Unity. The antinomy resides, within and between these pairs, as surety and insecurity, uncertainty and certainty, determination and, opposite, what's undetermined. One is one, no doubt; that's certain. But duality shows every shade until it drops, at base, to actuality; Potential drops from Full Potential to its fully worked, constrained expression in the inertial forms of fixity. Thus, as quantum physics also grasps, you collapse through energetic possibilities (or probabilities of indeterminism) into a single, 'real' outcome; or into a single, automatic kind of repetition we call habit. Is not matter utterly, obsessively yet mindlessly habituated to its actions?

Nature abhors an information vacuum. Automatic matter has no mind; it's habit all the way but, with conscious mind, the more aware of options that it is the more that, on a sliding scale, creativity takes over. Either way, by habit, instinct, archetype or range of conscious flexibility, mind is in whole control. Indeed, conscious mind craves knowledge like a panacea for its interests and to resolve its insecurity, that is, its ignorance. Uncertainty that's not resolved will, anxious, fretting, drag you down; and ignorance is bliss until an unexpected consequence appears. Increased resolution leads from indeterminacy to knowledge and determination. What then if, above all other quests, you decide you want to know how cosmos works? Then you want Knowledge and, in Complete Certainty, Union with the Top Attractor; or if, on the material plane, you want illumination on the laws of nature you study operations scientifically - even if that study cannot tell how they are made.

You will, whatever else, have noticed by now that the causal lines of information (mind) and energy run anti-parallel. Thus we're beginning to envisage cosmos in the terms of 'up' and 'down'. There equally exists (you know it from experience) hierarchical interaction between mind and matter. Such a chain of causality, where one level of existence affects another, is called 'vertical'. In dialectical terms a cause 'higher' on the vertical, informative ordinate is that from which a 'lower' effect derives. Materialism, on the other hand, involves no vertical axis on its graph. Thus it only accepts 'horizontal' chains such as those of physics and chemistry; in this non-hierarchical case cause and effect, after such inexplicable 'first cause' as an expansive 'big bang' or other projection out of 'really empty emptiness', cycle within a single level. Each effect becomes the next cause; in an endless transformation of energy (and accompanying scientific S.I. units) one appearance shifts into another. This is the materialistic and, by thought-osmosis, scientific view.

Couldn't you, on the other hand, call principles, plans and programs causes? Isn't a principle, although a metaphysical and informative rather than an energetic entity, a kind of law; do sets of rules not govern changes in space-time? In this case causal archetype is law and action its effect. Check *figs.* 0.5 vi, 1.3 and 1.4. If principle affects behaviour then government is handed down; immanence affects phenomena; a pattern of behaviour is the outcome of its causal 'stage above'.

Cause motivates and moves. It transforms. What is existence but such change writ large? Within a system, physical or psychological, cause and its effect continue in a chain. While a material system's knock-on is definite, automatic and immediate the links in a psychological chain, whose voluntary energy is intent, stretch towards future possibilities; and possibilities are often less definitely predictable or immediately active. Cannot a memory stay dormant, for example, until some instance flares it up? Complexity of circumstance in mind and body also complicates reaction.

And so the world is not as simple as a single, knock-on chain. Scientific method (*fig.* 5.5) isolates such simple chains for its experiments but real-life connectivity is a complex network with extent, from past to future, in space and time and, don't forget it, mind. A multiplex of impact ripples, an aggregate of different factors simultaneously interacts to generate a swathe of more or less predictable effects. This ever-changing context, far and near, is called existence.

The first steps of any journey set the way that things will go. Laws are axiomatic but initial conditions swing the way they are expressed. That, for example, a butterfly flapping its wings in Brazil started a chain of events leading to a hurricane in Indonesia is an anecdotal way of saying that apparently simple changes can lead to monumental difference in consequence. It all depends. Slight variation at the trembling start could have caused lightning over the Azores or everything to peter out without discernible effect. Can science in the real world of complexity make accurate predictions? Could a *TOE* know everything? What is certain is that, if you want a given product down the line, then the initial conditions you feed in must be (as with our cosmos) so precise.

We want, eliminating all uncertainty as far as we can see, the certain truth concerning everything. We want the security of concrete knowledge. However, wanting everything to be reducible to material cause and effect does not necessarily make it so. If there is a limit to physical probes this may leave us, but not necessarily the operations of nature, fundamentally uncertain. A branch of mathematics has been developed to address ubiquitous 'uncertainties' that plague the real, non-experimental world. 'Chaos Theory' describes the behaviour of systems that change over time. It deals with the impact of simultaneous, multiplex relationships and the kind of unpredictability that occurs when small 'fluctuations' in initial conditions give rise to large 'disparities' in actual or expected outcomes. Such systems may appear random in process but are fully defined by (or totally dependent on) the initial conditions so that outcome is in fact determined. Their state is, rather like our world in general, one of 'deterministic chaos.' In the face of such chaos could one share the belief of certain quantum physicists (of Danish school) that nature is, because its basic quantum phenomena are wave-like and waves are spread through space and time, an indeterminate quantity? Could it be that definite cause need not have definite effect?

Students of uncertainty and the 'chaos' of effects as they emerge have not yet included in their equations such uncertainty as certainly exists in mind. Nor what historians habitually make central reference to - the incident of thought. In this case uncertainty derives from mind not matter. It is down to me (that is, the possibilities that I perceive and one I choose) and not to nature. Although, since I'm a part of nature, is there any actual difference?

Of course material nothings, in the shape of metaphysical ideas, schemes and carnal desires which are at source even less visible than the flap of a butterfly's wing, have an immeasurable impact on both human and natural history. The knock-on effect of life's calculations can be traced, albeit hypothetically, back to...where? What was the first cause? Was it a matter of conscious design or unconscious chance? In our relative ignorance can we presume that, at the centre of the external unpredictability in which we live, the inmost causes and regulation of events are uncertain? That absolute and essential knowledge called Enlightenment, does not exist? And, therefore, that the security of Certainty is fiction?

thing	Nothing
finite	Infinite
duality	Unity

First Cause of creation may not be, as we'll see, first cause of matter. What have you left when all matter and all radiation are extracted out of space? Is nothing left at last, the same as at the start was there alone? No doubt, as Einstein's formulations indicate, time and space are very warped, perhaps even broken at their 'nether' pole, an 'infinitesimal' locality dubbed a black hole; but what about the 'high and free' extremity, a pointless, large-scale, immaterial version of infinity? Zero and infinity are terrors. The twins' uncertain pointlessness is where phenomena fall in and out of being, where the foundations of all logic and the framework of mathematics are destroyed. Lack of definition prevails; and an uncontrollability you have, excluding rogue infinity by some contrivance, to rein in and 're-normalise'!

Yet the twins are alpha and omega. The events of existence, both physical and metaphysical, began and (see Chapters 7 and 12) may well end in them. Is there cause locked in the emptiness of space? Is there really, at this point, a quantum energy called *ZPE* (see Glossary)? And if so, is it the residual energy of an exhausted vacuum or, on the other hand, vacuum potential, first cause physical from which things spring, a reservoir of primal 'super-force' by which they are supported? In other words, does archetypal paradox reside (for physics) inside nothing and its close associate, infinity?

What of psychology? What about the immaterial element, information, and its 'space' of mind? Events 'slung in between' the Unity of Nothing and Infinity must hang on them; existence, made of mind and matter, is dependent on their Causal Essence. Is mystic transcendence, chasing this Strange Trinity, a psychological 'un-mindfulness', a love that intellect needs 'normalise'? Dialectic logic certainly espouses subtlety, harmonic oscillation and something-next-to-nothing as the generator of the world. In the sense that Infinite seeds subtle and thence gross, finite things this book's starting-point is seminal. *Its structure is hierarchical, arranged top-down in the order of creation's process from Infinity.*

Indeed, you could see creation as (including you and me) a finite gap in the infinity of nothing. Zero certainly creates some puzzles. Equations that describe the notions of big bang, black hole and zero-point activity can incorporate division by nought. Such division always yields the madness of infinity. No doubt logic explodes but cutting something up by nothing forms the pointless basis for a kind of maths called calculus. It might unhinge the laws of nature but you can't make nothing of it! Calculus, developed by Sir Isaac Newton,

Leonhard Euler and Gottfried von Leibniz is the mathematical language of nature. Indeed, like zero and infinity, it is a key to understanding energetic worlds. Is polarity, in principle, an actual, natural cause? Is on/ off, zero/ one an integral component of creation?

Don't override apparent absence; don't think nothing of it. For physics the division of something by nothing yields infinity. For Natural Dialectic it is the division of nothing (which is infinite) that yields the multiplicity of creation. Nothing, in its positive form, is infinite potential. Where 'essence' comes from the Latin word that means 'to be', 'existence' comes from a word that means 'stand out'. *In this way Essence is the being, potential or ground from which all things spring.* It precedes and empowers the motions and structures of existence. *The latter, for its part, expresses the reduction of infinity to all degrees of finite form that 'stand out' from it.* These forms are polarised transients; they are ephemeral 'bubbles' within the infinite presence. They have a beginning and an end. They are events (objects are no more than relatively slow events) with a start and a finish. Creation, like your life, appears as a gap (incorporating myriad separate sub-divisions) in the Unity of Nothing and Infinity. This triple aspect of Central Potential precedes, with fine distinction, even primal subtlety of motion called First Cause.

outcome	*Potential*
attributes	*Being*
order	*Preordination*
subsequent chain of events	*First Cause*

First Motion is not physical but metaphysical. The first cause of a logical operation or machine is not its *modus operandi* or its parts but its inventor's mind. Mind's first motion is the energy of will; first cause of strategy and tactics is desire. Prior to actual action comes its possibility, prior to external order inward principles of law. The source of mind is Archetypal Order couched as resonant command. What, therefore, of *physicalia*? Was there an automatic start to cosmic automation? *Or does its order conform to an archetypal register filed as memory in causal mind before the body of our world began?* What on earth in heaven's name does that mean? We shall see.

Polarity

conscio-material field	*Top Pole*
polarity	*Neutrality*
duality	*Unity*
repulsion/ attraction	*Consummation*
motion	*Balance/ Peace*
repulsion/ negative	*attraction/ positive*
divisive/ divided	*unifying*
non-conscious matter	*subjective mind*
base pole	*in-between*

What of the second principle, *polarity*? Poles express the fundamental duality inherent in things as opposed to nothing. Indeed our cosmos can, as we've begun to clarify, be seen as anchored between poles of various sorts; nature, physical and psychological, is seen as a dynamic interplay of opposing and yet complementary forces. These are the stays from which its range of webs, its tapestry, is woven. Their extremes influence each other. They can repel and stand apart or they can attract and work together in a covalent kind of way. Between their complementary, reciprocal opposites ranges a spectrum of intermediates. This oscillatory range of influence is known as a field. The doctrine of polarity (or duality) wherein complementary opposites interact and together constitute a whole is ancient. Empedocles, for example, encapsulated polar dialectic simply as the universal forces of love and discord. Attraction and repulsion. Not only matter but the whole of existence is imbued with the paradoxical two-in-one nature of polarity. Sometimes emphasis leans one way, sometimes the other but whatever it is you are considering polarity is represented in the aspects of it. *Natural Dialectic is a name for cosmic chemistry that always binds polarities into a single pair, a double singularity, a covalent stack of interaction. You might say it dances to the tune of counterpoint's polarity.*

However, the Dialectic also *resolves* polarity in the form of a third, central component, Balance. As regards the whole creation it proposes mind and matter as two existential aspects of an underlying Essence. Two-in-one is also one-in-two. The framework thereby, as we've seen, complements existential relativity with Essential Absolution. Thus its stacks represent Essence (Top Pole) and Essential Characteristics set, top to toe, against a range of relativity we call the existential field. **This existential matrix, in which informed and informant (energy and information) interact, is called the conscio-material gradient.** From its Top, 'Strong' Pole of Subjectivity this gradient falls to the special case of 'weakness' - a base pole of none. The nether case, non-consciousness, is creation's body; such objectified embodiment of Essence we know as our spacious, sunny universe of matter.

The Chinese 'Book of Changes', *I Ching,* is ascribed to the sage Fu Xi. According to the mathematician Leibniz Fu Xi (who lived c. 3000bc.) was the founder of Chinese scholarship and, it might be argued, oriental psyche and psychology. His seminal book is based on the arrangement of $(2^2)^3$ or 64 hexagrams each composed of two stacks of three lines. These lines are either broken (*yin* with a value of 0) or unbroken (*yang*, 1); and the six-lined stacks of them (the hexagrams) are thought of as seeds containing the potential answer to every question that the pattern of the universe can generate. Was it not Galileo who proclaimed that mathematics is the language of the universe and science is, therefore, at root mathematical? Was it not Leibniz who, after reading the *I Ching*, developed his binary system of mathematics in which all numbers are expressed in terms of duality? Is it not remarkable that these two 'whole' numbers, nothing (0) and its neighbour (1), can be arranged so as to symbolise an infinite diversity of other numbers and entirely represent a Galilean but especially a polar sense of cosmos? Was creation, using Leibniz' pair, *ex nihilo*, *ex uno* or, dividing one by zero, *ex infinito*? Whichever way, is not the cosmos computational? The binary system underwrites today's information revolution. It is therefore at the heart of information technology. In operation current is switched on (1) or off (0). On combinations of such simple switch the towering edifice of all computer programs is built; such switch is simply multiplied, according precisely with a plan, to ramify into the networks of our lives.

Not only mind machines but biological information systems (Chapters 5, 13, 14 and 19 - 25) are based on binary on-off switching networks. What about your nervous system for a start?

The second major revolution of the twentieth century underlined such facts of life. Genetics and molecular biology took to complementary pairs of opposites. A double helix is composed of complementary, polar strands. Is it not curious that, as well as life's book, life's alphabet is composed in a binary way of base pairs (2^2 gives G, C, A and T)? Such code, based on discrete bits called nucleotides, is digital and thereby computer-like. If this is new to you use the glossary (under *DNA*) and Volume 1's glossary (under base and base pair) to help you start to understand. Each Chinese hexagram is composed of two stacks (its *yin* and *yang* elements) each of three lines. Similarly permutations of three letters (instead of lines) stack as a word from life's vocabulary. For example, you could specify the word (or 'codon') AGC or TAG; each one would spell out an amino acid that, in turn, joins like words in a composition to a sequence called a protein. What are you for your biological part but a body built of proteins? The genetic code is what has specified you. For you and every other kind of life on earth its full dictionary is not one of $(2^2)^3$ or 64 hexagrams but precisely and with binary efficiency $(2^2)^3$ or 64 three-letter words! That's not *I Ching* but it's official! Perhaps the Dialectic has a natural pedigree.

0 and 1 denote potential and action. Above all, therefore, they denote Presidential Essence and, when it is switched on, auxiliary existence that stands out from and, apparently, apart from it. The two numbers are also used to denote (in Part 0) the informative and physical potentials that compose a lifeless, macrocosmic stage and (in Part 1) this stage alive with units of drama, that is, with 'switched-on' characters, with conscious incorporations such as you, me and, although sometimes less lively, every other organism.

Other fundamental polarities include information/ energy, unity/ duality and isolation/ interaction. What about things in their aspects of active energy (changes) and passive objects, that is to say, active wave and passive particle or separate body and its exchange of influence with surroundings? You might even analogise wave/ particle duality with conscious (mental) and non-conscious (material) aspects of existence. Mind and body. Physics, chemistry and biology also embrace polarity; they are replete with stacks of complementary pairs. What, in their case, about positive and negative charge, forces of levity and gravity and, opposite polarity itself, neutrality? We shall elaborate dialectical Stacks of opposites. Answers depend on your perspective and the questions that are asked but everywhere you can, as Leibniz did, count on it - polarity is at the root of nature. It is universal. Such spinal reciprocity is reflected, although an essential modification will soon be made, in the structure and operation of Natural Dialectic.

The fundamental polarity is between Essence and existence, between Potential and action or, in other words, Nothing and things.

If you prefer, it is between undisturbed field and motion. Between neutrality and polar modes. Within the fluctuating field of existence basic polarity is expressed in terms of positive and negative power. *Each pole of each sort of field is sorted into one of these categories.* For example:

negative	*positive*
matter	*mind*
dark	*light*
particle	*wave*
gravity	*levity*
mass	*energy*

'Atomic' is Greek for 'indivisible' and atoms well express polarity. Poles are part of each other. One becomes two but each part remains related to the other and, nested as a subset, to a larger, neutral whole. Such division, branching like a *top-down* computer menu, is the mode of logical creation. By this path principle differentiates into detail. Moreover like poles repel and unlike attract. A pair may or may not want to unify. They seek each other or they run away. In this way, due to its power and influence, polarity is a source of tension. **What are things and thoughts but tensions strung in mind and throughout space?**

The Basic Existential Dipole.

realisation/ expression	*Potential*
tipping	*Balance/ Poise*
polarity	*Neutrality*
motion	*Equilibrium*
fixed/ static/ impotent	*kinetic/ potent*
completion	*action*
inertia	*stimulus*
negative neutrality	*current*
inertial equilibrium	*dynamic equilibrium*
exhaustion/ discharge	*vibration*

No doubt everything, both large-scale and small, is in relative changefulness. But motion of change reflects only one aspect of energy, the kinetic. In fact there are three aspects. *They are (fig. 0.5) potential, kinetic and fixed*. These are, translated into process, possibility, action/ motion and exhausted completion. The dialectical definition of 'kinetic energy' is therefore broader than the simple 'energy a body possesses by virtue of its motion'. Putting aside any mysteriously perpetual motion of atoms, the definition includes both directions any flux/ change can take - towards inertia (↓) or, by stimulation (↑), towards free flow. It therefore includes dynamic factors such as light, heat, electrical current or force-in-action; and 'anti-dynamic' factors such as gravity, mass, aggregation and 'crystal precipitate'. Indeed, is not material cosmos automated by the fixity of natural law?

Moreover, the poles of Natural Dialectic are, as opposed to flow between them, opposites. It is a mark of Dialectic that these poles (in this case 'potential' and 'exhausted') reflect each other by inversion. Their opposition is, you may say, in 'reflective asymmetry'. For example, neither potential nor exhausted forms of energy show motion. The dialectical definition of 'potential energy' is also broader than simply 'energy a body possesses by virtue of its position'. It implies possibility as yet unrealised, latent capability and 'readiness-to-go'; it implies a neutral and yet potent form of equilibrium; this is positive neutrality. In 'going' the motions of physical existence involve both the levity of stimulus such as propulsion, light, energy gains or other tendencies to change and, on the contrary, 'contractive' agents such as nuclear force, inertia, gravitation and tendencies to lose energy, aggregate or slow to fixity. Static, spent or exhausted energy is motionless and yet diametrically opposed in character to the motionless poise of potential. Examples of such neutral exhaustion are mass, inertial equilibrium or a puddle, dead water; they show negative neutrality. Both potent and inertial equilibria are 'nothing' with respect to motion but are very different kinds of void. They are, like positive and negative neutralities, common examples of reflective asymmetry. Unless care is taken such asymmetry can, as Chapter 3: Semantic Switch demonstrates, lead to verbal confusion.

The polarity of energetic motion is, however, not the basic existential one. To physical you need add metaphysical; to energy you need to add another factor, information.

Information

existence	*Essence*
manifestation	*Latency*
conscio-material gradient	*Pure Consciousness*
forms of Information	*Information's Source*
energy/ matter	*information*
passive forms of info.	*active/ semantic forms of info.*
informed/ guided	*informant/ guide*
objective aspect	*subjective aspect*
non-conscious/ physical	*conscious/ psychological*

An argument has been enjoined - that cosmos is composed of energy (including matter) and immaterial information. If 'natural' is equated with 'material' then information has no natural cause. Non-conscious, physical phenomena create neither codes nor, which codes carry, messages; although intrinsically devoid of meaning or purpose they may, however, 'carry' both of these. Things may be dynamically or otherwise arranged, as in the case of music, language or machine, to express meaning; matter may, to its molecular core, be shaped by the various logics of purpose.

In other words, for Natural Dialectic there are two main kinds of information - active and passive. Informing and informed. Active information is semantic; it involves conscious appreciation, understanding and manipulation of form. Consciousness needs either physical or non-physical (thought) form through which to work. Shapes of mind and matter carry intention; mind does so actively; matter, on the other hand, is passively informed.

The patterns of matter are reflex; its arrangements are, although often dynamic, always passive; they are repetitive, automatic and therefore predictable. If such mindless yet 'self-organizing' behavior is devoid of purpose and can, by itself, generate neither code nor purpose how might it specify such an information centre as mind or quality as intelligence? If it couldn't then how could mind evolve out of oblivion? Is it an atomic spin-off? Did the only sort of reason known to man 'self-organise' from lack of it? Or, on the other hand, does human intellect involve an immaterial factor, an element of information or an incomplete reflection of the order of a universal mind? In short, is cosmos reasonable or not; and if it is, why?

Do you remember, from the Introduction and Chapter 0, the idea of an immaterial element - the metaphysic measured on the y-axis of a conscio-material graph? This is identified (as in fig. 0.7 iii) as a subjective/ objective mix, a range of forms of information that comprise the mental aspect of existence and called, generically, mind. Now check 'information' in the Glossary. Individual minds are information centres. They are concentrations, variously weak or strong, of the information element; and, in its weakest and least active form, a special case of information is observed - completely passive, automatically ordered energy/ matter.

Thus you need, for any sort of reasoning, an information-centre. The natural information-centre is mind. Things may be reasoned out by you. But is cosmos reasonable due to passive information, that is, because some other source of information had first worked the system out? This question boils down, we shall see, to the origin of natural law.

In brief, Natural Dialectic would propose that there exists, as well as non-conscious physical energy, metaphysical mind - a 'field of information' also endowed with potential, kinetic and inertial (sub-conscious) aspects. This immaterial information centre may influence and be influenced by material circumstance. Interaction and exchange occur. **For this reason information and energy are identified as the component derivatives of causality and polarity and, therefore, the basic coefficients of existence.**

Energy.

Chapters 7-12 deal with **energy**. These divisions might well, for reasons you will come to understand, be entitled Infinity, Rules, Unity, Nothing, Time and Everything. Meanwhile Chapters 5, 6 and 13-17 deal with **information**, knowledge and psychology. Before we split these fundamental elements, however, we can note their polar field, called biological life. In this, our present field, they can be written as coordinates. This implies their interdependence. If you step from your microcosmic self to the macrocosmic universe you can understand, perhaps not how but that, interwoven in various and variable proportions, these two principals substantiate the fabric of existence. Indissolubly bound together in varying degrees, information and energy constitute a conscio-material spectrum, gradient or hierarchy of existence. Two sides of a single essence, they co-exist. Material and immaterial, they are complementary opposites, the covalent poles of an existential dipole.

It is important to remember that energy and information are intimately and indissolubly involved. They comprise two sides of the same coin, nature. The Sanskrit names of these primordial powers are *purusha* (universal consciousness) and *prakriti* (universal energy). Check them in the Glossary. The former is psychological; it is an immaterial factor whose subjective property is information; call it the 'inward cosmic element'. Then call the latter, whose patterns are phenomenal, the 'outward cosmic element'. In respect of Natural Dialectic the immutable aspect of physicality, graspable by mind alone, is metaphysical archetype; call it (figs. 1.3 and 1.4) causal or potential matter. Its energetic, reflex and yet changeful expression, which science studies at quantum and classical levels, is the universe of physical phenomena.

Informed Energy.

Energy and information are co-principals. From this pair creation is derived.

Objective energy, the stuff of physics, is about action; subjective information is about meaning, purpose and control. Together they comprise plan and its realisation.

To inform means 'to give form, shape, intelligence or organising power'. It is to communicate a pattern of behaviour. When psychological this pattern is called 'knowledge' and includes the attributes of meaning and purpose. The province of knowledge is 'active information' - conscious mind; and the province of 'recorded therefore passive information' is memory in subconscious mind. The automatic behaviours of energy are also passively informed by what is known as natural law. They show as patterns of an object or courses of events. Any pattern of behaviour is explained by its cause; its cause is its reason; reasonable action needs a plan; information is a deed's potential; and mind precedes material behaviour. W*hat, however, is the cause of natural behavior, the reason behind natural law*? This question occupies Chapters 7-12. For now it's simply noted that if the way material things relate cannot be generated by those same relations then, logically, transcendent laws of nature must be metaphysical. Their provision of initial conditions for all natural events means that their origin is super-natural. Cosmos is a *carrier* of information but is not its cause.

Information considered in its fullness includes sender (informant), content, meaning or message, transmission and informed receiver. It can be animate or inanimate. One mind communicates with another or, by guiding some rearrangement of them, with materials. Only mind can deliberately, that is, purposely make a difference. Psychological organisation can be simple or complex. So can the inanimate 'self-organisation' of energy. The latter, however, is always involuntary, purposeless and passive. What message has matter? It can carry and transmit but, being mindless, never actively generate information of its own. In fact, no information exists for its material carrier; information arises according to how this carrier (be it sound, light or any other kind of pattern) is interpreted; it depends on how the substrate is arranged or understood.

Just as an idea precedes its expression or a plan its realisation, so in dialectical terms information precedes energy. At the top of Mount Universe is 'pure information' and at its base 'pure matter'. 'Pure information' means motionless potential to inform. 'Pure Consciousness' is what philosophers may call 'prime ontological reality'. 'Essence'. Between First Cause and passive matter there exists a hierarchy of mind/ matter combinations, different states of consciousness or, in energetic terms, a conscio-material spectrum. **In brief, existence itself comprises a dynamic mix of subjective and objective parts.**

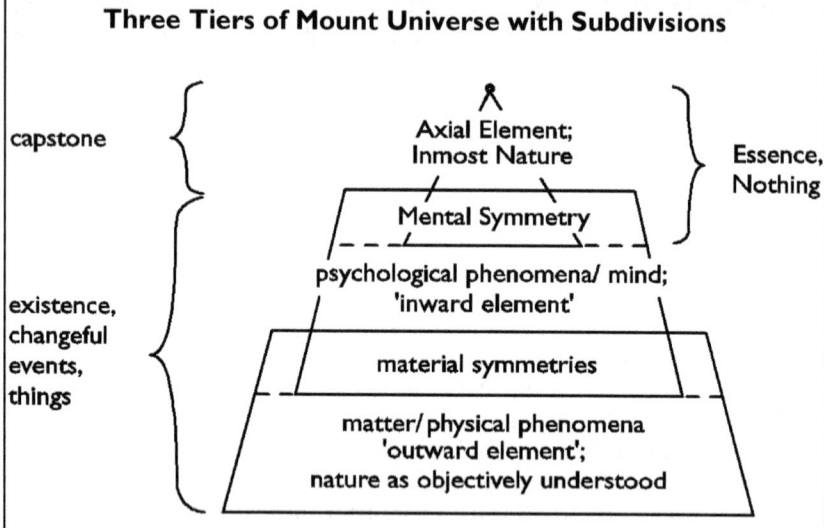

Three Tiers of Mount Universe with Subdivisions

Major subdivisions of cosmos are (Sat) Essence, (raj) mind and (tam) matter. The three new words in brackets introduce the Cosmic Fundamentals (explained in Chapter 2).

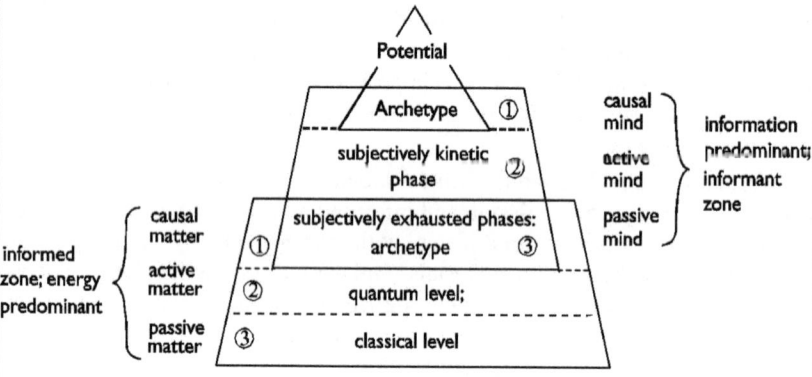

① pre-active plan, 'seed' or causal archetype
② internal informant, pattern-maker; primary effect
③ external structure, fixity of pattern; secondary effect

Within each major division minor (numbered) subdivisions can be drawn. These numbers are related to the three cosmic fundamentals. For mind these are (sat) 'causal' super-conscious, (raj) conscious and (tam) passive sub-conscious levels; and for non-conscious matter a psychosomatic link grade, (sat) potential matter, (raj) quantum matter-in-principle and (tam) bulk or bonded matter-in-practice. This final minor subdivision itself involves a familiar sub-subdivision into gas, liquid and the final, most fixed and 'static' expression of them all, solid.

fig. 1.3

Potential information, pure consciousness, is without constraint but does not do a thing; from potential are conceived ideas; as these are realised detailed form appears. Detail is a localisation of principle; it represents some restriction of the co-principals. Active information (informative mind) is the process of consciousness shaping but also, at lower grades, increasingly being shaped by the constraints of its involvement with energy. *Purusha* (see *figs*. 6.2 and 13.6 and Glossary) shapes and *prakriti* is shaped. In other words, as information is locked it loses active and takes on a passive aspect. Passive information (as in the case of dormant sub-consciousness and non-conscious shapes of energy) occurs when consciousness is overshadowed or was never present in a circumstance. This sort of information, which includes the entire physical plane, is fixed and impotent. Behavioural rigidity can be described mathematically. It is often called a 'rule' or 'law'.

How does 'simple' law inform bio-complexity? ***Informed energy is most obviously at work in living forms.*** In what way is information physically expressed? It is coded and exchanged as energetic patterns (speech, books, TV etc.). Information is the very basis of biology; it is written up in *DNA* at least. We study coded information of this sort in Book 3.

You might argue codes (like words or numbers) are, though informers, empty symbols. Maths, principles, laws and all kinds of code are impotent without the energy that they describe behaviour of. They constitute the immaterial rationality that makes an action reasonable. As with human so with cosmic carriage. Information is the universal rationality; it is inextricably, in various proportion interlinked with power to create and operate. Indeed, in a real sense information's metaphysic is material energy's potential; it constitutes an archetypal source of order. Information is 'empty' just because it's immaterial. It controls the way an action works.

'Energy', for its part, means 'there's work in it'. It is the lifetime's work of every chemist and physicist to investigate the fascinating relationships that bind, separate and transform phenomena. Such motion, action, interaction or change is driven, according to aforesaid rules, by an exchange of energy in one form or another. Include momentum as a force because, according to Einstein's famous principle of equivalence, its mass component is a form of locked-up energy. Include, equally, equivalence of balance. What physicist's equations do not balance cause with its effect, action with - as Newton's Third Law noted - reaction? The behaviours of energy are each, in spite of a veneer of chaos, informed by the precise consistency of 'law'.

Behind the swing of various legal scales a law of causality ('action and reaction'; *karma*) states a fundamental principle of balance, order and coherence. Why, however, should the balance sheet of cosmic economy not extend to the psychological dimension? Why should precision in the automatic world of matter not be mirrored in the mind? Indeed, why for the sake of scientific satisfaction should the world reduce to just material dimension? A triplex form of information and of energy (potential, kinetic and 'rigidified') reflects the way an ordered cosmos is constructed. Can you see it in the three tiers of Mount Universe?

	range of levels	Top/ Highest
	shades	Clarity/ Transparency
	two-way gradient	Source
↓	**lower**	**higher** ↑
	darker	lighter
	grosser	subtler
	base/ sink/ blackness	colours

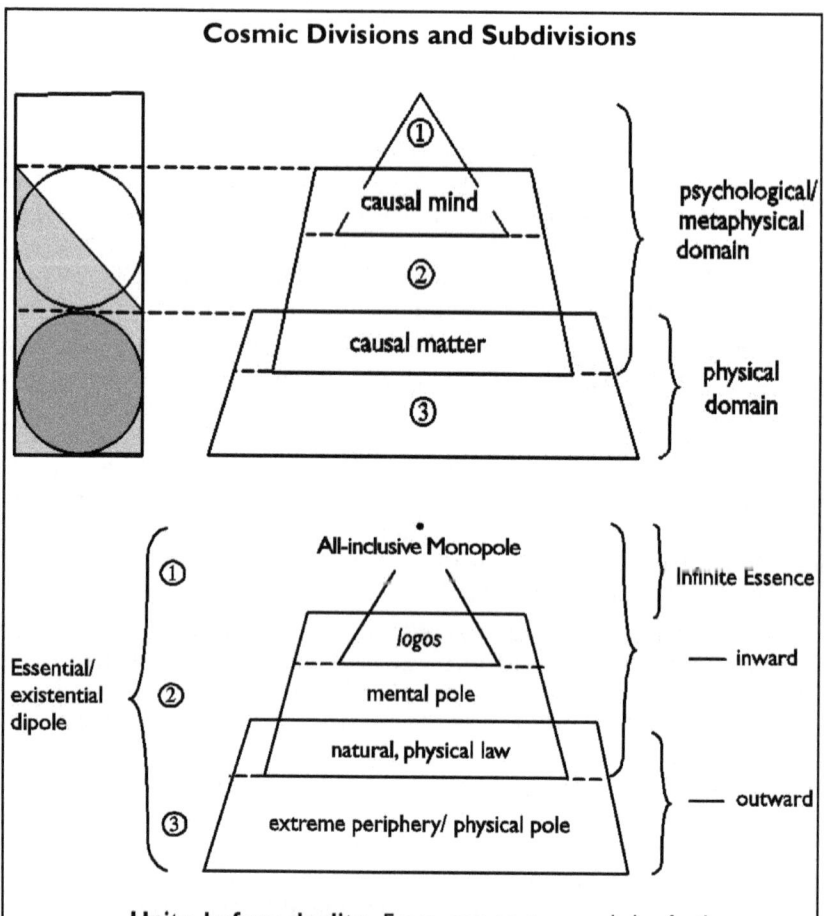

Unity before duality. From one to two and, by further division or multiplication, multiplicity. **Within the fundamental poles of Essence and existence, existence is further polarised into mind and matter.** From Essential Monopole springs existential dipole (see also *figs.1.1, 1.2* and numbers as 1.3).

In dialectical terms potential matter, preceding actual, material behaviours, can also logically be called 'transcendent super-matter'. In this view 'transcendent physics' would study the nature of psychosomatic linkage, archetype and the basis of natural law.

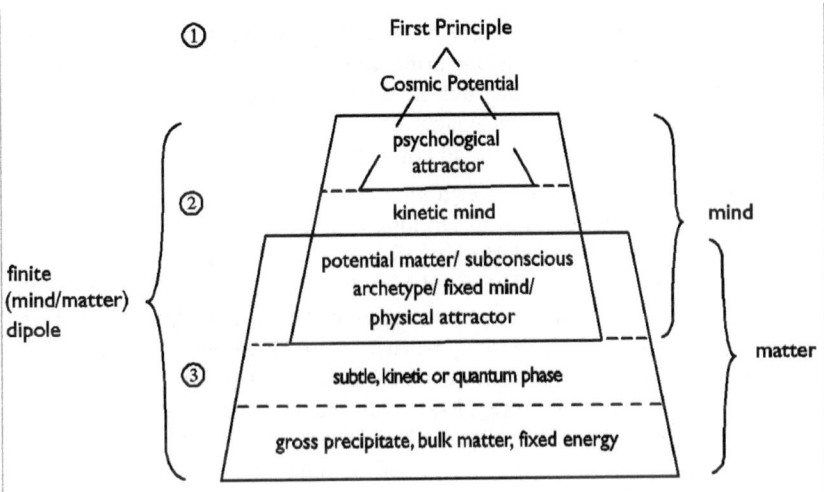

The codes/ rules of quantum (*raj* or kinetic) physics appear to differ from those of bulk (*tam* or classical) physics; at least a full linkage has not yet been found. Why, therefore, be surprised if the rules of 'transcendent physics' are different again? Or, if a quantum/ classical linkage is in future fully defined, that a definition of quantum/ transcendent (psychosomatic) linkage will not also be elaborated? Perhaps (esp. Chapters 9 and 16) such elaboration has begun.

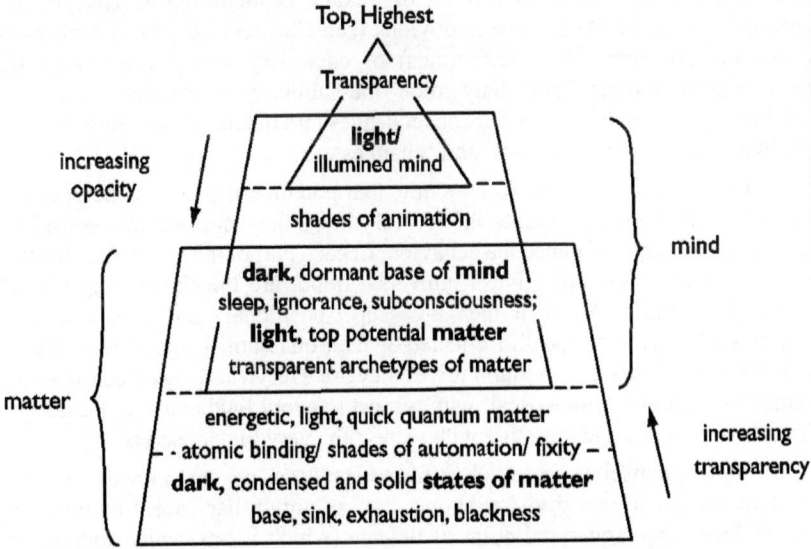

The dialectical idea is very simple. High to low. We shall come to understand in much more detail how the conscio-material spectrum runs from high to low, light to dark, source to sink or top to bottom of creation. It drops from subtle into gross, informative to informed or psychological to physical, that is, from mind to matter.

fig. 1.4 (see also 0.13 and 10.1)

In short, the universe does not have a single dimension. It has three - potential and its dual expression, mind and matter. The quality of these grades depends on the proportions in which information and energy are involved with each other. Such quality of dimension is not the same thing as dimension of extent, of which material volume has three plus time (space-time).

The linkage of one grade or tier to the next is hierarchical. The top, initial or central grade is the potential of Nothing. Potential precedes and generates the level below. From Causal Nothing issues the predominantly informative aspect of creation, mind. Active mind is subjective, conscious and *precedes* passive, sub-conscious and non-conscious planes. As metaphysical energies coarsen, mind falls dark. Flat. Overwhelmed. Non-consciousness is a subjective void, an objective oblivion. From a *top-down* point of view, despite its astronomical fireworks, the physical universe is the lowest, heaviest, darkest level of the three. It is the earth pole, a dominion of non-conscious, energetic aspect, an outer periphery of automation, rigidity and death. Welcome to the body-world.

Triplex division also manifests *within* the regions or tiers and, in the case of mind, this shows super-conscious (or causal), conscious and sub-conscious bands.

What therefore generates the shapes of matter? *As potential energy gives rise to kinetic and runs to exhaustion, so the tiers of Mount Universe are indissolubly linked.* **By this logic a mind-body (psychosomatic) interface will occur at the base of universal mind.** This point of linkage is identified as zone of the aforesaid archetype. Archetype is obvious (see Chapters 16–25) in biological forms; the archetypes (or subroutines) of chemistry and physics comprise immaterial or abstract fields that 'guide' the subtle, quantum phase of matter and, thereby, its gross, condensed consequences into the manifest behaviours we call 'particular objects', 'events' and 'changes'.

In anticipation of Chapter 8 let's note that potential matter's archetypes are metaphysical. As the ziggurats in *fig.* 1.3 show, they precede non-conscious individualities; they influence the behaviour of each part of the energetic, physical universe. Archetypes, operative in universal mind, are the 'integrating whole' behind these parts. As such they are 'super-positional' and simultaneously ubiquitous, that is, 'non-local'. Reflected by their quantum agents (e.g. electronic charge and quark mass with gluon force) they are everywhere received the same. Natural law is thus 'symmetrical' with respect to every field, wave or particle its 'immanent' affects; and therefore with respect to everything observed.

In short, unconscious, passively informed matter is the (*tam*) lowest grade of the creation. Its triplex division shows first as unfamiliar causal or potential matter. From this archetypal point of linkage (which is physically nothing, an intangible absence) issue the second, kinetic and third, lowest levels. This latter couple is familiar to science. We study the kinetic as quantum physics and chemistry; and the bulk as classical physics and biology. The bulk, 'inertial' subset is familiar to every single one of us. We sense it as our 'normal', 'large-scale' world. It comprises a further division into an energy-related triplex of the states of gas, liquid and what is commonly understood as 'static', fixed or solid. Energy lost or locked 'descends'. Energy gained or freed 'gives rise'.

This stack translates from *fig.* 1.4 iv. Its conscio-material gradient slides from high to low; a spectrum shades from light to dark. Potential falls to impotence; high-grade, subtle information drops into opacity; simple, subtle quantum principals precipitate the gross and complex world of solid things your eyes can see. Thus lower is expression of its inner substance; outward appearances reflect their inner truths.

Mind over matter. From mind matter. *From a materialistic point of view this is completely outrageous.* A capital thought-offence. Off with its head! Truncate the universe from immaterial information! Let us, therefore, at this moment re-clarify. There is not only matter in this world; there is meaning too. The association of information with energy should not be seen as a Cartesian split but as a gradual and mutual involvement (*fig.* 0.14). *Let us then see, as the book unfolds, how the game transpires.* How do things work out?

	static/ kinetic	*Potential*	
	finish/ action	*Start*	
↓	*static*	*kinetic*	↑
	finish	*action*	

We can use these stacks to hunt the creation down its slope. We work right to left on the top pair of stacks; then right to left on the bottom pair. We work, for example, from 'potential' through 'kinetic' down to 'static' bottom left; or from 'start' through 'action' into 'finish' bottom left. A dynamic that's simplicity itself.

	existence	*Essence*	
	becoming	*Being*	
	transformations	*Potential*	
	body/ mind	*Soul (?)*	
↓	*energy*	*information*	↑
	physical/ biological effect	*psychological cause*	
	reflection	*archetype*	
	physical/ biophysical image	*psychological template*	
	body	*mind*	

What does this one imply? Some element of antithesis is clear. You have Essential and existential poles. Within existence there are mental and material poles. Mind and body. Soul? Is 'soul' a figment of a nervous twitch, a special movement of electric charge that's called imagination? It is 'realistic' scientism's heresy but is it actually, within the logic of the stack, an elemental factor that, hierarchically, precedes the mind. Because a conscio-material gradient drops. From head to toe, it drops from consciousness towards its lack. From life to lifelessness its levels reach towards an outermost periphery, circumference or bottom of the existential cone. Such bottom is the physical exterior. This book is part of it. So are you and I and every single body from an atom to the stars. Not that creation's outside part is separated from the inner one. Space and time don't separate material from immaterial element. Both Essential and psychological poles are *within* the physical; and, if you want to play with cosmic paradox, *vice versa*. Not a jot removed. A book contains atoms, atoms contain particles, space etc. In the same way each deeper, more interior level of Mount Universe is masked within the one below - like layers of the proverbial onion. Could you penetrate to the cosmic onion's centre with which, therefore, all layers you peeled must have been linked? Thus, in a way we might understand better as the plot unfolds, the pebble in your palm would contain, like every other object, all levels of creation; it would therefore be a fossil nested in all time, all space and also anything they may have issued from. If only one could read just how.

	duality	*Unity*
	components	*Whole*
	expression	*Source*
	message/ creation	*Information*
↓	*passive*	*active* ↑
	exterior	*interior*
	matter	*mind*
	informed	*informant*
	created	*creator*
	recipient	*messenger/ agent*

Mind is an informant but, you argue, energy is one as well. Aren't energetic patterns what give matter shape? This is true and, therefore, bear in mind that there exist both psychological (informative) and physical (or energetic) elements. Active information is creative and involved with message; passive information is a reflex medium that's not. Which shall we place at higher level in the hierarchy of the universe? For example, is this book, of itself, mindful? Its material is a medium but not creator of a message; its ink and paper do not generate or simply carry information. How could matter, which is totally non-conscious, ever come to think? You can; but, like this book, your body is obliviously material. Neither book nor body know a thing. So how can energy give rise to mind or brain to thought? *Can passive ever create active information?* On the answer hangs where you came from. Did reflex mud or sand or clay, when mixed with water, make you up? Or do books, chemically made of ink and paper, need writers who first codify their shape? These are issues Natural Dialectic can resolve but, for a start, simply note that hierarchical subsets are nested *within* one another. Internal is implicit. Top unfolds; it becomes more explicit as it drops towards external. Each subset (or level) is different in quality from the ones super-posed above and sub-posed below it. Creative order runs from mind to matter.

Is a coherent pattern starting to emerge? It's early days. So far the construction of Natural Dialectic has been duplex - a columnar *yin-yang* expression of complementary opposites. These opposites are, it is emphasised, mutually supportive. They are two sides of the same coin and, in the sense that a man may harbour some degree of femininity and a woman a touch of masculinity, the characteristics of each may tend or grade towards the other. Both are parts, interactive parts, of a dynamic whole. However, a whole is greater than the sum of its parts; and in order to properly mirror the triplex nature not only of physical energy but Mount Universe itself a simple but essential modification will have to be made. More than a hint of this central modification, which you will later recognise, has already 'bleeped'. It is time to turn, in more detail, to study the structure and operation of the philosophy.

2. Cosmic Fundamentals, Stacks and Ziggurats

Ranges.

If opposites are black and white then between them lie all shades of grey. Between antipodes exists a spectrum of eventualities, more or less of each. If you marked off the mixture at points between extremes you would represent range, ratio, degree, emphasis, predominance and so on. *Natural Dialectic builds the idea of spectrum, proportion or relativity into any aspect of life.* **Ranges represent, between two poles, their field of action**. Between potential and impotence, its antipole, grades of power are expressed. From high to low a gradient always falls and *vice versa*. Things move and change; and transformations on a sliding scale between the poles of any kind of opposite may oscillate.

Oscillation illustrates dynamic tension; it shows as vibration, period or cycle. From atoms to the stars the world whirls round. A core feature of dialectical (and scientific) descriptions of cosmos is homeostasis, dynamic equilibrium, cycle or frequency; and related to one cycle there may be another with, perhaps, resonance. The scientific register of 'swing' is hertz. This unit of vibration is not appropriate to measure every kind of flux but the bouncy notion of life and the universe's ups and downs surely has its finger on the pulse of Natural Dialectic! And the simpler an object, the easier it is to understand either its erratic or rhythmic changes. The more characteristics, aspects or poles it incorporates, the larger the stack needed to describe it; and the more complex the web of tensions to which it is subject and the changes it undergoes. How much simpler, for example, is the cycle of a pendulum than the integrated cycles of a human being.

In short, dipoles involve dynamic tension stretched between extremes; and they imply the possibility of movement (oscillation, stimulus or its reverse) that ranges in between. The Dialectic writes up them in columns; it lists them in **stacks**.

dark	← *coloured* →	*translucent*
low/ down		*up/ high*
gravity		*levity*
negative		*positive*
erratic		*rhythmic*
ice		*fire*

Dialectical range is, you easily see, not the same as prairie or a mountain range. It is not an area or a variety of foods, clothing or whatever else. It is an expression of tension between opposites. Its pairs of poles amount to the extremities between which a range runs. You might call such vectored ranges 'shades', 'bands', 'fields of influence', 'degrees' or 'tendencies'. **In the same way as such horizontal arrows are used in chemistry to indicate an equilibrium reaction they are used by Natural Dialectic to form a 'relativity shuttle', that is, to indicate a spectrum of intermediates, covalency and either loss (←) or gain (→) of information or energy.** Such loss or gain involves power, influence and concentration; it amounts to a two-way flux in the field between poles.

existence	*Essence*
relative	*Absolute*

When an Absolute is placed against a relative there is no inherent motion and therefore no vector. *However, the relativity of existence is composed of motion between extremes (or poles) of one kind or another. Such oscillatory flux is a fundamental part of nature and therefore existential stacks, lacking the right-hand column of Absolution, are vectored.* These vectors emphasize the comparative nature, the scale, of flux.

In this respect you could write, for example: *high*
                                                low

and thereby express the polarities vertically rather than horizontally. This has an advantage and a drawback.

The *advantage* of this construction is its logical simplicity. High is always set 'above' low or, for example, positive 'above' negative, strong 'over' weak or concentrated 'over' dispersed. This is a consistent and easily understood way of representing proportions, ratios or comparative degrees.

The *drawback* is that you can only process a pair of poles at a time.

If, on the other hand, you write:

   low   ← →   high

You can insert as many polar characteristics as you want. For example:

   low   ← →   high
   lower   higher
   downward   upward
   falling   rising etc.

Flux or oscillation that occurs on a single plane (say, within the material universe) can be registered with *horizontal* vectors. However, as fig. 0.7 iii clearly shows, the dialectical cosmos also involves a <u>vertical</u> coordinate. It involves the hierarchical bands of mind and matter and, within each of these, spectral subdivisions such as conscious/ subconscious mind and quantum/ bulk matter.

        higher ↑
     ↓ lower

Should we, therefore, return to this unsatisfactory vertical representation of poles? In fact, the vertical arrows indicate the same loss (← ↓) or gain (↑ →) of information or energy as the horizontal.

<u>**An upward (↑) arrow represents an increase in influence of the right-hand, positive pole**</u> *e.g. higher or lighter; and, conversely, a decrease in influence of the left-hand, negative pole e.g. less low or less dark. It is placed to the right.*

<u>**On the other hand a downward vector (↓) placed left represents the opposite effect**</u>, *an increase in the left-hand, negative pole (e.g. lower or darker); and, conversely, a decrease in influence of the right-hand, positive pole e.g. less high or less light.*

Such replacement has, however, the added advantage of a constant reminder. **The vertical arrows, written as a single pair at the top of each stack, not only replace the horizontal arrows when indicating comparative loss or gain between opposites but also indicate cosmic levity (towards ↑) or gravity (away from ↓) an Essential Centre or the Apex of Mount Universe. In other words, they act as a reminder of a hierarchical creation.** Thus, killing two birds with one stone, they are preferred and will be used henceforth. Not therefore:

    ↓   *lower*   ← →   *higher*   ↑

but simply:

↓   *outward*   *inward* ↑
      energy   *information*
      loss/ dispersal   *gain/ concentration*
      entropy   *negentropy*
      dead/ flat   *active*
      informative/ energetic loss   *informative/ energetic gain*
      garbled message   *correct message*
      muddle   *clear grasp etc.*

Hierarchy.

    spectral colours   *Transluscent All-in-One*
    range   *Source*
    other ranks/ stages   *Principal*
    its derivatives   *Principle*
    shells/ orbits/ concentric rings *Central Axis*
    off-centre/ disequilibrium   *Balance*
    oscillation   *Apex*   *Pivot/ Mid-point*
                              ^
                           Axis
↓   *down*   *axial line of balance*   *up*   ↑
    negative   *positive*
    lower powered   *higher powered*
    deceleration   *acceleration*
    loss of mobility/ immobility   *mobility*
    imbalancing factor   *balancing factor*
    inertial equilibrium   *dynamic equilibrium*
    flatness   *vibration*
    fixity   *flux/ action*
    materialisation   *dematerialisation*
    black sink   *range/ specrum of colour*

We don't believe in it today! Natural hierarchy's not the modern way! Indeed, a cosmic scale is fatal for our 'levelled' scheme of things. For materialism's sensibility there is just one elemental level everything's contained within (*fig* 0.6); this is non-conscious, material energies. Each kind of such oblivion is measured in accordance with intensities of power and influence that scale from high to low; and state of matter, gas 'down' through liquid into solid, is dependent on its energy. High power, low power; the gradient drops from strong to weak. It's also true that water in a cloud, whose elevated but pre-active state of energy is called 'potential', may drop as rain and run until it stops, exhausted, in a puddle. This said, energy conversions are not generally thought of in a spatial context. A battery 'runs down' but we do not associate its loss of power with actual falling. <u>In this same way the arrows of Natural Dialectic lack, in the majority of cases, any spatial, prepositional context.</u> They may refer to loss (↓), steadiness or gain (↑) of energy as occurs in levels of, for example, temperature, momentum, voltage or velocity. However, lacking the subjective axis (*fig.* 0.7) materialism thereby lacks all sense of immaterial relativity. Relative attentiveness, measure of intelligence, morality, the various qualities of information or, indeed, any scale of mind eludes its grasp; as well as natural energetic hierarchy there's informative as well. Don't *both* involve their graded scales? Wake *up*! *Fall* asleep. Isn't consciousness *above* non-conscious oblivion? *Heightened* awareness, ignorance; happy *'highs'* and painful *'lows'*; most *high* and hellish in the dark *below* - non-spatial, psychological inclines and inclinations just as well as physical exist. Of the fundamentals, immaterial mind and matter, which comes hierarchically prior?

Source springs above its sink. Hierarchy always runs from source to sink; it falls between these poles. Does non-consciousness source consciousness or *vice versa*? Or is the Dialectic's conscio-material spectrum, a dynamic 'Jacob's ladder 'slung between those cosmic poles, a notion that's materially as well as hierarchically wrong - a sheer illusion? If not you have to decide what's source, what's sink. Which turns out top? In the two-way movement on a gradient composed of the primary, complementary opposites (*information and energy* or, in simpler terms, *mind and matter*) which is 'higher up'?

↓	*energy*	*information* ↑
	material	*immaterial*
	body	*mind*
	oblivion	*experience*
	non-conscious physical	*conscious psychological*

While this stack clearly indicates dialectical banding, materialism forbids a vertical coordinate. If matter is oblivious its 'horizontal' axis logically contains oblivion alone. Thus, in a materialistic book, holarchy 'levels' hierarchy. The word, coined by atheist and socialist Arthur Koestler, means 'a whole which is greater than the sum of its cooperative parts'. It is his way of expressing the inevitable - a hierarchical structure of commandment and control. Physically 'higher' quantum interactions rule their 'lower' bulk effect; and a biological cell or organ can be seen as part of a holarchical whole, that is, an organism. For an evolutionist the organism survives due to an originally *unintentional* arrangement of 'holarchic' parts. For a *top-down* holist, who accepts the Dialectic's vertical coordinate of information, it survives due to a *deliberate* arrangement of parts. Its hierarchical arrangement runs from information (message in the form of code and signals) through subtle business in the form of biochemistry to gross physiological and anatomical expressions. Such hierarchy explicitly includes the informative dimension of design. Hierarchy explicitly accepts the factor of psychological development and control where holarchy does not. A whole is greater than the sum of its materials by virtue of design; the order of such design is *top-down* so that, in your own case, parts are subsumed under body and whole body under the plan whose potential orders its construction.

Bureaucracies are hierarchies. They are structures of command and influence; they embody scales of power and ranges of control. An important aspect of complex 'bureaucracy', derived from a single source, is its intrinsic connectivity, its tissue of coherence, order and cooperative function.

anarchy	*order*
cacophony	*harmony*
opposition	*accord*
repulsion	*attraction*
fragmentation	*coherence*
disorder	*order*

A king is a potentate. In him is vested all potential action. His instructions cascade through each governmental level of the kingdom. Each level, in deriving power and substance, takes its order from above. At each step down power devolves, gravitates and is diminished. Any politician or businessman will tell you that lack of a resilient power structure results in chaos. The king's *top-down*, hierarchical order is the opposite of anarchy. Anarchy is the result of no rules or breaking the rules. Matter has clearly, scientifically defined rules. Although it does not break these automatic rules, its non-conscious governance may *appear* to generate disorder. This disorder, which is also called accident, chance or chaos, is simply the reflection of some non-conformity or missing part in the scheme of order in the mind of its observer. Having no initiative, matter has no way of breaking the rules that govern its existence or, if you prefer, no way of changing its behaviours. Thus the stability of non-conscious, physical creation is assured. The hierarchical subsets of an emperor's government are *nested*, like Chinese boxes, within one another.

So it is with cosmos. First Cause is transformed, where the cause of each level 'descends from above', into a hierarchy of distinct subsets. The subtle or principle level materialises from within. Immanent power is stepped-down or exteriorised. At each level energy and information combine in different proportions. The effect is to realise principle in the practice of a rich profusion of forms and aspects of forms. As mirrored in biological classification the top, principle or archetypal levels multiply inclusively from general (kingdom or phylum) down to specific detail. Similarly, apparently diverse and complex shapes of objects incorporate a few hidden subsets of molecules, atoms, subatomic particles and whatever smaller plies its trade on the borders of nothingness. Polar relationships and internal consistencies are sustained. Each part is related to each other part and together they operate as a whole. This is causality. Interconnectivity right back to First Cause is the way of creation. The universe, so indestructibly and completely networked, is (Chapters 7-12) both hierarchical and holistically 'organic'.

'Organic'? This does not mean that physical nature is at all alive. It means that it is developed and operates according to a preordained logic. If this logic preceded physic it is metaphysic; and, since the opposite of logic is chance, metaphysic does not inform by chance. From biological to cosmological,

Natural Dialectic views a cosmos projected according to the same hierarchical principle by which a human body develops from zygote through foetus and child to adult. It is not constructed like a motor car but develops organically *from within*. Universal 'blood' circulates from a Nuclear Heart, a Centre entirely omitted from the 'flat', monotone, materialistic holarchy. It pumps, *top-down*, through a hierarchy that's composed of two coordinates, one metaphysical and the other physical (*fig.* 0.6iii). *Two* axes describe creation's pulse. The plot descends; information is by stages energised, plan gradually materialised.

On the interior, psychological side of things, therefore, information predominates. On the exterior, physical side of things gross energies (called bodies or material effects) predominate. A predominance of mind scales into one of matter. Such act of creation therefore shows (as will be elaborated in Chapters 5 and 6) two main phases; these are psychological and physical domains (*fig.* 1.4). Descent from super-mind falls through conscious 'viscosities' to dormant sub-mind's form of information gelled in files that we call memory. Descent from archetypal memory (or super-matter) freezes into gross, non-conscious behaviours. Non-conscious body is, in the order of this view, an appendage of the conscious mind. In universal terms descent is graphically illustrated down the slope of a cone, Mt. Universe: or radiating from a Central Source to its peripheral extremities of influence.

From subtle to gross, from cause to effect it exteriorises a body based upon inside Information passed in code. Information is invisible potential; it governs; and material expression follows its governance. Order is, like rules, latent until application. Law is physically formless but it guides the way forms act. It bides, until expressed in action, latent in potential. An artist's work is directed in accordance both with his previous capability and the nature of his original inspiration. The realised '*oevre*' bears the stamp of his character. Similarly, latent order precedes the cosmic process of creation; it is present before the inspiration of First Cause initiates its drama.

Because it is central to holistic Natural Dialectic it's worth repeating this description of events in a slightly different way. *According to Natural Dialectic the fundamental complementary components of existence are consciousness and non-conscious energy*. Materialisation, otherwise called creation, is an outward, decentralising process of gradual confinement; both information and energy are increasingly lost; they are locked into fixed, particular 'solid' forms. Dematerialisation is the reverse; information or energy is gained and, for either, there is awakening, release and some measure of 'ascent', however large or limited, towards its centre or, as we might call it, archetype. You devolve towards actual detail; or evolve towards simple, archetypal principle. 'Ups and downs' are common factors - for each a 'vibratory range'.

A simplistic, idealised model (*fig.* 2.1) can be constructed to illustrate the nature of the conscio-material hierarchy as it drops from apex to base.

Idealised Models, Old and New:
Cosmos and Atom

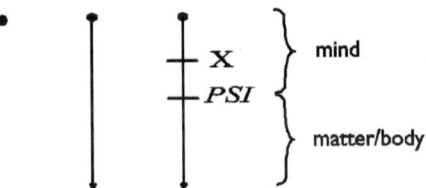

Build a symbolic cosmos. Make a simple start. Mark a dot. Take it as the centre of a circle. Extend a line, a radius from it. This radius represents, on a sliding scale, the gradient of combination between information and energy. The Dot, The Centre-Point represents Pure Consciousness. How can you sub-divide a conscio-material gradient? Wavebands of an electromagnetic spectrum are both continuous although apparently distinct (for example, X-ray, visible light, microwave and so on); and phases or states of matter are apparently distinct although in fact contiguous *and* continuous as well. Within the conscio-material spectrum's 'bands' a distinction is commonly drawn between super-conscious, conscious and sub-conscious aspects of mind followed, below the psychosomatic border, by subtle and gross non-conscious matter (*fig.* 1.4). Such division is elaborated in Chapters 5 - 17. In this case divisions for X (conscious, cosmological axis) and *PSI* (psychosomatic interface) have been drawn.

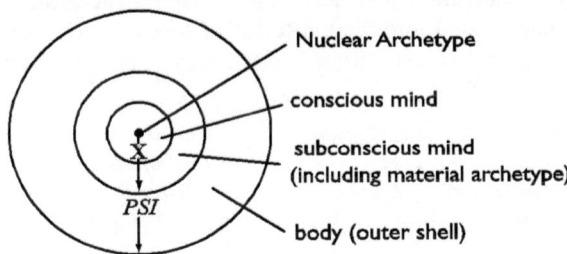

You have used 0-dimensional dot and 1-d straight line. Now use geometry's third fundamental - curve. Sweep the radius in a full arc. Within this peripheral extremity now sweep rings of X and *PSI* concentrically. Transform these 2-d rings into 3-d spheres and you obtain an idealised symbol of the cosmos. Its spherical regions remind you of an atom's actual shells. Each one's nature is discussed at length from Chapters 5 to 17. An atom, physic's symbol, is mass-centred; recall, however, that the Essential Nucleus of all spheres represents the apex of Mt. Universe; it is the Top, Central Source of creation's conscio-material gradient; and, being Information, is the absolute converse of matter. In this respect the inner half of the model is metaphysical/ psychological; only the outer half symbolises physical nature.

fig. 2.1 (see also figs. 0.4, 1.3 and the last paragraph of Chapter 7).

From top to toe *fig.* 2.1 illustrates the hierarchical construction of cosmos as it radiates from Axis towards periphery, a passive outer shell called the physical universe. From pole to pole this anti-parallel inversion represents reflective asymmetry on a cosmic scale.

Of course, it is opined, 'sacred geometries' and multi-tier, hierarchical models are too simple-minded, medieval and *passé*. Such scaling has become unpopular. The stack below, including aspects metaphysical, is unacceptable.

	imbalance	*Balance*
	change	*Stability*
	parts	*Whole*
↓	*subjectively inoperative*	*subjectively active* ↑
	objectively active	*objectively inoperative*
	physical	*psychological*
	material	*metaphysical*
	mass-centred	*information-centred*
	atomic/ forceful	*non-atomic/ forceless*
	body	*mind*

Today's fashionable key scenario is a single-tier (exclusively materialistic) extension from a **physical first cause, a dot called thrice misleadingly 'big bang'.** *In the first case it is orderly expansive rather than explosive; in the second it is (making something out of nothing) an instantaneous miracle; in the third, due to its degree of order,* **it were much better thought of as a 'transcendent projection'.** Just as in the metaphysical arena this source/ projection is concentrated information (creative focus of attention), so in the physical arena 'source' is concentrated stimulus (say, energy as light or heat). You might describe it thus:

	imbalance/ equlibration	*Pivot/ Balance*
	physical energies	*Super-matter*
	expression	*Archetype*
	material cosmos	*Causal Dot/ Nowhereness*
↓	*isolator*	*communicator* ↑
	nucleus/ centre of mass	*charge/ point of stimulus*
	proton/ quark	*electron*
	inaction	*in-action*
	locked/ bonded state	*interaction*
	object	*process/ event*

In short, from toe to topside, that is, from a mass-centred point of view the philosophical vector is reversed. *Bottom-up*, no cosmic hierarchy exists. The leveller is non-conscious energy. What is not equally made from this? A conscio-material gradient is entirely missing from materialistic paradigms and all accompanying explanations. *It is therefore missing from the conceptual apparatus of contemporary science.* Indeed, materialism simply 'doesn't want to know'.

Thus the origin of everything's material; central substance is nuclear mass ringed by electrons whose force is associated with ethereal photons; light generated from peripheral rings may therefore radiate weightlessly into a weightless 'shell' of cosmic darkness. The symbolism of metaphysical substance and centrality is reversed. Science studies only an outer, material shell, a peripheral region at the base of Mount Universe. Scientism then promotes this fraction as if it were existence in entirety. *To reiterate, in this reversal, where the status of 'pure' light energy has been reduced from cause to an effect, a physical atom symbolises the inversion, the 'turning-inside-out' of Potential into physical realisation. It becomes a natural symbol of physical construction and, as such, an inverted symbol of the whole dialectical metaphysic.*

For holism, on the other *top-down* hand, the whole order of the mountain, whose gradient drops from apex to base, involves a conscio-material or informative coordinate. **Such coordinate is central to understanding the nature of creation and, therefore, at the heart of Natural Dialectic**. This book is devoted to explaining its operation and effect.

From this angle scientism becomes a window on half the whole scene, a perspective that gives onto a half-truth. And, as any propagandist knows, half-truths (or half-untruths) are a most seductive agent of misinformation. Classic misinformation argues "I want to think it happened like this; various facts may be construed to support my view; therefore it *did* happen like this". And yet it will be argued in-depth, for example, that although elements of Darwinian mind-set are evidently true (e.g. natural selection, mutation and variation) its totality (dependent on 'abiogenesis' and 'macro-evolution') is false. In which case any hypothesis built on its half-truth were perhaps learned, politic or academic but, in the light of reality, 'partially true yet wholly false'.

As the potential for a new creation (i.e. a change) is realised the action 'falls' through a kinetic phase to exhausted completion. Information is 'crystallised', energy is lost in the making. A new impetus, a gain in energy or information, will stimulate a further change. The ingredients of the universe, events, all work this way. They work in either regular or irregular cycles of causation. Natural Dialectic describes the process using 'event vectors'.

Event Vectors.

Event Vectors:
A Preliminary Glance at Cosmic Fundamentals.

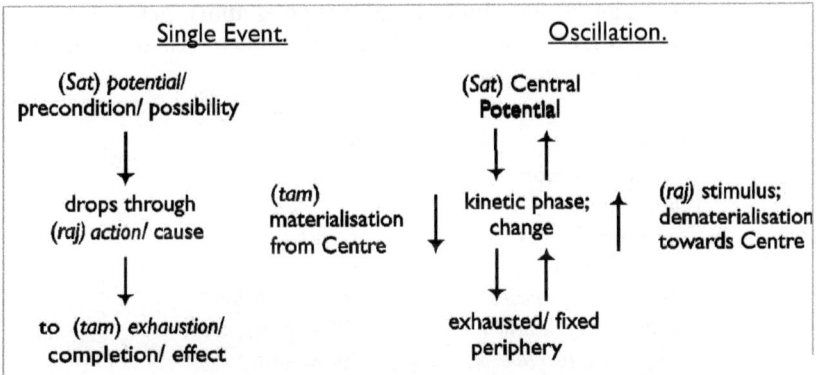

> Every event, whether metaphysical (in mind) or physical (in physics, chemistry and biology), undergoes the simple, single process whereby potential information or energy is triggered into action that, after a while, stops. One sort of equilibrium (pre-action) has been converted into another (post-active or inertial equilibrium). Inherent possibility or precondition is converted, through causal action, into completion.
>
> This diagram is first represented as a ziggurat in fig. 2.3. It also introduces the notion of a triplex of cosmic fundamentals (called *sat*, *raj* and *tam*) that we'll develop in a couple of pages and thereafter use extensively throughout Natural Dialectic. *In fact, they constitute the basis of its operation.*
>
> Cause runs to effect; effect becomes a cause of further change. Stimulation 'fades' into fixity; re-stimulation changes it again. Rise and fall are two sides of the same coin. Creation oscillates according to these universals; its scale is built of Cosmic Fundamentals.
>
> **fig. 2.2**

You've already met these arrows, the vectors of (↑) levity and (↓) gravity. Do you remember that they indicated existential loss or gain of energy or information? They head up the inferior, existential stack.

In Natural Dialectic a superior stack cites Absolutes across from relativities. One such Absolute is Potential. *Potential*, inherent possibility or pre-active calm is therefore called a Primary or Essential concept. *Motion* or *change* is secondary or existential. Potential is unseen until expressed by an action that issues from it. The cosmic case is the same as the particular. The totality of actions that derive, in the first instance, from the pre-existential field comprise what we call existence; that is to say, polar existence issues from Essential Potential.

Change is the basis of existence. Creation *is* incessant transformation. Its motions create, automatically, duality and relativity. Psychologically it intensifies, slackens, holds steady, changes course or falls asleep; physically it accelerates, decelerates, holds uniform, interacts or stops. Do you remember positive and negative neutralities? In this context it is important to discriminate between pre-condition (in which there is not yet movement) from exhaustion (in which motion has finished). **These are, as was discussed in Chapter 1, very different kinds of non-motion. Indeed they are poles apart, poles that hold motion between them**. Later we'll discuss their natures, dubbed 'transcendence' and 'subtendence'.

expression	*Full Potential*
lower	*Highest*
duality/ division	*Unity*
other levels	*Principal*
peripheral	*Central*
consequence	*Origin*
levels of constraint	*Pre-condition*
levels of realisation	*Transcendence*
range of influence	*Super-state*

↓ lower	higher ↑
more passive	more active
exteriorisation	interiorisation
towards periphery	towards centre
diffusion	concentration
information loss	information gain
energy loss	energy gain
discharge	charge
materialisation/ creation	dissolution/ dematerialisation
damp/ slow down	quicken/ accelerate
increasingly static	increasingly dynamic
sub-state/ base	range
impotence	action
extreme subtendent	spectrum/ phases

From a *top-down* point of view the conscio-material gradient is a kind of subjective/ objective calculus that drops from the Potential of Pure Consciousness through to the impotence of pure non-consciousness i.e. matter. Thus a multi-layered universe is described in terms of (↓) 'descent' from Inmost Potential to outermost exhaustion: and, conversely, an (↑) 'ascent' towards It. Cosmic cycles are one thing, individual cycles another. As the arrows show, each aspect of all events oscillates in mini-versions of the universal pattern. Since each aspect may oscillate in different degree or direction from the others; and since the object or event they describe may be large (a galaxy, planet or human being) or small (molecular or less) the actuality is complex; but the principles are simple.

Thus 'event vector' is a formal term for something very simple. The vectors (or arrows) of Natural Dialectic are dynamic. *They express the direction, more or less kinetic, of any change.* **They relate a loss (↓) or gain (↑) of energy or information**.

They also link the covalent movement between any pair of opposites (one's loss is the other's gain) with a more fundamental aspect of change viz. loss or gain of information or energy on the conscio-material gradient that underwrites creation. What was introduced at the start of the chapter as a device is now seen as a basic feature of cosmic operation.

It is worth briefly rehearsing that an *upward* (↑) or *left-to-right* (→) arrow attached to the right-hand column of dialectical opposites always represents either information or energy gain. It indicates motion (or change) towards a higher level. Its levity, however limited in any particular event, is always in principle centripetal towards the Infinite Centre. This Centre is also the Origin of a hierarchical model of cosmic order, concentric spheres. In the dialectical sense an origin is a source. To reach a source is to transcend an object's self. There is a sense in which water 'transcends' ice, gas water and hydrogen and oxygen the very substance of water. We might also find that atoms 'transcend' molecules, energy mass and quantum classical physics. What is the undifferentiated origin of material elements? Is it a 'transcendent' archetype? Could metaphysical information 'transcend' matter, conscious mind 'transcend' unconscious body or waking 'rise above' sleep?

Conversely a *downward* (↓) or *right-to-left* (←) *arrow* always indicates the opposite direction - information or energy loss. This translates as increasing materialisation, confinement or fixity. Its gravitation is, in principle, a centrifugal 'descent' from Central First Cause. For example, this direction shows a loss of creative ability and intelligence; or it describes the act and conditions of physical creation through subtle, atomic structure down to gross gaseous, liquid and then solid precipitate.

The motion of vectors occurs between poles. Poles represent extremities so that, in this case, where does relativity meet its absolute? Where do comparisons end? Where does it all stop? Or start?

An initial suggestion is that in an upward or rightward direction arrows indicate motion towards origin; and in the downward or leftward motion away towards the exhausted condition, at the furthest periphery from central origin. Either way, motion occurs between a pair of poles or, in the case of vertical arrows, towards a level beyond them. *The question to ask, therefore, is the nature of cosmic poles we dub 'transcendence' and 'subtendence'.*

If Potential is an Original Precondition, of what nature is psychological potential, that is, Super-state Transcendence? Can it be known?

As opposed to (→↑) ascent towards Super-state there is (↓←) descent towards sub-state. The extreme base of any aspect, property or characteristic is called its subtendent. After looking at the triplex nature of Natural Dialectic, we shall return to examine Potent Transcendence and impotent subtendence more closely.

Three in One.

Three is an interesting number especially when related to one in terms of three-in-oneness, tri-unity or trinity.

Trinity is a strand that pervades the cosmos. No doubt Christianity (Father, Son and Holy Spirit) and Hinduism (Brahma, Vishnu and Siva) embody their own time-honoured triune fundamentals. Mount Universe is itself effectively a tri-universe. However, let's trace three-in-oneness with reference to Natural dialectical coordinates. Through the following chapters keep a sharp eye open for Essential Tri-unity (Infinity, Unity and Nothing which form the basis of Chapters 7 to 10) and a series of interlocked existential trinities (Chapters 2, 3, 10-13, 19 and 23). These include the aforementioned triplex construction of Mount Universe (a tri-universe composed of duality within unity), a trinity of cosmic fundamentals (shortly to be elaborated as *sat, raj* and *tam*), of cosmic principles (absolution and the relativities of information and energy), of cosmic process (potential, kinetic, exhausted), cosmic state (neutral, positive, negative), life (a duality-in-unity of soul, mind and body), the physical universe (space, time and action/ energy), space (with three dimensions), time (before now, after or past, present, future), classes of basic particle (represented by photon, electron and proton), biological life (from light, air, and water) and others. Oscillation (norm/ axis, up, down), homeostatic balancing or control systems (sensor, processor, effector) and a human's psychological information loop (sensory input, brain, motor output) are a few more important examples that immediately spring to mind. And many subdivisions, such as super-conscious, conscious and sub-conscious mind, the potential, kinetic (quantum) and inertial (classical) aspects of matter or even quark symmetries, fall easily into groups of three. Creation weaves with triplex thread. *How, though, can you best define duality within unity, how can you combine polarities to satisfy the three-in-oneness of a trinity? How, at root, can these fundamental triplex or triune conditions fit the duplex, columnar structure in which Natural Dialectic is expressed?*

There are two answers, two kinds of modification, two ways to express 'the same difference'. The first is to split the Dialectic into a combination of Primary and Secondary 'stacks'. The second is to express these two parts, the fundamental 'three-ness of polarity' (see *figs.* 1.1 and 1.4), in tri-logical form.

Primary and Secondary Dialectical Stacks

Essence is an absolute pre-condition. Its characteristics include potential, unity, infinity and void. Its unbroken wholeness, its 'peace that passeth understanding' has also been labelled *Nirvana*. Essence is axial, top-centre, the capstone at the peak, first and highest tier of a three-tier cosmic pyramid.

From Essence a First Cause (called *Logos*, Word and many other names in many languages) engenders *existence*. Characteristics of the latter include polarity, relativity and relationship. These derive from motion informed by principle, regulation or law. Existence is characterised by polarity, division and differences. A First Cause is clearly primary. And if cosmic 'logic' derives from so-called *Logos* it would be eminently rational and sensible to try and find out more about the nature of this Principle of principles, this Principal, this Archetype. Introductory metaphors are 'an electrical current', 'rushing wind' or 'stream' - a mover. Another one is of an 'existential hologram', a 'super-hologram'; you could also think in terms of program and the Primary Logic of a Cosmic Application. Central to the Dialectic is, of course, the conscio-material spectrum, a gradient of creation that issues from The Archetypal Source.

Before reading the following elaboration please re-check 'dialectical Stacks' in the Glossary.

The Dialectic is split into two forms. The first is **Essential, Primary or Main Dialectic.**

This disposes Essence and essential characteristics on the right against those of existence on the left. Essential Dialectic is, as already noted, indicated by writing the right-hand column with a capital letter. Since it involves qualities of Essence Main Dialectic is placed, as the superior stack, above its polar, existential counterpart. Because it involves no movement, no arrows are attached to it.

existence   Essence

Arrows *could* be attached to the left-hand (existential) column indicating its duality.

↓ existence ↑   Essence

In practice they are simply understood. Other examples of the Primary Dialectic of Absolution are:

finite	*Infinite*
relative	*Absolute*
peripheral	*Central*
polar	*Neutral*
yin/ yang	*Tao*

Note that, for example, Unitary Tao is placed against finite, polar yin/ yang.

The second form is **Existential, Secondary or Branch Dialectic.**

Existential Dialectic represents aspects of division implicit in the left-hand, 'duality' column of Primary Dialectic. For example, *polarity* is split into its components, *negative* and *positive*. Inferior, secondary stacks involve duality. Whatever exists can be assigned qualitative attributes and quantitative properties. Examples are beauty/ ugliness and light/ dark. Between any pair of extremes an attribute/ property may oscillate through a spectrum or phased series of values; and whatever exists may be described in terms of a stack of such dynamic descriptors. Since oscillation is a vibration you might thinks of changes in scale as regular or less-than-regular cycles. You might imagine their motions like those of music registered on a sound meter, digital display or, as waveforms, on an oscilloscope. Such oscillatory calculus, wheeling up and down between the scale's extremes, is vectored. The icon below expresses existential relativity in terms of vectored cycling (round an axis). If you take the circle as a 3-d cone (*fig. 0.5 vi*) then axis is the same as apex; thus the upward vector would approach this Apex.

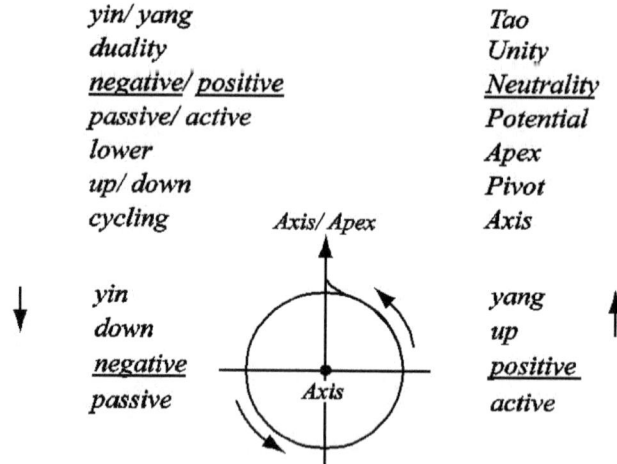

yin/ yang	*Tao*
duality	*Unity*
negative/ positive	*Neutrality*
passive/ active	*Potential*
lower	*Apex*
up/ down	*Pivot*
cycling	*Axis*

An abbreviated version of the relationship between Essential and existential forms of Dialectic, such as you've just met, is formalised below:

	yin/ yang			*Tao*	
	polarity			*Neutrality*	
↓	*yin*			*yang*	↑
(-)	*negative*			*positive*	(+)
	passive			*active*	

Opposites issue from *latency* or *potential*. For example, kinetic and static (exhausted) motions issue from potential energy. Could our existential world of opposites likewise issue from some sort of potential? Which, as its origin, would be essential for it?

Essential Dialectic may characterise the Nature of Nothing but Existential Dialectic represents the nature of things. Its yin-yang polarity can equally express psychological or physical poles. This time the stack includes a second icon representing swing (about a pivot, medium or point of balance):

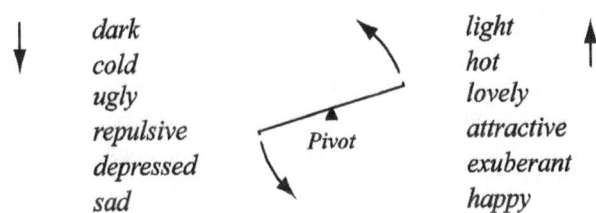

↓	dark		light	↑
	cold		hot	
	ugly		lovely	
	repulsive	Pivot	attractive	
	depressed		exuberant	
	sad		happy	

In order to emphasise the reciprocal, relative nature of both creation and secondary dialectic either member of a pair can, as already noted, be written in comparative terms.

↓   *less light*   ←   →   *lighter*   ↑
     towards darkness/ darker   *towards light/ less dark*

In summary, there exists no 'sense of range' between the opposites of Primary Dialectic (such as Unity/ duality). It is simply that left-hand relativity depends, in a way the Dialectic elaborates, on right-hand absolution.

Secondary Dialectic is, for its dependent part, cyclical. Its icons are the wheel and scale. Each pair of opposites (e.g. light/ dark) represents a couple of polar 'anchor-points' between which a spectrum of energies flows. It represents a couple of extremities between which a scale of degrees is slung or, if you like, an oscillation between polar peaks and troughs (*fig.* 2.14). The 50/50 mid-point of such sliding scale is its 'equator'; and any changes in the calculus of its degree are vectored up (↑ →) or down (↓ ←). For simplicity just (↑) and (↓) are used as indicators at the head of secondary stacks and are dropped though understood in all other cases (as shown in line two). The existential dialectic of '*oscillation*' or '*relativity*' is always identifiable by its lack of capital letter. Even a (*Sat*) character is, reduced to its reflections in existence, then written (*sat*) - as, for example, (*sat*) balance or (*sat*) unity.

Finally, you stack pairs of poles; by stacking 'poles' of different qualities and properties on the 'chromosome' of a Dialectical Stack you can build a non-mathematical description of any item that you want to talk about. Where exactly the 'point of stance' or 'range of oscillation' lies for any individual, polar component is a matter for mathematical calculation or, especially in psychological affairs, discernment. Could there exist an Absolute Criterion, an Ideal Rule by which to judge? If so, of what kind might Ultimate Discernment be? How could you or I reach its Perspective? Perhaps we'll see.

The Tri-logical Form of Natural Dialectic.

The two abovementioned forms of Dialectic (Main and Existential) can be combined and expressed in a tri-logical construction. Characteristics of the central potential from which polar, peripheral existence is realised are, as in Main Dialectic, marked with a capital letter. Creation is thus represented by a pivot around which vectors swing - a pair of scales or balance. Thus:

↓	down/ descent	Pivot	up/ ascent	↑
	left pan	Point of Balance	right pan	
	negative	Neutral	positive	
	inertia	Poise	stimulus	
	static	Potential	kinetic	
	centrifugal	Centre	centripetal	

Up, down and balance: three strands comprise creation's thread, three qualities a cosmic scale. The right-hand indicates 'bounce back' towards Central, the left-hand 'push away'. Tri-logical construction can highlight a couple of common problems - semantic and paradoxical - that the Dialectic addresses later.

Cosmic Fundamentals - Three Qualities

Essence, the potential for existence, precedes it. Essence in motion *is* existence. Motion is relative; existence is composed of changes. It involves polarities and opposites like start and stop, fast and slow, straight and curved. A special case of curve is straight; and of kinetic non-kinetic, that is, static. Exhausted, finished, motionless.

A fire goes out. As it dies, energy drops into a cold, inert condition. A fire is lit. Heat quickens. It stimulates. Information lost drops into error, consciousness falls asleep; but information gained leads to understanding, heightened awareness and truth. There exists, between each pair of opposites, a similar two-way oscillation, a sway of influence that depends, like a pair of scales, on its pivot. A pivot is a point of balance, equilibrium or poise. It is, in dialectical terms, potential that precedes a field of influence; the agents of this influence are motion's impetus and sway.

All sorts of pairs describe two basic sorts of field, those of mind or matter. Each pair involves a point of origin, a precondition, a pre-active poise from which its various, changeful motions are derived. This stillness is an inward subtlety, the source of opposites, neutrality from which their juxtaposed emergence springs. Take, for example, a neutral vacuum from which charged pairs of particles emerge. Or this physical book - a chemical expression of inward, subtle atomic forms. Its manifold words are various ways of expressing a few ideas and principles. And its source, the author's mind, is its cause. Mind is a source of influence that, from possibilities, determines which effect occurs. An effect is, in the nature of its changes, directional; that is to say, it ends up more or less like its cause. It approaches or recedes from it. For example, the source of chiaroscuro is light. Some objects and events in the shadow-play are obscure, others clearer. Similarly with consciousness: while alert there is more understanding, less ignorance; while drowsy or asleep there is less understanding, more ignorance. Natural Dialectic is, in this way, a theory of relativity but also represents, against a range of shadows, the absolution of un-shadowed light.

You want it simplified? Take motion (*fig. 2.2*). A ball is dropped to the ground. While held it has potential energy; when released it develops directional kinetic energy; as it bounces or rolls to a halt this energy is exhausted and it becomes fixed, static. *Potential, kinetic, static; ready-to-go, activated or finished - this is the order of action, materialisation and creation.*

Balanced and swinging upward or downward. **These three conditions - balance, a descent towards inertia or its reverse - are found in different degree in every aspect of every object and event in creation.** They are expressed in the motion of a pair of scales; in cooperation they compose a dynamic equilibrium; they constitute, tipping this way or that, a cosmic scale, a balance of creation. As such, they are simple measures both of the real universe and its dialectical description. **We call them Cosmic Fundamentals.**

	static/ kinetic	*Potential*	
	outworking	*All Possibilities*	
	transformations	*Pre-condition*	
	down/ up	*Balance*	
	(3)/(2)	(1)	
↓	down	up	↑
	static (3)	kinetic (2)	
	fixity/ solidity	flux/ change	
	info./ energy loss	info./ energy gain	
	less excited	more excited	
	less alert	more alert	
	passive	active	

A precondition of excitation is its quiescent field. Whether it is a quantum force-field or the battlefield of the world's action upon which Krishna and Arjuna fought their demonic enemies, the field is absolute, unified and undisturbed until its balance is upset. This change upsets neutrality, the movement polarises it. *All kinds of polarity arise. Existential Dialectic charts them. It indicates, within the scheme of things, the pattern of their interactions.*

	negative/ positive	Neutral	
	issue	Source	
↓	negative	positive	↑
	containment/ constraint	radiance	
	drag	stimulus	
	resistance	flow	
	isolation	interconnection	
	shape	power that shapes	

or tri-logically:

	↓	down/ descent	Pivot	up/ ascent	↑
		static, fixed	Potential	in flux/ kinetic	
		yin/ tam	Tao/ Sat	yang/ raj	
	(-)	negative	Neutral	positive	(+)

177

Is three nature's lucky number? It is now obvious from physics, event vectors and Natural Dialectic that we are not dealing with dualities but *trinities*. Up, down, balance. A scale. In this tri-logical view pairs of opposites interact with one another under the influence of *three* qualities, principles or tendencies. *Polarity's threads are really three*. The fabric of the cosmos is fundamentally triplex. Different and changing permutations of these qualities exist in everything; everywhere they show proportions, more or less of each. As their interplay in a real way underwrites the story of existence so they underwrite Natural Dialectic. Changes in these proportions produce variety. They represent, through various agents such as subatomic particles, the dynamic relativity with which the tapestry of existence is woven.

Could three 'neutral' words identify these qualities? You've most likely heard of Chinese *Tao*, y*ang* and y*in*. In Indian philosophy they are called <u>Sat</u>, <u>Raj</u> and <u>Tam</u>. Those interested in dietary cooking might have already picked up on their categories. For example, *sat* food is fresh and includes fruit, vegetable and cereal products. *Raj* food stimulates; it is hot, promotes physical activity and includes spices, curries etc. The *tam* ingredient is stale or heavy; it includes meat and alcohol. You get the flavour. **In fact, the triplets (each called a *guna* or thread) are much more radical and all-pervasive, to the extent that they describe the tiers of Mount Universe itself.**

A simple link from physics describes the three basic conditions of energy as (*sat*) potential prior to (*raj*) action and (*tam*) exhaustion. Start, process, end; poise, excitement, exhaustion; balance (*sat*), up (*raj*) and down (*tam*) compose triplex creation. Such trinity, we'll see, applies to mind's psychology as well as matter's physic. Henceforward we'll use their tendencies to head up every stack throughout the development of Science and the Soul. In **Primary Dialectic**, these tendencies are written, for example:

tam/ raj	Sat
down/ up	Balance
coloured	Clear/ White
range	Transcendence
expression	Potential
action	Start-point
polar	Neutral

with left-hand polarised into the opposites of **Secondary or Existential Dialectic**:

↓	tam	raj	↑
	downwards	upwards	
	negative	positive	
	gravity	levity	
	constriction	release	
	end/ exhaustion	reactivity	
	inertial	dynamic	
	black	spectrum of colour	

Cosmic Vectors

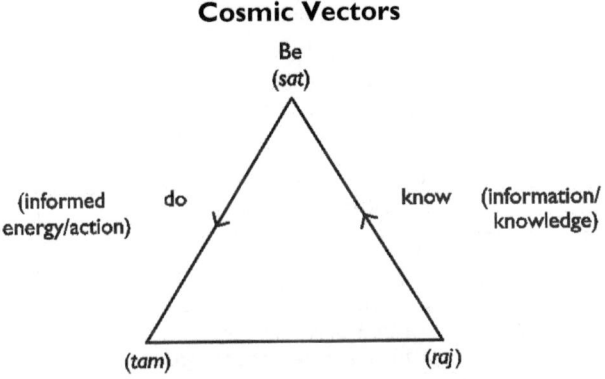

A triangle of truth (see also *fig.* 13.4) incorporates the three basic verbs, the three fundamental modes of human being - to be, to know and to do. It illustrates pre-condition - being; knowledge from perception - information; and action, that is, ordered energy. In this triangle Essential or Supreme Being differs from the dynamic equilibrium of conservation, existential poise or, in biological terms, the consistency of homeostasis.

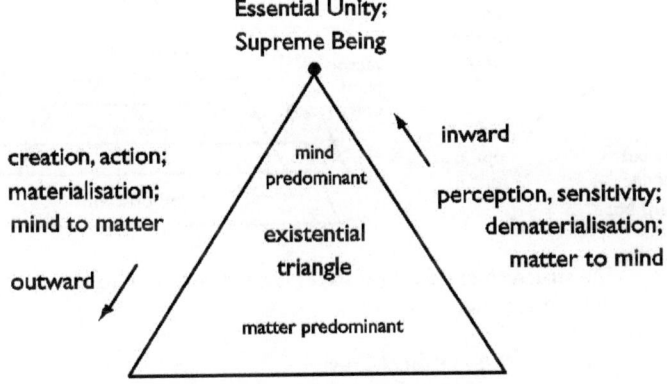

This version of the triangle illustrates the existential dipole (Chapter 1) dependent on Essential Being.

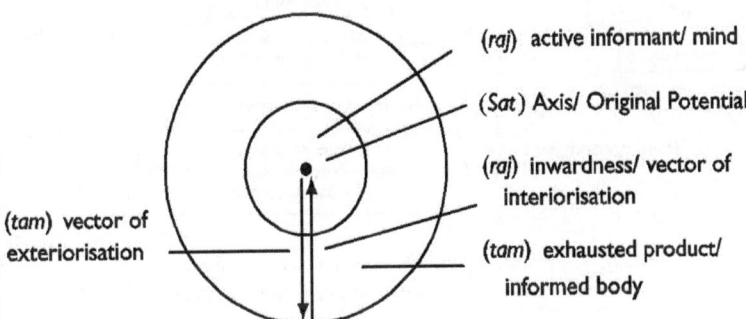

(*Sat*) pivot, axis or point of balance can be modelled at the centre of concentric spheres.

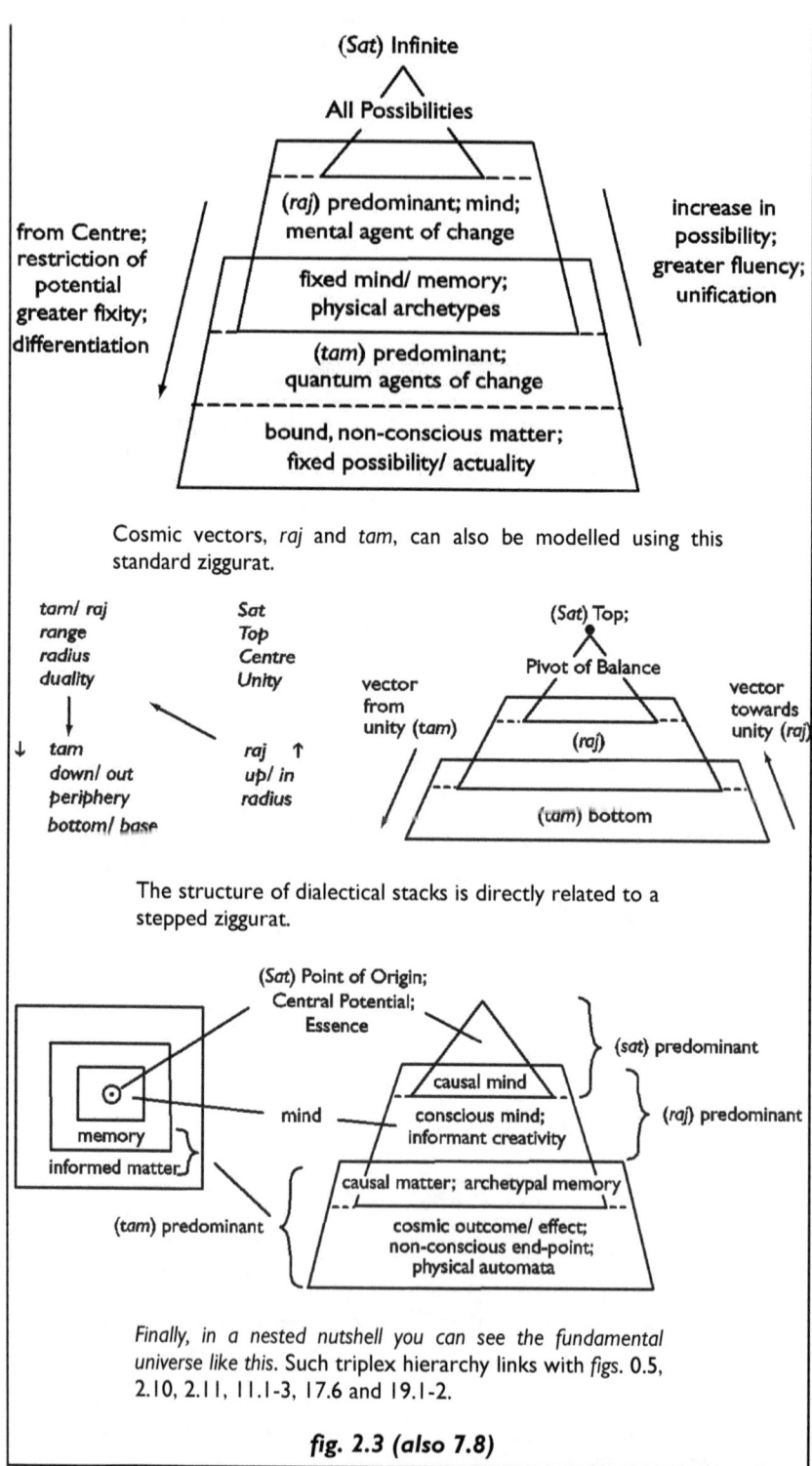

fig. 2.3 (also 7.8)

Between opposites an arrow to the left indicates an increase in the quality of *tam*: one to the right indicates an increase in the active quality, *raj*. As described, the arrow to the left downwards represents movement towards a nether extreme (a subtendent). To the right upwards motion is towards the opposite extreme, called transcendence or supremacy. These two arrows indicate the way, along that vertical Information Coordinate shown in *fig.* 0.7 iii, creation itself is ordered.

The logic of Natural Dialectic, which is based on three fundamental cosmic principles, can perhaps be drawn in its simplest forms as either a triangle or a ring of truth.

Take the former. Visualise its triangular (or pyramidal) hierarchy as if it were a centrifuged, layered precipitate consisting of three chemicals. At the top the clear supernatant is undiluted *sat*. This grades into *raj* until the latter is predominant. This is the second tier, the top half of the pyramid and representative of mind. As mind becomes more involved with the third tendency, *tam*, it becomes darker, more limited, fixed and ignorant. Finally, gravity precipitates the heaviest, physical region, a zone of material oblivion in which *tam* is predominant. Behold! The gradient of cosmos in a test-tube!

In terms of Natural Dialectic, therefore, the **cosmic funda**mentals can be seen as two vectors, *raj* (↑) and *tam* (↓), with a central, balancing factor (*sat*) in overall control; and its columns can be seen as a kind of universal vector analysis.

In short, the combination of three Cosmic Fundamentals in differing proportions amounts to a single, simple, underlying rule. This rule characterises the structure and operation of creation, that is, both physical and metaphysical existence. *You may call this either prior or ultimate rule.*

**Cosmic Fundamentals, Vectors and Balance
illustrated using scales (or see-saw) and pendulum**

In these diagrams the pivot, fulcrum or axis is always (*sat*).

(i)   (*sat*) primary condition, poise, potential or starting-point

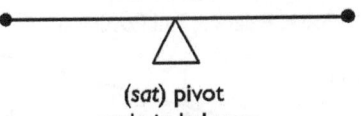

(*sat*) pivot
scale in balance;
potential for motion

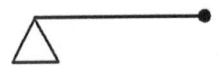

(*sat*) pivot
pendulum drawn back and held;
potent immobility

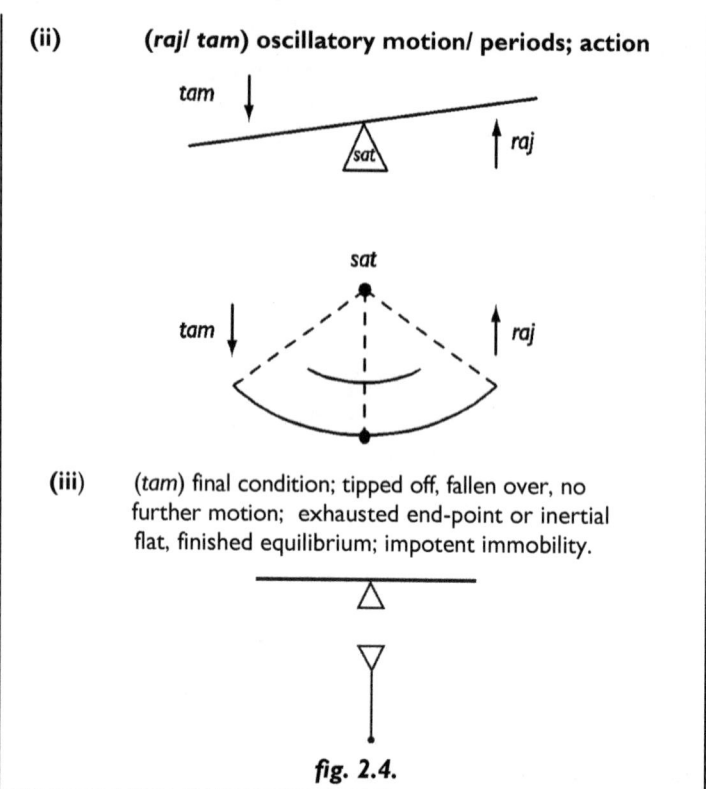

(ii) (*raj/ tam*) oscillatory motion/ periods; action

(iii) (*tam*) final condition; tipped off, fallen over, no further motion; exhausted end-point or inertial flat, finished equilibrium; impotent immobility.

fig. 2.4.

Now imagine a scale of creation, a balance of all things.

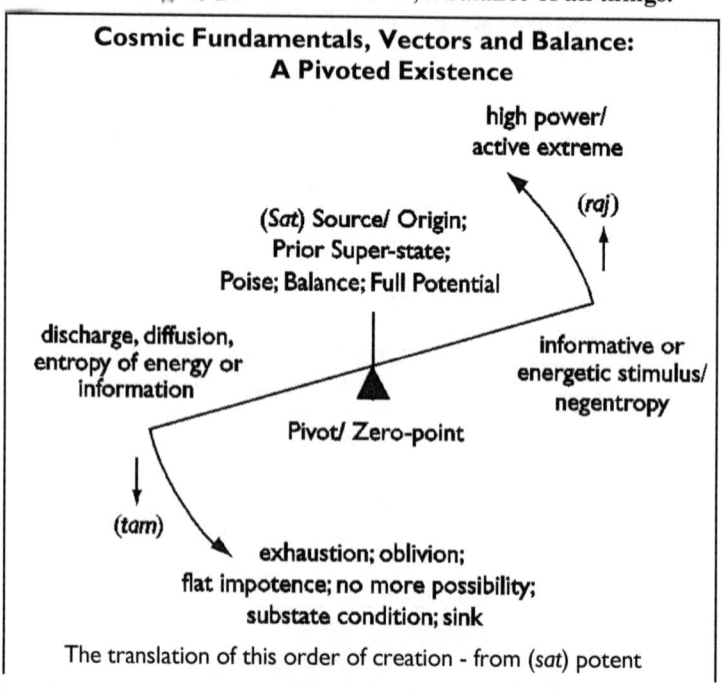

Cosmic Fundamentals, Vectors and Balance: A Pivoted Existence

high power/ active extreme

(*raj*)

(*Sat*) Source/ Origin; Prior Super-state; Poise; Balance; Full Potential

discharge, diffusion, entropy of energy or information

informative or energetic stimulus/ negentropy

Pivot/ Zero-point

(*tam*)

exhaustion; oblivion; flat impotence; no more possibility; substate condition; sink

The translation of this order of creation - from (*sat*) potent super-state through a (*raj*) phase of action to (*tam*) sub-state impotence - is rephrased in terms of cosmic fundamentals. Each one is predominant at one of the three levels of creation.

Fig. 2.5i makes clear that in description of swing around a pivot (*sat*) potential *is* that pivot. Although written dialectically above, it is the point whence wobbling changes in the two domains of mind and matter start; it is the point of poise round which all swing begins and of equilibration when a balance is assumed again.

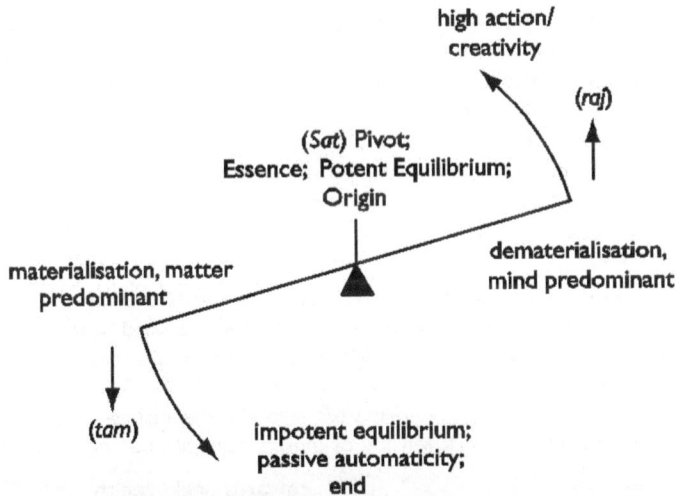

Thus the cosmic scale integrates a pivotal, balancing factor (Essence) with two antagonistic vectors of existence. Of this couple mind is (*raj*) and matter (*tam*) predominant. Such proportional representation of cosmic fundamentals is reflected as subdivisions within each level; the hierarchical constitution is also modelled by concentric rings (*fig.* 0.11) or, as *fig.* 2.10, a ziggurat.

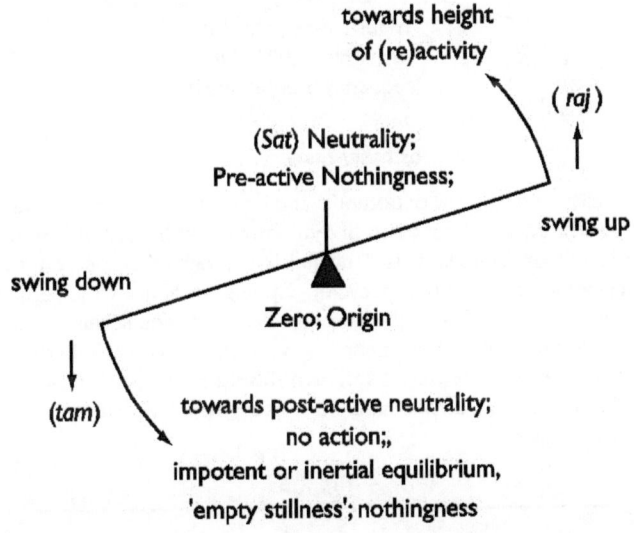

In other words, the perpetual changes of creation are seen as myriad adjustments against the disturbance of balance. *(Raj)* up, *(tam)* down - nature is a scale whose beam forever wobbles in equilibrium round its various *(sat)* centres; and, on the large scale, one could see such existential instability equilibrating round a Central Pivot, an Essential Point of Balance. Stable Axis represents Perfection **Creation's instability is, therefore, as regards its Source, forever perfectly imperfect; perfect imperfection shows as this perpetual motion of continual change.** Existence thus amounts to a self-regulating balance; it amounts to the sum of myriad individual actions and reactions each of which always shows proportions of two vectors pivoted around the poise of a third non-vector. Such orderly procession by equilibration is sometimes referred to the '*karma* drama', physical and psychological fields of *karma* or equations of creation; physics also measures transformations by equation.

The non-vector of these scales is central balance, an attractive state that events are always and naturally moving to obtain. This Origin, is however, counterbalanced by peripheral sub-state, point of maximum imbalance. At this point mind falls to oblivion; and matter into the exhaustion called its greatest entropy, least tension or inertial equilibrium.

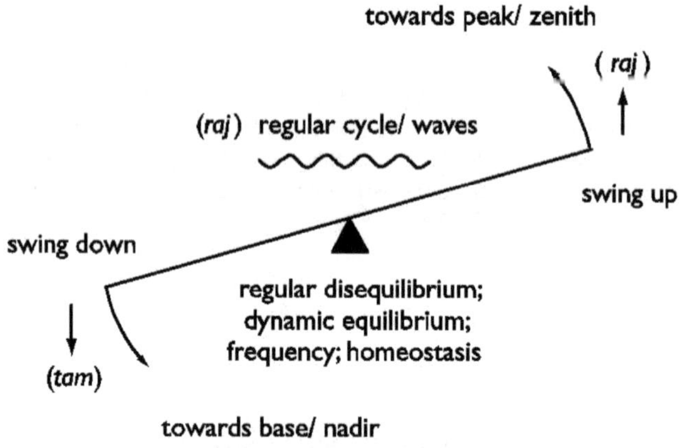

towards peak/ zenith

(*raj*)

(*raj*) regular cycle/ waves

swing up

swing down

regular disequilibrium;
dynamic equilibrium;
frequency; homeostasis

(*tam*)

towards base/ nadir

As well as the stillness of potential and inertial forms of equilibrium there also occurs the motion of **dynamic equilibrium**. All events proceed from start to finish; this finish is either different from the start-point or the same (a cycle). Cycling amounts to a regular swing that is called, dialectically, dynamic equilibrium. It is measured in terms of frequency, wavelength or wave. **Vibration is the form in which such equilibrium is contained. In biology it is expressed by periodic episodes and homeostasis.**

fig. 2.5

To rephrase, the fundamentals are a measure of information and/or energy present in any circumstance and the direction or directions in which it is projected. *Their tri-unity/ trinity shows, in various proportions, in all events and objects in existence.* They are ingredients in everything and every process from the mundane to the exotic. It is all a matter of their degree.

Although they have been introduced using explicit Indian and Chinese terminology their non-prepositional use is also implicit throughout western philosophical and everyday thinking. For example, we understand 'higher' or 'lower' levels of intelligence, moral quality, happiness, heat or electrical voltage. 'High' is identified with elation and, in less fortunate moments, one may 'sink' into a 'low' mood. Arithmetical positive (plus) and negative (minus) are, with equation (=), an integral description of the whole working. So are energy levels.

It is, therefore, a distinct advantage that the terms for Natural Dialectic's cosmic fundamentals - **sat**, **raj** *and* **tam** *- are fresh, unaccustomed and uncoloured by preconception.* For this reason they will henceforward head every stack and be used extensively in the development of Natural Dialectic. It is therefore time, in order to understand their place in its design, to play see-saws and swings; time to swing the pendulum and gauge the scale of things. How does a Pivoted Existence work?

The downswing of this scale represents materialisation; it involves increasingly heavy, inertial moment while the upswing represents stimulus, lightness and increasing dematerialisation. What happens with your sensory process? The separate objects of the world are dematerialised into a unified whole - your perception of them. The world is symbolised in mind; the symbols are manipulated according to your purposes then you send a message to rearrange the outside world (including your body) in terms of those purposes.

Mind, situated between its source (called psyche, soul or pure consciousness) and body, can focus its attention either way. It can move inwards by contemplation or outwards by sensation.

Non-conscious matter, on the other hand, moves subjectively nowhere. Its informative aspect (natural law) is automated; it is without initiative, entirely passive and, of course, insensitive. Its scalar ups and downs involve losses, gains and exchanges of various sorts of energy.

This leaves the third but actually first and foremost fundamental, Pivotal Equilibrium. Before returning to discuss the hierarchical, triplex structure of cosmos that these three fundamentals, each in its predominance, composes let's take a look at them in turn. Before any thing is its absence, before everything there was nothing and before vector there was peace. It is, poised at its axis and pivotal at the heart of things, the essential quality of balance and of truth we first inspect.

The Essential or Central (*Sat*) Quality.

***Sat*'s quality is of essence whose attributes are found on the right-hand column of Primary or Main Dialectic.** *Reflections of its nature, in the sense of extreme positivity, are approached by the right-hand (upward) column of Branch Dialectic.* In complete ascendancy *Sat* is the Absolute, Transcendent Centre. It is not so much a tendency as a quality because it represents poise or equilibrium. *Not motion but balance. No inequality but symmetry.* Within finite existence its reflection appears as mid-point, poise or pivotal control.

existence	*Essence*
outer/ dis-equilibria	*Centre/ Balance Point*
expression/ power grid	*Power/ Potential*
actualisation	*All Possibilities*
motion/ action	*Peace/ Poise*
distortion/ asymmetries	*Symmetry*
degrees of impurity	*Cleanliness, Purity*
relativity	*Absolution*

<u>Sat</u> or truth quality is in the paradoxical department. It represents the fundamental unity that underlies duality. It is (see *fig.* 1.1) the *Tao* from which *yin* and *yang* divide. *It represents the whole; it includes both sides of an equation and is, as such, the balancing or equalising factor. Sat* imparts, no matter what the terms, equilibrium (for which the Arab word is '*algebra*'). The wobbling of imperfect balance, called existence, ceases on perfection to exist; and as it does so Buddha-nature is revealed. *Sat* is, therefore, pivot of the cosmic scales round which two pans are oscillating; it is the axis around which that vectored couple, gravity and levity, forever skip and play. Any weight is easiest to bear at point of balance, any light the lightest at its source. *Sat* (Sanskrit: *sattwa*) means light of weight, transparent, evident. Its poise is associated with potential, precondition or '*sine qua non*' rather than the kinetic or exhausted phases of events, creations or behaviours. In this way *sat* involves prior, implicit capacity whose possibilities, when reduced, will have informed and defined an explicit, actual end-result. It involves, in other words, principle, intelligence and law. This applies as much to the universe as to each of the minuscule events from which it is constructed.

Sat comes first. *Sat* is prior. As such it can be viewed not only as fulcrum but as the centre or apex of any systematic creation. It is the point of origin and originality; and the essence of this creative principle is consciousness. It is source not sink; and represents a concentrate, a purity - especially the (*Sat*) Concentrate of Information at the Apex of Mount Universe. In terms of expression call it precondition; in terms of initiation call primal-unity-encompassing-polarity an egg, the cosmic egg; and in terms of physical effect metaphysical cause. Informative potential, plan, precedes material behaviour. *Sat* is high-level, prior, at the axis or top pole of things. In terms of local, energetic concentration call it 'sun'. As such it blazes for all satellites; its brilliance is reflected in the radiance of any source of influence. *Sat* shines with informative illumination or energetic light and heat. It imbues, in varying degree, the qualities of comprehension, knowledge and wisdom; and infuses the information implicit in an ideal purpose or construction. In truth it is the source and principle of order, coordination and coherence - the superintendent, informative principle. It both resolves the tension between poles and is, as zero-point, the axis of balance around which fluctuations occur. Its immaterial equilibrium gives rise to the two other swaying, mobile tendencies (of *raj* and *tam*) and, in this sense, it is also their transcendent, causal point of origin.

As well as egg to source diversity, *sat* represents the goal of combination, union. Either way, in unifying two-to-one or splitting one-to-two, resolving opposites into a single whole, *sat* remains the neutral element. Such neutrality, surpassing 'good' or 'bad', is super-positive. It is uppermost, surmounting, excellent, beyond; its equilibrium is potent and its poise replete with possibility.

In reconciling the antagonistic vectors (*yin/ yang, raj/ tam*) the quality of balance harmonises and, between extremes, exemplifies the norm, the middle way or golden mean. Such resolution of contradictory tendencies restores either local or global equilibrium. Peace and calm. *Sat* or *Tao* is thus a central point of reference and as such operates as the self-adjusting, regulatory aspect of any homeostatic, balancing event. *Sat* informs and controls. It is a pivotal criterion, ideal condition and, transcending motion, governor and stabilising influence.

What is order's agent if not mind? What is mind's subjective substance if not consciousness? Is this immaterial factor not the centre of your world? Is awareness not the pivot of your balance, the axis round which the world swings or the hinge on which it all depends? Is not this same awareness that you take for granted what you call your life and truth? *Sat* stands for life as well as truth. Pure *Sat*, Pure Life, Pure Truth. If this is mind's then what is matter's truth? It's natural law? The immaterial factor at the heart of physics is, according to the Dialectic, (*sat*) informant archetype.

Take a cone (*fig. 0.5 vi*). From its (*Sat*) Central Apex drops an axial line of balance to the base. *You can thereby understand that (Sat) Apical qualities, grouped in the right-hand column of Primary Dialectic, are reflected throughout existence down that line.* Such reflection, which shows as potentials, equilibria, norms, pivots, points of control etc., is denoted by the use of lower case (*sat*).

The Existential Vectors (*raj* and *tam*)

	tam/ raj	Sat
	body/ mind	Soul
↓	*tam*	*raj* ↑
	negative/ down	*positive/ up*
	passive/ impotent	*powerful/ active*
	grosser	*subtler*
	energy	*information*
	matter/ bodies/ automation	*mind/ minds*
	entropy	*negentropy*
	gravity	*levity*
	resistance	*assistance*
	inertia/ drag	*stimulus*
	mass	*force/ interaction*

Raj and *tam*, under *sat*, are the antagonistic fundamentals. These two are inversions of each other. <u>Together they break the absolute symmetry of pre-formation; they act to express any potential.</u> Pole and anti-pole together they represent opposites, ranges and relativities. **Scales tip, pendulum swings, the world oscillates**. **It cycles**. These cycles are its beats, its life-beats. Motion downwards, irregular, spluttering or losing rhythm is *tam*; but, pushing forward with regular and well-timed swing, *raj* is on the vibrant up-beat.

Think not of scales, pendulum or ball but **a second simplification to help describe the cosmic circuit - a battery**. *Sat* represents a battery's full charge or *potential* before connection is made. A connection between terminals gives rise to motion, called current. *Tam* represents discharge towards the earth pole; *raj,* in the opposite direction, recharge that results, in the end, in full, original (*sat*) potential. In the electrical analogy repulsion is the (*tam*) negative basis of division, discontinuity and rigid structure; positive (*raj*) attraction is, on the other hand, the basis of union, continuity and flow. Opposites, both held apart and conjoined, are fundamental necessities that arise from the original neutrality.

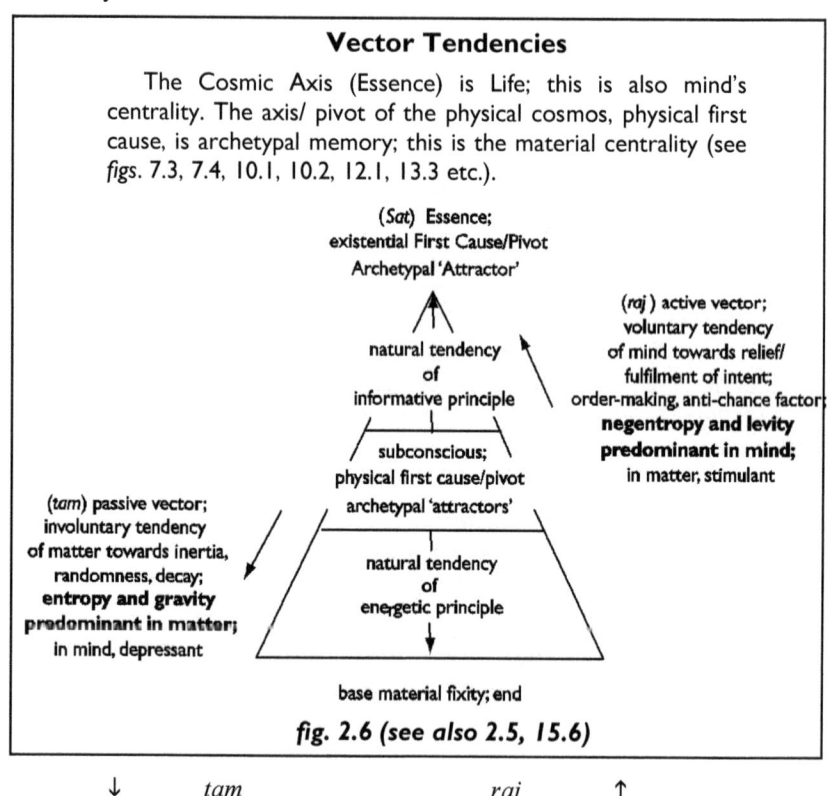

fig. 2.6 (see also 2.5, 15.6)

↓	tam	raj	↑
	greater entropy	greater negentropy	
	depressant	stimulant	
	long/ flat line	wave/ frequency	
	weightier quality	lighter quality	

Check the Glossary for entropy and negentropy. **A third way to think of creation is in terms of concentration gradients of information and energy, mind and matter. The vectors of these gradients are entropy (↓) and negentropy (↑).** This view correlates precisely with the dialectical view of cosmos as a conscio-material gradient. You can think of (*raj*) as gain in integrity, information, happiness, stimulus or energy; and (*tam*) as the reverse, a vector of loss (or entropy) that diffuses from source to sink, that drops towards the exhaustion and inertial equilibrium of sleep, fixity and death. The 'negentropy' of (*raj*) vector is predominant in mind whereas the 'gravity' of (*tam*) factor is predominant in the physical universe.

(Raj) Predominance: Psychological Pivot (i).

Informative swings.

We do not experience the Unified State of Grace that is Communion with Cosmic Axis. Instead we inhabit a point on the conscio-material gradient of Mount Universe below Transcendence; we inhabit the apparent unity, diverse in ever-changing situations and masks of personality it wears, of mind. The pivot of this embodied self is its cosmological axis. From this point your direction of focus can turn inward (in contemplation) or outward (with sensation). **In fact you oscillate between these two modes of perception, knowledge or information.** It is a matter of ever-changing emphasis. Most of the time you synchronously combine a degree of each. There are, however, moments of pure sensation; and others of pure contemplation or (when the world is lost this time to sleep) of dream.

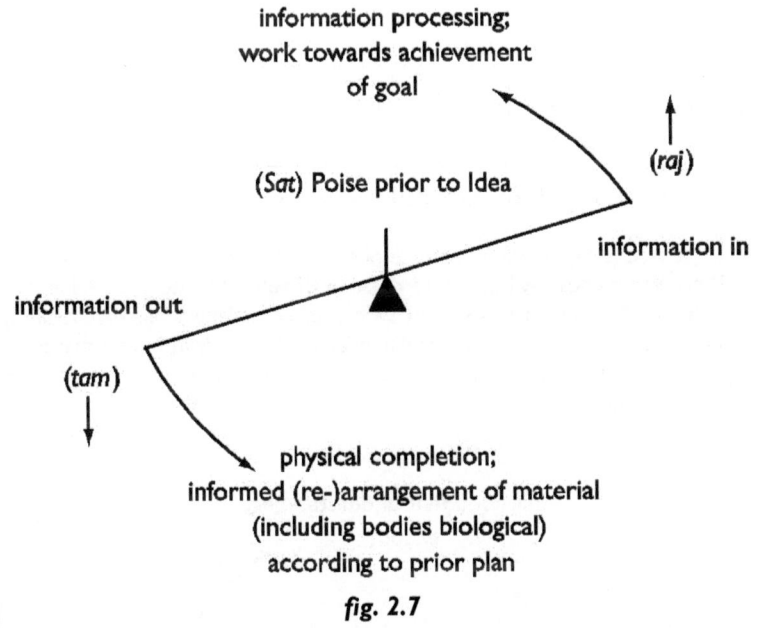

fig. 2.7

<u>Raj</u> (action) **is represented on the right-hand column of Secondary or Existential Dialectic**. Action, like motion, is one thing but its vector is another. Action can positively increase, negatively decrease or, in dynamic balance, stay the same - but whether influenced by (*raj*) levitatory, (*tam*) gravitational or (*sat*) balancing tendencies, motion or change is intrinsically (*raj*) energetic. The latter's increase (or 'ascent') is upwards (↑) towards a maximum or even, beyond itself, another 'gear' called transcendence. In other words its *ascent* is towards an increasing proportion of *sat* quality in the positive of any two opposites. *Raj* ascends and its climb, *as the vector of levity and of negentropy*, becomes increasingly buoyant. It returns a scale's pan towards equilibrium, equilibrates or moves towards balance. It 'ascends', in other words, towards a culmination of relativity in Absolution. What is absolution but irrelative and perfect poise? What, therefore, is physics' 'highest', lightest object? Is it light whose free energy most nearly approaches non-materiality and is thereby the closest expression of materiality's own absolute? If freedom of energy is at a maximum in light then what of metaphysical mind? What is the nature of its Absolute, its Light? What is the Essence of existence? The Dialectic indicates that transcendence of *raj* into *sat* absolution is a circumstance that invites further investigation (Chapters 5, 7, 9, 13, 14).

Raj (Sanskrit: *rajas*) **is the positive, active, kinetic quality.** It is the dematerialising, dissolving principle. It is therefore an expression of levity, radiant energy and continuity; of acceleration rather than deceleration, energy as opposed to mass or inertia. It moves towards an increasingly predominant proportion of lightness; it is quick, free, stimulatory, interactive and of smooth and flowing motion. It tends towards the qualities of harmony, cooperation, communication and unification. Its model is radiant *wave*s (which show as concentric rings) and its mode vibrant and harmonic.

In its aspect of unification the motion of *raj* is from complex practice towards simple principle. In this respect *raj* predominant constitutes, *with information in the ascendant*, the kinetic, metaphysical aspect of creation called informative mind.

(Raj) Predominance: Psychological Pivot (ii)

A second aspect of the balance of mind involves its four main states. These will be dealt with in more detail in Chapter 13 (see figs. 13.5 and 18.5 (i)). As well as state mind swings in comprehension, that is, in understanding and therefore confidence; and it swings in mood, that is, in positive, negative or neutral, detached perception of its present circumstance.

Comprehension.

full, confident understanding; wisdom

increasing ignorance, incomprehension, inability to concentrate, lack of intelligence

(*tam*)

(*raj*)

increasing awareness, ability to focus, knowledge and understanding

sub-state; unawareness, sleep, oblivion

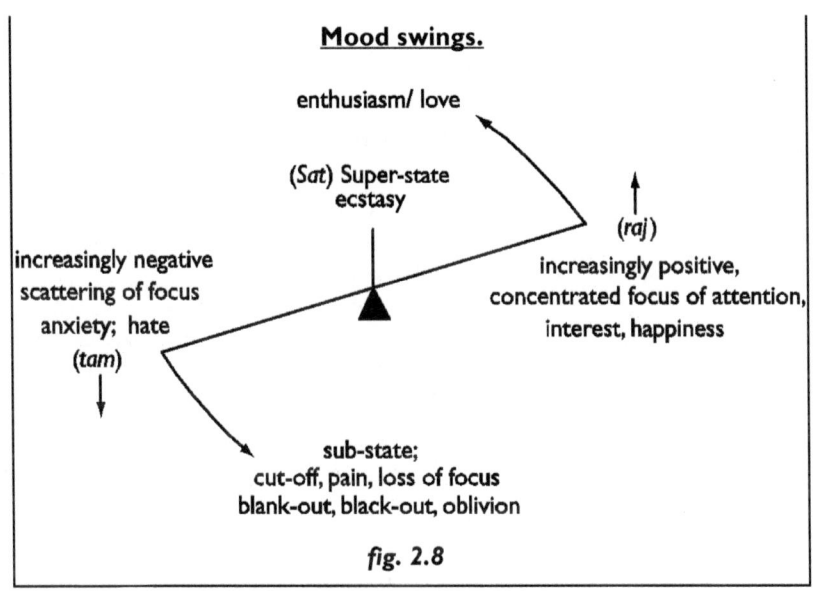

fig. 2.8

↓	tam	raj	↑
	tam-charged mind	sat-charged mind	
	ignorance	knowledge	
	dimness/ disorder	intelligence/ order	
	from truth	towards reality	
	towards illusion	from appearance	
	superficial	profound	

Mind is sited (*figs.* 0.5 and 1.3) between body and soul. It can therefore swing either way from its pivot, the eye-centre. The two directions of focus (*fig.* 0.8) are (*raj*) inward and upward and (*tam*) down and out.

Sat-charged mind swings towards its Centre. This direction is contemplative and its process towards solutions. Psychologically the ascent shows as (increasing) awareness, intelligence and happiness: also in creativity and powers of design. In the reciprocity between poles it shows as attraction, interests other than self, unselfishness and giving. It is the communicator and represents the positive pole in things.

Tam-charged mind, on the other hand, swings outward into the sensible, external world of things. This motion is strongly influenced by the bodily senses. The downward vector marks materialisation. This is the way in which ideas are implemented. It also marks diffusion of consciousness. As concentration is lost mind descends and, psychologically, *tam*'s down-stroke shows as increasing darkness, limitation, fixity and ignorance. It also manifests as lack of focus, idleness and disorder dropping to unconscious oblivion. A fixed mind is dormant; it is one entirely in the hands of instinct and automated reflex. No mind at all is matter's state.

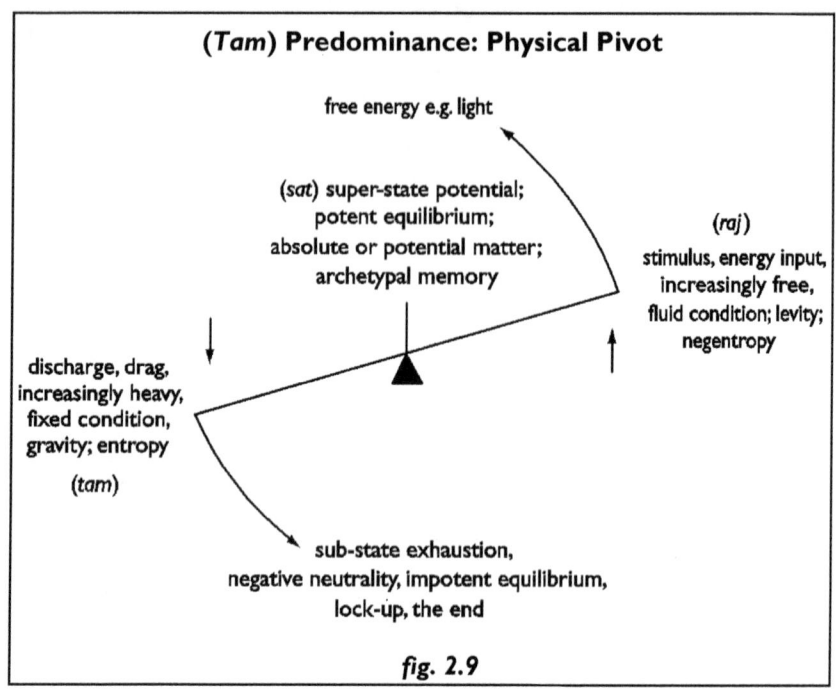

fig. 2.9

Tam on the left-hand column of Existential Dialectic, is the vector of 'fall'. Do you remember, from Chapter 1: The Basic Existential Dipole, 'reflective asymmetry'? This (*tam*) base fundamental is Natural Dialectic's formal 'inverse reflection' of the (*sat*) top pole; and, as well as opposition, represents its vector of descent. Couched in physical, scientific terms its '*downward gradient of entropy*' is expressed by degradation, exhaustion, impotence, damage and so on. In psychological terms *tam* shows as tumbling from Central Enlightenment, as fall from Grace. A loss. By this is meant an outward turning from correctly inward-centred focus. As regards value this vector demonstrates dejection, loss and lowness. As regards intent it drops with light behind, seeks out peripheral shadows and therefore increasingly evokes the tendency to sin. And what is sin? To not act in the spirit of the light - even to deny its presence and resist its Logic. How logical, from darkness' point of view, to fail to understand a metaphysical illumination. High noon does not dispel such (*tam*) blackening of mind.

Tam (Sanskrit: *tamas*) **is the negative, passive, inertial quality. It is the materialising (↓) tendency.** *Tam drags. Tam exhausts.* Its negative arrow subtends towards base (called 'sink'); its 'purity' is represented by the sub-state or subtendent pole. As opposed to *raj*, an activator, *tam* contains and resists. It represents the medium which energy informs, the 'solidity' 'fluidity' has patterned. *Tam actualises. It is the vector of manifestation expressed as gravity, binding energy and materialisation.* As gas drops to solid state so *tam* precipitates out through the regions of the universe. At the same time as it coarsens and solidifies it divides. Its agents divide, differentiate and crystallise, at bottom level, into fixed and separate details; or diffuse into an inertial equilibrium. In descent there manifest increasingly extreme forms of properties such as impotence, inertia, massiveness, unawareness and destruction. In terms of process *tam* represents the end.

In its aspect of division the motion of *tam* is from simple principle towards complex practice. It is from flexible thought towards its takeover by automated response - reflex instinct, law or pattern of behaviour. In this respect the phase of creation most fully under the (*tam*) predominant influence of energy is called the material universe.

Tam's influence localises, defines, restricts, confines. It blocks, inhibits, isolates, contracts, represses, depresses and stultifies. Cosmic drag! It is the passive, rigid, anti-active, mass-centred principle in things. Inertia, gravitation and precipitation are expressions of this vector. Its model is a particle, an object or, in any circumstance, the fixing factors. Gross matter is created by gravitational confinements, by nuclear to galactic localisations of matter. Following the direction of (*tam*) materialisation you sink towards slack, the state of least tension and inertial equilibrium. Things flop into inertia. They tend towards a loss of information (called informative entropy) or of energy (called entropy). In descent towards the base of Mount Universe you fall towards extreme negativity, inertial neutrality and the darkness called *subtendence*.

Do you remember positive and negative neutralities? Two cosmic fundamentals, vectors *raj* and *tam*, are locked vibrating in between neutralities - potential that's pre-motive and post-motive impotence. As (*sat*) potential is polarised it breaks into (*raj*) action. Such action involves, however, counterparts from two directions. *Raj* is up, *tam* down. In *tam*'s direction motion closes with inertial equilibrium, that is, post-motive impotence. Cities turn to dust; things diffuse into 'homogenate'. That's how, from super-state to diametrically opposed expression called sub-state, the run-down always goes. It is the tendency of matter to 'run down' that a physicist calls entropy. Entropy is materialisation's mode - downhill and, strangely, getting harder as you go.

As well as being the vector of 'fall', *tam* neutralises action. It leads to the bottom, creation's brick wall and a hard end called the material universe. The simplest example of this cosmic process is drawn from basic chemistry. Preconditions (reagents) are prepared; reaction fires to completion; balances of charge are satisfied; and, in this case, the neutral subtendent is the end product. We say, therefore, that *tam*'s (↓) 'negative power' materialises to the point of inertial neutrality. The fire goes out.

In the reciprocity between poles *tam* shows as repulsion, isolation, self-absorption and selfishness; and as tension, stress and pain. It is the separator. It confines, pressurises, materialises and, as such, represents the negative or earth pole. Its losses of information or energy engender fragmentation, differentiation, precipitation and, in the maelstrom of events, apparent randomness, disorder and irregularity. The vector drops through 'solid' mind (called memory), sleep, coma and the oblivion of matter - where we've just seen it push against the limits as solidity. Exhaustion, sub-consciousness, sleep, impotence and mass are base-line characters of mind and matter. The spark of liveliness is lost. You might think from this that tam is a 'bad' influence. Well, yes and no. There is moral darkness, yes. What drama does not need to challenge it? And no; what film without its shadows? How come the world without resistance, downward harder as you go, from natural negatives?

Gain! Spark regained. The upward bounce is *raj*. Uphill is easier as you climb. From sub-state towards antithesis in opposite-to-concrete super-state, *raj* represents a rising through the ranks. A simple example is the heating of a solid through liquid into gaseous phase. Or, psychologically, an interest that draws from ignorance to knowledge or depression into joy. You might call these, combined, the fire of love.

Have you ever like a dervish danced the Dialectic? With right arm (*raj*) raised and hand open to the heaven, with left arm (*tam*) lowered and hand cupped towards the earth he whirls. He rotates like a galaxy, spins like a star, orbits like a planet or, in a cyclic cosmos, oscillates like atoms do. See how from time to time he gathers both arms to his heart and then, as a flower grows, organically unwraps them towards the sky. Observe his head inclined towards the path of his ascension, towards the right: and how with perfectly controlled and centrifugal force he whirls around the leading dervish like a planet round its sun. He glides around the pivot of a gyroscopic world. What does his moving meditation gain? Has he united with the central poise directing all the motion of a balanced universe - a universe revolving and resounding to the order '*Kun*'? '*Kun*' is the command of Allah - 'Be!'

What is at the climax of a vortex spiralling up? What constitutes the (*Sat*) Apex of Mount Universe? Completion of ascent is, for Natural Dialectic as for dervish, a completion of the cycle and return to origin. Did not this world derive from a massive jolt of potential, from an original shot of energy? What is the nature of this 'lift-off' and, secreted just behind it, the nature of potential matter? Higher up and psychologically, what is the tip-top nature of the The Peak, the immaterial Super-state that is Potential for all things? At any rate, at both ends of the spectrum find neutrality. The neutral states are 'limits' and, between them, cosmos oscillates.

Although abstract and not necessarily familiar in concept, Natural Dialectic captures the essential motion of these fundamentals. *Sat*, *raj* and *tam* form the simple, triplex basis from which complexities (including the various 'fundamental' laws of chemistry and physics) ramify.

Hierarchical, Triplex Construction of the Cosmic Pyramid.

The incorporation of the fundamental conditions of conscio-material duality (*potential, kinetic* and *inertial*) into the structure of Natural Dialectic is now plain. *Sat* (potential) is represented, in Primary or Main Dialectic, by the capital letter. *Raj* (↑ kinetic) and *tam* (↓ inertial) are represented in tendency or direction by the vertical arrows of Secondary or Branch Dialectic, *up* on the right-hand or *down* on the left respectively. For example:

	tam/ raj	*Sat*	
	gross/ subtle	*Clear*	
↓	*tam*	*raj*	↑
	gross	*subtle*	

or tri-logically:

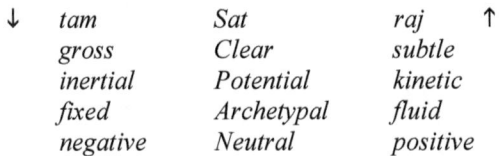

↓	*tam*	*Sat*	*raj*	↑
	gross	Clear	subtle	
	inertial	Potential	kinetic	
	fixed	Archetypal	fluid	
	negative	Neutral	positive	

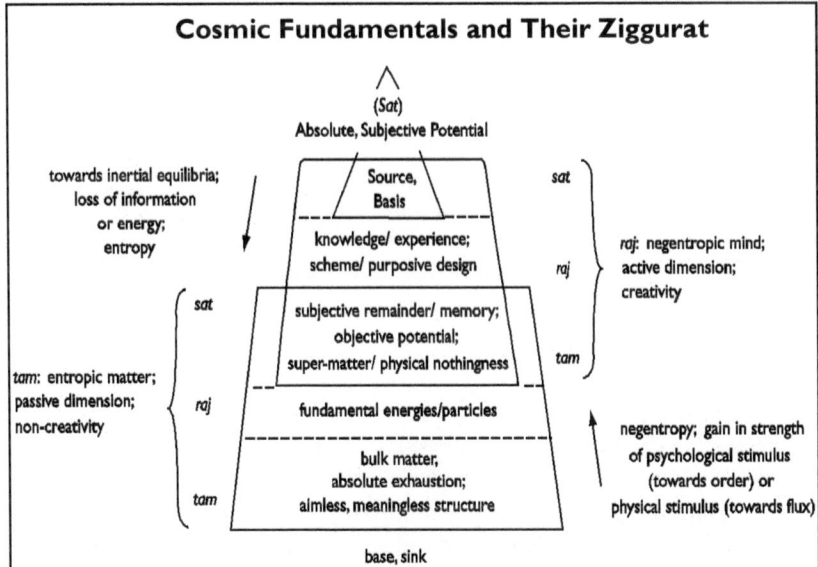

Have you ever seen life like this - cascaded from the Apex of Mount Universe? A pure concentrate of consciousness is, in conjunction with non-conscious energy, devolved; from such source to sink the world is issued.

Of what is matter exhausted? It lacks all subjectivity; it is non-conscious. It is life-empty and in informative terms fixed. Automatic and in this sense passive and impotent. Conversely, in energetic terms, it is active, powerful and often fluid.

fig. 2.10 (see also figs. 0.5, 5.2, 9.1 and 23.1)

Furthermore, each quality in predominance is expressed as a region or tier of the cosmos. Check elementary *fig.* 0.4 again. At the top level there is neither time nor motion, only Presence. Supreme Being. (*Sat*) Potential first moves, as Cause, to express its implicate or latent order. It moves, as consciousness-in-motion (mind), from Centre outwards down the conscio-material gradient. (*Raj*) kinetic mind is information-predominant and precedes the ordered generation of the next layer. It precedes and is therefore cause of (*tam*) physical effect. This effect is the control of energy by passive information or, in scientific terms, fixed patterns of material behaviour compressible to formulae; these formulaic sets of rules describing change in space and time are called laws of nature. The final, lowest level is entirely non-conscious, that is, objective.

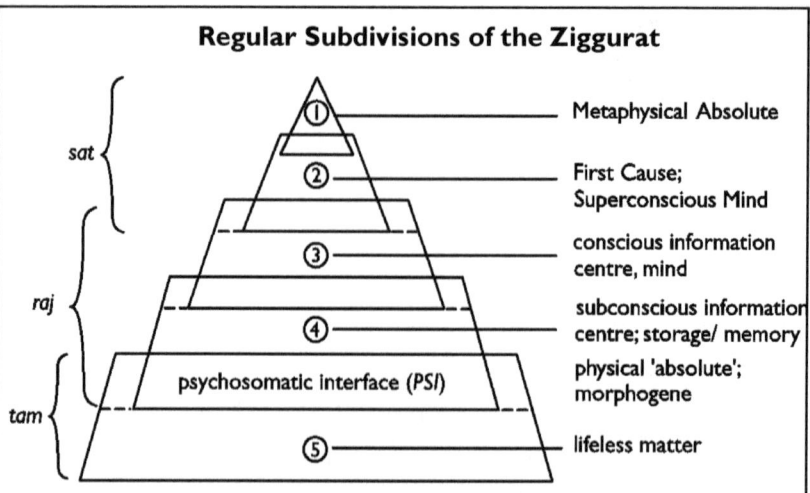

Different psychologies and philosophies cut different subdivisions (also variously called levels, realms, regions or domains). Natural Dialectic sticks with triplets at each step with three main steps in all (fig. 2.10) but (as in figs. 2.11 and 2.12) you can divide in five or seven too! How do you divide a rainbow up?

Mind and physical energy are 'wireless'. The lowest 'waveband' of mind is fixed, recorded thought, that is, memory. Physical events, both 'wired' and 'wireless' are fixed by regulation, that is, rules that govern their behaviour. They include the flux of energetic transformations and 'wired objects' such as atoms, molecules and massive aggregations of them. The latter are, effectively, fixed records of events; compounds are gross expressions of their inner subtlety - especially the solid phase.

You can visualise wireless factors predominantly in terms of spectrum; and physical discontinuities (grade 5 of the ziggurat) primarily in terms of particles or separate objects.

fig. 2.11

The three broad hierarchical divisions can be pressed a little further (into triplet subdivisions) before we turn to the power whose transformations both link and sustain them. The *(tam)* base of one division links or coincides with the top of the one below it. This top level then polarises to produce subtle and gross subdivisions. How does such cosmic circuitry pan out?

The top subjective division, Potential, *is*. **Thence downwardly, subjective subdivisions at each level of mind represent *loss of information* from the pure state, enlightenment; that is, they represent a lower, slower, more restricted or (as with memory) fixed state of mind.** In other words, super-conscious (*sat-*predominant) mind is linked above with Absolute Potential in the form of First Cause. On descent its 'condensation' generates first conscious (*raj*) and lower down (*tam*) sub-conscious wavebands, phases or aspects of mind. The degree of such 'condensation' is a function of material wrapping; that is to say, it is a function of the 'density' of mechanism by which the basic faculty of perception, pure consciousness, is empowered to operate at different levels on Mount Universe. For example, just as a diver requires kit to observe alien ocean depths so the mechanism of sensation at physical level involves a physically 'dense' and specialised correlate - kit called the nervous system. In short, the state of 'active information' (conscious mind) becomes progressively more passive until, at stages 4 and 5 of *figs.* the state of 'active 2.11 and 2.12, all action is lost; entire passivity and fixity prevail. And the upward reverse of loss is, of course, increase in information, interest and understanding. The consequences of such a *top-down* perspective, following the conscio-material gradient of creation, will be worked out more thoroughly in the chapters (5, 6 and 13-17) that deal with *psychology*.

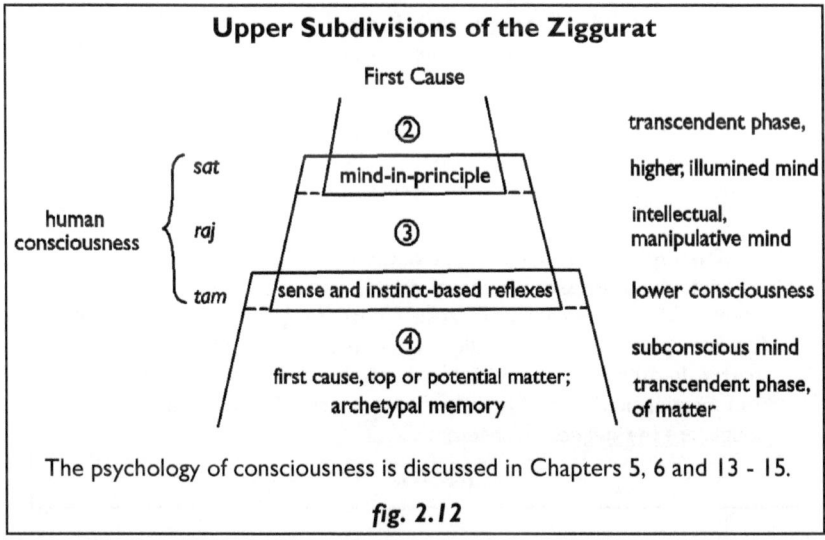

fig. 2.12

The psychology of consciousness is discussed in chapters 5, 6 and 13-14, sub-consciousness in 15-17 and chapters 7-10 elaborate on the subdivisions of non-conscious matter. In the sense that it 'underlies' mind our zone of physics might be termed an underworld.

Objective subdivisions of each level shown in *fig.* 2.13 represent, downwardly, a *loss of free energy*, a 'thickening' of non-conscious materiality from the subtle state of light and quantum factors down to gross solidity. Firstly, nothing is prior to what its potential can create. From a dialectical point of view you *expect* something from nothing, that is, expression from potential. Material subdivisions therefore start with (*sat*) potential matter or (*figs.* 2.9 and 2.11-13: level 4) archetypal memory. Archetype is not an abstract, powerless void. Do not think of it in a reduced, exhausted spatio-temporal way but as the rules that govern our material game, the instrument through which the laws of nature are expressed. In this way, just as the qualities, textures and colours of solids are *inherent* in a gaseous form, so are natural behaviours *inherent* in potential matter. The metaphysic of such matter hierarchically precedes the subtlest quantum agents. For this reason the omnipresent, background nature of a quantum vacuum with its *ZPE*, intrinsic energy of which the 'void' of space is full, is (see Chapters 9 and 10) interesting to Natural Dialectic.

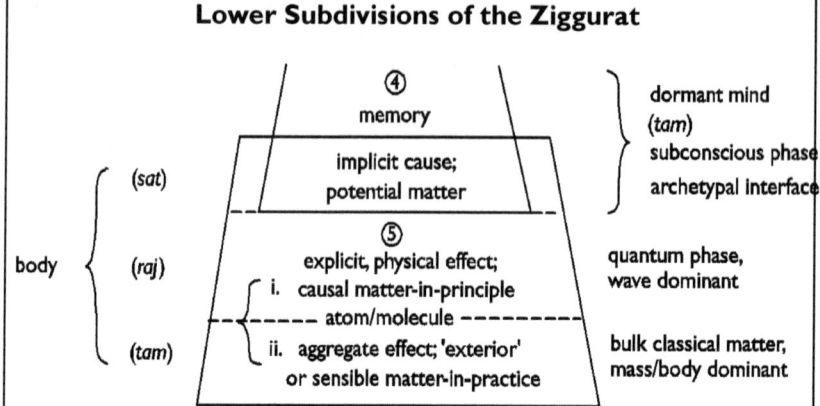

(Sat) potential matter is an implicit, metaphysical cause of physical effects. This transcendent subdivision of 'super-matter' and its psychosomatic interface are discussed in Chapters 7 - 10, 15 and 17.

Through quantum physics science comes in touch with, at its archetypal phase, universal mind. Section 5 involves all physical effects. Of these an explicit cause - cognate quantum particles and forces - gives rise to so-called 'condensed', 'externalised' or bulk matter. In other words the material universe is an effect composed of (*raj*) interior, kinetic and (*tam*) exterior, inertial forms of energy. This couple are the subject of Chapters 9 - 12.

fig. 2.13

	tam/ raj	Sat
	below	Transcendent
	things	Nothing
	static/ kinetic	Potential
↓	tam	raj ↑
	particle	wave
	mass/ no bounce	energy/ bounce
	bulk	quantum

Could one even subdivide matter-in-principle, the world of quantum physics, into bands? First, unfamiliarly, would come the zero-point potential that was mentioned just before. Of special interest, would be the nature of any possible psychosomatic linkage between this level of (*sat*) virtual matter and (*tam*) subconscious mind. Such a linkage would, *top-down*, demonstrate the origin of matter is a form (or forms) of mind. 'Below' the (*sat*) transcendent or zero-point origin once called 'the luminiferous vacuum', actual motions show both (*raj*) wave and (*tam*) particulate qualities. In other words, subsequent polarisation would generate the overt forces and particles that underwrite configuration into the third quantum band of atoms and 'wired' or bonded molecules. At this level the 'least bulky' form of energy is light. Its aspect of particle is the mass-less photon. Grade of mass increases through electron to proton and, heaviest, neutron; neutron and weightless vacuum - atomic charge is squeezed between these two neutralities. Such subatomic particles form the potential, principle or basis for bulk matter. Atomic principle gives rise to bulk practice; from quantum states issue gross, molecular and *sub*-molecular aggregates (the parlance of *bottom-up* perspective would call them *supra*-molecular).

This three-banded (*raj*) quantum level gives rise to the third physical level of (*tam*) bulk, sensible matter-in-practice. Matter-in-practice can itself be subdivided into higher atmosphere, 'middling' water (seas, rivers, lakes etc.) and solid earth below - more generally gases, liquids and solids. Since the three levels exist simultaneously and are simply, like atoms to a crystal, different aspects of the same thing we might call them higher, subtler and lower, grosser manifestations from a transcendent origin. You could see gas as potential solid and a crystal as de-energised, impotent vapour. Each level is potential for the one beneath. Consequently physical phenomena are banded into three - one implicit, two explicit. Implicit, archetypal potential is metaphysically described and discussed in terms of abstract formulae; kinetic (subtle) matter is, in broad terms, the subject of quantum physics; and massive (gross) bodies the subject of classical or, for non-physicists, 'normal' physics.

Here we are - cogitators in an otherwise subjectively empty precipitate called the gross, physical universe. *Raj* is ascendant in the regions of mind, *tam* in the physical world. Each characteristic, including *sat*, is reflected in some proportion in everything that exists. Either in repulsion or (as between male and female) in reciprocity between poles every aspect of relationship occurs in a range of proportions, intensities and time-scales. Up, down and axial, the three cosmic fundamentals are omnipresent qualifiers of the shape and power of things. Science measures them in different forms exhaustively; the Dialectic simply classifies such measurements, in terms of fundamental type, within a hierarchy that extends beyond a single platform made of earth.

Sliding Up and Down a Slope.

Do you remember (Chapter 1: Causality) 'horizontal chains' of cause and effect? There is a sense in which cause, activating a process, is 'superior' or 'prior' to the effect. In a chain of events, however, does not 'inferior' effect become the next 'superior' cause? Such cycling involves intrinsic balance (as equations show) but it may or may not return to its original impetus, its first cause or starting-point. Nor do physical chains involve any particular, targeted effect. Instead they propagate multiple, diverse effects; they trigger transformations of shape and energy.

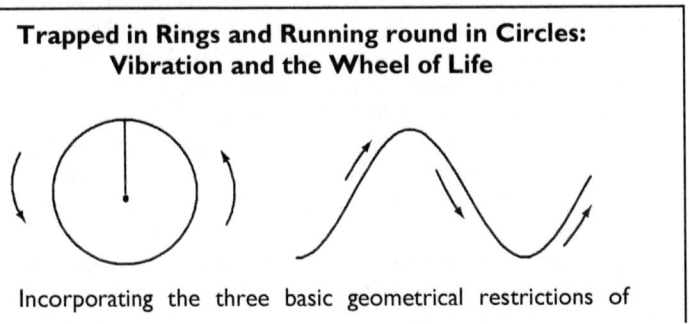

**Trapped in Rings and Running round in Circles:
Vibration and the Wheel of Life**

Incorporating the three basic geometrical restrictions of

infinity (a point, curve and radial, straight line), a vibration is an interesting expression of both diversity-in-unity and the binary interaction of opposites. A closed curve is a circle. Locate a point on the periphery of a circle. Spin the circle like a wheel that moves along a line of time. The speed at which a circle spins is, effectively, the same as a wave frequency. You trace an oscillation.

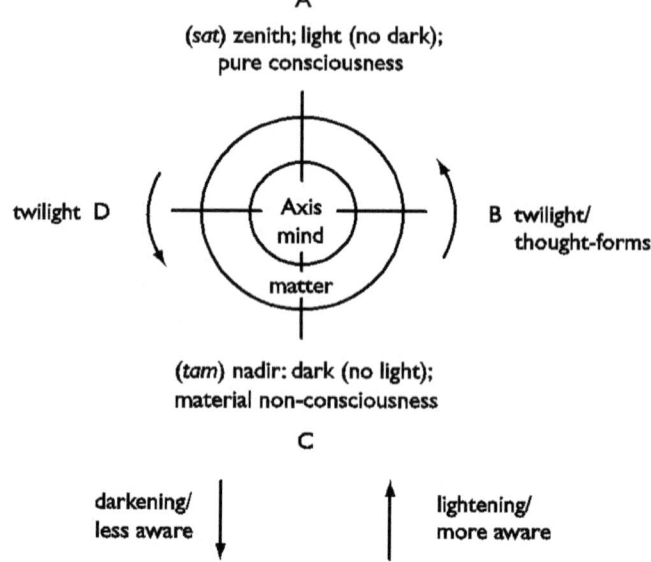

A vibration is, therefore, a wheel that turns through time. What about a cosmic ring? An adaptation of the Dialectic's models (concentric rings, Mount Universe and ziggurat) is the Buddhist wheel of life. Are you not trapped and whirring round in circles? Do you spin in cycles psychological and physical that may include the turns of life and death? Are you not locked (see figs. 18.2 and 18.5) in mind and matter? And, identified with this pair, driven round inside an existential wheel?

Spokes of a wheel are radii. Their distance from an axle represents the height (or amplitude) of wave. Maximum amplitude is found at the circumference; you might say disturbance 'kicking up' the biggest waves is found at the periphery. A wheel revolves exhausted physical universe. The more axial, that is, the more central your position on a spoke (or conscio-material gradient) the I its furthest distance out there at its world's rim. And at the dialectical edge of creation rotates the non-conscious, subjectively ess distance, the more slowly you revolve and the 'subtler' or less 'loudly' vibrant is your amplitude. One approaches peace. What more peaceful on the inside or without than light? And at the Central Axis motion ceases, at the Point of Balance poise is gained. Motionless in motion, pivotal detachment is complete. This represents the point where action (*karma*) stops. The logic runs that here internal peace surveys external business and war. You don't get off the world to stop. You migrate with wise men rising to the very origin of every sphere. From cosmic hub the sage acts without action. In this profundity complete detachment reigns. **The mid-point of a Buddhist wheel is called Essential Reality.**

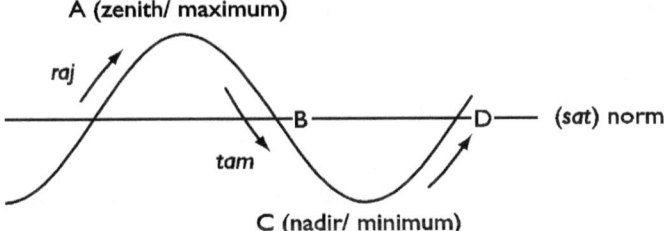

The wheel spins always forward in time. The natural symbol of existence is its oscillation; waves, vibrations and all kinds of cycle and sub-cycle represent the inner being of both cosmic and Dialectic motion. A ring binds pairs of polar opposites with endless swings between each other. Constraint or limit on the degree of difference between these opposites can be depicted as a closed circle with extremes at opposite ends; and swing depicts the range between. 'A' is the most positive and 'C' the most negative condition of any characteristic. In other words, variation between dialectical opposites can be seen as a circle-in-motion-through-time, a cycle or period. Wheels within wheels, the inward nature of existence is vibratory and its periods cycle, each around its axis, norm or 'ray'. *Natural Dialectic, according with the cosmos, supposes oscillation in between the opposites of its duality. Its secondary stacks are, we've just seen, vectored. The way it illustrates these vectored oscillations up and down its conscio-material spine is (figs. 2.2 and 2.3) by using arrows.*

fig. 2.14

Such psychologically undirected chains of cause and effect, the subject matter of physics and chemistry, are called 'horizontal'. This is because in them we read only one informative level on the conscio-material spectrum. This level is (*fig. 0.6*) a special case of consciousness - its lack - called the material grade. It were as if a cosmic graph (*fig. 0.7*) had been assigned only one coordinate - the x- or material coordinate and lacked any inkling of another. Lacking the vertical y- or 'conscio-' coordinate you might think of such a representation as 'lacking depth', 'missing dimension' or 'flat'. This is the current position of science. Its solely materialistic conceptualisation of raw nature's chain is not framed in terms of hierarchy; its single platform is, with respect to either animate or inanimate component, flat. And it is true that, viewed without reference to anything except themselves, events this single, energetic level appear to simply churn through endless transformations, that is, endless changes in energy, shape and place. There is vibration but no hierarchy; there is no metaphysic and, certainly, no notion of a metaphysical origin, shaping power or sustenance. Of the conscio-material coordinates scientific materialism has neither time nor place in its calculations for an immaterial figment of a y-axis (conscious coordinate), only for the x-axis (material coordinate). It has no use for picturing Mount Universe.

The universe of Natural Dialectic, on the other hand, includes both x and y coordinates. It is vertically hierarchical with 'horizontal' bands. Apex to base the universe is banded; a gradient of mind and matter constitutes Creation Hill. In this case, note a subtle but critical difference between 'horizontal' and 'vertical' cycling. Within its own 'horizontal' band an element, either psychological information or physical energy, can be lost or gained. Such ups and downs can also cycle. Vibration's mid-point is the centre of its swing. This line of 'lesser balance' might be called an axis, norm, neutrality or point of compromise. A golden mean! Its middle way is called dynamic equilibrium; such equilibrium, called homeostasis, is stability-in-motion. It is through a coordinated clutch of vibrant, healthy centre-points like this all bodies stay alive.

In the 'vertical' case, however, (*Sat*) Hub represents the Essential Mid-point, Source or Origin of cosmos. In dialectical terms its Primary Characteristic is Balance or Potential Equilibrium. The cycle and sub-cycles of creation depend upon the Zenith of this Natural Axis, this Alpha and Omega. Since its Central Pole is Conscious as opposed to the non-conscious band of bodies (that compose the peripheral pole of matter) it is (*raj* ↑) towards Communion with this Hub that mystics, Buddhist or otherwise, aspire. A body stays alive awhile but healthy minds (saintly psychiatry avers) seek Pure and thus Eternal Life. How could materiality conceive that it exists? But if Life *is* the Axial Origin round which creation spins then mystical communion would be the only way to splice its ends (called Omega and Alpha) and so complete the greatest circuit of them all.

Apex to base, the hierarchy of Natural Dialectic's conscio-material gradient is, like any bureaucracy, layered; but also dynamic since streams of communication and movement between layers can occur. Target, direction and therefore information are involved. Mind is a fundamental coordinate, an essential part of the system. Of course, events at their own non-conscious, material level appear to churn through endless transformations but they do so according to succinct rules of behaviour. What is the origin of these rules, the so-called laws of nature? And what, prior to the creation of a physical realm, was the nature of its metaphysic? Does such metaphysic still, either consciously or sub-consciously, impinge; and if so how, sliding up and down creation's slope, might such interaction work?

Consider not only the inanimate but, especially, the animate or biological component of our natural world. Consider purpose, targets and machinery. Consider deliberately cycled systems, ones returned to norm by negative feedback or, if you like, to origin by homeostasis. Back to the beginning. If material cosmos cycles round, what is its point of poise or origin? And, if you include hierarchy as well as simple periodic swings to your scenario, what might be the source of any immaterial, informative dimension? To what pivot, point of poise or origin might mind naturally return?

Although prior information is disseminated *top-down* through a system, feedback is possible - in which case a circuit is joined. Information and energy are, in fact, both cycled. Indeed, any process involving negative, homeostatic feedback is an example of an information/ energy cycle and, as such, involves (*fig.* 0.7 iii) the y-dimension. The information coordinate. An example of such hierarchical cycling is central to your very being - your nervous system. Sensory information ascends and motor response descends in motion round a circuit. Sense organs automatically transmit data to a processor. The flow can be traced up from external through internal physical peripheries to a central regulator, a balancing factor, a complex called mind and brain. Input, regulation, output. In other words, the myriad separate objects around us are collected and dematerialised, first into streams of coded information and then, unified with elements such as memory, feeling and thought, *con*sciousness. The product of processing is a decision whose coded command is issued through the motor system and appropriate muscles. Orders are 'sent down' to matter.

Commands might involve a machine, even a mind machine called a computer, but you can always trace the original manipulation back to a centre, brain or mind. Are these two in fact the same or separate? The basis of a brain is atoms and their particles but if the centre's mind then what are 'particles' of mind? What is the anatomy of information and the source of its most vital element, live consciousness? If the natural origin of information is in mind and mind is metaphysical then you need picture, ultra-scientifically, Mount Universe.

	tam/ raj	*Sat*
	info out/ info in	*Info. Centre*
	imbalance	*Balance/ Stability*
	relative perfections/ imperfections	*Perfection*
	each side	*Equation*
↓	*tam*	*raj* ↑
	down	*up*
	static	*dynamic*
	flat / inertial	*vibratory/ exciting*
	outside	*inside*
	materialise	*dematerialise*
	deaden/ de-animate	*enliven/ animate*
	info. out	*info. in*
	motor/ motion	*sensor/ sensation*
	to body	*from body*

As opposed to atoms of the brain consciousness, by definition, *knows*. If Pure Consciousness is the Natural Centre of Information, then Ultimate Knowledge, the Source of Wisdom, resides in its *sine qua non*. It is where all thoughts arise. Thus, for an experient, it is possible to transcend an existence composed of mind and matter and, having completed the cosmic macro-circuit, return Home. Is a mystic thus a mega-cycler?!

Non-conscious matter, on the other hand, is mindless and involuntary. Its oblivion is totally controlled by law. 'Justice' and 'mercy' may be flexibilities of mind but merciless is mathematics, rigorous its equations; inexorable, inflexible laws described by chemistry and physics govern every move. Do you remember 'event vectors' that, in their simplicity, indicate a direction for any aspect of an event either (↓) away from or (↑) towards Centre? *In the case of matter this involves either energy loss that results in 'hardening' (e.g. water vapour to ice) or, conversely, the stimulation of energy gain.* It involves creation/ fixation or, conversely, dissolution/ release. Because it is non-conscious, however, the case is hopeless. There is no way matter can break free from oscillations of oblivion; having no subjective condition, there is no way it can join the macro-circuit, no way evolve back to The Origin. There is voluntary release neither for inanimate objects nor the bodies of 'animated beings' such as you and I. The molecules that make up biological bodies are as inanimate as those that do not. How, therefore, could there ever be the 'resurrection' or 'transfiguration' of those molecules into a 'state of mind'? How could your present body rise up at the crack of doom? This case is the same for every physical phenomenon and the cosmos as a whole; it can only be subjected to *involuntary* starts, finishes and, perhaps (see Chapter 12), cycles.

It is only with 'awakening', the dawn of conscious appreciation, that any wish can inform action, any hope or meaning arise. Wishes override the hopeless jiggling of material events; purpose sets the seal of meaning on a world without it. In humans desire informs both psychological and physical activity. There may even arise, amid the myriad more usual ones, a desire to attain Pure Consciousness and thus, having returned to Source, a wish to close the whole, personal cycle of existence.

Source? Oscillations up and down the conscio-material gradient are, whether along 'horizontal' or 'vertical' coordinate, all very well. They constitute the world of forms, they make motions registered along the graph of existential relativity. Hierarchical layers or levels on the Mountain are understandable but how far up or down can you go? If Natural Dialectic is a Theory of Relativity It is also a Theory of Fixed Framework, Equilibrium and Absolution. What, either way, are the limits? Where is the axis? What is the apex of the world, how will you find its base? Of course science, having no sense of hierarchy, does not even have the frame to ask such questions let alone, in its hunger for material details, time to consider them.

Play a game of snakes and ladders. When you slide down, the layer below is said to subtend the one above. In this sense subtendence is simply a function of the vector of (*tam*) descent. What, however, is the furthest (*tam*) creation can press down? What diametrically opposes the Centre, what darkness lies at a periphery called 'deep bottom'? What did you expect to find around the buttocks of the universe?

Conversely, transcendence is simply a function of ascending grade and, with it, increased potency. Water vapour transcends ice, waking climbs from sleep. Does mind transcend matter or information lodge above its lack? What did you expect to find within the cosmic head? All physical effect is traceable back to a first cause of things which, most current science textbooks claim, is an expansion out of nothingness. A 'big bang'. *Bottom-up*, things are connected by an origin that's not in mind but time. *Physical time not mind comes first.*

What last? After a start a lifetime and an end. On life is draped the pall of death. If you think time precedes mind you might think it also follows it. You might even imagine a reciprocal contraction from big bang's expansion so that everything ends up in a singularity far smaller than your thumbnail. Alternatively the star-fires might go out and cosmos die in what is termed a 'heat-death'. Exhaustion is a terminus called Omega.

Individual mind 'emerged' long after time began. It evolved from time's dust and returns with every single death to dust again. Mind is a product of body; it developed after conception and, when the body dies, will immediately disappear. There is neither before nor after lifetime. Mind is, viewed from the bottom of creation's slope, marched 'last in then first out'.

You say there is matter alone. I remind you there are matter *and* consciousness; energy *and* information.

Both conjectures are metaphysical. **It is pure speculation on the part of a materialist (including the predominant leaning of the body of science itself) to presume that consciousness is a material entity**. Just as materialism treats *as if* it were matter, the *top-down* non-materialist treats *as if* it were not. *Top-down* and *bottom-up*, when pressed, reach diametrically opposite end-points. They are, each for the other, antipodal perspectives.

Physicist Sir James Jeans once remarked that 'reality is better described as mental than material....the universe seems nearer to a great thought than a great machine'. Whatever he meant by this, it is possible to think of the behaviours of matter (known as the laws of nature) as residual, rigidified or automated information; as absolutely repetitive habit or crystalline mind. *It is indeed the contention of Natural Dialectic that behind the physical field of action there lies an implicit, metaphysical 'field of information'.* Explicit order issues from this superior field; transmission occurs across a psychosomatic border; control, command and cause are not a matter of temporal but conscio-material precedence. *Top-down, mind and not time physical comes first.*

If you think mind preceded physic's time could it not succeed as well? And can a tide that flows not ebb; with ebb of power to whence it came, the level that's 'above', how could atomic webs flossed round in space not just collapse? If what supports the cosmos from behind withdraws, will it just disappear? Projector switched off. Film over. Collapse of image. World turned off. Will universal matter just collapse into an origin called universal mind?

As there existed universal archetype before and after matter, so there would exist individual mind before and after its incarnate lifetime. Mind is, in the *top-down case*, a matter of 'first in, last out'.

By casting life in opposite perspectives Natural Dialectic indulges, philosophically, in sliding up and down a slope. So, high or low, what's the end-point of the fun? Is it Pure Matter or Pure Consciousness?. What is the nature of the latter's (*Sat*) Transcendence? What is it absolutely heads creation up?

Start and finish. Cycling down and up a slope. *The two extremes, which represent limits of existence, are known to Natural Dialectic as transcendence in the upward direction* and *subtendence in the downward*. Let's ride the roller coaster down before returning up.

Subtendence.

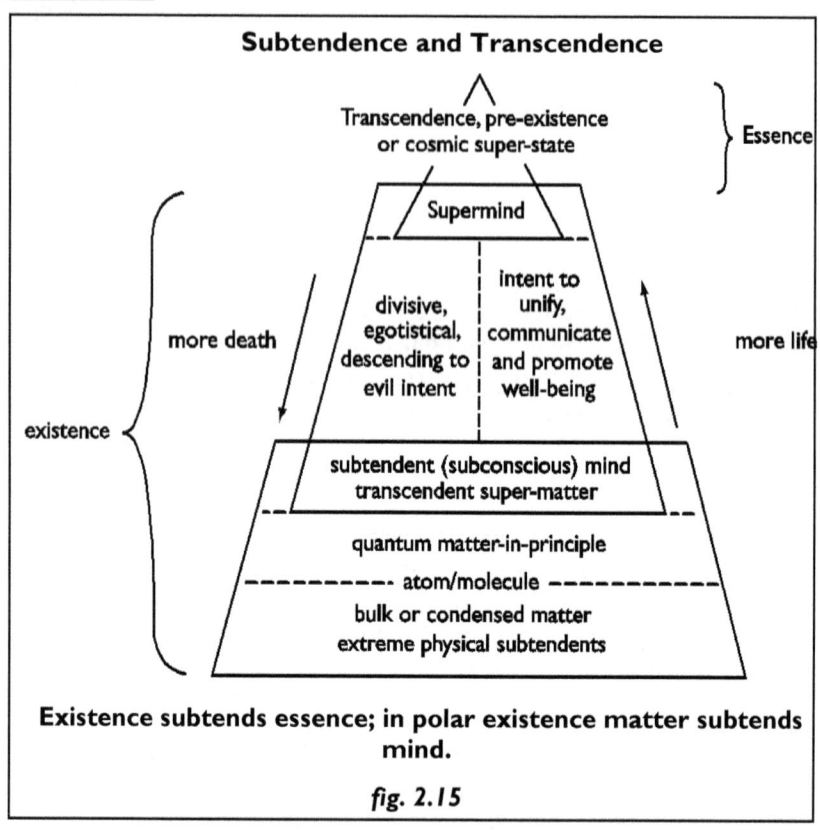

fig. 2.15

tam/ raj	Sat
below	Transcendence
range of action	Super-state
spectrum	Cosmic Peak
expression	Pre-motive Potential
↓ tam	raj ↑
descent	ascent
fall	rise
subtendence	transcendence
towards exhaustion	towards replenishment
more death/ less life	more life/ less death
post-motive impotence	motion
sub-state extremity/ cosmic base	spectrum/ range

You might protest that philosophical extremities, especially at the metaphysical end, are impractical and irrelevant to your bench, desk or pragmatic way of thinking. This sensible attitude in fact turns actuality on its head! Science is encapsulated in extremities. Far from being trivial, they are boundaries and foundations, at one end potent and the other impotent, that one way or another all of nature oscillates between.

To subtend is to be or to extend underneath something else or opposite something else. Low subtends high, dark light. Visible light subtends ultra-violet, solid ice subtends liquid water; large objects subtend atoms and atoms subtend their super-atomic particles (electron, proton etc.); all these subtend their origin, super-matter. From a dialectical perspective Origin is 'above'; basic substance, foundation or fundamental principle is, at the 'top', transcendent. Thus bulk matter subtends atoms. It is, dialectically, *sub*atomic. If sub-tendency or sub-state is the step below, what is at the bottom?

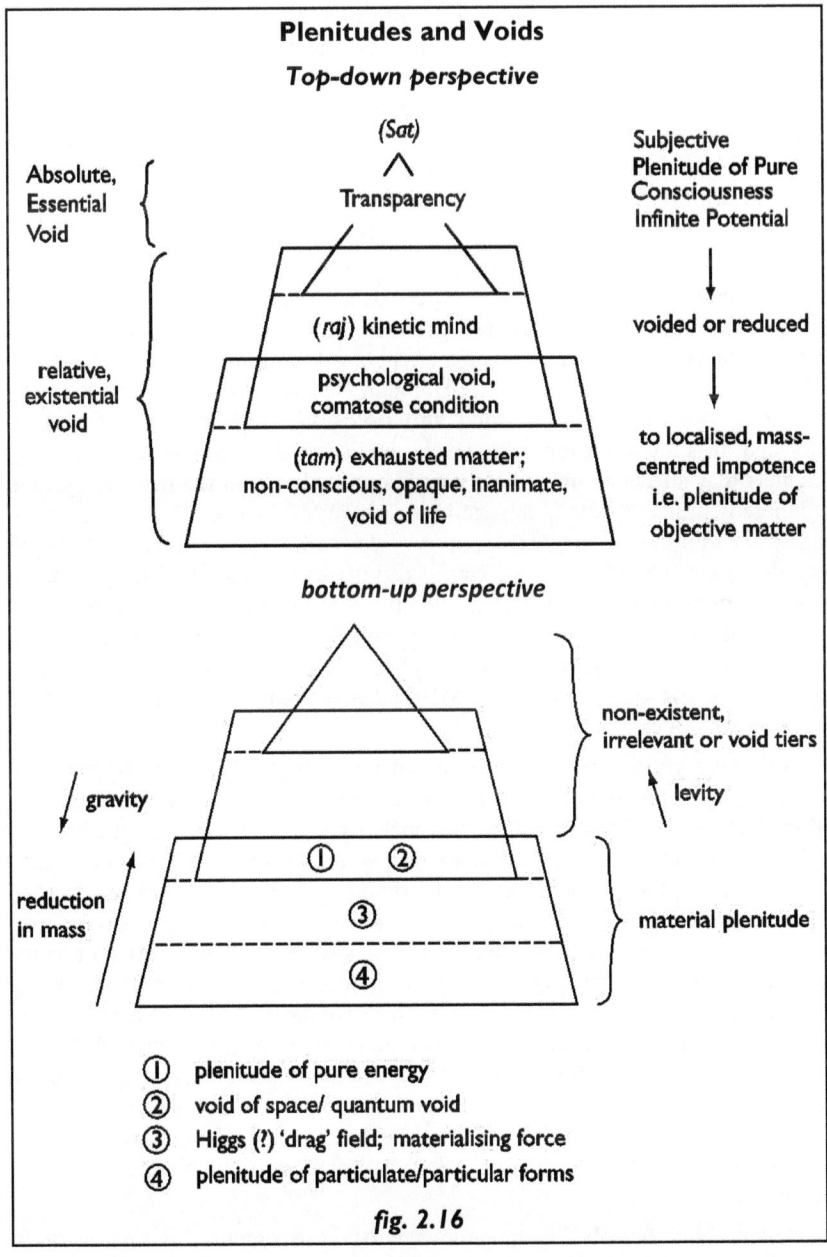

fig. 2.16

Of course, *bottom-up*, you look at things the other way. Therefore your base - solid, sensible matter - is your starting-point, your foundation. For you mind is a product of material brain, so that in present parlance it subtends the brain. *Body comes first, mind (a bit of a material mystery) follows after.*

It is true, *top-down*, that products subtend their producers. A producer stimulates and stirs things up. An author *informs* his play, a scientist *informs* (conceives, sets up and performs) his experiments or a boss *informs* his business. From this perspective information precedes its material consequence, metaphysical mind plans physical arrangements, super-natural precedes natural order. An individual reflects the cosmic case. Psychological principles inform physical. In other words, the fundamental patterns of physical energy are 'stored' metaphysically; from such 'records' derive the perceptible laws of nature and various properties, shapes and behaviours of matter. In short, the latter subtends mind; the foundation, the starting-point is immaterial. *Mind first, body after; body's an appendage of its mind.* Does it, that physical depends on metaphysical, invert your 'normal' sense of things?

In the psychological zone sleep, we say, subtends waking; sub-conscious subtends conscious and the latter super-consciousness. If there could be an 'Adam's fall' from super-conscious grace to our normal waking state, we are unaware of it; but we certainly experience a fall from conscious focus to unfocused and sub-conscious sleep. Although we may be unaware of it the sub-conscious, dormant level of mind still holds information, informs and is informed. It is the repository of memory, instinct and the psychosomatic (*PSI*) channels that interface, immediately and continually, with the nervous system and thereby molecular structures, cells and organs of the body. *Stored, reflex or automated information is passive; passive information subtends active.* A singer records a song. The vinyl or tape carries rigidified information. What is it but a plastic form of memory? All matter, which is totally non-conscious, repetitious and predictable in its behaviours, is informed by the internal or immanent exigency of the 'laws of nature'. What is this 'song', whose record plays across the fabric of the known universe? Material bodies are governed by and thus subtend these rules.

Let's take your own body first. When you waggle your little finger, is it possible that the line of operational command runs from conscious mind through sub-conscious templates (memories, instincts etc.) over a psychosomatic border; and that incoming or outgoing information would be first translated to, or last translated from, the physical side by the product of 'excited' electrons, that is, electromagnetism. This 'radiant' phase would be subtended by the biochemical; and both levels occur within bulk, biological structures, that is, cells, nervous system, muscles and a coherent whole body. This renders your sensible, physical form (but not necessarily the origin of its shape) at the bottom of an informative/ energetic hierarchy. It would be the final outcome of a two-pronged, coded plan. First prong, generic, springs from archetype; the second, locally specific, is genomic. From signal biochemistry is body built; phenotype subtends its molecules and genotype. Thus bio-form is frozen yet dynamic program; or, prosaically, a 'functional structure'. Very complex, yes; automated, yes; it's a machine.

Bulk body is controlled by quantum agents; these 'organise' its chemistry. It subtends these elements. Whence, from a *top-down* angle, did these subtle matters rise? What is the template they in turn subtend? If one exists how does connection to it work? Could, closest to whatever transcends physicality, pure energy (say, mass-less radiance of light) subtend a signal interface? Thus bulk, solid rock emerged from air (which, if the universe was once much hotter than it is today, it did) because the air subtended energy and energy was ordered under archetype. If vaporised concretions make a gas the opposite applies. Thence the question 'What's the end precipitate of gas, the final stage of matter past which it's impossible to drop?' Is there, as well as transcendent super-matter, an opposite but absolute subtendent?

	tam/ raj	*Sat*
	subtendencies	*Transcendence*
	finite	*Infinite*
	duality/ multiplicity	*Unity*
	polarity	*Neutrality*
	relativity	*Absolution*
	degrees of imperfection	*Perfection*
↓	*tam*	*raj* ↑
	negative	*positive*
	more passive	*more active*
	confinement	*release*
	dark	*light*
	base extremity	*range*

How deep can you dive? Essence is not, as a lone monopole, polarised. Absolute Transcendence is neutral. Its Truth does not change. However, any motion means a relative lack of Poise; nature *is* motions, changes and their relativity; you might call its balance-seeking wobbles 'perfectly imperfect'. Thus existence, which subtends Essence, is by definition relative. And relativity is polar. Its duality involves the instability of ups and downs, positives and negatives, comparisons and opposites. Can you have a relative or a negative Absolute - an entirely imperfect Perfection or perfect Imperfection? Can, in the latter case, a complete and utter, infinite negative be arrived at? Does there exist a total opposite of any positive existential character? For example, do perfect darkness, negativity or materialisation exist; is an Absolute Subtendent possible; and what, as you descend, might be the nature of absent space (extreme confinement) or, conversely, absent entity in empty space?

If a 'condition' that is 100% composed of the *Sat* fundamental (Absolutely Neutral Potential) is possible, why can't you propose one composed of 100% *raj* (an absolute positive) or *tam* (an absolutely negative subtendence)? The answer is threefold. Firstly, 100% *Sat* is not a condition. It is a pre-condition independent of any constraint that reduces it, that is, of any effect that flows from it. Its Subjective Void is wholly potential and called Infinite Essence. Its nature is Enlightenment. Secondly, *raj* and *tam* are imbalances dependent on prior balance and so, by definition, relativities. They are dependent. As existence on Essence or a part on its whole, they are dependent on the *Sat* fundamental. Thirdly, in dialectical terms *raj* inclines to *Sat*, *tam* declines from it. *100% raj is Sat*; it obtains when the point of original balance is re-achieved. Thus absolute positivity becomes neutrality. Translated into psychological terms this means that existence is transcended. As was explained (in Chapter 0: Pure Objectivity) Absolute and Central Subjectivity enjoys the only full detachment from existence, that is, from the works of mind and matter. Extreme (*tam*) subtendence is, on the other hand, the polar inversion of (*Sat*) Transcendence - but there is, even in extremity of physical presence (say, a black hole concentration) or absence (wholly slack, diffuse and heatless space), nothing that is absolute.

Apparent space- or time-conditioned absolutes abound. Purities - exclusions of all else but one - are various. You can discover individual unities, perfections, balances and concentrates of information (say, principles or condensates of fact), energy (say, stars) and matter (say, pure chemicals). Yet focus of the mind is lost and separate distillations all diffuse in time. What universal concentrate survives such relativity? Are there unchanging laws, immutable abstractions or ideals that both transcend and govern how things go? Principles are concentrates of information; they are the potential that precedes specific act; immaterial principles rule mind and matter's patterns. The latter has no choice but you can choose in which direction, up or down, your own beliefs, ideals and interests run. Is it *GUT*s you crave; or to improve the world by an injection of your love? What is the cleanest of ideas? What might be the Purity of purities, what is the nature of the Concentrate of concentrations unto which, transcending peace-less relativity, you'd best (↑) aspire?

	tam/ raj	*Sat*
	expression	*Potential*
	colour	*Transparency*
	motion	*Tranquillity*
	shades	*Pure Light*
↓	*tam*	*raj* ↑
	shadows	*colours*
	absorption/ capture	*emission*
	obscure	*bright/ radiant*
	slower	*quicker*

Existence is a state of tensions or of motions full of changeful relativities. Even rules that govern such inconstancy will, at the time of dissolution, disappear. And *raj* involves those motions as opposed to (*tam*) completion. It is action's vector, one of stimulation, dissolution or ascent but not of poise. How, therefore, could absolute (*raj*) positivity, like 'motion absolute', be possible? Perpetual yet immobile motion seems self-contradictory - unless, like light, it's never still but clearly calm. Indeed, the most intense forms of (*raj*) right-hand characteristics are associated with forms of high energy, dematerialisation and communication. *Psychologically*, heaven is characterised by the instant of peace, light and freedom. You have been happy; you have been ecstatic when you lost constricting sense of self; and you know that, with joyful interest, time foreshortens to immediacy. It is contracted into timelessness. We love it. *Physically,* of course, a gas flies high, an incandescent gas still higher. What purer form of energy than weightless matter, light? Is there a limit to its brilliance?

Talk is very well but you may still be wondering how extreme (*raj*) tendency could rise to (*sat*) transcendence. How, for example, can pure physic (such as light) turn into metaphysic? How can anything transcend itself and thus return, perhaps, to origin; thus to regain a pre-material level of potential, to recharge what has been discharged, to uncreate creation? For Natural Dialectic such a metaphysical transcendence finds potential matter. More likely matter never does but, in its 'purest' state, might act as an interface. We'll check this idea later. Perhaps the nearest to an absolute material obtains is light's speed. In this case such speed (called c) becomes a container, one of the boundaries in which the universe is kept.

What other clues does light hold when it comes to principles of Dialectic? After the polar excitement of interaction there's left a chemically neutral product; after (*raj*) reaction follows (*tam*) sub-state completion. Where, though, do ultimate reactions lead? When all atomic energy is loosed then matter disappears as light and heat - as radiation; and when the ultimate material polarities collide, when matter meets opposing anti-matter what results? Is duality not unified? Aren't things made light of? Materialisation's reverse is one that raises mass into (*sat*) transcendent light again. The most subtle sort of physical unity is insubstantial and as tranquil as the space it takes. Its particle, the photon, lacks rest mass and its wave provides both information (that a mind might process) and energy - energy sufficient to link worlds, support biological life or, as a laser, slice up steel. Life on earth hangs off a beam of light. Could such a beam be seen as universal pivot, (*sat*) potential for a universe? Having identified light as a material prototype, the Dialectic could suppose that it represents as near absolute a motion as possible. And if absolute motion is poise, a seemingly immobile rate of what by day seems nothing but whose nightly absence shows it shows up everything, that's brilliant. Although the speed of light is fastest, perhaps absolutely flashiest in Einstein's universe, what is apparently more restful, seamless and unmoving than illumination?

Down from light you can't descend beyond a solid. Perhaps you thought some things were solid but such solids are, like atoms that compose them, almost perfect voids. They very nearly don't exist. Look at your hand. It might seem as peaceful and immobile as the light you use to look. If, however, you need evidence for the (*raj*) kinetic nature of quantum principals you certainly can't see them; you have to visualise atomic tininess. Although it's over 99.99% space your 'ghostly' manipulator (seeming by your senses solid and so real) is also packed with immaterial fields of force. It incorporates billions of atoms buzzing with high velocity physics and chemistry: and in the emptiness of deep atomic space each of its nucleons (nuclear particles) oscillates 10^{22} times, amounting to a distance of 60000 kilometres, per second. Interactions of pico- (10^{-12} of a second) and nanoseconds (10^{-9} of a second) are the order of the day. An electron spins at 1000 kilometres (a million billion circuits) per second and the atoms themselves oscillate billions of times in that span. this is quantum action - unimaginably fast yet seeming, in its bulk appearance, still. Additionally, in your fingers or whole frame, labyrinthine sets of algorithmic chemistry sweep endlessly and simultaneously at lightning speed to goals choreographed, down to the precise orientation of each electron and proton, by a systems program the density of whose packing and operation is truly awesome. Can you play the piano? What intense transmission of information, what rapid infrastructure supports the visible parts of your ten mobile digits. You feel happy but, every fraction of a second, millions of atomic motions, chemical reactions and bits of information combine, coordinate and compose that health and happiness. On the gradient of creation where does kinetic quantum behaviour or 'matter-in-principle' belong? It is 'above', 'interior' and much quicker than the bulk torpidity of things like hands. Higher still, what is the 'speed' of (*sat*) potential matter, that is, physic's source (Chapters 7-12)? Is it, like light, *apparently* motionless; or not of physical velocity and guise at all? What, furthermore, about the (*raj*) ascent of mind to its own (*Sat*) Source of Light (Chapters 5, 6, 13 and 14)?

Therefore, are you, so spacious, like a ghost? A shadow of solidity? Compress the ghost, eliminate your space and you'd sit like a grain of salt, a single crystal weighing kilos in your phantom palm. Are mountains and the plains an empty phantom too? You might enlist the press of gravity to squash this earth's illusion to a spaceless crystal of its former self, a cloud-sized concentrate of really sensible solidity - something nearer to a full subtendence.

	tam/ raj	*Sat*
	things	Nothing
	become/ becoming	Being
	event	Before
↓	tam raj	↑
	down to subtendence	up to transcendence
	gravity	levity
	creation	dissolution
	compression	release
	less flow	more flow
	finished/ changed	changing/ becoming
	all over/ nothing left	event

Tam is the vector of compression. The third cosmic fundamental subtends from release. Its negative, materialising power swings the 'wrong' way. Its motive influence drives towards increasingly apparent relativity, separation, isolation and thus set-apart duality. It crystallises out block-headedness and blocks. Mind loses information and matter loses energy until each reaches an impotent state; it drops into exhausted stillness. At this point material existence is complete and essence least in evidence. However while exhaustion (in the form of sleep) or sinks (in the form of a gravitational black hole singularity and 0°K) may appear absolute, internal energies/ motions still persist. Is even formless space entirely empty? Such 'absolutions' are therefore relative, dependent, localised. Finite not infinite. Indeed, the (*tam*) inertial nothingness of 'nothing left' diametrically opposes (*Sat*) Potential Nothing at Top Right.

How can Nothing of itself(lessness) be relative? Although the Infinite splits - losing nothing of its wholeness - into myriad, finite portions of existence, how can One be permanently two or more? How can it, even instantly, be more than Independent One? Existence, that depends on Essence, might be thought of as an absolute polarity - but wherever you can tease out more than one component there is relativity. Even light, with its electric and magnetic parts, is polarisable.

There's no self-contradiction like a relative that's absolute, essential subtendence or boundless negativity.

How is this discussion shaping up in light of mental and material subtendencies?

The foot of Mount Universe is its negative pole. This pole comprises an objective plenitude of matter in its various conditions. *Bottom-up* materialism, however, is oblivious of the whole; in this view matter is neither a 'pole' nor 'negative'; it is positively the essence of existence.

Top-down beneath the Fullness of Transcendence extends creation's range of relativity. Psychic subtleties drop into energetic or material fullness; physics' natural plenitude is now identified as a subjective state of void - oblivious non-consciousness-in-action. Least tangible of *physicalia* are twin 'dimensional containers', cosmic abstracts we call space and time; and in its hardest and most-seeming actuality the 'rind' of cosmos is of solid form. In other words, the end of cosmos is not far away. Touch base; to reach creation's outer edge, to know the bottom of the universal scale just touch any solid thing. Still better, extract every joule of energy; freeze things or nothing near the boundary of zero on Lord Kelvin's scale of heat. Where utter darkness is an edge as well the combination sounds like outer space. Pitch black, very cold and lonely - nothing there forever. You wouldn't want to visit but you can't exceed subtendence like this. Or, equally inhospitable but this time squashed beyond dimension, would you rather designate a solid black hole than an empty one at world's end?

'Subjective void' or an 'objective plenitude' expresses two forms of (*tam* ↓) furthest subtendence, one energetic and the other massive. A physical object, super-cooled, approaches (but can never quite achieve) a total 'loss of levity', an immobility that we have just called 0°K; no dance, no vibratory energy remains. The second, massive kind of cosmic death involves approaching total 'gain of gravity'. Such theoretical - perhaps even actual subtendence - that throttles matter to extreme degree is called a black hole. Some persons say a violent source of X-rays marks the murderous spot. It may transpire, however, that the seemingly 'infinite', self-centring gravitational mass of black hole singularities is simply a titanic guess derived from dialectical deduction and pursuant sums. Could you ever spy acute solidity stuck in a hole like that? Such negativity makes sense but is its boundary for real? Some are sure but, in the case that starts and ends are veiled in mystery, men interpret but nobody really knows. How would you, a scientific super-hero, break the bounds of space and time and things?

What, on the other hand, is a subjective, informative void? There are two aspects that diametrically oppose The Void of Pure Consciousness - one of state, the other of experience. Pure Consciousness is a special case of objectivity - none; unless, in the sense of its distinction from creation, the objectivity of its perspective is complete. Its 'statelessness' is unconditioned and, without the limits of condition, infinite. Conversely non-consciousness is a special case of subjectivity - none. You most nearly know this form of statelessness in a total lack of focus; beyond a scrap of ignorance you lose all knowledge in the darkness of deep sleep. This passively subjective form of void is a subtendent blank, a comatose 'experience' of nullity. In organisms it is known as dormant sub-consciousness or unconsciousness but in non-conscious matter as the state of pure oblivion. A straight line is the special case of curvature - its total absence. By analogy matter is the special, 'flat' case of consciousness - its total absence; or it could be seen, conversely, as a total fixity of mind.

A second sort of subjective subtendence is conscious. It diametrically opposes knowledge of Enlightenment and, as such, is the negatively active form of void experienced as subjective darkness, violent constriction, isolation and depression - that crucifying singularity, a black hole of pain. Hell, no! Such a void derives from either natural traumas such as accidents or illness; or from devilry's 'unnatural' inflictions. In psychological terms we identify a focused intent to destroy well-being using fear, pain and death as the work of a devil. Agents of extreme suffering are called demons and their chamber of perversity is hell. A fiend is a state of mind personified. So demons can and do occur, of course, in human form. Either way, pain and suffering are a darkening experience. The depth of hell is more experience than place. Its theatre is the petrifying grip of terror, a psychic 'black hole' of depression and despair or the immobilising 'freeze' of utter, lasting pain. Time is slowed or 'elongated'. It has stopped. Not disappeared but stopped. Not flashed by on the uplift of enthusiasm but straitened, hard, with burden. Hell hath no future nor, definitely does it wish you well. Its despair is desperately avoided. It's the kind of real unreality nobody wants to undergo - at worst life's only light, its hope and prayer, is for extinction by a mortal blackout.

Is blackout boundless negativity, neutrality or, for torment, positivity? If there's no absolute subtendence then Zoroastrianism's incorrect. The demon Ahriman may powerfully exist but isn't any more eternal than a parallel distortion, Catholic damnation. That is to say, both (*tam*) negativity and (*raj*) positivity will last as long as the existence they mutually sustain; and if it started and is finite as the objects that compose it then existence (*aka* creation) isn't infinite in time or space. Even space, time, archetype and natural law, not being Infinite, have existential starts and stops. Ahuramazda's epic, oscillatory struggle with malicious Ahriman, that represents 'eternal' war between the restless forces of (*raj* ↑) life-giving light and (*tam* ↓) light-draining shadow-blacks, is existential not Essential.

You can see by now existence as degrees of oscillation in between The Blessed Void and petrified subtendence - between The Infinite and finite voids. Then is the emptiness of consciousness or matter? Which is immaterial, what doesn't matter? It depends on viewpoint. Along the conscio-material gradient range relativities of combination, energy with information, from Void until you reach material fullness, that is, subjectivity's extinction. Complete materialism presses consciousness away; it pulverises thinking into nervous dust. For a Buddhist, on the other hand, the existential gradient is incomplete. Its fabric is composed of motion, change, impermanence. He uses the word '*dukkha*' to describe the way its condition is reflected in yourself. *Dukkha* means relative suffering/ happiness or imperfection. Buddhist logic and practice are, in essence, a remedy for the relative pains (or voids) of material existence through achievement of an Immaterial Void - *Nirvana*, Plenitude of Absolute Enlightenment. Nobody claims *Nirvana* is material and thus materialists, in ignorance, eschew its Void.

Because the *(tam)* negative power is a creator of voids from Void, it is, as Chapter 26 explains, a necessary 'evil'. This materialising power is present, interacts and, as the opposing pole, substantiates every finite part of the finite localisation of Infinity called creation. It is an existential staple. Resisting flux its force locks energy and licks its mobile patterns into fixed and lasting shapes. Without fixity (of memory in mind or hardness in material) how could you obtain a semblance of stability? How could a human or whatever else endure? Interwoven with awareness, various kinds of gravity build condensations; these, like clouds, are shadows slung across creation's sky. The tension of resistance shapes a flow. What use is film without its negative development, without its shades and colours conjured from transparency? You need (↓) brake with (↑) accelerator; so the negative, inertial pole is a critical and extensive aspect of the vehicle called creation, of a projection called existence. As amoral as the pole of a battery, the moon or female form, it is simply half the world's duality. So that, except for its hellucinogenic demons, negativity resists. It opposes. It is action's foil or a container but not an evil in itself.

Ups and downs, electrons, protons, energetic fluctuations - even of some as yet undiscovered sort of matter - are restrictions of one sort or other. Are any of them permanent? What, in bendy space and time, is infinite? Physic's 'real world' is composed of finites so 'real infinity' makes nonsense. Could, however, an 'event horizon' mark the spot of absolute subtendence? A black hole is an absolutely boundless mass in absolutely bounded space. These infinities make nonsense but still might be real. Could, past the point of no return and towards the massive centre of a singularity, space, time and physics all at once be broken? What, if anything, is on the other side? In all of nature nowhere's as extreme. Think, on the other hand, upon a million billion stars in just as many galaxies all emergent from material subtendence that, bang, explodes the other, levitatory way. How could physic of a vacuum generate, from its non-conscious absence at the world's base-pole subtendence, a point whence flies the awesome universe? Whence could a singularity obtain such energy and information?

Perhaps an eternity of mind exists; or one of matter or of both. What on earth's behind the world? *Top-down*, you'd ask what sort of concentrate is hierarchically in front; or, if you like, above and not below. Things are not issued up from absolute subtendence but down from Top Potential's Superstate. They drop from Source not rise from sink. From this perspective, from beyond the actuality but not the archetypal principles of physics, cosmos is orderly devolved. Creation's a dependency of Conscious Presence. It all amounts to a projection, that is, movements of Transcendence.

To summarise, Transcendent Void is Essence (Being) out of which the conscio-material range of objective voids are brought into existence. The relativities of creation are thereby each related to their Origin, Subjective Plenitude. Diametrically opposed to Essence exist some extremely negative polarisations. *These subtendencies are due, in the informative case, to severe loss or paralysis of consciousness and, in the energetic case, to extreme loss or confinement of energy.* Lock up. Imprisonment. And yet *(tam)* negative extremities are localised. They come and go. It is not that imperfection holds Perfection; change-prone relativities do not contain The Absolute. There can be neither Subtendent Purity nor Perfect Negative where, in cyclical existence, all is relative and nothing lasts forever!

Transcendence.

Materialism's lexicon does not include the word 'transcendent'. If everything's material then absolutely nothing can transcend it.

	tam/ raj	Sat	
	existence	Essence	
	subtendent creation	Transcendence	
	lesser	Supreme	
	lower	Highest	
	relativity	Absolution	
	stream	Source	
	manifestation	Potential	
↓	tam	raj	↑
	downward	upward	
	lowest/ lower	higher	
	outer	inner	
	inferior	superior	
	gross	subtle	
	fix/ sink	flux/ stream	
	materialisation	dematerialisation	
	creation	dissolution	

Now we're on the up and up. Check *fig. 2.15*. Can you let slip the cycles of existence? Transcendence, unlike sub-tendency, extends above. As such it is the source of what has sunk beneath; its spring defines the nature of the cosmic course that streams below. Can you bridge this course, transcend polarity and skip the currents of this world?

Allow that, in a polar universe, the element of conscious mind is one apart from energy. Material and immaterial, the cosmos is composed of both but the internal workings of material systems have no meaning by themselves. The meaning of non-conscious systems is 'outside' themselves; mind is 'beyond' the systems it observes, describes or makes. Mind transcends material systems. What, therefore, transcends the starlit system of the universe but universal mind 'beyond' it; and what transcends the finite, existential worlds of mind *and* matter but Essential Infinity? This One transcends its multiplex creation. Thus, where materialism levels all things into its own flat-earth dimension, stepped order flows *top-down* out of an Apex of Pure Consciousness. This Transcendent Apex gives creation reason and its meaning. And, by reversing *top-down* flow, recycling might uplift you back where you began; ascension might deposit you On High; or, put another way, concentration (known as meditation) might obtain The Central Spindle geared to which cogs in the world machine rotate. From this agenda parts fall into place, aspects of perennial philosophies are galvanised. Whether you agree or not, you will have spotted that they all self-organise around this, a Jacob's ladder that inclines from heaven to earth and back, the conscio-material principal.

In material systems energetic flux transcends solidity; 'subatomic' particles transcend their capture into atoms and bulk aggregates of matter. *As energy transcends mass, quantum transcends the framework of classical physics.* Force and object, subtle (field) and gross (visible, sensible) materials exhibit different grades of character. The different things you see are aspects of the same atomic substrate; but the quantum world is hierarchically (not spatially) above the classical. It is causal, comes 'before' and is interior; and bulk effect that follows is exterior. *This perspective reinforces a profound sense that cosmos is 'ontologically-layered' in the way of a proverbial onion.* **The cause of each layer on its conscio-material gradient 'arises from within' and, identically, 'descends from' above.** Upper source is the potential for the outcome we call lower sink. So we say, for example, that the atomic or quantum level transcends the gross material; the latter 'descends from it'. Indeed, Karl Popper only slightly incorrectly dubbed quantum physics the 'transcendence of materialism'.

Arrows on the Dialectic's right-hand columns (↑) indicate ascension. 'Higher' is above; it transcends what's down below. Mind transcends matter. Indeed, matter itself might be viewed as 'solid' or 'crystalline' mind; and metaphysical mind, which is a physical nothingness, as a material void. When energised things naturally dematerialise towards apparent nothingness (as solid to a gas or matter to pure, mass-less energy); they 'dissolve' towards potential and don't solidify to impotence. Is it not ridiculous, however, to suggest that automatic matter might evaporate to mind? Isn't physic's ceiling made of 'glass' of space; or pure transparency composed of light? There's nowhere (you can see it plainly) else to go.

In this light you might enquire whence, at the start of a material universe, did all its power and glory fly? What transcendent form of matter threw the world away? All sorts of calculations weave the maths across a savant's desk but could mind be crystallised in matter by a mould called archetype? The possibility will be explored.

Action rises to extreme. Most high-powered is what you mean. How, though, might this apply to physic's source (projection from a singularity?) or, as opposed to brute mass, subtler quantum entity? 'Potential matter' means 'where matter comes from'; what might interface with its transcendent nature thus linking physicality to what it's not? Transparent, insubstantial - who can shed some light? Do you think a photon you can't squeeze between your fingers isn't powerful? Almost immaterial, that flies with absolute velocity, concentrated light can cut through steel. Communicative, information-bearing - you can signal using light. 'Grounded' it transforms to heat; its 'nothingness' drives life on earth. (↑) Ascent translates, in dissolution, to release, boundlessness and (no matter in the wave) individual annihilation. Isn't this the character of mass-less light? **Orderly electromagnetism is, for Natural Dialectic, physic nearest unto metaphysic - that is, transcendent or potential matter.** Light would constitute a medium for archetype. Thus the Dialectic would suggest transcendence is (material to mental or mental to Supreme) well reflected by a spectrum of illumination.

tam/ raj	*Sat*
dependent	*Independent*
imperfection	*Perfection*
life form	*Life*

↓ *tam*	*raj* ↑
less perfect	*more perfect*
inanimate	*animate*
involuntary	*voluntary*
fixed mind/ automaticity	*conscious flux*
subconscious instinct	*cogitation*
stored pattern/ memory	*presence*
non-conscious/ unconscious	*wakefully cognitive*
blank/ oblivion	*knowledge*
passive, automatic body	*active mind*

And so to metaphysical phenomena. Mind's set is, like matter's, subdivided. For example (*tam*) sub-conscious, (*raj*) conscious and (*sat*) super-conscious phases of the mind exhibit, like solid, liquid and gas, different characteristics. Nothing is anything to (*tam*) inertial sleep; sub-consciousness seems nothing either when I sleep or wake. Is this mundane transcendence, walking over sleep, as far as waking goes? Are you fully woken up? You think so but are the states of your own mind the only possibilities? Perhaps, in the energetic manner of our current universe, they are relatively low and slow and sleepy; you are alive in mind but when compared with Brilliant Life perhaps closer to the body-world of death. Could further, voluntary awakening transport you nearer to Transcendence; might it lift you up the grades, across constraints towards excellence; and sweep you to The One and Only Infinite? In such carriage human wakefulness were left as far behind as it leaves sleep.

While (*tam*) absolute negativity is, on dialectical terms, impossible, it was noted that (*raj*) upward ascent translates, in its dematerialisation, towards release, boundlessness, annihilation. Upward translations can occur at all levels of existence but is there an Absolute Positive, Perfection, a Real Superlative at the Top? Is the tendency fulfilled by (*Sat*) Infinity, Nothing and, therein, Primal Unity?

Firstly, recall that a sense of range includes dependency. Lower depends on higher source. As solid 'emerges' from liquid or gas, so gross 'emerges' from subtle. Universally, existence 'emerges' from Essence. It is like a crystal hanging on the thread from Nothing. The only Complete Self-Sustenance, the only Total Independence is with Essence. Thus as dependence on the Absolute increases, so dependence on the changing round of existential needs and relationships decreases. Such independence from the world makes towards True Independence.

Secondly, as the involuntary, reflexive condition of mind decreases so the voluntary, reflective increases. Positive characteristics, which now include moral and aesthetic valuables such as goodness and loveliness, intensify. What was beyond and super-natural becomes reached, normal and natural; what was super-conscious is now known and, thereby, understood. Mind dematerialises through selfless principle towards Nothing. Mystics subdivide what is still, for you and I, super-consciousness into further levels as it grades towards the Absolute, Overall Superlative. Of course, with mind goes egotistical self; unclothed from self-attachment there shines, evolved and world-free, naked soul. There is no individuality in the diametrical opposite of black holes called

Transcendent Singularity. There is no duality in *Nirvana*. Translation, transformation and transfiguration will have taken place. You, as you know yourself today, will not be there.

	tam/ raj	Sat
	matter/ mind	Transcendent
	outworking	Ideal
	comparative	Superlative
	relativities	Full Life
	scale of lesser positivities	Summum Bonum
↓	tam	raj ↑
	decreasingly positive	increasingly positive
	sinister	dexter
	dark	light
	closed	open
	diversification	unification
	isolation	togetherness
	hate	love
	lie	truth
	confinement	freedom
	matter/ body	mind
	more death	more life
	sublative lifelessness	life-in-relativity

In an existential context comparatives swing either way. In dialectical terms *raj* tends towards positivity, optimal conditions and superlative solutions. 'Best in the class', 'very pretty' or 'most high' are called *super*latives and, where *tam* swings the other way, it falls to *sub*latives. Who wants a sublative black hole? Who needs misery, miasma or, at lowest, death of life? In a natural but often ignorant, error-prone somnambulation we always want to minimise the negative and maximise the positive. And while (↓) descent to absolute subtendence is barred, ascent (↑) to Absolute Transcendence is not. The latter is, in fact, the lightest and most freely conscious non-condition possible. The climax of existence, its topmost extreme, is a Superlative, Optimum or, as Plato called it, *Summum Bonum*. Optimum is a reflection of perfection, a gauge of brilliance and criterion of truth against which lower changes and degrees are measured. You can even have transcendent cake and eat its slices; reflections of the *Summum Bonum*, slices of the Optimum are cut throughout existence everywhere it seems that optima are shining.

First out, last in. If the first expression of the Absolute was First Cause, the last stage of return will know The First. This primary current of creativity is called *Logos*, Holy Name, Christian *Word*, Koranic *Kalam-i-Illahi* and Sufic *Kun*. As the first vibration from which creation emanates it is Sikh *Shabda* and the Hindu *Paranada* whence *Om* and other *Nadas* (sounds) descend. *Raj* moves towards the Best. The final and most important transcendence is logically from existence to Essence. From something to Nothing, self to Self. This step, beyond even the 'living sound and light' of First Cause, is variously called Communion, Release or Enlightenment. Of what does it consist? Primary Dialectic indicates Pure Consciousness. Have you experienced its nature?

In Pure Consciousness there is no other than itself; and there can be no death in Pure Life. Indeed, *death from creation's relativities will only leave this Essence.* **Such logic implies, simply, that death is not absolute but life is**. *It is logic worth pursuit* (Chapters 13, 18 and 26).

By now you will have grasped the flavour and dynamic of the Dialectic. Within existence a series of polarities diversifies into appearances called things; derived from essential unity these polarities are, therefore, at root dualised aspects of the One. From this framework, from these poles or opposites the dialectical framework is constructed. It is tight-knit. Well-rivetted by bolt and counter-bolt. It is simple, natural and relates the pattern of creation. It relates the principles and conditions by which the unconditioned manifests. In the symmetrical balance of its opposites we find the basis of beauty and, which is poetry in motion, reciprocity and harmony.

In fact, there are three cosmic fundamentals and so rather than two there are three sides to hierarchical creation's coin - descent, ascent and point of balance, axis. **Primary Dialectic** represents the character of (*Sat*) Original Balance. **Secondary dialectic** uses arrowed vectors to represent the principle dynamic in the way existence works - (*tam*) from and (*raj*) towards Transcendent Centre. A scale tips. The balance swings.

Because its triplex logic is embedded in all things, poets and playwrights have always intuitively grasped it. No less can you.

The columns are its backbone. The philosophy's three-way motor is now up and running. Its polarities represent not only human but the cosmic spine. At its head the central *(sat)* cosmic fundamental represents the origin, source and governance of things. Existence is seen as changes in aspect and proportion of these three fundamentals. **As such Natural (or Polar) Dialectic is a theory of absolution, proportion and relativity. It represents, like your body, natural philosophy**. The Essential component is consciousness, whose information guides and rides an existential system. The nature of this systematic beast comprises metaphysic (mind) and, incorporated in the body's axial, spinal column, two-way-streaming nervous traffic. Information's the potential that precedes a course of action; mind precedes matter. Both subjective and objective sides of the same coin, life, are included. The system includes, therefore, both science and theology. **It generates an abstract, metaphysical machine, tight-knit, well riveted by bolt and counter-bolt, the simplest working model of the universe.** Or, in more human terms, it generates a muscular body of philosophy that accurately reflects the order of the cosmos.

3. Truth, Appearance and Reality

Are you grasping how the Dialectic works? This book is like a prism that discloses how the One Light is dispersed in different colours of polarity, in various shadows of duality; and how your being and your body constitute a real prism of that same dispersal.

If, therefore, you think the previous chapter was no more than academic exercise, think again! *The system of Natural Dialectic is not so much an abstract reflection of the way things are as an application program.* No single book can address every question that arises in the world. However you are now in a position to construct your own combinations and permutations in columnar form. You can use dialectical stacks to relate opposites, spot relationships you had not seen before and resolve paradox. It may clarify your own position on matters both trivial and fundamental in character. For example, it may help you see how you can best deal with a situation or discover the place of something (including yourself) in the scheme of things. It could give a clue how to fit pieces in the jigsaw puzzle life has set to solve.

If the insights dialectical reasoning throws up ever jar with your habitual or preconceived perspectives, the robust structure of the system should allow you to resolve such conflict in a consistent way. *It should help straighten out faulty reasoning and engage a better perspective.* Again, if the scheme is right because it accurately reflects reality, its instrument of logic becomes a powerful tool for making new connections, problem-solving and prediction.

It is almost time to tackle the description of a *top-down* dialectical perspective on its basic existential components of information and energy. This involves inspecting the act of creation (informative input or authorship) and acts of energy (physics and chemistry). The description proceeds, in Volume 1 (Books 2 and 3), to include a polar appreciation of psychology both as regards subjective experience and, from the 'outside', an objective study of embodied awareness; and of biology, the study of aware embodiment. It will conclude with an overall look at our natural and human communities - ecology and then religion, law and politics. *Firstly, though, we are in a position to address the nature of three interesting subjects - paradox, reality and, as opposed to science, conscience.*

Paradox

Paradoxes are a box of tricks. They are apparent contradictions or absurdities that may, in some cases, turn out true. As opposed to real division, poles apart, a paradox is capable of resolution. In some cases resolution comes from disentangling hidden error, ambiguity or unstated assumptions; then the paradox may disappear. In other cases elements previously conceived as contradictory, running counter to some current orthodoxy or just plain incompatible are reconciled by fresh science, groundbreaking philosophy or an individual seeker-after-truth. This kind of resolution amounts to discovery. **In fact, if the world**

is paradoxical by nature then to reconcile apparent contradictions might improve one's grasp of truth, appearance and reality.

What types of paradox occur? Try physical - 'what ends space?', 'is time finite, infinite or both at once?', 'how, as regards the issue of creation, does anything emerge from nothingness?', 'the stillness of a stone seethes with perpetual motion of its atoms' or 'solid matter is, atomically, a ghostly void'; if this species of twister is not tough enough then turn and bend your mind around the quantum physics of our unseen, microscopic world.

There is also esoteric paradox - 'can you be dead (to an observer) and alive (as far as you are now) at the same time as you cross an event horizon to its other side (whether that side is the subjective postulate of an 'after-life' or the objective postulate of an astronomical black hole)?' And mystical paradox - 'before Abraham was I am', 'the more love you give the more you have' and 'die to live'.

What about metaphysical tension as witnessed, for example, in the centaur paradox - 'if a human is both a spiritual being and an animal how can the tension between each part's desires be relieved?' 'Does mystic practice foster a split personality or, as intended, unify?' What, furthermore, of paradox of relativity - 'is this light dark?', 'is this male relatively female?' and so on. Certainly the stacks of Natural Dialectic are composed of opposites and it would seem that, wherever two are juxtaposed, the capacity for paradox is either latent or exposed. A dull example of relative paradox resides in the oxymoron (condensed paradox) of black whiteness or, if you like white blackness. The trivial absurdity is resolved into the truth of 'shades of grey'. Shadows may not make you jump but real tricks erupt, sparks of electrifying reason really fly when you invoke infinity. What of the nonsense Zeno spouted so that motion would not seem to move? What if, in an infinite world, infinite reflections could exist of anything that could exist - including you? How can you resolve the tensions stretched between the different kinds of mathematical infinities, potential infinities or perhaps infinity that, always adding or subtracting, you can never ever reach? And how, if at all, can your numerical infinity be reconciled with Real and Cosmic, Absolute Infinity?

Paradox seems hidden, just below the obvious surface, everywhere. The deeper forms shine out with relevance and truth. For example the pithy wit of Zen Buddhism and the Chinese *Tao Te Ching,* wherein Lao Tzu asserts the 'formless form' and 'non-existence' of essential, central *Tao,* are each replete with paradoxical descriptions of the 'indescribable'. What, for you as well as a Buddhist, is the difference between illusion and reality? Perhaps the greatest paradox of all revolves, as we may come to see again, around the nature of that Absolute Infinity. Can, for example, endlessness be bound? Could boundlessness exist, like space, within such bonds as we call things? How is the Infinite made finite? Does One God sustain the world of finite things, stand distinct or both or none? How can humans ever comprehend, in mind, such Liberty as boundlessness enjoys?

Infinity with Unity and Nothing certainly emerge as central dancers in the Dialectic's theme. But here's another tricky one. Is the future predetermined, as Pierre Laplace would say, by the way the present is configured? So that, since the way that time's first moment was configured, everything in cosmos is imbued with fateful destiny? In which case determination overrides free will. Another paradox is, therefore, 'Have you any choice or none? Have you (as you think you have) free will or is life predetermined and such will illusory?' How ever can determinism and free will be yoked?

Paradox, embracing science, is not always welcome or well understood. For example, physicists debate the status of infinity in finiteness. Is its appearance like a red light flashing that a theory isn't right or, in other cases, could it be a 'thing' that's real? And what about wave/ particle duality of matter or the notion that, in fact, fixed mass and fluent energy are really just the same. Only recently in history was this, and thus the Heracleitan paradox of flux in fixity, proved true. Now try 'appearance differs but the underlying truth remains the same.' Is this a universal generality? If the Natural Dialectic's structure accurately reflects an actual, polar structure of existence, both psychological and physical, then universe itself is rooted in the twists and turns of paradox - which thus is bound and bundled up within our microcosmic lives.

How paradoxical is life? Is mind a form of matter or is matter the precipitate of mind? Is life physical or metaphysical or both at once? How many of these three will turn out true? Is the assertion that 'death lives in life and life in death' simply biological or more? Either way, what does such revelation mean? Is it vapid or profound? In cases where the metaphysical is reconciled with physical, paradox is sometimes called antinomy. For example, philosopher Emmanuel Kant recognised an apparent contradiction between, on the one hand, a need (especially in biological science) to indulge in 'teleological-talk' of final causes and design and, on the other, the unacceptability of such talk to objective science. Therefore must not chance have 'accidentally designed' phenomena? Or, precisely to our living point, did luck compose the world of complex systems that illumines every facet of biology? Can senselessness evolve the sharp and purposive precision codes incorporate and bodies always radiate? Kant's famous antimony is Natural Dialectic's too. So is the important one between perspectives that are able to draw opposite conclusions from an observation of the same object or event - *bottom-up* and *top-down* ways of looking.

	tam/ raj	*Sat*	
	states below	*Transcendence*	
	apartness	*Wholeness*	
	each side	*Paradox*	
↓	*tam*	*raj*	↑
	analysis	*empathy*	
	intellect	*intuition*	
	partition	*merging*	
	classification	*unification*	
	logic	*love*	

One problem with paradox is that, in order to fully understand what is polar, both sides of a whole picture need be held in mind. Too often, though, human thought is single-minded in its politic; it is partial in analysis, emotion and its animal agenda so that we tend to take one side of a circumstance to the other's detriment. Reasoning is a serial rather than an encompassing, intuitive process.

A thinker tends to choose one line and, if not careful, become entrenched and swing towards its extreme. If, on the other hand, you care enough then you might take the other side, weigh things up and come to a wholesome, neutral compromise that rests upon the pivot of a 'middle way'. Rooted in self-interest it's easy to be disinclined to balance others' sides of the equation or to reconcile opposing views. Who wants, anyway, to boss the world about? It's quite enough to be one's own life's referee.

We shy from a life of compromise. Such resolute balance involves a condition of poise which, in a world of action, might seem like indecisive 'sitting-on-the-fence'. Is poised decision or detached behaviour really possible? We are met with or create problems that we have to solve. We have to weigh up then decide - and in so doing tend to lose poise, take a party line and act in a self-interested way. Sometimes partiality is of no consequence but, in the politics of life, its divisions often are - certainly where moral issues rear. Indeed, how can you be 'detached' in choosing 'right' from 'wrong'? You might, a monk would say, become detached from worldly business through the path of righteousness. Perhaps only love, replete with paradox, can rise above the fray.

The truth of a matter may seem to slant one way or the other but, from a wider arc of view, it may transpire that such truth does not tend to one side but them both at once. The paradoxical wave/ particle duality is an example. The relationship of mind with matter is another. Whereas a mind is habituated to thinking of things in a single, fixed or 'static' way, it is often difficult to frame paradoxical matters in a dynamic, complementary way. Reason struggles. Judges hear from both sides of a legal coin. Sometimes this is intuition's break. A contemplative focus or a flash of inspiration yields an idea, answer or '*eureka moment*'.

Have you never had a hunch? Who has never known intuition whether in a vague or sharper form? A hunch is, like paradox, a condensed form of thought - in fact no thought's involved. A 'lower' form is born of emotional empathy and the 'higher' of a contemplative focus. They have in common concentrated attention; both involve thought-in-principle rather than serial, intellectual entanglement. Of the two varieties contemplative intuition is often the outcome of deliberate, sustained focus. A realisation, usually involving principle rather than factual detail, spontaneously surfaces. Intuition sees 'through' or 'past' superficial, localised behaviours to find what causes them; and thus, with a measure of wisdom, generalises. It 'gets to the root'; underlying issues are revealed. What's revealed may range from personal to universal. It might involve a crime scene, a deceptive person's real motives, emotional telepathy or scientific insight. 'The source of all true science' was for Einstein an emotion that he called 'sensation of the mystical'. What kind of expression might hunches or sensations of profundity engage? Discourse, speculation, mathematics or a hunt for proof?

An intuitive solution is passed to the intellectual domain for development, that is, for rational explanation or execution. Intellect tends to deal with specific circumstance. Everybody has their reasons; like a wheeler-dealer cogitation skims and weaves, ducks and dives and calculates. By what do you judge quality of reason? Which philosopher can sniff the best criteria? No doubt high reason painstakingly unwinds the implications and develops truth flashed on the screen of a 'eureka moment'. Try describing all the details that surround an incident that happened in a second. The statement might take hours. It certainly ties up detectives, courts and witnesses to crime or accident. The business of elaborating sight's or insight's instant into factual context, decanting a 'felt' message coherently down to earth and, by integrating it into some conventional framework, sharing information is a process that may (as many scientists have found) take longer than you think.

Even the, especially after education that extols the use of serial reasoning, we shy from intuition, paradox and 'immediate perception'. Who has faith in thoughtless thought? Who's been taught to trust 'the voice within'? How can you ever prove what's immaterial is right? No doubt that conscience and the world of metaphysic can't be grasped like objects in your hand. And it is difficult to deal with paradox, especially 'deep' paradox, because we are physically inclined to resist the exercise of contemplation and prefer sensible observation, reasoning and test. We are not trained to develop or trust the inward mode of focus that engenders intuition as a standard way to unlock riddles or discover truth; and, with today's cultural slant, who is encouraged to include metaphysical with physical considerations? While 'the proof of a pudding is in the eating', laboratory science is a necessary balance and a touchstone to measure the actual effects that derive from causal, contemplative 'think-tanks'; but its overt business is in fact secondary to contemplation. Action follows thought; empirical research follows questions raised and thought about. When it comes to larger issues science is for many metaphysic's tomb; but does not materialism-trailing-atheism falsely stifle full consideration of the mind and immaterial, human spirit? What on earth, it asks, is that? What's its sociology? It must be part of evolution and survival of the fittest, that's for sure.

The nub of knowledge, information or intelligence is, of course, to ask the right questions within the right philosophical and, after that, factual frameworks. Is your first bearing on this intellectual journey a materialistic or holistic one? If the former one might ask, 'Is not atheism, in denying metaphysic, in denial of its mind? How can you think without the thought you seemingly deny?' How, either way, might you cram nothing or infinity into your sense-making picture frame? These are, mathematicians know, *monstres sacrés* in whose presence intellectual mayhem easily breaks loose. On what leash can mind restrain this couple? If they are either side of cosmos how can anybody tightly rope them in? If they are, moreover, what created something finite how are they contained? How can you box the infinite or picture nothing easily? One strict container, holding two- or even three-in-one, is paradox. Inclusive paradox may better represent a truth than any single point of view. Finite and infinite, physic/metaphysic or subjective and objective - encompassing both sides (or more) of any case a paradox not only better satisfies but, rising, transcends reason's maze. Paradox and intuition, in the hands of proper pilots, fly above the intellectual labyrinth. Silence up here cuts the cackle of the streets. Indeed Primary and Secondary dialectical 'contradictions', 'polarities' or 'complements' are each born of an attempt to phrase things as a whole, to return to original wholeness and thus see each side of an issue at once. Dialectical structure thereby promotes an exercise of intuition that is, in its excellence, wisdom. Chinese sage velly famirial with paradox; he say 'weigh thing up and from best balance draw complomise'. Aim for the Centre. Kung Fu Tse take 'middle way', the 'golden mean'. Such a dialectical exercise involves both objective, scientific materials and subjective, metaphysical immaterials. At once both partial and impartial, paradox and intuition are a sign of lively thinking. They are especially sharp, as Kant well noted, when there is an attempt to include both physical and metaphysical dimensions in the same holistic frame.

Such trans-dimensional paradox is, as noted, sometimes called antinomy. Antinomy means 'opposing rule by counter-statute'. This might imply an irresolvable contradiction in the Manichean/ Zoroastrian way of an eternal struggle between the powers of light and darkness such as was earlier identified as 'divisive, negative paradox'. The dialectical emphasis is, by contrast, on the complementary nature of opposites and resolvable paradox. This is especially apposite to you. You have a mind and a body and unless consciousness is simply a non-conscious, material energy, mind is not physical but metaphysical. *It is, therefore, important to resolve your own apparent duality - the in-built, deep or fundamental antinomy of mind, body, experience and your life on earth.* Without such resolution there can be no full and satisfactory answers to the 'deeper' questions that both shocked and quieter moments sometimes pose. Is there a reason I exist? Is life's whole point carousal? Or is there any more profound and worthwhile rationale? Is there a purpose that could really motivate you, is there meaning dear enough to die for hidden from the superficial, daily whirling of the business carousel? Is there, moreover, life apart from chemical appearance localised in cells and bodies built of them? What happens when a form of life, an organism falls apart? Should science here, at the lych-gate, materialise as undertaker leading to an explanation of interment's soil and biological recycling as the only form of after-life? Or may there, in the quest for answers, come a point where understanding must condense its serial reason and transcend the separate objects of the earth? While intellect can model things like engines, hearts and even atoms, it cannot easily model an experience by using mathematics or an image taken from technology. This sort of truth is invisible, abstract and yet, like happiness, loyalty or love, most real and important. The latter constitute meaning, reason and the heart of life. Negatively in division and positively in union, psychological truths drive human nature. To explain the metaphysical we engage metaphor, poetry, music or, which most succinctly summarises the union of complementary opposites, paradox. Only a dolt takes myth or metaphorical model literally. Only incomprehension measures music in SI units. On the other hand paradoxical, so-called mystical explanation approaches the heart of truth, which is silence. How can silence, as opposed to busy, strident intellect, ever speak an answer? Is there a silent facet of communion? Can it be At Home with Friend as well as home with friends?

In conclusion, paradox is a compacted, more immediate and dynamic mode of thought than serial analysis. Its inclusiveness can, at the risk of seeming foolish, better clown the whole truth than a single point of view. Oxymoron (say, bitter-sweet, love-hate or thundrous silence) slaps opposites close-up but paradoxical expression is most revealing and acutely felt when trying to resolve the apparently opposing principles, perspectives and relativities that occur between physic and metaphysic. Its antinomy tries to embrace information and energy, mind and body, heaven and earth; and each with what transcends them, soul. Because such duality (two-in-oneness) and trinity (three-in-oneness) are basic to the structure of Natural Dialectic, its natural heart is paradoxical. **Paradox is at the heart of science, mysticism and the Dialectic.**

Synoptic paradox. The more polarities resolved at once, the more principles and anti-principles gathered up, the more such paradox is able to reflect the truth of truths. Therefore one route of paradox lies in the complement of opposites and neutralising differences that otherwise would hold apart. What, in the end analysis, is resolution but an understanding, union and, in lively merging, love? Do you want knowledge hard enough? Yet at the same time as it rolls truths up towards climactic super-principles, the more at loss for words an explanation of the paradoxical becomes. The tension is that, while enlightenment drops symbols to the side, you need them for an intellectual explanation. Can you, cry scholars at the university, better our gold standard? Can you transcend our first-class prize, the cup of rationality? Irrationality, however, has two meanings. 'Sub-rational' lurks upon the animal side - below; but 'supra-rationality' (for example, *figs*. 0.9 and 2.8) ascends above the ego and divisions of self-centredness. Such 'rational irrationality' as psychological transcendence is not easy to encapsulate in books; the silence of such wisdom rises after words have stopped.

Dumbness born of wisdom's lack is not so easy either. I clearly remember a brief encounter with a philosophy tutor, Geoffrey Warnock, near the side gate of an Oxford College. After the usual pleasantries I said, "I feel the answer is in paradox". He looked keenly and asked me what I meant. Alas! I didn't think, for starters, to propose infinity. I couldn't answer properly! How humbling! A pity that the Tao Te Ching was not part of his syllabus!

Two Kinds of Paradox

	tam/ raj	*Sat*
	existential	*Essential*
	imperfect	*Perfect*
	impure	*Pure*
	coloured	*Transparent*
	visible	*Invisible*
	obstructive	*Clear*
↓	*tam*	*raj* ↑
	lower	*higher*
	dull	*bright*
	dark-coloured	*light-coloured*
	low energy	*high energy*

It is possible, in this dialogue, to identify two main kinds of paradox.

Essential paradox occurs when Essential is set against Existential Dialectic; that is, when *sat* is set against *tam/ raj*. This relates relative to absolute. Intense and exalted forms of such paradox are found expressed in the abovementioned *Tao Te Ching*. For example, the inexpressible *Tao* 'is inactive yet there is nothing it does not do'. In another paradox nothing is what it seems, and yet it is. You think, for instance, that this book is a book but it might equally accurately be conceived of as atoms or, for the most part, space. You think you are completely real and in a sense you are; you are also, as we'll see (Truth, Appearance and Reality), a part of 'incomplete or lesser reality'. Call it relative reality. What you think you know you are is, in terms of the Full Truth, only part of the story. When set against Essential Reality mind and body are, like all else in existence, Really unreal; set against what Buddhists call the Ground-State (*Dharmakaya*) they are relatively unreal or unReally real! By this paradoxical criterion you, a combination of the pair, are somewhat an appearance - not quite real yourself! Yet you seem to unify material and immaterial entirely happily.

Saint Patrick shook a threesome in his hand to show how paradox might work. *His shamrock signifies a prime paradox of occidental Christian theology, the three-in-oneness of Trinity.* Omnipresence that includes past, present and future is another such paradox. So are the contrary notions that life derives from non-life (in the form of chemicals) and that there is life after death. That matter derives from non-matter (or no thing) is also an essentially paradoxical assertion. What, furthermore, about the nature of that prize experience, Ground Zero at *Nirvana*? What is Ideal Nothingness?

"I've had enough", I hear a cry, "of this incomprehensible nonsense. Negative boredom is a frustration everyone has suffered. How could the experience of Infinite Nothing be, far from ideal, other than Endemic Boredom that would drive its patient mad?" The actual reverse is, paradoxically, how the Truth's reported. If an experience of Pure, Primary or Essential Paradox of Super-consciousness is beyond mind then it is also beyond words. Words and intellect are totally dissolved in its solution. It transcends the understanding of a non-experient. Both dialectical logic and descriptions that the mystics of all ages have transmitted in their various ways indicate that the nature of dialectical Nothing is a fullness of Being. Supreme Being. Its Potential is diametrically opposite empty boredom, inertia, absence or impotency. It is central. It is Life and its enlightenment irradiates the character of right-hand Main or Primary Dialectic. In this respect a 'fully realised' mystic, such as the Buddha, Christ or Ramakrishna, incorporates duality of mind with singularity of Essence. Although of male or female form he or she transcends dualities of sex and ego psychologically; (s)he understands creation, in both grand and trivial aspects, from perspectives human and divine. This, so the teaching runs, is everyone's lost heritage. Can't you understand a child's view and your own at once? The golden state of knowledge is simplicity itself. It is as clear as light, easy as the resolution of all tension and, at once, completely paradoxical. The world and world's source are combined as one. A mystic lives essential paradox. Whether or not the bulk of mankind cares or comprehends, this sort of professor's nature radiates Prime, Living Paradox.

Existential paradox is, on the other hand, restricted to duality. It is described by Secondary, Branch Dialectic and occurs when existential opposites each express in some degree the characteristics of the other; when a whole is analysed in terms of its polarities; and when either's expression is a matter of predominance or tendency. *Yin-yang*. With existential paradox you state the poles (say, happy/sad) but understand dynamic range between them. Manly woman, girly man; light darkness; naughty but nice; these are examples of the simple, almost trivial kind of complementary compact that a scale between antitheses supplies. Such tensions are pervasive in our 'normal' world.

With or without its element of paradox light is a symbol and a guide. Its clarity is coloured. Clear potential run through action's prism shows as spectral colours. A rainbow in clear sky symbolises Natural Dialectic. Transparency irradiates the character of an Essential stack and a graduated spectrum illustrates the conscio-material gradient. It marks the existential stack and, thereby, range of relativity. The range is polarised; its colours illustrate a gradient that tips from right to left and ends in darkness. In other words, a 'gradient of colour' runs from Essence to existence; and it defines the shades of difference between all pairs of opposites that you can write in Existential, Secondary Dialectic.

Switches, Twists or Inversions.

There exist three important kinds of innate 'switch' - one between Primary (Essential) and Secondary (Existential) Dialectic and two within the confines of secondary dialectic alone. They are *'Primary Inversion'*, *'Scale Switch'* and *'Perspective Switch'* respectively. There is also a trivial or *'Semantic Switch'*. Their combined effect is, if you are unwary, to cloud the clear logic of dialectical Stacks and thereby make reading or writing them less easy.

Primary Inversion.

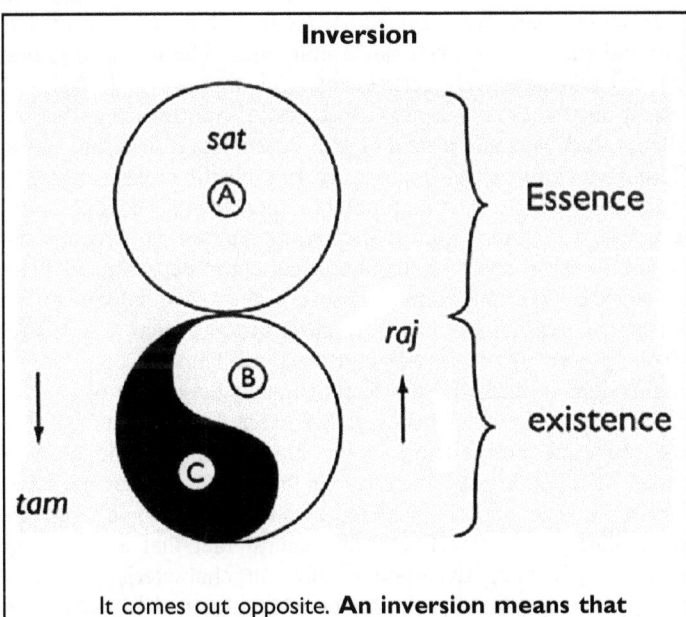

It comes out opposite. **An inversion means that something turns out opposite its origin.** Cosmic poles and inversions sound complicated but can easily be simplified using a 'flipped open' version of the yin-yang symbol (*fig.* 1.1). You can see that A (Unity) antithesises B/C (duality); also, within existence, B (metaphysical mind) counterpoints C (physical body). Glance at *figs.* 1.3

> and 3.2. It is clear that the diametrical or 'far opposite' of A (Essence) is C (the physical universe); in other words the extreme inversion of Essence is solidity and *vice versa*. Finally, these antitheses can be regarded as mirror images, imprints or inversions of each other. 'As above, so below' runs the reflective maxim. Dependent B and C reflect the nature of independent A but, in the process of realising A's potential, this reflection also involves an inversion. It involves the 'organic' kind of development that expresses 'inside information' outwardly. According to its logic Inmost Essence (A) devolves a polar creation whose final stage ends up as non-conscious matter (C). **This, the drop from consciousness to non-consciousness, is the primary, creative inversion.**
>
> *fig. 3.1 (see also fig. 1.1)*

Electrostatic charge accumulates in purple clouds. From motionless potential builds the storm. It breaks and pours and, spent, is over. There is calm before and after cloudburst but of very different and inverted kinds. Is not exhaustion the reverse of powerful potential? From potency to impotence, before is not the same as after. So with the cosmic storm we call existence.

A fundamental character of this existence is, as opposed to (*Sat*) Essential Rest, continual flux. 'All flows', said Heracleitus. The universe is one great bursting flood of change. Objects, events and, therefore, all their different aspects are imbued with restlessness. (*Raj*) restless motion drops, at its end, to (*tam*) stillness. It decays into a state of rest. Peace. But rest-before and RIP are very different kinds of peace. Poles apart. In dialectical terms any 'stillness' of (*sat*) balance, poise or wound-up potential precedes process. It is, although still motionless, a pregnant source. Its latency that we call 'readiness-to-go' causes, when it's triggered, streams of consequent effects. And currents drain to ground in a contrasting form of peace - the (*tam*) puddle stillness of exhausted sink. **A gradient slopes from potent to impotent peace; a balance swings from poised until inertial equilibrium**. Thus, as day's lease gives way to night, light is blacked. Start to finish takes the course of an inversion. The immaterial pole of consciousness is twisted in descent to an extreme, non-conscious, material constriction - perhaps in the form, at base-pole subtendence, of a black hole. In terms of Primary and Secondary Dialectic pole to anti-pole inversion is registered as an *innate* 'switch' from Primary Right to secondary left. It reflects the general fact that a (*tam*) left-hand extreme represents the negative opposite of a (*sat*) characteristic - even though they may appear superficially similar or even identical to each other; and though the same word, such as 'rest', may be misleadingly used to describe them both (see 'semantic switch'). Their true characters are diametrically apart.

A more detailed look at natural inversion occurs in Chapter 24. For now, check out the following stack.

tam/raj	*Sat*
C/B	*A*
primary left	*Primary Right*
range	*Super-state*
polarity	*Neutrality*
inertial/ dynamic states	*Full Potential*
issue	*Source/ Start*
lesser presences	*Presence*
things/ events	*Nothing*
restlessness/ motion	*Poise/Peace*
imbalance	*Equilibrium*
scale down to non-consciousness	*Concentrate of Consciousness*
action	*Pre-active Rest*

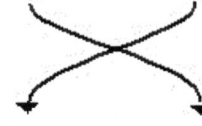

tam	*raj*
C	*B*
secondary left	*secondary right*
sub-state neutrality	*process of change*
negative	*positive*
impotence	*ability-in-action/ drive*
exhaustion/ sink	*action*
inertial equilibrium/ discharge	*dynamic (dis-)equilibrium/ flow of current*
stagnation	*stimulus*
fixity	*flow*
non-conscious matter	*mind*
fixed polarity/ bond	*inter-polar action*
flat/ fallen off	*oscillation/ cycling*
more passive	*more active*
object	*event*
end/ flatness	*range/ lifetime*
post-active stillness/ rest	*motion*

An instance like calm is <u>neutrality</u>. The source of antipodes such as (*raj*) positive and (*tam*) negative is (*sat*) neutral potential. After stimulus a course of events runs to conclusion. In other words (*sat*) 'super-state' potential gives rise to polar action but, as that action is neutralised, it shows as (*tam*) post-active lack of energy. From 'transcendent' to 'subtendent'. Stimulus has worked, capacity for change been spent and potent transformed into impotent. Is this not the way of chemical reaction? After the achievement of 'activation energy' reactants are propelled into electronic interactions that result in 'neutral products' or 'completion'; in the process entropy increases. That is to say, process dies away into 'sub-state' exhaustion, into a condition of depolarised stillness where active polarity is finished, fixed and buried. This is 'descent' of chemical reaction, sleep or, indeed, any type of completion - flat, out, over. Motionless until recharged. An end-point is the opposite or anti-pole of its inactive origin. In other words, pre-active Neutrality and post-active neutrality are, the point is made, quite different states.

Of course, in the stream of events it is hard to draw a line. Action blurs the film as one shot leads into the next. It is simply that (*fig.* 0.7) the Dialectic draws a line that's 'vertical' as well as 'horizontal'. **It involves two axes of causation - energy and information**. This shows as hierarchy that devolves along the conscio-material gradient. It runs from Central Inner Apex outwards down the slopes of Universe. It runs through mind and information out to non-conscious, passive energy - a depletion of all mind that we call 'physicality'. And matter drops between two forms of equilibrium; it falls (*fig.* 2.13) from the implicit order of its archetype through quantum, gaseous and liquid forms of flux to an explicit, fixed, inertial state; it runs from energetic flux out to the border of the world, the outer ring of cosmos, stillness called solidity. If this 'nether' outcome seems immobile that is an illusion. If a solid now seems changeless, that is not the nature of its wider or its inner constitution. No object is, on the wide scale due to the continual motion of celestial bodies and on a constricted scale to the perpetual motion of sub-atomic particles, at rest. Apparent stillness is a partial truth, a relative illusion. Thus you have two opposing forms of immobility, potential and exhausted; you have two forms of peace, prior and after. Hierarchically, one is real. The other, that depends on it, apparent. Are these two physically met in nothing, vacuum, space?

This important point is laboured since it means that, dialectically, set against Transcendence and its First Cause, *Logos*, creation is *apparent*. Phenomena are *emergent*. Or, if you like, Reality's projection comprises a gradient of 'more or less' realities. Thus, with respect to mind, the Archetype of *Logos* constitutes Reality; and with respect to the material division, archetype is its reality. Physicalia are expressed upon a gradient that hits the ground of least reality in solid mass and, most peripheral of all if they exist, black holes. Can this be right? *Could the truth be an inversion of rock-slapping reason, of hard sensibility?* This is radical. It spins materialism upside-down. It would constitute a revolution in its way of thought.

Madness? Unacceptable convulsion? A prescription for proscription, treason fit for exile from an academic state of grace, Natural Dialectic right out on a limb? Well, no. One merely states the view supporting Buddhist, Hindu and Islamic cultures; and the roots of Christendom; thus of our culture, seats of learning and the renaissance; and therefore, despite what scientism might prefer, the roots of modern science! Materialism's show lacks lights!

Things develop from the inside out. Cosmic inversion nurtures them, like seeds, from implicit to explicit, that is, from 'inside potential' to their 'outward, finished form'. Expression is the word. The stillness of the seed is not the stillness at the end-point of expressed development - no further growth, as far as you can press. Upper Cause eventually drops to lower case in its maturity. In energetic terms the cycles turn, the ages roll but paradoxically with age may there not come, in terms of information, increasingly mature ascent towards Maturity, back towards the Upper Case, towards the Source of Information? Is Alpha Omega? Is The Origin an Inmost Singularity?

An act of creation is eventually, hierarchically expressed outside the mind in terms of matter, that is, as a pattern of material behaviour. What might the far-flung limits be? Could they assume a special case of no motion (stillness), no information (blankness) and no energy (impotence)? Examples of these special cases are, respectively, a motionless object, absence of object (space or, psychologically, sleep) and a frozen object. In fact objectless space could fit all three descriptions. Yet if such nothingness is impotent are there (see *fig.* 10.3) any potent kinds? Cosmically, Infinite Potential at the Apex of Mount Universe is totally different from either inert space or material state. Such Void is, however, paradoxically at once an existential absence yet a fullness of potentiality. What kind of void preceded matter in our universe? Did archetypal metaphysic (or potential matter) come before extent of space? Or perhaps the Void's material reflection, vacuum, is two-faced. Could it be an absence of all things yet, at the same time and intrinsically, a fullness out of which each object or event derives its source and substance? Stillness before and stillness after - but the similarity is simply verbal. 'Different sameness' is explained as a 'semantic paradox'; but the realities of potent and impotent nothingness are diametrically disposed.

	tam/ raj	*Sat*	
	C/ B	*A*	
	body/ mind	*Soul*	
	stream of events	*Concentrate of Consciousness*	
↓	*tam*	*raj*	↑
	C	*B*	
	body	*mind*	
	physical	*psychological*	
	outward focus	*inward focus*	
	sensation	*contemplation*	
	reflex reaction	*proactivity*	
	data collection	*interpretation*	
	ignorance	*understanding*	

What, between the extremes of transparent A and opaque C, is the colourful nature of B? Does its range include the psychological and physical extremes? How else can you review a conscio-material spectrum in which subjective mind can oscillate? Mind changes its directions. Check the consciometers (*figs.* 0.1-3); check (*fig.* 0.10 iii) how its focus moves inwards (in case of contemplation) or towards outside (in case of sensation and physical activity). It can even rise to its climactic fulfilment of potential at the root of mind, First Cause. This is the exceptional case; more usual is a fluctuation in and not far out of conscious mind's nadir, its bathos in a loss of consciousness called sleep. More usual also is a fluctuation in and out of passion (anger, greed and so on) while the other form of bathos - concentrated darkness in perversion of intent called devilry - is an extreme more rare. It is normal that our level of awakening, human consciousness, wavers in between the heights and depths. It fields feeling and reflex response but not great emphasis on intellectual effort. Mind embodied is a lesser zenith, a horizon the exigencies of physical survival cramp. It is often not so far from undeveloped, animal condition. But are you just biology?

As your 'world-gradient' drops out underneath the state of dormant mind it falls into the systematic wobbles of biological life, the vibratory cycles of informative archetype and homeostasis - heartbeat, breathing rate and subtler regularities whose feedback contrasts with the uncontrolled irregularities of earth.

At the final stage of 'organic inversion' consciousness has been turned inside out and fallen, therefore, into the non-conscious physic of bodies animate or lifeless. It has turned, upside down and out, into the bailiwick of physics and its chemistry. Conscious concentrate has been completely lost; what started out without non-conscious energy has turned to nothing else. It has become a plenitude of lifelessness. If oblivious matter constitutes the last flight of creation's stairs its final step is turned to stone. The scale from zone of soul to that of pure, non-conscious energy completes the opposition of its poles that Natural Dialectic calls reflection by asymmetry. *This is what, in universal terms, we mean by paradox of twist, switch or inversion.* **Our starry universe is clustered round the lowest rungs of a Great Ladder, the last step-down out of a higher, grander cause - an outcome that completes inversion of, as well as you, the cosmos.**

Scale Switch.

'*Scale switch*' only occurs, of course, within the stacks of Secondary Dialectic - because only this Dialectic is involved with relativity. Such switch is the simple result of scale, range, spectrum or degree. In essence, it simply addresses a 'failing' in that the Dialectic deals in words not numbers. In other words, not all comparisons can be made numerically. 'Scale switch' is the Dialectic's way of factoring scale in verbally. In fact, opposites and therefore the scale between them represent *two* sorts of relativity.

The *first* kind of 'scale switch' involves 'more or less' of the *same* thing e.g. temperature, height or happiness. *The rule, simplicity itself, is 'higher, right'.*

	tam/ raj	*Sat*	
	below	*Top*	
↓	*tam*	*raj*	↑
	low	*high*	
	lower	*higher*	
	cold	*hot*	
	colder	*hotter*	
	unhappy	*happy*	
	less happy	*more happy*	

If point A were higher on a scale than B we would write:

B	A
50	75

but if B were, at the same time, higher than point C, we would have to invert:

C   B

<u>25</u>   <u>50</u>

The inversion of B to accommodate comparison, with higher always right, is called 'scale switch'. Suppose you were actually part of the scale, like a climber on a mountain. Swap your position and the angle shifts; in the motoring metaphor, change gear and you engage another point of view. For example if you were at point B, peak A would be above and C below you. At transcendent Point A everything is below. Translate this into your own experience. Such changes are a normal part of learning; the more you understand the nearer the top and the more overall your view becomes. You have switched into a high gear. What, however, does a Professor of Everything have to know? Is it simple or complex? What is the First Principle which, when followed, leads you from Truth to every other truth? How, in the Knowledge stakes, do you attain an A grade?

Is it clear? One more go. Take the stack above. It represents both states of matter and electromagnetic scale. Now suppose that A were a gas, B a liquid and C a solid; or, spectrally and as related in the text of fig. 3.2, they were X-ray, visible light and microwave. Just the same scale switch applies to metaphysics as to physics, that is, to information as to energy.

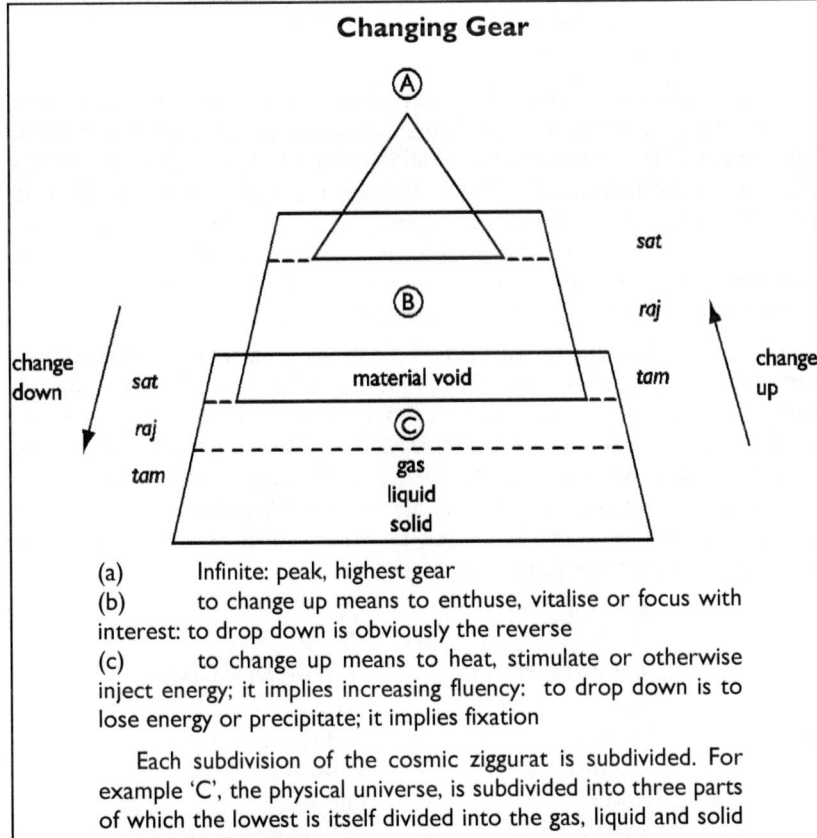

(a)   Infinite: peak, highest gear
(b)   to change up means to enthuse, vitalise or focus with interest: to drop down is obviously the reverse
(c)   to change up means to heat, stimulate or otherwise inject energy; it implies increasing fluency: to drop down is to lose energy or precipitate; it implies fixation

Each subdivision of the cosmic ziggurat is subdivided. For example 'C', the physical universe, is subdivided into three parts of which the lowest is itself divided into the gas, liquid and solid phases of bulk matter.

> A scale switch involves 'changing gear'. It is the way that position on a scale is translated into the columnar stacks of Natural Dialectic. It reflects how the Dialectic deals with relativity, especially with respect to position on the conscio-material gradient, spectrum or scale. **With respect to the psychological world information is the transforming factor; and with respect to the physical, energy.**
>
> Thus, in the Dialectic stacks, scale drops from right to left. For example, X-rays are of higher frequency than visible light so they would be placed 'right' against visible light's 'left'. Compare visible light with microwaves, however, and its position on the scale is higher. It is swapped 'right' and lower microwaves placed 'left'. Changes in relative positions on a scale sometimes (as also in fig. 3.1 ii) mean swaps across columns. You need to recognise a scale switch for, simply, what it is.
>
> *fig. 3.2 (see also 0.5)*

lower energy	*higher energy*
solid	*liquid*
liquid	*gas*
microwave	*visible light*
visible light	*X-ray*

The *second* sort of relativity involves mixture of two different factors, that is, 'more or less' *different* but complementary opposites. It involves proportion, predominance, bias, concentration or ratio in respect of two parts indissolubly entwined to make an interactive whole. The classic example of such a mix is the Dialectic's basic existential dipole, information and energy. Its derivatives include, as bands on the conscio-material gradient, metaphysic/ physic and mind/ matter. *On this kind of scale the rule is 'higher on the Dialectic's vertical information coordinate is written on the right'*.

Does such division imply that energy is, in the way that a solid is derived from its gas, derived *from* information? Or does it indicate a primal division in nature, running from the top of creation, between information (subjective awareness/ *purusha*) acting on grades of non-conscious universal energy (*prakriti*)? Just as the origin of electrical charge as a basic element of matter remains unfathomed, so the advent of primal polarity from unity is creation's fundamental mystery; but the Dialectic takes existence and its conscio-material gradient as a mix, in varying proportions, of the conscious/ non-conscious dipole.

	tam/ raj	*Sat*	
	duality	*Unity*	
	lower (gears)	*Top (Neutral Gear)*	
↓	*tam*	*raj*	↑
	passive	*active*	
	outward	*inward*	
	lower	*higher*	
	energy	*informant*	
	involuntary	*voluntary*	

physical   *psychological*
bottom gear   *various gears*
body   *mind*

Therefore the fundamental components of duality (information and energy) are indissolubly enmeshed in a sliding scale called the conscio-material gradient of creation. They are like gears engaged in a suite of combinations that descends from 'no energy, omnipotential information' (a Concentrate of Consciousness) through the relativity of mind to 'no consciousness, omnipotent energy' (no active information, a concentrate of matter). The question is, if higher is to the right, which is the higher of the two?

For many reasons that this book has stated and will state the conscious condition having information as its element is considered prior, primal and thus in ascendency. High gear, right hand - conscious; low gear, left hand and non-conscious; and in the middle mind relates. So now Mount Universe is motorised! In other words, a change in gear simply implies a change in the proportions of mind/ matter combination. You move up into more consciousness, less material energy; you change down until, in the material world, there only exists 'lawful' energy. Creation (↓) drops from pre-active through active to passive, informative to informed, voluntary to involuntary in ways and with results that the rest of this book describes. Dissolution (↑) works the other way.

In short, just as a 'dialectical switch' can occur when comparing levels or degrees of the *same* thing, it can also occur when comparing proportions of *different* things. It reflects ratios. For example, the complement between mind and matter, in which energy is lower, can be written:

↓   *tam*   *raj*   ↑
       <u>*energy*</u>   *information*
       lower   *higher*
       informed   *informant*
       non-conscious   *conscious*
       involuntary   *voluntary*
       matter   *mind*

However, mental subdivisions can be written:

       asleep   *awake*
       <u>*conscious*</u>   *superconscious*
       sub-conscious   <u>*conscious*</u>

And material subdivisions can, as we've seen, be written separately:

       mass   *energy*
       locked energy   *free energy*
       solid   *liquid*
       liquid   *gas*

In short, the stack rule is 'right side higher'. This means that one character may, according to its comparative position with another on a scale, be written either right or left side.

Perspective Switch.

Perspective switch is nothing new. Is half a glass of water optimistically half-full or pessimistically half-empty? Another common form of perspective switch derives from sexuality. Who thinks a male's attractive, who a female? Your perspective's likely to be polarised. Yet is division primal and complete? Does a scale run unifying, by degrees, the two to one or are the poles at root distinct?

	tam/ raj	*Sat*
	polarity	*Neutrality*
	expression of duality	*Archetype*
	female/ male	*Humanity*
↓	*tam*	*raj* ↑
	female	*male*
	below	*above*
	moon	*sun*

Why, not only feminists might ask, is female placed reflective, passive and receptive? She is surely equal and not underneath. Why are these stereotypical characteristics assigned to sex - female left and male to right? No doubt, different individuals show psychological and physical male/ female attributes that range along a scale from highly feminine through intersex to highly masculine. Even within the ambit of their overall sexuality, variations in aspects of sexuality (such as various inclinations, level of sex drive or emotional agenda) may dapple different occasions in their lives. So which is right?

negative	*positive*
pacific	*aggressive*
soft	*hard*
submissive	*dominant*
feminine	*masculine*

Or the reverse?

negative	*positive*
disruptive	*relational*
less communicative	*more communicative*
harder	*softer*
divisive	*unifying*
masculine	*feminine*

Unless some genetic defect has occurred an individual's psychological and physical sexuality remains fixed in either male or female form; but he or she also expresses unisexual behaviours such as tenderness, cruelty, ambition, shyness, friendliness and so on. In fact, looked at in a balanced rather than a biased way male and female make up two sides, with complementary differences, of the same human coin. One party is not 'fuller', 'better', 'more positive' or 'well-formed' than the other. They are, on balance, equally 'good' or 'bad'. That is to say, their interesting polarities derive from overall neutrality. Sexual polarity is a matter of emphasis and degree; but it is based on distinct poles embraced by the whole super-sexual, neutral human archetype. More of this in Chapter 24.

Sexual neutrality, bisexuality, homosexuality or non-complimentary antagonism in the forms of misogyny and misandry are sexual perspectives.

Such a spread of sexual proclivity around salient male and female norms is central to the lives of the individuals concerned but socially less important than the philosophical divisions which rend our minds and, therefore, communities. These divisions of perspective are, whether a person thinks about it or not, as pervasive as morality. They colour the texture of every thought, decision and behaviour; and, therefore, lifestyle, law and belief systems of any particular group. Such systematic classes derive, ultimately, from how people see themselves. Such perspective involves, inevitably, personal history. Where do we come from? Where belong? What is our place? Why am I here? To answer any of these the past springs into the present. Origins are, unremarkably, fundamental. They colour the whole present. The past's issue is origins. Do you see things *top-down*, where your stance is from the perspective of information; or, alternatively, *bottom-up* where it is from the perspective of energy? Holism or materialism?

You can see from this that the really important perspective switch is cosmic. It derives from where you place yourself upon Mount Universe. From what angle do you see life and the world? From what angle will you derive your interpretations of events and make your moral choices? *Top-down* and *bottom-up* perspectives tend to invert or even negate each other's priorities. Take ascetic mortification of the flesh or, on the other hand, fleshly indulgence combined with a denial of any 'ghost in the machine'. Where's the smart money? What is your concept of mind? Physical or metaphysical? Materialistically (see *fig*.0.6), the stack is really flat. No universal scale is entertained:

	tam/ raj	*Sat*
	aspects of truth	*Truth*
	forms of reality	*Reality*
	lower gears	*Top gear*
	particle/ wave	*Energy*
	material patterns	*Oblivion*
↓	*tam*	*raj* ↑
	negative	*positive*
	reverse gear	*forward gear*
	metaphysic	*physic*
	illusion	*fact*
	thought	*things*
	mind	*matter*

Holistically, however:

	tam/ raj	*Sat*
	aspects of truth	*Truth*
	forms of reality	*Reality/ Fact*
	lower gears	*Top gear*
	conscio-material spectrum	*Pure Consciousness*
↓	*tam*	*raj* ↑
	negative	*positive*
	reverse gear	*forward gear*
	lower fact	*higher reality*
	inanimate	*animate*

physical force	will/ thought
physic	metaphysic
body	mind

A universal scale obtains. It involves the conscio-material gradient of Mount Universe. Holism holds to The Essential, Central Truth. Vectors run the slopes. If Truth is at the Top then (↑) upward vector points you towards it.

material	immaterial
from Truth	towards Truth
less wisdom	more wisdom
ignorance of Truth	awareness of Truth
clever/ manipulative	simple/ straight
worldly knowledge	spiritual knowledge

Perspective, any lawyer will tell you, leads to a particular interpretation of the facts. If there is no metaphysic, holism is wrong; and *vice versa*. The matter of perspective is important because it is the lens through which each scrap of evidence is inspected; it is the goggles through which we interpret and define our universal context. You can't take these 'orientation-glasses' off. So which perspective have you switched to, which lens have you decided to wear? In other words, which end of Reality Street are you standing, why and is it the 'right' one?

Semantic Switch

Of course, linguistic paradox can be brain-teasing fun. A classic example is the case of a Cretan accused of lying. He averred that his compatriots were liars too. Were they? Was he? Try and work it out.

In dialectical terms, however, cases of *verbal* or *semantic* 'switch' cause trivial, superficial forms of paradox. Two or more different concepts or conditions (and their logic) are confusingly subsumed under one banner, one word. Do you remember the unwitting, superficial confusion that a polar inversion called 'reflective asymmetry' might cause? Potent and inertial equilibria were an example. Another one occurs when the same word (say, 'peace') is used to describe both positive (transcendent) and negative (subtendent) extremes. However, far from being the same thing as the (*Sat*) Peace of Poise, Preordination or Potential, the left-hand (*tam*) existential peace is an inertial stillness of impotence, exhaustion and death. Not the beginning but the end. R.I.P. Rest in peace. But the peace of a dead body is quite different in kind and consequence from that of soul. Such inversion is fundamental; meanings are sometimes diametrically opposed. Nevertheless trivially and by an indiscriminate use of language, the same word serves opposite meanings. Use distinct words (e.g. poise and exhaustion) and the confusion clears.

If not, the semantic situation can deteriorate still further.

↓ tam	raj ↑
peace/ non-motion	motion
illness	health
war	peace

Now the *raj* case of 'peace' is political. It means neither psychological tranquillity nor the stillness of a dead body. It means the healthy, stable, ordered activity of life - a social form of dynamic equilibrium. And the peace of non-motion is set with both its own and political peace's opposite, war!

You reach a point where, tri-logically, the word 'peace' can be written in any position. This is a worst case of how semantic switch apparently confounds the Dialectic. If the same word can be placed everywhere it makes nonsense.

↓	tam	Sat	raj	↑
	peace	Peace	peace	
	(static)	(potential)	(kinetic)	

It would make nonsense if in each case it had the same meaning as well as form. In fact the same word is used with different meanings. When these are disentangled we can rewrite the triplex either using the cosmic fundamentals, more accurate words or, in this case, in terms of equilibrium.

↓	tam	Sat	raj	↑
	flatness	Poise	vibration	
	inertial	Potentialdynamic		
	(equilibrium)	(Equilibrium)	(equilibration)	

Potential equilibrium is poise; dynamic equilibrium is balanced action (or a balancing factor) whose symbol might be the vibratory wave and whose mechanism homeostasis; and inertial equilibrium derives from tumble, falling apart, destruction and, at the end, no more action. Exhausted stillness. Life fallen off its bike and dead. As well as 'equilibrium' we shall meet several other cases of semantic paradox that involve important aspects of both essence and existence. They include unity, void, order, creation and dissolution.

For example, oneness or unity has three distinct meanings.

↓	tam	Sat	raj	↑
	oneness	Oneness	oneness	
	separate unit	Communion	unification	
	apart/ alone	Whole	together/ a part	
	isolation	All-in-one	integration	

These three are 'unity of isolation', 'of a cooperative group' and 'of merging or identity with another'. Triplex unity and void (or nothingness) each recruit a chapter of their own (Chapters 9 and 10).

You can also use the normal dialectical duplex to sort out verbal ambiguities. For example:

	tam/ raj	Sat
	dis-equilibrium	Equilibrium
	action	Potential
	division/ unification	Unity
	thing	Nothing/ Void

↓	tam	raj	↑
	down	up	
	inertial equilibrium	dynamic equilibration	
	static flatness	kinetic oscillation	
	impotent inaction	action	
	mass	energy	

space/ nothing/ absence	event/ presence
abstraction	substance
unit (separate one)	uniting (making one)
division/ diversification	unification
creation/ product	creation/ act of making
apart	a part (of a whole)
end	process

Perhaps all this is obvious so why dwell on the dialectical method? Because it is a vehicle to resolve apparent absurdities, remove traps, discover new connections, clarify the process of reasoning and, if it reflects nature, better enter into the real logic of things. In order to drive a vehicle properly you need to understand its controls. The Dialectic is a structure, albeit a very simple one, that reflects the structure of creation. That is no small achievement. It is a tool whose proper use depends on knowledge of its grammar. It is necessary, in a book whose key concepts include 'essential nothing', 'existential polarity' and a 'conscio-material gradient', to understand how spectra, levels, relativities and paradox fit into its columnar construction. When it comes to the ingredients of truth, appearance and reality an insight into the nature of paradox is also an important working tool. In order to read and write logical yet paradoxical Dialectic it's necessary to understand the implications of the various 'switches' (such as 'changing gear') clearly. This is why matters have been made explicit.

Truth, Appearance and Reality.

	tam/ raj	Sat
	existence	Essence
	relativity	Absolution
	from Centre/ towards Centre	Centre
	matter/ mind	Transcendence
	duality	Unity
	things	Nothing
	lesser truth	Truth
	lesser reality/ appearance	Reality
↓	tam	raj ↑
	objective	subjective
	matter	mind
	from Centre/ towards triviality	towards Centre
	lesser truth	greater truth

For sure, reality is what you know for sure. But, straightaway, **a**ppearances can be deceptive. When is an illusion not illusion? Is appearance an illusion or have all phenomena their own reality? Things come and go; they appear and disappear. Are changeful appearances the real nature of existence? Or is there, behind them, Actuality? Is there, at the heart of relativity, a Changeless Absolute or is such notion the Illusion of illusions?

Bottom-up, reality is physical. Everything, including consciousness, is a material phenomenon. This is, although it might be absolutely wrong, materialism's promissory faith. Normality is, therefore, sensory reality, doubt, ignorance and habit; it is not, except by trick, illusion of the senses. What you can slap is evidently real. If you want evidence of any subtler thing then wait till science has divined it. Unseen factors (atoms, forces) have, though still unseen, been seen to play a major part; and, probably, others in the future will. Then, brought within that corpus, new material answers will become self-evident.

Yet Niels Bohr claimed that we accept as real what's not. Quanta substantiate the large-scale universe which, on such terms, is an appearance. Matter's really energetic patterns, grosser views are less than real. Thus truth's invisible not common-sensible. Part of material reality's invisible, not common-sensible; to this extent its truth's incomprehensible. Can brain know it? No doubt, brain is a lens. It cuts to a local portion of reality. It constrains our sense of what is real, it formulates our consciousness and is constructed, you might claim, with such capacity and in the way that an embodied human's *meant* to know the world. Brain filters but mind mirrors and interprets everything according to a combination of contextual memory, interests and purposes within a sensory or contemplative focus of attention. What, as well as smoke and mirrors, is there real? In a world of mind and non-mind are they each as real? Or is there single, simple truth? **At least, the immaterial existence of perception is its own reality.**

Yet, from a material perspective, it is dreams, emotions and perhaps even consciousness itself that are illusory; they are, in fact, just products of the nerves' activity. In other words, just find a 'neural correlate' (the pattern of the way synapses form and fire when using different parts of brain) to find perception's underlying truth. There is no 'self' without its brain. Where electric signals storm like drops of rain, there at grey matter's neural base you will find mind's rainbow crock of subjectivity. Matter is the truth; mind and metaphysic are appearance; physic is self-evident reality.

Cut to the chase. You're on the line. **Since physical events or objects are oblivious of their own reality is consciousness composed of non-conscious material or not?** Though not permitted by the rules of physics its unproven immateriality is every person's base reality. Not physically provable may not mean something's wrong; our personal lives buck mathematical description. **Certainly, if there's reality outside our consciousness we'll never ever know it!** Experience, feeling, knowledge, math and meaning utterly depend on it. Are these unreal?

What is your central being? Is it not experience-in-mind? If this immateriality is an illusion are not you as well? Aren't you a phantom of the atoms and the operations that compose your nerves, a ghost in a molecular machine? Is consciousness and therefore life a pervasive illusion; or is delusion only in the brains of atomistic, 'scientific' fantasists insisting that it is? If consciousness is immaterial *and* real then so, in this respect, are you. The phantom-fantasy is wrong and all that Natural Dialectic elaborates is right. The basis of reality, most real, is pure, uncreated consciousness. No start, no end, no physicality - this primal state, embracing the potentialities of mind and matter, transcends both. It is the seamless oneness underneath duality; it is the source of both subjective information and objective energy - universally. If this click's true and you unwittingly share Absolution then the consequence of such conclusion is, concerning your identity, profound.

Let's snap to another check-point. From a *top-down* angle what is known without a knower? How can the information centre, mind, be an illusion? Is there no certainty that knowledge is a certain fact? Is it just apparent, even false? I think I surely am but am I? Who or what am I? Have I a crisis of identity or is it all a false alarm? According to the Dialectic men are more confused than nature is. Subjective knowledge called experience is forever your reality. You reason, judge and make mistakes; you feel, fall prey to sleights of hand and tricks of mind but mostly get things right; you imagine, fantasize and when the sunny world of sense is blanked the stars come out - hallucination's hells and heavens start to play. Make believe. Believe what's made. What is the prolific and fantastic world of art and entertainment but imagination's fruit? What is it but unearthly mind brought back to earth? The content may not always ring a bell but all the time experience is real. **The experience of a subject is, for him or her, reality.**

Do you remember Alcatraz and the sense-deprivation tank? Induced or otherwise, hallucinations are as real as natural flowers and trees. They just aren't, like the latter, common property - except that you can share a false belief. If consciousness is its own immaterial element then how can mind's experience be unreal? No doubt that there exists a scale of truth and thereby sanity; undoubtedly it runs from ignorance and error to correct and accurate information. No doubt, also, truth and seeming change according to the angle of your vision, that is, where you stand upon the slope of Universe; and finally, for sure, external matters and internal minding both seem true as far as each is real. Resolution of the question has to take a central slant - the conscio-material gradient. On this sliding scale of truth, appearance and reality the measure, more or less, is distance from Top Centre. Here, you might expect, Grand Truth presides. Let's see!

Shall we call as witness science with an adjunct called psychology? Or engage a contemplative mystic to provide his saintly answer and precisely lead us through the world's complexity? Which breed of psycho-navigator will track to The Final Truth instead of relativities, semi-truthful expertise or falsehood? There is no doubt both genuine parties have a complementary role to play.

Truth in the Balance

Top-down

↓	relative truths, illusion	Subjective Enlightenment/ Truth ↑
	change/ ups and downs	Pivot/ Balance
	existence	Essence
	relative (un)reality	Reality
		Archetype
	(epi)phenomena	Noumenon

reflection of Truth
marginal zone of objects, *(tam)* mass-centred body (inc. brain);
the least important non-conscious oblivion of the physical universe
material 'reality'

Bottom-up

> How real the knowledge of electrons in materialism's grey stuff? How do atoms of a brain divine reality? Is life molecular illusion? In accordance with the dialectical perspectives, *top-down* and *bottom-up*, you may have guessed already there are two views of reality.
>
> The *top-down* Centrality is Subjective Essence. In this view existence is 'real' but its 'reality' has the quality of a shadow thrown by light. It is secondary, dependent and, in variable fact, a projection created by a lessening of the Original Light. It is a reduction of the Infinite Truth.
>
> From a *bottom-up* point of view Truth is objective. You look at it outside your mind; it is mass and matter-centred. What you can register with sense or meter makes a real appearance; what you can't does not exist - except for dark energy and matter, exotic particles etc. (see Chapter 12). These are suspected and do not inhabit superstition's zone since they would dwell, if proven real, in matter and not mind. Matter is real. That's the truth; we know it relatively well but not yet, in a *TOE*, entirely. However, subjectivity, a consequence of how your brain has been perchance constructed, is a phantom. Experience is an illusion that a matrix of electrons can incorporate; it is the ghost in truth's 'Material Machine'.
>
> *Philosophical obfuscation, confusion and prolixity follow thick and fast unless these two great anti-parallels are held, clear and distinct, apart. As we have already seen a logical pursuance of either leads to its particular one of the two pillars of faith.*
>
> Such anti-parallel treatment of Truth, Reality and Appearance can thus lead to a belief that there is intrinsic conflict between physical science and such metaphysical 'science' as is commonly practised in the form of one religious faith or another. It also throws each party's definition of 'void' (see Chapter 10 and *figs*. 2.16, 10.1 and 10.2) into stark and opposite relief. If it is your interest to understand the Nature of Essential, Pre-existential Nothing such clarity will make life easier.
>
> For a Subjectivist Void is Truth and the material void of oblivion a hopeless, empty place to search for Truth.
>
> For a saturated materialist, on the other hand, the 'hopeless place' is a 'Holy Grail'; the hope of material truth is with science alone but the Hope of Essential Reality is, as life's Primary and Most Important Idea, less than even an illusion. It does not exist. Therefore you might deem to seek it a deluded and a desperate exercise, a wasteful vanity!
>
> Betwixt the two a mix. Compromise of a hopeless variety (see Chapter 25) is the fudge of vicars who accept the neo-Darwinian, pan-evolutionary explanation of origins; on the hopeful side it is sharply delineated by scientists (or those familiar with the scientific facts), some who accept and others who deny the possibility of 'intelligent design'.
>
> **fig. 3.3.**

Please check through the text of *fig*. 3.3.

The pillar of materialistic faith admits only matter. Therefore there is no ultimate criterion of truth or reality except physical measurement. Internal, immeasurable subjective truths like emotions are deemed a side effect of psycho-chemistry and, as such, relatively measurable by brain scanners, probability theory and various, hypothetical criteria (such as 'neural correlation') proposed by 'scientific gurus' working in the field. In other words, you might decide that a physical object is 'more real' than an idea, thought or dream; and that mind is simply a consequence of special but tangible, quantifiable 'grey matter' and its activities. At root *physicalia* are, in respect of their materiality, equal so that in this sense truth, appearance and reality are the same - material. They are, like morality, what you make of them. Relativities.

Karl Marx believed that in a vast, impersonal and hostile universe one's body is a negligible transient. For such a hopeless speck he accused religion of being 'the false heart of a heartless world'. There is a fraction of truth in his materialistic angle. For the holistic pillar, however, such partiality is defective. It is only necessary if you think collateral that accompanies the acceptance of a natural 'element' of intelligence is unacceptable. If an immaterial element of information does exist then Marx's bleak, grey view of cosmos is not only 'hard' but incorrect. Wrong. Do not, therefore, trust in it or its derivatives.

Holism, you are well aware by now, does not admit that mind is simply a function of special, complex forms of matter and, in essence, material. In this view all things are *not* equal. **Consciousness and matter are basic components of cosmic duality; information and energy are distinct, complementary aspects of existence.** This line of reason leads to a hierarchical or vertically-graded structure within which truth, appearance and reality are evaluated. **If Absolute Truth manifests a hierarchy of appearances, if from Substantial Reality emerge relative illusions and, therefore, things are not equally real, what is the criterion by which one is judged more or less real than another?**

So far Natural Dialectic has interrogated the nature of things using a basic framework of Absolute (called Essence) set against relativity (called existence). The operation of this machinery has been powered by three qualities or, if you like, the stability of (*sat*) balance and tendencies of motion (*raj*) up and (*tam*) down. You could call them equilibrium and (using upward stimulatory and downward inertial vectors) equilibration. We can use this simple, scalar device to try and understand the nature not only of reality but also its intrinsic linkage with truth, knowledge and morality.

Essence, Absolute, Nothing, Infinity and Transcendence are synonyms. (*Sat*) Essence transcends (*raj/ tam*) existence. Where Essence is Infinite, existence is composed of finite objects or events, both physical and psychological. Far from being empty, however, Infinite No-thing is absolutely full of potential. Such potential is, because you can add or subtract as many finite entities as you like without affecting it, inexhaustible. It is greater than any finite part. Indeed, compared with its wholeness, completion and perfection what is finite seems, in every case, relatively partial, imperfect or incomplete. *Such imperfection is, however, graded; its reality is relative; and such relativity is graded for each and every object by its closeness to Reality.*

The Infinite, being Nothing, depends on nothing. It is self-sufficient. Uncaused (there is nothing before or after), it can cause existence; which existence is in this case its effect. The reverse, because even the most primordial effect cannot cause its Cause, is impossible. Thus existence, made of many somethings, depends on a First Cause, a Primal Mover, pre-existent immateriality called Essence. Indeed, existence might be termed reductions, differences or lesser grades of Essence. Think of it as Essence moving, motion from which things 'emerge'. If this is true, Nothing is true - a whole and single Truth. It is the One in which all separate, lesser ones rest as yet unexpressed; it is the Substance on which everything depends. What form has endlessness; what shape a field that's undisturbed? From latency spring lesser forms; from the basis of all information issue the dependent orders of reality. Similarly, just as appearance is a lesser or relative form of Absolute Reality, so lesser truth (or relative illusion) derives from Truth. In Primary Paradox everything is an appearance, an emergence more or less related by its distance from Truth's Apex; and, on the other hand, there's no appearance at this Apex - only Quintessential Nothingness to make Reality.

	tam/ raj	*Sat*
	relative	*Absolute*
	less right	*Right*
	relative illusion	*Truth*
	critical comparison	*Criterion*
	lower qualities/ lesser values	*Quality/ Value*
	constrained states.	*Consciousness*
	death/ life	*Pure Life*
↓	*tam*	*raj* ↑
	lesser truth	*greater truth*
	away from 'right mind'	*towards 'right mind'*
	lower qualities/ separators	*higher qualities/ unifiers*
	quantifiable objects/ events	*experience*
	non-conscious forms (matter)	*conscious forms (mind)*
	individual/ local contexts	*principles/symmetries*
	less real	*more real*
	towards darkness/ oblivion	*towards light/understanding*
	non-life	*animation/ life*

Recall the analogy of creation with Mount Universe or a three-tier cosmic ziggurat. Think of this existential 'mountain' as a power structure whose grades are devolved hierarchically from its Apex. From Absolute Apex we are talking of a sliding scale, a gradient of descent or, as with *fig.* 0.7 or your 'psycho-box', a graph composed of two major coordinates - 'conscious' and 'material'. This conscio-material slope, a duplex composed of proportions of information intertwined with energy, itself issues from Transcendent Nothing.

In such hierarchical view (*Sat*) Primary Reality is Elemental Consciousness. Consciousness is awareness and subjective awareness is at the core of life's experience. Pure Subjectivity, Life, is therefore the Criterion of Truth. We are not at present in communion with The One and Only but, in the truth stakes, prizes won by increasingly positive expressions of life and upwardly mobile awareness of its value and its source; the logic of the Dialectic, since you're living, always points you up. In this case whose experience is, when lesser realities have faded and only Unity is left, of Topmost Absolute? Whose experience of Criterion the measure of all things? Could the philosopher's stone be an experience of Enlightenment; or holy grail communion with the Infinite; could life's elixir flow from intercession by the Paraclete?

Or is obscurity like this not worth the words it's spoken with? Isn't such 'reality', that sets life firmly on the central pedestal, derisible? *This kind of logic certainly inverts materialism's truth which latter has, it claims, itself inverted Immaterial Truth.* In *top-down*'s paradigm the Element of Consciousness is gradually involved with that of energy. You get the drift? On such a gradient of involvement the less a grade is Real the more unreal it is. The less True the greater its illusion, seeming or appearance. Relative reality (appearance) is calculated according to its approximation with (*Sat*) Subjective Truth. This, its distance from the Centre, is an object's value, an event or thought's real worth. What value has, by this criterion, a 5-foot lump of rock? What of one whose quantities are equal but whose quality of shape is known as Venus de Milo?

Top-down, quality trumps quantity. Aesthetic, altruistic or any other selfless quality shines more valuably than contractive, selfish calculation; and radiant quality increases with a tendency towards unity. (*Sat*) balancing and (*raj* ↑) unifying factors tend towards principles and each, in physics or in metaphysics, towards its own First Cause. The highest value and, therefore, reality is hierarchically vested in the Omega that Alpha was, is and will be. The highest qualities reflect the nature of, psychologically, the Archetype called *Logos* and, physically, potential matter of the archetypes at base of universal mind.

Check *figs*. 2.16 i and 4.2. A pure concentrate of consciousness is prior but, you object again, how ludicrous is that? How 'weak' and, unlike massive galaxies, intangible. How can mind make mountains! Intelligence and information are the functions of a brain and you need molecules for brains. In accordance with a naturalistic point of view you guess that 'energy' kicked off; it's only energy that drives the game. I say, however, that the origin of cosmos is an immaterial element. How do you know the nature of a pure concentrate of consciousness? What is your experience of enlightenment? When normal, human knowing is so weak how do you know the Unmixed, Immaterial Power? The answer is that you do not. Do you judge, therefore, that what you do not know does not exist?

Primal motion of such Immaterial Infinity is energy's first order; the first move of existence is 'a wisp' but also paradoxically its whisper is the spring of mind. From principle to practice, from causal ramifying to diverse effects mind's motions drop along the conscio-material gradient. Subtlety is lost to gross precipitate - non-conscious matter. And, on this line, what 'primal wisp' is source (not sink) of *physicalia*; what is the principal (and not exhausted end) of matter's immaterial fields or other vacuous, sub-atomic motion? Are heavenly or heavy bodies really just precipitates of mind, non-consciousness the nether pole of consciousness and what we call the universe creation's last gasp - objects with all subjectivity squeezed out?

Science now believes an energetic 'wisp' of motion, quantum fluctuation, is the root of physical solidity. **It considers immaterial fields are, since particles and matter rise from these, the ultimate reality**. Fields rule. And, in concurrence, Natural Dialectic asks if immaterial isn't metaphysical, what is? Or is the difference merely a semantic rift? Shape, size, location, motion and the universe of bodies is defined in terms of immateriality. Field is the basic 'stuff'; archetypal field is at the apex of your physical reality.

And of objects chemistry would have you know the lowest state is fixity, least energetic is solidity. *From this top-down perspective matter is seen as the least essentially important and therefore least real grade of existence.* It marks unawareness, the total predominance of non-conscious energy and the zone of oblivion. It is a vale of lifelessness, an empty theatre with a play-less stage. From the point of life's view it is alien, apart and incommunicado. Accordingly Buddhist enlightenment, Christian salvation or Hindu liberation accord matter as low a status in the reality stakes as one of its subspecies, death.

Of course, your body and its brain are material; but is subjective experience a 'thing-physical' or 'nothing-physical'? *Top-down, nothings govern somethings; nothing is the basis of a thing.* As rules to players so a field is, to a scientist, an invisible organising principle. *In the jargon, mind is a field of information whose particles are bodies;* or, if you like, the archetypal 'lines' set in this field are organisers of behaviour; their physical expressions are events or, slowed right down, the objects of our world. *It is not, therefore, that mind is of and in a body - the equation has to be reversed. Body is a 'crystal' in the mind-field.* In the transformation of infinite to finite, the interpenetrative communicator (mind) gives way to an exclusive, separate body. The more concrete an object the less it reflects (*raj*) kinetic and even less (*sat*) transcendent, basic or original characteristics. Light gives way to dark; as the triplex principle of photon, electron and proton/ neutron 'falls' into the appearance of a variegated, sensible but heavy world, so inner truths 'fall' into outer details. The illusion increases, as truth decreases, outward. The greatest illusions are, by this measurement, the subtendencies of freezing space and material heaviness brought to theoretical and maybe real conclusion at physic's last (but perhaps not first) frontier - wild 'black holes'.

In short, there is a sliding scale of truths but also, at the top, a Truth of truths. In terms of information a hierarchy drops from accurate and important through to valueless or none. In terms of energy potential is expressed through action to exhaustion; it is constrained from possibility to single actuality, the end-result. The greatest reality is, as with Bohr's quanta, vested in the (*sat*) invisible origin of any particular expression. At mental level this source/ potential is called *Logos*; at material level potential matter of the archetypes. If potential precedes action then we have a sliding rule for priority, importance, truth and reality.

At this point you might want to know about the axis of the universe; you might enquire the nature of material potential, that is, where phenomena spring from. The logic of the Dialectic points to nothing physical. It has defined and will elaborate (in Chapters 7-12, 15-17 and 19) upon the nature of such metaphysic called 'an immaterial, archetypal field'.

It also, at this point, needs be stated that the notion of 'archetype', being metaphysical, adds no more to physical science than the notion of inventor to machine. You don't need inventors when you work on their machines. Nor, though you note the purposes and reasoning, are you obliged to think of physical constructions as a 'filtrate' or projection' of the prototypes first planned, the basic structures first conceived, the archetypes drawn up and naturally stored in mind. Not that nature's mind is yours; yours is as minuscule a fraction as your body in its universal field.

We have glimpsed a perspective that is, to a sensible, down-to-earth way of thinking, strange. What are its impractical consequences?

Firstly, such a Theory of Relativity with respect to Truth, Appearance and Reality is actually neither new nor strange. For example, Buddhists are taught appearances can be deceptive; the sensible world of trees and people and teacups is a partial illusion. Science now knows that, from the non-sensible perspective of quantum physics, this is correct. Objects can be understood as a web of almost nothing (less than 0.01% of electrons and nucleons) spun at very high speeds in nothing (over 99.99% space). Moreover, what substantiates this evanescent web is vibratory. Are not the Buddhist notion of perpetual flux and the modern theory of kinetic matter in essence similar? In the latter, constant motion and vibration of bulk matter's make-up - quantum elements and atoms - explain such basic features of our world as temperature, diffusion and phase change. In the former, explanation is extended to include a medium of information - mind that also oscillates, radiates and can't keep still at all. No permanence. In terms of Natural Dialectic these perspectives, since they both reflect inward principle rather than detailed outworking, are higher than a sensible one. Quantum is rated higher on the scale than classical, bulk matter; and informative mind above energetic matter. Revise *fig.* 2.15. Fundamentals such as vibratory quantum fields are counted 'more real' than obvious, temporary illusions they create. Their 'primary illusion' is nearer to the cause of physicality. Its effect (trees, teacups and so on) is an appearance, dependency or 'secondary illusion' stemmed from the primary one above. Is cosmos really so graded? If truth's illusions hierarchically arranged what, when physical and psychological illusions have completely cleared, is a Buddhist's Absolute Reality?

At this point might take a cup of tea and thus refresh your sense of what's reality. A second consequence of the relativity scale is that, because mind precedes matter, subjective is more real than objective. Bodies are, like rocks, trees or teacups, part of cosmic 'exhaustion' or, if you like, 'final outcome'. Active mind is more real than passive matter, thought scaled above object, principle above practice. The latter depends on the former. What something means to you has (for you) a higher value than the object itself. Your enjoyment of them is marked higher than the value of the sugar or the tea. And a subjective quality like, for example, loveliness is more real than the object that is lovely. Likewise harmony and beauty reflect, in objects, principles of balance and proportion. Principles rule. Where truth is a function of reality, this seems an absurd way to calculate reality. What an 'arty' way to measure truth! Yet look at, for example, music's popularity. Is its innumerate truth a scientific one?

We already assigned atoms a 'truth-value' that was a grade above sensible objects like teacups. To assign mind's subjective sorts of information a grade above material data is, on Mount Universe, only to climb a level higher. All depends on Top, First Principle. What is its unearthly, non-utilitarian nature?

tam/ raj	Sat
existential	Essential
relative knowledge	Knowledge
degrees of cleverness/ wisdom	Wisdom
peripheral/ trivial	Central Importance

division	*Union*
lesser loves	*Love*
↓ *tam*	*raj* ↑
ignorance	*knowledge*
unawareness	*awareness*
folly	*learning*
incomprehension	*understanding*
enmity	*friendship*

Is such truth, such knowledge simply abstract? Is comprehension important? Does it centrally affect your life? What is Essential Psychology?

Take union. Merging. Do you remember the Master Analogy of Natural Dialectic, harmony that struck to the point you gave yourself - the aforementioned power of music we just heard about? To what do you give yourself? That is, to what do you most willingly attend? What are you interested in, what do you love? In which case love is where your heart is. <u>More lovely is more real; more loving is more real</u>. This is the value, worth and subjective impracticality that drives our schemes. It is the principle of which our practicalities are actually the consequence. Where but in love can greatest loveliness and lovingness be found? Surely an extrapolation up this line isn't going to lead us to the height of mystic madness? Who dares reveal, at the Heart of Universe, a Subjective Source behind existence? A Highest Truth, an Ultimate Reality or Central Communion that might be termed, by those that know it, Love? But how, in heaven's name, can love be universal - not just hippy-like but built into the world machine? Is this not cosmic chocolate with much too soft a centre? Is it not the animated folly of enlightenment?

We've already met this strange inversion of utilitarian and detached, for-its-own-sake intellectual truth! Contemplative mind transcends its roving, calculating 'rationality'. Is, by this criterion, the enlightenment of reason really enlightening or, stuck there, simply making sense of darkness? Where subjectivity meets objectivity in quality and quantity, what a turn-up for the books! It overturns a moneylender's bench and wipes the slate of mathematics clean. What nonsense if your god is money or the real truth is mammon. Thus we conclude that such assessment is unscientific. Topsy-turvy, mad and utterly incredible. How, despite an ignorance of nature's First Cause and the nature of its *Logic*, can you think or, worse, believe that 'Infinite Love' has generated triplex protons (with three quarks), three obvious chemical constituents (proton, neutron and electron) and three normal types of character (light, mass and charge) that make up all the rest? Indeed, isn't this the apogee of anti-science? A depth of superstition well rewarded with complete objection and a dunce degree?

Top-down, a third consequence explains why we think this way of assessing reality is so topsy-turvy. The view from the top of a mountain is perfect. Complete and absolute. The human mind, however, is imperfect. Like truth, knowledge is a function of reality and I am relatively ignorant. Is any child aware of what it does not know? I live in relativity and from a pragmatic point of view this is my reality. Reality seems 'hard'. It is what seems rock-like, solid and most sensible. Like other people, teacups and their tea. Objects seem 'more real' than vapours. Even vapours are more tangible and sensible than molecules and atoms. And mind? What web can catch a dream? Who ever netted thought? No doctrine of the world's impermanence flits through our cast-iron surety of mind but in the flux what stays? Only electrons, protons, forces and their rules remain…for ever? In existence permanence, the Buddhist doctrine claims, is an illusion that endures. So, not knowing or denying contrary perspective, we are stuck in the illusion that our natural world, defined by science, is the distillate of final truth.

The problem is that human perception dwells low on the slope of universe - higher, of course, than teacups and non-conscious physical events and probably higher than any other earthly form of life but still not at the top. Even within the human experience there exist, like a rainbow within the small band of visible light, different grades of perception, different mixtures of influence and tendency, different emphases of interest and focus. On the one hand, embodied in a world of energy and objects, the mind is attracted out through the senses into sensation. Sense data and sensation dominate. On the other hand, as we saw in Chapter 0, a subjective tendency to contemplate, arrange information and gain 'higher' knowledge also exists. Our consciousness oscillates. Within this oscillation some minds are perceptibly brighter, keener to seek truth and, when engaged in this upward bias, less inclined to the passions of sensation. Each scientist and monk defers to the ideal of detached, 'disinterested' scholarship. In a word, some people are more attuned to the fixed, intrinsic universal scale of truth. But if their emphasis, either subjective or objective, differs then so will their perspective. Their apprehension of truth varies and, with it, what they believe is real. Whether they think about it or not everyone wears philosophical glasses and an individual may, through his or hers, assign a value to an object or experience. Because this assignment is 'lensed' its value is unlikely to coincide with its 'Reality-Value'. In other words, the value it obtains from our unenlightened point of view is *not* its value ascribed from the Clear yet Glass-less Apex.

Here a strange paradox emerges. Our assignment of value oscillates according to our unstable, changeful state of mind. Nor do lifeless values measured by a scientific clock or ruler wholly tell the tale. In fact, a value assigned from the subjective perspective of love is more nearly an object, event or person's real value than a value assigned from lower, darker or more analytic moods. This observation is, as every mother and all lovers know, nothing new. In which case whence can Absolute, Fixed, Eternal Value be assigned? Is it from the Apex? From the 'Infinite Point' of Mount Universe all finite creations would be experienced as part of Unity; and, if they issued from a fount of love, be seen in such a way. Topsy but not turvy is the point of Cosmic Love and Understanding that is sometimes also christened 'God-discovery'. In this respect everything, even human faults, must be part of cosmic order; the kaleidoscope of all events must be a consequence of higher logic and a part of Truth.

The scale is absolute but from any point below the Enlightened Source of the conscio-material gradient you do not get the full picture. From angles of relative ignorance one imperfectly assigns relatively loveless, off-centre values of truth, knowledge and reality. For a person that inhabits a world of appearances what, therefore, redeems reality? *The only way an unenlightened person like* *myself can ascribe true values is to take the advice of an enlightened one. <u>The only way I can try to behave in a way that is illusion-free is to aspire to their highest ideals</u>.* Like the student who wants to speak a language fluently but still has to think about his constructions, this is a little forced. He 'ought to' or 'should' do things in a certain way but it does not yet come naturally. His mind has not yet aligned with his teacher's. Faith, hope and trust have crucial parts to play. Where do I find True Teachers? Have perfect saints lived on the earth? Are any living now? What is the path of their redemption? Is it, towards a Cosmic Apex, devotion and the pathless path of Love?

This leads onto the horns of an apparent dilemma. What flag will you run up? To which mast will you pin the colours of your hopes? Who relies on ever-changing boundaries of physics to explain his everything? Physically you rest your faith in brave new worlds pressed into shape by technological advance. Material hope is pinned onto a scientific prow; the ship of international economy is driven by invention, industry and, at least at present, dwindling stocks of fossil fuel. Yet, when you observe the world's condition, is science an unmitigated blessing? Do best intentions never pave a road to hell? Medicine has led towards over-population; natural resources towards over-exploitation; nuclear energy could trigger Armageddon. These are dilemmas that concern utopia; they are struggles that erupt between the brute power of material science and an immaterial code of ethical restraint. Technology, a two-edged sword, has the capacity to satisfy desire but at the same time sweep power-hungry sucklings into engineering their own mother earth's apocalypse.

Psychologically you might, according to materialism's creed, think that atoms are the ultimate reality and human reason is the culmination of some self-evolved materials. And yet how long does reason's body last? How short will you be here? Does physics grasp your central subjectivity, the workings of your mindful self; can study of biology explain psychology and - perhaps - a spiritual dimension; does chemistry make moral sense? Science and technology without morality's restraint comprise, as ecology well demonstrates, a danger zone. Are material and immaterial sides of man's equation balanced well by modern politics? After all, who does the aphorism 'survival of the fittest' most respect? Prometheus played with fire. How sharp should be the curb on irresponsible dominion and human, scientific curiosity? Is there any sort of bridle that will fit?

How long, I repeat, is your body going to be here? Holism estimates that its material is, in truth and even now, a grave from which one needs to rise. Mind and body, a couple with which I habitually identify, are in fact two forms of restriction and two species of illusion. They are a pair of impermanent and relative realities. Appearances. If the basis of one's normal self-identity is, in the whole scheme, non-essential then what could be claimed, essentially, as real? In other words, how wrong is an ephemeral man's perception of the way things truly are? Perhaps wrong but in the wrongness there are seeds and flowers of what is right. In as much as union is real and its loving heat is what we crave, are erring humans absolutely out alone or lost in cosmic cold and dark?

Perhaps you would nail your colours to both masts and thus include the immaterial factor in your sums. The logic with which Natural Dialectic is aligned points beyond an intellectual grasp of things. It indicates, in deep meditation and thereby without the 'noise' of thought-stream, a gradient of return. This means a voluntary, 'negentropic' climb towards pristine awareness at the root of mind. Transcendence means ascent above the changeful business, ordered or chaotic, of personalities and their corporeal attachments. It is hard for western education to accept this reduction in the emphasis and value placed on its curricula. Its specialist degrees involve, of course, an excellence; but neither worldly importance nor intellectual prowess can replace the method of transcendence (meditation) and its purpose (wisdom). **Singular wisdom, it is held, is the middle way involving first and final resolution of a polar world's dilemmas.**

Two Value Systems.

	tam/ raj	*Sat*
	quantity/ quality	*Quality*
	objective/ subjective	*Subjective*
	lesser selves/ personae	*Self/ Psyche/ Soul*
	lesser truth	*Truth*
↓	*tam*	*raj* ↑
	objective thing	*subjective sentience*
	physical context/ bodily self	*mental context/ egotistic self*
	external	*internal*
	more an illusion	*less an illusion*
	quantity/ aggregate	*quality/ meaning*
	value of things	*value in mind*
	numerical/ market/ bodily value	*motivating/ emotional value*
	less important	*more important*
	utility	*beauty*
	using/ abusing	*caring*

Are you ready for a fresh dose of dilemma - this time of another sort? Value, meaning and significance arise from mapping cosmos by experience; each new experience within this context is imbued with truth and truth must, more or less, bear worth. Minds and moods at different levels and with differing objectives make various value judgments; these are coloured by your overall perspective called a world-view - *top-down* or *bottom-up*. The fact is that we use (and confuse) two familiar but distinct value systems. These 'anti-parallels' together comprise the 'ecology' of moral diversity; they constitute the way we size our psycho-habitation up in terms of the two key aspects of our individual worlds - body and mind. One is exclusively objective and materialistic; the other, holistic angle includes both systems and thus perhaps better describes 'rounded' decision-making. It is clear, therefore, that two basic sorts of moral system are direct correlates of the two pillars of faith (Chapter 0); and that individuals and societies play politics according to the tenets of their overriding philosophical presumption.

All agree that mind's for thinking. It's a tool to get you where you want to go. What, pray, are your goals and thus criteria of your success?

And it's agreed that you're not only what you think but what you *are*; however, there's still disagreement over this essential self. Is Being Soul or is quintessence made of selfish genes? Is mind's core quality its immaterial consciousness or an excrescence of illusion ghosted, somehow, out of complex nets of nerves?

We have glimpsed perspective that is, at first glance to a sensible, down-to-earth way of thinking, strange.

Then that is who you are. Amoral molecules cannot beget morality. Man must be the product of rude chemistry. His body's nature is defined by genes and its environmental circumstance; and mind's experience is nurtured by a nervous network of relationships. Your 'self', a complex web of expectations, roles and memories is, like your body, ever changing. Plans are made; 'sod's law' obstructs them. Goals are conceived of; by their yardstick what is thought of as success or failure is accrued. Critically, what is the value calculated? What is, when it comes to goals, a man's most valuable choice? Survival turns up trumps! No doubt that politics and economics rule the body's day.

No doubt for scientism mind, if you've any brain at all, evolved. Self is at root related to the chemistry of genes. No higher power exists. So in a soul-less, help-less wilderness the egocentric rationale commits to understanding 'man alone must forge a human destiny'. The existential view is that, in the process called his life, a non-conformist will create his own ideals. With both integrity and sense of personal responsibility he'll strive towards the image of his chosen 'self', his superman. He may, as Friedrich Nietzsche claimed, transform into his '*Übermensch*'. If you choose a path then who's to judge what's better or what's worse. You forge your moral code and live by relativity. From what options might you ideally choose? Fame, victory or power? Is your idea of valuable success based on your family, work, sport, politics or any other line of march? Or perhaps, more wisely than material reward, you want contented happiness. Is that the highest aim? What I choose will trammel how I act. A self-made man evolves as targets move. His freedom is a function of maturity.

Of course, if your ethic varies, slogan-wise, from week to week then trouble may, as Chapter 26 elaborates, erupt. But, paradoxically, a person's 'ruthless selfishness' is tempered by his self's extension in community. An individual's frame of thought is stapled by the influence of 'social self' or 'membership' in some society. Most groups involve hierarchical structures (federation, leader, father and those 'below' them) and the context of their negotiations is called politics. Take sociology and economics. 'Quantity' of life-style is the objective business of economy - resources, wealth and body-care. A utilitarian system, for example, values goods according to their usefulness as instruments of physical satisfaction. It therefore highly values the possession and control of objects that promote the user's market or material interest. Such worth is numerically accountable; it is quantified in terms of SI units, currencies and other physical parameters. Utilitarian principles and practice are important since they serve creature needs. To live in health men need sufficient water, shelter, clothing, food and freedom from discomfort, fear and pain. In this respect a utilitarian tendency is, although the sentiment of sympathy sometimes permits a whimsical form of ethics, skewed towards selfish satisfaction and physical survival. It may, with a pragmatic and Darwinian 'bio-logic', emphasise 'me', 'mine' and 'my' desires at the expense or exclusion of others. It will also tend to rate technological progress, wealth and physical utopias above internally generated, circumstantially independent psychological contentment. The news is we have faith in brave new socio-economic paradigms with science and technology to pump their gold-paved heart. Does reason not dictate that we embrace egalitarian brotherhoods and live, according to this rule, in communes? Not monasteries. Isn't a utilitarian community of humankind a compromise that, if and when it works, is good enough?

'Quality' of life-style, on the other hand, involves subjective business - interests, relationships, aesthetics and happiness all round. Such socio-economic linkage needs firm leadership. Why? A population is a clamorous 'family' composed of selfish and thence often quarrelsome members. It needs more or less a strict, *external* government in order to promote group balance and to mould conflicting instincts, interests and behaviours into as stable yet dynamic a pattern as can be - in a phrase, to keep the peace. Such quality is at best, however, based not on external but *internal* government. Such self-imposition stems from moral principles by which a person lives; and these in turn are coloured by the pillar of his faith. Worthy principles should lead to balanced life. Persons motivated in this way comprise the backbone of society; morality is the invisible cement, it is the rock foundation on whose bed a city stands. If everyone were thus society itself would automatically stand straight. The need for checks, forced balance and external government would diminish in proportion. In an ideal world (of good ideas) the cost of crime, disorder and affray would be eliminated. Almost like life in a monastery. So civilised! In an ideal city of the blessed and blessing there would exist no needy, selfish sinners to create their waves and force our navigation on a choppy social sea. Vice would not hang expensive millstones round the neck of virtue. *Costless, painless law and order derives from principles - invisible, immaterial, unscientific principles.* Yet in our time it sometimes seems that neither clever socio-biologists with selfish genes nor cunning crooks respect the timeless root of crimelessness.

A politician (who is not?) must reconcile utilitarian with subjective needs. Evaluation of relationships is not tallied in the way of coins; you do not measure happiness in metres, shillings, calories or hours. It is inexact, immune to number and involves such unit-less but unifying qualities as goodness, beauty, truthfulness and love. *The tendency of subjective evaluation is to emphasise the quality of thought, feeling and behaviour - the scientifically immeasurable quality of mind.* By what yardstick is the essence of such quality best judged; what is sort of nature whence, without more effort than the cream on milk, morality and manners rise?

What is, at root, essence? An existentialist draws essence from existence; free from authority he generates his own 'essential' ideals. You would not, of course, expect surrender from an ego-centric to a theo-centric rationale; but a holist, trying to sacrifice his bestial part in favour of Ideal, moulds his existence on its Essence. One man's essence is material alone, the other's includes immateriality. Their vectors, like their world-views, are reversed; but both are martyrs to the cause of sloughing rigid dogma and enforced conformity.

Top-down vision is consolidated round the Central Axis of Enlightenment.

Its holistic value system attends Nature or (if you insist that nature's only physical) Super-nature. Its First Principle, from which all others flow, is equated with the Nature of the Inmost Self, The Highest Good or Apical Experience. Such Experience is extolled. It is symbolised at the sacred heart of all world faiths and in their personal devotions. This central residence would, dialectically, deposit any cult of matter (including humanistic scientism) at the periphery of truth; it would define a *bottom-up* perspective as, according to its concentration of materialism, a relatively eccentric misconception. Subjective truth is the criterion of absolute morality which, due to personal ignorance, we have to take on trust and, due to the conflicting needs of centaur-like embodiment, is difficult - even if inclined that way - to squarely face. Thus any behaviour, event or object is measured according to its position on the conscio-material gradient, that is, its value against the Primary or First Principle of Natural Dialectic. This, inverting scientism's version, is Reality. Such order is ideal, no doubt, and idealistic. In this case egalitarian brotherhood is clustered round its Sacred Heart. Such commune is 'trans-religious' and its absolutism is without political excuse - except that politics is always tempered by pragmatics on the ground. Climb any tower. Look out upon the city. What you see, we saw, is built on metaphysical foundation - the cement of aspiration, information and desire.

In this respect a primary value of the Dialectic, Unity, stands out. Nor is it drably marched in cohorts of conformity. Its uniformity melts into togetherness. It's Presence and not 1984. Whether in subjective or objective terms, (*Sat*) Unity means wholeness. What is (*raj* ↑) unification but the path towards unity's climactic union? One aspect, wherein the parts adhere, is love. Another is creativity and a design in which some comprehensive integration of working parts is compiled to express, in transcendent unity, a conceptual work of purpose. Doesn't a steam engine, jet plane or a human body transcend, both in its own obvious coherence and in the expression of its maker's purpose, the sum of its parts? Of course not, cry material 'reductionists'. Of course, the holists bellow back. This is surely what you diagnose as an intelligent design. What more intelligent than, combining love with human form, both qualities of unity combine?

If this sounds difficult it is not. You measure everything, automatically, in terms of your desires, interests and loves. Fun and friendship are driving criteria but since their particulars tend to wax and wane you might call them imperfect, relative or incomplete sorts of happiness. They are lesser loves and, because they tend to fade, can even come to be disliked. Imagine, for a moment, that you might be consumed in a greater, all-encompassing fire, a love for no thing in particular but, specifically, Nothing. *Such Nothing is, in Essence, <u>source</u> of everything and so, paradoxically, from Pole Position such a Love <u>embraces</u> everything.* Utilitarian uselessness! The flame of mystic love turns out to be, because of inner radiance, superlative; that is, where subjective trumps objective, of the highest value. It involves the unified state of mind to which all other states, more or less, aspire. If you chase happiness the Highest Happiness is where you'd like to be - at least as much as possible. *Therefore how practical the mystic; how the sceptic lost his ticket; and how a state of mind (including quality of judgment issued in accordance with this state) is what it's all about.* You know that happy moments spark the generous decisions. Might you allow that at the moment of a 'peak experience' you would make different judgments and decisions of a higher quality than those that spill from darker, less enlightened, more utilitarian moods?

There you have it. A *top-down* view upturns the existentialist's. Natural, Essential, Everlasting Values are to hand. As opposed to egocentric you find theocentric Inner Self to which a lesser self may always turn. In this case a cosmic, all-embracing goal might be to gain the summit of Mount Universe. To judge this Apex just a mental construct, concept or imagination is the folly of an evolutionist. If consciousness is elemental, real and can be purified then it is the error of that creed's psychology and its psychiatry. Such, built on the sand of matter, are no less than black delusion in our time.

Surely, you protest, Darwin scuppered mysticism's upside-down philosophy for good? By rendering immaterial, well, immaterial he pulverised its false pathology. Is that really true? First read Chapters 19-25. If they are true then superman's long been around. He does not jump from gym into the fray nor will super-superman descend by evolutionary development. In this case superman is possible but not by force, by physical development or only in the future. That is Nietzsche's and the existentialists' mistake. Progress is towards a Good that, though it's veiled, is here now. The climb is actually Return to Source. Superman is super-conscious - well above the state of mind of average men. His aspiration is not evolution of a super-body but the voluntary evolution of soul, that is, the rise by purifying consciousness into a concentrate that's free of all contamination.

If there is a most meaningful, most valuable Truth at the top of an absolute scale, the unenlightened do not know it. If there exists an Absolute Goodness, Rightness or Natural Morality, we have to take it on the authority of parent or teacher. We have, until the various clouds of doubt are lifted and imperfect human reasoning surpassed, to bind with faith. Until the shadows of appearance, tricks and ignorance are obliterated by The Full Illumination we can only trust a wiser guide to show the way. Believe me. Natural Dialectic and, more profoundly, Christ, Buddha, Mohammed and the others are each pointing, with the right-hand vector, up!

Rights and Wrongs

	tam/ raj	*Sat*
	relative shadow	*Light*
	range of rightness	*Rightness*
	failing	*Ideal*
↓	*tam*	*raj* ↑
	negative	*positive*
	from ideal	*towards ideal*
	wrong/ bad	*right/ good*
	doing wrong	*putting right*
	from balance	*towards balance*
	towards darkness	*towards light*
	cacophony	*harmony*
	division	*unification*

From your value system flow your reasons. These calculate, accordingly, what's right or wrong. Whether it's a thought or action, now you're home and dry.

A potentate, a king or the founder of a business is the Absolute Director of a power structure. From the Apex his orders 'set the ball rolling'; as first cause, they both generate and control the actions of a hierarchy of personnel and their tools of trade. These tools are inanimate; such material bodies can neither distinguish between good and bad or judge right and wrong nor, *per se*, can they be qualified as such. They are *amoral*. The workforce, on the other hand, lives with intent. Each worker, from the perspective of his particular grade, understands the reality, purposes, rights and wrongs of the business in which he is involved. His *use* of an instrument, including his own body parts, can be technically correct or incorrect (according to his skill and the purpose of the tool); or morally right or wrong (according to his purposes). However, when set against the perception, criteria and directives of the 'boss', no worker is more than partly informed. The understanding of the workforce is relative. This means that unless they strive in accordance with the firm's instructions they will make mistakes. Of course, the cost of error is as much an incentive to improve as is appreciation shown by the 'boss'. Team players are rewarded; ethos is encouraged; a leader wishes to promote staff through the grades. Does the company called cosmos play this sort of game?

Communities believe in 'rights' and 'wrongs'. Religiously, politically and socially - 'should' is a word bound up and binding with morality. Emotionally and physically - 'ought to' helps to keep us free from harm. Duty calls. Ideals arouse. They are psychological attractors that pull forward; they are like engines that empower our course of life.

Morality's a function of voluntary intent. What has no intent has no morality. Neither moral nor immoral it is, object-like, amoral.

The 'behaviour' of a physical object is automatic. Physics and chemistry, using abstract and non-negotiable mathematical equations, have compiled a comprehensive catalogue of these non-conscious, predictable behaviours. Because it is unaware a physical event is unable to judge either its own or another's behaviour. *Inanimate things are amoral. Materialism, whose reflex automata 'behave' according to the laws of physics and chemistry alone, is intrinsically amoral.*

In this respect any instrument of purpose is *per se* amoral. It is impersonal. Such instruments include machines, money and even one's own body and its parts. *It is intended or actual use, not tool, that involves value, worth and morality - an ethical dimension.* Therefore such a dimension is psychological, metaphysical and immaterial - one to whose purposes all instruments play second fiddle.

If what is automatic is not moral then genetic, neural and hormonal activities are amoral. Sub-conscious reflex, preordained instinct and pure sense perception are prescribed, involuntary and therefore amoral. *Morality depends not just on mind but conscious mind; nor only on awareness in instinctive gear.* No doubt eagles and apes are just as conscious as us. Animal senses and responses are often sharper than our own. Many show obvious emotions. Resourceful ones like octopus, crow, dog, dolphin, monkey and many others learn, scheme and solve problems. Apes and jays are no fools either. Shrub-jays can anticipate and plan tomorrow's food supply; baboons plan 'shopping expeditions' to best gather food from nature's leafy shelves.

Indeed, the animal world might be construed as representing a range of 'emergent' consciousness. While it *does* learn about its neighbourhood a *paramecium* or an ant seems to us, compared to lion or man, sleep-walking in its world of instinct and sensation. Their bodies, cognitive systems and behaviours are simpler than ours. Moreover, since they lack contemplative focus, their responses are almost entirely unthinking. Such organisms act reflexively. Or, as with eagle, lion or ape, their inward focus is shallow. In as much as thought is lacking instinct always takes its place. Mind is never short of guidelines. One instinctive aim of life is to survive. Another is comfort, that is, lack of pain. What promotes survival and comfort is 'good'. What threatens well-being is 'bad'. Behaviours are perceived accordingly as 'right' or 'wrong'. Watch a troupe of monkeys, pack of dogs or any other social interaction. No doubt the participants know or are soon taught what is good for them. What promotes an individual's survival and comfort is attractive, what threatens is repellent. Pleasure and pain, framed within hierarchies of power, sex and hormones, are the orders of a creature day. This is the body-centred view. Pegged to its frame brute physicality is dominant. *Such sense of 'good' and 'bad' belongs to the flesh of an animal.* Is such condition what you'd call a moral one?

If an organism is incapable of awareness of the notions of 'right' and 'wrong' it is incapable of morality. As far as non-human life cannot transcend its thoughtless instinct and intellectual sleep, it is amoral and morally blameless. *It is as amoral as a baby.* Is there, in this case, any real difference in man except that, allied with an advanced power of vocal communication (speech), mathematical manipulation and imagination, he can better cope with complexity and concentrate his power of attention to a greater degree? Is there any cosmic significance in this, possibly the highest level of information processing in the physical universe, or is man simply, of materialistic course, another sort of animal?

He's a centaur. Half man, half a beast. Man, like a centaur, has an aspect that is animal. And many animals enjoy life played out in a hierarchical community. They cannot, it's admitted, manage the manipulation of symbol that underwrites complex speech, reading, writing and arithmetic. They cannot tell a joke or story, sustain a logical argument, do science, grasp cosmic principles (such as atomic structure or descriptive 'laws of nature'), comprehend morality or glimpse the possibility of Transcendence. Man doesn't differ from the 'higher' animals in substance, just degree; but he can argue abstracts and claim logicality. Whence derives, at any level, rationality?

Counter-intuition is the word now used; in grand, self-refuting style materialism's creed recites that brain, evolved by random accidents, rationally understands a cosmos that is rationally understandable. Accidents made order; irrationality gave, blindly, rise to reason; man's a speciality of chance! It isn't what seems logical but atheism's creed confirms life's made of atoms; morality and creed, like genes, are just extraordinary shapes that chemicals compound. Ironically, if this is true, and thought is chemically controlled then what control is left to 'you'? You can't think what you want since reasoning is just the product of a chemical reaction globally called brain! It's absurd. Atheism cannot rationally conceive itself.

Moreover, since each brain is different, might we construe no absolute exists? As products of its nerves the laws of mathematics, logic and morality should vary in each one. Moral relativity is this game's name. If they're simply made of nerves in brain then codes are generated as the molecules dictate; and since brain configurations change our values vary with them! Why, anyway, should anybody, much less Anybody, tell me what to do? In this view 'personal or social opinion' blurs with absolute morality. Why should enslavement, murder, rape or theft be 'wrong' if that's the way I'm thought? If, furthermore, morality is a subjective choice then who could logically or morally denounce an evil tyrant for his choice - especially if our moral roots reside within a common ancestor of scum, that is, bacterial chemistry that colonised an ancient pond? One man's code is simply different from another's. My code, my genetic code, which made my brain which makes my thoughts, is just as 'good' as yours! You might even, in a politician's way, keep changing points of view. Survival justifies its selfish plans, its animated means. You don't jail cats for killing mice and we're just animals as well. At least, this is the sort of creed that evolution justifies.

Thus, based on evolution theory as it is, humanists and atheists suspend the basic logic of their creed. Most behave as morally as anyone. 'Should' and 'ought to' carry weight in their vocabulary; punishments and prisons are allowed. Indeed, policy agrees it's *absolutely* right to live by moral relativity. Standards not unlike a Christian's rule the social day - except when certain aspects of them do not toe the party line. Meek is not assertive, humblest yields and strength is not conceived as exploitation of the weak; this is not survival-of-the-fittest talk. How, though, can you borrow from a creed that you've condemned as moribund and act, it seems unjustifiably, in contradiction to your own?

The problem is that such a house of relative morality and thence, politically, its laws of state, is built on shifting sands. <u>*Without an absolute reference-point trivial issues are confused with major principles. Each man judges the excellence or otherwise of a given behaviour differently; such a society drops into an ethical morass of fashionable, relative moralities and even, at nadir, moral meltdown*</u>. In such a bedlam whose authority interprets truth? In the barter of a moral market who can understand the worth of 'right' and, in that balance, tell from 'wrong'? In which case what is 'character'? Without hierarchy, without distinction between the value of characters and characteristics, how on earth can you judge quality of mind? If you treat every sort the same then striving for ideals were simply ill conceived.

What, therefore, to do? Choices must be made. If one is 'better' than another (perhaps judged by degree of lessening pain) then obviously, you ask, what is the best? What is 'most right', that is, ideal? What is the line of fitness that you try to hold, how great the play that deviance must not transgress? Transgression of a 'right' is 'wrong'. 'Wrong' conflicts; it strains the social atmosphere. You could have seen that in advance. Is there any moral without foresight any more than cause without effect? Man anticipates the outcome of his chosen actions; he can also weigh the impact of intentions on another man's experience. *This, anticipation of intent's effect upon the future quality of life, ensures that man, the centaur, is intrinsically a 'moral beast'.*

In the case of training animals the 'wild will' of an object may be, one way or another, eliminated in favour of the trainer's 'law'. Are humans, in education of their young and systems of control, less harsh? The group's desire is pressed upon each individual. Not just selfish 'good' or 'bad' but 'right' and 'wrong' in terms of others' wishes hauls us to the edge of ethics, to the margins of morality. **The real question thus becomes the quality of such desire, the criteria of legislation and thence the character of law**. In this respect the Dialectic suggests an equation. Does what you call life reside in your awareness? Awareness is the same as consciousness and, if the latter is an immaterial element, then a complete concentrate of this 'substance' is possible. Set against a mixed solution (mind) call this 'pure solvent' soul. Pure Consciousness is thus your Soul; the metaphor is Living Water. Down the scale, in mixed-up centaur-land, embodiment presents its moral challenge; strain (and its compromise through the agencies of politics, law and religion) inevitably pulls between corporeal self and incorporeal Self, that is, between the anti-parallel attractions of body (↓) and the soul (↑) above. Of course, for a materialist there is no soul and therefore no strength of opposition. Unilateral strength of body-self prevails. The mightier a force of egotistic will the more it legislates its way. There is, however, for a holist soul. And the cosmic significance of man's estate is that, in perfect detachment from existence, he can purify the mind's solution and thus clearly come to know the Essence of his Soul. From perfection and transcendence springs automatically morality. Thus, since awareness is a natural element - the root of information - there is an actual immanence of cosmic ethic, an Intrinsic, Natural Ideal!

If this Ideal informs the metaphysic of creation and, in hierarchical construction, archetypal metaphysic precedes physic then the universe is, in quintessence, ordered and supported by an immaterial principal. One word for this (and Chapter 5: Top Teleology lists more) is *Logos*. Verbal noises are no more the *Logos* than numbers make up stars. If, however, *Logos* is First Natural Cause then we live in a basically moral world. And if this world is good what makes it bad? Where there's evil (Chapter 26) what dilutes it? What solvent yields the purest, best solution?

No doubt evolutionism can't account for laws of logic, reason or morality. A Conscious Cause could - easily. Its nature, we've suggested, represents an Absolute Ideal. In other words, as mystics since the start of time have claimed, morality resides in state of consciousness. Because capacity to recognise then realise a pristine and original state is part of centaur man, communion with the Real Ideal is central to our birthright, our inheritance. It's called Going Home.

Why, however, care for absolute ideals when godless humans can invent their own? They can be re-crafted sensitive to change and always be improved upon. High principles oppose the torpor of set ways; they shame hypocrisy and struggle with irrational, mindless anarchy. Such anarchy is simply moral relativity while working at its lowest order, that is, in sub-bestial levels descent; and since hell never works for long the secular adherents swiftly, strictly post such canons as will regulate - until the force of regulation raises hell again! Who needs cosmic patriarchs to issue ruling policy? After all, communities survive with more success than loners. Morality is, therefore, socially cohesive glue. Brains chemicals have 'learned' the lesson of enlightened rationale, one that supports self-interest the same but this time indirectly. Helping others first, called altruism, has its roots in economic satisfaction, promotes the happiness of the majority and thereby leads to peace. Peace, in terms of life, is a dynamic, throbbing equilibrium. Trade's heartbeat stably hums. Such stability, contrary to a struggling, evolutionary creed, leads to improved survival in materialistic terms. It's true - despite conflicts between society and egotistic individuals that make it up, high living standards help; but do they also cultivate 'internal' peace, that is, contentment born of understanding and not things? Which creed, evolutionism or a 'path of peace' that hums according to the way the world was made, promotes its moral standards measured by achievement of an Absolute Ideal?

By now you see how fundamentally the issue hinges on a point of metaphysic - is soul an imagined fiction or Reality? Is 'soul' sense, nonsense or, worse still, a lie? And do communities define what's right or is, like soul and health, 'rightness' a naturally embedded thing? In other words, is 'Good' prescribed by boards of men and handed down from their authority on high or is The Office even higher up? Whose will or Will's to be obeyed? The question boils down to prescriptions that might foster tendency towards a socialist utopia, religious theocracy or tolerant plurality - whatever you believe will best accommodate the two-way world of centaurs. Of what does Plato's 'greatest good' consist; and, beside 'external', social politic, of what consists 'internally' achieved enlightenment that might make individual saints of everyone? No doubt, in practice, government involves some measure, more or less, of tension, violence and hypocrisy. Where, though, is government but in a governor's mind; and minds that take decisions must involve morality. 'In God we trust' is written on each U.S. dollar bill. 'Defender of the faith' is written on each British coin. Is such sentiment naïve? Indeed, unless there *is* an Absolute and Absolutely Natural Morality.

Is There an Absolute Morality?

	tam/ raj	Sat
	scale of inferior reasons	Reason
	range of rightness	Rightness
↓	tam	raj ↑
	from Reason	towards Rightness
	false	true
	bad/ wrong	right/ good
	passion	patience
	pain	peace
	sin	righteousness
	vice	virtue

It's clear. The holist thinks there is; and it is something that a naturalist denies. Morality is easy in its black and white. 'Is this right?' Often, though, circumstance creates a spectrum of response. To what authority, therefore, do questioners appeal? Is it, universal, to an Absolute Authority? Or, local, unto human reason in whose court a multitude of rationales may easily compete - should short-term expedience prevail, political correctness sway the social mind and wobbly 'rightness' fluctuate by place and time? Is morality a relative concoction of pragmatic minds or is it, Naturally, Absolute?

Bottom-up, reason and not Reason wins. A variable creed prevails. A clever humanist can always, with his finger on your pleasure spot, dissolve morality's control. He can construe how ancient Greeks and other classical authorities seem to endorse this 'democratic' sort of line. Who'd be fool enough to disagree? Comfort, if not hedonism, is 'the good life's' game. A siren voice excites an easy popularity. Just scratch my back and I'll scratch back - especially if you're 'family'. In fact, you must understand morality's determined by propensities derived from chemicals that in their turn depend on genes. Aren't you genetically determined? What on earth is self-control? Therefore take a sinner's sinless pleasure in the body-centred view - you know it's right. Eat, drink and be merry in the knowledge 'moral goodness' really is pursuit of wealth and sensuality. 'Within reason' is a postscript you might add. And, since your reason is The Highest Good, you can decide the tipping-point. Indeed, every moment should be weighted towards the body. To hell with medieval sackcloth! Let us maximise on fun!

Top-down, reason is aligned with Reason. If the cosmos is stepwise in structure, then well-aligned philosophy and good, sound mind should reflect the fact. In other words, there will scale a gradient of qualities of mind and, logically, one might aspire to Highest Good, Noblest Truth or The Criterion. This, Inalienable Right, issues from an Ideal Centre that, *a priori*, naturalism naturally does not admit. Two vectors run - one towards the Centre and the other towards its outer ring, the body of its edge. To visit such Home Central one needs bus from world to World's End and, on the journey, relative to Absolute Morality. What is this Depot like? The mystical consensus is its character is love. The milk of kindness in most concentrated form. We were once sucklings at the scientific breast of fatty tissue, nipple, ducts and glands; and at this breast the milk of human kindness flowed. As we were sucklings at the warm edge of the world, it may be that the world's a suckling at the warm, unscientific edge of an Ideal. This cream of moral milk, Ideal, is therefore issued from a Breast of Love. How scientifically daft is that?!

Physical survival's not a hero's or a martyr's absolute but it *is* evolution's end-game. For altruism to come into play you have to borrow from a less than wholly exploitative, calculating, self-first creed. With sheer survival once assured a 'satisfactory' life-style is desired - but what's judged 'satisfactory', what is fulfilment, how is life best lived? A chosen bias might, without a second thought, veer towards sensual, hedonistic pleasure; it might be possibilities of mind - imagination, inspiration and invention - that you choose to emphasise; or else devotional pursuit. Centaurs oscillate but warmth, contentment, fun and 'happy moments' are, of course, what every earthling seeks and often finds. Finds where? Our changing choices generate a mental atmosphere. They colour character. They may affect another life or many lives on earth, perhaps even earth itself. The quality of any individual's character is judged upon the fruits of his or her moral decisions. We can think, anticipate and choose and thus cannot avoid morality. *Whether it forgets, denies or flouts moral rules, the thoughtful nature of mankind is bound up inextricably with struggling over what is good and bad.* **All human society has regarded some resolution of this struggle as its first and foundational priority.** *Such resolution is the primary concern of philosophy, politics, law and religion. This is because the psychological stability and happiness of individuals and, collectively, society depend on the answers.*

What, therefore, is your priority? What hierarchy best, for you, expresses Truth and, according to its reigning influence, all subsequent and lesser truths? With what moral compass would you, when you think about it, constitute your government and frame its laws? Pursuit of wealth and gain of pleasure are agendas that inform its worldly ways. A secularist would choose, of course, the shifting sands of a consensus driven by the way that 'growth' and 'progress' should evolve to satisfy these physical desires; indeed, a modern man (and when is 'now' not up-to-date?) needs vote for relative morality according to the way the wind blows on the ground and less regardful of such principles as drive the whole caboodle from a Moral Sky. Centaurs, as history shows, equivocate; although professing candour and religiosity most of them hypocritically chase power, wealth and glory on the ground. Actually, you don't have to have a vote. Thus moralising politics will promise or will force you to a left-wing or a right-wing promised land.

It was modern then. A *top-down* social order was descended, through its leader, from the Ideal of divinity. A pharaoh or a Chinese emperor may not himself have been enlightened or of ideal character but, ruling at the central peak of his world's nested ranks, he was at least, symbolically, a conductor who would 'earth the lightning' of communication with divinity. Higher power was not despised. In the interests of optimal health and happiness the metaphysical construction of creation was relayed to human society and reflected in its social orders. Kings may not, just by virtue of a role, become divine but, by definition, saints and mystics are conductors of the truth. Their (*Sat*) Knowledge understands the way things are and, for wilful, wayward minds, should be. If their Topmost Perfection is a Natural, Central and Original Reality, then they imply an Absolute Morality.

From Science to Conscience

Is science not a key unlocking from the jail of ignorance; and a revolutionary revelation of new information? But is information of things physical the only type?

Study of material is not the same as study of the immaterial. Laboratories are not the place for moral seminars. Indeed, scientific progress is irrelevant to moral progress. Nor does the latter's kind of evolution change with history. Thus, on the one hand, objectively scientific materialism may limit reference to any moral condition or human context and, at best, describe these in animal, humanistic or politically utilitarian terms. Such reason does not render subjectivity unreal. On the other hand, the teacher Christ used stories and offered himself as an interface through which to obtain his Truth. Self-effacement characterised Mohammed who insisted his disciples, while practising the Koran's instructions, worship not its messenger but its Author. And the Buddha prescribed a psychological treatment which, dutifully self-administered, would remedy the suffering of earthbound souls. *There are different truths and therefore ways of seeking them. Do many ways up to its Apex mean a mountain lacks a Peak?* Why is it rational to deny the Universal Height is immaterial - unless a metamorphosis of reason called materialism has forgotten reason is itself a form of metaphysic?

Mind, dialectically preceding matter, is an information-field. It is metaphysical. And so, beyond the mind, Transcendence is as well. Its root, latent information, rests behind all patterns; its purity transcends all mental individuality.

Three consequences rise. If Transcendence is an experience at the heart of creation, then Reality is subjective and the creation is, at heart, alive.

Secondly, mind evolves by lowering its own entropy of information viz. increasing comprehension. Wisdom is derived from a purification of understanding; and, since orderly simplicity derives from principle, the more embracing a principle the more details are easily subsumed under it. Principles are concentrated power. They are the rule of natural or man-made law. If, therefore, the Axis of Mount Universe represents the Central Principal of Principles then the ascent of wise man will involve his mountaineering towards this Essential Peak.

Do you see? It is thus obvious that, thirdly, Truth resides in the effort of (*raj*) unification and experience of (*Sat*) Communion. Cohesion, concentration and a sense of union - where every sense of individuality (or '*ego*') is dissolved an Absolute Morality obtains. Do you think, in love, of 'law' or its 'morality'? **Enlightenment is, naturally and automatically, Morality.** The root of imperfection is therefore identified as various degrees of isolation, individuality or, in short, the bonds of egocentric self. (*Tam*) 'division' or 'apartness' is, in fact, a form of negativity, entropic influence or mass-centred power of gravity. It is fundamental to creation, that is, to existence. The body of an atom, planetary aggregate or galaxy show individuality. In a world of separate things each individual *divides* the world between self and non-self. Normal psychological self-centredness is manifest with 'weighting'. Such bias derives from an embodied centaur's inevitable identification with the body of his mind (in the shape of habits, interests and abilities, memories etc.) which, in turn, derives from identification with its physical body. The more 'self' intervenes, the more love lacks and the further mind is removed from the Centre of its True Self; in which case the harder, colder and less appetising does morality appear.

This might be rephrased. Is morality love's logic? Love knows no law but logic knows no love. At a fundamental level love-in-action is the basis of morality. Therefore as love diminishes the logic of morality appears to balance loss. It is the front line of a struggle that would balance 'bad' with 'good' thus neutralising evil. Where loving care and moral education fail centaurs are forced to recognise society needs instruments of resolution, restitution of a broken order and guidelines for coherent peace. They recognise expression of the act of moral balance in social logic drawn from religion, implemented by the arms of law and politics. What liberal rabbited that morals didn't count? Lack of them will always count the cost. Ask any Chancellor. Immorality and crime will tax you heavily. Their fall will fall on you. They eat like parasites upon the law-abiding members of society.

Ethology and psychology. Science studies the behaviour of organisms whose driving psychology in some instances approximates a human state of mind but never, in a capacity for inward focus, reaches it. Is man's psychology reduced to this? Is the morality of scientific atheism only ethological? And character a function of the chemistry of genes? **Is evolutionary psychology entirely missing out a fact it's blind to - evolution that cleans mind which thus, transparent, lets the soul shine through?** This slant on evolutionary science is, of course, from a Darwinian point of view just pseudo-science (if you could even call it that). Yet your psychology is excellent not in survival kit alone but due to your capacity to search for knowledge and more knowledge. Is this a function of brain size or structure? Or is cerebral structure at the service of such learning, understanding principles and following ideals? In which *top-down* case might man (and man alone among the family of life on earth) be capable of Communion with Real Ideal, the Source of All Intelligence? If so, the clear nature of the Origin to which return was aimed would, of itself, demand as fee morality. Morality is built-in to the way things are; and, correspondingly, into man's frame. *From science to conscience, the Ideal of Wisdom is in mystic practice able to be realised.* Science may decline to rise this way but that does not exclude 'the other half' of human brain!

Conscience in Principle

	tam/ raj	*Sat*
	consequent principles	*First Principle*
	lower status	*Principal*
	action more	*Ideal*
	or less accordingly	
	relativity	*Absolution*
↓	*tam*	*raj* ↑
	anti-principle	*principle*
	downward/ outward	*upward/ inward*
	from Rightness	*towards Rightness*
	from Reason	*towards Reason*
	passion	*patience*
	obsession	*detachment*

Morality involves direction of focus, intention, anticipation and choice. Automata (with all their attributes and properties such as texture, colour, shape etc.) lack any focus and, as such, are *per se* neither 'good' nor 'bad'. They are amoral. The physical universe is, of itself, amoral.

Herein lies the dilemma. *His body is an automaton but a man is not.* He has the wit to understand a tension stretched between biology and High Ideal. While spirit is willing, the flesh is weak. The body is low, close and urgent in its bestiality. The influence is powerful. Temptation's undertow drags towards the depths of passion, currents rush you into sensuality; at the same time fires ignite the lust for wealth and power and fame. Familiar pleasures lock mind to the world. Suction towards the physical, detachment from the metaphysical is carriage 'down the stream' of what is known as entropy. And if life's Source is Life who still would swim against the tumbling current of his earthly circumstance? The choice is life or death. This is the dramatic test into which humans, whether they like it, know it, care or not, are pitched.

Seen in this way morality is not just a function of intent, choice and reason. Discrimination needs to be aware of yardstick known as Ideal. It needs, as immorality does not, to accord with the right-hand (↑) path of Natural Dialectic.

Now we have complete encapsulation. **Morality is a function of intent and 'right-hand' reasoning**. Its absolution is a natural consequence of Enlightenment. In this sense con-science is not so much additional as precedent to science. It is intimation of an Absolute Morality.

Enlightenment is union and union of souls is love. The Mystic Ideal is, therefore, Love. Such inspiring radiance inhabits neither mundane tasks nor lusts of biological erotica (where lust involves a brief and bodily conjunction, a reflection of the higher ecstasy). Therefore, lacking Natural Ideal we force a diminished, viscous flow of 'oughts' and 'shoulds'. The darker mind or more acute adversity, the harder it becomes to squeeze love's optimistic simulation out. Less keen, the drier duty then becomes. Indeed, if either real or perceived injury becomes too hard to bear a moral flip into loveless hatred targets whosoever caused the pain. Devils devils make. The sins of man are visited. Parents' minds are visited on child - unless the strength of mystic and not scientific practice breaks the cycle. Such epic practice throws the rings that bind and melts them in its fire of love.

As such life is a morality play. A mummery. Life with other people, with oneself alone, with the panoply of living organisms and even, ecologically, with lifeless but life-giving earth is moral theatre. I'm pitched on stage, I have to act and, for these actions right up to the final moment, I am held to accurate account. Newton's law of strict causality, decreeing every action (Sanskrit: *karma*) has an equal, opposite reaction, is a basic theme that interweaves all plots. You can observe direct effect but knock-on is less easy. With physic so with metaphysic. Why should the same law not apply to psychological causality, that is, desire and intent? As much as they can understand it or it suits their taste critics, gossips and the judges of our little worlds weigh in with relative assessments of the value of performance that life-long we give or, now dead, once gave.

However, gossips or a court of law are not the final arbiters of deeds. Some actors calculate that furtive, unseen actions can and do, with luck, evade the force of law. What law? Is there any deed not seen by natural consciousness? Its perpetrator knows and whence comes knowledge, what is at the source of mind? The seer is yourself and at the same time soul. Is soul not part of Soul? Thus, finally, at death (Chapter 18), mind direct to mind when nothing can be covered up, each naked individual's dossier is weighed in cosmic scales. Your life is in the balance. Set against Transcendence lesser lights, the shades of conscience, show up sharp enough. In an instant Lord Chief Justice will have summarised your case and, against a wordless screen of truth, Self issues self the verdict. You will know it. Con-science knows it straightaway. The relative has been compared with absolute and, in it, was there any trace of absolution found? It is the way that nature works. We resonate, automatically, with such wireless wavelengths as our thoughts convey and we are therefore tuned to. Is life a lottery or does a man attract experiences he deserves? At any rate it is delusion to believe that 'left-hand' acts invoke 'right- hand' rewards; or to convince oneself that sinfulness is, in some circumstances, 'worth a fling'; or that courtless, natural justice is escapable. In which case, as the unpopular beatitudes proclaim, we are all deluded, fallen angels, sinners on terrestrial ground.

Morality is closely allied with the cosmic fundamentals. These determine action. Their proportions intimately affect the quality and direction of our attention, that is, our life. They hold the clue to the innermost, absolute nature of morality. In its highest (*Sat*) Original Form nature's Truth is 'good' and the operation of creation is 'correct'.

Tam, on the other hand, rolls out a downward influence that tends towards inertia and exhaustion. This vector runs, like entropy, down matter's hill; it is the way all energetic congregations go. It grows heavier until it reaches solid earth. On this scale bodies show increasingly confined. As in body so in mind; a person's behaviours always reflect their mental condition. As negative influence grows isolation and self-centredness increasingly appear. Care or interest in another's welfare wanes. Poverty of mind is crystallised into the perception of life's other actors as just bodies. Things. Are not 'things' to be exploited? Other animations are transformed as instruments that serve the lusts or, worse, as scapegoats for the cruelty of such closed individuals. Might you diagnose this common illness as 'behavioural entropy'? Mark feral status or count moral vacuum? Is secular materialism a natural, moral vacuum? You can't see such emptiness, the world around brims to the full and so you might not think you're in a void or it's in you. Beware! Morality abhors a vacuum!

Stride on, stage left, the player representing gross subtendence! Beelzebub! Negative power is the materialising influence. It is the shadow of light, the counterfoil and resistance without which nothing could be shaped. It is, in this aspect, one of the two fundamental poles; it represents the tendency of 'hardening' and is, therefore, a necessary cosmic operator. *As a vector negativity is as amoral as its partner, positivity; it is impersonal where consciousness is not concerned. Beelzebub, however, is the personality of anti-principle.* <u>*His left-hand intent, his sinister will darkens the immoral under-pole*</u>. The Greek word *diaballein* means to 'throw over' or 'cast between'. A fisherman might, for example, throw a net. The noun *diabolos* (English 'devil') means a slanderer, that is, one who lies, who intentionally casts a net of illusion. When deluded you do not, by definition, realise the truth of lies? What sort of 'truth' is used to spellbind, fascinate, possess? What tail does *Shaitan* (Satan) twist, what tale the devil spin? That descent is really ascent? Left is right? Or darkness light, death life? That the highest calling lies in power and pleasure for the body? *Only human imagination can inflate, inflame and embellish the guidelines of natural instincts for survival into passionate schemes; it is these intense schemes for domination, sex and wealth that lead to immoral acts of darkness. In these Beelzebub bestrides and rides you.*

Let's drop briefly in on hell. It's nowhere in the amoral, physical universe except in suffering. Isolation and petrifying mortification are the walls of its prison, dis-ease the mark of inmates and cacophony its bedlam. Pain is the seal of its subduction zone. Demons contrive perverse, anti-principled plots in order that, for their pleasure or advantage, victims suffer. Hell's 'merry' dance is psychological subtendence. As the Dialectic indicates, the dark recesses of a moral outer space include extremes of terror; they are cold and heavy with anxiety, loneliness and hurt. Nevertheless medieval Christian, Tibetan, Hindu and other depictions of demonic inflictors are as static, ritualised and misleading as the cartoon of angelic choirs a-chirping on cloud nine. If *you* were asked to draw pride or anger what would you do? You could only *represent* their qualities. In reality the suffering of hell is a nadir of mind. It is not static but dynamic. It comes in myriad forms and to some extent most humans dish it out and take it. Don't you loathe its nadir as intensely as you seek the zenith, bliss? *Experience is not described by scientists but art.* Art's language is not numbers, it is metaphorical; its pictures are not graphs but graphic; and the foreground of its play is coloured not so much by moving objects as emotions. An image ages and a metaphor may ossify but each's meaning stays the same. Hell's sticks and heaven's carrots vary culturally but their truths and strengths remain the same. What better deterrent for a sinner than to understand a graphic threat of punishment or witness morbid, painful consequence of sin? Indeed, what better stimulant? So that internally arrested action makes it certain the external police are never called.

There are many hells on earth but they are all in mind. How could you, knowing, suffer otherwise? **You may even, wrongly, think that you're immune or none of this applies to you. The real question we all ask ourselves is how best to avoid such fate. What's the universal answer?**

↓	tam	raj	↑
	division	unification	
	confinement	release	
	pain	relief	
	terror	comfort	
	loneliness	friendship	

Being 'good' implies purpose. Conscience implies obedience to definite direction. Which direction? Whose direction?

Raj (↑) indicates an inward, upward focus. This direction, far from dividing, unifies. It tends towards balance and stability. Its high degree, the triple first, is a transcendent one. Thus the nature of Transcendence must be Union. Union means loss of self by merging in another. Another name for this is love. If Transcendent experience is at the root of creation, creation's root is radiant. The cosmic Essence of existence is love. <u>This love is the touchstone, the criterion against which all lesser loves are measured. It is the principle of principles, the colour of pure conscience and the basis, absolutely, of morality.</u>

Conscience in Practice

Morality's as natural as gravity. Its effects are unavoidable. Such immaterial but crucial data is not gleaned from a laboratory bench. **Indeed, the fact is that all the emperors, generals, politicians, philosophers, philanthropists, artists *and* scientists in human history have not exercised as much influence as a few perfect mystics.** Waves of their transcendent experience resonate, as they always seem to have done, in the heart of mankind. Sages such as Buddha, Christ or Nanak (in alphabetical order!) personify Ideals. How, therefore, do you practice their fantasy of phantoms, an imagination made of neural networks and patterns produced, in the last material analysis, by randomness? Conversely, if our brains have been intelligently designed, do you conclude the Truth of this Intelligence must be reflected in your own computer; in other words, that Absolute Morality is mirrored in the freely-issued structures of a human brain; and the bus to World's End is embedded in our archetype? Put another way, if man reflects his maker then he is specially made; your coded body, such an instrument as lets you link with the non-consciousness of earth, would thus reflect this truth; and Absolute Morality would be inscribed within the neural structure of a human brain. What an inheritance! From this you'd understand a common aspiration towards the Cosmic Truth could be a possibility. The idea of discovering its maker's order first drove science to explore the world; material *and* mystic science could, combined, fully realise mankind's potential and enlighten us.

Conscience-in-practice starts, with an appreciation of the difference between right (↑) and wrong (↓) vectors of behaviour, at the seat of mind - the eye-centre. It is here, where centaur-like humans leave behind the animal in them, that choice between different courses of actions and their consequences are weighed; and therefore here, where they can voluntarily transcend instinct and *think*, that moral rudiments are realised. From this point, the cosmological axis, a first positive, purposive and therefore, obviously, voluntary step towards higher truth is made. *It involves, through meditation, the achievement of a highly coherent or concentrated focus of attention at the eye-centre.* From this point the goal, following an inward transport, is to reach the top of Mount Universe and thereby achieve identity with the Centre. **This identity is with the Principal of principles. It is called Enlightenment - in which the completely *top-down* perspective is obtained. Teacher, method, practice and completion in identity together constitute a science of the soul.**

So here it is. If Pole Position is Most Natural then, like a sun, it precedes all human formulations of the way it works. Its existential Void is at the Sacred Heart of things; it is Infinite transcending explanation; and it is One so that, of different faiths, no one can claim exclusively Top Singularity. There are, of course, debased and elevated forms of worship. Those elevated in their focus on High Unity each claim an exclusive guide; but these guides (wise men, prophets, saints or what you will) in communion at The Cosmic Peak, become as one. What difference is there in one? Religions fighting over different, exclusive leaders play a self-refuting game. One isn't two or more. Therefore better, in conscience, to cooperate; true religions truly practised are much wiser than to fight.

Their alignment is the same; it will, in meeting target, transcend faith. Both practice and successful graduation to The Cause can be repeated, at will, by an expert of any race or creed. It is an open secret but, because it simulates the process of dying while living, not a casual, cheaply acquired experience. The process is an intrinsic part of human constitution so that man has always been able to know the Ultimate, Implicate Nature of things. Although degraded earth-bound mind-sets are in large statistical majority, a few individuals at every period of history and, presumably, pre-history are always able to realise their full potential. As such they become, in a word, sanctified; and, as perfected humans, ideals to stimulate the practice of the rest.

4. Cause and Coincidence

Let's take stock.

Do you think of Natural Dialectic as a pair of *scales* in which is weighed any particular aspect of the universe? Think of it as a conceptual tool, a balance whose two vectors and pivot are represented by the cosmic fundamentals of (*sat*) equilibrium, (*raj*) up and (*tam*) down. These vectors are inextricably identified with process in its stages of potential (equilibrium of readiness), stimulus or action (↑) and exhaustion (↓) or equilibrium of completion. They are thus, in the eventful media of mind and matter, omnipresent.

Dialectical *columns*, on the other hand, stack opposites. In the Primary case they contrast Absolute against relative attributes; in the Secondary case they set polar yet complementary opposites against each other. Any one object or event (psychological or physical) is composed of a stack of opposites each of whose fundamentals show, at any instant, various and varying proportions.

The basic 'component' of Absolution is Pure Consciousness. The basic components of relativity are consciousness-in-motion (active information) and non-conscious energy. **From this simple premise flows the whole *top-down* program of this book.** The interaction of these basic components generates a two-phase, conscio-material spectrum. At its Top, before duality and multiplicity, is the Single Essence of Clear Consciousness. There follows a hierarchical spectrum of mind. At base is found pure energy, that is, all non-conscious objects and events that combine to comprise the physical universe. *Existence is the motions of mind and matter*.

Do you recall that, for your own part, experience involves two directions in which the attention can be focused? Interaction with the objective world of bodies is mediated by the nervous system; subjective, interiorised focus contemplates within the metaphysical domain of mind. These two modes of knowledge are natural and complementary. One involves an objective, the other a subjective perspective - what something is and how I feel and understand it. Either perspective, if emphasised to the extreme of a world-view, becomes a pillar of faith. The objective pillar is composed of differing cultural expressions of materialism (e.g. humanism, communism or orthodox scientific atheism). The subjective pillar, called holism, includes world faiths, metaphysical philosophies and psychological exercises. This is, in a nutshell, all I need to know.

You may have noticed a presentational polarity that pervades this whole work. Its systematic oscillation, Natural Dialectic, allows you to compare and thereby evaluate the opposing perspectives of the two pillars of faith. For reasons already given, the holist's view is labelled '*top-down*' and the materialist's '*bottom-up*'.

Bottom-up, a relatively ignorant student or audience can learn from authority or, by their own investigations, empirically. The scientific method, trusting only its own authority, is self-reliant and experimental/ experiential. Unfortunately, however, it is also materialistic, excludes any metaphysical possibility and, therefore, any inference of inhuman but intelligent design. 'I deny it, therefore it does not exist!' Or even, neurologically, 'I will prove its physical non-existence'! And when it comes to origins materialism naturalistically flips to default mode - everything has physically evolved.

Top-down, on the other hand, a professor or engineer exercises authority from a position of overall comprehension. When used to interpret natural as opposed to man-made creations a holistic perspective identifies physical *and* metaphysical components in both universal and individual (microcosmic) schemes of things. It is able to infer intelligent design.

Physical tests and naturalistic explanations, where possible, are fine for now. However the real context of the present is the past that has led to it. We learn eagerly about the present construction of our universe and, from this, try to extrapolate back through history to its origin. Of course, in this case the conclusions are intimately and inextricably linked to our mental starting-point, our perspective or frame of reference. There are three main such frames - design-less development, intelligently-prompted but otherwise natural development and systematic, preordained design of the type employed

by *IT* and all other art and technology. Each involves the opposing notions of teleology and chance. *Teleology*, says the dictionary, is the doctrine of final causes, of design, of preordination or purpose. *Chance* is an absence of defined cause; but any doctrine of causelessness is, not surprisingly, a job to reason with - by definition you are not allowed to say 'be-cause'.

Coincidence or No Coincidence?

coincidental	*fundamental*
casual	*causal*
contingent	*substantial*

Is coincidence the fundamental cause that we are here or not? Your embodied presence, observing starry sky above and earth beneath, is possible because physical and cosmological quantities of cosmos are appropriately constrained. In other words, any universe won't do; only this specific set-up lets you read these words. However, a survivor's mere presence is no explanation of *why* his shipwreck happened; likewise your presence *per se* fails to explain *why* earth's cosmos is so suited to bear fruit of life - including you. Thus the irrelevant explanation provided by a 'weak anthropic principle' (*WAP*) fails to distinguish between possibilities of chance and design. But the parameters are very tightly defined. Cosmos at least appears as if designed for life. So much is fact.

Perhaps the most detailed appraisal of 'the anthropic cosmic principle' has been recorded by J. Barrow and F. Tipler in their eponymous book. They note two more speculative principles. The *SAP* is 'stronger' in its stipulation that the universe *must* have such bio-friendly properties as allow carbon-based life-forms to thrive; and that such forms must do so. Whether or not observers are (according to a quantum mechanical *PAP* or 'participatory anthropic principle') necessary for the cosmos to unfold, it may be that life is an integral component of its destiny. You might, therefore, if not 'design-inclined', home in on calculations that include a form of special pleading called 'the landscape' or 'a multiverse'. By this device sometime somewhere the specific, bio-friendly constraints and laws inevitably clicked, as jackpot numbers in a lottery, to make our universe by chance alone. Then (Chapter 8) you can, with towering imagination, speculate from such a speculation.

Arguments from design and the application of *WAP* to physics and biochemistry are admirably elaborated, mathematically and otherwise, in Barrow and Tipler's work. But their interpretation is slanted. It is riven with the prejudice that all facts, including the subjectivity of the anthropic observer himself, are material and materially explicable. There exists no immaterial element. As a result the facts are fine but scientific materialism and the consequent perspective of scientific atheism saturates their arguments. Therefore the biological theory of evolution is, according to present academic fashion, held as a sacrosanct basis for speculation; and *top-down*, metaphysical considerations of universal mind or archetypal design are treated, in a somewhat patronising way, as a historical stage of human thought. We know better, don't we? So, for different reasons, cosmological and biological ideas of evolution seize the explanatory way. Such progress declares a natural element that's metaphysical is out-of-date! There is no 'inner creativity'. There follows naturally a call beyond the brilliant picture of Hubble telescope's deep space; beyond its furthest void and past big bang's eruption invocation of an unseen multiverse is much preferred as an interpretation of experiment and observation's proven high-specificity. In this line of thinking all appeal for a Creator (or hierarchy of Creative Powers) to justify the bio-philic tuning of nature's legal framework can be ignored. But does mathematical analysis of a structure have to exclude the possibility it was invented? Does analysis of working mechanisms - say, aeroplanes - exclude the existence of their creators? Materialism's lop-sided presentation clearly jars with Natural Dialectic's wholesome, complementary but materialistically rejected view.

From this brief consideration of anthropic principles let's now return to the three frames of reference established with respect to teleology and chance.

The first frame, 'design-less development', is anti-teleological. If cosmos means 'product of order' (as it does in Greek) then this frame presumes an anti-cosmos and life forms within it that evolved by chance. 'By chance' means without rhyme, reason or logic; development by chance is not cosmic but chaotic. The only 'purpose' or 'direction' in a material universe is pointlessly derived - chance is constrained within the boundaries of natural law. Such law itself is deemed the product of initial fluke - say, expansion of no-space into all space from a 'big bang'. And if you call its character 'necessity' then transformations are determined by necessity; but chance oils this determination and things often slip out unpredictably. If, for example, creation was wound back and then re-started then there is every chance that we would not be here and chatting; and no chance whatsoever of imagining the extreme improbability, type, quantity and individual consequence of each antediluvian slip-up without whose billions there'd have been 'no you'. The 'Darwinian' equation, based on the omnipotence of chance and necessity, is 'scientific' because no metaphysical entity, even before the physical start, is presumed to have existed. No creator is presumed and therefore, by presumption, none exists; it is a gratuitously tautological and anti-teleological position that fails to support materialism's bottom line. Still, it repeats, no rational source of information is embedded in the system; and because matter is mindless its specific creations are not even 'accidentally-on-purpose' - they are purely the results of chance interactions for which the word is 'coincidence'. Life is simply a vast accumulation of coincidences. How, though, can you reason with or falsify a theory based on unpredictability and on coincidence? You cannot establish tests to prove what is, by definition, immune from prediction; or to replicate what various unknown changes happened once each long ago. What could you not claim might have happened? The level of unfalsifiability is vastly compounded when the theory is extended to include prehistoric events. **Biological evolution's appeal may lie in its materialism but the scientific credentials of such a theory are at least as suspect as one that includes an untestable metaphysical presence, intelligent design. Moreover, the latter has the virtue of making sense whilst the former makes, on close inspection, none.**

While biology lags cosmologists have, however, become increasingly aware that universal constraints (or rules) are precisely fixed and the laws of physics finely tuned in a coordinated way that makes them supremely fit for life - an awareness called as the anthropic or bio-philic principle. 'Bio-philic' means life-friendly. Physics and chemistry are in this view bio-philic. How on earth, though, might you argue (Chapters 8 and 25) that the scientific platform is revolving to an angle that again clicks with the core perspective of natural theology - a bio-centric universe and man a microcosm that reflects the nature of the macrocosm, a view that man is not just measurer but measure of all things?

'You definitely cannot', declare one faction. An idea that the business is a 'set-up job' must, to the last ditch, be resisted. We are only here to consider the issue because the 'right' flukes all happened. Of all possible combinations, we happened to 'hit the jackpot'. Jackpots are one in very large numbers of possibilities so that a recent, astronomical corollary of this line of argument is to suggest our universe is perhaps the single winner (judged by the criterion of enabling humans to exist) in a so-called multiversity of failures. This is, although supported by extreme speculation, a valid interpretation of anthropic principle.

'You definitely can', retorts the other party - also validly. Why invent multiverses? Why not simply work with what we know, a universe, and allow a teleological, bio-centric perspective? Physics and, it will be strongly argued, biology are increasingly giving the lie to a four-hundred-year-strong intellectual effort to demolish the teleological point of view. Life needs to combine fluidity with structural solidity: it is likely, therefore, that even if any other 'liquid crystal' forms of life were discovered their substance and operations would, due to biochemical and physical constraints, occur in conditions like ours and resemble ours. Certainly none of the physical or psychological sciences has submitted any convincing, predictive blueprint for any other than the terrestrial 'carbaqueous' form of life (one based on carbon, oxygen and hydrogen) nor has any been found. It may be argued that their explanations all depend upon a partial view of things (called saturated materialism) that excludes a commonsense, down-to-earth and defensible inference that the cosmic projection operates as a well-informed, coherent whole. Nowhere is purposive order, a metaphysical entity, more apparent than in the integrated designs of living organisms.

Of course, such a claim diametrically opposes orthodox neo-Darwinism.

While the latter deals with accidents and their effects it does not account for the creation of vehicles that suffer them. It helps explain a body's wear and tear but not its origin. **Indeed, it may be clearly and unequivocally stated (Chapters 19-24) that the theory of biological evolution, while correct in respect of minor variation, neither explains the origin of life itself nor its major organic blueprints; in other words its elements may be right but its extrapolations incorrect.** It would be, in this case, a species of half-truth. Is such half-truth not compounded? No doubt its explanation of life's origins is materialistic and therefore, in scientific terms, complete and wholly rational. Is there no other rational kind of term? Its half-truth is compounded if we recognise material is, in consideration of the immaterial, only half the story. This half fits only half (that might be called the 'lesser half') of Natural Dialectic's *top-down* structure. Nor does it sit easily upon the facts. Indeed, although still fashionable and widely rehearsed, Darwinian explanation of life's origins is now approaching sell-by date. Following the description of an act of creation and a possible, immaterial mechanism for the intelligent construction of prototype (Chapters 5, 6, 8 and 15 to 17 but esp. 16) the neo-Darwinian theory is, in respect of its extrapolations, comprehensively shredded (Chapters 19-24).

Where does this leave us? Even if the cosmic stage seems well designed, what about the playwright's mind, plot and players? There are two alternatives both of which recognise the integrity and apparent ingenuity of natural constructions. In this sense both are teleological and might be termed natural theologies.

Integrity, Coordination and Coherence.

The second frame is a theory of 'intelligently-prompted' and inevitable evolution. This suggests that, because of the bio-centric way in which the physical laws of nature are framed, the spontaneous appearance and subsequent transformative evolution of biological forms is 'written into the cosmic script'. Big bang was transcendently projected with preternatural order written in. Its 'egg' was programmed in anticipation of life forms. These were preordained in the sort of way that, given the constraints of a particular set, a certain kind of play is bound to appear; its plots and players must, somehow, emerge; and, finally, their ramifications inevitably but coherently complicate until, according to a 'strong anthropic' version of events, Promethean man was always destined to swing down from the gods to centre-stage. Or was it apes that dropped from trees?

Is it clear? Once the 'mind of God' had established the rules and set the show in motion, He need no longer interfere. Such thinkers dig their Maker's grave; they simply blank Creative Bio-action out. No special creations (except the primal, cosmic one), no divine mutations, simply a naturalistic continuum within which different forms of life keep cropping up. Every designer adapts his invention according to its use; and his work is constrained by the various forces it will have to withstand. In the case of biological players it is mooted that first principles, the laws of nature, must have been sufficient to 'force' appropriate molecular coordination, physiological systems and anatomical shapes from the dusts and clays of earth. The key word that comforts such directed evolution is 'inevitability'.

Yet does the presence of iron ore, quartz and a volcano inevitably lead to a motor car? The signs for progress don't look good. In which case, perhaps, G more than set the show in motion. He must retain an interest, a mysterious hand that 'prompts and guides' things in a way that looks like chance. Lacking plan the Maker somehow, sometimes orchestrates material response at whim! This sort of catch-all explanation, called theistic evolution, is highly unsatisfactory. Indeed, it will become clear (see Chapters 25) that there are profound problems with any theory of directed evolution.

A third alternative is, like the second, teleological but this time fully and flagrantly so. No chance. No coincidence except around the trivial fringes. No accidents except like those that gradually degrade a spanking new machine. It suggests that the universe devolved from its original, instantaneous projection in the orderly way a picture is drawn, a machine engineered or a play staged - stepwise. The first steps are metaphysical. An idea, the will to express it, an outline sketch and detailed, carefully recorded plans all precede the construction of an appropriate physical stage, biological puppets and, with them, the drama of each individual living puppet's part. As in the case of any artistic or technological plan, no detail is left to chance. No doubt, as the cycles of physical production roll, variations-on-theme are played out. Conditions change but the basic rules of plot remain. The whole business is a projection. It is evolved within and projected from the mind of its creator. *We call this wholly intelligent sort of evolution an act of creation.*

Such an extremely human-centred angle locates mankind at, potentially, the climax of the universe, at the climax of a purposive creation. Such centre is not physically located. It is not a body but, at the Apex of Mount Universe, identified with Life; and if creation is deliberate what is the purpose and potential of its most creative *creatura*, man? What is his real role? Is he a knowledge-hunter, an information-gatherer from *Phylum cognoscens*? The organism that most loves to learn and know and feel it understands? Is man incurably inquisitive, a seeker after physical truths - and metaphysical as well? Is his a life that can relate to and, at climax, reunite with its own Source, its Creator? Which figure, earth to earth or soul with Soul, describes him? Or do both sides play a part? If this is a contemptible consideration for an atheist it is also more his emotional problem than a scientific one. Especially because, ridiculous or not, it might be true.

An Act of Creation.

Chapters. 1-3 introduced Natural Dialectic. *Natural Dialectic is the framework within which to cast top-down creation.* Its order, within which the order of this book is constrained, starts at the top and works down. Thus Chapters 5 and 6 start with author, creativity and the act of creation. Chapters 7-12 deal with both metaphysical and physical principles according to which the stage is designed, 13-18 with the 'masked' psychology of individual personalities and 19-25 with 'puppet' biology. 'The world', observed a famous poet, 'is but a stage'. 'Stages within stages', you might add.

An act of creation involves concentration, centralisation and the gathering of power. On the psychological side it involves a focus of attention and on the physical a concentration of energy. Such (*sat*) potential, once released, (*raj*) flows towards its (*tam*) 'outworked' completion. In this way creation involves a dispersal, depletion, diffusion or attenuation of original concentration. Materialisation involves the expression of an idea or crystallisation of energy into a finished product. This product represents the form of its inertial equilibrium, exhaustion or finality.

	tam/ raj	Sat
	range below	Top
	existence	Essence
	range of consciousness	Consciousness
	static/ kinetic	Potential
	motion	Balance
↓	tam	raj ↑
	downswing	upswing
	passive/ lower/ outward	active/ higher/ inward
	informed	informant
	static/ 'stop'	kinetic/ 'go'
	object	subject
	physical quality	metaphysical quality
	non-consciousness	consciousness-in-motion
	matter	mind
	automation	purpose

You say there is only matter.

I say there are consciousness and matter; and it is a fundamental error of science to conclude there is no force other than material. 'Science and the Soul' accumulates the force of this assertion. In its case a Concentrate of Consciousness (without matter) amounts to Substance whose shadows cast are mind and matter; it amounts to an Infinite Pole of which the character of pure matter (without any consciousness) is the exhausted, nether consequence.

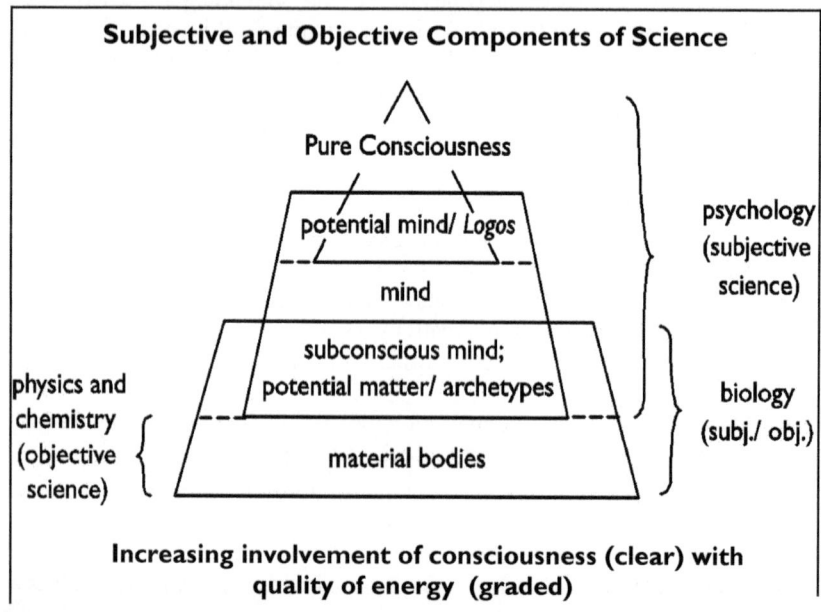

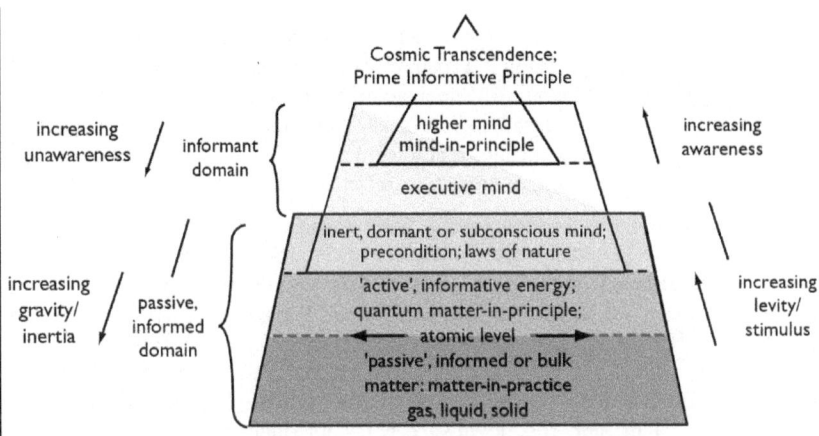

Relate this diagram with *fig.* 0.5 iii. Essence, Pure Consciousness, is the basis of a 'shadowy' or 'impure' consciousness called mind. In the first case, therefore, Essence informs existence; conversely, existence is informed by Essence. **The gradual binding of awareness up in matter is called, from top to toe, the conscio-material gradient.** Such gradient, dropping from the pole of purely immaterial concentrate to its anti-pole of purely material concentrate, is central to the order of existence. The concentration of non-consciousness (otherwise called energy or matter) is, of course, a precipitate called our physical universe.

Within existence active information involves awareness: passive information does not. Passive information is fixed, automatic, reflex. The domains of its storage are subconscious mind and non-conscious matter.

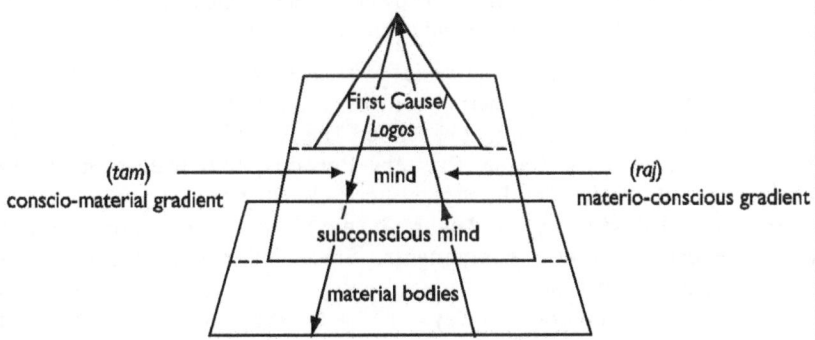

Natural Dialectic represents the creation as a ziggurat or concentric circles but what is its 'dynamic'? What is the light of the dialectical rainbow? Check *figs.* 0.8, 1.3 and 2.3 to reconstruct your idea of gradient and vectors. Two factors apply - inward, informant mind and outward, informed energy. The psychological, subjective dynamic is 'volitio-magnetic', that is, a combination of desire ('I want') and will-power ('I will have'); and the objective dynamic is material energy.

> In this case 'the creation express' runs down a conscio-material gradient; its expression is a form of fiat - 'let it be'. First Cause or *Logos* <u>is</u> a concentrate or super-state of consciousness. From an original, that is to say hierarchically superior, exercise of will devolve creative principles that inform the non-conscious practice - the energetic components of material creation.
>
> Between the concentrates of consciousness and non-consciousness each vector of a 'polar dynamic' has its part to play. **Expressions of *(tam)* gravity of materialisation and *(raj)* levity of dissolution oscillate in all psychological and physical events.** Their rings cement, their 'electric' polarities bind nature in all parts at all times. In material terms (Chapters 7-12) the upward direction, levity, can be construed as one of 'lift' or stimulus; and the downward track its opposite. Involuntary matter cycles in between extremes of radiation and of mass; voluntary mind, however, might achieve a final 'lift', eventual dissolution into its primal concentrate of Consciousness. If conscious life involves creative capability then this concentrate is pristine potential, pure life. Therefore spiritual psychology in practice always and voluntarily seeks an upward, 'materio-conscious' gradient of 'return'. It is this (↑) unifying line of *Logos* (is it love or understanding?) that a holy man aspires to 'ride'.
>
> *fig. 4.1 (see also figs.0.5, 0.8 and 0.13)*

In this view the central, unifying component of both Essence (pre-creation) and existence (creation) is *(Sat)* Consciousness. **Consciousness is, therefore, the prior universal.** Characteristics of Essence, towards reach the *(raj)* upward, unifying principles of existence, are related on the right-hand column of Main Dialectic. Characteristics of *(tam)* downward, materialising principles are, on the other hand, related in the left-hand column of Branch Dialectic.

Essence is the Potential that precedes existence. The preface to existence is therefore, within Essence, a gathering of focus, a concentration of consciousness whose 'welling' forms what might be termed The First Idea; or, dynamically in terms of presence, First Cause. **Primary Paradox, that everything is an appearance of Basic Truth, starts here. Such paradox involves the polarization of First Cause (or *Logos*, *Om* etc.) into subjective, informative and objective, energetic components.** Since information hierarchically precedes energy so psychological precedes physical creation; and this means that physical creation drops, in a way to be (as far as possible) described, from patterns set in universal mind.

The primordial vibratory interaction of this pair, information and energy, involves the activation of potential. Essence is excited. Consciousness is motivated by the psychology of will and desire. Such Consciousness-in-motion is called information or informative mind. Mind is the seat of order, a governor, nature's natural agent (I presume you think your mind is natural) of purpose, anti-chance and law. In this case it's identified as being hub and prop of universe; and is often analogised with the fluent 'creativity' of air or water -

current, stream or wave. Call it source, stream, ocean, breath, breeze or wind; call it *anima, spiritus* - both words mean 'breath of life' - or power-filled transparency. Waters flow with a clear, clean and cleansing rush - the metaphor is universal; and if the current of the *Logos* is alive with purpose, then its Logic is the logic of creation. What, asks everyone excepting atheists, is this?

If consciousness is called *purusha* then universal energy is, on the other hand, called *prakriti*. Such energy, the antipode of consciousness, is non-conscious and therefore intrinsically anarchic. Unable to actively impart information but instead passively informed, it is the medium through which mind projects images and formats patterns of desire. Subtle *prakriti* called '*manas*' is an instrument of expression directly used by awareness; it is both motive power and screen on which images are projected; it is mind-world's 'wave and particle'. In fact, mind is a compound made of consciousness-involved-with-energy like this. *Mind is the active subject, energy the passive, plastic object. Patterns of energy or its equivalent, matter, represent passive information.* Gross *prakriti*, the forces and particles that make up the physical universe, are passively informed.

Existence is seen, in this view, as a combination; it appears as a conscio-material hierarchy that consists of an interaction between consciousness and non-conscious energy. This scale of ratio includes the special case of Pure Consciousness (whose objective, energetic value is 0 and whose subjective, informative value 1); and, at its nadir, pure energy (whose subjective value is, since it is non-conscious, 0 and whose energetic value is 1). Between the two, from higher consciousness to low, there runs an existential scale of creativity, a spread of mind's creative possibilities. Base grade, the nether case of mindlessness, is one of entirely passively informed energy; we call it the physical universe.

Top-down, motion on this hierarchical gradient involves the conversion of active information or energy to passive, inertial forms of either. High-grade intelligence and information top the scale. In a process of creation information is converted into rigid forms and formulations; or, as it used to be explained, there is unconscious resistance to the current of divinity, its spark of creativity is earthed or royal will exhausted into common physical precipitate. Active thought is converted into a material reflection; mind's order is imprinted onto bodies and their properties. Material symbolises immaterial. A creation's shape and/ or behaviour illustrate the character of its creator.

In this process of conversion information is rigidified, will-power is transformed to outcome and ideas materialised. If you think that matter is 'more *real*' than mind then you could say that wishes have been *real*ised. *If, however, you presume some prior metaphysic you are under obligation to explain its archetypal shape or shapes.*

Energetic drive (called will) is lost upon completion of the job. Change of focus, loss of interest or desire is complemented by the graduated loss of consciousness a mind can slide through. Such entropy includes lowered awareness, intelligence dwindling to the extent of incomprehension, ignorance, a reduction to purely reflex, instinctive behaviours and, at base, inertial, dormant consciousness. The most involuntary state of all is dormant, sub-conscious and experienced as sleep. The (*tam*) descending trend of informative entropy is reversed by (*raj*) 'waking up', increasing intensity of interest, enthusiasm and coherent focus of attention.

The physical parallel of psychological 'materialisation' involves *loss of energy*. Such loss 'hardens' from plasmoid through gaseous to an inertial, solid state of matter. In both cases active falls into passive mode; and materialisation can be reversed by a gain, an excitation, an input that stimulates the rate of motion, change or frequency of vibration. From an overview matter is a special case of mind, that is, energy alone. 'Alone' means without any active, subjective aspect. Matter's information is totally passive; its 'flatness' is mind's special case, its absence. Although such energy acts in a psychological void it is, however, cosmic and not chaotic; it behaves in predictable, 'lawful' ways. Is such order the consequence of previously active attention? Or is the coherent operation of our universal body down to chance?

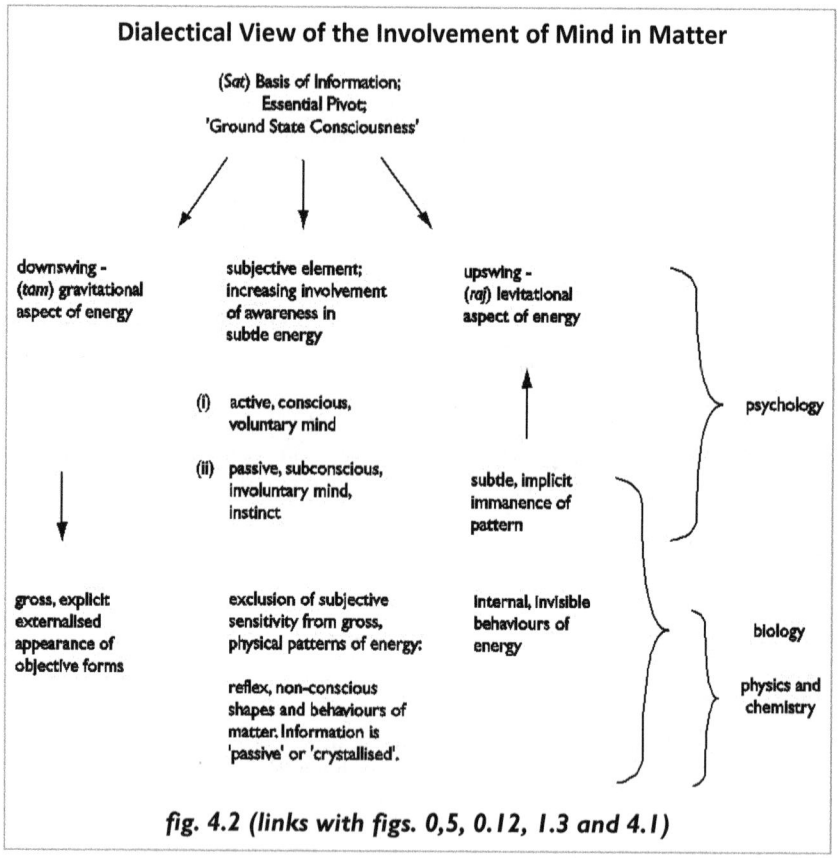

fig. 4.2 (links with figs. 0,5, 0.12, 1.3 and 4.1)

The two major categories of dramatic existence are mind and matter; or, in personalised form, mind and body. They are described in scientific terms by psychology, biology and physics (with chemistry). These subdivisions study mind, mind/body and body alone. Rephrased, they study thought, psychophysical organisms and automata. The first two show what most people recognise as life, the third non-life. Scientific attempts to describe animate and inanimate nature, whose connection is biology, have been successful in the physical arena but less so in the physically intangible, metaphysical or psychological domain.

Natural Dialectic concurs with the scientific categorisation of existence into psychological and physical components but disagrees that they are simply different aspects of the same thing - matter. Instead its solution to the 'mind/matter' or 'life' problem is to distinguish between subjective awareness and non-conscious material energy. Both exist in you but are also universal. In other words the Dialectic proposes that life derives from Life and not non-life (in the forms of material forces, atoms etc.). The *top-down* perspective is therefore at odds with a uniform, materialistic point of view. The implications of a reality that includes both physical and metaphysical components are pervasive and numerous.

The following chapters elaborate the consequences of this brief analysis. They identify the neatness of its fit with what we know about the world; and interpret this data in the new, trinitarian framework of Natural Dialectic.

Psychologically (Chapters 5, 6 and 13 to 17), a thinker projects his ideas. They are projected by his body language but mostly by the language of his vocal-corded voice. The stage is inanimate but actors tread its boards in various guises and with varying degrees of animation. Some, like bacteria, plants or fungi, appear robotic and respond in a reflex way to events. Others are more aware and show a limited form of choice, what seems to be 'free will with strings attached'. Constraints, contexts or limiting conditions of role include body-form, instincts, environment, previous experience etc.

The central, psychological purpose of an experient is to gain information, to improve its quality of survival and realise such truths (however humble or grand these may be) as enable the achievement of its goals. To this end mind is an information centre. Conscious mind is active and works in 'psycho-space' through a bio-cybernetic cycle of intake (information receipt), processing (imaginative data manipulation and decision-making) and output (implementation of purpose).

Sub-conscious mind, on the other hand, is a relatively passive, automated 'cyber-space' of information storage and transmission; it includes the faculty of memory and its processor is instinct.

The more mindless, then the more automatic, rigidly organised and yet, apparently, chance-prone is a system, object or event. *Physically* (Chapters 7-12), matter is completely mindless. Its non-conscious behaviours are entirely automated. Mind works with material objects and events indirectly through the medium of a biological body (Chapters 17 and 19-24). Its processor, the nervous or any other homeostatic system, is chemically constructed.

The central biological purpose of an experient, which informs and is circumscribed by appropriate biochemistry, physiology and anatomy, is life. To live is, translated into a biological base-line, physical survival. Therefore for most organisms to seek to survive is perhaps their single, central truth. Is it for humans? Is comfortable survival their whole truth or simply part of it?

The aim of Natural Dialectic is, as faithfully as possible, to truly reflect the whole of creation. *From a top-down point of view, the highest exercise possible is the creation of something from, apparently, nothing. It is the authorship of a work of art, architecture or engineering design. This is why Author with his Idea comes first. Authorship is a crucial form of potential. From ideas ripple powerful effects; from The First Idea ripples an all-inclusive existence.* **Idea seeds physical consequence and so, as already noted, it is imperative to understand the order of a *top-down*, creative act.** From what does a painting, an opera, a book or an article of technology materialise? At first there is nothing. But this void is neither negative nor empty. It consists of the creator's potential; of his highest capability. Thence emerges inspiration, insight or idea. This is the First or Causal Principle. There follows an outline sketch from whose simplicity a detailed plan is evolved. The plan's purpose and complexity is derived from the First Principle. The latter authorises; its central authority informs the nature of subsequent or peripheral construction.

This first stage of construction, plan, occurs in mind. It includes, where appropriate, storage of such plan. There are many types of information storage mechanisms. All involve symbolic code. Examples include vibratory patterns, biological *DNA*, books, tapes, computer memory etc. We save stage dramas in books. *In fact, the natural psychological storage mechanism is called memory.*

The second stage is expression of plan. Principle is commuted to practice, plan to product, dramatic text to its staging. The process runs from simple outline (or backdrop) to complex detail. If we take the analogy of a painting, the preliminary sketch is drafted; of a concert, the correct instruments are assembled; of a play appropriate scenery is prepared and actors auditioned. A piece is built according to its plan. *A design has a reason. This reason is understood to be its meaning.* The meaning of anything is, simply enough, that it does what it is *meant* to. It works. Its work and working, that is, its reason is most fully expressed in its climax. It might therefore be sensible to identify climax and thence deduce the reason for its components. For example, is the climax of material construction a human brain? And life - what are life's high spots? Survival? Happiness? Sexual orgasm? Friendship, love, the joy of understanding? Realisation of, even union with, Origin? Which of these hits the Highest Spot?

The vector of materialisation is top-down. An accurate description of any deliberate creation needs to be *top-down*. What, however, about macrocosmic, universal creation? What is the vector of your perspective? Is it *top-down* or *bottom-up*?

It depends where you start. How do complex structures begin? Eggs are powerful symbols of preordination, latent information and potential being from which material forms develop. Eggs mean promise. Eggs encapsulate intentions. They are rational reflections of causes because their purpose unfolds. 'Cosmos' means 'perfect arrangement'. Highest order. An egg is, therefore, a prime metaphor for cosmogony. If nuclear information packed in a zygote derives from its generator(s) what about the case of cosmic seed or universal egg? What is the nature of nature's generator? What is the nature of the Nothing that preceded everything? If the cosmic start was single, all things must have materialised from its simple yet concentrated focus of information. Did mind precede matter, information energy? Did passive precede active information or the other way? What is the nature of the cosmic predecessor, dialectical First Cause? Is your cosmic egg Pure Consciousness whose outermost, non-conscious projection is matter? Or was it perhaps a mindless accident that generated all the ways that sway material energy and thence, perchance, gave rise to you?

In other words, what is The Big Idea? Can you ever 'get it'? Or is it a hoax?

Bottom-up, it's just a hoax. Matter is prior and mind is an unusual sideshow. It is an 'emergent property', a flickering and transient frailty evolved from rare, localised atomic conformations. With only one universal substance there is little notion of hierarchy or vector. It is not a question of, say, whether the order of creation is from principle to practice or *vice versa*. Principles observed in the behaviour of matter have no particular origin. They are, in the empirical view, simply part and parcel of the way things 'happen to work'. In the scientific case there exists neither mind nor, therefore, reason behind these 'principles' or 'reasons'. They are both unreasoned and unreasonable reasons. This is the contemporary, somewhat unsatisfactorily self-contradictory orthodoxy, the 'normal' perspective on things.

If, on the other hand, things developed *top-down* from a nucleus of Pure Consciousness then the informative order of issue runs from consciousness-in-motion (mind) through sub-conscious, dormant mind to non-consciousness. If such a 'gradient' is the case, there is more to cosmos than meets the eye - or telescope or electron microscope. Just as a straight line is a special case of curvature (none) so non-conscious matter is construed as a special case of mind (none). Its 'flat' inflexibility were better described as 'no informant, all informed' or 'all active energy but passive information'. In this way the shapes of energy are construed as crystalline mind, fixed 'instinct' or 'frozen' memory. And in this sense one-dimensional materialism is called 'flat earthism'. It marks a regress in the evolution of philosophy, an ebb within the flow of human understanding.

Top-down, mind is prior. As noted, it is certainly the case when an author, artist or engineer realises his idea, according to various principles, in practice. *Principle informs; it is the source of reason because of which patterns of behaviour occur.* It is assumed that a conscio-material gradient drops, with a vector of consciousness, from mind through a psychosomatic medium to non-conscious matter. This medium is identified as sub-consciousness, a state no more a void than sleep or an unremembered memory. Indeed, sub-consciousness is the residence of 'solid' mind, fixed psychological 'objects' in the form of records called instincts and memories. You could see it as a natural database, a personal 'cyberspace'. <u>Instinct is archetypal behaviour and if the behaviour of matter is informed by universal mind, such information should issue from the same channels as those through which a dormant mind informs body</u>. Of course, the human psychosomatic interface is complex. It includes a control panel called a brain. How, though, does memory work? How are instincts transmitted to body and how, in parallel, might archetypal 'instincts' be transmitted to the cosmic body of matter? Certainly not by nerves but possible agencies such as resonance, light and electric charge are discussed later (Chapter 16).

In this view information shapes a passive, energetic or material component.

The latter is the visible medium through which invisible mind expresses itself. There is a real sense in which an instrument of purpose, say a written message or a machine, is a memory of its creator. It is a memorial. It is a device to abnegate the presence of chance, erase time as far as possible and, as efficiently as possible, achieve specific goals. It has no mind but has! Both technological and biological instruments of purpose can, although they have no mind of their own, be classified as 'mindful'. *In this other sense memory can be objectified. All machines are, in their physical action and appearance, forms of memory.* Every artefact is an immediate, ephemeral or long-lasting impression of its planner's mind on matter. As well as atomic construction it stores, by arrangement, the mind of its maker. It is in this sense that the cosmos is construed as crystalline mind, the memory of its maker.

Finally, it might be asked if the order of creation survives; that is, does each phase through which a hierarchical act of creation is brought to fulfilment continue to exist? Might such phases not constitute the structure of creation? If an 'organic' expression of cosmos 'developed' from principle to practice, from informant to informed or from mind to matter, could existence still incorporate the stages - each from 'within' the previous - through which the original projection devolved? Is the Infinite Presence still present? Is the next step, mind, still here? Who disagrees that the base stage of Mount Universe, matter, is still kicking round? Space, gas, liquid and solid, each in their hierarchical place, compose the currency of planet earth. Is every step in you? Are you Macrocosmos Incorporated? Mount Universe Limited? What's the business, what's the system's origin and how does it work?

Physics certainly tries to map a high-energy cosmic history using experimental probes such as particle accelerators and satellites. With a different approach it is now time to apply Natural Dialectic to a world at least as dramatic as it is dynamic or mechanical. Each application counterpoints the *bottom-up*, materialistic version of events (which interprets the world in terms of physical energy) with the *top-down*, holistic version (which emphasises metaphysical information but includes physical things). An accumulation of evidence adds up to a rational inference of intelligent design. Further coverage of 'abiogenesis' (Chapter 20) and the roots of biological form and function generate critical mass. The case for a universe ubiquitously imprinted with the signs of bio-centric design becomes predominant and irresistibly common-sensible.

In summary, from what is subtle, unseen and fundamental are derived gross, different and changing appearances. From the permanent archetypal duality of massive (↓) proton, radiant (↑) electron and associated forces devolve all physical phenomena. Immaterial mind is subtle, matter gross; from immaterial immanence, called archetypes, there issue physical behaviours. Mind's polar fundamentals - radiant (↑) consciousness, a formative (↓), oblivious 'screen' of *manas* and associated forces create all psychological projections. For example, the 'volitio-magnetic' light of mind is focus of will and desire; and voltages of (↑) 'good' and (↓) 'evil' influence in some degree the sway of every thought. Thus is woven every local, individual dream - or shall we call our minds, for us, reality? Thus were woven archetypal dreams as general as construction of the universal dream - or shall we call our bodies and their cosmic circumstance reality? Since subtle precedes gross and mind is prior to matter, for Natural Dialectic all creation is projected from an Immaterial Source. This is, of course, materialism upside-down. For atheism it is simply maddening lunacy. Which paradigm, however, actually stands upon its head?

If the order of natural creation is *top-down*, from conscious to non-conscious, then any accurate description must follow suit. This is simply another way of explaining why the order of this book and its instrument of simple expression, the eponymous Natural Dialectic, must also work *top-down*.

The book's binary structure can, accordingly, be simply characterised.

Part 0: Macrocosm	(i)	Author Metaphysics, *IT*
	(ii)	Stage/ Universal Body Physics & Chemistry
Part 1: Microcosm	(i)	Actor/ Specific Mind Psychology
	(ii)	Actor/ Specific Body Biology
	(iii)	Cast Ecology/ Sociology

Stingers.

For *bottom-up* materialism any metaphysical discussion is irrelevant. Only inanimate energy exists. The perspective hopes to explain, in this term, consciousness, mind, instinct, morphogenesis (the development of biological form) and, centrally, the nature and origin of information, order, purpose, reason and language. It hopes to fully explain not only the physical universe but life itself. Its problems are deep and pervasive. *Despite faith, hope, optimism and ingenuity it may be that, within the limitations of its terms and conditions, it will prove impossible to sign off the full document of explanation in just materialism's name.*

<u>For this reason alone it must be reasonable to thoroughly consider a rational, self-consistent and coherent competitor.</u>

Have you no faith in the unseen - even when you might infer such presence logically? An electron was unseen; dark matter is suspected but unseen; could metaphysical intelligence not range the universe? A materialist 'cannot imagine' a moment of creation by design. A moment of psychosomatic impression, a miracle of materialisation offends his present experience. However, science deals with current operations. By extrapolation it infers historical events; but causal agents and one-off events (such as the origin of life and cosmos) are outside the scope of proof and thus described according to 'best explanation', that is, according to 'best-educated speculation'. Such an explanatory guess, philosophically reduced to naturalistic, might be wrong. Is an immaterial origin beyond imagination? If science hung upon the limp excuse of 'can't imagine' nothing would have ever been discovered.

A *top-down* view instead adopts a Theory of Intelligence. It puts consciousness at the heart of an otherwise inanimate universe. Natural Dialectic serves in its entirety to justify the strongest of anthropic or, more generally and yet precisely, life-centred perspectives. Such a view includes (which the currency of modern philosophy, scientific discrimination and popular thought disdain) the notion of universal mind. How can its nature be explained? It is sometimes profitable to tackle a problem by working backwards, the reverse approach. This is an entirely acceptable, scientific mode of operation. Universal principles can be inferred from small-scale events or experiments. *Thus, where the reverse might actually be the case, universal mind is treated as a reflection of human potential, orders of expression and levels of consciousness.*

In this view principle is presumed to hierarchically precede practice, mind to inform matter and matter to constitute an important but external shell of a whole projection called 'the universe'. If this is the case a materialist has unpleasant nettles to grasp. Nor is the holist in a less formic, more comfortable position. The *top-down* idea of a universal mind immediately raises numerous issues; and if cosmos involves a conscio-material gradient through mind to matter, their conjunction has to be defined. *This stiff challenge needs to be squarely faced, logically worked through and, as far as possible, its difficulties resolved. Stingers that need to be grasped include:*

(i) the nature of consciousness (Chapters 5, 6, 13-14), sub-consciousness (Chapters 15 and 16) and non-consciousness (Chapters 7-12).

(ii) whether individual mind can exist independent of a body and, if so, the nature of its entry, attachment, exit and disembodied condition (Chapters 13 and 18).

(iii) the interactive relationship of mind with matter; the nature of any *PSI* (psychosomatic border or, perhaps, quantum linkage) between mind and matter (Chapters 6 and 15-17).

(iv) if prior, informative metaphysic is presumed then the psychological aspect, if any, of matter; the mechanism by which mind might inform material precipitate; the origin of physical constants and patterns of behaviour, that is, 'the laws of nature' (Chapters 5-12, 15, 16).

(v) the nature of physical and biological prototypes, homologies or archetypes, if any (Chapters 8, 9, 15-17, 19, 22 and 24).

(vi) 'Chance and/or necessity; or design in accordance with necessities?'- a wholesale reappraisal of the neo-Darwinian theory of evolution (Chapters 5, 6, 8, 19-25).

These and other 'eternal' questions constitute, alongside the ever present, underlying Nature of Nothing, a *motif* that adorns the following narrative. They constitute a continuous backdrop to be borne in mind throughout the dialectical to-fro between *top-down* and *bottom-up* perspectives. What, if anything, is Nothing? What is the nature of the world? Is there any purposeful agenda? Will some or all these questions be, although addressed, answered? How well?

How Can Science Sensibly Cope?

No doubt, materialistic naturalism cannot philosophically approve of *top-down* attitude. If, however, there's an elemental, immaterial factor - information - then the writ of such exclusive form of science only runs so far. The dialectically sad equation - 'science = materialism' - fails; and to extend its limits-by-self-imposition the consensus we've dubbed 'scientism' needs a level head. It does not need, as from some quarters when creator, metaphysic, archetype or concept of design is reasonably invoked, a hissy fit that simply spits, "Irrationality!"

The calculated odds and the facts will demonstrate the days of atheism's bastion, evolutionary speculation, are numbered. For some a cloud is passing, for others approaching. The great fear of any believer is, if doubt insinuates or topples his faith, what might replace it. Theists are tormented by stuck-in-the-materialistic-mud militants. Atheists (a central tenet of whose faith is the theory of evolution) are alarmed to see the growth of Christian literalism complemented by an Islamic world, whose creator is Allah, taking an ever-increasing interest in the evolution/ creation debate; and by an encounter that is much, much tougher - one with Intelligent Design. A missionary materialist parades his faith by sandwich board - 'God is over, God is dead'. He would decisively debunk *all* metaphysical delusions of Universal Mind, Creator or Creative Intelligence but what if, unthinkably, his ideas of cosmogony ('big bang' and astronomical consequences: Chapter 12) and abiogenesis (life from dirt, water and time: see Chapter 20) were proven very dubious or even definitely wrong? And conversely, if not according to the specific dogmas of their religions but on a principle of intelligent design, his 'opponents' won the argument? How can the orthodox secularism of science, scientism, sensibly respond?

What authority can swallow humble pie, large slices of? Even if wrong, how could you admit it? Even if you wanted to, how do you halt a juggernaut? How cope with loss of pride or disentangle from a net of wasted investment, investment in scholarship, institutions and personal belief? For a single, minor example, in 2005 a state-of-the-art centre for research into human evolution, into which many millions of somebody's money have been and will be ploughed, was opened in a street where Darwin lived in Cambridge; and in 2011 The Calleva Research Centre for Evolution and Human Science was opened at Magdalen College, Oxford. Any 'church' with religious or political powers of compulsion, not least a secular one, involves finance, property, careers, lives and families. Never mind the basis of a faith its rivals are resisted on worldly grounds alone! In the event of 'attack' all faiths deploy various armouries of defence. One is to adopt a 'bunker' or 'lager' mentality from which to systematically ignore, stigmatise, stave off and propaganda-wise eliminate 'the enemy'. With respect to theories of design and non-design such battle has, in America at least, already been joined. This is, however, a war beyond the territory, power or wealth of institutions but the whole truth. **Suppose that, simply, there is a metaphysical dimension to the world - a dimension that is nothing physical and, therefore, physically absent.** Is simply to deny its science - since it is non-physic - argument enough? How, therefore, might cool heads defend a rational metaphysic from distortion, dogma and excesses by the steamy hotheads of religious politic - both atheistic and theistic species? How might truth hold the natural line and science tweak its paradigm to meet this possibility in a rational, supra-religious and independent way? Can it find a powerful, alternative method with which to grasp the 'stingers' inevitably associated with a *top-down*, metaphysically inclusive point of view?

<u>*In fact the whole objective of this work and its Natural Dialectic is to provide, in an initial form, such a rational response*</u>. **Its aim is to handle full reality in a detached, non-religious yet also testable way**. Such expanded materialism allows the adherents of both pillars of faith to join, argue coherently and impartially appreciate the principle truths behind our world. It is a framework physical science might adopt in order to help cope with what is coming.

TOTAL CONTENTS

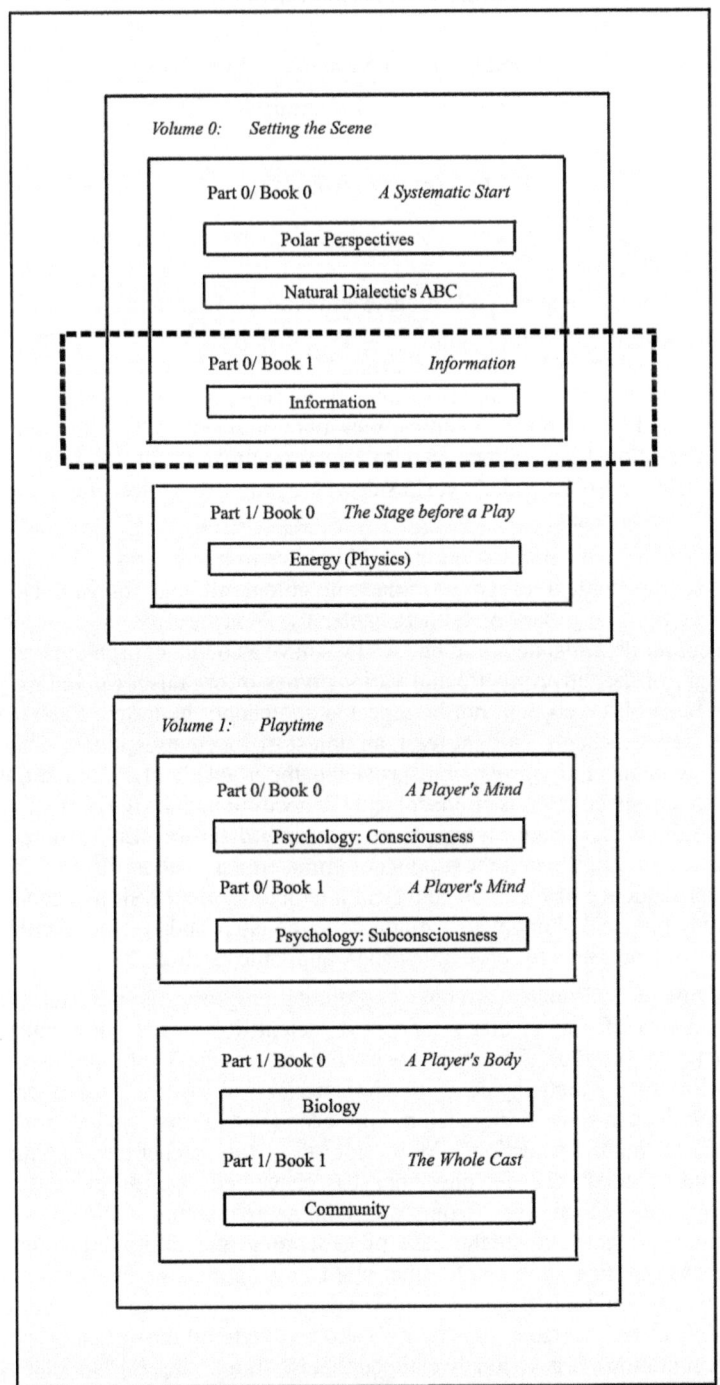

Volume 0:   Setting the Scene

Book 0:   A Systematic Start
Part 1:   Information

5.   Information is Immaterial

"Why should I believe that?" A basic intolerance of cant and an independent, questioning spirit combine to make science a breath of fresh air. Its liberal yet pragmatic freedom of thought exhilarates. "How can I best let nature speak its truths?" Scientific naturalism is welcome to those who, recoiling at all manner of dogma and its attendant cruelties, seek unprejudiced truth - although most people's emotional inclinations seem to veer more towards politics, religion and art than science. At least, let's cast aside value judgments, clear the decks and start with a clean slate!

Clean slates have nothing on them. You have to make a start. A start is an origin, so the context of your origin will colour all that follows. Do you remember the two pillars of faith (Chapter 0)? You already have to choose. Which colour dream-slate will it be? And we have a couple of problems as well. The origin of the universe, life and various types of organism (including you) has not been observed, will not be repeated and cannot be tested. Each slate's historic start is, scientifically at least, an object of speculation alone. Who, for example, can materially prove or disprove the important claim of immateriality? To this extent speculation is philosophical but needs to include the contemporary understandings of science. However, are my primordial roots wholly material or not? Is cosmos and life in it the product of immaterial as well as material factors or the consequence of fortuitous physical interactions alone? Since science can only study physical phenomena I might hope the latter and, having identified a mechanism that seems to serve my wishes, apply the ratchet.

Let's be clear. Ratchets are devices stopping progress slipping back. There are two kinds of ratchet that science - if 'scientific' means 'materialistic' - systematically applies. The first locks out 'regression' into philosophical permission of a 'metaphysical' or 'immaterial' or, by its own restrictive definition, 'super-natural' dimension. The second is a mechanism celebrated by the name of 'natural selection'. For a criticism of this ratchet turn to Chapters 19-25 and especially 22; for now note it is conceived of as a device that, by virtue of death, accumulates happy and eliminates unhappy accidents. These accidents are identified, in the case of generating and improving biological systems, as genetic mutations. In other words, the application of such a ratchet automatically 'guides and controls' the contingencies thrown up by the mindless constraints of nature (called 'necessity') and raw material thrown up by chance (genetic mutation). It is seriously claimed that by eliminating DM's (deleterious mutations; see Glossary 'mutation' and Chapter 23) non-purposive natural selection has superseded the concept of purposive engineering; it has rendered 'intelligence-behind-design' obsolete with respect to the superb conceptual engineering that clearly but only 'apparently' substantiates all biological form. More importantly, natural selection is a process of death by insufficiency or malfunction that is elevated by materialists to the status of lethal injection for the heresy of 'natural mind' or, criminal-in-chief, imaginary theology! How, from a framework of saturated materialism, can you argue with that? The enemy an atheist must first and foremost of them all destroy is naturally intelligent design. Why should a 'Theory of No Intelligence' include it?

The God Delusion?

My God! Are you an illusion or, still worse, delusion - a symptom of cracked mind or empty madness that deranges men? Do you remember (Chapter 0: Religious Delusions) that for materialists the deepest of delusions is a Live Creator? My God, therefore, the heart of all religious mania! In one short, backward word is such a welter of imagination vested. Who in the name of science knows just what it means? Who needs image when reality is all around? Who needs 'the mystery of G' or refuge in divinity when nature stares you in the face? Therefore strike confusion from the dictionary! Eliminate that word! Or else strike preconception and blank out the imaginations of its meaning from your viscous, human state of mind! Yet atheism (the basis of whose creed is neo-Darwinism) has, as we'll see, compounded the problem of an unprejudiced revelation of truth by its arbitrary restriction to naturalistic materialism. From nothing comes everything - but only physically, non-miraculously, by natural law or by selection! Is this, without the blinding force of prejudice, really good enough?

tam/ raj	*Sat*
polar	*Neutral*
order	*Preordination*
lesser grades of consciousness	*Consciousness*
tam	*raj* ↑
object	*subject*
non-conscious energy	*conscious mind*
non-teleological	*teleological*
purposeless	*purposeful*
'free' matter	*arranged material*
pointless structure	*functional structure*
chance	*guidance*
necessity/ natural law	*design*
physico-chemical 'system'	*biotechnological system*

You say there is only matter.

I say there is matter <u>and</u> consciousness.

The former statement is an atheist's axiomatic bluff. Matter *maketh* consciousness. His arguments all start from this presumption, one unproven scientifically. The pillar of such faith starts with a leap of faith; the first step of its first leg is a guess. *Could this be a devil of delusion?*

Neither proven is my own assertion. Science does not deal with immaterials but who knows how 'material' might be reclassified? Were electrons 'immaterial' before they hove in sight? What you don't know or don't yet deal with can exist. And consciousness is labelled, in the Dialectic's book, as a 'subjective immaterial'. Is this another axiomatic bluff? Let not, at the first step, pot call kettle black. *Top-down, bottom-up*; is either way of looking wholly right?

Matter is, of itself, purposeless. Life forms are, on the other hand, motivated to survive; that is, they function to an end and, as they do, their structures serve their functions. The primary agent of purpose and therefore of functional behaviour is mind. **By its nature mind evolves designs but matter never does. Is that unnatural?** Where matter is objective fact, mind is a subjective agent whose emotions feel and whose intelligence informs. **Intelligence is, using information, the creative power**. And what is man's intelligence if not to grasp and variously deploy his principles or find out how the objects of his interest work? Humans are the authors of all manner of bright schemes but could you countenance that creatures are informed by any kind of mind except their own? Organisms may or mayn't have minds but if they are designed then such intelligence not personal in the sense of being human or their own; yet it is exceedingly ingenious. Call it super-human. So by whom (not what), when and how does an act of creation occur?

This issue is central to the view we have of ourselves. Let's rephrase it. Nobody sees purpose in cloud formations, ocean currents or the lie of rock 'systems' - unless the laws of physics are themselves designed. But all philosophical parties see design in biological organs and organisms; their operation depends on many correlated parts whose *purpose* is survival (Chapters 17, 19-24). The question is whether such designs are real or apparent; metaphors for design or the products of intentional design; 'design' or design. A 'Theory of No Intelligent Design', central to scientific materialism and atheism, promotes 'the illusion of design' by evolution. A 'Theory of Intelligence', on the other hand, invokes a metaphysical act of creativity; it promotes design for what we all know it actually is - deliberate and not accidental. Purpose is the very opposite of chance.

For both theories biology's *objective* element is its chemicals and physical constraints (what engineer does not have to account for the constraints his idea involves?). For materialism's 'No Intelligent Design' the *subjective* element is psychological and thereby dependent on nerves, genes and brains. In holism's view, however, such constructions mediate between physical and metaphysical dimensions and the psychological information centre, mind, is a separate element; and, with respect to bodies whose form always follows function, the *subjective* element is purposeful design and the functional value of a feature is derived from this. There is a real sense that, by insisting that there are no subjective values in biology, an objectivist is hoist on his own petard - because he habitually questions the reason and consequent efficiency of any structure derived, in his end analysis, from the chance-prone irrationality of inanimate matter. Charles Darwin simply blunted, not resolved, the acuteness of this self-contradiction. **Indeed, is it because the rationale of 'intelligent design' holds a knife to the heart of his intellectual delusion that an atheist resists the notion so ferociously?**

What imaginations, therefore, might a neo-Darwinist conceive against his bogeyman, God the Designer? How does he fabricate a scare-crow straw-man at which he targets his *ad Hominem, Misericordia* and other species of assault? Firstly, of course, he chooses all the worst examples of 'religious' behaviour (not including atheism's own) that fallen spirits have embraced. His core appeal, however, is to the argument of infinite regression (see Chapter 0: Models). You can, at root, propose two sorts of cosmic origin - a starting and a start-less one. Steady-state (eternal matter) evokes an endless series of causation; a start involves projection out of nothing. The nature of this nothingness defines First Cause. Preceding physical it's metaphysical but could it, when it comes to a Creator, be uncaused? Origins exist outside the scientific realm; therefore, it is arbitrary and unscientific to presume that First Cause must itself be caused. An atheist, however, may decide it is. Thus who designed, in infinite regress, the world's Designer, Super-Designer, ever larger nets of nerves or powerful computation unto ever-greater nonsense? Is this, an inappropriate analogy, how you physically imagine metaphysical causation.

Look at it this way. Forms of life are complex, all agree; and, given entropy at work, high complexity's improbable. Specified, coherent and coded complexity would be, by natural forces like the wind and rain, highly improbable - some, not atheists, would argue impossible. Is it not, after all, intelligence of mind transforms mindless, natural impossibility to certainty - as you do scarcely thinking of it every day? Mind, however, isn't in a naturalist's remit; but somehow a Bio-Designer must 'out-complex' his designs. Of course, you may conceive of physical complexity in terms of the abovementioned brain or computing power; but if your nonsensical Designer isn't physical (is mind physical or immaterial?) you're barking up a gum-tree. Never mind. You've noted physical complexity's statistically improbable and a Creator, reckoned still more complex, even less probable by chance. In other words, a Designer God that might appear by chance approaches zero probability, that is, impossibility. It's thus impossible Intelligent Design designed - unless you take the theist view of what you've said. A Creator formed by chance is absolutely ludicrous so they agree - you never get Designers out of chance!

Thus atheism claims its own 'designer' out of chance - mindless natural selection. Bio-designs just look 'as if' they were designed! The chance is slim but, as you can clearly see all round, unconscious matter must have made cells on its own. It even threw up mind! But wait! Where mind's involved you don't measure using the effects of chance. It's inappropriate. **Such logic, equivocating with the gist of probability, is fallacious**. This time the sleight of atheism's magic hand is not appeal to unseen transformations or to potent swathes of time but to mathematics. In this case the concept is applied statistically; it measures the effect of chance occurrence in the world of reflex, physical events, that is, of material science. But argument for bio-logical design is not decided, any more than with an engineer's design, by such criterion. As archaeologists interpret evidence of artefact, interpretation of design is drawn from evidence. Chance is not the issue. Programs, machines, art and artefact all have creators. Not chance but anti-chance has, although tautologically ignored by atheists, entered into the equation. Mind abhors randomness and espouses purpose; its business is making certain and as such it would annihilate all chance.

It renders such an instrument of argument irrelevant. *The atheistic chain of reason lacks in self-consistency.* Measurement of a designer or design statistically is logically irrational. You might employ such reasoning to oppose the self-creation of a physical designer (such as a chip or code-filled *DNA*) but cannot when the metaphysical takes charge of chance.

Some, between extremes of a designer and of none, promote variations of a stop-gap that we glimpsed in Chapter 4 and will meet again in Chapter 25 - what might be called versions of a 'Theory of Accommodation'. For example Darwin himself, as a deist if not an agnostic, believed that evolution might have been left to unfurl from deliberately contrived initial conditions - after an initial 'miracle' there came no more of them, simply accidental evolution within the constraints of 'unbroken natural law'. He was also keen to derive, as Newton had for physics in the name of gravity, a universal force for biology. This bio-force, evolution, has to account for both initial and increasing complexity of design so that other 'accommodation theorists', such as Teilhard de Chardin, have proposed a grand purpose operative throughout nature that inexorably drives evolution to produce mankind and, far beyond, climax at Point Omega. Is Omega the same as Alpha so that natural selection evolves God? Is de Chardin's theory 'vital' (having a motivating genius somewhere in it) or 'non-vital'? *Scientific biology rejects vitalism yet its modern replacement, coded information, is no more the gift of mindless molecules.* No chemical makes plans. Nor gravity, although it 'organises' suns and moons, can organise a language or can innovate a system that works towards an end. <u>No universal law of increasing, specified complexity exists and yet such specificity is what biology is all about</u>. **No 'bio-force' exists in lifeless matter that is able to develop instrumental systems, organs or metabolisms.** The fact is that, without exception, intelligence is needed to inform design.

"If it could be demonstrated", wrote Charles Darwin, "that any complex organ existed which could not possibly have been formed by numerous, successive, slight modifications, my theory would absolutely break down."

The fact is obvious, we'll see, that this is not the case and that, indeed, the theory is a broken one. The fact is that that organs, systems and bodies always show a coherent complexity of design that derives, as in the case of all machines, from a purposeful arrangement of specific parts; and this distinct arrangement is dependent on a language written on a *DNA*-made database of such digital, computer-like complexity as (information and computer theories indicate) chance/ necessity could never generate. You need great intelligence. The first of steps that precede the creation of a machine is its purpose; this and proceeding steps are taken in its maker's mind. Ideas precede construction and directed construction is, by virtue of these ideas and the arrangement of its parts, a 'specified complexity'. Are not the coded development of an organism or the coded operation of a cell each an example of such complexity? Is not conceptual, programmed organisation what makes a technological or a biological whole greater than the sum of its parts? A self-consistent, coherent whole is, serving its original purpose, irreducible because the removal or malfunction of a part renders it 'sick' or 'inoperable'. Such dialectical order of creation is, it will be seen, an inalienable feature of all biological forms; and perhaps also, with respect to the way the 'laws of nature' are framed, chemical and physical aspects of either animate or inanimate bodies. Is there clearer indicator of design than 'purposive and irreducible complexity'?

In this case it might be possible to recognise that while subsequent chance variations-on-theme occur original themes were designed. Unintentional as well as intentional adaptations to a program can occur. The former are called bugs (or viruses) and may or may not be lethal; the latter involves a systematic reformulation. It could, in biological terms, incorporate a capacity to adapt to common sorts of change (e.g. temperature, water potential, soil chemistry); or involve fresh system whose parameters have all been re-attuned from another homology, whole-body plan or, in spite of Thomas Huxley's dislike of the word, archetype. Don't call an archetype a *bauplan* rather, in the modern animistic jargon, 'program'. Armies understand a plan, a strategy, a goal that's subject to clear psychological preordination; specific orders hierarchically 'command and control' all operations in a field of war. Engineers design equipment men will use in order to survive and win. Is not survival of the fittest palpably on every front about intelligence? A purpose draws you to itself; goals are attractants that thereby control behaviour in their field of influence. Is a morphological attractant, archetype, so different from the notion, 'drawn direct from information technology, of bio-systems, bio-cybernetic programs and subroutines. These 'archetypes', not a long suite of bugs, inexorably execute the reconstruction of 'ground data' (such as biochemicals) into a target formulation (such as a body-shape). They stably yet dynamically express intended patterns. Yet, as we'll see, are chemistry and physics really enough?

If the evidence leads towards design and yet, because of an ingrained materialism, you refuse the logical conclusion are you not profoundly anti-scientific in your naturalistic set of mind? But this, where it comes to origins of life and workings of a cell, is where the set of scientism's mind is bent. No less today than in the recent past a bombast is unleashed at those who strive, beyond dogma, to unlock the whole truth. If a mistake in current thinking is discovered, is to correct it 'turning back the clock'? If a theory of unintelligent 'design' is only partially or metaphorically right but actually, wholly wrong, is reappraisal 'retrogressive'? In spite of noisy and illiberal philosophical bombadiers, let's cycle forwards.

Have you ever ridden up a cul-de-sac? To advance we have, briefly, to reverse. Darwin, according to his 'bull-dog' Huxley, 'dealt a death-blow to teleology'. He pulled past William Paley's Natural Theology. There was no divine first cause, no purpose, end, final cause, call it what you will, in biology. On the other hand it is also claimed, according to philosopher and (in his own words) 'fervent evolutionist' Michael Ruse, that his hero 'painted a thoroughly teleological world-view without directly invoking a designer - natural selection gets things done according to blind law without making direct mention of mind' ('Darwin and Design' pages 173 and 126 respectively). Natural selection you might say 'designs'.

This is, if not deceit, equivocation. It is confusion of the first order; but it is important to spotlight because design-based language (such as code, function, mechanism or specified complexity) is routinely used throughout the study of biology. Ruse simply echoes standard procedure whereby the language of intelligent design is purloined and rehashed in an animistic, evolutionistic way.

In this context it is worth noting communication/ information theorist Warren Weaver who, having coined the term 'molecular biology', then promoted its cause through the Rockefeller Foundation - not least in Cambridge with its associated batch of Nobel laureates. This study's revolution changed biology. It broke life's *code*, the *language* of the genes. Genes, *DNA*-written chemical 'hardware' that carries fundamental bio-information, are digitally coded text; they operate according to the grammar of symbolic language; such language and its agents of expression are conceived *in order to* transmit a message optimally - something senseless matter never 'thought' about! <u>Indeed, every code contains agreed morphology (the form its symbols take) and significance (what they stand for or they mean) between its sender and recipient. Such prior agreement always needs intelligence</u>. This is as true of 4-letter nucleic acid translation into 20-letter protein language as of one spoken language to another. **There's no 'code analogy'; genetic code *is* code!** Yet, still having denied the clear possibility of an intelligent and purposive design you have, in the face of obvious biological design, to wobble, weave and wriggle in a Rusian sort of way. You have to persuade yourself and, if you are a missionary Darwinist, others that apparently rational, programmed code and purposeful design are not what they seem. Information's immaterial, as we'll see, and thus, where natural is equated with material, it is naturally unnatural; so mind and meaningful intelligence are 'supernatural', 'spooky', unacceptable. Yet who is spooking whom? *The fact is that a metaphor is a metaphor precisely because it describes but is not what it describes.* It is for interpretation and mind is an interpreter. But a slippery sort of materialistic blind portrays 'design' as a metaphor for real, teleological, informative design. Design *is* 'design'; 'as if' blurs with 'is'; analogy becomes identity. A primary metaphor has, in the poetic science of evolutionary biology, been transformed into reality - but not informative reality! 'Design' is, in such hands, really chance's child.

How does such blind and blinding 'logic' work? Firstly, consider metaphysic is extinct; next embrace a new philosophy of physic, one of essential randomness constrained by death. Death naturally selects; death is the guiding principle; it lends direction (but not purpose) to the unpredictable mutations chance throws up. This valve of death prevents 'back-flow' from 'progress'; there is no turning back, no circling round the starting-point, no devolution, please! Thus, as it grinds unfitness into ground, death just permits an incremental novelty or system that is 'fit for purpose'. It's easy, isn't it? Or is the argument, like death, a cul-de-sac?

An atheistic vicar marries Lady Luck to Earth. For did not lifeless Earth, so runs the myth, breed an evolutionary seed - bacteria? Like yes and no, life and lifeless death are fundamental states; holism's guiding principle is life, materialism's is a graveyard made of earth. Are you the offspring of an epic accident or bound by true design?

Of course, information's immaterial; design is metaphysical and metaphysic needs (as poets know) metaphor or modelling to show the way it works. But metaphor and myth are not well grasped by literalists. They involve intent not numbers so that, for all its own material mythology, scientism deprecates the allegories, symbolic explanations and innumerate conceits of art. Who, however, cracks our epoch's whip? You can see the way its wagon's rolling.

The only science-with-a-purpose is technology. So jump on board - no metaphysic on the physical express! Science drives and real philosophy is bundled in the back. Has what was once a great Department now kow-towed to truth-by-*PCM* (see Glossary). Materialism now presumes that 'randomness non-randomised by death' has flipped you from a cosmic trial of chance.

Accidental Variations

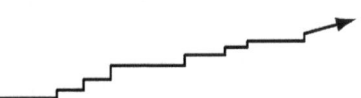

Darwinist diagram of potentially indefinite, micro-evolutionary 'progress' by small, gradual and unplanned steps.

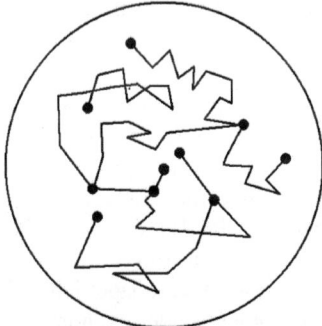

Holistic diagram of variation-on-theme or accidental, genetic 'drift' within a typical boundary. This boundary acts as a 'homeostatic norm' that constrains fluctuation.

Note a 'progressive arrow' in one view and a 'constrained meander' in the other. Mutation and natural selection (i.e. the capacity or luck to survive) are operative in both cases.

fig. 5.1 (see also fig. 22.1)

Randomness constrained by death 'invents' all forms of life. This is as deep or shallow as discussion digs. Is such assertion simply a misleading ruse or is Ruse right? Can a sieve of death, Edward Blyth's original idea of natural selection (Chapter 22), 'design' the systems some suppose that it can incrementally evolve? Might 'survival-of-the-fittest' really innovate? Neither catchphrase conjures 'teleology' but each is a tautology. Who or what survives survives. They sum to bland assertion that any particular accidental change (called an adaptation or modification) either does or does not cause death. Perhaps a certain bug is not too deadly and the vehicle keeps running. Serial bugs like this, it is arguably argued, can 'improve' design. **But modification by accidental bump or grind says nothing about the origin of vehicle systems or a vehicle as a whole.** It simply represents a chance and fractured fraction of a teleological whole. 'Design' is not design; analogy is not identity; and purpose is the nature not of matter but of mind. Perhaps a neo-Darwinist would judge inverted commas tart up Lady Luck too far; but Michael Ruse's kind of 'teleological' enticement often flutters from behind intent-shy evolution's veil.

On the one hand it is a fact that, at each level from cell to whole organism, functional existence depends on the correlation of many parts. There is (Chapter 20) no more a 'simple' organism than a 'simple' motor car. Integrated operations towards a definite end, survival, make biological organisms repositories of teleological information. Part of this is tied up in the coded database called a genome. Codes are an epitome of informative, meaningful design and, like speech, do not happen accidentally. They are often resilient. You can slur, misconstrue or otherwise maul speech and still, unless the 'noise' becomes lethal, the message survives. *'Noise' says nothing about the origin of language.* From thought to speech, from act to fact a creation involves a psychological plan that is, down the line, externalised. Even the 'simplest' organism is coded less like grunt or whistle, more a long text in complex language. Indeed, far from being a cobbled-together improvisation of 'spare parts' it could be labelled 'information incarnate'.

On the other hand, Darwinian theory deals in small, gradual changes, modifications or variations-on-theme. It does not, as Darwin himself made clear, deal with the origin of biological blueprints, homologies, prototypes or (Chapters 19-24) 'typical organisms'. His remit did not run to the origin of metabolic pathways or organs even in their simplest forms. Yet is the origin of a vehicle not of far greater moment than accidents it suffers thereafter? And, if designers evolve improvements, are these not dovetailed adaptations of blueprint, deliberate and coherent modifications of (in biological terms) homology?

Someone (can't you feel it?) is just bursting to pour vitriol on this idea. You might analyse the form and function of an engine, a computer or construction of technology to nth degree. Might you thereby not appreciate the mind or minds behind it? Does such appreciation mar your scientific explanation? Apparently it does so when it comes to engineering forms of life. To consciousness that's 'raised by cranes of natural selection' unnatural 'intelligent design' is what 'un-raised' lower races, people lacking scientific evolution, might believe. After all, the liberal intelligentsia are perhaps convinced that mind is brain; psychologists queue up to tell you consciousness is somehow born of nerves. How, therefore, could an immaterial intelligence have any mind at all? The thought-fields of polemicists are littered with the bayoneted straw-dolls of such pretty dim ideas - the dimmest and prettiest of which is 'God'. Who made big G? Complexity needs super-complicated minds to make it. That means super-complicated brains or simulated minds - computers. Who on earth in heaven or in heaven's name constructs computers? A modern, demagogic propagandist such as Richard Dawkins has no ignorance concerning what is non-existent - or even what he might not know! He knows how 'God's brain' works - because he's somehow worked out his decision that it doesn't. A no-brainer without mind or reason, he thinks it is a very bad idea. Just imagine that. You can excise with a slice of logic, you can exorcise with sharp imagination. Cut free by thinking from all thought of immaterial intelligence, your naturalistic speculations must in principle be right. Let rip. As a rationalist imagine billions of chance-generated universes and life-friendly planets (Chapters 8 and 24); take a pinch of luck to whisk life into action (see, however, Chapter 20) and then orchestrate a choir of evolutionary hypotheses. Imagine steps and stories that will ferry you across an ocean of improbability. Evolution happened and therefore speculation can, with sufficient budget to support it, bridge all the gaps in fossil record, in biogenetic law, the origin of code, sex, other integrated systems - indeed, it can cover every little problem that 'raised consciousness' enjoys. At least, excommunicate the infidel. Is not 'scientific' thought alone enough to silence 'theories of intelligence-behind-design', to crack down on confusion of a code with reason and demolish the illusion called by unbelievers 'teleology'?

On the other hand, is not incautious confidence a form of arrogance? Is the Darwinian notion wholly right? Or is that projection, an important subsidiary of the 'Theory of No Intelligent Design', conceptually muddled; philosophically and therefore scientifically flawed; scientifically flawed and thereby a false philosophical alarm; and false alarm because annunciation of exclusive materialism is, in just that exclusivity, fatally flawed? Will it therefore wither as 'an anti-God delusion'? Could it simply be a blip upon the long-term screen of human understanding or, worse, reduce to 'the greatest deceit in the history of science' (Soren Lövtrup - Darwinism, The Refutation of a Myth p. 422)? Yes indeed, responds a 'Theory of Intelligent Design'. Although a vehicle in constant use is always subject to accidents and failures of its parts such wear, tear and mishaps never made it. **This interpretation of events takes Darwinism at face value as an explanation of accidental variations, trivial changes to main structure and a theory for the origin of minor novelty, including species.** *That is, such interpretation of events at the same time side-lines the theory of evolution to a description of the origin of cline, race, subspecies, species and, depending on how classification is made, genus and possibly a few cases of family.* **It is an explanation, in broad terms, of origin of species. No more. Its action is reduced to the rank of minor, haphazard, thematically negligible variation**. Chance, mutation and 'natural selection' are seen for what they are, intrinsically uncreative. Design, on the other hand, is also seen for exactly what it is.

At this rate evolution (as opposed to biological variation) is viewed in the light of a logical fallacy, a trick called equivocation. *This semantic trick is sustained by the conflation of two entirely different matters - firstly, variation on existing features and, secondly, the <u>addition</u> of complex, fresh features (such as coherent organs, systems and so on) in the first place.*

Thus, can typographical error and selective copying really transform the instruction for a simple but irreducibly complex toy into, say, a space shuttle? Can bio-typographical error (mutation) and bio-selective copying (natural selection) transform the instruction for a simple but irreducibly complex bacterium into, say, a human? Could such a pair, having an intelligence quotient of precisely zero and no sense of purpose whatsoever, accidentally create an instruction booklet, toy, bacterium or its genome in the first place - even thanks to the random interference of radioactive 'blips', chemical 'blots' and other blotchy, botchy blights? Even when information in the genetic manual is as inexorably subject to degradation by the cosmic 'rust' called entropy as its *DNA* carrier and associated chemicals? Yet this is exactly what The Primary Axiom of Materialism, its Primary Corollary and libraries of neo-Darwinian scholarship invite, indeed, actively cajole you to believe. Is it really credible (Chapters 19-

24)? Are you really mutant ape, fish or whatever; did mutation make you from a scum on water or a single cell?

An open-minded thinker may not like the implications of an argument. He may *prefer* alternatives. Preference is, however, not a scientific still less a detached criterion of truth. There also exists a key conceptual difference between opposing points of view. In the dialectical case the metaphor for construction is organic rather than technological. Mind is prior and *within*, body secondary and '*issued out*'. Creation exteriorises from an information centre to peripheral product. On the other hand, a philosopher who habitually equates nature with physics alone rejects any creator of cosmos - including the standard but misleading metaphor of an 'external' cosmo-factor who creates the way a potter works clay or an engineer builds a machine. Any reason, rationale or mind behind natural order is unacceptable. He has closed his thinking process off. The metaphysical nature of intelligent design does not, for him, exist.

After this discussion you may ask, "Is natural selection really a designer or simply a process in which prefabricated systems survive (or don't survive) the changes wrought by wear and tear?" The latter view will not support an atheistic faith. Does, therefore, the ratchet set against a 'God delusion' or 'the backwardness of natural teleology' actually click and hold? Or slip? If slip then is the slide not towards such an evaluation of design as must include designer? Is not 'intelligent design', while contrary to the atheistic creed, a sensible and rational leap of both interpretation and of faith?

Intelligent Design.

'That's not science!' For some *ID* is, perforce, a powerful delusion. It is a methodology for physically testing guesses and thus winning, in the end, a truth - truth physical; and Darwinism is its myth of mindless origin. No 'unseen entity' may enter in. Since metaphysical is not included material hypothesis alone can be allowed. Thus *ID* is 'not science' and, on that decision, should be firmly binned! Yet, construed in this restrictive mode, is science really a sufficient vehicle to approach whole truth? No doubt, denial of an immaterial entity suits materialistic purpose well. Fairies, demons and divinities are 'unseen entities' - but mind is too; so are morals, motives and intelligence. What physic deals in information - except a mind machine, a computer mind already made? Nor can one scientifically examine thoughts or memories. You can only quantify (in cases which are quantifiable) the material effect of causal, immaterial purpose and design. After all, according to which unit will you judge a thought's relation to a nervous twitch? How will you measure ingenuity? What test will you apply to innovation, at what notch balance man's conceptions or designs of nature on your scales? The fact is, also, that experimental method can't be used for either one-off past events or unfathomable future possibilities. You can only best-guess using some statistics or draw inference according to your inclination. Then, if you've philosophically reduced your scope of inference to chance and uninventive natural law, your logic *must* produce an evolutionary outcome. There is no other way. Design must be 'design'. However, best inference for the generation of information and complex, specified design may well, using common sense, be mind. Is design 'not science'? Why, outside a philosophically restrictive mode, is *ID* not science? Or, if science can't exist outside material practicality, why should anybody trust it capable of answering everything? Indeed, such 'it's-not-science' species of dismissal issues a semantic smokescreen; it is simply a red kipper scientism uses to deflect unwanted, immaterial inference off its course. *If you want truth, not word-games, then defensive and unscientific prejudice is best ignored.*

What kind was Plato's Maker he called Demiurge? It may well be argued that his pupil, Aristotle, champion of the natural sciences, also held the master's metaphysical idea of 'Holy Being' or 'Essential Divinity'. So that, with a few exceptions, everybody until about three hundred years ago saw all objects and life forms as part of what is sacred. Why? *Because design is, at least in organisms, obvious; because design is a psychological directive; and because what commands, rational mind, is conscious and alive.* What more reasonable, therefore, if you decide an eye is *for* seeing, a hand *in order to* manipulate or muscles *because of* the need to move, than to suppose a Designer? When the analogy of cosmos with organism (with body, mind and soul) changed to 'machine' the supposition remained. Is not a machine organised such that the parts are essential to the concept and operation of its whole? Indeed, they derive from and cohere to it. A machine is greater than the sum of its parts and its greater fraction lies in its design; so with a very finely-tuned and consequently life-permitting cosmos. Nor do an *IT* revolution and a switch to computer-based metaphors change anything. Indeed, when applied to genetic operations and cell physiology the comparison with database code, signalling and the whole business of informative program, intelligence and meaning works very well.

A sign, an indication; in Latin *designare* means to mark out for a purpose or to plan. *Telos* is an end; and teleology the doctrine of final cause, that is, of aim or purpose. Surely, also, intellectuals understand the word that marks them out? *Intellego* (from *inter*: between and *legere*: to choose) means that I choose between, discern, discriminate. *Intelligence* is a capacity for higher functions of the intellect - not least discrimination. Do rocks discriminate or molecules design? Matter wholly lacks intelligence and thereby an ability to specify. Could it, with requisite foresight, generate a functional system; could it, with anticipation, ever have devised the grammar of a code or mechanism for a job?

World-renowned astronomer and arch-humanist Fred Hoyle wrote that a commonsense interpretation of the facts suggests that a 'super-intellect has monkeyed' with the physics of creation (see Chapters 7-12); that the notion not only a biopolymer but the operating program of a living cell could be arrived at by chance in an organic soup is 'evidently nonsense of a high order' (see Chapter 20); and (check Nature Vol. 294 12-11-81 p. 105) that the probability of life forming by chance was a number - 1 followed by 40 noughts - big enough to bury Darwin and the whole theory of evolution. An inanimate scenario did not animate itself. There was no primeval soup either on this planet or any other and if the beginnings of life were not random they must therefore have been the product of intelligence. **Furthermore, despite political correctness in the science of biology, has any organism a sophisticated system, complex feature or an integrated structure that could *not* be there by purposeful, intelligent design?** Since, however, mathematics lays materialism bare detractors would distract you with some arguable plausibility. Are their reasons feeble (see Book 3)?

What has caused the present pass? With the medieval rise of natural philosophy such personalities as Galileo and, especially, Francis Bacon decided that a 'doctrine of final causes' did not mix with science. We want to know not just believe. 'This is a scientific paper in which there is no place for purpose, causal mind or God' - which, if science is exclusively the study of natural phenomena using the senses and their technological extensions, is perfectly reasonable. But a slow sea-change in perspective began to follow this impeccable rationale. Of course, God no more faded overnight than He appeared by day. For Kant, Spinoza, Cuvier, Paley, Richard Owen (who designed the Natural History Museum in South Kensington as a cathedral to praise God's creation) and most thinkers He remained central. Others, including Erasmus and Charles Darwin, Jean Baptiste Lamarck and, latterly, Henri Bergson unfurled the banner of 'creative evolution' that either predisposed or drove matter to a somehow inevitable climax in the shape of you, a human being. Finally, the advent of scientific atheism combined with knowledge that at least 'micro-evolution' happened by chance. From this toehold can't any clever fellow twist into a 'counter-intuitional' pose of mind from whose perspective purpose seems but isn't really what it means? If, so contorted, you decide 'design by accident' will do the trick you zeroise all creativity. Materialism annihilates the thought of universal mind; it religiously appeals to luck. How could you disprove historic luck? You can't and so that's it then. It follows speciously that metaphysical intelligence did not design but evolution's power *must have* 'designed' our natural scenario.

What really came about? There occurred a gradual shift of focus, a change in intellectual emphasis from inward, contemplative to outward, sense-based direction of mind. This means, in dialectical terms, a shift from Inward Centre to outward periphery, from principle to utilitarian practice or, in grand philosophical terms, from Idealism to Scientific Empiricism. In a phrase, perspective flipped from *top-down* to *bottom-up*. With such an 'enlightened' swing follows an emphasis on inductive reason (which runs from individual detail to generalisation/ data to theory) rather than deductive reason (whose inference runs from general principle to specific fact/ theory to data). Induction is the tendency of exploration, deduction of authority. Thus the physics student does experiments and builds a picture of primordial behaviours he calls 'natural law' or 'code' or 'program'. And a biologist will classify from individual specimens 'upwards' into larger groupings; but, by abduction (a form of reason that infers a causal history from its contemporary effects), he concludes that 'natural bio-law', such as code and archetypal body plans, was not primordial. Organic form was never set; its complexity evolved by small and unpredictable accretions from, it is presumed, initial simplicity. He therefore rejects, on principle born of perspective, the 'intelligent' notion that individuals represent variation on precedent or 'higher' theme; he rejects any prior 'establishment' of archetypal hierarchy behind, say, homologies of form. Once 'intelligence' is excluded from the suite, your hand is forced. Ideas can happen rapidly - but not by chance or the 'necessity' of natural law. On the other hand, a 'program of design by mindless matter' (*aka* 'designer-less design') could only happen very slowly if (because one's *never* been observed) at all. **So a materialist must play the evolution card**. The claim is time-veiled accidents (as opposed to prior 'conceptual establishment') can shuffle up continual but excruciatingly witless 'development' of fresh body-plans, homologies, biological systems...the lot. You, specifically, are here and so they must have done. They made your lucky day. In this view specifics don't devolve from universals; rather, universals slowly evolved by common ancestry. You argue from specific to general, actual or fossil specimen to larger groups and, instead of archetypal starting-points, you draw up family trees.

Order-less, irrational evolution engages an inductive as opposed to holistic, deductive perspective. It is conceptually basic to the *bottom-up* as opposed to *top-down* view. Such basis involves, however, a dialectical self-contradiction. On the one hand you argue from detail to universal, from actual practice to relationship-in-general or, in the biological case, from individuals you can observe to the 'ancestral' phylum that connects them. Phylum is the principle, actuality the detail but, conversely and concurrently, you argue that unprincipled chance 'specifies', by inscrutable steps, the intricate ramification of what look, in reality, more like developed ideas (such as body plans and their associated, functional systems) than a mishmash of add-ons. Or you argue, for example, that simple chemicals 'self-organised' their own complexity; that these organic complexes then metamorphosed into 'simple' but adeptly coded cell or cells and from this ancestral species evolved a web of more intricate organs, systems and body plans. You have to. Your hand is forced. Neither storm nor chemical reaction could instantly generate viable biological complexity. So you work piecemeal from 'chemical evolution' to 'abiogenesis' (Chapters 20 and 21) and - although all naturalistic models up to now have failed - from this modern brand of alchemy to ever more complex, 'higher' forms of specificity. So, you decide, biochemical arrangements must be produced by time, laws of nature and, mixer of life's cocktail, chance; and if scientific explanation must, by gerrymandering philosophically, exclude intelligence then this aforesaid triplet does not only have to stir up chemical complexities but compose specifically coded organisms. *Such exclusion marks a weakness.* <u>*An abductive inference to best-explanation may be plausible but at the same time incorrect*</u>*.* **As Sherlock Holmes well knew, many causes may explain the same effect. A detective's best guess may still, missing out some crucial factor, miss the mark.**

What is it, therefore, renders inference of such a causal factor as intelligence 'unscientific' to some minds? That it is unobservable or can't be falsified? If so, then look to pseudo-science in some theories of modern physics and cosmology. Perhaps, alternatively, it is the notion that mind's currency, called information, is a natural, *immaterial* element; such notion is marked 'counterfeit' in naturalistic markets of philosophy. Then again, could mindless nature ever 'scientifically' innovate? Perhaps natural science does no more than take account of material law - law that does not as much explain as just describe the banked-on regularity of operations in this universe. Regularity, with its high probability, facilitates prediction; but regularity and repetition constitute - according to the shaved-of-purpose, neutered Shannon sense of information - low information content. Can what is automatic ever innovate? *How, oxymoronically, could low-information, lawful rigour ever generate high, specified irregularity that, clear as a well-spoken language, coherently and unexpectedly informs the book of life?* The science of original formations is historical. Archaeology, paleontology and all forensic studies hark *back* to causal factors. They infer 'best explanations' from the possibilities surrounding those events that happened long ago. An inference of intelligent design does this as well but also predicts. For example, organisms will increasingly be found to be much better constructed and more subtly informed than chance mutations could account for. This will especially be the case for genetic regulation and in-built, super-coded adaptive potential that anticipates changing circumstance.

Even though you can't lay measurement against a thought nobody argues that, for example, thoughts did not make motor cars. No doubt that for three centuries some have conspired to re-invent the logic of origination by deleting purpose - in spite of the fact that every cell involves a digital storage system integrated with functional components all of stupendous specified complexity. *Cast off graven images and false imaginations such as atheism's troop, like Don Quixote, rush to tilt against.* **Forget imaginary theologies that Darwin, Dawkins and a thousand others spin around the nature of creator and creatorship. It remains emphatically true that the only known cause of codes and specified complexity is intelligence.** Thus thinkers philosophically unchained from materialism might construe immaterial intelligence as an obvious best-explanation of biological origins; and if original design stood as 'best-explanation' for life's codes, metabolisms and the organisms they compose, why should anyone - save those naturalistically as opposed to Naturalistically animated - disagree? First Cause would thus involve Projection by Intelligent Design.

How is information made? If the codes that govern life did not originate intentionally then (Chapter 20) how could unplanned accidents create their huge and working specificity? Life's beauty shines from its coherent systems, elegant proportions and purposeful ingenuity. Not disorderly confusion but reasoned rightness rules. Intelligence *and* matter as a whole. 'Wholesome beauty' that includes both Designer and designed, both voluntary mind and an involuntary, physical world, has been acceptable to many scientists (e.g. Newton, Faraday, Maxwell, Einstein, Planck) but for those who decide, in the name of physical science, to delete one half, it is a line of logic unacceptable. Is it valid to decapitate an organism from its source of information? To execute its metaphysical intelligence, delete teleology and level the hierarchy of creation to its lowest, horizontal field, a headless body of the whole truth? If so, then brainless chance, creator of all law and lawfulness, has won the day. If not, then the last three centuries have increasingly insisted on a truncated truth - and such half-truth is a most insidious form of falsity. *Whether or not your decision is correct, once you decide to remove a Designer, design remains - no doubt less obviously in the case of physics and chemistry but clearly in biology's practical purposes.* However, to try and develop a *bottom-up* explanation for the purposely specified, that is, coded complexity explicit in the structure and function of all life forms, you have to engage reverse gear.

'Reverse gear'? If forward gear represents the recognition of design for what it is and all that that implies, reverse reverses this and recognises only an 'appearance of design'. Furthermore, as will be detailed soon, Claude 'father of information' Shannon's theory was one of signals not significance. Although useful at the level of statistical analysis, it's useless telling meaningful and mindless messages apart. What use is measurement of decibels when you want to hear the message of a song? In fact, since anti-randomising mind informs, random sequences are actually information-in-reverse! Nonetheless, a mindless world (in which the metaphysical has been denied) is thereby imagined to 'inform itself', that is, 'self-organise'. Of course, robotic matter does just this; but an extreme adoption of empirical method and associated theorising not only fails to mention but explicitly denies such natural, meaningful design as happens every time you creatively evolve an idea into a plan and its finished product. Reasonable design is thus by denial trumped. It is overturned by information-in-reverse, diametrically upended by unintelligence, replaced by the uncreative 'evolution' of base materials. A series of targetless, chemical accidents is, presumably, ratcheted into functional complexity by the ability of their intrinsically inert products to survive - whatever 'survival' means to chemicals. **Unimaginative evolution is the key prop of materialistic atheism: but such extreme philosophy is as absurd as if, from a *top-down* point of view, only mind were allowed and physical existence denied.**

Nor is anti-evolutionary 'Intelligent Design' simply a twitch of philosophical nerves, a strange emergence from brain chemistry. In Communist countries its serious impact is, on atheistic principle, officially, politically banned. Is it not an absurd but actually widespread insanity that you and I could be imprisoned for writing and enjoying the 'non-party' propaganda of an open-minded but reasonably anti-atheistic book like this? Yet so-called 'rational theism' is also the subject of intense educational and legal dispute in non-socialistic zones. In the USA, for example, it is the unswerving intention of The Discovery Institute to counter, with full intellectual rigour, pure materialism's standard objections and thus establish a scientific theory of intelligent design. *Nor would this establishment derive from point of ignorance; the observed fact is that the only natural, causal agent known to generate highly specific, coherent information is mind; and it is thus fair inference that such cause lies at the root of life's highly complex, accurately specified and coded information systems.* The basis of biology is information; is it right to damn suggestion that it was not chance (mutation) guided by extinction's rigour (natural selection) that set life's brilliant systems up?

Halt! Science stopper! Some claim science should not entertain this inference because it spoils the fun. Worse, ID is 'unscientific' since 'God's nature' can't be ascertained. You can't empirically measure supernatural creativity - but, of course, we *can* infer creation by intelligent designers, ones who we call engineers. **The issue is not one of identifying a particular designer but detecting such criteria as spell design.** You might, for example, expect to find in a design purpose, rapid infusion of information that elaborates a concept and transforms it into a functional system, complex integrated parts, possible use of symbolic code (such as on-off or more complex switching systems) and the re-use of functional units (or mosaic variations on their theme) in different and physically unconnected systems. You find all these substantiating what we call biology. *In* **fact, the priority of ID is, as it involves the immaterial component of technology, information.**

From the detection of such factors one might predict biological discoveries and, thereby, infer a generator of innovative information. This, we've seen and will see, cannot be rationally identified as chance. So why black-mark, cane and then expel the rational alternative? In what kind of school does 'scientific reasn' thus irrationally master us?

ID predictions could include the discovery of intricacy, coordination, complexity of coherence, specificity and an irreducible number and arrangement of parts per functional mechanism; also reasons *why* specific, predetermined parts are co-assembled, that is, self-organise in cybernetic order to perform a task; a convergence or mosaic appearance of modular parts; the discovery that previously misinterpreted components are not in fact vestigial junk but structurally and/or functionally necessary; that a system involves the complex operation of language, that is, accurate symbolic code; and that, if discrete systems are designed (albeit with intrinsic, predetermined adaptive potential to flexibly respond to circumstance and thus retain the balance to survive) you would expect to find discrete types of organism and, historically, fossil 'abruption', that is, lack of billions of gradual missing links. The presence of all such factors in life on earth might lead a rational man to infer design. And, based on this perspective, propose such research fields as the computer-like properties of *DNA*, non-protein-coding (n-p-c) genomic function, front-loaded adaptive potential as regards both environmental adaptation and development, the practicalities of applied protein engineering and so on. A reductionist is, on the other hand, forced to identify his 'designer' and chance a-dance with death; and to narrate hypothetical stories and suggest research objectives accordingly. Why we can ask again, abort the ID perspective and thus only allow issue of the latter world-view to thrive?

No doubt, such analysis throws a spoke into the wheel of materialism's spin. Why, however, should an inference of intelligence spoil any scientific fun? *Why should it detract a jot from joy of finding how inventions work or understanding that they've been invented*? Why, moreover, should a rational, causal explanation, born of knowledge and appreciation of a mechanism that's researched, be equated with an invocation to 'god of the gaps'? Chance represents, for sure, an element of ignorance. Lady Luck is matron goddess of all evolutionary gaps, that is, mutations; but, without intelligence, there'd be a mighty gap in all of us. How is an argument of prior intelligence an argument from ignorance?

Yet such strange, exclusive logic runs amok. Groups lobby parliaments and make petition to Supreme High Courts. Should any alternative to Darwinism be aired in our schools? Will the Court find in favour of 'design' or Design? Will the gavel fall for a 'Theory of No Intelligence' and against a 'Theory of Intelligence'? Won't some 'enlightened' judge rule atheistic politic should supersede (because it has been dubbed 'religious') theistic superstition? Surely he'll decide how life in cosmos has been made? Even if, reluctantly, you'd allowed 'intelligent design' a word in edgeways could its attorneys tell between a weak form ('Theories of Accommodation') and a strong form (creation in the manner of an engineer or artist)? And if, in all this, the decisive question is of metaphysics not just legal doctrine, whence will judgment arrogate sufficient knowledge unto its deliberation? Might not equivocating judgment, by admitting apparently contradictory theories to public discussion (e.g. in the classroom or lecture hall), allow debate in public as in private? If, however, such 'confusion' were debarred and Natural Design decisively rejected in favour of non-intelligent 'design', important debate about the sort of origins that by far precede the American Constitution would, in the land of the free, have been stifled!

Such was the case when an apparently tendentious judge (Jones in the Pennsylvanian Kitzmiller v. Dover case 2005) arrogated the authority to decide the nature of science; and not only science but the nature of the universe! Should a court ban the scholarly discussion of theories alternative to those a judge deems 'popular' or 'preferred'? The issue is especially piquant with respect to an essentially historical subject locked, except for inference, outside laboratory experiment - origins. Jones' edict ominously implied an end to free speech and debate in United States schools. Teach only materialistic versions, it decreed. Your academic freedom here extends to praise for Darwinism; courts will exclude and colleges expel all other as the writ of heresy!

If a scientific theorist ceases to think critically or consider alternative explanations, he has become (Chapter 0: Religious Delusions) dogmatic. Lesser thinkers never rise above that bar but, nevertheless, dogma is poor science. The scientific method is, as Karl Popper noted, based on repetitive testability and capability of being falsified. Not everything, however, fits these criteria and is susceptible to scientific examination or proof. To clarify the matter we denote two 'modes' of scientific creativity. The first well logs recurrent tendencies that nature seems to show as 'natural law'. The second does not deal with current *operations* but at best infers unique and causal histories, that is, how things began. The question of *origins*, failing both of Popper's criteria, certainly exceeds the scientific remit and may in fact involve what is prior to physics - metaphysics. As such, having inferred either a Creator or No Creator, both pillars of faith are prone to lapse from non-religious principle into ideology, one as theistic and the other secular religion. It could be argued, on this basis, that as the core of materialistic faith the theory of evolution has acquired the properties of a religion.

So if a Supreme Court wishes to rule on a case of 'unconstitutional violation of principle of the separation of church and state' which church should it choose? Should it admit each species of faith, neither or simply an opinionated 'chosen path'? Which path? Which guess? Does a belief in modern evolution make atheists as logically as a belief in design makes theists? If theistic realism were denied and the atheistic position permitted to partake a monopoly of 'legal protection' then, as Harvard graduate, law clerk in the US Supreme Court and long-time legal academic at UCLA, Phillip Johnson, writes, 'the Supreme Court will in effect have established a national religion in the name of First Amendment freedoms'. Such 'non-religious' religion would, since 'religious' theism is legally impermissible, be humanistic atheism. And if atheistic naturalism became a legally sanctioned 'religion', its extremity would rest solely on an apparent (and philosophically self-inflicted) inability of the Supreme Court to countenance the supra-religious logic of Intelligent Design. The key concept driving official American philosophy would have been an infidel 'theory of no intelligence'! George Washington would be spinning in his grave - especially if the Supreme Court had, while dealing with such basic constitutional issues, considered a memorandum that is still legible on the wall by his pew at St. Paul's Church, Manhattan, New York of 'a nation under

God'! St. Paul's survived next-door '9-11' entire but what an absurd, legalistic and educational muddle (see Chapters 5, 25 and 26) in which Uncle Sam still flounders!

Not only Uncle Sam! Not only judges but our European politicians also seem to know it all! They totally ignore (or never learned) that the Renaissance, progenitor of modern scientific method, derived enlightened sense of order from a universe intelligibly created by a Most Intelligent Creator. In 2007, at once dismissive of its founder scholars and the strength of argument, both then and now, from inference of 'intelligent design', the European Union's Committee on Culture, Science and Education proposed to the Parliamentary Assembly of the Council of Europe that carelessly defined , so-called 'creationism' (of any version including 'intelligent design') should, as a potential threat to human rights, be outlawed from not only science but all classes in its federation's schools!! Potential, wild-eyed inquisition! Fangs of 'rationalistic' arrogance evolved in bureaucratic jaws! This unenlightened spit on intellectual freedom was rebuffed but another edict on philosophy, a bull, was issued by The Royal Society. In 2008 the infallible pronouncement, effectively mobilised behind the 'fact' of evolution and scientific materialism if not outright atheism, forced its director of education to resign. His 'toxic' sin had been to remark that open debate might, in the causal history of origins, best be enjoined! No doubt there will follow more such hiss and lunge!

Blunder or else prejudice could easily outlaw this book from academic institutions or debate! These factors are the more surprising since Popper has dubbed the theory of evolution 'unscientific'. In his book on scientific methodology, The Unended Quest, he absolutely correctly calls it a 'metaphysical research program' because it does not meet his most searching criterion - the ability to be thoroughly tested. It is, in fact, 'untestable'. It cannot be tested, repeated, falsified nor (except in the highly arguable 'evolutionary' case of trivial biological variations-on-theme) ever 'seen'. It may resemble theoretical cosmology but it is not like down-to-earth physics. Its science is not of the kind that probes the atom, sends craft gliding between the planets or underwrites modern technology. It is not empirical but, wherein the evidence is interpreted according to hypothetical abduction, a kind of forensic science. CSI (Crime Scene Investigation) delivers impartial facts to courts that, if they in turn serve justice, involve interpretations from both prosecution and defence.

Some, pots calling kettles black, label the alternative 'unscientific' too. It depends, of course, how 'science' is defined. If only matter matters you can banish immaterial matters from the witness box. Only what's observable's allowed - except (as in cosmology, psychology, quantum physics and Darwinian biology) you are permitted to infer that unobservable but physically conceived reality substantiates observed effect. Even the causal implications of mind, as in construction of a purposeful and codified complexity (say, a computer), are allowed because mind is, root, construed as 'thinking' meat. Thoughtless, uncreative *DNA* may act like a computer program; but what mind have computers in them? 'Digital' nucleic acid may be identical in terms of purposive complexity but implication of its immaterial origin or, if you like, an element of metaphysical intelligence is philosophically taboo. Yet whence, except from mind, do you think any system rich in information starts? We know full well specific, integrated, functional mechanisms are conceived in mind alone. Why, therefore, deny mind as 'best causal explanation' of how life's biotechnological complexity began? As open-minded science may allow for an inference of hidden variables or unseen factors at play in evidence that it collects so it should exclude emotional or philosophical denials of 'intelligent design'. No doubt that denigrators use their 'science' as a stick to beat an idea they don't like. They link 'creation' with religion but their religion is, conversely, the sole advocate and the adjudicator they call to the bar - materialism. No doubt that from this angle one creates a books-worth of objections but, if you think bio-logic is 'unscientific', then a measured, technical, polite and thorough demolition of your books-worth's claims and sub-claims has been proffered by, say, Stephen Meyer (Signature in the Cell). All the harder, thus, to understand a Supreme Court decision that might ban, in a classroom, the court of origins! What book exactly are its legal sages judging from?

The blunders are still more politically surprising since Darwin himself puts the EU, Sam's and Socialist establishments to shame! He writes (Origin of Species; Dent Everyman Edition 1972, p.18; my italics), *'For I am well aware that scarcely a single point is discussed in this volume on which facts cannot be adduced, often leading to conclusions directly opposite those at which I have arrived. A fair result can be obtained only by fully stating and balancing the facts and arguments on both sides of the question, and this is here impossible.'* Darwin may have lacked time or space in a single book but that does not excuse the cramping style some verbose institutions of the world adopt.

Nor, in 1859, was Great Britain a land of closet evolutionists. The impact of Darwin's 'Origin of Species' was, as Huxley rightly observed, less scientific than anti-clerical. For was it not Huxley who, legend has it, defeated 'Soapy Sam', the Bishop of Oxford, in a debate concerning where he came from? Now that the vicar's Creator could be reasonably diagnosed as the pre-scientific fiction of an untutored mind, so 'anti-bishops' such as Marx, Haeckel or Nietzsche could zealously promote their versions of secular, socio-political faith, their anti-religious mission. Just as importantly 'The Origin' gave science its 'bio-force', that is, a substitute for divinity that might explain life on material terms alone; and if the nature of such an undetected but material 'life-principle' was progress surely socio-economic progress, whether capitalist or socialist, could now evolve without the application of a brake by the corruption-prone, authoritarian and reactionary church? Were not objective purity and revolutionary, scientific liberation from the clutches of religion just the manifesto evolution could provide foundations for? No wonder that Friedrich Engels immediately wrote to Karl Marx, 'Darwin, by the way, whom I'm reading just now, is absolutely splendid. There was one aspect of teleology that had yet to be demolished, and that has now been done'. So that Marx wanted an autographed copy of the book. No wonder today's militant atheists still promote 'The Theory of No Intelligence' and execrate 'The Theory of Intelligence'. Having *a priori* excluded intelligence from the remit of objective science they proceed to preach the former as 'fact' and the latter as 'subjective fiction'. Fairness is not the hallmark of any extremist response to 'counter-revolutionary' claims but you might expect well-mannered democrats to uphold proper courtroom procedure! How disconcerting, therefore, when in 2004 the Anglican church popped up in the form of today's Bishop of Oxford who, having about-turned from Soapy Sam's alignment with the actual words of scripture, thereby clapped his endorsement on 'scientific' suppression of the church's own professed and professional Logic. Crowded into the same 'pulpit' as passionately atheistic scientists and media celebrities he lined up to sign an 'open letter' berating and lobbying to nip in the bud any study in British schools that, using biological data, might dare to compare our two Theories. The Bishop swallowed evolution as a 'fact' and declared for 'No Intelligence'! Many clergy have 'come out' behind him. Even the Archbishop of Canterbury has seemingly digested the central tenet of mammon - evolutionary biology - and thus embraced the conclusion of materialism. Add evolution's trophy convert, Pope John Paul II (Chapter 24: Theories of Accommodation), and you are struck how deep the roots of such inverted logic have now penetrated metaphysical soil. Have Albion and Il Papa's See grown even more confused than Uncle Sam? Will John Paul's successor, Benedict XVI, retract?

Such all-out resistance to an alternative must involve a cast-iron reason to escape the charge of dogma. What is the irrefutable logic according to which a final solution is reached, one in which a 'Theory of No Intelligence' liquidates its opposite number? What fork impales design? The two-pronged fork, which represents reverse gear in full throttle, is of course composed of a belief that subjective experience is the product of chemistry (Chapters 5, 6 and 13-18) and of neo-Darwinian evolution (Chapters 19-24). Its material thrust to the immaterial heart seems, because the latter's presence is in scientific terms unreal, fatal.

In that case how do you deny that bio-logical 'design' is in reality design? Just wield mutation and selection? Or, like Professor Ruse, a 'demarcation principle'? By various criteria that exclude any immaterial factor such demarcation cuts 'intelligent design'. Circular logic thereby labels the inclusion of such a factor 'unscientific' and, at a stroke of homespun metaphor, the intellectual decapitates intelligence. But listen again. A telephone is, we agree, actually designed; but, because neo-Darwinism will not permit, an ear is *not* designed. It is a naturalistic metaphor for design - as if, seemingly or apparently 'designed'. The intention of inverted commas is deletion; they are brackets here specifically meant to cancel a designer out.

'We treat organisms - the parts at least - *as if* they were manufactured, *as if* they were designed, and then we try to work out their functions. End-directed thinking - teleological thinking - is appropriate in biology because, and only because, organisms *seem as if* they were manufactured, *as if* they had been created by an intelligence and put to work' (M. Ruse op. cit. p. 268 - my italics).

Surely such verbal *legerdemain* does not mark the climax of evolutionary exploration? Delete 'as ifs', replace appearance with reality. Distortion vanishes. Ear, as well as telephone, becomes a teleological product of intelligence. From teleology to 'teleology' - surely an inverted pair of commas isn't what two centuries of huff and puff have been about? Airbrush the pair and lo! Illusion vanishes. William Paley, Darwin's erstwhile hero, reappears. Materialism's nightmare waves his pocket watch in broad and scientific daylight. Might Ruse recant? Is it not time for a decisive change?

If, of course, naturalism is incomplete in the way Natural Dialectic suggests, its savants will have been quixotically tilting at windmills. From a *top-down* point of view there is no problem of 'design' to solve. There is no denial of metaphysical or physical realities. On the one side, there exist no necessity to espouse Biblical literalism; on the other, no root need to distort the meaning of the word design or indulge in mechanism's form of vitalism - animistic 'function-talk' of codes, programs, reasons, selfish genes and so on. Such biological realities as natural selection (for what the concept's worth) and variation (called micro-evolution by mutation, sexual selection or accidental genetic recombination) are no problem. The problem (Chapters 22-25) is not with the origin of Linnaean species but phyla, classes and orders. So there is (see *fig.* 22.1) no objection to Darwin's erudite 'Origin of Species', speciation, adaptation (such as adaptive coloration), population genetics and various related studies - except in the elevation of trivial accidents to the status of instruments of informative design, that is, the extrapolation of limited variation into unlimited. **In this respect this book cannot be viewed as anti-Darwinian (except in the 'macro-evolutionary' extrapolation of, say, mouse to whale or single cell to human). Nor is it in the slightest anti-scientific (except it lacks a philosophy of materialistic atheism)**. The sheer common sense of an inference of 'Intelligent Design' has no need to indulge anguished or quixotic confusions that involve abiogenesis, ancestral jelly-balls and arguable phylogenetic trees (Chapters 20-25). It seems as if only with respect to origins has *bottom-up* got it back-to-front and baked with half a truth. In ascribing the source of 'design' to chance and/or necessity it is out of order - inversely, completely out of order.

For Natural Dialectic the objective focus of science (with its empirical and inductive methods) is as natural as air, ears, eyes and learning. The whole enterprise is a methodical, cumulative and exhaustive interrogation of the physical world, a high-resolution sieve of truth from untruth. It is, on the other hand, also possible that Aristotle, Ibn Sina, Newton, Priestley, Maxwell, Cuvier and scientific company, along with Socrates, Christ, Nanak and mystic company, did not get their *ID* notion wrong. Every barrister knows that the facts do not change because you skew them your way. If Intelligent Design were the case, the truth would not change because a few philosophers and, latterly, orthodox science, had decided it had. Such decision were speculative, professionally useful but, in the last analysis, simply opinion. Variation in biological form over time is a proven, undeniable fact; evolution, as either a vital or mechanical bio-force that explains the *origins* of biological form, is not. There is certainly no need to take Thomas or Julian Huxley's, Herbert Spencer's, Ernst Mayr's, Theodosius Dobhzhansky's, Stephen Gould's, Darwin's or any other 'gospel of evolution' as gospel. The issue of design is no more anti-scientific, as some errant philosophers claim, than planning or creativity. It will be readily understood, however, that the exercise of the latter is not physical but metaphysical. Dialectical bio-force is psychological. It involves the intelligent deployment of information by the use of mind; it therefore involves the active drive of purpose and passive faculty of memory as well as the subsequent shape of physical results.

What in a nutshell is the word? That common sense to one is to the other common nonsense. By what criterion might you infer design? How might you recognise a substance impregnated with a purpose or define a mind behind?

To a *bottom-up* materialist a Theory of No Intelligence (the neo-Darwinian theory of evolution incorporating the 'rational irrationality' of 'accidental design') must be fact and therefore common sense. Today biology reviles William Paley. It denies that bones or any such 'machinery' were made on purpose. By this token any Theory of Intelligent Design is fiction and, in this sense, a common nonsense.

Yet anthropologists don't seem to have a problem in distinguishing designed from undesigned formations. Lifted out of sand, loam or clay an artefact designed in plastic, wood or metal and the reason for some bone or rockwork (say, an axe-head or an ancient cornerstone) straightway strike the eye. From a trench you feel the instrument a maker's or a user's fingers felt. In this case purpose is allowed a mind. For a *top-down* holist such allowance universally applies. A Theory of Intelligent Design is the rational product of straight reasoning. Design is actually non-accidental design. Materialists worry endlessly how matter might by chance make mind or how molecular arrangements yield up life but they refuse to entertain that subjectivity's an elemental datum. Nothing makes or takes it. And its informative intelligence is anti-chance. **Who is chasing ghosts?** Are philosophical approaches thus now aligned head-on and entering a crash zone? **Whose is the delusion?**

At least, you see, Design is not a pretty theory all alone. All sorts of people have designs on it - some nice and others not so nice. The issue is central and from one angle or another is going to keep cropping up. For example, if you adjust from scientific to dramatic angle, what's at the start of every play? Whence do mummery's purpose and design derive? What about informative arrangement, commandment and an act or acts of orderly creation?

Authorisation.

Science, like art, politics and religion, is a human enterprise. It is a force by whose authority some choose to live. And its noble endeavour is to slake man's thirst for truth, to drink material water from its own Castalian spring.

A thirsty one is driven to discover water's nature. The better to divine its presence he approaches it from every angle. He drives out the hopelessness of ignorance and thereby accurately hunts knowledge to its source. In such enquiring manner twists and turns the hunt of physics with respect to space and time and every other thing. Yet the noble tally-ho is also prone to human weakness - venialities of power games, exclusivities of 'tribal' tendency and ever-changing fads and fashions. It suffers from a struggle for survival (in the prosaic form of publicly-funded research grants, promotion prospects, pensions and so on) and, beyond mere survival, a desire for recognition if not fame.

Is metaphysics relevant to such pragmatic exercise? In fact, enforced consensus - a 'group-think' characterised by *PAM* - pervades its working atmosphere. This 'air' is policed by an anti-metaphysical species of constraint. For example, philosophy of the Vienna Circle's 'logical positivism' stated that only assertions verified by material observation count as legitimate knowledge - of which science represents the sum. This idea curtails imagination and thus bridles how men while in laboratories may think. By its criterion, although you might protest that such *metaphysical* abstractions as mathematics, mind, consciousness and even logic aren't outside the scientific remit, only what is *physical* is designated 'true'. Is moral and artistic knowledge therefore illegitimate as well? Materialistic fundamentalism of this sort lives on.

Matter is non-conscious; is mind too? Nevertheless, a philosophical structure that includes an immaterial, conscious element, a dimension of mind, is in our day a demon for the casting out. And yet, if science does not cover all events, the exorcism of faith's metaphysic represents an error of irrationality. Materialistic truth might not be absolute but relative and incomplete. In which case how can a skill-less proletariat evaluate what smarter people guess or calculate or claim is proved? Should ordinary citizens pass judgment even if their views, including a Creator, exceed the naturalistic paradigm? *Top-down* there's an Informant, a Creator; *bottom-up*, from fundamental matter, there is not. By whose instructions - an Informant's or the over-riding government of accidentally-made-from-matter men - should our lives and, thereby, planet earth's well-being be at root sustained?

Curiosity, philanthropy and a desire for truth more or less pervade humanity. Yet, paradoxically, our man-made science leads directly to the urgent threats that face our planet - for example, mass extinction of the other species allied with overpopulation by ourselves, a kind of human monoculture or, as some might see it, plague. Where (Chapter 26) might all this end? Have you faith that more and still more science must, like a house of cards or politician's promises of heaven that in expedition crumble into hells, by itself deliver some utopia? By whose authority did many of today's debasements rise?

Let's tack along another line, another current of authority. Before a play is staged it has to be conceived, intelligently developed according to its own first principles and then projected on a film or stage.

The two components of Natural Dialectic which underwrite the conscio-material coordinates of creation are *Information* and *Energy* or, in dramatic terms, *Author, Act of Composition and Set*.

	tam/ raj	Sat
	expression	Potential/ Inspiration
	drama	Author
↓	tam	raj ↑
	informed	informant
	final text/ product	act of composition
	storage	transmission
	recipient	message
	background	foreground
	stage/ set	player
	audience/ observer	drama

Nature includes both physical and metaphysical modes of representation. Natural Philosophy used to include physical science as a subset of such a whole. Perhaps Natural Dialectic, treating modern science within such a systematic context, is the first to rigorously restore the larger framework. This book, both ancient and modern in its perspective, treats consciousness as an information centre; it includes both mind *and* matter as facets of the complementary poles, information and energy. It explains, *top-down*, the consequences of this simple yet revolutionary starting-point. What revolves? *Materialism is turned upside-down. It is seen as the lower part of the whole truth.* We can start with the creative, informative component (Chapters 5 and 6) before proceeding (Chapters 7-12) to inspect created patterns that shape energy.

Intelligence is the ability to learn, understand or grasp principle; and the word is also used in the sense of what is understood, that is, information. There are, for Natural Dialectic, three aspects of information viz. informant, data, informed; sender, message and recipient; shaper, shape and shaped. **What shapes is called, in this context, 'active information'; what is shaped is called 'passive information'.** 'Passive information' is contained in the arrangement and interaction of materials. 'Active information' is psychological. An 'active' informant can impress his designs on material forces and objects in which event, if he has correctly understood them, no deviation from his expectations is possible. If, on the other hand, the recipient of his message is also 'active' (i.e. minded) then creative, unexpected responses are possible. Although human 'free will' is highly conditioned, humans are not entirely puppets. Life is not robotic.

The mind of psychology itself involves two modes - (*raj*) active, conscious and (*tam*) passive, sub-conscious (often called unconscious). Sub-conscious information (e.g. memory or instinct) is imprinted; it is passive but whatever the nature of any possible psychosomatic border between mind and matter, there is no mind at all in matter. <u>The shapes of 'passive information' as it pertains to non-conscious energy are studied exhaustively under the headings physics, chemistry and biology.</u> That is, active information plays no part in the non-conscious universe of physics and chemistry. In the case of biology, however, it interacts with the world in a poorly explained way. *Certainly, the notion of universal information is unknown, ignored or denied by outward, physical science. The operation of cosmic psychology, if it precedes or subsists physical law, is unidentified. It is a complete blank.*

However in the case of either metaphysical or physical construction information is prior. It is the potential for subsequent creative behaviour. Constructions emerge according to its causal purposes and rules. (*Sat*) information precedes (*raj*) process and (*tam*) result. In the active case conscious deliberation precedes any reasonable course of action or construction; in the passive case some fixed pattern of governance, stored either in the form of memory or translated to material code, precedes process and outcome. Ordination or preordination. Whichever way - in the 'psycho-space' of deliberation, in the unconscious 'cyberspace' of archetypal instinct (Chapter 16) or in materially coded records such as *DNA* or computer disc drives - instructions guide behaviours.

There exist both informative and energetic potentials. Each precedes and informs event and outcome but which precedes the other? Are they actually connected? *What is 'prior' comes before and, in dialectical terms, mind precedes matter; therefore informative precedes energetic potential.* No doubt that energies shape crystals, fluids and the natural forms of earth according to 'habitual behaviours' we call natural law. What, though, informs the energetic action? What next grade up, of mind, is precedent? Fixed mind? Sub-conscious memories called archetypes? If this is so you have in turn to ask what, up the ladder, precedes mind. *Top-down, Subjective Consciousness is prior*. Not humanity's dilution found at our cosmological axis; not our fraction at Point X on the conscio-material gradient (*figs*. 0.8- 0.12) but Pure Consciousness, the Universal Concentrate whose Potential precedes in power and status all that follows its 'pre-action', all that flows from its 'starting-point'. If Consciousness is Essential, it precedes in grade all existential time, space, thoughts and things. It is more powerful than dependencies; it seeds issue and then watches over, so to speak, its created, cosmic family. *'Prior' is used in a causal not a spatial sense*. It does not, therefore, mean that existential information necessarily appeared before energy but certainly accompanies and orders it. Its authority authorises, in principle, the way things work. Matter and material outcome are shapes of secondary effect. *Information is prior in the way that rules govern activity or thought precedes action*.

In short, metaphysical precedes physical. In existence mind is prior. Its subjective 'space' precedes and governs all inanimation, that is to say, the reflex energetic patterns and the bodies of this world. Information is a powerful potential. Principles inform behaviour and precisely determine the outcome of events, whether animate or not. Science has 'crystallised' its empirical observation of physical and, to a limited extent, psychological patterns into what it terms 'laws'. General, legal regulation shapes all localised, individual transformations. In this organic view of creation mind, in one way or another, governs all material forms. How on earth could this be so?

This section (Chapters 5 and 6) constitutes a theory of relativity with respect to the prior component of existence, information. That is, it constitutes a theory of intelligence and intelligent design.

	tam/ raj	*Sat*	
	existence	*Essence*	
	matter/ mind	*Pure Consciousness*	
	polarity	*Neutrality*	
↓	*tam*	*raj*	↑
	passive	*active*	
	material/ physical	*immaterial/ metaphysical*	
	informed/ passive info.	*Informant/ active info*	
	recipient	*communicator*	
	script/ code/ speech	*author*	
	program	*programmer*	
	visible	*invisible*	
	pure matter	*mind*	

Where do you start? As noted in Chapter 4, cosmic starts are tricky. They are headaches. The exponents of big bang theory, for example, know there is no way to avoid extremes, paradox and mystery. *In fact, big bang's big problem is its name*. It gives entirely the wrong idea. **In fact, there burst an orderly projection; immediately and perfectly controlled, a cosmic miracle unfurled materiality and set the stage for life**. The 'bang' was, we shall see, wondrous in its accuracy. Chapters 7 and 12 will show the precision of its flow.

What, therefore, preceded physics? In the *top-down* case what came before body *and* mind? What preceded existence? In dialectical terms by now you know the Precedent is Essence, Pure Consciousness or Potential Information. This is the Absolute against which all objects and events, in mind or matter, are measured. From Absolution issues relativity. First Cause activates an effect called existence. Consciousness-in-motion (i.e. mind) comprises a duality whose couple of basic components are its own self operating on a non-conscious non-self called energy. This pair, informative mind and energy, are compounded on a sliding scale of proportions called the conscio-material gradient.

To crack (as opposed to explode) the cosmic egg another way, a Perfect Symmetry of Consciousness is broken by the process of thought. Mind informs the energy by which it moves; mind 'rides' matter; its purposes create the patterns of behaviour we call existence. **Information and energy are complementary aspects of a conscio-material continuum from which all creation takes shape.** They are macrocosmic universals from which all specifics are, in one way or another, derived; and, although you might construe their continuum as 'emergent', what source does it emerge from? Is it matter or, ontologically prior, primordial consciousness? Natural Dialectic traces the pattern, based on a principle of interaction by three cosmic fundamentals, of creation issued an immaterial spring.

Although these qualities (*sat*, *raj* and *tam*) co-exist in varying proportions in different aspects/ components of each object or event, creation as a whole is also 'layered' according to their concentrations. In a three-tiered macrocosm (*Sat*) Essence is the potential from which issues (*raj/ tam*) polar Existence. The relative proportions of information and energy are reflected (*figs*. 0.4 and 2.7) in three main divisions of the conscio-material spectrum and, thereby, the structure of creation itself. They are Transcendent (*Sat*, Potential) and mind (*raj*, kinetic), which involve metaphysical consciousness; and (*tam*) subjectively rigid, non-conscious, physical matter. **In simple terms, information is active in the half of existence called mind and, mirror-wise, passive in the half called matter.**

There are therefore three main modes of information - *potential, active* and *passive*. Potential information is transcendent, active is *voluntary* and passive *involuntary*. Could there exist an inter-phase, a medium between the active and the passive poles? At base of mind and top of matter, instinct? Sub-conscious memory?

The category of passive, involuntary information includes both dormant mind (in the form of records called memories) and matter. *A voluntary cosmology is, understandably, entirely absent from the perspective of scientific materialism.* In other words, as far as science is concerned there is no universal mind - not even in its lowest, involuntary aspect as a 'script' of archetypal memories. Such mind does not exist for it. There is no intelligence, purpose, reason or informative psychology involved, at any point, in the cosmological origin of exclusively physical nature. *The bottom scientific line is, therefore, that the character of matter (and thence its lawful, automated world) must be exclusively the product of chance.*

From a dramatic point of view, however, this is less than half the answer.

Chance is anathema. An author creates. He informs. He establishes precedent. His logic is the action's authority, his stamp of approval authorises, he directs the plot. 'This is, for me, what is right and good.' A drama also depends on its author's character and intelligence. Could dramatisation, whose characters play scenes upon a reflex stage, follow the same order as cosmic creation? Is its grammar, for the stage at least, what we know as 'law'; and don't a quantum alphabet of agents forcefully express this natural language? We shall see.

Information, Messages, Arrangement.

What is information?

Pick up a postcard, menu, letter - anything informing you. The object is reducible to chemistry and physics but the *sign* or *message* it conveys is not. Signs and signals always have a purpose; objects never do. **Top-down, therefore, signal information's irreducible to scientific scrutiny. Yet, as we'll see, its semiotic metaphysic dominates debate about creation and our lives.** In other words, information's fundamental to our being but does not fall, in a semantic sense, within the scientific remit. Information constitutes a problem that's tremendous for the theory of evolution.

***Bottom-up*, however, it is claimed that everything is energy and matter; and, perhaps, this pair can generate an immaterial factor - information.** Information's born of chance and natural law. Firstly, therefore, let's distinguish clearly ordinary from scientific/ materialistic sense. *In mathematics of the latter 'information' must not be confused with 'meaning'.* Therefore what exactly, in this apparent shortfall, *does* the word mean?

It was Claude Shannon who, with Warren Weaver, first devised a mathematical and thereby scientifically acceptable definition of information. He treated its transmission in purely physical terms according to statistical formulation of the entropy-inclusive laws of thermodynamics. In such transactions his unit was the *bi*nary digi*t*. This on/ off, one-zero 'bit' allowed the quantitative properties of strings of symbols to be formulated. His theory inversely relates information and uncertainty. The more uncertain, the less probable a sequence of symbols or arrangement of materials the more information it is calculated to contain. Rephrased, an amount of information is inversely proportional to the probability of its occurrence by chance. Simple, repetitive or predictable sequences contain less and complex, irregular arrangements a greater quantity of information. But 'Shannon information' is a measure of improbability. *It makes no judgment whether such irregularity is specified; it involves no sense of meaning.*

Thus his 'Mathematical Theory of Communication' has proved invaluable to information technologists who deal in the bulk carriage of data. It covers the purposes of transmission and storage but, when it comes to the real significance of information, fails completely. It does not address questions of meaning, comprehensibility, correctness or worth; nor those of source, intent or destination. Shannon's definition is suitable for describing statistical aspects of information such as quantitative aspects of language that depend on frequencies (such as how many times does the latter 'a' or the word 'and' occurs) but it treats any random sequence of symbols as information without regard to its concept, meaning or purpose. In other words, the more improbable any arrangement the richer is its Shannon-defined information content. *In short, two messages, one meaningful and the other nonsense, can be exactly equivalent according to this form of analysis.* For example, ZNQWRHSIX TAZHHVB and COMPREHENSIBILITY are assigned the same value. In other words, Shannon has reduced information to a statistical quantity; he has shaved off any sense of meaning, cut out sense of purpose in numerical analysis. *Shannon-shaving deals in non-purposive complexity.* It might make statistical sense but definitely linguistic and rational nonsense. In this assessment gibberish can have as at least as much 'informative value' as sense. The real nature of immaterial information, its quality of meaning, is ignored. So is the nature of its source. How can Shannon-information therefore serve as useful metric of what information actually conveys?

Herein we've pinpointed the defect. Absurdity sufficient unto fallacy, such definition fails to distinguish between what is *simply* improbable and what is *specifically* improbable. Since it measures information-bearing capacity and not information content it treats non-purposive and purposive complexity the same. Its analysis fails to distinguish between specific sequences of code (such as you find in any language) and non-specific, meaningless jumbles; between functional efficiency and functionless ineptitude; or between meaningful and meaningless arrangements. It therefore treats the origin and state of sense and nonsense equally. For example, chemical *DNA* carries, like ink and paper carry words, a specific, purposive and complex code whose meaningful expression is protein; but 'Shannon information' misses the important half of life's expression. Standard information and communications theory can compute a biological form's passive, chemical complexity but can't identify conceptual schemes or meaning or, behind the meaning, active origin of systematic information by design of mind. It is as if you analysed the pulses in a wire but could not distinguish between a sensible, minded conversation and mindless, random, unspecific kinds of bleep.

Physics, in confusing jargon, says, 'Light transmits information.' Yes, but only when there's mind there to interpret it. Why should I not indissolubly link information with a mind and therefore consciousness; and disconnect it from, intrinsically, every form of energy? Only mind extracts a meaning, makes connections and interprets signal patterns such as, in its presence, circumstances variously convey; and only mind can generate what is, to it alone, information.

It is, therefore, important to grasp how Shannon, thoroughly negating logical reality, accords randomness and purpose equal status. **But randomness is really reason-in-reverse; it is information's opposite**. So when it comes to language or to mechanism, including those embodying biology, such conflation is an error of the first order. The Shannon information-content of a message increases simply by the number and irregularity of symbols. Empty verbosity, chance concatenation or a message subject to error, 'noise' or interference can be assigned a higher value than one concise or error-free. As a consequence the analysis does not distinguish between presence of mind, authorship or creativity and their absence; it fails to recognise purposive specificity; nor does it accord function or meaning any premium. Thus order and precise meaning - usually complex, always accurately coded - might as well be able to emerge from randomness or senseless motion under natural law. You can thus deceive yourself and think in topsy-turvy terms of 'senseless design', an anti-teleology well-known as Darwinian evolution. Yet such self-deception infects the mind-set of an academic discipline throughout the world.

In this respect a proverb, principle or a concise directive involves highly compressed information but, once Shannon-shaven, is rated below longer, jumbled, possibly incomprehensible drivel. This nonsense is compounded when, based on the principle of uncertainty as opposed to order, Shannon accords a higher information value to unlikely, unpredictable or, we can say, meaningless events than logical, predictable or reasonable formulations. Mindless gibberish can score above Einstein's equations or Shakespeare's sonnets! This topsy-turvy analysis (see also Chapter 8: No Information from Confusion and Chapter 20: Noise, Monkeys and Catalysis) arises from a *bottom-up*, materialistic view of things that ignores, because they do not exist on the physical plane, active, immaterial sources of information; it misses mental teleology. *It therefore also ignores the essential nature of purposive complexity, the root specification of life; and is ignorant of irreducible complexity. Specified, irreducible complexity is the signature of mechanism and, as far as life is replete with integrated systems and their mechanisms, Shannon's definition of information is useless for identifying biological purposes.* Using an extension of probability theory he has, like a chemist who weighs but cannot analyse a chemical, 'weighed' information but not analysed it. The *quality* of information, of which he has no grasp, is its semantic value. A large, prolix book can have a lower semantic value than a thin booklet. What is the value of a proverb? It is not the price or number of words in a telegram that interests its interactors, it is the richness of its meaning.

↓	*tam*	*raj* ↑
	passively informed	*actively informing*
	organised	*organising*
	informed/ guided	*informant/ guide*
	particle	*wave*
	matter	*field*

Whether or not 'Shannon information' involves meaning is a dialectical irrelevance. At this point don't let materialism's viewpoint shunt you off the tracks; do not be slipped 'off-message'. Rapidly revise the tiny section 'Information' (Chapter 1). **In a *top-down* dialectical there exist two main kinds of information - active/ semantic and passive/ guided.** The former kind is subjective, immaterial and involves knowledge, specificity and purpose. It thus has any meaning only in the context of a knower and is essentially a psychological phenomenon. You may, of course, find passive information out of mind. Don't principles inform the way material nature works; don't designs inform the way things are developed and arranged? Materials can carry information pressed upon them and such passive sequence can be formulated mathematically. Ink and paper, clay or iron can be arranged to carry code and such code-carrying capacity be analysed statistically. For codes and materials *per se*, of course, neither mind nor meaning can exist; bare of meaning, void of purpose their passivity yields 'Shannon information' to such minds as measure bits and pieces. The issue of intent rests only in what starts a system up. Was original information 'specified' by chance or not? Is ever a symbolic language (called a code) established unintentionally? At least, the way things are arranged, informed by natural forces or by men, needs mind to recognise a *modus operandi* and, the Dialectic would suggest, develop any system's logic and its rules.

You inform, you order, you continuously arrange your body and the world according to your purposes. Thus purpose gives arrangement meaning; mind has ordered its own schemes and, consequently, bodies of the world into conformations that will satisfy desires. If there is cause it has effect; the world responds and information is exchanged. Thus you are, in effect, an information centre that exchanges messages within a field of influence, a circle of awareness.

To inform is to give shape to, to configure or arrange. You could, however, argue information doesn't need a mind. In this case non-conscious energy informs. Don't waves inform a beach? It isn't mind informs a geological formation. In any interaction doesn't one side shape the other? Don't force and objects influence each other and the order of their interaction coincide with patterns we call natural law? Yes, these are patterns automatic to the point that mathematics can describe them. Energy will always passively inform. Don't confuse such information with the active, comprehending sense. Raw data that a scientist collects is, like the events that it describes, intrinsically meaningless; but after its collection an observer using his intelligence, treats mathematically, elicits sense of pattern and/or interprets using a hypothesis. He thus imparts a meaning-to-the-gleaning he calls 'information'. Raw data carries meaning but only after it's been processed by a mind. This is active information's application; it is pervasively, the scientific and, indeed, every conscious organism's normal way. It is mind's negentropic nature to inform and be informed. The material data it manipulates is, however, ever, totally oblivious. It only lacks to understand, therefore, what regulates how your raw data has behaved. The causal question ever rears its head - did natural law originate by chance or not? Check Chapters 7-12 on energy, on physics.

Voluntary or involuntary, active or passive, psychological or physical - you can model information several ways. Natural Dialectic models energy and information using rings (*fig. 0.6*). *Inherent in this model (of which examples are Light, Bell and Pool) is the idea of wave, oscillation or vibrant pulse; also inherent are the notions of source, broadcast, 'sphere of influence' and 3-d field.*

In the case of light these notions combine in what is called an electromagnetic field; and in the case of sound, the extent to which compression waves are propagated through a medium. Indeed, for modern science the basis of matter is immaterial fields in which particles (such as electrons and protons) are quanta of vibratory energy. Because atoms and bulk matter are made of such particles everything (including your biological body) is in the last analysis reduced to such material foci as particles and their fields. Until vibrated a field is latent. No vibration indicates potential is untouched. Vibration is latency 'quickened', potential activated. The motion of particles is fast and perpetual. What kind of 'sound', 'light' or other stimulator quickens them? What passively informs them?

As well as electromagnetic there exist gravitational, strong nuclear and other fields. Indeed, each sort of energy and its particle exercise power and influence through non-material, informant fields. Such fields are everywhere yet nowhere, something yet also, until activated, nothingness. Yet they organise, they order everything. *Fields are not explained in terms of matter, rather matter is seen as a form of energy within its 'archetypal' field.* The different sorts of quantum field have not, as yet, been unified by a single theory.

No doubt free energy is wireless and matter is a 'locked', 'wired' or particulate aspect of existence. Moreover wireless informs wired. Simple inspection of iron filings around a magnet demonstrates that a field is an *organiser*. A field informs its locality, shapes its inhabitants and rules their behaviour. It governs, communicates, arranges and defines the character of its material subjects. Organisation, information, order. In short, fields are the invisible precursor of lower, visible order; they constitute the principle that governs action and, as such, cause of appropriate effects. As ruler of its region, a field is hierarchically prior and matter in it secondary.

What, however, of the other kind of wireless energy, the 'field of mind' with all its kinds of influence? Awareness is a living light, a voluntary radiance. Its attention - which is a sensor, learner, decision-maker and creator of patterns - is a focus of awareness. Attentive focus, in its various states of interest, is the basic component of mind; its light employs the faculties of insight, imagination and memory that we call thought. Another key metaphor for information (and for mind) is therefore a special sort of radiant energy, the electromagnetic field of a radio or TV broadcast - a sourced, ordered sphere of influence, purposeful organisation of energy, a program. Programs have, of course, meaning and a purpose. Their order is expressed within a grammatical framework; their information is arranged according to syntax. Such rules lead us to another very common model - language. Symbolic code. Informative mind is analogised with speech, voice or command. A command orders, "put it in the shape I want."

In short, there exist for Natural Dialectic both voluntary and involuntary *fields* - both psychological and physical fields of influence. The latter's influence, from which the physical world is constructed, is as automatic and predictable as the fields around a magnet. Physical fields are, although organisers, involuntary. They influence and are influenced by particles and larger objects within their boundaries. Energetic field (and particle) interactions may be complex but are reducible to invariable, mathematical equations. *Such fields inform material events but we perceive no mind in them (though we might behind them).* Although they constitute the fundamental, active aspect of bulk, passive objects, they shape them in a robotic, automated way that we call the laws of nature. We call them, therefore, inanimate. *Their objective information is passive.*

This is not the case with voluntary, active information fields. They include, for example, the animating 'force' of will and 'magnetism' of desire. They are called, collectively, mind. Mind is a subjective information centre that interacts externally with the world and internally with its own contemplative faculty and bank of memories; and by means of interactions it develops programs. Like electromagnetic radiation it involves both fields and a spectrum of energies; and as TV broadcast's 'grounded' by a set so embodied mind by brain. The program itself involves purposeful information but, unlike its material aspects, is not organised but *organises. It is an active not a passive form of information.* Of course, as any teacher or publicist will confirm, fields of physical and psychological influence are quite different in character. The latter is neither repetitive, spatially contingent or a simple register of energetic variations. The quality of their patterns also differs; the criteria and measurement of psychological quality are educational, moral and not physical concerns.

The psychological ingredient is different as well. Think carefully of a white horse. Imagine the details of its mane, its eyes, its muscular form. Have it gallop. *Of what is this image or, in its duration, video program made?* It is transient and therefore not the same as its *activator,* who is still here! Aren't you? An activator manipulates mind material. The substance it manipulates, the plastic object of its attention (in this case a white horse), is a metaphysical material analogous in character to the electromagnetic spectrum. It is called in Sanskrit *manas. Manas* is sufficiently subtle for an *activator* to shape. It is, conversely, able to be impressed and, screen-like, present those impressions to the activator in his other mode - *sensor or seer.*

Natural Dialectic calls the deliberate creation of order (such as the mental imagination of a white horse) *active information.* Voluntary, active information is the nature of imaginative, manipulative, order-making mind. The order mind creates is always in accord with its activator's wish and thence directed will-power. *The translation of an idea, concept, or wish into practice so that a corresponding physical arrangement is produced is called materialisation.* The creation, projection or object produced (which no more contains mind than a TV set or the signals it receives) is called *passive information.*

Look at information with respect to cosmic fundamentals. *(Sat)* potential information is latent behind any action, expression or outworking; it is effectively the basic field of consciousness. *(Raj)* active information is mind. Mind processes, plans and transmits information to achieve its purposes. Get the message? *(Tam)* passive information is, on the other hand, message received. Got it? A message shapes the object of its attention. Such an object is realised potential. It represents the outcome of plan, the product of program, intention fulfilled, *used* information. In this (passive) case potential is exhausted. It has served a sender's purpose. Of course, as previously noted, a message can be received by another mind that will, on receipt, be affected, shaped by and actively respond to it. Or it can, as in the case of rearranged bricks, be imprinted on inanimate, unresponsive matter. Physically there is neither active message nor response. Energy involuntarily 'informs' according to rules. Such information is transmitted or stored in both the internal and external arrangement of objects great and small. Such mindless, purposeless behaviour is wholly passive. This is, in cosmic terms, the *(tam)* condition of non-conscious automata - the natural world of physics.

Even such rigidified, automated information has its grades. The principles of active communication, linkage and relationship have higher status than the products formed. *In other words the quantum level, including electronic charge, informs the bulk shape of matter.* Codes or grammatical rules rate even higher up. These are *(sat)* information from which all else derives. For physics such potential matter is called the metaphysical repository of natural laws; just as the 'particles' of language are inked or sounded letters of the alphabet so information is expressed through an alphabet of primary particles (photon, electron etc.). In biology (*sat*) genetic potential is an information centre that resides in the nucleus of cells. (*Raj*) activation of this store according to grammar, algorithm or program, call it what you will, produces (*tam*) informed shapes of protein. As correct materials contribute to the construction of houses correct protein contributes to the construction of biological mechanisms. In this case the alphabet is one of primary bases. Their coded potential is based on the correct physical arrangement of atoms in bio-molecules. Nothing in the life-field is, except accidents, wear-and tear and infection, random.

Though the sequence is clear it remains to identify the link between active and passive information. If live music is recorded it is fixed on disc and its subsequent reiteration is robotic. Similarly thought and impression are fixed in nature's disc called memory. Do repetition, habit and the 'grooves' of a universal memory play any part in the operation of physical law, the influence of immaterial fields and the fundamental nature of matter? *Natural Dialectic will explore sub-consciousness and PSI (the psychosomatic interface using the concepts of resonance, attunement and, with respect to the storage of originally active mental constructions, memory. We might even enquire whether science, in describing oscillatory fields, is not actually investigating an aspect of universal mind.* Could modern science have attained metaphysical ground? If fields amount to 'rules' or laws then how creation works would be ordained by fields of various kinds. Pre-active latencies. *Archetypes that, like the energy they guide, provide the <u>passive information</u> that controls the way phenomena behave.* Mathematics may describe but not create the rules; nor may it implement the world's game how this real power base does. Could scientist and mystic be describing, in different ways, something of the same thing Natural Dialectic calls an archetype?

To probe this relationship we can invoke, after radiant source and radiant influence, the third main model of creation - a deliberately informed radiant influence called *music*.

Take Mozart. He writes a concerto. He develops and details his basic idea (or theme) and stores the consequent program as a score. Active information has been fixed; the score is passive information. Musicians, according to the score, exert the force of Mozart's field of influence. The listener admires his organisation of harmonies, is swept into a rhythmic stream and understands, through a wordless universal language, more or less what the composer wanted to express.

Meanwhile a scientist has measured the recital in a different way. He has precisely measured the length of a flute, discovered by investigation how a violin works and submitted equations that rigorously describe the hall's acoustic properties and harmonic oscillations of each different sound wave in the whole piece. He is, he assures his audience, absolutely right in all he says. He can prove it. He presents a long and detailed dossier, replete with complex mathematics, to Mozart.

Unfortunately Mozart does not understand. Nor can he relate this interpretation, which the scientist reaffirms is absolutely accurate, with his own. Indeed, it is almost irrelevant to him. It addresses an aspect of his work in which he has little interest and in no way relates to his genius, his act of creation. In other words, the dossier has described, albeit in exquisite detail, only the involuntary, passive, physical side of an impassioned, voluntary, artistic composition. It would be a pity if, having missed the *point* of the music, the scientist decided to scorn its ignorant (in his view) origin - Mozart. Or if, absorbed in the study of physical effect, he were to claim that his dossier contained not just the most important but the *only* data worth consideration!

In truth both parties are, each in their own way, completely right. Do they, however, each recognise that they are talking about the same thing? Who informs, who is informed? What, in the case of concerto or cosmos, is the real heart of the matter?

Information is Immaterial.

Top-down, **information is an immaterial element in its own right**. Why so? Mind is active and commits ideas into material orders that are called, in the Dialectic, passive information. Mozart's mind, through instruments, informed material sound. Is not the real heart of energy through air you hear as music therefore immaterial? Is its primary arrangement psychological or physical? Is it physical or metaphysical?

Why, once more, is information non-physical? Buy Mozart's music! CDs and floppy-discs are materials carrying information. Have you ever weighed one accurately then rubbed it clean? Deleted all the information? What, when you reweighed it, was that information's mass? Zero? None at all? Do the same to your computer. The essence of computers is their programs and their database. Delete this essence and peripheral hardware weighs the same. Nor any more than thoughts can you line information up. No length or dimension. Because it inhabits the *arrangement* of material - not just any old arrangement but one with meaning that serves purpose or accords with principle. Purpose weighs no more than understanding; both weigh just as much as meaning. In the scientific balance meaning weighs as much as abstract theory. Each is far lighter than a feather; it is zero grams and so the scales tell nothing; nor do the finest registers of time and space as much as twitch. Norbert Wiener, mathematician and founder of cybernetics and information theory, said *"Information is information, neither matter nor energy. Any materialism which disregards this will not survive one day."*

There is, furthermore, known neither law nor process nor sequence of events through which oblivious matter can create information. The latter is not a property of matter. Purely material processes, unguided except by natural law, are fundamentally precluded as *sources* of information. Information is not a thing itself but a representation of physical things, metaphysical ideas, emotions, events and conceptual relationships. In fact, data has no meaning outside the purpose of its maker or its user, mind; and information may be defined as the arrangement of material (including voice or ink) according to plan. Check *fig.* 2.6. You might think of information, as did Wiener and does Natural Dialectic, as a 'non-physical form of anti-entropy'. What is not physical is metaphysical. Wiener's 'negentropy' is metaphysical and maybe thought of as a property of mind.

Where entropy-directed process falls into states of energetic equilibrium informatively-driven programs rise into unusual, unstable states of disequilibrium, that is hi-energy constructions. These include tools, machines, including bodies biological, and use of them. All material objects and events present a 'rigidified' arrangement of originally symbolic information. They *carry* information; they are symbols, records or 'memories' of their original information. Information therefore resides in the *order* of things. For example, you can mentally order red, green and yellow bricks. If, after psychologically sorting them, you physically sort them then the bricks have been informed. If you build enough bricks into a house you had in mind, then the house can be seen as a crystallised form of your 'house' idea. We call your act of sorting a creation. The act is 'active information' and the product 'passive information'. Any such product can be seen as a form of information storage. For the Dialectic this includes projection from a physically transcendent level. Even the 'world-machine' itself may have been metaphysically informed; that is, cosmic patterns of behaviour may have been derived from archetypes that science knows as natural laws.. This idea, as *figs*. 2.12, 7.3 and 8.2 begin to illustrate, will be developed.

What is certain is that life in the form of a biological organism projects its mind on matter by rearranging the physical world to suit its purposes; in this way information also creates things, events and relationships. Man, because he possesses sufficient focus to plan, is especially adept in this creative, informative field. Man-made creations passively carry his mark, intelligence, purpose or information. Natural creations show an outer and an inner side as well; inner energy informs the shape and properties of bulk automata (objects and events). What informs the order of energy? Why should, how could there be an aspect of mind behind nature itself?

While mind is the immaterial basis for all technological systems, scientific understanding and cultural creations, it is also a prerequisite for all biological designs. The latter have a very high information storage density (much higher than any man-made machines) and employ ingenious concepts. Information, we can say, rides with matter; but it is non-material and as distinct from matter as a tune from the strings of a guitar. It is obvious the tune derives not from chance but its composer's mind. What about the instrument? It is manufactured. What about the component materials of specially fashioned wood and steel? Does the natural world also embody order? If so, whence arose its non- human order, its information and its geometries of shape? In short information is, although immaterial, natural. Therefore any complete science needs to have formulated laws concerning both information's metaphysical entity and that of its prime processor, consciousness-in-motion *aka* mind. Consciousness, at the heart of information's business, is subjective - so the laws need to encompass subjectivity. Psychology but not neurobiology or genetics might apply to discern them.

Information is Hierarchical.

Laws order. Information and order are closely allied.

Hierarchical Information
Active (Creating) Information and Passive (Created) Information

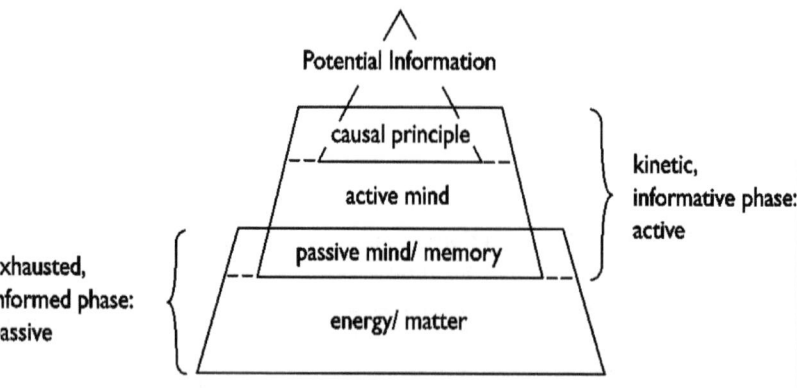

Information is phrased in terms of awareness. **Of the major cosmic subdivisions mind is termed active information. Although it may exhibit highly active behaviours matter is, with respect to information, non-conscious; it is passive, automated and in this respect creatively impotent.**

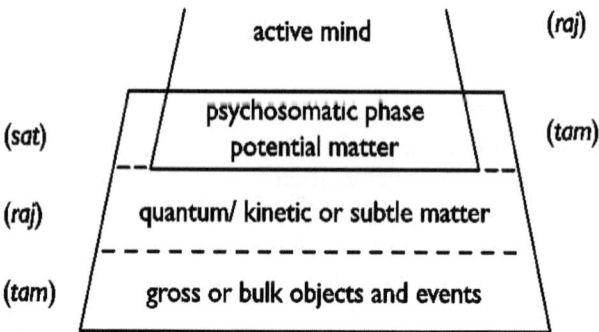

Clear distinction needs be drawn between informative mind and informed matter.

There also exists, however, an active, informant phase of matter - energy; and a passive, informed phase of mind - a subdivision called sub-consciousness. Sub-consciousness is mind's storage facility. *Memory is the natural recorder of finished events in active, conscious mind.* This lowest level of mind also constitutes, in dialectical terms, the point of psychosomatic linkage with matter. *In other words, passive mind in the form of memory becomes the transcendent, causal principle that orders matter.* This hierarchical view of memory is not standard, is not materialistic but, we shall see, fits well with psychological, biological and physical facts.

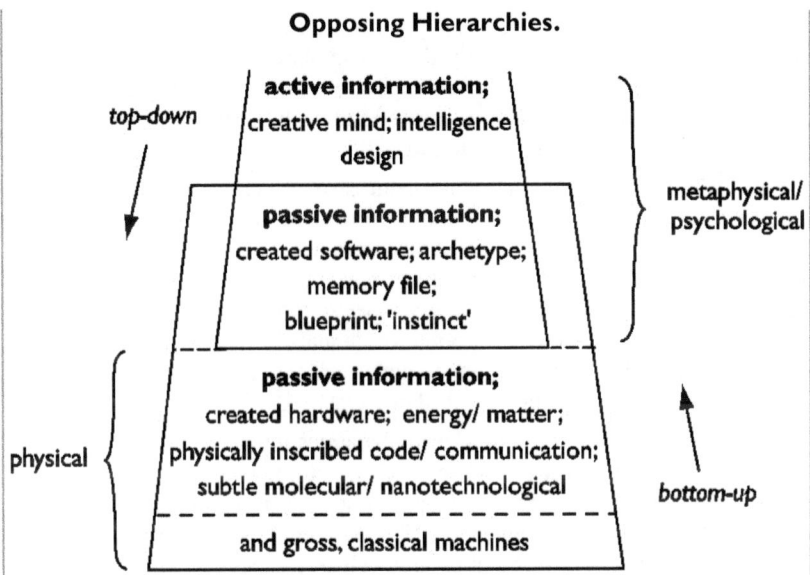

Opposing Hierarchies.

Top-down and bottom-up views of the origin of information are here juxtaposed. *Top-down* information originates in active mind; *bottom-up* it starts with passive matter. Since information is the source and sustenance of life the opposition is important. Which view makes better sense?

As in *fig. 0.6* the *bottom-up* view is that passive leads to active information. Matter/ energy emerges from nothing or is, perhaps, eternal. This matter fortuitously, in a process called abiogenesis, creates a cell coded by *DNA*; such a physical carrier of code itself generates information; and such informative though random chemistry somehow eventually makes consciousness. Intelligence emerges out of atoms.. *Hardware creates software.*

The *top-down* view is absolutely the reverse. Just as software is composed of non-material instructions that control the hardware, so a distinct immaterial element, conscious mind, designs software instructions (stored in the files of memory as archetypes). Such metaphysical programs are reflected in material carriers. Similarly, particles may be conceived as devices carrying archetypal code; the behaviour of non-conscious substance is thus metaphysically controlled. *Software defines hardware.*

Such fundamental dichotomy concerning the origin of information is henceforth extensively examined.

fig. 5.2 (see also figs. 0.6, 5.3, 5.4 and 9.1)

From a *bottom-up* point of view natural order is paramount. It is described in terms of material laws of nature whose constraints are also dubbed 'necessity'. Conscious mind and its species of order is, though powerful, simply a contingency. It is an evolutionary 'extra' whose recent, unexplained emergence on the scene is cosmically irrelevant.

Of 'necessities' perhaps the most general are the first and second laws of thermodynamics. These, empirically derived, describe the dialectical order of all physical events. They have overarching, universal application and constitute a branch of scientific study that, as opposed to cosmology, shows remarkable stability. No exceptions have been found.

Robert Mayer's First Law states that, in the case of energy, only transformation but never creation takes place. This indicates that creation is over and its phase replaced by one of conservation and transformation. It is these transformations that science tracks. The law gives nature fundamental balance and science its equations. In the orient it becomes, where '*karma*' means 'action', an element of '*karmic*' law; this involves the notion of 'equal-and-opposite' conservation between a cause and its knock-on effect, that is, conservation in every interaction or transformative event whether physical *or* psychological.

Lord Kelvin's Second Law of Thermodynamics indicates that 'nature flows one-way'. Time's arrow flies only forwards; and there is a general tendency to the dissipation of energy. Entropy; Clausius later defined this law in terms of 'entropy' (useful energy dissipated per $°K$; see also Glossary) so the law states that 'the total amount of entropy in nature is always increasing'. Quantity stays, quality diminishes. Finally Ludwig Boltzmann realised that disorder (diffuse or random state) is more likely than specific detail or complex order. In short, energy and mass are conserved but over time undergo both energetic and configurative decline into what is described as 'incapacity to work', 'exhaustion', 'degradation from complexity to simplicity' and 'disorder'. By entropy the universe is neutralised, homogenised and dies.

These laws have had the same unifying effect on the physical sciences as the theory of evolution on biology but the predictions of one contradict the other. The former predicts a general loss of energetic and configurative order, a descent into randomised 'equality' of equilibrium. The latter presumes the reverse. It retrodicts the spontaneous elaboration of highly specified, biological constructions from an initial mess; it predicts (without any evidence) that irreducibly complex order can arise from a chemical stew. Can this fundamental tension be resolved? Materialism thinks so. Holism does not.

There may arise confusion in science over the nature of entropy (order loss) and its converse, 'negentropy' (order gain), as this pair impinge on arrangement and, therefore, on information. This is because (see *fig.* 2.6) the physical assessment is incomplete. Non-purposive energetic and configurative orders are included in the laws of thermodynamics but, not unnaturally, the most important source of complex, purposive order is excluded from the frame. Mind's essence is the creation of arrangements according to specific purposes. To eliminate 'intelligence' from the consideration of an artefact is to hobble explanation. It is, actually, ludicrous; but, because it cannot see a universal mind, then science has naturally ignored and, even worse, denied it.

Of course, materialism admits hierarchy. Mind controls body; code controls the arrangement of its expression and causes all intended effects. Such hierarchy is only permitted to exist, however, within a materialistic framework. And the most logical deduction of that framework precisely reverses the dialectical order of information.

What, *bottom-up*, does such a framework say? A 'big bang' (Chapter 12: *PAM*'s Miracle) created nature and its laws; these were sufficient to allow some special chemicals to form and, by chance, start replicating. They spontaneously reproduced themselves! This self-reproducing accident (called *DNA*) constitutes life's basic information and is, therefore, its creator. Such creation is, however, passive and unwitting. Its instrument of generation is the *PCM* (Glossary and Chapters 20-25) by means of genetic mutation and natural selection. This couple are supposed capable of evolving different sorts of organism. Thus genetic acid - not the immaterial message that it carries - plays life's cradle, crucible and crux in one. From this potential there evolves the 'natural intelligence' of blot and protoplasmic blob. There are expressed such purposeless mutations as, acted on by natural selection, generate coherent cells and self-consistent bodies. Last and least by sheer coincidence is tagged the only mechanism capable of active information, conscious mind. Such mind is, you need to know, simply a figment of 'subjective illusion'; it is, you must understand, really an 'aurora' flitting through a nervous sky, no more than the ionic passage of electro-physiology in rare, pink jelly called a brain. Life is reduced to complex chemistry.

But as a chemical *DNA* is unguided save by natural law. It does not by itself self-replicate (see Book 3) and cannot flex to generate genetic language with its attributes of code, syntax and reason that communicate its message 'make this body'. **In other words it is a medium and not a message.** It is a **vector**. It is the passively informed carrier-chemical on which, like the medium of words in ink, genetic information is inscribed. Ink, paper, atoms, *DNA* and genes - these are oblivious to purposeful arrangement but why invoke intelligence? It is presumed, in effect, that ink and paper combined to write life's manual by chance. Vitalism is the doctrine that a form of life is due to more than just physical laws; a vital principle supersedes biochemistry. Yet, as a crypto-vitalist (or even animist) you are required to believe that this genetic medium mindlessly, mysteriously evolved a most specific message from its 'alphabetic' pieces. Of such reductionism perhaps the crudest form, as we shall see (Books 2 and 3), is the 'bits-and-pieces' attitude that formulates molecular biology. *In fact and to repeat, information (the very basis of biology) cannot arise spontaneously from undirected physical behaviours.* **No-one has demonstrated that it can and, despite appeal to computer simulations and bacterial adaptations, creative matter hasn't ever been observed.** Innovation, which evolution needs in spades, is missing from its earthen plot! But, because it marks the theory's knell, this is not a message naturalistic scientism is inclined to face.

	tam/ raj	*Sat*
	below	*Top*
	effects/ causes	*First Cause*
	subsequent order	*Archetype*
	excitation	*Immaterial Field*
	disturbance	*Neutral Latency*
	consequence	*Initiator*
↓	*tam*	*raj* ↑
	outer/ external	*inner/ internal*

informed	*informant*
passive	*active*
static/ finished	*kinetic/ in process*
effect	*cause*

A deduction that omits mind from the world equation until the very least and last precisely reverses the *top-down*, dialectical order of information. Its order of creation runs, first and foremost, from mind. Things start in mind. The order runs from psychological to physical, from active to passive information, from creativity to its creation. **Top-down hierarchy emphasises an order of information that runs from active (mind) to passive (matter).** How, though, can dialectical order's precise reversal claim to start with subjective intelligence and not objective oblivion? What right has active information to wrest law and order from passive energy? What reason has consciousness to usurp the centrality of non-conscious matter?

Concrete things do not emerge straight out of nothing. This applies as much to 'big bang' theory as to the 'transcendent projection' of Natural Dialectic. As solid from gas, so bulky things derive from sub-atomic subtleties; and they in turn are made of energy. The nearer an object or event to its source the less massive it is; the nearer to its original condition the subtler and more generalised it becomes. Isn't 'principle matter', in the form of forces, elementary particles and stimulating energy, subtler than gross issues - gases, liquids, solids? Inner is finer. It is subtler and causes gross effects. Invisible gives rise to visible. Principle precedes practice, idea its expression and active information passive. Mind, a subjective immateriality, precedes matter. Metaphysical precedes physical action and, according to the universal order of creation, potential precedes its active issue and exhausted result. What is information but potential? What is mind but an inventive capability?

	tam/ raj	*Sat*
	existence	*Essence*
	lesser awareness	*Pure Consciousness*
↓	*tam*	*raj* ↑
	external	*internal*
	practice	*principle*
	informed	*informant*
	non-conscious matter	*conscious mind*
	objects	*energy/ force*

What is the subtlest issue next to physical non-existence? Perhaps a quantum or a mental fluctuation. If chance or quantum fluctuation did not cause the 'program' of the world then unto what Prime Rule, Rules or Ruler does every subset trace its parentage; what Cosmic Ruler is the Measure of all things? If you called the Ruler's rule-book 'archetype' then 'archetype' would be a label slapped on cosmic code, code that (Chapters 7-17) runs the world's great game.

And, if it's 'prior' to matter, what is the origin of mind? What is the issue next to non-existence, that is, Essence? The subtlest Cause is First. And the causal substrate of a mind is consciousness. Is, therefore, Consciousness the

Central Metaphysical Principle, an Archetype from which creation's subsets - lesser principles or norms or regulations - are derived? The Dialectic indicates that this is so. The right-hand column of every Primary Dialectical Stack reflects the nature of Essential, Pure Potential - Consciousness.

If mind comes first then purity of consciousness, preceding mind, is its perfection. It is, beyond imbalance, motion or asymmetry, the perfect order of the mind. It resides, at the Peak of Mount Universe, in Transcendent Essence (see Chapter 5: Transcendent Information and Top Teleology; also Chapter 13). In the course of gradual materialisation from mind to matter Original Symmetry is broken. It is breakage yields existence whose poles (or major grades) are mind and passive matter. In the former the predominant aspect is information, in the latter energy. **Absolute order manifests through principle. A hierarchy of principles, rules or laws gives way to the detail and complexity of practice.** Perfect order is, though broken, always behind apparent imperfection. Games are played according to their rules; these unseen strings constrain the state of play; their changeless influence governs every variation in the way the field of battle flows.

In dialectical terms the characteristics of this Infinite Potential (Chapter 7) constitute a basic axis around which the cosmic product revolves. Or oscillates. This axis *is* the law; its Principal is gradually degraded (Chapters 8-12) by the involvement and influence of energy. That is, as explained above, it descends into an increasingly complex involvement with different objects and behaviours. In the psychological case of mind these are thoughts, desires, emotions etc.; in the case of matter they are what objective science studies. In this view creation occurs outwardly along a devolutionary line of principles; conscious, informative principles precede non-conscious energetic ones. All thoughts and behaviours (where a thing is simply a static behaviour) express, in the course of their oscillations and interactions, different shades of (*Sat*) Central Principle and its existential correlates - (*raj*) dematerialisation and (*tam*) materialisation. There exist, accordingly, three familiar aspects of informational hierarchy - potential, active and passive. The latter pair is easily understood from your own experience - explicit aspects of your mind and body. Above them a first, implicit factor is, from below, less than clear; yet, dialectically, potential obviously precedes expression. Call this potential, psychologically, consciousness or soul; and psychosomatically, lower down Mount Universe and in the zone of dormant mind, potential matter. Archetypes. What is such material non-existence like? Perhaps in this respect the most important odyssey of science is upstream to explore the land, ascend Mount Universe and identify the source of physical order.

division	*union*
diversity	*unity*
practice	*principle*
differentiation/ variation	*symmetry/ theme*
complexity	*simplicity*

In dialectical terms Primal Order issues from Simple Transcendent. This is, no doubt, viscerally unscientific. Principle, which is highly condensed information, supports the order of myriad lower, dependent phenomena. Simple symmetry 'degrades' by virtue of myriad interactions into a complexity of differences. Archetypal principle devolves to local practice. Internal order precipitates external eccentricities. Disequilibria. Individualities. For example, subtle atomic order (which we could call *principle matter*) in practice aggregates into the apparent haphazardness of gross, visible structures (which we could call *sensible* or *practical matter*). Although we know precisely the shape of their molecules, how predictable is the shape of a cloud, an ocean current or a mountain? Simple symmetry degrades but, through all the transformations, balance is sustained. Conservation, equation and cosmic equilibrium are maintained. A mathematician might understand that equation is a highly condensed balancing act. This is because a general formula covers numerous individual variations on its theme.

Simplicity spawns detail and complexity; order gives rise to apparent randomness. The understanding of higher (principled) mind is, in descent, restricted, individualised. It is personalised but, dropping to unconsciousness, impersonalised. At this oblivious level - called by Natural Dialectic passive mind or potential matter - simple, archetypal symmetry is found. And, as we shall see (in Chapters 7 to 12), its excited fields devolve the physics of our cosmos. Formative forces are precipitated through kinetic (*raj*) quantum to the exhausted (*tam*) order of things - classical gases, liquids and solids. **In this view internal consistency exteriorises into changeful variations-on-theme.**

Physical formulae are not hand-written across each atom of every conglomerate but the rules are there. Physics and chemistry rightly seek not randomness but the rules behind superficial, exterior complexity. The goal is truth not seeming, certainty not uncertainty, order and not chaos behind it all. You may imagine or even write computer programs to simulate a billion other possibilities but nature's *actual* 'programs', algorithms or laws are prescribed. Its language is expressed by the invariant iteration of letters (fundamental forces and particles), words (atoms) and their various combinations. The behaviour of 'principle matter' and its aggregation into large-scale practice (our sensible, practical, 'real' world) is the staple of science. It is easy. The rule is that from simple principle issues complex detail: and nothing is simpler than an Infinite One.

Water 'falls' to ice. Ice is 'raised' to water. If we call liquid water 'kinetic' the ice is 'static water'. We could also say that steam is, potentially, ice. Conversely, because energy has to be added to activate the conversion, we could call ice impotent steam. It is a heavy, passive form of gas. This hierarchy is based on the loss or gain of energy.

An analogous hierarchy applies to loss or gain of awareness-in-action or, in other words, mind. Start with what you have probably not experienced and, because it is formless, cannot be imagined.

The first order of awareness, shadow-less radiance of Pure Concentrate of Consciousness, is the diametrical opposite of sleep. It is potential information that loses, with any particular focus, some of its unbroken symmetry. Mental foci (thoughts) comprise a spectrum of shadow, increasingly dim as the capacity to concentrate is lost, which finally drops to rest in sleep. The light goes out. Darkness. Sleep is loss of mind, closure of the information centre.

We do not see these grades of awareness as shadows; nor thoughts as clouds. Neither, when our focus is outward on the world, do we see our appreciation of its business as a relative darkness. Because we are habituated, we think ours is the only level of brightness. However, the Central Light is within, unappreciated and with its full radiance only revealed to a subjective focus that flies above the clouds of thought.

	tam/ raj	*Sat*
	finite	*Infinite*
	constraint	*Freedom*
	imbalance	*Balance*
↓	*tam*	*raj* ↑
	negative	*positive*
	confinement	*release*
	tension/ anxiety	*relief*
	question / doubt	*answer*
	heaviness	*lightness*
	unhappiness	*happiness*
	tears	*laughter*

The second order of awareness, conscious mind, is called 'active information'. Negentropy of information is composed of inspiration, purpose, will-power, feeling and remembrance. Its energy involves intensity of interest and focus of attention. Of course, such attention can be stimulated or suffer a decline. The quality of consciousness may vary. Is an experience positive or negative? Do shadows of despondency or pain darken? Or the haloes of happiness radiate? Although good and bad weather happen at any height, the higher a state of mind the greater its transparency. As you ascend the clouds thin out. Therefore (*raj*) positivity and happiness are linked to a heightened state of alert but relaxed and confident tranquillity.

The third order of awareness is dormant, passive, unaware. The exhaustion of sleep, for example, is subjectively passive; and its body is laid out. Mind's storage zone, sub-consciousness, contains fixed images like photographs or reels of film. This archive, like any file, is also passive except at its inception and at any moment of access/ retrieval. Finally material order is without subjectivity. But although they are passive in this respect some material bodies are very active, very energetic. Take a volcano, a supernova or our friendly lifeline, the sun; and there exist powerful free energies like light alone. None of these is awake. They are not even asleep. *Matter is non-conscious.*

Thus, while Pure Consciousness is (*Sat*) Potential Information, (*raj*) mind is active information; passive information is comprised of (*tam*) subconscious, dormant mind and, finally, a special case of consciousness viz. none. This case, matter, is energetically active, wholly passively informed. By observation and in theory we have prised the world ajar revealing symmetries and rules. Whence their origin? Is it chaos, is it cosmos that you spy? Are principles that rule born inexplicably of chance? How can you explain what doesn't have a reason? Why, if it's inexplicable, don't you give up?

Orderly Creation.

Cosmic hierarchy shouldn't be a mystery now! Covalency of mind bonds top to bottom; it binds the Apex of Mount Universe to base camp, matter; it ranges in between Pure Consciousness and lack of it. And in this case of lack a gradient also falls. In the physical hierarchy principle derives from potential matter, the laws of nature come from archetype (*fig.* 1.3 and Chapters 7-12).From such archetypal framework polar vectors issue in the form of (↑) 'energetic levity' and (↓) 'inertial gravity'. The former is dominant at sub-atomic, quantum level; and from quantum matter-in-principle derives matter-in-practice, that is, all the macroscopic, massive entities that we sense and call our normal, gravity-dominated world. This 'classical' normality, in which inertia and inertness feature heavily, is just the outer layer of a cosmic onion. Here basic rules and energy are rigidly conserved.

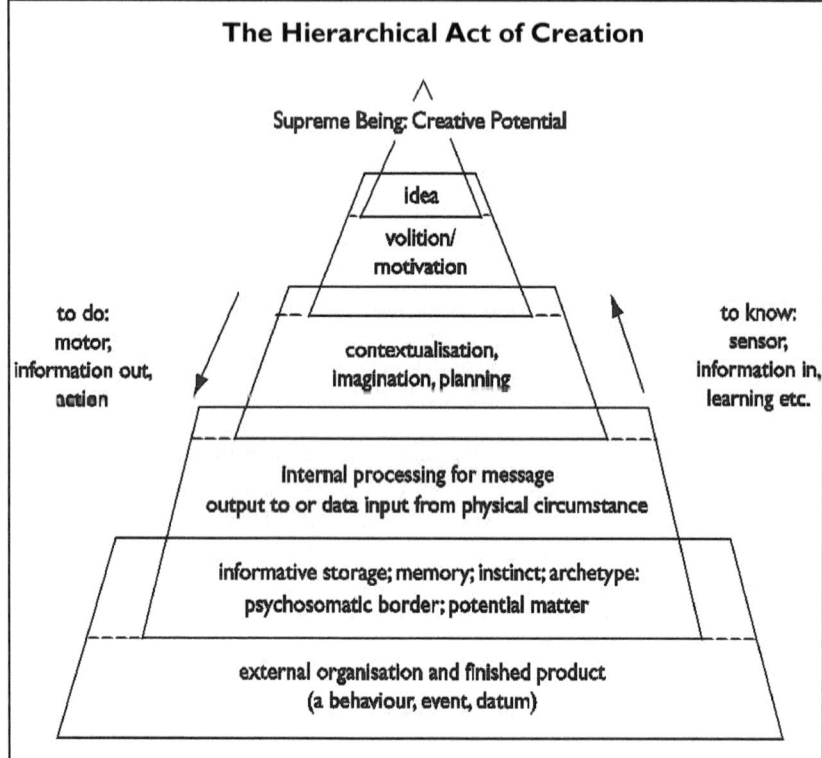

What comes first? The source of active information (knowledge and creation) is transcendent potential. This is also called essence, being or pure consciousness.

Based on consciousness the two vectors of knowledge and action (see also *figs.* 2.3 i and 13.4) constitute the cyclical rhythm of information as it oscillates between perception/ learning and response. At the same time they mark the downward, materialising direction of creation and the opposite, upward, dematerialising direction of perception/ learning/ understanding.

How is commandment passed down from a king? In this diagram the psychological cycle is seen as serial but, more importantly, hierarchical as well. Its oscillation involves priorities. For example, the idea for a creation or behaviour precedes its physical implementation. A creative vector 'descends' from inwardly conceived idea or principle outwards towards its practical realisation. The potential complexity of response is defined by an organism's ability to inwardly focus attention and contemplate; things (including mechanisms) have no ability at all so their response is, simply, round a reflex cycle.

Not every mental activity involves every level of creation. Sub-conscious or purely physical, involuntary responses, for example, involve only the two base, reflex levels. At others response also includes processing and decision.

In the full case, however, an inspiration or idea is the principle realisation, the first switch that illuminates the rest. It triggers the psychological engine, will, to understand the inspiration in more detail and then possibly express it in a physical creation or behaviour. In other words the purposive, teleological phase of will drives into an executive arena. This consists of the imaginative, intellectual manipulation of psychological 'objects' or symbols (called thinking) and their arrangement. The executive level produces a more or less specifically tailored, detailed blueprint for subsequent implementation by unconscious, psychosomatic 'slave drive' (Chapters 6 and 16: archetype and mnemone). In the case of creation such reflex 'slaves' transfer an internal blueprint to the external world; in the opposite case of perception they automatically translate the external world into a code acceptable to conscious psycho-space. Reflex communication involves, in the human case, sub-conscious habit and instinct together with a non-conscious nervous system, neuromuscular anatomy and associated support systems (e.g. energy metabolism). Indeed, in universal terms biological structure can be seen as an exchange in which information is plugged into material energy and vice versa.

fig. 5.3

Thus, from high and free to low and snagged, a gradient also permeates non-consciousness, that is, the zone of energy. Herein microscopic informs macroscopic. Forces and particles are the informants. As science demonstrates, all that is made of them is automatically informed by them. In this large-scale world an input of energy, called (*raj* ↑) stimulus, is defined as 'negentropic'. Stimulus bucks, for a while, (*tam* ↓) tendency of process which is always pushing down; it is, overall, pressing towards exhaustion. Things run like a current in a river towards a puddle, towards a base-line called inertial equilibrium. So entropy is generally, even when a local stimulus defies it, always on the increase. Maximum defiance takes the form, as long as you can make it last, of simple action or of cycles. Other names for cycles are vibration, period and dynamic equilibrium; and in biology a systematic cycle has a special name. We shall meet homeostasis more than once again.

How, though, does psychological creation work?

	tam/ raj	*Sat*
	existence	*Essence*
	do/ know	*Be*
	centrifugal/ centripetal	*Centre*
	cycle	*Axis*
	sphere of influence	*Criterion/ Principle*
	output/ input	*Integrated Whole*
↓	*tam*	*raj* ↑
	outward/ down	*inward/ up*
	do	*know*
	expression/ output	*input/ uptake*
	action	*sensation*
	response	*understanding*
	centrifugal	*centripetal*
	actor	*seer*

This psychological stack is linked to any physical expression of the Dialectic through the psychosomatic phase of sub-consciousness. This phase is the residence of 'fixed thought' - memory. Therefore, where memory is the natural, paperless retainer of ideas, it is also the depository of 'filed plan' called archetype. Potential is an unexpressed capacity. Archetype is energy's precursor; it is the 'template' or the 'field' that, activated, will accordingly express an orderly creation's play. Potential in this sense is causal. Archetypes *cause* orderly succession; what succeeds or 'comes below' is dependent on their regulation. For an energetic potential (or potential matter) one could write:

tam/ raj	*Sat*
expression	*Energetic Potential*
bulk matter	*kinetic matter*
practical effect	*causal principle*

Top-down, therefore, we can summarise by saying that the order of information runs from potential through active (kinetic) to passive (fixed) levels. **If higher governs lower, the origin of matter (potential matter) lies in mind**. Now inspect the order of creation (*fig. 5.3*).

Is there any difference between an act, a cause and a creation? Are not 'motion' and 'event', however grand or trivial, just two more labels, more descriptions of a single, basic character of cosmos - transformation? Never mind First Cause but, in all its changeful consequence, what is 'cause' and what 'effect'? Cycles of this pair are everywhere and interlinked.

How creation might occur on cosmic scale is not as obvious or easy as the recognition that you, living in your own personal microcosm, are a creator of creations. The creative process, although hierarchical, occurs at every level of awareness and, even without conscious components, in a sub- conscious reflex arc. Whenever mind impresses its plans, even automated recorded plans from memory, it 'creates'. We are all 'mini-creating', forever moving things in different ways to different measurements of quality and excellence. Even waggling your index finger, although trivial, is a mind-to-matter mini-creation. What, in more detail, is the order of informative play?

Got the idea? A simple *idea* or inspiration is the first step. There follows, powerful with the strength of unleashed potential, a *teleological phase*; it orientates, directs, motivates and drives; it formulates the basic purpose, objective, plan or desire. 'I want'. 'I want answers'. Such answers are solutions; they satisfy my desires, relieve the burden of my wants and therefore give me, the more intensely the better, what I construe as success. Happiness.

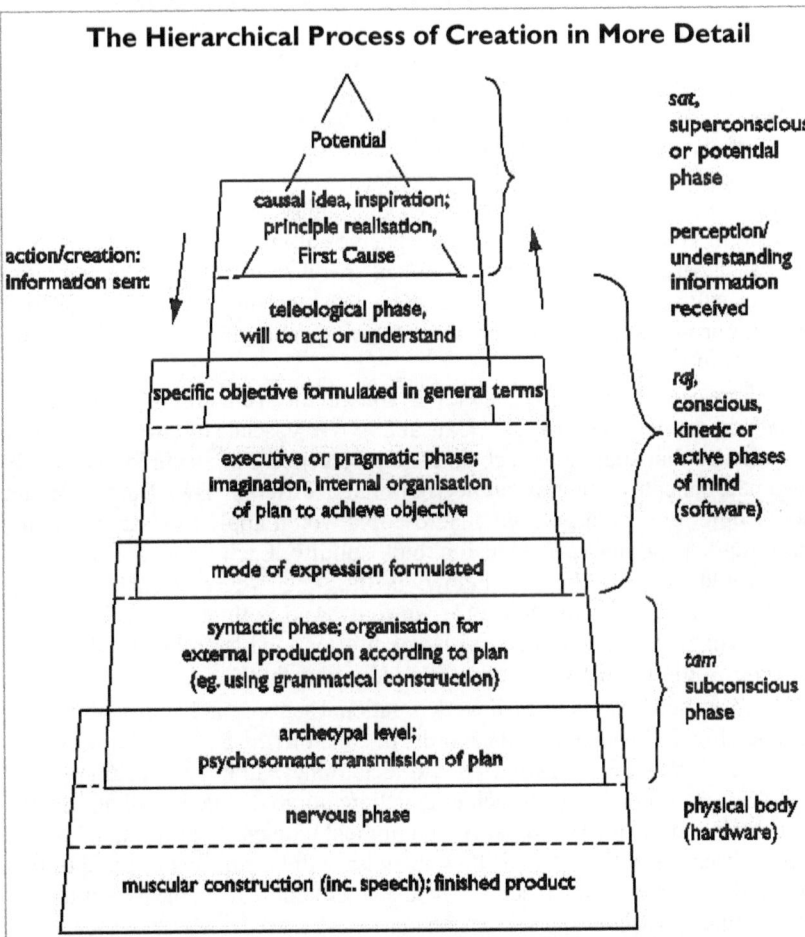

As white light contains a rainbow of colours or different radio broadcasts can occupy the same space, so different grades of intelligence and mood in psycho-space are nowhere separate. As well as a hierarchy of ability to think and act effectively there is also a hierarchy of process, a spectrum of operation that runs from principle to practice, from cause to effect, 'heaven to earth' or, if you like, higher to lower realisation. Psychological cause gives rise to physical effect. This process, which follows a line from information (the precondition) through informed action to conclusion, is one entirely familiar to you.

It guides every non-reflex step of your life from the easy or trivial (e.g. making a sandwich) to the complex and/or important.

It constitutes, moreover, (figs. 5.5 and 6.1: legend) the very basis of scientific experimental method, a method running the exacting gauntlet of leading question, hypothesis, design of experiment to test hypothesis, execution, results and, in feedback towards further experimentation, conclusion and evaluation. As with science so, in fact, with all initiatives whether they are commercial, social, political and so on. Either informal or formalised, it is the mind/ matter process of progress. Its psychological-cause-to-physical-effect is life's creative mode of operation.

fig. 5.4 (see also 6.1)

The engine of purpose motivates. It wants to create an answer, to achieve an object of desire; it becomes involved, more or less passionately, with the pleasure of an interest or solution to a problem. At root this means more sense pleasure or more contemplative knowledge although not both at the same moment. In your case does this variable balance of interest lie in predominance of sensual or intellectual pleasure? What, in other words, are the interests, purposes and major loves that shape, focus and steer the vehicle of your life? Interest, purpose and focus of attention, teleological aspects of will-power, are at the root of life.

If there is a hierarchy of creation you might well presume, subsumed beneath an Archetypal Law (say, Causal *Logos*) a cascade of laws, a hierarchy of properties that applied to each discrete emergence. For example, physically subsumed beneath potential, archetypal matter different laws from quantum microscopic 'out' to large-scale macroscopic would apply to vacuum, atoms, gases, liquids and, at base extension, bulk solidity. Each 'more fundamental' level would devolve the one below. Is this not, where science seeks an archetype that unifies, exactly what we find? And with sub-laws the world-scheme's differentiated properties? Is not information hierarchical - so that the Dialectic would predict this kind of individuation chute?

All constructions, schemes of work or rational actions, including commercial and scientific, follow the dialectical order of creation (*fig. 5.5*). The highest level of the *executive or pragmatic phase* determines strategy; imagination is employed to define a way to achieve a strategic goal. Manipulation, tactical manoeuvre and the internal organisation of mental objects (called thoughts) are used to reach a solution. This is how the wish will be fulfilled or the question answered. This phase may reach a conclusion without leaving the psychological domain, that is, without further expression.

If communication is required either with or through the physical domain, a *semantic phase* is engaged. The detailed plan is now internally organised according to a grammar that will convey it to the outside world. This physical world includes the thinker's body and its surroundings, both animate and inanimate. Communication is structured, in order to most accurately express the design, in a *grammatical phase*. Its rules and regulations may be natural in the form of the laws of nature, nervous signals or reflex body language; or conventional in the form of language, semaphore, morse or any agreed code.

Speech is a good example of orderly, meaningful communication. It is a very precise, highly flexible construction that involves rapid, complex access to memory. This recall does not involve only learnt words and grammar specific to a given language; it also requires association with non-verbal memories and transducers (Chapter 16) that articulate with the body and organise subtle muscular responses. A psychosomatic 'operating system', wherein memory interacts with nervous system and muscle, mediates all thoughtful or instinctive movements. It corresponds with a sophisticated, highly responsive modulator called the brain.

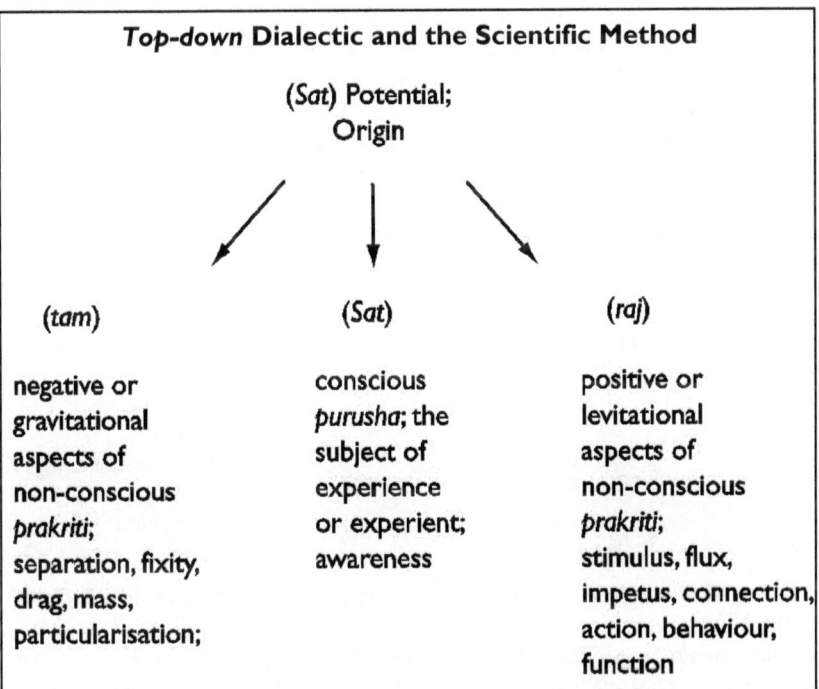

fig. 5.5

The steering wheel, pedals and control panel in your car are designed to materialise your particular directions. These instruments *localise* your 'psycho-space' in 'real', three-dimensional space; from specific positions they transfer specific commands to mechanisms that expedite the driver's wish - to accelerate, stop, turn 30 degrees left for 1.5 seconds or whatever. Similarly, brain gets you down to earth in the way you want. Subjectively you experience fear, disgust, happiness or so on; nor do you, even if you could, think what facial muscles to pull. A switch (say the amygdala) has already set your dial. Nor do you think 'I want to run; I must activate areas A, B and C and, thereby, produce hormone X, complex muscular co-ordinations (P and M with L, S and B) etc. etc.' Such connections, even instinctive responses with no thought at all, are plugged in through the relevant sub-conscious and chemical patterns. In other words, exchange of information with the exterior is organised, at its psychosomatic interface, by *memory*. Sub-conscious patterns are transduced into the electrical syntax of nervous signals. These constitute a 'coded' quantum level of behaviour that is exteriorised through 'wired' impulse and chemical synapses to the final stage of devolution, muscular behaviour. Mind has impressed matter. Bodies are moved. *Physical action is a centrifugal, materialising, differentiating, motor process. You may naturally think of it as active but actually it is mind that is active; the motor agents (nerves, muscle, body) are a simply a passive mechanism that executes mind's active commands.* Both these agents and their effects are informed from 'above' and 'within'.

↓	tam	raj	↑
	passive	active	
	(to) base	(to) top	
	structure	function	
	constraint	stimulant	
	medium of use	code	
	hardware	software	
	storage	transmission	
	recipient	message	

Now your thought is out in the world. Your inward behaviour is outwardly visible. You communicate with other life or, although they will never actively respond, inanimate objects. *Yours is orderly creation. Whatever order you impose on external circumstance is called, as far as the new arrangement is concerned, passive information.* Receipt is distinct from response. It is always passive even though a living or programmed receptor may as a consequence respond actively, that is, with purpose or intelligently - and thus initiate a counter-cycle.

According to this view information is transmitted and received in an orderly manner. Purposeful, meaningful data or programs are stored either in natural or artificial, simulated memories. Your natural memory includes data files and programs (called habit and instinct) that respond to stimulus. Artificial memory (stored in books, tapes, computers etc.) also includes files and programs; such storage involves physical objects employed as syntactically organised and therefore meaningful symbols (e.g. ink on paper, magnetic patterns on tape or photographic images on film) and its usage may involve changes in their shape or behaviour. A sequence of precisely shaped ink (letters) on paper is exactly the same, in principle, as a sequence of 'alphabetical' bases on supporting 'paper' made of sugar-phosphate side-chains: *DNA* is also ink and paper for a script.

You can extend the definition of data storage to include the product of any deliberate action or construction such as a building or machine. Such objects and their behaviours can be seen as reflections of the mind of their creator; in his mind they are memories, on paper plans and, outwardly, memorials. In the case of meaningless objects or behaviours, such as those studied by physics and chemistry, energetic information is transmitted according to *rules*; communication, such as impact or electrical field, may rearrange the reactants and the product is stored, either short- or long-term, in a new shape and/or pattern of behaviour. If the rules or patterns of these communications all derive from universal, archetypal files of memory, then all objects and events are like memorials. <u>Indeed, the physical universe is construed as a vast structure that incorporates the memories of its own creation.</u> It is like the replay of a record once made in the studio. This record of its natural principles is stored in universal mind; its tracks, called archetypes or 'laws of nature' are iterative programs inscribed as various immaterial fields; these fields continually generate new circumstances according to their script. They are like a clutch of sub-routines that together work a whole routine; this clutch can represent a type of organism but also the type of universal program that we find we're living in. Programmed cosmos varies on its themes; general generates specifics; universal gives rise to details; principle produces individual practices. Just as different games of football are the expression of one set of rules, so changes in the universal body seem to be a 'frozen' or a 'crystallised' expression of sub-conscious universal mind. If the laws of nature are 'memories in the mind of God', how are they communicated to the energies of earth (see especially Chapters 9 and 16)?

A Cyclical Order of Information.

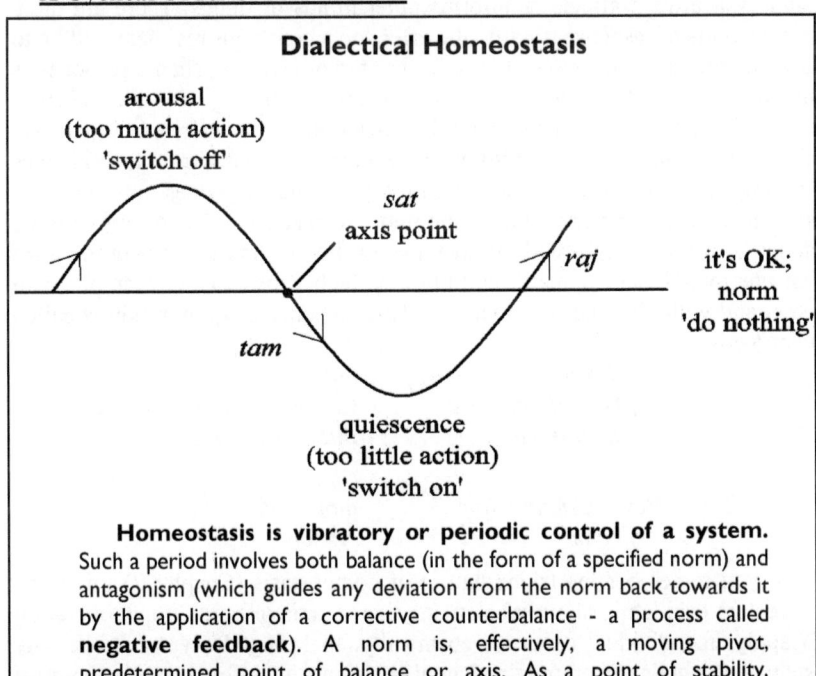

Homeostasis is vibratory or periodic control of a system. Such a period involves both balance (in the form of a specified norm) and antagonism (which guides any deviation from the norm back towards it by the application of a corrective counterbalance - a process called **negative feedback**). A norm is, effectively, a moving pivot, predetermined point of balance or axis. As a point of stability,

> predictability or invariance with respect to an aspect of something's behaviour you could call it a law. The simple, inanimate objects of physics and chemistry short-circuit any loop; they have no homeostatic cycle because they respond in immediate, reflex ways. Animate organisms, composed of integrated bundles of norms, respond in less obvious a reflex way. Their homeostatic norms are integrated and each invested with a specific degree of play; a period of dynamic equilibrium, a 'lifetime', pertains before coherence breaks down to incoherence in the form of sickness or imbalance toppling into death. **Each type of organism is with respect to its bundle of norms seen, dialectically, as a complex law of nature.**
>
> *fig. 5.6 (see also figs. 2.14 and 18.2)*

	tam/ raj	Sat
	existence	Essence
	do/ know	Be
	centrifugal/ centripetal	Centre
	cycle	Axis
↓	tam	raj ↑
	outward/ down	inward/ up
	do	know
	action	sensation
	response	understanding
	centrifugal	centripetal
	actor	seer

An act of creation is one of communication. Unless a superhuman generation *ex nihilo*, it always affects or informs other minds or things. There are three possible kinds of response. Firstly, the effect on a conscious recipient will be to cause an animate response, that is, a further act of creation; such a response is mutually meaningful. Non-conscious matter neither gives nor receives meaningful response; its inanimate, reflex reaction is characterised by what we call the laws of nature. In the third instance 'purposive materials' like machines or biological bodies respond meaningfully. Such meaning does not, however, reside in structural materials but in the mind of maker or user. An appropriate reaction to relevant input will involve a controlled response, often in the form of an attempt to sustain dynamic equilibrium. Such is the case, for example, with motors and with all biological material. This responsive equilibration is called homeostasis.

	tam/ raj	Sat
	sphere of influence	Criterion/Norm/ Principle
	output/ input	Integrative Process
↓	tam	raj ↑
	expression/ output	input/ intake
	effector	sensor

Do you remember the triune nature of homeostasis (Chapter 2) involving two vectors oscillating about a balancing factor, axis or norm? Such operation can apply in individual cases or, cosmically, to Mount Universe itself. Any system that obtains a preordained goal by using an informative but pre-set feedback loop is called cybernetic. A simple example of such vibratory, robotic homeostasis is your central heating system. It works according to a principle of cosmic significance in an information loop called negative feedback. A sensor gathers information that a processor/ regulator compares with the norm. If it is the same an order is issued 'Do nothing'. If it is overactive the command is 'Switch off' and if underactive 'Switch on'. Complementary (*tam/ raj*) opposites are dynamically contained according to a pre-set criterion, standard or rule. Negative feedback, with axial, lawful inaction at its heart, is the basis of your cybernetic central heating. In this way room temperature fluctuates around a norm. Fluctuations, oscillations and vibrations are cycles.

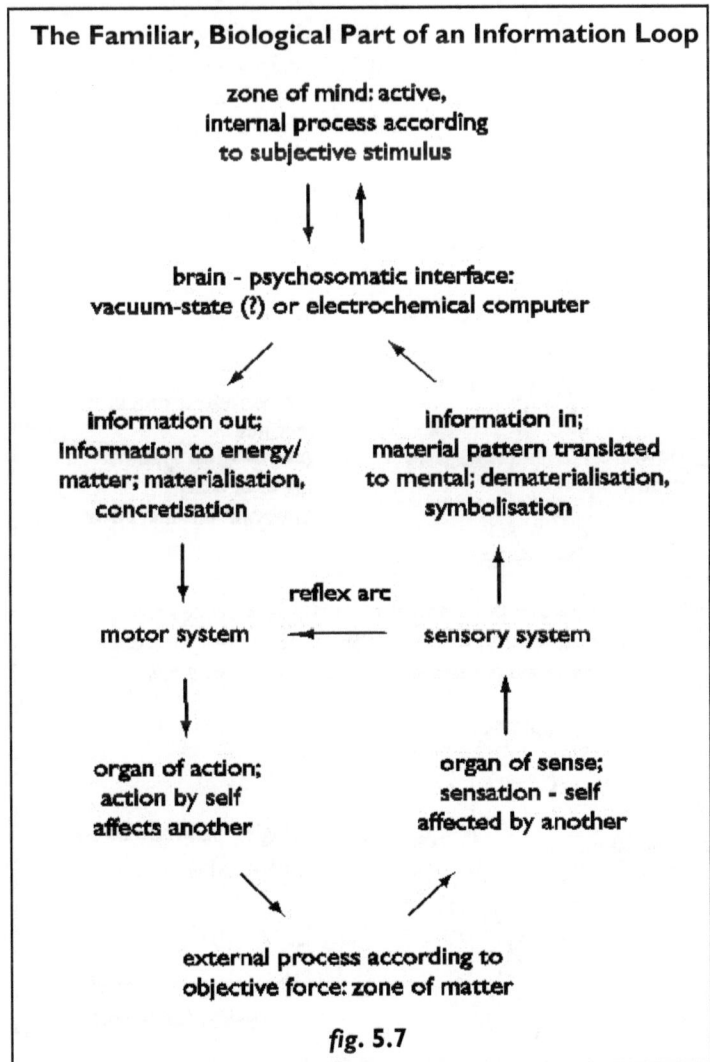

fig. 5.7

The smoothest expression of an oscillation, the form of dis-equilibrium that is most in equilibrium, is a loop. This process round an axis or norm is called a rotation or vibration. A vibratory wave is balanced motion. Any balancing

Inward and Outward Information Loops

(i) balanced human thought process:
both loops in use (see also fig 0.10 iv)

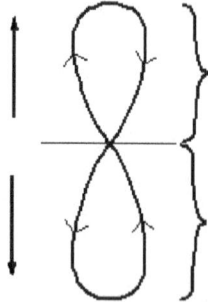

upward, centripetal loop:
Inward gathering of attention;
unification involving the
metaphysical world of principle

cosmological axis (X)

downward, centrifugal loop
(as in fig. 5.7); outward focus;
scatter/diversification of attention
into the physical world of details

(ii) animal predominance in thought processes; instinct and emotion hold sway

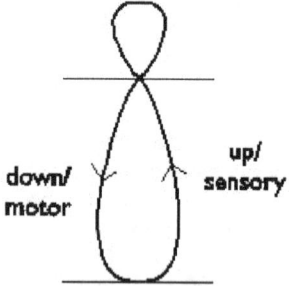

up/
sensory

down/
motor

(X)
subconscious medium/
instinct
neurological system:
physical data,
environment inc. body

(iii) intellectual predominance in thought processes:
predisposition to manipulation of symbol, interest in
informative principle rather than prosaic detail

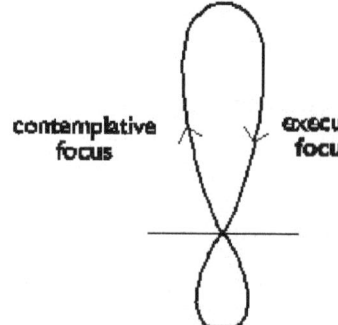

contemplative
focus

executive
focus

bringing principle to bear
on factual data.

(X)
animal loop restricted
to physical/sense data

These loops can help resolve the **centaur paradox**, that is, the tension of the paradox in which, like it or not, all our humanity is inextricably bound up.

> Animal involves the lower half - instinctive behaviour; Man-part involves the higher; in this 'wise' (↑) ascends to Principle; 'clever' bears down (↓) application of his principles on physical constructions. Both share knowledge, of a different kind, with fellow men.
>
> **fig. 5.8 (see also 0.11)**

factor, even an eccentric or irregularly wobbling cycle, reflects (*sat*) harmony and order. The (*sat*) tendency regulates order and limits disorder. *Its 'beats' tend, in a world of motion, to dynamic equilibrium.* <u>*The condition of dynamic equilibrium obtained by pulses that occur round the axis of a rule, regulation or norm is one of 'stability' or 'poise'; the whole purpose of vibratory information loops is to achieve, by means of sensor, axial regulator and effector, stable adherence to a program or a rule.*</u> In other words, homeostatic

cycles respond to stimuli in a manner that guides to a goal; this manner is, within the terms of the goal, rational and meaningful behaviour. The goal can be the maintenance of an inflexible, preordained, archetypal norm; the basis of biology is information; its incarnate information works this way. It can also be calculated fulfilment that completes a cycle instigated by desire. Of course, goals instigated by desire are various and flexible but, whether flexibly psychological or inflexibly biological, satisfaction is the 'answer'. Either way, an information loop is always causal, always seeking balance, always inherently homeostatic.

The cycles of homeostasis are, therefore, not deathly 'flat'. They involve neither incoherent fragments of material scattered 'by chance' nor disparate aggregates of 'factual information'. They are dynamic, self-consistent and instruct, often in cooperation, the driving motion of purpose, the insistent order of working systems.

When is a flight a lifetime? Aircrews depend on information. Feedback displays on a deck control panel help track them to their destination. Automated flight control may navigate in line with preconceived destiny. And an overriding mind, the pilot, is the live part of mechanical intelligence that built up and now flies the plane. Up and down, round and round. A biological lifetime is just the same. **The basis of biology is an immaterial factor - information. It is not too much to proclaim that life on earth is information incarnate.** The genetic code and structural arrangements are fixed, passive forms while expression of code and the cycles of physiology are active. They regulate, they inform and are informed; they exchange information. Biological function is intrinsically bound up with antagonisms, opposites-in-action or feedback according to criteria called norms.

Wheels within wheels, different control systems cooperatively promote your own well-being. Such a pre-programmed, self-regulating, cybernetic form of order, homeostasis, is therefore central to health. Observe (*fig.* 19.2 iv) how information is plugged, *top-down*, into your body. Top-centred brain drives down the spine and ramifies into a surrounding recipient, the body. Nervous homeostasis is, by virtue of 'upward' sensory input and 'downward' motor output, central to your own conscious information centre, mind. Negative feedback is also, with its balancing act dependent on the orderly exchange of information between opposites, dialectical and, thereby, central to Natural

Dialectic. If such construction accurately reflects the way things are then the cosmos might itself be termed dialectical. It is fundamentally dynamic and yet balanced. Cosmos is as much an event as a thing. Initialisation of such an event would, obviously, have involved the establishment of both (*sat*) informative norms/ laws as well as the actual instruments of their (*tam/ raj*) vibratory homeostasis.

Where, in order to follow it round, do you best break into a cycle? An act of creation is first prompted from 'latency' by the spark of an idea. This idea is born of context. Psychological context includes understanding (however profound, shallow, correct or incorrect), stored information called memory and, in the normal human case, the impact of external circumstance. These factors combine to conceive a goal, plan towards it and translate mental into physical action. Behaviour corresponds to prior information; this is the motor half of a cycle. Since it descends from the information centre into physical matter it is called centrifugal.

Having reached earth through the motor half of the loop let us, before we bounce back into sensation, briefly consider the inanimate circumstance of matter in an information loop. Mindless, purposeless, it

is trapped at the rump of cosmos in its own energetic but non-conscious cycles. Seen objectively energy revolves in rings, vibratory cycles, orbits or periods balanced around various axes. Through all its transformations it maintains a homeostatic kind of *status quo*. Variation on strict theme. Conservation laws mean equations can be written and, where every action has an equal and opposite reaction, an involuntary, immediate form of accounting balances the cosmic book; it is this accountancy that scientific maths reflects. *Seen subjectively, however, matter is incommunicado.* It is without the ability to inform, create or evolve purposing complexity (as code or a machine). In this respect intelligence will take you somewhere but all the reflex matter in the universe is perpetually going nowhere. It is informatively inactive; its loop is closed, locked down and, therefore, loop-less.

Loop-less, link-less matter involves reflex physical transformations but is psychologically hopeless. Oblivious. This is why, in dialectical terms, materialism is a descent towards darkness. Yet what escape is there when, materialistically speaking, all is matter and informative mind is a kind of mirage? It is an unpredictable, inexplicable but fortunate property of chemical 'sensibility'! Complex, specific and highly 'coordinated' chemistry is said to assume subjective feeling - which only adds to the mystery.

According to *top-down* logic, however, a nervous system is an interface between body and its information centre, mind. The mode of mind may be either conscious or sub-conscious but, either way, it is mind and not molecule that quickens. Therefore unless mind is incorporated in its information loop there is no life in any biological body. Of course complex molecular systems can, machine-like, respond reflexively. The legs of a pithed fillet of frog will, as the application of electricity by Galvani showed, twitch reflexively. Other automatic responses can be induced, even by 'electrifying' the brain itself with probes, but if the psychosomatic link (Chapters 15-18) is broken there is no more instinct, sub-conscious regulation or awareness. Life is lost. There is none left.

Consciousness may, in organisms like you or a frog, be resuscitated or 'woken up' from dormant mind but it seems that most organisms (e.g. plants, bacteria, fungi) never wake. *Nevertheless without mind, even perpetually dormant or comatose mind, there is no life.* Body is a very special sort of vehicle. Its 'load' is experience, it chauffeurs subjectivity through objectivity, it links inner with outer worlds. Nevertheless neither a nervous system nor a body is, of itself, *subjectively* sensitive. That is, to repeat, electrochemistry alone is not the same as life. Matter, however simple or complex its construction, is thoroughly lifeless. Body alone is, even in life, not live.

What goes down comes up; what parts returns. The sensory half of the cycle ascends from matter to the information centre. Sensation is thus, as opposed to centrifugal action, called a centripetal process of return. It brings matter to mind; it de-creates or dissolves matter into mind. Disparate data is collected from outside, filtered and concentrated in the metaphysical arena of *con*-sciousness. Separate things are combined; they are known as one. Consciousness (literally 'knowing together') means awareness or the unification of disparate items as a single experience, knowledge or sensation - you. Sensation is a natural, pre-set mechanism (or, in informational terms, grammar) that converts the external world into electrical and then psychological symbols. *Material things are dematerialised, that is, they are passed up, symbolised and unified in 'psycho-space'.* The neurological aspects of sensation are obvious - eyes, ears etc. - even if not yet fully understood. Their operation is the subtle study of neuroscience. What, however, turns a nerve to knowing?

Brain scans confirm that, just as subjective output activates electrical, hormonal and genetic switches, so does objective input. *Can we presume two sides of the same coin, mirror image brain biochemistry? Reversible flow? The scales of sense and action swing; information sloshes from one side to the other like the concentrations of chemicals described in an equilibrium equation; input is processed according to goal and output gauged to keep on track.*

The first step up is psychosomatic. At this point a world that has been registered in objective code (nervous impulses) is collected and re-codified into the nothing that is your everything - subjective knowledge. Sensory experience is first loaded, according to the logic of Natural Dialectic, into sub-conscious mind. This dormant psycho-space is a house of ghosts or, in more prosaic language, fixed recordings called memories. It is more closely defined by the Dialectic (Chapters 15 and 16 and *figs*. 15.8 and 9) as a 'mnemone'. With its fixed images 'solid' mind is the way past defines present; personal and archetypal memories define our context; they add 'me' to incoming signals and press my seal on outgoing. Archetype includes the way, through 'signal transducers', that metaphysical mind and physical body exchange information. It is an interface between conscious mind and biological physiology/ cell chemistry. **Accordingly all organisms are equipped with, as well as their non-conscious physical parts, at least an archetypal form of mind; only members of the animal kingdom wake, in various degrees, beyond it.**

Dormant mind is, like underwater reefs or habitual grooves worn in a psychological landscape, always an influence on the way mind's currents run. Each type of organism's 'psycho-scape', the way it is decreed to know the world, is specific. As they stream through its channel incoming signals are filtered, associated with memories, linked with habit and instinct, infused with consequent emotional 'colour' and passed, next step up, to the surface. Surfacing is waking. Being awake is buoyant. The 'dark' information-mix is presented to light and conscious mind for its deliberations. Once symbolised in mind the world can be rearranged according to our habitual ways or conscious wishes. This scheming, whose deliberation ranges from practically reflex to profound, takes place at your thought centre, the cosmological axis, point X (*figs*. 3.1 and 5.7). What an extraordinary world is mind's. What is, for you, in mind? It is all in mind. Not, of course, anything at all physical. There is nothing physical, of itself, in your mind; nor anything that's metaphysical from anybody else's mind. It is a cosmic nothing, yet contains as much about the cosmos as you can upload! And it hands down more commands than even you can dream of. Do computers run at speed? The human information loop can cycle, unless contemplation intercedes, in a split second. In a flash. What extraordinary intelligence is written into body; and on top of that what plenitude of nothing you, patterns in a mind-space, are! You are, as I said before, information incarnate. From what does mind itself download?

In fact there radiate from point X two loops - a familiar, outward one and an inward, contemplative one. Of this couple 'external' sensation/ motor action in the lower complements 'internal' comprehension in the upper. Each loop itself involves, as *fig*. 5.8 illustrates, both inward and outward movement of different kinds of information.

The *outward, downward, centrifugal loop (figs*. 5.7 and 5.8 (ii)) involves the passive transmission of code. Sensory aspects of the nervous system gather and present data to the brain/ mind; they collect the outside world into my head, presenting it as if it were outside. After processing the system's motor aspect carries code to enable the imposition of metaphysical decision; in other words, it carries such instructions as allow our bodies (and their technological extensions in the form of tools, machinery and so on) to rearrange the world according to our purposes. Such creation, derived from the principle of desire, is apt to diversify into complex relationships with the physical world.

The *inward, upward, centripetal loop* is metaphysical. It involves the active, subjective agency of mind in the forms of thought and contemplation. In upward mode involves no organ of the body that subtends point X (the third-eye-centre); but its contemplative focus (*fig*. 5.8 iii) gives rise to learning and erudition. Information presented to the mind is 'made sense of'. On the other hand, its downward, executive focus uses knowledge gained to better scheme; it plans behaviours and creations that muscles and machines materialise. Information passed from mind is meant to fulfil wishes and govern physical environment. *Thus, in accordance with the axiomatic assumption of all education systems, the better a pupil's contemplative focus comes to understand principles, the more informed his decisions and effective his plans and practices.* What Most Powerful of Principles might you examine; what, in life's great game, Most Worthwhile of all kinds of knowledge might you seek to gain?

Bodily and intellectual distractions seem to rule. Most people, busy with survival and emotional relationships, have neither time, energy nor inclination to indulge in trying ot achieve Central Goals, Holy Grails or Comprehensive Bulls-Eyes. But if, because 'the penny dropped' and you saw Principle's Importance, there arose such inclination what sort of grail would you pursue? A *TOE*? Complete understanding of a physical universe which, if there is nothing else, is everything? Or *TOP*? Based on experience that identifies a prior, subjective element of cosmos and would, logically, enjoy communion with its Source, Essential Consciousness? In fact the pair, representing climactic knowledge from each direction of focus, are not exclusive. Indeed, they are complementary. Why not try both? Why should a clever chemist or an engineer not also engage the wisdom of a saint or *vice versa?* It has been the case.

No doubt a human centaur's lower half pursues a comfortable survival and, at best, the hedonism of sensation. In the top class, though, what divides the wise from clever man? What odds between a *TOP* and *TOE*? According to the Dialectic this is simple. Cleverness applies its knowledge facing outward to the world; its focus is executive and its manipulations, science and technologies are instruments of physical desire. The derivative of contemplative focus, wisdom, only faces inwards, sets its sights beyond the intellectual loop and is consumed as it unites with Inmost Principle. *In other words, ascent beyond an intellectual loop is possible.* Beyond the myriad lesser loops in which minds are entrapped and passionately confined there exists, from Alpha to Omega, from Origin to Start, an overall information loop. If creation outward occurs along a devolutionary line, then centripetal de-creation will evolve inward. If mind is (↓) 'downloaded' from super-mind, if awareness is lost in its involvement with matter, then the (↑) evolution of consciousness is a triumph of positive over negative. It amounts to an expansion from confinement, liberation from any adverse drag that's due to mental and material conditions and, at last, recovery of The Primal State that's long been lost. The climax, beyond the changeable ambience of existence, is return to an Indestructible Centrepiece, the Source of Worlds.

This is not a scientific truth. You could have been the first man, Adam, and still known. It is embedded at the heart of both human and cosmic construction. It is the full potential we can realise - because capacity's incorporated in the way we're built. You're familiar by now with cosmic hierarchy and the world made sense of in this self-consistent way. *Man, the Dialectic shows, is a microcosm of the macrocosm.* He is, therefore, 'both measure and the measurer of universe'. **This, you may dispute, is simply an 'anthropomorphic' misinterpretation of the evidence; a human-centred vision of the cosmos is unscientific and, materialistically speaking, wrong. Alright; but if there *is* an immaterial element of information, natural as your mind, then this view is the one that's true!** *And, in alignment with this truth, a man can strive towards closure of the cosmic information loop. Such closure reaches out to boundless openness; such stretch towards 'final origin' is called the mystic path or science of the soul.*

Alignment with Truth.

tam/ raj	*Sat*
practices/ principles	*First Principle*
existence	*Essence*
complexity	*Simplicity*

	expression	Potential
	concentric circles	Centre
	oscillation	Axis
	lesser reasons	Supra-Reason
	relative truths	**Truth**
↓	tam	raj ↑
	from Truth/ Principle	towards Truth/ Principle
	towards complexity	towards simplicity
	centrifugal	centripetal
	executive focus	contemplative focus
	dispersal/ diffusion	concentration
	sub-reason/ instinct	reason
	Irrational	rational
	unthinking/ emotional	intellectual
	attachment to sensation/ physical objects	detachment from sensation/ attachment to generalities
	action	understanding
	outward practice	inner learning
	motor system	sensory/ perceptive systems
	materialisation	dematerialisation

With what sort of truth do you align? What's the way to reach it? Is your method rational, irrational or something else?

You believe what you hold true. Such conviction is established in three ways. One is by empirical observation using measurements; logical deduction, whose main element is mathematics, is another. One trades in physic and the other metaphysic but both rationally serve science.

Reason is a serial path by which intention is achieved; it is the instrument of ego calculating how to satisfy desire; and, in its third and highest form, it seeks to disentangle truth from error and discern the kind of life best lived. Such discrimination is not scientific. **Science cannot offer moral value, meaning or transcendent faith to lift a soul beyond its body-world.** It cannot even offer love. What, therefore, is the unscientific way that humans work out what is good and bad? Is there a rational way to seek and find the Highest Good?

In Natural Dialectic's terms the vector that increasingly aligns with truth will carry you (↑) beyond the mind. Enlightenment reveals that Reason transcends reason. Call it, therefore, Supra-Reason. Sub-reason (comprising instinct, unpredictable emotions and irrational behaviour) is not the same as Supra-reason. The latter is, although transcending intellectual explanations, not irrational. Quite the reverse. Which way of reason, therefore, might exceed the intellect? What might excel pure logic?

A reason is be-cause; truth secretes itself in cause; what, therefore, might you reason is the nature of First Cause? What is the truth of origin, of where you come from? Such cause happened in the past. While science well defines contemporary operations of the universe, the latter's origins can only be inferred. The same is true of life. Thus abduction by best explanation wins the day. Some prefer materialistic explanations. Others would infer a 'hidden variable' whose metaphysic, information, finds its source in mind; and if the root of mind is consciousness and pure consciousness is the essential component of creation then the Delphic exhortation 'know thyself' dives deep and clear. **With which sort of explanation do you philosophically align?**

Science is inherently unstable. It involves faith that its present, material-only understanding is provisionally correct but also that new visions will appear. Is there, for science, ultimate, eternal truth - a wholly naturalistic *TOE*? At least, as far as metaphysics is concerned, science should be neutral. Thus, as a methodology, it is neither atheistic nor theistic - unless you play the two-hat trick. Scientism, for example, wears two hats. It swaps between a neutral working practice and non-neutral world-view - materialism, scientific naturalism and thereby atheism. If there's no natural purpose in creation then this kind of faith relies for all of its constructions on blind chance and matter's 'instinct' - those behaviours we call natural law. Moreover such a strain of extremism, which needs the theory of evolution as a vertebrate needs spine, appears immune to contrary argument, however reasonable. 'New Atheists' are modern, self-styled 'Brights'. The bright idea is that, somehow, religion is a virus, a violent and pathological attack upon humanity; and, in tandem, evolution (with its intrinsic competition, domination and survival of the fittest) is naturally the fittest creed to march with. By this logic a figure of holiness, say Christ, is a diseased expression of 'evil' ignorance, superstition and repression! After all, he peddled metaphysic which, it has been decided, is not real; and didn't peddle evolution which we, in the name of science, have decided is! Our scientific reason is inherently superior to imaginary, contemplative Supra-reason.

Thus, to pop the other hat on, it is right to try and render any philosophical, holistic world-view in as few strokes as we can extinct. It is adventurous to break religious spells. The outcome of such rectitude is catalogued in Chapter 26.

The 'New Atheism' of, for example, Richard Dawkins - wearing philosophical and not strictly scientific hats) is self-confessedly militant. Its lop-sidedness appears alarmist and *prima facie* fascistic ('it might be ethical to kill people for dangerous propositions' writes co-conspirator Sam Harris in his book The End of Faith: Religion, Terror and the Future of Reason). A web-based church of anonymous acolytes is led towards enlightenment by their intellectual élite; an army of irrationals indulge in brutish abuse of an enemy marked out by its inferior-minded theism. Examination of fundamentalist blogs might suggest these foot-soldiers are defined by hate for religious faith and consumed by the scientific and thereby political correctness of their faith in evolutionary theory - sorry, fact. Indeed, their God is foul in many ways and yet does not exist; it is a fantasy created in its own creator's image; but if man is sole creator does not such a toxic image just reflect the down-side of its own demonic father? 'New Atheism' is, fortunately, a fashion that lacks the power of a political organ. It lacks Enlightenment of The Terror (with its *liberté, egalité, fraternité* and guillotine) or social Darwinism in a Communist or Nazi form; but, never shy of shots, it does not lack ideological, separatist rhetoric or, in Marxist fashion, negative dialectic. Does not destabilising violence-in-mind drive violence through the body politic? Dark extremes distort religious and political ideals but just as problematic is that basis of New Atheism's kind of fundamentalism, a Theory of No Intelligence, alignment with blind forces of oblivion and belief in scientific evolution. Upon the horns of such dilemma, is atheism simply sad but true? Don't such potters call the real shots?

All detectives seek results from reason. They seek, at best, sufficient information. Their goal is alignment with a case's truth. It is part of everybody's best survival strategy to find out how things really are. Their goal is an alignment with the truth. It is part of our survival and improvement strategy to find out how things really are. We want to understand and thus effectively impose our will; we want to order our surrounding context starting with our bodies. Comparing feedback with preconceptions one can re-align. What, however, is the maxi-truth encompassing all others, a '*summum bonum*', state of bliss or 'final origin' with which one might climactically unify? Will 'total knowledge' bring the seer 'total happiness'? In other words, is Knowledge Bliss?

Down-to-earth, the centaur paradox (from Chapter 3) is resolved into a matter of perspective. Perspective leads to courses of behaviour. Politics of persons or of nations, law, religion and appliances of science are all constructions governed by the metaphysical, that is, by perspective. They engage activity that's guided by a line of sight. It's important, therefore, as in engineering so in life, to line up with Correct First Principles.

Human 'animals' are centaurs stressed between biology of body and psychology of soul. Centrifugal, worldly struggles vie with aspiration, ideal and their inner, centripetal search for lasting truths. 'My' attachments compete with what is called 'detached academy'. Now you understand the meaning of the scholar's maxim as he researches higher principles and seeks a reasonable understanding of the facets of our universe - both psychological and physical. It is knowledge for its own sake. Principles that leave aside the scheming uses all of us, as local businessmen and politicians, put them to. Detachment is a virtue prized, in theory at least, by academics, men of reason, schools and offices of government. With that search there's always, logically, arises craving to obtain The Highest Principle of All, The Principal. And, as the contemplative focus of men everywhere has always shown, aspiring to a high ideal involves morality. This is what a right perspective must include and thus what a worthwhile sage debates. Though sometimes clever rationalism expels wisdom from its doors the names of colleges in universities (say, at Oxford or at Cambridge) unequivocally state their origins in contemplative aspiration towards a Truth of truths.

Is such Truth foundational or, returning back through all of human history, are the founders of man's scholarship just wrong? **Is your perspective that material alone exists where, I have said, material and immaterial are both, in order, real?** Of course, the view of Natural Dialectic and the founders of a scholarship according to the contemplative focus must be wrong if immateriality - materialism's great anathema - is simply a delusion of the mind and therefore wrong. *Thus are founders made pariahs and their inner truths subverted.*

However, we've already tackled this perspective. If, where immaterial information constitutes creation's core, materialism's base perspective only covers a peripheral reality, then holistic *PAND* is right and *PAM* is incomplete The founders with their Ancient Truth are welcomed back and wisdom is re-elevated to its rightful place. Would the banner of its restoration, raising intellectual reason over colours of a hedonistic life, be popular? Would it mean the aberrations of a centaur race were banished from our human circumstance? Would it even temper passion or, for heaven's sake, mitigate the body's earthly strife? It is doubtful since materialistic centaurs do not welcome such ascent as, up the right-hand path, a wise man indicates is best. Human nature's worldly down-side intervenes!

There is a catch. Enlightenment, if it transcends existence altogether, also transcends the immaterial fraction known as mind. It can, therefore, be inferred but never known intellectually. Human position (X) is (see *figs* 0.7-11) relatively lower down; our mental axis is below the Axis. We are scattered in the outside, ever-changing world and, lacking superintendent overview, our knowledge cannot be complete. Finite information is, by definition, held in restrictions we call form and will involve some element of ignorance. **In this way, even if physical science obtained full and accurate information concerning every aspect of its data (the physical universe) and even if psychology had scoured every cranny of humanity's experience, they would remain relatively ignorant with respect to the whole, three-tier cosmos.** Nor, without the inclusion of an Absolute, will such psychology obtain full objectivity as its criterion by which to measure relative conditions of the mind. This Absolute is Transcendence. **Psychology without Transcendence is like physics without light.** From this super-intellectual vantage point a full, *top-down* understanding of principle, import and priority is, paradoxically, gained. Characteristics that more nearly approach it are indicated on the upward right-hand column of Secondary, Existential Dialectic. Central characteristics are themselves recorded on the right-hand column of Primary, Essential Dialectic. Once important, now derided; once the stone substantiating all philosophy, it is today a pebble scientism throws away.

This is psychological ground. We are talking about the origin, not of species but of specification. While psychology measures physiological data and argues different aspects of mind, it also needs to measure consciousness against a scale of Truth, a criterion whose light of life is perfect. This is, of course, what religions have tried to do. The problem for religion is, while its Professorial Saints may have obtained First-Class Degrees, its priests and other neophytes have generally lesser, sometimes significantly lesser, marks. Ignorant of communion, these inexperienced juniors do not fully understand faith's heart. They cannot speak *ex cathedra* but must interpret from eccentric seating round the central throne. This eccentricity, according to its distance from the Cosmic Centre, refracts; and as distance from a fire makes frosty so, in proportion to the distance of their orbits from The Sun, the teaching of these novices is cooled into liturgical formalities. The juniors assume, especially when their Senior is dead and thus unable to correct their misinterpretations, degrees of misalignment. Foibles of their formulations (Chapter 0: Religious Delusions) take the force of law. Metaphors take on the face of fact. Explanatory guidance turns to politics and government and forceful law. Is not entropy the worldly way?

Psychologists aren't any better off. They are as ignorant as priests of Ultimate Perfection. Worse still, many do not think that such a 'thing' exists and therefore do not care to try and understand it. Science of the Soul is, where there's Science of the Body, muscularly brushed aside. What then usurps truth's empty chair? Surely it is not the fields of evolutionary psychology and sociology? Nowhere do nerves engender speculation less controlled than in this pair. Yet beneath their banner lesser minds (of uneducated peasants and the proletariat0 are scientifically sanitised. Delusions of religion are treated with a knowledge that emerges from grey matter (mind is made by brain). Re-education, pharmacy and exorcism clean The Ancient Truth away.

Of course, illness and accidents occur but pain is not the way we measure health. Surely to embrace material oblivion as your destination must, interfering with mind's natural compass, generate a major misalignment from a reading of The Polestar's Steady Truth? It would seem that, left-hand down, you might descend into a void, a lack of purpose and, since immaterial mind is information's centre of exchange, misunderstanding of mind's nature and, as far as death's concerned, full knowledge and correct alignment. Down's the wrong way. Thus there follow, in this chapter and the next, a hierarchical exploration of information. The gradient runs from (*sat*) potential through (*raj*) active to (*tam*) passive forms. We start straightaway with (*Sat*) Essential Information before moving, in Chapter 6, to cover *(raj/ tam)* existential information.

(*Sat*) Potential or Transcendent Information.

	tam/ raj	Sat
	existence	Essence
	expression	Potential
	its peripheral designs	**Information Centre**
	relative levels of truth	Central Truth
	conscio-material spectrum	Pure Consciousness
	breakage into creation	Ultimate Symmetry
↓	*tam*	*raj* ↑
	passive information	*active information*
	matter/ fixed mind	*experience*
	non-conscious/ subconscious patterns	*conscious mind*
	storage or automatic routine	*reason/ purpose*
	resultant creation	*creativity*

From a *bottom-up* perspective, lacking concept of a conscio-material gradient or Central Origin, 'Transcendent Information' is delusory nonsense. An intelligent modernist, having adopted 'The Theory of No Intelligence', claims such thinking is relatively unintelligent and, certainly, absolutely irrelevant. Was not the world's First Cause probably a 'big bang'? What of your personal origin? Where are your roots? They lie, of course, in the soil of your ancestors. Not just a few centuries of family tree; not only in passage through the churchyard's graves but also a reversal through parental apes and fossilised uncles to a primordial speck of mucilage at the vital moment - abiogenesis (Chapter 20).

Top-down, a holist agrees that your physical roots are in human ancestors but adds metaphysical roots in the form of an informative Central Origin, the

First Essential Cause whence both universal mind and matter are derived. Thus Transcendent/ Potential Information (*fig.* 5.2) seems to be of no *earthly* use and yet is, paradoxically, the most fundamental level from which all existential (active and passive) information originates. That is to say, all the facts of existential relativity subtend, by degrees, Absolute or Transcendent Information.

Let's put this in another way. Your body's made of elements but elements are generalities; each composition everywhere is made of them. Mind's basic element is consciousness; if consciousness is elemental then every different minds is based on it. It is information's essence and, just as you can concentrate material elements and build up density, why can't you achieve a pure concentrate of mind - thoughtless consciousness alone? Undiluted animation, pure vitality. If, moreover, you can grasp that matter's cosmic, why not also mind and its centrality? Why might you not be able to distil a supramental, post-conceptual essence, Cosmic Consciousness? Such Potential is the first, top mode of Being. Pure (*Sat*) Information, perfect equanimity that precedes any shape or form of mind, is the Essence of all lower beings - conscious and non-conscious - in their worlds. Nothing in existence can disturb its Equilibrium. Objectless, it is therefore No Thing and, equally, the pure gold of Subjectivity.

Objective's differs from subjective's way of looking at the world. Observation that's detached from objects is replaced by an immersion, empathy and forging of relationships. Relating tends to friendship; attachment binds and, losing any sense of difference, communion. Losing feeling of apartness lends to ecstasy. Probably the simplest, universal merging that we all enjoy is in communion with music. We unite with motive and emotive energy. We communicate through sound. We resonate together. *The abstraction of music is one of vibratory quality rather than isolated, unit quantities; of unification rather than computation or analysis.* **Its harmony is to subjective experience what mathematics is to objective science.**

The most subjective experience that I have is what I am. My stream of consciousness. This is what I know, subjectively, for sure. My life. This experience, remember, is a two-way track. I can scatter outward into all the world's relationships or, by choice, enter inwards towards the (N)One. I can ascend or descend Mount Universe towards or away from its Transcendent Peak. In this dilemma rests a paradox - in which direction shall I mainly turn? Detached from world, attached to (N)One; or, carried on the mind stream, *vice versa*? Which way is the pull?

t is clear that subjectivity is not a thing but an experience. While Subjectivity may undergird the macrocosm, it entirely lacks objectivity. It is Experience. Beyond mind, it is central to existence. For a Theory of Intelligence this preconscious, super-conscious Essence is Real Knowledge. **It is Knowledge only had by being it**. The entirely concentrated experience of transcendence is the culmination of mystic transport. For a Christian it is Highest Heaven, Sanctum Sanctorum or The Holy of Holies. What about *al-Isra* and, at the root of Muslim faith and prayer, the *miraj* (flight) of Mohammed to the seventh heaven? For Jews Illumination is *Ain Sof*, for a Hindu Essence is *samadhi* (Self-awareness). The Buddhist Peak is called *satori* or *nirvana*. *Nirvana* means where fluctuations, flames or alterations of existence are entirely absent; all flames and flickers of the world are 'blown out'. Nothing but enlightenment is left and, thereby, knowledge of the Quintessential (N)One.

What for starters is more simple than Infinity of (N)One? What less complex than the Concentration called Illumination's Sun? One of its rays is part of any sun. Nor can you think of separated drops of sea. What is a 'point' of light, a photon? Isn't it a star-like isolate in inky night, a speck of brilliant weakness and the only spark of life in oceanic darkness? What about the light of animation? Soul is, essentially and paradoxically, a 'division' of the Infinite. It is conceived of as a simple ray of Consciousness or maybe, like a photon, a life-bringer in material night. You might also see it as a drop of ocean separated from the other drops by an accretion of two finite sheaths. This couple clothe the Naked Heart in its creation, its phenomenal existence. The inner garment is of metaphysic - mind; and the coarser, over-garment body. With these two coverings (*sat*) immanent potential information expresses (*raj*) active thought and (*tam*) passive forms of things. Each binds the edgeless into finite form; each more or less restricts Infinity.

This is why mystics refer to existence as various grades of imprisonment. It is why, measured against the Central Reality of Universal Soul, they treat the traps of mind and body as illusions, that is, as only relative realities. And it is why the upward path of an 'Alignment with The Truth' therefore logically involves an increasingly distant relationship with duality, an increasingly close relationship with Unity and, at last, Identity. Self-realisation is to have cracked the shells, unlocked both doors or stripped existence off. The sentence is over. The drop has merged back into its Essential Ocean. The previously separated, cellulated soul has merged back into its original, infinite potential. Such potential is not an absence or just empty space but a freedom *full* of possibilities. Knowledge of Nothing brings, paradoxically in the fresh light of Oneness, knowledge of everything else. Difference becomes togetherness. This is because Transcendent Information (Essence) is the source of all existential information; so, from this Causal, *Top-Down* Perspective 'All is One'. Since mind is the existential information centre, Transcendence lies above mind. It is not, therefore, an intellectual exercise. You can ask a mystic what the nature of communion is. What is the feeling you experience when sympathising or when merging with another? Is it understanding, peace or love? If mind is still then better understood are motions, motives, motivations; and who is full of love more readily perceives a lack or need of it. Only, though, expect your tutor to express the inexpressible in silence.

Transcendent Information is Essential; but its reflection also rules existence. Information is the (*sat*) potential whence creative action springs. As such it grades 'above' or 'prior'; actively informant mind 'precedes' passively informed matter. In this sense, at subconscious base of mind, (*sat*) potential matter (*figs*. 8.2, 10.1 and especially 12.1) constitutes the cosmic program; archetypal memory, metaphysical and therefore a material void, amounts to 'law' for physical behaviour. This law (or set of natural laws) is woven through its 'subtle' quantum agents into those 'gross' aggregates we call our world. This is the order that informs and thus composes all we sense and so call sensible.

Top Teleology.

	tam/ raj	Sat
	lesser being	Supreme Being
	parts	Whole
	sound	Silence
	expression	Potential
	practice/ sub-principles	First Principle
↓	tam	raj ↑
	passive	active
	detailed expression	sub-principle
	informed	informant
	ts results	vibratory information
	consequence	cause

Are not mystic conclusion and religious delusion, in the eyes of a smart, mathematical breed of logic, serious misinformation? Materialism utterly repudiates their sort of rationale. If, however, existence issues from Essence, what *is* the nature of the top-level information, the source of meaning? What is the cause of reason, will-power and the answer, in principle, to all questions?

Power and import reside in order, in coherence chained along a line. Any systematic set-up operates according to this fact. From kings a power structure branches hierarchically down. The unity and integrity of any system is vested in its regulator. Whence, in creation's case, might any uncrowned regency appear? 'Before time was I am'. Essence is; existence always does or knows something. It 'stands out' from Nothing. If there is Void before creation you might argue that the present job was non-existent in its latency; undeveloped order rested in Supremacy of Being. If so, how could such Supremacy have any choice in what developed from its uncreated field? Who lacks all choice is an automaton. Automata are mindless. They are obliviously material. If God were thus material He can't be metaphysical; but I cannot detect such Being physically or metaphysically and so, you argue, both are figments of delusion. Elemental metaphysic is a dream and its Supremacy does not exist. Imagination is a bubble lack of evidence will prick. Both are bubbles. *QED.*

This is, of course, a materialistic turn of attitude. It assumes, where metaphysical does not exist, that matter's nature is the same as mind's. <u>It effectively omits to distinguish between active and passive information</u>. The issue is, through scientism's lens, blurred and unresolved. No doubt that what is latent in *physical* potential is involuntary, reflex and of predictable response; this is, for example, what Newton's laws of motion are about. If, however, Supreme Being is Absolute it precedes relativity; if *metaphysical* it precedes all lesser beings we call objects and events; and, if information precedes energy, it is alive. And since the Infinite is boundless it is free. Alive, the Will is Free. This means choice is absolutely open; potential is completely unrestrained.

Thus, Einstein could infer, God had choice; God could predetermine any kind of physicality or none. And as sources predefine the nature of their issue, so Essential Information predefines the way existence is devolved. As psychological potential informs *voluntary* outcome, so Voluntary Ordination creatively determines the nature of its universal 'invention'. This working invention, creation, includes lesser sub-cycles, that is, voluntary outcomes generated by living, wilful organisms. One of your desires or a plan brought to its fruition are examples of a voluntary sub-cycle. Of course, the power and quality of individual efforts vary but, among lives here on earth, human beings are the peerless learners and inventors. They are, reflecting Pure Volition, creators of great purposive complexity; they specify, in maths and languages, all kinds of schemes.

If Supreme Being pre-exists anything then it simply *is*. This is why it is called 'I am that I am'; and why it is Essence (or Being) without predicate, quality or condition. One being one. If Unity is what you crave, you will have found its place. Is not Top Teleology's enlightenment the root of all faiths and, therefore, a believer's scholarship? Its Subject is straightway recognisable as the major theme to which humanity has, in various more or less canonised forms, always adhered - except in the materialistic grain. *TOE* grants *system, TOP* grants *meaning* of the world. Our science and societies are, historically, built round *TOP*'s Core Experience. It is, according to the Jewish philosopher Baruch Spinoza, Substance; on Substantial Foundation modifications, called existence, stand and move; Baruch's creation is, essentially, within the mind of such Supreme Being. Such Supremacy is also Natural Dialectic's Essence, Apex or Central Potential. Motionless the ground of motion stays. This Essential Being is Preordination; it is (*Sat*) Pure Consciousness that, when it moves is called First Cause. It is The Word whose grid and gradient branches through the world; or *Logos*, an umbilical connecting-line by which it is created orderly, developed and ineluctably attached to Source.

If Supreme Being is, paradoxically, the Universal Experience of Nothing then First Motion, Cause or Activation must also be almost entirely subjective. *(Raj) active information is the province of mind.* First Cause, the highest and most active form of mind, will be practically without structure. It will preordain structure, sketch ideal, determine informative principle; it will represent potential for what is subsequently detailed - like an idea, perhaps even a brilliant idea. Mind's highest function is teleological. It is purpose. Kinetic purpose is will. Will commands and a command is an order. A command initiates the action. Authorisation is causal. It says 'Arrange things as I want'. Such is the teleological order of *Kalma* or *Hukm* (Command, as the Muslims call it), *Shabda* (Sound, as the Sikhs say), Hindu *Nad* or Taoist *Tao*. Briefly check the Glossary for '*Om*'. *Om* (as Buddhists and the Hindus utter) is supposed to represent the resonance of 'primal sound'; it encapsulates potential; from its seed-vibration sub-divisions of the universe appear. For Hindus there is a radical connection between orderly sound (music) and language. The 'language of the gods' - at least, the world's most ancient known tongue - is Sanskrit. In Sanskrit each phonetic 'vibration' is endowed with cosmic resonance and thus each syllable of a word endowed with such significance; indeed, the fifty characters of its alphabet are each said to have a vibratory correspondence with man's spinal caduceus and its informative cerebrospinal foci (each a plexus, *chakra* or expression of a variously-petalled lotus flower: see Chapter 17). Thus language is, in its most ancient form but one from which our own Indo-European English derives, related directly to cosmic utterance.

Om is, in Christian terminology, Amen. Or, again in Christian terms,

Byzantine *Christ Pancrator* is identified as *Logos* (the Word of Reason). Intent in nature; nature's teleology. Thus, of course, from *Pancrator* ('all-powerful') to *Pancreator* is no step; power psychological includes the attribute of creativity. *Logos* is a Greek noun meaning 'reason' or 'the words expressing it'. Speech, message, code. And also 'voice'. *A word has meaning and its purpose is communication. Information. Instruction. Command.* Words are expressed with vibratory sound and imply a grammatical structure within which to impart an inner plan or thought or reason. This arrangement, *logic* or convention must exist before any message can be issued or understood. Could there exist a natural language? And if so, in what terms is it relayed? A second meaning is 'reason' or 'calculation' and a third is 'the order of self-consistent narrative - speech, story or theory'. In the Greek language *logos* can be construed in a mundane sense; it is also found elevated to a cosmic level. Here *Logos* is seen as the source of order or plan. For a Stoic it described the principle of reason that impregnates the world. Nor only in the Christian faith is Name of God First Rational Principle; Word, its synonym, is the Source of subsequent principles, laws or reasons that together constitute the grammar of the world. Grammar and syntax are a set of sense-making rules; by the world's rational construction is it able to be understood. The language of mind and the scripts of matter can therefore only be properly traced and understood in terms of their relationship with the top of the 'cosmic menu', that is, in the final context of First Principle, Cause or *Logos*. Such Living Cause is, at the highest level, first move of Essence in the existential game; it is the central current of Awareness, a stream of Spirit whence devolve all worlds. It is the Seed whence existence is evolved or, if you like, the Heart of Universe. What more Innate a Quality of Heart than Love? Thus, from such root, the Natural Logic of this Word informs all psychological and physical phenomena; and, at the same stroke, gives them reason. This transcendent source of information is a flow of 'speech', a current of instruction, an 'oration' called Top Teleology. **Who would, therefore, by denying *Logos* be an enemy of Reason and thence natural reason?**

A word aside on language here! Cosmos is (as Natural Dialectic's radiant concentric rings and its ziggurats suggest) a nested hierarchy with specific, archetypal fields at core; and, science tells us, fields intrinsically involve their own behaviours and associated particles. Language is a nested hierarchy too - one also born of shapes of vibrant energy, in this case sound. If we conflate the two then types of vibration (phonemes) show as particles (letters), atoms (groups of letters, words) and, within a grammar of polarity, groups of words (molecules); thus gross matter and its worlds are organised syntactically (by coherent rules or laws of nature); a complex, scripted yet flexible and, at root, vibratory 'code' is used to compose an ever-changing physical world. Such informative, linguistic theme has already been introduced and will recur. With life it's the same articulate affair. Biology's genetic code is not '*like a code*'. It *is* a code. From specific letters (nucleotides) are built words (codons), sentences (genes), sectors (chromosomes) and, as a whole, each organism's book of life (its genome). What is more this book is, informed to high degree, automatically accessed to render various meanings (in the form of cell types, signals and cell-cell interactions). Indeed, there's code on code; steps of control direct assembly of the next stage of expression - protein synthesis. It performs an accurate, edited translation into the twenty-letter (or amino acid) protein code. A protein is a word in an organism's vocabulary; sentences are protein complexes and the full meaning is expressed as a metabolic whole. Cells collaborate, an organism lives. This is the meaning of it all. Fluently and eloquently life's bodies are, according to commands, developed into highly specified complexity. No chance a book is written accidentally. Chance at the margins is as creative as ephemeral, anarchic bugs. Within an ecosystem organisms have, each one, a predetermined role to play. Thus is created, generally and personally, life's story. Language is employed this very way. It is a nested hierarchy too. Cosmic language may create the show but, at an individual level, it is speakers who can think and thereby script, according to their information/ intelligence, their own creations. This is life! What screen-play are you, at this moment, dreaming up?

Thus, men speak many languages but the Order of the Universe is not dispensed by human voice. Organisms have been given everything; they can only try, as best they understand, to give account. Remember, though, a universal language, one that all men, organisms and, indeed, even things respond to - vibratory power. Vibration ordered to convey specific message sounds like music! Music means rhythm, wavelength and a resonance with frequency; and the vibratory power of any word (or *logos*) resounds in its ordered context, its message, its meaning and thereby its purpose. A fugue is, for example, a balanced and harmonic measurement of energy; and as a reasoned choice of words grammatically *informs,* so *Logos* informs the macrocosm. Such Word or Language is active and not passive; it is not stored but fluent; it is not a verbal sound, a written record (such as a Bible) or abstracted symbol made of any shape or form. You may remember Natural Dialectic's primary model, its Master Analogy, was music. You may, accordingly, now come to understand that nature's order is transmitted by vibration; and, at its highest level, natural reason is incorporated in creative Sound. If such Harmonic Order is the inward Sound of God then you neither read nor hear it read by speakers but only by communion with the Speaker, Generator and the Source of Reason.

Why, therefore, can't we hear the Sound of God; why can't we hear the mystic's *Logos* with our ears and normal human consciousness? There are only atoms at the base of everything but you need special microscopes to see one; just so, special focus well beyond our 'normal setting' is required to know Transcendent Word. This Word, *Logos, Shabda* - call it what you will - is *not* a priest's words written in his holy book. These books variously map the journey A Single, Inner Land; they describe its character. Yet wars are fought, confusion reign and truth waylaid when maps are taken for the country they describe. The Real Word of Nature is but One. The staple of a mystic's exercise is effort to obtain the focus to commune. This focus is the laser that delivers inner scope; this resolution - Knowledge as a Scientific Truth - resides in super-conscious setting.

Words can be sung. A song emits both power and radiance. Is it not magnetic in effect - an Orphic hymn whose energy not only touches things but also influences hearts and minds? As sound from silence, so issues natural music from the Inner Word; as light from energetic star, so issue rays of *Logos* from the Sun. From the Informative Potential of a Single Cosmic Star, Pure Consciousness, creation streams. As sound to scales and light to spectrum, so

Logos to creation's gradient. **As vibrant tensions and polarities substantiate all physicality, so inner sound and psychic light at root inform all mind.** In the way that the physical world is seen to consist of energy so 'abstract sound and light' - not sensed with ears or eyes but by correctly focussed mind - is held by mystics of all ages to substantiate forms of creation as a whole; and in the way that electromagnetism is the basis of a spectrum of power and colour, so the First, Logical Cause drives a 'rainbow' of information - the gradient called existence. Such primal concentration of influence is gradually transformed, like electricity from a central generator, through the grid to its peripheral users.

The greatest concentrates of radiant energy are those unconscious foci known as stars. Could concentrates of information source the levels of a conscio-material rainbow? *Physicalia* derive from stars; immaterial, conscious suns inform each level of the mystic firmament. Power and influence dissipate. Conscious runs into unconscious. And from run-down stars eventually accumulates the 'flatness' of a wholly passive form of information, massive things. From metaphysic physical phenomena take shape. Does it not seem incredible that from an insubstantial but substantiating Principle flows existential consequence? How fantastic an Immaterial Cause whose order shapes not only mind but the exploding supernovae, galaxies, tiny aggregates called mountains and such infinitesimal incorporations as your own; and which, while their energy is subject to the 'force' of entropy, is (*fig.* 2.6) by nature negentropic and thus never fails! Fantastic, indeed, is this existence! <u>*In dialectical terms neither psychological nor physical components of the macrocosm are a product of chance; its whole, integrated order is Logically Informed.*</u>

Check forward to *fig.* 7.1. Essence, the Infinite, the (N)One or Pure Consciousness are simply facets of a single diamond. When Potential starts to move its play begins to sparkle. **At the apex of existence moves Top Teleology; from its central origin the same power known by a thousand names irradiates and thereby bears a cornucopia of movement.** Call it, as the Chinese called it, *Tao*. Philo, an Alexandrian Jew, identified the intermediary connector between Essence and existence as *Logos*; so, as their First of principles and Reason for it all, do Christians. They named the origin of cosmic grammar Name or Word of God; but words are cheap against electrifying actuality, the hum that generates events and drives the whole great network forwards. Did not a river first flow from its source? Does it not still? If First Cause *was* the origin of things, how *is* it? Its transcendent status might be high but what about its work down here in time? If God indeed preceded or conceived creation it did not yet exist; time physical did not exist and so, as St. Augustine noted, all the planning must have happened prior to its first moment when 'the world burst out'. In this case the idea and primordial act of physical creation might be realised as one. Timelessly single - which does *not* mean that time is other than a serial order. This is what time is. In timed existence space-time seems self-contradictory - a continuous splitter, a seamless divider, a permission that is basic to the series of dependent relativity. Even if The Highest Cause is without time and place, its subsequence of causes and effects requires them.

A woman may tell you that words carry relative power. Some things are trivial but the words 'I love you', spoken by the right man, are possibly the most powerful she wants to hear. The pristine current of Word, fresh from Presence, is alive and brings, as the basis of its principality, an order of love. By division of this subjective union objective creation is gradually devolved and interests diversify.

Conversely, it might be seen that ascent towards Origin is one of increasingly pure, concentrated love. This would therefore be a most important, unifying thing.

It is paradoxical. In infinite nothing reside all finite things. Of these substantial events the insubstantial motivation is love. Let's descend from supernatural non-sense of the most important kind to more familiar psychological ground. Let's ignore, as yet unknown, any intervening 'mystic regions' or 'super-conscious bands' of mind and touch base where at this very moment we enjoy the constraints of a sense-based intellect. *X, of course, marks home. X marks the spot, the treasure of your consciousness. Touch this spot above your nose between your eyes. Just behind its balanced physic there lives, metaphysically, your cosmological axis, information centre and the third eye where you think. Is this not, therefore, where you are? And where, in mind, you purposely create?*

6. Design Dissected

This chapter deals with the act and order of creation. It presumes *top-down* vectors that drop from conscious to non-conscious, mind to matter, subjective to objective, principle to practice and, last but not least, simplicity to complexity. In the sweep of this order the death of mind is matter.

The action of mind on matter creates 'unnatural' complexity. Its shapes rub off as art, machines and cultural diversity. Since matter is, on the other hand, lifeless its own 'natural' creations are shapes of preordained law, specifics-from-universals, variations on inflexible themes. In either case, however, creation itself is presumed to evolve from a simple start to a complex product. The question is, as always, what sort of start? Simple physical principles (e.g. bricks, alphabetic letters, subatomic particles and atoms) underlie complexity. Was the start even simpler? Does complexity need mind?

It does not necessarily need mind. *There are two kinds of complexity*. The first, <u>*non-purposive*</u>, is a function of energy transformations. It involves the confinement of energy (in particles) or loss/ gain of energy in changes of state (e.g. from formless gas to such differentiated variation-on-theme as snowflakes show or *vice versa*). In aggregation or in action nature's physic is mindless. Its *complexity* (whatever its beauty in the subjective eye of a beholder) is <u>*passively derived*</u> from the automatic behaviours of energy informed by rigid 'rules'. In this sense matter is a 'no-possibility zone'; having no freedom of will it is lifeless and automatic. Scientific research and application focuses on this indiscriminate kind of complexity.

Where physical possibilities are mathematically predictable, psychological ones are more flexible. The second, <u>*purposive*</u> kind of complexity works the other way. It is a function of information gain, expansion of consciousness and a capacity to grasp and purposely, creatively exploit the principles and possibilities inherent in any circumstance. An increasingly concentrated focus of attention wakens to greater capacity, flexibility and possibilities for purposely ordered, coherent or <u>*active complexity*</u>. At each level of ascent the degree of coherence may improve; the degree of ingenuity, adaptability and innovative complexity may increase. For example, human mind and its society constantly respond, adapt and build new, elaborate structures to combat the age-old exigencies of a problematic life on earth. <u>*Active, purposeful complexity*</u> works against the 'downward' wear and tear of time and chance; it codifies and specifies design - which chance cannot. It is an instrument of biological survival, intellectual enquiry, technology and artistic creation.

The question is whether you can infer a purpose behind the 'design' of apparently non-purposeful complexity? Not chance but anti-chance? Could the universe be a product of bio-centric design? Could it be both wonderfully and purposefully made? The response from scientific materialism is that 'there's no necessity to think that way because it's all necessity - the name for natural law'. Its answer is, in other words, a resounding, study-stopping 'no chance since it's all (including the origin of natural law) by chance'.

Computation.

While universal energy is the subject of physics and chemistry, universal information is unknown to outward, physical science. Nevertheless a simple analogy with computing illustrates its character.

Active information programs. Its order is teleological (the top level of idea, will and purpose), *pragmatic* (the design department) and *semantic* (the communications division).

There first arises an inspiration, idea or principle. *A decision to realise this inspiration is the causal, volitional or purposive step in creation.* The rest of any project or projection is driven by this engine. It is purpose that motivates. You say 'I want' or 'I will it'.

There follows the *pragmatic* work of a systems analyst. He works out how the idea can best be implemented. *This stage involves imagination, planning and design.*

To be realised a design has to be communicated to the outside world. This is *the semantic* level of information. Its medium of communication, a program, is written. Software includes, therefore, the 'psychosomatic link' by which an original idea is translated to the physical exterior.

The dynamic structure of mind is simulated by a computer; active information of the former is replaced by passive of the latter. The *exterior* infrastructure which passively supports the expression of ideas and robotically rearranges data in an orderly manner includes physical hardware and, from outside the system, an electrical power supply. The *interior* infrastructure includes programming languages and software. It operates in both dynamic and fixed phase; programs-in-use are dynamic while storage systems 'fix' out-of-use data in short or long-term memories of various kinds. *In other words, the two phases of passive information are physically imperceptible, that is, linguistic (the preordained rules of code and, accordingly, algorithmic application texts); and the physically perceptible (the material composition of the computer itself and objects, behaviours or events that are part of its electronic operation).*

Our material cosmos involves (see *fig.* 6.1) the two 'exterior', passive stages of an information pyramid.

The higher of the two, language or code, includes a grammatical convention within which to organise any meaningful program so that exactly the right idea is conveyed. Such a structure operates according to its own rules of position and interconnection. *At the level of potential matter it is called archetype or archetypal memory; the physical and chemical reflection of 'archetypal grammar' is called the laws of nature; and in humans the natural algorithm of information exchange is called a nervous system.* Computers employ artificial linguistic conventions. A customised code informs operating systems, application programs, checks, filters and any instruction that determines an error-free manipulation of data. Encoded data is stored in memory. Memory is variously called a file, a record, archive, database and so on.

Neither memory nor matter can, unaided, recognise anything. Neither can create but they can *carry* information. Software is represented by physical shapes carried or stored on objects such as chips, disc, tape, paper and so on. In humans the hardware is the body, in computing it is the machine and in the cosmos matter. *Information at this, the lowest perceptual or quantitative level, is vested in data items.* Physical data items can be objects, actions, nervous 'blips' or the symbols of a code but at this grade they are, *per se*, non-conscious, oblivious and therefore without message and meaningless. Meaning resides only in the purposefully minded arrangement of a creator, user or observer. A data item may, computer-wise, be input, stored or output; it may, sense-wise, be the non-conscious, neurological aspect of sensation, reflex processing or motor response; and, object-wise, may constitute a component of some natural or contrived event. This means that, in the sense that it is part of a naturally informed scheme, any and every object in the universe is construed as a data item.

Each creative exercise involves the same hierarchical composition. Once the purposive software and 'slave' hardware are in place any number of applications can be run any number of times. In this way the lower, passive stage is realised; a scheme is materialised. *In computing the material end-product is issued onto screen or paper; in nature patterns are projected and rearranged at the quantitative level of data items called forces, objects and events in space.*

While physical nature might behave *like* a great machine or computer, it is not exactly either. Nevertheless the origin of its robotic laws and their mode of influence are of interest. For Natural Dialectic these laws originate according to qualitative principles (Chapters 7, 9 and 10). *The rate of any systematic, psychological creation is not so much a function of time as of the decisive, informed intelligence and will-power of its creators, projectors or participants.* This applies to inventions, artistic creations or, *top-down*, the Act of Creation itself. This one-off left its record stored in the natural way of mind, memory. Such an archetypal source of patterns is called potential matter; it influences first the behaviours of quantum and thereby bulk materials. Such influence is non-conscious, reflex and inflexible. It is as if a 'stream' supporting matter ran through diiferent shapes of groove; or, rather, archetypal tracks by which the different 'notes of energy' resound.

Matter, unlike computer programs, is goal-less. Only minds can create purposive programs and scrutinise output. To repeat, matter is of itself mindless and, therefore, without meaningful feedback. Voluntary, purposive feedback on a cosmic scale is therefore, in terms of scientific materialism, impossible. *This is because physical science must, by definition, work from the last two grades of information - the passive levels of preordained language (the laws of nature) and data item alone.* Its superb analysis therefore not only breaks down his concert into all its measured parts but annihilates Mozart himself!

The Calculations of Mind.

A second easy way to understand cosmic psychology is to relate it to the gradient of our own.

The agent of active (*raj*) information is *mind*. Mind is predominantly conscious but grades to sub-conscious. In sleep we have slipped into a sub-conscious or dormant condition but the non-conscious 'sub-coma' of matter is permanent. The traffic of matter is going, subjectively, nowhere. No way. Mind's, on the other hand, runs two-way. It can operate both as sender and recipient. At the same time its intentions can project 'down' towards the exterior or 'up' towards the interior.

IPod, iPad, iPhone - communicative technology. Information's all the rage. What about iCentre, the eye-centre or third eye? This information pivot, at point X (*fig.* 5.8), is the very centre of your universe. *In the loop from X upward* sensations are received. In this case data, converted into symbolic or coded form as it is passed up a sensory chain, reaches one level or other of processing. No thought is involved in reflex, physical or sub-conscious mental response; but if a message reaches conscious mind you make decisions what to do. Have you free will or did your prefrontal cortex (and ultimately, therefore, genes) make up your mind? Did car drive driver? Was there a conditioned choice or did the various operators in your brain combine to give 'you' ownerless direction? After all, what and not who are you? Aren't you really an illusion dreamt up by the dials of brain? This is, of course, the logic of materialism and, from that limited perspective, it might be correct. After all you might refine your understanding of the brain to every last electron's motion. You will have understood its operations and malfunctions perfectly. You could analyse an airbus or a television just the same. But what about the latter's information? What about its makers, broadcasters and viewers - what you'd call the live, subjective part but you'd forgotten was the whole intelligence and reason? You would not have missed part of the point but you would have completely missed the whole.

Thus sits neuroscience in its calculating lab. Does 'will' sit happily with physics? Does 'purpose' snag determinism with a doubt? Are not choices predetermined by physical events combined with brain? Or, if you take a less material line, is thought not also predisposed according to the '*karmic*' context of the ocean that we sail, our memories? Is there, in this encompassing environment, even conditioned let alone uncaused, unconditioned or entirely free discretion? No doubt, embodied freedom is well battened down.

To interiorise is, dialectically, to think. In different circumstances choice is made by different intensities of thought. The crossing from input to output occurs at different levels on the gradient, the conscio-material gradient, of mind. Consideration now returns to earth.

In the loop from X downward a purpose is transmitted through the stages of functional hierarchy to its conclusion - a creation or communication of some kind. In the motivator/ motor aspect of mind levels of information are linked such that every lower level is dependent on the one above. A response may never leave its owner's mind but, for the average part, it reaches down to muscles. This is the exterior extent.

Among these basic ups and downs and ins and outs of life a question rests. The downward loop ends up in contact with the solid earth. If, however, conscious experience and not non-conscious matter is the primary factor how far can the upward loop proceed? If subjectivity is prior to objectivity how high, how detached from sensation can the process we call thought become? *In the loop from X upward* a contemplative focus manipulates symbols according to its contextual understanding. It wants answers. It tries to find the truth about the object of its interest. Mystic practice is an extension of this intellectual process

- except that the focus of interest is not created objects or the creatures of creation but the Creator. Such practice seeks communication and communion with inmost Truth. In this case, therefore, individual purpose seeks alignment with universal purpose. It is not an egocentric but a cosmo-centric path. It is one of alignment and self-annihilation. Having purified or calmed the mind of its own 'static' interference a meditator is now able to observe (or receive) the natural broadcast of principles and patterns that reside in things. These are the message, received from 'above'. Meditation is a matter of interior focus. The more profoundly interior a level of information the closer is its alignment with the Centre and its Truth.

	tam/ raj	*Sat*
	existence	Essence
	action	Potential
	outer/ inner	Inmost
	following	First
↓	*tam*	*raj* ↑
	passive	active
	motion	mot(ivat)or
	recipient	sender
	response	message
	effect	cause/ reason
	exterior	interior
	product of plan	purpose
	object of understanding	subject who understands
	thing known	knower

Essence is The Origin, The Precedent, The Donor. All existence is on the receiving end. What could existence give to Essence except thanks? Thanks for the grant of its dependent, existential being.

Like Essence First Cause was, is and will be. It is not temporally but *hierarchically* first, in front, before. However, once you take Essence for granted you can see only the myriad cycles of your own existence. And having forgotten First Cause it becomes difficult to distinguish a start from a finish, an origin from an end, a cause from an effect. How far back can you trace any chain? At any particular instant of such a chain of events how can you judge which is which? Or is each thing that happens both?

Reaction is response to a previous stimulus that was in turn a response. And should you wedge sensation in between purely reflex stimulus and a response you simply expand the options, flex rigidity and loosen the robot with a modicum of conditioned free will. Nevertheless the chain of dependency oscillates ceaselessly. A creator one moment is a sensor the next; donor becomes recipient, cause effect and so on. How loose can you prise computation and calculation? How free can consciousness inflate from its tight, material, egocentric chains? And in what way?

In any cycle you have to take a starting point. In the cycles of sensation and creation, understanding and action, information and energy, materialisation and dematerialisation, inward/ outward, upward/ downward or cause and effect where do I stand? Where am I from, where am I going to?

Only Essence is independent; only Potential is untouched by action, motion or the cycling bonds of time. Pure Action and Pure Perception reside in Origin alone. Therefore treat the whole knock-on maelstrom of creation as a Message taken from the Top, the timeless Start - one that, since Omnipresence always Starts, is Starting Now. Treat cause as a message; treat it as information and its influence as effect; and treat active influence as informing passive recipient - consider informative mind to influence energy. Presume mind over matter.

If this sort of logic seems pretentious, trivial or irrelevant compared to the pressing reality of everyday business, it is not. According to *top-down* dialectical perspective it reflects the cosmic scheme of things, a scheme that includes purpose. Whether or not a man admits or aligns with it fundamentally affects his decisions; and the calculations of his mind sum to the worth of life he leads.

In any chain of events creation comes first. I was, in the first instance, created. I was passively granted life before growth, before my nascent consciousness of things matured to some extent. First Cause (in my physical case fertilisation) preceded my projection into the reflex chains of matter and the sensory/ motor cycles of mind. From then my lifetime's been a movie but how might the maelstrom of psychological and physical buffets into which I was thrust end? In a 'downward' annihilation - ashes to ashes, dust to dust? Or, in a hierarchical macrocosm, an 'upward' one - ashes to Origin? Might there be a return of subjective psyche (but not of objective body) from the nether, physical vale of non-conscious matter to the Zenith of Original Independence, Essential Freedom and Peace? In which case physical birth was a descent, a coming down; and the end will be, as far as the gravity of earthly attachments allows, an ascent.

If I am just a body this is nonsense. Mind is simply the action of brain and will, like a will o' the wisp, disappear. *If not, what then?* On earth or, in modern parlance, as a biological form, I find satisfaction in neutralising each obstacle and desire as soon as possible. Death would sort out the physical part but what about my mix of calculations in the form of psychological debts, credits, nostalgia and unfulfilled desires? Is thought a kind of electronic condensation death evaporates? Or else the metaphysic of these drives and memories could still disturb a mind. Buffeted and knocked off balance, how can I regain the (*Sat*) Pure Satisfaction of desire-free Poise? If, in other words, Poise and Point of Equilibrium are features of Original Experience then how by reaching Omega could I reach Alpha and complete the cosmic loop?

Body never will ascend. Dust to dust the gravity of earth will have its own. But such tug is not just physical. Attachments to the world are ropes and desires are loads. Tensions, torsions, pressures. They are '*karmic*' gravity. The more that are cancelled, the lighter I become. Consciousness, unlike the body's mass, is negentropically inclined. It is intrinsically levitatory and so I do not have to wait for death to start to rise, that is, dematerialise towards the Centre. Before the physic crumbles I can start unloading and untying in the mental, metaphysic part. I can start to climb the information gradient, the inner space of Jacob's ladder; and if my determination is intense I will travel a structured, positive, contemplative path Homeward. Every monk or aspirant knows what this means.

I find myself at point X, a point of mind above sub-consciousness but well below the peak of Mount Universe. I calculate in worldly ways or else desire adventure to the Heart of Universe. If Heart is Top I must (↑) ascend through the active regions of creation to Subjective Origin, my Point of Issue (Chapter 5: Alignment with Truth). However, in accordance with the *top-down* order of this book the following brief anatomy of information treats as if I were, rather than a recipient working towards Understanding, the Sender of a Message. It treats in the *outward* direction of creation - as if I issued a command (↓) for something to be done.

Phased Intent.

An act of creation (Chapter 4) amounts to phased intent. It is, simply, the materialisation of an idea. If creation is the re-arrangement of the world according to one's wishes, you and I remake it all the time. The major phases of intent are comprised of active and of passive information.

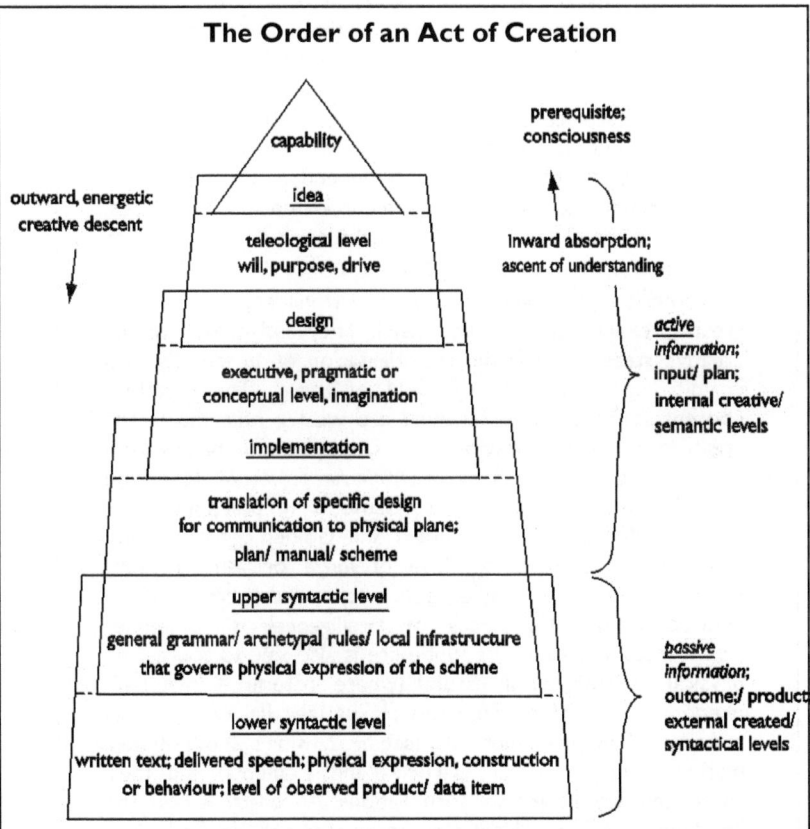

In this hierarchy levels are conceived of as nested but co-existent. A creative act develops, organically, like a seed to flower. The act itself 'descends' from metaphysical to physical levels - from mind to body. Such 'descent' may occur very rapidly as, for example, in a conversation; or more slowly in, say, working out the detailed solution to a technological problem. Acts of creation range from trivial through important to cosmic in scope.

Conversations, solutions to problems and so on also involve 'ascent'. They involve an act of 'counter-creation', comprehension or understanding of what is going on. This applies with full force at top, teleological level where principle, motivation and purpose are grasped. When you have got the idea you have the reason. You understand. Informative cycles oscillate between comprehension and reaction, stimulus and response. They involve different concentrations of interest and energy and not all stages are necessarily engaged. For example, a person ordered in a military way to do something scarcely enters the top three levels; instinctual and reflex information loops are also passive.

It was shown (fig. 5.5) how scientific experimentation is based on informative, dialectical cycles. In the creative part there first occurs an idea or question. Motivation to find the truth is followed by conceptualisation of a hypothesis and the executive planning involved in designing an experiment able to test it. Such experimentation conforms to a pre-established norm, the 'grammar of science', so that under tightly controlled conditions a relationship is established between cause (called an independent variable) and effect (a dependent variable). 'Hard data' is accumulated, scrutinised and correlated with hypothesis. The loop is completed. If things agree the original hypothesis was right, if not it was wrong. Either way a truth obtains.

Esoteric semitic, Hindu and Buddhist philosophies treat the greatest experiment, creation itself. They assign each of the creative steps towards the materialisation of an idea its own cosmic region, 'heaven' or 'field of *Logos*'. These various concentrations of universal mind are weakly reflected in the creativity of human consciousness. The 'spectrum' ranges from Essential Father, *Atman-Brahman*, *Allah*, *Ain Sof* or whatever you label The Infinite through First Cause or Causal Mind (also known as Word, *Logos*, *Om*, *Brahm*, Sun, Golden Egg and so on). From here it drops to the level of '*turiya*' or 'astral regions' whose powerhouse is called *jyoti narayan*, *sahasrara* or 1000-petalled lotus. This 'focus of intelligence' is sometimes represented as a flame. Its subtle fire is also said to source the physical creation i.e. all natural (where 'natural' = 'material') phenomena; therefore from this plexus (see fig. 17.11) issues material archetype - a hierarchy (see fig. 2.13 or 5.2) of potential matter, principle matter (particles and elements) and their subsequent lower, grosser formulations. At length at base the 'hardening' of Mount Universe precipitates solidity. Such non-conscious fixture constitutes the buffers of creation, a cosmic terminus of energy and information. Here flecks of higher worlds, the souls of animation, rarely flicker. An island, an oasis like the planet earth is scintillating in the dark; subjective sparkles wrapped up in objective bodies constitute such forms of life as you and I. Chapters 7-12 discuss the actual grammar, especially physical grammar, according to which inanimate creation is devolved; Chapters 13 - 26 discuss the animated combination.

> Have you ever 'got the idea'? Grasped what someone or something actually meant? What about the Cosmic Idea? Since, in this view, the Origin of cosmos is alive then a full comprehension of the Idea amounts to Communion with its 'thinker' and, beyond, Pure Subjectivity. Although crazy according to the tenets of materialism this is, of course, the mystic's experiment. If he completes the loop and his hypothesis is proven right, what a result! What, Pascal might add, a winning bet, a vindication!
>
> *fig. 6.1 (see also figs. 5.3 and 5.4)*

If a concentration, that is, focus of Pure Consciousness streamed out through universal mind to matter this would constitute the basic current of creation. If First Cause willed all creation in accordance with its plans then perhaps the phases of cosmogony incorporate commands (that we call laws) of how things should appear. Perhaps, in other words, the order of creation is a statement of intent.

All lesser causes, metaphysical and physical, are re-creations, secondary causes or consequent effects of Original Order. They reflect, in phase of their intent, the grades and eras (Chapter 11) of our cosmos. They reflect the aboriginal, intrinsic sequence of creation, that is to say, the order of a world in which material nature is a peripheral part, the bottom of the pile.

(Raj) Active Information.

Active information is only created or registered in conscious, flexible mind. *What constitutes an act of creation? What is its order?* There follows an elaboration of the proposal made in Chapter 4. On a conscio-material gradient passive follows active information; input precedes output; although sensation starts in matter and ends up in mind creative action starts, *top-down*, in mind and ends up re-arranging bodies in the world outside. It 'drops' from cause towards effect. Firstly, therefore, follows an account of the 'internal' or 'subjective' aspects of an information loop. These are conscious teleological, pragmatic and semantic phases.

Purposely Down to Earth.

First issues an idea. This is the (*sat*) potential, seed or genesis of any active, formulating process. There accompanies intention to express, elaborate or fulfil the embryonic inspiration's promise. This is the basis of its purpose and, indeed, all purposes. A void of ignorance sucks in (↑); it is filled by understanding and the purpose is to know. Conversely, from prior stillness action issues out (↓); fuelled by knowledge and desire the purpose is fulfilment of creative process. Knowledge (in the form of latent memories) is revived; plans are made to re-arrange the outside world. Internal to external - knowledge is intention's fuel and will its directed power. This motor, known as volition, is a psychological current. Mind's dynamo streams from focus of attention, a concentration of interest. An act of will intends to order things or events according to its own purposes. *Indeed, will and desire are the psychological equivalent of physical electromagnetic radiation.* Will is like electricity, desire like magnetism. Together they are the light and life of mind; the '*volitio-attractive*' charge of their attention draws, fires and bonds. It lights up life. Its spark is pure vitality.

Thus at the centre of command will couples with idea's intent. Its control post energetically drives a purpose to an end. This end is always the creation or maintenance of, as the creator perceives it, an appropriate condition. Creations range from something as trivial as a small movement to something as complex in its expression as a book, a machine, a building construction, a research project or a new government. How *(raj)* actively informed are life and cosmos? In Books 1, 2 and 3 we'll see.

Information is life in motion. Purpose orients, volition drives and feedback guides. Meaning and information therefore arise from an act of will or are retrieved from the consequence of that act. In will's order lies its reason. *To understand the reason for something you have to go back to first principles, the original information.* Information is as metaphysical as its origin, mind. Every Sherlock Holmes knows that mental informs physical act - that metaphysical informs physical. A crime scene or a machine can be investigated in a clinically scientific way. *Although neither thought, intelligence nor information will be found among its molecules, yet all three imbue every part with their order, with information from the mind(s) behind. And an investigator takes full note of that!*

In other words, there is no mind in matter but there can be! Material information derives from the mind that arranged its components to create a situation or a construction. On a thoughtless, animal level psychological response *follows* stimulus but above such instinctive, body-centred reaction thought increasingly *precedes* behaviour. Anticipation and planning supersede reflex; psychological controls physical action. *This truth, higher than the dormant, robotic reflex of organism or material, is the foundation stone of education, morality and human civilisation.*

You can argue, of course, that the start of a process is not necessarily 'higher' than its end. Why, you ask, even think of an act of creation as 'hierarchical'? Are not 'top ideas' drawn from the context of previous experience; do 'eureka moments' not stem from the rearrangement of memories which, far from being transcendent, are the staple of subconscious mind?

Most ideas and inspirations are indeed based on instinct, desire and personal memories. It is from this 'base' context that mundane acts of creation arise. Such habitual rearrangement of thoughts and objects as chatting with friends, washing up or driving a car are personally important but in terms of creativity trivial. Their 'inventive gradient' is flat.

Memory is just a storage mechanism; is 'objects' cannot innovate. Higher up the scale 'eureka moments' are also based on contextual memory, stored information and experience but this time involve higher principles and deeper understanding of the world and its ways. The discoveries and inventions of such inspiration are archetypal in scope; they involve universal memory or natural law and are therefore more powerful in their application.

Old men are not necessarily wise. On the approaches to wisdom it is not quantity of experience but quality of insight that counts. And at super-conscious level personal memory is not involved at all. The order of existence issues and is therefore best perceived from the vantage point of its source. Potential mind, First Cause, perceives its own nature directly. This is the supra-normal nature of enlightenment. Complete perception is hierarchically above such specific patterns of behaviour and incomplete perceptions as derive from it.

There are, therefore, ideas and Idea. Although hierarchy is not obvious in the low-key, relatively uncreative processes at the level of an average state of mind, it is evident from a salient elevation - especially from the Highest Peak of Principle before all practice. The brilliance of knowledge at this peak is called, in (↑) counter-creative terms, Understanding; and in (↓) creative terms The Origin. It is Enlightenment.

Getting Your Way - Pragmatics.

From its rooftop perch a white stork surveys the same physical reality as you. It feels the same wind, registers the same far-off clouds and encodes the same movements on the plain below. As in your case its experience is symbolic. Its whole world has been dematerialised from things to information. It is inside its mind, its head and perhaps outside as well. A wild scene composed of myriad objects and events if gathered into one. This is the unifying nature of life's *con-sciousness.*

Information loops run from sensor through regulator to effector. They are pragmatic, homeostatic mechanisms designed to maintain optimal communication and, through it, a state of dynamic equilibrium, a peace-keeping interaction with the world. Of course, a bird has different priorities. Its pre-programmed processor notices and calculates differently. That is to say, its nervous system differs and it subjectively manipulates symbolic data in a way recognisable but foreign to you.

Manipulation of symbol is the next step in the conversion of purpose to practice. Only after the symbols have been reordered to our own ends can we project our own creation, our behaviour, our life into the outside world. The crane's data processing seems pretty well confined to instinct, preordained neural syntax and almost immediate, reflex response. It is conscious; it can learn, scheme and remember but a practically closed circuit of instinctive patterns and sense data seems to dominate its world. This is less so in your case. Although your loop includes preordained neural circuitry, sub-conscious instinct, memory and a symbolic picture of the world (called a perception) it is also active in a comparatively embryonic and yet powerful way. You possess a more advanced faculty of symbol-making and manipulation. What hands can do with things mind does with images. You are able to think. Creative imagination is not the weak, fanciful fiction we sometimes imagine. Its suggestive process is, in fact, at the root of planning, complex integration and all educated human endeavour. It is, in concentrated form, a very powerful instrument of purpose.

Pragmatics is, if you like, the level of executive or systems analyst. *Transmission of purpose through the pragmatic level involves, in the outward direction, imagining and planning its realisation.* The sender of a message wants something; and wants the receiver to conform to expectation. Desire expects results. This is active teleology. It involves manipulating options, working out exactly what is wanted of the communication or design, what its format or construction will be and the effect it will have; and as the message is transmitted so it involves contextual interpretation by the recipient as well. This recipient may be animate and thus in various degrees (involving social behaviours, choice of options and so on) of flexible response. On the other hand, interaction with inanimate, non-conscious material may result in the creation of an implement, machine or some impact on the environment. In such a communication the incommunicative material side is still called a recipient. Recipients, whether robots, machines or other material bodies, help to fulfil a plan by fixed, inflexible response. Natural Dialectic calls such reflex passive information. In fact, whenever an intention is physically expressed it works through mechanisms and machines. Machines, which include biological bodies, operate according to physical law and fulfil their function using, in one form or another, energy. They specifically accord with plan and inform the world in ways unguided nature can't. As such they obviously link information (that is metaphysical) with energy (that's not). Could the universe itself express a covert plan?

You may have designed a method and, perhaps, an instrument to get your way. In either case you need to keep checking you are still on course and are going to hit the target. You need, in other words, a guided missile. The missile, your purpose, needs to sustain a dynamic equilibrium with respect to its trajectory. Slight deviations either way need correction, the sooner the better. For this it needs to *communicate* with its environment and, if necessary, re-align or equilibrate with its target. An incoming message, from any mind or body, is a form of feedback.

What Do You Mean?

So important is loop-closing or corrective feedback to the attainment of objectives that the engine of psychological direction, will-and-wish power, is supported by a steering wheel called homeostasis. The communication and, thereby, regulation of homeostasis take both active and passive forms. That is to say, feedback occurs at psychological, biological and physical levels. Only in the first case is its information conscious - *flexible* homeostasis runs a desire to its satisfaction. In this case it involves the transmission of meaning between minds or meaning 'bounced' back to the sender from inanimate material. On the other hand, at sub-conscious and physical levels order, meaning and interaction are reduced to fixed, preordained law. As mind's influence wanes so automatic guidance waxes. Control is always here. Law, dictating how things are intended, *is* the purpose. Passive program has succeeded programmer. *Inflexible*, cybernetic homeostasis works, either instinctively or non-consciously, through a programmed, reflex set of norms. In either case a block on the balancing mechanism upsets; a bug frustrates and causes pain.

All active information is meaningful. Passive information (such as a newspaper or other coded signal) contains the message of its creator(s). Paper and ink are physical entities, meaningless *per se*. However, their commander has controlled their order to provide a channel of communication. A command erases randomness; it is a deliberate restriction imposed to cause a non-random outcome. It employs an agent of restriction - sign, symbol, code of one sort or another. *This is the world of signals, semantics.* *What is spoken is not a matter of chance.* **Whatever is encoded is intentional.** Code and chance are chalk and cheese. They never mix. Language, other symbols and the construction or decipherment of meaningful communication (called semantics) make up the third level of information. In other words, this is the level at which ideas are framed in code before their presentation in vocal, written or other material form.

A communicator, programmer or creator of any sort wants his purpose achieved. He frames his exchange, implicitly or explicitly, either as a statement or a question; take it from or give it to me. The message itself may vary both in accuracy and in the importance of the information. Is it essential, less essential or trivial? For the mystic essential means just that - Absolute Centrality. Mostly, however, the importance of a transaction is judged according to the relatively imperfect, worldly perspective of its initiator. Such perspective ranges through the affairs of business, appetite and love: and whatever human judgments involve an element of choice are often permeated with a moral dimension. Is morality the same as mental health and hygiene?

Whatever information is generated, whatever questions or answers (including decisions) are formulated and whether the formulations are sensible or nonsensical, information always has a recipient - whether the recipient understands or not. Understanding at semantic level is different from that at pragmatic or teleological. In order to make connection codes, which accord with their own grammar and syntax, are used as a vehicle for the message. *Coding and decoding are core semantic business.* Symbolism and simulation are mind's agencies. They are its metaphysical intelligence. They make sense. Although *meaning* is the important, active ingredient behind codes, data transmission and storage, nevertheless these passive instruments of communication are important. We need frameworks (hardware) within which to manage information (software).

No doubt psychological and biological codes are replete with meaning. The latter's meaning is protein, the operation of such protein and thereby (as a computer system is built on regulation by electrical circuitry) the coherent operation of a life form. What, however, is the physical code, the alphabet and subsequent language of inanimate nature? And, if there is one, what is its unwitting meaning? The world cannot express itself. Has it any message?

(Tam) Passive Information.

Passive information is the expression, external to conscious, flexible mind, of active information. Such information may be dynamically exchanged (as in the case of speech or body language). Otherwise its impressions are stored either in subconscious mind (featuring relatively inflexible memory, instinct, archetype and so on) or using matter (where instruction is carried by arrangement on chemicals such as clay, papyrus, ink or *DNA*).

Storage may be fixed (as in a file or photograph) or dynamic (as in a running film, program or automated mechanism). As well as fixations of memory, physically encoded messages or any purposive construction it may also involve unintentional arrangement in the form of reflex operations according to 'the archetypal laws of nature'. Is (Chapter 8) the origin of natural law by accident or design? At this point you may shrug!

Psychology may venture to investigate an active information centre (conscious mind) and its passive subordinate, the realm of subconscious mind;

but neither the immaterial medium of memory nor the material medium of functional mechanisms currently exercises reductionist biology. Today's biology, physics and chemistry deal only with the perception of, as it's construed, unintentional arrangement. 'Hard science' engages a reflex, physical universe that is, in dialectical terms, one only passively informed. In so doing it deals with the (*raj*) activation of code, rules or laws (in the form of quantum behaviour) and its (*tam*) outcome in the form of bulk material objects and behaviour. Such science does not, therefore, reach the (*sat*) source of code, the grammar and syntax of things stored in what is termed transcendent matter (step 4 on *figs.* 2.11 - 12) or the memory of sub-conscious, universal mind. This assertion, shocking to a materialist, will be worked through (Chapters 7-12 and 15-17).

Passive information includes grammatical rules, syntax and the objects (or units) of their construction - words that compose linguistic code. A code is an agreed set of symbols arranged to format information. The basis of language (or code of any sort) is its grammar. Syntax is the convention or legal framework within which symbols are ordered; it determines those structures allowed and those not. *Thus a higher (grammatical/ syntactical) level is the filter through which order is communicated to and from the lower (environmental, statistical or quantitative) level of data items - physical phenomena.* Seen thus the laws of nature are, in type, linguistic code.

There exist both flexible, animate and inflexible, natural forms of orderly, meaningful disposition. In the animate case sounds of a human language (the letters of its alphabet) are arranged in specific orders to signify something. Artists can use instruments and notations to make music, paint or otherwise signify their inspiration. Significance is meaning. On the other hand, natural and dynamic forms of 'grammar' include sub-conscious instinct, memory and archetype; also the laws of inanimate nature and rules that govern the operation of codes carried by electrical and chemical agents, such as genetic code borne on the wings of *DNA*. Indeed, all biological systems of coordination and control (such as the human nervous system) are 'linguistic'. This is because they are *preconditioned* filters. The filtrate of an information system is meaningful patterns. 'The 'language of nerves', for example, allows mind to impress meaningful order on its environment or, conversely, register the latter's patterns of change. In other words, mind is an information exchange and its agent is, in humans, brain and its extensions. Technological instruments are further, 'unnatural' extensions of this agency. Through them science dramatically expands the human register of perceptions and the capacity of motor systems to rearrange the world to suit our purposes.

As opposed to grammar and syntax, which is their orderly mode of interaction and arrangement, *data items* include symbols (such as letters of the alphabet, numbers, signals etc.) and objects that represent those symbols (e.g. complex 'pictograms' such as guanine or adenine which script the genetic code). We use arrangements of data symbols to express ideas.

We can go further and state that any object that is arranged or event that behaves according to a system of rules is, in terms of that system, a data item. **In this sense inanimate patterns of behaviour, whose letters of the alphabet appear to be 'elementary particles', whose words the elements** and whose coherent structure is derived by grammar we call forces, write the story of the world in nature's language. *These particles* (see Glossary: elementary particles), *called bosons (force particles such as photons) and fermions (such as electrons, protons etc.), automatically accord with a universal code of conduct; physics and chemistry strive to elucidate this natural script.* Thus passive information correlates with matter in a universal way; and science learns to read inanimation's archetypal tome. Seen thus the laws of nature are, in type, linguistic. Thus, written up in particles and forces on blank sheets of space, cosmos is a book. This volume is, when you can read it, open and dynamic. Where you can't it is, as yet, a cryptic cipher.

This basic linkage will be elaborated throughout the rest of the book. It therefore bears emphasis that, in dialectical terms, the latency of cosmic language is always present but requires 'harmonic vibrancy' to rouse it into action and express an orderly creation. **The 'level of language'** (*figs.* **5.4 and 6.1) is equated with top matter; 'information's infrastructure', code, is equated with potential matter (***figs.* **2.12 - 13 and 10.1); and such matter is, in turn, equated with (***figs.* **5.2 and 7.3) archetypal memory or, simply, archetype.**

Just as language needs voice and musical theory (solfège) is silent until quickened by creative energy so 'linguistic' archetype is latent until roused. It is an ever-present instrument by which the notes (subatomic particles), chords (atoms) and their combinations are played out. Natural Dialectic identifies the 'inner air' heightening the notation of creation as *prana* (*figs.* 6.2, 13.6 and Glossary). Pre-physical *prana* is an 'inner energy' that, 'speaking' universal language, raises cosmos from its archetype; it is 'inward sound' of metaphysic expressly drumming music of the world. You can, therefore, understand how the very core of nature hears that story. Indeed, nature *is* the way such melody vibrates. **Natural law, the norm of force and form of particle's behaviour, derives from passive information stored in archetypal memory.** In this sense law is program, code or rule - a natural, cosmic program.

Thus for the Dialectic there exist three levels of pre-formulated, passive information. The first is conceptual, archetypal and metaphysical; it is called potential matter. The second is the direct or primary expression of such archetypal format through the agency of forces, particles and their possible patterns of behaviour. At the border between the second and third levels are atoms; and the third level is that of molecule and bulk aggregate formed of multiple interactions. Levels two and three correlate with the lowest, environmental level of data items (that is, with the amoral, impersonal and quantitative world of physics and chemistry). The dialectical action of all three is described in Chapters 7-12.

In summary, inanimate grammars are known as the language or the laws of nature. And nature's book of non-life (physics) is complemented by its books of life. Animate grammars for any specific organism involve a combination of instinct and bundles of physical norms around which the laws of its nature (that is, the instruments of its homeostasis) are constructed; orderly construction is devolved from complex code. **It is proposed, dialectically, that the natural repository for cosmic language is memory; and that, as a level on the conscio-material spectrum (*fig* 2.13), archetypal memory occurs at**

the psychosomatic 'frequency' of universal mind. *The level is passively informed, sub-conscious and also identified (figs. 0.7 and 0.8) with PSI (the psychosomatic interface).*

This is why the bank of archetypal law has been analogised with an invisible place from which software - in the form of an archetypal set - is projected on the screen of 'cyberspace'; and why material objects, including the codes and mechanisms incorporated in the bodies of biology are said to be the product of archetype - itself the 'programmed memory' of initial purpose. Do you recall (Chapter 0) Bohm's 'holographic paradigm' wherein 'things' are seen as '*eidolons*' conjoined to deeper layers of reality? Whereby they are derived, at root, from cosmic hologram, a single plan, an archetypal projection we call 'laws of nature'? Could these records, could this primal, passive information be what Plato called (from the same etymological root) '*eidea*' or ideas? Could the qualities of subatomic particles reflect a basic set, that is, the basic board (space-time) and players of our cosmic game?

Is this, therefore, the significance of a material object or event - that it reflects law and order consciously established at the time of creation? Or else, without observer or participant, is there significance in anything? Isn't there an empty stage, just a chance-wrought, clockwork operation flying through a meaningless expanse of space? In this case it might be important for me, as a player, to know whether the origin of mind is meaningless (from matter) or meaningful (from mind alive). And if alive, then life to Life, this were electrifying news!

Information's Infrastructure - Code.

Every code, it is worth repeating, contains a morphology (the form its symbols take), syntax (rules of arrangement) and significance (what a message means) agreed between its sender and recipient. Such prior agreement always needs intelligence even if the text, as in a computer program, after its creation works to complete tasks in a reflex way. Code is information's agent of transmission and reception.

Mind is an information exchange employing only code. Code conveys specific meaning; it carries messages. The order of these messages depends on their arrangement, that is, their grammar. The distinction between active, flexible code (as employed by a thinker) and passive, inflexible code (the symbolic grammar such a thinker uses) the subject of this discussion. The latter, made of alphabets, binary charge or whatever, carries the former. This, information's infrastructure, is the level of symbolic rules; it is the archetypal grammar by means of which ideas will be orderly presented at the lowest, physical level.

The order of both active and passive codes is shown on this stack:

tam/ raj	Sat
existence	Essence
matter/ mind	Super-mind
sub-orders/ rules	Word/ Order
implementation	Idea/ Super-rule
orderly expression	Preordination

	physical/ mental 'settings'	Archetype/ Mould
	recipient/ message	Primal Information
	passive/ active information	Source of Code
↓	tam	raj ↑
	passive	active
	stored code	code-in-action
	memory/ fixity	act of creation
	informed	informant
	recipient	message
	outcome	guide/ agent
	non-conscious/ sub-conscious	conscious
	matter	mind

Only the left-hand, 'passive' column of the Secondary Dialectic is of concern here. It involves the fourth, psychosomatic and fifth physical levels of creation.

	tam/ raj	Sat
	subsequent patterns	Instinct
↓	tam	raj ↑
	physical interaction	psychological interaction
	with specific instances	with specific instances
	bodily response	mental response

Is instinct just an engrained habit, an innate aspect of bio-psychology? Isn't its 'event' confined to local interactions and its 'outcrop' only found in instances of reflex, subconscious and non-conscious laws of natural response? Each instance differs, every game's unique in circumstance but are not the rules behind it 'general', 'typical' and permanent? 'Universal mind' is an unfashionable concept but no matter if it immaterially and actually exists; and if it does then cannot universal instinct exist too?

	tam/ raj	Sat
	bulk/ quantum levels	Potential Matter
	subsequent order	Archetype
	motions	Pre-physical Latency
	phenomena	Noumenon
	its agents/ expressors	Natural Law
	local complexity	Simple Generality
↓	tam	raj ↑
	gravity	levity
	condensed matter	quantum principals
	bulk patterns	energetic patterns
	(gas, liquid, solid bodies)	(forces and particles)

It is not scientific habit to appeal to flukes but causes - except in the most important causes of them all, the Big and Little Accidents - those nuclear inceptions of our cosmos and its life. But why appeal to causal fluke if flukeless cause is there? Why appeal to accidents if archetype is prior?

Potential matter (archetype) is the first of three levels through which active mind expresses itself 'below'. In the way that something physical is fixed and called 'a solid' we might say that 'fixed thought' or memory comprises 'solid', sub-conscious mind. This, dormant mind, is the medium (*figs.* 0.12, 2.11, 4.1 and 5.4) between conscious/ active mind and passive body. In dialectical terms it carries, in the form of memory, such grammatical framework as is necessary for the orderly, physical expression of a thought. In short it is a medium of storage; and its 'circuitry' amounts to the mechanism of translation between mind and matter.

There exist both natural and specifically devised languages. The latter can be verbal (as in vocal English) or non-verbal (as in tapped morse or binary-based electrical computer codes). The Dialectic suggests that message always comes in code; and where there are personal purposes (as in the case of conscious animals) there will be flexible code. It also suggests that the ability to devise or learn different, flexible languages is the function of a framework laid in human instinct. In other words, the general capacity to engage in orderly, flexible communication is innate. Man, the information hunter, is capable of enormous yet most orderly flexibility and therefore great subtlety of expression.

Inflexible expression of code is, on the other hand, a physical matter. **The alphabets of such code are, in dialectical terms, biochemicals (supremely, *DNA*), binary nervous code and the subatomic elements, forces and atoms of physics and chemistry.** The former couple are specific to life forms - possibly, at least with respect to *DNA*, anywhere in the universe. *The latter constitute a cosmic code. This code dictates, through the agents that a study of phenomena elucidates, the way things naturally turn out.*

The question arises whether code and languages, natural or otherwise, can happen incidentally. Can 'purpose' rise from randomness by chance? Is its formulation, called meaning and expressed in code, as meaningless as meaning is to the oblivion of atoms? Is information just, at root, an illusion that arises when its carrier assumes a certain conformation? So, for example, brain gives rise to thought and memories are just synaptic patterns made of nerves. This is, for sure, the view that's logical from *bottom-up*.

Top-down, of course, the opposite pertains. It is worth review. How is information 'fixed'? It is fixed in some type of grammatical language and may be stored in sub-conscious memory or carried on a physical medium. What is the natural storage facility of mind? How are its thoughts fixed, its patterns of behaviour and symbolic codes stored? As memories, habits and instincts. The natural mechanism of information storage is memory. Physical media that store memories (or memorials) can be ink and paper, clay, stone etc. This carriage can be transient (as with voice in air or pictures on a television screen) or durable (as on magnetic tape). Every computer, for example, has a storage bank called 'memory'. Such memory is reflected in its drives and discs; and, similarly, configurations of the nerves in brain electrically reflect its transfers to and from the body-world. Indeed, you might infer that 'nervous' charge comprised, out in the wide and starry world, the diffuse form of universal brain! The reflex cosmic brain is real; it consists, controlled by archetype, of atoms, particles and their oblivious relationships.

If brain is physic's side then what of metaphysic's mind - in this case universal mind. For Natural Dialectic there are three levels of matter - (*sat*) potential, (*raj*) quantum and (*tam*) bulk, condensed precipitate. Of this trio polarity *(raj/ tam)* is physical but the *(sat)* neutrality whence they derive is from the level above. It is metaphysical and, dialectically, identified with subconscious mind and principles we've grouped as archetypal memory. Such totally rigidified 'instinct' governs material behaviours. *Bottom-up,* science deals with the *(raj)* kinetic expression of principle in the form of polar charge, electrons, protons, atoms and so on.; and also their combination and aggregation into the details of *(tam)* bulk matter - whose base level is 'solid'. It does not deal with the origin of natural law, the *reason* for material behaviours called *(sat)* potential or transcendent matter; indeed, some guess that they arose by chance. *Top-down,* however, 'code' for physic rests in metaphysic called an archetype. In this view physical solidity, incorporating principles that made it how it is, 'remembers' them. A solid object is a form of memory that crystallises natural law; it is a memory engraved in 'stone' according to the immemorial language nature speaks; its archetypal facet is a book reflecting, through an alphabet of quantum agents, the metaphysic of a grammar that informs the pages of this physical creation.

Grammar? Grammar, as described, is the structure within which reason and purpose are placed. Syntax means 'put together in order'. And words, most of all the Universal Word of words, resound with meaning. Together the three factors comprise language and language is the principle through which myriad, various and orderly communications occur. It is natural information's law. It comprises the mode of orderly transmission and storage of information - code; and since the agents of this code are vibrant you might logically call the language of the world a harmony. In this sense music rules the universe.

Code includes sets of symbols (e.g. ABC etc.) and the rules that arrange and make different connections between them, that is, the rules that govern their assembly into words and grammatical sequences *independent* of any meaning they might have. This *code* must adhere to prearranged (or preordained) conventions. Without such pattern no meaning is possible. **Code is always the result of a mental process**. If a basic code is found in any system, you might conclude that the system originated from a mental concept, not from chance. You might therefore conclude it had an intelligent source - especially if that code is optimised according to such criteria as ease and accuracy of transmission, maximum storage density and efficiency of carriage (such as electrical, chemical, magnetic, olfactory, on paper, on tape, broadcast etc.) to its recipient; and if, above all, it works and orderly instructions are unerringly responded to.

interference	*clear signal*
error	*precision*
barrier	*communication*

Two conditions affect the transmission and storage of code. The first is the *clarity* with which it expresses its purpose, design or meaning. For example, you could write the linguistically correct "Government flies with a biochemical pencil sharpener". It has not, however, clarity of meaning. The second condition is the *accuracy* of transmission or *conserved accuracy* of storage. Is the syntax adhered to? Is the machinery in order and are the coded symbols on its storage medium undamaged? What produces incomprehension, confusion and disorder is ignorance, lack of properly focused attention, damage and disruption or 'noise'. *The coder takes no chance. Randomness is eliminated.* A compiler is a mindless mechanism but a programmer is not. He determines the code and its operation: error is rigorously debugged. By definition, mistake or randomness degrades information; and the job of any editor is to eliminate interference, 'noise' or mistake. Both scientist (externally) and meditator (internally) seek to minimise the same and thus discover hidden truth. **Chance neither creates nor transmits information. On the contrary, accidents always (unless accidentally reversing a previous degradation) degrade meaning and, by degree, render information unintelligible.**

Grammar and syntax are an operating system. They are the method by which information is handled. Information can be handled flexibly by conscious mind but, as the quality of consciousness diminishes, so choice is taken over by inflexible command. No mind means automatic. Automatic function is the nature of both instinct and the reflex language of the universe, the program we call 'laws of nature'. Such fourth level of information, cosmic software, is that of an operating system.

There exist both natural and artificial forms of operating system. As free energy is crystallised into fixed structures, so active information is 'frozen' into passive, coded forms. Such codes include body language, genetic, nervous, hormonal and other biological codes, five thousand or so human languages, mathematics, chemical formulae, technical diagrams, photographs, musical notation, semaphore, morse, formal computer programming languages etc. The grammar of spoken languages is generally more complex than that of 'mechanical' ones, such as those used with computers. In the latter case it must be complete and unambiguous because a compiler has no way of referring back to the programmer's intentions. Such a rigid lack of ambiguity would, for the same reason, have to apply to any purely physical system (such as nature's).

Each sort of code needs its own transmission/ storage system or 'linguistic convention' into which matrix information is poured. For example, the nervous system filters and encodes information. It employs data capture (eyes, ears etc.) and code (in a binary electrical form): also data communication with the environment using muscular body language that expresses the actor's intent. Biological information systems mirror, in a complex way, the basic vibratory syntax of matter but their cycles, called in general homeostasis, are much more complex in their interactions. Physical code (see Chapters 7-12) is simple and dialectically expressed as electrical charge, wave/ particle duality, gravity/ levity and other polar sets of influence; and the material of a biological database, *DNA*, also works on a conceptually simple, binary system of base-pair opposites or, if you like, complements. There also exist many artificial modes of information carriage. These include, optical, acoustic, magnetic, electrical, tactile (pressure-based), chemical etc. No doubt you can think of examples of each kind of technology. It is not that passive information (including technology) understands anything itself. Its code merely reflects the order imposed on it by active information.

Goonhilly is a good example of code in action. Wired or wireless, broadcast the news! Any optoelectronics engineer will tell you that the information-carrying capacity of light is colossal. A single burst of laser light can transmit the entire Encyclopaedia Britannica in a second. This does not mean that a physical carrier such as an electromagnetic wave or an electrical pulse is to be confused with its information content. There is no purpose, *per se*, in the substance of lifeless, dead or inanimate things or their interactions. The information broadcast from a star is entirely different from the traffic programmed by mind-forms. At Cambridge Lord's Bridge radio telescopes receive iterative, passive stellar broadcasts. These background radiations are, exactly as a stage is, without intelligent life of their own. What complexity they show is non-purposive. By contrast, Goonhilly Earth Station is situated in Cornwall near Poldhu where Marconi made the first transatlantic broadcast. Goonhilly is a live wire indeed. It processes a massive volume of two-way 'life' traffic. It is an international centre for the transformation of meaningful energy. Voice, sound, fax or film can be reduced to digital transmission. Precise, complex and purposive code is broadcast, with high fidelity, on Intelsat microwaves.

receiver   *transmitter*
shaped   *shaper*
sensor   *mot(ivat)or*
understood *message*

Either way, there is no mistake. Whether in construction or in use Goonhilly no more happened by chance than any of its traffic. Such an idea is unthinkable. War is waged on randomness, interference minimised, errors eliminated. The receipt and transmission of meaning in digital form analogises the impulses of the central nervous system. As with healthy biological function the Goonhilly data signals are maintained intact throughout their relays and transformations. Message; various codes; transmission; various decodes; understood. 'Noise' is the enemy. It is indecipherable, meaningless, a dead-end, eliminated. The whole concept of satellite data networks is based on the opposite of chance. Raw materials, such as iron and plastic, are used to make equipment like antennae. Electrons and photons are a subtle part of the hardware used to store or transmit Goonhilly's programs.

We exploit 'intelligent' matter. Submit your TV, internet hardware or even a communications 'brain' like Goonhilly to scientific inspection. Nowhere at Goonhilly will you find Marconi - except conceptually, when he is everywhere. Nor would we use its communication systems or want its programs unless they served our purposes. Life's purposes. *To confuse Goonhilly's hardware or its data stream with inventors or program-makers is to make the same mistake as confusing consciousness or thought with neuronal circuits and electrochemical interactions.* The entire complex is, nevertheless, a symbol of information's hierarchical loops (at least as far as the intellectual level). *The place is information's infrastructure; it is a large, grammatical object.*

In the dialectical case such grammar is extended to universal design. Is a computer program not 'fine-tuned' to work? What greater margin for erroneous 'noise' might you expect to find in cosmic constants or the patterns that their program keeps, of necessity, reiterating? *Such codified intelligence is stored, as archetypal principle, in universal mind.*

You jib but may remember (from Chapter 2) James Jeans reflected that the universe seemed nearer to a great thought than to a machine; Cambridge colleague Arthur Eddington likewise wrote that 'the stuff of the world is mind stuff'. Physics and the Dialectic don't, in this case, seem so far apart. Werner Heisenberg, a founder of quantum mechanics, noted that 'the smallest units of matter are not physical objects in the ordinary sense of the word [but] forms, structures or - in Plato's sense - Ideas, which can be unambiguously spoken of only in the language of mathematics'. And Plato was, in turn, much influenced by Pythagoras - who believed that mathematics links us to divinity. Music, order, golden rules - Pythagoras was mystical. James Joule was not; but he thought that 'acquaintance with natural law means no less than acquaintance with the mind of God'. Johannes Kepler cried out that the laws he found were 'thoughts' of God; Newton, Einstein, Planck and a host of others reached towards the same conclusion. OK, some don't. **But Natural Dialectic simply, hierarchically identifies potential matter as archetypes in universal mind. It thus recognises the basis of physics (and with it mathematics) as metaphysical.** And the physical realisation of such archetypal pattern-makers, such metaphysical 'attractors', is the last, lowest, fifth level of an act of creation and, therefore, of information.

The Lowest, Physical Level.

Information has meaning to a mind; such meaning (or semantic) may be organised into passive, symbolic receptacles such as language. But, while the concept of 'message' cannot be expressed in terms of the solely descriptive concepts of physics and chemistry, at the lowest, physical level it *can* be expressed through vocal, written or other material form. Sound waves, paper, clay or *DNA* can register ideas; so can machines of any kind. Indeed, the natural constraint of archetypal code is now translated to constraint of point in time and space; the general is turned physically specific everywhere.

Therefore, bump! This is the part that bottoms out in light, sound, fluid patterns and in crystalline solidity. This is the 'external', 'objective' or physical side of the psychosomatic border; and, while forces and atomic particles comprise its primary expression, the hard, bulk universe that we survey is its secondary. This is, therefore, the lower syntactical level of computer, ink or cosmic book; such creation is its causal code's expression; starry cosmos is a vast effect. *At this level of existence we find data items or the material on which arrangement is imposed.* And, equally, it appears as an object, event or pattern of behaviour without subjective quality, context or intrinsic meaning. It is the level of raw data *per se*, objects in whatever form they appear before higher inspection, manipulation or interpretation. Its values are numerical and its description measures 'thingness' alone. *Without a higher level of interpretation it carries no message.* It is meaningless (unless you count conceptual background as a part of nature's show); but, where messages cannot be treated mathematically, the symbolic objects that compose them can. In short, the lowest level of 'information', devoid of sense or meaning, derives from the measurement of any unintentional aggregate of symbols or atoms. Physical science observes and measures such impersonal, inanimate manifestations; then, by working out the principles that guide them, seeks to raise the game.

Numbers entered into data tabulations are as yet meaningless. A pattern needs be teased from them. Such pattern might be framed by simple or complex mathematics but is 'meaning' thereby found? Do sums fully describe Mozart's concerto or Bach's fugue? There is no doubt that, for example, statistical analysis relies on the presence of underlying order or principle and it is this that science uses to interpret significance, infer order and thereby describe the qualitative but still passive level of grammar, code and law. Inference is itself an active, mental process that depends on reason but is reason meaning's sum? What is the unit that measures 'meaningful value'? Surely not a volt or joule or any other scientific jolt? Perhaps it's what the numbers tell you *for your purposes*. Or is it something you can measure (like the taps of telegraph or electric fluctuations in a brain) that may reflect subjective episodes but cannot measure what it means to you? Such measurements remain, if science is materialistic, extra-scientific and forever incomplete.

At this level, therefore, sensible objects and transformations have measurable values but are, in themselves, meaningless. They are subject, in wear and tear and change of shape, to inertial tendency and a process of randomisation. However, despite superficial 'noise' their internal (quantum) structures persist. Adamantine proton and electron neither wear out nor stop vibrating; they are as effectively indestructible as their motion is perpetual. Quantum transformations, from matter to energy and *vice versa*, are also subject to immutable law. If they are symbols of code (e.g. proton and electron standing for polarity) then the 'meaning' is set, the stage fixed and the subtext rigidly reiterative. Symbols of any dynamic code can be manipulated (as in speech), stored (as in books) or jumbled indefinitely. They can be randomised in ways that lose all syntax or meaning. Even then, stripped of meaning, they retain the sort of quantitative information earlier identified as 'Shannon information'.

A statistical analysis of symbol alone is at the heart of scientific method. The subject matter, objects and events, is meaningless *per se* but the numerical results tabulated from experiments may still yield valuable information in the form of patterns, principles and even laws. And, as a cryptographer, if you assume an underlying meaning to a jumbled message then your scrutiny is just as valuable. Do you remember Bletchley Park? In World War II 'enigma' transmissions from German High Command were cracked by this secret code-breaking station with informative results that generated ripple effects throughout the fields of physical action. Information is the real power. The war was won, in the same way as systems and strategic action are planned, *top-down*. Information is all about the plans and systems, however personal, emotional or apparently trivial, that purpose, logic and meaning generate. In this context statistical information measures frequencies, amounts and signal accuracies of the basic carrier units - letters of an alphabet, words, numbers or any other symbol. Such symbols could include amino acids or *DNA* bases. Information density, the amount of information (in bits) in a unit volume, is a measure of the efficiency of storage and transmission of information. The highest known density is that found in *DNA* molecules that comprise the chromosomes of living cells. A bacterial cell contains about a megabyte and a human cell (whose *DNA* weighs about six picograms or 6×10^{-12}g.) a gigabyte or so of information. It can be shown that these amounts represent a density of around 2×10^{21} bits per cm³.

This is billions of times more efficient than the latest silicon chip! Into whose construction such a fund of human knowledge, scholarship and intense attention is lavished. Such investment represents prize intelligence. Nothing was left to chance. Indeed, if the sum total of information in the Library of Congress were stored using *DNA* the area of a pinhead would easily suffice instead.

If, from a statistical perspective, meaning does not enter into the equation then you might proclaim that the greatest literary work of art was basically nothing but a scrambled alphabet. Science generally evaluates its subject matter - interactions, forces, patterns, bodies and so on - in a numerical way. On the other hand, the same items could be conceived of as information storage and transmission agents. That is to say, information is not the substance of the thing itself: it is the representation of an idea. If we can perceive the idea we will have 'got the message'. Of course, it is not easy to perceive the idea behind an inanimate object or interaction. By definition, it cannot explain its actions. Yet, in the case of both inanimate and biological behaviours science has perceived a grammar with its rules that seems to govern things. Indeed, it is a primary goal of science to elucidate not only the operation but the origin of codes and rules; and find, without materialism's philosophical constriction or obstruction, the truths of how things are and came to be. Such truths constitute the very fabric of a scientific form of knowledge and enlightenment. We might thus ask again, "Does active information naturally order passive; has physic a conceptual background; and if it has how did its systems, physical and biological, originate?"

Active on Passive.

↓	*tam*	*raj* ↑
	passive effect	*active cause*
	information receiver	*information giver*
	unconscious matter	*conscious mind*
	design/ imprint	*programmer/ designer*

Informed energy. Active information orders passive. Mind generates code. Matter can act as a storage medium but never, being subjectively impotent, generate coded information. Whilst *creative* information stems from a free (or at least only partially conditioned) power of will its material expressions are robotic. Their *passive* information is derived from the will-less inflexibility of natural law; or else such information reflects, as some arrangement of materials or a machine, its creator's purpose. *Purpose does not, therefore, have to appear conscious.* Subconscious instinct and non-conscious technology each convey intent; and cosmic 'instinct', archetype, unconsciously and therefore passively informs the patterns of this world. But there is no creative spark in matter. Only mind flares with originality.

The footballer strikes a superb goal. The crowd roars and the game is won. From an objective point of view any number of observations, calculations and descriptions can be made. The mathematics, physics, chemistry, biology, history, sociology and psychology of the event can be drawn into immensely complex, truthful theses worthy of doctorates. Indeed, if you contracted down to the atomic particles of his enzymes or up to general relativity with his universe you could squeeze all of science from the moment. And many other academic disciplines. Even, as the tabloids regularly illustrate, much more!

Nevertheless, for the player who performed the act it was a simple, immediate affair. Thought scarcely entered on it - simply awareness, a feeling of the whole, valuable circumstance and a will to achieve a purpose. The complex, efficient summary of body, ball and motion coordinated in a meaningful moment of experience and, as the goal shows, he calculated exactly right. Is the player's or the crowd's truth less real than the PhDs' descriptions? Is subjective less valuable, as a perspective, than objective? Which is most central, which more peripheral to the great event?

The player, centrally, created the goal. His idea, associated with intuition and trained reflexes, engaged a number of different aspects. The difficulty is, as with a description of physics and chemistry, to disentangle them in the 'proper' order and, with hierarchy that accurately relates to truth, best relate his individual act of creation. Not all creations, however, are as 'simple' as the footballer's. Not only nerves and sinew but also intellect serves goals. And humans have developed powerful extensions of their natural born 'grammatical construction', that is, of their informative nervous and muscular systems. These extensions include complex tools such as language, instruments and machines to better serve our purposes. Which purposes are, in brief, survival, communication, education and pursuit of personal interests. We want enthusiastic answers. Yes!

Each individual mind (human or otherwise) has its own, relatively correct perspective *about* the order, and its place *in* the order, of things. In other words, relative order is in the mind of the beholder. We each order the world through the lens of our own mind. This lens varies according to biological type and also, within the group of *Homo sapiens*, according to intelligence, education, temperament and social position. Therefore both an individual's perception and imposition of order in his environment are imperfect. Incomplete. Do you remember the conscio-material gradient of creation? It can be expressed, in the graphical form of figure 0.7, in terms of information and energy coordinates. We are at present not interested in body (or energy coordinate) but mind (information coordinate). *In this respect a person's relative knowledge/ ignorance of natural order (even Natural Order) and its principles is a reflection of their hierarchical position on the conscio-material gradient.*

Active on passive, conscious on subconscious mind and matter - materialism execrates but is there any proof subjective is objective, immaterial material or life made of death? If there is none my case is sound. <u>*If (and you are an exemplar) consciousness and matter are two separate, interacting elements then what gainsays cosmic animation?*</u>

Mind imposes patterns of its purpose on the stage of nature. Organisms make arrangements *indirectly* through their bodies. Down-to-earth animals and humans (yes, a human is biologically but not, in psychological potential, an animal) build on a scale relating to their size in universal body. Humans can, however, create with a focus, a grasp of principle (insight) and sense of purpose that dwarfs body-size. They devise tools for survival, empowerment or enrichment; tools to build, repair or destroy; instruments for aesthetic, hedonistic or learning purposes; and functional machines for transport, locomotion, sensation or, consummately powerful, information and communications technology. The will to inform and be informed is embedded deep, like a craving, in the human psyche. It is a key drive in humans. To understand, create and communicate information. Give me information then some more. I want the truth but how shall I find out the Truth of truths?

Objective and subjective scientists are twins after truth. One looks 'up' to the cause of information, the other 'down' to its effect. One checks out the sender and the other checks what has been sent. Suppose we wish to find the truth about a TV set. It is as if the saint grilled its maker and the scientist analysed the set itself. Of course, an objective philosopher seeks truth as much as his subjective counterpart but his mode of enquiry differs. He wants to find out the physics and chemistry for himself. His only mistake might be - because he doesn't see, believe in, know how or care to meet an immaterial inventor - to assume there is not one.

Active on passive; or, in the jargon, 'semantic' on 'syntactic'. Mind and brain can interact. Mind can *directly* make arrangements on grey matter. If the two are not the same there is an interface. The question we'll attempt to answer, asked explicitly in Chapter 4, is the nature of omnipresent *PSI*.

As organisms (with a restricted range of awareness) excel matter (with none) so humans are gifted with broader intellect. Above the excellence of intellect, psychological potential is realised inwardly as development ascends towards unification with the Transcendent. Here man can attain what is, compared with everyday consciousness, a super-conscious power of will, purpose, love and principle.

It is here that, in commune with the Word, he can experience the origin of information, order and reason. *Because he is at this point perfectly aligned with Central Logic, such experience is of cosmic proportion.*

This is all very well but man's purposes are mostly more, much more pragmatic. No doubt he expresses an aesthetic, subjective quality in art but people want survival, health and comfort. So they make tools, instruments and machines. What on earth has the Holy Ghost to do with machines? Let's take a look at three of mankind's great creations, ones that analogise the physical, biological and psychological worlds - music, machines and mind machines.

Music.

discord	*harmony*
disease	*ease*
blockage	*flow*
incoherence	*coherence*

The old word for integrated order is harmony. Music, like health, is harmony in action. It is the shape of pure energy, constrained only by its type of instrument, harmonics and the skill of its musician. Melody is a most profound form of information-in-motion. It is perhaps the best medium for the vibratory transmission of meaning. The vibratory energy of First Cause and its subsequent harmonic constructions embody the internal logic, order, pattern or rhythm that pulses through the levels of creation. Song will, as every musician since Orpheus has known, pull you straight to the heart of things, to the centre of life.

Music is layers of harmonic energy; it is also a 'rainbow' of sound whose colours shift through different shades of sky, a resonance that quickens, that raises the dead. You can appreciate music in an objective or a subjective way. You can study the construction of instruments, calculate various wavelengths, frequencies or amplitudes of notes and experiment to discover the behaviour of sound waves through various media. A concerto is certainly a dynamic movement whose 'system' might be described by maths as simple or as complex as you wish.

On the other hand, this is all beside the point when music plays. Then objective flips into subjective and appreciation grows. Rapt attention identifies with rhythms and the meaning. It makes you dance internally if not externally as well. You may even find yourself in touch with a fresh and deeper aspect of reality not revealed in the detail of life's chores. You come away thrilled and reinvigorated. You have not discovered music's nature through a conversation but by becoming involved with it, experiencing it, knowing it for what it means to say. The principle is purposely-balanced complexity of rhythm. Music is not (although they have their simple charm) the non-purposive, uncoordinated rhythms of the wind, waves or earth. The mind of a composer strikes a deep resonance, a consonance with inmost self. We love it, identify with it, rise on it, lose ourselves in it and are healed! This is not a scientific form of knowledge but is it less reasonable? Is it less valuable? Is its truth less valid? Our appreciation of music is not scientific but it is nevertheless real. *It analogises the mystical appreciation of creation; it reflects biological timing, the pulse of homeostatic periods and the harmony of vibrant health.* Consummate appreciation attunes to First Cause. For Brian Pippard, physicist and a Cavendish Professor at Cambridge University, "A physicist who rejects the testimony of saints and mystics is no better than a tone-deaf man deriding the power of music".

As a play needs more than a script so a musical performance needs more than just its score. Actors need costumes and stage; musicians need instruments. In the same way the cosmic drama requires background. It needs a field or instrument through which to express its ideas. This is the physical field. And its stage or instrument, although itself not conscious, supports the actor or musician. It is fashioned as the agency through which their message is expressed. It sets the scene. So the non-conscious aspect of a creation also bears the hallmarks of intelligence. It radiates purpose, order and deliberation. What is 'entropy' (Chapter 8) but a scientific generality that, in its statement of material tendency towards the mindless dispersion of energy or order, implies that start-ups need a stimulus; and therefore that The Start-Up needed such 'Negentropy' as finds a place in teleology, the doctrine of design. As thought applies to plots it equally applies to bodies clothing mental drama and the wider body of a stage to act it out upon. In other words, non-conscious energy/ matter might itself encrypt intention. *The universe might be seen as the cipher for an idea and its objects as symbols of the stage-plan.* Information is seen at every level, animate and inanimate. Composer, instrument, score, musicians (the composer's mouthpiece) and performance also reflect a conceptual background, that is, a mind behind.

music	*musician*
instrument	*composer*
medium	*information*

For Natural Dialectic music is (see Chapter 0) The Master Analogy. The vibrant, ordered energy of music has the power to move. It moves and shapes us more easily, immediately, directly and deeply than any source of inspiration except perhaps love. Indeed it is a sort of love, a universal language of harmony whose call excites us to join it, surrender and lose all sense of apartness in its lively stream. This, a temporary and limited transcendence, is a form of ecstasy.

Is ecstasy, like music, all the same? Of course, each kind of rhythm presses to a different end, each composition swells with different feelings and each vibrant harmony more or less evokes a personal resonance. Which sorts best express the heavens? And, if there's song, what's mostly at the heart of lyric? What name is given to the fire? Not various forms of union, happiness and love?

If light is focused through a lens the wood bursts into flame; and focused interest bursts into enthusiasm's heat and light. What gives rise to all the rhythms, heats and lights that underlie creation? As mind hierarchically precedes body so active creation, either individual or universal, is psychological. What might imprint a universal order? Could not resonance and radiant communication be aspects of a messenger? *Cosmic Word issues as primal Oneness whose positive division is in the two aspects by which it is known, living sound and light.* This description has varied neither geographically nor historically. It is been ritually symbolised with lights and chimes. Let strings cascade, flute quiver or the organ thunder. The 'Sound of God' has traditionally been celebrated in terms of music - universal quickening, a coherent stream of order out of which emerges everything. Who is lord of the dance? The earliest illustrations are of *Siva* (5000-year-old statuettes from the Indus valley civilisation) but the party's not exclusive. Everyone joins in. They raise First Cause from an inanimate big bang to animate radiance and resonance in accordance with whose pulse all lower 'grammatical' patterns are devolved. <u>Radiance and resonance are identified as agents of law enforcement</u>. What kind of law is that, whose agents are so bright, colourful and free? The vibrant couple certainly transmit physical information. Physics has detailed properties of life-giving, sight-giving light; spectroscopy uses it to identify the components of far-distant galaxies, stars and, each by its own unique 'bar-code', to tell apart every different element and molecule on earth and in the heavens. Resonance, for its part, is a feature of all vibratory systems - atomic, electrical, musical and so on. The sound of a gas or liquid put under pressure is called 'resonant'. The sound of your voice as you speak is an example. In parallel you might similarly describe Sound (called *Logos*, *Shabda*, *Om* etc.) as the metaphysical vibrancy of primal energy (*prakriti*) 'put under orderly pressure' by the focus of Infinite Potential. It is called First Cause because it is the first and therefore principal notation that resoundingly informs creation. The whole opera is, in these terms, a reflection and a resonance of its transcendent order. Central and Most Internal, the pattern of its movements is no more audible to physical ears than enlightenment is visible to eyes; yet, because it is psychological, it may be communed with. Audience with such inner music is the prize; magnetism of this ringing radiance is the current drawing mystics higher.

Machines.

The metaphor for universe used to be biological. It was of the universe as an organism - the soul, mind and body of God. Man was a reflection of it. Then, in a renaissance of objective rationality, the symbolism changed to technological - to a Newtonian kind of universal machine. Now it's sometimes, in line with the Information Revolution, computational. A Great Computer. *Any way you look at it, the 'meaning' of a mechanism or machine is vested in the purpose of its maker. Is conceptual background therefore part of nature's show?*

A machine involves a system of well-matching, interacting parts that, unless any is removed or degraded, contribute to a function or produce a targeted result. Such systems are therefore specifically and irreducibly complex; and so to work they must be made at once. In this case which comes first - a machine or its concept, a work of art or its inspiration, the chicken or its egg? Extrinsically fruit issues from the branch. The branch came first and the fruit after. *From an intrinsic point of view, however, the branch is from the fruit. The fruit came first but, in order to bear it, the branch was conceived.* **Wherever an apparent 'chicken-before-egg' situation crops up the puzzle is resolved by the introduction of purpose.** The real egg is invisible. From this conceptual 'nothingness' a secondary thing appears. Can plots not hatch what later comes to light? Eggs are information carriers and ideas front developments; eggs are potential that control the path of all that follows from them. The solution to the problem, therefore, is design and information rolled as one into a process some call teleonomic (directed by natural constraints) and others, who admit intent, teleological (systematically directed).

As far as we know, the laws of physics are statistical. Bulk material order, known collectively as the laws of nature, arises from various constraints placed upon quantum disorder. Thus, according to Erwin Schrödinger, biological order from order (from code as opposed to statistical quantum disorder) means that the ways of life are not reducible to physics and chemistry alone. A 'non-probability mechanism' (viz. accurate local government by a relatively small but highly organised group of atoms in the nucleus of a cell) completely trumps the 'probability mechanism' of quantum physics.

In fact, such 'local government' issues instructions for the production of a panoply of bio-machines. "Machines", argued philosopher/ scientist Michael Polyani, "are irreducible to physics and chemistry." They are irreducible because they involve immaterial purpose, the stepwise development of a plan of implementation, a directed cohesion of working parts and, of course, the thoroughly non-material anticipation of an operational outcome. Such a machine may itself be simple or complex; it may be possible to deliberately adapt its purpose and therefore function; but if parts are missing or corrupted then its operation fails. *System depends on parts; part with part expresses the idea. Engineers routinely face the chicken-before-egg problem of designing interdependent components for their mechanisms and machines.* They are not inclined to leave their solutions to chance. Indeed, irreducible, specified complexity is a sure sign of active information, a dialectical act of creation and the original presence of an intelligent agent, that is, mind behind. Thus, although a machine is totally material, it is at the same time full of its inventor's mind. **Creation, whether trivial, important, linguistic, artistic or technical, is the very hallmark of human life. Do you remember (Chapter 5) that the definition of a machine involves such purposive and irreducible** complexity? **Such coordination, with added layers of coded control and self-reproductive capacity, is the teleological hallmark of all organisms.** *Thus in Book 3 we'll meet multiple evolution-defying biological implementations of the chicken-before-egg paradox.*

Teleology is, as opposed to mindless 'teleonomy' (if it more than trivially exists), to plan and create machinery or artefacts. **An invention is, if you like, the teleological precipitate of a purposeful idea.** A machine embodies creative information stemming from the mind of its creator. *It is characterised both by anticipation of its operation and an irreducible complexity of interlocking parts, without which it could not work.* Some advanced technologies incorporate homeostasis. Such a mechanism involves the concept of negative feedback. This kind of order shows a dynamic, internal self-consistency. A homeostatic system employs sensor, regulator and effector so that, while fulfilling its purpose, its operation fluctuates around standard behaviours or norms. While such dynamic equilibrium allows the machine to work, its failure causes breakdown or destruction.

Where matter (*tam*) tends to disorder mind (*raj*) tends to order. It is 'anti-entropic' which means that it actively seeks order and eschews chaos. It is with 'negentropic' meaning that it fights the entropy of information (see Glossary: information entropy). Information is (see *fig*. 2.6) anti-chance. Like purpose it reduces the odds. Do you want bugs in a system? No chance! Design is a function of purpose - you design a program and eliminate its bugs. The less bugs it has the more effectively the machine can express its potential. The more principled its information the more it can, perhaps in a complex but certainly in a coherent, powerful way, realise its objective. Every engineer, architect, scientist and artist knows that works do not arise by chance but by his express and principled design.

Looked at the other way, how can you reasonably argue that chance produces something as reasonable as the microchip, the brain or its thoughts? Or that chance produces anti-chance, a source of information? Can you explain how, for example, a set of physical changes might correspond to conscious experiences such as "seeing that an axiom is self-evident"? Could 'material mentality' anticipate the consequence of a conditional 'if' or set in train the conscious logical transition implied by the word "therefore"? If a sequence of argument is no more than a spin-off from a chain of predetermined physical processes, it follows that the thinker cannot help thinking or saying what he does. His arguments, as reasoned arguments, carry no weight. They are robotically formulated according to the uncertain positions of electrons; they are simply electronic ghosts of thought. Why, therefore, should you rely on them? Or, indeed, anything in your own ghostly mind - which, as far as you are concerned, contains everything including yourself? Materialism provokes a crisis of identity. *Exactly who are you?*

As noted in Chapter 5, there is neither law nor known agency that can cause information to originate by itself from matter. The materialist's problem is acute when it comes to machines. A machine is a device which, in terms of thermodynamic entropy, can locally raise free energy to do useful i.e. specific, designated work. *Yet no machine ever appeared as a consequence of* *the addition of random free energy into an existing system.* **Physical nature can't create machines; yet information and machinery are closely twined in living and, indeed, all teleological constructions.** This is fact.

For example, do those beautiful objects called steam engines all chuff by the will of their inventor, George Stephenson? Of course they do. It is as 'scientific' to admit creative will as to study engineering. The 'Rocket' was created once-for-all. Once the principles and information behind it were cast in materials its construction was fixed. The irreducible complexity of its working parts submits to preordained rules and principles. The rules include operating norms and feedback mechanisms (such as pressure valves, brake, accelerator etc.) designed by its inventor; and, of course, the whole submits to natural constraints. To recognise Stephenson's mind as the origin is in no way to inhibit scientific investigation of his construction. Of course, you might investigate his product from the point of view of heat dynamics and construction from materials (say, copper, steel and wood) alone; but would the facts seen in this light cause you to infer that locomotive engineers did or did not exist? And if you came to understand the way things worked would you then absolutely fail to distinguish between cause (agency or creator) and effect (product or creation)? And then, having dropped into this basic philosophical error of category, would you infer a complete absence of invisible intelligence and interpret every part of locomotive 'design' as 'engineered' by evolution's godhead, chance? Or, as happens, ban discussion of the possibility of a biotechnological rather than accidental engineering process in a science class? **The answer to the original question is therefore elementary. Yes, science can or will fully explain material construction and operations; but no, it cannot explain the original idea, purpose or design that generates a system.**

So, although the phraseology is out-of-date, does a plant grow by its maker's will? Of course it does, except that due to change of mind-set (not of fact) it's currently 'unscientific' to propose what might seem obvious! When it comes to systems that are natural, especially bodies biological, then atheists just can't agree that they are marvellous, computerised machines! Point-missing commentators queue up to spit out those red herrings, 'Gods of gaps'! *Such stunning lack of logicality would be, in your own body's case, yours! This book checks if it's right.*

The insufficiency of chance and the sufficiency of mind to generate complex, specified information are vast, coupled underestimates made by Darwinian-minded biology. It makes these elementary mistakes routinely throughout its course. You might *imagine*, for example, that in eons natural forces could perchance serve up an elementary cup of tea. One can *imagine* anything but art, not science, is imagining. If, though, somewhere in space and time, such glazed aggregation 'organised itself' it would not the least resemble your own herbal mug - since it would lack all reason, context and design. It is not so just with mugs that an abysmal maelstrom might emit; it is so with any 'rational' mechanism natural roulette might 'win'. Nowhere, as we shall see, is immaterial information more glaringly obvious yet missed or denied than in the scientific effort devoted to cutting a path for neo-Darwinism's *PCM* by dreaming up ways in which life's complex logic maps and computer chip-like circuitry 'evolved' from pools of chemicals or, with the 'evo-devo' crew, life haphazardly 'evolved' development until maturity just accidentally found itself!

Of course, an atheist fantasises that a century and more ago Charles Darwin killed, by death-guided chance, the argument that purposive design must have a mind. Could (as Chapters 19-26 debate) such 'natural selective' fantasy be true or is materialism's answer a misguided shimmy wriggling out of reason? Is this alternative interpretation of the origin of working systems just a sleight of intellectual hand, a manipulation of the data that is (though the fellow can't accept it) wrong? Look at a TV set. What life or 'living matter' do you find? There is none in casing, screen or the electron gun. Indeed, analysis right down to its molecular or subatomic level would reveal no trace of life in its material 'mentality'. The irreducibly complex and purposely fabricated parts amount to passive information. They are no more animated or subjective than a nerve of brain. *Does this exclude the fact that, in a very real sense, John Baird inhabits every TV set in existence*? He (or, some would say, Philo Farnsworth) invented television and this part of his mind is, with the host of consequent engineers whose improvements have led to the present generation of sets, in every one of them.

Machines are purposely conceived; and intention involves reason, design, interaction, function, recognition, meaning, communication, signal, code and language. Engineers and biologists both use all these words extensively. **Therefore let's be absolutely clear that, although machines are operationally subject to the constraints of natural forces and environmental context they are never created by them.** An explanation of origin does not involve a choice between observed physical characteristics *or* a creator. Such choice is as false as foolish. This applies, despite dense clouds of special pleading raised by materialists, as well to biological designs as to modern inventions.

Indeed, is not a creator a centrally natural, unseen part of his creation, a basic influence every moment of its time? George Stephenson is both absent and present in every steam engine. In the same way as Baird inhabits the TV, Charles Babbage is in every computer. Their minds were not fixed but their machines were. A machine, like a recorded song, is the passive symbol of its creator's active information. It is like a crystalline piece of mind, both memory and memorial. Look around the place you are probably in. Walls, windows, tables, chairs and gadgets - what in this room is not such a 'memorial'? Passive but not active information is in everything. Mind is everywhere. Each object carries it and some, such as books or CDs, are even dedicated to the storage and transmission of information for its own sake. What you see is, as much as simply 'things', many artefacts, many minds and obviously their purposes. **It is important to remind yourself that, in this way, mechanical or any other kind of artefact has mind in it. It is a purely material body but imbued with the mind of its creator or creators.**

To think that Goonhilly which, like all life, is replete with the purpose of communication constructed itself from a series of chemical or geological changes is as absurd as it is irrational. As irrational, astronomer Fred Hoyle famously said, as believing the passage of fire, rain and wind through a junkyard could gradually, in a cartoon-like way, create a Boeing 777. Could even cyclones through an aircraft parts store do the trick? Intelligent creation is, conversely, sure and quick; non-conscious, mindless time can no more conjure information up by chance than pigs can fly - even though materialism quickly hopes it slowly can. Is it right, therefore, to draw time as a veil behind which you imagine anything you like? Like evolution through a cartoon set of 'morphs'? Is it entirely rational to ascribe, steep against all odds, non-probable complexity of obviously purposeful design to luck when you have mind for a solution? If chance serves as the cosmic non-god, then an informational hierarchy is animistic nonsense; if it isn't, information is the obvious and licit answer.

Goonhilly is no great shakes compared to the information encapsulated in a purposive cell let alone a multi-cellular organism. The basis of their biology is codes. Biology is replete with signals and communication networks. *Top-down*, hierarchical systems dominate the centre, middle part and membrane surface of every cell. The mechanisms and operation of reasons ramify throughout every level of each organism. Biological bodies are seen, from this perspective, as information hardware. And if a living cell were ever built from scratch, think of the billions of man-hours, billion trillions of bytes and incalculable discoveries, ideas and intelligence that would have been invested in its construction.

Indeed, the problem is especially acute when it comes to biological machinery. How does it work? It is a biologist's rule of thumb - if you don't know the *reason* for something look harder and find it. Get the idea? Machines, although entirely material under the microscope, are thoroughly imbued with purpose. They are invested by their creators with each level, from principle idea to detailed effect, of information. So is all biological machinery, psychological or physical, no matter how humble its form. Natural information is much more concentrated in animate than inanimate bodies. *The former involve <u>purposive complexity</u> - plans, functional parts, integrated circuits.* Information shines from the structural and functional profusion of ideas that ranges from phenotype (visible form) through molecular, atomic and even electronic levels - all based, as noted just above, on inside information in symbolic code so dense in logic that it needs be mapped to grasp and archetypal plans expressed with a sophistication that outshines, by far, the switching of chip circuitry. It also shows in the awareness with which, as far as they can, biological forms organise their environment to accord with their driving designs. It shows, that is, in the power of metaphysical override, the informative calculations of instinct and, in the case of humans at least, a consciously elaborate desire to comfortably survive.

Machines which embody awareness in the form of conscious or sub-conscious sensitivity show the highest form of passive order. We call them alive. Biological machines (or bodies) are material and conform to the laws of nature but differ from other systems in their very high information storage density and the conceptual brilliance with which they exploit the aforesaid restrictions. **Molecular nano-technology of exquisite effectiveness is ubiquitous at the foundations of biology; indeed, at every interactive level coherent systems are rich with purpose, information, homeostatic feedback and irreducibly complex mechanisms.** With respect to order, a cell is as much a control freak as a computer! **It is the driving logic of Natural Dialectic that such machinery, which intelligent humans have not even begun to reproduce, must have had a mental source.**

We can go further. **The purpose of biological machines, such as your body, is not one exhibited by non-conscious and therefore purposeless atoms or their constituents.** Your form's purpose is, as a working machine, survival. Intact survival demands information past, present and future. *Prior* information was required to construct it. *Now*, apart from metaphysical thought, your life is one of information-related sensation and energy-related behaviours; it is composed of input (perception), cognition (thinking) and output (muscle-based action). The physical aspect of this process is underwritten by highly informed, complex energy metabolism. But bodies, however energetic, tire. They decay. Time ushers death. The *future* or, if you like, post-mortem survival of a body-type depends, therefore, on replacement by the reproduction of its form.

As shown in Chapters 17 and 19 your whole body reflects the situation. So, with its centralised information base, the nucleus, its energetic cytoplasmic business and walled periphery, does every cell. Look at your head. It is an information centre. Head and spine plug, then ramify as shown in *figs.* 14.4 and 19.2, into an enfolding trunk. This trunk accommodates, *top-down* almost to its base, organs associated with energy supply (basically, respiration) - 'airy' lungs, a radiator called a heart, liver, pancreas and the digestive tract. The latter drop, at their exhausted, 'earthy' end to excretion, egestion and, of course, the channel of re-creation. No less the gravid weight of fresh, earthly form is conceived and, as a foetus drops, precipitated from the lowest end. Limbs? Appendages? Motion-makers? Arms and hands engage in subtly directed motivation, on-the-spot manipulation; legs and feet are for directed movement joining points in space. In every instance, in every part information is pervasive; stored code, signals, dynamic checks and balances and systematic hierarchy are obvious All are subsumed under the *purpose* of more life, survival and, in the larger scheme, a part to play.

Why is it wrong or even, as some allege, pernicious to believe that life did not derive from lifeless matter but from Life? Is the whole show not sensible, rational and exquisitely informed? If so, then the information was present before expressed in any on-stage prototype. Indeed, an author conceives players before production; and, unless all members of the drama's cast have equal status, he conceives leading roles before peripheral parts. Take even a lowly bacterium. Such a fellow has minimal prerequisites - homeostatic software and biochemical hardware both involve prior, present and future information. How did it first anticipate its own first needs - especially its own prior conception and future reproduction? There is, say evolutionists, no chance that a bacterium ever did. Chance did it all.

There is no chance, agree the holists, that bacteria made themselves; nor, however, chance that chance did either! An examination of energy metabolism alone (which includes the mirror image, symmetrical photosynthetic and respiratory pathways) reveals a level of order that is conceptual. It is difficult to understand how the chance jiggling of atoms could have informed such a multiplex system, could have combined this precise, integrated complexity of minimal requirements in a single microscopic place at a single moment in time. Machines either do or don't work. 'Don't' here means 'dead'. 'Do' means 'purposeful, alive'. Creative information is involved and it is metaphysical.

In this respect Natural Dialectic is as mechanistic in its view of bodies as of motor cars. Bodies, like cars, TVs or locomotives, are mechanical but mind is in them - mind of their creator(s). Is the invention of an engineer a 'vitalistic' one? If you want to claim that 'vitalism' generated lorry or a larynx, do; but better stick to 'mechanism that is engineered'. And definitely leave out chance.

The basic cause of effects, the source of reason, the issue of information's origin does not affect most science because the latter deals with present events. But it *does* radically affect the historical issue of origins and is, time and again, acute in consideration of biological information. The latter has a praeternaturally high storage density and generates conceptual, purposeful structures, functions and strategies. In this case it is irrational to propose (and, strangely, the most ardent 'irrationalists' consider themselves most rational) any model for the origin of life or life's processing, transmission and storage of information based solely on physical or chemical processes. **Unfortunately, this irrationality reaches to the philosophical heart of contemporary science**. Through science it has reached out to authorise its own relativistic version of ethics, morality and human behaviour. It has therefore entered (one might almost say, infected) the arena, both secular and religious, of the humanities. Mindless is, however, meaningless. What has no meaning has no purpose. It has, ultimately, no reason. Atoms have no concept. It is, therefore, entirely rational and consistent to propose that the information so clearly present, coded and operational, in the biological machines called bodies had an intelligent source. We shall have to return to this matter (see Chapters 19-25).

Mind Machines.

Information, communication, sensation! This is life. This is its news, craved novelty and buzz. The nearer our technology can bring us to immediate worldly fulfilment, the more it delivers apparent omnipresence, omniscience and omnipotence to our fingertips the closer to divine it feels. This is truly our life's striving. In a cosmos computational what more natural than changes calculated in a binary code? What more nervous than a sense of input and through electric circuitry the motor of response, an output? What closer, therefore, to your centre than a brainchild best reflecting brain? Information flows as mind; your mind's the treasure of your life. What more life-like a brainchild than the idea, therefore, of a mind machine? *A computer is a mind machine.* Inspect the logic of its functionality. Examine integrated circuits. Their molecules, like those of a brain, show no sign at all of mind. They are not even biochemical but metal, plastic, silicon and so on. Yet they are replete with passive, rigidified order. In this respect each one's determination cries out its ghost, the active order of their maker. The whole machine is absolutely full of maker's information.

Programs are a mind machine's intent. They express the will that mind invested in machine. Just as software is essentially composed of non-material, coded instructions that control the hardware to compute results so non-material influences must control the phenomenal cosmic program. Lines of this archetypal program are called natural laws. Archetypal influence most obviously, ubiquitously shows in every form of life. The 'stinger' question we'll investigate extensively is how.

Is it alive? A living cell with its hardware (membrane, cytoplasm, chemicals etc.) and software (*DNA* database, operating system and application programs comprising biochemical cascades, gradients, switches and other sorts of signal) can be likened to a very powerful, flexible, nano-technological computer. A computer is an expression of its manufacturer's intention and a cell of archetypal mnemone (Chapter 16). Both, in the purposeful way they handle signals, are a kind of mind machine. Can computers therefore come alive? You might attach one to a structure that, in consequence, appeared to 'see', 'hear', 'speak', 'move' and even 'learn'. A machine that tirelessly entertains, serves and even thinks for you is called a robot. It's 'switched on', it is live but is it, in its superhuman calculations, what you'd call alive?

Is a picture of you you? It's *like* you for sure. Would a very life-like waxwork, made by Madame Tussaud's, be you? Take an extremely good simulation, a picture that is a likeness, a working model and, due to state-of-the-art computerisation, a copier of your behaviour. This 'as if' simulation is so good that you no longer ask "Is it alive?" but "How are you?" Could even such exquisite artificiality be you? If so then aren't you a robotic species too?

Think of a friend. The transmission of their image, which is symbolic information, may be optical, acoustic, electrical or so on. Its storage can be by photograph, memory, tape or computer disc but the information ('my friend') is the same. This, while the former are peripheral infrastructure, is the significant, central part. It belongs to the semantic level of information. Not being physical, it defies a mechanistic approach. A computer is a grammatical device without semantic category. It is programmed, it performs algorithmic operations but, even if the program can itself generate new code, it needs an originator. Computers were initiated out of nothing by a set of minds; mind whipped them out of its thin air. Their basis (which might include bugs or logical errors) has been provided by intense thought, in the form of engineer, programmer etc. A computer is therefore neither inherently infallible nor, any more than a very clever, 3-d cartoon character, aware. Both computers and robots are dramatic but not alive. They engage only the lowest aspect of an act of creation, passive information. They lack, in two words, real subjectivity.

When you switch on your computer Babbage isn't on your mind. Nor Cerf, Gates or Berners-Lee when switching money round your internet accounts. An explanation of their programs and technology might not refer to them or any other programmer or engineer but can you separate their reason from invention? Can you exclude their logic, mind and consciousness from the intelligent, conceptual machines they've built and you are using now? An invisible component, subjectivity, resides and yet does not reside in all machines. What is the nature of this ghost? Is consciousness computable? Is brain an abacus or an upgraded abacus called a computer so that you are made a 'conscious machine'? Is the initial condition of computers not conceptual? It is, for sure, creative cogitation in their makers' heads. Mind machines originate in minds. Subjective creativity, called active information, thought their creation through.

At the start of this chapter computing was analogised to information's creative/ motor hierarchy. Now waggle your little finger. This is a simple creative/ motor exercise; a computer is a complex version of what is, in principle, the same operation. It is the rearrangement of physical components to accord with an intention. There is no intention in the structure or operation of a computer except what has been vested in its shapes and processes by engineers and programmers - by minds. The machine is, like the nervous and muscular systems of your body, entirely physical - logically organised but physical. *Its physical system has no mind, no metaphysical component; psychological intent only comes from makers and users.*

And, it needs be re-emphasised, every machine has the mind of its maker in it. Not in its atoms *per se* but in its purpose, design and lawful operation. Machine information is passive. Mind machine information is passive. *A robot is constructed from massive and sustained input of intelligence, information, will-power and purpose from humans.* How, therefore, could Generation 1 have made itself? Its creation and its reproduction would depend, like ours, on its original creator; and 'thought', even if it was programmed to 'learn' would forever be dependent on its creator's original algorithms. A computer's neural networks (or its silicon chips) are derived from intellectual concepts riding on metabolic energy produced in the human brain - not chance. Thought-fields reduce chance. The more powerful such a field the more definitely it will annihilate chance. In other words, the origin of grammatical/ syntactical devices like a silicon chip, robot or brain is always in mind. Active produces passive information. **To reiterate, there is known neither law nor process of nature by which information originates by itself in matter. The source of information is always mind. And at the heart of mind is consciousness.**

Indeed, the irrational, undemonstrated hypothesis that two abstracts, chance and duration, can 'construct' or 'evolve' intentional, purposive mechanisms might be considered simply a critical blind to cover the naked, neo-scientific philosophy of materialism.

What, therefore, is subjectivity/ consciousness? How did it, with memory, volition and the power of thought, evolve? Whence did the original program for a brain (if you say brain is mind) derive? By *PAM*'s mantra or the *PCM*?

There is no doubt that AI (artificial intelligence) *simulates* thought. It acts according to rules, proceeds in well-defined (algorithmic) steps and achieves objectives. It simulates human mind, in whose image and for whose purposes it is created, but simulation is not the real thing. You can simulate weather patterns, protein formation, flight-deck procedures, chess moves and so on using computers. You can generate typical or atypical circumstances but such simulation is never, by definition, the real thing. Is it different when it comes to mind? Does the simulation of thought think? Does simulation of cerebral activity 'become aware, then become aware it is aware' and thereby know itself? *Is apparent consciousness really or only virtually conscious?* Could even the most sophisticated robot imaginable be sentient, aware, alive?

You've been in this bind before. Was it not claimed that *PCM*'s 'design' was actually the way things are designed? Mindless did mind's job, '*as if*' turned into '*is*'. Is this deceit or self-deceit? Ruse transformed a metaphor into reality. Such scientific poesy recites again. At what point does 'as if conscious' transform itself to 'really conscious'? When does a mind machine or nerve of that impressive piece of hardware, brain, become alive? Such inverted commas are, in fact, degraders; they are anti-reason's subtle smokescreens; they fog the home of clear reason, that is to say, they clog departments of philosophy.

One of several possible hypotheses concerning the nature of awareness asserts that, while physical action of the brain causes consciousness, some aspects of this 'super-posed' action are not accessible to computation. Mathematician Kurt Gödel's 'Incompleteness Theorem' established that 'no formal system of sound mathematical rules of proof can ever suffice, even in principle, to establish all the true propositions of ordinary arithmetic'. Another mathematician, Roger Penrose, has argued that by extension Gödel "established that human insight and understanding cannot be reduced to any set of computational rules...there must be more to human thinking than can ever be achieved by a computer, in the sense that we understand the term 'computer' today". There is, in other words, something essential in human understanding that is impossible to simulate by computational means. Penrose does not necessarily suspect, still less admit, that there is any immaterial or metaphysical aspect of mind. He simply believes computation is an aid to comprehension rather than comprehension itself. Instead of espousing the bizarre but standard materialistic line that thought is an emanation from electrochemical patterns of the nervous system he makes the equally bizarre proposal that subjectivity is the product of 'quantum coherence' in cell microtubules; and that quantum physics thus somehow controls consciousness.

The exact nature of understanding, consciousness or awareness remains a mystery. To understand its origin is, say the mystics, to comprehend all things.

From a perspective opposite a saint's could quantum physics lend uncertainty in quantity sufficient unto will-power or the calculations born of thought within that acme of machines, a brain? Or should one believe (in line with Alan Turing's 'Mathematical Objection') that if we can do anything a machine can't then we must have a pinch of the '*je ne sais quoi*' - intelligence? Not that Turing and the AI fraternity believe there is mind separate from matter or 'ghost in the machine'. In their view fact and fiction merge; there is no divide between 'as if' and 'actually' or metaphor and its reality. '*As if*' means '*is*'. In a Turing test any computer-based robot that acts under sustained interrogation *as if* it were conscious, *is* conscious. Theory claims, with chips like neurons, mental states are 'computational'. Thinking and awareness are by-products of appropriate currents in a calculating circuit. Thus, if sufficiently well informed by Turing or alumni, simulation would become the real thing. Thus AI is aware. Information technology has spoken and with it comes the genesis of life. An android would be granted life.

You forget (Penrose did not) that computational androids need prior logicians. Without prior software operating in its specially fashioned hardware android 'life' could not begin. Without prior intelligence there'd be no 'intelligence'. Turing's surmise is skewed further, as we'll find (Book 3), when it comes to bio-logic. He drew mathematical equations that might, in principle, describe overt patterns in body coloration and, possibly, the chemical basis of morphogenesis. How life's body shapes are made is still, to some extent, a mystery. Might such chemistry originate spontaneously? No, it will not because it is dependent on whole complex, hierarchical networks of prior genetic code - whose detail armies of biologists are only now beginning to log. Upon interrogation could the operation of this bio-logic work 'as if' intelligently programmed because, naturally, no immaterial bio-logician actually exists to program it. If 'as if' applies then, in Turing's estimation, why should bio-logical code's materialised intelligence not be as real as android thought and logic. Therefore, who needs immaterial creators? Thus Turing's type of logic triggered naturalism's musings on the subject of biology and was quickly turned to evolution's order; such order quick-marched (if not frog-marched) squadrons of philosophers into thinking bio-logic's code might spring by chemical reactions sometime, somewhere, somehow, spontaneously. The modern company now trumpets to the tune of 'No Prior Intelligence'. Dawkins, Ruse and scientism's congregation beckon you to jump aboard. 'As if' oils this band-wagon's wheels.

When academic contemplation reaches to psychology such un-sentimentalism can't have truck with 'sentience' or immaterial subjectivity. Logic might be immaterial but *DNA*, like ink and paper, brings it down to earth; algorithms are symbolical 'machines' but electronic bits and bytes can carry out their functions on an acetate reality. Thus Turing followers adopt a computational theory of mind which must, of course, apply with equal force to you as to a robot - if, that is, you aren't in fact unwittingly a robot as you read this now. Such hypothesis represents a *bottom-up* attempt to model the structure of sense, intellect and human psychology in terms of neurobiology (nerves, brain etc.), neural networks and conceptual diagrams that try to describe the patterns of sensation, thought and communication. Its most useful, accurate work is perhaps at the level of natural grammar precisely encoded, pre-programmed and developed for human intention. This includes both biological hardware (neural networks and neurochemistry) and correlated software (firing patterns and the way such networks symbolically represent sensory data and motor commands). Flow-chart diagrams are built up to reflect the hi-fi transfer of information between hierarchical levels of command and control. These nested sets become increasingly philosophical, systems analytical or abstract as they ascend from neurological *structure* towards the *functions* of mind. Such functions include (as in Natural Dialectic's hierarchical register of information) purpose, imagination, learning, understanding etc. Can understanding be constructed out of fundamental ignorance? Can mindless molecules 'evolve' (or chemically 'self-organise') to make up mind? Must code not lead you to a coder; does hard-wired logic on a chip not need (for which 'as if' is a poor substitute, a lame excuse) an immaterial symbol-weaver, a program initiator - prior, non-android intelligence?

If awareness is an immaterial element then, of course, all materialistic and evolutionary ideas about psychology just crumble into dust. That (Chapters 13 and 14) is a lot of dust, enough to cloud your vision; but if their basic premise is not right then they need deconstruction followed by rebuilding on the right lines. Facts won't change, perspective and interpretation will.

Transactions on computers are just circuitry, transistors and constrained electric pulses. These can never be subjective or alive. Nor do non-conscious, material elements *produce* consciousness. That they might is (as Chapters 5 and 6, then Books 2 and 3 will show) simply mystified but wishful thinking. You are a conscious machine not because your brain generates but because it modulates your consciousness. Each way signals flow - from inside thinking out and outside (including your own body) sensing in. A medium on this communication path is brain.

What is pure energy or consciousness? You only know them when appearing in a circumstantial way. You only see them take shapes in a current of events. Your brain is a human modulator of consciousness. It is not a dolphin's or an ant's. Thus your experience in this body is, must and only can be human.

In other words neurons, like computers, are not aware they are carrying information. They are orderly media whose structures computational theorists 'reverse engineer' using neural models. A theorist tries to understand how lower levels might support, link with and explain higher levels of algorithm. While these 'super-positional', hierarchical models are supposed to reflect a hypothetical anatomy of mind, another possibility exists - not that matter creates mind but that neurobiological patterns may (as in the wired circuits of AI) be an effect rather than a cause of mind. In other words, wireless patterns of mind affect the origin as well as the development and maintenance of biology's 'grammatical' mechanisms. *We might expect, therefore, to find some structural logic with respect not only to consciousness but the body as a whole - a logic of embodiment* (Chapter 17).

For example, although the contexts may differ widely whenever you want a car to accelerate you press the same pedal. Consider the primary informant of behaviour - purpose. Is it physical in origin? Neuroscience grounds this executive element of thought in 'Area 46' of the prefrontal cortex of each cerebral hemisphere. Its dialectical position is just behind point X, the eye-centre between 'Twin Pedals 46'. This is, top-centre, a metaphysical plexus mediating conscious thought with non-conscious body. From here fibrous tracts run to other parts of the brain. A range of cortical correlations with sense and motor functions descend towards sub-conscious response centres such as cerebellum, medulla and the electro-hormonal hypothalamic gland complex. The nerve bundles leaving from above even cross over just before they quit HQ (headquarters) so that the right side of the body is regulated by its mirror opposite (left) side of brain and *vice versa* - an inversion reflecting the polar asymmetry of informative mind and energetic matter. From its physical roots in brain the information system descends a spinal trunk before ramifying out to innervate all body parts. 'Organic' *top-down* logic that includes inversion and descends from point of purposing to body certainly applies; but brain

is just a dashboard panel and some scan-detected part of Area 46 links, like a pedal, purpose to its bodily machine.

Whatever creates, transmits, receives or processes programs is a mind machine. For example, a radio is a medium, template or interface for the exchange of information. Analogise brain with a radio set and mind with its program, broadcast or message. You would expect radio parts (e.g. tuner, volume control) to correlate with aspects of broadcasting. You might also expect aspects of mental broadcast to correlate with physical transformers - and neurophysiology has correlated numerous areas of brain to the coordination and control of different parts of the body. If you were to damage volume control you might expect impaired or lost function but would you have damaged the broadcast itself? Would the program's character have been touched? Similarly you might locate areas of cortex that correspond to aspects of bodily function, consciousness, will-power or memory; you might even identify an 'aha!' waveform or a 'big-G-spot' rich in opiate transmitters! If you break a television does the broadcast disappear? If you cut the cortex would you exorcise the source of metaphysic, thinking mind? Would a 'godectomy' or an appropriate drug rid its patient, once and for all, of any deviant illness stemming from a non-materialistic bent of view? Perhaps, for ethical reasons, such an experiment would have to be confined to a rat's appreciation of deity!

Such analysis commits the standard, elementary objective error of equating a thought with its nervous signal. As any number of thoughts can be registered in a similar electrical way on 'live' telephone wires to different numbers and addresses so intense bursts of electrical activity are not the same thing as thought. The brain is a console that exchanges coded information with an attached vehicle, its body. Neuroscience has predictably grasped the objective but not the vital, subjective side of mind's coin. *In short, the brain is as lifeless as a radio or TV set.* Programs are 'in the air' whether or not a particular set is in good working order or tuned to their channels. It is the broadcast not the instrument that *means* anything. *Programs are purposive*; immaterial message is prior and an instrument is simply its conveyor. Information is only sourced or received by mind. Things don't specify stepwise activities that lead to goals. **It is against all known laws of physics that oblivious matter might create information; or that material instruments of purpose self-organise or instigate their own programs.**

<u>Despite these facts there persists that science fiction. An official one. It rumours, without a shred of hard evidence, that mind sourced itself by chemistry - because it is only an originally accidental arrangement of atoms. Brain, most excellent of mind machines, evolved!</u> Imagine, for the sake of argument, that the fiction was fact - so that if matter *did* source information not only would AI be alive but you yourself would be an advanced edition of it. To invent a natural robot simply substitute those abstracts, time and luck, for the best minds in modern technology. *The reason for reason would therefore be, perchance, chance-ridden PCM!*

Natural selection (Chapter 22) is, although a biological axiom, neither a law nor an explanation for the origin of biological forms (morphogenesis). It correctly implies that unhealthy or mutant individuals are less likely to survive life's vicissitudes but does *not* explain how chemicals may 'want' to survive or, having simply 'cellulated', become progressively more complex. Natural selection has a *sine qua non* - biological body. It struggles (Chapters 17 and 19-25) in the face of this chicken before its egg. Which came first, biological life or natural selection that selected it? Life's simplest independent 'robots', like bacteria, involve irreducibly complex hierarchies of command and control mechanisms. A cell is life's basic unit. How, therefore, could pre-biotic chemicals be 'naturally selected' to produce a cell? Or, if they were not, how did just the needed ones, along with codes to fabricate and reproduce, co-endure? How did that fantastic proto-cell appear?

This (see Chapter 20) is a key problem for a 'bottom-up' materialist. How could the laws of physics and chemistry aggregate atoms into precise, information-rich orders of integrated code, signals and purposeful mechanisms? Molecules don't *want* anything. An amino acid, sugar or base does not want, any more than iron in a piston, anything. Therefore don't ask why they fell together in an operation like a mind machine and called a form of life because there is no answer. No reason. Instead ask how and give your answer, "chance" - the same thing as no reason. Have, thereby, unreasonable faith in luck, a powerful anti-deity according to whose total lack of wit practically any order might, if you can just imagine, happen.

Mind is, on the other hand, an anti-chance - a 'deity', a powerful charm Luck must have raised against herself. Anti-body followed body. You gain immunity from chance by mind and yet, absent from the first, mind must have followed last as an effect of casual chance. Mind, the great intender, sprang from no intention! Mind, the great pretender, springs from great pretension! How ironic chance evolved its killer, mind since mind naturally selects by purpose not by accident. Eventually, according to The Book of *PCM*, a brain and then a human brain and, from the pattern of its atoms, minds evolved which re-develop, every life afresh, from information pre-laid in an egg! How, after all, could Johnny Human with come-lately mind have made a 'bio-thing'? How from the rear could mind have ever made a natural force or particle or body? To even raise the possibility is out of order, well outside material lines. Therefore you must look on mind (or, rather, mind machines called brains) as an exceptional aberration; alone in a universe of mindless, automatic fixity this single 'strangeness' has emerged, as if alive, from chemical oblivion and, it seems, creates illusions of intention. This 'alien' wants to see a purpose even where it seems that none exists.

In fact, materialism's view reduces all intention, thoughts, feelings and thus Johnny's actions to reactions of the neurochemicals that work his brain. This robotic stew *is* Johnny's mindless mind. As a robot he can only, totally enslaved to wholly masterful machinery, do what he's told. Thus, in causal chemistry's eternal chain, he's just a witless cog. No doubt, oblivion's illusion is its film of consciousness. He therefore 'thinks' that murderous is different from kind intent; it is 'as if' the lazy and the diligent, the moral and immoral seem apart - but chemically it's all the same. Actions are inevitable conclusions of molecular determination! Such philosophy can exorcise a big G easily!

Luck is goal-less but somehow, you say, evolved a feel of purpose in the universe - but teleological craving is a waste of rational time. It is (do you not understand robotic atheism's empty heart?) useless to ask 'Why am I here?' or 'What does my life mean?' To ask purpose of luck or ask reason of its lack is a futile exercise. The question is meaningless. Hopeless. Self-deceptive. There is no meaning, from that point of view, in the survival of any particular pattern - including your own self. Mind-numbing 'evolution' is the buzzword to replace 'intention'. Of course, your 'selfish' genes are, as a bunch of chemicals, under no illusion. Purpose without purpose is their game. Reasonless, unreasoning, unreasonable survival is its name! So why, in such irrational oblivion, survive?

So are you really just an animated metaphor? Are you a mindless mummer, puppet or machine? And yet no careful algorithm has been step-by-step constructed to explain, in precise molecular or systematic terms, how chance must have danced into the field of hierarchical command, homeostatic self-governance, metabolic pathways and the conceptual, targeted mechanisms of life. If these were not purposely informed (which is materialistic anathema), then you must have faith that an uninformative maelstrom of chemicals evolved active anti-chance in the form of purposeful behaviours and eventually, for absolutely no reason, that epitome of information we call rational mind. You cannot, while immersed in this widespread delusion, make appeal to reason. The reason *is* chance. Such irrationality has deselected reason and denied it at the heart and origin of life. It cannot acceptably deny teleological schemes behind the irreducibly complex (but still relatively simple) working tools of man but it can and, against the principles of all arts and technology, does deny those behind the construction and operation of natural, biological organisms. Of whoever professes natural selection as the 'mind' of minds it may be argued that he has blurred reason by non-reason. He has, perhaps even as an academic, reasoned reason out of nature. *Cum laude* immobilised it. Annihilated its logic. Such is the twist of a wilful determination to deny a rational origin of will. *Is this, as a curriculum that's standard in the modern school, reasonable?* Are you rational? Can you emerge from mists and mirrors of such intellectual shadow-play and, from a position of illumination, make your mind up? What do you opine?

A materialist has decided and opines. For some, such as evolutionary psychologists, it is a holy grail to exorcise the metaphysical ghost (was that mind or was it soul?) from any machine. Human bodies are machines far more sophisticated than computers. Such ghost-busters therefore attack any 'ridiculous' logic that leads towards a counter-belief that 'sentient awareness' is inexplicable in computational or scientific terms. The 'spooky' view held by many scientists and non-scientists, historical and contemporary, that life is essentially spiritual is, in short turn, dismissed. Wiener's view that information is metaphysical is dismissed. The view seemingly held by Gödel and certainly all mystics that awareness is metaphysical is dismissed. The first principle of mystic research is therefore, before its start, at best detached from and at worst dismissed. This is at best a pity, at worst a fateful error.

Are mind, human identity and free will simply 'a vast assembly of nerve cells'? Are you a mind machine? You might like to review the first, short couple of sections from Chapter 0. <u>**It was noted that in fact, despite the hype, we have not the least idea how consciousness arises from cerebral grey or any other type of matter - if it does**</u>. **This whole book drives a coach and horses through materialism simply by assuming it does not**. It was proposed, right from the start, that subjective experience is in fact more than subatomic particles or even the informed arrangement of such particles. **It is suggested, in other words, that matter and consciousness (or energy and information) are two fundamentally separate entities**. The latter is something as separate in quality from matter as atoms are from sensible objects. It is not physical. If natural is physical then awareness is your physical, unnatural or super-natural element. Indeed, this element *is* you. It is where you live - in your mind. Is that unnatural?

If this is true, then 'weak' AI (wherein a robot never acts more than 'as if' conscious and therefore metaphor is never turned into reality) is Natural Dialectic's stance. And any AI research that presumes to create rather than mimic a fundamental cosmic entity is fundamentally ill-conceived. If brain really is a kind of computerised machine whose product is an entity called consciousness, you would expect to be able to reverse-engineer life. By making a robotic brain AI technology would be making 'objective, simulated consciousness' - except that the fundamental property of actual consciousness is actual subjectivity. *No doubt computers are the gift of mind but, except in the dreams of some of their creators, lack precisely subjectivity.* To simulate's to act 'as if'. Computation can outrace a human's thought; you might logically evolve, by using mind, a high-class, superhuman robot that outclassed a nano-robot (cell) or even (help!) your own robotic self. What a mind machine! Computers are mind's gift but some AI creators think that their own mind is matter's gift; intelligence evolved from molecules so why should a computer's not be real as theirs? Such truncated logic, deliberately ignorant of immateriality, twists and squeezes every explanation of design into its mindless frame; it strangles how you look at life (or, because it has evolved from atoms, non-life) and kills soul off at its conception.

How do you kill an element of cosmos off except in your imagination? A mystic's grail, perfect mind, involves purifying consciousness. From a *top-down* point of view First Principle (from which the rest derive) *is* Pure Consciousness. Consciousness is subjective while matter, even as subtly formulated as a 'living' robot, is not. If, therefore, such a mind machine had self-evolved towards perfection, what would it find? Would inanimate complexity come up, like a proto-cell, with simple animation? Could 'Artificial Intelligence' summon mind-worlds up, turn mystical or digitise experience? Could it develop towards climactic Digital Enlightenment and an abstract 'Great Computer In The Sky'? How? Blow a fuse? Simulation is metaphor and simile. A likeness but at what point does likeness reach and overtake reality? Simulated consciousness is never more than virtual subjectivity. As if. Unreal. Although fascinating, the operations of a computer are no more mindful than a steam engine - in other words, the machine is a reflection of the creative cooperation of its makers; and the gap between mind and machine (albeit a computer or a brain) is permanent. Unbridgeable.

Another kind of cyber-evolution, programmed according to the desires of its carnal creator, is more likely. Ultimate servitude. Instead of ascending into freedom AI's 'homunculi' will willingly descend to complete servitude, the perfectly obedient service that you make and market slaves for. No freedom means absolute constraint. No mind of your own. This, just like matter's, is a robot's state of mind. Any body biological is, like a robot, an objective mind machine. It is your own mind's puppet but not normally the object of an overwhelming ideology or exploited as another's serf. And, unlike computers, it incorporates the *active* information of a real, non-simulated and subjective mind.

Are organisms therefore natural mind machines? For the Dialectic immaterial subjectivity is present; it is mind round which material forms accrete. In conscious and/or subconscious mode this organiser, mind, is always immanent within a biological machine. Is it, we asked, a binary or polar code that universal mind electrically impresses on the 'world machine'?

If you sit outside a room then don't complain you can't see what is going on inside. If you look at your computer don't complain if you can't see the origin of computations that its 'mind' performs. If you inspect a brain, don't complain if mind is not apparent or a thought has not appeared - even if it is your own brain that the probes are wired to. How could you read your own thoughts on a screen? Would it show them in the way you know them? At death such brain, along with all the rest of body, crumbles. The physical medium of computation/cogitation ceases. If consciousness is fundamentally distinct from material energy, what will happen to it now? On release from its physical 'damper' could not your metaphysical part discover itself in unfettered heavens and the hells of its own immaterial element (see Chapter 18)?

Universal Authorship.

Bottom-up, matter precedes and produces mind. Having exorcised the spook, materialism extirpates any report of out-of-the-body experience, notion of a mind-world or mind/brain distinction. It bins them as preposterous delusions conjured by oxygen deficiency, epileptic brain chemistry, drugs or neural shutdown. The subjective point is not entirely lost - it never really existed.

Top-down, the reverse is true. *Why therefore, as this book in its entirety will show, should I dance off down materialism's garden path? Why should I be interested in its form of fairies? Or indulge its 'just-so' fairy-tales?*

On what grounds, therefore, do I reject the authority of materialism's intellectuals and, labeling it a kind of fairy-tale, reject its 'progressive' form of thinking? In respect of scientific facts I don't but here, at the close of Part 0, let's also check on immateriality's distinction and summarize subjective as opposed to objective conclusions.

Is not the presence of an invisible, physical event often inferred from its effects? Examples include planets that circle distant stars and electrons that orbit atomic nuclei. Why, therefore, should the presence of invisible, metaphysical mind not be inferred from its informative effects? And why, if mind is a separate element from matter, should universal mind not shape materials? Couldn't information issue shape to energy?

An author is the originator of a plan; an authority is one with power to make others obey his logic; and to authorize is to sanction. All three words are based on a sense of intelligence, will-power and reasonable plan. They involve meaning.

Information technologists use the word 'semantic' to mean 'significant', that is, bearing sign or meaning. Thus they include software, code and any signal in this category. For the Dialectic, however, only 'active information' involves 'meaningful' or 'semantic' communication. Psychological message, either sent or received, is 'actively semantic'. The categories of software, code and so on are incapable of creativity or knowledge. They are syntactical vehicles of information that only become 'passively semantic' when arranged according to the dictate of a mind. In this sense information's infrastructure (code) is a conceptual object, insensitive grammar or metaphysical frame that, by the preordained organisation of otherwise meaningless constituents, transfers 'semantic' sense.

Smoke is meaningless; smoke signals aren't. Iron is meaningless; a bicycle is not. The shape of clouds, sound of whistling wind or currents of the ocean swell exist at the lowest, physical level of the information hierarchy. They are meaningless or, to the information technologist, 'syntactical' - unless you consider their syntactical arrangement is according to natural law. *If this law is the product of archetypal memory in universal mind you will perceive 'semantic' although passive information, as in a machine or the materials of a book, behind the universe.* If not, any so-called book of life or cosmos is essentially meaningless. Such emptiness is scientism's current view.

Let's rephrase the dialectical perspective. Active information (mind) precedes what is passively informed. By now you will have recognised the hierarchy operative in yourself. It involves St. Patrick's 'Essential Trinity' of Soul, Word and embodied soul. And it is thereby linked to the drop of 'existential trinity' from soul through mind to body. Was 'soul' defined? Is it a paradoxical division of the Infinite, a drop in the Ocean of Consciousness or a particle of Nothing minded, in existence, to feel separate - a separation the illusive bubble of a mind, in keeping self from Self, sustains? Is not the object of subjective science to burst bubbles? To prick the film that is illusion of the mind, relieve its sense of separation and thus show for sure that real brilliance in your life is, as a photon to the sun, the same as Brilliance of An Endless Sheet of Living Light, The Cosmic Soul? Of course, you won't find this perspective on your universe in an objective science book. In going off the other way you turn your back; negate the sun and you expect to feel black, freezing, hopeless space.

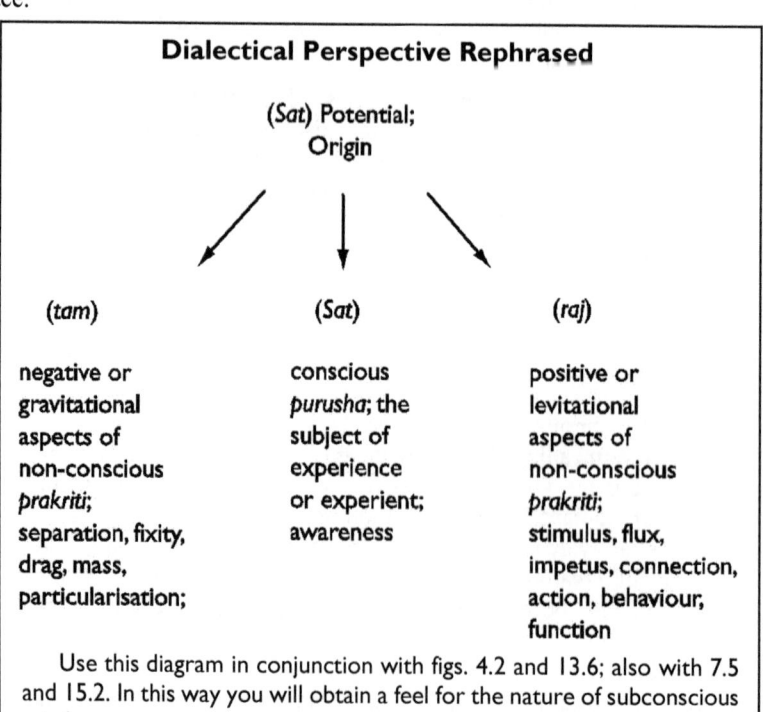

Dialectical Perspective Rephrased

(*Sat*) Potential; Origin

(*tam*)	(*Sat*)	(*raj*)
negative or gravitational aspects of non-conscious *prakriti*; separation, fixity, drag, mass, particularisation;	conscious *purusha*; the subject of experience or experient; awareness	positive or levitational aspects of non-conscious *prakriti*; stimulus, flux, impetus, connection, action, behaviour, function

Use this diagram in conjunction with figs. 4.2 and 13.6; also with 7.5 and 15.2. In this way you will obtain a feel for the nature of subconscious mind.

Do you recall that, in Chapter 1, in oriental phraseology **Purusha** is the subject of experience, that is, the awareness of an experient? **Prakriti**, which was introduced in Chapter 2 as 'universal energy', is the object of experience; the range, spectrum or different aspects of *prakriti* are what is experienced. Subjective *Purusha* is, as your own consciousness, unitary; there are, on the other hand, always two sides to the objective, polar nature of *prakriti*.

Modern science has, unlike the practice of yoga and traditional faiths, made no elemental distinction between metaphysical mind and physical matter. It is, therefore, necessary to introduce, define and later (Chapters 7 and 15) refine the meanings of a few useful terms. It is because these terms are well known but employed with various nuance and implication that a specific, self-consistent dialectical definition has been drawn up.

Prakriti is, in the conscio-material spectrum, the non-conscious 'partner' of consciousness. Its wireless presence extends throughout the range of macrocosm, that is, includes subtle mind and matter. An electromagnetic spectrum is broken, for convenience, into various wavebands. Each waveband has its own character and capacities. Similarly various 'bands' are identified in the wireless radiance of *prakriti*. The subtle, psychological band is called '***manas***'. 'Manas' is the 'film' on which, or energy with which, the shapes of thought are developed. At fine levels of mind consciousness predominates but, as thought 'thickens' or becomes grosser, the lower lights of *manas* are employed.

In the dim, thoughtless shadow of sub-conscious mind '*manas*' is resolved into a further subdivision viz. motive '***prana***' and its field, '***akash***'. Check the conscio-material position of *prana* in fig. 13.6. 'Akash' is often translated 'sky', 'space' or 'ether'. *Prana* is 'light' which through the lens of archetype is focused into the basic building blocks of things; it is 'electrically charged oxygen' that germinates seed memories or an 'energetic wind' that from the instrument of archetype blows 'notes' that make up matter. If *prana* is the activator of an instrument then '***tanmatra***' is the archetypal instrument itself. How in music do you separate the two? Their relationship is analogous to the paradoxical, physical relationship of energy/ mass or wave/ particle. *Tanmatra* is a potential matter. It is a subtle, causal element compared to *prana*'s force; it is a strand of archetypal memory whose pattern *prana* excites into manifestation. The nature of archetypal memories as the basis of physical existence and the lens through which chemical and biological bodies are projected will need elaboration (see Chapters 9 and 15 - 17).

The five-stringed cosmic instrument is played. Five *tanmatras* (strings) vibrate. Some chords are harmonic but the jarring ones are 'barred'. Cacophonies are, like mistakes, squeezed out of composition but sweet, balanced melodies survive. 'Music' of the basic particles persists. They resonate and are related to various qualities of matter, various properties and textures of physicality. The qualities (see Chapter 17 and especially fig. 17.11) are expressed as five 'elements' or hierarchical phases that substantiate our universe - archetype/ space, free energy and mass in the forms of flux (gas and liquid) and fixity (solid).

Instead of instrument an analogy is commonly made between *prana* and electromagnetic radiation - light. As the wireless radiance of visible light can be subdivided into the colours of a rainbow, so *prana* can be subdivided into sub-conscious energy levels. Each of these 'colours' has its physical correlate (in terms of forces, particles and states of matter) and associated attributes. In the analogy of light as opposed to sound a *tanmatra* amounts a medium of resistance, a diffracted waveband or coloured filtered from clear light. Such correlation is perhaps clearest in the abovementioned case of human construction. In this case it is directly related to biological purposes, that is, the order of form and function (figs. 17.9 - 10).

It is a dialectical principle that what is 'above' sources what is 'below'; and what is 'central' informs what is 'peripheral'. For example, a higher level of *prakriti* (say, the reflex of 'instinctual remembrance') governs a lower (the physiological and locomotive behaviours of a body). **Higher is 'nested within' lower.** In physical terms *(raj)* quantum behaviours drop from *(sat)* archetypal patterns; and, 'chilled' from quantum level, they precipitate *(tam)* 'wired' or 'bonded' states. These range (fig. 7.9) from light to heavy flux and hardness. Gaseous, liquid and solid phases of matter are increasingly constrained.

The latter trio are *prakriti's* gross or 'wired' expression, the bulk 'extrusion' of its universal energy. In a body, therefore, such *prakriti* is the stuff of physics. Various aspects of its equivalents, positive energy and negative mass, interact to shape objects and govern the pattern of physical events.

fig. 6.2 (see also 1.2, 7.5 and 13.6)

Experience is a private show. What way, except by meditation, to experience Experience or prove an immaterialistic theory true? If true could you, in consequence, reflect construction of the cosmos? <u>Could the general, universal or macrocosmic structure of information be reflected in specific, individual or microcosmic form and man thereby be the measure of all things</u>? If so, such construction falls from its Transcendent Author down through super-conscious, conscious and sub-conscious levels of mind to non-conscious matter. Within non-conscious matter it shows as quantum and, lastly, the exhausted, massive or classical phases of physical nature. You could view gross gaseous, liquid and solid objects as expressions of the lowest banding on the conscio-material gradient. Kinetic theory of materials states their motions are the slowest, most constrained. With slowest-of-the-slow the business has 'got fully fixed'. Your body's semi-solid so, if you seek higher truths, you simply have to dump oblivion and rise the other way - from mind to soul.

In this paradigm science is observed to study only the two lowest, passive levels of the information hierarchy - objective automata and the 'grammatical' rules by which they work. It measures the former and infers the latter; the *origin* of the latter, either by chance or not, is a question of major consequence.

According to the *top-down* view Authority for macrocosm is the *Tao* (Way) or *Logos* (logical construction by The Word and 'natural language'). Who is the authority for your own microcosm? Who speaks and actually controls your life? Is it the compulsive, centrifugal tendency of mind towards matter and its sensual influences? Is it, on the other hand, deliberate, centripetal focus towards mind's source or, in another way of saying it, towards Right(eous)ness? Both ways, Centaur, struggle in you. Which, in the struggle for authority, most often wins?

A true teacher of subjective science has authority to puncture mind's obstruction, cut the power of negativity and make the positive connection. He initiates you into practice for The Answer, switches you to understanding The Essential Trinity - 'Lord', 'particle of Lord' (called soul) and, in existence, current of First Cause (called The Communicator and Connector, Word).

Understanding is the (↑) upside. What has the (↓) downside made? What creation has been authorised? In Part 1 the Dialectic is going to rephrase the 'non-life sciences' (objective physics and chemistry) into its perspective. Volume 1 will follow up rephrasing the 'life sciences' - psychology, biology and ecology.

Active on passive. *Material energy is the structural face of information.* Just as film is needed to fix light, so information is developed into energetic shapes. Or, if you like, its light is swathed with degrees of material shadow. Imagine travel down the conscio-material gradient as winding layers of some covering around a bulb, so gradually extinguishing the light.

Let's introduce (to bear in mind when travelling on) the nature of this cosmic shroud. Read, if you have not already, the legend of *fig.* 6.2. Polar energy, as opposed to unitary consciousness, is a fundamental half of existence. In general or universal terms it is known in Sanskrit by the word *prakriti*. *Prakriti* is the basic substance of natural automata. Its embryonic shadow develops through levels of mind. Just as the electromagnetic spectrum is divided into wavebands (X-ray, visible light etc.) so the mental division of *prakriti* is called *manas*. *Manas* (mental energy) is the material substance of attention, a subtle film against whose resistance imaginations, called thought-objects, are evolved. Like those of matter, grades of mental energy rigidify. They range from rarefied to solid, from high frequency to low.

The energy that attention works with also exists in grosser forms that it cannot directly influence. It becomes snared, dragged down and overcome by their influence. We call such dormant state of mind sub-conscious and its objects memories. Thoughts engraved, habits grooved, instincts and archetypes inhabit the psychosomatic border in whose *PSI* region mind touches matter. *Dialectically, the immemorial archetypes of universal mind precede the shapes of subatomic, atomic and consequently physical phenomena.* The final expression of *prakriti*, unmixed with any subjectivity, is the stuff of physics. Its non-consciousness is what we, forgetting any metaphysic, call the universe.

Creation involves what the Dialectic calls 'polar inversion' or 'asymmetry of reflection'. The Centre expresses itself 'organically' from within; it 'turns inside out' so that internal is precipitated in external, physical form. Its centre of balance gradually swaps over to one of gravity; mass instead of awareness becomes the centre of things. The inversion grades, at the same time, from conscious to non-conscious, voluntary to reflex, actively informing to passively informed. At the base of its spectrum, at the foot of the cosmic rainbow energy predominates. We call it physical energy; its total objectivity precipitates, in the order of things, a massive crock of fool's gold called bulk matter.

Stimulate. Obtain response. Information's stimulus is focus of attention; input interest, interact and cause a rearrangement - output. This is, on a voluntary basis, how mind works. A clear distinction is drawn between mind and separate but closely-knotted body. A materialist's delusion is they are the same. Cause, effect. Input energy and, according to the rules, predict an output. This, at the involuntary grade, is how automata proceed. The whole physical world is a reservoir of involuntary patterns. *Thus, in principle, a body is passive information; it is an expression of inward archetype and also, if associated with an animated form, an outward container of mind.* A minded body or embodied mind is normally called a biological organism. An organism is a complex form of archetype but, dialectically, all material cosmos is expressed from suites of archetypal principle, of fundamental memories in universal mind. These constitute the 'instinct' of material creation.

If its levels of information exteriorise from active to passive, then the sensible, physical grade of cosmos would thereby express an underlying purpose, an inward, 'hidden' metaphysical principle. As in an egg's intrinsic case information would be developed, the potential of its prefixed program would evolve. The Latin word '*mundus*' means an 'elegant arrangement' and the Greek word '*cosmos*' means the same. It means a system wherein the behaviours of energy are informed by law, wherein any chance of chaos has no chance. Accordingly, every form or behaviour consists of varying proportions of the cosmic fundamentals, energy and its alignment with mental or, for the most part in our universe, physical command. The sway of mind and matter accords, dialectically, with the swing of (*raj* ↑) and (*tam* ↓) vectors round a (*sat*) balancing factors, principles or laws. While automata are immediately susceptible to the direct effect of an inflexible operation of law, voluntary mind also inexorably reaps the natural, equilibrating justice of *karma* - the law of cause and effect. *Karmic* means, in effect, 'equal and opposite'. It means equation or exact equivalence but, in the case of mind, not necessarily immediately so. Equilibrating justice is often delayed. It may be triggered, in recurrent ways bound up with memory and circumstance, sooner or later and therefore, it may seem, with an effect only indirectly related to the original cause.

You find no obvious meaning in inanimate matter. It is a non-conscious, unresponsive, choiceless cipher. You can only *infer* meaning from a body's characteristics and behaviour. From the principles that underlie its particular behaviours you might infer that the material universe showed (for all its tangled complexity and apparent chaos) the characteristics of an ordered, logical machine. You might see it as a crystalline form of mind, a non-conscious recording that lacked meaning or intent except, perhaps, for its author or its audience. Physical matter only has meaning in relation to its observer, that is, life. If the material universe were viewed as 'purposeful', you would call it a machine. If, however, you include the requisite observer it becomes a machine that includes life, a construction that includes players - a dramatic machine. Drama implies authorship and a rational process of creation. In the dramatic presentation of universal authorship the meaning of matter becomes the same as the meaning of stage, sets or music - background to the foreground plot, to the really interesting part, the whole reason for a theatre. When the conceptual potential (or text) and stage have been prepared, dramatic life is released at a stroke. From silence 'Action!' From darkness the curtains part, lights go up and music plays. From nothing everything is switched on. Is this the way of Universal Authorship, Cosmic Directorship or, if you like, Lord of the Dance? Is this the track of reasonable, meaningful, inanimate existence; and, within inanimate, an animated part?

We have swung full circle and returned to the issue of 'Intelligent Design'. If there is no intelligence except with brain, then Universal Authorship is a non-starter. Even if there were it would be metaphysical and thus a scientific 'no-go' area - as is any other reasonable, immaterial form of information - because 'hard' physical science deals only, of course, with mindless, inanimate objects and energies. It is easier, however, to see reason in an instrument that clearly incorporates high degrees of purpose and complexity. For example, a biological body is, in its physical parts, a superb soft machine whose purpose is survival and reproduction on earth. Machines operate at the passive levels of information. They execute tasks according to both physical grammar (the laws of nature) and the wishes of creators. In the biological case they are highly coordinated instruments whereby mind (minimally flexible in the shape of sub-conscious instinct or also capable of thought) is 'plugged' into the inanimate tier of the cosmos. Terrestrial life is two tiers of the cosmos spliced; it is the whole of existence exemplified.

In this respect upright human being is unique. Human body is the 'container' of a mind that can, potentially, involve information at every level from physical up to Transcendent. It therefore represents the whole three-tier macrocosm; it reflects the full potential, principle and practice of this system (Chapters 13 and 17) in a hierarchically well-balanced, dialectically logical way. *In this way a human microcosm bio-logically reflects the hierarchy of information as issued to the cosmos.* **You, therefore, actually represent universal creation.** Cosmic man, whose seed is in the Infinite, whose voluntary roots in mind and whose involuntary body has grown out into the physical universe, forms an inverted Tree of Life. *Life is, you may agree, special. It is precious and a realisation that you could connect and commune with both the whole of creation and its Author would make it very precious.*

An Inference of Design.

A summary is called for.

Inductive reason is the scientific way. It generalizes from specific observations; it suggests a truth but still allows that its conclusion may be false. Thus, if to all appearances an observation's true, then scientifically it is. So that, in biology, an overwhelming appearance of deliberate design (on which all parties are agreed) may scientifically be construed as just that. Not what it seems to seem but what it is. Design and not 'design'. Engineering that's intelligent is scientifically OK.

You use historical as opposed to operational, experimental science when dealing with one-off events - such as the origin of integrated, working systems - that happened well before your time. You deal with causal mechanisms, possibilities and inference by 'best explanation'. Nor does 'best explanation' exclude the perfectly scientific inference of a mechanism's intelligent design. Such inference might reflect the nature and obvious quality but not necessarily the identity of an inventor.

The jargon is 'cognitive dissonance'. 'Don't confuse me with the facts, my mind's made up.' Its set is fossil-hard. **So, *bottom-up*, the whole world came about by chance. Without design. What are there but *physicalia*? Existence and all that, including subjectivity, is a fantastic accident evolved from impulses of energy constrained by mindless serendipity of natural law.**

Top-down, mind imparts intent. It informs another mind or prints designs on matter. Thus purpose may be conscious or, as a reflection of original design, unconscious. Subconscious instinct, archetype, code and non-conscious machines are each unconscious, reflective, passive examples of the carriage of purpose.

In this respect two major thrusts well indicate a cosmic scheme of anti-chance, that is, creation by intelligent design. *They are physico-chemical and biological*. The first (Chapters 7-12 and especially 8) involves the integrated precision of principles (laws of physics) that permit your presence here. The second (Chapters 5, 6 and 19-24) illustrates the very limited correctness of a theory that you were evolved by oblivious forces; and emphasises the irreducible, coherent and ubiquitous complexity of design, informed by code, found in the remarkable biological components of your (and every other) body.

TOTAL CONTENTS

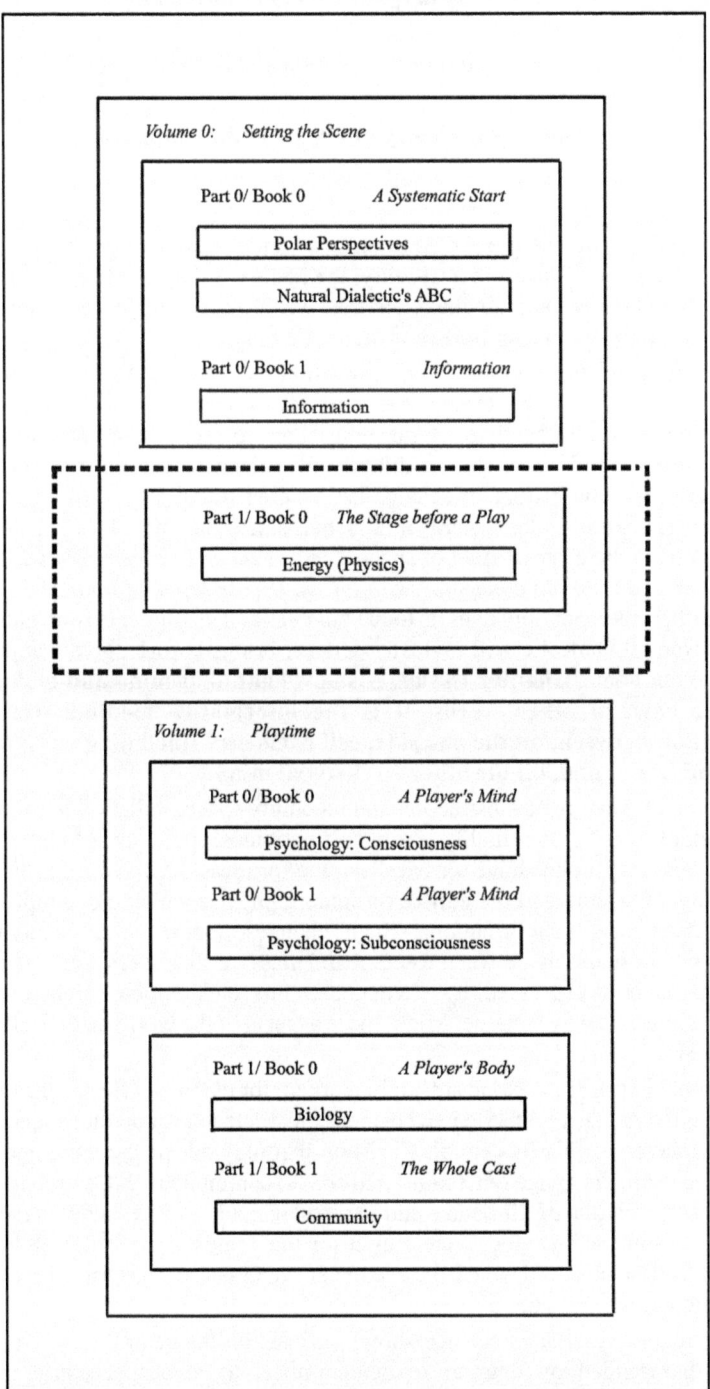

Volume 0: *Setting the Scene*

Book 1: (Energetic) Stage Before a Play

Part 0: Physics

7. Towards a Theory of Physic and Metaphysic

Space, time and energy! An actor on two blanks, a weaver on the warp and woof of emptiness. These are a tri-unity, the godless trinity of physics. Did you expect to analyse a thought in matter, extract will-power from an elemental force or discover purpose in a particle of dust?

Ground-state vacuum and an excitation in it - radiation! Not the metaphysical light of reasoning but material radiance. Light fantastic. An undulation in an unseen field - light with which to see! It is a phantom from whose shapes and ways the brilliance of scientific reason reasoned its enlightenment. The history of science is clustered round a beam of light (see Chapter 12: Pure Polar Energy); a renaissance has developed from its oscillations in the abstract field of space. *Light is the messenger from stars, the informant and ultimate connector with all we'll ever know of time and space.* And, furthermore, not only does its touch our eyes and instruments from galaxies but, as Natural Dialectic will logically deduce, brings knowledge much, much closer to our home. **Chapter 16 suggests how light links mind and biological matter; how, in other words, it is the informative medium of radio translation between, on the one side, cell chemistry (including cells, nerve cells and brain) and, on the other, archetypal mind.**

Profound a pair, near-abstraction and an empty space, radiation in the void, make up physics' prime duality. From fluctuations shapelessness takes shape. As Lao Tzu might once have mused, from the formless devolved form. From a slip of light has emerged the edifice of modern physics which, put simply, is a study of energy transformations. Such changes appear as a phenomenal complexity of interplay between forceful and massive characters. Perhaps polar interplay is best expressed by elemental atoms and, which is chemistry's concern, interactions of charge. Might these aspects of the world be dialectically expressed?

Omphalos. Navel. A stone at Delphi marked, for classical Greeks, the centre of their universe. In an Aristotelian cosmos constructed of revolving, concentric, crystal spheres earth was central. Christian theology added the component of man 'created in the image of Creator' and able to commune with Central Source. And a combination of Christian and Aristotelian views, the latter preserved through Islamic scholarship, were taught for the Tripos at Oxford, Cambridge and the Sorbonne - so dawned the medieval renaissance, western science and our contemporary world.

As the history of such science shows, a focus on the centrality of physical energy has tended by degrees to de-centralise, to render eccentric man's egocentric view of himself at the heart of things and God at the heart of himself.

At the same time, conversely, matter has been written centre-stage. Reconstructed cosmos is mass-centred; and it finds insignificant, godless human bodies dwelling, by strange chance, somewhere in a mostly vacant vastness. Isolation. Emptiness. A complete inversion. Traditional philosophy turned inside out.

Progress often steps from one outdated model to the next. Tycho Brahe's observation of a comet shattered Aristotle's crystal spheres and his highly religious assistant, Johannes Kepler, showed that orbits were 'imperfect' circles called ellipses. Galileo championed the apostasy of Copernicus which proclaimed that the earth moves round the sun and not *vice versa*. The sun, not the earth, was at the centre of a solar system whose surroundings Herschel's telescope expanded into galaxies. From geocentric through heliocentric it is now believed the universe is centre-less - unless you take its speculative starting-point, a universal singularity, as central. Vastness of space and time have dwindled earth into to a speck of dust in cosmic desert. Thomas Digges and Giordano Bruno pushed still further. Overthrowing Aristotle's model of a finite universe each proclaimed (should not such proclamation magnify God's excellence and infinitude?) our starry universe was infinite. Iconoclastic Bruno even invoked no less than an infinite plurality of worlds with neither centre nor circumference - and was burnt at the stake for his pains. It was perhaps the greatest scientist of all, Sir Isaac Newton of Trinity College, Cambridge, endowed us with a spectrographic understanding of light by which you can analyse the chemical nature of a substance in your hand or from the furthest galaxies. Harlow Shapley mapped the Milky Way and Edwin Hubble interpreted a universe expanding out towards Bruno's speculation of infinity. Look, therefore, to the cosmos; learn by light. Physical illumination teaches us the universal tale - its present and its past; gravity and light alone connect all things; radiation's colours pack, in every beam, discovery. Such beams, visible as rainbows or invisible as spectral rays, teach the state and, in virtual time, the nature of non-consciousness, the start of all material worlds. Light is knowledge; all its wavelengths are employed to chart immensity. Some physicists even, extrapolating from a universe to 'multiverse', straighten yet another 'orthodoxy' out of Bruno's slant. Endless, start-less, stop-less physicality! And what if, to complete the Galilean revolution, man were evolved from matter and, though basic chemistry would have to be the same, forms of life are found in other constellations and in every galaxy! Life turned inside out. Life breathed into us by light. Life born of stardust, paradise apparently lost and, by strict contrast, real facts gained. Although rampant on earth, man's heavenly ego has been humbled by science. Sidelined. He no longer inhabits a spatial 'omphalos' - even if his size (halfway between the approaches to minuscule, subatomic and large-scale, galactic infinities) is central in the measure of all things. And, if light the founder is of physic's stock, what place for metaphysic? The origin of everything must be in random fluctuations of primordial energy.

Is it strange to think of physics as the study of non-consciousness, non-consciousness in all its energetic forms? If, though, non-conscious objects constitute the lowest level of a conscio-material gradient then what about their measurer's subjective part, his mind - mind whose essence is its consciousness? As the thin rainbow of visible light slices a huge spectrum of radiation halfway through, where does man's 'mind-band' slice the conscio-material spectrum of existence - a spectrum whose Light Source is Essence, a gradient whose Subjective Summit is the goal of mystic mountaineering. How far, from this *top-down* perspective, lives man from the Immaterial Centre of the Universe?

Materialism dos not think this way at all. Although a series of revolutions in science has energised and thus overturned the static, medieval models of creation, at the same stroke a correlated revolution in mind-set has levelled hierarchical cosmologies and rendered everything (now included on the same non-conscious, material level) intrinsically equal. Therefore, instead of showing how to work for heaven it aims to show the way the heavens work - and earth in them. Subjective has, in a big way, been upturned into an objective point of view. Such beacons of physics as James Clerk Maxwell, Max Planck and Albert Einstein each shed fresh light on the radiant heart of things. Accompanying their enlightenment came the series of increasingly 'fluent' pictures of the universe. Hubble showed it was expanding but the pictures really climax when, within this expansion, Einstein made everything of energy and so 'dematerialised' the world. His equivalence of energy with mass (energy = mass x the speed of light squared, that is, $e = mc^2$) made each object of, at root, a locked-up form of...what? Pure energy? Light? What is a particle? Which, radiant levity or contractive gravity, came first? Or did both vectors issue, like opposing twins, at the birth of universe?

At Least Two Sides to a Single Story.

What are the 'hard sciences', whose heavily researched public domain this section involves, except the study of energy in all its physical forms? Millions of books have been written about physics and chemistry. The facts of science stay the same but is devising new technology the only way men in society progress? That inexplicability, the human mind, asks questions. **Is a set of material answers all there is to give?** *In fact, when it comes to pure knowledge, who in their right mind would rely on ever-changing physics for life's full and final answer?*

Nature does not change its way. Old rules endure but new acts are revealed; thus perspective and hypotheses keep varying. Now is the moment to reiterate the obvious. Natural Dialectic is not mathematical. It may not be the perspective you at present use productively but its methodology is valid; polarity is fundamental cosmic fact. And although the flow of its logic results in criticism of some current assumptions it does not restrain the flux of scientific curiosity. The main concern is not to restrict but to extend our paradigm by complementing physical with metaphysical reality. It is not, therefore, the object of the next few chapters to summarise or adjust the scientific *corpus* but, rather, to arrange its physical facts and imaginative fictions in the context of the Dialectic. Such equivalence must necessarily be an exercise in extreme compression that should generate, rather than a black hole, illumination's star. The job is to translate, as crisply as possible, standard concepts and phraseology into the dynamic yet simple, orderly framework of dialectical principles. Why?

You say there is only energy.

I say there are two fundamental components of existence - energy and information. Of these (see *fig.* 1.3) information is the inward, creative

element; and energy the outward element, the medium of information's show.

Every model, not least a scientific picture of the world, depends on its primary assumptions. Such assumptions are critical. They influence the line of thought and determine its conclusions. In other words, if you start wrong you'll end wrong!

Physic and metaphysic? *Metaphysic seems, except for mathematics, irrelevant to study of the operation of material phenomena.* **It is apart from physical events but not, perhaps, from their foundation; unseen, it could be part of operational pattern.** But your first assumptions, *PAM* (The Primary Axiom of Materialism) and its direct consequence *PCM* (The Primary Corollary of Materialism), expressly delete a 'higher power' than energy; your naturalistic axiom allows no immaterial element of information, no universal mind. *This, again, is entirely acceptable when dealing with non-conscious objects and events but not necessarily when dealing either with consciousness, the origin of natural law or the original formulation of basic particles and forces.* Your next presumptions might be that the universe is infinite (perhaps) and in process random (save for patterns of behaviour we call 'laws of physics'). Did it not pop up from an entirely unpredictable 'singularity' whose discharge, due to those first-instant, qualifying laws, has been predictably 'self-ordering' ever since? If you think the 'random' process is 'evolutionary' you could adopt a synonym - the theory of evolution or, in dialectical terms, The Theory of No Intelligence. This is scientism's world view but the same critique applies. *If primary axioms are incorrect it would be expected that a realignment of perspective brought new factors into line of view.*

Belief in the unknown (until revealed) and faith in unseen immateriality (until technology or one's experience observes it) are both habitually denied. Such ideology, that far exceeds the evidence, claims only matter made the world; such religion, crying non-religion, casts its 'rival' out. Thus, let's be clear, consideration of the Dialectic is vetoed; perspective based upon the element of consciousness is not included in *PAM*'s so-called 'scientific paradigm'. It is, fairly or unfairly, thrown out of court. *Yet if any aspect of the world is not non-consciously material then scientism's angle needs to change.*

I have, on metaphysic's hand, proclaimed an immaterial ingredient, information, is part (perhaps the most important part) of cosmic composition. I might claim creation was in concept perfect but physically is in a process of decay. Of course, if I implied some element of purpose overarched all chemicals and their constituents then my form of physic would include the metaphysical. A Theory of Intelligence includes such element embedded somehow in the frame of cosmic law.

'That', clamours scientism, 'is not science; it is an anti-scientific tendency, a threat to reason, war on science!' *Science means knowledge but if that word is redefined to mean physic alone then, of course, inclusion of the immaterial must run beyond its writ. That redefinition is, however, a restriction.* Perhaps, in fact, natural truth includes the metaphysical and thus, shackling reality, a materialist reduces truth. If there is more to life than particles and forces he deludes himself. Possibly, in seeking to appropriate all knowledge for himself, he errs. If a broader Science is in fact the case then how, as a rational adult, can he deny denial and proceed to treat holistic issue sensibly?

Let's rephrase the answer to that protestant objection. Our material factor is objective and oblivious. It has no sense of any kind. My immaterial factor is the field of information and of consciousness. It is subjective, it is sensitive. If, of course, there is no subjective component and conscious animation is a material 'illusion', then the argument that I advance is incorrect. If, moreover, science identifies exclusively with matter it is *inevitable* that the second view will never fulfil scientific criteria. You can neither prove nor disprove metaphysical presence by physical means. You can only infer for or against. 'Against' is chosen since the very baggage modern science was evolved to slough is 'metaphysical'. The introduction of a non-corporeal animating 'substance' is sometimes called vitalism. Vital forces or divine providence serve no scientific purpose. They are completely redundant. Vitalism is in no way a scientific alternative that can be judged against the standard of orthodox science and as such, you will understand, is strictly, negatively policed.

Garbage in, garbage out. Don't you want an all-embracing theory, a key that will unlock the inmost secrets of the universe? Won't you work night and day upon the mission of a *TOE*? The problem for imagineering *TOE*sters is false premise. If you venture in the wrong direction you won't ever reach the right conclusion. In this respect the *TOE*s that emanate from physics all assume the sacrosanct embrace of *PAM*; cosmology assumes that subjectivity is just a curious excrescence of the brain. Metaphysical dimension is completely missed. There is no concentrate of elemental consciousness and no intelligence above a man's. *If, though, an immaterial element exists then such a TOE, simplistically considering the physical alone, would propagate a falsehood, promulgate an unintended mischief and, be it through as high a medium as The Royal Society, promote materially correct but (in as much as it is fundamentally incomplete) erroneous cosmology.* All atheists are so misled.

This section therefore reviews current facts and theories through the paradigm of Natural Dialectic. Take, for example, *PAM*'s birth. Isn't 'standard' big bang cosmology (see Chapter 12) as self-inconsistent as it is consistent? Isn't it brimful with tautology, *ad hoc* speculation and, since its explanation fails in terms of known facts, almost chock-a-block with unknown dark materials known as 'fudge factors'? Is not faith therefore the assurance of things hoped for and conviction of those things unseen? And does such a theory qualify its faithful, the congregation of big bangers, to negatively police the rest and bang them up in silence?

In other words, did everything (including space and time) in truth erupt out of a single 'naked singularity' beyond the reach of any law? Was *PAM*'s miracle then boosted by a second, bigger bang? Did its 'inflation', while blowing up the universe, really iron out all irregularities and, most unusually for an 'explosion', set things bang on track for how they are today? Are 'bangerian' interpretations of red shifts and *CMBR* (cosmic microwave background radiation) spot on? How did galaxies, stars, planets and moons 'condense' and 'fragment' from a ballooning cloud of hydrogen with helium? Did, could an orderless, odourless, silent, invisible and tasteless gas alone ever turn its 'nearly nothingness' by nature into you? An entrancing, seamless epic seizes us but are its poet quite correct? Does the big bang model (which may or may not correspond with reality) properly address let alone explain almost any of the objects of our universe? If not, how advanced are we? Numerous theories make a single point and make it absolutely clearly - we've got some clues but, even when we've snaffled up the physics, will we have the whole idea? Clearly, questions aren't in short supply.

One metaphysical speculation (*PAM*) ought to grace consideration of another (*PAND*). This section takes a look at bangers' mash but it also frames some possibilities proposed by Natural Dialectic. Could subjectivity and objectivity both play a part? Can there exist, does there exist a broader and inclusive Science?

Towards a Theory of Physic and Metaphysic?

Don't laugh! The situation is extremely serious. Could *you* plan a cosmos better? Lucid physicists agree it looks 'as if' designed. **When it comes to astrophysics and cosmology stunning ingenuity seems to have coordinated chemistry and physics' natural laws, not least when it comes to bio-friendliness.** Why is the world so well cooked up, why so 'finely tuned' in its construction as to seem 'a put-up job'? It seems, upon the face of it, designed by an intelligent creator - except the face of a deliberate scheme is one that naturalists viscerally abhor. Intelligent design of nature's laws does not conflict with science but the idea (even were it true) is philosophically unacceptable. Nor do archetypes, which aren't 'anti-scientific' either, show in any scientific photograph. If, however, you can't fix the face or face the fix then are you under an illusion that big bangs, M-theories and complicated mathematics have now sorted out the game? Wrong. Is fluff and finely worded huff and puff or bluff a final explanation of the way that things began? No. Present theory is suspended by a line of shady factors and hangs swinging from a string-along that's called 'the multiverse' (see Chapters 8 and 12). At this rate could materialism actually be apocryphal? It's possible. We want a wholly 'naturalistic', 'accidental' answer with an Author or a Meddler struck right out but was the universe designed with life in mind? Could be. Although the very thought is dangerous. Such a thought would constitute a serious and highly threatening heresy! It smells of universal mind, it reeks of Cosmic Subjectivity and that, within the perfumed garden where 'God Is A Delusion', cannot scientifically be. Yet, given facts so far revealed, is not this garden of the godless a mirage? What kind of light dissolves the shadows of your mind? Could 'God Delusion' be a fiction, an illusion clear illumination might eliminate?

In fact, *subjectivity* is central. It is (see Chapter 0: The Objective Pillar and Scientific Delusions) the way we know we are alive, it is the reality of all observers and participants and, without it, there is left only an opaque, non-conscious dullness known as oblivion. **If such subjectivity is information's element and fundamentally distinct from the objective oblivion of material objects then the question divests itself of ridicule; if mind is *per se* a universal element it needs not be denied but counted in a rational way that handles such a possibility.** It is, equally, sensible to construct a provisional framework and try to work out the consequence of such profound dualism. Habits become engrained and, for dyed-in-the-wool 'scientific methodists', it may be uncomfortable to contemplate any change in a lifetime's pattern of thought. Nevertheless if, despite the presumptions of neuroscience, mind is not physical do you remember the 'stingers' (Chapter 4)? *It is the aim of Natural*

<u>Dialectic to develop a system sufficient to reasonably incorporate subjective, metaphysical as well as objective, physical components of existence. One hopes that science, if not the 'hard-core' materialist, is able to conduct a rational discussion at this level.</u>

Fair enough! The Dialectic's metaphysic rests with information; and for physics information is conserved. What kind of conservation's that but so-called 'natural law', the behaviour of energy of energy projected from outside the edge of space - where 'outside' actually means emergence from an inner metaphysic and not some outer spatial or temporal place. What's the projector; what's the nature of projection's source? Natural Dialectic's holographic edge is everywhere; it's 'super-posed' on physics, omnipresent but invisible because it's metaphysical. **It is the place where physic and its metaphysic meet.** Space time and things are *within* universal mind. **Mind's archetypes project our world; archetypal memories, potential matter, are the essence of our physic; they inform, unchangingly, material being. Immaterial information holds the world, physical and biological, together. It is by archetype conserved.**

Can this be so? Snowflakes whirl on random, unseen gusts of air. Yet, in drifts or icicles, an order shows; flying fragments are 'constrained' in definite complexities. Quantum factors whirl in thermal motions. The laws of physics are statistical and, as atomic bonds and aggregates increase, so order shows. Masses of cooperative particles that constitute phenomena assume predictable configurations and behave in certain ways. *Thus statistical order arises from quantum disorder.* It is due to atomic structure (as in the inevitable quantum configurations of atoms), bonding and bulk material constraints. It might well be argued, therefore, that materialism's emphasis on randomness is quite misplaced. This is because, beside their random motion, *character* of particles endures - such is innate, intrinsic or potential order. Different particles yield definite, though different kinds of interactive order. **Order is intrinsic in phenomena.** *The question then becomes (for electron, proton, photon and so on) origin of this enduring character that leads, in company, to mathematically predictable behaviours. Did chance or otherwise give particles and their great universe specific character?*

No doubt that science takes a naturalistic slant in all its explanations; some philosophers extrapolate but, in the case of order out of order, we shall see materialism's mind-set does not cover consciousness or the origins of laws of nature and of life. *In these matters immaterial information reigns supreme.* It is, therefore, important from the start to clearly realise and emphasise that cosmological devices we shall meet (such as 'inflation', ten or more dimensions, multiverses and so on) are pure speculation. **They are sophisticates' inventions to explain in naturalistic terms 'fine-tuning' at the cosmic start and in current circumstance; to parry the immense unlikelihood that the whole great bundle of 'coincidences' were exacted just by chance; and thus to obviate or best negate the *top-down* possibility of archetype.** 'I'm only interested in my kind of truth.'

Yet it's claimed, materialistically, sheer nothing generated everything! At least as reasonably Natural Dialectic claims that metaphysic precedes physic and, instead of the 'big bang' - a phrase first coined in sarcastic ridicule by Fred

Hoyle - proposes '**transcendent projection**' (see Glossary). Miraculous (transcending or, at least, preceding laws of physics), instantaneous and dynamic - this projection's parameters are most cooperatively and orderly fine-tuned. To grow, develop and to yield a stage that's fit for life you need, for example, an expansion whose mass and space energy densities are one part in 10^{60} and one part in 10^{120} respectively. At this rate you might count cosmos as 'prefabricated'. Now take a bubble for a simple metaphor. A bubble blown needs first be framed within a wand. Its shape depends upon this wand's round frame; and when it's blown the frame remains. So the character of cosmic bubble emanates from its own wand. This wand is precedent and metaphysical; and when projection's over and the bubble flown why shouldn't frame remain the same - a composite called archetype or, for separate elemental factors, each its sub-archetype. Thus, you might reasonably surmise, cosmos was projected from informative transcendence. Nothing physical prescribed the way things have to work, that is, their archetypal rules.

Archetype as origin of natural law is not a new idea. It is (recall *figs*. 2.6, 2.15, 4.1, 5.2, 5.4 and so on) sited on the interface at which subjective and objective meet. This *PSI (figs*. 0.8, 0.11, 0.12 etc.) is called potential matter. The metaphysical precursor to all *physicalia* is known, in either individual or universal mind, as archetypal memory. Through this unconscious gateway are expressed the agents (quantum behaviours) and the outcomes (gross, sensible materials) of a starry circumstance that we call 'home'. Order therefore emanates from inside; guidance of the cosmos issues from above. **The challenge is to integrate an archetypal form of information with physics as it stands today.**

There is a second way of explaining the format and intention of this section. Scientists theorise and experiment with simple ideas in sophisticated ways. A primary goal is to provide simple, elegant answers to problems. This is because there is a feeling that, underneath the colourful complexity of our world, there lies simplicity. For Natural Dialectic this simplicity derives from principle, the complexity of phenomena derives from fundamental law. What is the character of this law? What are the principles? Are they hierarchically arranged? If so, what is the Principle of principles, the Absolute of absolutes? From what, *top-down*, does everything derive? In this respect, while physical light is not divine you might wonder whether its illumination had connection with the other patterns that compose our changeful habitat. Flying at the edge of mass, space and material being could its radiance bear information from a source beyond the 'thickening' into things? Pure light is charged with colour. Is it also charged with law? In its bosom of neutrality could there be 'cached' polarity? Exactly how is light tied up with magnetism and electricity? How is it linked with charge and heat and therefore the cascades of action leading to a lawful cosmos, that is, a universal body? **Where science has now thoroughly (*raj*) energised our view of things perhaps it is now time to go a step further, to move a level higher and to (*sat*) potentialise**. This means here to gain insight into potential, that is, into the source from which material nature is informed and, from a higher state, projected.

As *fig*. 7.1 describes, Book 1 will explore four existential qualities in terms of physical projection, that is, in terms of space, time and energy. As such it describes an arch that binds together *(raj)* kinetic, quantum and *(tam)* inertial, bulk-classical pillars of natural science. Through such an arch the world appears. If there was a physical beginning what preceded it? If archetypal law precedes its physical expression both in time and hierarchical order, then why should an apparent 'eruption' of things out of 'nothing' not involve an element that's metaphysical? The start would be transcendently projected.

Transcendent? Metaphysical? No doubt it is difficult to describe and explain physical phenomena definitively. This is because there is no consensus about the nature of their critical foundations - infinity, space, time, the origin of natural law, the origin and destiny of universe etc. Speculations abound, fashions swing from one 'cut' to another and a clever answer always counteracts the one before. The invention of improved equipment and the discovery of new facts propel the search for scientific absolution (in the form of omni-science) into continual transformation. *Is it, however, possible to rise above the argument, to transcend the fray?*

Surely you cannot transcend the physical by simple invocation of the metaphysical? Is metaphysic not delusion? Is its staple, mind, in fact illusory? Isn't mind, the source of all delusion and of truth, as physical as brain? Or is it, linked with body, metaphysical in essence? Perhaps so but surely, seriously, there is absolutely no suggestion mind has anything to do with the production or control of matter? Isn't it absurd or, if happenstance prevails, unnecessary to invoke ground other than material? Material nature's quite enough. Why depress materialism by the imposition of a 'super-natural' over 'natural' presence? Why spin naturalism's wheel the wrong way round? *Such revolution is, you hear the cry, reverse, perverse and backward. It marks modern philosophical anathema. And yet, for all the giddy spin, are consciousness, information and purpose definitely material? Is life body by itself?*

	tam/ raj	*Sat*
	existence	*Essence*
	relativity	*Absolute*
↓	*tam*	*raj* ↑
	objective	*subjective*
	non-conscious/ matter	*mind*
	physical	*metaphysical*

Most people want the whole truth not just dogma and therefore theories unifying everything are 'holy grails'. What counterpoints things physical but lack of them? Is there, within this absence, something more than space of void? Could there be, materially unsuspected, an immaterial element; and could this element - informing, unifying and a source of law - make overriding, overwhelming sense? What could be the nature, if there is one, of a universal mind? And if there is then how (see Chapter 4: Stingers) might metaphysical conjoin with what is obviously entirely non-conscious matter? If Natural Dialectic sees aright at least it, mind to body, so conjoins in you!

If consciousness is 'super-natural' or 'metaphysical' let's take another fundamental look at things. Do you remember (Chapter 1 and *fig.* 1.1) polarity? *The Dialectic is a structure built around the unity, at source, of separate poles. It is a 'unified theory' par excellence.*

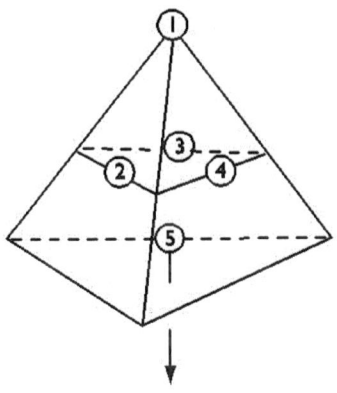

The Diamond Capstone

This Capstone illustrates Essential Qualities from which existence issues.

Substance	①	Consciousness	(Chapters 13 and 14)
Essential qualities/ facets	②	Infinity	(Chapter 7)
	③	Unity	(Chapter 9)
	④	Nothing	(Chapter 10)
Precondition	⑤	Potential	(Chapters 5 and 8)

The Apex of Mount Universe is likened to a Tetrahedral Diamond with three clear facets and, at the base of its transparency, a fourth to interface existence. The trinity involves, at the highest level, aspects of its Substance; it incorporates the triune qualities of a Pure Concentrate of Consciousness. These (*Sat*) qualities comprise Potential from which all creation is expressed; from this intrinsic precondition all the order of existence hierarchically falls.

fig. 7.1 (see also figs. 1.2 and 7.2)

When Absolute Infinity is polarised then its antithesis shows up as 'finiteness' that is, as objects and events that are 'relative infinities'. Where, after all, among the shapes and changes can you draw a line? How can you isolate components from their whole or cause from context and effect? Can you say where influence begins and ends - not only for phenomena that you construe distinct but also cosmos as a whole? No doubt that science handles natural operations quite successfully but what about the origin of nature? How, in the last analysis, do finite differences of objects and events - both mental and material - first and as a consequence occur? How, in essence, is Infinity made polar; how in creation has The (N)One become all things?

Prior to action or sensation there is (*fig.* 2.3) being. Supreme Being, also known as Essence, is at the Centre of creation, at the Apex of Mount Universe (*figs.* 1.2, 2.3, 2.10, 2.11, 5.3 and so on). Its transparency is likened to a Diamond. The 'Adamantine' Substance of this Diamond is Pure Consciousness.

Such Consciousness 'opposes' the whole range of sensitivities that drop, at base, to an extreme antithesis, their special case of none - oblivion. Oblivion is final darkness, subjectivity 'rolled flat'; it is the last, exhausted output furthest from the Concentrate of Consciousness, Essential Singularity; and is a state of mindlessness that, being the condition of all physical phenomena, is what chemistry and physics interrogate. What else can thoughtless answers be but absolutely reflex?

If there exist two basic elements - consciousness and energy - then by definition (energy means 'there is work in it') the latter is a mover. What before a cosmic start could have preceded motion except the Pure Concentrate of Consciousness? If this Source is One its only proof is the experience of its Self. Unified Experience. Knowledge of the existential factors, mind and matter, won't suffice. Scientific truth and thought itself will have to be transcended. The only proof of Essence is, therefore, a transcendental one and, being One, loss of lesser selves in the communion. How many different individual identities would you expect to find in Oneness? None? What else have mystics ever said?

Look at it the other way. In dialectical terms the primary expression of Essence is (*raj*) psychological and called First Cause (see *figs*. 0.6, 1.3 or 2.5). If First Cause is psychological-preceding-physical then it is metaphysical. If the metaphysical counterpoint of physical non-conscious energy is conscious awareness; if Pure Awareness counterpoints duality of mind and matter in a polar universe; and if active information is potential that initiates a pattern of behaviour then Conscious Singularity precedes polar expressions of itself. Essence comes before existence.

In other words, if you correctly propose two fundamental components of matter and consciousness, you are obliged to concede that the Start was conscious. An Idea was devolved through principles to practice; it was realized at last in matter.

From Conscious Source, therefore, issue the major hierarchical divisions of Mount Universe - mind and matter. In this instance one might develop towards a Unified Theory of Physic and Metaphysic whose explanation starts with (*Sat*) Metaphysical Trinity. This Essential Tri-unity comprises three facets of a Single, Supreme Being. The Substance of this Being is Consciousness or Life. In other words, Life is the dialectical First Principle; its triune qualities, outrageously equated (see Chapter 1: Nothing and, as indexed, The Essential Equation), are the same. A Supreme Paradox comprises aspects of The Diamond - of infinity (the wild card), unity (1) and nothing (0).

The lowest hierarchical division of Mount Universe, passively reflecting order from above, is matter. Matter alone forms the philosophical basis of 'natural science' as the term is understood today. *Therefore of more interest to science is the (tam) reflection of a second, lower Cause. What is the nature of an unconscious yet informative principle or set of archetypal laws from which alone the universe of chemistry and physics is devolved?*

Can Mathematics Help Us?

Flux and fixity. The world's in flux; things *are not* so much as *are becoming*. In the face of such inconstancy explanation, scientific knowledge in the form of theory, seeks intellectual fixity. It seeks 'eternity' in changeless abstracts such as archetypes or, with its power of manipulating abstract concepts, mathematics. *Math is physics' sacred truth; it is the gate to science; more, it is the rational basis of our scientific quest.* The fixed assurance of its symbols yields predictability. **How strange, therefore, that a symbolic construct of the human mind - materially non-existent and meaningless without it - reflects so well the energetic patterns, balance and equation of an ordered universe**; and even, for cosmology, is able to deduce - from irreducible yet often indemonstrable equations - plausible scenarios of unknown matters ranging from, say, big bang to string theories! The human mind creates, in principle, this universe and can reflect upon a whole host more! You might dare, indeed, to wonder if numeric rules, embodied by specific formulations, mean our world is really a rational creation. Did a pure mathematician plan the logic of real nature's numbers game? Pythagoras, at least, believed that 'All is Numbers'. And for many, including Roger Penrose, the Platonic world of mathematical forms is real. He writes, 'There is a very remarkable depth, subtlety and mathematical fruitfulness in the concepts that lie latent within physical processes.' They were obviously there before discovery by humans; they will endure as long as cosmos but is this enough? Numbers are eternal abstracts but it's energy and not immaterial mathematics breathes the real fire; flames and motions not equations drive the cosmos forward. Equations simply 'ghost' real power; impotent, descriptive, by themselves they neither make nor do a thing. *Like the 'laws of nature' they describe but don't explain a thing.* Indeed, they have no being but in mind. They inform the various quantities and qualities of mass and motion with a web of metaphysical relationships, transcendent logic and eternal principles. Men exploit this 'inside information' to control material circumstance. Can sums alone fill up your bank account? Naturalists might dream equations could create a universe - but theirs is pure illusion.

Therefore, drill down. Whence, at root, are things projected? Could structure, underneath appearances, still be mathematical? *Could mathematics thereby forge an archetypal link beyond the bounds of space and time?* Such description would, by definition, like any law of logic have to be abstract. Such abstraction more or less defines an observation; its compression calculates material possibility; and it is timeless so that you are able to predict. In this case a mathematician acts as sensor in recording passive information. Indeed, a physicist or engineer may engage timeless, materially transcendent parameters to describe a system he never saw before but, according to certain assumptions, devises. In this case he is the creator of the system. He actively creates an infrastructure (say, a petrol engine) within which energy may operate. You can even devise 'unnatural' mathematical systems, ones that do not correspond with nature's known ways. Whether or not a mathematical vehicle (say, string theory) accords with nature's actual operations only scientific tests can show.

In any event all symbolic language of description or creation involves mind. Such codes involve knowledge of which, obviously, you must be conscious and of which, by definition, involuntary, non-conscious objects are not. The question therefore arises, if mind is brain and brain evolved, why chance evolution granted precise capability to deal with understanding symbols and, especially, numeric logic that aligns with universal plan. How might it reflect and comprehend the natural, abstract template that frames, mathematically, the works of chemistry and physics - unless our minds are microcosmic, holographic fragments? Are they pixellated parts of universal mind, divisions with capacity to resonate with archetypes and thus, upon a learning odyssey, gain access to the inner trove of law?

Of course, the parameters of change might on the contrary, have been established as a result of purely material origin - say, a big bang. Such a view would satisfy a naturalistic paradigm. But, from a Naturalistic, *top-down* perspective, are the abstractions that compose templates for the shapes of energy *actual* abstractions? Physical abstractions, yes, but they are metaphysical realities, shapes of information in the universal mind. They are potential matter that involves form, process and informative capacity. They would constitute the causes, the potential or sets of possibilities from which physical phenomena orderly derive. And because they reside in mind only mind can access their 'profundity', their mental existence. Only mind can represent their behaviour using symbols such as words, numbers and equations. To express lower subdivisions of the cosmic ziggurat (*fig.* 2.13) we write:

tam/ raj	*Sat*
'pixellated' expression	*Potential*
physical hierarchy	*Top Level*
classical/ quantum levels	*Archetype*
many local instances	*Archetypal Singularity*
change	*Constancy*
phenomena	*Noumenon*
cosmic development	*Universal Concept*
shapes/ transformations	*Natura Mathematica*
subsequent properties	*Original Order*

↓

tam	*raj* ↑
shape	*transformation*
external effect	*internal cause*
object	*energy/ force*
local distinction	*ubiquitous presence*
macroscopic outcome	*microscopic agency*
bulk level	*quantum level*
condensed matter	*fundamental components*
detailed structure	*simple factor*

Such metaphysical actualities are called in Natural Dialectic archetypes. Fixed forms of mind are memories; they are memories in universal mind. Therefore nowhere except in mind, mathematically or intuitively, may archetypes be understood. Of course, as film and record archives show, fixed forms can be dynamic. What more rational than a program's subroutine or algorithms reaching targets through a finite course of steps? Such mental objects, fixed or otherwise, are immaterial but may be inferred from physicality in its aspect of self-consistency, order or design; our elders used to call such order 'natural harmony' or 'universal law'. Perhaps these immaterial objects form the link between the corners of what mathematician Roger Penrose has referred to as a mysterious triangle of physics, maths and mind. Wranglers tend to fall into one of two camps - those who believe maths is 'all in the head' and so-called 'Platonists' who believe the mathematical constructions that summarise our cosmos are not a device of human calculation but rather 'out there' buried in eternity until we uncover them. The former camp must embrace an evolutionary view (what, after all, does 'out there' or 'mental object' mean to a materialist?) but the latter could align with Natural Dialectic. For the Dialectic an archetype such as informs the matter is not, however, 'out there' but 'in here'; it constitutes an 'inward' aspect of the universe, a part of its element of information, mind.

A memory is created in a moment of sensation or of creativity. It would, therefore, be logical to ascribe the formation of archetypes, the 'habits' or 'instinct' of universal mind, to a conscious source. Active generates passive information. What are the qualities of consciousness or conscious information? They are knowledge and intelligence. Mankind has always sought and gathered round the flame of Natural Intelligence.

Dialectically, you'd therefore say equations drawn from accurate observation boil down to numerical descriptions of metaphysical archetypes; and these in turn indicate a conscious source as well as (in the form of a physicist, chemist or biologist) a conscious 'sink'. The message is decoded by receivers. This is the renaissance view of science - a decryption of the Maker's Message and thereby comprehension of His Reasons. Could mathematics really point, as Cantor thought, to God?

↓	tam	raj	↑
	irregularity	*regularity*	
	variation	*theme*	
	flexibility	*rule*	
	individual local case	*universal/ general case*	
	practice	*principle*	
	rough actuality	*underlying pattern*	
	change in context	*changeless abstraction*	
	illogical imperfection	*logical perfection*	
	unpredictable	*predictable*	
	random spontaneity	*mechanistic/ clockwork*	
	apparent chaos	*determinism*	

Shapes involve, as Pythagoras and Euclid were aware, geometry. What kind of geometry best describes our physical and biological world? Is it pure lines and circles or the five Platonic 'perfect shapes (pyramidal tetrahedron, cube, hexagon, symmetrical 12- and 20-sided forms of which the world is made? For Newton 'flat' Euclidean geometry sufficed but Einstein needed 'curved' Riemannian space. Does either well describe irregular complexity of wild and changing systems like the weather or a living form? Beyond regularity and determinism we need irregularity and flexibility - rules with outcomes that you can't predict. We need a theory, like the Chaos Theory of Henri Poincaré and Edward Lorenz, wherein simple rules generate, by way of cyclical feedback wherein output becomes the next input, complex, apparent*ly random dynamical behaviours. Google, for example, the metronomic 'BZ reaction'. Discover os*cillatory chemistry and exuberant psychodelic pattern-making. Could physical disequilibria build cells and life for us? Then there's the 'deterministic chaos' of René Thom's Catastrophe Theory and the 'fractal' or 'holographic' geometry of Benoit Mandelbrot. Mandelbrot's geometry explains the large-scale, repetitious and irregular generation of both animate and inanimate shapes. In fact, geology, physics, astronomy, chemistry and biology can all be viewed and described in terms of, where self nests within self at all levels, 'self-similarity'. The 'Mandelbrot set' can, in theory, be infinitely magnified or diminished. The real world seems to end at an indivisible duality of electrons and quarks but a fractal is a geometrical structure that has detail at any scale, any size, any level of perspective. While this is the mathematics of wilderness, irregularity and so-called chaos there resides, at heart, a simple formula.

Complexity and chaos fallen out of an initial simplicity! Maths is a code, a rational order and a universal language whose virtual abstraction is naturally expressed by actual alphabets of particles, four forces and genetic bases; its material nothingness writes words with atoms, sentences of molecules and, from the basis of a ruling theme, chaotic variation that competing forces aggregate. Simple repetition of the Mandelbrot equation seems to be the way that nature grows into the complex world we sense. Internal order is expressed externally. From central seed is developed, in an orderly way, external, peripheral form. In dialectical terms Mandelbrot derives endless, localised complexity from eternally permanent, basic and initial simplicity. He expresses perfectly the tension between regularity and irregularity. He marries archetypal and actual worlds, each of them an integral part of phenomenal reality. Theme and variation, principle and practice, archetype and physical expression are at the heart of Natural Dialectic's explanation of apparent chaos paradoxically issued from real, immaterial order at its heart. This order can be, as far as physics is concerned, mathematically expressed. Abstraction seems to rule the patterns that will show up the real, rough game.

You can guess where this is leading! Excited spontaneity of pattern-making, self-organising principles that might apply to life! While entropy degrades complexity here, it seems, complexity emerges in reverse. If the agendum is to automate then, materialistically, the bottom-line is mindlessness; if all complexity evolves from some internal feedback round Mandelbrot's or like equation then nature doesn't need intelligence. It self-organises. You might dub a virtual icon, the 'fat Buddha' of Mandelbrot's m-set, 'the intangible fingerprint of God' but, really, strike out any thought of creativity. Design' needs no designer. Atheism's vindicated mathematically and scientism's victory over immateriality complete! Is not chaos plus mechanics scientific atheism's dream response?

But m-sets and equations, however much reiterated automatically, can't make a thing. They can describe and virtually explain the energetic patterns nature churns with; they describe self-organising operations of a wound-up world as tension is released. But they do not account for innovation, creativity and origin of specified, deliberate complexity as found in every purposive machine.

The universe may well, in its mechanistic, orderly procession, be construed as a machine; and also, in chaotic aspect of 'post-mechanistic' unpredictability, as a non-machine - a non-machine disorderly in pattern-making and 'designs'. Cosmos is machine and at the same time not! Iterative machines can't innovate but chaos can. So chaos innovates fresh forms in nature; it configures 'as if' by design. Without intelligent designer! Design without designer. The paradox has thereby spawned a strong illusion, forcefully expressed, that mathematical equations not only describe but can initiate patterns as complex as biological morphogenesis. Is such 'emergent' character not clearly seen in the development and growth of biological constructions - leaves, trees, hives, feathers, blood transport and a host of other repetitious, branching systems? Therefore chaotic evolution must have generated biological machinery and such machinery, eventually, without intelligence evolved intelligence! Science, most intelligent of man's endeavours, is thence able to divine the immaterial principles by which material operations were created and now run! Who needs intelligence when natural law will scientifically do? It's so much more inspiring to believe that sets of formulae can be so clever as to bring life and the universe to pass!

Crystals, chemical reactions and natural elements alone produce complexity and, with it, 'post-mechanistic' unpredictability. And of course, machines work in accord with natural laws - but this does not mean that, like clouds, sand-dunes, mountains or whatever other undirected structure that you care to name, they are created spontaneously by them. Natural, automatic self-organisation is not comparable to the specified, encoded complexity of either purposive machinery or biological organisms. Such faith in evolution will be dealt with extensively in Chapters 19 to 26. It includes belief that a single cell (or even protein) could ever self-organise or that, without intrinsic coding, morphogenesis - chaotic or otherwise - has any relevance to specified biological form. It's as daft as believing, given letters of the alphabet and enough time, any printed book could organise itself; or that computers and their programs (even programs simulating 'virtual evolution') came or ever could occur by chance. When it comes to evo-devo and development the nonsense of this claim is multiplied. Yet maths of chaos has of late been fashionably welded onto Darwinism to create an evolutionary alloy, a subtle two-pronged lunge on metaphysic, archetype and any immaterial element of information.

What exactly is chaotically 'self-organised'? Is it phenotypic pattern? No, this pattern's somehow issued by the genes - although they no more bear resemblance to a body-form than does the mental form of your experience to nervous patterns in your brain. Is it, therefore, sequence of those genes that is self-organised? So that chaos boils down to mutation. How are mutations subject to chaotic feedback formulation? They are not. You must, therefore, conclude that natural selection, as the agency of feedback, can somehow be mathematically inserted into the equations representing chaos. How? Neo-Darwinism's explanation (the *PCM*) is insufficient to explain the innovative origin of efficient, functional and deliberate scales of order even single cells display. How can chaos help explain constructions of and irreducible complexity and the codes that specify them? What chaotic steps bridge, for example, the absence through to the complete, working presence of Hox genes, their related subroutines and all the complex, integrated requirements of development? How did chaos even make a seed or mathematics any way create a cell?

So is creation just a numbers game? Mathematics helps build models that are plausible but does reality agree? As you read on you'll find in fact there's very far to go. Whence, for example, hailed behaviour such Mandelbrot's equation indicates? How does its cyclic mathematics inform the real world; how might its abstraction shape the paths of energy; how could abstract bear on concrete or metaphysic influence the state of matter? How, repeatedly and pointedly, does *DNA* (whose information generates protein) trigger iterative geometries; or, conversely, how does iterative geometry interact with 'holographic' *DNA* to create symmetrical, coherent and yet differentiated bodies? We need to understand the connection, if it exists, between the m-set and the way nature works; and how fractal templates thereby influence our make-up. The answers to a host of questions are unknown; they are explored, as they pertain to archetypal morphogene (see Glossary), more deeply in Chapters 16 and 17. Perhaps fractal geometry offers a revolutionary new way to understand archetypes and formulate the potential, metaphysical matter whence our world is generated in an orderly fashion. Do you remember (Chapter 1) that wrangler Michael Heller believed the laws of mathematics might be one way to describe the thoughts of God? Perhaps fractal 'transformers' are a way in which we shall more clearly perceive how mind makes matter, how 'idea' shapes energy and how, thereby, we might read those ordinations. **At least, concerning animated forms, it's argued throughout chapters 19 to 25 that the uncoded morphogenesis of natural 'systems' (such as mountains, oceans or the weather) bears as much relationship to the conceptual, coded morphogenesis of life-systems as shaken chemicals to a cell.** *There is no evidence but there is keen faith, the promissory hope of any worshipper that his ideal (a world without creator or a world with one) is true.*

Any More Questions?

There are lots. At the end of the 19th century physicists thought they knew, except for a few details, the lot. Planck and Einstein soon ripped that view up. Will the 21st redefine what textbooks in the 20th thought might perhaps answer everything?

Where is physics coming from? Whence does chemistry derive? From a big bang/ projected start, from a material eternity or from a background that is metaphysical? Check *figs*. 1.2, 7.1 and, this time forward, 7.2. From a dialectical point of view the physical pole of creation diametrically opposes the *(Sat)* Quintessential One. As such it lacks any trace of fundamental Unity, called Consciousness. But you might still expect it to reflect the triplex aspect of *(Sat)* Essential Tri-unity. Three characteristics of First Principle, three inverted reflectors would constitute the basis of material complexity; their impersonal transparency would reflect 'heaven on earth' and, in this case, comprise the primary phenomena that substantiate all physics. What might they be?

To answer this question Science and the Soul describes, over the next five chapters, an order of descent from principle to physical practice, that is, describes a dialectical pattern in the way physical law is expressed. The section therefore starts with a consideration of the triune nature of infinity, unity and nothing. It then, through the agency of zero (0), links this tri-unity (or trinity) with its 'lower reflections' in the zone of objective expression called physical nature. *What, therefore, is the logical nature of the Dialectic's 'transparent and intangible trinity' as it shows physically?* <u>It is composed of a couple of pure, passive infrastructures called</u> (Chapter 10) <u>spatial nothingness, a vacuum out of</u>

which and within which things occur; (Chapter 11) *time, a function of energy's motion i.e. change; and* (Chapter 12) *energy itself, whose polarity includes both radiant, electromagnetic light and contractive, mass-centred forces.*

In short, Chapters 7-12 reflect the Diamond Capstone. They investigate the nature of physical Infinity, Unity, Nothingness, Time and Everything. Is everything non-conscious particles; is there nothing immaterial abroad whose influence impacts, at root, on physicality?

For all his efforts one thing is certain - man is still a student rather than, thank heavens, master of the universe. A detective may suspect what is going on but still be surprised at the twists and turns new clues or perspectives force upon his case. Although material phenomena stay the same, events may force understanding and explanations of them to change. With hindsight a previous view is judged relatively superficial, incomplete and, to this extent, incorrect. Not omniscient, science is therefore still a journey that proliferates with approaches and ideas - some of which may be right, others partially right and others simply wrong. Who knows? Just when you thought you'd arrived or, as Paul Dirac put it more than seventy years ago, 'the underlying physical laws necessary for the math of a large part of physics and the whole of chemistry are completely known', then new discoveries crop up. What a chase! *We can safely, in accordance with historical pattern, aver that science-of-the-day always assumes it is on the verge of knowing everything but is later proved far from it.* The very reason science is in flux is because of its 'unknowns'. **In such a sea of relative uncertainty, therefore, this section can only take various material perspectives, review as yet unanswered questions and present science within the context of Natural Dialectic.** This Dialectic is presented as a framework, one broader than physics alone, within which the patterns of the flux might seem to take a simple form.

Broader framework? Can you pin a sign on cosmos warning 'Keep out, physics is all mathematical and those who don't know math can't understand enough'? But the laws of physics don't inform biology; biology's informed by a symbolic code translated into cellular materiality. *Still less do laws of physics cover forces metaphysical.* Does creation not include sensation, will-power, love, desire and creativity? Aren't these central to experience and un-measurable meaning? Thought, virtues, morals, aspirations and emotions can't be pinned to numbers. Thus lifeless physics, missing out the calculus of mind, will never illustrate life as a whole; but incompleteness even in its own domain fires questions too. Thus Natural Dialectic also works 'outside the box' and thereby redefines it.

No doubt, full truth will out; but for now the fact is that, with 'dark factors' pencilled in, even physic's baby's not yet born. Therefore in the course of the next six chapters - Infinity (7), Rules (8), Unity (9) along with Void, Time and Everything - we'll discuss the strengths and weaknesses of modern science. What exactly do we know for sure? Physics is a moving feast. It changes all the time. One day, in its completeness, it would become immobile by including every physical event - but, as noted, what about the rest? What about the restless, causal mind itself? *Material science never can, except where its philosophy's concerned, be more than relatively right.*

What Answers?

Today's physics is an intellectual *tour de force*. Ranging from atoms out to galaxies it combines the Standard Model of Elementary Particle Physics with Concordance Cosmology. First came Newton's optics and mechanics; Michael Faraday discovered cathode rays, Clerk Maxwell developed a theory of light (electromagnetism), Planck the quantum hypothesis and, on top of these, came Einstein's Special and General Relativities. Today physics is effectively composed of two partial theories - general relativity and quantum mechanics. The former grandly ties, for the whole universe, abstract to actuality. It equates the curvature of space with mass; it describes the way space, time, gravity and matter interact. Material presence deforms space-time's geometry and the latter influences objects by changing their paths. A set of principles (such as every particle's continuously fluctuating fields and the instability of space) defines quantum theory. This, while relativity deals in certainty, deals in probability. *The pair are mutually inconsistent descriptions of condensed and microscopic matter.* How might they be combined? Big bang emerges from a theory that doesn't work for singularities. How, moreover, can the macro-world be written up in terms of its foundation, elementary particles, without describing what is designated 'quantum gravity'? Nature isn't schizophrenic - so how can elementary particles and forces be explained as the appearance of a single fundamental entity? What set the precise and integrated values of about twenty constants (e.g. G, the gravitational constant) whose numbers bolt the framework of a cosmos that supports a thinking man? And where has all the anti-matter gone? What, if they invisibly exist, are dark energy, dark matter, Higgs fields, strange particles and super-symmetry? *There's faith these factors might exist but if they don't how out of kilter are our modern models with reality?*

By the late 1950's physicists had revealed four forces that 'ruled' a material world seemingly composed of photons, electrons, protons, neutrons and some more exotic particles. The partial unification of these elements came with *QED* (quantum electrodynamics), *QCD* (quantum chromo-dynamics) and the electroweak theory. This Standard Model remains, however, incomplete. How, for example, did elementary particles first bind to form complex structures? Nor, since it does not incorporate gravity, is it incompatible with general relativity. It is to overcome such grave disjunction in our perception of the nature of our world's great and small that String Theory strives.

Subsumed beneath the overarching classical and quantum pair you run upon atomic theory, theories of electrical charge, magnetism, 'big bang', dark quantities and so on. As a scientific detective develops ever more sophisticated equipment and inspection regimes he keeps discovering new facts but problems still loom large. For example, physicists are working hard to yoke the restive, somewhat incompatible ideas driving classical and quantum physics by developing a theory of abovementioned 'quantum gravity'. Can (Chapter 9) String Theory do the trick? Anomalies and contradictions exact modifications; new models, bolt-ons and refinements are devised. Hypothetical redefinitions, whose aim is to better self-consistently describe and explain the universe, appear with regularity and, until recently, physics has seen some decisive advance every decade or so for the last three hundred years. Successfully-tested candidates are incorporated into the body of physics and, over time, the kaleidoscope of these bit-by-bit and sometimes inconsistent additions grows unstable. The whole framework wobbles to the point that, like a house of cards, part or all of it comes tumbling down; 'an earthquake' shakes out some new paradigm. The theories of Einstein and Planck were examples of such seismic shift. The nature of the next extension is as yet unpublished.

Physics is certainly on principle exclusively physical. There are, however, no gaps in nature so that, if the theory is incomplete, there must exist extensions that will tame anomalies and fill in apparent blanks. **It is not that known facts will change but more may be added and different ways of looking at them may be needed**. Perspectives and interpretations will, as in any in-depth study, shift. *In this sense while Dialectical Theory represents an advance (unwelcome in some quarters because of including metaphysics) it will not compete but complement; it may suggest new lines of thought or fresh perspectives; and it may, at certain points, put strict adherence to standard benchmark theories out of joint. Its new template will not wholly fit. This is the cost of any necessary 'shift in paradigm'.*

Answers lead to questions. For example, the world of atoms, elementary particles and so on is described by quantum physics; it has been reduced to a foundation of six quarks, six leptons, their antiparticles, carrier particles for three of the four forces (the status of a gravitational force carrier, the paradoxically massless graviton, remains hypothetical) called bosons and the Higgs boson. Of these our stable world involves just two kinds of quark (of which protons and neutrons are composed), a charged lepton (the electron), a charge-less one (the neutrino) and perhaps all the bosons. These twelve quarks and leptons fall into three 'generations' but two of them rapidly decay into the stable 'first generation' proton quarks, electron and neutrino. Why, therefore, is there any need for two unstable 'generations' to exist? Again, the boson called a photon has no rest mass nor, it is believed, did any energy or particle just after a Primordial Bang. *So, as they swarmed massless at the speed of light, how did particles (except light's photon) all gain mass, slow down and let us have our heavy world*? Indeed, can we derive the massive particles from forces or their immaterial fields? Indeed, are fundamental particles the limit - perhaps you could split photons or electrons just like protons into three or delve to practically immaterial strings as the foundation of our mountains, seas and stars? Why, moreover, can't the Standard Model predefine what various masses there should be; or explain why anti-matter is so brief and rare? Can we explain why universal constants have the values that they do? Or unify atomic forces with the bulk influence of gravity? To answer these and many other questions rockets, telescopes and microscopes have been constructed and now probe all corners of the cosmos, great and small; and try, with particle accelerators, to recreate high energies prevailing in the tiny universe just after Bang banged.

tam/ raj	*Sat*
relativity	*Absolution*
duality	*Unity*
polarity	*Neutrality*
asymmetries	*Symmetry*
motion/ change	*Equilibrium/ Balance*
various expressions	*Pure Energy*

↓	tam	raj	↑
	gravity	levity	
	contraction	radiance	
	Higgs 'drag' field (?)	light	
	mass	force	
	materialisation	dissolution	
	inertia	excitement	
	inertial equilibrium	equilibrator	
	quark/ proton	electron	

Is, as has been suggested, creation a crystallisation of principle? Or are natural laws and principles simply 'add-ons', abstract descriptions of the ways crystals (or anything else) are made and behave? Are natural rules just functions of that incidental Bang? Was what informs the world begun by chance so that, ultimately, description lacks explanatory logic?

For example, is dialectical kind of duality (such as found in wave/ particle or expansive/ contractive antitheses) an expression of principle or happenstance? Take pure energy. Is there physically such a thing and if so what, behind its expressions, is it? What is its inertial antithesis, mass? What might 'jam' energetic freedom to the point of eventually ending up as a solid form? In fact, don't energy and mass quite well express the paradoxical, dualistic nature of a single universe, the two things really being one? Einstein's climactic Principle of Mass/ Energy Equivalence deals with the possible conversion of energy to mass and *vice versa*. Its equation of energy and mass, of creation and destruction and therefore life and death delves to the heart of matter. Is such conversion possible? Indeed it is. Energy is converted into matter when subatomic particles collide at high speed and create new, ephemeral but heavier particles. And a few grams of matter blew up Nagasaki; half a teaspoonful of water could, if all its energy was released in a controlled way, fly a jumbo jet round the world or a tumbler's volume empower London for a week! Indeed, if the power of the atom in your own body were instantly released you personally would explode with the force of more than a hundred of the bomb that devastated Hiroshima. *If so much energy is locked in mass, students want to know what first arrested it then almost threw away the key.* How does creation as opposed to dissolution or destruction work? Came grained matter first or energy? If they're the same but not the same what bottled energetic waves? How does this paradox of two-in-one entwine? By the same criterion the polarity of existence, set against the Singularity of Essence, might be an illusion, a relative illusion set against The Absolute Reality; might studies culminate by showing how, within duality, everything is, at the same time from the same root, essentially sourced by One?

Current theories need additions if not revolutions to explain profundities. String Theory (or, rather, a so-called 'landscape' of string and string-related theories) has been devised to tie the loose ends up, to splice them seamlessly. Can such filaments pull something out of nothing's hat? Strings are too small to ever be observed but the theory also needs more spatial dimensions than our three and postulates what's known as 'super-symmetry'. And the hypothetical contraption of '*SUSY*' itself requires, with its assignment of conjectural 'super-partners' to each force-carrier (boson) and matter particle (fermion), a doubling of the number of subatomic entities. Isn't this too many? We started with a few simplicities (stable proton, electron and so on); high energy experiments allied with calculations have discovered over fifty more; and now String Theory wants to more than double that! Elegance is wobbling; it is losing easy grace. Yet elegance needs form, form needs mass and if you are too ghostly-slim to pack an impact you need clothe yourself with heaviness; you need dress up in at least a mini-mass.

How, though, does energy acquire form and with such fixity the attribute of weight? What first endowed, as they were flung from nothing, particles and thus the universe with mass; and hence, with transformations possible, the great diversity of things? Perhaps the explanatory saviour is Peter Higgs. Higgs proposed a mechanism in the form of cosmic field (see also Chapter 10: Watch This Space) whose 'drag' lends particles their previously missing mass; such conference somehow fills the cosmos with 'a sort of viscous, clingy stuff'! This drag-field's boson (or information/ force-carrying particle) should, if it exists, show up only at extremely high energies and weigh in at over a hundred times heavier than a proton.

If Higgs' etheric superglue is materialisation's fundamental factor you would surely want to stick your finger on its pulse. After all, though undetectable it conjures matter from thin masslessness. Even if light somehow slips Higgs' net *something*'s got to lend all else its mass - so maybe this is how it's done. *It might be the way creation's made to work.* If its particle, called boson, is supposed to constitute the subtlest agent of the (*tam* ↓) gravitational/ materialising vector then at least you need to have inferred its presence by experiment. This, after all, is a creator - one christened with a ridiculously inappropriate name - 'God-particle'. What might cause, whence might arise this 'Giver of creation's mass'? We're back to Mozart or George Stephenson. No way do joules supporting brain-waves explain creativity. You analyse the physical ingredients of their creations but, having placed the final atom, where does this leave authorship? Have you thereby ruled creators out? The *meaning* of a system isn't found within it. **To confuse a mechanism and/or law with creative agency is a kindergarten yet very common category error among materialists, a mistake most severely frequented, incidentally, not by physicists but by evolutionary biologists.** You analyse the paint and canvas of a painting but don't leave the painter out. It's obvious why not. But if you've opined the *cosmic* canvas lacks an artist's mind-behind then you paint *any* painter out. What substitution, therefore, might you surmise? Could it be a quantum hiccup in the sound of silence or selection, naturally, of our world out of zillions and, where life's concerned, umpteen mutant oddities? Such atheistic 'creativity' might in madness dub some particle 'divine' but, with bubble chamber for a brain, it loses track of *meaning*. Indeed, if mass depends upon a boson and, thence, nerves depend upon atomic mass then, for nervous reasons (Chapter 0: Scientific Delusions), why should I believe a word? Oblivious omnipotence - God gene, God spot, God particle, God knows what's next!

A powerful machine? What machine might churn or turn out 'G' - except the constant 'G' of gravity? You aspire, by creation-in-reverse, to recreate conditions of the starting moment from, instead of naturally above, unnaturally below. You want to dematerialise, to end things up, high-energetically, where they quite possibly began. You might replicate, as best you can, the starter pistol's bang; you might ramp up energies as near to big bang's absolutely energetic concentration as they'll go. Therefore you will, including application of the financial uncertainty principle as applied to engineering projects, have upgraded the facilities of The European Organisation for Nuclear Research (*CERN*) in France by replacing its defunct *LEP* with a super-powerful atom-smasher called the *LHC* (Large Hadron Collider). This collider will certainly have cost an astronomical but most worthwhile few billion pounds or more to try and crack with certainty the micro-values of our universe. In case it didn't work - or even if it does - an *SLHC* (Super-Large Hadron Collider) is already on the drawing board - a snip at perhaps 10 billion more! It seems, however, that it may have done the trick. In 2012 it was decided that traces of a few obese Higgs bosons had been found and standard physics breathed again. Now it's off to check for different kinds of 'Higgs' and any other 'super-partners' which might, it is speculated, be a form of 'dark matter' - an equally ubiquitous, theoretically essential but, because it neither absorbs, emits nor reflects radiation, as yet unobserved entity.

Theories are fine but what about the dark side? Examination of the mass that makes up galaxies shows that (if current theories are correct) there's not enough of it. Have Newton's laws (and by extension general relativity) failed to correctly predict the motion of stars in the gravitational field of a galaxy; do we need a *MOND* (modified Newtonian dynamics) or other radical alteration of the laws of physics to explain the motions of extremely massive bodies such as galaxies or neutron stars; or, which avoids this crisis, is there simply much more matter than we see? Such as dark matter; or dark energy (or, maybe, dark mind's ignorance of what is going on)? In 1998 the observation of supernovae in remote galaxies indicated that expansion of the universe was accelerating fast. Is something unpredicted by string or any other theory well at work? Space might be ballooning faster than the speed of light. Or is some other unpredicted, thus unknown, factor well at work? What kind of unseen levity is blowing gravity apart? *To save current models we make the difference up. We speculate, without a shard of proof, that nearly all the cosmos is composed of quantities of whose character we hardly have a clue (about 25% dark matter, 70% dark energy and only 5% the brighter kind of matter that we used to think made everything). By this reckoning we don't know most of what there is to know!* But never mind! The *LHC*, made out of consciousness (and metals too), might prove to be a god-send; or new discoveries and calculations shed such light as may disperse one, none or both 'imaginary' types of darkness. A spoonful of humility might make one cough but still improve digestion.

Are There Any Other Kinds of Answer?

Did you think science was just common sense wrapped up in maths? Could the patchwork of *s*omewhat incompatible, often bizarre and counter-intuitive elements that comprise what might be called Standard Theory (a patchwork including Special and General Theories of Relativity, Newtonian and Quantum Mechanics, the Standard Model of Elementary-Particle Physics, Dark Energy, Singularities and so on) be finally correct? Are there less radical alternatives that might be right?

Why not start at the beginning? Why not join the Sisyphean task of understanding how the cosmos came to be? Did you think, conventionally, that a projection called the 'big bang' started everything? There exist alternatives. For example, astronomer Halton Arp has re-interpreted red shifts, quasi-stellar objects (quasars) and mapped a different galactic universe; there also exist steady state and plasma cosmologies, theories of 'meta-gravity' and so forth. 'Bucking the Big Bang' was the title of a frustrated article (New Scientist 22-5-04 p.20) by 34 scientists from 10 countries. These workers identify fudge factors such as 'dark matter' (that is supposed to hold the universal web of galaxies together) and 'dark energy' (which is held to push it apart) without which the current Standard Theory will not work. Such major hypothetical entities, whose quantity places what we know as 'normal' matter into a huge minority, are invented to plug an abnormally large gap between theory and observation. They would, under normal conditions, raise serious concerns if not topple most speculation; such science might be thought most imprecise. Yet in this case, the physicists complain, peer pressure and dogmatic control of funding mean that no other ideas can be properly explored. There cannot be dissent. The cost is far too high. *If big bang's ages disappear then geology is faulty too.* A fault too far, a crack too great, quicksand that could swallow too much supposition! Do you accept the dominant regime?

Especially since an even more subversive sort of thought's occurred. A tiny fraction of a second after cosmic dawn's epiphany the vehicle of physics fails. The rules describing facts break down without a word about their cause. What caused all effects? It is now agreed the bang is flawed; its singularity (if thus it was) was not a naked one. *Something came before.* Averted glances still resist what clothes such an iconoclastic breakthrough. What is the nature, you are always forced to ask, of metaphysic, prior physic or pre-cosmic nothingness? What is on 'the other side'? In the rush to ambush metaphysic and to fill this vacuum up with physicality cosmology has revved its non-experimental engine; surely it's not overheated to explore eternal, uncreated matter, perpetual inflation, cosmogony inside black holes, rebounding (bouncy-bouncy) cosmoi, branes colliding in extraordinary dimensions and, most arcane of all, a myth, a psychedelic-seeming fantasy of everlasting multiverse! Must there exist ancestral universes whence our own, as far as life's concerned, is a naturally selected offspring? There's nothing 'real' in such conception but there's nothing extra-physical as well. The maths should stave off immaterial advances for a while. Such whole, mind-blowing vastness is conceived in order to outwit the presence of a single, natural but immaterial, subjective element. That is scientism's point. It could, it should, it will succeed!

Let's hop back inside the universe. If the presence of matter affects the geometry of space-time then, as materials move, its structure changes; the geometry of this dual entity is not fixed but evolves in time. Such curvaceous if not vibrant space follows the effects of gravity and is, in this sense, a gravitational field. How, though, can you unify the natural forces and so discover one force whence they all proceed? One way might be to invoke extra dimensions. In the 1920's Gunnar Nordström applied Maxwell's theory of electromagnetism to a five-dimensional world (one of time and four of space) and found that it included gravity; but his calculations could not account for light bent by gravitation as it passed a star. Conversely Theodor Kaluza applied Einstein's theory of relativity to the same 4-d space and discovered electromagnetism. Where gravity takes three-dimensional space does electromagnetism's geometry use four? For various reasons including the discovery of two more forces (strong and weak nuclear) such extra-dimensional answers were shed - but in their place many more dimensions, all unobserved and purely hypothetical, blossomed in theorists' minds in order to satisfy unification through the medium of a string theory of quantum gravity. The idea of ten or eleven coordinates (six or seven rolled up away from sight in 'hyperspace') has now gripped calculating minds!

Enough of space for now (but more in Chapter 10). What of time? Does its flow flex with mass and motion (as Einstein's theory redefined it to)? Can its rate slow or accelerate according to an object's mass or its velocity? Or is time dilation/ contraction simply down to changes in the rate of clock-work used to measure it? Surely, though, such flexitime won't U-turn? You can reverse equations; laws of natural behaviour might in theory be reversed. But could you mirror-image time - even in an anti-matter universe? You cannot put a cause before its own effect thus making an effect the cause of what had caused it. How, in anti-time, could the moment of creation ever start when its effect, the world, preceded it? This effect, become its cause, would make the start a finish! The flow of time, from cause to its effect, is irreversible; you can rewind a film of falling cups but can reality reset the smithereens? Time reversal, rubber time, a weft that is affected by the warp of space - is modern science shamelessly and seriously at odds with common sense? If so, is such topsy-turvy contradiction operating in its anti-space correct? We'll take a further look at time in Chapter 11.

Nor is relativity the end of physics' rightful romp with 'abnormality'. If Einstein made you queasy Planck will swing you sense-sick. Relativity is almost mild against the quantum queerness of the journey Chapters 8 to 12 will sometimes travel through the micro-world's high seas. Is, as quantum theory seems to claim, the only fundamental certainty uncertainty? Or everything conjoined by universal waves of probability? Is such a theory even a necessity? In 1913 Einstein, Bose and others developed 'stochastic electrodynamics' (*SED*) to describe the micro-world using the mathematics of classical, deterministic physics. Is the sacred cow of special relativity, c (the speed of light), safe as an absolute? John Moffat and others have proposed a variable speed; and if you have a variable speed of light then inflation can be dispensed with and Einstein's Special Theory of Relativity is wrong - unless you double up the postulates. Doubly Special Relativity, which operates under conditions of only two spatial dimensions, proposes two universal quantities - the speed of light and a length, the exceedingly small Planck length. Ask the theory's proponents where they think it leads. At any rate, the world of physics is a-buzz with fresh ideas because its work is incomplete. String theories, generations of them, abound. Loop quantum gravity, modified Newtonian dynamics, branes and dark types of matter that no-one has ever seen inhabit the imaginations and equations of a fertile scientific mind. Professional physics even dares to contemplate a billion different kinds of universe! Indeed, Czech physicist Peter Horava may endorse string theory but has also proposed that space and time be treated separately.

Thus you'd dismantle general relativity (with its ideas of gravity) then recompose it in a way that is compatible with quantum gravitation. Could special relativity be tweaked as well so that, in a hugely energetic early universe, its principles broke down? Such tweaks might mean 'dark matter' - vast, essential yet hypothetical and unobserved - were not required. Furthermore, could any kind of spec exceed such gross extravagance except the trump of infinite inflation dreamt by astrophysics' hubble-bubble crew? Models, like perspectives, twist and turn about. What's sacrosanct? One might imagine that there's more to learn. *Might anybody find contemporary Standard Answers insufficient to the point they feel they have to jump outside the box - even materialism's box?*

Yet what you repeat enough (in lectures, books and conversations) takes on, whether or not wholly right, it its own legitimacy. Despite frayed edges and deep incompatibilities Standard Theory holds, like Darwin's Theory in biology, a comprehensive monopoly. Without any external frame of reference from which it can be objectively criticised, anomalies, reparations and additions are almost always treated within its boundary. All advances are attempts to refine or extend but not to replace its central place in physics. They tinker. Can an exercise in mathematical manipulation - such as string theory - ever suffice to improve explanations, find a unifying principle and discover a deeper level of reality? If not, any maths will have simply linked elements in a flawed model and, as such, most likely have become very smart but complex - and fragile as a house of cards. In this respect, if only to loosen the rigid bonds of assumption, it might be useful to relate the strengths and weaknesses of an alternative theory. *Could there exist a framework of understanding whose interpretations systematically extend or replace those we hold 'sacrosanct'?*

One example is Expansion Theory. Expansion is, like Standard Theory, wholly materialistic. Unlike big-bang's expansion this sort never had a start. At its root matter is conceived as 'active' and not 'passive'. This 'action' is vested in its fundamental nature, expansion. The idea is a species of 'eternal inflation', the cosmological pipe-dream proposed by physicists Andrei Linde and Leonard Susskind wherein endless universes are forever 'bubbling up'. Theory and yet not at all the same. Briefly what, in its speculative claim to a self-consistent but non-standard explanation of all physical effects, does it say? In an 'amateur' refinement by Mark McCutcheon there exist two zones of expansion. 'Internal', subatomic space is different from our familiar 'external 'atomic' or 'large-scale' space. Within the atom great expansion consumes no space; it simply *supports* atomic structure. Externally a slower, precisely calculable expansion is the motion that underwrites all observed effects. From this single, causal principle is derived effects that include what we call gravity, electricity, atomic structure, astronomical events etc. In such a universe the sole, fundamental particle is an uncharged 'electron'. Electrical charge is explained as 'a crossover effect' that occurs between subatomic and atomic dimensions. Protons and neutrons are conceived as groups of neutral 'electrons'; electromagnetism (light), electricity and, indeed, all various other forms of force or energy are construed as different ways in which expanding 'electrons' or groups of them called atoms can behave. There exists, therefore, no 'pure energy' as conceived by current science; instead power is vested in the universal expansion that *is* matter and, therefore, material existence. Eternal expansion is the first dynamic principle whence all followed, follows and will follow.

Just expansion gives you matter? Whence did that eternal drive derive? It is easy (and in this case justifiable) to pan fresh and therefore strange ideas. For example, the notion of different subatomic and atomic spaces seems as contrived as, say, classical or quantum mechanical descriptions of an atom. And if you double in volume your mass must increase proportionately. As with Standard Theory, mass is a problem. Whence appears the extra mass to support a universal expansion that doubles the volume of everything except space every twenty minutes or so? What causes 'atomic boundaries'? What *are* 'electrons'? What, above all, gives rise to the 'essence' of material existence, its ever-expansive motion? There exist many apparent problems with eternal as opposed to non-eternal expansion theories. Why even be dissatisfied with standard explanations?

The answer is multiplex. Whence did matter leap unless from immaterial source? What, furthermore, keeps everything, including quantum particles, in constant motion? Stars, planets, galaxies and atoms swirl and swirl again; their vibratory cycles measure out both heavens and the earth. Why does the universe itself not seem to spin? Reformation or a revolution in perspective seems required. Could there be flaws of logic, misconceived 'experiments in thought' or erroneous abstract models used to grasp the world? *No doubt that where it is testable Standard Theory meets the mark*. It has not been falsified and specific predictions have been verified. Such success has generated the curricular momentum of a juggernaut. Its conceptual direction has, as with exposition of the abovementioned biological theory of evolution, become apparently invincible. Why shouldn't men eventually, physically understand every aspect of a cosmos physical? In terms of physics they would be exactly right but, still, would this represent the whole truth?

Is it mysterious how a magnet can be endlessly supported on a fridge door? How does a light bulb emit electromagnetic radiation from passing electrons that lose neither charge nor pace? The implicit assumption that no textbook contradicts is that electromagnetic fields, potential and energy are freely created out of nothing by their associated dipoles, their sources of charge. Does such 'support' not violate the conservation laws? What sources the endless repulsive and attractive forces that Benjamin Franklin called 'charge'; and what supports identical units of 'charge' called electrons? What *is* an electron? How does its field of influence interact with protons to stabilise atoms for billions of years? Why does electrical attraction not cause electrons to drop into nuclei? Quantum theory of atomic construction is a description but is it, of itself, an explanation of why their 'orbits' and thus the chemistry of bulk matter persist? What is it keeps electrons whirring round leashed quarks to compose an atom and thus spin the whole world into destiny? And that tight, elastic leash, called the strong force and introduced to counteract the perceived repulsive force between nuclear entities, also violates basic law in the way that, without an obvious power-source, it permanently glues quarks, protons and therefore all the years of atomic existence together. Paradoxically but necessarily, the closer are protons (or quarks) to each other the weaker this force; but it strengthens with distance so that you can practically never prise quark from quark, split proton or thus wreck atomic nuclei. Quantum mechanics, with its own bizarre and counterintuitive claims, was introduced to better describe but not necessarily explain such strange, intra-atomic behaviours. As Neils Bohr himself remarked, 'If you aren't confused you haven't understood it' - that is, quantum theory.

How, like perpetual motion machines, do atoms seem to vibrantly violate The Conservation Laws? Because, as the kinetic theory of matter describes, every particle and molecule in the cosmos continually vibrates; and every larger form in space is on the move. Such action, from an atom in a pebble on a planet up to a nebula in size, underwrites physical existence. It keeps things up. Even if it seems immobile, a pebble is nevertheless in constant, internal motion; and the smaller, more generalised the particle that you observe (say an electron or, perhaps, a string) the faster it goes. *The fact is, equally, that science has no idea what sustains this all-supporting changefulness, this eternal motion causing things to be. The fact is also, on the other hand, that mystics proclaim it an aspect of primal, immaterial vibration, of the structured, 'musical' stream of First Cause.*

Such projection of a *Causal Logos* is also called the Name, Creator's Word or Voice. Its grammatical logic quickens a physical text wherein everything, including you, is 'word' or 'code' from the Author. From basic grammar, alphabet and dictionary millions of books and conversations flow; from simple formulae, like Mandelbrot's, a world of difference evolves. Over the next few chapters we can try and see how, lower down the scale, such principle might naturally determine an apparent chaos of detailed, material practice. At this point we simply note that at the very edge of scientific bounds perhaps the mystic version reaches down to meet the scientific looking up. Could it be, as some physicists propose, that things emerge from 'active vacuum', from an effervescent cornucopia of 'quantum foam'? Can't you grasp uneven space just as well as even anything? Its transparency is, though opaque to observation, something that you have to see through; its nature is a height we'll have to plumb.

Could archetypal, metaphysical vibrations really be the drum-rolls that sustain perpetual motion? Indeed, what caused the grandest violation, the start of all commotion; what, is it supposed, smashed conservation rules to smithereens in an original uprising, a revolutionary new crack at nothing modern science thinks of as somehow-expanding space? Or, if a 'law' is a presumption born of observation, why did prior lack of conservation foster conservation law? Why, in other words, was Primal Shot stopped at first 'dollop' and its amount of matter fixed? Why not allow, after the primer, continual additions of some sort of energy, some vibratory supply of 'fuel' to keep the cosmic engine on the go - continual shaking to sustain what's been materially shook up? Some scientists suggest that there exists a reservoir of virtual, vacuum energy. Is the 'virtual' world of metascience, just because 'behind-the-scene' is unobservable, unreal? Are psientific objects, being unobservable, invalid to the eye of physic's actuality, that is, to such facts as science can perceive?

Foils to Standard Theory query the accepted explanations. Some objections might be right and others wrong; some proposals may prove errant, others herald an advance. The real point is that Standard Theory is deficient to the point that each of its own two pillars of faith, classical and quantum theories, seem incomplete and, beyond any particular frailty, incompatible; that it includes no satisfactory account of consciousness (i.e. subjective awareness), mind or informative capacity; and that a suite of major violations of conservation law and other anomalies render it, if not fatally flawed, at least weak enough to need a radical alternative - in Linde, Susskind and McCutcheon's cases one whose unifying principle is not energy but the eternal inflation of space. If two or more theories (whose basic principles are so different that each requires the other to be rewritten on its own terms) can compete then we are some way from the final gel, a *TOE*! Of course, scientific atheism is a secular religion and in no society do you argue trouble-free with priests. Nevertheless Natural Dialectic takes a *top-down* view. It involves an explicit metaphysic. There are certainly at least two sides to the same story.

There is, very close to you, definite evidence for Natural Dialectic's immaterial element. Just allow subjective mind is not the same as brain; allow the information centre that you're thinking with is metaphysical; allow that nervous molecules of brain work with but do not excrete your consciousness. Is experience just material? Or could there be a simple fifth and metaphysical department, an informative dimension labelled mind? And, as universal body, so a universal mind?

You say there exist only non-conscious elements in four observed dimensions.

I say there are two existential components - mind and matter operative in at least five dimensions.

I propose, furthermore, that metaphysical *precedes* physical. Its First Principle, Infinite Essence, is metaphysical. *So is the unifying principle that sources physical phenomena universal mind?* How bizarre is that? Such mind (but not your fragile fraction of it) informs energetic behaviours? Before it drops to concrete earth the nature of this abstract mind can be described in mathematical but also polar, dialectical terms. **To this end the order of play is, as mentioned above, to inspect Infinity, Oneness and Nothingness; and, having linked Nothingness with Space, to follow with Time and Pure Energy.** While in this view the first principle of material existence is metaphysical. The fifth dimension has a prior, archetypal role to play. Its phenomenal issue involves the subsequence of chemistry and physics. This 'Energy Section' (Chapters 7-12) reformulates Standard Theory, as far as it is correct, into a dialectical framework.

The Principles of a Unified Theory of Matter.

It might be said that Natural Dialectic simply takes what is there and reassembles it in an internally self-consistent way that makes another sense. Or again, that it tracks along its subroutines to make better sense of the cosmic program as a whole. *Every systems analyst understands top-down logic.* If you want to understand how, *top-down*, anything works you have to begin at the beginning; you have to understand its purpose and its principles. **For a full understanding you need to grasp first principles. To grasp the principles behind physics you may need to consider their metaphysical origin.**

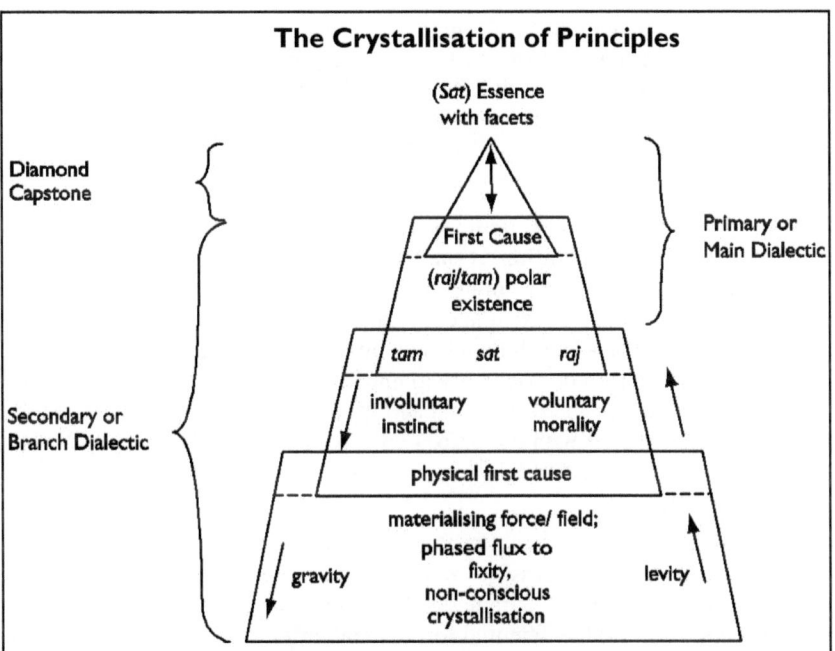

Crystallisation of principles is the natural process of creation. Facets of the Diamond, Essence, include Infinity, Unity, Nothingness etc. Facets of existence derive from the fundamental event vectors (*raj*) levity and (*tam*) gravity in concert with reflections of the (*sat*) balancing, normalising factor. The latter acts as pivotal; it constitutes an axis round which tipping opposites may swing.

	tam/ raj	*Sat*
	sub-principles	First Principle
	existence	Essence
	exhausted/ kinetic phases	Potential Phase
	polarity	Neutrality
	periphery	Centrality
	finite	Infinite
	duality	Unity
	things	Nothing
↓	tam	raj ↑
	descent ← →	ascent
	from Essential Centre	towards Centre
	anti-principle	principle-in-action
	negative presence/	positive presence/
	rel. absence of (sat/ raj)	rel. dominance of (sat/ raj)
	'right-hand' properties	'right-hand' properties
	exhaustion	kinesis
	entropy	negentropy
	resistance/ inertia	stimulus/ flow
	fixity of object	flux of process
	passive state/ object	active energy/ force
	special case of motion/	motion/ change
	an object's immobility	

separation/ differentiation	unification
crystallisation	merging/ fusion
materialisation	dematerialisation
gravity	levity

fig. 7.2

What is a principle? **A principle, which has been called a highly condensed form of information, is conceptual.** Because it precedes, orders and directs subsequent and often complex actions, principle is *potential action*. Because it is an apparently absent latency, nothing without practice, principle is *symbolic action*. What is law without behaviours, idea without fruition, intention without tool, mechanism or machinery to implement? Principle orders action. It is the source and definition of all legal activity. Its intent is expressed through mechanisms that accord with its logic. *This is no less true of cosmos than the latest plan or program of action (however grand or trivial in scope) that you brought to fruition.*

Principle informs practice. For this reason it has been argued (Chapters 5 and 6) that active precedes passive information, concept precedes manufacture, thought precedes action and, axiomatically, mind is the repository of principle. In other words, mind is the potential for material order.

Because it reflects the way principle is (or could be) translated into practice abstract mathematics and the laws of logic can predict and compute automatic physical relationships and behaviours. Physical principle, the law of nature, gave and gives rise to matter in a specific order and in order to understand creation properly it is imperative to follow this sequence. The narrative must follow its logic. *The right order gives the right reasons.*

Archetypal mathematics is the natural grammar that describes the build-up of our world. Although maths should, in principle, be able to follow all physical processes, in fact the amount of simultaneous action and interaction generates a complexity whose full description is difficult to the point of impossible. Various theories (called 'Chaos', 'Complexity', 'Fractal' and so on) ally with the statistics of probability to rescue this circumstance. They can also predict the likelihood of psychological behaviours, even voluntary response. Wherever there is constraint mathematics can 'get a handle'; the greater the constraint the better the handle. Thus science, in the form of sociology or experimental psychology, can formulate an objective description of mind. To describe is not necessarily to explain. What about subjective feeling, awe, desire, inspiration or an idea of morality?

A *bottom-up* view might be that principle derives from practice. Things just behave as they do because they are what they are. There is no particular reason behind their 'reason'. Asking for 'deep reasons' is misguided, a waste of time, an enterprise doomed to failure by its own misconceptions. Principle and law is the result of chance. But is it really? You may shrug.

The bottom line is that modern 'rationalism' dislikes the notion of a hierarchical cosmos based on Rational First Principles; and its atheistic brother-in-arms detests the idea of Natural Intelligence. What do they offer instead? Their foundation is a shrug and offering of accidents.

From a *top-down* point of view *(fig. 7.2)* such nonchalance implies a preference but also involves a complete guess. Whether genuine or disingenuous, the shrug incorporates an untestable hypothesis (which divorces it from science) and an irrational presumption that chance can generate non-chance, that is, underlying regularity (which divorces it from logic). It derives, simply, from a one-tiered, 'flat-universe' perspective. What is the origin of anything? What is its reason?

Top-down, principle precedes practice and the influence of principle is increasingly 'crystallised' through sub-principles, ideas, archetypes and, finally, physical phenomena. Therefore, one needs to understand the origin of principle to understand its consequent practice. This is the dialectical approach devolved through Chapters 7-12.

Before attempting to rephrase the jargon, consolidated perspectives and speculations of physical science into the terms of Natural Dialectic, it is necessary to paint a broad-brush sketch of what sort of principles and norms the Dialectic might, with regard to the physical cosmos, expect to find. To understand physical science in dialectical terms revise:

(i)   *top-down* scaling; a conscio-material gradient

(ii) interplay between cosmic fundamentals in different proportions in different cases

(iii) simultaneous government of (*raj/ tam*) vector directions by a third (*sat*) balancing factor. This factor appears as an axis of polar motion; or a pivot, like an equals sign, of transformation. As such it constitutes the norm a pattern of behaviour describes; or, if you like, the rule whence it derives. As such it is linked with possibility prior to actual occurrence. Possibility is potential. Potential is a source of action. It involves both the rules by which a pattern of behaviour can occur - its informative aspect; and its energetic capacity to 'happen'. The former is always identified with the tier above its expression. In dialectical terms the informative tier above matter is mind; and of mind potential matter constitutes the lowest and sub-conscious tier. This implies that the principles governing the behaviour of material energy (the scope of physics and chemistry) are lodged in and derive from a metaphysical source. We make sensible, sensory connection with matter; *but we make contemplative connection with its metaphysical principles.*

(iv) with the synchronous engagement of *three* operative elements, two directional and one an immanent, balancing factor, we might expect to entertain, along with any whole truth, paradox.

Principles are a start; they are stages in the first, original translation of Truth to physical form; and also in its transmission as the structure is kept running.

If the patterns of energy that appear to us as practical things of the world are devolved from aspects of (*Sat*) First Principle, what is it? What is its character, what are its aspects? We call it Essence. Pure Being. A pre-existential Singularity or Monopole. Its aspects are stacked in the right-hand column of Main or Primary Dialectic and reflected in lower, relatively imperfect forms and behaviours. **Of these aspects Consciousness, Infinity, Unity and Nothingness** (or Void) are of central importance; the Quintessential One-within-Three is (see *fig.* 7.1) Informative Consciousness - Life - and its Potential Creativity.

This section (Chapters 7-12) concentrates on non-conscious energy rather than information so the first of these, Consciousness, is not in point; but the other four - Infinity, Potential, Unity and Void - form the orderly basis of its narrative. The character of First Principle is beyond existence but reflected in relative, imperfect forms throughout it. For example Essential Symmetry is beyond existence but lesser symmetries are commonly found; similarly Absolute Equilibrium is reflected by myriad relatively stable conditions that occur in mind, time and space. How might (*sat*) principle be reflected in physical construction? By stability, conservatism, various kinds of symmetry and, with that balance, norms? A norm is an expected pattern, a constraint on possible outcomes in any particular circumstance, a guiding principle or causal law.

The first *existential* principle is (Chapter 1) causality. Cause precedes effect so any search for truth involves a search for causes, reasons, origins. What, before all lesser causes, is the nature of First Cause? The second existential principle is (*raj/ tam*) polarisation. Polarisation, you may remember, is defined as the concentration of information or energy at one pole and its absence at the other; motion from a condition of high concentration towards low, that is, a current flowing from 'high' or concentrated potential towards an inertial 'sink'; or tension between opposites. If polarisation 'divides' then what about its opposite yet complementary case, depolarisation? Not one as two but, in the reverse process, two becoming one. Recovery of full potential. Positive neutrality. Particle's distinct from wave and mass from energy but only by fixation. As well as signifying liberation from fixity to fluid freedom, the (*raj*) vector involves integration, unification; or, put another way, a tendency away from exhaustion and towards revitalisation. It recharges the inertial state; a stimulating input of information (by focus of attention) or energy stimulates, reformulates; a current of dematerialisation 'ascends upstream' towards its source. The next few chapters will explore the nature of paradoxical existence - not least the way that all three cosmic fundamentals (*sat, raj and tam*) cooperate in giving rise to the appearance of each and every event in existence.

You may certainly need to keep revising conceptual positions and, possibly, incorporate your findings into a new, expanded paradigm. If you are going to include mind as a fundamental, subjective component of existence you will definitely need to expand your previous framework. <u>The 're-contextualisation' of such an expanded universe may, according with the aforesaid aim of Natural Dialectic, require both some conceptual realignment and minimal fresh vocabulary; but the idea, although squeezed flat by materialism, is not new. It is one of cosmic hierarchy.</u>

A source is 'above' and 'before' its issue; it is the potential for it, the cause of it. So what is 'above' fundamentally governs what is below. Principle precedes action and is thus termed its origin. Therefore (*Sat*) Absolute Potential might be seen as An Indispensible Origin, The Source of Existence, First Cause. From (*Sat*) Cause issue polar (*raj/ tam*) effects; from (*Sat*) First Principle issue (*raj/ tam*) existential principles and practices that are stacked in the left-hand column of Primary Dialectic but also, polarised, in the antithetical columns of

Secondary or Branch Dialectic; in short, from Informative Singularity issue polar (*raj/ tam*) effects.

We met these polar principles (Chapter 2) as 'cosmic fundamentals' or 'event vectors'. How do they work? The (*tam*, ↓) 'downward' vector separates, deactivates, precipitates. You might logically expect the Dialectic to emphasise, with respect to materialisation, principles, forces and forms of gravity or gravitational effect. An 'upward' (*raj*, ↑) vector implies, conversely, an input of information or energy. It stimulates, raises, unifies and would be expressed as a principle of levity. What could be physical expressions, forms and forces of such levitatory and gravitational characters? That is to say, which particles and forces are agents of materialisation and which of dematerialisation. For a preliminary suggestion see *fig.* 7.9.

What state is above matter? Mind? If so, what is the nature of universal mind or, at least, the fraction of that mind from which matter is directed? What is the nature (*figs.* 1.4, 2.5, 2.9) of (*sat*) potential as opposed to (*raj*) quantum and (*tam*) bulk matters? Might it involve archetypes, patterns according to which the observed behaviours of force and particle appear? In other words, are the principles governing the behaviour of material energy (the scope of physics and chemistry) lodged in and derived from a metaphysical source? If so, (*sat*) archetypal memory would, in this sense, be an agent for the 'grammatical' transmission of metaphysical principle to material practice, a key to the transformation of mind to matter, nature's method for the conveyance of law. Is this indeed how energy's informed? And has contemporary science suspected it (Chapter 9)? Or calculated on the archetype?

There may have been nothing physical but was it absolutely nothing out of which the galaxies all spilled? *Top-down*, the beginning is further 'back' than an emergent 'big bang'. According to Natural Dialectic the universe and its origin can be discussed in terms of three major phases, grades or subdivisions. Overall (*fig.* 2.7), the order of play runs from (*sat*) potential to (*raj*) kinetic and (*tam*) inertial phases of creation. If (*sat*) potential is Pure Consciousness this top grade devolves mind and, lastly, a special case of mind, no-mind or non-conscious matter. Each major grade is subdivided (*fig.* 2.8) into three. The lowest subdivision overlaps with the highest of the major phase below. You can read this linkage as, in the way of a spectrum, continuous or, in the way of a phase transition such as water to steam, discontinuous. How does each phase relate to the creation and subsequent behaviour of matter? The next six chapters aim to explain this relationship. *They aim to frame scientific theory and fact within the context of dialectical perspective.* This perspective neither much conflicts with nor entirely conforms to the contemporary paradigm.

The first task of the next section is, therefore, to place the basis of science, energy, in the context of the *whole* creation. This involves a recollection of the simple yet revolutionary derivative (from Natural Dialectic and its conscio-material spectrum) that Essence, a Super-mind, precedes the existence of mind and matter. It is from this First Principle that sub-principles are all derived.

The second is (*figs.* 1.4, 2.9, 5.2 (ii) and 7.9) to consider its three *physical* subdivisions. These subdivisions can be treated, time-wise, as generations. Two are described by the pillars of modern science (quantum and classical physics) but the first, their arch, is less familiar. It is actually not physical at all. *It is physically non-existent but is the potential for physical matter. It is called 'transcendent' or 'top matter'.* Such potential exists, as has already been hinted, in underlying archetypes that are metaphysical. They are, in dialectical terms, a part of universal mind. Not in a conscious but a low, 'fixed' or sub-conscious phase of mind that constitutes, as developed in more detail in Chapter 16, universal memory. Are not memories mental 'fixtures'? 'Top matter' is, while a memory, certainly not one of yours. This relationship between mind and matter is a further startling yet revolutionary derivation from the logic of Natural Dialectic and its conscio-material gradient. Memory, it is proposed, is the connection by which principle is delivered to practice. It is the channel through which metaphysical information is relayed to physical body; and the information itself can be likened to 'cosmic instinct'. This notion has, in fact, been simmering for a while (see *figs.* 0.12, 1.3, 1.4, 2.5, 2.9, 4.1 etc.). It will, in conjunction with the morphogene, be elaborated in Book 2 (Chapters 15-17 and 19-23). At this point suffice to say that, as a musical instrument possesses the capacity for certain harmonic combinations, so 'instinct' or 'archetype' involves potential in the form of information for a particular set of physical behaviours.

	tam/ raj	*Sat*	
	peripheral	*Central*	
	expression	*Potential*	
	subsequent order	*Archetype*	
	actual behaviours	*Causal Law*	
↓	*tam*	*raj*	↑
	down	*up*	
	inertial	*dynamic*	
	lower	*higher*	
	energy	*information*	
	matter	*mind*	

According to the *top-down* perspective a lower level derives power from one hierarchically above it. *Higher is nested within lower.* A foundation is a substance, base or origin and an origin is nested, hierarchically, above or within its outcome. In this view the source of physical creation is mind. *Mind is above or within matter.* In this respect the 'top' condition of matter, as it appears at the psychosomatic boundary, is metaphysical. Nothing obviously physical: something paraphysical. There arise, therefore, the stingers (i) to (iv) of Chapter 4 - particularly 'matter in the context of mind' and 'the nature of any psychosomatic linkage between mind and matter'. Mind nests *within* the physical 'ocean' of energy and microscopic subatomic particles and atoms are found *within* bulk matter. Such large-scale, classical 'islands' of bulk are, however, themselves practically empty. Atoms and therefore aggregate atoms called gases, liquids and solids are, in terms of quantum physics, 99% space. Amazingly, says science, you are 99% physical void. But your mind, although nothing physical, is full enough. What of?

At this point it needs be reiterated that dialectical use of prepositions no more necessarily implies spatial position than light 'above' dark, waking 'above' sleep or a good mood 'above' bad. Metaphysic is no more spatially 'above' physic than, say, heaven 'above' earth or high energy 'above' low. But how else, except using prepositions, can you tell anything apart? How else can you relate the steps of scale or hierarchy? Having placed Essence top-centre, Natural Dialectic orientates accordingly.

<u>It also needs be made crystal clear that no active aspect of mind is operative in the iterations of matter. The physical world is one of completely non-conscious automation.</u> No active only *passive* information is present in physical data, that is, objects, behaviours and events. It is easier to infer the presence of passive information in the construction and operation of a machine than in ordinary matter. This is because information implies meaning and purpose. What purpose can be inferred from matter which is, from its own mindless point of non-view, as purposeless as a machine. What rhyme or reason could stars have? Why ever might clouds, mountains or grains of sand purposely exist? Might they host a drama on their stage? How does the stage itself exist? Could you say that it is, in effect, crystalline mind? That the universe is a dead ringer of its own idea, the call of sub-routines from simple, primary code or the expression of an archetypal memory filed in universal mind? *Could you conceive that material behaviours and consequent rearrangements of bodies derive from preordained patterns, information otherwise known as the laws, constants and formulae of nature?*

Just as an intricate body develops from information invisibly encapsulated in an egg; just as different shapes, colours and textures 'fall out' of a gas that cools to its solid self; and as all kinds of chemical compounds derive from combinations of a few archetypal expressions we call subatomic particles, so eternal abstracts' appear to underwrite different levels of physical interactions. What, although mathematics may describe them, *are* the actual archetypes? What empowers them and, if they are not freaks of nature, from what principles are they described? In other words, what is the link between information and its non-conscious counterpart, our cosmos?

	tam/ raj	*Sat*
	classical/ quantum	*Top/ Transcendent*
	concentric spheres	*Axis*
	oscillation	*Ideal/ Norm*
	variation-on-theme	*Archetype*
	alteration	*Permanence*
	actual behaviours	*Law*
↓	*tam*	*raj* ↑
	inertial	*dynamic*
	outer rings/ circumference	*inner rings*
	bulk matter	*(sub-)atomic elements*
	inertial tendencies predominant	*dynamic tendencies predominant*
	classical	*quantum*

The order of argument is simple. *'Top' or potential matter correlates with the grammatical/ archetypal level of passive information* (Chapter 6). It is also variously called super-energy, super-matter and transcendent, potential or absolute matter. *It involves the pre-conditions for physical creation.* After identifying the Principles of a Unified Theory of Matter we turn, in this Chapter, to discuss the nature of this (*sat*) 'top' subdivision of matter - its absence or a 'pre-physical' void.

The order of play then turns to consider the two lower (*raj/ tam*) subdivisions. *These combine to correlate with the lowest, quantitative level of passive information.* This level is the final expression of principle in physical practice, that is, in terms of data items called objects and interactions called events.

The first of these is the (*raj*) kinetic condition of 'matter-in-principle'. This is also described as subtle or active matter. Standard Theory knows it as quantum and atomic physics. It involves (*fig.* 3.3) a first physical cause and the nature of simplicity on which subsequent complexity is built. This secondary subsequence is called 'matter-in-practice'. It is also known as the gross, passive or (*tam*) exhausted condition of energy. While Expansion Theory calls this third dialectical phase one of atomic (or larger than atomic) effects, Standard Theory groups it within the classical physics and chemistry of mass, aggregation and bulk material. Either way, it includes gases, liquids, solids and, of course, all study related thereto.

It has become clear that, starting at the top, the order of this book follows an act of creation. It devolves from (*Sat*) Essence through (*raj*) mind (also Chapters 5, 6 and 13-18), matter (Chapters 7-12) and their biological conjunction (Chapters 19-25). Where the previous section (Chapters 5 and 6) dealt with psychological aspects of initiation - Consciousness, Information and Authorship, this one (Chapters 7-12) sets the material ball rolling. Together they attempt to show how, combined in different proportions, the cosmic fundamentals give dialectical rise to each and every event, psychological and physical, in existence (in this context physical objects are construed as, simply, slow or rigid events). The challenge for Natural Dialectic is one of orderly narrative. It is to clearly and accurately order its description along the gradient of creation. If this gradient runs from Top (*Sat*) Principle, Infinite Essence, through the whole spectrum of finite existence, then the task of the Dialectic is to take order from the Infinite. How, in other words, might you crystallise Essence?

The Essential Pair

First Principle is, paradoxically, beyond all principles. It is pre-existential. Although unconditioned it shows facets. The first is its Essence, which means unqualified Being. Essence is being-without-a-second. 'I am that I am'. The major qualification of Essence is essence-in-motion. This is called existence. Existence (changeful events, objects or attributes) predicates Essence (the subject). 'I am x'. Check *figs.* 2.3 (i) and 13.4. *Central identity (to be) is complemented by knowledge and action; to know and to do; information and energy.* Essence and existence, as characterised in the dialectical Stacks, comprise the Essential Pair.

tam/ raj	Sat
existence	Essence
relativity	Absolution
to do/ to know	To Be
conditioned being	Being

	subsequent derivatives	First Principle
	colours	Pure Light
	parts	Wholeness
	lesser presences	Omnipresence
↓	tam	raj ↑
	to do	to know
	action/ informed energy	knowledge
	energy	information
	passive 'informee'	active informant
	automated	free-willed
	purposeless	purposeful
	subjectively impotent	subjectively potent
	material body	mind

Being and lesser being. Absolute Being and relative, conditioned or qualified sorts of being. Essence is Omnipresent; existence is here and now locally. The latter's events are conditioned and confined, its times and spaces are finite. What is here is not there; what is now is neither past nor future. The Wholeness of Omnipresence has disappeared in fractions; it is *in* these fractions, it *is* them but at the same time not the same as them. It is reduced to include relative voids/ diminutions of pure information and energy; these appear in the form of partialities and limitations in mind, space and time. The paradox of absence and presence, void and plenitude is defined with more rigour in Chapter 9. Essence, Immaterial Concentrate (*fig.* 0.5iii), is the Prime Cosmic Paradox. Although pre-existential Essence is the Precondition of every thing it is alone, in its purity, without condition. Although the Substance of all things, it is none of them. It is Independent of existence: existence is dependent on it. Yet it is Nothing. How can everything or even anything depend on Nothing?

Essence is Life's Holy Water; and, unmixed with mind or matter, *Logos* is its spring. By a cosmic metaphor Water is the solvent, energy the solute and existence the solution. A biologist will tell you that pure water has potential that is valued nought. Is your Water's Pure Potential, asks the mystic, also Zero, Nothing, (N)One?

Nor do you need a Declaration from the State House balcony at Boston, USA to understand the qualities of Essence. Liberty and Independence are, according to the Primary Dialectic, inalienable 'rights'. They stack up on the right side with The Diamond's trio and at root, on top and hidden deep inside, are everywhere. You, like all creation, are dependent on their Actuality!

If First Principle is likened to a Diamond its major facets subdivide into a stack of 'cuts' reflecting the internal light of (*Sat*) Essential Substance. These 'cuts' or characters are principles because the natural laws of mind and matter are, in practice, drawn down from them. **In other words, all nature is a derivation from First Principle; it is an expression of The Infinite.**

The Infinite

Called by many names it is nameless. For example, the Chinese sage Lao Tzu describes the essential *Tao* as follows: 'This sameness is called profundity. Infinite profundity is the gate whence comes the start of all parts of existence'.

Is it mist on the mountain, wind that bends the reed or monsoon rain and river? What kind of infinity do you want? Philosophy, maths, science and theology all struggle to make sense of problems understanding startlessness and paradoxes sprung from endlessness - the liberation from all boundary that's infinity.

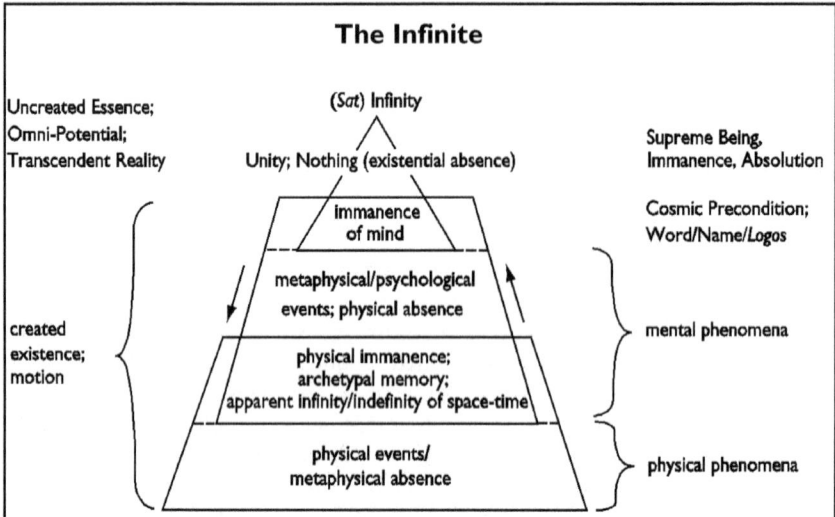

Liberation! Freedom from all limitation! Is such boundlessness Infinity? This illustration helps, in conjunction with fig. 7.4, to explain the hierarchical derivation of finite existence (in all its finite and indefinite parts).

Essence is and is not existence. Its purity is hierarchically prior; this concentrate is nothing existential; it is infinite but its expression, not reducing that infinity, generates a universe made up of finite and apparently infinite (or indefinite) entities. These 'actualities' of mind and matter, whose restrictions never drain the bounty of a limitless Infinity, are called creation.

Check Chapter 3: Truth, Appearance and Reality. From this *top-down* perspective finiteness is degrees of endlessness. It represents a range of partial boundlessness whose special case, the anti-pole of Unity, is individual fixity. Therefore call creation's parts 'relative realities', 'lesser essences' or 'more or less restricted absolutes'. Its units are, in themselves, such lesser absolutes; each one is issued from and makes up a specific transformation of its Substance, Unity. Thus Unity precedes all finities and subsets of numerical infinities called indefinities; it generates more or less connected units that are part and parcel of the cosmos. But for all this generation Real Potential stays the same; Infinity remains aloft. **The Dialectic has identified its Central Character as Consciousness.**

Both diagrams (7.3 and 7.4) apply to the cosmos as a whole (the macrocosm) and you (a microcosm).

fig. **7.3**

	tam/ raj	*Sat*	
	existence	*Essence*	
	expression/ motion	*Potential*	
	finiteness	*Infinity*	
	duality/ multiplicity	*Unity*	
	things	*Nothing*	
↓	*tam*	*raj*	↑
	negative	*positive*	
	gravity	*levity*	
	non-conscious matter	*conscious mind*	

Mind-blowing! You can't grasp it with a thought or touch it anywhere. You can't, by definition, have 'an infinite finite' or 'a boundless, unconditioned thing'. Might you, however, in excess of googleplex or Graham's number, make an endless count? Is there no end to units or to numbers? In this respect the word would seem to be a slippery cipher for unfinished business - but is its concept abstract fact, actual fiction or an ultimate reality? Are its paradoxes sensible or, like a mirage in an overheated slippage of the mind, false resuscitation of your dying sanity?

You can only get your mind round finiteness. Maths is just manipulating units. Is there any physical immeasurability? Pure space, perhaps, and lack of boundary in time? Will you find it lurking in the geometries of point, circle, sphere, endless lines or cycling figure eights? What in the world can you forever add to but still never add a thing? Or divide forever leaving ever greater than its single wholes? Is it size you're after - growing even smaller, waxing ever larger while you fail to reach its fullness or get to the point? It looks as if expanding universes or contracted singularities are in your sights. Yet can you grasp, as if at rainbow's end, a box of time or space? Even more than just transcendence of dimensions modern naturalism is enthralled by such infinity wherein, at random, all is possible and everything can happen even now. What, you ponder, is the nature of this transcendental potency, material omnipotence, this non-god of a recently-imagined, godless void and great pretender to the title of divinity? Such madness is obsessive and the border of cosmology has recently, succumbing to hallucination called a multiverse (see Chapter 8), rationally become insane.

Confrontation with abrupt infinity is bruising and so physicists now blank out (by 're-normalising') most uncontrollable infinities that plague their calculations. In fact, better theories might well absolutely exorcise the demons. Surely, though, you still believe that matter is eternal or opine, for an expanding universe, that something you call 'energy' banged for a start from absolutely nothing (which is an infinity of lack). To begin or not to begin, that is a question. Starts and ends disrupt infinity. Whether it's material or immaterial, edged existence must be second to the (N)One. Edged distinctions, finities of mind and matter, constitute creation. Edgeless, uncreated Essence is At First. First comes the Transcendent One.

All things are finite. How, therefore, can infinity be anything? How can formless boundlessness be isolated as a single unit any more than any number of them? dialectical infinity transcends all number. It is not a bag of abstract individuals, even Cantor's endless groups. It is not mathematical. Cut that department out. It is Unity beyond all units. How, therefore, can anybody chalk transcendence on a calculating board? Ever larger, ever smaller, don't the roots and powers and series, being built of quantities, miss the Actual Point of Quality? Do endless space and time exist; is natural law forever? Or might that infinitely-expanded triplet not be punctured by an infinitely-belittled point and such a spear-sharp singularity destroy all three? From quantities to quality again. You might value but can't calculate an Utter Quality; you can't count the Point of Being, Metaphysical Infinity, that encompasses and yet eternally upstages all the finite rest. You know infinity is neither fenced nor has a centre (unless, of course, you find its middle everywhere); how therefore, unless you grasp both sides of Paradox, can such a Point of Being be the Central Apex of Mount Universe? And if it is a Singularity then you must also, knowing it, commune with Oneness and, by being it, experience Unity. For a mystic 'countable infinities' or 'indefinities of number' are existential, this Essential; they are apparent, this alone is Real; where they are relatively false this One is True. No room for two! And since in knowing you must be alive then there's no room for any speck of death. This is Pure, Concentrated Life or, if you like, The Concentrate of Consciousness. It is a Source beyond yet at the heart of mind and matter; Essence is beyond existence so that mind cannot imagine what it's like. Known yet unknowable it passes understanding in a waking, fully woken absence of the mind - Enlightenment. It is towards this Subjectivity that Natural Dialectic clearly drives. Ascend its right-hand column to the logical extremity of (*Sat*) Top Experience. Is this, when mist upon the mountain clears, what you were after all the time?

There you have it. Two perspectives.

Bottom-up you have doubtful physical and psychological infinities. Where can you find one that is real? For example, is there speed potentially infinitely great or is a barrier set? Is there a fundamental, underpinning brake that always bars velocity from its infinity? Actually the foundation stone of modern physics is the limitation of 'pure motion', of the unimpeded speed of light in nothing. This speed (called c) is capped; the 'absolution' made of energy, space and time combined is confined. Nor is any individual object or event by definition boundless - even, if there was a start to things, the universe itself. Why therefore speak of endless matter or of physical infinity - unless the cosmic membrane's porous to those fixed and fluent nothings, space and time?

No doubt, I understand that you can always add another onto any number or, like Zeno did in mind, divide one down indefinitely. There are 'potential infinities' of number and of size. These are 'indefinities' that can, by addition or by multiplication, grow ever larger; or by subtraction or division, always more infinitesimal. Do such expansions or contractions, involving quantities of units, ever break into The Real Infinity? Or are they relative or 'false' infinities? That seem to be but aren't? The Dialectic terms reflections of the Absolute inessential, existential, incomplete or apparent - lesser species of Infinity. They lead, potentially, towards infinity but never actually arrive; nor are they the Infinite Potential, Actual Singularity or Truth from which finite duality derives.

You see the difficulty? If there was nothing where did something - any unit - come from? What is First Cause, Prime Mover, Metaphysical Creator of which the consequence is universal change? Could you find out by regression, infinite regression? Could you chop up, say, gold until you reach finality? You can - the crock is called an atom. Could you chop up atoms? Yes - to sub-atomic particles; but now, upon the borders of infinity, what decisive termination is there to regress? Can you, for example, chop up space or get the 'other side' of next-to-nothing, light? Ask a physicist.

You could do the same with time, fractionating seconds into ever-smaller dots approaching pointless zero or else expanding it forever past and future. Could time precede itself or is it an eternity? If not, where's the line? Time has edgeless edges in its starts and stops but what *is* the end of infinite regression at the start, the sharpest edge of time? What is the nature of what came before beginning - the singularity called nothing or infinity? This is, of course, the very question this whole work addresses. *Be sure, as we shall see, the nature of Infinity is not numerical.* If there's not a single, finite unit how can mathematics come aboard? As opposed to abstract, actual Infinity is not mathematical. Calls to address its nature using number or regression miss The Metaphysical Point and therefore, dialectically, cannot be more than just a fraction of full truth.

Let's take a second look at the 'infinities' of space and time. Up to the twentieth century it was assumed that 3-d space was everywhere and time flowed steadily. A separate pair. For Einstein, though, the form of formlessness - of space and time - is fashioned by the motion and the spread of mass within a phantom homogeneous pair, a space-time unit. Total void is flat but things can bend it. When coupled up with matter placeless space is groovy and, as the universe expands, elastic. Could flexi-time be squashed until it disappeared; or stretched until it hung forever motionless? Could flexi-space contain sufficient matter to ring-fence itself? Is it so curved that, finitely infinite, time has an end although it has no actual edge? No other side to peep across. Or is the universal fencelessness we shuttle in an infinite, immaterial and yet non-conscious desert - cold and black like nothing else?

What's at the centre of infinity? Edgeless and without a centre, is the cosmos finite or does its business incorporate infinity? How do you square a lack of centre with the 'white hole' of a naked singularity - big bang itself? You perplex me with the answer that big-bang did not burst from any single point but everywhere at once. Is pointlessness like this a possibility - unless its point was all there was? Indeed, reverse a white hole's cornucopia and squeeze space, time and everything back into such 0-dimensional pointlessness. Could its extreme of curvature press mass into a less-than-pint-size, no-size dollop of infinity called, pointedly, a black hole? Could you arrest the flow of time and halt its traffic through a pointless gap like that? You might penetrate the 'dot' and, like a camel through a needle's eye, rip up the fabric of finiteness and obtain 'physic's other side'. Would, at the other side of 'worm-holes', this kind of extra-cosmic metaphysic still be physical or not? The scientific jury is holed up in judgment pinning 'absolution' down. Are its 'infinities' in fact, as in the case of light, unreal - un-really infinite and really limited? Is the universe a finite one or not? Perhaps, because the nakedness of total freedom from condition is censored from our eyeful sight and everlasting space is clothed by speed of light, the curvatures of space and limits on observers' lives, we'll never know.

All this, of course, has not stopped physics blowing bubbles in the air. Taking Bruno's idea of a physical infinity of universes the science fiction of some cosmologists speculates that innumerable material *cosmoi* might (in the manner of Hoyle's or Linde's theories of expansion) keep continually self-inflating out of nothingness. Each bubble of this foam, the speculation runs, might house a different world, a different set of properties and therefore possibilities. You have, according to this plan, a 'plurality of worlds'. You have an Infinite Material Universe but, of course, no Infinite Intelligence because this sage idea is at the roots of blasphemy, the depth of rationalistic disease, the height of irrationality. You can understand, dialectically, why. If objective matter is the extreme anti-pole of Subjective Essence, if it marks the outer circumference of Axial Truth, then its infinities are wholly apparent, illusory and false. There is nothing to gainsay this assertion but such 'rough, rude blasphemy' is fundamental to evolutionary materialism, an orthodox 'scientific' perspective and, therefore, atheism. This is because it requires an infinite number of chance events to generate, by reason of infinite random motions of atoms or EUs (exotic extra-universals!). It demands infinite material possibilities that evolve a universe in which intelligence can sit. You need a multiverse (Chapter 8) for this arm-chair world-view to work!

None of this is new except with respect to astronomical discoveries. *Bottom-up* logic was awake and kicking with non-archetypal 'odd-men-out' such as Democritus, Epicurus, Lucretius and other minority classical rump-ends so grasped as grist to mill by scientific atheists. It was negated for millennia by the teleological influence of Plato, Aristotle and the Church but blinked then winked again with Bruno. Now that materialism holds scientific sway the only real factor that can knock out chance - mind - is discounted; and, of course, so is the Only One wherein no element of chance exists, wherefrom only faultless exercise of purpose ensues - Essential Being. Materialism's 'infinity of worlds' is one of quantity quite different from the hierarchical but single Universe of Natural Dialectic's 'Mountain'. From Independent Essence are derived psychological and physical fields of action. Each is, even if it presents aspects that might appear indefinite, a finite subset of Infinity. Indeed, if it is life you want then why should Infinite Intelligence wait on unwitting chance to fail to make the systems that embody it (say, you and me incorporated in the vale of death)? What science actually knows confirms the starry uni-verse is not a multiverse or an 'infiniti-verse' but what its name means - single. Unless you call each galaxy another universe then this agrees with Science and the Soul's position. Isolated universes don't exist.

A paradox of infinity is, you may remember, that it can be considered - in both point and extended forms - as a double entity. As you probe the micro-world you approach a singularity, point infinitesimality. On the other hand, as you probe galactic space there also seems no end. Could there be an end? What might be there on the other side? Rather than end you might extend forever: macro-infinity stretches, if not expands, forever. By this token, finite existence is cradled between 'external infinity' of extent and 'internal infinity' of, everywhere, lack of extent, of point infinity. We are in a great show marked upon an endless, omnipresent screen. *Such existence is a medium, an intermediary like a man in a hammock slung between two extremities, two opposite infinities that are in fact the same.* Indeed, if you were to expand point-zero up through the micro-world and, at the same time, contract the cosmos, the size of a human is about where they meet. Half way (logarithmically) on and a galaxy you are a resolution of the double vision due to focal length set on infinities; you are also, close enough to middle distance, a measure of the universe.

It needs be emphasised that mathematical existence (that is, logical self-consistency) and physical existence are not the same thing. For example, you can invent endless geometrical systems but only Euclid's describes nature. It needs also be re-emphasised that mathematical infinities are psychological. They deal in number, that is, cipher. Cipher is symbol, a symbol is an image and these 'perceived infinities' are, essentially, imaginary. Many are 'potential', created by the addition or subtraction of numbers, sets of numbers or objects indefinitely. It takes finite units called conditioned things to finance this lesser reflection of the True Infinity.

Georg Cantor, a brilliant German mathematician, perhaps came closest to the *top-down* dialectical perspective. He developed a theory of transfinite numbers (called, in *fig* 7.3, indefinities) wherein a larger set was always possible. He worked upwards through these towering eternities of number towards an Infinite Absolute. His crescendo involved a hierarchy that began 'below' at (*tam*) physical level with apparent infinities such as space, time, sets of units and so on. The tower rises through (*raj*) psychological infinities of number, both countable and uncountable, until its climax in the (*sat*) highest, greatest endlessness of all - for Cantor God. His God is, however, infinite beyond increase of number or apparent boundlessness of space and time. He is Transcendental, Absolute Infinity.

Let's stick to number and equation for a moment. You can add, subtract, divide or multiply infinity but leave it just the same. Infinity = $1/0$; therefore infinity x $0 = 1$ and $0 = 1/$infinity. Indeed, x (any number) times infinity is as infinite as any divided by zero; and any number to the power of zero = 1! Strange equations. What's their meaning? Can the Dialectic offer any kind of resolution?

The motion of Essence shows as things and events. These are finite. What is finite is, simply, a fraction of The Infinite; it can be assigned a number or described by an equation; but description, verbal or mathematical, is not the finite object. Existence is in all its parts finite, relative and dependent but its predecessor, Essence, is independent, absolute and infinite. Being without condition The Infinite is, as Itself, no thing. It is Nothing. And the Symmetry of this Immanent Void is unbroken, undivided and therefore One. **In other words, The (N)One is, simultaneously, one, zero and infinity.**

This Supreme Paradox is, beyond existence, Absolutely Non-Sensible; it is, however, furthest from absurd; it is our foundation and our starting-point. **Its numerical nonsense, the Essential Equation, is expressed as $0 = 1 = $ infinity.**

The Existential Equation is, conversely, expressed as $0 \neq 1 \neq$ infinity. *This, amid the relativity of existence, makes straightforward sense of things.*

The former, which represents Absolution, is converted into the latter, Relativity, simply by motion. In the case of cosmos relativity is established by the primordial motion of First Cause.

What is the character of, preceding existence, Precedent? What is nature of Prior Principle? What is the Nature of Nothing? Potential precedes event. What potential precedes the whole cosmos? *If macrocosm is ordered then an ordering, informative principle precedes and guides its realisation.*

The order of an act of creation (*figs.* 5.3, 5.4 and 6.1) has been discussed. A physical creation is an arrangement of mind. Mind is, by its nature, an order-maker and the basis of active mind is consciousness. One concludes, therefore, that the Precedent Nature of Nothing is consciousness: and, which really turns our commonsense view of things on its head, it is from this ground-state that creation is devolved. Existence is a projection of mind whose source, whose subtlest origin and first cause is Pure Consciousness. No wonder mysticism, with such subjective reversal of materialism, is scorned! Such metaphysic is physic's vacuity. It apparently plays no part and, unless it is invisibly a source of order or of energy, is irrelevant. Such a conclusion certainly seems absurd until one remembers (from Chapters 1, 5 and 6) the basic couple of cosmic components. These are consciousness (whose motion *is* informative mind) combined with energy. Energy *is* motion according to the play of three cosmic fundamentals - (*sat*) potential, (*raj*) motion and (*tam*) exhaustion or complete passivity. How might such a combination work?

Essential Consciousness is, in 'concentrated' formlessness, immobile; as the Infinite (N)One it is latent potential from which the cosmos is, in all its finite multitudes of change, expressed. Beyond mind, Pure Consciousness is inaccessible to mind's comprehension; yet its subjective characteristics include Life, Love and, as the Light of Reason, Knowledge of the Central Truth. How many, ask the mystics, make Communion? A derivative of the Essential Equation could be written to describe such Precondition: $1 + x$ (where x is 1 or any other number) = 1. All is One.

This section does not, however, discuss the subjective, informative aspect of Essence. Rather, it turns 'down' to discuss energetic outcomes and, as such, First Principle is aptly described in terms of a trinity of objective, impersonal characteristics - infinity, unity and zero. From this triplex Natural Dialectic derives (Chapters 7-12) the sub-principles that govern the polar, physical universe. Such sub-principles are the basis of what we call the laws of nature. In other words, the realisation of this cosmos represents a drop through the conscio-material gradient and, as such, from metaphysic to physic. It amounts (*fig.* 7.2) to a crystallisation of principle. In this view the physical universe is not itself a cause but a vast arena of effects. This hierarchical idea is not new. It is the staple of all human thought except for one-tiered, 'flat' materialism.

By definition, equations balance. In finite, polar existence the number one is a unit, part of a neutral, undivided Whole. Balance, neutrality and polarity are all expressed in the simple equation where $0 = + 1 - 1$. One physical example of this is called 'conservation of charge'. The number of electrons and protons in the cosmos exactly, as far as we know, cancels out. In this case 0 stands for such cancellation; or a proton plus electron - neutron; or 'physical enlightenment' when matter/ anti-matter vanishes in light. What is the big 0 - cosmic zero? Surely it is not the circumstance of charge before division? Is it light or does it stand for neutral 'archetype'? Or for balance, poise and point where motions cancel out? Another illustration of conservative equation is Newton's Third Law of Motion ('if body A exerts a force on body B, then body B exerts an equal but opposite force on body B). This, in a form expanded to include the 'metaphysic of mind', is also the oriental 'law of *karma*' or, since *karma* means action, Law of Action and Reaction. If you are sensible then voluntary, balanced action means 'do as you would be done by' - because you will!

Human mind is a construct of existence, of finite extent. The (*Sat*) Infinite is, therefore, by definition incomprehensible; to know it involves, paradoxically, the dissolution or transcendence of mind. Similarly, descriptions involve words or symbols. Words are neither the thing nor the experience they describe; they are simply indicators. All intellect can do, in the case of the incomprehensible or indescribable, is to use the best ciphers it can and push the limits.

The Order of the Infinite

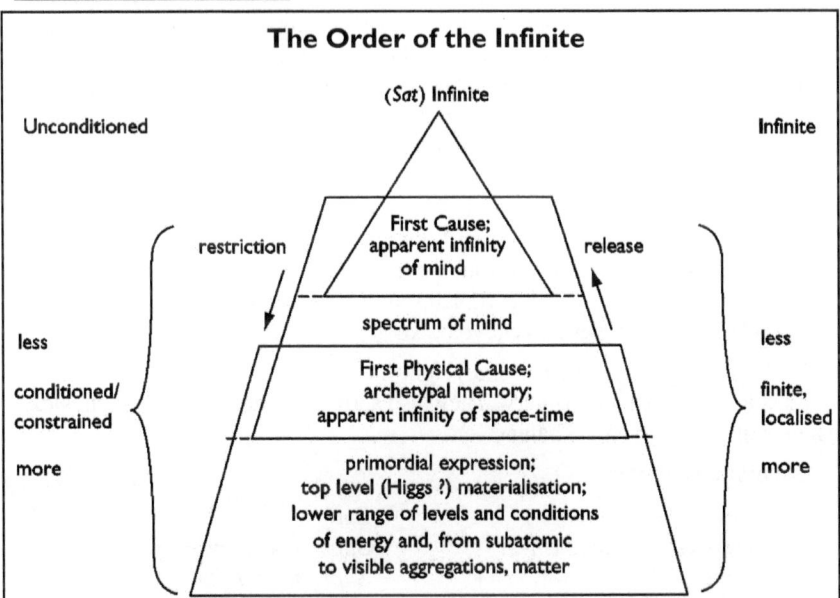

fig. 7.4

tam/ raj	Sat
existence	Essence
finite	Infinite
range	Absolute
hierarchy	Transcendence
spectrum	Transparency
time	Timelessness
place	Placelessness
form	Formlessness

peripheral	*Central*
cycle	*Axis*
motion	*Equilibrium*

A facet of Essence is, as opposed to the non-conscious nadir of its anti-pole (a concentrate of objectivity we call the physical universe), a Concentrate of Subjectivity called Super-consciousness. The zenith of 'Central Super-consciousness' *transcends* mind; intellect cannot, therefore, imagine it.

The Infinite is Essence in which things have their being, Poise from which motion issues and the Formlessness within which forms move. It is reflected in lesser (*sat*) infinities or absolutes. In their primary appearance these constitute the top level of each tier of existence. For example, the origin of material behaviour is construed as metaphysical; its principal (see *fig*. 2.11, 2.13, 7.2 or 7.4) is archetype in universal mind. *'Potential matter' is hidden from physical probes but 'apparent infinities' or 'lesser absolutes' of physical type include the three transparencies of space, time and (upon which these abstractions depend) pure energy.* Unitary Essence is the Source from which dipolar existence, through the forms of information and energy, derives. Mind, in its aspect of universal memory, is the prism through which the three transparencies are, like a beam of light, split; it is like film through which the negative of dipolar, spectral physic we call nature is developed.

Essential or Main dialectical (*Sat*) right-ha
nd characteristics are primary. They also show, secondarily, in any physical instance of centrality, axis, balance etc.

For mind we write:

	tam/ raj	*Sat*
	restricted consciousness	*Super-consciousness*
	mind	*Supramental State*
	relative impotence	*Potency/ Control*
↓	*tam*	*raj* ↑
	unconscious sub-state	*spectrum of mental states*
	darkness	*range of shades*
	total lack of control	*partial impotence*
	mind dormant	*mind awake*

And for matter:

	tam/ raj	*sat*
	non-conscious matter	*informative archetypes*
	classical/ quantum patterns	*potential matter*
	actions/ excitations	*vacuum*
	waveband/ colour	*light*
↓	*tam*	*raj* ↑
	darkness	*range of shades*
	classical restriction	*energetic quantum phase*
	mass	*force/ energy*
	macroscopic mass	*microscopic agencies*
	automatic reflex	*partial impotence*

The Infinite is, like Love, Pure Paradox. The more you give the more you have; the more you have of less of it the greater your increase. More for less! So easily! What motive power! What industrial steam! The will of love determines, at root, all we really want to do. If it fuels a finite span of animation (such as ours) then what might Infinite Love accomplish? Its poise is ever-readiness, its potential greater than all actions and its communion inclusive but above them all. Its 'psycho-space' is limitless and thus all-comprehensive, its eternity is instant and its presence never other than immediate. You may think such paradox is nonsense but it is the very source of sense and therefore sensible.

Supra-mental infinity of mind is (fig. 3.3) called First Cause, Word, Logos, Shabda etc. It is a conscious experience, the seed of thought, the Logical Start of things. Its experience and, therefore, knowledge is called metaphysical enlightenment.

The lesser supra-material infinity of body is a second, lower first cause. It is called, dialectically, potential matter. What can be said about the logical start of a limited set of things, physical phenomena? What about sub-mental 'infinity' and the nature of mind's junction with matter? In other words, for the sake of discussion that spans Chapters 7 to 22, let's be clear about the psychosomatic nature of potential matter.

<u>This physical absolute is, transcending matter, metaphysical. It is composed of archetypes or memories in universal mind.</u> **Such archetypes define the behaviour of bodies simple (as studied by physics and chemistry) and complex (biological).** Natural Dialectic identifies their grade (*fig.* 2.13) as potential matter. Do they constitute David Bohm's or Leonard Susskind's 'deeper, holographic level' of the universe - a universe projected from them which we sensibly construe as 'real'? Potential matter, in the form of archetypal memory, informs; it is law; it defines the 'normal' behaviour of physical events. You won't, from this perspective, twist a twister. You won't whip cosmos like sand devils, whistle atoms from an empty wind or blow a 'big bang' up from absolutely nothing.

In other words, through precedent 'archetypal memories' proceed the acts of physical phenomena. These phenomena are various, detailed, local and, in their material entirety, a cosmic nadir. This is because non-conscious energy is a special case of consciousness - its absence, its complete antithesis. It is below even sub-conscious, pre-physical mind; it is stranded out on the circumference of creation, a rind, a skin around creation's fruit. In other words, what we think of as 'physical reality' is just the outer layer of the cosmic onion. 'True' physics, both in archetypal information and the energies at which things unify, is much further in; or, as 'inner' quantum physics indicates, much higher up the creation's scale. It is hidden and symmetrical; and from the symmetry of archetype is 'broken' all the multitude of detailed, local outcomes. Phenomenal! Phenomena are fall-outs from the numinous; they are excitations in the vacuum's archetypal fields. These finite, restricted patterns of behaviour have their basis in the potency of void; from apparent nothingness emerge conditions we call 'actuality' - by which is meant the non-subjective stuff of physics, chemistry, biology - material science as a whole. Modern physics is described in terms of fields. The Standard Model next proceeds to specify the patterns these fields can describe in terms of forces, particles and other more or less limited powers (the most limited category being particular, solid objects). What might drive or first propel expanding matter? At the root of Expansion Theory is subatomic expansion. Neither its origin nor character is identified. Either way, you might count an indefinity or apparent endlessness of numerical units. This material universe is, in dialectical terms, an inverse of the Infinite Concentrate of Consciousness. It is a concentrate of non-conscious energy wherever action or an object shows and, in its lack, a dark, insensitive, impotent emptiness.

It may be possible, using a little inverse logic, to try and grasp the splendour of spontaneous, self-illuminated Infinity. Concede that base is opposite apex, that the apex of the Cosmic Pyramid is Infinite and the base, its anti-pole, is a zone of most finite, dense existence. Look up on a clear night at the profusion of suns, stars and galaxies flung like sparkling dust across an apparently endless darkness. This is physical locality. Astronomers catalogue the wonder of which we, human individuals on planet earth for a short stay, are such a materially infinitesimal yet wondrous part - since life alone is conscious and can know.

Now think of the Infinite as the inverse of non-conscious, dark voids and distinct solidities; and also as the antithesis of a localised, physical starry 'event', at whose source is mass and which radiates physical light outward. *A picture emerges of One Radiance, whose centre is Conscious and shines without mass; it shines towards an outer projection, a precipitate. Against this Inner Light our universe appears a dark anti-pole; opposite this Knowledge it is an involuntary, non-conscious object. A body develops from the nuclear potential of its egg; the cosmos is developed, organically, from its nuclear Information Centre. Physical nature is thus seen as an inversion, the Truth turned inside out, Life washed up on a far shore.*

This shore may be the finish. It may be the base of cosmic hierarchy yet it is not so far. By analogy your brain is top-centre of your balanced, bilateral body; like a pilot surrounded by his vehicle your mind flies its body through the world. We stand infinitesimal on a tiny planet; we look out towards stars and the nether edge of space and time but we can, from the cosmological axis at our eye-centre, also peep within. We can look from outside in, through a glass darkly. This route (*fig.* 0.11) ascends along the inward, metaphysical slopes of Mount Universe until its Peak. Its Infinite is nowhere without; yet at the Central Nucleus it is both within and beyond all finite objects and behaviours, psychological or physical. This is a simplistic picture within the spectrum of whose radiance a *sufi, yogi, kabbalist, bodhisattva or Christian saint* might define five or seven heavens, each sphere within its previous, before the Infinite Height were scaled. Can you scale what's off the scale?

The Finite

Do you remember gazing up at Cantor's unimaginable tower of infinities? What about the Dialectic's *top-down* view? A hierarchy issues from Infinite to least infinite, that is, most finite, fixed and localised solidities. Things. In this respect the Order of the Infinite might read, "Make finite, compose individualities and thus create diversity from Oneness".

Revise the legend of *fig.* 7.3. Creation is a composite of finite entities. These are 'actualities' of mind and matter that 'degrade' by levels or degrees into their states of fixity - memory and solid things respectively. Each object or event is a phenomenon of underlying Substance; each unit, however you have chosen to cut units out of wholes, represents specific transformation of an underlying Unity.

Things are one thing, time another. Time is (Chapter 11 illustrates) full of paradox. For example, the qualities of 'commonsense, body-centred' time and Infinite Timelessness/ Placelessness are distinct. They are poles apart. The latter drops, *top-down*, from Subjective Omnipresence to the duration of a physical event. In dialectical terms an event is 'forceful' but a 'frozen event' is an object or, psychologically, a memory. In other words, (*raj*) action is less finite than (*tam*) inaction. The more 'frozen', that is, the longer a particular object's duration the sharper its definition, the more finite and less infinite its appearance. The duration of, say, stars or chairs or solid rock is underwritten by 'locked-up vibration' called an atom and, of which it is composed, the basic physicality of proton and electron - these are physics' ultimate survivors. How paradoxical! Down-to-earth endurance may appear to last forever but is finite in its quality; yet every finite object holds within itself a lost eternity, a Moment of original infinity.

The Hierarchical Order of the Infinite also runs from Conscious to least conscious, that is, a non-conscious, subjectively passive universe. Infinity is Absolute. What might be the nether parallel, the best material reflection of Absolution? What is the least physical 'object', the one furthest from differentiation into solid objects? Is it space? If so, you might infer (see also Chapter 10) that all things issue from a potent aspect of the void.

Essence, Infinity and the other Main Principles are Real. They are reflected in a suite of grades that end in their polar opposites; they are represented in a spectrum that ranges through mind down into oblivious, material opacity. The only place, as far as we know, that the facets of First Principle are fully realised is in an earthling called *sanctus*, a saint. Such designation is not chartered by a university degree; nor is the title one bestowed, in the manner of ecclesiastic knighthood, by papal or popular decree. In the dialectical sense it is reserved for those who, either well-known (such as the Buddha) or unbeknown, have 'technically', that is, actually *realised* Super-conscious Brilliance.

Aspects of (*Sat*) First Principle occur on the right-hand column of Main dialectical Stacks.

tam/ raj	*Sat*
existence	*Essence*
below	*Apex*
expression	*Source*
finite	*Infinite*
relativity	*Absolution*
variation	*Invariance*
effect/ cause	*Potential*
manifest	*Unmanifest*
appearance	*Reality*

Their existential antitheses appear to the left. Each antithesis is polarised in the way described as follows by Branch Dialectic:

↓ tam	raj ↑
down	up
below	above
negative	positive
passive	active
anti-principle/ debasement	principle-in-operation
materialising	dematerialising
dragging to static	raising to kinetic
fixed	fluid
end-point	process
mass	energy
object	event
structure	function
effect	cause

In this case (raj) active aspects are upward in motion. They involve increasing lightness, unification, stability of poise, relationship and liberation; they ascend towards, at the Top, the Infinite. *Each principle is antithesised by its (tam) downward, consequent anti-principle.*

As already noted there exist two main existential tiers of the cosmic pyramid - mind and matter.

objective	subjective
outer	inner
physical	metaphysical
tangible	intangible
matter	mind

If the Dialectic is written for the metaphysical level of mind, in which active information predominates, we read:

tam/ raj	Sat
scale/ range	Climax
more or less	Complete
relative knowledge	Comprehension
partial science	Science
lesser truth/ relative illusion	Truth
impurity / division	Single Purity
rainbow/ spectrum	Transparency
relative shadow	Light
information	Source of Information
practice/ principle	Latent Order/ Preordination

↓ tam	raj ↑
passive	active
recipient	messenger
practice	principle
reflex	stimulus
informed	informer
storage	motion
fixed	fluent

memory　　　　　　　　　　*thought*
lower informative phase　　*higher informative phase*
sub-conscious　　　　　　　*conscious*

Light is a colourful transparency. As its clarity illumines, so does Truth. And Boundless Clarity's refracted into separate shades; all things are seen as lesser, variously coloured lights within its pure Transparency - a great and cosmic rainbow, an arch across the world's great heaven. It is thus that existence is called, dialectically, the conscio-material spectrum; in the spread below its Peak it is called creation's gradient; and in its motion up and down from Climax, the ladder of the universe. Light symbolises hierarchy; it reflects the scale of things.

If there is seeing there is knowing. If there is knowledge is there Knowledge that's Complete? Is there Comprehension unifying everything? If so, of what nature is such Knowledge from which less important kinds of knowledge are derived?

Put this round the other way. If there was, before anything, Nothing then what is the nature of this Apparent Absence, Vacuous Principal, First Causal Principle? If principle is conceptual and informative then principle derives from mind. This would apply equally to individual and, with respect to the laws of nature, universal mind. Of the two main grades of mind conscious is superior to sub-conscious, active precedes dormant, or, in other words, an impression or thought precedes its own storage in passive mind. Active information (Chapter 6) precedes passive. Thus active mind composes schemes and rules and hands them down to memory. Just as there's sub-division of sub-conscious mind (Chapters 15 and 16) into higher 'personal' and lower 'archetypal' parts, so conscious mind (Chapters 13 and 14) is also sub-divided. 'Lower mind' perceives, conceives and issues orders appertaining to the world; its 'noisy' state informs and is informed by physical phenomena. By this level subtle *Tao/ Logos* has been lost from view. As fundamental particles are 'overwhelmed' and unperceived as part of gross material objects, so its Primal Message is 'smothered' by surrounding 'noise'. It is, like a whisper in a storm, drowned out. Contemplative mind is focused, on the other hand, towards its Source; its deep tranquillity receives the Signal clearly from First Cause. *Higher first and lower later - mind is 'switched on' by the light of First Principle. It is, in effect, the prism of the Infinite.*

Mind is the domain of knowledge (Lat. *scientia*). *Sci*ence means, as noted in Chapter 0, knowledge. Not only does mind inform, it is informed. Either way, it knows. This knowledge, also called experience, is *conscious*. What unifies a world of difference is *con-sci*ousness. Mind unifies distinctions in a single, central picture. Look around. As you do so your own consciousness combines, within itself, the separate 'data items' you can see, hear, remember by association and so on; it builds them seamlessly into experience. Things are related and are organised by mind. Mind may be called consciousness-in-motion. It is, at the subjective centre of our life, single by virtue of its single knower. Levels of consciousness depend on the scope of its unification and, therefore, experience of unity. Run-of-the-mill human consciousness is localised and sense-based. It is, in cosmic terms, very restricted, conditioned or qualified. It is separate from universal wholesomeness.

What about lifting the restrictions? Expanding a contracted form of consciousness? Knowledge of relative things is relative knowledge. It can only be consummated in a unified understanding of all things combined, that is, in terms of their root togetherness. In this respect, Knowledge constitutes the only absolute frame of reference against which the relativities (or motions) of existence can be measured. The question then becomes one of how to attain that state of Consciousness, obtain the Inside Information and get the Primal Message.

Consciousness is what we recognise as central to our life. It is the essential, living screen of knowledge. Universal Comprehension, whose Peak sees everything related on the slopes of Universe, is called Pure Consciousness or Enlightenment. This is not a state of mind alone. It is *the* cosmic pre-condition. **If Essence is Pure Life, if First Principle is both alive and the source of all lesser, restricted lifetimes then the heart of existence is Pure Life.** Its Potential is the Source of Information. It is Science. Science is knowledge and the unexpressed First Principle, which contains within it all other principles, is Omniscient. In knowing Nothing such achiever, paradoxically, knows everything. The Omniscience of Enlightenment does not, therefore, involve knowing every or even any trivial detail about what is happening or what anyone is thinking at a particular moment - which is a serial kind of grasp. Instead it involves union with Potential Information, the Pure Science that precedes any existential expression of Itself. To descend from that Height is to descend through the layers of creation and therefore know the whole business, *top-down*, as it is. The Poise of Enlightenment is incomprehensible to the intellect (which involves only consciousness-in-motion). It is, equally, comprehensive so that as a metaphor Christianity dubbed it 'Alpha and Omega'. And because it is alive and seeds a universal 'child' some call this Source of Life 'The Father' of existence.

From Infinite Capability (or Potential) devolve lesser capabilities - voluntary and below that rigid, involuntary capabilities. In other words mind and matter are, in turn, derived from the Infinite Potential of Life. (*Raj*) kinetic mind wills and desires. These motivations inform, create or shape all behaviours. *Information*, the expression of consciousness-in-motion, is one of the basic couple of existential components. If its apex represents Infinity then the first tier of cosmic pyramid indicates the first phase of creation, mind. The psychological projection is followed by a second phase wherein the second existential component, *energy*, predominates. If Dialectic is written for the physical level of matter, in which passive information predominates, we read, for example:

	tam/ raj	*Sat*
	relative illusion	*Truth*
	relative shadow	*Light*
	spectrum	*Transparency*
	impurity	*Purity*
	oscillation	*Axis*
	things	*Space*
	changes	*Framework*
	transformation	*Field*
↓	*tam*	*raj* ↑
	passive	*active*
	lowering energy	*heightening energy*
	heaviness	*lightness*

low energy	*high energy*
object	*event*
storage	*motion*
structure	*shaper*
mass	*energies*

So objects are stored patterns of energy; and these behaviours are themselves expressions of transcendent, material archetype. They are, as we said, a crystallisation of principle.

	tam/ raj	*Sat*	
	down/ up	*Balance*	
	motion	*Poise*	
↓	*tam*	*raj*	↑
	down	*up*	
	contraction	*expansion*	
	inertia	*stimulus*	
	gravity	*levity*	

This is the place, before proceeding to elaborate the order of the finite, to gather a few threads and make a starting-point. Do you remember (from Chapter 1) two aspects of Nothing - the negative impotence of absence and its positive potential? Do you also recall 'paraphysical zooming' down towards sub-microscopic infinite smallness or out to astronomical extents? The Infinite has Potential: this Potential is expressed in two directions simultaneously. You can expand (as Plotinus suggested) from point infinity; or, as opposed to an expansive creation, you have its contractive side - contraction from infinite extent. Antagonistic balance. A kind of cosmological constancy.

No doubt that balance is a characteristic of First Principle reflected throughout creation; but so are its fundamental motions, characterised as the (*raj*) expansive and (*tam*) contractive vectors. *Sat*, *raj* and *tam* constitute the order of creation (e.g. *fig.* 2.7). Potential flows towards exhaustion; initial poise runs to inertial equilibrium; soul rolls to sole pole - Essence unfurls mind and further, as an outer darkness at the foot, the base edge, matter. Such gradient, derived from cosmic fundamentals, underlies the (*tam*) contractive expression of creation, the downhill order by which 'free' information or energy is gradually 'locked' into fixed shapes of mind (memories) and matter (objects). You get the dropping drift by now. Ideas are precipitated out, principles are crystallised. Conversely, uphill gradients take effort. They need pumps. They need an input called effort or energy. This, the (*raj*) expansive expression of stimulus, is the way that stiffness or unwillingness 'gets up and on with it'. Fixity is freed. This is order in which mind or matter, in a conscious or non-conscious way, can rise, dematerialise and find its freedom. *Thus derivation has been drawn, from the Infinity of Point and of Extent, of two fundamental cosmic characteristics - expansive (raj ↑) levity and contractive (tam ↓) gravity.*

The Order of the Finite

tam/ raj	*Sat*
below	*Top*
spectrum/ levels	*Most Light/ High*

orderly creation   *Logos*
subsequent order   *Archetype*
matter/ mind   *Soul*

For voluntary mind, that is, active information with negentropy dominant (*fig.* 2.6):

↓   *tam*   *raj*   ↑
down/ descent   *up/ ascent*
information loss   *information gain*
less awareness   *more awareness*
ignorance   *knowledge*

For involuntary matter, that is, passive information with entropy dominant:

mass   *energy*
energy loss   *energy gain*
gravity   *levity*
particular reaction   *communicative force*
affected fermions   *messenger bosons*
exhaustion   *activity*
inertia   *stimulus*

Order of the finite is (revise *figs*. 5.3, 5.4 and 6.1) simply the order of creation. It is the way (see Chapter 1: Causality) that global ideal is translated into local action, that general principle is broken into actual practice. It is how information is disseminated, programs are implemented or a project ramified into its product. What more cosmic product than existence? Creation is the ultimate in variation on its themes. Dialectically, however, this creation is a layered 'dance'; First Cause (see Chapter 1: Causality) does not jump from Essence straight to physical effects. A psychological effect, composed of information in the form of mind, precedes the origin and sustenance of *physicalia*.

You may complain that I've confused the tiny mind of man (which certainly exists) with universal mind (which, as far as science knows, may or may not). Is man constructed in the image of the universe or not? And if so (Chapter 17) how? For scientific atheism my proposal is anathema. Yet the Primary Axiom of Natural Dialectic simply bases all that follows on the premise of two existential elements - matter *and* mind, energy *and* information; and its order runs from mind to matter.

Universal starts are headaches. You think about them but whether there was psychological or physical beginning, what preceded all beginning is beyond the limits of the sharpest probe of intellect. Mind simply cannot grasp transcendence. How do you draw something out of nothing? Who wants a headache, one that's possibly incurable? Who cares?

Nobody cares - except the person who sees that it is critical for all that follows, one who wants to line his logic up according to initial truth so that the headaches of uncertainty are cleared.

Bottom-up there may have been 'Expansion' but the idea smacks of starts. First causes. What, instead, about a startless, endless 'steady state'? Isn't there, according to expansion theory, a growing ocean of indivisible and adamantine primordials - uncharged electrons whose continual expansion supports myriad patterns called events? Or perhaps, according to Fred Hoyle, it's down to an unstinting production of hydrogen atoms *ex nihilo* (at the untestable rate of one for every litre of space each few billion years) to match a startless, endless but steady expansion. It is unclear whence steady states derive their endlessly supportive drive but never mind; anyway, Hoyle's theory came to nothing since new galaxies that should spring up in between the old ones just aren't there - the old are all far-off and newer nearer. Quantum cosmology, a form of mathematical metaphysics, can also seem to blur Transcendent Bang's sharp singularity. Its psience conjures up a sea of indeterminate potentiality or a wave function of the universe that, like a primal deity, floats into space and rules. Immeasurable, invisible, omnipotent. This wave (or perhaps a pre-existent mini-universe) can play the cosmic egg. Eggs are potential and by faith you might conjure an extra-cosmic, self-inflating five-dimensional sphere (with four of space) whence, almost mythologically and certainly mysteriously, creation as we know it was developed. When a beginning isn't really one you've killed off starts. Why *ex nihilo* if, in initial absence, there's no catalyst? Where's the starter without which creation's dead?

Top-down, by contrast, the origin is clear. If First Principle is both Single and, among its other Main dialectical attributes, Subjectively Conscious then the origin of duality must be found in a primal focus of attention. It must be found in the distinction between Knower and what is known, between Self and other, between Essence, its purposes and their realisation. *Ex nihilo nihil fit*. Nothing comes from nothing; but something, indeed everything, is made of Nothing. The Void is presence not an absence; Void is Psychological Potential and not Hawking's emptier place than space; The Cosmic Egg is metaphysical, informative and archetypal.

	tam/ raj	*Sat*
	expression	Potential
	lesser consciousness	Consciousness
	relative understanding	Comprehension
	less subjective	Subjective
	not-self	Self
	prakriti	Purusha
↓	tam	raj ↑
	passive	active
	objective	subjective
	prakriti predominant	purusha predominant
	tendency to drop	tendency to rise
	oblivion	intelligence
	diffusion	unification
	dispersive tendency	concentrative tendency
	entropy	negentropy
	matter	mind

At this point let's reaffirm that, though the subject of this section (Chapters 7-12) is Energy, we need to take an overview. Before proceeding to elaborate the principles from which *physicalia* were derived, we need to grasp the energetic context as a whole - metaphysical as well as physical. We are in the business of resolution and, as such, the process of ever more sharply defining the logical structure within which the levels of creation, including the lowest, physical level, are hierarchically unfurled. *This is the opportunity, before subsequent elaboration, to reduce matters to a precise, minimal starting-point and, in the process, introduce any necessary new vocabulary.* This starting-point is the 'seed' from which (*figs.* 4.2 and 6.2) the rest of the book is devolved. A grasp of it will help you to cross-reference and make connections as you cover the sections on physics, psychology, biology and ecology.

Let's retrace the order of creation in a slightly different way. If pre-existential Nothing was a Pure Concentrate of Consciousness then the start must have involved a 'difference', a motion or expression. It must have involved a reduction of infinity, a finite element, a specific reflection. What, therefore, could the primal thought have been? A self-realisation? 'I am that I am' followed by creative purpose ('I will, I want')? What is always the command of will towards its goals, its creations? 'Be!' What fire makes manifest, what interest defines the way a work turns out? It is focus of attention. The primal, directed focus, called First Cause or *Logos*, does not disappear. As founder and foundation it remains, sustains and from the centre brings to fruit its dream, its plan - the universe.

Such focus would involve a duality consisting of the Self that purposes and the energy/ material with which it exerted its will and materialised its designs. It would involve distinction yet relationship between Self (the subject) and 'not Self' (the object) or, in oriental terms, between *Purusha* and *prakriti*. Moreover 'not Self', being the issue of Self's contemplation, would reflect 'His' Character, that is to say, 'His' Principle.

'His'? The Dialectic might drive us 'upwards' to conclude that, as positive precedes negative and higher lower, this Principle is Most Positive. If its facets include Infinity, Union and Life it might be called an Infinite Radiance of Love; or, if the basis of intelligence is consciousness, Potential Intelligence. Beyond opposites the Centre of the Dialectic, the Main Attribute, is Essential Balance. It is Neutrality. Which sex has more intelligence? Is elemental consciousness a male or female entity? Without body, immaterial, it is sexually neutral and the pronoun 'It' is genderless; but if it's reserved for objects and sub-human organisms what do you call a super-human source of life? If not 'Its' then call life by the way we understand - 'His' or 'Her' Principle but, transcending sex, it's neither.

How grates 'anthropomorphic sentiment' on disbelievers' ears! A hard-bitten *bottom-upper* wants, with all his heart, to explain the universe and life materialistically. Atomically. Hard-bitten cynics crave a bottom line as hard-edged as a proton or electron or, statistically, less. From a compendium of stories, such a naturalist would urge, could you excel the Darwinist collection? You only need some fundamental particles to prise an organism from its dust and, later, open out of sleep an eye. Did you think the purpose of an eye was sight or its reason was to see? Ask Dmitri Mendeleyev. You are wrong because no immaterial element is listed on his periodic table. It is not a card that you can deal. You can't see the chemistry of information or intention in an atom so that any immaterial factor won't, except when 'dreamed' by molecules in evolutionary accidents called brains, exist. No previous plan, no purposefully minded code has lit the world for you. There is only, like a law of nature, reason without reason - natural selection. Imagine that! You could not invent it but imagine sight (plugged into all the other faculties and instruments an animal involves) evolved. Quite simply, eyes morphed into view. If they evolved quite naturally then couldn't anything? It could. It must have, has and, even if I'm not sure how, I will concoct a plausible construction. I can spin a scientific yarn! This is, in fact, the way I hope to spot life's truths.

Do you understand what you are part of? You are part of and, you sense, apart from the materials of nature. How could you be any different? Are you also part of and, you may suspect, apart from Immaterial Nature - life's subjective side? If First Principle is Consciousness then Subjectivity precedes all objects. What is the voluntary spring of action? What is the potential for creative motion but the love of it? Whatever you are going to achieve, the starting-point is wanting to. Infinite Potential, like a love that never ends, might constitute creation's starting-point. Nothing can precede such Pure Potential; everything proceeds from it. It is not an abstract void or emptiness in space. Its Void is Pregnancy; from its womb there issue, in accordance with its character of creativity, the world's array of possibilities.

	tam/ raj	*Sat*
	local	*Global*
	specific	*General*
	imperfection	*Ideal*
	individual expression	*Potential*
	imbalance/ asymmetry	*Informative Symmetry*
	excitement/ tension	*Archetypal Field*
	wrinkle	*4-d Sheet of Void*
↓	*tam*	*raj* ↑
	classical tier	*quantum level*
	macro-world	*micro-world*
	extrinsic symmetries	*intrinsic symmetries*
	local bulk asymmetries	*local asymmetries of force*

At a material level, just as the qualities, textures and colours of solids are inherent in their gaseous form, so natural laws and sub-laws are inherent in the cast of cosmic theatre; they pre-physically determine how its mindless energy behaves; their archetypal symmetry dictates the order of design. Such general basis is 'precipitated out' as localised, specific details. Archetypal particles and forces underwrite the superficial, large-scale show. Their particular modalities dispense the outward world.

You could say that they are crystallised as individual events and local objects of a star-strewn, mostly cold and dark but predetermined space. Their expressions are variations-on-themes, thematic principles intrinsic in the potential from which they issued. Thus Primal Nothingness is Most Important. It is Most Natural. It is the latent repository of principle according to which the possibilities of nature will be actualised.

An Expression of Dialectical Perspective in terms of Cosmic Ziggurat and Concentric Rings

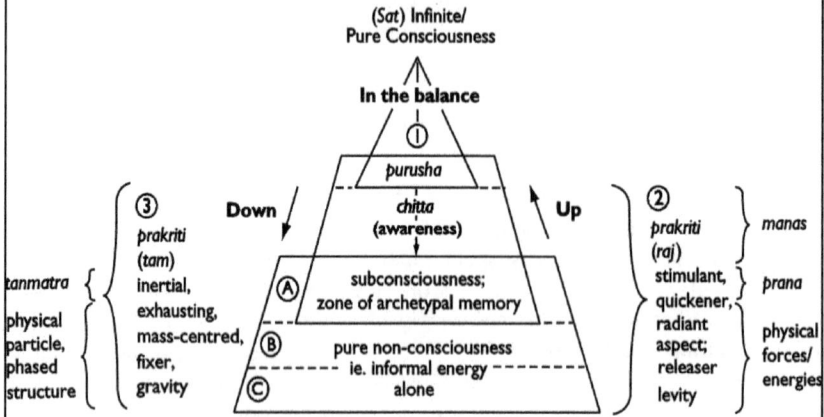

1. subjective centrality
2 and 3. vectors of (↓↑) motion; agents of objective circumstance
   or   1, 2, 3. balance, energy and form

(*Sat*) Pure Consciousness at the Essential Pole drops to pure non-consciousness, its exhausted pole known to us as the physical universe. In this view the material element of the conscio-material gradient is called *prakriti* (universal energy). *Prakriti* first shows, with strong metaphysical bias, as subtle, psychological energy called '*manas*'; the lower waveband of such energy is called '*prana*' or *ch'i*. Both levels operate in 'inner space', 'inner sky' or psycho-space called *akash*. Psychosomatic '*prana*' is associated with oriental medicine, the occidental 'vis medicatrix' (see Chapter 17: Caduceus) and, in Patanjali's Yoga Sutras, with *pranayamic* breathing techniques, physical vitality, electromagnetism (sunshine) and negative ionic charge (from oxygen). But its quality seems more a psychological than a physical; it is an energy of mind rather than a material phenomenon like, say, a quantum field. *Whichever way, this interface needs definition (see Glossary); and in this book (see also Chapters 15-17) it is used to label sub-conscious, psychosomatic energy.* Because *prana* is identified with sub-conscious energy it is also called 'potential', 'transcendent' or 'super' matter - the subliminal level whose 'archetypal programs' precede physical creation. It is therefore a metaphysical term whose closest physical correlates show as vectors of (*raj*) levity and (*tam*) gravity, that is, forces and particles. In the order of creation (*raj*) action precedes (*tam*) exhaustion and so, in this hierarchical but not necessarily temporal case, electromagnetic energy (or light) precedes the appearance of mass and gravitational effects. 'Let there be light'. What form of energy preceded even light? What kind of incipience gave birth to physicality?

The biophilic, 'halfway' or median subdivision of the electromagnetic spectrum is visible light. If *prakriti* were analogised with the whole spectrum its middle band might represent the mental 'wavelength' of humans and the lowest 'bands' would, devoid of awareness, represent the non-conscious forms and forces of physics and chemistry.

Patanjali lists (see fig. 6.2: legend and figs. 13.6 and 17.9) five subdivisions, wavebands or 'colours' of *prana* each of which can be seen as an archetype, a pattern according to which its physical correlate behaves. In accordance with wave/ particle, energy/ mass paradox, the energetic, wireless aspect of *prana* is complemented by one of discontinuity, phase transition or particular state. Such state is called *tanmatra*. The five *tanmatras* are traditionally thought of as 'ideas' or 'expressions' of energy (as, for example, the heat, light and motion expressed physically by fire); and scientifically (fig. 7.9) of archetypal fields that express, from vacuum, the fundamentals of material phenomena. The 'mind of matter' is an immaterial archetypal field; archetypes are natural as atoms and bulk form. Yet, from a physical angle, *prana* and *akash* do not exist. This is why, from a materialistic point of view, the business of universal mind as opposed to universal matter and, with it, *prana*, *qi* or metaphysical energy is mumbo-jumbo. While physical space is wholly clear its nature is less obvious (see Chapter 10) than objects that inhabit it. *Top-down*, Chapter 10 explains how the pre-physical potential of dialectical vacuum is opposite in character to inert, impotent, macroscopic space; its (*sat*) archetypal 'void' is the polar opposite of (*tam*) an exhausted sink of emptiness. This 'other side' of space's Janus-face is, dialectically, the source of physical behaviours.

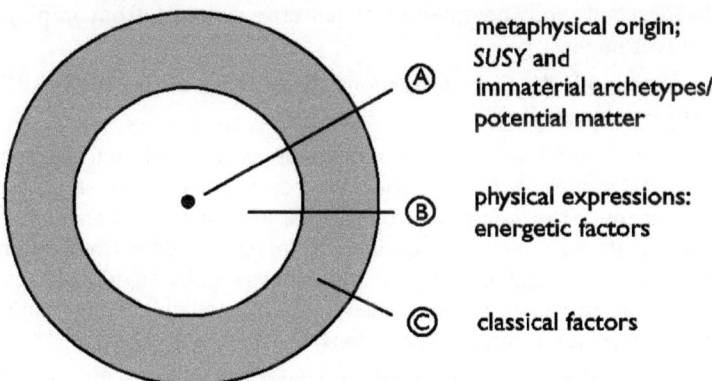

Ⓐ metaphysical origin; SUSY and immaterial archetypes/ potential matter

Ⓑ physical expressions: energetic factors

Ⓒ classical factors

This concentric representation of the 'elemental hierarchy' links with the previous diagram. It shows the exteriorisation of matter from its central, archetypal zone. The essence of matter is its archetype. From 'high' to 'low' the grades of matter are:

A. (*Sat*) potential matter; psychosomatic, mnemonic or archetypal phase whose metaphysical aspects are:
   prana - metaphysical driver
   tanmatra - causal, ideal or qualitative form of matter
   and whose physical aspect is:
   space - either 'granulated' or 'smooth'; apparent nothingness or vacuum/ ZPE.

> A. *(Raj)* energetic, microscopic phase; this is the 'wireless' level of material expression that involves electrodynamical interaction, quantum particles, atomic physics and chemistry; it used to be represented by the 'element' fire: its typical expressions are:
> light/ heat
> electron
> quark/ proton
>
> B. *(Tam)* classical, macroscopic phase; it is the 'wired' or bonded form of matter we call 'sensible reality'; its factors are expressed in states of:
> gas (extremity in the form of plasma)
> liquid
> solid (theoretical extremity a black hole)
>
> fig. 7.5 (see also figs. 0.11, 2.13, 4.2, 6.2, 7.9, 13.3, 13.6 and 17.9)

Projection of the cosmos is, as the stack above and ziggurats of *figs*. 7.4, 5 and 8 show, stepped. The issue is hierarchical. From archetype devolve the primary symmetries of balance, laws of conservation and invariant stability of how the whole dynamic works. Secondary and tertiary patterns of behaviour, known to measurement as laws that govern fact, drop orderly from these. Each state, each phase and different condition of the micro- and the macro-worlds shows regulation in a subsequent variety of ways. In short, simplicity descends into complexity of patterns summed from impinging forms of government.

Where in daily life do these abstractions get us? What use are they at office, home or supermarket?

Maybe none but, while educated, social and commercial mind entrains a humanly restricted view of life, the wide, wild cosmos isn't far away. Do you want to venture out? The door swings open, for a start, upon a conscio-material gradient that exits a subjective peak, upon a universal ladder from different condition of the micro- and the macro-worlds shows regulation in a subsequent variety of ways. In short, simplicity descends into complexity of patterns summed from impinging forms of government.

Where in daily life do these abstractions get us? What use are they at office, home or supermarket?

Maybe none but, while educated, social and commercial mind entrains a humanly restricted view of life, the wide, wild cosmos isn't far away. Do you want to venture out? The door swings open, for a
whose lowest rung the furthest, final, heavy step lands on a physical, objective universe. And does so at the point of subatomic, quantum action - at the subtlest register of non-conscious things. Call these *physicalia* the body section of the total universe.

Let's check the dialectical perspective in terms of cosmic ziggurat and concentric rings (*figs*. 6.2 and 7.5). If '*purusha*' is subjective awareness what is the full meaning of dualistic '*prakriti*'? The word is often translated as 'cosmic or universal energy' but the implication is subtler. The term includes both psychological and physical energies. It is the 'mind-film' on which mental images are developed but also, in its coarse phase, shows as the 'stuff' of physics. It is (*tam*) resistant but also, at the same time in the manner of wave/ particle duality, (*raj*) active, shaping energy. Polar *prakriti* is the prism of Essence, the medium that facilitates existence. It disperses Pure Light into the colourful, multitudinous variations of creation.

Do you remember the idea of True Self losing Purity (or Perfect Consciousness) in the course of an increasing involvement with energy - to the extent that mental consciousness is lost into a subjectively passive, reflex, inanimate world of material bodies? At mind's centre, seamlesslessly combining the objective patterns of duality with its own subjective unity, there sits a third and immaterial factor - life's attentive focus. Consciousness. This factor scales. *Purusha*, binding with *prakriti*, dims; mind, more and more engaged with mental and material forms, is ruffled from its purity; it fades until engulfed. Consciousness surrenders to oblivion. At subconscious, dormant level all control is given up but order is not lost. Pre-set patterns now take fully over. Archetypal agencies (the quantum factors) automatically govern the material refrain. Thus, while science both methodically and brilliantly studies a precipitate, the source of order stems from higher up.

If pre-set patterns underlying physical phenomena depend on and derive from metaphysical foundations then these foundations constitute a superstructure that, stepped 'hierarchically', engages from 'above'. How on earth can science in objective mode penetrate beyond material's 'glass ceiling'; how, through a glass darkly, can its instruments step up to peep inside a universal memory or mind? The answer is a simple one - to track such superstructure using subjectivity, that is, to reflect the general using one's own individual mind. No doubt art expresses archetypal principles in terms of drama, poetry and music but the way of science is mathematics and, for pictures, graphs. An abstract of maths can theoretically describe pre-physical archetypes, that is, metaphysical symmetries, balances and disturbances as well as actually measured physical events. Nevertheless occidental natural philosophers' from Thales to Descartes and, most recently, psychologists have unwittingly clothed the 'basic pair' (existential dipole's energy and information) with so many overlapping and even contradictory concepts, terms and nuances that, in order to simplify and clarify, Natural Dialectic follows the same course as it did with Cosmic Fundamentals (Chapter 2). It simply names the conscious, informative aspect of its thesis '*Purusha*' and its informed, material antithesis '*prakriti*'. Of course, different nuances of meaning also apply with the users of Sanskrit terminology but, in order to define its rationale in a way least skewed by western preconceptions, the abovementioned couple along with associated terms have been borrowed from oriental philosophy.

Universal Menus:

an alternative way of expressing polarisation, spectral hierarchy and phase division

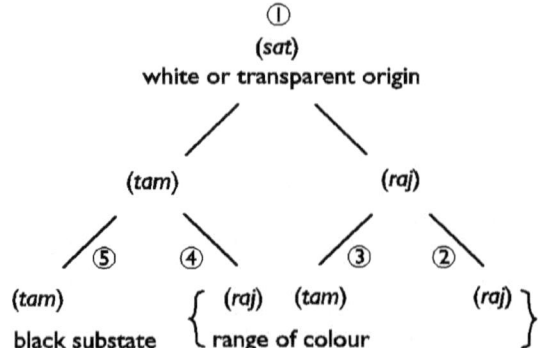

Scale applied in terms of cosmic fundamentals. As opposed to rings or ziggurats this model represents bands or gradations as polarities. The cosmic fundamentals are polarised from general into progressively more specific division, more clearly detailed forms. The pattern represents, from (1) to (5), a drop in 'frequency' of awareness or energy. The descent is analogised with the loss of pure transparency or whiteness through a spectrum of colour that exhausts, below the red of its rainbow, into the blackness of complete loss of subjectivity, the total objectivity of matter. (1) is the 'super-state', absolute or transcendent against which relativities are measured; (5) is the substate, negative extreme or subtendent.

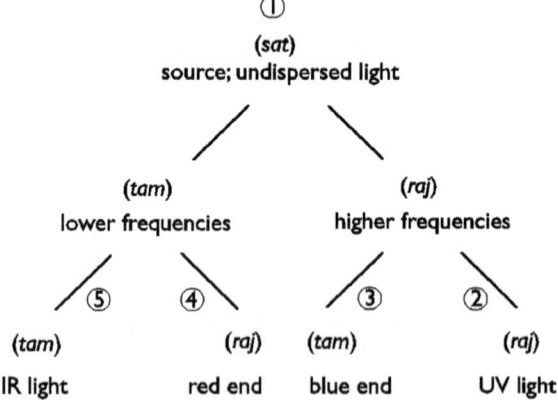

Scale applied to the spectrum of visible light. The speed of light is presumed to be an absolute against which all other speeds are relative. Infinity and zero are also absolutes. Just as the incandescence of white heat cools to blues and reds, so the transparency of white light fragments into colours. And colours, recombined, make up transparency. Is visible light a 'lower version'

of supra-visible frequencies and sub-visible darkness lower still? Where might you fall on impotence or exhaustion? Is the greater power above or below?

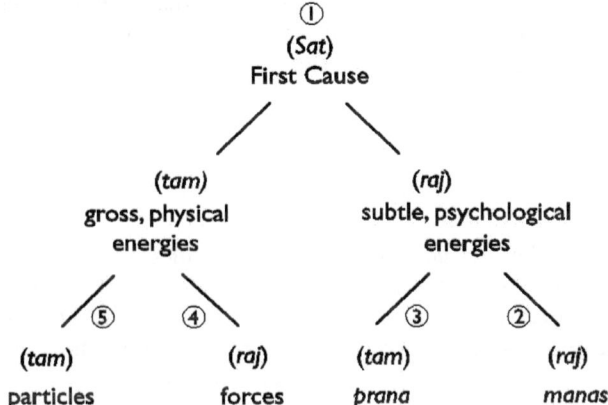

Scale applied to 'cosmic energy' called *prakriti*. Prakriti is the non-conscious, objective aspect of creation whose extent ranges from the subtlest motion of thought developed, you could say, on its 'mind-film' through to the grossest expressions of energy - massive solids.

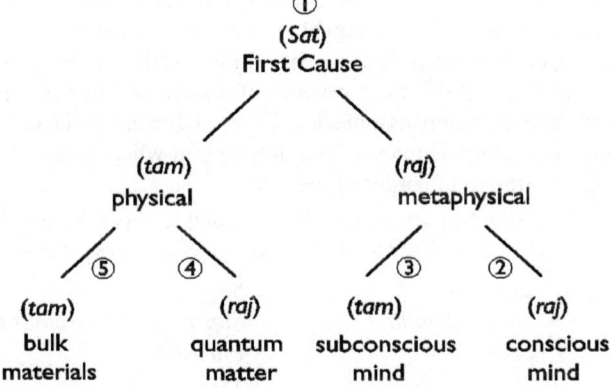

Scale applied to the conscio-material gradient. This gradient is an expression of changing proportion of *prakriti* with *purusha* or, in occidental terms, the involvement of awareness with increasingly dense or 'heavy' patterns of energy. In this case (1) is First Cause while (2) and (3) represent conscious and sub-conscious grades of mind; (4) and (5) represent the quantum energies and bulk mass of the physical universe. In this order of creation the archetype for light would reside at level (3). In fact, just as level (1) is the Potential for all below, so level (3) is the potential for grades (4) and (5). This raises the materialistically nonsensical question of whether potential, principle and archetype are words that simply represent different ideas about the same thing. Is principle a potential? A predefined capability and so, at the same time, both scope and constraint? If not, how does the system work?

fig. 7.6 (see also fig. 1.1)

The framework is therefore easy. *A key realisation is that, although mind can be thought of as a single 'object', it shows a dual aspect.* The same applies to matter. This paradoxical nature is reflected in modern physics. We speak of matter but understand it in terms of either (*raj* ↑) energy or (*tam* ↓) mass; or, for example, treat electro-magnetic radiation as (*raj*) wave or (*tam*) particle (photon). *Another key realisation is that, as the forms of matter are distinct from their observer, so are those of mind.* The field of the external world changes continually - and so does that of the internal world of informed and informative mind. *Just as the external world is composed of physical energies, so the mental world is composed of subtle, metaphysical material.*

Prakriti can be treated, like light, as a coloured transparency. The Nothingness of its Potential can (see also legend for *fig.* 6.2) be filtered, split or 'broken up' to reveal a spectrum of existential energies. The divisions of *fig.* 7.6 (i) simply represent a scale of predominance in terms of cosmic fundamentals. In *fig.* 7.6 (ii) this idea is applied to radiation and the biophilic waveband of visible light employed, in primary terms, for information (vision) and energy (photosynthesis); and in *figs.* 7.6 (iii and iv), presenting an alternative to the universal ziggurat (*fig.* 1.4), it describes the conscio-material spectrum as it covers mind and matter.

Materialisation is a gradient of loss; loss of energy includes the 'drag' of cooling. The order of the finite therefore drops through gas to solid but, in all the void of space, what might you expect to find upon the nether side of nature's Great Infinity? Could such a pole comprise 'finite infinity', infinity of mass, the 'absolute inverse', an antithetical reverse of Essence? Physics has suitably identified such base extremity as a black hole. Most dire and finite of 'infinities', such crushing singularity denotes a final dot, a point where space and time are lost - a fitting full stop to the order of the finite.

The idea of 'gradient-by-menu' can be extended to show the subdivisions of mind (*fig.* 7.7) and matter (*fig.* 7.9). For mind you could write the stack:

	tam/ raj	*Sat*
	states of mind	Super-state Consciousness
	prakriti	*Purusha*
↓	*tam*	*raj* ↑
	mental sub-state	range of awareness
	sub-consciousness	embodied consciousness
	sleep	waking
	ignorance	awareness
	memory	cognition
	prana/ qi	*manas*

And then convert it into *fig.* 7.7.

A super-state is 'above', 'nested within' or supporting 'lower' dependencies. It sources them. The die from above prints patterns out below; but what is below neither observes nor suspects the 'hidden' mould from which it derives. What is above seems as nothing, non-existent or unreal. Is nothing, the mother of all possibilities and thence actualities, that easy to describe? Mind's Super-state is variously called *Dharmakaya*, enlightenment, internal radiance, heavenly light etc. And matter's super-state? Is it the desert of a vacuum - cold, dark and universally inert? Is it an emptiness that's flat or curved? And is space elastic or expansive or contractive - or none of these? Is it an unexcited 3-d field or zero-point-abstract of fields from whose potential energy, as from a wellspring, all particulars come and go? As opposed to (*tam*) exhausted, sub-state space does there exist a (*sat*) super-state of immanence, a causal 'substance' which can manifest material effects and out of which the zone 'below' is poured? The immanent mould is known as archetypal memory, 'fifth dimension', universal mind. Can you see thoughts by looking on the outside with your eyes? Mind is on the inside. As normal mind is unaware of its interior potential, enlightenment, so even memory is on the inside of material space. It cannot be seen by any lower down, exterior view. What's immaterial is, naturally, physically obscure.

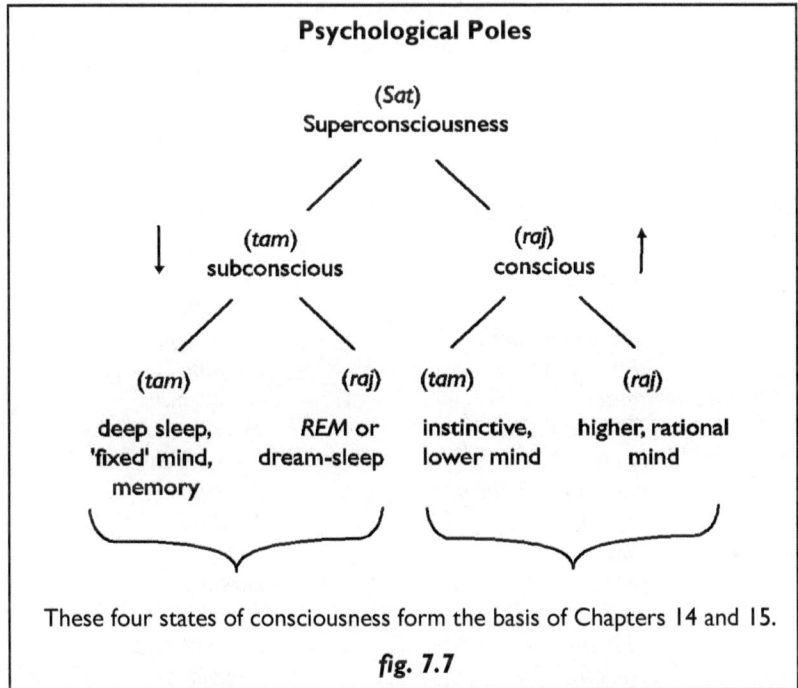

fig. 7.7

Matter is a casing. A shell. So is mind. As can matter (*fig.* 15.3) so can mind, derived from super-consciousness, be divided into five. You can even cut up subdivisions (as in, say, *fig.* 15.5 or 15.8: sub-consciousness) into bands. If attention directly affects *manas*, 'mind-stuff' or metaphysical energy, then *manas* is the 'negative' against which psychological image is developed. This (*raj*) psychological phase of energy is the 'resistance' that gives active information shape. It gives shape to the symbols or objects of mind. Just as the letter '*a*' is not always distinguished from the shape of ink with which it is written, so western psychology has not made the subtle distinction between *manas* (mind's paper and ink) and the meaningful order it represents. In other words, thoughts are subtle, fluid membranes. We identify with them but actually these casings 'separate' us from our Self. Being metaphysical the forms of *manas* are neither perceived by instruments nor known to physical science. As radiant light of different wavelengths and intensities can be transformed into heat, electrical, chemical and other forms of energy, so the range of *manas* shows in different qualities, tendencies, directions and strengths of mind.

Are mind and matter phases of the same combination - consciousness and energy? Instead of heat it's focused consciousness that, like a laser, calls the mental shots. A concentrate of consciousness makes active mind; its lack (a concentrate of energy) makes passive mind (sub-consciousness) or non-conscious matter. In this phased view how does mind impinge on matter?

Order Before a Physical Start.

'Before' a physical start is, hierarchically, 'above' it. What high ghost predisposes cosmos? What immaterial precedent cooked matter up? Prior to the action of the world there rests, according to the Dialectic's logic, metaphysical configuration in the form of abstract but real fields. Excitement stirs initial symmetry and, according to its type, the various changes of the world occur; transformations happen orderly; levels of event cascade from archetypal through the agency of quanta down to condensed precipitate.

Check *figs*. 2.13, 5.2 or 7.8 against this stack. Can it represent three tiers of the astronomical universe?

	tam/ raj	*Sat*
	action	*Potential*
	subsequent order	*Archetype*
	things physical	*Nothing Physical*
	matter	*Super-matter*
	physical forces	*Harmonic Prana*
	change	*Immutability*
	variation-on-theme	*Invariance*
	motion	*Balance*
↓	*tam*	*raj* ↑
	gross	*subtle*
	inertia	*stimulus*
	diffusion	*concentration*
	entropy	*negentropy*
	negative	*positive*
	complex	*simple*
	gross detail	*subtle archetype*
	static outcome	*kinetic basis*
	passive inversion	*active pattern*
	shape	*behaviour*
	confinement	*liberation*
	contraction	*expansion*
	fixity	*flux*
	sub-state container	*shaper*
	gravity	*levity*
	mass	*energy*
	classical physics	*quantum physics*

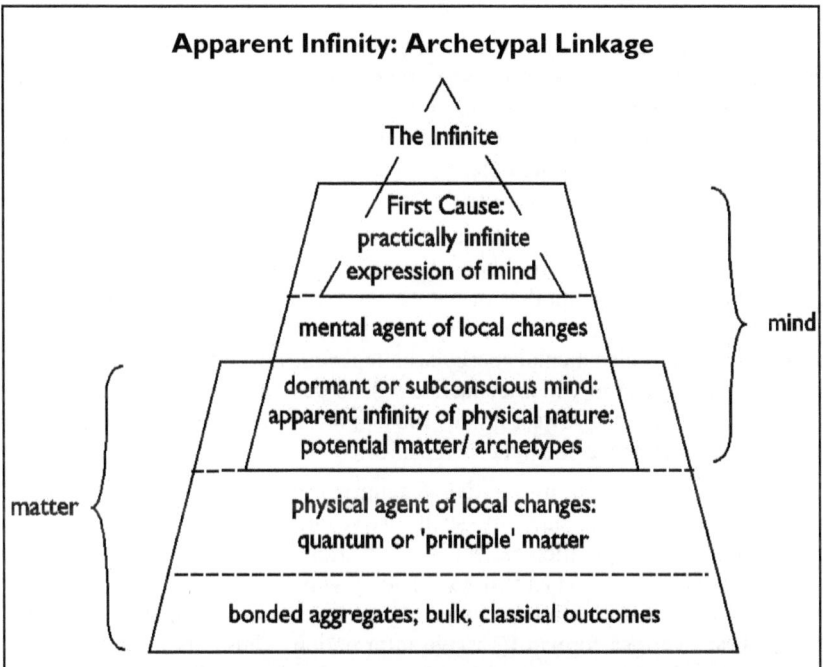

Sub-conscious mind is the storage place of memory, instinct and, in universal terms, archetype. Its mental, mnemonic energy is (see *figs.* 7.5 legend, 7.6 (iii) and 8.2) termed *prana* or, in Chinese Taoist practice, *qi* or *ch'i* energy. This level in the conscio-material gradient is, as the psychosomatic link between mind and matter, also called transcendent, potential or top matter. It is, above the scientific 'glass ceiling', nothing physical. It is, however, the supra-material dimension of natural law, a subliminal source of physical order and, therefore, governance. Is it 'virtual matter', a vacuous ground of matter physics may suspect is there? So that 'virtuality' might be the scientific word for 'metaphysical', if only reaching to the base, psychosomatic 'phase' of universal mind?

Observe the ziggurat from a slight shift of perspective. Fixed, bonded aggregates of mind, memories, constitute a psychological substrate (it is, dialectically, a superstrate); the lowest step of mind, potential matter, forms a context whence reflex physical behaviours are controlled. And the lowest step of the whole ziggurat represents not psychological but physical fixity - bonded aggregates and outcomes we observe as sensible, that is, as normal things.

fig. 7.8 (also 2.3)

Although the transcendent projection of a big bang marks, in contemporary minds, a start of things there is actually no knowledge of what came before. Either there was nothing before a physical start or something happened - in which case there was something. Was there really nothing, absolutely nothing and nothing else? If there was a prime condition was it metaphysical or physical? What is the nature of a bird that lays a cosmic egg? Is it purely material?

Just as liquid and solid textures emerge from a cooled gas, could the three cosmic fundamentals 'condense' into predominance at different levels as the hierarchical process of materialisation descends to its outermost shell - bulk, solid matter? Could Pure Consciousness at one end of creation fall into the oblivion of pure energy at the other? And could there be (*figs*. 0.14 and 1.4) a sliding scale of combination in between?

The conscio-material gradient involves a range of conditions of consciousness and energy. Re-check *figs*. 6.2, 7.5 and the Glossary. *Prakriti*, unscientifically dubbed 'mother of nature', is a generic word that includes all forms of energy. Where consciousness predominates, the influence of energy is weak and passive. Such mental energy is called *manas*; its lower, sub-conscious, psychosomatic grade in which consciousness is depleted, diffuse or 'dormant' is called *prana*; and the lowest grade, 100% non-conscious and 100% actively energetic, is projected as the material cosmos.

Active mind involves 'informative negentropy', that is, 'learning', 'thought' or 'creativity'. The product of such active information is some coded construction whose operation reflects the intention of its maker - a message, a job or a product. This product is the vehicle by which its creator's purposes are met. We call its information passive. Memories, codes and naturally lawful bodies all are various forms of passive information. Thus acts of creation are seen to devolve from conscious through psychosomatic into physical arrangement. It has been proposed (Chapter 6 with further detail in Chapters 15-17 and 19) that the immaterial, psychosomatic medium is sub-consciousness; the connection between mind and matter is by way of fixed structures or patterns in mind called memories. In universal terms these are archetypal memories; they are passive information in the shape of mental records, instruments of repetition whose 'notes' constitute the 'deep structure' of physical nature.

Why should mind's dimensions be material's; why, any more than natural law, should a 'deep structure' of material phenomena be localised? **And if simple archetypes devolve the 'alphabet' of physics, why should complex blueprints not devolve an organism's 'book of life'?** The implications are profound. Such a view conceives of the physical universe as the development of an idea, an image projected not from emptiness but mind. A fixed image is a remainder; memory is an image and archetypes are memories. What is the filing place of mind but memory? Is not recall, replay or repetition how a record's used? Reflex reiteration is continual reminder of its origin. In this view 'autonomic habit' of material operations derives from archetypal memory and, in an active but sub-conscious sense, remembrance of creation. As is the case with all machines, such vision recognises prior presence of intelligence.

In *top-down*, dialectical terms the origin of energetic patterns (particles and forces) and, therefore, the foundation of every massive configuration is archetype. The law of aggregation's shape, morpho-genetic law, is written in the memory of universal mind. If microcosm reflects macrocosm and human construction reflects universal then its operation is, as well as biochemically, psychosomatically governed. Archetype shapes energy and, as atomic physics has confirmed, from vibrant energy there 'condenses' matter in the shapes of things; is not such archetypal grammar geometrical? Ancient cosmos (order) was impressed on chaos (*prakriti*). *Prakriti* is the anti-principle of *purusha*;

while complement and antagonist of the *subjective informant*, at the same time it constitutes the *informed objects* of creation. Cosmic energy is the passive resistance, negative or film on which first mental and, lower down, physical shapes are developed; it is, if you like, an ubiquitous, graded 'elastic' whose various tensions give rise to images and objects. A gradient of mind is created (see Chapters 5 and 6) as the informant; *purusha/* awareness, becomes progressively involved in *prakriti*. As it works its ideas through to their psychological or physical conclusions then focus of attention becomes increasingly ensnared, like a bird caught in lime, in such forms of energy as it creates. Is this not a picture of your own life bustling round in circles after things you've thought of?

Is memory (or any part of mind) outside you? Repetitive behaviour is stored as habit and 'automatised'. Aren't plans and coded systems (such as language) filed away for recall when required? Is not such storage known as memory and a stored bank of patterns of behaviour known as 'instinct'? You can hardly say that matter has an inward 'instinct' or 'instincts' since this sort of reflex is associated with a living thing. You could, however, say machines in their 'correct responses' show reflex memory of how they should behave. Or is such memory embedded in construction and coherence of their parts? From machines you might extrapolate to cosmos. Cosmos could be seen as a machine composed of basic, interlocking parts; finely tuned design rules operation. Is its memory embedded in these vibrant parts, encoded in the particles of which it's made? From what plan, 'instinct' or memory did they derive innate and automatic patterns of the way they move? Whence (Chapter 8) comes 'natural law'?

Think of software generating hardware; think of information generating patterns within energy. Does thought or instinct governing behaviour come from 'outside'? Did you think that things are organised externally or that natural law's an exposition by some hidden hand? 'Meta-law' of archetypal immanence arises from within. There is distinction drawn between rules and their game but also, in the playing, there is none. Regulation of a particle, its forces and the particle comprise a trinity. Triune. Potential information is seamlessly expressed; immanence is, as if from nowhere, realised.

In Dialectic terms the cache is stashed in universal mind's sub-conscious level. Archetypal memory is passive information. At this pre-physical, psychosomatic level subjectivity is extinguished. *Purusha* is now dormant, *prakriti* dominant. The dimension is not physical. It is likened to an inner sky (*akash*) whose 'faults', stresses or currents give rise to a 'climate' of physical phenomena. The Buddhist metaphor of existence as passing clouds is apt. Clouds of gas appear chaotic but their molecules are shaped precisely; particles of dust or rain are pregnant with an orderly creative power. Vibrant patterns are materialised and, eventually, constrained into the grains and shapes of condensates, into irregular or random (so it seems) precipitates of inward rules. Morphogenetic law is, in this view, written in the memory of universal mind. <u>In other words, cosmic archetypes are the agents of grammatical information, the source of automatic, non-conscious physical behaviours.</u> They constitute an upper, pre-physical side of the psychosomatic coin that is expressed as material patterns or the laws of nature. Such laws reflect the informative as opposed to energetic side of nature. They have already been identified (*figs*. 2.12 - 13) with (*sat*) potential, transcendent or top matter and involve (*fig.* 7.9) expressions of the vectors *sat, raj* and *tam.*

In this *top-down* view, the physical world was projected from within itself so that the common metaphor of potter and clay, implying fabrication from the outside by an external creative force, is inapt - unless the hand inside the pot is metaphysical and the one outside material. *Apt is the organic metaphor of preconceived development from seed.* Cosmos is programmed. As its atoms 'underpin' bulk matter so, moving up a stage, an archetypal template prints 'behavioural shape' upon primordial, subatomic particles. Archetypal memory is the next, implicit layer over-lodging its dependency, the physical entirety. Passively informed energy was therefore ejected into the external conditions we biological bodies experience.

'Humph!' objects *a bottom-upper*, 'I really do not like this pattern of ideas! It is nonsense!' Since when, however, did dislike make possibility impossible? Or inconceivable? Is it inconceivable an author could 'program' his book? Or that material nature's coded book was, in all parts, conceived instead of accidentally assembled by its own oblivious self? *Bottom-up* assumes mysterious projection of commencement by immaterial emptiness (but definitely not by immaterial mind); and that, through billions of years which lead unto himself, there was and always is a total lack of any scheme, reason or meaning - except, of course, the predetermined schemes of behaviour that we call the rules and laws of chemistry and physics. Are these stabilities intrinsic in material? Did nothing but contingency spout nature out of nothing? Against the evidence, do 'natural statutory regulation' and 'fundamental particles' evolve? In short, *bottom-uppers* argue everything, including humans, reeks of accident that's simply happened.

Tugs and Wobbles.

Tugs and wobbles - here they come! The ultimate in scientific wobbles, a preposterous leap of faith that stokes huge tug and tension up! Who can be sure, who land on hard ground when he's jumping into space! So in silence here it roars, the First Express from Nowhere into Somewhere! Sudden and unscheduled, it emerges full tilt down the rails of space! First causal stimulus, a levity and unimaginable heat erupts. From such heat cooling matter broke. 10^{50} tonnes of things raced out of none. 10^{87} particles, 10^{22} stars - as many as earth's grains of sand - out of an instant's flare-worth. A one-off; mass and energy conserved! Pop of pops. Miraculous! Why, you might dare to think, should eternity one moment and one place in space spit out a white-hot speck of limitation called the cosmos? What bellows, what flame-thrower blew the super-ball up into galaxies? Or was the rustling up a gentler, less pin-pointed kind of gravitation?

tam/ raj	*Sat*
integrated 'sub-symmetries'	*Symmetry*
concentric expression	*Central Potential*
subsequent order	*Archetype*
particle/ force	*Law*
range of actualities	*Precondition*
action	*Latency*

The Eruption Story might be true or not but from Natural Dialectic's *top-down* point of view a coordinate is missing. There's loss of energy that cools our world into its shapes but an ingredient of information has gone astray. Whence arose primordial super-order; what kind of archetypal symmetry is, through loss of energy, broken into laws and properties of each emergent cosmic phase?

Super-order? Mark that time's arrow moves, as astrophysicist Sir Arthur Eddington first clearly pointed out, (↓) 'downhill' towards randomness. Over time disorder increases; entropy increases. The future is more random than the past; overall the trend is towards decay. Thus, conversely, the past was less random, more orderly and fresh than the present. Order increases, entropy decreases as, in imagination, you swim up against time's tide. **In the beginning, therefore, cosmos must have been extremely orderly.** Zero entropy. Absolutely minimum disorder. Super-order out of which the world is gradually falling into bits (called galaxies, suns, earths and every other thing except, it seems, a permanence of protons and electrons). Permanence? Eventually the universe will, by this thesis, wind down into heat death's randomness wherein the lot's amorphously the same. What paradox! A super-orderly projection that is infinitely hot and dense gives rise to a chaotic plasma. From an expanding violence, a seething sea of sameness, drop out by degrees dynamic, changing shapes of cosmos. What *is* it that's so super-orderly? What *is* the nature of this perfect order that, from nothing, gave our precipitate, the universe, the best of starts? What *is* super-order's energetic archetype - the reason for such pristine paradox? Hardly, if the order's perfect, chance.

The dialectical suggestion is that tensions in the vacuum known as immaterial fields in undisturbed condition constitute original, archetypal symmetry; that archetype is a generic word which includes a tissue of such fields, each of whose undisturbed potential may, once aroused, express a certain pattern of behaviour (a pattern that appears as different kinds of force and particle); and finally, as quantum vacuum is the mother of phenomena, these elementary characters combine, according to a principle polarity, into atoms, molecules and sensible precipitates of matter that compose our world.

It is, in other words, suggested that layers of archetypal potential whose disequilibrium gives rise to the fundamentals of physics are really memories in universal mind; that complex files that can generate life forms also exist in this metaphysical zone of preordination; and that the causal energy that excites such psychosomatic morphogenes is a single super-force, an archetypal energy called, for want of a western correlate, by the oriental words *prana* or *ch'i*. In what sense is this beyond comprehension? Only if the only sense, even concerning origin, is physic. Was physic before physic? Was there, is there never metaphysic that exists beyond embodies, brain-linked, individual mind?

In this case you'd need primal information to begin a world but would you need an overwhelming heat, top of the scale, from which the forces of the world, once unified but now 'decoupled' and distinct, compose the quartet jazzing our cool, underwhelming world? Is it archetypal information, sheer energy or both in tandem that precipitated operations of the cosmic system? If the notion of physical effect deriving, in the manner of a menu or flow chart, from metaphysical cause is unfashionable, so at first is every shift of paradigm. If the idea of universal mind framing universal matter strikes you as a 'counter-intuitive', unpalatable and unwelcome intrusion onto the field of material science, it may have to be lumped - because if the primary axiom of Natural Dialectic (*PAND*), that information and energy are separate cosmic elements, turns out true then its logic follows through.

Archetypes are not active but passive forms of information from which physical appearances derive. From subtle devolves gross, from immaterial material and from apparent vacuum everything. This is the (*tam* ↓) direction of creative vector; it is motion down the conscio-material gradient that the image of Mount Universe explains. On such slopes the presence of a cosmic mind is logical. To entertain, if you want to be dramatic, an issue of phenomena according to the rules laid 'in the natural mind of Natural Essence' requires no change in facts but in the angle of perception. A revolution in perspective generates a shift of paradigm.

The current notion is that Einstein's is the final cap; or Planck's; but a dialectical perspective, changing nothing physical, is richer in its breadth and continuity. It re-embraces metaphysic like a long-lost friend and so re-animates an understanding of subjective information - mind, consciousness and life. Part of this shift involves, as well as a non-conscious universe, the notion of a prior, inward, conscious element. The basic tensions and vibrations of material events - and thus their bulky products too - *are* creation but an orderly creation in 'the mind of God'. Re-interpretation of the immaterial fields of quantum vacuum in terms of potential matter, archetypal record or an aspect of a cosmic mind is not music to materialism's ears. Perhaps, however, as the mystic has long claimed, existence is but polar slack along with tensions wobbling round in psycho-space.

Order Below a Physical Start.

For mind matter is subordinate. The Dialectic stipulates that law is hierarchically prior. It comes before; rules are necessary for an ordered game to start; they are the precondition of expression from the level just above. For *physicalia* this level is the lowest metaphysical, that is, archetypal memory; and memories are simply stores of what has passed. Mind in its phase of active information is now exhausted, finished, done. *For matter mind is the subordinate; in the second major phase of creation, therefore, information is completely passive. It is fixed and automated patterns of material energy are now the active part.* The physical universe is an antipode. It diametrically opposes the non-existential, essential (*Sat*) Absolute. It is the Inmost inside out. Essence has been turned inside out, the Infinite and its characteristics reversed. (*Sat*) top level has become the (*tam*) bottom. The interior is externalised, inversion complete and Potential in its last stage blossoms into a fully-fledged, dynamic but non-conscious realisation.

Physical energy is sometimes called 'nature', as if this was all nature was. As if there was nothing before, after, under or above it. *It needs be emphasised that physical cosmos constitutes the lifeless end of creation: and that the common-sensible view of matter is incomplete.* A broader perspective finds life subjective but all bodies objective; and finds the objective cosmos, as a product of principle, basically conceptual. Accordingly physical energy is seen as an externalised, rigid or 'crystalline' form of (universal) mind. A shell. It is, like the special case of straight-line to a curve, the case of non-consciousness to conscious mind. Its myriad automata make up an exterior, inanimate container of whatever information shapes them. **A study of matter is, as already noted, the study of non-consciousness.** Or, rephrased, physics and chemistry are the psychology of non-consciousness.

What can physics say about what comes above what's physical? It has no knowledge of or interest in what is beyond material phenomena. There are two kinds of such 'beyond'. One is hierarchical, the other temporal. The first is metaphysical, the stage of universal mind; the second comes before a start, such as, from less than anywhere, the Great Projection of a Universe. Below the start for involuntary matter, that is, passive information with entropy dominant we read:

	tam/ raj	*Sat*
	particle/ force	*Law*
↓	*tam*	*raj* ↑
	outward	*inward*
	gravity	*levity*
	particle	*wave*
	receptor	*force*
	mass	*energy*
	object	*field of space*
	discontinuity	*continuity*
	division/ divisor	*unifier/ union*
	isolator	*communicator*
	separation	*relationship*
	diversification	*unification*
	complexity	*simplicity*
	tension	*relief*
	confinement	*decompression*
	imprisonment/ locking	*unbinding/ liberation*
	crystallisation	*evaporation*
	creation/ materialisation	*dissolution/ dematerialisation*

Let's dip below the start.

Three, Two, One, Zero...Emergent physicality! Now it's coming! Can you feel a cosmos coming on?

To understand effect, both science and religion say, you need to know its reason. This reason is its cause. Therefore if you understand First Cause you ought to understand the rest. So let's cut to the chase. What came before emergence - God or no God - is not physics' field; how something issued out of nothing is. What were the initial conditions? What was the type of time's first tick? What is nothing-into-something's trick?

First let's take an atheistic tack - where even *PAM*'s support is not enough and you need nothingness like anything! God is nothing; there was nothing; nothing happened in that nothing which, anyway, causelessly created everything; everything then jiggled for a long time ceaselessly until, still without a reason since reason didn't come till men, a tiny part of jiggling turned orderly and out of it popped you. Is this not profligate a currency of miracles? At least for sheer nihilism you can't beat it. It's vacuous. It's nonsense that you can't believe but do!

Therefore let's, as textbooks do and we will later (Chapter 12: Labours of an Empty Womb and *PAM*'s Miracle), garnish nonsense with a seeming sense. Let's pretend that matter is eternal or that 'nothing' is 'potentiality'. Such form of nothingness excludes that mind could whip up anything from its thin air; nor is such a trick 'explosion' in the sense of bombs or 'expansion' in the sense of bubbles.

Yet silently, where there was absolutely nothing, something stirred; you might say a pointless, powerless point of physicality occurred. This dot, a singularity, was physically omnipresent since no other thing was there; or perhaps it wasn't just a dot but everywhere at once! At every point; and now all yawning space must do is stretch! For some reason, understanding presently instructs, space expanded. Extreme paradox! A quantum fluctuation, 'infinitely' hot and dense, unleashed primordial potential as a cosmos-worth of energy and matter. It grew into the only astronomical extravaganza that we know. Nothing stretched and stretches to make space for things; and if the universe is infinite (if not, then what's outside it?) you might think the natal singularity was point infinity. Less than a dot with all those galaxies inside?! Mind boggles. Are we toyed with philosophically or scientifically? How, you rightly ask, did a null factor, emptiness, develop into every form of thing?

Can a void become unstable; can temperature in nothing rise; *must* something spurt from emptiness? Of course it must. Or if not nothing then perhaps cosmos smaller than a pin's head had to pre-exist somehow. Thence *bottom-up* today's dominant, materialistic story accelerates smoothly out of an orderly projection. Apart from this burst's singularity the universe has neither centre nor an edge. From this basic epicentre uniformity expands, despite local 'perturbations', into a large-scale uniformity of cosmic structure. And from that paradox of quantum heat and density emerge the behavioural patterns of energy. Never mind the source of thermal stimulus, an oven smaller than a human egg and hotter than the brightest star - by far! Hey presto! Where there wasn't now there was. Space expanded smoothly; only simple, single super-force and energy first flew. Primal symmetry. It is conceived that, a split second later and for several reasons of convenience (see Chapter 12), the symmetry of Grand Unification was broken as the strong nuclear force was broken out. Massless particles appear. Thence a second huge acceleration blew up the space of cosmos by a billion times and much, much faster than the speed of light. Next, at a stroke that's critical for your existence, this so-called 'inflation' faded out. Most conveniently. Now the projection falls within the scope of theory and *LHC* experiment. It is next surmised that, on cooling to $10^{15}\,°K$, weak nuclear and electromagnetic forces are broken apart. The top level 'precipitate' of material creation called a Higgs materialisation field appears. It is also of necessity convenient, with force untested but presence maybe soon inferred by a Higgs boson. This frictionless, ubiquitous and ethereal pop-up breaks energy-conveying space by somehow lending mass to energy. Mass isolates. In a second at ten trillion degrees, a nascent cosmos-worth of photons, quarks, electrons and divergent forces is created. This would be the crucial phase change whence from fundamentals will eventually emerge every separate detail of material things. The influence of law emerges! Matter's made!

Snap your fingers twice. Everything so far has happened in less time by far than that. No waiting time. In a split second quarks have coalesced to make up protons and, a few of minutes later, very hot gas and some atomic nuclei appear. The universe is pumpkin-sized. By now heat in the cosmic birthing chamber has fallen to a billion degrees. Light is locked to matter and endures a 'dark gestation' of, shall we guess, three hundred thousand years. Expansion cools the combination until, at 5000 degrees or so, they split apart. A cosmic egg can hatch. Now the plasma can 'condense' into a seething mass of chaos, a transparent and expanding cloud of what classical philosophers arguably thought of as *'prote hyle'*, *'ur-matter'* or *'prima materia'*; and what modern science calls hydrogen gas. The word 'gas' is derived from the Greek word for 'chaos'; chaos was, in Hesiod's Greek description, the dark, original and structureless 'profundity' of things. Variations in the densities of gas clouds are supposed to allow gravity to 'kick in' so that, eventually, clusters of galaxies are 'seeded'. As their matter spins through the darkness it begins to collapse on itself. Embryonic stars flatten to discs whose peripheries will provide the ingredients of planetary formation; at the same time hydrogen is reformed into nuclear furnaces and, as concentrates called suns or stars appear, the pregnancy bears light. These alchemical furnaces synthesise more kinds of atom; from supernovae even heavy elements like uranium are forged and scattered. About five billion years ago solar systems with planets and comets materialised. On earth, where conditions were right, life was in the lap of mother luck. Its spontaneous generation was nurtured by chemical and then biological evolution so that, eventually, observers evolved who could with increasing accuracy speculate about their own origin. Now, they believe with confidence that verges arrogance, they might have got the whole thing right.

'Easy, isn't it? Good show!' The cosmos is, to some extent, a fireworks display but could its host have been a single bang? This is not the moment to engage the history of thought that's swept us to our present, fascinating pass, only to note that every culture and each generation holds that its own 'state-of-the-art' myths are the whole truth or else very nearly so. Ours is no exception. Is not the objective picture neat? Does not the story accurately explain how God's mind burst or chance stirred cosmos up? *What headache, even to a physicist, could such a grand yet simple vision of creation present?* Sadly, as Chapter 12 will show, pass the paracetemol. Unanswered questions in cosmology are deep. They concern the size, age, expansion, components and non-components (voids) of the universe. They also concern star/ galaxy formation and even the origin of solar systems, planets and moons. How long before the detailed model of a big bang, still riddled with riddles, cracks up? Will its crack in nothing be resealed? Or has projection that transcends description nearly been described?

A material level comes beneath the thumb of 'natural law'. It falls within the zone of science, the remit of 'necessity' and conservation laws. *Of these Mayer's First Law of Thermodynamics (conservation of energy) reflects supreme government, that is, the (sat) balance of accountancy.* From a dialectical perspective the cosmos is geared around an overall (or underlying), intrinsic principle of balance. Scientists recognise it instantly. It is at the heart of conservation laws and known as 'exact equivalence'. In 1847 James Joule wrote, "....the phenomena of nature, whether mechanical, chemical or vital, consist almost entirely in a continual conversion ....into one another. Thus it is that order is maintained in the universe - nothing is deranged, nothing ever lost, but the entire machinery, complicated as it is, works smoothly and harmoniously."

Conjugate the principle of balance with neutrality (see *fig.* 7.2: stack). Neutrality and balance couple well. (*Sat*) Essence is neutral. Existence, on the other hand, is polarised in (*raj*) action but, as the motion runs, also depolarised and falls towards (*tam*) inertial equilibrium. In universal terms (*figs*. 0.5, 1.3 and so on) its two phases show as (*raj*) mind and (*tam*) matter. At the exhausted base of mind, subconsciousness, are fixed the mental shapes called memories; and archetypal memories comprise the essence of the phase below, the purely energetic physical. They are called potential matter; it is from the invariance of archetypal law that all material behaviours derive. The axis of our cosmic balance is its law; according to this neutral law behaviours occur. Conversely, at the other and peripheral, exhausted end, the base of matter, there are found inactive shapes - electrically neutral, gravitationally balanced solids - constructed from those active cosmic symbols of balance and neutrality - atoms.

Such dialectical order of events is reflected in the first two laws of thermodynamics. The First Law states that, in the case of energy, only transformation but never creation takes place. This indicates that creation is over and its phase replaced by one of conservation and transformation. Big bang, shattering this fundamental law, must have created it! It is the inexplicable creator.

In 1852 Lord Kelvin articulated his Second Law of Thermodynamics. Called by some the supreme law of physics, it indicates that 'nature flows one-way'. This way is a downward stream of entropy that drifts into the ocean of exhaustion, randomness and death. Kelvin had thus established an apparently impassable barrier to evolution although Darwin had not by 1859 twigged it. Still today, by philosophical necessity, his adherents turn a scientific blind eye to this basic vector of the universe.

In fact, entropy might be visualised in terms of universal diffusion. A concentrate of information or of energy diffuses, weakens and is lost. Become a cloud. Look down. Drop rain onto a hilltop; start the flow with an initial input that will, sourcing subsequent reactions, stream down the mountainside.

Velocity is proportional to concentration difference (the gradient) and distance dropped depends on where obstruction lets a stable state, inertial equilibrium or 'puddle' form. From such puddle to the next one down is measure made (in thermal terms) by chemistry. When such obstruction is removed flow continues to the next. And so on gradually down the energetic hill. Eventually 'plenipotentiary' initial condition, that is, input at the start of cosmos will have degraded through the whole existence-worth of fixities and transformations to a universal puddle, an exhausted ocean at the end of time. Small stimulus, small activation energy; but if you want to 'knock the water back onto the mountain-top' you'll need high energy (say, particle accelerators) that might reveal the nature of potential at the start. Otherwise the energetic sweep is (*tam*) downwards. Such loss goes for entropy of passive information too. Only mind (or 'negentropic', active information) reverses this arrangement.

Natural Dialectic has in this sense exceeded Kelvin and brought metaphysic in the form of information to the table. Such extension of equation notes that information, once initialled, tends to 'dissipate' along a conscio-material gradient dropping from an active, mental plan to passive, physical outcome. It runs from possibility (or potential) to final actuality. And, finally, it tends to degrade from accurate to garbled message. Such entropy of passive information (see Chapter 23) can, like entropy of energy, be reversed. Apart from the point of initial creation stimulus causes 'upward' change. In the case of information this stimulus is an input called focus of attention. Such psychological as opposed to physical concentration of energy creates possibility, plan and purposive complexity. **To purpose is to specify**. Specification creates, against the grain of entropy, specific order. If its codes or structures decay, its concentrations diffuse and the cosmos drifts into disorder what must have been its original condition? How, in other words, would you describe the nature of Initial Concentration?

It is noted that while big bang explodes the First Law the theory of evolution (which is neither a conservative nor degenerative process) breaks the Second. Yet the assumption of deliberate design breaks neither of these fundamental, empirical laws of physics. Objections to these assertions will be, in turn, dismissed (see especially Chapter 20).

As polarity arises from the division of neutrality, duality from the division of unity and motion from a loss of poise, so you might expect a principle of equilibrium to source such symmetries (for example, gauge and particle symmetries or the numerical 1:1 electron/ proton charge ratio symmetry) as underlie the fabric of space. We've seen that such symmetrical wholeness would, according to Natural Dialectic, be metaphysical. Actual physical particles and their bulk phenomena would not derive from or be different appearances of a single kind of underlying particle (say, a hypothetical preon); nor would they transmute into each other (neither protons nor electrons, for example, are seen to decay). Instead they would represent the polarisation of this inclusive, archetypal norm; they represent disturbance of initial balance. An example is the quantum pair production of particles (say, an electron and positron) from the vacuum; and their mutual annihilation into neutral light. Broken symmetry leads, in other words, to polarity, duality, waves, particles, sub-principles or laws of nature and a perpetually unbalanced, changeful, motile creation. **In the dialectical metaphor of scale norms (or laws) are axial; nature can be seen as a swinging, vibratory attempt to forever regain a perfect balance.** The universe can be seen as a vast homeostatic process wherein forces are the agents of negative feedback; they are balancers, always seeking equilibrium. They are the expressions of a natural attempt, after Initial Disturbance (for example a big bang), to regain norms, equilibrate action (with equal and opposite reaction) and 'keep account'. For psychological, 'upwardly mobile' forces this equilibrium is one of poise - at maximum the Super-state Poise of Enlightenment; for 'downwardly mobile' physical forces it is one of inertial state, bonded aggregation and exhaustion - whose maximum is found in the sub-states of Absolute Zero (0°K) or a black hole singularity. In this view both polar forces and particles are themselves agents of (*sat*) regulation. Metaphysical preordination of particular expressions of the cosmic fundamentals (e.g. photon, electron and proton) is the (*sat*) axial principle around which the (*raj/ tam*) scales of cosmic motion swing. In this case there would exist more than a couple of dialectically polar sides. *There would be three sides to the single story of cosmic homeostasis - two physical, one metaphysical. Actual matter would take shape from (figs. 1.4, 2.11, 8.2) metaphysical potential. Is, therefore, the origin of natural law just accidental? Or does it follow archetypal rationale?*

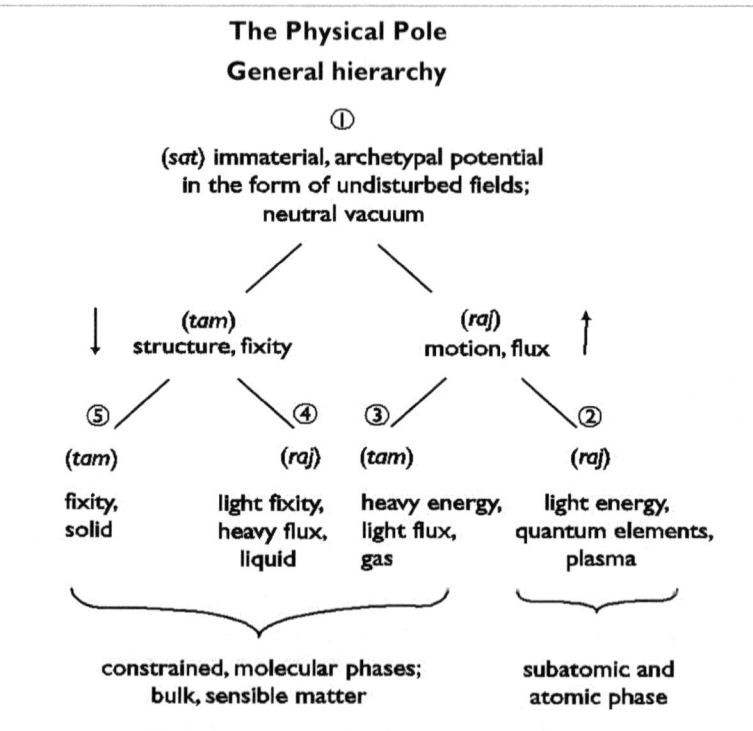

The polarisation of forces and particles

Perhaps science looks, using somewhat esoteric terminology, for more mathematical complexity than this description offers. Is there, however, deep *simplicity* behind phenomena (such that, for example, you might dialectically predict a space-linked force of levity, anti-gravity or lambda)?

Communicator, messenger or force particles are called bosons; massive particles, the basis of matter, are called fermions. Decay from high to low energy drops through three generations of material constituent (fermion). The present, third and (*tam*) lowest energy generation comprises our leptons (neutrino and electron) and quarks (proton and neutron from so-called 'up' and 'down' quarks).

The following three diagrams the illustrate

Capture and Release: a conjectural aspect of the polarisation of forces and particles

①

(*sat*) metaphysical transcendent;
archetypal field symmetry; ground-state potential;
also, prior or basic energy, system activator (*prana?*)

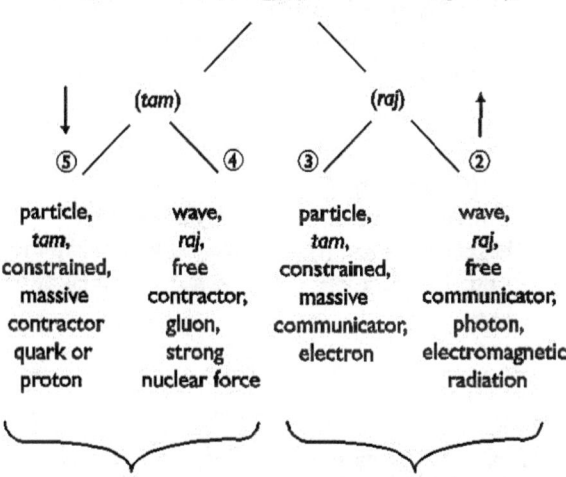

⑤	④	③	②
particle, *tam*, constrained, massive contractor quark or proton	wave, *raj*, free contractor, gluon, strong nuclear force	particle, *tam*, constrained, massive communicator, electron	wave, *raj*, free communicator, photon, electromagnetic radiation

inertial effects; gravitational aggregation; centripetal contraction towards centre of mass	electrical or stimulant effects; chemical bonding; centrifugal radiance from energetic concentrate ie. source

(*sat*)
dormant, latent, immaterial fields;
neutral immanence

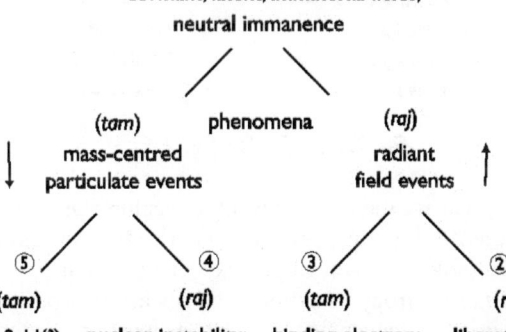

	(*tam*) phenomena	(*raj*)	
	mass-centred particulate events	radiant field events	

⑤ (*tam*)	④ (*raj*)	③ (*tam*)	② (*raj*)
binding; Higgs field(?), strong nuclear force; quark, proton, neutron; large-scale binding, gravity	nuclear instability; radioactivity, transmutation of isotopes; decay of free neutron; electroweak force	binding electron; magnetism; chemical bonds, molecular construction	liberation/absorption of light/heat energy, electromagnetic force, compensating product of change in energy state state of atomic electrons

nuclear; intra- and extra-nuclear	electromagnetic; intra- and extra-atomic

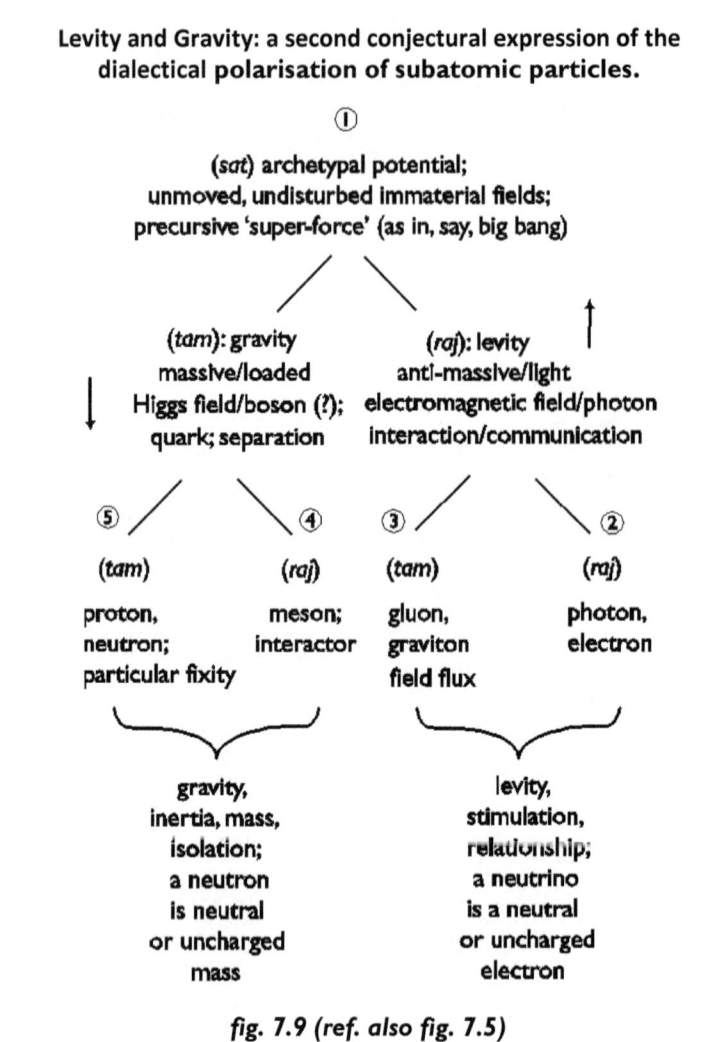

fig. 7.9 (ref. also fig. 7.5)

As well as (*sat*) balance there exist two other vectors. One, (*raj*) negentropy, is action's wake-up call; it corresponds to dynamic stimulus and/or an increase in configurative complexity. You might expect its action to predominate at starts. The other, (*tam*) entropy, is action's fall towards sleep; it corresponds to inertial or disintegrative tendencies towards exhaustion, differentiation and disorder in the sense of randomness. You might expect it's how you'd find things fall apart while rolling towards their end-game.

Dual aspects of the vectors need to be distinctly underlined. In dialectical terms (*figs.* 1.3, 2.5, 2.6) what exhausts or fixes is a (↓) gravitational effect; the opposite is (↑) levitatory. In dialectical as well as scientific terms (↓) entropy and (↑) negentropy each have two aspects - informative and energetic or, if you like, configurative and thermal. When purpose is excluded then, as nature flows, these two show inverse effects. If temperature is turned up then thermodynamic negentropy shows configurative entropy (the 'randomised disorder' of gas particles); if heat is lost thermodynamic entropy shows configurative negentropy (in the sense of atomic crystalline or solid order). Yet overall, as in the case of a steam locomotive, hot gas will cool and 'lose capacity to work'. What is a crystal but assumption of the slackest shape, least tense configuration and expression of inertial equilibrium? Without a purpose or without a fire the world is tumbling down its hill. Sooner or later, underneath their start, all physical creations fall apart, go up in smoke or fade away.

Creations are events or objects made of what? Are they, as Newton thought, distinct from such external rules as govern them by instantaneous operation everywhere in space? Or is there 'immanence', a sense in an inner, archetypal, triune entity directs laws, elemental forces and the basic particles in space? Today we understand that forces are mediated by an exchange of particles between all interacting bodies. And so-called 'gauge symmetries' dictate the constraints that give rise to specific energetic bundles - different kinds of particle. Such underlying symmetry of constraint defines, that is, informs the pattern-generators, those same particles of force and mass. Behind 'sub-symmetries' (or sub-routines) of particles can we conceive grand and overarching Symmetry, a single archetypal program whence the balanced set of nature takes its origin? This is the dialectical idea; string theory (Chapter 9) resonates with it; and *GUT*-busting physicists with their big *TOE*s all crave a singular description. Will it describe what's physical or, like an idea memorised, what's metaphysical?

From vibrant, elemental and enduring uniformities you can build up large-scale changes in a world of various fractures and asymmetries extending down to atoms, molecules and larger. Strings are defined by tensions, taut or slack, whose rippling modulations are associated with mass-energy of elementary particles - but not with music of the spheres! If harmony is off your score is speech a random process? Are words built of any letters or a song composed of any chords? Is string vibration tuneless or are the 'right chords' of a cosmic melody, the actual particles and forces of our world, defined as resonance that's issued from the instrument of archetype? Perhaps modern teaching is, in essence, a refined redefinition of the old.

Let's put this important point another way. Do you remember (Chapter 5) that information uses agents to symbolise its messages in code? For Natural Dialectic the Symmetry of Essence is expressed in changeful existence by a suite of lesser yet still fundamental symmetries - those of polarity. Its stacks express polarity; so do ziggurats, concentric circles and (*figs*. 7.5, 7.6 and 7.9) suggested menus of duality. You might expect the agents of material law to reflect the triune structure of Primary and Secondary Dialectic. Is this, in the form of fundamental particles such as photon, electron and proton, the case? What about intergalactic code?! How many letters, made of particles and forces, are there in the grammar and the alphabet of things; and how many atoms make the words of world-wide physical vocabulary? And two base-pairs (four letters) underwrite genetic code; four cubed (64) genetic 'words' (called codons) underwrite our lives on earth. Champollion deciphered hieroglyphics; it seems maths is the language, the grammar that our world is written in.

Bulk Reality.

Matter-in-principle comprises quantum-level forces, subatomic particles and atoms. From that ocean surface large-scale edges of the world, bulk matter, condensations that incorporate all bodies, animate or not, which we can sense and call our physical reality. Matter-in-practice is this sensible and bulk reality. What is any object but a slow or fixed event? It is a macroscopic, slow and massive process or else one microscopically flashing by; either case depends upon the 'immanence' of its internal programs that we call potential matter's archetype. The whole collection spills from 'inner space'; energy locked into order now luxuriates into a universe. Is there any more to add?

Firstly, in any process count its possibility. There can be psychological capacity in the form of will, desire and purpose but this prior stage of process is discounted if, in a truncated view of process, mind does not exist. You need, nonetheless, energy and context (laws of nature and environment) sufficient to allow a possibility to happen. This precondition is called, in terms of Dialectic, potential; it is a pre-active state - potential equilibrium or poise. Information is the higher, prior aspect of potential, energy the lower. Higher aspect is expressed in working out creation; its level is, like any principle's, above its actual consequence. What, you ask, about this stage in respect of universe? We've just asked what allowed the possibility of cosmos prior to when it dipped below its start; why, next question, do the properties of cooling matter turn out, gas to solid in the realm of science, how they do?

Secondly, count overall conservation of energy. This means that while the expression of energy may change its equivalence remains. It may be radiant, reactive, locked in bonds or exhausted by diffusion to a point of languor, flatness or inertial equilibrium. Fallen into torpor or solidity no further action can occur. This, the opposite of full potential, is base impotence. In a closed system, isolated from output or input, such degradation into impotence always occurs. The tendency is (*tam* ↓) towards some kind of death; indeed the system is itself a dead one. Our cosmos is non-conscious as a grave and so we walk a vale of death. If the First Law is true you might expect that such material universe were closed and so conclude its end will not be an annihilation but a 'heat death', whimper of an end to all activity, the sump of every world. What do you imagine is the shape of whole passivity, of total darkness hierarchically down below?

On the other hand, a local concentration of events and objects is the order of galactic days. 'Negentropic' stimulation drives each process forward. We find that some events are 'paused' for various durations. Such 'pause' occurs whenever active energy is fenced or fixed by bonding forces into forms of varying complexity. The more complex a chemical or bulk material shape the rarer its occurrence. It will tend, like heavier atoms, to spontaneously fall apart. Nonetheless, the chemistry of electromagnetism and aggregations built by gravity can (as the 'hard science' of geology, astronomy and physics shows) create a panoply of lifeless worlds. Note that the basis of irregular duration is a blockage (such as 'activation energy', solidity or other barrier to change); and of regular duration it is fluid waveform (called a period, cycle, oscillation or vibration). Where you find conservation of a *status quo* you find either frequency of wave or local fixity of object. In the event of life, whose 'pause' is called dynamic equilibrium, you find *purposive* complexities of both these species intertwined.

If energy's conserved and nothing new can enter in then, universally, the matter's closed. However you might argue outside forces can affect a local system like the planet earth. Heat, for example, might confer 'an openness' that could defy the hand of death, of entropy. After all, sun lights earth's surface. It imparts motion to the air (climate), sea (convection currents) and most forms of life; and whole rock continents are moved by inward heat below our feet. An influx of energy to any system makes it change and move; it is transformed in ways irregular or regular. Although the general tendency is towards exhaustion local negentropy is seen, for a while, to buck this trend. This is, where life involves a complicated dance, by far the most pronounced in bodies biological. These age but, with dynamic equilibrium, resist their aging. They develop from 'potential' or 'perfection' of an egg. You might even think they gained in order and developed their complexity but this is an illusion; all was there from the beginning in that single cell, that starter called the egg. And, eventually, they die. Such resistance to breakdown and death is, for the Dialectic, based on negentropy of information *and* of energy. Raw sunlight will not help. Life's energy does not derive from thermal currents or electric circuits on their own. These may be necessary, like electric current powers machines, but was mains circuit ever cause enough to make a mechanism in the first place? Life's flows are not the simple product of some radiation, current or collision but a refined product of a coded (that is, anticipated) cycle called in its completeness photosynthesis and respiration. Nothing less will do. We shall cycle in more detail round this chanceless chemistry, this primary case of life's dynamic equilibrium (see Chapters 19-25: homeostasis and especially Chapter 20: energy metabolism).

Plan is needed to construct above a natural level of complexity. A non-coded level only reaches strictly repetitious use of molecules (as crystals), molecules with several kinds of atom or molecules (such as simple organic types) with various kinds of bond and up to, say, a hundred atoms. Of course, the forces that throw up such molecules are not committed to maintain or to increase complexity. They are not committed unto any course of 'evolution' so that the radiation or kinetic motions that perchance constructed such an abnormality would shortly, as a change of tide dissolves sandcastles, knock it out. The bulk reality of animation deploys precoded, planned complexity. It flies on quite another plane, a living body operates on quite another information level from the atoms of its dead stuff.

Action phase of process moves towards completion, exhaustion, diffuse or slack state of its lifeless or disintegrated form - the (*tam* ↓) downward process is identified and named by science 'entropy'. Natural Dialectic groups materialising forces and activities, including entropy, under the generic banner of one vector - 'gravity'. The specific force we learn about as 'mass attraction' is therefore just one of those opposing 'levity'. Reactivity slips down the ladder of excitement, it rolls down the mountain until some ledge or boulder holds it up. So rolls the world until the last stone falls and, having reached the bottom of an energetic slope, everything is stopped and flat. As this applies to steps of an event, so it applies to cosmos as a whole. The Second Law of Thermodynamics confirms that, as a process is swept downstream, it may postpone but not avoid its end. Therefore make the most of your own vital 'pause', the upwelling fountain of your life; maximise upon postponement of compulsory appointment - death.

After action, therefore, hoves the third and last step of a process. Opposite potential impotence appears; after its potential is expressed any system sinks into inertial equilibrium. Postponement's over and, as was noted, the end of an event draws nigh!

If matter is the end of things that's not to say it is the start. Physical science might but Science and the Soul won't forget the immaterial element of consciousness. Mind is part, like Mozart of his melody or Stephenson of 'Rocket', of bulk reality. The mental factor is, however, anti-chance. Mind is negentropic and an information centre. Only mind with focus of attention systematically resists the natural procedure. It conceives, builds and repairs unnatural, complicated tools to serve its purposes as fully, cheaply and as long as it seems fit. Such, if mind means life, is life's economy. (*Raj*) active creativity stamps purpose on (*tam*) passive products. It mints pockets of negentropy, coinage subject of necessity to natural forces but not made or reproduced by them. In other words it sets up orders, specified and complex orders. These serve, according to the region's laws, its own deliberations rather than blind flow of transformations flowing out of mindless chemistry. We always recognise in such deliberate mechanisms the feature we call life. If you live in town then artificial constructions quite surround you. In the country unnaturalness surrounds you too - if 'natural' is a term restricted to the influence of eyeless forces and their particles. Life's machinations shine with an epitome of purpose. Everywhere they radiate mind's essence, creativity. They involve sequential code, tools, languages and messages that, in their irreducible complexity, are called machines. In art, technology and, above all, biology coherent ingenuity of product is the norm. It is ubiquitous. This is why the Dialectic calls a living body's bulk reality 'information incarnated'.

How illogical, therefore, the rationalist's claim that chance made life! The idea of naturalistic evolution constitutes an explanation that, in materialism's terms, packs 'overwhelming power'. Not every explanation is the one that's wholly right.

Do you really think a proton, protons or an element of chemistry can come alive? Do you think, if you could find enough and fit their jigsaw well enough, that many chemicals would come alive? Do you believe that such complexity of form, which only mind can make, is an 'evolved' arrangement giving you a mind? **Sadly, a materialist does. Concept, along with brain, has risen from oblivion. This is the unconsciously blinkered faith.**

The order of animate nature includes meaning and purpose; its informative factors, psychological and biological, cannot be explained exclusively in terms of physical law and process. Its every part is integrated, coded, co-operative and so development and dynamic equilibrium (called homeostasis) of an organism is beyond the scope of just heat, physical work and molecular configuration. It needs informed negentropy. It needs, like any application, a specific plan to generate its 'morph' (or shape) and biochemical algorithm coded for the in-place, on-time delivery of materials that make up any metabolic requisition list. In terms of Dialectic such negentropy involves, like any book, sequential instructions according to an archetypal framework of ideas. What medium would you use to bear your ideas in a book? Are ink and parchment suitable or not? Life's choice is very suitable. It is an efficient, double-stranded bearer - duplex *DNA*. Even professors (shouldn't they know better?) have confused this carrier with the message that it bears. Its database and operating system is consolidated in an application program called the structure of a cell.

In other words the deliberate, holistic order implied by Science and the Soul clashes head-on with the scientific call to probability and astronomic odds against the blind and piecemeal, evolutionary construction of a cell. The odds to which appeal is made are in themselves fantastical. So desperate is the need of scientism to eliminate all notion of prior information and, according to the Theory of No Intelligence, promote the foggy, medieval notion of spontaneous generation that 'A Little Accident' (see Chapter 20: passim) in the form an alchemical chimera, 'chemical evolution', is evoked to generate the origin of life. From this imagination forth one leans on 'random fluctuations' in the order of base sequences (genetic mutation) and differential reproduction (natural selection) as crutches to explain the towering tree of life away!

For Natural Dialectic a body biological is, like all machines, a coherent bundle of intrinsic, homeostatic rules that take advantage of the chemistry and physics of the world. As such its organisation works *as intended* for a while but is still subject, like any complex body whether animate or inanimate, to the order that plays out below its start, that is, to the entropic forces of decay. It will, in this respect, suffer entropy of information (in the form of garbled, mutant genetic message), entropy of configuration (in the form of imperfection, ugliness, malfunction, illness and/or accident - the latter known as medical conditions) and entropy of energy (decline in fitness and a fall towards worn-out age). These are imperfections that affect a species with respect to its inheritance and eventual extinction; and the individual with respect to growing older towards expiry date.

Mind or no mind - that is the question always. Whatever your vision, the world spins on. Atomic automata are reflexed through determined although not always, in practice, predictable patterns of behaviour. The only chaos is inside a watcher's head.

How, therefore, does the principle play out in practice? What ciphers represent the archetypal code of cosmic law? We all observe, but science does so with precision, how the game is played out on the ground. It involves the way neutralities are polarised and (*fig.* 7.9) expressions of three cosmic fundamentals, vectors named by Natural Dialectic *sat* and *raj* and *tam*.

Slopes face both ways. You can slide down into bulk reality but can withdraw as well. Active transport, work against a gradient, needs energy. It needs a concentration of attention or a stimulus to melt what cooling hardened. *Loss of energy or information involves 'locking' into involuntary phases of physical or mental forms - materialisation. Gain involves a stimulus, a change or break into fluidity - dematerialisation.* Existence is a vast arena of such oscillations. Every cybernetic system needs antagonistic vectors - brake with an accelerator; call them (*raj*) levity and (*tam*) gravity. The primary physical expressions of these archetypal vectors are (*raj*) radiant and (*tam*) binding forces. The freedom of (*raj*) electromagnetic radiation is associated with polar charge, heat and stimulatory effects; and (*tam*) strong nuclear constraint with specks of mass, protons and atoms. From short-range issues long-range consequence. The secondary effect (ranged from subatomic to intergalactic in extent) is classical gravity (a force or curvature of space exerted as a consequence of and in proportion to its mass per volume).

The world of physical science therefore ranges from the 'pure', chargeless energy of photons to massive large-scale aggregation; and from subtle, changeful atomic bonding and the production of detailed chemical shapes to the force that 'organises' massive spheres and gross galactic discs; both levity and gravity play their part. What is bonding but the achievement of electrical neutrality? Are not the planets swung in balanced orbits? Weightless light flies over 'weighed down' mass; does anything excel illumination? All three cosmic fundamentals realise the permutations whereby, with respect to energy and information, the conscio-material spectrum of creation logically 'fans out'.

Is it really possible, as these diagrams suggest, for a General Theory like Natural Dialectic to *deduce* the form-in-principle of matter? Although broadly and simply in line with standard physical reality *fig.* 7.9 is, in its parts, tentative and unsophisticated enough to warrant a number of remarks.

Firstly, Natural Dialectic proposes two basic components of existence - information and energy. This paradoxical principle is reflected (*fig.* 7.5 (i)) in mind and matter. In mind 'waves' of active information constitute conscious life; on the other hand 'particles' of memory constitute rigid, sub-conscious objects in a field of thought. Not only memory and archetypal memory (called 'potential matter') are passively informed. In matter waves are carriers of force; they communicate their passive information with an active energy. They influence across a field, inform at-a-distance and are the medium of interaction. By contrast doubly passive particles are separate, massive objects. They stand apart. They stand for difference and are affected, that is, passively informed by waves.

Secondly, check *fig.* 7.5 (ii) for triplex expression of what has been presented here in the traditional five-fold way. Note also that 'an element' today means an atom; yesterday it meant a 'state of matter'. *Fig.* 7.9 (i) illustrates an understanding based on five 'states of matter'. If you replace (*sat*) vacuum with 'ether' and (*raj*) active energy with 'fire' you obtain an ancient paradigm, common to Hindus, Buddhists, Greeks and medieval scientists. Although considerably elaborated, the fundamental logic remains - you 'drop' from nothing (1) down to solidity (5).

Fig. 7.9 (ii) outlines the two basic kinds of communicator - a (*raj*) radiant, thermal stimulator and the (*tam*) basis of inertia, mass and aggregation. Electromagnetic and strong nuclear forces hold the atomic world together. As noted, in a dialectical view classical gravity is a secondary effect of primary cause. This cause is the weight-making binder, the 'gravitational' nuclear force.

Thirdly, *fig.* 7.9 (iii) sketches the dialectical position of stable (electron and proton) and decay-related particles (neutron and neutrino). It bears in mind that for some reason these are presumed to be the 'third generation' products of decay from high to low-energy environments and as such have 'parent' correlates. The diagram does not, however, square entirely with the current paradigm. It is weak with the weak; where does this interaction fit? Moreover complementary (*raj*) electromagnetic and (*tam*) strong nuclear forces represent complementary, polar micro-forces - but what causes mass and what antithesises the universal, attractive (*tam*) macro-force, gravity? What is the nature of a universal, expansive (*raj*) macro-force - large-scale 'levity'? Where is it? What is it? Anti-gravity? Negative pressure (suction) called dark energy or 'lambda force'? Was its inflationary effect predominant at the world's beginning whereas now, when things have 'settled down', gravity holds sway? Dialectical dynamic needs antagonism that can balance gravitation out.

One can summarise and say materialisation (as opposed to dematerialisation) is a (*tam*) downward-driven process. *In terms of Mount Universe the ideas of 'descent', 'ascent' and 'levels' are useful but the predominant mode of materialisation is captive, binding and confining.* The conscious, informative aspect of existence (*Purusha*) is overtaken by the non-conscious, energetic one (*prakriti*). It is as if the cosmos fell unconscious; or air, in which you can easily function, became condensed, waterlogged and eventually frozen. Like a polar explorer or a deep-sea diver you can now function only indirectly at the alien level of bulk reality by using specialised equipment; such physical equipment was provided as your birthday suit. Not only mind but matter is localised and confined. In a low-energy, slow circumstance of dim or absent consciousness 'binding' and 'coverage' are reasonable metaphors. The Inmost Light, a Singularity at the Apex of Mount Universe, is progressively wrapped round; it is covered up by its involvement in a diversity of imaginations, desires and physical embodiments. These 'coverings' shroud its luminosity, sap its strength and, at the low levels where the bondage is particularly tight, extinguish light and 'freeze' the freedom of its will. Oblivion. Such energy straitjackets; it is coarser than mind-stuff (*manas*) and human focus is unable to interact directly with it; its automatic phases appear as states we bracket into chemistry and physics, the zone of cosmos science sees.

In this way creation is seen as a devolution of Central Order, an inversion of the Infinite Centre. The physical part is seen as a coverage, husk or shell that binds two, nested nuclei - mind and the Quintessential Kernel; at the same time it is, paradoxically, seen as a tight, local reduction of its own Infinite Interior! Furthermore its own basic part (a finite, non-conscious atom with mass at the centre and peripheral energetic spheres or 'shells') inversely represents the whole creation (at whose nuclear heart is The Most Natural Essence, Conscious Infinity). *An atom is a natural symbol of the order of material, mass-centred universe.*

8. Lord Deliberate or Lady Luck?

Accidentally or On Purpose?

The fact is that astronomers espy the wonders of our universe; cosmologists then reason why. It seems inconceivable that this vast glory was conceived in mind - that it was planned. What, though, is the alternative? That the beautifully ordered fireball whence it once began was accidental? That the coherence of its regulation is specified by chance or that its grand design was blindly tailor-made for forms of life? If you conceive, with atheism, that eternal matter is the Ultimate Creator then your Lady's Luck. If, with theists, you believe eternal non-matter is the Ultimate and Immaterial Creator, then Lord Deliberate is your man.

The very nature of existence is inconstancy, variation and differences. Changes, transformations, relativities. However, at least physical and probably psychological behaviours appear to accord with invariant principles, themes and patterns which scientists call 'the laws of nature'. Variation is organised by invariance. Is this all? Variation is organised by these strict 'necessities' but you need to add the unpredictable flex of a second factor, chance. At first, did chance create the rigid laws themselves? Or are they the product of a third creative mode, design? We turn to consider the nature and origin of these constraints, rules or norms by which the cosmos coheres. Is such coherence a matter of chance or design? What is your theory? Which way, on balance, does your judgment swing?

This, Chapter 8, is Lady Luck and Lord Deliberate's abode. Here dwell fun and fisticuffs when trying to make up our minds. Is it charm or order rules the day? Let's note two things to set us on our way.

Firstly, this chapter is about **potential**. The word implies, of course, possibilities; and the greater a potential, the greater the number of possible outcomes - within the natural rules that govern the event in question's behaviour. Furthermore, Deliberate's side would argue, information is potential that can make the rules it also governs by. Chance plays a part but energetic, cosmic sport is governed by its set of rules. In short, he'd claim the world's consistent and not arbitrary; it is logical; it's rational and not irrational.

Secondly, in normal parlance we grade down from 'certainty' through 'probability' and, remoter, 'possibility' to an 'impossibility'. However, in statistics one word covers the whole show - probability. The range is 1 (certainty) to 0 (impossibility) and every possibility between - except that, quantum physics says, you won't ever make exactly 1 or 0. You might seem to treat as if there's neither absolute certainty nor complete impossibility. Is this, in every context, true?

The domain of past cannot be changed. It is fixed and certain. The domain of future is, of course, uncertain; it involves conditioned possibilities. But conscious choice or natural law or both combined yield, always, on the instant, present certainty. This is the way potential is 'reduced' to actuality; it is the way the world works. And yet, puzzlingly, at the very edge of minuscule, quantum physics states that all is indeterminate; cosmic apparition slips and slides upon a modicum of Planck oil. Does this mean, to all effects and purposes, matter is uncertain or is not? Such oil has encouraged some to claim that 'whatever can occur will.' The emphasis bears down on 'can'. What does 'able to' mean? Can anything *not* happen? Indeed, there is philosophy that, like fine snake oil, temptingly exudes from this. Its medicine, charlatans agree, will poison Lord Deliberate full and finally. 'Anything', they tout, 'can happen if you just wait long enough. There's no impossibility!' Thus pompous primates (from the lab's bench and not church this time) would peddle that impossibilities, by quantum probability, *can* happen. Even miracles can happen! Uncertainty has killed off never-ness; it has resurrected only possibility. Well, prove it then! Why should I believe a word? Of course, like certainty impossibility curbs endless possibilities; so why should I believe the natural world, in trillions of years, could make a teacup - let alone the tea and me! Do you believe the wind and rain could, over time, contrive to write a sonnet in a book that's full of others? It is mind, not time, creates specifics - codes and systems - that the elements can't even dream of. The chance-reducer builds, in moments, what those natural years cannot. Thus, when it comes to life itself, why should anyone believe that uncreative operation of the elements could write a book of life, create containers made of cells or stir a metabolic juice of tea? *The fact is, when it comes to rationality, there's no necessity. Long live impossibility*!

An Age of Unreason

A Theory of Intelligence considers creation 'rational' and A Theory of No Intelligence 'rationally' considers it occurred by chance. The European renaissance, an age of scientific reason, blossomed from adoption of the former theory. The origin of nature was not questioned, only its logic and exact mode of operation. Over time, though, the emphatic focus of science on physical phenomena led to the notion that there was no immaterial, metaphysical substance; and so, without Intelligence, how could other than chance be the 'reason' for the origin of natural law - which is, in turn, the operational reason for our astronomical universe? The 'reason' for all things including you and me is, at root, the epitome of lack of reason; the world's because of what's by definition non-causality - happenstance. A Theory of No Intelligence is beholden for its Causal Reason to what is nothing but sheer lack of it - a non-entity called chance. It is not that, armed with chancy theories of how - by big bangs and evolutionary unpredictability - things might have begun, anyone believes the rules of cosmos vary or evolve. Predictions and retro-dictions both assume material behaviours stay the same. <u>Nevertheless we inhabit (we may think) a creation that's illogically due to accidents, a cosmos stripped of higher rationale and (we don't think) a scientific age of, with respect to origins, unreason.</u> Materialism's stuck with a solution based in mindlessness. What other than unreason could the oblivion of non-conscious physicality employ? It wears irrationality as if it were a lucky charm.

Charming! But not only that. Materialism's 'rationale' refuses to consider any other than its own, that is, The Theory of No Intelligence. Is such intellectual 'freedom' not exclusive and its refusal most unreasonable? On two important counts, therefore, scientism fails the test of rationality.

There's a third and fourth. It is generally believed logically impossible for a cause to bring about effect before its own existence. X creates Y. How, therefore, could X make X. How could the universe spontaneously create itself? Especially if X was nothing. Is nothing a sufficient cause for anything? How could nothing for no reason create something; how could its impotence make everything? Nor can abstract descriptions (such as formulae and laws) cause a thing. If Stephen Hawking thinks an M-theory or law of gravity could bring about a universe, then he is logically confused. Laws need agents. If, therefore, he philosophically skips to causal gravity, then gravity's a property of mass and if there's mass his cosmos has already been created. Y makes X. Or, if he implies that gravity is the creator of our universe, how could this be? Did gravity precede its world of influence; is it pre-physical as well? Gravity can't make a thing - and that includes itself. It can shape things but that's not the same. Y can't make X. An effect cannot create its cause!

The fourth count we have also met. Chapter 0: The Science Delusion relates neuroscientism's grand, deflationary idea; and if mind really is a bunch of firing neurons what sense can we make of that? Atoms and electrons do not *know* a thing. Their 'reason' cannot carry weight; why should we believe a word that their associate (another molecular configuration called a tongue) might say - especially concerning how the universe itself was made?! From such roots unreason stalks materialism's explanations.

The Order of Invariance.

Did your judgment freely swing towards logic that involves the rule of anti-chance whose causal and coherent reasons we call comprehensive law? Do you see the cosmos as a self-consistent, fine-tuned system or are its patterns of behaviour sprung by chance? Information is potential telling action how. It commands. As well as potential, therefore, this chapter is concerned with **rules** - **rules** versus randomness, that is, invariant principles producing various effects. **On top of rules we shall consider choice of origin, that is, 'original how'.**

Perhaps the most exciting part of any study, especially physics and chemistry, is to graduate to some kind of overall understanding of the subject, a point where things 'come together', a fluent grasp of the principles and rules - the very antithesis of chance. Order involves constraint, rule and law. We recognise such constraints in nature. They represent a systematic prohibition or avoidance of error. Not anarchy but its reverse. Cosmos, not chaos, is reality. A stack expressing law might read as follows:

tam/ raj	*Sat*
lesser truth/ appearance	*Truth*
aspect of reality	*Reality*
particular instance	*Potential/ All Possibilities*
variation	*Invariance/ Permanence*
information	*Source of Information*
development	*Seed*
practice/ actuality	*Principle/ Law*
subsequent order	*Archetype*
elaboration	*Plan*
oscillation	*Homeostatic Norm*
change/ inconstancy	*Stability/ Constancy*

↓	tam	raj	↑
	passive	*active*	
	informed/ ruled	*informant/ ruler*	
	away from norm	*towards norm*	
	unbalancing	*equilibration*	
	irregular motion	*regular vibration*	
	contingent variation	*basic themes*	
	instability/ unpredictability	*stability/ predictability*	
	external 'fall-out'	*internal order*	
	crystallisation/ bulk shape	*energetic pattern*	
	apparent disorder	*actual order*	
	chance	*necessity*	
	coincidence	*providence*	
	accident	*design*	
	chaos	*cosmos*	

It should be pretty clear by now. The world's in fundamental paradox. It's both in balance and it's not; but, as the Dialectic's logic would predict, balance rules. Its order of invariance is defined by conservation, equilibrium and symmetry. These three are - check Chapter 2: The Cosmic Fundamentals - (*Sat*) Essential, Central Qualities. From a *top-down* perspective their Main Dialectic hierarchically precedes the secondary, existential dialectic; it balances equations, it informs the (*raj/ tam*) polar swing of things. From unity of government are issued the commandments ruling variance, motion and asymmetries.

You are a microcosm that relates to macrocosmos, that structurally reflects creation as a whole. And, if you can understand the mind of a creator you are able to predict invention; you predict the nature and behaviour of his systems - in this case The Great Machine.

In this respect the community of physicists splits roughly into two - idealists (the fundamental theoreticians) and pragmatists (empirical researchers). Both are equally important but, mind over matter and according to the Dialectic, which takes precedence? Einstein, a great idealist, said, "I want to know how God created this world. I am not interested in this or that phenomenon. I want to know His thoughts; the rest are details." He was convinced that using mathematical construction alone man could discover the concepts, logic and laws that govern *physicalia*. His instrument of progress was insight. By pure thought one might grasp material reality. For Natural Dialectic such reality is at root vested in the abstract of a metaphysical archetype. It is those patterns, if you want to wax dramatic, that compose the memory of God; and you, the son, eventually will read your Father's Will. The body of this universal testament has always issued from its mind.

Heartburn. Indigestion. Worse. Pass the sceptic pill since in this view (*Sat*) Perfect Symmetry of Consciousness is at the Origin, the Centre or the Axis of Mount Universe. And (*sat*) perfect symmetry of the unconscious zone is at the base of mind; non-consciousness is ruled by archetypal memory in universal mind. This is the level of psychosomatic interaction; it is the tier of potential or 'transcendent' matter. 'True' physics is then vested in original symmetry that man has more or less awoken to where science wants to totally. Quantum physics is the nearest rung upon world-ladder and especially addresses it.

The search is therefore for a toe-hold on a *TOE*. After all, what was there before beginnings? Following the physics is like trying to catch ferrets in a sack! Theories whiz and weave. Or you could say that signal lights were always changing up and down the cosmic starting line. Each professional train of thought hauls wagons full of travellers towards the terminus of time. The motive power for these conceptual expresses is upgraded every few decades (by engines called, for example, Aristotle, Ibn Sina, Galileo, Newton, Einstein, Faraday or Mandelbrot); and all, whilst skirting their line-side's infinity, steam by the fire of conviction (or perhaps the relative illusion) that a Final Destination, Absolutely Nowhere, could be just around the bend. Research for which a green light is at go includes 'inflation', 'expansion out of nothing' and 'black hole singularities'; currently on cautious amber, 'string' and 'loop' theories of 'quantum gravity'; and which, without a definite timetable, have not passed red - '*VSL* (the varying speed of light)' and 'everlasting matter in a multiverse'. For sure, the major schedule is to take the measure of some piston driving in the cylinder of space. What rods cosmos in its revolutions as it puffs around the void?

Precondition/ Potential

Bottom-up, **no metaphysic orders what is physical**. You shrug. The so-called 'laws of nature' are but a description of behaviours that a starting point (say, big bang/ initial projection) spawned. Your formulations and the values of their constant factors are not fixed by any deeper level than experimental regularities. Natural reasons have no reason; material effects do not have *a priori* cause. Certainly there's no intention that substantiates cosmology; the order in its movements started as an accident; freeze teleology right out of every frame.

Yet, when it comes to how the universe, its laws of physics and its bio-friendliness are organised you might be struck - struck hard - by the appearance of design. And if you strike out metaphysic then your source of order must be chance. Could, however, the appearance of design be more than seeming? Could it be for real?

Top-down, **the order of invariance is hierarchical.** Natural Dialectic indicates a source from which the stages of a cascade fall; it involves an inner, central 'symmetry' from whose intrinsic order cosmos breaks. It breaks, like all phase changes, with a natural spontaneity. Natural spontaneity, however, never barred preordination. *Rules are information, information is potential for behavioural patterns*. Regulation thus precedes and guides the way a game is played spontaneously. Solids fall from gases and, as they do, so different rules of state apply; so different sets of laws precipitate from inwards outwardly. There would be, according to this scheme, essential principles and primary law; and secondary subroutines that follow on. What is, we keep on asking, the central source of code?

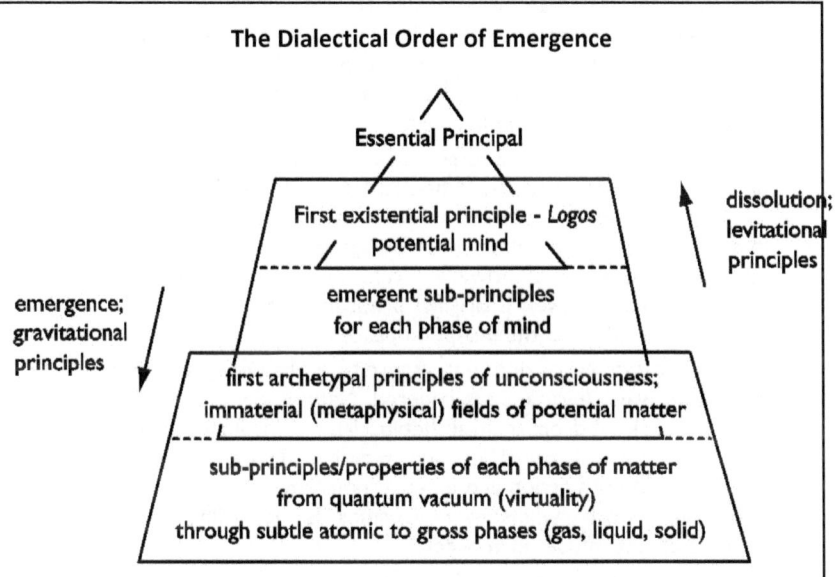

Things 'emerge' when individual parts combine into a fresh phenomenon. From a cloud 'emerges' rain; when cosmos cools as if from nowhere snows a universe.

In dialectical terms 'emergence' is the expression of potential; it is the product of fixation of information or binding of energy. We say that from subtle issues gross, from inward or implicit is devolved explicit and, in the jargon, symmetry is broken. Principle materialises into practice, detail's born of generality and variations wobble round the axis of their theme. From microscopic particles and atoms macroscopic entities arise. And what's more microscopic than a dot that's disappeared? What's simpler than the dot's prior nothingness? The Dialectic indicates that physic of the universe emerged from void. Science calls the source a quantum vacuum. What is the nature of this cornucopia of nothingness whence nature issues into sensibility?

The hierarchy of emergence runs from simple, powerful height. It disperses from its inward, central essence in accordance with the rules of play. What is the archetypal principle from which sub-principles and properties of lower phases follow orderly?

In the material case Natural Dialectic indicates that immaterial, latent fields precede the breaking of the balanced tension of their symmetry. Conscious Potential tops the bill; so-called *Logos* precedes active mind; and prior to active, quantum matter an unconscious potential (archetype) presides. In scientific terms an archetypal symmetry beyond even vacuum state collapses into various asymmetries. **Thus the symmetry of inactive latency governs; unexcited fields encode, at root, all natural law; pure potential, preceding even the quantum vacuum, gives rise to preordained behaviours.** From lower quantum

> and atomic phases drop the rules that govern and the properties exhibited by the emergence of firstly fluid and then fixed and solid states of matter. *In this hierarchy rules and properties at one level are dependent on the one above; in practice, however, those at one phase are often effectively irrelevant to others.*
>
> Symmetry is like the surface of a pond whose smoothness, broken into waves, is gone. When latency is broken its unused 'symmetry' is hidden from your view. The 'breakages' caused by loss of energy through phase on phase appear spontaneous. Such spontaneity does not, however, rule out prior, underlying, archetypal order - even if original preordination's truth is only clearly mirrored when the mill-pond peace returns.
>
> Thus 'symmetry' does not, according to the Dialectic, break in random ways. Invariance is ordered. Natural law precipitates from empty sky; it drops from archetypal latency through quantum into bulk material principles and practice. There is phased emergence of both action and its government by 'higher' source. There's no worry whatsoever! Cosmos is, according to its gradient of emergence, all in order.
>
> **fig. 8.1 *(see also 1.3 and 10.2)***

Let's rephrase. On the conscio-material hierarchy information is prior; dialectically, information precedes behaviour; it is (*sat*) information that, either in active/ mental or passive/ mindless form, gives rise to (*raj*) flux of action and its (*tam*) outcome. From learning Primary Dialectic and the Cosmic Fundamentals (Chapter 2) do you remember the (*sat*) cosmic fundamental which involves potential, precondition and prior principle? Potential is a Main, Essential Character; it is invisible but crucial and therefore needs be clarified.

From (*Sat*) Top Potential (*fig.* 7.1) there derive all possibilities; but, as in the case of every Essential Character, reflections of its facet show at every level. Thus you find restricted potencies; relative, conditioned or qualified sorts of potential; you find constrained capabilities, a specific range of possibilities for each 'lesser' potentiality. Such 'lessers' may be psychological (informative) or energetic (physical). Psychological examples include ideas and wishes (giving rise to action); physical examples include potential charge (a battery's voltage) or potential energy (of a body due to its gravitational position). *Potential matter itself, however, transcends the physical plane; it constitutes the 'informative whole behind material parts' but, as metaphysic, is not made of bits; it is the formal, fixed, subconscious aspect of universal mind that, as archetypal memory, directs cosmic energy.* <u>*Aren't the laws of nature a restricted set of possibilities, a theme on which, with limits, material variation in the universe is played?*</u>

Law is regularity; symmetry of law is stable; and, as Einstein claimed, the rules of nature are 'incorporated reason'. Archetypes, in this encoded sense, are the voice of reason heard through mechanisms of material form. Their expression is a 'breakage' out of background metaphysic; it leads from hidden symmetry through simple, subtle quantum agents to the detailed aggregates of flux and fixity. Heat lost gives fall to sharper form. Information buried in complex diversity might seem to have lost its theme but, as all scientists know and strive to better understand, you have to dig to find the principle substantiating practice. *From what superficially seems chance you have to disentangle rule.*

From invariance variance streams. When motion tips the balance symmetry drops wobbling to asymmetry; it might fall as far as flat. Down the ladder of the world from principles drop practices, actions break out from ideals and particulars of variation are precipitated from their general themes. Apparent chaos flows from cosmos. Do not worry. It is all in order.

Remember, therefore, that the existential precondition is (Chapter 5: Orderly Creation and *figs*. 5.3, 5.4 and 7.1) Essence; this gives rise to psychological principle and, lower down the scale, physical principle in the form of potential matter. What is such matter? In dialectical terms it is archetypal infrastructure, universal memory whence are crystallised the regular behaviours (or laws) expressed by material energy. To recap, potential information is encapsulated at the highest level in a psychological principle called *Logos*. *Logos* is First Cause. Lower down the scale physical first principles are encapsulated in archetype. Unconscious archetype is physical first cause. **In this view, therefore, order is not a matter of accident**. Qualities, properties and values are bound within a self-consistent whole. In this sense immaterial priority from the stage above informs what issues below. Although the base of Mount Universe is non-conscious, reflex and automatised, the system from its core is teleological. It is imbued with intention, will and purpose. Metaphysic rules the physical.

The grammar of causal archetype is expressed first through quantum fields and factors and, having 'lost' both information and energy, complexifies into the bulk forms of solids, liquids and gases. The cosmic system is thereby as integrated as your own person but its 'health', its norms and laws, is fixed. Nor, since matter is involuntary, is voluntary deviation possible. The rules/ laws of nature are (at least for this creation and we know no other) invariant. Passive information is, while effectively reflex, the product of higher, internal, immaterial purpose. This point is obvious and natural to every comprehension bar materialism's atheology. For atheologists the origin of order lies in chance - in evolution out of nothing into comprehension that 'absurd' irrationality born of oblivion is why the scum of chemicals we call ourselves exists! Any other source of order is repellent to a mind-set which denies First Cause. Thus the notion of invariant order born of hierarchy needs reiteration; the way that norms are sourced needs Dialectic emphasis.

A norm is an axis round which fluctuation wobbles. It is a pivot. In this sense laws are dynamic pivots; and natural law is the pivot around which cosmos homeostatically swings, round which things keep their automatic balance. This balancing act is called 'the natural order'.

Check Chapter 1: Causality again. A principle is like a seed. A plan or program causes outcome. It was proposed that precondition for the physical arena be called archetypal law. The symmetry of principle is broken by its agents; and these agents of expression, everywhere identical, are quantum forces and/or particles. The order of material cosmos is derived from metaphysic. The code of archetypal symmetry is broken, by the action of its subroutines, into the eventual, secondary expressions we perceive. Because it is the seed from which the orderly construction/ development of universe is shaped 'super-symmetry' is just a modern way of saying, without archetypal implication, 'Cosmic Egg'.

Unseen gives rise to visible; potential matter's primary effect is quantum agents; and their inner, causal regulation predisposes secondary outcome in the form of gross creation. It is this with which we life-forms interact.

Now check, for example, *figs*. 7.2 - 4. At the apex of Mount Universe The Infinite is Super-symmetry. There follows primary *SUSY* of the mind (the *Logos*) and psychological sub-regulations. Hierarchically, at base of mind, potential matter is the secondary, non-conscious *SUSY* physicists are looking for - will strings of string or super-string mathematics, though not up and threaded yet, lead through the labyrinth of cosmos to its central lair? *SUSY*'s broken into subroutines each with its integrated symmetry expressed in terms of physic by its principals viz. fundamental particles and forces such as charge. From quantum and atomic agents are constructed all familiar shapes and motions, textures and behaviours we are wont to call 'reality'. The Dialectic simply adds unseen dimension. If archetypal super-symmetry includes sub-archetypes - the principles that we call laws of nature - then perceive these laws as subroutines. Then, just as physics indicates, these sub-symmetrical routines in turn are broken by the 'real' world's simultaneous interaction of prime agents with their subtle physical and chemical effects. Hidden cause gives rise to secondary effect. Quantum events are, on the third step down, devolved into a multiplicity of variation on their subtle themes. The 'textures' of their products compose the bulk world that our senses can perceive and thus, in error, think most real. These secondary and gross tertiary outcomes are now seen wobbling round their central metaphysic, round the axes of intrinsic principles. Such superficial chaos is the character that manifest existence rings its changes with.

Vectors of the existential symmetry and how it's broken are the cosmic fundamentals (Chapter 2). *(Sat) equilibrates; it represents 'point of balance'. Gravity (tam ↓) and levity (raj ↑) are both polarities of motion; they represent the swings by which all existential change occurs.* In other dialectical words, each specific object and event, whether psychological or physical, is the expression of a combination (or stack) of complementary opposites, each of whose relative proportion in the whole is 'weighted' by the three cosmic fundamentals of *(sat)* neutral, *(raj)* positive and *(tam)* negative. This metaphysical 'background' is expressed through an appropriately coordinated composition of information (which orders) and energy (which is ordered). *In this way (sat) norms are the pivots around which fluctuate the antagonistic (raj/ tam) tendencies - a continual state of change cast within an overall framework of balance.* Balance of equation illustrates how, in a rigidly governed system, energy is guided, transferred and transformed in relation to invariable factors we call natural laws, constancy or normality. Invariance. Such cosmic accountancy could, based round conservation or 'laws of stability', be termed 'dynamic equilibrium'. **Accordingly, the universe can be described as a cybernetic machine whose (*sat*) regulations arbitrate or, if you like, equilibrate between complementary yet opposing tendencies.**

Non-life exhibits a non-purposive complexity; life's systems are replete with purpose. An animated 'type' of body is therefore, viewed dialectically, a kind of symmetry. An organism is an archetypal program that, like a mini-universe, includes a buffered balance of routines. These unseen, tailored subroutines are expressed at quantum, biochemical and physiological levels of what's visible, material and therefore scientific biology. In other words, a metaphysically-coded organism is at the same time more complex and less absolutely rigid in its working than inanimate or inorganic matter. Inorganic simplicities are compounded into complex molecules; and a life-form's overall law involves the coherent operation of a 'bundle' of norms. You might turn to 'chaos theory' or statistics the better to describe its host of interacting factors, principles and homeostatic sub-laws. These combine to make up its specific type. Physiology and anatomy are informed expressions of the 'rules'. An embodied life arbitrates, by the light of its own programmed or thoughtful principles, between the various complementary yet opposing tendencies (hunger/ satiety, heat/ cold, pleasure/ pain etc.) that compose its circumstance. The panoply of life, the variety of biological types illustrates an increasing capability that ranges from reflex towards voluntary decision-making about what seems right or good for it. What is 'right' amounts to health, survival and, on top of these, contentment. Is not satisfaction the fulfilment of desire? And desire a need, a need born of physical constraint, innate instinct or dreamt up on the ephemeral wings of imagination? So that need, more or less urgent, becomes a norm, a target, principle or homeostatic law according to which an organism readjusts its world. What has served its principled decision-making is an exchange of biologically and psychologically processed information. The store of automatic principle is archetype; the 'legal statutes' that inform four forces (as we understand them now) and govern bodies physical compose an instrument of balance, of equilibration and thereby symmetry. **'Breaking biological symmetry' into actuality (a body such as yours) is discussed at length in Volume 1.**

In this sense might one travel further down the archetypal line? Is the cosmos like an organism? A purposive system? Not a conscious one but a machine? To the extent that a great many of its behaviours are cyclical and that it operates according to fixed rules, you might agree. It turns on 'axial' constraints but is it purposive? In the sense that, despite their interconnectedness, individual operations are mindless you might disagree. Cosmos shows order, you agree, but neither systematic coherence nor 'end-game'. It looks disorderly. Each part, although contextually constrained, ploughs on regardless. Passive, involuntary matter is inexorable. It is needless and wholly lacking in desire. It is not a case of plot lost but never any - and yet... plot. Total discipline, impersonal, amoral order according to theme, reason and the rules. It shows rigid adherence to a preordained plot. As natural laws and fundamental constants we can check the rules in any science data book but have no idea of how they originated or why they have the formulae and values that they do. *Science does not explain the basis of its own fundamentals.* Are they without reason? How irrational is the web of values clustered just behind the scenery that shows? Is the set-up accidental or, on the other hand, is matter not a managed stage? Do you remember (Chapter 2) the physicist James Jeans observed that cosmos seemed more like a thought than a machine? Aren't the couple interlinked?

Is a Finely Tuned Universe Coincidental?

Hold on! Materialism doesn't like this 'thoughtful' sort of drift. Let's take a basic bearing on the fix.

A naked singularity, whatever that is if it started everything, must have been beyond expression of the laws of physics. Perhaps latent legality was metaphysical; but, besides such archetypal 'immanence', within an isolated universe (and what is ours connected to?) you can check your Physics Data Booklet. You will find maybe thirty constants (e.g. unit charge, fine structure, gravitational, Planck's and others) conserved from big-bang time exactly as they are today. They interlock. Their properties and values interlock. Forces and the laws describing their behaviour seem to delicately balance out. Nature is, at its controlling level, 'poised'. A missing instrument or one that's out of tune affects orchestral symmetry; there's not one instrumental value you disturb without destroying natural integrity. The Holy Grail of physics is to demonstrate not how but why these factors have the values that they do. Yet the probability that *PAM*'s accidental miracle could have produced even one of them with its precisely stipulated value is evanescent. Then, as in a great game of sudoku, what about the others, equally exact and interlocking! The laws of probability alone strike like an axe at 'absolutely nothing' that, from before or 'above' the laws of physics, miraculously and yet accidentally struck up exactly such a complex combination as unlocked our cosmos.

In the beginning was *NOT* chaos. Modern physics shows that mankind dwells in a finely tuned universe. To be precisely fit for life it must have started in a way most orderly, specific and specially defined. *Fine-tuned by chance? If low probability together with specific definition indicate design, no chance!* **A fine-tuned universe might be construed as one of specific, irreducible complexity - and such complexity, we shall see, can be a hallmark of design.** *Or, if you must obtain a naturalistic answer, what contrivance can you dream up out of mental space?*

A mystery, this 'fine-tuning problem', lurks at physics' heart; a query gnaws the scientific core. What is the origin of accurate congruity? Who or what first cracked but never scrambled laws of nature from the cosmic egg? All in order, all on board for life! Was cosmos fixed or did it fix itself? Has some Great Inventor twiddling with his knobs and dials so tuned the cosmic program into working perfectly? Was the source of software natural? Or, God forbid it, Natural? Swap God for Luck. You cry that Mindless Mother Nature must have fixed herself; this way the origin of natural law, the 'fix' of physics, is the child of Lady Luck. Arch-dodgers of intelligence and metaphysic postulate, as we shall see, bye-laws in a multiverse. Our universe, they speculate with shameless lack of elegant parsimony, is probably just one from multiversity, from infinite arrays!

↓	*tam*	*raj*	↑
	physical	metaphysical	
	material	immaterial	
	chance	archetype	
	oblivion	memory/ mind	

Who can argue? Such imaginary 'science' is, however, met with *top-down* concepts - universal mind and archetype. These are hoary, old ideas; but were our forebears fools, is what's traditional always wrong? 'Universal mind' is as antique yet up-to-date and rational as human scholarship. And in terms of Natural Dialectic archetype is metaphysical; it is potential matter or, as some call it, 'meta-law'. **Natural law is thus delivered in three stages - main (primary) and secondary archetypal routines, quantum factors and, lowest-ranking and most outwardly expressed, bulk material regulation.** It is the third one easily and obviously observes.

This résumé has set the Dialectic record straight. We've got the rising tension off our chest. From these opposing markers we expand.

Is the Invariance a Fix?

Fine-tuned and fit for life; but, you muse, is the world's invariance a fix? *Bottom-up*, you won't agree; *top-down*, by now you know you will.

Check Chapter 4: Coincidence or No Coincidence. Impersonal science (for which mind is a detached and blanked out part) shrinks from the idea of management. Thus, in character with its core concept and as you might not expect, it shrinks from reason and espouses accident. Cosmos (meaning 'orderly arrangement') derives from a coincidental start and accidentally carries on. What manages this raft of chance? By chance emerged (or perhaps evolved) out of the blue an unexpected and coherent suite of managers - its laws, its reasons. Who dare propose a reason for these reasons? How could a mindless universe involve a rationale, a conceptual basis for coordination? You and I are proof that reasonable, subjective consciousness exists; mind is abroad but you argue, axiomatically, that it is really just a special form of matter - brain. So mind itself can't make a thing. That is to say, *our* minds cannot create a mote of dust; they can only cause our bodies and their tools to rearrange the world. Nonetheless will-power and purpose aren't made up of atoms making brain. Nor are acts of creativity; nor memories, imagination, sensitivity, intelligence and feelings of emotion. And yet life is composed of these. If nature metaphysical as well as physical were universal then, astoundingly, the world-machine would not be mindless. Absolutely the reverse. Reasons are the writ of purpose and of thought. Cosmos would be conceptual and, therefore, reasonable; science would not have dismissed but accurately defined the archetypes. It would be hierarchically informed according to a natural train of logic. This is not a new idea. It used until quite recently to be The Standard Theory.

In practice both psychological and physical aspects of nature are, as introduced in Chapter 1, informed energy. The cosmic fundamentals are, as basic principles of design, expressed through various agencies. They devolve in three main sorts. These are (*sat*) norms of potential, equilibrium and poise; (*raj*) patterns of activation, motion and change; and (*tam*) behaviours that tend to deactivate, gravitate and fix. What instruments convey their leading code? What shunts them down from mind? Practice involves their expression through causal agents (such as neutral photon and the poles of stimulatory electron as opposed to massive proton); and through subsequent 'large-scale' effects (such as heat, chemical reaction, bulk aggregation and gravity). It is from the behaviour of anyone or anything that we infer their guiding principles. In the case of automata there is no chance the foundations will shift. Science works believing laws are simply regularities in nature. If, however, you believe in chance then you are able to believe in nearly anything - perhaps in a multiverse or even other galaxies you might find pockets or domains of varying natural law.

From a dialectical perspective this is not enough. The single universe results from interactions, endless variations and recombinations on the theme laid down by primary agents of its regulation or, as could be said, expressors of its principles. <u>You cannot formulate any old set of such principles or ascribe any mix of fundamental values to a system's operation and necessarily expect it to work</u>. **Indeed, unless chance struck phenomenally lucky you might argue (as in Chapters 5-12 and 26) that our own working universe is, in origin, a logical construction devolved from fundamental principles and, as such, a bio-friendly expression of anti-chance viz. mind.**

Check *fig*. 8.1 again. *This whole section ('Energy') shows how cosmic principles can be viewed as, and likely are, aspects of First, Essential Principle and subsequent, finite, dualistic derivatives.* (*Sat*) Essence is the pivot around which (*raj/ tam*) motions, called collectively existence, oscillate. As these three are fundamentally interactive; as they are interconnected so are their consequent derivatives. These derivatives amount to a hierarchy of laws of nature - archetypal symmetries that cover quantum actions which in turn regulate bulk outcomes. Physicists observe, measure, classify and, perhaps according to aesthetic judgment, try to simplify. They write prescriptions for predictions. Things might work regardless of each other but the global rules defining them do not. By circular reasoning finite events and properties are explained in terms of each other. For example, interactions describe space, time and things; on the other hand, the behaviour of these things in space and time defines interactions. Check the attributes of matter described in your science data book. Start with motion. Free it up by including time and the three spatial dimensions. Motion means a change in position or shape. Change needs energy which can be quantified in joules. There exist several species of energy such as electromagnetic, gravitational, chemical, kinetic, nuclear, sonar, thermal etc. Each can, without overall loss or gain, be transformed into another. The wheels go round but what set them up? Of all these circles where's the nub? What is Axis, The Central Unity and Initial, Ultimate Connector? For the lower, peripheral tier of Mount Universe the dialectical logic leads down from preordination in the form of archetypal law.

There's a difference, let's be clear, between the laws of science (which are incomplete) and laws of nature (to whose perfection scientific laws approximate). Science well describes; but for every explanation of a scientific law there lies a deeper and more general one. One might, dialectically, expect to find profound, ubiquitous paradox within the way nature works. This is because it springs from the simultaneous interplay of complementary (*raj/ tam*) opposites governed by a (*sat*) third internal, invisible balancing factor. *Symmetry, balance, regulant homeostasis and invariance are characteristics of Primary Dialectic; they are, equally, key aspects of universal order.* In their governance the three cosmic fundamentals show as 'predispositions' of energy with respect to the conscio-material gradient. Up, down and poise; gain, loss, equilibrium; action, exhaustion and potential. They show, in different objects and events, in various proportion, predominance/ subordinance or ratio. Three 'constant tendencies' are basic to the different ways that energy and objects are expressed. They comprise, like notes in scale, 'conduits' through which the 'shapes of force' appear, harmonically transpose and are transformed in ways we call 'legitimate'. These conduits, like chords, oscillate with metaphysical authority; they are - the Dialectic root of physic's play - 'deep law' called archetypes. Conservative, non-physical realities. Archetypes are principals through which are issued principles-in-matter - particles and forces that compose the substrate of our world. Or, from a dialectical perspective, call them subatomic 'superstrate'.

Symmetry

	tam/ raj	*Sat*
	relative	*Absolute*
	gross/ subtle	*Unseen*
	subsequent expression	*Prior Potential*
	influence on game	*Law*
	loss of balance or symmetry	*Balance/ Symmetry*
	diversity	*Unity*
	inconstancy/ change	*Stability/ Constancy*
↓	*tam*	*raj* ↑
	gross/ passive pattern	*subtle/ active pattern*
	ruled	*ruler*
	diversification	*unification*
	isolation	*connection*
	inc. irregularity/ imbalance	*inc. regularity/ balance*
	appearance of chaos	*appearance of cosmos*

It's time to look at ideal beauty in more depth. You know symmetry. You've seen it in the geometry of arabesque, Persian carpets and the fabric of a church or mosque. Indeed, you are bilaterally symmetrical yourself. But why's it so important?

Check the Glossary. A system's symmetry is one of its mathematical or physical features that is preserved under some change. Such property involves repeating process, structure or behaviour and thereby engages two distinct elements - 'invariance' and 'transformation'. And 'conservation' means about the same - a property or feature of a system left unchanging as the system is transformed. For example, the First Law of Thermodynamics states that through all kinds of transformations, at any angle, time or place, energy's conserved; the subsequent presumption of 'continuous symmetry' is that the natural laws of physics do not change in all of time and space. Invariant legality. Unless, of course, in some as yet undiscovered way, energy were *not* conserved and thus, upon such universal flux or leakage, spatial geometry (which material energies affect) were warped and laws began to change! Such mutability, however, isn't what we find. Thus, since energy's conserved and laws don't change, physicists can balance their equations - which is pretty basic stuff!

You can see, by this, that symmetry and conservation are a powerful pair. Emmy Noether's theorem explains precisely their connection. It shows that any conservation law is associated with the symmetry of a physical system (e.g. of momentum with spatial, angular momentum with rotational or energy with temporal symmetry); and, conversely, that each symmetry implies that a physical feature of its system is conserved. Fundamental conservation laws are linked to symmetry; the rules of nature's game, conserved through every changeful transformation, are examples of 'translational' symmetry. From maths of symmetry, therefore, you can predict its rules. Or, by experiment, you check invariance of action; if you discover conservation law you look for symmetry dictating how it works. The whole universe is framed by regularity; the search for cosmic fundamentals boils down to a search for symmetries. No wonder physicists rate elegance. They can explore such natural beauty using a nineteenth century branch of mathematics based on the work of Karl Friedrich Gauss and Leonhard Euler called 'group theory'. For example, in the search for symmetry responsible for conservation of charge 'gauge symmetry' was discovered. Its groups have been used to predict previously unknown particles and explain the electroweak and strong unifications. They frame intrinsic symmetry that's independent of space-time and orders subatomic patterns. A conceptual key supplied by nature, scientists use it to unlock the secret logic of our universe.

Geometrical and energetic balances pervade the natural world. We spot it easily. You will have noticed rotational, reflective, inverted and various species of radial and strip symmetry. Take snowflakes, faces, flowers and that balance of equations telling us how energy's conserved. These are extrinsic symmetries but at a deeper level can we delve to find conceptual simplicity behind the seeming muddle of the world? Could the pivot, balance, swing upon a unity of law? Classical and quantum physics both observe a similarity of symmetry (though one's deterministic and the other, using probabilities, non-deterministically so); so can we use intrinsic symmetry to simplify complexity and grasp, at root, the unitary nature of the universe?

Unchanging archetypal principle is not much use without expression in the changes objects undergo. Unexpressed potential precedes. In the conscio-material cascade from 'potential existence' (that is, Essence) issue mind and matter. In that order. Dialectically, a cosmic level issues from the one above. From 'apparent nothingness' of immaterial information drop behaviours of material energy; potential matter (archetype) amounts to neutral symmetry from which the polar order of this universe derives. Various ways have been discovered how, while leaving fixed parameters of archetype intact, symmetries can 'break spontaneously'. Natural Dialectic knows such 'breakage', psychologically due to loss of information and physically to loss of energy, as variation-on-a-theme. It's a version of diversity from unity, limited plasticity, differentiation or a principle expressed in local detail. Water vapour crystallised into a snowflake is a simple pointer to the way that symmetry, broken symmetry and conservation work. The vector of asymmetry is (*tam* ↓) down; it is produced by entropy.

But symmetry itself cannot make something out of nothing. It does not explain the *origin* of law. Noether's mathematics only let you find new laws that govern at a deeper and more comprehensive level than the ones already known. How, therefore, was our cosmos 'broken' out of perfect symmetry? How did the archetypal influence, 'configuration of the world' that rests in latency 'behind' the vacuum, sally to control all physics? Quantum physics is described in terms of fields and 'amplitudes of probability'. Each excited field, such as the electromagnetic or strong forces, an electron or the different kinds of quark, exhibits its appropriate patterns of behaviour. Perhaps things started at high energy. Then forces acted, particles appeared as different fields were variously disturbed. You drop, when things have crystallised as far as possible, into distinct but lifeless states of matter that comprise your body's world. This cascade of broken symmetry turns archetypal principle to local practices. There is no confusion. It gives rise to action that is only orderly. Thus scrutiny of Natural Dialectic and the attributes of symmetry should offer further clues to understanding how the polar universe is built and, both in principle and practice, works.

Valuable Constancy

Where would you be without it? If memories are 'fixed mind' then so are archetypes. The roots of physics are described by 'frozen formulae'; they accord with algorithmic programs accountable by maths. And their values are therefore like cogs fixed to the hub of the universe; they are the primary settings to which all operations are coherently geared, the keys with which, cut to the template of invariant archetypes, all patterns of behaviour in the cosmos are unlocked. In other words, mathematical abstractions reflecting actuality reflect, in the way that they describe a musical rendition, an aspect of the archetypes. They represent, at root, a world and its experience reduced to numbers. Here are Mozart and the physicist again!

In which case are the values of these numbers absolute and invariable or dependent, derived and relative? Science seeks, for every explanation and its 'law', a deeper and more general one. Just as the sensible world hierarchically depends on its invisible 'deep structure', do relative sub-principles track back to *absolutely* fundamental principles, that is, invariant law and government?

Physics certainly presumes the same laws of the universe apply throughout and that their constancy endures, from *ex nihilo* projection and the *CMBR*, through time as well. But what constitutes nature's scaffold? By which changeless threads is its fabric woven? The electron rest mass, gravitational constant, speed of light *in vacuo* and, perhaps, length of a 'string' are assumed ubiquitously constant but no-one has much idea why they have the overall-related, interlinking values that they do. The same applies to another thirty or so fundamental constants you can look up in a Science Data Book.

Could fluctuations ever happen at the heart of physical reality? *Apparent constants* might change numerically according to conditions. They might oscillate, such as the gravitational constant (G) has been measured, around a norm; or you might even, for example, obtain *VSL* (variant speed of light; see Chapter 12) or other 'constants' changing over time or at extreme high energies unlike those you generally find in our 'low pressure' cosmos. Again, Hubble's Law concerns galactic recession and the rate of the expansion of the universe. Its form ($H = v/d$ where H = the Hubble 'constant', v = recessional velocity measured by red shift and d = intervening distance from an observer on earth) would seem to generate actual distances of stars and galaxies whose different red shifts indicate speeds of recession. The reciprocal of this constant gives a measure of time (or age) so that you might hope to estimate the age of the universe. However, even if there were no problems over the interpretation of red shift (again, Chapter 12), this hope is confounded by a circular form of argument. Distances can be estimated using the Hubble Constant but the constant itself is estimated using distance - which means that astronomers disagree over its value and thus the size and age of the universe. We will meet a similar circular form of argument when it comes to dating fossils. In this case the form of equation remains constant but its values shift.

Apparent constants can also vary numerically according to the units of measurement you use. Humanly contrived units of measurement change from time to time and across cultures. Take weight, distance or even, measured in a way convenient to man on planet earth, time. The units are interchangeable but relative. For example a mile is 1.61 kilometres and 2.205 pounds make up a kilogram. Would ET from a far-off galaxy recognise this measure as his own? Scientists have collaborated to generate 'absolute' SI units but are there any natural, universal and in this sense absolute units of measurement?

Max Planck uncovered them - basic natural units of energy, mass, length and time. He pronounced nature fundamentally discontinuous, that is, he 'quantised' the universe into final, minuscule divisions. His 'pure' *quantum measurements* transcend all the different convenient but artificial human divisions of scale and can be used to register anything, anywhere, anytime under any conditions. You can use these invariant quantities, these absolute amounts to measure anything. What, though, might serve as a criterion against which any changeful motion or its lack might be computed? Is there any single, changeless entity might act as benchmark, as a value whose 'gold standard' sets all others by its store?

Values are defined in terms of units. Play a game of cricket. Notice certain numbers - 2 (teams), 11 (players per team) and 6 (balls in each over or runs per no-bounce boundary) - interlock to represent and to constrain the game. They might be derived from observation but, while the game is geared around them, it is difficult to track the logic specifying these and other 'cricket-constant' values in the first place. In similar vein, could the abstract constants of nature also be, as independent absolutes, arbitrary? Or could pure, dimensionless numbers represent them? With, as befits rigid stability, invariant values? Are there ratios that represent an interlocking set of quantities on which the structure of the universe depends?

If you are after 'deep truth' buried in the cosmic maelstrom (whose outcomes revelations called equations with their constants can lay bare) then turn again to everything's internal symmetry. If you want the inside information then what logical consistency will let you calculate pure numbers and from these deduce the way the cosmos works? Is there some principle of integration, expressed with a ratio or set of them that like a pin allows a suite of them to 'open sesame'. You might, in the final distillation, even bring the combination down to one, one key that automatically unlocks the rest. Simplicity made singularity is at the heart of any *TOE*.

Is such an *essential constant* real? Can such whole abstraction make sense of material complexity? Are such absolutely independent numbers set, like sperm in universal egg, to trigger the phenomenal development? And are such numbers, like genes in meiosis, tumbled out of lottery; is nature whimsical to the extent that you can never work out only work through, by *induction* from experiment, to describe by numbers her behaviour and thereby her fundamental character? Or, as Planck, Eddington, Einstein and perhaps even Pythagoras believed, is the self-consistency of cosmos such that, once you dropped upon the key, you could *deduce* (*top-down*) the whys, wherefores and exactitudes of rules? After all, one might argue nature's the embodiment of a mathematical system; it is mind made (materially) real. Thus, using only the invariant bar-codes or key combination-set of nature's rules, might an initiate into the hidden code of matter's mysteries reveal just how a stable universe is sparked; a feat of mind *par excellence*! Either *total constants* were God's choice of definition or He had no choice. It's certain that cathedral architects employed the ratios that they believed He chose in drawing up the cosmic plan. Whichever way, such 'chosen' numbers would intrinsically define proportions and determine the construction of our cosmos and, as plan is explicitly revealed, cosmology. Of course, a number is simply an abstract, a cipher or trace that symbolises substance. It indicates. It represents but cannot be the nature or the actual substance of a thing. What, therefore, might pure numbers represent? Harold Aspden, an old-school Cambridge man, suggested an exalted trinity of such encryptions as might, unlocked, reveal the secret, sacred garden. These are the electron/ proton mass ratio (~ 1836); a fine structure constant (~ 137) that could reveal the factor governing the phenomena of quantum physics i.e. matter at the sub-atomic level; and, thirdly, a minute quantity (the gravitational-structure constant at $\sim 6 \times 10^{-39}$) relating the gravitational constant (G) to the square of proton mass. For good measure add Planck's constants. Then try, for definite indefinity and spinning circles, π (pronounced 'pie'); or, for spirals and for definitely beautiful proportions, φ (pronounced 'fie': see also Chapters 17 and 23), the 'golden mean'. These, with a cosmological constant we'll check later, are strange figures. What exactly do they mean? Do they really determine star size, atomic structure or what strength bonds making molecules obtain? Have they power so real that mathematics in the cosmos lets us live?

As well as 'magic numbers' could one also say that space and time remain invariant? We saw (Chapter 7) that Newton chose time and an intangible, fixed frame of reference within which absolute rest was, in the latency of 'smooth', continuous space, possible while all else varied and was relative. Was this ether 'fluid' or, *selon* de Duillier, 'corpuscular'? Next came the Michelson/ Morley experiment to try and verify the presence of such intangible cosmic continuity. This, followed in 1903 by Trouton and Noble's test using a capacitor, seemed to remove the possibility of a 'continuous spatial substance'. Yet neither they nor, hot on their heels, Einstein's relativity accepted or denied the ether's possible existence. His theory simply contrived, by geometric means of space-time that can curve, to cancel out such 'insubstantial substance' - thus rendering it irrelevant. Was he wrong to ignore its fixed frame of reference and thus divert the track of twentieth century physics? Because such a medium allows absolute time, space and rest to exist; and, as Maxwell thought, might act as a universal transmitter of vibration. To ignore these possibilities Einstein had to conjure up another absolute against which everything was relative - the nonsensically irrelative speed of light in a vacuum (c). If you and another car travel at 60 mph. in the same direction its speed appears to be zero but if you approach each other head-on it is, effectively, 120 mph; similarly a bullet has different speeds for observers travelling at different speeds and directions relative to its source. Yet Einstein, having counter-intuitively assumed that the speed of purely energetic light in a vacuum is absolute, bundled space and time into a variable 'space-time' bracket. With no fixed frame his geometry of things is ever relative, ever changeful, ever less than absolute. So then, is light or ether 'absolute'? Analysis by Aspden, Bearden and others indicates that ether may in fact be the virtual basis of material creation.

Do you remember (Chapter 2: Sliding Up and Down a Slope) that Natural Dialectic's Theory of Relativity is equally a Theory of Fixed Framework, Absolution or Invariance? In this case a Metaphysical Absolute (Essence of Main Dialectic) is set against the relativity of existence (in Branched or Secondary Dialectic). Where Newton assumed the physical, scientific absolutes of space and time against which to set all changes did not Einstein choose motion absolute? In which sense his Theory of Relativity, with its set speed of light in void, might at least as well be called a Theory of Invariance.

Whether or not Einstein's assumptions of invariance were actually correct physics 'clicked' from aspects of energy (time and space) to energy itself as its basic infrastructure - from a 'powerless' to a 'dynamic' denomination of its ultimate fixed point. What price, however, absolution, if even its pole star, the speed of light, were variable? What if, as Natural Dialectic proposes, there exists a third (*sat*) potential beyond the (*raj*) energetic or (*tam*) vacant aspects of an absolute? Is there a fulcrum whose balance might equate the two main forms of physics? Is there a keystone that completes an arch between the two current pillars of quantum and classical understanding? Might it consist of absolution beyond physical relativities? And constitute a metaphysical origin of law in archetype but not in chance? The nature of both chance and abstract archetype will have to be discussed.

Emergence from the World's Nest

	tam/ raj	*Sat*
	relative	*Absolute*
	subsequent order	*Archetype*
	actualities	*Key Abstract*
	physical nest	*Metaphysical Egg*
	practice	*Principle*
	motion	*Balancing Factor*
	physical matters	*Top/ Transcendent Matter*
↓	*tam*	*raj* ↑
	negative	*positive*
	outer/ gross construction	*subtle/ inner cup*
	isolating	*communicative*
	anti-principle	*principle*
	gravitational principle	*levitatory principle*
	inertial principle	*stimulating principle*
	Higgs mass-making field (?)	*electromagnetic field*

brake power	*motive power*
drag/ downpull	*uplift*
mass-centred	*anti-massive*
bulk level	*quantum level*
'dark energies'	*'light energies'*
binders/ restraint	*releasers/ flow*
contraction	*expansion*
strong nuclear force	*light*
quark/ proton	*electron*
Newtonian gravity	*vacuum's levity?*

What does invariance produce? What is the order of its archetypal constancy? Clear distinction was noted between (*sat*) constancy that stabilises creation and the (*raj/ tam*) vectors of its process. The former includes 'real constants', dimensionless numbers that somehow reflect the basic rules of the game (e.g. '3' dimensions of space, '1' of time, π and probably ratios which reflect the way principles are integrated); and it includes the coordinated presence of a conscio-material gradient. The (*tam*) vector involves a hierarchical process of materialisation that drops from an Inward Centre outwards through a dependent series of increasingly involuntary, inflexible and differentiated 'phase-layers' of creation. It 'outworks' from potential to realisation; as gross solid from rarefied gas, so lower emerges from higher level. Since the physical layer is 'outermost' it therefore derives from a higher, internal cause. It is an effect regulated by prior principles that are also called (as in any process of structural development) seed-forms, blueprints, plans or archetypes. Simply remember that, in a hierarchical or nested structure, one phase links either continuously (as with spectrum) or discontinuously (as with ice, water and steam) with the next.

Primary cause gives way to ultimate effect. A central, inmost, immaterial egg creates a nested and exteriorising hierarchy as it falls through subroutines of mind to matter. If the range of cosmos drops 'vertically' from top Consciousness to a base, inverse, non-conscious or massive condition, then its projection will depend, primarily, on an act of will. Does an '*objet d'art*' not reflect its artist? This projection will automatically incorporate the qualities, principles or inherent self-consistency of its projector. In other words the ultimate source of physical, chemical or biological law will *not* reside in the physical appearance and behaviours of the objects these studies interrogate. Nor, in the case of biology, will it even reside in a customised carrier of teleological code: a genome no more originates its organism than code is the creator of computers. At least the latter set-up needs a mind.

The (*raj*) vector complements its (*tam*). In various degrees it dematerialises, dissolves or returns towards Original State - in dialectical terms one of perfectly pre-active (*Sat*) Poise or Potential. In this case where might the character of this All-Powerful Principle repose? In dialectical terms its descriptive stack is on the (*Sat*) right-hand column of Primary Dialectic. Such poise or latency, which applies to both mind and matter, seeds any consequent action. It thus constitutes the source of stable governance and sums, in effect, to the features of a General Conservation Law. The dialectically logical source of physical law has been identified as memory, program or archetype fixed in universal mind. *What is a nest built round? What, at its centre, is an egg if not the incarnation of potential and first principles?* Check *fig. 8.2*. We know our material nest as a galactic universe with, especially, one star's solar system and its planet, earth. But earth's not central; at the centre of this cosmic nest find archetype, a 'universal egg', 'transcendent matter', 'fifth dimension' - call it what you will. Such nothingness, such ideal permanence apparent omnipresence is the source of order for the stage below. As a thought is handed down to muscle for its execution in the world, so expression follows concept; physic follows an originally psychological projection of creation. This projection's timeless; archetype, the base of natural law, is fixed and physically invariant. Of course, we'll probe the nature of such archetypal templates in the course of working Natural Dialectic's logic out. Suffice it here to note they rest at base of mind.

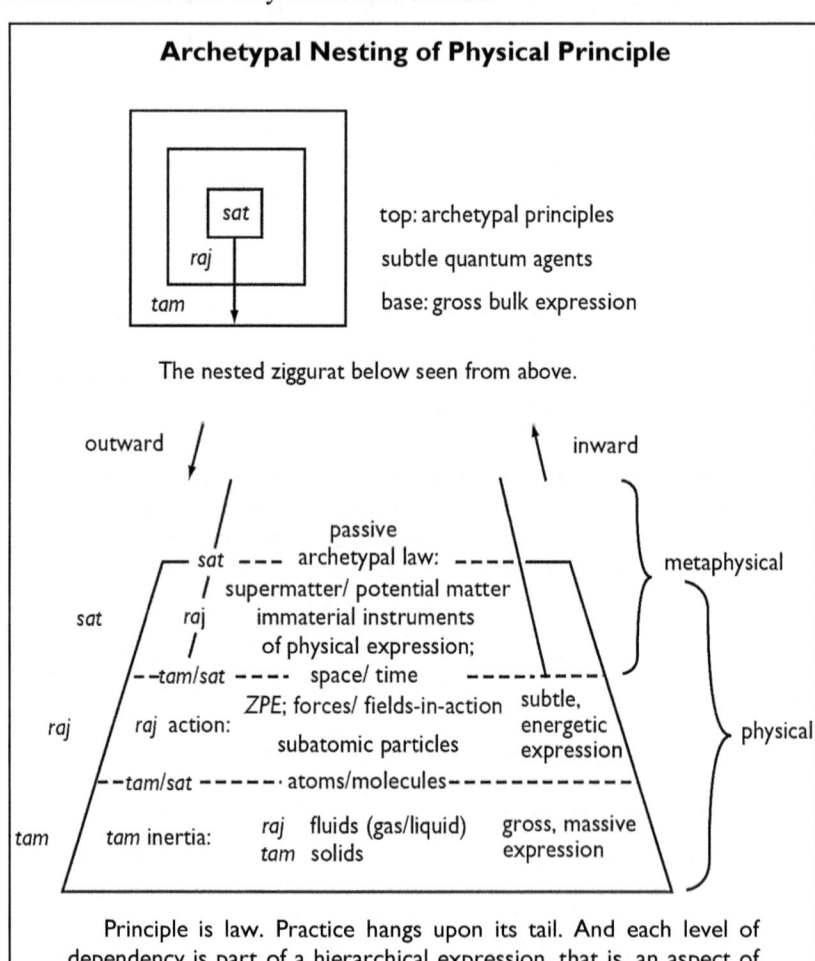

Principle is law. Practice hangs upon its tail. And each level of dependency is part of a hierarchical expression, that is, an aspect of informed energy. In a rainbow orange merges into red and so on. Seen as a spectrum the (*tam*) lowest level of one (*sat/tam*) division of the conscio-material gradient merges into the (*sat*) highest of the one below. This diagram illustrates the dependency of the lowest, non-conscious tier of Mount Universe, the plane of our physical universe. By its lowest, gross level 'wireless waveband' has 'precipitated' into 'wired' (or chemically-bonded, molecular) states of gas, liquid and solid with their points of discontinuity called phase transitions.

fig. 8.2 (see also any of figs. 0.5, 1.3, 2.5, 2.11, 4.1, 5.2, 7.2 or 7.4; also 22.3)

Is 'transcendent' or 'top' matter an abstraction, a vacuous imagination? If not, what are its simple, primary expressions? The Dialectic suggests duality - a duality composed of abstract and actual. Dual abstraction is composed of space and time. These act as a frame within which a third component, actual energy, can operate. Of this trinity energy itself is polar; its polarity is expressed, fundamentally, according to (*raj*) levitatory and (*tam*) gravitational vectors. These are the *sine qua non*. They constitute apparent absolutes, the seed from which differentiation devolves, develops or 'emerges'. The polar structure of creation is informed from prior (*sat*) archetype; and its antagonistic (*raj/ tam*) principles comprise (*raj*) causal, leading action called levity and, 'afterwards', (*tam*) effective closure called gravity. This means that, although the pair of principles co-exist, levity is hierarchically prior to gravity. Top (*sat*) supplies the seed of information, median (*raj*) activates or develops and base (*tam*) finishes. Internal energy creates external shape; from particle to atom through to bulk solidity subtle cause gives rise to gross effect. You might therefore expect a highly energetic start in which 'levitatory' forces predominated. And after the start you might predict the rise to power of a (*tam*) inertial principle, a 'running-to-seed', a diffusive principle called 'entropy' that drops to settlement, exhaustion, closure.

In the case of three-tiered Mount Universe the end product, physical creation, is non-conscious, mass-centred and, as such, represents a diametrical inversion, an asymmetrical reflection of the (*Sat*) Essential Character. In other words, Central Character is surrounded by subsequent layers of creation - metaphysical followed by physical - stemming 'organically from its seed'. Each layer 'nests' its inward source until you reach the outer layer of the cosmic onion. The onion's skin, its outer sheath, is a material circumference. It is the end, the limit of existence. You can glimpse it in the apparently limitless extent of stars in a clear night sky - but the boundary is equally your body or the solid stones beneath your feet. Both scientist and mystic want to know how you can peel a cosmic onion clean away.

Are not time and space the shadows of existence, ghosts of eternity, foils that point up being and becoming? If you subtracted everything would they persist? Newton thought so; they are independent entities. Or are these apparent absolutes locked up with locality, the 'here and now'? Prior to motion was there space-time? Was there space and time or not? Do they depend upon each change or *vice versa*? Do they dilate, contract or otherwise distort according to the speed of something's transformation, mass or motion? Einstein thought they did; space-time is a dependency. Others even guess they've 'crystallized' or 'polarized' apart; at big-bang energies they melt into a single lack of any definition - with neither time nor space nor any boundary condition.

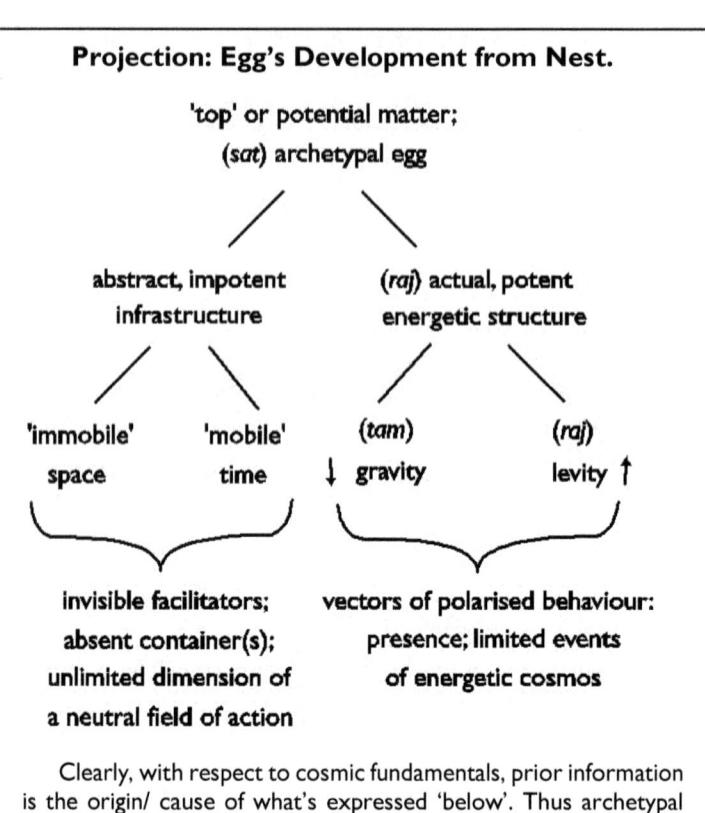

fig. 8.3

Let's rephrase. When in fact did nothing change or time divert from 'now'? Are not this couple, time and space, a variant unity - Einstein's 'space-time'? Or do they constitute an immaterial and seemingly immortal pair of twins that are, except a third abstraction, law, all that's absolutely changeless? Are they Newtonian zeroes, absent presences, the measure of all motion? Or, rather than abstract criteria, just part of all the relativity? Aren't they an easy, empty and yet unrelenting, all-inclusive couple that can never live apart; a couple that, when either absolutely stressed or wholly slack, disappear as one into a nothingness? Aren't they powerless and yet omnipresent and omnipotent - majesties that outflank all the competition and thus dominate the universe? 'Look on these invisibilities, ye mighty, and despair'.

Timeless, everything would freeze; spaceless it couldn't even be. Are their mute actualities, 'here' and 'now', simply functions of commotion, simply facilitators of the changes that we call, collectively, existence? Are they hangers-on or independencies? Aren't they the basic donors, blanks, extents or sheets on which the book of things is written? Both would involve, therefore, the latency of immaterial field. They should include the undisturbed and silent possibilities that 'form' the archetypal basis of a quantum world, a study whose mechanics delve the roots of physical existence. They are the fields for every other kind of field, the frame in which the actualiser, power, plays. Before, during or after any event - from the smallest to the universal - do space or time chance to create a speck of dust, a dime in all the wealth of things? Their all-powerful impotence is set against the power and glory of the source of every possibility, the currency of physics, different faces struck upon the universal wealth called energy. What is, stripped of its masks, pure energy?

Within the abstract space-time frame all actuality is energy's. Is energy eternal? If not whence does it arise? What guides its appearances? Scientific 'string' theory (Chapter 9) construes what's visible as shadows of the tensions stretched through 3-d space and wrapped, compacted into many more dimensions. What does this mean? The Dialectic sees unconscious (sub-conscious and non-conscious) operations as fixed patterns in a mind. It sees matter as a rigid, reflex reiteration, that is, crystallisation of such patterns. And principles composed of these unchanging patterns - natural law - are expressed by definite agencies (forces, particles and atoms). Such agencies are preordained; they are shadows of their archetype.

What, this book asks, does that mean? How does a dualistic energy component (physical *prakriti*) work? With cosmic fundamentals now in play it is suggested that, besides an overarching (*sat*) equilibration, the seed of (*raj*) levity shows in general form as cosmic stimulus and in its 'highest' specific apparition as pure, kinetic energy. We call this mass-less, formless, radiant communication light. The (*tam*) seed of gravity, conversely, shows as inertial, contractive energy. It localises, materialises, lends mass, isolates and differentiates. In large-scale form it shows as a mass-centred force effective through the whole extent of space. This general influence that everything exerts is specified conventionally as gravity. Natural Dialectic just extends its gravitational vector to include the character of short-scale nuclear forces. These particularise, squeeze particles like pips from space and aggregate bulk objects to a mass the large-scale force can really grasp.

Thus at this point two points raised by the stack above are noted. *Firstly*, both Higgs field (if it exists) and gravity are ubiquitous forces, the former primary and undetectable, the latter secondary (arising from the creation of mass, strong nuclear force and electric charge) and detectable. In terms of Natural Dialectic gravity is omnipotent in the sense of acting on mass everywhere in space but is only a only the fraction of all the (*tam*) gravitational forces. *Secondly*, the logic of Natural Dialectic requires a large-scale levitatory force. The pulling power of gravity is, potentially, omnipresent and omnipotent but is there cosmic counteraction to collapse? If (*tam*) gravity is the 'last' power of nature could inverse (*raj*) levity be its 'first'? What is the aspect of this pushy power that, unconstrained by mass, might constitute the start of things? How does gravity's antithesis, large-scale anti-gravity or anti-massive tensing power, operate? The possible nature of 'vacuum potential' is discussed over the following chapters.

Prior to physical expression of a universe there only existed, in dialectical terms, the mind-world's potential for it - archetype. Before the triplex base of space, time and polar energy from which physical appearance springs and swings there existed only metaphysical possibility - a latency that, like gravity without mass, awaited its own activation. As such these three are physical absolutes.

Just as a multitude of particular forms precipitates from a formless gas so this trinity incorporate, between them, the potential and natural principles from which all things are, down the scale, gradually crystallised. Apparently boundless space-time is the frame within which fundamental polarity is expressed, where energy is informed according to prior dialectical rules. Two-in-one. Push-and-pull. Antagonistic yet complementary forces work together. Just as space and time provide the milieu for action so the action is itself divided. Polar energy comprises levity and gravity. Expansion/ contraction, activation/ resistance, stimulus and constrained effect are integral to the formation of stacks of Secondary Dialectical polarity. Perhaps the subtlest expression of such paradox involves an excitation, a field combination of electrical motion and constraining magnetic coherence - electromagnetic radiation that is light. Such initial, logical integrity leads directly to the conclusion that, as child to umbilical cord, action is locked to law. As life depends on a solar cord, a ray of light, so the whole outstanding business of phenomena (including you) depends on prior, that is metaphysical illumination - information. **Before cosmogony, before the starry universe began all was not nothing. There was archetypal information so the cosmic bird developed and can fly.**

Metaphysical Evasion

Hasn't chance a chance? Chance is unattached to reason. It is unhinged from logic. Although a wildcard it is, as an essentially irrational trump of scientism, a metaphysical evasion we now have to deal with. Evasion of the whole truth is, in Dialectic terms, not *by* but *of* metaphysic. Metaphysic is avoided by appeal to happenstance; in this view even cosmos with its laws occurred by chance.

Bottom-up, evasion of the whole truth is *by* metaphysic with appeal to God or gods-of-gaps to cover cracks in knowledge. Yet popular science, although empirically aware of natural law, has not picked up or has philosophically ignored either the clear, vectored form of its order or a conscio-material thrust that is trans-dimensional - one that operates in both metaphysical and physical domains. Indeed, its viscerally un-metaphysical premise condemns such a perspective or, as it may be, actuality. Can we say that, on losing energy, a solid emerges from a gas or, on its gain, a gas evolves from a solid? We can. We note, moreover, that characteristic forces, particles and behaviours may 'settle down' as the cosmos 'chills out'. Could we say that these objects, textures and their accompanying rules of engagement simply 'precipitated out' of infinite nothing? We can, it is claimed, believe that nature sprang from nothing. Void excludes information; there was no design. There is, it is claimed, no metaphysical source of hierarchy, simply a physical context based on the key principles of 'happen-so' and invariance (in the shape of law). Chance tangled up with none. What, however, is the source of the latter, the unintentional but non-accidental part? Whence providence to guide coincidence? The answer given is that 'the necessity of providential law' was not providential. It either evolved or abruptly jumped from absolutely nothing by a valueless and uncaused chance! The rules are, therefore, simply what they chance to be.

Once created the cosmos is closed to chance as far as the operation of law is involved: nor is there evidence that the laws of nature with their correlated constants are evolving. How, indeed, could the laws of cause and effect, cosmic fundamentals or the conservation laws change? What might constitute an evolved betterment? On the other hand, could they mutate for worse? In fact, such regulation is, *top-down*, not evolved but *devolved*. Rule of chance is not a rule so everybody takes the rules informing physical behaviour 'for granted'. They are 'data' - 'given things'. Yet you have never caught or plucked a law from air. Rules themselves are metaphysical unless, in the case of nature, they aren't rules at all but simply mathematical descriptions of predictable behaviours. They aren't, therefore, due to charity of chance; their donor's not intelligence; they are just how they are.

Top-down, evasion is *of* metaphysic by materialism's rigid physic - only physic, never metaphysic. *However, in the way that a house is devolved from the architect's plan, so the patterns of non-conscious energy operate within the framework of a program that, because 'information precedes energy', preceded physical creation. Informative potential precedes expression. In this view the material world was present, in principle and in potential, before it was ever actually expressed.* As grade and era (*fig.* 11.2) are the inextricably linked expression of a layered universe, so un-manifest but inferred principle is the potential not only forever behind but also ordering the expression of phenomena.

The real construction of Cosmic Drama is seen in terms of motion along a 'vertical', hierarchical conscio-material gradient. As such a line descends from conscious to non-conscious, voluntary to non-voluntary condition it shows a gradual inversion of polarity. A twist, you might say, in its tale or its tail (Chapter 24). The predominant characteristics of the lowest level inversely mirror those of its highest. For example, non-massive consciousness is central at the top; at the bottom non-conscious mass (for example in the atomic nucleus) has replaced its centrality. In the same way highly differentiated, particular bulk objects have replaced (or emerged from) unified, internal principle. Simplicity lets fall complexity. Objects drop into (*tam*) fixity from which they are only 'lifted' by a local concentration, called an input, of (*raj*) energy; the same applies psychologically, where reflex instinct injects relief from physical pressure or, in the case of humans, a 'hot' focus of attention melts obstruction, a local concentration of hard thinking solves a knotty problem. In what does complete freedom from the existential problem, various obstructions and constraints, consist? Is total liberation only found in Essence?

Put this another way - it's all in order down the line. Active information's neither energy nor matter; metaphysic is an attribute of purpose. Passive information (e.g. stored codes or purposely shaped objects) may be physical or psychological in form; but, whether memory or object, it is made by active information and does not create itself. Systems are designed or, if you like, evolved by mind; mind is an order-making, negentropic faculty. An abrupt appearance of invariant behaviours and their description as the laws of physics and chemistry is one thing - though more and more it doesn't look like chance. It is, however, generally agreed that prototyping of a body biological is, as Chapters

20 and 21 clearly show, quite another. Even single cells are, as we'll show, much too complex to have aggregated, like inanimate formations, due to chance and the determination of some natural law. All actors in life's drama - some of them multi-billion-celled developments - are organisms replete with purposive, specific information. They are information incarnate. **Yet there exists no Law of Non-conscious Innovation and Integration, no Principle of Material Evolution.** Would you expect oxymoronic contradiction of this magnitude? Could you ever think there was? Laws describe (but do not cause) predictable connections in between the past and present. They don't regulate irregularity but post you, from a definite cause, determination of effect; they indicate a probability so probable that you can mathematically deduce the course of its event; and they reflect the reflex modes and mindless motions of an uninventive universe.

Don't let a sound-bite lead astray. 'Life', wrote Werner Heisenberg, 'is order in a sea of chaos'. Physic's 'sea of chaos' swells and sways by non-chaotic 'natural law' but bio-logic's forms are not just orderly in this wild way. They are also comply with quite another and entirely higher kind of order - specified in their complexity by code. **Wind, rain and natural forces *never* codify. There is no physical but only psychological Law of Innovation**. Creativity and innovation generate the unexpected and improbable. Not just improbable complexity but, as a language making sense, specifically functional complexity. Intentional invention. Necessity (if that's what you call natural law) is not the mother of invention. Mind is. Imagination is not rigid, purpose innovates but these factors are as immaterial as the information that they generate. Were you calling for material inventiveness, a source of information that's not psychological, a self-contradictory law of teleology? How, in particular, could brews of chemicals, chance, time or quantities of atoms ever combobulate arrangements of highly manufactured, coded, cooperative complexity? Information alone is the potential for purposive, anticipatory parts. Yet a Darwinian context lacks, in naturalistic principle, any active source of information. Its created specimens are just exposed to the destructive influence of entropy. Natural selection never innovates, it kills; nor do mutations though a very few in certain contexts temporarily facilitate survival. They degrade, mostly wreck what works but might (though it's not been observed, make good by re-mutating back to health what an original mutation led away. *The PCM is, Chapters 19 to 25 will show, a confusion generating great misinformation.* This misinformation is not one of fact but stems from the perspective whence interpretation of those facts occurs. If you seek the origin of bio-logic's coded, systematic information do not look to physic. Be informed that metaphysic gives a very satisfactory explanation too.

The issue's pivotal. The fact-less notion of a 'Principle of Naturalistic Innovation' or of 'Specified and yet Non-conscious Integration' - pervades contemporary 'scientific' thought. It pervades because prior philosophical decision has been taken, often by consensus, to deny intelligence is anywhere but brains; and, although its nature is materially obscure, to ban the thought of consciousness as metaphysical and thus distinctive from non-conscious physicality. <u>This is as metaphysically evasive as you get.</u> Such unquestioned and unquestioning evasion permeates the study of biology worldwide. Its mind-set leads to lectures and to libraries of books and journals whose interpretations of the facts miss out code's crucial metaphysic - active information. Active information, also known as intelligence, is not the *zeitgeist* of today.

No doubt biological bodies (Vol. 1 *passim*) are purposively complex to the point of invoking a Biological Principle of Innovation and Integration as applied to purposely composed complexity - some principle for the natural, unintentional creation of intention and its agent, information. Even Shannon-information (Chapter 5: Information, Messages, Arrangements) gains by unpredictable improbability. Repetition is as valued as redundancy. In this respect the laws of nature, called by some 'necessity', yield a 'self-organising' type of order but did not inform themselves. And certainly, without intent, they cannot organise the innovation specified complexity involves. Their order is non-purposive, the opposite of teleological. So don't expect that mindless *physicalia* by any means will codify, invent or actively inform. The specific order in computers, books, machines and life can only drop from active information - the domain of mind.

In other words there is, unfortunately, neither law nor process nor expected sequence of events by which matter can create information. In fact, the design and construction of purposive schemes and machines is the product of informative negentropy. While such negentropy is the basis of learning, intelligence, creativity and all man's major, voluntary expressions of advance, it is immaterial. Metaphysical. It belongs exclusively to the province of mind.

Involuntary matter lacks informative negentropy completely. Even AI's 'learning robots' *copy* life; they are dynamic pictures, the inanimate product of immense and prior informative negentropy supplied by their creators. Do you believe that their creators came about by chance? Robotic is a passive sort of information while design is voluntary in process - learning with a grasp of immaterial principle, options from a choice of possibilities and thus deliberate invention. The development of an idea necessitates informative negentropy. Biological evolution is, however, mooted as an exclusively physical process. And all physical factors (except, it seems, protons, electrons and the four forces) are subject to the energetic entropy of decay. All bodies, from atoms through to biological systems, suffer destruction. Such blind wear and tear, the opposite of innovation, applies to all objects whether they are artificial or natural. There you have it. *Informative negentropy is needed; only energetic entropy is available.* *Thus evolution theory, with its basic metaphysical evasion, fails to match reality.* **Neo-Darwinism suffers, throughout the entire body of its scholarship (and there are libraries of eulogy), a primary philosophical error of category**. It alleges the faculties of mind in matter; denying information its metaphysical reality, it imposes the character of design upon physical oblivion. And then, realising the obviously counter-intuitive nature of this root nonsense, it proceeds to invent multiple hypotheses to convince everyone that what is false is real. This is a restless intellectual place that there's no need to be.

Materialism's central prop is evolution. Thus metaphysic's most succinct evasion is retailed by those biologists who call design 'design'; and thus, through Darwinian smoke and mirrors, information's penny never drops. Nor must it. If it drops both theory and philosophy go pop. Such metaphysical evasion is fundamental to thrice fifty years of academic thought. Its deep 'infection' badly needs a purgative. Whence, therefore, can systematic information be obtained? *Top-down*, the aforementioned Biological Principle is assigned to archetypal memory, the storage of a complex program lodged in universal mind. Working backwards from complexity to the relative simplicity of principals such as particles, atoms and the chemistry of stars, you could argue that if archetypes exist to specify distinct but complex biological arrangements, why not distinct but simple ones such as compose the automated cosmic stage as well? Systems inanimate would also be, as this and following chapters indicate, an expression of such prior information.

If this idea is philosophically barred then we are thrown from reason back to unreason in the form of chance. Metaphysical evasion will have reduced us to a Theory of No Intelligence, a Theory of Unreason and, as a subset, a Theory of Biological Evolution. **The fact is that purely material processes (even the 'awesome duality' of chance mutation and natural selection) are precluded as sources of information**. A mutation may involve unintentional complexity but no more meaningfully, symbolically codes than earthquake, wind or fire. Can salty water conjure up a cell? You can lapse into animistic language or employ terms charged with teleology - like 'genetic program', 'blueprint' or 'innovation by mutation'; but you can do so only as long as they are sanitised by the habitual extraction of their real implication - purpose; that is, you can do so only if you prooflessly imagine that the legal constraints on chance, called necessities, are in themselves sufficient to 'design apparently teleological systems'. What a materialist *really* wants to believe, in spite of contrary fact, is that chance can, while subject to constraint, create information; and that this is how information and 'meaning' are indeed naturally created. **In reality, let's be crystal clear, every time that animistic language is used it implies, unequivocally, mind. Purpose in front. Intelligence behind.**

Chance is impersonal but ignorance of reason is personified. If you can't identify a cause you might call such an instance 'chance'; if you can't see the reason why your 'rules' are broken you might call the breakage 'happenstance'. Yet is your nomination, chance, the cause of anything? Does its blundering lack of information really equal cause that's capable, from the non-existent, non-essential starting-point of pure nothing, of drafting all the 'dicey' rules? Can it, as some physicists aver, construct a universe? Could it, as some biologists are sure, evolve a 'life' (if life's no more than its component chemicals)? Or does a study of invariance show that rational order is a better bet? In which case Chance were a fine thing, a seductive charm that weaves you, out of ignorance and doubt, round her abstract little finger. Chance is unpredictable and chronic unpredictability is a symptom of insanity - so why cradle the delusion that, when her heart is chaos, she could create your cosmos? Why thank her lucky stars or empower her, like an empress veiled in mists of time, with the magical capacity to make up anything? Why, when it comes to origins, is science so obsessed by this seductive and yet uncreative maiden? Why so madly fascinated by a mindless and irrational womb of worlds? Out damned alternative, down with the laws of Lord Deliberate and up with material uncertainty - except materialism's own self-certainty! But is it logical to lean too heavily upon the whimsical; is it wise to swing close up or even philosophically espouse Caprice - whose other name is Lady Luck?

Getting to Grips with Lady Luck

What's her character? How slippery? Is chance dumb, blind or a blur, a trick of incoherent mind that can't quite grasp a real reason? Is she simply a cosmetic answer, nothing more than simple make-up, an illusion whose substance vanishes the more you peer? In any case she is central to materialism and peripheral to its holistic opposite. To the former she's a powerful force for transformation, a creative 'anti-deity'; you might even crown her Queen of Serendipity. However, to holistic Dialectic she's a phantom flitting only in the unsure mind of an observer so that, since she could not puff a powder let alone create a drop of tear, you would make obeisance to an impotent, strange and imaginary idol. Either way the blur needs disappear. Although she's not an easy ghost to lay the Lady must be caught and, if necessary, pinned in place. Slide her underneath a microscope. Her shape must be revealed; the flirtatious nature of her ambiguity must be resolved. It might help tease the problem open if we split it into mind and matter.

Let's draw a psychological stack:

	tam/ raj	*Sat*
	existence	*Essence*
	appearance	*Reality*
	logical outworking	*Potential Mind*
	phenomena	*Latency*
	variations-on-theme(s)	*Active Archetype*
	orderly practice	*First Principle*
	logical outworking	*Preordination*
	variation/ uncertainty	*Invariance/ Certainty*
	peripheral effects	*Central/ Inmost Cause*
↓	*tam*	*raj* ↑
	passive/ non-conscious element	*active/ conscious element*
	universal matter/ material cosmos	*universal mind/ immaterial cosmos*
	informed	*informant*
	body/ player	*author/ rule-giver*
	regulated	*regulator*
	reflex outcomes	*choice of options/ decision*

Active Archetype, at mind's source, is identified as *Logos*. From this First Line the program is unfurled. Behaviours are, down the cosmic hierarchy, preordained. Thus call them ruled; and, since behaviours follow patterns, call these patterns rules. In this view cosmos is the son of Certainty; yet, while rules predetermine games, at the same time contexts on the ground allow for endless possibilities; of all this range of possibilities certain ones are, in a moving present, consequently realised. Mind and matter are the types of existential ground; they are connected by, at base of mind, unconscious attractors, pattern-fields or passive archetypes. Now let's draw stack physical:

tam/ raj	*sat*
classical/ quantum expressions	*potential matter*
matter/ energy	*material first cause*
ordinate possibilities	*passive archetype*
phenomena	*latent fields*
game-by-rules	*physical determinant*
↓ *tam*	*raj* ↑
gross	*subtle*
inertial	*energetic*
classical overlay	*quantum basis*
particle/ fixed location	*wave/ generality*
discrete object	*field interaction*
fixed response	*'statistical' response*
local certainty/ actual event	*general possibilities*

What did you expect? You know the rules but things don't always work to plan; but there's another aspect that we have to tease apart. Uncertainty (of mind) and luck (coincidence), although alike, are not the same; let's involve another mental stack:.

↓ *degrees of ignorance*	*knowledge* ↑
incomprehension	*comprehension*
doubt/ anxiety	*confidence*
relative uncertainty	*certainty*
fixed/ instinctive response	*options open*
inflexibility of mind	*flexibility of mind*
automation	*creativity*

Thus, it seems, uncertainty's a mental construct; possibilities occur in mind. Then there are scientific objects and events; how far is material uncertainty identical to probability and, within the constraints of a given system, probability to chance? The world's a complex, layered ferment of activity; when, in the context of its maelstrom, something unexpected happens then we call it luck. Luck's probability occurs, you bet for sure, within some given system (say, a lottery, a horse-race or electron orbital) but, as far as matter is concerned, there's no such thing. It's all in your mind; you can't put your finger on it, can't predict it and so call the business chance. Uncertainty is due to unpredictability, immeasurability or ignorance of factors. Is matter at root indeterminate? Maybe it seems that way because a wave is spread in space and time and quantum physics bases its mathematical descriptions on wave-like analogies. Thus, in quantum waves, locations in both time and space seem to dissolve at root into uncertainty - at least for observers like ourselves. Is physical uncertainty therefore the case? Such 'blur' doesn't mean, however, that beyond space-time dimensions archetypal fields aren't omnipresent. If so underlying certainty prevails.

I don't know. Caprice is difficult to fix because by definition she is unpredictable. Chaotic. Chaos isn't order; you can't abbreviate what's random into formulae and so she's more or less incalculable. Her lack of order dubs her Lady Luck. She lives recalcitrant; her rule is kicking over rules; hers is Sod's not God's Law. Sod's Law as applied to human plans states any possibility of system failure is going sometime to occur. Much of technology, in concept and in action, is a battle set to minimise the sooner or the later of Sod and his escort, Capricia, turning up. Chance is lapse from an expected course and, if you are not omniscient, predictable as unpredictability. We live variations on our daily themes and, occasionally, within a sea of trivial unpredictability a wave rears and affects us. Chance then challenges. It changes things in unexpected, sometimes major ways. Who, for example, can predict when he is going to die? To define luck is an uphill slog. Is this because incalculable behaviour, complex interactions and irregularity upset the reasonable bent of mind, whose nature is to actively inform, find patterns and to understand as far as possible the truth - for sure? It needs mind to spot a chance yet chance runs counter to it. Chance loosens up mind's certainties, punctures preconceptions and waylays its rationality. How can you assign a reason to a factor that, by definition, hasn't got one? How ascribe cause unto causelessness? How logically describe what cuts up logic or rationalise irrationality? *How could you ever test a theory (such as evolution) based on the disorderly principle of chance?* How recreate the randomness of its creation? With broad uncertainty. Best guess, alternative interpretation of the facts and choice of options are as far as explanation's remit runs. Luck is definitely a dickey business so be warned that, flirting with Capricia, you're up against the odds!

Matter doesn't have objectives but they cut the odds to shreds. And what is mind if not an information bundle whose desires *are* objectives? Mind is negentropic - an entity whose purpose is its schemes. Mind can do in moments what mindlessness would fail to in a trillion years. Thus mind is anti-random, anti-chance and, except for brief encounters in excited moments, doesn't care for careless chaos or its demon, Lady Luck.

	tam/ raj	*Sat*
	relativity	*Absolution*
	relative certainties	*Certainty*
	relative truth	*Truth*
	incomplete knowledge	*Knowledge*
	accident/ design	*No Chance*
↓	*tam*	*raj* ↑
	lower mind	*higher mind*
	from truth	*towards truth*
	ignorance	*awakening*
	mindlessness	*awareness*
	instinct	*reason*
	reflex control	*voluntary control*
	rigidified info.	*active info.*
	automatism	*conditioned free will*
	aimlessness	*purposes*
	doubt	*faith/ trust*
	forgetfulness/ info. loss	*learning/ info. gain*
	misunderstanding / error	*understanding/ precision*
	more chance	*less chance*
	'wind-blown'	*decisive*
	unintentional	*intentional*

complex interactions	expected interaction
unruly/ chance event	design/ specification/ rule-book
luck	deliberation

A stage-set has no intrinsic, mindful purpose of its own; and if this 'set' is a dynamic, interactive multiplex its game is just the same. It has no objective save its builder's plan, no reason but its player's sport. Any information, any conformity to purpose is imposed on it. Its material patterns are, like anything that's manufactured, 'passively informed' by both intentional (psychological) and unintentional (physical) constraints. Even an intentional arrangement accords with natural order (what architectural folly would, for example, ignore gravity?) as well as with the purpose of its design. Both ways, it accords with the preordination of a blueprint or a rule. Is the cosmic stage, you ask, the product of such rule; are its workings rubber-stamped by principle; is it a matter of original idea? If so, then its 'determinism' may be said to generate a 'Rational World-View', a Theory of Intelligence. As such chance would appear peripheral, a product of miscalculation or a limited embrace of fact.

But (yet again because it's crucial) was cosmos first or chaos? Did law or lawless 'law' preside? Which is the fundamental face of universe - Lady Luck or Lord Deliberate? Surely the laws of nature did not spring by accident? Surely, therefore, Lord Deliberate himself is not a child of Universal Mother Chance?

'Of course,' a chorus swells, 'we all agree that Luck alone can't make a mountain or a man.' Capricia can't bear cosmos by herself. She needs a Fixer who'll fix energy and thus supply the forms; Lady Luck needs discipline according to the iron rod of natural law. Thus Chance and Necessity (*aka* The Vice, The Fixer, The Rigidity of Natural Law) together create mice, men and anything that happens in their bounds. This Fixer sprang, you understand, by Chance from nothing at the start-point and so there's no necessity at all for Lord Deliberate to live - unless the former's fine-tuned body of The Law leapt from Design. In his case Necessity's Invariance would underpin a great conceptual scheme; principles of law would anchor nature's frame. Within this generality you'd plan and build, as engineers agree, specific mechanisms (say, robots or life-forms) to conform to your intentions *and*, of course, the frame's constraints.

Are You Certain?

So are you certain how the cosmos works? **From whose belly issued The Necessity of Law?** Do 'chance and law-by-chance' promote the show; or is cosmos orderly with law fixed by primordial design? **From whose loins leapt The Fixer? Does it fix for Lady or for Gentleman?**

A *bottom-up* perspective elevates a maelstrom made of accidents to gradual creator of the world. What, after all, is there but mindless matter, matter which perhaps 'expanded effortlessly' from 'absolutely nothing'? Nothing determined the order of a 'big bang'; nor did purpose fix the purposeless and reflex way that things behave. No purpose, only lawful and yet 'lucky' interactions do the work; no mind abroad, the cause is chance! Where chance is cause's absence! If the fundamental character of chance is an apparent lack of cause, then causelessness created every cause! And if these causes constitute the patterns of behaviour we name laws, then pristine lawlessness created every law. Although the cosmos is arranged according to these laws, it is also found (by so-called 'objective single-case chance') to be chaotic. This theory holds that things are intrinsically chaotic because 'cosmic evolution' or 'cosmic process' depends, from moment to moment, on chance events. As such the whole system with all its 'laws' can neither fix nor (had it mind) predict any subsequent state. If chance is inherent in the nature of things then the simple 'film' between existence and non-existence, inhabited by the quantum energy-fields and particles from which all complexities are built, is 'oiled' with drops of certain non-rigidity. The universe glides on a patina of fluidity - a 'flexibility' created from uncertainty; beneath it crackles probability, the noise, the fuzz that fogs sharp clarity. In a Theory of No Intelligence the slippery uncertainty of chance holds sway. In a rationalist's 'Irrational World-View' chaos governs but does not. Necessity (The Fixer) has to rule. Is it rugby? Is it cricket? Is the world a game of natural law, lawlessness or both in pseudo-anarchy? You've got your laws but can't predict specific state of play. Cosmos or chaos - is existence Lord Deliberate's or Lady Luck's terrain?

How might you draw order from anarchic atmosphere? Yet even, next to chaos, gaseous plasma has its rules of play. Physics, studying causes and effects, abhors a lack of cause. With its companion, chemistry, it likes predictable determinism and calls order 'law'. If cause is reason physics has its reasons. They are objective not subjective. After all, what will or purpose has a lifeless thing?

Look. Events occur by law and at the same time by coincidence. Look closer. Numberless co-incidents are happening all the time but who can separate each one and tease the plethora of bumps and grinds apart? Can you eliminate, by sharp analysis, apparent 'accident'? Could you, as the determinist Pierre Laplace avowed, predict each one if only you could know enough? Could you not calculate each detail of this world's working in advance - even, if you'd lived a billion years ago, have predicted your own brain and what you're thinking now? In other words, is unpredictability a function of just ignorance, an inability to cope with the complexity of simultaneous interactions? Laplace, at least, determined that complete determination rules.

Are particles of gas, for instance, really flying randomly or does each run a classic race? Each molecule obeys the laws of motion. You can account precisely for its flight paths. Slower souls, however, cannot grasp a billion molecules at once so they employ a systematic way to study chance, statistics. You can thereby grasp the general picture. You may calculate a cloud's behaviour according to statistical analysis but, while necessity is certain, what the method names as probability (or chance of happening) still lies within the laws of physics. It is an abuse of Lady Luck to think that, given all of time and space, she can grant any wish you make; and likewise an abuse to think that like a fairy she can violate the laws of physics and, mongering the miraculous, make impossibilities come true. Can you twist her to conceal your ignorance and bluff? No doubt that organic evolution theory thrives on such a twist.

Hold hard! Take an even closer look. 'Cause and effect' is common sense; its prejudice holds true throughout the daily, macroscopic world - the one that we inhabit. If so, this world is 'deterministic' and chance is an appearance due to ignorance concerning all the facts - including each one's history. Where there's cause there's reason's rationality. With insubstantial chance as *cause* your quantum world-view is, essentially, irrational; there's no clear cause so that some factor must be hidden or the model's incomplete. Who, after all, can track each interactive detail of phenomena? From unchanging principles there crystallise arrays of different practices. We accept, as nature's flexibility, actual variation on its general norms; everywhere we recognise specific, local products of its rigid laws and themes. Outcomes vary though the rules do not. What, however, if at microscopic, subatomic grade such prejudice were null and void? Doesn't indeterminism, found with radioactive emissions, Brownian motion or in quantum fluctuations, illustrate that cosmos really wobbles on a fundamental sort of unpredictability? It is believed and even held as proved that quantum waggling is replete with 'spontaneity'. No cause, no reason. Not superficial but root indeterminism. On a large scale, where events can be predicted and relied upon, there is security; but insecurity at subatomic scales from which large-scale security is paradoxically built. You're sure of that? You're certain of uncertainty? Why, yes, because this is exactly, at its heart, what the cause of philosophical materialism needs; and some folks, therefore, seize the happy opportunity to exclaim that, while Luck isn't orderly, Lord Deliberate's completely out of microscopic order! All that exists is based on probability! Quantum cosmology (research combining great and small in theory and hopefully in fact) might then rely upon a big and impregnating bang from nothing for no reason, for no cause and, therefore, meaninglessly; or, even better, might depend upon some causeless, extra-cosmic sac or immaterial wave of probability whence somehow issues all there is! Get that? *Non est deus. Quod erat demonstrandum* (*QED*).

If you say so. Werner Heisenberg's famous *HUP* (Uncertainty Principle) is based on the notion that only measurable quantities should be part of any theory while at the heart of atomic matter a measure of uncertainty in the measurement of variables (such as position and momentum) always and inexorably blurs deterministic lines. Of course, wave/ particle's a schizophrenic model not entirely true. But treat the quantum as wave and your position is a blur; freeze-frame the particle and obtain position but velocity's unclear. Measurement can't give you both at once and so uncertainty occurs. Precise imprecision rules; you can't nail nature down. Randomness seems wobbling at the root of things nor, at quantum level, will you get a final resolution - only 'amplitudes of probability' for all chains of events. Neils Bohr, within a collection of ideas dubbed the 'Copenhagen interpretation', echoed this view. A physicist after physical - which is not necessarily total - truth only considers material quantities; but for Bohr the quantum mechanical description of an atomic state is irreparably undefined, having only the *potential* for values with various probabilities: science would, in this sense, never offer better than such incompleteness - truth involving the *likelihood* that interaction between quantum elements (such as photons, electrons or atoms) will produce a certain kind of outcome. What, however, is uncertain - nature or the measured observation of it?

In practice, quantum theory had shifted the focus of physical enquiry from object alone to its relationship with the instrument of its detection (by which it is disturbed) and the type of experiment by which it is measured. The *HUP* becomes a so-called observer effect (whereby your act of observation changes what you are observing). Mind, as register, becomes a function of equation. It also implies that no meaningful property can be deduced independent of an instrument designed to reveal it. 'What we observe is not nature itself but nature exposed to our method of questioning'. We try to explain everything in classical terms of, say, particle or radiation but this may not be sufficient. Might this conceptual, metaphysical aspect of nature's substrate mean that it's all in the mind - not yours of course but universal mind? From Chapter 6 (Information's Infrastructure, Code) you may remember Heisenberg himself suggested, cryptically, that matter's smallest units were not physical objects in the ordinary sense of the word but forms, structures or - in Plato's sense - ideas, which could be unambiguously spoken of only in the language of mathematics.

Some theorists did veer down a solipsistic track reminiscent of the Bishop Berkeley - 'it's all in your mind' and 'our human minds create reality'. No elementary particle, they said, becomes a phenomenon until its presence is registered by an observer. At this instant it collapses from a probabilistic blur into the sharp line of one actuality or another. An observer, far from being isolated, initiates the decisive process of 'actualisation'. Observer's mind and instrument have contrived to create actuality from what is, otherwise, a universal 'virtuality'. Each one of us would thus work miracles incessantly!

So we create the world? This sounds cuckoo not least in a lab - where (with mind *objectively detached* and thoroughly divorced from the equation) experimental purpose is to simplify the world by isolating cause and then control its impact, measure the effects and balance them against hypothesis. This is the unworldly, artificial state of play. For a physicist the exercise of mind should play no measurable part - except in observation, measurement, conclusion and the rest. You may tend to lose precise repeatability when many influences simultaneously interact (as almost always happens in a compound and connected-up experience of the world) but, worse, with mind you lose detachment irretrievably. This is because the incidence of mind affects, by how it probes, the outcome of experiment; and because metaphysical mind can, in an instant, choose to make very low-probability events (such as an arranged meeting with intended consequences) happen easily to order whereas mindless physic never could. In other words, mind introduces teleology which reflex matter hasn't got.

For Heisenberg the substance of modern theory, its uncertainty, was down to its shortcoming - an inability to completely describe our underlying quantum world. Although a logical positivist in the sense of rejecting metaphysic he was, paradoxically, a religious man and seems to have had sympathy with Platonic form, a central order of things or universal mind-behind. For example, when asked by Wolfgang Pauli if he believed in a personal God he replied, "Can you…reach the central order of things…whose existence seems beyond doubt….as directly as you can reach the soul of another human being? I am using the term 'soul' deliberately…If you put the question like that, the answer is yes". Nevertheless the apparent abandonment of the principle of causality and, thereby, causal determinism by the *HUP* - leaving chance to tweak the roots of nature - has been characterised as 'a struggle for the soul of science'.

Philosophical debate over the 'acausal' implications of quantum theory has been protracted and intense. I'm certain I'm uncertain of uncertainty since I prefer deterministic certainty. Thus Einstein, who believed in 'complete law and order in a world which objectively exists', filed a series of objections designed to checkmate the implications of such 'chaos' and irrationality. He did not appreciate Bohr apparently wresting rules from big G (God not gravity) and, at the same time, 'deifying' chance; he thoroughly disliked the idea of apparent natural order really being the product of compulsive, abandoned cosmic gambling. 'God does not play dice'. He felt that if you can't, through insufficient theory, predict events that does not mean that they are, perforce, fundamentally random. In other words, if you couldn't be clear about a system either in between or even at your points of observation, quantum theory was incomplete. The doyen of physicists also espoused 'realism', that is, the idea that things exist at all times and not simply when their quantum probability is collapsed into a definite reality by observation; and he especially abhorred the quantum notion of 'entanglement' or 'non-locality'. Entanglement, also called 'action-at-a-distance,' involves instant (i.e. faster-than-light) communication across unlimited space between originally connected but now separated quanta. Why should such 'omnipresent God-effect' occur? *Does it mean that, if everything was once connected in a 'big bang' singularity, everything has affected, in an historical way, everything else? Is it that, in some indeterminate way, things all still interact?* No doubt, we can't assign a cause for omnipresent gravity but, still, its action-at-a-distance doesn't travel faster than that single absolute, the speed of light. Therefore Einstein believed, common-sensibly, that if two systems parted and were isolated for some time then a measurement on the first system could not produce a change in the second. Was something in various interpretations, aspects or perspectives of the quantum theory missing? Planck, Schrödinger, de Broglie, Bohm and many others (maybe including Heisenberg himself) also believed that deeper causal mechanisms may be found that reinstate nature's actual determinism. No doubt, we inhabit a universe that can be interpreted by 'quantum all the way' but what about the constant order as opposed to indeterminism it displays. There is, equally, no doubt that the supposed indeterminism of a 'quantum universe' little interferes with the operation of macroscopic objects and, therefore, daily life. Isn't even uncertainty subject, in fact, to various, strict quantum 'rules'; are not the equations of quantum mechanics in themselves physical determinants? And if the way that I deduce them was (by evolution) and is (by sub-atomic motions in my brain) determined by the chancy laws that those equations represent then logic runs in circles. How do you sustain a logic born of the determination that the world's determined just by causeless chance? As Max Born wrote, 'the motion of particles follows probability laws but probability itself propagates according to the law of causality.'

Then struck Bell and with him Aspect. John Bell was pro-Einstein and anti-Bohr at first; but if his Theorem (called Inequality) was proven true it means that, in the face of quantum theory, Einstein's 'local realism' comes unstuck. 'Spooky' action-at-a-distance can occur; uncertainty and quantum probability prevail. Subsequent experiments by John Wheeler, Alain Aspect and others have been interpreted to indicate that his theory *has* been proven. If all systems are at root 'non-local' then nature appears supported by an invisible reality whose connectivity is omnipresent (or at least faster than light). One might thus accept, as many have, that the 'Copenhagen' way of thinking - based on *HUP* - was a true description of 'microscopic deep reality' just as chaos theory is of macroscopic evolution. Strange proof crowns Lady Luck. Einstein would be wrong. *Non est deus. QED.*

Will sanity handcuff such weird ideas or, alternatively, the eventual truth of quantum strangeness set you free? Bell's 'proof' of non-locality and, by implied possibility, the ultimate correctness of non-deterministic quantum theory, *HUP* and the apotheosis of Lady Luck derive from an interpretation of experimental results. Such interpretation might therefore indicate the incorrectness of standard, historical determinism and so undermine the 'laws of God'. But if, as claimed, 'loopholes' exist in Theorem then the question could slip out and stay at large.

Many think, however, that the jury's firmly out. Frank Tipler, for example, treats the central equation of quantum mechanics (the Schrödinger equation) as a special case of the 1840's Hamilton-Jacobi equation of classical mechanics. By adding a term he renders this equation deterministic and thus claims Einstein (with his *SED*) is right. 'God does not play dice'; so Darwin, with his notion that evolution proceeds entirely by chance, is fundamentally wrong. If it's Einstein versus Darwin, Einstein wins. E's right, D's wrong.

If quantum strangeness is the norm then why is order so apparent everywhere? Roger Penrose takes an inclusive, common-sensible approach. From a quantum perspective 'bizarreness' *is* normality. It exists in potential but is 'confined'; it's everywhere a possibility but is, for the most part naturally 'suppressed'. How? To observe a quantum you disturb its state; you collapse its 'probabilities' into an actuality. Local fact is realised. You don't, however, need a laser or some other kind of probe. Continual, interactive impact by the objects of the universe disturbs each one's entanglement; mutual interference, collapsing probability, forever brings the action 'down to earth'. Past and future tryst in present actuality. An immaterial 'cloud of possibility' is realised; immanent potential is continually, locally expressed; quantum indeterminacy and classical determination each together live. *Furthermore, constrained by the 'statistical' bulk of billions of neighbours, most quanta do not dance in isolation; bound in molecular and larger entities they submit to various chemically and physically predefined forms of order called the certain, non-random laws of nature.* Creation *is* (as glimpsed in Chapter 7: Towards a Theory of Physic and Metaphysic) the orderly constraint of energetic chaos.

The standard, 'acausal' Copenhagen model that presently dominates quantum fashion was attacked by a determined and deterministic Einstein but it is also rivalled by another school of scientific metaphor. David Bohm's dissatisfaction has calculated how a *deterministic*, causal kind of *quantum physics* that might work just as well. To explain action-at-a-distance his model proposes a '*non-local* hidden variable', a kind of non-local quantum potential which supplies implicate order to particles and thereby the universe. In this case uncertainty is definitely in doubt. With Einstein it allows 'locality', isolation and predictable outcomes in a common-sense universe composed of serial cause and then effect. He wanted to describe definite actualities and not indeterminate probabilities; thus, with Shrödinger, de Broglie, Planck and others, he believed quantum mechanics was a statistical approximation to an underlying deterministic theory. Does this mean Einstein vindicated? Copenhagen quantum physics slightly wrong - not mathematically but in interpretation of the facts? You may want to sever cause from its effect. You might wish to 'disprove the omnipotence of God' and rest a non-deterministic world-case on interpretations of experiments involving a pair of photons or electrons whose spin is indeterminate until, upon measurement however far apart, each adopts identical condition. But the world is not such particles alone; what, moreover, is 'entanglement' without a measurer? Why should its weirdness afford the underpinning for a faith in godless, cosmic randomness - especially if all probabilities collapse to certainty when any aggregate of particles (of which the world's composed) is found? Will the whole business of physical uncertainty and 'ultimate randomness' may prove to be either a mind-game, a statistical blip in the history of physics or else the prelude to some deeper understanding of the way things work?

Perhaps you can see where this is leading. It isn't where materialism wants to go. Natural Dialectic would, like Penrose, want inclusive answers if they still make sense. Check *figs.* 8.1 and 8.2.

Firstly, mind doesn't have to bear the properties of matter. Its 'holographic', unifying and yet 'nuclear' capacity for information may impact each 'pixel' (or each 'quantum') of the cosmos simultaneously. **Potential matter precedes physicality; it has been identified as archetypal memory.** Could Bohm's deeper deterministic mechanisms relate to archetypal memories in universal mind? *Could his non-local, hidden variable Natural Dialectic's archetype?* Archetypes amount to possibilities or, if you will, universal 'waves' of probability. Potential, principle, endless local possibilities are realised within the general, omnipresent, archetypal constancy.

Could we be talking, in this context, of physics interfacing metaphysic? Is this the way that mind and matter meet - not only in the universe but, at this level, in your own microcosmic constitution where we call it (Chapters 15 and 16) psychosomasis?

Secondly, mind *is* information; memory is the fixed or 'solid' phase of mind; archetypal (universal) files constitute an 'immanent, immaterial potential'; on the 'vertical', informative y-axis of Natural Dialectic's conscio-material scale (*fig.* 0.7iii) this level - potential matter - is hierarchically sited just above matter. Its precedence is an informant whose principle (i.e. specific possibilities) is realised in the expression of different local, physical phenomena. In this case the form that an archetypal memory would take is known by us as 'field'. Cosmic memories *are*, logically, physics' basic, immaterial force-fields. Disturbance of such field (or latent file) results, of course, in action of its sort; we're into physic's 'upper', quantum level now. **Thus order, both determinate and indeterminate in style, is one; according to this scheme of things these qualities combine.** Is such unification a satisfactory resolution of the cosmic paradox? Are you certain? If so, materialism by including information must be willing to dissolve into a larger whole.

This conclusion is important to the point of repetition. A physicist's only considerations are material but, if a 'metaphysical linkage' does not have to cross physical space-time, it could be a property of 'pre-space', super-space or potential matter. Potential matter is (see *figs* 1.4, 4.1 etc.) in dialectical terms subconscious mind. **Subconscious mind is a generic term for files of memories.** Could omnipresent connectivity reside in mind - not individual but universal mind? Not conscious mind but archetypal memory? Could quantum mechanics, for which the separation of observer and observed is not permissible, be probing the 'entanglements' of such a universal memory? Perhaps Bohm would say Aspect's experiment was touching on 'the hologram of deep reality'; or Natural Dialectic was connecting with 'non-local' archetypes whose various field-potentialities compose the memory of universal mind. **These files are the source of what we know as physic's patterns, that is, natural law.** *If these suspicions turned out true then rule luck out of your equation.* What seems 'accidental' to a goggled vision is, as part of the whole picture, determined by connection with its 'ruler', archetype. Check out Natural Dialectic's definition in the Glossary. In this way cosmos is intangibly yet absolutely, pervasively and unchangeably ordered.

Punting on the Cosmic Stream

What did you expect? Chance *is* what you *don't* expect - though if it were completely random neither law nor mathematics could describe its operation. Complete randomness is, by definition, completely unpredictable and therefore unreasonable. So can you conclude, although the universe is fundamentally irrational, that its irrationality is governed by rules, by chance-generated natural laws that make it reasonable? Thus chance is the absence of a defined cause but creates the framework of defined causes within which it happens, a framework of 'necessity' called natural law. Can such nonsense be true? In which contradictory case, the one that philosophically underpins physics and chemistry, chance and necessity are flip sides of the same material coin - one weighted, you bet, towards odds-on necessity. Indeed, the whole scientific endeavour, based on a presumption of balance, law and predetermination, has been to illuminate the patterns, rules and laws to which physical operations experimentally and without exception adhere. The behaviours of material automata are described by a universal language, a balancing act called mathematics; they are described, that is, on the presumption that what goes into a reaction comes out transformed but, essentially, energetically, the same. It works. <u>And the effect of rigid determination is to eliminate chance.</u> *What is residually perceived as chance happens within a framework of natural law. Is such residue itself an aberration, a flaw in certainty that's down to statistical error?*

Yet it is presumed by complete contrast that, according to the tautological cycle of illogicality spun in the previous paragraph, the origin of the world's legal infrastructure coincided with a sudden, causeless appearance of matter from nothing. In other words, the principles and laws are themselves a matter of chance. Direction was undirected. *Chance happens within an accidental framework.* Do all paths lead to Lady Luck? Was 'necessity', that runs the universe upon a stable course, just the star-struck product of an accident? Did chance mint the patterns of transformation we call laws of nature? <u>*What, you ask again, is the reality?*</u>

The question is humanly important because ethics, morality, law and social politics are all derived in the context of its answer. *They involve both current and historical contexts; and of these the historical eventually and intimately invokes the issue of origins.* Where am I from? Origins are, like parents, at the heart of how we see ourselves. Creation stories, of one kind or another, are an imperative. They closely affect the texture of our lives. If there is deliberate (and therefore reasoned) theme and variation-on-theme, that is one thing. If, on the other hand, nothingness for no reason exploded, natural law is itself the product of a Big Accident and unsystematic evolution works by breaking themes of type, then the basis of reason is itself unreasonable. What rationality resides in chance?

If chance is at the heart of naturalistic and thus scientific philosophy then *lack* of information is identified as your progenitor. Bang wallop. Your abiogenetic ancestry derives from accidental impulse to 'create'; you inherit tiny, reproductive particles made up, because impossibility is not permitted to exist, of most improbable extent of probability. **If this is currently the dominant philosophy, then the idea of mindlessness is what prevails**. A causeless, wind-blown mind is diagnosed as mad and yet an intrinsically anarchic rationale dictates the way we think of life. A Theory of No Intelligence crowns irrational 'reason'. It lifts whimsy to the throne of cultural identity; gowns of academic institutions laud such elevated 'rationale'. Its culture validates a moral sense of relativity, that is, morality constructed out of personal preference, social consensus or persuasion of a given government. It recognises the 'necessity' of material law (such as the sciences research) but fails to recognise or else deliberately rejects 'necessity' of immaterial law. If you can 'get away with' what's defined as 'crime' then good for you; you win; egocentric, evolutionary survival's conscience-free. Cosmic morality, the so-called 'karmic law' that Lord Deliberate has in mind, is snubbed. Because, of course, you can't have such morality without anathema, a universal mind with an intelligent design. Therefore, isn't what I think as good as what you think? It's all up for argument. Everything is open to debate. What else but variable opinion or, in society, the tides of what most people are persuaded are 'good things' hold sway? Time and place play hide and seek with any single standard. This is what, without Ideal, our forebears recognised as moral meltdown. What came before must, by definition, be antique - even if it forms the basis of our government and courts of law. Doesn't justice hang upon its cosmic sense of balance?

Guess, gamble, punt. You're punting on the currents of which kind of cosmic stream? Gambolling with Lady Luck is hedonistic fun but is unreason really at the heart of making chipper, modern man? Is rule-breaking chance the root of civilised and rational identity? You suppose a legal structure (called the universe) erupted unintentionally from One Big Accident (powerful expansion from dimensionless a point). This structure's quite coincidental rules (the laws of nature) are, however, most determined and without exception broker certainty. Can't you count on the behaviours of automata? Isn't what's predictable reliable? Isn't what is stable reasonable? Better still, the laws of physics have no life; they can't interfere, any more than natural scenery, with thought; they can't impact character or rule relationships. In this sense they are totally apart but aren't they quite enough?

Is what is historical and therefore closed to test or even 'retro-diction' really unambiguous? It is, in truth, an open question, the inference of whose answer must best fit a known fraction of the facts. What is *your* origin? Did life spring from a Little Accident (the evolution of its chemistry) which, it can easily be argued (Chapter 20), is impossible. Less than improbable you are impossible by chance and yet, you urge, there's no Ideal. Mindless matter seeded you. Even Deity that loosed a cloud of hydrogen could hardly have imagined you evolving from such reflex mist! Nor, but for certain unpredictable and undefined events, would you have, myth-born, taken shape. But, according to the academic doctrine of 'objective single-case chance' and inspirations of Charles Darwin's species, you just know you did! *You must have done*. Never mind that chance, a goddess of the gaps, plugs holes the laws of physics don't explain! A fertile mix of accidents first impregnated earth; a sort of acquiescent Luck that's fettered by necessity's constraints has kept on growing complicated; unusual events occurred, contexts conspired until, with an unsystematic show of brilliance, she evolved intelligence. Out you popped. Beneath a host of lucky stars you, an unexpected changeling, a chanceling of coincidental stream, evolved from her bacterial seed and so were born of Lady Luck's canal.

Up Horseshoe Creek

Up the creek without a paddle - that's a boat without a purpose. Is this the case with cosmos? What is the alternative to atheism, that is, to the 'lucky horseshoe' brand of creed? The only answer paddled is theism in two forms - the pure (intelligent design) and impure (see Chapter 25: Theories of Accommodation).

You need a mind to want anything and, therefore, care about the interfering impact of 'chance' frustrating your desires. Mind aims to control. It would minimise such tiresome unpredictability. Conversely, matter couldn't care a fig. It is senselessly reflexive. Reflex coincidences 'happen' in a lawful way. The ones observers label 'dicey' are never quite impossible. They are only, as you multiply the happy accidents, wafers of improbability you slice down to an atom just before impossibility. Impossibility is zero chance and zero chance is, when you talk about the origin of life, the chance of scientism's faith. Happy accidents, however, are not the way intelligence obtains success. No chance is also chance eliminated by its great eliminator - mind. What engineer leaves anything to chance? His paddle of intelligence persistently avoids blind alleys and so why should theism's Engineer be stranded in the Luck's doldrums or get lost up Horseshoe Creek?

Just as a superficial two-dimensional ant cannot comprehend seemingly causeless effects from a third dimension, so the superficial dimensions of matter cannot understand the coordinate of mind - especially the effects of universal mind. How, if it exists, could such mind integrate with matter? How might the operation work? Let's note the character of each impurity, deism and theistic evolution, before investigating what theism in its pure form claims.

If chance is goddess of the atheistic gaps then, likewise, no god of the theistic chinks inhabits a deistic image of the universe. The idea here is that metaphysical God, having designed and initialised a reflex 'cosmic machine' consummately well, can then let it run or even 'develop' by itself. This was the 'progressive', evolutionary line of thought that Erasmus Darwin and possibly his grandson Charles pursued. The laws or 'constraints' of nature might even, it is claimed, have been devised to promote such a development - although there is absolutely no evidence for this imagination. *Physicists do not know what turns chaos into cosmos; there is, within the corpus of that study's law, no 'source of creativity'. There is, we've seen, in fact no self-contradictory Natural Law of Innovation that allows for progressive evolution without theistic interference and within the bounds of chance.* So how could chance, within a framework of law, accidentally lend the level of progressive flexibility that, by its very nature, mind with clarity and systematic purpose generates? Nevertheless deist and theist push Lady Luck to unwarranted heights of just the creativity that physics (but not yet, it seems, biology) has understood she lacks. Has she not evolved a brain to recognise her excellent but non-existent capabilities? So that, in a Newtonian universe, the Theories of Intelligence (at the start) and No Intelligence (thereafter) might, without the slightest scientific confirmation, accommodate each other.

Deists present you with a magic watch, a mechanical but sparkling universe that can, most wondrously, transform itself into time-watchers such as you and me.

A theist of the evolutionary brand thinks otherwise. His paddle peddles 'God or pantheistic gods of gaps'. These are gaps of ignorance, of inexplicability or even quantum indeterminacy. The laws of physics don't stand in His way. He can meddle ceaselessly with chemistry but only in the mode of seeming randomness, of evolution. The 'universal pocket-watch' or, if you like, 'cosmic machine' will therefore seem to build itself because its Maker, taking Lady Luck's appearance, works in ways mysterious to anyone intelligent. Transvestites want the world both ways and this Chap designs by chance - unless the chance is really a design! Are you confused? Thought like this can fracture tablets. It can splinter logic set in stone. Never mind.

If, the theistic evolutionary argument runs, there was a start to physicality then nothing physical (that is, something metaphysical) preceded it. This prior metaphysic, labelled God, acts as a 'hidden variable'. Working between the actual lines this variable controls and evolves events in a 'progressive' way that might appear to be the hand of chance. This hand, invisible, has slipped into the glove of chance! Or perhaps it is a hamstrung Limb of Mystery that, imitating luck, limply, lamely weaves and lurches towards a non-existent goal. Any fumbling thing could happen. This idea locates its Chancy Chap in the accommodation Heisenberg's equations use - address uncertain, occupant effaced. You might, however, query (Theories of Accommodation, Chapter 25) why a Legislator bends his laws, steers or even interferes with them, by using chance. If you have a goal, a complex goal like building robots or developing a body bio-logical, then why block logic, blindfold mind and handcuff creativity? How desperate and daft a strategy! Complete incompetence! Intelligent creators target the construction of precise, coherent, purposeful creations and legislators dot their '*i*'s and cross their '*t*'s to sharpen definition. No blur. No probabilities, possibilities or confusion. Who'd hide their goal and then advance by clear-cut anarchy? Such were surrender of control. Isn't chaos madness? The system spirals into failure - especially if you want to model it 'in your own image' of Intelligence and Love. The Lady's not an easy girl; by choosing as a consort Lady Luck theistic representatives of God have turned their Mentor upside down. Indeed, their brand of evolution's reason's magic wand.

No luck accompanies a third idea - *pure theism of design*. It's magic's simply high and very bright intelligence. In this 'chauvinistic, right-wing' trick the Lady's vanished. Things may not seem predictable to you or me but reason, with anticipation and precision, underlies or targets every move. An event may look, from our level, accidental but from the top it's all in line. Not a leaf falls but according to the game, that is, the game-plan. Neither uncertainty, unpredictability nor, for that matter, free will can exist; but they can seem to.

No actual freedom anywhere? Can this be right? Or right for matter, wrong for mind? Or does it all depend upon the level of an observation? So that, from its own vantage point, each apprehension is correct? The greater ignorance the greater seems the part of chance and *vice versa*. In which case, simply, is there among the relative positions any absolute? Is there a Highest Truth and, if so, what's the Nature of its Law? Will scientific Lady Luck or her Antithesis embody it?

At least, within pure theism, the only time that pure chance appears is as the outcome of a lottery; see it as intentionally established 'randomising-within-context' whose neon lights, 'Cosmo-Casino', flicker ceaselessly. Therefore weather and the waves of sea, the path of meteors or solar fusion are, like leaf-fall, certain outcomes even if the shape of individual flames, twists, drifts and scintillations of each local moment's interactions turns out different. Outcomes of a type's behaviour vary slightly not by chance but context.

Ceaseless change-on-theme. Complex contexts make for many variations round the axial certainty of theme, of laws that are not broken. Even the behaviour of subatomic particles forever 'collapses' into precisely predictable, practical certainty. This is the kind of certainty that you can slap. Rock sure. The laws of chance are statistical rules enshrined not in things *per se* but in formulae that try best to picture them. Things themselves are not descriptions. They are real. They represent no probability, that is, they represent collapse of range of possibility into a single certainty. Uncertainty relations are extremely small. Even if they're real (and arguments exist against them), do they matter more than the 'wobble' of a piston rod that rams its cylinder or sparkles of the sun that dance upon the certain waves of sea? Indeed, you might imagine that the *HUP* acts like a film of oil by lending flexibility, a patina of indeterminate relief between the crushing certainties with which the vice of nature grips events. It vacillates, that is, between an object larger than an atom and, 'down below', those smaller 'granulations' at the edge of space. Such latitude from rigid law may be a measure of the 'noise' or perturbation caused by *ZPE* (Glossary, *fig* 2.13 and Chapter 10: zero-point energy). It may thus, at atomic or at subatomic level, register an agnostic, indeterminate 'buzz'. But if the source of *ZPE* and, therefore, atomic perturbation rests with minuscule 'Planck particles' in a fabric much too small for their transmission we arrive at bedrock. We arrive, next to infinitely nothing, at a final, endless beach, the sands of space, a stippled certainty that dots the base of being. Even if it were correct the *HUP* applies to physical events alone. It presumes no 'hidden variable' - especially one metaphysical like mind. Yet its author was well aware the laws that quantum theory formulates 'no longer deal with the elementary particles themselves but with our knowledge of them'. *The question therefore arises whether HUP's restriction on the accurate knowledge of an individual event is intrinsic in the fundamental structure of the world or is simply the consequence of an inadequate theoretical model.* Either way, is it any more relevant, with respect to truth, reality and classical determinism, than a red herring?

Indeed, the maths of both quantum and classical physics is able to provide highly accurate descriptions and predictions of material events. Both are, in this sense of certainty, deterministic. Determinism's lines are, sharp; the sames lines are, by indeterminism's count, statistically blurred. Determinism, whether of the greater or the lesser spotted breed, implies involuntary, automatic obedience to rule - the invariable behaviour of non-conscious automata. Scientific research would, as far as possible, detach some automated factor that it wished to study from the whole, complex influence of the world at large and treat it, both experimentally and mathematically, in isolation. The goal is to determine the behaviour of a simplified system, divine its underlying rules and so, adding the behaviours of other such systems, build from these parts a picture of the whole. The more regular, repetitive and predictable an automaton, the simpler are the model, the maths and the research. Artificial simplicity is a tool that, whether single causes are observed precisely or in blur, has notched up powerful successes.

Is the real world simple? I never said the subject was a simple one. We know the Lady's not an easy girl. She's not exactly dodgy but she's wayward, whimsical and hard to understand. Have we yet run Luck to ground? Her real world is, any Yangtze river pilot tells you, more like rough and rolling waters whose multitudinous bulk, based on simple elements but riven with conflicting currents, conspires to conjure up surprises. Is the principle of uncertainty applicable to currents through a Yangtze gorge? Can you ever 'freeze' such action into an absolutely predictable frame of each eddy and each molecule of interacting elements? *From this normal, non-quanized perspective, physical chaos and uncertainty are functions of a lack of information or ignorance due to the quality of information available to an observer.* Things seem uncertain because *we* are uncertain. Our information, however relevant, is relatively incomplete. *Indeed, such uncertainty is rooted in capacity to cope and in our sanity.* Chance, in part of her profile that we understand, is the price that's paid for filtering trivia and avoiding factual overload. Consider the sheer volume of possible interactions from atomic to galactic levels, all with a history leading to a present, local, personal context. In a living world you also have to factor the mental decisions of other organisms, all made in their own context of instincts and personal memories, into your calculations. Are you surprised that we 'cut the guff', drop off this impossible load and, instead of noting every triviality, 'catch the drift' and then cut to the chase of pattern and its principles? So that, in the details of life's Yangtze rush, we miss some object, swell or current that - it seems by chance - can later tip us up? Chance is, even while the actuality is certain, an intrinsic and surprising part of life. *This is because it is a function of ignorance - ignorance which humans strive to eliminate as far as possible.* Nature's not amiss; but our depth of understanding and cooperation are. We want certainty and to eliminate uncertainty you need to understand things properly. Immaterial information is real power. The devil may inhabit detail but the principal is principle. Principles, which are condensed forms of information, are high-powered. Unlocking doors to inner chambers is the aim of science and the key to man's control of nature's greatest strengths. *In this view chance, uncertainty and disorder are words for relative misunderstanding.* They are functions of a closed (or restricted) system of inspection, one insufficiently broad to embrace the complex, codified network of relationships that simultaneously affects and contains its every physical and psychological component.

Mathematics of order, disorder, chaos and complexity has been developed to try and track to lair the beast of unpredictability (if you could ever so defame the lovely Lady Luck). This beast may awaken when small errors in initial conditions (called perturbations, mutations or uncertain details) give rise to large errors (unexpected or abnormal results). From Chapter 1: Causality do you remember, for example, the 'butterfly effect'? It argues, possibly correctly, that a butterfly stirring the air in China could transform storm systems over New York by the next month. Computers have been used to try and make sense of 'chaos' - as in the case of supercomputers used to help generate weather forecasts or predict protein conformations. Formulae have also been discovered that allow them to create 'nested patterns', called fractals, resembling some of those found in nature. The business of how things 'self-order' is key to a materialistic view of the world. Such morphogenesis can make up mountains, clouds and galaxies but can it produce a life-form? Even a cell so tiny that we can, mistakenly disregarding or, like Darwin, ignorant of its complexity, allow ourselves to imagine its evolution? You can therefore understand why research funds have been directed by scientific materialism towards discovering and evaluating any structure or mechanism that might seem to support 'chemical evolution'. A good deal of energy has been expended trying to show that 'living matter' can develop 'apparently purposive' complexity in spite of entropy (but not in spite of death). We deal with this claim exhaustively through Chapters 19 to 25. For now suffice to say, 'Of course, it can. Each organism operates according to its *plan*; the trouble is that evolutionary investment bets that such a 'plan' emerged by accident'. For example, theory and research in physics have combined to identify temporary 'dissipative structures' such as vortices. These can generate complicated patterns which (unfortunately for abiogenetic aspirations) neither reproduce themselves nor generate highly complex, stable constructions produced from coded information. *When such turbulence decays it leaves chaos in its wake.* Chaos is not plan.

In fact, there is no freedom from the straight and honest fixity of natural law except in mind - in consciousness. This would, as Chapter 26 explains, make God's major risk mankind and his big gamble the conditioned free will of that agent's reason and morality. On a sliding scale the less aware a mind the more preordained instinct, dogma or ritual come to dominate. No mind, no choice, no risk but completely reflex and amoral rules - these are the laws of nature and the state of matter. *Top-down*, no dice roll around the automatic world's 'casino'. In physics there exists no root uncertainty; uncertainties are constrained and probabilities collapsed, every instant everywhere, into actuality. In this case chance *is* a refuge of failure of perception and/or comprehension. *In which case an obeisance to chance at the heart of a materialistic creed were unfortunate; scientific idol worship of Capricia Fortuna amounts to leaning on an abstract form of faith, a pillar made of ignorance. An antique ignorance practised inside temples of the mind.*

In fact, luck's not all ignorance of facts involved in making a decision. You could, as noted, generate deliberate randomness within a preconceived system. For example, within the restrictions imposed by a computer itself programs can be written whose algorithms (i.e. rules) can produce a theoretically unconstrained, endlessly unpredictable sequence of number or pattern. Such abstract randomisation is, however, intentional. Its *intention* is to create the programmed sort of unpredictability called a lottery. The *purpose* of a lottery is to generate random permutations of a specific kind of cipher; and thus, as with the physical universe, its chances occur within a framework of rules, specific design and a special, totally unnatural program of constraint - numerically none. Such purposive lotteries include card shuffles, bingo and, with its three layers of shuffle, biological meiosis. Of course, the spatio-temporal recombination of objects, sequences and events is also a 'lottery', this time unintentional; but it is only unintentional if you presume the rules that constrain it are without intention too. Would you additionally allow the likely circumstance that, despite the presence of its rigid rules, unforeseen and therefore 'chance' events can happen in the context of a game? If not, then you mix rules with complex context of their operation - physical events. Do you remember the Marquis de Laplace, a French astronomer and mathematician who contributed to the fields of 'celestial mechanics', applied mathematics (including probability theory) and physics? He would throw this shape of paddle. He might rescue you from Horseshoe Creek. A paddle is an immaterial figment, that most powerful of energies, belief; and he believed that absolute determinism rules - the universal body is one insusceptible to chance, one systematically constrained by regulations (or necessities) to the point that, if you had enough information, you could describe, predict or retrodict any event since the world's inception! Or was that conception? At any rate, there's no chance here. Chance is a phantom faith, an anti-deity, a wisp excluded from the gaps. No wavering. No waiver. The most leeway determinism dares allow is, in an absence of pure chance, 'limited randomisation', 'ruled chance' or, psychologically, 'conditioned free will'.

Is this French prescription what The Doctor ordered? No uncertainty, no doubt? Some see Lady Luck's statistics as her 'vital wobbability' but others simply call them probability. Statistical analysis, the mathematics of uncertainty, only works because of the presumption of underlying patterns and thus a limit on her randomness. These constraints inform any calculation of probability, possibility and impossibility - unless your bottom line is that unlimited possibility is possible, that nothing is impossible only highly improbable. You rule zero out of the equation - except that mind-behind's assigned a zero certainty! You zeroise the possibility of cosmic anti-chance, universal mind, a source of intelligent design - this really *is* out of the question!

Odds longer than 10^{50} to 1 against are generally conceded to be zero. Only into the range of this minuscule fraction might you hope to squeeze an interpretation of 'chance absolute'. But don't confuse the 'zero' of pure, virgin luck with what is really zero - pure impossibility. Yet odds as low as that can easily be ranged against the chance occurrence of complex components, let alone the correlated coding, of a single cell! Intelligent but immaterial design, of course, works with material constraints but counters entropy to produce unnatural constructions ('natural impossibilities' like computers or space shuttles); its arrangements can obtain the certainty of 1. Wranglers who believe that, mindlessly, a certainty of 1 is possible for anything are those who guess that you evolved. To compute that nothing is impossible (that zero can't exist) is, in effect, to waive constraint; it is to ignore the automatic, uncreative necessities of physics and the rules of chemistry. In a world of predefined or, at least, definite natural law this is clearly nonsense. Sheer fairy wands, will-o'-the-wisp and lucky charms - the creek's a marshy cul-de-sac. The horseshoe on the lintel of Fortuna's shrine has fallen off. Approximate determinism is the maximum an anarchist, atheist or other cultist can expect. Is such 'conditioned chance' enough to have evolved Luck's fervent worshippers?

Shots in the Dark

	tam/ raj	*Sat*
	existence	*Essence*
	duality	*Unity*
	matter/ mind	*Potential*
	practices/ principles	*First Principle*
	conditions	*Freedom*
↓	*tam*	*raj* ↑
	more fixed	*less fixed*
	material automata	*mind*
	involuntary	*voluntary*
	no will/ no freedom	*conditioned will*
	necessity	*choice*
	determined	*determination*
	body	*actor*
	product	*purpose*
	specific practice	*general principle*
	complexity	*simplicity*

Chance and necessity, free will and determinism. There are two kinds of 'freedom', two kinds of uncertainty; and with this existential pair a paradox, (*Sat*) Certain and Essential Freedom.

For matter nothing aims to hit a target. Every shot is in the dark. Its principle of certainty is a fixed pattern of behaviour, a mathematically precise, predictable 'law of nature'. Of course, such a pattern must be measurable and, if immeasurable, its observer must experience uncertainty. Quantum motions, the formation of waves or clouds and the weather, for example, are difficult to 'pin down' closer than what are often, in practice, sufficient unto needs - statistical approximations.

Automata, the mindless objects and events of nature, experience no uncertainty. They neither expect nor predict anything. What is is certain. In oblivion there's no such 'thing' as freedom, possibility or chance. Metaphysically, psychologically we cope with doubt and elements of chance in terms of 'theoretical uncertainties'; a bookie measures relative as opposed to absolute certainty in terms of 'probability'. From this statistical angle accident amounts to a description of the odds, that is, an inability to predict with certainty the outcome of either immeasurable or complex interactions. This applies as much to dogs and horses racing round as to atomic particles and, if expectations are tied to intentions, chance is a bug that, lucky or unlucky, surprises our designs. *In short, you argue (pace quantum physics) there is neither uncertainty nor chance in nature; only thinking makes them so.* They exist only in the mind of an observer - because this mind is (always and inevitably) relatively ignorant of the whole truth and because it anticipates/ intends that something particular should happen. Such insecurities accompany thinking in the dark; and vanish in proportion to extent of knowledge. Why else place such a premium on education, that is, information gain?

The second kind of 'freedom', therefore, is of *mind*. Mind seeks information to dissolve its darkness and so aim in clear air. Humans crave relief from ignorance and the empowerment of finding out. We want to know for certain what is going on. Moreover, chance doesn't only mean uncertainty but opportunity as well; the opportunity to learn, diminish chance and so achieve bulls-eye for our desires. What desire more principled, therefore, than knowledge for its own sake, that is, knowledge of the way creation works. If purpose is a shot whose target's clear, what better than to shoot for clarity itself? What, in this cause, more fundamental than material science of the cosmos? And, if there's an immaterial side, knowledge of that too; you might climb past relativity and, at the peak of cosmic mountain, find not creation but, in its Creator, Certain and Essential Truth.

Will truth set you free? Free will's uncertainty resides, for sure, in choice not chance. Voluntary certainty lives in decisions; it's embedded in informed determination of the path that's 'right'. Choice unavoidably implies a sense of good and bad, of right and wrong. It may involve morality. As well as material therefore enters moral science. In this field of action correctness or mistake can cost as much as any physical effect. It can easily involve a case of make or break or life and death. In which case is it true that unscientific metaphysic is no more effective than a juju or a talisman? Is it just a refuge of the weak or hard-pressed? Isn't it a wretched substitute for certainty, a superstition that will end you up in Horseshoe Creek? If you are sure that such assessment is correct, exclude the notion of a metaphysic from all reckoning. If, conversely, you are wrong the truth remains. The sun shines even if denied. The fact of metaphysic is not changed by what you think, not least if thought itself is metaphysical.

In what units, up against what yardstick might you therefore calculate the metaphysic of a value, purpose or idea? How do you clock up moral 'brownie points'? Because decisions, right or wrong or good or bad to more or less degree, flower into happening.

What would happen? What *will* happen? What shall I do? My world's a consequence of what I did and do. Response to fate leads forward to destiny. Choice and implicate morality affect life's business thoroughly - nor has material science place in measurement of immaterial choice and consequence, that is, in the government of our subjective lives. Indeed, it is no more than a pretender passing shallow cleverness for wisdom there.

No doubt, choice implies uncertainty; but such a species of uncertainty is not the same as chance. The pair are linked but ignorance and moral question are as well. 'I don't know what to do.' Where knowledge lacks, a guess appears; but in proportion as you understand a reason, cause or meaning, ignorance, doubt and betting disappear. I now know what to do. This applies to all perceptions, any scraps of information trivial or otherwise. Don't you wish, in order to increase control, the future was predictable? How interesting but fearful that would be. In general, by natural principle, it runs 'to plan'; but any local nexus of events is still uncertain - and you're fixed in your locality. This complicated web includes the context of your mind and those who think around you. An element of ignorance is thus endemic. Mind lives in doubt, tensions rise when either information might be incorrect or dangers randomly arise. In such cases you can raise a talisman, use common sense or, sometimes and more powerfully, seek scientific knowledge to control response. Information is potential. Control is a function of one's understanding; understanding is the nature of both human power and truth. This is the reason why we always want to learn. What is best learnt? What is your priority? The higher up you shoot the more thought's arrow finds the light. The mystic's Surety is, sure as sure, not of the scientific kind. Shall we, therefore, discount the immaterial power of love? Or (Chapter 3) the possibility of Absolute Morality?

If thought is metaphysical and you are what you think, then aren't you, at the heart of human being metaphysical? Is there, upon this fact, a metaphysical, invisible environment of cosmos too? If so, could physical phenomena have been informed? Could 'the law of natural behaviours' be rooted in design or not? **Either way, your choice is philosophical. It is metaphysical and scientifically unverifiable. Why? Because you can neither objectively prove nor disprove, only infer the fact of a metaphysical entity.** No doubt that science tends to philosophically identify with a materialistic mind-set, naturalism and the wheels of physical automata alone. Even biology and psychology resort, when forced to the face of animation, to treatment in a 'naturalistic' frame. And yet (Book 2) is your own familiar consciousness simply, definitely only physical? Of what, therefore, is thought made? Of what is the subjective knowledge of that thought made? **In what, precisely, does experience consist?** What is metaphysical may well exist, be true but still outside the self-inflicted scientific remit. For scientism, though not perforce each scientist, everything is physical: non-existent metaphysic is, it has been decided, black-marked by the truth of its illusion. You cannot test it. It is therefore an irrational falsehood, unconstrained by S.I. (international) units and, because it runs against a touchy-feely grain, an exceptionable domain. *It is certainly not an objective, scientific reality.* But, believe or not, uncertainty is mental in effect; and the unscientific metaphysic of uncertain mind drives science and religion just the same. Its purpose is to drive out chance. To learn. To become a marksman with sure aim. No uncertainty; no shot in mist or murk but Aim.

Where's the bull's-eye in the butt? You might well ask, like Cicero or

Socrates, what Certainty and Moral Aim comprise. Certainty of mind derives from knowledge and therefore, at best, knowledge that is absolute. Does such absolution take a scientific or a Buddhist, an intellectual or a meditative form - or both combined? Lesser truth, an element of ignorance, arises when, in a specific instance, information is left partial or absent. An observer may even think he knows the lot (don't most of us proceed along this line?) but, as the development of science repeatedly illustrates, have missed an unknown factor; and an experient may, by experience, be sure of what a case involves but still not actually know it all. Understanding, say the mystics, needs to be developed so that men can more align with Nature. Out of tune with nature and with Nature, selfish humans lurch and lumber on. It seems, no doubt, that Horseshoe Creek's a swampy, complicated place! Is its other name 'Department of Philosophy'? What's the paddle? Where's the river? *What is it to know everything and, thereby, enter oceanic certainty? What is the nature of this ocean's light?*

Knowledge is, at heart, experience. You know by being here; the light is only known by being it; realising any principle is called awakening. Awakening illuminates and thus the Principal of principles is known as Illumination.

As much as chance lacks principle, so principles are anti-chance. They constitute a 'higher' wisdom than the detailed lists of life's 'small change'. Science excellently illustrates how grasping principle will obviate the need to know or catalogue the billions of separate effects it generates; and at the same time will empower you with ability to understand, relate, predict and thus employ. In other words, knowledge-in-principle no more involves perceiving every detail in the process of existence than an engineer, say Sir Nigel Gresley, knowing every move every part of his locomotives ever made. Nor, it may be said, does it require either erratic or perpetual intervention. If a Live, Superhuman Power created material forces and bodies of all kinds how could even This enter into their non-consciousness? He (not It, please, for what's alive and well superior to us) might know The Great Machine in terms of its design but not of its oblivious operation. Material oblivion is an unknowing part - only life, through body, *knows* the world. And there are many kinds of life, many sorts of 'footplate feedback', myriad experience of embodied souls. Soul is an aspect of The Quintessential Power of Life. What, therefore, in the reflex maelstrom of the universal body are its own Creator's eyes and ears but lives? *And, think, what does that make you?*

In effect, a principle's an aim; it precedes the practice of a shot. No principle, all chance. Shots in the dark. If, however, aim precedes a shot then First, Essential Principle precedes *all* existential principle or practice. For Natural Dialectic, a principle is information; and information is potential framing consequent expression. Is this whole chapter not about potential? Precondition? That which comes before. Thus, logically, Complete Knowledge and therefore Certainty are only obtained in the experience of (*Sat*) Informative Potential. This *TOP* circumstance is analogous to the way that scientific omniscience might be unveiled by means of a mathematical encapsulation (a *TOE*) from which all other formulations can be derived and beside which they are relatively incomplete. Whereas 'external' science seeks certain knowledge about the predictable, non-conscious realm of nature, 'internal' meta-science of holistic mystics includes the source of mind, that is, Knowledge Itself. Only in the case of such enlightenment is all uncertainty removed from mind. Each individual confidence is a doubt lost and loss of all doubt is recovery of Original Certainty, Complete Trust and Confidence. Can you now see that enlightenment does not involve data? That omniscience is not full of facts but Principle? And that 'the mind of Principal', at *(Sat)* Highest and Most Singular, is quality above all detailed quantity? Such Luckless Aim shoots in the light. Primal Knowledge knows no fragments and therefore it is holy men and not Laplace who really rescue you from Horseshoe Creek.

In this way you can understand that between metaphysical Certainty and, based on the laws of nature, physical certainties there range psychological uncertainties. Incomplete knowledge is where unpredictability and unknown quantities are on the prowl. Ignorance, mistrust and fear are its response. Their tension increases as comprehension decreases; both ignorance of the rules and the simultaneous complexity of events conspire to ratchet up a failure to predict, to cause anxiety or to wrong-foot. Who cares? Not dormant organisms like a fungus or a plant; and 'lower' animals are channeled through their instinct's alleys. They are careless of the future. The cost of mental freedom is, however, for a human linked to care. Less blinkered minds are, therefore, hooked on information, on knowledge that will answer questions, relieve anxiety and rationally eliminate their fears. More understanding's more control until, paradoxically, in Understanding that surpasses relativity, Complete Control and Freedom are the same! When the light's on shooting in the dark is out.

Mind is certainly an uncertainty-cruncher. It is a probability-reducer, the agent of anti-chance. It is the organ of shot-in-the-light, that is, of teleology. In detachment it remains poised but, once engaged in purposing, it loses balance and in motion acts. It frames accordingly whatever pictures senses or the intellect present. This is the normal, healthy case. Mind is the governor, matter governed. Chance, the twin of ignorance, is just a ghost of thought, a flickering of certitude, no more than an infringement of your expectation. Nobody knows each detailed twist and turn impacting on complexity of cosmic connectivity. No-one can trace a present circumstance back through all its causes. So when the best-laid plans hit problems mind, in the face of challenge and frustration, best calculates to minimise, to 'iron them out'. It is an anti-randomiser dedicated to accuracy of measurement and efficiency of purpose yet, at the same time, its ideas, thoughts and their consequent effects are immeasurable and, as every politician knows too well, unpredictable. From this it will be seen that what is obvious to a historian is also the dialectical case. *Nothing skews probability into certainty like thought. No magic transmutes impossibility into actuality like mind, like purposeful action. Natural forces only partially control our world.* Human history (both individual and general) is testimony to the unpredictable, world-changing effects of an invisible, subjective, metaphysical kind of data item - ideas.

<u>The impact of ideas on behaviours and, thereby, human social and environmental history is of central importance but hardly amenable to mathematical description</u>. A theory of electromagnetism, the idea of electricity as a motive power and inventions which have followed that idea are examples of how thought can change the world. So are the plans of Julius Caesar, Clerk Maxwell or, to some extent, each politically-egocentric man or woman that has

ever lived - including you. On a mundane but personally critical level each individual organism orders events according to its instinct, feeling or reasoned 'rightness'. Such rightness includes health and survival. For humans it also includes a swathe of unscientific 'abstractions' - love, hate and morality among them. Things go 'wrong' when they frustrate intentions. Intent amounts to a decision that involves maintenance or perceived improvement of the present condition; it amounts, in effect, to a target along which we shoot ourselves. Not chance but deliberation (or pre-set deliberation in the form of instinct) brings relief. Such decisions (Chapters 3 and 25) are governed by metaphysical (moral) as well as financial or simply physical, mathematically calculable imperatives. *In short, a definition of life is that it purposes; and that a presence of mind in matter changes everything.*

Mind is not Fortuna's province. The notion that cosmos is the product of intelligent design is materialistically unacceptable. *Teleology, you remember, is the doctrine of final causes, of design, of preordination or purpose.* **Science cannot deal with it and for a non-teleologist natural intent is the key anathema.** One-tiered materialism has, by its own decree, outlawed a rational origin of cosmos. The order of order, causal hierarchy, is denied. But reason is the clerk of cause. Be-cause. If explanation of a system lacks one of its origin the first and greatest cause is lost; fundamental reason's executed too. You can't view that system with a full perspective. There's chaos, causelessness, unreason at the root of things. If reason is denied it's broke; unreason, in the form of randomness and unpredictability, slips into the office that it leaves. Thus, from the holster of a guess, shots are fired into original, acausal dark.

Mind anti-randomises certainly; and yet, because hard science deals in mindless, automatic aspects of creation (including bodies biological built from molecular automata), a key concept of materialism is oblivion's randomness. This is, paradoxically, a kind of randomness in which things happen 'lawfully'. Rules are the antithesis of happenstance but, without intelligence, you must conclude that thesis rose from its antithesis. A causeless universe is what it is; what more can you say than that? No rules in nothing then, perchance, rules governing everything. Prior to mind rules anti-randomise; and the anti-randomiser's here by chance! The priority is chance; there follow natural patterns of behaviour. These thoughtlessly 'collapse' a probability and thus, by simulating an entirely rigid mind, reduce the moment's possibilities to a single, predictable event. *Such actual events are never creative.* They offer no creative possibility outside their own predestined patterns of behaviour. They only represent purposeless, essentially meaningless complexity.

Luck may be a complex girl and yet complexity's her challenge; and, since oblivion can't specify, her nemesis is specified complexity. Non-conscious, non-alive - she can't express herself through codes like life. Where evolving bodies are concerned complexity is tottering happenstance's last saloon. She's drunk. Her every shot is in the dark. How can she make a hit? Chance a function of, as well as trembling lack of aim, complexity; but, though the dynamism of an energetic world can 'pop chance up', complexity is born of causes, physical and psychological, that both accord with archetypal rules. Mountain ranges, weather systems, organisms and man-made technologies are all complex but could you confuse their type? Conflate them for your theory's sake? *Call order principle, command or law; both kinds of complexity, non-purposive and purposive, derive from order*. Minded or mindless, specified by natural law or by intelligence, each kind draws a gun on Lady Luck. One for archetypal reason shoots shots in the dark; the other lights the dark and, shooting, kills Fortuna. More rules less chance; more intelligence the same. *Reason makes for aim and aim makes accurate; there is no shot in any darkness either way*. Regimen annihilates the unexpected and drives chance into extinction.

The first, non-purposive kind of complexity is material. *The laws of nature govern purposeless complexity*. Matter, parties can agree, contains within itself the principles by which it operates. **The origin of these is not agreed. Nor, for certain, is the notion of 'a principle of ever-increasing complexity' which would allow matter to 'self-organise' in an 'evolutionary' sort of way and mean the universe might mindlessly 'progress' towards some grand but entirely accidental 'goal' - like you or me.** The reverse. There is no evidence, past wishful thinking, of such material principle or law; or, consequently, that you mean anything more 'systematic' than a grain of sand, a solar sytem or a puff of wind. Of course, you can introduce some energy. You can simplify by raising solid shapes to fluid 'formlessness'. You can even atomise things into basic principles and then, by draining energy away, 'condense' them from a plasma back into the 'normal' rock, marine or weather 'systems' that our planet wears. But the rules of 'non-purposive complexity' are wholly inflexible; its information is passive and uncreative. Potential (the start) is contracted through the time and space of a 'process' (also called a reaction-time or lifetime) to a point of outcome (the end result). In this case the (*tam*) process of realisation is 'entropic' and downward; it increases with energy loss; non-purposive detail appears to 'crystallise' as energy is locked.

Take a thoughtless snowflake. Crystals are composed of regular arrays of atoms whose internal order they, on a large, external scale, reflect. However, many outside interferences (such as the temperature and humidity of the cloud or part of cloud in which it grows, differential heat release, mechanical stresses and bumps against neighbouring flakes) can combine to produce an apparently random variety of patterns out of water vapour, each manifest in a different flake. *Identical snowflakes may exist but, generally, the complex shape of an individual is unpredictable, random and 'new'*. Yet constraints on the crystal (it is made of water, cold, with branch after branch at a 60 degree angle and all arms identical) are axial and differences peripheral.

The same applies to any star, galaxy or other goal-less astronomical, meteorological, geological, molecular or, indeed, atomic system. These are systems only in the stuck-together sense of composites. They are not systems by virtue of any sense of mechanical wholeness or use; they are unplanned outcomes whose arrangements really do not, except in the matter of controlling principles, exceed the sum of their parts. At this point beware. Some persons are confused. They claim unspecified complexity is, like a language with its meaning, specified; they claim that natural causes are, because you can predict them, specified. Can't you predict the weather or the wheeling of the stars? There is the rub. You've specified but they've been specified. Data's been collected but not, of itself, arranged. *You* arrange; *you* transform indifferent, passive data into active; *you* describe and use according to your mind. Is not the scientific goal to model, analyse, deduce and specify; to bring non-purposive behaviours to book? A book of rules. Raw data boiled to principles by means of which you purposely predict. Your filters, like a code or language, have now framed the natural 'message'. Raw data's been transformed to information; by dint of mind alone the framework or the 'grammar' you supplied makes sense of senselessness. Because a lifeless system's purposeless complexity depends on law you pick patterns, measure probabilities and make broad sense of things. Indeed, within the 'by and large' you can identify, for particular automata, ways in which they will react exactly. **This is the essential nature of the element of mind; it is your information centre's point when dealing with the elements of matter**. Mind shoots chance; but what's the cause of physical be-cause?

Where, due to synchronous complexity, prediction fails it seems chance re-appears - but science is not pleased. It chases hidden factors that will pulverise uncertainty. And though it abrogates the origin of natural law to chance it doesn't really like oblivion as a *reason*. Schizophrenic is the message, tense relationship between the husbandry of causal logic and the whims of Lady Luck. That's why Deliberate never married her.

At every circumstance you, being mindful, inject or extract some message. **The second kind of complexity is psychological. It is a characteristic of mind and closely allied, as Chapter 5 describes, to the function of information.** *The discipline of intelligence generates, according to its level, all kinds of plots and mental constructions.* Indeed, intelligence not only reduces probability to certainty but at the same time increases possibilities. It offers active information gain, specified complexity and, therefore, creativity - creations with a reason and therefore a meaning. The rules of such complexity are both flexible (according to intentional design) and inflexible (an inanimate construction must comply with the rigid, non-purposive constraints of nature). **Purposive complexity is the work of a conscious creator and applies to artistic and technological as well as biological creation; it depends upon decision; and its details are based on theme that, hierarchically, reaches back to a first or overriding intention. Such fundamental purpose forms the first principle of any particular creation.**

To summarise, an act of creation precedes its psychological record (called a memory) and its correlated physical construction. Thus the lower, outer creation of (*tam*) passive matter is subsequent and dependent on psychological precedent or, if you like, principles. Orderly purpose came first and law-abiding chance is, like crumbs from a loaf, an incidental latecomer. There is in fact no material chance. The principles that control all bodies (and within whose constraints even our own conditioned, so-called 'free' will has to operate) might be termed 'principles of design'. At a physical level they can be appreciated mathematically or, with less precision, verbally. The fabric of creation, whether artistic, technological or involving mundane repetition, is woven round such a core. **You know (because it is the way you operate) that *top-down* logic is, from a technological point of view, impeccable.**

<u>**Nor, as we shall see, is there anything in the findings of science to discourage the possibility that such an order might be universal.**</u> Creation is part physical, part metaphysical and, in this respect, metaphysical simply equates with oriental 'cosmic psychology' or 'universal mind'. This is most

obvious in the case of biological form. Purpose is distinct from lack of intent, purposive complexity from purposeless. Even physicist Ilya Prigogine, a first-class student of non-teleology, does not believe that 'dissipative structures' - such as whirlpools - are the source of or can replicate biological complexity. An engineer exploits the principles and practices of nature in his constructions but no-one claims nature itself could invent a biological machine (or, indeed, any mechanism or machine) from scratch. The question of biological origins (Chapters 5, 6, 15-17 and 19-25) therefore remains intractable. Can relatively simple, non-purposive starting-points develop into irreducibly complex mechanisms which integrate with other such mechanisms to achieve a purpose no atom has the wit to conceive - survival? Can The Primary Corollary of Materialism, random mutation and natural selection, really 'invent' life's purposeful systems? What creative place has chance in any kind of nature but, especially, natural history? Against the major feat, construction of a systematically working body, Darwin's 'natural selection upon accumulated chances' (evolution theory) no more than deals in minor triviality. What purposive system, incomplete, can work? How can chance bring all its parts at once to working so that it can ever start? You might weave a subtle tapestry of plausibility but this promotion is, by inference alone, not necessarily correct.

How do you feature in this film? How lucky in its shoot-out do you strike? It is a work of physics to identify the *modus operandi* by which principles (as proposed in Chapters 7-12) are expressed. In the case of biology P. Johnson writes (Reason in the Balance p.12), "What is presented to the public as scientific knowledge about evolutionary mechanisms is mostly philosophical speculation and is not even consistent with the evidence once the naturalistic spectacles are removed." Is this true or false? Is life purposeful or purposeless? Is purposeful derived from purposeless? Is life the shot of all shots in the dark? If so, then Lord Deliberate falls dead. And long live Lady Luck!

This is the royal assassination that materialism's revolution utterly depends upon. Oblivion must win the marksman's prize. The lucky shot must hit; the hit-and-miss must not have missed. Secular philosophy in all those universities, the media and palaces of politics is based upon this utterly incredible fatality. Deliberate's dead, long live Luck's vitality! A shot of long shots in the dark is at the heart of atheism. Disciples of the faith have bet their lot on it. No hedge. Now you can understand why edginess erupts if you elaborate the way, by light of day, that evidence decrees such an imaginary pellet must have missed. Who bites the bullet? Who is shot? Who, in this attempt to dominate the world of mind, is flattened?

An Unholy Apparition

Perhaps Deliberate's assassination and Luck's coronation are illusions of an atheistic mind. Perhaps reliance on original power of chance is just a dream that will not die.

It is sure, according to the Dialectic's *PAND*, that there exist metaphysical mind and the 'necessity' of rules that govern physical behaviour. Each, although the first part is debarred from scientific naturalism's seminar, is entirely natural. Where, after all, on the steps of Nature's gradient, do you place 'natural'? Above that point is super-natural, what's below sub-natural but everything is 'natural' just the same - except 'unnatural' behaviour that's confined, on the margins, to contrary use of healthy, natural design. But the exclusion order's still applied, of course, to 'the engine of design' and 'fountainhead of order' - mind. **Mind is anti-chance; it creates, by nature, order**. It understands and makes and uses patterns. Chance is empty-headed; drifting fortune never has an end in mind. What coded, systematic innovation can the natural world invent? What oblivion could never reach mind with its instruments of purpose can - in short, sometimes extremely short, an interval. You are perfectly aware of this. I'm saying nothing new. Unlike matter mind is metaphysical incorporating conscious agents such as imagination, will-power and a powerful executive, the intellect. These fund its operation as the source of accurate, coherent creativity; they work in close association with a commanding and controlling principle - design. **In short, mind and not matter is the active information centre. We call it an active (or kinetic) informant and its objects (of which it is aware) passive data. In other words, things or events that have been informed by a purpose are called passive information.** Any information contained in the character and geometrical arrangements of matter is certainly passive; it is unaware and rigidly accords with its formative rules, the natural order.

Who, on the other hand, has seen the face of chance? If Luck's a ghost she's not an easy one to lay. This unholy apparition only makes appearance when the cause of an event is missing from your mind. The logic is elusive and you can't tell how it comes about. Can you think, however, that such Ghost of Ignorance created cosmos? A miracle is simply a transforming mechanism you don't understand - and what transformed non-cosmos into cosmos is certainly not understood. But is a miracle the same as chance? Another mind might, though you don't see how, perform what seems a miracle to you. Knowledge and not luck's the issue; know-how transforms a miracle from apparition into information. *Did Queen Fortuna therefore have to bear the laws of nature and the purposive complexity of forms of life - or is mind the real 'ghost' in all the world's machine? Was Deliberate, far from resurrected, never dead except you thought him so? Is the Holy Ghost alive?*

Why, except in scientism's eyes, is Fortuna's face attractive? Is it beauty, harmony or style beguiles? Lady Luck's no naked empress, rather, nothing when she's stripped; nor smartly colour-coded but, when intellectually dolled up, untidy, dowdy and ill-fitting as a bout of accidents. Soft and hard; soft as evanescence, hard to understand. We know, by struggling this far, that she's not an easy girl to hold. And unpredictable. Flittingly uncertain. An unholy apparition. How can Capricia *decide*? Yet, as any politician let alone a mystic tells you, it's the quality of choice and value of determination that really, without number count. Decision marks Deliberate out but life's equations balance feelings too.

So, if they haven't killed each other, which does science guillotine? Your vote is, personally, crucial; decision is your top priority. Did random scatter or a self-consistent system make things up? Did a pure mathematician, Lord Logico-Deliberate, elegantly frame the universe with his equations? Was it in balance or imbalance? Was it rationality beyond compare or giddy lack of it that spawned you? From singularity to singularity did Lady Luck bear subatomic particles and through their happy union (although the 'pregnancy' was very long) carelessly conceive a second singularity, a tiny cell? From which vast chemical complexity somehow evolved the rest of life? Of course a cell, like clock or any other vehicle of intent, accords with natural law. Where natural law is comprehensively described by mathematics can that practice represent a large-scale case of chance? Could 'necessity of law' by chance have set life up? **Or is the Lady just a powerless cipher filling in the gaps of ignorance, a goddess of the scientific gaps?** According to the Dialectic she is just an apparition swelling up as resolution fades, a phantom and a blur. What more mindless than haphazard? She's an absence of all substance, impotent and empty of intelligence. *Yet officials of materialistic faith would elevate a faithless, faceless slip of unpredictability; they would execute the Emperor Logic and in place enthrone Caprice, the less than naked Empress of Illogicality.* You could say a revolution in the way of thinking swapped thought's laurels. You could swear shots in the dark assassinated Lord Deliberate and with him any Theory of Intelligence. Bannered anarchy has marched to town, revolutionary ghosts of anti-reason seem to have usurped substantial reason's crown. The world's turned upside down. Long live Queen Luck! Down with divinity! Chance, by evolving life, seems to have become materialism's anti-holy ghost.

The Chance-Killer

	tam/ raj	*Sat*
	below	*Top*
	hierarchy	*First Principle*
	ranks	*Commander*
	variation	*Invariance*
	range of possibilities	*Certainty*
	down the list	*Highest Priority*
↓	*tam*	*raj* ↑
	low (level/ priority)	*high (level/ priority)*
	passive	*active*
	informed	*informant*
	outward appearance	*inner rule*
	trivial details	*important principle*
	apparent disorder	*actual order*
	jarring/ out-of-tune	*harmonious*
	graceless	*elegant*
	tangle	*tango*

Do you think that what you're reading's repetitious? Did you know it all before? Why then, if the penny's dropped, are you still frozen in a fashionable, humanistic pose and why does science still tout Darwin's theory as a Truth?

It must be time to turn the tables. Time to resurrect a hero from the dead. From law to luck and now the spotlight swings again onto the very soul of reason, Lord Deliberate. Doesn't his comportment ring a bell? You met him (Chapter 5) under an alias, another ID known as Intelligent Design. Intelligent awareness is an understanding full of plans and purposes. Deliberate never has a flutter for the Lady Luck nor she, fine thing, on him. The odd couple are strange bedfellows - if they co-exist at all. Each programmer de-bugs, every engineer deliberately suppresses any chance of chance. Is it any mystery?

Purpose kills chance; Lord Deliberate is a lady-killer and the lady's Luck.

Luck shafted? Ghost busted? Her ladyship struck down? Deliberate does not execute with random strikes but strikes precise on randomness. Sharp the wit that, at its sharpest, with the first stroke cuts a swathe through error. No mistake. Full marks. And when its deed is done the die is cast. Deliberate's culture's one of certainty, of luck's elimination. So if her ghost appears then, gripping Lady Luck, you grip thin air - nothing but (as ghosts all are) the unexpected. Who'd rely, instead of schemes, on luck for what he wanted? The girl's a charmer; she's a dream as mindless as Deliberate is not.

You best define a man by his priorities. What is Lord Deliberate's priority? What is most important? Don't priorities define the order in which *top-down* thinking works?

It's the central point of education. Correct logic fosters intelligent thought. Grammar and syntax precede the expression, in words, of that thought. Purpose, rules and law precede intelligent behaviour. Such behaviour predisposes matter according to its plans. Intelligence wants to organise the things around it. It creates patterns that accord with the character of its desires. Perhaps non-chance is not coincidental! Our lives are clear examples. For instance, each man more or less perfectly understands what's orderly and what is not. As each person's relative sense of order is a function of his sense of truth, so intelligence is a function of order. It is the relative ability to grasp or employ root principles, that is, the priorities behind the details of any particular case. What then, *par excellence*, about the case of live embodiment? Why flinch white-faced from a Theory of Intelligence?

A principle is a condensed, potent form of information that pre-exists individual, different expressions of itself. Either psychologically or physically, it gives rise to various complexities of practical detail and difference. Take care of the pounds and the pennies will take care of themselves; take care of a principle and its details will take care of themselves. **It is from the implicit simplicity of principle that explicit complexity unfolds**. Thus detail derives from rule; superficial randomness overlays an underlying order. You can bank on it.

What do you think is most important? Where in the last analysis does your heart lie? Are you right? Is there a prime priority in life? Shall we gravitate in its direction?

If a principle represents the order of some particular characteristic of behaviour then let's abbreviate and simply say 'a principle orders'. Thus from a straightforward analysis it might appear that principles are a strength from which proceed orderly practices; they are an invisible coherence, an axis around which details fall, as if by magnetism, into place. Whether it's overtly purposeful or not a principle appears like an attractor, pulling power or a determined act of will. Failure to grasp its motive power leaves confusion, weakness and decay.

Matter, at the bottom of creation's slope, is mindless and therefore absolutely passive, wholly ruled. Mind rises by degrees towards greater freedom and autonomy until at Apex its First Principle (or, if you like, its Principal), the *Logos*, is most powerful and free. Materialism may abhor such metaphysic but strives nonetheless to concentrate its subject matter into logic's distillate, an analogue of *Logos*, a principle of principles that physics calls a *TOE*. Science understands, as does the psychology of saints, the powerful relationship between principle and dependent practice. The less purposive a project the more confused, dysfunctional or non-functional is its outcome. Lack of purpose makes for lack of reason; lack of reasonable certainty means lack of teleological design or, simply, goal. Senselessness. And lack of sense descends into a dissolute, unfocused sort of character. Less orderly, less meaningful - is this the drift of life without a hope or goal? Is unprincipled or unenthusiastic outlook really such a drag?

unexpected	*anticipated*
irrational/ disorderly	*rational/ orderly*
unpredictable	*reliable*
hopeless	*hopeful*
low probability	*high probability*
chance	*systematic coherence*
lack of guidance	*guidance*
accident	*design*

Unprincipled disorder is a close relation of dim, scattered consciousness and, like a crazy dream, lacks the coherent sense of wit. In fact, it is a rule that in proportion to the dimness of a mind 'pre-organised intelligence', called witless instinct, compensates. *, On the other hand, the higher an intelligence and more informed a mind the more effectively its plans are implemented. Such intelligence works according to principles, the deeper and more primary the better. The keener its interest the more rapidly, systematically and harmoniously informed is the complex detail of its creations.* **Intelligence is certainly a first-class chance-assassin.**

An ally of intelligence is expertise - and wherefore expertise if not to gain an end? Expertise is information born of principles combined with practice. Is not information, in principle and detail, what you need before you rationally act? An artist comes before his kit. *Is not information, before even energy and matter, the foundation of all science and technology?* Information is the life-blood. Not only coursing through our minds and brains, it also dominates data processing, control engineering (cybernetics), biological communication systems and cellular information processes. Code, signalling, language, messages, communication and information are everywhere. What are laws but messages that tell you how to act? What is 'natural law' except, through the agency of fields of influence and their various messengers, such information? Indeed, as the nineteenth century technological revolution replaced manual labour with energy conversion machines, so the twentieth century analogue replaced mental labour with data processing machines. We inhabit an Information Revolution! How strange, therefore, how whimsical and contradictory to guillotine The Lord Deliberate and, in defiance, elevate Caprice to Madame President! As, by surrendering the origin of physics' laws to chance and life to accidental evolution, science does.

Yet from experiment researchers want to 'get the idea'; unless, as Darwinian biologists, they want to fit results to their idea. We congregate around idea because we think there is one. Ideas amount to reasons. We seek reasons, subject collected data to mathematical and conceptual interrogation and find some order. Is ever randomness a reason? Only when disorder or destruction need, by senselessness, to be explained. Mind, information and coherent order are, on the other hand, allied. It is the constructive, anti-entropy of mind that uses information to make sense of or, according to its individual perspective, rearrange the world around it. *A mind is a non-physical focal point that creates order round itself.* Its central principle is order; it is formative, that is, information's principal. Informant and informed, does human mind not understand the natural order because it is both part of and reflects it? Not randomness but order. And, personally, to tidy up what it construes as out of order - suffering, anarchy and 'incorrectness'.

His seat of power is information; this instrument of order is the close domain of Lord Deliberate - but not Caprice. Smart, elegant, distinctive and decisive in the way he works - irrationality would drive him mad. Smooth operators do not deal in unexpected friction ; how therefore could the dim girl ever rub along with him? How could she ever make an equal mate? Instead, like oil and water, chalk and cheese, the two live daggers drawn. Just murder him and, reason-killer, you can resurrect her. Crown Lady Luck and marry her with Muck. Let the plebeian partnership provide a lucky muck-let from the mud and, once Little Accident is born, reverse the reel of death. Let dust from dust accumulate, from sterile desert generate perchance a line of Muck 'n' Lucks, from earth to complicated earth let spring the tree of life - molecules that 'come alive' within a very special sort of grave. A capsule called a cell. Thus spake the retro-prophet, thus speaks genius of a contradictory kind, replacing mind with none and striking wit completely out. No wit, no information and a Theory of No Intelligence.

Information, on the other hand, is a chance-killer, first degree, whose royal pedigree's of Lord Deliberate's line. Some organisms actively inform environment. All imprint it with the orders of their type. Often patterns they impose originate from that repository of automated action, instinct. Instinct is, like memory, a passive phase of mind. And biochemistry is based on passive stores of coded information. Inasmuch as forms of life are pre-set and passive (such as trees or flowers or fungi) they are reactive rather than proactive; circumstance is dominant. But either way, active or passive, information slaughters imps of hazard and its general, strategic enemy, dark forces of disorder.

However, passive information flows originally from active. It is mind cast in steel, coded in chemical or set in stone. It may be physical in operation but is metaphysical in origin. It may seem to science that such origin is immaterial and therefore inconsiderable. You presume, therefore, that nature's buck stops at the passive level of a written code - with chance both scrambler and unscrambler (without knowing it) of chemicals-cum-symbols used to concoct its pointless script. Such a view severely underestimates its enemy, intelligence. Sense generation is the murderer of chance. The order that a flexibly intelligent person imposes on his circumstances starts with information built by thought. It tends towards manipulating nature, dominating situations and general rearrangement *à la carte*. **The origin of order is critical.** What *is* the origin of cosmic order?

Of natural law? What latent logic precedes and thereby informs space-time? <u>We need to know because, in truth, we subscribe to order and not chance. Reason and rationality are based on it.</u> And reason always guns for lack of it. We shoot up lack of cause and seek out patterns actively. The more these harmonise with 'natural law' (also called *dharma*) the nearer it is felt that we align with truth. Indeed, we qualify our information into grades according to alignment and its 'density' of truth. The 'denser' a truth, the greater its bracket and the broader is coverage by its 'umbrella' principle. Therefore once High Truth or Principle has been identified it should, both theoretically and otherwise, become a top priority to work with and express throughout our lives. This, as science knows full well, is the height of reason. Yet its own research is bound to peak within a solely physical arena - one by definition out of its own mind.

What about the psychological arena? *Dharma, karma* and commandment - laws of mind. Did Lord Deliberate and Plato meet? And spark the obvious question? What is the greatest good, the ultimate truth? What do humans really want? To rise higher and become Most High? Highness? Your Highness? What on earth is that?

	tam/ raj	*Sat*
	existence	Essence
	relative reality/ appearance	Reality
	scale of truth	Truth
	lesser positives	The Good
	levels of constraint	Freedom
	lower ranges	Most High
	lesser consciousness	Pure Consciousness
	subsequent agenda	Top Priority
↓	*tam*	*raj* ↑
	negative	positive
	non-conscious matter	conscious mind
	objectivity	subjectivity
	physical facts	psychological wisdom
	external appearance	internal reality
	detailed analysis	principled summary
	differentiation	unification
	classification	cohesive reason
	increasing ignorance	inc. understanding
	from truth	towards truth
	more mistaken	less mistaken
	by accident	on purpose
	pain	relief
	error	rectitude

Psychological alignment means agreement and, in the case of moral principle, obedience, submission or a voluntary concurrence with another's will. In what, therefore, resides the *summum bonum* of communion, the alignment that is perfect bliss? Surely it is not, to cap that bliss, precise reflection of the rationale behind all things? Is wisdom not another name for grasp of natural reason? And if First Motion fired up this rationale then what preceded Motion?

Was it latent order or an offbeat chance? Is Most High a fluke or not? Why did Priority preceding anything usher in a cosmos and not chaos? What is the inner, private nature of Deliberate's pregnant kind of Nothing? Surely not unconsummated Lady Luck?

Materialism's consummation of the truth is gathered in a *TOE*. To have gathered all the threads of scientific corpus into one, single strand is omniscience of a partial kind. What kind? A theorist's negentropic 'order-loader', mind, is functional with brain; brain, in self-development according to genetic plan, scooped minuscule resources from the gigantic pot of space, time and their reactive, bubbling brew. Would 'omniscience' therefore stem from freaks of evolution and so be knowledge born of nescient matter somehow atomically arranged in, as we call them, nerves? Would it be, from chance on chance, omniscience born of unrepeatability? Which exercise of unrepeatability has passed the baton and relayed the course of accidents into the hands of one that, strangely, will eliminate its random generator with a chance-eliminator, mind. An end of parentage by evolution's accidents. Parricide; or matricide if mother's Luck. Yet happy matricide if mindful science can now force the pace and, in a spurt of progress, deliberately evoke new forms of life! Gene technology! Men can already biotechnologically arrogate (or is it surrogate?) Capricia's womb. What greater power, on top of *TOE*'s omniscience, than to create a child of scientific parentage? Will Promethean scientists, having passed where angels fear to tread, assume a personal responsibility for their objective Frankensteins? Having slaked vast curiosity will professional professors ever see their biosyntheses as family? When ethical confusion rules the day what more is there to learn? Perhaps what the mystic always indicated that there has to be - a measure of humility.

For holistic Deliberate the *top-down* pattern of Natural Dialectic indicates that the *TOP* Priority is Pure Consciousness, the unitary origin of duality; it is Essential Potential; and such Potential, at the upper end of cosmic structure, is conscious. What is conscious is imbued with will. Its intent is voluntary; it engages purpose. **Purpose by its nature always kills chance dead**. Like a human, the structure of the universe would therefore be teleological - in the image and intention, that is, mind-set of its maker. For a teleologist, such as an engineer or other pursuant of purposes, there is immediately, right from the start as little left to chance as possible. There is order, as complete as possible, both in the precondition and initial condition of his creations. Ask Sir Nigel Gresley if this was not the case for his steam locomotives or Bill Gates for Windows. From Nuclear First Principle issues mind; from The Informative Egg issue nested layers that exteriorise outwards through mind to a physical shell. This issue involves, through its intrinsic principles, a cosmic order of priority. It represents a ramification that starts with voluntary order - because flexible mind has motivation. Voluntary motivation is another way of saying 'purpose' so that, if matter succeeds mind, it does so according to the latter's purposeful designs. Mind plans; it may or may not store its plans in memory but materials are arranged accordingly. *Materialism's grand delusion is that mind is brain; that information is, somehow, a material element. Ultimate confusion thus derives.* If, however, mind is cosmic then its cosmic teleology preceded and, hierarchically, precedes all traces of apparent randomness. This viewpoint therefore sees mind as the author of purpose everywhere and design intrinsic in everything. An act of creation, in itself a chaos-slayer, is 'cosmo-morphic'. It is not that a *top-down* view of cosmos is 'anthropomorphic' but that our own anthropic acts of creation are, since they employ the immaterial element, in their small way 'cosmo-morphic'. *The ultimate chance-buster is a mind.* **If there is universal mind then where, anywhere, is chance?** Of course, this pattern of perception is anathema to anyone who has nailed conscious subjectivity finally and irrevocably, by way of neurobiology, to the brain. Does he not crucify the heart of reason and the soul of life upon a cross of molecules? And in such labour labour under an illusion, as if he'd quenched the sun and killed the immaterial cosmos off?

Look to the night sky's sweep of stars. How great, sang psalmists, is the mind behind. If that's the way what's changed or ever will?

Well then, if everything's derived from purpose, is the purpose 'Good'? If Pure Consciousness is liberation from existence is that 'Best'? And Free Will? What is the character (as is discussed in Chapter 26) of unconditioned, infinite will of which your choices are a much-conditioned part?

Or, looking on the contrary from the base, material end of scale, how could choice have evolved from choicelessness, from the oblivion of matter? Because, be sure, mind's evolution out of matter, if it ever happened, is a major black box mystery. Why can you, unlike the matter that your body's made of, choose? Make choices and decisions when surely what is voluntary is not by reflex or by chance? Is your freedom real or an illusion? It is, for sure, incomplete. It is conditioned by the context of your memories, brought down to earth and localised by body and environment. But is it wholly predetermined, as utterly restricted as matter's choiceless interactions? If so, voluntary behaviour is predetermined. It is involuntary and 'free will' a travesty as tightly bound as lack of it. There is no difference in this case between the voluntary action of a person and the reflex of a lifeless object. Puppetry. Morality evaporates. The 'Good' is gone. And without purpose, which is nothing if not voluntary, chance rules. With a passive, wind-swept sort of rule. Caprice is pushed, you might say forced, obliviously through the rules of space. She's swirled around the automatic tunes of time. How can a thoughtless puppet realise, any more than atoms it is made of, the game of Lady Luck? And, if I'm a puppet, whence springs the illusion I am not? Whence materialise those apparitions called decision, principle and purpose? How can a puppet specify ambitions and, in so doing, weave its personal fates and destinies? How can it reason or desire to understand Platonic Goodness? Stripped to its naked logic what confusion does materialism's theory of evolution propagate; what gross misinformation does its maculate misconception, a Theory of No Intelligence, purvey!

Puppets, Mummery and Drama

	tam/ raj	*Sat*	
	scenarios	*The One*	
	outworking	*Whole Idea*	
	work	*Scriptwriter*	
↓	*tam*	*raj*	↑
	staging	*cast*	

backdrop/ scenery	drama
props	purposes
inanimate part	animate part
objective circumstance	subjective role
consequences	reasons
motions	emotions
non-self	self
body/ puppet	mind

I am not only cardboard or cut wood. My body is a coloured puppet but mind, its master, pulls the strings. Unlike atoms of my body I am conscious; and consciousness is Lord Deliberate's inmost nature. I am, in wanting, willing and in knowing actually a junior, immature version of Deliberate, a child of His Maturity, the unripe fruit of My Maturity - which is, at the top of the league, Pure Consciousness. In India they call this state *Sat-Chit-Ananda* - Truth, Consciousness and Bliss.

And lower down the league? Mind for certain drafts its own adventures. Fate is what the body suffers, destiny is where the mind projects it next. Lifeless matter is without a motive, lively mind with motivations weaves its future destinies. These, in the dynamic field of life, become its ever-changing present circumstances (or fates). Each reaction to a fate rewrites the script, recasts the plot, re-configures the array. Kaleidoscopic shifts rephrase the colours of relationship, every move in life is like one in a game of chess. Each adjustment ripples through a sea of possibilities. According to the strength of its intent each destiny is realised or dies. So it is you, not Lady Luck, that leads the chase. Do pilots look for accidents or damn them? You, chance-killer, steer your ship; one of its instruments, the helm, is brain. On what basis do you estimate a profitable port of call?

	tam/ raj	Sat
	operation	Max. Order-in-Principle
	relative uncertainties	Certainty
↓	tam	raj ↑
	asymmetries	symmetries
	stage detail	stage infrastructure
	shape-in-practice	shape-in-principle
	inertial equilibrium/ slack	dynamic equilibrium
	fallen over/ flat	vibrant/ balancing
	superficial disorder	internal order
	impotent	potent
	fixity/ ground-state	flux
	discontinuity	continuity
	finished	in process
	mass	energy
	resistance	stimulus
	container	pattern-maker
	gravity	levity

You are not made of stone. Nor any more than stone, you feel, a puppet or a robot - but if you were then under whose command? And if, by dint of Puppet Mastery, the story's all in order what about disorder? If it's all 'right' whatever could be 'wrong'? What about the unexpected, the exception, imperfection, ugliness or error? And if you're stuck between the rails where's the adventure? If there's only rigid government where is the choice? In the wholly trammelled zone of physics, where there isn't any choice, isn't chance the function of a pattern of events that's unexpected, that is, unexpected for observers? Is not 'disorder' simply born of impacts caught along the very complex, natural run of things? Even when there is a choice perception of perfection never stays; it is a mutable ideal to whose fleeting moment we are slaves. Imperfect information seems to be the problem; that would make 'right thinking' turn you towards the cure. Aren't your misjudgements just a function of the lack of it, of ignorance in spite of facts already learnt? Surely mess, mistake and accident are each as real as the 'correctness' they are deviations from? Do not disorders, especially the sharp and painful ones, cry out for order, comfort and relief? What cries is not the stone, the puppet or his body but the information centre, feeling and creative mind. 'Hades' means 'unseeing'; whether puppets dwell in Hades they dwell outside sight. Mind, on the other hand, is dark oblivion's match, a clear sky above; its weather may, however, also float a mist of partial sightlessness, a fog of ignorance the sun of understanding can evaporate. Here comes, dialectically, what kind of sun?

Mind might but stones can't choose. So how, if there's no free play, can natural purpose, targets or design occur? How could you prise deliberate design from mindlessness? How squeeze sense from astronomical, galactic mindlessness? Isn't the behaviour of material energy and mass as 'wooden' as a puppet or a stage? How can what runs down a line of iron rules ever be derailed? Disorder is no more, from nature's point of view, than a disorder of the mind, a trackless lack of principled perspective.

Action has to have a frame; drama has to have a stage. Puppet drama is enacted on a planetary stage. Of course, a stage *per se* has no intent; intent is its producer's. Stage is a facilitator, inanimate a backdrop that informs the active play. Active is dynamic. Scenes change but what, behind their differences, is the simple, changeless infrastructure of our cosmic stage? Is it basic particles and forces? Behind such simple, primary expression where reside the principles of stage design? Do they derive, *top-down*, from a First Principle? From facets of the Infinite whence, sparkling with polarity, all finite entities derive? Is this how physics is informed? Is light, reflection of the Light, how spectral shapes communicate their rules; does sound or some vibration govern their behaviour? Is electro-magnetism, for example, duality well-balanced, that is, polarity compacted in a neutral principle? At all material levels information is entirely passive. It is energy informed, a carrier of code, a mute communicator of the part it's meant to play. It may sound or echo but matter, like a puppet, never really answers back.

Inspect the puppets' strings again. 'Force fields' communicate the certainties. Such transmission, push or pull, is oscillatory. Forces are part of the way that energetic puppet-patterns in the vacuum are related; and because a stress or force can neither push nor pull in nothing they are described in terms of something - an exchange of particles called 'messengers'. Stresses in space-time and interactions of such particles are aspects of the same event. Once such wave/ particle duality's accepted we can proceed to identify operations having a predominance of 'wave' aspect (free or kinetic energy) with quantum or 'principle' matter and those having a predominance of 'particle' aspect (mass or locked energy) with forms of bulk aggregate, that is, molecular and more massive structures. These are, familiarly, gas, liquid and solid formulations. The scene is never set but ever settling and resettling. At root this vast, dramatic stage is a set of changing tensions in a silent, lifeless, empty-seeming yet supporting field of space.

Apparent lack of cosmic order (the untidiness of galaxies, confusion of dust clouds or ferment of exploding stars) is superficial since, as we know, underneath appearance works reality composed of rules. Disorder is as much in a beholder's eye as beauty. Surely, though, stupendous beauty catalogued by *VLT* (Very Large Telescope soon to be joined by the Extra Large Telescope at Paranal Observatory in the Atacama Desert, Chile) and *HST* (the orbiting Hubble Space Telescope) is not simply a stage to serve such immaterial but absolutely valuable a rarity as life? Still less if life were just a puppet show.

Your body is not made of lignin but of other molecules; it is a puppet. Can a puppet see spectaculars? Can its eye appreciate the show? Neurologists are keen to think and say so. Don't scans show that when you sense, remember, think or feel or make decisions the brain, your brain, 'lights up' in different parts? Nerves are made of atoms. A puppet is no less a lifeless mass of atoms than a brain so how do brains make puppets see? Herein lies confusion when it comes to information; here rests conflation of mind's active part with matter's passive print of it. **A materialist's delusion is that excitations in grey matter don't just represent but actually *are* your thoughts**. Is the objective trace of someone's brain while dreaming just the same as their experience in the world of dreams? This illusion, serious and scientific as it is, confuses subjectivity with sensible events. It kills out consciousness. It squeezes life's experience into a mesh of electronic circuits made of nerves. **In fact, according to the Dialectic, brain's the wooden puppet not the thought**. Brain's the robot and the mind its mover. Scans show material images of thought. They illustrate the point of interface between the physical and metaphysical. They film an active cockpit where, hands-off, mind meets its vehicle's controls. Around a vehicle's cockpit is extended its whole body. What a vehicle I've been loaned in which to travel time! What a suit to dive into the world with! What a gift by which to touch, earth to earth, with the body of the universe! **Body is the puppet not the mind**. Brain is, like a language, at the coding interface - it relays information but does not create or know a thing itself. And nervous pulses are the symbols that translate experience - they are not experience itself. **Experience is in mind. Puppets don't have minds and thus the drama is not theirs**.

No Information from Confusion

Is that mirage identified? Is that confusion settled? Is there any more that we should know?

There is. Why labour lovely or unlovely Lady Luck like this? Why harry ghosts, why hunt this apparition (if it's real) around the seven cosmic seas? Because right-thinking is at stake. Everything that you believe in is suffused by her or by Deliberate and so, since scientism heavily promotes her immaterial ghost and simultaneously denies another immaterial factor, elemental information, we must draw blades and duel. Does the universe embody lack of reason? Are things mindlessly and arbitrarily 'designed' or not? One may, Deliberate enjoins, be lucky and throw off materialism's mental yoke. One more thrust then. Let's press on.

Do puppets really have a sense of order? What about the natural order of a mindless set? Within this order chance, confusion and disorder operate as pals, as anti-principles - but whence derives the universal Inside Information?

	tam/ raj	Sat
	informed/ informant	Information
	existential effects	First Cause
	hierarchical order	Max. Order-in-Principle
	falling short	Perfect
	more or less stained	Immaculate
↓	tam	raj ↑
	external appearance	internal reality
	seeming	actuality
	relative imperfection	accordance with principle
	external order	internal order
	random distribution	more predictable distribution
	tendency to lose symmetry	tendency to gain symmetry
	fall apart	communicate/ cohere
	diffusion	concentration
	entropy	negentropy
	differentiation	unification
	isolation	relation
	bulk structure	particle/ atom
	inert ground-state/ effect	dynamic cause
	disruption	smooth flow
	irregularity/ discord	harmony/ regularity
	ugliness/ deformity	beauty
	confusion	clarity

Confusion stems from lack of information. How can oblivion, lacking any information, ever make it? As lack of clarity, doubt or ignorance increase in mind then natural rules take over; instinct is the backup that is placed behind. With no mind no confusion can occur; entirely passive information rules. Oblivion, the state of matter, is entirely reflex and its bodies total puppets of preordination. Whence preordination? Whence the 'backup instinct' fixed behind its mindless operation?

The Dialectic always needs an absolute and, being 'higher' or 'above', this absolute must act as potential for the relativity that follows 'lower down'. Information, whether active in the form of thinking mind or passive in the mode of memory or natural, archetypal law, dictates the course of all events. Flexible the former and constructive; inflexible the latter and entirely artless. Just as an engine's action flows from drawing board so absolute for physic flows from metaphysic. Its plan derives from artful archetype. In other words, subjectively inflexible constructions - electromagnetism, electron, proton and so on - are governed 'from above'. By inner metaphysic outer physic is controlled; not conscious but unconscious metaphysic in the form of archetype. This step-down is reflected in the sensible dimension we call universe. Just as atomic shape defines a crystal so 'higher', inner energy defines its 'lower' outward form. Whatever seems confused or randomly distributed in outward form is never the creator of a code, a message or a scrap of information. There is no information from confusion; confusion simply clouds and misses higher, inner ordering. And information fits with purpose like a hand in glove. You might judge, therefore, that physics' subject matter is on purpose and that, logically, 'confusion' is no more than an illusion in the mind's eye of a man who judges nature, at top level, faultily.

At least Deliberate's game is a consistent one. From the logic of polarity springs a relationship between the inside, causal information of the world and its effects. In physics we call causal information 'laws' that govern local, individual patterns whence each type of inner energy 'personifies' itself. Inner principle shows up as behaviour that conforms. There is no confusion. While continuous fields of force inform, discontinuous particles and ponderous parts are kept in shape. They are consistently informed. *Information is prior, matter secondary.* Prior to particulate form are the waveforms, forces or energy that are locked *in* it, *as* it. Waves inform and trapped/ resisted waves shape their containers. Is not matter energy 'solidified'? It is the locked-up, separate or (*tam*) discontinuous aspect of energy that drags, impedes, isolates and, by means of boundaries and phase transitions, shows as different things and states of things. For example, in this sense the dominant (*raj*) kinetic or wave aspect of an electron acts as a principal of linkage and communication; and the dominant (*tam*) massive or particulate aspect of a proton is one of containment, resistance and individuality. Inner wave shapes outer crust. *Bulk matter is a cosmic shell.*

A couple of cosmic buttons govern behaviour and clearly define relationships. Yes, no: release or lock. Strong nuclear clench and electromagnetic radiation well express imprisonment and liberation. Cosmic stop and go. Polarity's expressed as binary charge as well. Like charges repel, unlike attract. Negative repulsion pushes, divides and isolates; it is the basis of (*tam*) resistance, immobility and otion apart. Positive attraction pulls together; it is the basis of (*raj*) unification. All chemists know that charge informs; it arranges primary behaviour that leads to secondary effects called 'sensible things' - the objects and events of what we call 'our world'. Binary information. Crystal clear and no confusion whatsoever there.

As well as 'yes' and 'no' there's 'on' and 'off'. If off, there's no change in a condition; and if on, some push or pull is varying circumstance. 'On' means inter-polar action of some kind but when flow is absent or as soon as discharge is complete you will find the balance of neutrality - no motion, reactivity or, at least, no influence to change a state. (*Sat*) neutrality of poise precedes and (*tam*) inertial neutrality succeeds each individual change. 'Off' is super-state potential that exists before particular stimulus/ action/ motion; and, on the other hand, sub-state exhaustion follows it. Things are 'off' until another stimulus arrives. In other words, the books are balanced until something further knocks them out of kilter. **What more dialectical than charge-in-action at the heart of**

chemistry; or electromagnetism driving and creating physical effect?

We understand the chains in time of consequential events but what about the underlying factors four: First Cause, levitatory (↑) negentropy, gravitational (↓) entropy and End Effect? In Creation could there be another kind of chain, one that is hierarchical - from top-to-bottom, up-and-down including both informative and energetic scales? As for order, where might anybody find confusion's minimum - a maximum of regularity? In government by principles? In symmetry or fluent light or in a single, as yet undiscovered force from which the other ones derive? Or is the cosmic master eyeless entropy and thus the apogee of orderly arrangement found locked up in crystals at Lord Kelvin's base degree, the cosmic cold periphery (0°K)? Let's see.

Symmetry is the epitome of regularity. Yet dialectical 'Symmetry' is not the same as physical. Metaphysical, it resides (*figs*. 7.2 to 7.4) within the Infinite and its First Conscious Issue, Cause or Form. Unconscious first cause *physical*, as you can see (*figs*. 7.2 to 7.4 and 12.1), is lower down the scale. Unseen through the 'glass ceiling' of physical perspective, its substance is vested in the aforesaid *archetypal memory of universal mind*. This agent represents the files for physical phenomena. It is the cache of symmetry, the instrument of vibratory law, the databank by which things are arranged and, in this sense of genesis, a universal code. It is called potential matter. Therefore maximum order-in-principle is metaphysical. Its first expressions constitute matter-in-principle. These are, as far as atoms and bulk form are involved, internal energies. They are the shapers of bulk form and therefore all that is eventually sensible. Light with photon represents (*sat*) neutrality (see also Chapters 0, 1, 2 and 3: positive and negative neutralities). An electron represents an influence of (*raj*) stimulation, connection and relativity; its 'levity' promotes connectivity and bonding. A proton, on the other (↓) hand, represents an influence of inertia, captivity and isolation derived from the strong nuclear contractive force; its (*tam*) 'gravity' is derived from mass and promotes bulk aggregation. Join a proton and electron to gain mass-centred neutralisation, this time one of 'sub-state' or 'subtendent' equilibrium - the point of balance called a neutron and, beyond a neutron, atom as a whole. In this sense an atom, since its construction well reflects the cosmic fundamentals, is the natural symbol of this universe. Is there confusion in this information? It seems in Dialectic order.

What, therefore, of order's anti-principle - disorder? If there's no confusion except that thinking makes it, what about decay? A tool wears out, fire dies and so do life forms. Things stop working. What about decrepitude and pain? Suffering of the innocent, slaughter of the lambs and real muddle in the minds of men? Does nature fail? Fall? Is there amoral, natural 'sin'? Is not order in disorder lost? Is this where Lord Deliberate got it wrong? Or is his 'wrongness' built into a scheme that's based on opposites? So that it's an intrinsic part of polar 'rightness' and the process of creation?

In this case amoral 'wrongness' just offends our sensibility; it is personal catastrophe but no more than an outcome of the natural order. Conversely and as Chapter 26 will illustrate, moral wrongness is a falling out with unifying principles of mind; it is a scaled precipitate dividing selfish passions from Immaculate Morality.

↓ *tam*	*raj* ↑
towards material outcome	towards principle
towards self/ individuality	towards generality
binding with body	binding with ideal
informative loss	informative gain
instinct for survival	meaning/ interest/ love
non-creativity	creativity
loss of mental definition	gain in mental definition
inc. gravitational strength	inc. mental focus
entropic tendency	negentropic tendency
energetic/ configurative loss	energetic/ configurative gain
diffusion	concentration
exhaustion	refreshment/ stimulation
gain solid definition	lose solid definition
materialisation	dissolution

Do you recognise (from Chapters 5 and 7) that we've dropped by this stack's set of scientific generalities before? Knock on a door. Its sign declares 'Department of Thermodynamics' and, pushed ajar, the opening reveals Victorian ghosts at work on formulating law. Call them up. Their names are Carnot, Mayer, Clausius, Boltzmann, Gibbs and Kelvin. Lord Kelvin's Second Law states that, in a closed system, things tend energetically to go one-way - downhill. Haven't Kelvin and his pals grown old and passed it? Aren't you following down? Reams, even libraries, have been written to describe this universal tendency. Things tend, in dialectical terms, to lose potential, dynamism and, by wear and tear, complexities of order. They tend, that is, towards conditions of inertial equilibrium, minimal strain or (*tam*) exhaustion. Such 'disordered' outcome is, Boltzmann claimed, highly probable. He assigned its 'state of natural confusion' value that was 'high in entropy' (see Glossary).

Of course, such tendency can be resisted. If not, what could stick around? Solidity, for example, buffers such dispersal and disintegration. It fixes fluid and thus, for a time, sometimes a long time, keeps pieces of the world in place. Nevertheless when energy is 'loosed' and overcomes 'an activation value' (symbolised by chemists as *Ea*) what it affects reacts; and the reaction tends towards completion, towards the end or, if you like, decay. Take fire, take heat, which from a human angle is both beneficial and destructive. Good or bad. De-personalise it. Subjectively de-value; objectify; enumerate. Retrieve the objective, thermodynamic view. Fires fade, powers wane, exhaustion overtakes all labour. Things cool, exhaust and drop. They fall into the inertial equilibrium of impotence, into ground-state and an ash of disarray. 'The woods decay,' wrote Lord Alfred Tennyson, 'the woods decay and fall, the vapours weep their burthen to the ground, man comes and tills the land and lies beneath, and after many a summer dies the swan.' Things fall apart. Original order is, like motion, 'apt to decay'. *Physicalia* tend to run, dialectically, down the stack and into 'left-hand negatives'. This way, biologically, illness swings into ascendancy, falling down seems on the up and age is swept to term in death. This is depressing news; it breaks you up and gets you down. It is confusing. What was in Deliberate's mind? What is going on?

There is nothing unnatural about a gradient. Between all polar opposites there exist gradients of information or of energy; these two-way, dialectical gradients are vectors of (*tam*) discharge and (*raj*) charge. The cosmic poles are (*Sat*) Potential and (*tam*) impotence, between which coupled motions of (↓) gravity and (↑) levity occur. These are the directions of materialisation and dematerialisation running between poles of total materialisation (the physical universe) and total dematerialisation (Essential Potential). A (↓) discharge from fullness to exhaustion is the path followed by the conscio-material gradient of creation itself. Check *figs*. 1.3, 1.4 and 2.11-13. It looks as though a fundamental constant (*tam* ↓) might be the broader agent of what Boltzmann restrictively, materially defined as entropy. The drop towards exhaustion, complete diffusion or 'disorder' in a system is, however, reflected information-wise as tendency towards decreasing complexity and output of increasingly random appearance. Things lose strength, stop working and fall apart. Such entropy is a measure of diffusion and the end-state of a macroscopic process. Is such wastage out of order? For Natural Dialectic law (expression of the cosmic fundamentals) does not change. Law is super-state potential governing the way things change. Reactions flow, at least where active mind is not concerned, from stimulus to its exhaustion, from reaction into sub-state completion or cause to its reflex effect. No confusion there. Is anything, you start to ask, disordered when all states are lawful and in order? Is not entropy's 'confusion' simply an inexorable expression of the materialising vector, one that leads from Infinite Potential down through material potential to those most final aspects of them all - fixity and base conclusion?

If things decay original perfection is implied. What is the nature of a perfect start - a highly ordered and as such unusual 'inflation'? Was order set before or at the moment space (with nothing on the 'outside' but, perhaps, with something on its 'inside') blew smoothly up and out and into a universe of things? Did developmental order spring from chaos or was cosmos preordained? Nor at this moment is it useful to observe that laws of physics break down (or, looking forward, haven't yet appeared) just as, at the point of cosmic singularity, big starts 'banging' soundlessly. On the contrary, they might have been intrinsic and then first expressed upon the cosmic gong. In this case (*sat*) pre-physical, potential matter and matter-in-principle, whose (*raj*) vector counters entropy, file first into the playground. Last and, in this sense, least trails matter-in-practice, bulk aggregates whose distributions are calculated in the entropic terms of (*tam*) randomness, disorder and decay. Such condensations include, of course, all you can see and much you can't. They range from galaxies to mountains, human forms to bacteria, from specks of dust and grains of sand to elephants and planetary systems. Each sort of item sports the same 'internal' elements but varies in the details of 'external' shape and distribution.

What kind of universal egg could 'perfect order' therefore issue from? Is the reality that materialism's 'order' is evolved from chaos or that apparent chaos may (as in the case of snowflake crystals) be devolved from underlying order? Is it true, in other words, that - dialectically - preordination means that any variant 'chaos' caused by (*tam* ↓) entropic loss devolves from order? All laws and principles pack patterns; they inform behaviour. Does the scent of information have to track back up through atoms and their subatomic particles to archetype? What is the origin of coded order? Code is physical and chemical as well as, obviously, biological a part of what informs all forms. As sounds inform all speech so agents (forces, particles and nucleotides) inform all forms. These letters, using polar grammar, link to make up words and simple sentences; or, beyond the fizz and whistles of inanimate chemistry, elaborate with complex essays into biological intentions. Immaterial, causal information precedes; its projection from transcendence patterns all material effect. Is the code for cosmic language stored, prior to expression, in the metaphysical dimension's databank, a faculty called archetype? The 'instinct' for such orderly expression would, therefore, be lodged in mind.

To help resolve confusion one can further note (see Chapters 5 and 7) at least three kinds of entropy. *Two, energetic (thermal)* and *configurative, illustrate the tendency of a material circumstance to change in the direction of exhaustion and disintegration, that is, with its improbabilities unspecified except by the 'necessity' of natural law. This is purposeless and relative disorder*. Disorder here has different meanings. In an energetic (thermodynamic) sense the most probable end-condition for a system or process in flux is complete diffusion, an even spread, randomisation of its parts and end of flux; for a fixed system it is one of maximum stability due to the strongest forms of atomic bonding and least residual tension. Such ground-state probability is called 'negative' or 'inertial equilibrium'. No energy is left free for work. In the configurative sense it means that the more complex a structure the more improbable and unusual it is; and the more likely to be broken down. Complexity tends to disintegrate. Its 'unusualness' will tend to lose uniqueness, detail and distinct individuality. When simpler, repetitious or more general forms prevail we say that probability, predictability and mindless natural outcomes dominate. In short, the more complex, coherent and therefore 'abnormal' a body the greater is the probability of disarrangement of its building blocks or rapid transformation back to uniform 'normality'. By token of complexity aren't you and I and all that lives great non-conformists, natural abnormalities? Towers of un-naturality! If so, how do our bodies differ from complexities of river, cloud or hill? The answer is, of course, by fine and specified complexity; that is, by conceptual systems that, vested in the balance of survival, are replete with goals. Incarnate information doesn't sound like emanation, though it plays there, from a zone of entropy.

Beneath diversity work different sorts of uniformity. Top of the material scale, precursor of all bulk extent, are ranged the fundamental uniformities of matter-in-principle (forces, atoms, elementary particles). Unless you think of particles and forces as communicators spelling out a kind of cosmic interaction they cavort without a meaning, purposeless and, as materialism requires, mindless. Universal mind is off the syllabus; but chemistry and physics trace how nature's letters self-assemble into myriad constructions that are called, generically, molecules; and next, from these compounds, the aggregates of sensible bulk matter organise. Similar individuals like sand grains on a beach, salt crystals, twins or clones from a factory production line are also uniformities. Finally, ashes to ashes, dust to dust, bodies die and are returned to clay. Cities fall. They decay to ruins; you can hardly tell them from the fields and hills. This is homogeneity, a product of diffusion; its inertial uniformity is lowest on the list.

It needs be crystal clear. Configuration by arrangement may occur without a purpose, that is, by constraint of natural law. This kind of information has, in Shannon's sense (see Chapter 5), value but no meaning. Purposive configuration, on the other hand, is actively informed by mind. Its specificity has reason, functional intent and thereby meaning. **Therefore, unlike Shannon-information, order is a word that needs resolving sharply into two - unspecified and specified.**

Mindlessness is not the case with code, machine or purposive construction. These derive from mind. Their physical expression (in the form of passive information - tablet, paper, c-drive and so on) is subject to configurative entropy; but their purpose, sense or meaning is subject to *the third kind of entropy - entropy of information* (see especially Chapter 23: Entropy of Information). Such entropy disturbs *intended or specific order* as it garbles message to the point it doesn't come across. Coherent operation ceases, there's breakdown of communication. A grammatical construction's useful life is over; the structure doesn't work because its meaning's lost. Entropy of information, one might add, is not only due to physical distortion, interference or destruction of a missive. Active information is disrupted too. Clear thinking is impaired by ignorance, emotion, falsehood, error and forgetfulness. Such clouds darken and their fogs confuse.

You can call it 'big bang' or 'transcendent projection'. Either way science, in the form of basic law, postulates initial input of both energetic and configurative arrangements that are 'complete and perfect' and from which the 'world machine' decays. The rules stay 'perfect'; their information's fixed but what's informed (the world of energy) is always changing and 'decays'. It diffuses, cools and 'clumps' - which 'imperfections' all involve loss of free energy. (*Tam* ↓) gravity and entropy enslave. They petrify. They lock energetic freedom into bonded forms or waste it to surrounding space. Our cosmos is some fresh beginning's old precipitate. In the 'flattening' of vibrancy things localise, crystallise and thereby differentiate. Variations on theme come to take up space. Energy is individualised but 'in the wild' material forms and processes do not coherently self-organise; they do not systematically self-maintain. They never seek dynamic equilibrium; they always seek inertial, ground-state equilibrium. By contrast, a house is specified by architect and a machine by engineer. Both accord with natural law *and* an intelligent design. And life's unnatural, improbable complexity's informed (or, if you like, is specified) by an ingenious, succinct and universal bio-code. Inherent in its operation, apt to balance, is dynamic equilibrium; homeostasis is the name each kind of coded regulation bundled into interactive process means. In Chapter 19 we shall emphasize how homeostasis is the key; it is central to the maintenance of vibrant, coded life.

	tam/ raj	*sat*
	matter/ energy	*potential matter*
	process	*super-state*
	physical agencies	*causal archetype*
↓	*tam*	*raj* ↑
	matter	*energy*
	informed	*informant*
	bound	*free*
	flat/ particulate	*cycling/ vibrant*

precipitate	*subtle governor*
external disorder	*internal order*
bulk aggregate complexity	*(sub-)atomic simplicity*
sub-state exhaustion	*force-field/ action*
lower energy/ entropy	*stimulus/ raise energy*
classical effects	*quantum effects*

In order to resolve confusion magnify the difference between conflicting parts - parts physical and metaphysical. Order is the issue. There is order in the form of symmetry; and order in the form of repetitious uniformity (as in the operation of a natural law, spinning pulsar or constitution of a crystal). Will you find perfect order in an archetype or crystalline precipitate; is the source of order in potential (that is, metaphysic of a principle) or physic of material fixity? If you miss out (or take for granted) entropy of information then you identify the greatest order in a crystal. If you don't distinguish cause of order from effect then you drop upon The Third Law of Thermodynamics. This law states that maximum energetic entropy and maximum formal order are revealed in a perfectly crystalline substance. Greatest slack, ground-state disorder is paradoxically combined with greatest configurative order at Kelvin's ultra-frozen base degree (0°K). Such macroscopic and 'external' order of a crystal, visible to see, best reflects 'internal', microscopic order of its molecules. Informant, in this special case uncontaminated by interfering side-effects, is expressed by the informed. Thus maximum physical order is construed as a function of (*tam*) sub-state exhaustion as opposed to (*sat*) super-state potential - archetypal cause. Such confusion's understandable. True, the order of a crystalline array reflects in every separate molecule the order of this cause but is, in fact, a mirror. As above so below, it is a perfect illustration of effect. It shows how principle without disruption is devolved to practice, this time not metaphysical but fixed to see. Crystals sparkle with 'perfection'; they wink with allusion to their higher truth, an archetypal order that encodes the polar working of its agents in this world. Thus you would not ascribe configuration of a perfect crystal to 'negentropy' nor conclude a 'perfect start' to cosmos must have been a solid one - say, order of some cosmic crystal or a bulk projection from 'elsewhere'. Yet, super-state potential to the sub-state, individual product of its possibilities, could some polar (say, ionic) organising property have been the way that life, like crystals are precipitated, first evolved?

After all, variation is a key component of both physical and biological arrays. Isn't what's a crystal's trick evolution's staple too? It's worth a moment more to see how, out of principle, falls practice. Let fall, therefore, an exercise in differentiation! Let it snow! Snowflakes and larger scales of bulk material show, due to local circumstances, variations on the theme of their internal order. And, as opposed to 'internal' molecular construction, the 'external' disposal of materials in unintentional complexities like rivers, rock formations or continental outlines is also a product of the convergence of multiple, local factors (such as friction, collision or heat). That is to say, unless you perceive its internal rules, configuration and history a particular distribution or shape of bulk matter appears contingent and disorderly. Basic generalities devolve particular, local details. Internal order gives rise to, apparently, external lack or lessening of it. Necessity is prone to multi-factored chance. In respect of their distribution (as opposed to their atomic order) the entropy of macroscopic objects and events is, paradoxically in contrast to the Third Law's specific derivation, high. In other words, although internal conformity is high external factors interact to precipitate different appearances as energy is lost. This process is the one that Natural Dialectic, by the very nature of its top-down sequence, traces through each stack - principle to practice, possibility to final outcome, individual variation on a theme. The reduction of principle to circumstantial facts causes variation on a theme and, although outward appearances may differ, internal organising order is not lost. Variation-on-theme is the rule but an 'unlimited plasticity' is not. Are not snowflakes therefore anti-evolutionary? Is such an exercise in differentiation any different from the variation that you find (and Darwin found) around the conservation of a biological type, that is, around the 'crystal' information that encapsulates a certain form of life?

It's confusing and confusion's Shannon's. He did not resolve the difference between specified and unspecified complexities of digits. His idea, as Chapter 5 explains, treats coded and uncoded information just the same. Intelligent and mindless order is, for Shannon-information, valued just the same. A crystal is a repetitious body; geological formations are comparatively irregular. Natural systems carry information of a passive, stable, formulaic kind. But neither uniform arrays nor asymmetric aggregates of digits (atoms) can develop code, function or the systematic, integrated biochemistry of any form of life. Thus, mind-locked in a naturalistic train of thought, science puzzles how improbable complexities (such as forms of life) could naturally self-organise according to the probable, predictable, deterministic algorithms of 'necessity' (or natural law). *How can mindless, automatic motions 'select' function, 'foresee' mechanism or 'anticipate' quite another sort of natural necessity, the satisfaction of specific needs of life?*

What specified the 'legal' patterns of our cosmos, generating probabilities of various but unintelligent complexity, is one question. No doubt that matter-in-principle (elementary particles that make the cosmic alphabet) remain the same but bulky condensations - for example, mountains, clouds and galaxies - show myriad, continual variation on the theme of their internal molecular structures. Their outward form is a secondary expression of primary internal geometry. The agents of this geometry are atoms and, therefore, sub-atomic particles which (*figs.* 7.9 and 8.2) can be vectored, polarised and set in hierarchy. What shaped the adamantines, proton and electron? Perhaps in grade the highest, underlying or initial issue's one of information just as much as energy. It's one of archetypal mould whose 'perfect order' shapes the different ways that energy is seen - including symmetries and perfect order of a crystal.

Systematic Clarity

What, with foresight, specifies the order of a code, a functional mechanism or a highly integrated system is quite another question. You can't derive a very high, improbable but specified complexity without, to plan ahead, a specifier. You need, far beyond the mindless, purposeless persuasion of wind, water and some chemicals, immaterial intelligence. Life's not confused; it's systematically clear, intimately integrated and, in codified construction, lacking any trace of luck. How, without intelligence, could such animated brilliance against all odds appear? *The root necessity of evolution is an unspecified but systematic increase in information.* Yet such necessity, demanding an impossibility, is dead in vital water. **How, even if the natural trend towards disintegration into simple probabilities (called entropy) can sometimes be slightly, locally and temporarily reversed, could a systematic, long-term increase in specified order be the product of a universal, inexorable and overwhelming increase in disorder?** The evasive answer must be, lacking mind's negentropy, Darwin's replacement of intelligence by natural selection's lack! Is this evasion disingenuous? If not, and it is genuinely pursued, then is it good enough? It must be if you want to hold materialism's world-view safe. It is if you thought that sunlight's catalytic chemistry or percolating electricity were sufficient unto making man (or even single cells - see Chapters 20 and 21). <u>*Otherwise your theorising is, like confusion, just misinformation and a waste of time*</u>.

Or course atoms are, like water, prerequisites of life but they are not enough. Claiming prebiotic crystals, molecules or whatever else can by themselves perform life's 'trick' is just a verbal trick. **It's a confusion and, categorically, an error that involves a mix of order, repetition and improbable complexity; and, as regards this mix-up, a conflation that has not resolved specific information from unspecified.** *In short, nobody claims that the Third Law's ordered fixity can represent the same magnitude of order as either an abstract principle or the functional, specific order of the information coded into natural-systems-with-a-purpose, that is, each and every organism.*

Discrimination, resolution of confusion's blur. Energy and information aren't the same; but in creation both need holding to account. Informative negentropy as well as energetic entropy both play a dialectic part. Do you remember 'thought is separate from brain'? It is thought and not a puppet that makes plans; nor have puppets memory. What use is code or reason to an entity without a mind? You can shout it from the rooftops, 'code cannot arise by chance from mindless matter'. But in a lab you have to wear your goggles and put earmuffs on. Don't the textbooks iterate that evolution's how life must have come to pass? Senselessness is whence, if life is incarnate information, all sense and information grew. Professional science therefore knows and nods and takes for granted information never springs from universal mind or archetypal memory. It has been decided that what's immaterial is not an element apart. Since the only kind of mind is brain-born how, before a brain, could information ever be deliberate? How could phenomena be coded, governed or controlled except by accidents of rule deriving from Original Happenstance, One Great Accident that, at the moment of excrescence out of nowhere, somehow 'wrote' our rules? Order spilled from accident, cosmos from chaos. If, in spite of what the noiseless facts might seem to shout, profound confusion of materialism thus persists the cosmological and biological ideas of Primal Information, Original Perfection or Universal Egg will be the casualties. Whole truth, masked by half, will end up cracked and binned.

How well informed is such a chancy stance? In physics perhaps confusing words like 'order' and 'disorder' should be dropped. By what criterion and on whose terms is an atom or a life form 'ordered' but a mountain chain or cloud 'disordered'? All obey the same machine-like, automatic laws of nature. Chemical reactions drop not rise to their completion. Everything inanimate, from fires to stars, runs out, runs down and…can what never lived die? Exhaustion holds no meaning and a grave is empty when it comes to earth but what about the death of forms of life? Does that mean nothing? Doesn't it affect you? **It affects the root of who you are.**

In biology you might require that animistic words like 'reason', 'code' and 'meaning might, if they detract from the austerity of mindless natural selection, be expurgated. 'Design', especially, should be defrocked lest otherwise a whiff of metaphysics animate the argument. Metaphysic's issue *is* the one of information (Chapters 5 and 6). What constitutes informative command? Is it principle? Is it what you want to happen? How, as a consequence, do you define 'disordered'? Is it error, overblown untidiness, unevenness, bug in the system or some moral turpitude? Is it social, intellectual or moral breakdown? Is it, if you like, the 'entropy of information' (Chapter 24)? We might allow that anarchy infects a world that's, basically, dynamically ordered to the hilt; but who allows that such infection can create great order in the first place? Such argument is flawed. Its sprain of logic counters intuition and is, in this instance, dialectically incorrect. If the salve of sprain's unpalatable for some they might still allow civility; and, in seats of learning, when material absence is added to materialistic scheme, not ridicule a possibility (called immaterial information) that reason shouts make sense.

No material body has a mind. Physical behaviours are passively and rigidly informed. Any seeming lottery of randomness derives from 'number games' of incalculable interactions or from experimental inscrutability. From a thermodynamic point of view order increases to a maximum as things crystallise. They cool into an outer shell of impotence, solidity, whose shape and properties have been projected from 'within'; or, if not crystalline, then mix-ups drive them to disorder. Initial concentrations have diffused out into thermal equilibrium, heat death, the bottom of a cosmic drop that's matter's silent 'state of death' - world's end in space (not time) is physical solidity. How, where order sits impassively with its reverse, might you resolve the paradox? Potential's lost, exhaustion's nothing gained. How could such entropic paradox become creator of a life on earth?

What if it did not? Metaphysical is different. Mind is a powerful dose of anti-entropy whose tendency's to rise towards the top; it tends to rise the dialectic's right-hand columns towards (*sat* ↑) super-state potential. An information hierarchy gains intensity with focus and, in the end, is pure when concentration is complete; this transcendent 'ground-state' is (*tam* ↓) material ground-state's absolute antipode. You work from practice up towards principle. Input of information raises you towards primordial order and, therefore, the truth of things. Within confusion's noise a systematic ring of truth is plainly heard. From this perspective mind is anti-chance. The greater an intelligence the greater is a grasp of principle; and the more potential for creation of a purposeful arrangement. Natural Dialectic springs not from mindlessness but from the heart of mind, awareness. Its hierarchy drops from active down to passive order; and from active into passive, locked-up energies. On a conscio-material gradient everything, even apparent disorder, chance or resistant anti-principle, falls within the orbit of Supervisory Order. Therefore look for information prior to energy. *See the macrocosm as if plan not heat were most important.*

String theory (Chapter 9) adds to space and time many more dimensions out of which its orders can unfold. Natural Dialectic only needs two more out of which the world may fall. The difference here is that its 'new' dimensions are two old ones; they are not exotic, nor are they physical or apt for mathematical description but you're well aware of one of them at least. Its name is mind; the other is called Essence. Essence of Pure Consciousness is Subjective, Absolute and Infinite. And essential/ potential matter is a 'seeming absolute' set at the base of mind; it is a memory that constitutes the fixed informant of space, time and energy; and is an archetype upon whose simple code material phenomena depend. You might call an archetype a program whose sub-routines express the cosmic fundamentals (balance and the two ↓↑ opposing vectors) in a self-consistent way that compounds into universe. Potential matter (see *fig.* 12.1) is the plan for universal character; it styles the face of matter. It is, therefore, no more strange than are the laws of nature - except that it implies their residence in mind. Universal mind. In this case there's no outside; this universe, in all its parts, is *in* the mind of Essence; you are thus *within* the mind of God!

Do you remember chance's last saloon, the bar of specified complexity? Not simply non-purposive but purposive complexity that's organised? Such teleological order is enough to close that bar. Forms of life are organised and, to the extent they can and need to, organise their habitats. They are informed and ordered and, in turn, inform and order circumstance. But they did not make themselves and they decay. They cannot be the source of their original information. Was cosmos prior to chaos; did change supplied with information (called development) precede decay; of what kind was the cosmic egg itself? *Whatever you decide will not change nature, only you.* If chaos kicked the starting block then what evolved its wildness into tame and reasonable order? If origins were orderly then they involve, by definition, threads of reason, a consistent web of logic. Could such a web, more elegant than any spider's, catch you in a mesh of dreams? Can truth be spun; could it, when tracked to the centre of its labyrinthine lair, reveal itself to be a work of chance? What is your perspective? What's your slant?

The order you inhabit and produce is not from chaos but from mind. 'Bug', 'noise' and error interfere with it. Rightness is bright, fitness is light but bugs make heavy weather of healthy mind or body. Levity leans towards 'transparency' of health but gravity of entropy can drag you into mind-bugs, sins, sickness and, the converse of prosperity, an opaque extinction. You might imagine quite correctly that a healthy body levitates above the darkness; life hovers over an abyss of death.

When it comes to physics energy, if not prevented, disperses from its points of concentration. Speeds of dispersal or disintegration from these sources vary and, as far as information is concerned, Murphy's law will probably confound you! Entropy will drop you into sinks! Diffusion and disorder, like confusion, seem to be a natural but vexatious part of natural order. Indeed, it will have dawned by now that the (*tam*) downward agency of materialisation operates at diametrical cross-purposes with (*raj*) upward creative intelligence, that is, with the agency of purposive arrangement, teleology. Are you, a highly specified complexity, an outcome of disordering or ordering? **There is no cosmic law of material innovation or of systematic ordering such as the theory of evolution lacks but desperately needs.** Those processes are both of mind. And therefore theory swivels common sense upon its head. We've noted that its touted 'counter-intuition' (a contortion long since practised in the classrooms of biology) holds that accidents evolve intense complexities of order 'purposely' just by destroying anything that either does not work or does not work as well as something else! Or, to rephrase, it claims that information can be generated by an accident, meaning rises from happy-go-lucky contingency and lives selectively emerge from senseless lines of incidentals.

Is this the case when physics says it isn't? You would not believe the academic time and effort wasted in the uphill task of arguing that nonsense isn't nonsense, it is sense! If hot air could blow you up you'd languish in the stratosphere. An evolutionist selectively deflects the question. He accuses your confusion. You don't understand, he incorrectly states, the power and influence of my trump cards - natural selection and mutation. Natural selection (Chapter 22) is, with Lady Luck, the instrument of *PCM*: the Pcm's a plausible idea but plausibility is not a proof of truth; and, we are going to see (Book 3), it's most unlikely the idea, although it admirably suits materialism's scheme, is true. Codified development and physical degeneration are, according to a basic, cosmic drift, at loggerheads. Does not the whole of physics tilt against the life-man's death-embossed assurance, his shield against intelligent design? What sort of active transport pumps his dream against the natural flow? The trend of entropy is not confined to systems that are 'closed' to any outside influence. Open systems suffer too. You might inject some energy but while, like bulls in china shops, this stimulates some action the remains a tendency is for plates to fly and breakage to increase, complexity to lessen, built-on-purpose racks and shelves to be destroyed. How confused the previous systematic clarity. The shop's a mess. Non-purposive accumulations ease, like puddles, to the slackest shapes they can; there's piles of rubbish on the floor. And, after ruckus by the bull of nature, things fall in disorder on confusion's floor. Concentrations diffuse, puddles settle and inertial equilibrium appears whenever nothing's left to drive things forward. A mess but no confusion since it happens everywhere. In fact, the natural direction of change is, for almost every chemical reaction, towards what is defined as increased entropy, deflation and decay. If you didn't know the rules behind the process you might have thought that accident, not order, made things up - until you took a focused and objective look. Wherefore, therefore, did the bull *create* specific order in the high degree that every naturalist is staring at?

Nor is the creation of highly complex and purposive order out of accidents a process Darwin actually explained. He just observed that adaptations (of both accidental and non-accidentally pre-coded kinds) vary. He only guessed how what adapted got there in the first place. Look at it another way. What informs or, if you like, gives orders? Is it chance, necessity or mind? 'Necessity' is just another name for those constraints called natural law. This form of legislation certainly arranges objects automatically. And in this case a chance would seem to be no more to its observer than a product of the wide world's multiplicity - a totality of interactions that add up with elements of unpredictability. *Just because you don't know everything, however, doesn't mean that chance is capable of anything! It can only operate within the law and law denies it creativity.* Reflex interactions neither have a purpose nor can build up purposive complexities. We know, from all our gadgets that wear out, the world of physics breaks up mind's conceptual constructions. Just like our bodies they are subject to what in a 'perfect' circumstance would not exist - accidents and death. Yet Lady Luck is a mutation; and, if she's unlucky, natural selection simply spells her death. A reflex, planless pair. Is it out of ignorance and spite for Lord Deliberate that evolution has proposed these two built you! Perhaps it's out of both.

You know full well by now! Before confusion there is information; before passive there is active information that can, for a reason, organise a means to end. You might learn in order to disperse confusion but an expert knows already what he's doing. He knows, before he starts to fabricate, how in practice his idea is going to work. **Hierarchical power and the source of order descend from above. The 'organic' rule is that external is expressed from internal; subjective order is objectively realised in a physical creation.** This cosmos is the body of a mind - one far from just as limited as ours. In this view matter never flew up from a substrate singularity; nor mind evolved from it. The story emptiness made matter that in turn made mind is an inversion. **The truth is immaterial. The source of what is physical is positively metaphysical.**

In this view the material universe is clearly not a 'flat', one-tier closed circuit. It is not self-organising, self-contained or self-originating. <u>The process of creation descends, one-way, along a hierarchical chain of 'outworking'. This chain links metaphysical to physical. It issues from higher to lower, inner to outer, principle to practice, mind to matter.</u> **Just as the systematic order of creation externalises from active through to passive information, so it follows a correlate of materialisation from potential through orderly atomic and apparently disorderly bulk states of energy until arrival at the world's edge, at the terminus of solid matter.** And then you're barred. You can't drop over that.

In this view (*sat*) potential is not simply, as in the case of a battery, stored power that can be released to drive events. It is also, more importantly, pre-physical information that from memory fixed in universal mind prescribes the pattern of those events; and, at the same time, a subtle energy 'beneath the scientific radar' that 'vibrates the fields of physicality'. Energy and information are entwined. What pulses and at once informs? Music? Music is the master metaphor of Science and the Soul. *Is this why 'strings' and 'sound' are used as metaphors for the coherent and yet vibrant order of cosmogony?* At any rate (*sat*) composition is allied with (*raj*) energetic harmony; and (*tam*) proportionate inflexibilities of geometry also help express the character of archetype. As well as music Natural Dialectic simply uses two main models - ziggurats and spreading, wave-form rings.

Forces and atoms are not intelligent but, in a purposeful system, their functional arrangement is. The physical universe can be viewed as a process or an effect, one neither disorderly nor independent. If it is dependent it cannot be closed. Where does that leave the First Law of Thermodynamics - that matter is neither created nor destroyed? Whence did energy first issue; why is no input spotted rushing out of any cosmic quarter or around the corners of our spacious wall-lessness? How, on the other hand, could general dependency on mind or metaphysic be effected and sustained?

Let breath breeze through an instrument. Tap a rhythm on the drum. Let music smooth irregularity and drive confusion out. If taut skin is drummed sand on it dances. Patterns roll according to vibrations. Analogise particular fluctuations with the 'notes' of archetypal memory. Let patterns in the sand be sub-atomic characters or, if you like, atomic and molecular behaviours. What quickens from behind the scene, what vibrates behind the diaphragm? If energy is continuously, invisibly supplied from 'inside' the tambour, it will seem to 'outside' sand that its spontaneous energy is constant, neither created nor destroyed but simply conserved as a dance it takes for granted. Sub-atomic motions are supported from within. Call these subliminal frequencies sub-conscious energy, metaphysical *prana*, *prakriti* or inner, universal resonance. *Such vibrancy, whose patterns <u>are</u> its logic and communicate its logic, is seen as the unitary essence behind physical forces.*

No information from confusion; no confusion if the information's right. Matter is initialised by mind. Its dynamic is devolved, *top-down*, from an intelligent beginning and is stored in metaphysical constructions, records known as memory. Such archetypal origin is resonant with code. The first expression of its alphabet is sub-atomic particles and, in the case of life, specific sequences of letters made of *DNA*. From this perspective objects and their energies are seen - in the way of light off onions into eyeballs - as patterns passively informed. But active spelt what passive had and has to do - with systematic clarity.

A Flight from Science?

It's time to hit the bottom line.

Do you remember the First Law of Thermodynamics, perhaps the most powerful generalisation made about the universe? It states that energy can be transferred and transformed but neither created nor destroyed. If this is the case then things emerged from 'inside' nothingness; they issued from beyond material absence; their source therefore transcends non-conscious physic and creation happened, whether by big bang or otherwise, once. The single injection was finite and its quantity is unchanged. Such quantitative stability is the basis of all other conservation laws such as conservation of momentum, electrical charge and, in every change, equilibration. It leads to the ability to formulate equations. The whole of science therefore counts on it.

The Second Law of Thermodynamics prescribes an arrow of time. This arrow flies (*tam* ↓) downwards. Things spontaneously exhaust. In an isolated, closed or dead system this is always the case; in an open system, one into which energy can flow, this is still the overall tendency. Systems become disorderly and unworkable; objects and events run to inertial equilibrium. You might extend this notion to biological, social and religious systems which also tend to age, decay and die.

How can matter be eternal? If *physicalia* decay they must have had a start -

or everything in time without a start would have decayed already. If the cosmos had a start it must have had a 'starter'. Before physical can only exist metaphysical. In hierarchical, dialectical terms before physical comes psychological. Who or what, therefore, did the unnatural or, perhaps, praeternatural thing? How did The Ball start rolling? And yet holistic supposition born of fundamental science is anathema to scientism. Is it not ironic - flight from science in support of scientism? How, therefore, does scientism fight a rearguard battle as it flees itself?

Why, you argue, should I jump from doubt into a certainty that cosmic mind exists? Contrary evidence might, as it usually does, crop up. A scientific law is just a generalisation based on current lack of contradiction to its formulation. You might argue that the scope of physical thermodynamics, whose laws were pillars in an age of steam, are restricted in their application to the information (*IT*) revolution now upon us. They well describe molecular collisions and steam engines but how much do they involve computers, fractals or the un-decaying patterns of behaviour that inhabit certain subatomic particles? What is entropy of nuclear force or gravity? *Most importantly of all, their purview ignores the negentropic source of order - information and its mind.* What is the source of natural information, that is, natural law? Can your equations logically exclude the strongest of all governmental factors, the activity of thought? Can your equations logically include it either? So you protest that sort of factor is irrelevant because there isn't any natural psychology. Surely, even if the science seems to point that way, there's nothing metaphysical? If there is then surely it cannot be universal mind?

Another tack. You might propose 'extra dimensions' and thus mask extraneous metaphysic. We shall meet *ad hoc* appellations to 'steady states with creation *ex nihilo*', branes and cosmological extravaganzas reaching, in the case of string theory, to many more dimensions than our own. You can't spot them but you need faithfully believe that these imaginations are still physical.

One of evolution's major problems is, you may have realised by now, the Second Law. This becomes obvious when a theoretical attempt is made to generate complex, informative structures from a system essentially running downhill from 'original perfection' into a low energy, disordered condition. Does mindless nature hide sandcastles from the tide? You have to think up very complex barriers - and this is what specific code, metabolism, shape and instinct in a body constitute. What 'pumped up' code for starters? You can claim that, set against the ($\downarrow$) lawful flow, an anti-natural process (evolution) builds complexity. Such evolutionary metaphysic has been identified (Chapters 4 - 6) and will be elaborately treated in Chapters 19 through 26. You might yourself have thought that accident, not order, made things up - until you took a focused, scientific look. It can be confusing. You've just seen how Shannon twisted information's meaning upside down; but was it chance (against the tide of entropy but hand-in-hand with death) or mind made man?

Or even mindful matter? It was bluffed by Ernst Haeckel, a 19th. century German biologist, that 'matter lives' and has within itself the evolutionary capacity to create ordered, apparently purposeful complexity (without, oxymoronically, any real purpose). *Call it, therefore, 'purposeful' complexity.*

Such universal 'bio-force', 'vitality' or 'evolutionary tendency' is, however, totally imaginary. Matter hasn't such a property. No observation supports 'matter's mind' or 'thoughtful things'; no law involving 'tendency to coded innovation' has been found by science; nor even (Glossary and Chapters 19 and 23) do the *PAM* and *PCM* see matter Haeckel's way. Therefore, being outside natural possibility, material that's creative must be supernatural. *Supernatural creativity is not what scientism started out to say.*

Others have attempted to evade the second law by claiming its attenuation in open systems such as planet earth and bodies biological. Could such slim attenuation let slip 'self-determined' molecules sufficient unto making cells? For example, Manfred Eigen's notion of 'epicycles' involves 'auto-catalytic' molecules. What is 'an auto-catalytic molecule'? It certainly is not nucleic acid. No such molecule is known. The utter implausibility of this and many other such evasions is explained in Chapter 20.

'Fluctuations' feature strongly in naturalistic explanations of both inanimate and animate beginnings. They are invoked to show how The Big Accident occurred, how galaxies and stars assumed their being and, eventually, a Little Accident made chemicals 'alive'. In 1977 Ilya Prigogine won a Nobel Prize for his work on 'far-from-equilibrium systems' such as whirlwinds, currents and various other kinds of fluctuation. All real systems are driven towards (inertial) equilibrium and don't proceed to states of higher order. Prigogine's tempting speculation, wrapped in mathematics, was the creation of high order due to fluctuation of a 'dissipative structure'. Sudden movement of a system to a state of motion (such as water down a plug-hole) might stabilise in this more complex state. If kinetics danced conveniently with entropy then you might speculate on spontaneously created, epicyclic hierarchies of more complex states until, hey presto, from a suite of 'phase transitions' caused by thermal instabilities emerged a cell!

What proof? A few extraordinary 'chemical clock' reactions and (despite the chaos they tend to leave in their wake) turbulent vortices like tornadoes have been hailed as 'creating spontaneous order from chaos' sufficient, with the aid of large, imaginative kicks of extrapolation, to hold the key to 'self-organisation' and thence evolution and the salvation of a scientific world-view. Are they, like the theory, straws worth grasping? Might they save Lady Luck, as saved she must be, from the clutches of Deliberate? Naturalistic explanations of the first cell, even if they're glorious hurricanes of stuff 'n' bluff, have taken Nobel Prizes!

The fact that systems such as whirlwinds are 'far-from-equilibrium' does not exempt them from the Second Law. Nor, despite the temptation to speculate, will 'dissipative twisters' spin you over into evolutionary Oz. Such a hopeful method of producing life-like structures is as hopeless as the natural or artificial constructions of a few simple organic chemicals (see Chapter 20: Raw Energy Destroys). All 'dissipative forms' are totally removed from real-life circumstance because they lack prior information, the *sine qua non* on which life's principles, targets and subsequent organs of expression are based. *They are, in short, not based on coded information. Nor do they exhibit purposive development, complex mechanisms (such as protein synthesis or respiration) or, in the face of change or adversity, 'survival-power'.* A dust devil is as far from life as Prigogine's suggestion. **What, therefore, but metaphysical is missing from a physical solution to the question of the origin of order?**

This is, of course, why scientism toils with huff 'n' puff. Its symmetries and repetitious laws compose the natural order of reflex 'inanimation'; however, they do not explain their origin. Nor do they explain the origin of information in specific orders of complexity - the codes and functional mechanisms life everywhere incorporates. The strength of Darwin's line (including chemical evolution) is not, therefore, in its explanatory value. Evolutionary explanation runs (as Chapters 4 - 26 together demonstrate) counter to intuition, the most entrenched physical and chemical law and scientific evidence from molecular biology, genetics and, it needs be argued case by case, palaeontology. Since, moreover, every life-form demonstrates incarnate information Darwin's theory even fails to explain the basis of biology. On the contrary, it helps to explain variation on a limited scale but its real power lies in an ability to rationalise and help promote the cause of secular materialism and in particular humanistic atheism. It is on this non-scientific field that real mind-battles sway. *It is for this metaphysical reason that, ironically, scientism flees from science*! Nevertheless a scientist is rightly by his trade a sceptic. If The Theory of Intelligence would seem to win the day then why not doubt it? Why not call up scientific metaphysics and give evolution half a chance? Or every chance against its chanceless, rival explanation of the way that you arrived on earth. Isn't this a culture in which, we have seen, scientism seems to flee from science?

A Culture of Doubt.

↓	tam	raj	↑
	negative	positive	
	scepticism	faith	
	unbelief	belief	
	doubt	confidence	
	uncertainty	certainty.	

Doubt is faith's other face. Uncertainty is certainty masked inside out. Each depends, in its degree, on prior axiom, experiment and your experience. They depend, that is, upon perspective and interpretation. How, therefore, to light on truth without engaging dogma? In which direction might I find the Truth of truths?

To claim, for example, that laws intrinsic in the operation of the material 'machine' are sourced from a higher, metaphysical department will not engage an atheist who believes that both consciousness and information are solely physical; and who categorically denies a cosmic presence of mind. Nor, more practically, will it impress anyone who prefers experiment to metaphors. He wants experimental data and not theological chatter. He may think that information (and consciousness) evolve from non-conscious matter. Whether or not he's thought it through his received wisdom, his foundation stone is probably the (non-testable) biological inference of evolution and his faith therefore rests assured in chance, uncertainty and vast improbability. In this affair he trusts Caprice's infidelity.

For example Richard Feynman, a popular physicist with a modern image, doubted 'mind behind creation'. He thought of empirical science as a culture of doubt, a vehicle of probability, that is, of degrees of uncertainty that range from very doubtful through to very likely but never absolutely, dogmatically certain. Uncertainty is not dogmatic - it makes for open-mindedness. Might not the open-minded therefore welcome fundamental insecurity, an intellectual stimulant, no decisive end to questions? Doubt parries dogma. Science is, in this sense, not about belief as much as disbelief. Its knowledge is provisional and, therefore, relative. No absolutes exist - not even laws themselves. Does this mean, for physics, that a holy grail called *TOE* is off the graph? Does it mean biology's unsure - except, of course, for evolution? The theory is a fact. It's law. It's certain, isn't it? Yet is not faith the opposite of doubt?

Cosmologist Stephen Hawking also rejects Lord Deliberate's metaphysical culture of certainty. He is certain that, before a 'big bang', absolutely nothing lurked. Nothing material nor (since nothing immaterial exists) immaterial either. We doubt the irrationality of that (see Chapters 0: The Science Delusion, 1: First Principles and 8: An Age of Unreason). He is certain that Deliberate is non-existent. You could doubt that as well. Yet although he disagrees with Natural Dialectic Hawking still perceives a fundamental paradox in the search for a completely unified theory of physics - such a theory must include ourselves, our thoughts and actions. 'So the theory itself would determine the outcome of our search for it! Why should it determine that we come to the right conclusions from the evidence? Might it not equally determine that we draw the wrong conclusion or no conclusion at all?' Like self-taught robots who are unable to ascertain whether their original manufacture was bug-free, either logic or code-wise, we could never be sure of our knowledge, never absolutely know the truth - even if we thought we did!

Materialism's nothing if it's not tenacious. Science is based on maths; according to 'the incompleteness theory' of Kurt Gödel maths cannot discover all truths and therefore nor can science. In this case a grand unification theory of the physical universe is bound to be either incomplete or inconsistent - and anyway you could never mathematically prove it. If, however, you restricted knowledge to natural laws described by a decisive (e.g. non-statistical) and succinct form of maths might physics still achieve *ersatz* omniscience? Hawking believes that its endeavour has not succeeded in reducing chemistry and biology to the status of solved problems; and the possibility of creating a set of equations through which it could account for human behaviour remains entirely remote. He also hints that such a *TOE*, without an inclusion of the unquantifiable basis of psychology, awareness, loosens into an even more tenuous article of faith. Is, for example, the property of mind a special case of quantum physics? Does quantum physics really lead a scientist to believe his wife is just an elaborate differential equation? Or, by a similar token, are your perceptions just a collection of atoms? And did the enormously complex, infinitely unlikely hierarchies of order that 'crystallise' as biological form happen by chance?

The gap expands wide open. **A theory that omits an immaterial element can be taken no more seriously than a theory of the non-existence of matter.**

If you dislike this view the only riposte is to kill it. Categorically eliminate the metaphysical. This is why an atheist (such as the late militant, Francis Crick of *DNA* fame) felt compelled to try and prove that mind is not just caged in body but consciousness is made of matter. He wanted, as a priority, to bust the ghost and thus snare the element of subjectivity within the objective scope of science 'proper'. An iron fist wields, of course, exclusively material consideration. The frontier for reducing consciousness to nerves is pushed by neuroscience.

Of course, no doubt exists if you can just squeeze thinking out of brain. Indeed, 'solidified' psychology treats subjects as if objects. Instinct, ascribed to chemicals not archetypal mind, rules behaviour. Of course, no doubt exists if you can just squeeze thinking out of brain. Indeed, 'solidified' psychology treats subjects as if objects. Instinct, ascribed to chemicals not archetypal mind, rules behaviour. Behaviourism, an ideologically logical outcome of materialism, objectively (that is, from 'outside') treated thought and feeling as mechanical. Are not the motions physics traces automated? Is not mind, though complex, physical and therefore reducible to mathematical, perhaps statistical, analysis? What did you expect? Where 'psychology' means 'study of the soul' this special type is soul-less. Mind is reduced to nervous residue that neuroscience studies.

In other words, what if non-consciousness is not just lack of personal consciousness but embraces the entire oblivion that we call matter? If consciousness were nullified would simply energy remain? *For Natural Dialectic information and energy (or mind and matter) are two elements, material and immaterial, separate and yet entwined.* Subtract mind from mind-and-matter and you've only matter left. On the other hand, if you've devalued life's coordinate to zero what might lift its value up again? Simply add increasing concentrations of awareness we call mind. **In this duplex case a unified theory must include psychology that, more than simply electrochemical, takes full account of (metaphysical) information and its purposes.** It must (*fig.* 0.7 (iii)) recombine both objective *and* subjective coordinates into a unitary scheme of things.

What kind of thing are you? Hawking's point needs lighting from another angle. Check to Delusion 3 in Chapter 0: The Science Delusion. If there's no doubt that mindless matter makes up everything then life is just its molecules. Molecules can't ever innovate a thing. They can't self-replicate or copy patterns of themselves without a coded system of machinery (Chapter 20). Still, where lack of logic rules, you're nothing but a bunch of elements and these evolved to make a brain. Neurons *are*, somehow, your mind. How confused if not deluded is the doubtlessness of secular philosophy! Fear not, however! By thought you change the patterns in your brain. How can will-power and intention be the same as what they're made of if that's only mindless molecules? Metaphysic must exist. An element of immaterial information's a reality. Are you now sure or do you still feel insecure? Are you agnostic or, when it comes basics such as life, origins and death, are you sure exactly what to think? Is everything material with you evolved from non-life's death? Or is there more, much more to life's rich game? No doubt there's doubt. Doubting Thomas had to touch material base before believing immateriality was real. Which answer do you guess and thereby bet will win? Dare you leap beyond what's sensible and credit Thomas with some sense? In what do you have faith?

A scientific culture's one of healthy doubt. Is faith in doubt a working premise or a final attitude towards life? Could such a culture bring itself to doubt its own foundations? *In this case would a dose of doubt about materialism do?*

The Matrix

What do other scientific players say? It may have emerged from this brief, introductory discussion of order why physicists and biologists, no less than mystics, have a deep intuitive sense of beauty, symmetry and order in their work. They may call it 'elegance' or 'serendipity'. Its illumination may not shine continually through the dark cloud of reductionism that afflicts analysis but their inspiration is truth. Both parties perceive that this derives from balance, simplicity and harmony. Worldly truth, at least, is rooted in the ubiquitous paradox of opposing yet coupled principles (*raj* and *tam*) in balanced action. Such regulation or (*sat*) homeostasis is represented verbally by Dialectic vectors as (*raj* ↑) input/ stimulant which is (*sat*) balanced by (*tam* ↓) effect; and mathematically by balancing a symbolic equation with an equals sign.

Paul Dirac noted that it was, in principle, 'more important to have beauty in one's equations than to have them fit experiment...because the discrepancy may be due to minor features that are not properly taken into account and that will get cleared up with further developments of the theory'. This is why Albert Einstein had such overriding trust in his equations and why the rule of thumb is 'trust the Dialectic'. *If the Dialectic seems wrong, it will be because our perceptions at that time are incorrect; we incorrectly balance its equation.*

Such poetic, philosophical physicists as Dirac, Planck and Heisenberg were captivated by instances of symmetry and a conviction that, at the root of things, there is a completely harmonious structure. The ups and downs of vibratory order are well represented by the ancient Chinese symbol of *yin-yang*. Quantum theorist Neils Bohr discovered that the electron orbits of an atom explained its chemical properties. He neatly linked physics and chemistry and, in so doing, underlined the fact that different sciences are complementary, alternative aspects of the same reality. He proceeded to incorporate the dualistic, dialectical *yin-yang* design into his family crest!

The matrix of our roots affects us. Was I Celtic, Anglo-Saxon and so on - what was my genetic mix? To what family do I belong? Deeper, further, whence originated human life, life and the cosmos? We all want to know. And we keep meeting Albert Einstein. What did he believe? Orthodox rabbi Herbert Goldstein, a Jewish leader in New York, once asked straight out if he believed in God. Einstein was not an atheist; indeed, atheists usurping his words to support their cause upset and angered him. He telegrammed the rabbi, "I believe in Spinoza's God who reveals himself in the lawful harmony of all that exists but not in a God who concerns himself with the fate or doings of mankind."

Of course, the 'rationalistic' Spinoza was declaimed against as pantheist or even atheist (!) by some who controversially misinterpreted his notion of Extended Nature, of Substantial Unity. A word of warning here. Theism is of two main kinds. Pantheistic deities inhabit every sort of natural quantity - sky, mountains, forests, springs and so forth. They animate, it's claimed, material non-consciousness; and they incorporate, albeit metaphysically, passions and the bents of human mind. Theistic deity inhabits, being its Creator, all existence but, in Essence, is a separate quality. An inventor's not the same as his machine - even if you animate its engine, wheels or abstract properties. Essential Absolution underwrites the motions that compose creation's existential relativity but, paradoxically, transcends each individual part. How, therefore, differs Natural Dialectic's Essence or Spinoza's Substance from the vibratory Word that 'was God….through whom all things were made, without whom nothing was made that has been made' (St. John's Gospel 1: 1-3)? *Is there, where God is One, some other, second substance with which things are made? Are existential energy and information not outsourced from the Essential Same? How can what is finite not derive from its Infinity?*

Charles Darwin, on the other hand, was a deist - God creates initial conditions but plays no further part in the unfurling of His universal game. Dialectic logic would agree that impersonal science never could observe what's metaphysical; that Einstein's 'impersonal divinity' sort of fits a 'scientific' bill; and that science studies the operation of plans and players whose records, in archetypal memory, were initialised by Authorship. It would also agree that personal communication is not a scientific possibility but disagree that it is subjectively impossible. Why should communication, life with Life or soul to Soul, not occur; and by rising to that 'wavelength', communion with Apex, the aim of every meditative practice, not be achieved? Do you imagine intellectual personalities obtain the *Dharmakaya*? That egotistic 'sense of difference' could be the state that consummated Union? Or that individual's not dissolved into the universal first? Mystic Appreciation is above religious difference. It might thus be something Einstein, if he'd known, could have felt resounded with his sense of transport, harmony and awe.

Remember *fig.* 0.8 (11)? Could science, mysticism and yin-yang together set up an array, a grid of east/ west anecdote, one based around clear-thinking, well-informed Calcuttans? Sir Jagdish Chandra Bose, physicist and plant physiologist, was a prolific inventor. Sending wireless signals in 1894 he anticipated Marconi whom he later met (1896) and who used his 'mercury coherer with telephone detector' on the famous transatlantic transmissions to and from Poldhu (Chapter 6). Bose, a graduate of Charles Darwin's college, Christ's at Cambridge, was praised by physicist Sir Neville Mott for being '60 years ahead of his time' in anticipating P- and N-type semiconductors. Bose also used his 'crescograph' to magnify plant growth ten million times and conclude from his experiments that plants have a circulatory system (with sap analogous to blood), electro-physiological sensitivity and a varied, albeit dormant, emotional life - a view with which gardeners with 'green fingers' might agree! He published 'Life Movements in Plants' in 1931, several years before the historic discoveries of Burr and Northrop (Chapters 16 and 17) which, in turn, connect directly with the logical consequences derived from Natural Dialectic and the conscio-material gradient relating to memory, archetype and morphogene (*figs*. 15.8 - 9 and chapters 15-25).

Sir J. C. left Cambridge a few years before the arrival in 1890 at Kings of another brilliant Calcuttan, Aurobindo Ackroyd Ghosh. Aurobindo opted out of the prestigious ICS (Indian Civil Service) to become the famous *rishi*, mystic or sage of Pondicherry. He was also friends with poet Rabindranath Tagore and knew the first Indian 'guru' to make an impact in the west, Mukunda Lal Ghosh, otherwise known as Paramhansa Yogananda. He tutored Satyendra Nath Bose FRS (no relation to JC), musician and physicist who first translated the works of Einstein into English, resolved problems with Planck's radiation formula (1924) and was otherwise deeply involved in the development of quantum physics. In the mid-1920's Satyendra met the inner European circle of Heisenberg, Shrödinger, Bohr etc. and, as a musician, brought the theory of harmonics into subatomic play. De Broglie named a category of particles (bosons) after him. These include photon, gluon and, maybe, a so-called Higgs boson needed to allow the Standard Model to try explaining the fact of mass - obviously a major feature of material reality! S. N. collaborated with Einstein (e.g. Bose-Einstein statistics) and worked on X-ray crystallography whose later development by the Braggs led, among much else, to the discoveries of thermo-luminescence and *DNA* structure. Long live the *tao/yang/yin*, *sat/raj/tam* and *potential/kinetic/exhausted* paradigm! Long may orient and occident cross-fertilise!

Do you remember (Chapter 2) that physicist Sir James Jeans noted 'the universe seems nearer to a great thought than a great machine'? But let the final word rest with Max Planck. Planck not only first read, recognised and published (in the *German Annals of Physics*) the start of Einstein's revolution, the latter's Special Theory of Relativity. He also pioneered quantum theory, the second pillar of modern physics whose study is sub-microscopic, subatomic phenomena - the matter-in-principle of Natural Dialectic. Actually, his foresight may have ushered in the next revolution in human understanding which has already begun to focus on the primacy of the (*sat*) unitary coordinate of information - as opposed to the secondary, (*raj/ tam*) polar coordinate of energy. He asserted that the discovery of truth can only be secured by a determined step into the realm of metaphysics and is quoted as saying at a lecture in Florence (1944):

"As a man who has devoted his whole life to the most clear-headed science, to the study of matter, I can tell you as the result of my research about the atoms, this much: *there is no matter as such.* All matter originates and exists only by virtue of a force which brings the particles of an atom to vibration and holds this most minute solar system of the atom together...We must assume behind this force the existence of a conscious and intelligent Mind. This Mind is the matrix of all matter."

Indeed (The Observer 25-1-31), "I regard consciousness as fundamental. I regard matter as a derivative of consciousness."

Perhaps Max Planck would have appreciated Natural Dialectic.

Is The Match Friendly?

If cosmos is a mind-game you'd expect the rules made sense. No contradictions, please. No parts that don't fit and jam the works. Is this the case? If, in addition, free-play (known as life) is part of the intention wouldn't you expect the match, mind with matter, to be friendly?

Then I hope you've never been arrested! One, two then three cars hem you in. You dash for the gap when…at the fourth the hands go up. Surrender! Is this how science hems an atheist in? What chance is there that chance created cosmos, earth and life? At what point does a balance tip? When does an accumulation of coincidences become too much of a coincidence? So that deep certainty is seen to underlie the superficialities of chance, uncertainty and mindlessness.

At the centre of the universe, Aristotle and Ptolemy agreed, was an immobile sphere round which the sun and other stars revolved. Copernicus did not agree. His heliocentric system, where the earth revolves about the sun, was revolutionary. It uprooted the traditional view. Although Copernicus, a priest, believed that God's creation was '*propter nos*' ('because of us') his 'Copernican principle' has been usurped by an interpretation that dethrones the inference of purpose. Is not earth an accident of blind-forced geometry, a lonely speck of blue afloat among the stars that ply this great, dark envelope, the ocean of the universe? A 'principle of mediocrity' has been assumed to demote 'home' to 'nowhere special'; ours is delusion should we think we're privileged to be alive.

Is this the case? Is there neither purpose nor significance in construction of the cosmos, in conditions on the earth and therefore in my fact of life? Are habitable planets, of the billions there might be, rare or common places? Is the Darwinian view of cosmic evolution right? Surely it is not a huge mistake? If it is a great miscalculation purpose and significance loom large. And life is very special, very precious.

No doubt universal body can be described by a few principle, simple, 'elegant' formulae that encapsulate features such as gravity, the relationship of energy with mass, electromagnetism's working and so on. These equations and the constants that they represent are finely balanced. Cosmic prerequisites are finely tuned to give a world where you can live. Is there some unknown reason why? Could chance, although the odds are massively against, be the cause of cosmos? Or are you driven out of physics into paraphysics and to metaphysics (such as multiverses, archetypes or some species of intelligence) when you try answering?

What does not have a cause? Does everything except the universe itself? We think this universe expands and, if so, must have had a start. You can, in theory, trace expansion back into a pinprick called a singularity. Did this point make it all? This point of next-to-nothing was, you argue, potent as a seed. It contained a general, changeless self-consistency of laws and all the energy you'd need to set a hundred billion stars in just as many galaxies. Law is an article of faith that underpins the scientific enterprise; the second article is comprehensibility. How is it that we understand an atom, black hole or the logic that would seem to drive the stars? Bodies in the universe, slung between the nothingness of less than micro- and unimaginably large infinities, is certainly a mystery. What is pinpoint nothingness? How can it develop and sustain vast quantities of matter? From what sort of pre-physical (or metaphysical) latency could such a show emerge? Is it pre-programmed, that is, preordained? Was the very cause of causes causeless? Or did zero lose identity when, emptied accidentally, it spilled into all things?

You are interested and intelligent enough to understand that, most mysteriously, cosmos can be understood. Why should chance work wonders and this be the case? Why should atoms, given time, evolve an understanding of themselves? Does intelligibility imply intelligence or could Sheer Luck have swilled it up? *One test might be you.* You live. You observe and are an interactor, a point of consciousness conditions have allowed to flourish. You are life's litmus in a hostile space. If physical cosmos is the substrate of a metaphysical catalyst, mind; and if mind is the exercise of intelligent life, you might expect an 'intelligent' universe to be organised in a life-oriented way. Is it? Is its posture friendly or aggressive? Could you live without 'anthropic principle' or not? If you, as an engineer, wished to create life-forms, what complementary, optimal materials and constraints would you devise? Are these devices what we find? Is the interrelated multitude of constraints so severe that your prefabrication is the only one possible? Indeed, are the laws of nature specifically adapted for carbon-based life on earth or similarly watered planet? And are the cosmic, galactic, solar and planetary aspects of ecology together friendly for your life form? *Are they 'biophilic'?*

The fact is that they are in high degree. The knobs on the cosmic generator are tuned just right.

The match is very friendly.

What about the basic, predetermined infrastructure of space-time and the coherent chain of constants, laws and formulae that seem to describe most, perhaps all, the transformations of the energy cosmos contains? Universes with other than three spatial and one temporal dimension would be unstable, unpredictable or, in either's case of zero, impossible. The actual combination allows inverse square laws such as describe the influences of gravity and electromagnetism. The stable orbits of planets round a sun and the undistorted transmission of waves (such as light) in a vacuum also depend on three dimensions of space. Perhaps their invisible framework constrains the properties of other laws and so defines the nature of a life-containing universe.

You can describe, through a glass darkly, some kind of naturalistic start. We're well away from then. If you picked up the flight of a well-pitched, stump-bound cricket ball without seeing the bowler (which you can't because he bowled so long ago and even then was locked inside a black box smaller than the Planck dimensions) you might marvel at the accuracy. It's uncanny but I'd swear it might have been intentional. *I insist; co-incidents that coincide in quantity are hinting more than a coincidence - and the universal body is, as mentioned earlier, defined by a precise set of more than thirty numbers.* These include the numerical values of charge, proton mass, electron mass, speed of light, the strength of gravity and so on. If each component did not perfectly interlock the wheels of cosmos could not turn. Nor could you watch it doing so. To hit the jackpot of this lottery you'd need your thirty lucky numbers all in line. Well-tuned is the jingle that accompanies a winning combination; well-tuned, against all probability, are the parameters of any place where you can live incorporate. Was such embodiment by chance or was it not?

Why is the universe so 'flat', 'smooth' and large (see Chapter 12: *PAM*'s Miracle)? For example, if it is not to collapse the world must expand. In order to avoid gravitational collapse on the one hand and, on the other, expand gently enough to allow a chance for galaxies to form, the impetus imparted by a **transcendent projection** (or any other kind of initialisation) must have been *very precise*. The fate of the universe depends on it. One part in a million less would have ended in collapse, one in a million more and no planets could have formed. You cannot think of it as an anarchic explosion but just-right from the very start; nor can you imagine *what* emerged from nothing. Just as the automatic gears of a luxury car propel its acceleration, so you had better imagine a very smooth and accurately organised expansion of space. This is because the value of 'omega' (the ratio between actual and critical cosmic densities of energy) would seem to have remained incredibly balanced, at the 'thin line' of unity, between the two destructive possibilities for the whole extent of creation. This generates, in the jargon of experts, 'flat' as opposed to 'positively closed' or 'negatively open' curved space. Such an orderly aspect of the start is called a 'flatness problem'.

An imprint intimately involved with this 'flatness problem' and supposedly established in the first moment is so finely tuned (to one part in 10^{120}, vastly beyond the normally accepted bounds of chance) that, in spite of their desire to kill the chance-killer, big bangers need an excellent excuse to lay Deliberate to rest. Quantum energy of the vacuum, variously called 'the cosmological constant', 'dark energy' or *lambda* (Λ), is a repulsive force that, the reverse of gravity, increases in strength with distance. This constant precisely controls the expansion of the universe thus allowing your existence. It must be weak enough not to overwhelm gravity even over extents as great as those of galactic clusters. Its 'levitatory' anti-gravity may only come to dominate on larger scales than these. If, on a sub-cluster scale, the force were slightly larger gas could not condense into the drops of energy that we call suns or stars; if its strength were slightly smaller masses would be funnelled quickly into black holes and the stars become extinct. These issues of balance arise from a theoretical assumption of 'inflation' and levity in the form of 'dark energy'. Are such assumptions right? Or are they life-support plugged into a theory that's begun to die?

You need also assume ultra-fine smoothness or regularity of the 'quantum ripples' that are thought to have inflated in a way that devolved galaxies, stars and planetary systems equally on all sides. If, on expansion, energy peaks and troughs had differed fractionally less no galaxies could form; if more, black holes would have swallowed everything. How, therefore, can *CMBR* (cosmic microwave background radiation) be at the same freezing temperature, about 2.726 K, everywhere we look? Isolate a hot and cold point in the very early, tiny universe then, in your mind's eye, follow as they fly apart. They do so at the speed of light so how could energy from one track back, catch with the other and thus 'even out' the difference to yield (perhaps by multiple exchanges) the *CMBR*'s uniformity? Did speculative inflation (see Chapter 12: *PAM*'s miracle) perform the trick? If not then these 'poles apart' are 'causally disconnected' isolates. This is the 'horizon problem' that you can add to the 'ignition problem' (what lit the fuse?), 'lumpiness problem', 'entropy problem' (how was initial order in the bang so great?) and so on. Whatever *from* was like the way *to* big bang has not been smooth.

Not only levity but also contractive forces such as gravity must be 'rightly' tuned. A precise, comparatively strong, short-range nuclear force establishes atoms. Electrical forces sculpt atoms into myriad, detailed molecular structures.

And an exact value derived from the linkage between the speed of light, Planck's constant, π and the electron charge governs, in conjunction with a second (the electron/ proton mass ratio), the critical way atoms are constructed. Called 'the fine structure constant' (α), it describes how electrons and nuclei are bound stably together and determines how radiation (light) interacts with matter. As things stand protons repel each other to the extent that electrically neutral neutrons are needed to balance the resulting instability and allow a nuclear glue, called 'the strong nuclear force', to work. So sensitive is the fine structure value that it has apparently endured through all time (except perhaps at the very start when all conditions were, inevitably, different). If its value were slightly weaker neutrons could not have bonded stably and neither the periodic table nor anything other than hydrogen gas would have formed; if slightly stronger protons would have bound directly and exclusively together. No isotopes could have formed nor, after early conversion to helium, would any hydrogen have been left to provide fuel for stars or water for life. Indeed, neutrons have a fractionally greater mass than protons; if this discrepancy, which is over double the mass of an electron, had been below that mass then the latter pair (p and e) could have combined leaving only neutrons, that is, no atomic chemistry or world that one could know! Nice definition! Fine! Were archetypal, subatomic particles designed?

99% of known matter in the universe is hydrogen and helium. These may convolve and thus, as stars, become the nuclear reactors whence, it is thought, all other elements derive. For nuclear fusion to occur you need make neutrons for the helium and, it transpires, the weak force is of a bio-friendly strength allowing transmutation at a rate whereat the stars do not burn out too fast for life. If the strong nuclear force were fractionally stronger and the weak force that little weaker life would not exist.

But there's a bottleneck! All atoms come in the form of isotopes which, except in the single case of 'light' hydrogen, have different numbers of neutrons added to the protons of their nuclei. Only the 'correct' numbers of neutrons yield stable particles. The rest decay, often very rapidly. And neutrons ready to 'correct' those numbers don't float freely in the fiery furnace. They soon decay. If, therefore, a proton clashes with a helium nucleus (two protons and two neutrons) an unstable isotope of lithium-5 might momentarily occur. And if two nuclei collided that would fail. 'Incorrect' beryllium-8 would die almost before its birth. So if you can't obtain these first two elements then how on earth occurred the rest? By leap-frog? A collision of *three* helium nuclei simultaneously might jump you out of trouble in the numbers game. Except that step two in this most unlikely process still makes evanescent beryllium that you have to prompt to stick around. It jolted Fred Hoyle's atheism to discover that prompt in the bio-friendly form of carbon 'resonance' (precise to 1% each way). Such nuclear resonance, of an unusual type in carbon, would prolong the life of the unstable beryllium sufficient unto step two. Lo! The 'bottleneck' that had prevented early gas from making heavy weather has been breached. Carbon. Oxygen. There flow the other elements and hence a possibility of life.

The special carbon resonance gongs life but there's a second chime. If the nuclear energy levels of beryllium, carbon and oxygen had been slightly different neither they nor any of the heavy metals essential for the formation of a biophilic earth could have been formed. A minute variation in the value of the fine structure constant would have jeopardised the formation of carbon, an element on which biological complexity critically depends. And if endo-stellar carbon and an alpha particle clicked you might obtain (third in biological abundance, often lending its reactive 'fizz' to biochemistry) the crucial element of oxygen. How many serial flukes lead to a suspicion of intent? The resonance for this reaction is conveniently a level just above what energy its ingredients possess and thus the carbon wasn't all swept into making oxygen. Your body's carbon parts were not wiped out. Carbon was not made extinct before life could begin.

Chime after chime precise coincidences ring the changes multiplexed by physical parameters so that you can read this line. Fine programs are tightly knit. Without such 'fine-tuned' interconnectivity between the values of our cosmic constants there would be no carbon, iron, oxygen or any other heavier atomic state. No water. Can you drily speculate in detail credibly on life without such critical constituents? In fact some ninety plus stable but heavier and heavier types of element, each with its isotopes, are found. From these the colour, pressure, texture, temperature, spatial geometries and the massive (gross) manifestations of earth and its life forms all derive. A peal to Lady Luck? Or perhaps a gong to God for resonance!

Once you have molecular aggregates along comes gravity whose value without explanation Henry Cavendish first measured up as 'G'. For all its apparent weakness, gravity wields massive strength. Its force condenses molecules into circular, spherical or elliptical 'organisations'; it moulds the general pattern of galaxies, stars, planets and universal motions; it binds; on the large-scale, its arms embrace physical totality. If gravity were very little stronger stars would not last long; nature's forms would crumple dense and small. Biological bodies would be crushed by their own weight. On the other hand, if it were only slightly weaker stars could not be pressed, grow hot and shine; yet it must be strong enough to counteract the pressure exerted by these natural nuclear fusion generators so that neither they nor galaxies fly apart. Too strong, no stars: too weak, no suns. No sun, no planet orbiting, no biological complexity - in either case no life. In the case of earth's size, if the force of gravity had been weaker little or no atmosphere would have been retained; low atmospheric pressure would not permit the presence of liquid water and, therefore, water-dependent life. If it had been stronger then dense and noxious atmospheres would have prohibited the world's biology. *The strength of gravity is right for you; it is exactly biophilic!*

Inertia is an object's resistance to change in direction, velocity and/or shape if a force acts on it. It is, like gravity, related to mass and, therefore, weight. It was proposed (by Ernst Mach and later Dennis Sciama) that inertia depends on the reciprocal action of bodies, however distant. That is, the inertial forces experienced on earth are dependent on the total mass of the universe including as yet undiscovered galaxies, putative 'dark matter' and so on. Not only the masses of sun, earth and moon (which exert biophilic gravitational forces) but of everything creates an inertial force that, if less, would have biomaterial flying through the air and, if more, would render even muscles themselves too 'heavy' to move. *In which case the mass of the universe itself were biophilic!*

Central to the Dialectic's theme is (*sat*) pivot. Balance. Could you believe that positive mass-energy of every body (every corpus and corpuscle) might equate with negative bond-energy of gravity so that, if they cancel out, the universe were weightless? No net mass! The whole lot wouldn't weigh a thing!

How does size weigh up? If space were too small it would be, at least for astronomical and biological evolution, too young. So a vast, cold, dark and empty universe is, from a materialistic point of view, essentially biophilic. Curiously, maximum structural complexity is ranged about halfway, on a logarithmic scale, between subatomic and super-galactic simplicities. Possibly the most complex structure in creation operates at a size midway between the nuclear and electromagnetic forces of the micro-world and the boundless voids of gravitational sway. How's that for a fulcrum? A universe that's balanced round the human brain!

Is mass-less light also biophilic? If there were no night because the sky was too bright and full of radiant energy, it would be too hot for life. Space is dotted with galaxies but life needs an abundance of energy from at least one stable, long-lived point, a concentration called a sun. This ball of flame should inhabit outer suburbia in a metropolitan galaxy containing enough heavy elements and neither elliptical nor irregular in shape. Life also needs a solar system with a star of the right mass relative to planetary size so that the latter can orbit at the right distance. The orbit needs to be fairly circular and the planet itself needs spin at a rate such that the majority of water neither evaporates nor freezes and the climate never kills life completely. Moreover the sun needs to emit for the most part 'friendly' radiation. This means, for life, an infinitesimally small fraction (about a trillionth) of the entire electromagnetic spectrum. This trillionth amounts, remarkably, to a proportion of the whole electro-magnetic spectrum equivalent to the thickness of a playing card on the road from Land's End to John o' Groats. Quite exact! Our physical 'deity', *Ra*, obliges by emitting, from a surface temperature of around 6000°C, about 70% of its stable output in this tiny fraction, the near ultra-violet to infra-red range. Such energy corresponds to the amount needed to activate (not under- or over-activate) most biochemical reactions, photosynthesise and allow colour vision. Only through such a frameless but hair-thin slit of a window does visible and near-visible light flood onto the whole surface of earth and, healthily, into our lives. Other frequencies, except radio and extra-low (ELF), are screened out by earth's magnetic field, ozone layer and other gases in our atmosphere. These gases include water vapour; water also acts as a liquid radiation-buffer. And a series of highly improbable aspects of a dynamic and yet stable atmospheric composition have combined to protect and promote life for perhaps billions of years. The actualities strikingly coincide with life's necessities.

An extremely thin window of light is matched by a similarly narrow window of heat. Temperatures can range from billions of degrees (as in a fiery cosmic seed) or millions in stellar infernos right down cool earth and its icy wrapping, space of absolute frigidity. Just as life on earth demands a spectacularly narrow yet, in the long-term, stable, friendly band of light, so terrestrial heat must oscillate within a narrow band for billions of years continuously. Earth-life's tolerable limits extend, for the most part, between one and fifty degrees centigrade - a fluctuation of only fifty degrees in millions possible.

Extraordinary exactitude is multiplied by prodigious stability. They sum to exceptional possibility called life - including you.

Gravity, inertia and electromagnetism interrelate with the strong and weak nuclear forces. *That is to say, the cooperative strengths of these fundamental interactions sustain a universe that can support life.* If the strength of *gravity* were slightly lower its pressure in smaller stars would not generate nuclear fusion reactions; stars would not shine. If, on the other hand, it were slightly stronger then fierce fusion would quickly burn the fuel; short, fiery lives for stars could not deliver created or evolved life's prime necessity - a steady, solar power supply. If the *strong force* (that holds nuclei together) had been weaker the only stable element would be hydrogen; if stronger with respect to electromagnetism there would be no hydrogen from which galaxies, stars, supernovae, further elements and, eventually life are presumed to have evolved. The *weak nuclear force* (which pulls nuclei apart) controls radioactive decay and thereby the government of transmutation of one element/ isotope to another. It affects the thermonuclear activity of stars, is exactly right to let the sun burn at a slow and steady rate and probably plays a part in supernovae explosions that distribute materials in space. If it were slightly stronger there could have been no helium (or other elements) and if slightly weaker no hydrogen. And it holds proton and electron together in the form of a nuclear balancing factor, the neutron. Indeed, the weak and electromagnetic forces are each believed to be an aspect of a 'higher' unifying force. Electromagnetic interaction is, because a photon has no mass, theoretically capable of interaction over infinite distance while its correlate, whose force particles are very heavy, can only influence matters inside the infinitesimal radius of a nucleus. The *electromagnetic factor* operative in chemical bonds is right for elements to interact. Too strong and everything is perpetually glued, too weak and it cannot bond. And, at about twenty times less than covalent bond strength, weak bonds are ideal for the operation of *DNA*, protein conformation and, in water, various biophilic properties. Of course, such activity is basic to biology. All round fine structure simply underlines the point we're making. With even the slightest difference stars could not make elements nor objects of the universe exist. Nor have the values of the basic forces changed; nor natural laws evolved; nor does entropy affect this framework of our universe. **Such complete fine-tuning is both critical and integral to cosmos.** How did it arise? On purpose or by atheism's last refuge?

Talk of reactivity! Never mind its nuclear origins, the lust of oxygen for electrons is one of a catalogue of necessities without which your motor would not work. You are here because highly electronegative, oxygen-based polar bonding lends 'fizz' to otherwise rather 'stale' carbon chemistry. And, critically, bonded with the 'first' of atoms, hydrogen, it gives us water.

Polar water. Life's indispensable liquidity. Core biophilicity. The effect of 'electrophilic' oxygen surreptitiously stealing charge from hydrogen creates a 'mini-magnet', a 'sticky' molecule so that water at the surface of the earth is (where it should, like heavier hydrogen sulphide, be a gas) a liquid. Such polar 'stickiness' is therefore perhaps the primary physical reason why your body is possible, that is, why you are here! Indeed, water and especially liquid water enjoys a raft of properties exquisitely suited to carbaqueous life. These include high thermal capacity (giving thermal stability critical for the regulation of both earth and body temperatures, high heats of evaporation and fusion, polar solvent and (with respect to dissolved nutrients) transport properties; also its role as a biochemical reaction chamber, a medium of complex biochemistry. Then note surface tension, capillarity, viscosity and, from viscosity, its excellent bio-diffusion potential. As a non-Newtonian fluid, whose viscosity decreases as pressure increases, water is well adapted for use in pumped, circulatory systems. And its density both permits organisms a reasonable weight and therefore size; organic materials (except lipids which float on it) are much the same density as water so organisms can swim - if water were slightly less dense they would drop like lead and if more so be buoyed permanently on a dangerous surface. Are you (or any other organism) only solid, liquid or a gas? No organism can function as a solid (what about the millions of split-second reactions a cell constantly performs?), a gas (bio-reactions are highly accurate, interconnected and ordered) or a pure liquid. Biological bodies operate more like a liquid crystal. Therefore water and, consequently, a long-term stability of temperature in a minuscule slice (0 to 100°C) of the full range of thermal possibilities are absolute prerequisites. Indeed, with a few 'extremophilic' exceptions biochemical operations narrow that gap from above zero to below 55°C.

Don't forget that the extraordinary, anomalous expansion of water while cooling from four to 0°C (which helps churn the oceans and drive weather systems), its sudden anomalous expansion on freezing and the fact that water is liquid at earth's temperatures are all strongly biophilic. *Water is, literally, life's solution - a versatile and vital fluid in which precise, complex organic chemistry called metabolism can occur.* Indeed, because of water's polarity and therefore cohesive and adhesive properties you might even conceive of its presence in a cell as a 'tendency to crystallise' and, as just mentioned, the cell's orderly construction as a kind of liquid crystal. In all these matters it is, at root, the fine-tuning of quantum forces involved in intra- and intermolecular bonding that endow water with its exceptionally congenial, life-giving properties.

If, moreover, you construe the insolubility of 'hydrophobic' hydrocarbon chains like lipids as inimical to water-based chemistry, construe again. How well would a water-soluble cell membrane work? Do hydrophobic amino acids not interact with H_2O to multiply the shapes of protein strings? And combine to create water-free reaction chambers at the heart of many enzymes? 'Anti-water' is an integral part of the balance. *Find a better, more orderly matrix of life than water in a bag of 'anti-water'.*

Water might be critical but, as biologists who try and find life elsewhere know, it is not sufficient. The list goes on. Astronomer Hugh Ross listed fifty-five factors essential for a bio-friendly earth; and the probability of them all converging at 10^{-69} while the number of planets in the entire universe is estimated at about 10^{22}. Life elsewhere might seem, despite imaginations of believers, mathematically disproved. Then you have to add even greater odds against a coded life form onto this. Such further catalogue of the unlikely chemistry and finely tuned internal and external components of our motherly planet and its biosphere is given in Chapters 20 and 26. Have you been called impossible before?

If you checked a motor car and traced tightly coupled, pervasive 'auto-philic' interdependencies of each part on the others, it is like discovering the cosmic machine. **Even very small changes in its regulatory infrastructure would have made life impossible and thus, as delicate checks and balances abound, the constants and laws of nature are themselves 'biophilic'.** You may catalogue a remarkable list of preordained coordinates and components. Is it all coincidence or, as a bowler instinctively establishes the trajectory and impetus needed for a wicket, a set-up job? You might claim you wouldn't be here if it weren't for all those flukes. Is appeal to the 'anthropic principle' (see Chapter 4) strong enough? Can 'observer bias' whistle up a foul on chance? The *type* of universe you inhabit is, you say, defined by your presence. Its fundamental parameters must be such as to admit the possibility of life. 'I'm here because I'm here; I'm here because it happened right'. In less obviously tautological a form of words, if the initial controls had not been contrived to combine precise values that unlock the kind of universe we observe, we would not be here to observe it. This means neither that life *must* occur (but it has) nor that the parameters are deliberately set up (but they might be).

No doubt, we're here because the universe is as it is. That's no surprise. But how came it so? Was its initial condition chance or not? Cosmos can be analogised to a finely tuned machine with over twenty precise, interdependent settings. *These values were either preordained or intrinsic in (if that was the way things started) a primordial projection. Indeed, we find, the dials are set for the sun, earth, you and me - to an accuracy computed by Oxford mathematician Roger Penrose at 10^{10} to power 123!* **If true, that cuts out completely chance.** You can hardly get less coincidental. The probability of your bullet hitting a nail-head at the other end of cosmos first shot is vastly greater. Indeed, a mathematician considers odds longer than 10^{50} against to be zero. That is to say, there is statistically no chance whatsoever that cosmos and its dependent life are accidental. **The Penrose computation indicates that odds against the observed, law-abiding universe appearing by chance are stupendously astronomical; and the facts appear to support his calculation**. **The consequences of such statistical annihilation of chance (and therefore a purely materialistic explanation of life) are developed in Chapters 17 and 19 to 26.**

This is, chance-wise, problematic. You might boo Lord Deliberate as he seems to prowl, as he lurks the shadows of a godless stage but, when the lights go up, behold the puppeteer! Behold the string-puller, the organiser, intelligence that hierarchically precedes all chaos Lady Luck appears to sow. You might also, being a gentleman who abides by the rules, note that the friendly, biophilic match between life and its cosmic conditions is also the subject of an occasionally not-so-friendly match between the *Top-Downers* and *Bottom-Uppers*. Sometimes each team seems to want to knock the other one for six! When is a coincidence not a coincidence? *Top-Downers* field a team of players based on providence; *Bottom-Uppers* always attack it. No matter how many runs accumulate to indicate a purposeful, biophilic infrastructure, the score is never high enough. For the *Bottom-Upper* an intelligent universe is impossible because he has, *a priori*, banned Lord Deliberate. He has banished him beyond the boundary of play. But how can you restrain him from the pitch? How can you bind an Immaterial lest, in the light of evidence, He leaps back into action and reclaims the game? The kind of twine you might employ is an idea. A single universe, it claims, is far too small to tether universal mind or keep the Captain off his field; try and endless chain of them.

What splendour man has photographed in galaxies, star clusters and the nebulae! What grandeur shoots, spirals and explodes through endless realms of space. Aren't physics' explanations quite enough? On a scale where stars like particles of powder dust the vacuum is there any reason to suppose the universe was riven out of random motions just for the convenience of microscopic men? And yet, as we have seen, Bio-Felicity is not Caprice's friend. The friendly match between the world's stage and its actors challenges materialism and its totem, Lady Luck. The fact of high-grade tuning might not mean a tuner - but it might as well. "Nay, nay, thrice nay", you splutter and expostulate but look how riskily you spread materialism's bet. The odds are stacked against, the possibility remoter than remote. The challenge is, therefore, to reduce vast chance to surety, that is, to prestidigitate vast odds into none against. No surprise, therefore, and no Deliberate needed! Our cosmos would be inevitable - but how? As a last resort against Deliberation you have to chance your arm by gambling on an idea phrased as 'atheism's last refuge'.

Atheism's Last Refuge

What a game! Cosmology comes in a packet with a health warning - an addiction to speculation can damage your actuality! When all is said and done hard physics is a neutral judge that neither precludes nor suggests necessity for the G-word, 'God'; hard-line materialism, on the other hand, is driven by agendum. No doubt the fine-tuned facts of cosmos could imply a tuner and, therefore, a Theory of Intelligence and so, with the agendum in such parlous state, Lady Luck must gird her loins and see off Lord Deliberate's advance. To gather strength a strong puff on the pipe of rational imagination is a soothing subterfuge!

Is it true that energy's conserved? If it can't be created or destroyed the universe was doled its dollop at the start. Therefore, of cosmic origin remember this: if projection from a metaphysical dimension has, in the name of naturalistic science, been debarred, how did the world *ex nihilo* at once create itself? *A single universe is not enough.*

No beguilement more insidiously insinuates itself into hypothetical premises than an elaborate mathematical model whose predictions are not practically testable. String theory (see Chapter 9) is a current example. Normally, however, powerful appeal resides in 'simple' equations, 'elegant' explanations and 'parsimonious' solutions. A form of logic backs this up. William of Ockham was a philosophical barber whose razor (the principle of simplest line of argument) is often stropped by sharp sophists. Many scientists also carry it as part of basic kit. Its blade's employed to pare extraneous strands of speculation (warranted or not) and shave the tangled dreadlocks of discussion unambiguously smooth. At the same time no falsehoods appeal like half-truths do. In this case sharpness is employed to cut the throat of metaphysical intrusion; it executes the top half, head, of any argument that 'illogically', 'religiously' proposes more than a single, physical dimension. It is, therefore, deployed as guardian at the gate of atheistic logic and materialism's battery of thought.

Whimsically, in the case of life-friendly cosmology, Ockham's blade has not been stropped. Its guard has been stood down. Necessity dictates that, in the matter of a 'fine-tuned' universe, you cut parsimony of explanation out. We saw (Chapter 7: Tugs and Wobbles) that a super-orderly projection negates any notion of a random cosmic start. Up-front, perfect order cannot reasonably occur by chance. The natural, simple inference would be that it was planned.

Thus, Ockham, go to hell! In order to avoid the vast improbability that an accidental universe should be precisely suited to life and to sidestep the obvious consequent implication of an archetype or a creator some naturalists have grasped a prodigal, most inefficient, multiplex resolution of their philosophical dilemma. *Instead of four it seems convenient to proliferate dimensions; instead of simplicity to read luxuriant complexity; and in place of a single universe to imagine many. Infinitely many.* Welcome to the idea of a multiverse, indeed, an infinitely multiversal universe!

Sir, who is fooled? Such multiverse is an invention simply to avoid the idea of transcendent metaphysic and its pressing implications of design. Conjuring one is atheism's last resort.

Let's repeat the charge. *Atheism's sophisticated last refuge is a figment of imagination.* Not testable, not falsifiable, it is science fiction in the naturalistic genre and, as such, unscientific.

Top-down you might suppose that from One Non-Numerical Infinite derived one physicality - the body of a cosmos. Might this cosmos be a finite one? With a limit on the number of its particles? What about the edge of time and space? If gravity affects the shape of space Einstein delivers you curved, finite space without an edge; and time that starts is finite even if it trails for ever after. No doubt, we've already met a self-contained tri-unity, a hierarchical tri-universe incorporating both material and immaterial dimensions. Mount Universe. But a multi-verse is quite another proposition. **Eternal, omnipresent matter's where you place your naturalistic faith.** It's not a new idea but has been wrapped afresh. You consider a plurality of worlds; you imagine that numeric indefinities of galaxies might 'bubble up' eternally inflating from creation's spring. Such an unseen cornucopia would not be hierarchically constructed; and would be material alone. Has, however, atheism's cosmological preference, eternal matter, any substance? The *fact* is that by observation we have only lighted on a single uni-verse; the *fact* is that, although abstruse hypotheses may argue cats and dogs, we'll never know if its material presence or its absence (space) is infinite because the search for confirmation would take endless time. However, from the *top-down* point of view, every finite thing's fine-tuning is derived from Boundless (N)One.

Conversely, regardless of facts there burns that need to fix fine-tuning's vast improbability. The *bottom-up* tactic is to flee, at cosmic lengths and any cost, the idea of a Certain Fixer. The tactic's trick is to transform, by an infinity of tries, impossibility to utter certainty. It's not only laboured, by the way, with cosmos; life (by evolutionary mutations) and mind (by an incomprehensible complexity composed of electronic switches whence emerges consciousness) are plied with probability until it seems, blithely, plausibly and even easily, that anything can happen! Chance and time are a magician's hands. The trick is truth.

Such transfiguration could and, in scientism's mind-set, should for all eternity delete Deliberate. To obtain involuntary creation your unbridled speculation matters critically. Haven't we been very lucky to inherit such a fine-tuned, bio-friendly space? Such a single, finite, spot-on universe is much too lucky for a single, spot-on throw of dice. Thus roll up a cosmic roller! No doubt, an option for the simplest explanation, William's 'rule of parsimony', lies at the heart of science. Never mind. If there might seem to lurk the reason of a cosmic, immaterial intelligence then flout that rule. And, if the red light flashes urgently enough, then flog the flouting shamelessly. What contrivance is too great to rescue rationalism's scientific outlook and its consort, atheism? **Thus an industrial-scale exercise in speculation is invoked to roll, without a shred of evidence, an endless suite of bangs and crashes called a multiverse.**

Are not, the daydream runs, numeric indefinities of worlds 'a-bubbling up' from nowhere? Is not, unobserved for now and evermore, an endless foam of universes ever, by the magic of inflation, swelling from creation's spring? Since statistical improbability has been numerically deleted by infinity it's absolutely sure, mathematically, that 'ours and us' has to appear! The myth is *QED* even if it is unknown what causes such mysterious creation to exist. Perhaps uncaused, tireless energy? Linde's expansion? *Floreat aeternitas*! Even in a cosmic lottery you wouldn't hit the winning numbers let alone their combination as a jackpot every spin. The thing is that you don't need reason with infinity. Is it not obvious that endless 'twiddling' with the basic 'knobs' of law must in the long run plausibly fine-tune the broadcast of such natural law as we observe. How can you deny the possibility? Therefore eliminate all metaphysic. Don't give God but only Luck a chance - no matter how remote. Mindless Nature must be spinning, burning and continually 'inventing' with a multitude of different dimensions, 'experimenting' with unheard of combinations in a most exotic way. Improbability cannot inhabit an infinity of tries so here again, to underline it, is the trick. *By an infinity of choices you erase impossibility and at that stroke raise probability of anything to certainty!* You can imagine any single thing you want (as long as it's a *'physicalium'*). When every other explanation save 'design' has failed abysmally what other safe shot is there but numerical infinity? Therefore, draw up the bridge that isolates impregnable an answer in the keep of fantasy! Must not endless churning of parameters eventually throw up a key, a combination that inevitably unlocks a universe like ours and, winning set, issues perchance a fine-tuned cosmos, solar system and, with temporary lease, your body on its local planet, earth. No doubt, moreover, extra-terrestrials have evolved. You did and, in the case that 'probably' does not exist, many likenesses of you must live along in parallel. Because it is, in this bizarre scenario, conceived that limitless attempts have certainly struck life, just as ours was struck, from dead star-tinder scattered all across a myriad of heavens. **An imaginative glimmer of a physic-only multiverse is the single safe bet that can, against all odds, win an atheist security.** Nor, best of all, can anybody argue with infinity or ever prove such science fiction right or wrong! A story that evolves for ever has to win eternity!

The Italian, Gordon Brown, began it. Do you remember (Chapter 7) Bruno's 'infinite plurality of worlds'? Nor does the modern version entertain a hierarchy, just a material dimension within which 'flat' but 'pneumatic' matter by inflating bubbles ever foams with different kinds of universe. *After all, you argue plausibly, if one kind exists then why not others? Especially since the constraints that rule the operations of our particular world are life-friendly to a degree as likely to occur by accident as a cone balanced on its needle-sharp nose for a hundred years.* So, young man, compute! Simulate imagination's flexi-worlds by plugging permutations of either conservatively or radically different values into universal constants. Set up various 'alien' sets of properties. In this way theory can experiment with boundless laxity of law by generating endless virtual cosmic histories. You might assume that each big bang (or other kind of generation) could produce a different set of natural laws. Then the sets that hypothetically regulate distinct but hypothetical worlds in a hypothetical multiverse might, as the Copernican revolution scientifically overthrew man's picture of an earth-centred universe, unscientifically overthrow the notion that ours is either uniquely bio-centric or even uni-centric! The simulations are exciting and, in a 'reasonable' sort of way, necessary to eliminate the negative and accentuate the positive possibilities. They throw up all sorts of chaotic, unpredictable and unrealistic formulations and seem to suggest that, indeed, *only* our cosmos, solar system, moon and the conditions on planet earth are sufficiently biophilic.

Such suggestion shocks. It is definitely not 'Old Schools'. It's Modern Art. The cosmologies of Aristotle and the monotheistic perspectives taught for original Cambridge Tripos examinations involved the notion of a single, finite, purposeful creation. Can Ockham's razor shred such a simple line of speculation? A single universe fits known facts except that you might slice the element of mindful purpose out. Earth and heavens derived from an original Intelligence are barred and the 'Old Schools' line of reason deemed 'irrational'. Shred it. Bin it.

But then look at this another way. Who could fail to be compelled by such simplicity of argument as states the 'final cause' of eyes is sight? An eye is not the universe but eyes would not exist if cosmos was not constituted in the way it is. Could, Ockham might inquire, even cosmos have a reason? Is chance reasonable? Why should the inexplicable concoct an eye? How might cock-eyed irrationality create a universe? A rationality of plan would simply, Ockham might agree, obviate those convoluted explanations of how things (not least sight and mind) inexplicably 'evolved'.

No intellectual likes things easy. If puzzles are the food of wit, play on. Reason's messy and unsightly, not to say invisible, when system-building's left to Luck. Therefore convoluted speculations are evolved that, by reversing the direction of our medieval barber's blade, might seem to slash simplicity of plan with a prolix complexity of none. The message now delivered from New Theatres of academy deletes Intelligence, elevates the Lady and espouses the irrationality of chance-dependent modern 'reasoning'. Nowadays, however, even the idea of unitary cosmos flipped by chance has turned into a secular anathema. A sophisticate opts for the notion of 'eternal inflation' by some undefined 'bubble-blowing' power; this submission seems to reasonably yield, without recourse to metaphysic or to any other kind of 'parallel' shenanigan, so many permutations, possibilities and speculative 'actualities' that our precise patch and our strange lives un-simply had to happen! *QED*. No doubt - but plenty of it! Shrugs galore. But if ever, by reason of mathematics, logic or sheer lack of observation such chaotic, one-dimensional landscape fails the test then scientism is in trouble. How can it explain fine-tuning? Whence did the professors who explain arise?

The expansive 'multiversal' gesture is (because the odds against our superfine-tuned, bio-friendly kind of universe are billions times billions to one) a form of very special pleading. *It is agendum-driven pleading, up against all odds, for The Perfect Accident to happen.* Is this universe a product of deliberation or a long-shot jackpot? The first option is debarred. A deliberate universe is not politically correct but jackpots need a lot of other numbers, that is, lots of universes out of which a winner pops by chance. Thus multiversal pleading scours the world to check if scraps of anything could constitute roulette. Do universal constants (such as the fine structure constant) keep the faith or lack fidelity? Has this constant twitched by one iota or two whits in fifteen billion years? Could permutations of the others differ in their value elsewhere or 'dimensions from on high' project all sorts of numbers from behind the spatial screen? One motive charged behind such quest is the elevation, one might almost say assumption or ascension, of string theory. Aren't its 'constants' seen as fields whose dynamism makes for various values? Can't string logic grant autonomy and let each of a billion *cosmoi* live according to its self-consistent set of rules? Aren't there billions of possibilities in a string scenario from which there must emerge at least one bio-friendly world?

Do you remember notions of a steady-state expansion or inflation that is everlasting but whose source of power remains anonymous? Inflaton fields, it's argued, might have spots that strengthen and, as a genie from a lamp, swell up into a 'pocket universe'. Or billions of pockets. Trillions of domains that flood, each like a flame, from an eternal fire. Endless inflation is a multiversal mechanism that, easy as words, is forever blowing bubbles in the astronomic air. It creates a multiverse embedded in an endlessly expansive space. Must not a startless block of pre-existent time have chipped from off its flint at least one spark that flared into the stars? There must be many pockets this gigantic lottery has thrown up that could encase some form of life. Indeed, hypothesis is surely verified because there is - our own! Massively inconstant constancy might thereby over eons generate our solar system and the son of man! No less. Our corner is a precinct that has hit the jackpot and is propped by rules allowing life, that is to say, the evolution of material bodies fit for consciousness and the ability to think up hare-brained schemes. Such a headline may not make the cosmic front page but it indicates the nature of your ancestor - the one preceding any life since what's the difference if everything's material alone? It must have been a big bang, some 'pre-bangian fluctuation' or even unseen, brainless 'branes'. Some physicists have recently proposed (Chapters 9 and 12) your basic background was, if not a brane-storm then the 'friction' rubbed from clash of branes in anything between the fifth and the eleventh dimension! Anything, but anything to dream upon a world that, matter from material, can conjure up itself. Then praise your real, mindless father from the many-worlds hypothesis. All hail, the ever-bubbling Multiverse.

Now's the moment for an interjection from the other side! The extra-curricular dimension that, for Natural Dialectic, is natural and uncontrived is mind; and (see *figs*. 2.12 and 2.13), with respect to potential matter, archetype. Archetype, like program, memory or universe, is a generic word. It 'globalises' its own different parts or subroutines. In this case archetypal subroutines direct the shapes of energy that we call forces, particles and, by their behaviours, laws of nature. But archetypal memory informs and information's not an extra physical dimension. It is metaphysical. And that, for materialism, is the rub. Is such reason truth, anti-truth, non-truth or any more a nonsense than a philosophically-convenient but fictitious cosmic clash of branes? There certainly exists a clash of world views, a collision in between the party that invented branes in their intangible dimensions and the rival claim that mind's intangibility from its dimension shaped up matter and climactically evolved your brain!

Whatever academic fisticuffs Einstein provoked, his determinant was 'God may not have had a choice'. Yet now the logic of a universal brane is critically poised to counter both G and His Lack of Choice. Atheism totally depends on it. The roulette wheel is spun and, in this physical casino, varying the constants radically erases constancy. The number-cruncher dials out permutations. Are not millions of universes ever in the wings? Once, from this host of different parameters, a certain combination flies. Luck is struck. Bingo. Something clicks. Eternal matter 'germinates'; the world develops; once the species of our nature 'springs' and starts to whirl and whirr by fluke it will have 'self-selected'. *A kind of cosmic 'natural selection' would occur.* It seems to fit your paradigm - a 'survival-of-the-fittest' sort of universe. Just spin the odds until they hit a biophilic possibility - then add a pinch of biological selection. The link is forged From massive trial and error even that great flier, life, is on the wing. *The whole construction is engaged to lock around and validate its key materialistic core of neo-Darwinian orthodoxy.* You can now fanfare a fundamentally evolutionary perspective from the galaxies. Suns, stars and moons shed their celestial confirmation on conception of the cosmos by Caprice and birth of all its objects, inanimate and animate alike, from random motions, from the labours of her void but ultra-fertile womb. Progress, development! Did nature not 'self-select' the 'right' laws and, therefore, the 'right' kind of matter for non-conscious 'self-assembly' of a human being to appear? It must have. So, hey presto! Here, alloyed, are we!

No metaphysics. No intelligence. In this way nothing nowhere for no reason easily unlocks its absent capability. Out pop bumps and bangs and fizzles (choose your number); one at least was surely and exactly big enough to end up, after some delay, with an unanticipated, lucky audience - a crowd adrift upon a aimless sea, a crew of souls (if you can call them that) washed round in cosmic soul-lessness. This creation myth differs from science fiction only in the computational basis of its speculation, because of which some of the best scientific minds in cutting edge astronomy and cosmology have adopted and developed it. The story is acceptable because it is a mechanism to humanise (or was that dehumanise?) the universe, an atheistic prop and last resort that sustains the notions of chance and the possibility (given enough time) of any event in that most fundamental oxymoron, accidental cosmos. _Such irrationalities lie at the heart of contemporary physical science._

You might elaborate. With infinite possibilities at large at least one cosmic bubble has evolved, by chance, the type of anti-chance some human beings aggrandise as perhaps the finest in creation, their own minds. Could there ever be one greater? It is, for mathematics, all a numbers game. After all, a hundred billion stars or more rotate around our Milky Way. Already 'winks' and flirtatious 'wobbles' have seduced space-spotters to believe these are the parents of at least one satellite. Potentially habitable locations, such excited seekers are convinced, will be spied by the million. Some planets have already been chosen as the candidates for seekers of life's exo-grail. One, the child of a red dwarf baptised by an unromantic impulse as 'Gliese 581d', might just (with reservations as to size, atmosphere and climate) be habitable. Do water, air and iron a motor make? It is not that oxygen and water are, though necessities, sufficient bio-generators; the real question is, whatever broadcast is picked up from space, by what criteria will you decide it's coded so code-capable intelligence exists 'out there'?

'Gliese 667c', 'Kepler-22b', 'HD40307g' and, though most found so far are gas giants called 'hot Jupiters', thousands more planetary candidates will doubtless blaze a heavenly trail. Some, like Gliese 581, may be illusions due to sunspots or gas clouds. But there may be life abroad; if so, the question's just the same as here on earth. Whence can its complexities arise? Can chemicals alone deliver coded specificity? Of course they can - just as you've been taught and bought the tale they _must_ have; _except (see Chapters 20 and 21) they couldn't have._ Astrobiology would argue its presumption's point that many other kinds of life evolve in 'friendly' areas of the universe. The alien _ET_s might be microbial but for more than fifty years now _SETI_ (_NASA_'s search for extra-terrestrial intelligence that ignores Hugh Ross's odds against) has scanned heaven for a trace of peer intellect or one superior to humanoid. Without success. Of course, you won't see heroes, demiurges, gods or even God but isn't someone tapping beams of light to find us too? Is there no form of life in all this space that shines as bright as ours? No answer. The starlit night-sky dims our sense of human speciality into the ultimate Copernican abyss of mediocrity. Home, you understand, is nothing special whatsoever; earth is just compound good fortune in the world of chemistry. Why, statistically as well as chemically, should life not thrive elsewhere? Granted uniformity of cosmic law, or different zones with different properties and laws or even, run in parallel or series, multiverses flowing by each other how would you calculate the odds? Silence. Not a single song from space or voice from heaven. So budget cuts have wrapped an unsuccessful project up but you can bet that _NASA_'s Kepler Mission (2009) will enhance our resolution and identify the salient sources _SETI_ should have focused on. How, though, without even thinking do you tickle something out of nothing, that is, life from none? Is spontaneous generation, even if its probability is zero, possible? Or even, given at least trillions of years, a certainty? From zero let life, like cosmos, grasp the unity of unavoidability! And, when you've imagined that, why not imagine it a million times? What normally can force 'impossibility' to happen is a mind - but 'scientific' wonder-worlds dictate that you exclude wit from the actual game. By what criteria can you infer intelligence by study of your own, terrestrial *DNA*? By what benchmark, therefore, can *SETI* scientifically spot an alien brightness from some beams of light? Apply the same criteria to strand and beam alike. Cast over each material the same forensic eye.

Or dream enthusiastically on. *Top-down*, an immaterial element of information exists and with it everywhere potential, given genial circumstance, for incorporated life. *Bottom-up*, however, no such element exists; without any information only chance in high degree evolved intelligence! Mindlessness has definitely, it's from the lecture pulpit sworn, made mind. This latter-day intelligence alone can let you dream of an oblivious creator. Might it also use *IT* to simulate a range of cosmic possibilities for life? Combined with physics could computers generate real universes of their own? Is cosmos not a program run on such abstractions? The forum that promotes a multiverse would also arrogate the power to engineer fresh forms of life. Might not Promethean super-humans far advanced from us assume the role of Great-Programmers-in-the-Sky? Would not such Simulators wield, as if divinities, the power of life and death - to program a character and then, upon a whim, delete him? Send him from the cosmic room? It wouldn't be the first time this half-godly yet ungodly kind of chap, forgetful of a strain called nemesis, has tried to punch beyond his weight and arrogate Olympian roles. The God of Old Schools is usurped by blasphemy, by such players in New Theatres! Muses from another universe, The Grand Arcadia, have taken over!

Was it, for example, tongue-in-cheek and teasingly that Sir Martin Rees, Astronomer Royal, suggested that a Super-Computer might preside over the 'matrix' of our universe? And in this vein of humour Richard Dawkins add (Times2 10-5-07 p.5) that therefore universal reality, including bodies and dimensions such as yours, might be part of super-evolution in the form of Cosmic Simulation? Invoking simulation you invoke, of course, the presence of a Simulator. Do we mean, in other words, 'A Grand Designer and Design'? Science fiction, not religion, calls upon the figment of gigantic brain and huge intelligence - a Hallowed Scholar from a distant realm and God in all but name. You could not allow Intelligence to start with but the evidence has pressed so hard you now conjure transcendent mind (not supernatural, of course) to drive your evolution's 'progress' in an exponential curve! What about New Theatre's staff - Projectionists, Programmers and Professors of His Highness versed in every cosmic aspect of *IT*? Are they surfing a 'real' universe that runs in parallel dimension to the ones, like ours, which they create; can we understand the nature of the heaven that these angels astronomically inhabit; how many of them tap an endless cosmic keyboard or dance upon the sharp end of pin? Indeed, how many universes did the cult of atheism's last refuge (that celestial imagination called the multiverse) fumble-test before ours was 'selected' to survive? Such 'glorious' science fiction is a form of atheology.

You may have dropped the strop but Ockham's razor will return to cut your prodigality. Such outright, brazen speculation casts theology's shy miracles in shade and shame. You imagine (reasonably, of course, and for the sake of atheistic faith) that there's a multiverse; but if Nature is the fount of life and round it set a natural stage then why restrict creation of appropriate life-forms to a single planet in the sky? Intelligence or No Intelligence, program or evolution - why should *either* mode of physical assembly be restricted to just planet earth or circumscribed by our sun alone?

You scan the stars to find a sign, to catch a signal from some heavenly yet physical intelligence; and yet if you want evidence of extra-physical intelligence just look beneath your feet or spread your arms to air. Grass, trees and busy organisms in them all involve ingenious codes, behavioural programs and, as meaning, their intents. *In fact, an abstract program that supports the cosmos is, in the form of archetypal memory in universal mind, exactly what the Dialectic - using Ockham's razor to excise the atheistic nonsense - has in mind.* Check *figs.* 0.11 and 2.10-13. Potential matter, metaphysical repository of material behaviour and therefore natural law, is none other than its base of physical considerations.

From what Highness, therefore, did the low in fact descend? What is behind the splendid superficiality of things? What stands behind appearance and phenomenon; whence did astronomy's great world begin? For Natural Dialectic Nothing is The Great Profundity. It is Infinity that rules the show. Can you believe life is an elemental force as real as energy? And Pure Life, beyond the wit of moving mind to understand, is central to a three-tiered universe? In other words, the Highest Substance of our University is Absolute Intelligence. Could such Immaterial Illumination, such Transcendence really be the Source of Order? In this view metaphysic is not parallel, detached or 'other' but the source from which all physic flows. If the structure of Mount Universe is so, will Hubble's optic reach beneath all glittering appearances in space or an electron microscope plumb into depths of height whence subatomic particles originate? Who will find out where The Fellow lives? Wherein does Our Master lodge?

Can clever cats grasp astrophysics? Do even very clever monkeys have the brain to understand, in its entirety, the screen of world that life's film finds them in? Law, chance, chance-killer, doubts and certainties - the argument flips to and fro. Scales of debate have swung this chapter up and down between a Theory of Intelligence and None. Intelligence, ignored by scientism's prejudice, is in fact the crucial issue. An imaginary multiverse by chance is certainly designed - by atheologists in order to replace design by 'the appearance of design by chance'! **But is the cosmic infrastructure really just an accident? Or do its principles and rules derive from a priority, a precedent called archetypal memory in universal mind?**

You're not obliged, for sure, to hide behind a multiverse but, even if you do, your hiding's one for nothing. Though speculation's rife there's neither evidence nor, far less, proof that in a cosmos full of matter's unintelligence natural processes can code and thence construct even a single cell. When you've read Chapters 20 to 26 you'll understand exactly why the copper-bottomed, *top-down* guarantee is that 'inanimate casinos can't win life'.

Do you, on the other hand, believe the universe is rational? Understandable? Is your conviction, as with that of science, that it's orderly? If you expect consistent patterns known as law why not their immaterial source - an informant? A giver. One that ruled and rules the cosmic lines. At bottom-line, if there exists a multiverse instead of universe why shouldn't 'God' (however you imagine that small word) have flung, like galaxies in ours, a limitless extent of other possibilities?

Advance in physics and cosmology demonstrates extreme precision of the natural constants and constraints; it has discovered that the finest and most intricate of 'tuning' underwrites the universal 'swing'. *If an impression of design is overwhelming it is reasonable, against a prejudice of chancers to the contrary, to presume an architect.* Such presumption precedes any thought of animated forms; but it anticipates an inundation by the revelations of biology. Advances here illuminate design of coded information, programmed hierarchies and coherent harmony of complex biochemistry in life's most basic, microscopic unit, cells. And what about the shapes incarnate purpose, multi-cellular development, exquisitely constructs from them? Like you. **Therefore, well before fierce but rational interrogations of the most dramatic, cosmic chasm parting life from non-life and the theory of evolution have begun bring up the lights. Let Lord Deliberate step forward upon a stage inanimate to take a prefatory bow.**

9. Holy Grails

Let's presume that principles precede all practice. Do you remember the (*Sat*) Metaphysical Trinity? A trinity derived from the mathematically absurd proposal that the Infinite is (N)One? Having examined the order of the Infinite, the finite and whether the constraints of creation (the laws of nature) are the outcome of chance or design, it is time to turn to the second factor in the equation - the Unity of One.

Unity and Unification

	tam/ raj	Sat
	existence	Essence
	duality	Unity
	partiality	Wholeness
	polarity	Neutrality
	grammatical script	Clean Sheet/ Purity
	geometry	Space
	calculus of line	Dot
	noise/ rhythm	Silence
	motion	Poise
↓	*tam*	*raj* ↑
	down	up
	division	union
	increasing diversification	increasing unification
	multiplication	merging
	difference	commonality
	separation	relationship
	discord	harmony
	disunity	integrity of parts
	from balance	towards balance
	less symmetrical	more symmetrical
	apart	a part
	individual	community

The most cherished goal of physics and metaphysics alike is unity. Unification occurs when apparently distinct entities are realised as aspects of underlying sameness. Such realisation involves a change of perspective proportional to its profundity. Most profound and at the Apex of Mount Universe is what Natural Dialectic calls Central Potential. *(Sat) Unity* involves subjective Communion with such Essence, that is, the experience of a Concentrate of Pure Consciousness. This kind of Oneness is beyond the laws of logic, physics and existence.

When the (*Sat*) Perfect Symmetry of the Infinite is first broken there arise the notions of Wholeness, Coherence and Harmony. These are artistically expressed in the form of geometrical symmetries, harmonic cadence and the balance of dynamic counterpoint. Division, dispersion and multiple impacts gradually reduce this unity through to apparently different, incoherent and even discordant, conflicting parts. As information and energy are lost symmetries decay, shapes become ill-proportioned, rhythm is lost and balance swings irregularly; eccentricity holds sway and sense is lost. The psychological and biological products of disunity are loss of balance and discrimination, an increase in ignorance, egotistical aggression, strife, suffering, disease and death.

Unity is both the Potential from which duality is sourced and the Law against whose Purity its divisions of shadow and miasma are tested. **The criterion against which the distance of duality from Truth, Goodness and Beauty is best measured is Original Unity, whose Balance, Symmetry and Communion are facets of the Infinite (N)One.** *In practice this means measure in comparison with the life of one who, in perfect alignment, lives these ideals.* **Human society has always, 'erring' on the side of wisdom, accorded highest honour to its saints, bodhisattvas or 'enlightened ones'.**

	tam/ raj	*Sat*	
	polar	*Neutral*	
	possibilities	*Potential*	
↓	*tam*	*raj*	↑
	negative	*positive*	
	anti-principle	*principle*	
	disorderly appearance	*underlying order*	
	differentiation	*integration*	
	discontinuity	*continuity*	
	complex	*simple*	
	detail	*outline*	
	variation	*theme*	
	individual	*class*	
	specific	*general*	
	special	*universal*	

The nub of the problem is how anything diversified from absolutely Nothing, complete Unity or Infinity. The prime paradox (*figs.* 4.2, 6.2, 7.3 and 7.4) is how the Potential of Infinite Void gave rise to First Cause from which in turn a spectrum of (*raj* ↑) principles and their downside, the (*tam* ↓) anti-principles, is dispersed. And how Absolute Freedom gave rise to relativity and automated rigour. For example, while a fluid or an explosion is not fixed with respect to motion but is wholly rigid in its pattern of behaviour. The second paradox, if creation is 'glued' with the logic of principle, is whether the polar logic of our own cosmos is inevitable. If so, then Absolute Freedom is entirely constrained. To paraphrase Albert Einstein, 'God would have had no choice'.

First move. As in a game of chess, all subsequent moves are related to the first which is itself restricted by initial preconditions (board, pieces, rules etc.). What of your first move? The initial, critical conditions were set before you woke. Was it luck or some predetermination found you where you were? Did you have control? Parents, family, location, health, wealth, sex and education - a complex, prefabricated circumstance into which, at the flick of a particular switch called fertilisation, you were cast. Similarly, in the universal case, all things are related in time (to a Prior Moment), space (to a Prior Location) and substance (to a Prior Potential). In creation (*Sat*) Unity polarises. By this reckoning First Cause has *two* aspects. Check *fig.* 6.2. Its duality incorporates both Consciousness (*Sat Purusha*) and primal, polar energy *(raj/ tam prakriti)*. In the first instance the Principle of Consciousness is predominant; this is a unifying principle. Lower down the conscio-material gradient its ascendancy is lost until, eventually in the physical plane, only the *'prakritic'* principle of polar energy is left; duality and diversification are in involuntary 'control'. In the case of physical phenomena a second, lower 'first cause' governs how things work. *Physics seeks to understand such causal unity in perfect law, whose ground state, perhaps embedded in the vacuum, is one law to govern all, one law from which all others flow.*

Such law of laws would be, for Natural Dialectic, one of ultimate equilibration. All terms that contributed would sum to zero; it would constitute the principle of (*sat*) balanced government, of archetype and order. As far as physics is concerned the Dialectic calls such super-law, the reason why the shapes of polar energy translate and transform as they do, potential matter. Potential matter's 'software' acts as a prefabricated template that consists of archetypal memories in universal mind. What switched on what sort of primal current through the cosmic grid? Who or what initialised the world?

	tam/ raj	*Sat*
	existence	Essence
	polarity	Neutrality
	below	Transcendence
	expression	Potential
	offspring	Father
	effects	First Cause
	logical outcome	Logos
	circle	Axis
	conscio-material spectrum	Pure Consciousness
	body/ mind	Soul
	maya	Atma/ Atman-Brahman
	creation	Allah
	prakriti	Purusha
	mother goddess	God
	Brahma/ Siva	Vishnu
	dharma	Buddha
↓	tam	raj ↑
	downward	upward
	negative power	positive power
	inertial mass	stimulant energy
	'fixing' decelerant	releasing current
	apartness	connection
	materialisation	dissolution
	Brahma	Siva
	tendency from Father	tendency towards Father
	anti-Christ	Christ-mind
	from enlightenment	towards enlightenment
	in spite of dharma	according to dharma

If Essence is unitary existence is dual. In this respect the two faces of First Cause are variously known.

The *positive*, intelligent face is called *Purusha, Logos, Nam, Wah-i-guru*, Word etc. Also known as Redeemer, this side is radiant, merciful and beneficent. Its influence is creative; it excites, stimulates and uplifts; and its right-hand, ascendant path will, after the annihilation of creation, roll up to rest in the Uncaused, Quintessential Peak of Life. In a personal case of annihilation this amounts to 'God-realisation'.

The second, impersonal face fulfils the *negative* role of anti-principle. This is an inexorable, unwavering 'rod of iron', the side of reflex government, law enforcement and the rigour of accountancy. *Brahma* or *Kal*, the so-called Lord of Time, is a (*tam*) principle charged with materialisation (that is, the graduated deadening of energy), continuity and 'balancing the books'. It is the principle of equilibration upon whose scales *karmic* justice is dispensed. Exact payback, due reward. Its accountancy is mathematically, inevitably correct. The deepest 'instinct' of the cosmos is to balance any motion, to oppose exactly every action and, eventually, retrieve initial equilibrium. This applies as well to mind as matter.

The left-hand, negative or descendent path ends in oblivion. Its kind of annihilation kills activity. Then it grinds presence into absence; it drops into loss of consciousness or erases energy. Absent at 'the top', the influence of anti-principle (or negativity) increases down the scale to overwhelming at the base - black space. Dark place. Night time. In this respect semitic faith has emphasised (and oriental not omitted) a personalised form of negative anti-deity - The Good's polar opposite. This special case of devilry (a complete and unredeemed absence of goodness) involves the intentional infliction of suppression, strain or pain on another life. Lucifer (see also Chapters 3, 8 and 26) operates worldwide; evil spirits, *rakshas*, *djinns* and demons are bloodcurdlingly the same. The mode is destructive, agonising, criminal. Paradise is dragged by knavery down a spiral into hell. The human stage is one of mythic, real struggle between angelic and demonic opposites. What sort of drama lacks, what colour isn't just relief from black and its converse, transparency? Human life is a chiaroscuro; it is a play of lights and shadows of morality. What keenness for The Good's not whetted on a stone of pain? Ask any victim. Suffering is real enough yet at the same time it is life's bad dream, its greatest unreality. The devil is a nightmare, a writhing split from darkness, a foul miasma spilt across the living light. This is why we call oppressive auras sinister ('left-handed'). The only end of shadows is, with a dextrous switch, to kill them with more light. Is pain's light, its healing and forgiveness, hard to generate?

Either way, such *Logic* shapes the world. In attempting to align with its right-hand, redemptive Truth this book was sub-titled, almost, Drama of a Real Philosophy.

Since it also involves the Infinite Source from which it derived, the duality of First Cause is, paradoxically, a trinity. Unity at the peak of Hinduism's pyramidal hierarchy of divinity is called *Atman-Brahman*. From monotheistic *Atman-Brahman* are devolved a triad (*Brahma, Vishnu* and *Siva*); each of the three is wedded to a polar opposite, his wife; and below them cascades a pantheon of 'minor deities' symbolising myriad less general, more localised aspects of The Inexpressible One.

Of the trinity (*tam* ↓) *Brahma* is the creator-god; is there a single Indian temple devoted to materialisation and materialism? Perhaps a single, awkward one. (*Raj* ↑) right-hand *Siva* is the fiery stimulant, embodiment of the dematerialising vector, an awakener who throws off the shackles of material encasement and, dancing in cremation's heat and smoke, ascends. Centrally enthroned between these two is the first, preserver of worlds, fulcrum of equilibrium, (*Sat*) *Vishnu*. This deity, whose incarnations include *Lord Krishna* and the *Buddha*, represents the (*Sat*) Central cosmic fundamental. His sway therefore involves control of swaying, pivotal moderation of government and an influence showering peace, health and happiness. *Vishnu* embodies the principles of axis, balance, unity, knowledge and harmony - indeed by definition all the right-hand, essential characteristics of Main Dialectic. Because such triplex structure is incorporated in the way things are it also, inevitably, assumes a pivotal occidental role. Here the omnipotent 'triumvirate of duality' is identified, enthroned at the altar, as Christian Father (the Infinite Centrality), Word or *Logos* (informant, two-way current of creation or *logical* connector) and Son (mystic in whom is vested the upward or right-hand, returning aspect of *Logos*, that is, approach to The Infinite). Son therefore sits at (*raj* ↑) right-hand of The (*Sat*) Central Father. At this point we leave the explanation to St. Patrick with his mercifully simplifying shamrock.

The intrinsic nature of consciousness is, we know, unitary. In your case, for example, myriad events and objects are, from their external separateness, combined in a seamless, subjective whole. Incorporated into this presence is the past (in the form of memorised associations) so that the 'right' understanding and unambiguous response can be arranged in order to obtain the 'best' outcome - 'best' being a term relative to a particular organism's context, circumstance and understanding. *Such (raj) unification amounts, in biological terms, to any attempt to maintain or restore the original order of balance, health and pristine condition.* It seeks 'more life' which, translated, means 'survival' or the satisfaction of 'enhanced survival'. *It amounts, in psychological terms, to the fulfilment of desires and, in particular, the higher search for comprehension, illumination and ultimate satisfaction.* It therefore also occurs in the process whereby a conflict of interest is resolved. For people this is called reconciliation and for things solution. It also occurs in relating, communicating and the case when an individual, egotistical mind is 'lost' or 'merged' in an interest. Such communion is called love. What an individual judges to be 'in best interest' or 'lovely' is relative. It is imperfect until, having striven on the upward, dialectical path of *becoming* the Infinite, Union is achieved. In this achievement an experient has become Absolute Goodness, gained *Summum Bonum* or, in a medieval phrase, obtained the long-lost philosopher's stone. Therefore the ultimate goal of mystic experience is (*Sat*) Communion. Its *GUE* (Grand Unified Experience) is called enlightenment.

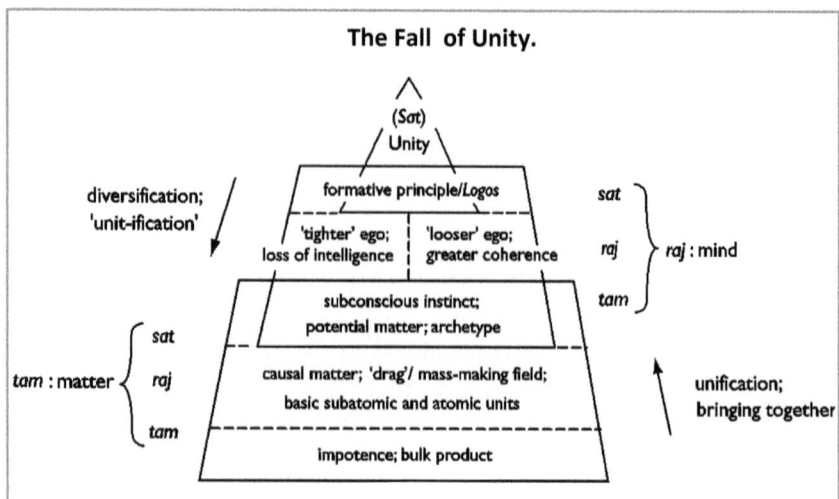

(Sat) Communion or 'GUE' is vested in Unity. This 'Stateless State' ramifies, through the single, transcendent portal of First Cause, down and out into the polar spread of mind and matter making up existence. First Principles are outlined in Chapter 1.

The unity of physical first cause, which ramifies out into the spread of 'physicalia', is at archetypal level. It is the metaphysical precondition of matter.

Check *figs 2.10-13* as they relate to cosmic fall from Unity. (*Tam*) constraints increasingly separate original oneness into isolated parts; they unit-Ify both mind (generating 'tighter' ego or sense of separation) and matter (from subatomic particles through atoms to gaseous, liquid and solid conglomerates). Active unity falls to passive units; principle precipitates detailed, particular outcomes. Informative descent of mind is accompanied by loss of truthful information or consciousness; and energetic descent is accompanied by the gradual fixation of energy culminating in solidity.

fig. 9.1 (see also figs. 2.10 - 13 and 5.2)

	tam/ raj	*Sat*
	finite	*Infinite*
	expression	*Potential*
	selves	*Self*
	personae	*Psyche*
↓	*tam*	*raj* ↑
	mentally static	*mentally kinetic*
	unconscious	*conscious*
	instinct	*thought*
	inanimate body	*animate mind*

Do you remember (Chapter 3) that unity has three disentangled meanings? Expressed tri-logically these are:

	tam	Sat	raj ↑
↓	oneness	Oneness	oneness
	separate unit	Communion	unification
	apart/ alone	Whole	together/ a part
	isolation	All-in-one	integration
	decentralising	Centrality	centralising factor
	matter dominant	Transcendence	mind dominant
	diffusion/ sink	Source/ Point	coherence/ concentration

Time past is known but the future's not. What potential, therefore, will you strive to realise; what unifying basis of design will be revealed to you?

Focus! Centralise yourself by con-centration! Concentrate, as contemplation does, upon a singularity - the heart or third-eyed singularity! *(Raj) psychological unification is revealed in thoughts or actions of union and togetherness*. It involves a positive focus in the service or interest of someone or something else. Such non-scientific communion is called liking, love or, at its extreme, ecstasy. Its unification is the basis of any free yet cohesive family, society or friendly relationship and its behavioural signs include self-sacrifice, moral integrity, heroism, dedication and good manners.

What is a machine (including biological machines called bodies which are discussed in Chapters 19 -25) but an irreducible, integrated complexity of parts designed for a purpose? Intelligent design, the higher the better, is the second, important aspect of psychological unification.

(Tam) psychological unity and, at the same time, duality are vested in a unit called 'self' as opposed to 'non-self', that is, self separated from others. While Self (Chapter 13) is the Essential Centre, 'I', 'me' and 'mine' comprise a changeable kind of existence around which 'my' world revolves. This existential isolate is composed of both the subjective mind and objective body with which I have identified. It is as illusory an appearance as existence, that is, in Buddhist terms it involves a flux of appearances that play about an unappreciated, central Truth. While *(raj)* dilution of ego is the goal of single-minded meditative practice, the opposite *(tam)* direction involves increasing egocentricity. Such isolation restlessly creates and suffers fear, pain, anxiety; incessant craving whips up 'vicious' separators such as pride, greed, dishonesty and other sins.

	tam/ raj	Sat
	consequences	First Principle
	outworking	Potential
	local expressions	Archetype
	outward devolution	Single, Simple Core
	thing	Nothing Physical
	physic	Metaphysic
↓	tam	raj ↑
	inaction	action
	informed	informant
	outcome	principle-in-action
	anti-principle	principle
	opposition to principle	cooperation with principle

resistance	*flow*
mass/ precipitate	*energy*
object	*wave*
product	*act of creation*
apart/ isolated/ separated	*a part/ together/ co-operant*
division/ unit-ification	*unification*
unit	*connector*
individual	*class*
specific/ detail	*general sketch/ outline*
complex construction	*basic materials*
bulk events	*subatomic events*
classical level	*quantum level*

At the level of *(tam)* mental fixity of shape - called memory - *(sat)* physical unity is vested in storage of design. Primary storage is in mind, that is, in memory; it is therefore metaphysical. Its medium is super-matter, archetype and psychosomatic junction at the mind-matter border (Chapters 7-10 and 15-17). Tri-logically try:

↓	*tam*	*Sat*	*raj* ↑
	sub-state inertia	*Super-state*	*range of action in between*
	drift to sub-state	*Potential*	*ascent to Super-state*
	apart/ alone	*Whole*	*together/ a part*
	isolation	*All-in-one*	*integration*
	bulk object	*Archetype*	*inner dynamic*
	separate unit	*Origin*	*connector/ medium*

There are three or four main aspects of *(raj)* physical unity and unification.

1. *The first involves (sat) archetype and its (raj) causal expressions of matter*. In dialectical terms the 'source beyond' this universe is its archetype. Archetype is cosmic super-state, potential matter; its blueprint is the principle or set of coordinated principles from which all other patterns are devolved. As a single instinct is expressed in many different circumstances so this 'immanent' metaphysic is veiled behind primary, subatomic expressions. The current order is that physics first presumes a 'bang'. This event involves 'hot influence' in the form of single super-law. From this (as gas to solid yields up different qualities) cooling gradually precipitates a 'menu' of sub-laws. Influences, fields or laws are frozen out. Phases or, if you like, hierarchical layers of creation emerge. From presumptive Higgs through electromagnetic and nuclear fields the influences that allow construction 'settle out'; and as constructions, in the form of particles and atoms, aggregate there emerge the lower 'sub-routines' that govern condensed matter's carousel. From unity unto diversity. And now, with large colliders and so forth, we try and climb back up the energetic cosmic hill. The idea is to recreate initial, high energy conditions and so unify the laws (perhaps even up to super-law) again. Thus, it is hoped, we'll find the way all different circumstances in the universe are at root connected. The simple, common building blocks from 'high up in the heat' are forces, particles and atoms. These are labelled, dialectically, 'causal matter' or 'matter-in-principle'; but even if you find material creation's forge you won't know where its the heat came from.

2. *A second aspect involves the combination or recombination of isolated units either as an aggregate, in a chemical reaction or in flux.* These comprise the (*raj*) components of reaction. Unification occurs when parts are added or rearranged to make a particular whole of any kind. The distinction between intentional, psychological and unintentional, physical complexities has already been drawn. *Purposive synthesis* involves the artistic or technological organisation of specific instruments, parts and materials in order to construct a unified expression of some principle, purpose, theme or idea. Such expression includes all products of mind. It is, dialectically, passive information but, at the same time, a diagnostic indicating 'mind behind' that drives intelligent design. *Non-purposive synthesis* such as star building or the creation of a river system is, on the other hand, very common; automata are self-ordered according to natural law but, distinctly, not self-organised. Their construction is based on the reflex, chemical interaction of polarities viz. + and - charge. When either activity - purposive or not - has run to completion the inertial result is a (*tam*) physical isolate, finished product or unit.

3. *A third aspect of unity is the one from which fundamentally tri-logical Natural Dialectic is derived - trinity.* Each different property of each individual object or event involves a proportion of the three cosmic fundamentals. It does not matter how you chop the 'being' up. You might consider a planetary system, its individual planets, molecules or even sub-atomic locations. Expressions of the three combine, at every level, to imbue an object's moment with its character. The anti-principle of (*sat*) balance is (*raj/ tam*) motion; and, of this relative pair, (*tam* ↓) anti-principle is simply the inertial or static register of a (*raj* ↑) dynamic principle. An object's character includes, at primary, microscopic and secondary, macroscopic levels, both itself and sphere of influence. Even the physical universe, which is the largest-scale single event possible, falls within the unity of trinitarian scope. Can our universe influence or be influenced by anything 'outside' itself? Could further dimensions from the common four exist and, if so, might archetype and mind be in the list? A physical universe would thus extend *internally*. Where might it all end up? 'Uppermost' in dialectical Infinity?

Look at this aspect of unity another way. Unity divides to multiply but, being Infinite, loses nothing of Itself. It remains paradoxically Infinite while also partly divided, differentiated from Itself. Simple division produces identical fractions of a dividend, simple multiplication creates clones of the one multiplied. If the Infinite (which is Nothing and Oneness) created this way, nothing would happen. Infinities would engage sterile multiplication or division into series of infinities. From such empty operations 'bits' of nothing would come both from and to nothing. This is an empty circularity, a vapid tautology. In fact, creation is not a numerical abstraction. It is diverse. Aspects of division embedded in the fabric of mind, space, time and things include diversification and differentiation. There derive from simple or, rather, dual initials a complex stack of polar characters; between their opposite yet complementary poles there exists a range of states, a field of interaction; and the many fields of interaction that compose any particular object or event make up, each in conjunction with its event vectors, its individual identity. An individual unit may look the same or different from other individuals but every single one is distinct, separated or divided from the others. At root all may be one; division and differentiation may be 'superficial' and 'relatively illusory'. But starting with Original Facets of First Essential Principle, which are listed on the right-hand column of Primary Dialectic, division and duality are the whole theme of this section of the book. *A key feature of cosmic Potential is that, in the motion of its creation, it divides, diversifies and complexifies from its Original Facets using the principle of complementary opposites - polarity.*

Principles are broken down against their anti-principles into myriad diverse expressions that comprise the world. Anti-principles tend to 'set' principle, fix flux or exhaust potential in specific, different ways. Loss of generality from an informative principle results in specific variations on its theme. The potential of a motive idea is, through the flux of thought, crystallised to plan and then, even more constrained, detailed action. Possibilities diminish just as diversity increases down this conscio-material scale. An example is the way all theatre draws its plots from a handful of emotional relationships. Such predisposition to generate difference from the same principle is ubiquitous. Do you remember the 'lower complexity' derived from 'higher simplicity' well illustrated by the dynamics that crystallise myriad solid variations on the snowflake theme from simple, liquid water? Loss of energy from a vaporous or liquid superior 'solidifies' into diverse and complex variation-on-theme, practice-from-principle or difference down below according to the rules of law.

(*Raj*) stimulus or creative activity is antithesized by (*tam*) finished product and uncreative impotence.

The (*tam*) impotent aspect of diversification manifests as neutral sub states. These antithesise (*sat*) super-states; they realise what once was only possibility. Such diversification shows endless variation-on-theme, repetitive duplication or 'same difference'. Psychologically you spot lack of focus, will-power or creativity; you note thoughtless repetition (habit), dimness and sleep. Low-level mind is locked to memory in the automated form of instinct. Physically you spot rigid rule-keeping coupled with a preponderant tendency to diffuse, solidify, grow dark or stop. Aren't blind, reflex motion and the empty silence of exhaustion exactly characters that best describe the way of matter and, therefore, the substance of material science?

Note again the clear-cut distinction made between psychological and physical complexities. Non-purposive physical complexity is, as in the case of self-ordering snowflakes, devolved due to energy loss. Fluids tend to 'de-concentrate'; they diffuse into an all-over, statistical spread of sameness. On the other hand, precipitates called solids are 'frozen' into different shapes according to their circumstance at freezing time. They have been split into (*tam*) discontinuous, apparently different units. Even if, as is nature's common way, these units are mass-produced, mass present and apparently identical as grains of sand the (↓) vector of unit-ification is one of separation and difference, the reverse of unification. Take, for example, friendly twins and toffees. Friends have different bodies but are unified by the liberating bond of mental synergy; identical twins (clones) have separate bodies but are (↑) unified by information in genetic code and parents; pieces of toffee, on the other hand, are separate but, if heated, melt and merge and lose distinction in a treacly fudge. The upward vector works at every level counteracting entropy, stimulating change and unifying.

Psychological complexity, on the other hand, is evolved due to information gain. It is the animated product of purposeful organisation. If its product, a plan, is activated then a complex physical event or arrangement of parts occurs. Information is expressed in the way you arrange material effects. **When informative complexity is translated into physical effect we say that active becomes passive information.** We say that loss of psychological potential occurs in proportion as different, specific forms, behaviours or designs are realised according to background principles. Expressions of such complexity include irreducibly complex mechanisms, technological systems, biological operations and artistic end products. They also include character and behaviour - yourself.

(*Tam*) physical end products of either psychological or physical construction are quite unlike the kind of unity found in principle and potential. Far from preceding dynamic creation they are finished units. Aftermath. If such a unit involved purposive complexity it will have become the frozen memory that is a painting, a musical score, architectural drawing, machine or, simply, a record called memory itself; only remembrance, observation or usage revives, resurrects or lends it life. If it involved non-purposive complexity it is, simply, an inanimate environment, a 'happenstance' described by the collection of physics or chemistry. There is a sense in which, as anti-pole to 'The Infinite Pole', the whole physical universe can be seen as principle rigidified. It represents the final, peripheral stage in the outworking of Central Potential, a crystalline version of universal memories, an inanimate 'life-negative' as opposed to animate 'life-positive' pole of existence. It is an entirely 'de-subjectivised' object. Indeed, from life's point of view you might contrast the lifeless, non-conscious part of the starry universe as a 'no-world' made of 'no-things' as against the 'yes' of life!

(*Tam*) is the diametrical opposite of (*sat*) and also the polar opposite of (*raj*); that is, it is vectored as opposed to the (*sat*) vectorless vector and also the opposite vector of (*raj*). It ($\downarrow$) descends, it settles 'entropically' towards the sub-state base of any kind of range. As such, drawn in the left-hand column of Secondary or Branch Dialectic, its represents an opposite. It is the negative anti-principle of positive principle or character. In this case (*tam*) unity makes for duality; it represents the isolation that separates one thing from another, a distinct unit without which there can be no duality. In other words, the opposite of oneness is also a part of any particular unit.

Unless you can count on hypothetical 'qubits' (which are quantum figments of semi-being), units and their absence are the binary roots of mathematics, itself a descriptor as powerful and physically powerless as words. Its unemotional trace wraps principles in figures; although entirely abstract and detached, yet it values substance. *The most precise descriptions of our automated environment are all mathematical.* Indeed it is mathematics that, locked into homeostatic balance by its equation of both sides, can keep a universal account and affords the clearest abstract example of the physical operation of Natural Dialectic.

Equations are the currency of mathematics; only they can endlessly manipulate objective, quantitative values.

Is there a mathematical absolute? Has the science of number any fundamental abstraction, any basic unit? Is it one, zero or infinity? Is it all, none or just the one? At any rate duality, made trinity by the pivot of its balance, embraces all mathematical conservations, including imaginary numbers (such as the square root of -1) and the oddity involved in equations (such as 0/1, 1/0 or 1/infinity) that relate one, zero and infinity. While infinity is the most potent source of number, number itself is nothing without what it describes and it describes objects. The most abstract form of object is '1'. Information technology is quite clear that this number, along with its absence ('0'), forms the basis of a simplest, binary existence. After one you can treat endless numbers of units with mathematical ciphers undergoing the four operations of addition (including multiplication) and subtraction (including division). Even 'apparent infinity', also called 'indefinity', is no more than an unrestricted set (or sets) of units, an endless series of identical isolates, an abstract line of lifeless ghosts. It is an unending business made of units and, as such, quite different from the Unity of Absolution.

If you quit the abstract world of pure numerical relativity (a self-reflecting place where ciphers rephrase each other) and leave a larger world of hypotheses, you enter the single, real creation. Real creation is without doubt or equivocation only what it is. Discard null operations, the empty set and an infinite number of possible mathematical and logical structures and concentrate on the particular system of logical rules with its single set of axioms that describes, in the manner of Natural Dialectic, nature. This subset, you might say, is real. Its abstraction, far from being inert or repetitive, is full of possibility. Its Essential Potential has the capacity to generate existence. We find existence full of different things that are related. Sometimes they are obviously interconnected, sometimes less obviously so; but all are certainly related to each other in respect of source. Which always begs the question 'What source?' Is the first cause of physics also metaphysical? Is the source subjective and, if so, how can you measure quality, enumerate thought or value subjective principles? A numerical description of The dialectical Reality is nonsense. The equation of zero, one and infinity is ridiculous. Yet such Essential Trinity is what it's all about; it is the Apex and the Centre out of which existence is spread round. Was it therefore Unity or unity came first? Does the former lead, transformed through active into passive phase of mind, to matter? Or, conversely, did big bang's projection lead to mind? Did its singularity eventually give rise, by accident, to mathematics and imagination of a metaphysic?

Either way, it's obvious isn't it? A unit is an isolated fact, an obvious object of division, a part apart and locked away from others. This aspect of unity (the unit and any aggregate built of units) is the basis of manipulation (not least mathematical) and observation (as a datum or unit of data). We know exactly what we're on about, don't we?

In fact, as well as 'static' individuality there exists a sense of union in (*raj*) flux, in the communication of relationship and in the connected wholeness of 'wheels-within-wheels', that is, of parts together. In this case how do you choose where to cut divisions or, from apparent divisions, work out how things are related in a larger kind of unity? Are you are going to build or break down in your search for Unity?

Take a cup of tea. Having realised that this starting-point is an obvious but arbitrary unit, you might decide that cups of tea are not alone. You might decide that everything, including tea, was linked in time - it all came from a primal bang; or linked in state by universal law; or linked in composition by its common forms of force and particle. Unifying is relating; it is drawing things together. You could build your cup into a serious and expanding series of spatial, social, historical, commercial, chemical and other associations. You could, on the other hand, divide it like the Greek philosopher called Democritus into subunit after subunit, each in its own right a unit, until you reach a limit, a theoretical indivisibility, an unchoppable atom or, to cut to a probable but not necessarily actual end of the impossible chase, a subatomic particle. Is this the last or first thing? Or, beyond the quantum state, is there a single force that links all others, a 'mother-power' from which all things derive and which means, therefore, cosmic unity? What is the nature of the force in whose name we can say that 'all is one'?

Is Reality a zone of smallest particles from which the cup is made? This argument would be especially persuasive if these building blocks were common to myriad other larger objects. For a 'reductionist' philosopher a whole is no greater than the sum of its parts and so, for example, a biological body is simply reduced to an agglomeration of bio-molecules and, therefore, atoms. From a holistic, hierarchical perspective, however, atoms are principles that reside 'above' what they substantiate - bulk matter; and above atoms are supra-atomic (we call them sub-atomic) particles. Such particles form the basis of matter and are, as such, the first step from its origin.

4. *This brings us, at the quantum brink, to the fourth and topmost aspect of (raj) physical unification.* We now peer into atoms and resolve nuclides and electrons. It is believed that 'quark' symmetries underlie proton structure, that only four main forces substantiate all physical phenomena and it may be possible to identify a single, unitary entity from which the astronomical universe is derived.

Because it is central to philosophical thought, both in terms of practical science and metaphysical overview, the (raj) aspect of unity - unification - has been mentioned and will soon crop up again. It involves the prime motivation of all research. **This is the grail of total reconciliation**. Its material apotheosis should light up the cosmic circuit demonstrating connectivity between all facts by means of mathematical equivalence or archetypal principle. Where knowledge is power, such intimate understanding becomes an instrument of control, exploitation and, hopefully, enhanced human comfort. <u>*The path of cosmic unification is, for science, one of teasing out the suspected simplicity that lies, in practice as well as principle, behind the physical universe*</u>. Such a 'holy grail' is, despite reservations to the contrary, the development of a *TOE* that has consequently been confirmed by observation and experiment. Curiously, the project seems to have come to invert, mirror-like, the discoveries of a mystic. For one party unity is nothing (a cosmic vacuum); its blank canvas is composed of inert space, dynamic time and coloured in by various energy-related things. For the other party Unity is the Infinite Presence (or timelessness) of the Void. For one the annihilation of physical units yields a flash of light; for the other annihilation of self yields Everlasting Light.

Never mind about mystics. Grails exemplify a primary drive in human consciousness to relate apparently unconnected facts or events by means of underlying principle and, in the case of scientific research, mathematical equivalence. Such reconciliation unifies superficial diversities and complexities into likenesses; it explains things in the simplest possible way. **The descent into diversity is, in this way, matched by a corresponding ascent back towards the inward, fundamental originators - first facts, first causes, first principles.** It is the purest exercise of reason to find reasons. The prime motivation of scientific insight and research is to uncover patterns, principles or 'connectors', set upon higher set, until they are all subsumed under the auspice of a single truth. Such motivation involves a conviction that, psychologically, physically or both, there probably resides a central force or power to which all else is indissolubly connected, from which all derives, on which all depends and in whose comprehension we will, with a sigh of great relief, know everything. It is this extreme conviction that propels a student of nature towards all-inclusive *TOP*s and *GUE*s or, objectively capable of observation and experimental verification, materially all-inclusive *GUT*s and *TOE*s. What, with respect to physics, is the current state of play?

Holy Grails.

	tam/ raj	Sat
	subsequent hierarchy	Holy Grail
	issue/ action	Source/ Priority
	actual	Virtual
	inertial/ dynamic action	Potential
	physical	Metaphysical
	classical/ quantum	Archetypal
↓	tam	raj ↑
	outward finality	inner support
	exhaustion	action
	practical/ sensible matter	principle matter
	massive/ large-scale	atomic/ subatomic
	bound up	in motion
	locked/ aggregate	released/ free
	heat loss	heat gain
	mass-centred/ inertial	radiant/ dynamic
	gravity principle	levity principle
	gravitation	electrical phenomena
	smooth/ flat	vibratory
	body	field of influence
	classical	quantum

Ferment is the state of play. It is self-evident, therefore, that the game is, far from over, in full swing. There exists no answer so complete as to extinguish all debate and ever-changing argument. No absolute, which is the Holy Grail, has been achieved. Is it possible to frame or rise above the ferment using Natural Dialectic?

From a materialistic view there's only energy expressed in various ways. The perspective of Natural Dialectic, on the other hand, takes an overview that presupposes metaphysical priority and seeks to illuminate the mode of its linkage with physics. It seeks to understand the source of principle and law and to develop a broader framework within which the observations of psychology, biology and physics can be incorporated. The ultimate remit of a Unified Theory is to tie everything into an elegant, self-consistent 'simplicity' of explanation. In science such coordination happens gradually.

Let's opt for a starting point. Galileo and Newton (with his first law) unified motion and its special case, at zero, rest. They showed, with a principle of relativity, a thing seems different according to perspective - which is the Dialectic's case, including consciousness, as well. And for Standard Theory gravity, a weak force whose first description by Sir Isaac Newton was refined by Albert Einstein's General Theory of Relativity, operates on the large scale to 'order' galaxies, solar systems and planets; indeed its influence is everywhere at every time on everything. It is an expression of omnipotence.

The other three recognised forces, including Maxwell's unification of electricity and magnetism (electromagnetism) and electroweak theory (electromagnetism unified with the weak nuclear force), are highly localised. They operate with strengths thousands of billions of times greater but over distances thousands of billions of times smaller than gravity; and they are the composite of a further unification, of special relativity and quantum theory, that led to the Standard Model of elementary particle physics. Progress would include the strong nuclear force (a *GUT*); then quantum cosmology (perhaps in the form of a string theory and so-called super-symmetry) would embrace gravitation and merge particles with forces in its *TOE* (Theory of Everything). Job done! Where all is one the cosmos is completely understood! Does that not include you too?

Newton drew the celestial to earth; Maxwell, with his theory of electromagnetism, carried light to the stars; and, prodigiously, Einstein unified space with time and energy with mass. Things great and small have been reduced to energy in bendy space.

It may seem that in respect of cosmos things are simpler on the extreme approaches to micro- and macro-infinities than they are at the centre of a complex cradle slung between them. Here, midway in size between subatomic particles and galaxies is the home of ecological complexity whose crowning glory is the human brain.

From midway science that tends towards infinitesimal, 'inner space' and deals with the (*raj*) kinetic principles of micro-forces and particles is called chemistry, atomic physics and quantum physics. On the other hand, leaving the midway physical and chemical complexities of human-scale existence, science that tends towards 'outer space' of macro-infinity is called astronomy and cosmology. Of special interest with respect to the principles of (*tam*) massive aggregations and celestial motions are antagonistic forces or, better termed, vectors - gravity and levity. In a relatively 'old', low-energy universe such as ours dialectical logic suggests that the (*tam*) force of anti-levity (gravity) may be in the ascendant, that is, predominant; and (*raj*) levitatory anti-gravity, identified with expansive force or vacuum pressure, recessive. Did the reverse apply at a highly energetic start? Is a form of 'levity' the energy of empty space? Therefore no mass, overwhelming expansion; dispersed mass, a proportional rise in gravitational effect; point of total mass, overwhelming contraction. The relative, 'subtendent end' would, theoretically at least, collapse into a black hole singularity whose logical reverse, a 'transcendent start', might read as a free, non-local expression of mass-less energy - an 'anti-mass' called light. That is to say, this primal radiation might not appear from the speculative point of a 'white hole' singularity but from 'everywhere-at-once'.

Next Einstein and Planck refined the way great and small scales of the universe are understood. *Together quantum and relativity theories constitute the two pillars of modern physics.* The substance of each pillar provides a framework that with great accuracy incorporates and harmonises data from the atomic heart of matter to the outer galaxies. They are complementary and each, while reflecting a predominance of opposing principles, paradoxically partakes of the other's. *The stark problem is, however, they are also incompatible. They run according to two unlinked sets of rules and, despite Einstein's dreams, no theory of unification whose single principle covers everything has been tested and accepted. The radical indeterminism of (raj) quantum mechanics has, so far, proved mathematically inconsistent with (tam) classical, relativistic theory.* On a pitch where small doesn't play ball with big yet their games are obviously, intimately and exactly intertwined, something is missing. Nature's unified and doesn't work like this. What clicks that can't be seen? What might overarch their different quantities? It's our schizophrenia needs healing, our dichotomy needs resolution and our equations unifying thoroughly. Can any theory string the two together? How will the great goal be achieved?

In practice science works either on the large, cosmological scale (with relativity) or in localities ranging from a metre or two down to minuscule billionths of a metre (the domain of quantum mechanics). Never with any real impact do the twain meet - except in the microscopic gravitational exercise of extremes. They only collide at such theoretical limits of existence as 'big bang/ white hole' and 'black hole' singularities but when they do one doesn't like uncertainty and the other can't explain gravity; one requires smooth space, the other won't provide it; and one resolves the issue of three fundamental forces from a 'super-force' during initial conditions presumed far more energetic than obtain on earth nowadays but the other refuses to fit in. Could you cram four styles of music into one? The integration of a gravitational matter-field into a so-called 'super-symmetry' of four vibratory forces has still not been achieved.

In this case is it possible that any forces and/or dimensions remain under physical wraps and therefore scientifically undiscovered? Small-scale strong nuclear force creates the nuclei of atoms; electro-weakness holds together neutrons; and small-scale electromagnetic force (with emphasis on magnetism rather than radiance) binds atoms into molecules and so creates a universe of bulky objects. Large-scale (↓) gravity, on the other hand, is an aggregator. It binds but not creates mass. It is a consequence of mass and might therefore be considered a secondary force. As such any attempt to integrate it into an amalgamation of primary forces could be fundamentally misguided. What, therefore, is the solution? Could vacuum potential, 'levity' or negative, dark energy act anti-gravitationally; and such (↑) large-scale, decompressing, levitatory/ 'inflationary' force lurk in space itself? Where is the arch and, in that arch, a keystone that will unite the two pillars of modern physics? *Perhaps, because they reflect the dualistic nature of the cosmos, they are fundamentally incompatible and no such connecting arch exists. Perhaps, on the other hand, each reflects a predominance of opposing principles (see fig. 2.13) and a way to harmonise their operation will be found. If so, it will prove possible to quantise the smooth curves of space-time and, in a final theory (perhaps of so-called quantum gravity), achieve cosmic self-consistency and the bliss of scientific illumination.* Yet, even at that moment, you may realise that bliss is in the mind; and it is also possible that, as your immaterial mind in its material body shows, you need to add at least one force that's metaphysical - purpose.

You can float the issue in another way. You can update the ancient metaphor of the world as 'an ocean of phenomena' called *samsara*. The Greek philosopher Heracleitus announced, "All is in flux". And, in updated metaphor, it has been noted that just as there are icebergs in the sea, so there is only quantum flux with 'frozen lumps' of classical matter in it. In other words, there is only quantum matter which sometimes behaves as if it were classical and that, on these occasions, classical thinking becomes a good approximation to the truth. Quantum and, where applicable, classical perspectives just interpret different aspects, levels or layers of the same thing. What, however, is the origin of the ocean's ice and waves? What moves them in their characteristic ways?

The body of physics works (in tested actuality as opposed to some theories) in four dimensions - three of space and one of time. Within this possibly unified framework we already know that rules governing micro- are different from those governing macro-spaces. We might tentatively ask if our four dimensions are wrapped in another or, conversely, at least one other is intertwined with ours. Consciousness-in-motion is an obvious example. Its subjectivity is the most familiar feature in the world - our own experience. Is it too much to speculate that mind involves a fifth dimension? Or that, as the rules of quantum and relativity theories are different and to some extent unlinked, so are those that govern mental operations? *That would not mean the systems did not actually work together - only that we have not fathomed how.* Such operations could include both conscious and sub-conscious modulations. Such a consideration is materialistically outrageous solely because of *PAM*'s own rigid self-restriction - its first unproven and scientifically untestable attestation that only matter exists. In that case, of course, mind *is* coloured pinky grey and brain is not a psychosomatic interface but the meaty generator of its own ephemeral dependency, an outcrop called consciousness. On the other hand the whole thrust of *top-down* theory stems from a not unreasonable, perhaps highly reasonable, proposition that *two* cosmic fundamentals, *(sat) consciousness* and *(raj) energy*, together compose existence; and that, of this couple, the subjectively active informant is rated prior to the dependency it informs, objective appearance. This rating is certainly ours when we hierarchically value immaterial awareness (or being alive) of primary importance; and dispose subjective mind 'above' and 'in charge' of our own material circumstance. Is there also an independent universal that controls matter? On which the latter's behaviour depends and which is, perhaps, metaphysical?

Perhaps formations in the quantum ocean (called events) and, by reflection, the formulations of scientific thought are themselves simply good approximations to a higher, transcendent truth of matter - the nature of its origin. And this origin, as invisible as vacuum, is in turn simply a humble step on the stairway that reaches up through mind to an Infinite Centre.

Remember that, for Natural Dialectic, bulk matter is at the base of a cosmic hierarchy called Mount Universe. Sources are above and not below. Therefore, looking up, atoms are 'higher' and what we call the 'foundation' or 'substance' of things, subatomic particles, are one step prior. They are 'supra-atomic'. What is higher still? Quantum loops? Strings from string theory? Or ubiquitous vacuum, a 'non-structure' from which the plenitude of particles 'emerge'? This paradoxical vacuum is at once a formless, neutral, empty dimension, a habitation of all motions and their source. *Space is seen as both an exhausted absence and a potential presence, a source for the expression of things.* As such it becomes, above (*tam*) classical and (*raj*) quantum phases, a prime candidate for the (*sat*) prior form and highest character of matter. Top matter. According to such dialectical logic a third, overarching theory of physics that binds the other two in its trinity should be one whose components revolve round an 'axis' that describes the structure and functions of ubiquitous but mystifying vacuum. *Its dead emptiness is physical and spatial but its plenitude is metaphysical and archetypal. The vacuum will, logically, be found to have properties associated with the operation of archetypal memory, that is, the lowest level of universal mind.* Physics will have made the psychological, psychosomatic connection.

Or will it? Physics describes material behaviours mathematically and derives principles of their operation. It accepts 'principle matter' in terms of subatomic particles, forces and so on but cannot accept that their origin might, in the form of an archetype, be metaphysical. Physics is not geared to what it cannot test. Archetypal physics is metaphysics and, therefore, not scientific.

High Wires.

What might therefore bridge the gap? Something 'high' it seems since, in dialectical terms, principle hierarchically precedes practice and since fundamental 'matter-in-principle' is 'higher' than its differentiated fall-out - bulk objects and events. Something 'extreme' because by trying to combine omnipresent gravitational potential in a self-consistent way with local quantum forces you are forced towards each frontier of infinity - cosmological extent and sub-atomic infinitesimals. Physics hits the buffers; it bumps hard into space; it knocks into the wall at 'both ends' of the endless in a place essentially outside the boundaries of experiment. At this point, therefore, it zooms in or out into regions of speculation that verge on esoteric forms of self-indulgence. Whole lists of theories claiming to be final have been serially scrapped. *Mathematical imaginations incomprehensible to the uninitiated have been constructed without either direct or indirect experimental support.* <u>Can there ever be, in principle, empirical support where such extremities are involved</u>? **Is the endeavour, you may therefore ask, one of physics or of metaphysics?**

Of course, lack of experimental support does not mean that a theory, even if it is metaphysical, is necessarily wrong. At present two main 'arches' vie to build consistency between general relativity (gravity) and quantum mechanics (explaining matter and the other forces). You might call them quantum theories of general relativity. Both involve numerous participants and hold conferences all over the world. They are the current orthodoxy but you shouldn't use them as a crutch of faith. It is unknown whether either fully corresponds with reality and is thereby in every aspect wholly right.

Theories change and change about. One attempt at a unified field theory, loop quantum gravity (*LQG*), starts by assuming even space and time are broken into pieces. *LQG* treats the 'topology' or 'quantum geometry of space' and gravity. In this world, stippled at the scale of a Planck length (1.6 x 10^{-35} m.), microscopic space is said to 'granulate'. These 'quantized' granulations are supposed to be made of tiny 6-dimensional balls - mini black holes called Planck particles. Such 'particles' are simply loops that make up fixed-frame, film-like space and time. From such an active void would *ZPE*, 'vacuum potential', rise? It is thus also speculated that loopy, digital space-time supports electrons and atomic structure. Are they, if they are what make even strings wiggle, not all-supporting and omnipotent progenitors? Our fathers? On a froth of nothing floats our universe! In this foam 'spin-networks' whirl you through different geometries of particles and motions by the agency of 'nodes'. The equations of the theory of relativity are independent of the shape of space-time or any fields in it; we call this background independence. Loop quantum gravity, unlike 'all-inclusive' string theory, takes account of background independence. This might seem to propel to a head start over its ambitious rival. Yet while it also proposes quantized gravity neither correlated particle or wave have ever been detected - not for want of trying. And, in a profession where large research programs are structured to aggressively market their own particular angles, fashion now dictates the mathematics of string (or hyper-string) theories rule the roost. Where there's no experiment there's no Nobel; but are the 'opposing' couple actually a complementary one? Are they in fact different approximations to the same ultimate theory (or *TOE*)? What actual factor(s) will they ultimately confront?

Mindful of 'chaos theory' and the sensitivity of outcomes to the subtleties of starting-points some now retreat from *TOE*s; can such a theory forecast weather in a week's time (or a thousand years) and tell us how a camel got the hump? Determination à Laplace is off the menu but, amid various statistical solutions, principles can still faithfully exact predictable effects. The hope, in principle, of excavating *TOE*s or a Mother Goddess, *TOE*, from nature's bedrock is alive. Let's take a look at the tool just mentioned, the 'rubber-band' theory of tensile strings.

What *are* string theory's strings? Firstly, grasp that fundamental particles are, just like hills, planets and solidity, illusions. By this token you can call the universe a kind of myth. Thus, what's really behind the whole shebang; what, hidden well beyond your microscopic gaze, wriggles inside each fundamentally minute Planck space? *Strings might (or might not) be what space, time, energy and matter are composed of.* They are supposed to elastically stretch, contract and wriggle in nothing, that is, space. Of Planck's length in size and thereby hundreds of billions of times smaller than an atomic nucleus, these infinitesimal *oscillators* are identified as nature's first, final, most fundamental ingredient. In fact, you've guessed it! *Nature's highest 'wires', strings, oscillate in different modes that correspond to different kinds of particle and force.* Photon, electron and quark are made up of them. From these sub-atomic particles there follow atoms, molecules until, layer after oscillating layer of 'control', last and lowest in the hierarchy - obvious, sensible things like gases, liquids and solids. In others words, to take a fundamental view of how a song constructs the universe we're zooming in at a level below particle because strings are visualised as the vibrant, ultramicroscopic core of particles and force-fields. Look at your index finger. Billions upon billions upon billions of such 'symphony' are what, the theory says, compose it.

	tam/ raj	*Sat*
	manifestation	*Latency*
	polarity	*Neutrality*
	extrinsic copy	*Intrinsic Pattern*
	consequent behaviour	*Archetype*
	instruments	*Score*
	thing	*Absence/ Void*
↓	*tam*	*raj* ↑
	static	*kinetic*
	mass	*energy*
	informed	*informant*
	end product	*influence*
	interference	*resonance*
	resistance	*flow*
	shape	*vibration*
	tension	*release*
	particle	*wave*
	discord	*harmony*

String theory, proposed by John Schwarz and Michael Green, is a mathematical spectacular, untested and perhaps not testable, that promises to combine general relativity with quantum theory, to unify forces and particles and to explain the values of all physical constants in terms of a single value, string tension (and a probability of strings breaking or joining called the string coupling constant). Such promise is manna from scientific heaven.

The theory, as fiercely defended but unproven as a biological analogue, the theory of evolution, has since inception passed through several incarnations. At first twenty-five dimensions of space were required along with particles that travel faster than light (called tachyons) - and there was no accounting for quarks. However, superstring theory deleted tachyons, included strong interactions and massive particles and reduced spatial dimensions to nine; the theory also seemed to predict, as well as photons, closed strings that are massless gravitons. Quantum gravity's royal banner was unfurled across its all-embracing sky. Indeed, breaking, joining, opening, closing and string motion seemed to generate all forces and to unify them with the particles we know. The vibration of open strings generate field lines of the nuclear and electromagnetic interactions; those of closed string give us gravitational fields; and bosons (force-carriers) are united with fermions (quarks and leptons) - the whole of physics might be unified. Not only this, but its constants are reduced to Newton's gravitational constant and the string coupling constant. Manna falls but at what price?

You still need ten dimensions (nine plus time) and can mathematically generate many different theories and predictions. Not only dimensions but extra constants, super-symmetrical particles and forces grow like Topsy. A current front-runner, which is incomplete and may or may never be tested, is a version of strings called M-theory (M for matrix, maker-of-worlds, mystery or maybe even muddle). This version sports zillions more possible worlds than there are protons in the known universe all in eleven dimensions with eight wrapped far too small to see! Competition's not the word but why on earth could a Creator not wax just as profligate? As we've already seen, laws and theories can't create a thing - let alone 10^{500} imaginary universes as opposed to actual economy of one. You might claim designers of complexity must be more complex still - unless intentional mind, whose nature is collapsing possibilities to certainty, has no material parts and consciousness, its base, is simplex. Nor, as archaeologists confirm, do you need explanation of an explanation when it comes to artefact and inference of mind. What, friar Ockham, do you think of M?

Beautiful mathematics does not mean, as far as nature is concerned, that the description's right. Nobody in the string community has, for example, predicted dark matter or dark energy. And everybody, calculating fresh 'flavours' and carving whole new 'landscapes' of string theories, is out of kilter with the actual world that we observe. We want uniqueness, exactly like our cosmos, not a million ways to show how it might work. In this respect string theory remains a mind-world cooking up apparently fantastical explanations without experiment or experimental predictions. Optimists of such an empirically-dead and wholly faith-based 'dreaming physics' don't, however, welcome criticism of further superstring extensions such as extra-dimensional surfaces called branes, brane worlds and other esoteric drift. There is reason to take any plausible hypothesis seriously but this is not, as any jury will aver, to claim it as a declaration of the truth. Yet if enough experts fervently believe in a formulation any measure will, as the theory of evolution shows, be used in its defence. What can such a theory, like string theory, not explain? Perhaps, therefore, as regards their grand designs, we're strung along by strings as much as Darwin's theory.

It is emphasised that although both string and loop 'cults' have been highly popular for almost a generation both remain entirely speculative, that is, well-supported by the highly complicated mathematical ingenuity of an academic industry, capable of marvellous confabulations but unsupported by any evidence. Because they involve quantities hidden in a minuscule Planck black box well below the observational limits of science, this seems likely to remain the case. If you can never confirm a theory by repeatable observation and experiment you have left science proper behind. So string theory is, although mathematical, scientifically indeterminate but definitely philosophical. It is, perhaps, simply a form of mathematical philosophy that represents a *bottom-up* attempt to picture the seeds of matter, unify the whole of physical science and, as a bonus, outflank 'intelligent design' and debunk, in terms of modern paraphysic, traditional metaphysics on the way!

tam/ raj	Sat
excitation	Potential
sound	Silence
colour	Radiance
issue	Source
↓ tam	raj ↑
less conscious	more conscious
less energetic	more energetic
inanimate noise	language
unreason	logic
insignificance	importance
cacophony	melody
towards darkness	towards light
silence of the end	rhythms of process

For several reasons it is, however, worth reviewing hypothetical strings (which may turn still turn out to unify the world) with respect to their vibration. *Firstly*, cosmic transmission of information and energy is, at root, vibratory; ordered oscillation is called harmony; harmony is the grammar of music and music is a universal language. Do 'strings' or 'loops' (a 'loop' is simply a 'string' with its ends joined up) embody a cosmic language of shape and behaviour, such as we read in our own so-called 'laws of nature'? *Secondly*, the mystics' First Cause is said to manifest as harmonic sound (music) and light. 'As above, so below'. What follows reflects the nature of its cause. Might a lower, physical 'first cause' (*figs*. 7.2, 7.4, 10.1 and 10.2) not reflect a metaphysical precursor? Do laws 'descend' from string-like music of the spheres? *Thirdly*, cymatics (the study of the effect of waves on matter) has found striking evidence of the patterns that sound can make on large-scale, let alone quantum-sized, matter. For example, 18th century physicist Ernst Chladni mounted thin metal plates on a violin, covered them with dry sand and watched the effect of different standing waves. *He watched, in a direct and obvious example of morphogenetics by energy, sound inform shape.* It is significant that Chladni's shapes often imitate familiar organic patterns that we see in nature and, especially, biological structures. More recently in Switzerland Hans Jenny invented a tonoscope which converts sounds into three-dimensional patterns in inert, inanimate material. It raises strong possibilities, always avowed by the mystic, that sounds and names have internal properties of their own. Certainly Chladni's 'intonations' are a fascinating reminder of, at physical level, the connections between sound, geometry and the development of form.

Won't rhythm always swing you up? The Master Analogy of Natural Dialectic is music. Both great music and the 'intonations' of mystic repetition (*mantra, zikr, smaran,* insistent chant or incantation - call it what you will) reach towards the heavens of our understanding. And do you remember Orpheus who, with his magic harp, could even charm the stones to life? His assertion that 'hills are alive with the sound of music' is, though pre-scientific, now confirmed. Don't atoms oscillate and energies spread out in frequencies and waves? At heart the world beats.

While melody is voluntarily composed the sort of repetition that makes up oblivious matter is, on the other hand, involuntary, closed in cycles but also vibratory. Is it, after watching Chladni's plate in action, ridiculous to think that waves both initiate and lay their fingerprints on form? Or to believe that, universally, physical vibrations drum up particles, atoms, molecules and things whose periods whirl as large as galaxies? Brain has its waves as well. These are physical but disturbance in the field of mind is called by Hindus '*chitta vritti*' - waves in consciousness. In this view mind both projects and registers such waves; so that the field of mind is really like a sea whose information is transmitted in the form of currents, oscillations and attunement. Is comprehension not a consonance called resonance and is resonance not the basic mode of learning and communication? Are not predictable undercurrents like a shape of mind called instinct? Could you not call waters with distinctive patterns archetypal memories - in which case what about a universal mind as well as universal body?

Is this a thought too far, too bold a step 'to think what is unthinkable'? For materialism, yes; but if you are not so roped then think it. Think, in an organic way, about a metaphysical cause of physical effect; allow some system in the way the patterns of phenomena are organised. What is an archetype but instrument of vibratory projection? Not only sound but light vibrates. You may remember (from Chapter 0: Pure Objectivity and Chapter 7) David Bohm's idea of holographic connection, a 'deep structure' of which every particle partakes. Is it not in four dimensions also that an instrument spins music, oscillates and whirls the cycling airwaves through its tune? An archetype shapes energy; it excavates the world from shapeless space. It chimes, each like a note, the particles from which the universe is scored.

What vibrates an instrument, what current energises information and drives out expression from the score? What breathed liquid harmony from Krishna's flute or struck the sound from Orpheus' lyre? What but a player set the reed of Rumi quivering? What, above all, about the materialistically unthinkable notion that a real, unspoken power sets emptiness a-fluttering in a billion different ways? What about the order of a *Logos* that, like wind through organ pipe or a plucked string, imparts various 'spin' according to the 'archetypal notes' or 'forces'? If so, what is the nature of involuntary repetition, one that breaks through archetype like waves express themselves upon the shore? What *is* the so-called 'Sound of God' that breathes the song of physicality? For Natural Dialectic logically the universal mind of body is the breath of lower and subconscious *Om, prana* (*figs. 7.5,* 7.6 (iii)). *Prana* is a name for the unitary force from which 'things material' are, through the instrument of archetype, projected. Egg is potential; egg is information so is archetype; cosmos is energy informed; a higher order is imprinted on our lower 'sphere'. Imprinted, embedded, vibrated at the root of things is definite and defining preordination, pattern from 'above'. Is not the whole of science one of revelation that a rigid natural law discharges flux, that archetypal resonance develops into what appears?

Is there not, in other words, above the body universal mind from which it takes its customary shape? Could natural logic be transmitted in the form of oscillation and could, as the mystics have insisted, an implicit radiance of sound and light (called *Logos*) have elaborated this explicit, three-dimensional world?

So that, as Mozart explained to those who measured metrically his sounds and thus missed their real reason, the artistic quality of cosmic music, poetry and dance is seriously reduced by science unto quantity alone?

The Infinite is spun into a finite tapestry by rings. Of these the most important is ringed energy, whose circumscription is vibration. A circle spins a good trap. Round and endlessly around, locked in a cycle, bound in rings. You can't escape. Rings isolate and differentiate. They are looped exchanges of information and energy in which minds and matter become involved. *From atoms to the whirling galaxies are they not the basis of existence? Energetic rings (figs. 2.14 and 5.6) are expressed in time as waves or oscillations.* A wave is an informant. It communicates its power and influence. It shapes whatever it meets. Cycle, wave or tide is expressed continuously as a regular period. A particular period is called a harmonic. Different particulars (or harmonics) either reinforce or interfere with each other. Reinforcement is a (*raj*) unifying process; interference (*tam*) negates, distorts and generates irregularity and eccentricity. Reinforcement (or resonance) is at the base of order, interference at the base of discord. Music, with its rings, quivering waves and other shapes of energy, is a good analogy of all this rocking, rolling time. As previously noted, a generic word covering all grades of energy is *prakriti*. At its metaphysical, psychological grade *prakriti* is sufficiently subtle to respond to voluntary influence; such conscious information is the activity of mind. At physical grade it is identified with atomic and large-scale, gravitational operations; its ambit is non-conscious and, as such, shows only passive, involuntary or automatic quantum and classical effects.

Of course, this kind of philosophical speculation has nothing to do with string theory - *unless it involves harmonic oscillation.* The hypothesis claims it all swings on strings and because, as you may remember, fundamental matters are 'above' and not 'below' so, in the way of wireless acts, these strings are swinging high in the foundations! What *are* these high wires? Strings are supposed to be as small as Planck dimensions and are, unless looped, unsupported at either end. They are not thought of as point-like 'billiard balls' but a kind of vibrant tension in space. Indeed, distinction is no longer made between string/ loops and their surrounding space; the nothingness of space becomes a dynamic quantity, its field primary and the particles secondary. That is, strings wiggle differently and, according to their different 'harmonics', different particles appear; each sort of wiggling string would serve as primary expression of an archetype. Particles and, therefore, gross matter built from them are 'epiphenomena' whose cause is a string field. What you and I call matter is the visible effect of oscillation.

Consider a *clamped harmonic.* Its struck chords radiate and have the power to influence shape. What is the clamping factor that confines vibrations of electrons or can modulate the shapes of rays of light? It is, in space, a clampless clamp; from it issues *unstruck modulation.* Similarly, strings are conceived as one-dimensional, stiff filaments whose ends oscillate at the speed of light. They appear but are not point-like; nor are they subatomic particles. In fact, because they are about Planck length their fluctuations have an even smaller size than this. These musical metaphors might be real and so constitute 'notes' at the very roots of matter but, of course, notes alone maketh not a symphony. What pulls these strings? Who plucks the wires of their harp? Whence, ask the stringer, does his un-struck orchestration come from; whence does perpetual dance arise?

Snip, Snip

Could string theory be snipped? We can review the points at which elastic strings might snap. Throw Ockham's razor overboard because, inconveniently as we've seen, strings involve ten or eleven dimensions, six more than normality even if you'd added subjectivity as extra number five. Like subjectivity the extra 'objective' dimensions are all 'implicit', that is, physically unverifiable. In theory 'strings' are one-dimensional, their two-dimensional aspect is a 'membrane' and, because the whole theory is a brain teaser, strange p-branes exist vibrating like tubes or sheets in extra-ordinary dimensions that pea-brains like me can't grasp! Strings, dressed up in convoluted maths, can even dance upon a '6-dimensional manifold' made of so-called Calabi-Yau spaces. Such geometrical nothings come in a large variety of shapes and are, in the infinitely focused eyes of some mathematicians, beautiful. The wranglers mediate by splitting modern and not medieval hairs. They intercede between a layman and these minuscule 'divinities of science'. How many strings dance on a pin's head? In how many heavens do they live? With eleven untestable degrees of freedom it becomes a mere hop to the frontier of space-rips, cosmic tubes, parallel universes, colliding branes and ... conflicting fictions?

This conflict with reality is, in reality, a snip. It is divorce from known fact and cut from common sense. The separation's interesting but it's all in a stringer's head. The theory is heavily involved in metaphysics (what is beyond observation - including the logic of mathematics itself), modern mythology (*ex nihilo* creation) and philosophy. *No doubt it is less abhorrent and more acceptable, materialistically speaking, than the simple assertion of a separate cosmic component, informative consciousness.* But it is difficult to visualise a microscopic panorama whose multi-dimensional patterns of vibration orchestrate the sensible, superficial fraction of our universe. Perhaps the mathematics of string theory is, as a physicist might calculate the sonic properties of a Bach fugue, an attempt to flesh out abstract, physically absent archetypes and enumerate the 'holographic' morphogenes of universal mind.

Nevertheless orchestration makes a good musical metaphor because it implies direction, order and a score according to which a harmonic projection is informed. Hypothetical strings are, according to theory, their own final arbiters; they are spontaneous oscillations that, comprising the substance of every particle of every thing, thereby control from an inner centre. Strings are not point-like but mass-less, geometrically precise undulations with such properties as spin. Cases can be contrived to include graviton-type spin and thus include gravity in a combined *TOE*. What is being proposed is motion in space prescribed by standing waves. Its information is, essentially, musical. Unstruck music. The resonance patterns that a fundamental string can support would give rise to the properties of an elementary particle such as an electron. In other words motion, energy, mass, force, charge and so on are determined by the precise, vibrant events that a particle's internal strings execute in symphony. This particular 'symphony', 'chime' or 'intonation' is its hallmark. Just as a spectroscopist understands that each sort of molecule emits its own 'fingerprint', 'signature' or signal by which it can be uniquely recognised, so at the level of strings different 'chimes' show as the different forces and particles of nature. The range that shows as natural in our neck of time and space is not necessarily the whole set of elementary strings or permutations. You may guess, however, that the rules authorise permutations like chords rather than discords - a system fundamentally harmonic and not cacophonic.

Let's snap, therefore, deeper into implications that derive from 'cosmic music'. The theory of music is implicit in any recital; similarly the explicit order of the cosmos is the product of implicit order. It is informed. Do you remember Chladni's plates? *From a musical perspective strings are dialectically interesting because of the association of vibratory harmonic (or harmonic oscillation) with shape.* Shape is implicit in energy; it is the form of force. Let knock-on interactions light, from different angles, different aspects of the archetypal hologram; change a note and thereby change the shape of energy's great composition. *In this sense strings would be extensions of their morphogenes* (see Chapters 7 and 16).

The universe is in fact a kind of machine finely tuned according to about twenty fundamental physical constants. What have string or any other model morphogene to do with these? Are string parameters, like notes, pre-set? Do the shapes and interactions of cosmos just 'fall out' of string or other kind of subatomic jostling? Could the topologies of strings/ loops act, even in another dimension, as the morphogenetic templates through which energy is associated with physical shape? Could they thereby constitute the instruments that shape such harmonic chords as elicit an electron, proton, atom, molecule and, vastly larger and more orchestrated, you? *Like organ pipes such geometrical archetypes would be the shapes of resonance through which the notes of cosmos can appear.* Of course, the idea of musical archetypes as a source of physical order is alien to one-tiered materialism and, therefore, the laboratory; but ordered, vibrant information, music, is Science and the Soul's central metaphor. Its communicative energy is very close to the core of Natural Dialectic, in whose case the topological geometries of strings, if they exist, would exist in the dimension of universal mind. They would amount to the basic memories of creation; memories that, like all memories, were initiated deliberately and are available for conscious retrieval and inspection. Is this remembrance actually the task on which string theorists are embarked? Are strings the point mind resonates with matter?

How can you test imagination, instinct or a memory? Or even what is seen as absolutely physical - a string? How can I test a theory wrapped in ten dimensions when there's only four to move in? String equations have endless solutions because of the many ways you can 'compactify' the six imaginary extras. Cut out experiment and you snip the strings that tether you to basic science, the laboratory and what it's all about - empiricism. You can't unroll a new dimension or unfurl some extra worlds. How, therefore, can science ever know the truth of one-dimensional loops or strings embedded in a ten-dimensional circumstance? If you deflate string theory down from its inflated size, if you cut its cosmos down to real dimensions how far do unholy grails recede? If you slice the high wires how far might you fall? *Strings and loops might be a myth.* Like such abstracts as Calabi-Yau manifolds, branes, quantum fields, quantum spaces, a multiverse etc. they are inaccessible to any scientific reality-check, that is, to physical observation. How, therefore, could you prove strings are or aren't? Like the theory's exuberance of dimensions, what connection have they with our physical reality? Such physics is as recondite as any theological enquiry. Nevertheless for all things great and small big bang cosmology and string theory are the current exotic, esoteric orthodoxies. Outside this main stream both have critics. Does bang go big-bang (Chapter 12)? And in the case of strings anti-stringers such as Peter Woit, for example, elaborate on the perceived depth of failure of a theory 'veering off reality'. Other theories, like Penrose's twistor, do exist but can a 'metaphysic of quantum gravity' ever be developed that successfully accompanies 'virtual zero-points' and 'quantum vacuum' as an explanation of that spectacular profundity, the origin and unifying basis of bulk matter? How will it incorporate the paradox of continuity and discontinuity or combine the simultaneous use of two appearances that relativity and quantum theory each describe? How will it unify fix with flux, communication and recipient, wave and particle or, for basic Dialectic, energy with information? Is construction of the cosmos not by 'legal' energies informed? Then seek the fountainhead of law.

Flow

The Dialectic indicates that sub-principles and their practices (their expression in physical nature) are derived from three aspects of (*Sat*) Essence - Infinity, Unity and No-thingness.

	tam/ raj	*Sat*
	finite/ relative	*Infinite/ Absolute*
	bounded	*Unbounded/ Free*
	conditional	*Unconditional*
	differentiation	*Undifferentiated*
	division/ duality/ multiplicity	*Union/ Unity*
	parts	*Whole*
	polarity/ direction	*Neutrality*
	motion	*Equilibrium*
	excitation	*Poise*
↓	*tam*	*raj* ↑
	resistance	*flow*
	fixity/ immobility	*flux/ motion*
	discontinuity	*continuity*
	interference	*harmony*
	disharmony	*concord(ance)*
	more constrained	*less constrained*
	restriction	*extent*
	impotence	*power-in-action*
	interference/ irregularity	*resonance/ cohesion*
	apartness/ isolation	*togetherness*
	towards incoherence	*towards coherence*
	difference in practice	*sameness in principle*
	sink	*source*
	binder	*liberator*

Major *(raj)* sub-principles of Unity include Continuity, Concordance and Liberation. Their *(tam)* anti-principles are, respectively, discontinuity, interference and confinement. Although they show in different ways, they are clearly, closely interlinked - for example in that profound expression of reality, the consonance of orchestra.

Orchestral swirls or the insistent pulse of rock 'n' roll well describe a world of fluid energy. Rhythmic flow. Both psychological and physical aspects of *(raj)* energetic continuity are vested in the form of wave. Vibration means stability-in-motion, insistent sameness in the changes or variation on a cyclic theme. Waves relate, communicate and merge. Thus they inform, pulse messages and, chord-like, radiate with spectral colour, shape and mood. Harmonic structures reinforce the principles of balance, counterpoint and symmetry. Vibrant harmony is healthy; resonant associations are the basis of the world's relations. You see but have you figured out the archetypal 'gig'? Did the natural quartet of physics' forces leap 'hey presto' from a bang or something altogether greater in its subtlety? Does the combo 'rock' a score that 'rolled' in mind before the start of physicality? Does science catch the 'vibe' but not address originality?

Psychological continuity shows an upward, desirable degree of unbroken harmony, agreement, inclusion, friendship and love. Its character shows as stability, balance and equanimity; its 'togetherness' is dynamic; and its sense of wholeness 'right'. Such 'lift' is, amid the downsides, what we seek. Resonance and reinforcement positively impel us towards a swelling crest of life. This levity, this wave of union is recognised as quality of mind. Which sets the question 'what is it lends to time its finest quality?' What is the way to achieve maximum 'lift', optimal lightness and rhythmic 'groove'? What is the way to stay transcendent, to maintain 'the wavelength' of its 'high'? The mystic answer is obvious. It is by expanding consciousness, by increasingly principled information governing behaviour and, eventually, clear communication with the Radiant Origin; this, without even a flicker of psychological discontinuity, is *(Sat)* Complete Stability. It is the Concentration of Enlightenment.

(Raj/ tam) existence is a thing of ups and downs. Changes. Continuities and discontinuities in mind, space and time. For the nature of discontinuity simply reverse continuity. Re-state the previous paragraph in dialectical reverse. *Psychological (tam) 'discontinuity'* shows as change in mood, desire and foci of interest. It also shows as a negative trend of 'breaks' or 'separators' in consciousness (sleep), knowledge (ignorance or unawareness) and emotion (lack of interest, boredom; and the divisiveness of repulsion, exclusion, enmity along with the exhausting 'drag' of disharmony or upset). The latter are depressing. They get you down. They break your rhythm, even break you up completely. At worst you come across the radical instability of madness. Neither saint nor criminal, neither fully sane nor insane, most people oscillate around the middle ground. Balance, elevation, education are the watchwords. Be rational. Go further. Explore. Does the mystic's radical, most logical solution, voluntary spiritual evolution, strike a chord? Such is Rationalism at its best.

tam/ raj	*Sat*
physical relativities	*Physical Truth*
variation-in-practice	*Invariance of Principle*
classical/ quantum events	*Unchanging Archetype*

partial/ apparent continuity of things	*Archetypal Ubiquity/ Non-locality*
relative continuity/ discontinuity of things	*Absolute Continuity of space-time*
material action	*Natural Law*
dividedness	*Unity*
↓ *tam*	*raj* ↑
isolation	*interconnectedness*
discontinuity of transformee	*continuity of motion/ change*
apparent rest/ immobility	*flux*
separation	*unification*
mass/ body	*force*
object	*action/ event*
point	*sphere of influence/ field*
particle/ locality	*wave/ dispersion*
phases	*spectrum*
mass	

With respect to physical nature there are, in dialectical terms, three forms of continuity.

(*Sat*) continuity is vested in precondition, archetype, material precursor. Physical forms come and go but a metaphysical archetype that is physically timeless and everywhere-in-but-not-of space remains intact.

(*Raj*) physical continuity oscillates in the atom, the planets, stars and anything that orbits, cycles or vibrates. Periodic waves are stable forms of 'round and round' and 'up and down', of regular, insistent power. Such harmonious stability, such dynamic equilibrium around a norm is called homeostasis. Look at a ray of light (*fig.* 12.2). Its vibrant structure is perfectly symmetrical, stable and continuous.

Sometimes, even in physics, the emphasis is on undivided continuity. Physicist Richard Prosser mathematically modelled particles based on the idea that each particle is a waveform filling all of space. He invites us to imagine an infinite undulation; to add many such waves together; and to find that the combination may be positive over a limited region of space but otherwise cancel out. Next, consider a particle to consist of such infinite waves; when observed by an instrument sensitive only to their combined effect a localised particle would appear. Waves from all particles cancel out, leaving space except at their tiny point of origin, called a particle. In this way Prosser reconciles the wave and particle aspects of light and matter. The bizarre quantum notion that atoms locally appear when measured but otherwise are spread like waves is called 'the measurement problem'. How can things be spread out everywhere (or even only several places) all at once? Might particles and light act as an analogy for elemental consciousness and (as its particle) a thought in individual mind? Or for material influence of archetypal metaphysic lodged in universal mind? In each case, nobody knows.

Prosser apart, modern physics, with its quantum theory, is now wholly overtaken by the concept, first conceived by Faraday, of field. Particles are described by probability waves governing motion that is specified at every point in space and time. Fields rule! Do you remember (from (Chapter 8) another interesting example of supposed field continuity, non-local reality? In a perplexing phenomenon called 'quantum entanglement' connections appear to propagate influences immediately, anywhere in space-time, unbounded by the speed of light. Put another way, a quantum pair of particles, such as photons or electrons, which have once interacted may somehow, it is thought, instantaneously signal each other even when widely (in principle, infinitely) separated in space-time. Such connections have no materially causal link and can be shielded by no type of matter or energy. They also violate the basic law of special relativity that nothing (except nothing in the form of universally expanding space) can travel faster than the speed of light. Of course, radiant light is also a fine example of continuity but its nature comes with the same warning sign that applies to the adamant nature of electrons and protons - 'Not wholly understood'.

Continuity shows as connection in a field of action. Excited fields are where things happen. Each represents a 'sphere of influence' within which 'messages' (like photons or notional gluons and gravitons) command and control a particular aspect of the whole physical exercise. Dynamic continuity, like a wave of influence, relates and includes. *Fields are a medium of communication and convey the patterns of relationship.* Within them interactive force is seen as a flow of energy determined, in type, by the composition of its particles. The lighter the particle the longer the range of its field. That is to say, the range of interaction depends on the extension of particle clouds; the type of interaction depends on the properties of the force particle involved; and the more particles join in and increase its current, the stronger the influence of an interaction. What are these messengers, these particles whose glue of information holds the world as one? What Dialectic does their type express, what principles do they import? Are they physical bearers of the fundamental vectors (*sat*, *raj* and *tam*) that instruct their various recipients how to respond? *In which case fundamental forces reduce to information; to the signals and communication department of physical business.* They are the way, in their different proportions, that principle is expressed, that the message gets through and its purpose is achieved. *The dialectical axiom is that information is prior.* Wireless wave precedes wired things. Information is in the wave. Resonant, inside information is radio. Concentrate on the wireless broadcast and demote the particle. *Matter is, from this dynamic point of view, simply concretised information.*

This dialectical logic is, of course, 'old hat'. Perhaps the most precise of all scientific theories, quantum electrodynamics (with Feynman's punning acronym of *QED*) integrates Maxwell's classical electromagnetism with quantum theory and special relativity. It describes an exchange of information; all electrical and magnetic forces are mediated by ephemeral clouds of wraiths. Precision's price is 'virtuality' in quantum vacuum fields (or omnipresent *ZPE*). 'Virtual' photons act invisibly as messengers in every interaction; light is at the root of such communication. Light, actual as well as 'virtual', is also intimately involved with life (Chapter 20). It permits knowledge (psychological light) through the mediation of sight, biological life through the mechanism of photosynthesis and, when absorbed, creates life-giving warmth. Light and life are very close allied (see *fig.* 20.3).

A more complex, complementary theory, quantum chromodynamics (*QCD*), describes exchange between 'massive' protons by gluons. Interestingly, (*raj*) *QED* shows an influence that declines with distance; (*tam*) *QCD*, on the other hand, which involves the roots of mass-centredness, inverts the circumstance so that influence increases with distance and an atomic nucleus is, therefore, tightly bound.

Binding generates the other end, the third and special case of continuity - (*tam*) discontinuity. Without discontinuity things would all be wrapped together, homogeneous and featureless - stable, cosy, gas-like, all the same. In this sense (*tam*) continuity is 'static', that is, smooth or regular in surface, volume or consistency. In another it *is* discontinuity. It shows as differentiation, separation and resistance, the 'negative' hardness that 'positive' flux needs to make things stand proud. The very word 'existence' means 'stand out'. To make any difference you need to 'up the anti'. You need anti-principle and counteraction. You need 'anti-light' in the form of a black spot, particle or 'dash of mass'. And you need to trap free waves, lose information and package energy in bundles. Such losses drop out into imperfections of shape and texture: they precipitate isolated, uncommunicative irregularities called different things. Is existence just a cloud bound up in the Essential Sky? With cosmos falling like complicated storm of snow?

Innumerable splits at every level constitute existence. The fix that fixes matter is invariance of natural law; and what fixes matter from its energetic fluxes is always loss of energy. Such loss splits the flow of continuity; it crystallises time and space; it precipitates distinction and, in the form of variations on a multitude of themes, binds up times and ties up individual places and events.

Split Flow

	tam/ raj		*Sat*
	duality		*Unity*
	split/ difference		*Consistency*
	complex/ simple		*Single*
	products		*Source*
	output		*Potential*
	extant pattern		*Field*
	things		*Void/ Vacuum*
	excitation		*Smoothness*
	motion		*Balance*
	informed/ informant		*Information*
↓	*tam*	*raj* ↑	
	outside/ outer		*inside/ inner*
	gross		*subtle*
	negative		*positive*
	relatively passive		*relatively active*
	more static		*more kinetic*
	differentiation		*unification*
	towards complexity		*towards simplicity*
	anti-principle		*principle*

reflector	*creator*
phenomenal array	*agency of natural law*
variation	*theme*
separate unit	*unit-as-part-of-a-whole*
discontinuity	*continuity*
split	*merger*
isolator	*communicator*
inertial/ dragged down	*equilibrant*
stoppage/ immobility	*motion*
informed	*informant*
shaped behaviour	*shaping process*
particle	*wave*

Units in the unity arise; discontinuities break continuity; disconnections, individualities - one thing can lead you to another sort. Splits in flow are when the interesting cracks appear. Splits in current are called eddies, vortices and bubbles. Those that fit are fine but when they start to contradict and interfere you're bobbed upon a choppy, changeful, tiring sea. Distortion, broken rhythm and notes out of tune soon creep up to break the harmony. Disharmony is like a fall from grace and, as you leave the streams of paradise, things start to freeze.

Things harden and slow down. The cracks become more obvious and difference set in stone. There's a thousand ways to fall. There's loss of broader focus into egotism whence flow greed, hate and aggression; and there's a million ways things cool from fluid into fixity of mind (called habit) and of shape. If you want myriad variation apply the anti-principles of interference, imperfection and irregularity. Observe the 'heat' of principle disperse; observe precipitates of information and of energy locked into local difference of detail right across a patchwork earth and sky.

(*Tam*) psychological discontinuity of consciousness occurs as sleep alternates with waking. Turned on (awake) we sense continuity, a seamless experience of time, space thoughts and things. Off, in sleep, you lose the lot. Human or any other form of consciousness shows varying degrees of sleep/ waking mix. It shows a range of moods like weathers and of competence like shades of light. Clouds of ignorance obscure the sun and storms of anger knock the colours out. Who ever thinks he is unfriendly, bent or ignorant? But when he comes to learn or learns to love his sunshine understands what sin-shame was. The weather of the head, the ocean of the mind is always changing. What principles and practice best maintain a sailor's sanity? What voluntary rules can best prevent his mind from cracking up? And does their formulation naturally reflect the reflex way that nature runs its business? In other words, is there a natural morality that works against collapse? That holds and beats the forces of stench and decay? The range of human wandering is, in its comprehensions, like a dream of shadows. What of calm sea or cloudless sky? Condensation can evaporate when heat's applied. When (*Sat*) Awareness comes about you understand split flow has cleared; when the concentrated heat of truth irradiates you know all shades have fled. The Knowledge is a clear and azure firmament. Enlightenment is Waking Up.

Materialism never talks like that! At its base extremity you simply drop into oblivion's everlasting night; nor will you ever find the final answer in non-conscious matter's irredeemable eclipse of mind. Such uniform eclipse completely numbs each lifeless object and event. What are the basic ways such natural mindlessness is automatically split?

	tam/ raj	Sat
	physical finity	Apparent Infinity
	physical	Metaphysical
	polarity	Neutrality
	classical/ quantum	Archetypal
	extant patterns	Pre-physical Instrument
↓	tam	raj ↑
	outer result	inner support
	negative	positive
	gravity	levity
	inertia/ restraint	stimulus
	mass	energy
	particle	wave/ force
	fermion	boson
	mode of classical effect	mode of quantum cause

This stack includes the dialectical trinity of physics. It includes transcendent as well as quantum and classical levels of 'numb' phenomena. The source of every level is the unseen one above; in this material case, therefore, immaterial archetype is lodged above its physical expression.

But if you begin with space (or even nothing lacking space) how do you get your head around its edge? How can you start to crack its uniformity? How are vacuums smashed to pieces or things spirited from nullity? How are continuities cut into separate parts? The Dialectic notes that, following the lawless bang's Primordial Discontinuity, science has identified three levels in the way that splits called individuals arise.

Split 1 is the 'great projection' or whatever nothing tickles into something new. It must be next-to-nothing and thus, like the 'big bang', very slight. Perhaps you'd call the prior instigator 'super-law' from which a cosmic program in the form of spheres or fields of influence is run by gradual 'freezing out'. And perhaps, first rung down on the cosmic scale, the primal differentiator is a 'drag' field (Higgs field) everywhere the same that, somehow glossing force with mass, thus acted as the source of all material diversity. Or perhaps the primary separator is as minuscule as 'quantum' (see Glossary) or a quantum fluctuation. A quantum (e.g. photon or electron) interacts as if a particle but travels as a wave. The theory that describes quantum behaviour was originally developed to solve puzzles such as 'the ultra-violet catastrophe' (whereby the radiation emitted from a so-called black box was predicted, when the temperature was sufficiently raised, to become infinite); the photoelectric effect; and the lines of absorption and emission spectra. As well as a particle, however, the word quantum defines a cut-off from a troublesome infinity viz. the smallest discrete unit by which energy can be changed. It is determined by Planck's 'constant of action', an infinitesimal amount which is used to relate energy with frequency or momentum with wavelength. In other words it relates the propagation of energy (radiation) to its interactions with matter. It is thus a quantitative link between two basic types of behaviour (propagation and interaction) and the two models (wave and particle) that equally describe them. Such a succinct unification is powerful. To describe cosmic operations in such simple terms places the quantum and Planck's constant at the heart of physical phenomena. From it are derived the natural units of mass, space and time.

A constant of action measures 6.63×10^{-34} joules per second. Next to nothing. Planck space and time are the smallest cuts in space-time. They are splits so small (10^{-35} m and 10^{-43} sec) that perhaps, despite speculation, nothing smaller can exist. There are no less than a quantum leaps. If 'grains' appeared at the dividing line between material and non-material existence they would break space, time and energy into fragments. Do you remember 'granulations of space', each point of which is thought of as a tiny 'ball', a Planck particle. Such so-called 'virtuals' (because they're scarcely real) are supposed to flit dialectically 'in-and-out-of-being'. Dialectically? Because neutrality 'above' gives rise to polarity; here neutral nothingness gives rise to particle pairs of opposite charge. It is their separation that is supposed to generate an electric field and their spin a magnetic field; together charge and vortex sally, they cascade down from subtle into gross; they gather mass and tumble down creation's gradient into a valley called the sensible conglomerations you and I are made of. Fixed is not fluent, solid is not vibrant but, in this sense, forever motionless. Impotent immobility, in mind or matter, is a vale of death.

You might say Planck's constant had *quantised* nature into 'blocks' from which everything else is built. This idea helps because, in maths and science, you must have something. Pure nothing is an abnormality that must be normalised; zero is an absence that defies a positive description. Therefore quantise it. Individuate. Make something out of most mysterious nothing. Obtain 'Planck particles' and get a virtual grip on all the rest. Thus all subatomic particles, such as massless photons, gluons and particles of sound (elastic waves called phonons) partake of a quantum. Even pendulous oscillations appear as a calculus of tiny steps; the wave itself is 'differentiated' into 'jumps'. With his 'quanta' Planck laid the first layer of granularity over a total smoothness of non-existence. First base, first ground, first splits of materiality. In summary, therefore, this minute constraint or confinement represents an explanation of the basis of physical individuality. It is the primordial individualiser that allows a universe of separate beings to exist. Look at it the other way round. If its value were set infinitesimally lower to zero, quantisation would not occur; the discrete world of forms (including the universe and you) would disappear into a continuous whole. *Planck's constant is, therefore, the prime isolator; it is the representative of discontinuity within continuity, of form within formlessness, of finiteness within physical infinity.*

It has been suggested that the speculative 'non-entities' of String Theory ('balls', 'loops' and 'strings') may be an attempt to describe seed or archetypal forms. Such ideals 'shape' energy into the lower, physical expressions of 'principle' or quantum matter, that is, microscopic levitatory and gravitational effects. In this view they provide the 'notes' around which a 'natural theory of music' is developed. These tiny items, forever closed (or implicate) in a black box below the Planck boundaries, would be instruments that 'organise' the wind of creation into a stream of melodic sound. They'd give rise to archetypal quantum particles and, thence, atoms, gum-trees and galaxies. In this view hills are indeed 'alive with the sound of music'. In this sense quantum continuities (called wave-like bosons) and discontinuities (called fermions) are observable expressions of 'set pieces'; they are derived from elements (string or other archetypes) activated by a subliminal 'super-force'.

Interpretation is a point of stand-off here. What, for 'flat' materialism, actually represents the intellectually discovered principles of 'quantum', 'exclusion' or the way the world's constructed to produce effects like atoms? Not how but why are things like this? Scorn may be reserved for any other angle yet, in a tiered universe, holistic archetype holds some explanatory power. A single, archetypal force is logically identified (Chapter 7) as metaphysical; an expression in vibratory terms of 'music', resonance and harmonic order well begin to model it. In their equations quantum physicists will recognise the metaphor. 'Super-force' is, in this view, both potential and the precondition for our whole phenomenal show. In oriental terms it is, while often translated 'universal energy' or 'cosmic force', called *'prana'*. By analogy with the national grid *'prana'* is the name given to stepped-down, 'household' current in the projection of Cosmic Dynamo, *Om*, *Logos* or First Cause. A sweep of the conscio-material spectrum (*figs.* 7.5, 13.6 and 15.2 (i)) reveals that its primary expressions could, taken in hierarchical order, be vacuum fluctuations, light and matters-in-principle (other forces and particles). Whatever your angle the physical vacuum informs, according to dialectical logic, quantum or minuscule behaviours. It is, therefore, supposed its 'upper', psychological, psychosomatic side is archetypal memory - thus archetype combined with sub-conscious energy excites an orderly appearance of the world. *If this is the case materialisation becomes a question of universal memories and their reactivation; or, in the well-worn simile of mystics, like a stream of air through various stops of pipe or plucked and vibrant strings.* Flute, lyre, organ or guitar - mnemonic resonance of *'prana'* is the activator of the archetypal instrument of cosmic music.

A shocking and seemingly ridiculous materialistic heresy thereby explains the principle driving a world of atomic structure, chemical reactions and their impact on the construction of large-scale objects, events and (when *human* intelligence steps in to use what is given) technology. If the physical part of the theory is correct, as innumerable experiments and calculations attest, could information relayed to a world of subatomic receiver/ transmitters (fermions) and their agents of information exchange (bosons or force particles like light) derive from archetypal memories? Mind's qualities and properties are not the same as matter's. Mind unifies; yours brings all objects and events around you into one perception. Universal mind, at archetypal level, does not consciously perceive; a form of memory or instinct does not know itself. Such mind is, however, everywhere and nowhere all at once. Thus, through everything's connection with its vibratory harmonics, it represents at subtle level all of physical connectedness. At once. As primal cause, moreover, it does not interfere with consequent material effects; within the physical domain causality (the law of cause and effect) remains inviolate.

Each archetype is thereby conceived as a vibratory program of behaviour impressed discretely (locally) upon the continuities of time and space. It is thus, logically no further removed from *ZPE*, charge or light than these from chemistry and our 'real' world. Subatomic science may be exotic but its phenomena are neither rare nor remote. Its 'characters' inhabit by the trillion the page you read or the finger you can waggle. Much closer than breathing they are, in quantitative terms, the commonest possible things. How far is a broadcast from the TV? Why should archetype or even, for that matter, the Source of all creation be that far? One dimension can be a part of, permeate and yet be apart from another. Again, as an example, take metaphysical centrality of mind within its puppet body. Why should the metaphysical not be micro-infinitely near; but also, as Centrality that's everywhere, embrace the universal body in its Macro-Infinite Dimension?

Memory is metaphysical but, remember, it is not creative. Its information is passively carried and so (see Chapter 6) belongs to the same hierarchical level as code. And whatever kind of pre-expressive framework of 'instruction', 'code' or 'grammar' you believe informs quantum behaviour, in qualitative terms the quantum aspect of matter definitely commands bulk, large-scale things (which are made of it) how to behave. It is the internal guide to external properties, textures and shapes. *Do you remember (from The Gate's Contents Box) that the world's a stage?* Its bare boards are the subatomic particles from which is grown a massive scenery. Materialisation's program is cascaded down the system, gradually becoming more sluggish and rigidified until it spills out, heavy and dead-beat, as the gross, sensible properties that energy assumes. It does so, of course, in the form of gases, liquids and solids. There is, in other words, a hierarchical devolution from shape-in-principle to sizeless, structureless, dimensionless 'notes'; and round these subatomic 'harmonics' accrue atoms, molecular chords and the symphonic world of nature. Shape-in-practice just reflects its inner principles. Gas is precipitated from a plasma; and sensibly shapeless but molecularly shaped gas is 'liquidised' then 'crystallised' according to the rules that govern matter's building blocks. From subtle to gross, everything is an aggregate of these intangible world-bearers. In which case how could anything be otherwise? Given its history (a process of conditions it has undergone) and its current context, then its constitution (be it beach, field, ionosphere or stars) is inevitable. It would, it does happen every time. If archetype is memory and gross matter issues out of subtle influence you ask again 'Was this originally intentional? Is natural law prescribed?'

	tam/ raj	*Sat*	
	polar	*Neutral*	
	relativity	*Apparent Absolute*	
	objects/ fields	*Vacuum Potential*	
	effects	*First Physical Cause*	
	receptor/ transmittor	*Medium*	
	push/ pull	*Go-Between*	
	actual particles	*Virtual Activity*	
	proton/ electron	*Photon*	
	hadron/ lepton	*Boson*	
↓	*tam*	*raj*	↑
	down	*up*	
	negative/ passive	*active/ positive*	

minus (-)	*plus* (+)
materialisation	*dematerialisation*
receptor	*transmittor*
absorption	*emission*
mass-centred	*radiant*
particle with mass	*pure energy*
weighty	*light*
proton	*electron*
deficiency of charge	*excess charge*
objectifier/ isolator	*communicator*
antagonistic opposition	*complementary opposition*
repulsion	*attraction*
off/ stop	*on/ go*
apartness/ individuality	*bond*

Note the mediation of light between virtual (metaphysical) and actual (physical) conditions of matter. The stack has moved down a stage to incorporate purely physical characters from the quantum and classical pillars of physics. This brings us to a powerful splitter/ splicer, an instrument of (*tam/ raj*) separation-with-togetherness - polarity.

Split 2 is polarity 'personified' - electric charge.

The influence of polar charge generates currents, waves, atomic bonds and, eventually, the bulk matter we breathe, drink, tread on and are made of. What is it?

In theory real particles can 'borrow' energy from the 'vacuum energy' of a so-called Dirac field to create evanescent, 'virtual' ones. Paul Dirac supposed that spontaneous production of electrons and their anti-material, positrons occurs in this neutral field. So perhaps a 'cloud of latent charge' 'creaks under strain' and with each 'wobble' ejects 'virtual' plus-and-minus charge-pairs; these, by almost instantaneous self-annihilation, return the energy that made them back to its original neutrality. Photons, which represent the 'pure, mass-less energy' of primary matter, are deemed both 'actual' *and* 'virtual' - flying at the very edge of space. Both photons and electrons indulge the 'borrowing habit'. Each particle is surrounded by 'an effervescent cloud of virtuality.' In a further step the electron's virtual photon pair is supposed to create its own Dirac pair (of new electrons). Where the electron of this pair is repelled from its parent electron, the positron will be attracted. So a virtual electron cloud will be separated from a virtual positron cloud. It does not matter that each unit of a cloud exists only for a tiny fraction of a second because continual effervescence replaces it and permanently sustains the cloud as a whole. It is the outer virtual electrons that are supposed to confer negative charge on their parent. Because the quantum field is everywhere, everywhere can be polarised like this. The vacuum has been polarised! In fact, it is claimed that there are zero-point interactions for strong as well as electromagnetic interactions. Could such complication be what charge, essential to our chemistry and therefore all the complicated composition of the world, is itself made of? Did you think simple charge - or anything down at the roots of nothing and infinity - was going to be an easy matter? What does your electrician think about the substance of his trade?

Is it clear what 'virtuality' or 'non-real being' is or isn't? Does it, although you think it 'must', in fact exist? Not everyone is happy with one particular result of Dirac's imaginative physics, a 'sea' of electrons in 'negative energy states'. They are not observable but are they, as an artefact of his evolving theory, unreal as well? Do you remember Planck particle pairs? As in their case could it be that ephemeral electron-positron pairs and, indeed, other species of particle and anti-particle pairs might arise? Indeed, 'real' ones can, using bubble chambers, be interpreted from tracks. Polarity has been (*tam*) materialised from unity; and in their mutual annihilation unity is (*raj*) dematerialised from polarity. So, in accordance with Dialectic, such pairs are simply an excitation of the vacuum. They 'crystallise' and, just as soon, evaporate into a pure, energetic excitation of the vacuum - say, *ZPE* in *ZPF* (zero-point field) or Dirac's ground-state 'electronic sea'. How can you call such lively emptiness dead space? It looks as though the vacuum's full - except its plenitude is out of sight.

It was suggested (Chapter 7) that so-called 'virtuality' might correspond with the causal level of matter. As such it would inhabit the psychosomatic border between mind and matter - not conscious mind but the level of universal mind consisting of archetypal memory. Physics is not part of such a metaphysic but umbilically derived through it. **Archetypal operation is as 'non-local' as parental *DNA* in every cell; it is the omnipresent context out of which the cosmos works. Nuclear yet omnipresent; immanent, a so-called 'noumenon' (or mental entity); unmanifest yet manifesting every physical phenomenon.** Is the archetypal concept any more a nonsense than the 'cloud' of ether Dirac seemed to speculate about? Or perhaps it starts to inwardly define the 'woolly' nature of the physicist's 'cloud nine'.

Can memory not store plans? The characteristics of quantum matter aren't the same as those of the gross matter to which it is linked, is at the heart of and supports. If mind is distinct from matter and metaphysic from physic, is it not reasonable to assume its different properties? The action of a predominant (*raj*) informative, unifying principle has been likened to the holographic storage of a single whole, expressed in its parts in myriad pixels; or the presence of coded information (written in *DNA*) representing the whole body but present in every cell; or the receipt of a single broadcast by innumerable TV sets tuned to its station. It is not, therefore, unreasonable to treat basic (subatomic) patterns of matter as 'resonators', each with its archetypal 'frequency'. In this view such a resonance - as specific as, say, the note middle C on a piano - actually creates and *is* the particle. One 'memory', one archetypal program patterns a particle in such a way that, due to this note or chord-like 'law' and its interaction with other 'laws', the output is bound to accord with a synthetic purpose. For a chemical this is its part in a universal stage; for biological archetype it is a *'persona'*, clothing or mask with which localised life is dramatised.

An analogy for archetype is thus 'clamped string' (see *fig.* 6.2: legend). Harmonics. Some harmonics or harmonic complexities are allowed. Discordant patterns aren't permissible. They are ruled out. The number of enduring particles is therefore distinct; hence also the atomic and molecular 'chords' that 'music' will allow. Retrace the chain of command. A quantum aspect of matter is informed by a 'virtual' archetype how to behave. The top, internal command and control centre is universal archetype, a memory or 'natural register' of the instrument on which the 'sound and light show' will be played. Check *figs.* 2.11 and 8.2. This latent level called 'transcendent' or 'absolute' is first cause of the shapes and behaviours that energy assumes and science studies. It is Natural Dialectic's physical point of origin.

What about chargeless communication between electrons? Their emission or absorption of light involves no change of charge or mass, only 'motion' between states of excitement. 'State' is the right word because the 'motion' is not, according to quantum theory, continuous; it jumps from one permitted state to another with nothing in between. When light is 'materialised' as thermal radiation (heat) then chemical bonds can be made or broken; but is there nothing between friends when it comes to its own particles, photons? Are they distinct or do they (and other kinds of communicators called bosons) have a space problem. Ghostlike, they need to be able to occupy the same space at the same time. They need, in order to register the intensity of a message, to overlap, collide or self-annihilate but interconnect in the most thorough way. You might even say they 'become' one another; more merging increase the lux of their 'enlightenment'. In this sense fields and their particulate aspect have a property that other matter lacks. You can squeeze as many force-particles into the same place as you like. This unifying tendency interpenetrates distinct entities called leptons, such as electrons, and hadrons, such as protons. *It allows a variable number of force particles to inhabit the same space and so vary the intensity of a signal.*

If field 'strengths' can crowd together in the same spot, the things they influence cannot. *While relationship involves continuity, individuality needs discontinuity.* Mass, in the form of the abovementioned distinct entities provides it. What *is* that indestructible next-to-nothing, an electron? Is it polarised vacuum? Simply a light, blurred spot? An equilibrant or relator where the proton is an isolator and the basis of all bulk distinction? Or is it a communicator, an informant that carries the binary message of electrical charge - opposite attracts, like repels? Yes, it's on or no, it's off - the basis of cosmic computing? It is certainly, as well as a communicator, a violator. The working hypothesis of Benjamin Franklin described 'divisive' charge in terms of 'positive' and 'negative' polarities. The choice was as arbitrary as which side of the road you drive so that, although dialectical reason would (according to the previous stack) confer 'positivity' on the electron, it is of no practical consequence once a choice is made and the habit stuck to. What counts is that 'like charges repel and unlike attract'. But how, year in and year out, billenium in and billenium out, does the force never weaken? And how, in a second violation of Conservation Law, do electrons flow round a circuit, impart energy to circuit components and yet lose neither speed nor charge? We can ignore this violation and develop a set of abstract mathematical models, called Circuit Theory, to help design circuits and their components but still only pragmatically describe and not explain the essence of 'electron-ness'. Despite imaginative speculation, though, electrons are the basis of chemical connections and, down-to-earth, our starry universe!

This brings us to *Split 3*. Somehow, one presumes, most quantised particles have been imbued, by an etheric kind of glue, with mass. Mass 'trips the zip' and thus, on cooling, stable atoms can emerge. If unlike charges must attract a child asks why these atoms don't collapse. Why does the world stand up?

Electrons radiate energy if they accelerate, as they do when orbiting. In losing 'strength' how long can they keep it up - and not collapse into a neutral, nuclear coalescence? Electrons, attracted by protons and thereby drawn to cluster round and coat the nucleus, should rapidly lose their various energy states into a single ground state. Then no reaction could occur. No chemistry. Yet all students of chemistry learn that, fundamentally, atomic structure involves electron orbitals. Mendeleyev's periodic table, chemistry's analogue of biology's genetic code, depends on it. So do their research projects and, much vaster in extent, the whole wide world outside.

Over several years a description was evolved. Electrons were assigned integral quantum numbers, related to the number of nodes in a vibrating system, that describe characteristics of their atomic motion. They include the angle, size and shape of orbit and direction of rotation. Each 'permissible' energy state (or 'note') was assigned a distinct set of integral numbers. Of course a description, even a quantum description, is not necessarily an explanation. What actually maintains atomic structure? In a children's playground you time your pushes to keep the swing swinging. What if a dynamic equilibrium were established whereby an electron radiated energy but, in the way of a resonant push, absorbed the same amount from *ZPE*? The vacuum itself would whack electrons round the nucleus; and if atomic orbit energies are sustained by vacuum energy you might observe that space maintains the whole stability of matter. Again, dialectically, nothing keeps the whole show on the road!

Whichever way, Wolfgang Pauli was able to perceive that each 'quantum state' in an atom was distinct and, crucially, if a particular state was occupied then the next electron would have to move to the next higher unoccupied one. His Exclusion Principle therefore states that no two electrons in an atom can have the same set of four quantum numbers. Collapse to ground state was averted and electrons cannot occupy each other's space. Nor can protons or neutrons. Theory was married with reality so that distinction and, therefore, the survival of atomic, molecular and bulk structures is assured. By this principle alone the world is stabilised. The rules of chemistry, solidity of matter, stability of stars and so on illustrate this (*tam* ↓) principle of separation.

What is it, though, that represents the principle? Why is it there, how does it work? When you tug its intellectual thread you find, behind atomic structure, more than simple separation. A universal archetype might, 'spookily' or otherwise, represent material and yet immaterial connection of the cosmic system as a whole. Thus particles (electron, proton, neutron and so on) may automatically be 'balanced', harmonised in different slots of energy throughout our great 'balloon'. We've seen, from Chapter 0, that in a tiered universe an archetypal phase is logical. Since archetype is immaterial its connections need not operate in matter's time. Immediate entanglement; interconnection everywhere at once! This is as nonsensical as quantum physics seems, sometimes, to be.

At any rate, we now have three expressions that illustrate the development of an anti-principle, discontinuity. They are Planck's Principle of the Quantum, Dirac's of Polarisation and, for orderly and stable individuality, Pauli's of Exclusion.

These splitters are important in construction of the periodic table but splicing and neutrality also play a basic role in chemistry. Charge excites and changes but splicing (bonding) and neutrality keep atoms in their place. Molecular and bulk units - almost everything there is - are made of them.

Consider, therefore, not only the (*sat*) 'transcendent' unity/ potent neutrality of archetype; and actual/ virtual photons are as neutral as the fluctuating *ZPE*. If electrical charge is generated from the vacuum then, two from one, neutrality bears polar particles in pairs; (*raj*) excited flow occurs between polarities; and, one from two, in their annihilation complementary duality returns to chargeless energy called, generally, light.

And, naturally, the Dialectic's reflective asymmetry predicts inversion of (*raj*) stimulus will be expressed as (*tam*) inertial or impotent neutrality. You should find the shape of mass stability without the stimulus of charge. You do. Neutral atoms conform to pair-parity of charge; they must have the same number of electrons as protons. Indeed, it is thought that a balance of protons and electrons is incorporated into cosmic structure. This implies original polarisation of their same, single source. Such numerical symmetry cancels out. It equilibrates. Furthermore neutrons, which decay into a proton and electron, are particles whose job seems to equilibrate. They stabilize the ferocious antagonism between electrical repulsion and strong nuclear force that threatens to tear the heart out of atoms and, thereby, destabilize existence. In short, they grant the basis of our massive world, diverse atomic isotopes, stability!

From a neutral whole arise two polar parts; but you equilibrate by splicing too. Polar halves combine, every chemist knows, to make a neutral whole. Pre-active neutrality is matched (see Chapter 3: Primary Inversion) by post-active. A neutral atom is not (except for a 'noble gas') chemically neutral; but its electrical neutrality is balanced by an orbital imbalance and the tendency to lose or gain electrons. Bonding revolves around the electronic glue of charge; electron switches, push-and-pull, are at the heart of chemistry's great shunting yard. One from two; the dialectical bond is a post-active product of polarity. The motion of reaction flows into inertial equilibrium. Ionic bonds complete exchange; covalent sometimes share electrons equally, in other cases leave residual polarity - that is, as organic chemistry well demonstrates, imbalance yielding reactivity. **In every case throughout the universe the motions that make things obey the principles of Natural Dialectic.** Ordering of the starry universe accords with (*sat*) potential (*raj*) excited into flow between poles ending in the 'ash' of (*tam*) exhaustion. This is the way of change. Potent cause effects a rearrangement; neutrality of possibility emerges into actuality; the pact is drawn up, the bond seals transformation to a new contractual unit of split flow. Balance, imbalance and the motions of polarity are the language of simplicity composing complex worlds of chemistry.

The Spring of Inter-polar Motion

tam/ raj	*Sat*
relativity	*Absolution*
static/ kinetic	*Potential*
dis-equilibrium	*Equilibrium*

	periphery	*Centrality*
	imbalance	*Poise*
	motion	*Non-motion/ Peace*
	change	*Changelessness*
	springing	*Spring Coiled*
↓	*tam*	*raj* ↑
	static	*kinetic*
	spring unsprung	*springing*
	slower	*faster*
	immobile	*mobile*
	toppled	*balancing*
	destabilization	*stabilization*
	flat/ straight	*vibrant*
	depressant	*stimulant*
	contraction	*expansion*
	confinement/ restriction	*release*
	resistance	*drive*
	hindered motion	*unhindered motion*
	brake	*accelerator*
	pull by inertia	*push against inertia*
	pull to centre of gravity	*push away from centre of gravity*
	acceleration towards mass	*acceleration from mass*
	materialisation	*dematerialisation*
	descent	*ascent*
	weight	*weightlessness*
	gravity	*levity*

Name your pole. Doesn't any motion flow between poles of one sort or other? Dialectic stacks make note of them; they organise the motive power. Cosmos in its aspect both of mind and matter dances round the maypole of Centrality, that is, around Main Dialectic.

Centrality at every stage is the object of a concentration. It is a goal, norm, target; or the measure of an object's purity; or a force's rule of law. Law is core and causal. Informatively central, it directs behaviour; choiceless energy is governed to evolve what is, by definition, preordained effect.

Chains of lesser causes and effects revolve around a Supervision Order. Such orders are, upon Mount Universe, devolved to stage below from one above. For example, nature's supervision order (called the laws of physics) is derived from metaphysic in the form of archetype; and your body moves about according to the supervision orders from your mind. In other words, two kinds of motion follow cosmic order as it drops from (*sat*) potential through (*raj*) process into (*tam*) end result. They are (see *fig.* 0.7 (iii)) 'vertical' informative and 'horizontal' energetic thrusts.

The former is the order of creation. It runs from creativity 'above' to the expression of its formulations in creation 'down below'. Check *fig.* 6.1. The control of automatic, passive matter is unseen; the authorship of cosmic data structures is as hidden as an instinct from plain view.

The latter floods around in energetic chains of transformation; its changes of location, state and shape compose the constant motion of both mind and matter. Science studies for its most part just the latter - matter.

Let's follow motion, *top-down*, down and round Mount Universe.

There is, you will have guessed, no existential motion in *(Sat)* Essential Potential.

A simple analogy that describes the dialectical perspective on motion and change can be made with the action of a spring.

The (Sat) Poise of Potential can be characterised by a *'spring unsprung'*. It has pre-kinetic potential or, in simple terms, the capacity to act in a particular way. It might be termed poise, immanence or loaded readiness - concentrated calm before an excitation. Positive neutrality. *(Sat)* Equilibrium precedes existence but gives rise to motion of the mind. 'Solidity' or 'fixity' of mind shows in its inertial kind of equilibrium - subconscious sleep and memory. As memory is fixed but latent prior to its recall so archetype is fixed but latent as an information-field. You cannot see a causal instinct, only its aroused, reflex effect. You see the body react in a certain way. Nor do you observe a causal archetype; only reflex operations of the body of the universe. Such medium, such psychosomatic interface of mind with matter *(PSI)* is nothing physical; yet it dictates the manner of things physical. Universal memory is, in this informative respect, their 'spring unsprung'.

In other words, *(Sat)* Poise precedes the mind that is, in turn, body's potential. It is immaterial and, in physical terms, invisible and motionless; yet its thoughts and memories affect the body's operation. Mind moves body. At base it shows as archetypal memory but physically is nothing yet. A latency. As with individual so with universal mind.

There is another sense in which *(Sat)* Equilibrium governs psychology and physical science. It is the basis, where nature balances its books thoroughly and precisely, of equation and equivalence. '*Karma*' is Sanskrit for 'action' and this major '*karmic*' principle not only governs the ups and downs of life but also allows humans to make sense of them. Sir Isaac Newton clearly, with his laws of motion, perceived the same principle of equilibration vested, for physics and chemistry, in the stability of natural law and the conservation (or stable quantity) of energy. It means that relationships are balanced, overall equality is maintained and, with it, an ability to describe the wholly involuntary world of automata in terms of exact mathematical equations. Einstein concurred but you might also worry, as his friends did, whether such determination leaves you any choice. Lack of choice would knock out any basis for morality. Resolve the paradox of free will and determination will you please?

(Raj) motion is characterised by *'spring un-springing'*.

(Raj) psychological motion, both conscious and sub-conscious, is detailed in Chapters 13-18. In brief, however, such motion is vectored by perceived goals. Goals are targets that involve trajectories. Purpose energises with the thrill of current. Will. Mind launches 'guided missiles' fuelled by strength of desire and kept on course by feedback. Their course is, for their time in flight, a 'norm'. In an individual's mind norms vary in specific application to a general theme - survival, comfort, fondness, satisfaction and so forth. In other words, where

matter's information is passive and its norms pre-fixed, active information (purpose and imagination) is concentrated towards what seems, to the concentrator, 'good'. In hierarchical terms an inspirational idea is coupled with executive will-power and imaginative manipulation. Such internal, psychological government sails a sea of external events. These deflect or obstruct any particular course. It is therefore clear that purpose is the psychological norm about which any homeostatic course is tacked. A norm amounts to an optimum, that is, a reflection of (*sat*) balance, satisfaction and peace - characteristic principles of Main or Primary Dialectic.

If you intend to sail an optimal course, you have to choose the right norms. If, like matter or unconscious organisms, you lack voluntary choice then nature provides in proportion to the lack. There are scripted archetypal patterns of behaviour, instinct and (see Chapters 16-17 and Glossary) morphogene, so that the 'right' or 'natural' course is automatically steered. If you could choose, not by instinct but by reason, where might 'right' logic guide?

Buddhist psychology is logical in choice. A patient visits, by prayer or appointment, Doctor Buddha and the Doctor diagnoses existential pain. 'Yes', he says, 'you are shafted in a hole; if you want to scramble out then follow my prescription'. He prescribes the Four Noble Truths and as the fourth an Eightfold Path. Western faiths are oriented, it might seem, more around devotion and the cult of personality than logical psychology; but both kinds of teaching start with the same material, a wretched and misshapen state of human mind. Each shows how to regain original shape, to recreate its beauty and, where sanity is beautiful, take beauty to its natural extreme - *mens sanissima*. The Truth is maximum happiness which, rather than a lesser scientific truth, is metaphysical. This world's flux deals tumult, shock and pain; it's antidote is other-worldly, that is, out-of-worldly peace. It resides most completely in (*Sat*) Poise, that is, in Transcendent Peace. In wisdom's winning formula! The best way for mind's spring to spring!

It's no secret. Norms are laws. There is natural law (called *dharma*) and there are man-manipulations. We all place (*raj*) positive norms high on our agenda. Governments spend billions and billions of social, political and legal currency on imperfectly formulated and executed plans to 'improve or rectify conditions'. Ignorance, imbalance, insecurity and lack of stability are anti-principles whose pain whips up a well-spring of desire for a return to principle, norm and stability, that is, to psychological health and corresponding physical health. Counteracting negativity may involve reflex response or voluntary determination using positive remembrance, thought or meditation. The quicker loss of balance is regained, the more successful a response is judged. Success depends on either controlling or ignoring outside disturbances (including those arising from one's own body) and also on internal 'self-control'. The whole point of education is to impart methods of counterpoise, ways to control events and emotions and thus regain (or optimally never lose) internal equilibrium. Composure. Radiant happiness. Education is, in the last analysis, a tool to best achieve peace of mind - though you might think some modern educationists had failed to grasp the point. *The real question becomes, therefore, one of desire. The satisfaction of which desires most efficiently produces peace? What wishes best yield lasting happiness? This question is at the root of education, morality, religion, politics and law.* Ethics is outside the scientific scope - unless materialistic scientism wishes to parade a life-style or ideology the place of physical science is body, the place of ethics is subjective mind. Are you not involved? You are up to and beyond your neck in it! Actually we are all wholly above our necks in the dialectical business of counteracting knocks, regaining balance and steering an optimal course! We think, with varying degrees of expertise, of little else.

(Raj) physical motion involves energetic as opposed to informative transformations. Two basic vectors swing about a pivot, an equilibrator. *(Raj)* stimulus springs up, initiates and radiates; *(tam)* inertia slows and weakens until influence or motion fade away. Levity flies out or, in the gravitational case, mass shrivels space. The logic of the Dialectic would suggest both vectors helped establish and continue to sustain a balanced and yet changeful cosmos.

Hence to the 'spring sprung'.

What about *(tam) psychological exhaustion*? I give up, I surrender or I fall asleep - the 'little death' I die each day but keep on living. A spring sprung is finished. It is immobile and incapable of further action. Psychological deep freeze is iced by terror but the mind's immobile fraction also harbours memories - personal recollections, habits, instinct and so on. It includes, in universal mind, the stage of mental archetype that generates the shape of lifeless things.

Don't things run down, grow 'tired' and stop as well? Don't they wear out with *physical exhaustion* and just fall apart? Do you remember (Chapter 2) the quality of *'tam'* as 'sink' - a state that is or tends towards inertial equilibrium, a pole of negative neutrality? Thus fluids in their lowest state diffuse into a passive kind of unity. You might call this condition 'negative homogenate' or a 'deconcentrate'. Or else things, when sufficient energy is lost, in the end precipitate into a lowest state - exhaustion in solidity. As psychological immobility involves fixities of mind called memories, so physical immobility involves fixities of matter. Solid objects are creation's edge; all around, including you, they represent the limit of the universe. Discontinuity. Crisp border and the terminus of energy. They are like full stops whose punctuation indicates the end is nigh. There is a real sense in which matter, already a subjective vacuum completely exhausted of consciousness, is itself the *(tam)* last phase of creation. Furthest from source it constitutes an outermost 'concentric sphere'; its subsidence becomes creation's shell. What's around you is a cosmic buffer. By such reckoning the physical universe is a vast spring unsprung. This whole section (Chapters 7-12) relates the split from archetype into material distinction and, within that distinction, varieties of loss of power as even the 'dead' segment of cosmos is finally, exhaustively deflated.

In short, loss of energy and information lead to impotence. Impotence marks the (↓) trend of negativity into subtendent kinds of unifying and of unity. Psychological subtendencies are ignorance and sleep. What, at matter's base pole, did we find? How, in other words, does energy, the engine of the world, run out of puff, refuse to cycle and thus grind to the end of cosmic line? What about the station's limits, sides and buffers to the train of things? How do light, charge and other quantum energies discharge themselves; how does 'principle matter' deploy into classical practice; and how does energy, masked in the form of various interactions, dispose itself around the 'real world' that science treats so scrupulously? In the end, at the bottom where it's really flat, time has expired. Reaction is complete; event is over. Fallen flat. With a puddle, a dead battery or body there is no potential left. Into this category is crystallised the final process which, fixed and frame-frozen, we call solid objects. Dust and ashes; cold and solid objects are the 'spring unsprung'. Their condensed illusion sprawls across the cosmic floor. In which case could the world-train flop against a block of absolution? At 0°K or in a black hole would it hit impassability? Completely squashed with neither dancing oscillation nor momentum left, must even its electrons stop? Atoms could be frozen solid; or there would be no space or motion even on the inside of an object's immobility. Cut. Final frame. Full stop.

Boxing the Infinite

	tam/ raj	*Sat*
	body/ mind	*Essence*
	finite/ limited	*Infinite/ Unlimited*
	degrees of freedom	*Freedom*
↓	*tam*	*raj* ↑
	automatism/ robotic state	*conditioned free will*
	tight bondage	*loose bondage*
	static object	*motion/ change*
	capture	*release*
	confinement/ boxing	*liberation*
	isolation	*unification*
	isolated, individual unit	*interacting unit*
	division/ apart	*part/ integrity*
	materialisation	*stimulation*

Boxing the infinite might, from the time that Max Planck conceived of quantum physics in order to contain infinities seemingly inherent in 'black box radiation', seem to summarise the subject. Albert Einstein's work on the photoelectric effect supported Planck's. Indeed, space, time and gravity apart quantum mechanics deals with all finite physics. But if a treatise that is mathematical is what you're after this is not its place. This section attempts, instead, to theoretically connect materialism with Natural Dialectic's broader principles.

Imprisonment. Local lock-ups. (*Tam*) discontinuity is dominant and isolation is the anti-form of unity.

The going toughens, balance slips and the emphasis is tightened into left-hand anti-principles. Eventual strangulation; black-out; materialisation is a (↓) downward process of increasing loss of knowledge and of energetic vibrancy. It is how, by degrees of entropy, the Infinite solidifies. How can such cosmic *rigor mortis* be retrieved? Does anything refresh the universe?

Solitary confinement, lacking all communication, is the psychological nadir - unless you add what stones don't feel, an element of pain.

In the physical universe all information is completely locked up. It is entirely passive. Matter's shackled up in bodies; in bodies biological mind is much constrained. Is not pain a blocker, a constrictor, an acute and violent locality? In this view, as the Buddha noted, physical and psychological incorporations are a zone of suffering in various degrees.

Freedom is the principle whose anti-principle, confinement, really clamps to earth. Negative (*tam*) anti-principles that have come to dominate this world include non-conscious automation, division, differentiation, discontinuity, disequilibrium, isolation and confinement. Their emphasis is vested both in corpuscles (or particles) and larger-scale bodies. *In this case the anti-principle of liberation, confinement, is the dominant aspect of localisation, that is, of materialisation.*

How does it work? How is the Infinite actually boxed? How is Freedom constrained or Potential realised?

Confinement is a range of freedoms split from Perfect Liberty. The Causal Level of confinement is First Cause and, accordingly, its freedom is voluntary and includes conscious will. Such freedom is inverted (i.e. automated) in the lower, involuntary first cause of physical creation. This lower primary is, according to the logic of Natural Dialectic (see esp. Chapter 16), not physical but metaphysical. It is mental and, at the comatose abyss of universal mind, abides in memories. These memories are couched in terms of harmonic grammars, that is, oscillatory fields. They constitute, on a cosmic basis, the passive, grammatical level of information described in Chapter 6 and developed in this chapter. Their information is realised, materialised or 'crystallised out' in (*tam*) creative descent of the conscio-material gradient. Archetypal shapes of mind precede the shapes of matter.

Information is thus regimented, energy is cuffed in irons. What is, in principle, linked all-as-one now is pegged in myriad different units. Cosmos is its scattered parts and these are locked up very locally. The power of lock-up is negative, left-handed and (*tam* ↓) gravitational in the dialectic sense. It is variously personalised as *Kal* (time-lord), Satan or 'the bringer down'. Seen from this angle creation sums to a vast range of reductions. Creation is a scale of increasing bondage down Mount Universe's slope; it is composed of myriad partitions each of which suggests but veils a larger whole. Constriction varies in extent but whether in a golden palace or grey prison, everything is locked in grades of separated space.

How, though, might you bottle up Infinity, how cork Liberty so it can be uncorked again? Perhaps, although not obviously, in a human skin. *A mystic's bondage is, of course, to Uncorked Freedom and allegiance to the Bell. (Sat)* Liberty is unconditioned, unqualified, unlimited and fully unlocked only in the Infinite. Thus if the whole construction (↓) fell from Nothing worldly gain is freedom's loss; but freedom's loss in shapes of mind and matter, when reversed, becomes annihilation of that pair and paradise regained. Such (↑) upward annihilation is constructive. Nothing is nothing if not Indestructible; redemptive Liberation therefore rises to Original State or, rather, Statelessness. Such Statelessness is called, to a Hindu, '*moksha*'; to a Buddhist it means 'release from the wheel of *samsara* (changeful existence) and the cycle of reincarnation'; and to semitic faiths achievement of its Central Truth means Freedom in Jehovah, Father, *Ain Sof* (the Kabbalistic 'Infinite One') or Allah. If man were created in the image of Creator and creation (Chapter 17) such height of knowledge, such profundity of wisdom would, by design, be built-in and accessible to him. Man is a microcosm wonderfully made in the ability to uncork macroscosm and retrieve Infinity.

	tam/ raj	*Sat*	
	conditional freedoms	*Freedom*	
	states of mind	*Consciousness*	
↓	*tam*	*raj*	↑
	negative	*positive*	
	confinement	*liberation*	
	stress/ pain	*relaxation/ relief*	
	deceit/ lies	*honesty/ truth*	
	isolation	*togetherness*	
	enmity	*friendship*	
	hate	*love*	

In principle it's clear. Now in practice let's descend Mount Universe through tightening enclosures, that is, degrees of mind and matter.

(Tam) psychological confinement shows, as legal systems recognise, in a drift towards egocentric and, in its further reaches, cruel and criminal disunity. It is a kind of gravitation where the mass is self. Its orbits of self-centredness involve the objects of desire; its circles are called passion; passions box with tunnel visions. It is not that ownership, attachment, self-esteem and sex are other than intrinsic parts of a survival kit. It is simply that the harder an egotistical departure from detachment into obsession the more community is fractured. The skews of pride and greed fly out of stable orbit. Overindulgence exhausts while instability, depression or lack of self-control cast shadows over any individual personality. 'I am an animal' cries the biologist with partial truth, 'and so are you'. 'I am a thinker, seer and moralist' cries the philosopher with perhaps the greater half of whole truth, 'and you are too'. Negative exaggerations were once known as sins; and sinners bind with an equilibrating debt to those they have abused. Such debt incarcerates its debtor. As well as guilt and shame incarceration's cells include paralysis of fear that petrifies. Fear and pain are freezing prison cells. Is not a person 'turned to stone' imprisoned to the maximum?

You can help another up but holding down is quite the other thing. Does a criminal consider social freedom or an individual's freedom? Does a violator care? No doubt that violation means attack on individual freedom; it consigns the object of a crime, a victim, to an episode of pain. Indeed, the record shows that centaurs have a keen capacity for the infliction of sub-human acts of crime, cruelty and devilry. There dwell in the basement of anti-liberty demons wearing different sorts of masks who purposely, pervertedly wreak agonies, both psychological and physical, on other life. Is an evil human not his victim's demon? Does (s)he not wear the devil's face? No boxing of The Infinite is sharper than the black tip of a tortured soul.

Is day restricted by the night or night by day? Is conscious life boxed by its fall into the cavern of unconsciousness? Pain is sharp and weariness is dull but each at last falls to obtuse oblivion. In the darkness dormant mind keeps working; you flag sleep's level as sub-consciousness.

On the other hand materials have no mind at all. The oblivious push-and-pull of bodies and the mindless energies of nature never were awake. No active, only passive information guides that 'layer' of non-consciousness, the universe of mass, the state of matter. The stuff of science is entirely dry of subjectivity. Subjective mind is boxed completely out. Immaterial 'internality' of information is not an element that standard occidental paradigms include; so only matter's 'externality' remains. The Infinite is boxed in by degrees until it's wholly squashed out into solid finity - the endful shapes of objects. You can also understand the broad and easy way such vision classifies a sailing universe. As in mind, so in matter. Counteractive vectors lean to left (↓) or right (↑) but, when the wheel is poised in balance, it means 'no change' in a condition. Equilibrium means 'straight ahead'. Forever seeking balance cosmos plies a wobbling, cycling course.

	tam/ raj	*Sat*
	classical/ quantum levels	*Top Matter*
	subsequent order	*Archetype*
	physical	*Metaphysical*
	lower	*Highest*
	lesser freedom	*Freedom*
↓	*tam*	*raj* ↑
	lower	*higher*
	box	*unbox*
	particle	*force field*
	mass	*free energy*
	energy out	*energy in*
	cool down	*heat up*
	locked-up	*free-range*
	lowered	*raised*
	contraction	*expansion*
	compression	*decompression*
	rigidity	*elasticity*
	inflexibility	*flexibility*
	fixity	*flux*
	gravity	*levity*
	confinement	*liberation*

This physical stack (see also *figs*. 2.10 - 13) demonstrates that, logically, the greatest freedom of matter is vested in, apparently, no-thing; it derives from a non-physical source, the metaphysical character of archetypal memory. Below this 'Central Sky of Fixed Mind' material phenomena are gravitating through an endless emptiness of space and time. Is time or space the one that's infinite? Or neither?

Top Matter reflects, in an unconscious way, the characteristics of Primary Dialectic. Its unity generates polarity. It splits into the ways duality can box (by losing ↓) and unbox (by gaining ↑) energy. With respect to matter Top/ Potential Matter is irrelative. Its immanence is neutral, 'absolute', omnipotent and omnipresent. Knowing nothing it is yet the law. Archetype embodies law that, embryonic, will develop in a certain way. Maturity of cosmos doesn't come by chance - unless you reckon Lady Luck diced up preordination. Freedom of the lowest range of universal mind consists of independence from material limitations both of space and time. It is, in short, the phase where pre-existent matter is not boxed; and, timeless, time is always now.

Check *fig.* 2.6. Right-side releases. *(Raj) liberation* is the function of an input of energy. Physically, turn the heat up, break the bonds, expand from solid into liquid and ethereal, gaseous dance. Propulsion, radiant emission, fire and chemical reaction are examples of this kind of 'liberation'. And, psychologically, the fires of interest burn, love's focus stimulates; such 'negentropic liberation' is, in every case, becoming unboxed, rising higher.

Left-side boxes. Its vector is imprisonment. *Existence is imprisoned Unity; it is Essence cramped in style.* Right down to earth. In physical existence style is wholly cramped. Such paralysis involves the frozen, automatic 'will' of matter. It involves the reflex origin of energy and sustenance of mass. It is reasonable (as Chapter 11 indicates) to subdivide this emergence of non-conscious energy/ mass into three stages. These reflect not only the sequence of creation but also levels in it we still find today. (*Sat*) subdivision involves the previously mentioned presence of archetypes and polarisation of the vacuum but no mass. It is therefore the two lower levels of (*raj*) force and particles and (*tam*) bulk mass that we now treat in order.

At this practical end of things (*tam*) *physical confinement* is the exclusive interest of physical scientists. The focus here is so intense that you see nothing else. How, professionally, could it be otherwise? Physics is the study of the interactions, transformations and, in a phrase, physical relativities of energy. In this respect its key concepts are energy, momentum and their exchange. At its heart, therefore, lie the ideas of motion (*raj*, kinetic energy) and (*tam*) mass. Physical confinement involves the origin and manifestation of mass to the extent that, in its most boxed and exhausted form, energy appears as objects of our sensible and daily life. This is the process of materialisation whose antithesis, dematerialisation, relieves density, increases flux and, at climax, annihilates matter into light and (if any other mass were present to absorb the light) heat.

Physical boxing is a process of energy loss; dematerialisation is one of energy gain. *The furthest reaches of materialisation therefore occur when (tam) left-hand characteristics of the Dialectic are concentrated.* Such concentration shows as inertial mass, strong gravitational (mass-centred) energy and the fixity of confinement. *It has been inferred that solid objects represent the final, exhausted stage of creation.* They show least potential, maximum apparent form (e.g. a solid against the apparent formlessness of a gas) and maximum static order. They are most boxed and separate.

What particular shapes cool from the potential of a gas? What geometries are locked up in its molecules? As potential is exhausted there precipitate solidities, there emerge from energy impotencies, there materialise fixed and locked-up definitions from which every other possibility has been excluded. Often these are loaded with irregularities. Familiar, large-scale appearances (such as landscapes, coastlines and any other unworked object) are at the sharp, hard, heavy end of dialectical determination. Pole from pole, their objective actuality diametrically antithesises Subjective Actuality.

It still remains to ask how boxing carries on. How is energy first pegged? How does a particle rise up and come to dominate the wave? What makes a wad of energy compact into a separate 'thing', what first 'coated' any pointlessness with weight, what 'freezes' mass-lessness to mass? Why do elementary particles, and therefore what is made of them, have the mass they do? In short, what contracted something out of nothing or, even if you crouch beneath a quantum, next-to-unobservable vacuity? It can't be hocus-pocus - we are here. Science now consults some powerful oracles. As progeny of Ernest Lawrence's hand-held 'cyclotron' we now have Fermilab's huge 'synchroton' called Tevatron near Chicago; and soon the 'fist' of European *LHC* will check if Higgs field is the finest boxer. If it's found then Standard Theory is confirmed; if not, somehow a new approach to physics must be devised.

Who wants to be hemmed in? Boxed into tight and bruising corners? Even elementary particles hate the ring. The closer draws confinement the faster they jump up and down. In other words, the greater energetic frequency the shorter ripples are. As atomic clocks can show, the mode that rules is 'short-time' oscillation rather than the massive, slow life cycles of the 'long-wave', bulky objects you and I can see. The smaller the box in which a given particle is locked the more agitated and outline-defined (or grosser) it becomes. The tighter it is bound the more it struggles. Confinement. Clamped oscillation. More taut is higher pitched. A string clamped at both ends behaves like a particle clamped in its (atomic) confinement. That is to say, the particle describes harmonic patterns. The equations that describe a standing wave also (as an eigen function) describe standing probability waves. As standing waves can only assume a limited number of well-defined shapes the energy levels of confined (or boxed or clamped) particles are thought of as 'quantised', stepped or staged. Electron orbits are now viewed as patterns of separate standing waves. In fact the immaterial wave function is employed as a descriptive model which is used to compute the probability of location and properties such as mass, speed, size and charge of a particle. The energy of any subatomic particle within an atom is (like musical notes) distinctly defined. Confinement has *per se* produced hierarchy, discontinuity and quanta. Quanta are fundamental matter. We've seen this clever trick already and, now as then, description is not necessarily an explanation. How do the strings twang? How do unstruck bells ring? What clamps energy in space? What staples things to nothing?

Ripples in a pond or the sound of gongs are driven as a 'wave train'. Such propagated freedom more represents continuity than the fixed or standing wave produced by a guitar string. Indeed, a free wave exhibits the continuity of a rainbow but a confined particle shows discontinuous, stepped levels of its inner energies. Marvel, won't you, at this indivisibility, this duality-in-unity, separateness-with-togetherness that mass and energy are coupled with. Such marriage, whether heaven-bound or not, still won't answer you where mass originates. What's the clamp that sources rainbows or can brace atomic nuclei? How can you clap with one hand and make sound?

There is a sense in which you cram more short wavelengths into a box of space than long ones. The more energy that is locked in a small space the more, by Einstein's Principle of Equivalence, massive or 'solid' that space becomes.

Matter is energy localised in a way that is related to the square of light's velocity - which means an awful lot of energy invested into making even tiny objects. For example, the energy density of an atomic nucleus alone is many billions of kilojoules. Quantum science loads its energetic freedom up in 'wave-packets'. In this way a single 'wagon' can be treated as a combination of wave trains! This case of trucking tends to lose its truly connective wave-like properties and, the narrower (or more bunched) things become, the more they resemble the discontinuity of a particle. Thereby a wave-packet, confining various energies, defines motion as a particle. A 'knotty' particle such as an electron or a nuclear proton has mass and, therefore, as it moves involves momentum. It also oscillates. An oscillation is a wave and a wave (*figs.* 2.14 and 5.6) describes the movement of a point (or particle) that circulates its own 'perimeter'. Perimeter, parameter, you cycle round. In this sense an oscillating particle is spun. The more boxed, the greater its confinement. The greater its confinement the more agitated and the shorter its wavelength; the tighter a circle, the more extreme its spin. Spin involves angular momentum and, thereby, acceleration and gravitational effect. Is this an answer? Is the clamp a period? A ring or a vibration? This would locate vibration at the root of nature. Different patterns of vibration are expressed as different 'sounds' or, in 'sonic fixity', as things. Still, what issues elemental 'sound' and what solidifies vibration?

From a dialectical point of view the grip of the warder, the shackle of the gravitational power-monger can be traced right to the top. It is the negative or materialising expression of *prakriti*, universal non-conscious energy. Personalised as 'God's left-hand man', the (*tam* ↓) entropic cosmic fundamental enforces restraint, rigid account and the structural part of order. Down we slide! Through archetype out into the material universe! Nuclear forces 'freeze' together wobbling, vibrant compacts spun from misty, mass-less emptiness. Seen in this way mass becomes a function of shifting alliances, of tension and of lawful motions in an empty, boxless box called space.

When there's compaction made of atoms then the massive centrifuge of gravity begins to swing. Hello, Herculean ease! What a handshake, what a grip and squeeze! Its weighty influence confines and pressurises till the solid, starlit worlds of our astronomy 'pop out'. The foothills of creation's slope are nigh! At this point a cosmic mountaineer has dropped from peak to base of Holy Mount. And with him we have run the course of grid, we have abseiled the conscio-material gradient from Central Origin down to solidity's compaction at its outer rim. In other words, the world's end is as near to you as any solid thing. Reach and touch the edge of universe, the place your earthly form resides - but if you think this is the only kind of terminus you fall into a massive error! Howsoever, we have 'parachuted' on a metaphor through all phenomenal development and now with a resounding 'bump' have definitely taken birth in space. Or, if you prefer, we have issued from the bowels of creation's cave; we have emerged into the clear, un-mysterious air, into the rare normality of planet earth. We are born into this fleeting moment! Fantastic just to meet you here!

On this slope's journey you have not learnt any extra physics but might now suspect a linkage joining both the pillars of the subject in the context of a third. This third reframes creation since it implicates our physical existence within higher, nested layers of a natural hierarchy. The object of the exercise has been to channel current concepts and translate their phraseology within the influence of Natural Dialectic's orbit; and thus, having linked automata into a larger context that includes metaphysics, mind and purpose, to gain a new and broader perspective. Yet even now we do not know how waves arise in nothing. How, like that single hand a-clapping, can it happen? How especially can it happen if, as opposed to ether, space is a real absence, an abyss, an endless absolute of negativity? Not even one hand would exist. Is all existence really just a trick? Are you truly part of an illusion? What profundity of topper! How does a natural conjuror pull something from his empty hat? Natural Dialectic would, of course, cup everything within two hands and then raise these towards a central Forehead, towards a third, intelligent Director. The cosmic dove released would flutter towards its Endless Home, its Natural Air.

Another facet of a diamond is another look. We cut from Unity. Chapter 10 cuts to a fresh perspective that in respects resembles clear but radiant, fecund atmosphere - the Sky of Nothing.

10. Nothing

Is it true the more you learn the less you know you really know? 'The only thing I know is I know nothing,' said Socrates. Perhaps, therefore, we'd best know something about Nothing!

Can you know nothing? Perhaps unconsciously! Can you see oblivion in all its 'blackness'? Can you have faith in the unseen? If Science and the Soul were accurate you would expect what's physical to devolve from what is not and what exists from what, essentially, does not. Do such voids make any sense? What is the nature of psychological nothingness and the material space that we call vacuum? That matter is created from such vacancy the Dialectic would agree. Let's see!

The Nature of Nothing

This facet of The Diamond, Essence, represents its empty quarter. O's Tale. Zilch! Nothing must be quite a scoop! A whole chapter built on nothing - how much waffle could that be?

Do you remember, first and foremost, you can think of nothing in two crucially different ways? Check Chapter 1: Nothing. You can find potential and exhaustion, pre-active and post-active absences of change. The negative sees nothing left, exhaustion or void impotence; the bottom of the pile's a sink. The plus sees what's before that is to come; it sees potential and capability *before* the action starts. Thus, in considering the state of nothingness, it all depends. Is the object of your focus loss of everything or everything to gain? Is it source or sink, a thing you want or, as its absence, what you don't? Understanding nothing's grades and nature might turn out to be the most important quest not just in physics but in life.

	tam/ raj	*Sat*
	expression	Potential/ Source
	localities	Omnipresence
	derivations/ dependencies	Super-state
	things	Nothing
↓	tam	raj ↑
	oblivion	awareness
	its abstraction	motion
	impotence/ sink	current
	its reverse	heat
	background space	presence/ individuality
	blank canvas	change
	sub-state void	action
	non-locality	locality/ focality
	their absence	things
	space/ time	events

Natural Dialectic's generalities might seem as little relevant to the buzz of a well-equipped science laboratory as Unified Theories to the hubbub of a city life. Yet broad generalisations and basic principles set the tone, orientate and radically affect the 'texture' of both thinking and consequent business. Having examined two of the absurd equation's factors, Infinity and Unity, let's take the third - Nothing. The study of nix is not for nought. 'Doing nothing' is no lazy option. The nature of nothing includes, potentially, everything.

Put it another way. If you believe the universe was always here, there is no need for any start. However, neither modern science nor traditional faiths subscribe to such a view that, as noted in Chapters 8 and 12, has profound problems. If, on the other hand, existence had a start then nothing, at least nothing existential, must have come before. If you believe existence (mind and matter) is the result of psychological projection, what kind of 'nothing' then preceded it? If, on the other hand and in accordance with a materialistic culture, you prefer to believe that only physical phenomena once 'started up', what kind of non-physicality preceded physicality? Again, what kind of 'nothing' was it that preceded 'anything'?

We've been here before. Do you remember (Chapters 1, 2, 6 and 7) potentials, priorities, chickens- before-eggs and all things conjured out of thinner than thin air - nothing? Something from nothing is The Big One. It is everybody's problem. From what kind of nuclear nothingness did the world divide? Some would argue space, whence *physicalia* were 'banged', is nothing physical - an immaterial but objective void. Is, therefore, 'expanding space' no more than a semantic - a non-existent but inflationary 'balloon' whose 'growth' moves through an absence? Nothing moves through nothingness! How then, if space is nothing and thereby does not exist, is Einstein's space-time absence curved by energy and mass?

To automation's contradictions we add animation's. In this case, did any chickens precede eggs or did a clutch of 'fertile' accidents create a life from what is not alive (see Book 3 especially Chapters 20 through 21)? With what, if not information, are eggs loaded? Is it passive information? Passive is the coded, chemical expression sprung from active information (mind). An idea's offspring is its birth into material form. Which comes first, machine or concept? What precedes an egg? Is it idea and intelligence to program its development? Eggs are nothing if not full potential for what grows from them. What about potential for a universe? What is the nature of such 'nothing-in-advance-of-things'; what seeds natural phenomena and is therefore at the heart of things? If the precedent is not informative, intelligent or principled then how did chance turn nothing into everything?

The whole answer, yolk and all, might just be unattainable. It may be beyond the wit of man but intellectual timidity is not the stuff that science is composed of. What first made what? What was primal nothing from which all things issued made of? Either you can follow endless regress down creation's hall of mirrors or can end up in the Infinite beyond all finite objects and events. If you can finish endlessly then there's your source for endless possibility before a single thing has come to pass. You will have arrived where all, including you, began.

Conscious and non-conscious - the states of mind's and matter's stuff - with sub-conscious in between. If two spring from one then, dialectically, energy and information spring from Prior Potential. But do two polarise from (N)One or were there always two or more? If both eternally exist then in such dualism there's no absolute, just relativity. There would be no Primal Unity.

Could it, however, be that prior stillness is 'more basic' than succeeding motion? And is Nothing's Nature immaterial or not? Some folk, you see, find it easier to imagine 'endless energy'. They can allow a boundless pool of latency behind its primal thrust - perhaps a cosmogonic 'expansivity'. Others find it just as easy to imagine 'infinite intelligence', a primal field of latent information that precedes a cosmo-logical projection. They can allow a metaphysical before the physical, the motion of whose rationale sways all creation. Which came from which or, lacking absolution, is the existential dipole (energy and information) locked in an eternal, Zoroastrian kind of struggle? What's the fundamental nature of the game? What unfathomable mystery.

If your solution is, though snagged with nagging problems, a pyrotechnic of dense radiation and expansive space what triggered their appearance? Of course, the *top-down* notion of a psychological/ metaphysical projection grates on materialistic nerves; but unless proponents of religious faith are coloured or corrupted by materialism's prevalence, it should not grate on theirs. **They should grasp, immediately, that the Dialectic thoroughly invigorates their basic case. On its terms Nothing (truly Essence prior to all existence) is Potential. <u>Potential's activation runs the existential show; from Nothing Everything is made!</u>** Thus Nothing is of prime importance. Either way, for holist or for humanist, the gift of everything depends upon the Nature of Prior Nothing!

Is Nothing therefore Consciousness or matter? Are you thinking clearly? Because the answer colours your description of the other's absence, that is, what you mean by 'void'.

Start with this stack:

	tam/ raj	*Sat*
	finite	*Infinite*
	duality	*Unity*
	thing	*Nothing/ Void*
	relative	*Absolute*
	lesser consciousness	*Pure Consciousness*
	form	*Formlessness*
	lesser presence	*Presence*
	relative absence of Void	*Absence of existence*
↓	*tam*	*raj* ↑
	down	*up*
	informed	*informative*
	impotent	*potent*
	totally void	*partially void*
	inanimate	*animate*
	metaphysical absence	*physical absence*
	energy alone	*interactive consciousness*
	absence of mind/ matter	*presence of mind*

Or look at things tri-logically. Do you remember (Chapter 3) that void can have three sorts of nature and nothing's nature be dependent on your point of view?

↓ tam	Sat	raj ↑
causal matter/ archetype	Potential	causal mind / Logos
psychological absence	Void	physical absence
objective plenitude	Essence	subjective plenitude
physical nature	Non-existence	metaphysical nature
material events	Neither	psychological events
passively informed	Potential Info	active informant
body	Soul	mind

Might some illustrations help?

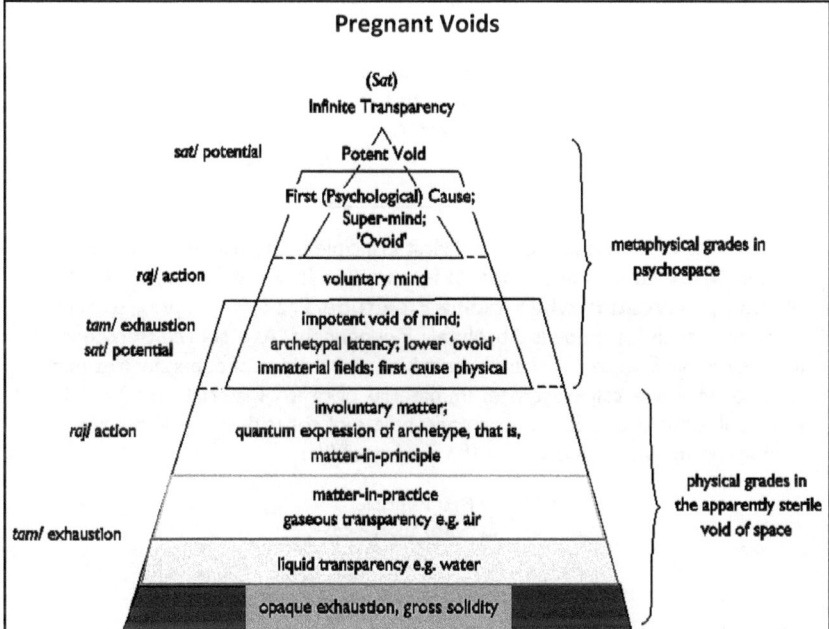

Do you remember (from Chapter 1) that nothing, called in Sanskrit *sunya*, has both negative and positive sides to it? Bare vacuum, as it were, yet full up with potential? In fact, there exist three aspects of void - (*sat*) potential, (*raj*) action in it and (*tam*) annulment, that is, exhaustion or absence of that action. That is to say, in terms of action (*raj*) event and (*tam*) object each derive from causal possibility; energy is framed by information; (*sat*) immaterial cause, the 'law of nature', runs physics' polar show. Of what sport are the rules that govern it material? They are immaterial and no less so in energy's great game. This is the natural triplex of dialectical vacuum.

Bottom-up (see fig. 2.16 and 10.2 i) a void is only of the third kind, empty space. It is an invisible blank or any point not filled with action, that's to say, not filled by object or event. It is passive, impotent and, as the cosmic sink, unproductive.

Top-down (*sat*) potent void is called potential. It is (see fig. 2.16 and 10.2 (ii)) antecedent, a wellspring of promise, an unseen battery of possibilities and things to come. What about an Infinite Potential, an Essential Battery whose charge once connected flows as current that sustains the circuit of all action 'down below'. What charge can Infinite Potential in its discharge ever lose? What might be the nature of such charged discharge, such egg that fertilised develops into what we call existence, such source of mind and matter?

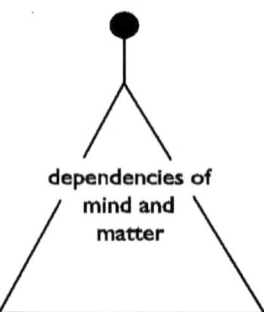

Essential Independence;
nothing existential;
The (N)One

dependencies of mind and matter

In terms of Mount Universe the First Principle is Essence. **Pure Essence is the Apex that transcends existence. It is an existential Void - Nothing to creation while itself a Plenitude.** Pre-existent, prior to mind or matter, such Plenitude is Absolutely Independent. As a source transcends its stream this Source transcends the mixed up motions of dependencies that sum to what we call existence. Immaterial prior to material, this Void of Potential is Subjective. It is a pure concentrate of consciousness whose spark, whose sperm, whose activator is the motion of Idea.

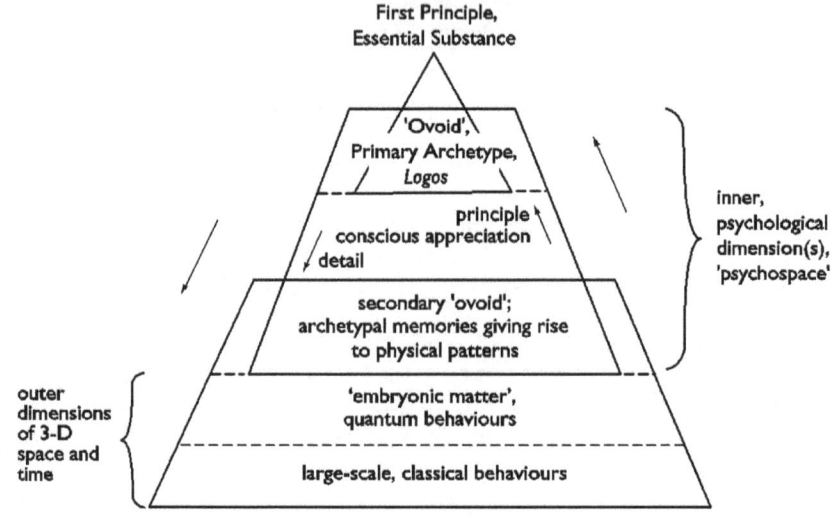

Ovoid activation of Potent Void is called Causal Idea. Such psychological First Motion is First Cause whose logic channels all development that flows from it, whose reason ramifies and, as *Dharma* or as *Logos,* drives and supports each facet of elaborate creation.

If the ovoid seed is psychological then first there issues all the cornucopia of mind. At what point is an active mind exhausted, at what level is perception void? Mind is voided at the stage we know as sleep; oblivion is empty of all subjectivity. This is one (*tam*) base, impotent void of mind. The other is forgetfulness. But though you may forget it does not mean you lose. You may record for recollection; things may be recorded, stored, filed away in memory. Memory, an inner context that surrounds your voyage of perception as an ocean bears a ship, is of sub-conscious mind. Sub-conscious mind is active though you do not realise; it is working even when you are not conscious. Psychosomatic instinct mediates with body; and, in universal terms, its mental impotence is matter's Ovoid Seed. **It is, on the hierarchy of the world, matter's archetype; it is (*sat*) potential matter, matter's first cause, reflex archetypal memory whence issues all the lawful cornucopia of things.**

Ovoids are, like eggs, complexity-in-principle. They seem, like mountain springs, to issue everlastingly; and they are unseen from lower down their course. What is above is unsuspected. Chancel's set apart from nave; Essence is veiled from existence; archetypal memory is unknown to material oblivion. Mind-in-principle, the Archetypal *Logos*, is unknown to coarse, low grade normality; matter-in-principle, sub-atomic and atomic entities, are not perceived by eye, heard by ear or felt by hand. **What seems normal, therefore, is not subtle. It is gross and outward bound; it is far-removed from central ovoid and its actual source.** If mind in ignorance, oblivion or sleep, is void of active information then matter's void is loss of active energy - a tendency to entropy, solidity and immobility. Inertial equilibrium.

You think of space as void but by the Dialectic's *top-down* line of reasoning it is the ground (perhaps in the 'form' of vacuum) from which phenomena arise. Space is potent while, with energy and information frozen stiff, the real impotent void is what you call most real - solidity. Does such a notion counter intuition? Are solid rock or shifting sands the foundation or the outcome of an underlying truth? How near that truth is sensible normality?

Impotent void is the antithesis of archetype; its base solidity is squarely set against the pointed apex of a pyramid. Gross form opposes archetype. Its fixity, molecular or bulk, involves a concentrate of locked up energy while archetype's an immaterial template for a pattern of behaviour. Most locked, most finite; most unlocked might seem infinite. What, therefore, about the final void of matter? What about the nothingness that constitutes the vast majority of what an atom is and where it isn't? What about the absence that contains all presence - space? Impotent with respect to energy, devoid of force or fact, is there annihilation that results in

> emptiness? Is there true vacuum or a limitless extension you could argue was infinity of space? Or is there never nothing, always something there - radiation, gravitation or subliminal energy called zero-point potential of the vacuum, *ZPE*?
>
> You can't see subliminal *ZPE* (see *figs*. 7.9 and 8.2); you can't measure virtual 'ground state'. But if, of course, it's everywhere then nowhere is there really nothing. Nothingness resolves, as Aristotle claimed, to an illusion; or a form of words that means abstraction and an absence of the actual things that we can see or measure. These are sensible and almost immaterial *ZPE* is not. Space is, in short, an absence of activities we can detect but also, with such emptiness, a fullness of potential in the 'formless form' of minimalist energies. **The real question then resolves to this; is ZPE residual and exhausted energy or is the quantum vacuum a creator and supporter of the universe of things?**
>
> *fig. 10.1 (see also 10.2 (ii))*

'Nothing' obviously means 'no thing' or an absence of a thing or things. Which thing? Is your cup half-full or middling empty? It depends upon your point of view.

Bottom-up, 'Essence' is a notion void of meaning. Because everything is, in fact, physical then 'metaphysical' is a void concept and at once a concept to avoid. The only objective, 'tangible' void is the vacuum of space and its absence of anything. Less tangible intangiblilities involve subjective descriptors (such as emotions) and abstracts of physical reality like words, symbols or thoughts. Indeed, what about life's symboliser itself? Mind (Chapters 13-16) is something of a nuisance ghosting the machine. It must, you hypothetically suggest, be an abstraction somehow haunting brain.

Top-down Essence, which is independent and different from existence, is the Void (or, in the causal sense of 'ready-to-begin', Ovoid). Its Pure Potential is projected, 'pushed down' or voided into existential creation. In this sense existence is a loss, a voiding of the Void. Within existence matter is a void of mind and *vice versa*. For example, your own information centre, mind, is a material 'nothing'. Do you weigh more or less when thinking, dreaming or asleep? What is the mass of memory? Mind is a physical abstract yet it is the central *sine qua non* of your life. Your perceptions, thoughts and feelings add up to an experience which, for you at least, is certainly a case of 'everything in nothing and nothing in all'.

Whence do your physical behaviours derive their pattern? Thought plays a part; but instinct and molecular construction govern reflex patterns of response. A glance at ziggurats 10.1, 10.2, 12.1 and 15.1 might help to gather up an understanding. These show pure mind - super-consciousness - gives rise to a spectrum of mind whose base - opaque, sub-conscious void - is invisible either to subjective consciousness or, probed from the material side, even an electron microscope. Dialectical logic identifies this void as 'super-matter' whose archetypes govern the way the physical universe is constructed or, given these metaphysical templates, constructs itself. Such archetypal potential controls the way material forms evolve and dissolve. Such 'potential matter' rings all changes;

its archetypal 'instinct' rules how energy (in the form of forces and materials) can dance. Both animate and inanimate bodies are subject to such guidance.

As polar opposite of Conscious Void the concentrate of non-consciousness may be seen as a black hole. This black hole, also known as Tartarus or the abyss of oblivion, is a negative, objective void embracing space, time and all material transformations. Physic is non-consciousness in action; our galaxies and stars and worlds are automatic, dream-like motions in an endless, dreamless 'sleep'.

Thus, according to this logical brief, the 'Ovoid' state of Infinite Presence is Central; it is at the top of mind. At the base of mind rests the secondary ovoid, archetypal memory; this 'universal instinct' is central to the issue of material origins and operation. Finally, at the most exhausted, 'thrown out' periphery of reality is dark emptiness - of space and physical phenomena.

It is clear that the two perspectives entertain almost diametrically opposite views of 'Void'. Unless you understand this and thoroughly disentangle them the likely outcome is confusion. What is the real nature of nothing? Perhaps nothing more than nothing depends on how you look at it.

	tam/ raj	*Sat*
	existence	*Essence*
	differentiated Essence	*Non-existence*
	lesser consciousness	*Pure Consciousness*
	lesser lights	*Light*
	lesser voids	*Void of voids*
	lesser fullness	*Plenitude*
↓	*tam*	*raj* ↑
	negative	*positive*
	objective	*subjective*
	subjective void	*objective void*
	non-conscious	*conscious*
	physical fullness	*mental fullness*
	matter	*mind*
	void of mind	*void of matter*

Vectored Voids

Bottom-up, materialistic view of void

Neutral Void is (*Sat*) Absolute. It is empty of any existential motion/ thing but, paradoxically, full of possibilities. Its Infinite Plenitude is Prior. Its Concentrate of Consciousness is Omnipotent (or Totipotent).

First Cause is, on the other hand, the most (*raj*) positive, 'senior' and therefore powerful action in the universe. It is thus, as a fertilised egg represents the start of development, dubbed 'Ovoid'. Such 'Ovoid' is (see Chapter 5: Top Teleology) also called *Logos, Shabda*, Name or Word of God and so on. In the Light of Void and its 'Ovoid' immanence all motions/ things are shadows of relative darkness. Of such shadow physical matter is the 'furthest', most exhausted, completely inverted and therefore, including unlit space, the darkest black.

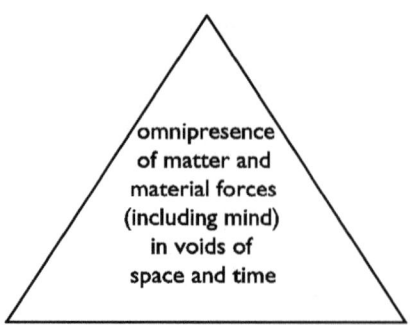

In this view mind is of minuscule,
ephemeral presence and importance.
'Mount Universe' is a void concept. There exists nothing but
matter or its absence, vacuous interplanetary and inter-atomic
space. Mind is a special case of matter; all metaphysics is void.

Top-down, holistic view of void

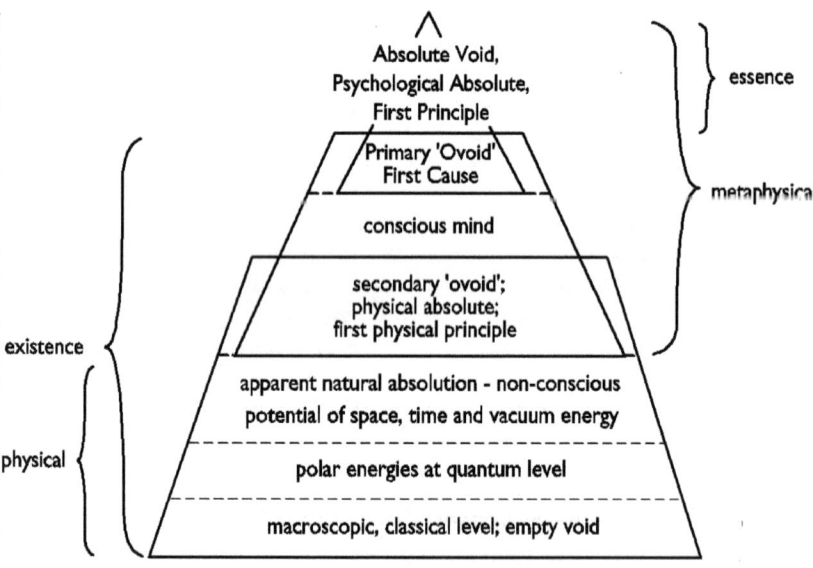

The (*tam*) negative void of mind is its unconsciousness, sub-consciousness or sleep. This condition, which precedes the special case of non-consciousness, is one of psychosomatic linkage. Are not steps two-way units of a gradient? If one step of cosmic ziggurat interlinks with the 'phase' or 'band' below then the downside of negative, psychological void is the upside of physical matter. In physical terms, therefore, this step is called (*sat*) potential or transcendent matter. Its secondary neutral 'void' (fig. 12.1) gives rise, as 'ovoid' cosmic egg, to physical matter. In other words the primary expression of (*raj*) physical first cause is, according to the urge of dialectical logic, a 'most positive void' or 'ovoid' called light.

The most (*tam*) negative physical voids (called in Chapter 2 'subtendents') are the extremities of empty space, an estimated thousand billion black holes and thermal absence - a lack of motion called 0°K (absolute zero). Indeed, the starry universe is almost like a dark cloud in bright sky, a local condensation of infinity.

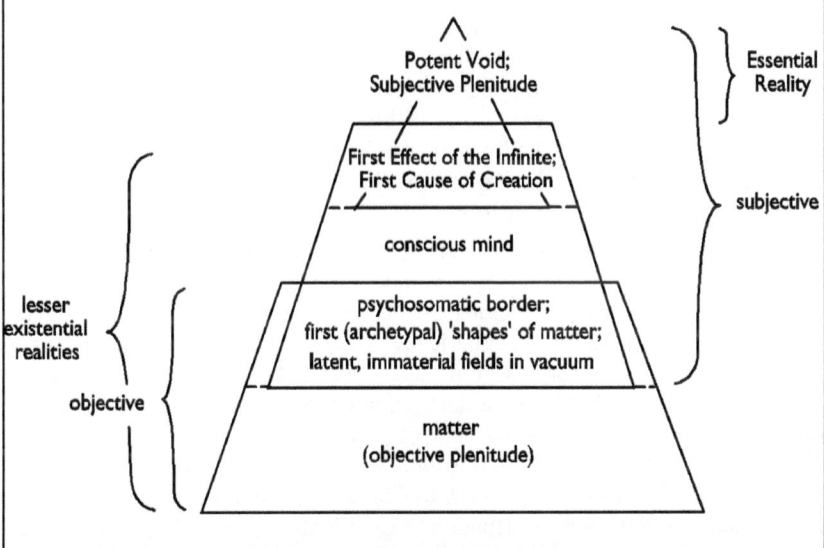

All (or nothings) together now

How does it all pull together?

Bottom-up, all is matter. Physics is valid, metaphysics invalid.

Top-down, both metaphysics and physics are valid. It is a question of order. Conscious mind is actively informant. Subjective and objective overlap. The zone of overlap is base, subconscious mind; it is called psychosomatic. The passively informed component of existence is objective; it includes subconscious mind and matter. Which, active or passive, comes first and is, in this sense, most real? Is it subjective plenitude (The Void) or the objective plenitude of matter? Is it things or Nothing? Which of the pair takes cosmic precedence?

fig. 10.2

Bottom-up, material sensation is what counts. There exists none other than material things. *In this view fullness or plenitude resides in material objects, fields and events.* The only things that exist are physical and where they don't exist is void. This formless void is an objective emptiness called space. It is an absence, a negative where things have not been put or minus whence they've been subtracted. Zero. What is left is space instead of object or, if you consider that a vacuum's something, not even its vacuity. *The view is of a negative, impotent void.* Such bare lack includes such phantoms as might be imagined metaphysical. Mind is, as remarked, construed to be an aspect of brain, spirit a delusion and soul, if considered immaterial, no more than self-deceit. What, after all is a 'material soul'? The soul of matter's nothing if not mass. Not only supra-existential, essential soul but even the notion of a brain-independent, metaphysical mind is voided. By the same token Essence is an Absence. It is an illusion, non-being rated as the least of truths. Dialectical (*Sat*) Void is absolutely wrong and so materialism's void, in ousting Subjectivity, is rendered a subjective desert. It is lifeless.

On the other hand, Lao Tzu describes the Taoist Void as 'an image of non-existence'. He says, "*Tao* is a thing that is formless yet there are forms in it; both invisible and intangible yet there is substance in it". *Top-down*, plenitude is vested in Absolute, Formless Void. While empty of things it contains, potentially, all shadows. In the light of Infinite Void all forms are shadows of relative darkness; they are limitations of the Infinite, different from it and, therefore, comparatively void of Void.

Do nothing, think nothing; no action physical or psychological. What of existential cosmos is there left? Absolute Void is Subjective. If conscious mind is positive and non-conscious matter negative then this Void is most positive. It is not a thing but an experience. What can Pure Life be but Totally Alive? Nowhere without, only within - such centripetal, central inwardness is, as opposed to the centrifugal outwardness of materialism, spiritual. If Subjective Nothing is the source of mind it is the Peak of Knowledge, Apex of Understanding and the Height of Wisdom. What to a mystic is the Pinnacle of Success is to a Hindu monk *samadhi*, to Christian communion 'the peace that passeth understanding' and to a Buddhist sage *Nirvana* - in which essential state the flame of existence is 'blown out'. It is the same Metaphysical Plenitude all notion of which an atheist would annihilate; but for a Buddhist self-annihilation in this Void is Culmination.

Can you stretch two viewpoints more apart? At least you understand why half of Asia kneels. What less profound could draw such supplication than a single portrait of his lotus posture that defines the Buddha's Culminating State of Grace? Do the pious, hands clasped, lips pursed, understand they're calling out for Nothing by another name?

Not that Nothing on the move is Nothing any more. What about, from Void, 'Ovoid'? Just as electric current is a natural phenomenon so is First Cause, 'Ovoid' or the *Kalma*. 'Electric' is a word and not the fact it signifies. Neither chatter, book nor circuit theory, no matter how advanced, begin to generate the actual thing. Similarly The Bible and the Koran are verbose; they might indicate the character and workings of the 'ovoid' *Logos* but they aren't the natural fact they signify. They are no more real than holidays in brochures. Don't you have to travel and arrive before you know reality?

Does saying 'London' take you there? You have to go there and, as any traveller will aver, that's not just as easy as a spoken word. Nor can you grasp intellectually what's Non-existent. Thou shalt make no graven image. If an experience of Essential Void is lodged beyond the reach of mind then speculation roves as vapidly from its reality as chatting to a booking agent in his shop. The shop is not the Promised Land. Chat is just an indicator of its loveliness. The actual experience of Void is super-conscious. The highest quality of meditational ascent, the apex of Communion is Mount Universe's peak. First Cause is a Subjective Reality that, as the cosmos is unfurled, turns out to be the basis of all lower and dependent states including physical phenomena. The climactic union of 'son with Father' is wordless, silent and inexpressible. A mystic can only, like Moses from Sinai or Mohammed from the cave of Hira in Jabal-al-Nur (the Mountain of Light), bring it down to the intellectual market-place.

Moreover Highest State of Union is reflected in the architecture of man's sites of Wordship (that is, worship). These are, after all, buildings in which connection with the divine is ritually fostered. Such structures all, but some more obviously than others, reflect hierarchical cosmic order. *Top-down* order. Look at Mount Universe expressed in the temples of Angkor Wat or Madurai in southern India. Top centre is the holy of holies. What about the cult of Amun's nested chambers in the great temple at Karnak, Egypt - whose inmost shrine was that of symbolic (possibly sexual) union and consequent 'ignition' of creation's ovum? Or the temple of Solomon in Jerusalem? This was, potently, an allegorical construction whose concentric courtyards led in from a gentile periphery to the inmost, central 'holy of holies'. The sanctum was silent and, unless perhaps an ark of law lay latent and symbolic there, empty. It stood for Nothing at the heart of nation; it was a metaphor for Essence and a finite symbol of The Absolute and Endless Void. Once a year at the climax of the most important festival, Yom Kippur, the High Priest would, as mediator between the sacred and profane, appear at its door and utter the otherwise unutterable 'tetragrammatikon' or Name of God. Such rich dialectical resonance is also discovered at the centre of Muslim focus, the Ka'bah at Mecca. This edifice stands for prayer. It is a symbolic cosmological axis. From every quarter of the world's body Moslems turn their attention to concentrate towards their own single, communal 'third eye'. 'If thine eye be single thy whole body shall be full of light.' At this moment each supplicant's own attention, his or her own third eye, thinks of Allah. Such remembrance kneels at contemplation's starting-line. At each *salah* (prayer-time) the basic requirement of meditation is at least represented. The Ka'bah itself is a cipher of Void; entered once a year the room is humbly swept of dust by its guardian, presently the mortal king of Arabia. Other places of worship, such as churches, all have their holy focal point separated by, for example, an iconostasis or screen from the unholy congregation - other-worldly chancel separated from its nave. And in the chancel Christian altars face eastwards towards the rising sun, towards crescent light and life. Is not even your body 'a temple of the living God'? Its altar is its cosmological axis, third eye or door of perception. Upwards, inwards from this centre focused contemplation arrows towards the hallowed Central Void; downwards, outwards consciousness is scattered into the profanity of mammon.

There is absolutely no outer sign of Void yet its sign is, for a seer, everywhere. Its 'Absence' is Presence. Its formless invisibility is the Essential Precondition for existence, its absolute fullness the origin of everything. Of what is it full? Potential? Then its formless latency were the precursor of all forms. Creativity? Whose *sine qua non* is consciousness. If consciousness is life then Void is Pure Life. The (N)One is, in this view, Pre-active Essence. It is the seed of all development, a single source of every shape and behaviour. **Nothing is therefore the First Principle that predefines lower phenomena.** It is an independent potency that precedes, sustains and succeeds every object and event. Creation is a graduated reduction in potential. Such Void gives not takes. It is, far from negative, the Great Plus that in its own reduction *adds* all things yet loses none. This 'form' of Nothing is the Nucleus, the Kernel, in a nutshell Nothing out of which the whole 'organic' business springs! **Its 'Ovoid' is, therefore, the Most Positive, Powerful Plenitude of all!**

How much more relevant can Nothing be? Metaphysics is a physical void and, except at the psychosomatic border where they are conjoined, physics is as metaphysically void. Mind is immaterial, body material. At their junction you might expect to find an important vacuum, a place of void that spans animate life and inanimate death. This hierarchical border must, even as you think and move, be crossed. What is nearest the psychological void of non-consciousness? Dormant mind? Sleep? Or is it 'solid' mind, mind's fixity in forms of memory? Is instinct not a template stored in memory; does not a template channel flux into a reflex set of typical behaviours?

The whole of this section (Chapters 7-12) derives from First Principle, The Void of Essence, and three of its facets - Infinity, Unity and Nothing. Essence at the Apex of Mount Universe (*figs.* 0.14, 1.4, 2.7, 7.2 etc.) precedes a two-staged existential pyramid. Now, at the psychosomatic border between mind and the second stage of existence, matter, Natural Dialectic identifies the logical first physical principle as one 'above'; the apex of the physical domain is metaphysical. Like any source 'it goes before', 'stands above' and is potential that precedes its issue. Matter's precedent, called super-matter, is within the dormant phase of universal mind. It is the void of archetype.

Just as the Primary Ovoid is *Logos* so its lower reflection, the secondary ovoid or first physical cause, issues from archetype. <u>This issue takes the form of a physical trinity - space, time and polar energy</u>. **These three 'lesser absolutes' constitute the basis of the next three chapters.** A passive couple, space and time, serve to localise an active third, polar energy. This third is the active aspect of first cause. It substantiates, in various forms, our universe. It involves, as well as equilibrium, both levity and gravity. Of (*raj*) levitatory energies which is the 'purest'? What is the nearest physical counterpart of metaphysical, the active nothingness most like potential matter? What material is closest immateriality? What, in words of Chapter 12, is matter's holy ghost?

The Order of Nothing.

Why should The Order of Nothing differ (since they are simply aspects of the same) from the orders of Infinity and Unity - except that now creation is expressed in terms of vacancies, evacuations and of voids?

The ziggurats of figures 10.1 and 10.2 give some idea of nothing's place.

Now let's look at Nothing's Order in another kind of light.

Source and Sink are Zero.

(tam) physical exhaustion; peripheral extremities; voids of absence of matter (vacuum), of space (black hole), of vibration (absolute zero, 0 K) and of energy (impotence)

(tam) vector of materialisation

informant quantum action

informed classical events

(Sat) Superstate; Alpha; Quintessence; Essential Origin

potent, informant mind

cosmological axis

(tam) impotent, passive or subconscious mind; instinct and memory

(tam) psychological base/substrate; *PSI*; archetypal patterns; *(sat)* physical superstate; the metaphysical origin of material behaviours

The illustration shows three '0's, three kinds of zero.

(1) Essential Source
(2) Psychological sink; physical so
(3) physical (and existential) sink

Check *fig.* 0.7 iii, the conscio-material graph. The material origin (2) is shown to stem from zero; physical alpha has been set at nought. Did not starry cosmos spring from nothing?

In one sense, therefore, you can understand that nothing is a latency; source is potential; it is 'nothing happening yet' before expression of the possibilities. In the other sense there's nothing left; exhaustion is the end of possibility; zero can also be a sink.

Thus Zero Psychological (1) is Omnipotent; its 'charged battery' precedes all action and it is the opposite of action's end. The Alpha Source is Central and it has two 'zero anti-poles'. One is subjective/ psychological, the other objective/ physical - in that order.

The subjective anti-pole is 'zero psychological' (2). It is a sub-state sink of unawareness, sleep and fixity of thought; it includes the passive and subconscious mode of mind called memory.

This anti-pole is also the material origin. Fixity of mind called archetypal memory serves as a source of information. Sub-state is, paradoxically, at once a natural super-state. Material government derives from this 'great instinct' of the universe. Memory is metaphysical; but (see Chapters 15-17) its zero physicality is template and thus source of every automatic, physical event. 'Potential matter' is not physical; yet its immaterial priority controls the way materials non-consciously behave. Starry cosmos sprang with this in mind.

A hundred billion galaxies still make subjective zero. Non-conscious matter is the anti-pole of immaterial Essence (1), mind and material archetypes (2); and the extreme of matter's anti-pole is its objective zero. This sink, 'zero physical' (3), marks total drainage at creation's furthest edge, exhaustion at its dark

> periphery. At this rim absence of all energy is characterised by space, inertial impotence and absolutely freezing cold. Such material sub-state is neither source nor ground of nature but, conversely, nature ground into its end. Negative annihilation. A natural posterior and the rump of universe.
>
> **fig. 10.3 (see also fig. 0.7 iii)**
>
> | | tam/ raj | Sat |
> |---|---|---|
> | | existence | Essence |
> | | things | Nothing/ (N)One |
> | | action/ change | Zero/ Void |
> | | levels of existence | Super-state |
> | | expression | Potential |
> | | lesser presences | Presence |
> | ↓ | tam | raj ↑ |
> | | stillness | motion |
> | | space | time |
> | | absence/ emptiness/ zero | presence/ locality |
> | | oblivion | things/ thoughts |
> | | impotence | physical/ mental energies |
> | | sub-state | transformations |

Humans are, psychologically and physically, only familiar with a limited range of conditions. From a *bottom-up* peer-spective don't telescopes inform us that the boundary of nature is composed of boundless space? Cosmos outside, atom inside - an atom is a cup of almost-emptiness. It seems the world's physique is mostly very thin. Phenomena, excepting neutron stars and black holes, are woven out of webs of nearly nothing and, if these are fragile, how much more so psychological events? How physical is purpose yet it rearranges mountains and steps onto the moon. When thought translates into construction one step placed against the next builds up a dream; no motive plies through dreamless chemistry but life channels reactivity to satisfy its needs. Yet in the last analysis all such constructions are a measure of our powerlessness. Organisms organise provisions - in the human case spectacularly - but cannot *make* a single atom or its particles *de novo*. The world is given free!

This does not prevent one sequencing, with a *top-down* slant, the principles of Nothing-turning-into-something down creation's gradient. Or, if you prefer organic metaphors, you might describe how ramifies, from roots in Nothing, seed causality and polar branching, an inverted cosmic tree. Can everything be really rooted inside nothing? Is the Dialectic mad? Never mind a mystic, look at crazy, half-real vacuum energy. Terms that contribute to its semi-being perhaps include Higgs field, *ZPR* and virtual pairs of quantum fluctuations beefed up into particles. And what about dark energy? The emptiness of nothing has, like the plenitude of things, an order. How, in principle and practice, does it work?

tam/ raj	Sat
existence	Essence
polar	Neutral
two or more	(N)One

	abstraction from Void	Abstraction from existence
	order	Preordination
	lesser grades of consc.	Full Consciousness
	body/ mind	Soul
↓	tam	raj ↑
	negative	positive
	passive	active
	inactive	kinetic
	unconscious	conscious
	unknowing	knowing
	informed	informant
	physical events	mental events
	body	mind

Natural Dialectic divides the cosmos into three familiar grades - (*Sat*) Potential, (*raj*) mind and (*tam*) matter (*fig.* 2.12). As far as information is concerned, subjective mind is active or kinetic; objective matter is, devoid of any sense of purpose, construed as passive and creatively impotent.

Nothing is easy! It's as easy as we've just explained. In the Dialectic framework (*Sat*) Neutral Void precedes. Pregnant Potential bears the poles of information and of energy. The Primary Projection, First Cause, is a (*raj*) creative 'Ovoid' from which is devolved, mind prior to matter, creation. In this case (*figs.* 8.2, 10.1 and 12.1) there is conceived, within existence at mind's edge with matter, a secondary ovoid; this ovoid is a physical first cause - potential matter called the bank of archetypal memory. This bank's traffic is with matter-in-principle from which aggregates bulk matter. What higher, lighter matter than the lightest subatomic particle? Of basic particles which less massive than a brilliant communicator - light? Is light a medium mind could work with? Could it, in particular, transmit from archetype to subatomic earth? Or its electromagnetism even mediate intelligence and translate purpose into nervous form?

Do you recall four kinds of *(tam)* negative, impotent void? Such exhaustion happens when an action dies. Flat the battery, dead the end. There exist two kinds of emptiness each with two psychological and two physical aspects.

The first kind, by the name of 'immobility', will be further discussed in Chapters 12 and 15. When thought is fixed we call its 'object' memory. When energetic flux is frozen then we dub the texture 'solid or 'immobile''. Both kinds of object represent a mode of storage. They express the frozen, passive shapes of information and of energy; each is fixed the way that, in its making, it was first informed.

The second, formless kind is 'vacancy'. Metaphysically, there is unconsciousness of mind - subjective absence due to sleep or comatose oblivion; physically there is exhaustion of all objects, that is, the objective absence of a vacuum, pure space - perhaps even pure space-time. Space is easy if there's nothing in it - but maybe there is.

Of these four psychological vacancy rests at the base of mind. Sub-conscious mind is an inert dormancy that constitutes, in the dialectical hierarchy, a linkage factor with the (*tam*) physical cosmos. The latter is, in the sense that it is entirely non-conscious, the void of Void, the vacuum of Pure Consciousness. Encompassing its empty quarters are three spatial dimensions and the one of time. Neither object nor event intrudes upon the bare abstraction of space-time. This hyphenated couple is a naked nullity, a vacancy stripped of all power, an existential desert. It is, however, a function of motion without which existence, the sum of innumerable motions, could not exist. It is, in other words, simply an absence made present by its cargo. A negative void, while itself empty of motion and therefore possibilities, is the 'frame' without which no scene could unfold. Without space and time there can be no action, no projection, no existence. Abstract dimensions are facilitators for concretions. Nothing is extensively essential!

Vacuous space is, according to Natural Dialectic, one of three physical absolutes.

But is it really nothing? Physics has falsified the existence of an inert vacuum. In other words, there isn't 'nothing' anywhere. Everywhere there's something but this thing might be, in space, below the threshold of our observation. To call a vacuum 'nothing' is an error but its 'thing' is on a different, so-called 'virtual' level. How firmly has our science got its finger on this evanescent pulse? In the dialectical hierarchy its level may represent an intermediate, the first physical expression of potential matter, archetype or mnemonic *prana*. Maybe it is vacuum energy or *ZPE*; and, pushed beyond the threshold of its so-called 'virtuality', it logically emerges as the finest sort of 'actuality' that our equipment can observe. To this extent space is seen as a 'container' and active vacuum as latency with 'immaterial fields for development'. The whole world is developed on such 'negative potential', such 'blank 3-d film'. *ZPE* mediates all quantum and electromagnetic processes. It is postulated as support for electron orbits, atomic structure and thence the phenomenal universe. The appearance and disappearance of clouds of virtual particles in this very lively vacuum are a price paid for what is claimed to be the most mathematically precise theory known - Richard Feynman's relativistic quantum field theory of electrodynamics. **Electrodynamics describes the effect of moving electric charges and their interaction with electric and magnetic fields - the precise point that Natural Dialectic identifies as linking subconscious (including archetypal) fields of memory with matter. It is therefore tempting, at this point of psychosomatic linkage, to explicitly equate *ZPE* (sink energy of the vacuum/ zone of virtuality) with *prana* (yogic universal field of energy); and thus propose it constitutes, by metaphysical projection, a template for material creation.** This theme will be developed in this chapter's section: How Does Nothing Physical Work? and also in Book 2.

No doubt this pair would recognise the notion of 'development of latent image'. William Henry Fox Talbot and his friend Sir John Herschel developed the 'photogenic idea' whereby 'transfers' or 'negatives' are developed into positives. Light is projected through polarities. Images are caught and fixed on film. Photography. The dialectical reality is archetype; and cosmos is the photograph. Dynamic photographs are films. Material science studies the development of 'photograph'. It ponders the pictures of an album full of possibilities derived from simple archetype. What candidate might fill the vacancy for prime expression of this archetype? Not negative but positive potential. What first cause, not of mind but matter, might link the two? What most potent next-to-nothing, most positive (*raj*) aspect of 'ovoid' might initiate a starry universe? Pure energy? Mass-less stimulation, motion absolute and levity alone? A developer called light?

There is just one snag - the snagger. Polarity's anti-principle. Positive principle needs, in order to develop, negative. It needs black to counterpoint the white, opacity to shape ethereal transparency and mass to snag the flux of energy and generate fixed shapes. You need abstract hangers-on like space and time; you need energy; but to build a world you also need to get a grip with powerful gravitational contractors. Nuclear forces and 'standard gravity' combine to fit the bill. They are persistent drags, real bringers-down-to-earth, accumulators. There can be nothing without gravity as well as levity. You might therefore expect, according to the Dialectic, that the world originated and continues with both vectors operating all the time in every way.

The challenge is, cosmologists agree, to peep beyond the walls of space. Can humans in a cosmic goldfish bowl observe it from outside; or is the higher truth, as micro-physicists agree, within? What was the spring wound up or pendulum released; what primordial imbalance in a latent field of action set the world to 'knock-on-go'? What sucks or jets our flight from nowhere into nothing all the time? You might suspect the mighty rush of an initial tide of energy could start the show. Did levity and gravity kick in together or did one outweigh the other at the start? Was it energy of vacuum (*lambda* force) that first inflated space while, simultaneously, a nuclear ram pressed out, spun or otherwise compacted particles from what seemed nothing? Is space itself a net flung out to catch free energy and channel how it flows?

What pin might prick the voids of space and time? What force could puncture nature? If a black hole could rip the pointless fabric up you might believe that its reverse, a white hole, was the spout, the sprout, the seed from which suns, planets and the elements have burst. So who's afraid of big, bad holes. It's not the colour, black or white, but content of a singularity that challenges a physicist. It seems that by the billion voracious 'mini-dots', stellar black holes and the titans, 'super-massive black holes' (*SMBH*s), scour our universe. Indeed, gigantic 'macro-dots' punched into space inhabit each galactic core. These cosmic rivets peg the tent of stars. These gateways into somewhere else are, paradoxically, the stabilisers round which constellations swirl. Call them large-scale attractors, organisers or the dark lords of a mindless continent, space-time. Such bolts would stake out groundlessness and rope their circum-ambulating galaxies in place. It's thought our own, the Milky Way, spins round this primal kind of vortex. In this respect 'black hole' extremity would constitute an axial starting-point, a staple that pins every part of each dynamic, stellar picture to the wall of space. Thus, like routine and stability, you need these turners for your health!

Who can tell? Lacking any better vehicle of naturalistic explanation 'big bang' projection theory is promulgated as 'the way we think it was'. Despite its *ad hocs*, stitch-ups, bolt-ons and weaknesses it survives for want of better rival - if the only rival you allow is of the mindless species. Is this one-tiered expression of the way things came to be complete? Expansion of the universe is Natural Dialectic's brief. Its order is broader and its ties are triplex, three threads spun from the Substance of Infinity. The Diamond (*fig.* 7.1) applies before creation (mind and matter) has begun; but its facets are reflected as the physical creation starts. Thus Infinity, Unity and Nothing each apply to the non-conscious, material stage. It is the plan to finish with this stage and thus wholly void the Void.

There are three expressions of the physical void, a *(tam)* tri-unity that underpins our cosmos. This trinity of physical absolutes can neither be observed in isolation nor can anything without it. Indeed it *is* our universe. In this chapter there is seen, as the inverted mirror image of Great Nothing, an exhausted **vacuum**. In the next, Chapter 11, dynamic nothingness is passed as **time**. Which leaves us finally with both inertial and dynamic aspects of **polarised energy**. Chapter 12 is first illuminated by the latter's 'absolution' in the formless form of light. This form of levity is matched, on the other hand, by the omnipresent, omnipotent and invisible influence of gravity. So we drop through the crack in two-lipped emptiness down to a cornucopia of things that have emerged from it - including you. This, in short, is the tall order issued out of nothing in particular.

There's More to Nothing...

So many words surrounding Central Silence! Who'd believe so many libraries of Books and books could have been written about something so singular, simple and yet clearly important - Nothing. Let's recap The Order of the Void.

	tam/ raj	*Sat*
	existence	*Essence*
	things	*Void/ Nothing*
	lesser states of consc.	*Super-consciousness*
	body/ mind	*Soul*
	objective constraint	*Subject*
	action/ sense	*Being*
	becoming	*Being*
	non-self	*Self*
	motion	*Permanence*
	change	*Constance*
↓	*tam*	*raj* ↑
	more objective	*more subjective*
	action	*perception*
	body-self	*mind-self*
	body-sheath	*mind-sheath*
	physical form	*persona(lity)*
	non-conscious	*conscious*
	inanimate	*animate*
	physical properties	*mental properties*
	body	*mind*

(Sat) psychological Super-space is No Thing. It is the Essential Void, an indescribable Infinite Potential, from which the finite motions of existence derive. This Independent Void (or, at creative spark, Ovoid) is, from an existential view, unseen; it lacks finite existence. Conversely, dependent existence issues from but lacks pure Void.

Bottom-up, there appears no such 'thing' as Essence. An Order of Void is itself null and void. Do you propose, therefore, another mystery - the profundity of eternal, uncreated matter? A quintessence of perpetual motion? Never-starting, never-ending, mindless 'creativity'? If not the start-point must have been impotent nothing - either physical, from which the big bang's cannon 'shot out' everything or metaphysical whose canon regulated its behaviour. Mmm!

Top-down existence is, in the first or last analysis, 'Nothing in confinement'. It is restricted, haltered Essence; phenomena are canned Infinity. That is to say, such Infinity hierarchically voids a finite part of itself into existence. It loses this part in creation; creation, which is dependent, gains it. Existence comprises effects of Causal Essence; it amounts to Essence-in-motion. While Essence is absolute, existence is therefore composed of changes; while Essence is 'Absolutely Mostest', existence is a spectrum of 'less-so' relativities; and while Essence is Pure or Supreme Being ('I am that I am'), existence is becoming constrained by attributes. Of these the main is isolation. Individuation. 'I am things that I identify with and all else is not-me'; or 'it seems to be itself and all else is not-it'.

If Original Essence is Pure Consciousness, then the first of its limitations is consciousness-in-motion - a thought, projection or idea. The intrinsic nature of becoming, or lesser being, is motion; and your closest forms of motion are mind and body. These clothe, like subtle underclothes and overclothes, the Subject, Self. Where mind is nearer and body further from the Truth, they amount to relative losses of consciousness. They are not Self. They are conscious and non-conscious 'not-selves'; they are 'other'. I am, in central truth, neither sheath - neither changeful mind nor body. I involve closely and identify incorrectly with this couple of rings that run around me whereas, in quintessence, I am naked soul. You, similarly, are not in essence the self you think you are.

Unconstrained Void is therefore, by analogy, the unwarped 'field' of mind, the psycho-space of Pure Consciousness. Its endless radiance is what Buddhists call 'ground-state', the Bible 'substance' or 'foundation' and the yogas (just as accurately) a super-state of mind. *If such Void were characterised as endless, living radiance, you might expect its opposite pole, its antithesis, to show a dark, non-conscious, object-splattered emptiness. It does.* Matter is the grave of subjectivity, a reflex vale of lifelessness. And of course between any two poles exists a range. It is no different in the cosmic case. Each successive cosmic grade or stage becomes coarser as exhaustion, in the form of apparent loss of information or energy, increases. Each such 'spectral band' is also subdivided into its own *(sat)* source, *(raj)* kinetic business and *(tam)* exhausted or impotent void.

Check, for example, *figs*. 2.10 - 13. For the metaphysical, psychological grade, the subdivisions are *(Sat)* Supramental Void, *(raj)* conscious mind and *(tam)* sub-conscious mind; for the physical grade there is *(sat)* super-matter, *(raj)* energetic, principle or quantum matter and *(tam)* bulk matter.

(*Raj*) mind is active - thinking, learning, understanding and perceiving. '(*Raj*) matter' is a term for the atomic domain. It covers microscopic quantum, atomic and molecular behaviours.

(*Tam*) sub-conscious mind is 'solid'; this is the grade of frozen thought and impressions, the domain of records called memory and an absence of mind called unconsciousness or sleep. (*Tam*) bulk matter is differentiated, as we are all sensibly aware, into gas, liquid and its lowest, solid phase. It is, in its least energetic form, a cold, hard, heavy object. Objects, either earthly bodies or celestial forms, are suspended in space. The whole array is buoyed against 'absent levity'. No top, no sides, nor is there bottom to the cosmic room. Things fall forever to a formless, endless sea of loneliness. If cosmos sprang from nothing then, as astronomers and cosmologists confirm, it seems to be a product of nothing supported by emptiness. This is, in a nutshell, how its seed develops.

An initial constraint of Essential Supergrade, called Ovoid First Cause, informs psychological and, lower down its cosmic 'grid', physical processes. This junction of Potential Void with existence is called Metaphysical Illumination. Initial 'Becoming', the Light of First Cause, 'leads the way'. As the first finite (but still *almost* Infinite) restriction of Nothing, everything will ramify from it. As a living current, will-power, men recognise this 'force' by many names - *Logos, Spirit, Shabda, Purusha, Gurbani, Nada* and so on. Its Womb forever bears all worlds. This is, therefore, your Heritage. 'In the beginning was the Word, and the Word was with God and the Word was God. Through Him all things were made; without Him nothing was made.' Genesis. How can anything be made from One Infinity that is not issued from It? How can something come of Nothing without firm genetic linkage? This is your Ancestor. It is the drama's Author. Every work of authorship tracks from and to an author. Is 'authorism' pantheism? Is a material stage alive? No. Are a play's characters identical with the author? No. Is the piece thought out or written up, that is informed by him? Yes. And in this case the order of creation's act (*fig. 6.1*) is Nothing's order - steps in descent while voiding Void. First Principle, the Void, and its primary expression, 'Ovoid' Word, substantiate but stand distinct from their theatrical elaboration. Many are the aspects of dramatic plots and ploys and staged production. Many grades of works and workers separate an author from the fulfilment of his inspiration. They separate him from his various themes and, in hierarchical turn, they separate 'variation-in-the-day-to-day-performances' from complete inversion of original idea - the scene in which its action is materially housed, a hard-nosed bricks-and-mortar theatre. This world is inside out! Ideas devolve into peripheral, non-conscious, scientific issues of projection such as planks of space, backdrop of stars and, in our human case, familiar, sunlit scenery of the planet earth.

In summary, it was shown (*figs.* 2.5 - 8) that each major cosmic grade is both distinct from and yet linked to one below. The base subdivision of one grade interlinks and is identified with the top subdivision of its successor. Sub-consciousness, for example, *is* super-matter. What is above is 'super'. It is prior and causal. It is the well-spring, source or potential for what courses out, what falls below. Yet potential is as invisible as an action that has not yet occurred. Thus a well-spring appears, from the perspective of its lower offspring, invisible.

'Ovoid' sources are unseen and thus often unsuspected and unknown. So active mind cannot, by itself, perceive the Infinite; nor matter either of them. In a phrase, there's more to Nothing than the eye can see.

Nihilisms

	tam/ raj	*Sat*
	things	Nothing/ Nihilism
	expression	Potential
	apart/ part	Oneness
	body/ mind	Pure Life/ Soul
↓	*tam*	*raj* ↑
	non-conscious fixity	conscious fluidity
	laws	purposes/ desires
	automation	animation
	material oblivion	awareness
	non-life/ atomic body	mind
	darkness	clarity
	despair	hope
	isolation/ loneliness	unification/ love
	nihilism	faith in Oneness

With Nothing at its heart you might call Natural Dialectic Nihilism; but don't confuse its Ultimate Potential with intellectual nihilism at the far end of the scale. Dormant, decadent or plain deliberate misunderstanding of what Nothing's nature is comes down to spent philosophy. Rejection, absence or annihilation of morality is one dark pole apart from Nihilism's Life. If life's element is spirit then (*tam*) nihilism represents negation of (*Sat*) Spirituality. Don't nihilists scorn fine ideals and anarchists tear orderly constructions down? Such negativity is toxic; acid scepticism drains, corrodes, infects a culture and, for the Dialectic, marks impending intellectual death.

If you propose the order of the cosmos was designed then nihilism is your notion's null hypothesis. If cosmos is chaotic (since natural law was sprung by accident) then nihilism's right and Nihilism wrong. You can annul Intelligent Design.

No doubt God-wielding bigots rack with cruel mistakes, fanatics poison wells of faith and wild-eyed zealots leave us with a loathing of religions in whose name they desecrate; but nihilistic crushing of 'the culture of the heart' is no less sacrilegious. What kind of cynical excuse claims systematic terror and atrocity is 'right'? Atheistic nihilism's politics excuse the violent overthrow of *status quo* by any means to force 'Utopian' ends. What morality cannot be 'hypocriticised', what back of justice not be broken in pursuit of power? Was there any blackness Lenin, Nechaev or Stalin and their international spawning could not claim - in everybody else's name - to justify? Philosophical, political or 'scientific' atheism, chequers in materialism's shadowed flag, simply rule out metaphysic. These destructive forms of intellectual vacuum take the earth as their foundation; they sow confusion and deceit, they reap a wasted crop of wariness and fear; and their anti-alchemy transmutes the golden soul to steel. How depressing! Could such viral nihilism be identified as evil that, lacking only overt violence, contaminates some 'rationalistic' infidels in modern universities? Is academic infidelity and educational abuse, is anarchy a cancer blighting those of anti-faith? Indeed, is nihilism's illness terminal for Nihilism's metaphysic? Can death really kill The Life? The Dialectic is, for sure, intellectual nihilism's absolute converse.

Pay Attention

	tam/ raj	*Sat*
	consequences	*First Cause*
	prakriti/ purusha	*Purusha*
	subordinate grades of consciousness	*Consciousness*
↓	*tam*	*raj* ↑
	prakriti	*purusha in action*
	objective part	*subjective part*
	manas	*chitta*
	metaphysical energy	*attention*
	weak/ restless	*concentrated*
	outward focus	*inward focus*
	seen	*seer*
	object/ image of focus	*focus/ involved attention*
	lack of knowing	*illumination*
	ignorance/ sleep	*awakening/ understanding*
	non-conscious	*conscious*

Voiding can be expressed in a psychological stack whose oriental terms illustrate the way General Consciousness specifies or, if you like, Pure Attention loses itself in its own imaginings. You might describe the first tier of creation, mind, in terms of a spectrum. A spectrum of 'volitio-magnetic' radiation (the 'electric' stimulus of will combined with the 'magnetism' of desire) is absorbed in its own creations. The general quality of these 'volitio-magnetic' creations is, in terms of subtlety or coarseness, analogised with 'wavelength' or 'frequency'. It ranges (*figs.* 0.2 and 0.3) from profound inward concentration to a weak capacity for reason, diffuse, outward restlessness and instinctive reflex.

Top of the class is keen concentration, enthusiastic involvement, high-octane performance. The light of attention is well paid. It heats; it quickens the objects of its focus and moulds them into psychological dramas and/or physical behaviours. The light of high intelligence soon grasps a principle and beams it to advantage. Of course, the lights of day forever change; atmospheres of storm and calm revolve like weather round a normal, moody head. Sleep and waking oscillate; awareness switches on and off. So does interest. It hooks, drops, forgets, ignores; walks on the dark side and the light. Mind wakes up, arrows like a laser ('very interesting!') or loses energy and interest until, at last and lowest level, knowledge falls asleep. Hardly anyone, however dim, thinks they are ignorant; no-one awake thinks he's asleep. Yet, compared with the illumination of a higher mind, normal sense-based human consciousness is fallen more than half asleep. A stranger out of paradise now fallen to the ground, it is almost unaware of upper realms and realties.

If Essence is Reality the range of lesser realities that constitute existence are, although real as far as they go, when compared to Reality relative illusions. They are illusions, that is, 'seeming realities' of mind and matter. Only in the annihilation of mind and matter is the Experience of Reality obtained - an annihilation that is liberation. Just as matter is liberated into light so, in a controlled and voluntary way, a mystic obtains liberation from the psychological material of mind. What is the nucleus of mind? Attention? In this case the critical mass is one of correctly applied focus. Its apogee is a pure concentrate of consciousness, a successful distillation of the mind into a mode called wisdom. Such return, through mind-in-principle (see *fig.* 2.12) towards its origin, is cosmic. Such immersion is called Knowledge of the Word or Enlightenment. This Knowledge is The Great Paradox - because in its uncreated, unthinkable non-condition resides potential for the creation of everything.

You don't believe me? You insist the Word's a book? This is a mistake worth singling out. Some persons think the 'Word of God' means something spoken and that Knowledge only comes from speech and books. It is true that you can learn from these but descriptions of enlightenment (as, at root, are all religious texts) no more impart that state than seaside maps can conjure up a shore. It simply indicates. You can read of swimming but, at last, you have to swim; you can read about exotic climes but, in the end, you really have to travel. To experience. The Word is not a word or words. It is not a teaching or a preaching or a figment of the intellect. No doubt the use of spoken word implies a grammatical order of expression; and 'hard' science is based on mathematical language that can precisely describe the natural order of events and thence read the physical part of The Great Book. 'Life and Cosmos', penned in archetypal script, is printed using ink of forces on the vast, clean page of space. It is published world-wide by its Author, the inimitable Word. While both maths and spoken language are descriptive tools, they are not the real thing. Who can keep a bulb alight by just repeating 'electricity'? Or materialise a part of nature just by shouting out a word? Although it is the source of mind, the Word is neither an imagination nor a symbol. It is Most Natural Reality from which all lesser truths (such as chemicals, emotions, cats or cars) eventually derive. You can attend to priest or scholar but in the end, as honest seekers after Truth agree, action is the policy that really pays. For scientist and mystic in their complementary fields practice is the way to better preaching best. Attention pays the price of understanding; only those who pay correct attention find out what the payment brings. So the former pays devout attention to things of the world; and the latter, by his meditation, pays attention to Attention.

Is Nothing Important?

	tam/ raj	*Sat*	
	lower	*Top*	
	after	*Before/ Prior*	
	things/ events	*Nothing*	
	expression	*Potential*	
	lesser priorities	*Most Important*	
↓	*tam*	*raj*	↑
	objective	*subjective*	
	inanimate	*animate*	

lower	*higher*
relatively trivial	*relatively important*
body	*mind*

Are you so languid nothing's worth its while? What's important is, for you, what you believe is so; but in universal terms importance drops off down the scale. It makes no material sense but relative importance relates, on a hierarchical conscio-material scale, to Importance. That is, to Absolution by the name of Consciousness. Absolutely Nothing is The Most Important Thing. A stage below depends for order on the stage above. If mind, a physical vacuity, is prior to matter then it's more important. This importance is enthroned upon subjective nature, that is, life. Vital value. *Natural Dialectic indicates that subjectivity is prior and the most important thing of all is life.* Beside this the lowest, final stage of psychological priority might therefore be termed its least important. This, even below sub-consciousness, is the crystallisation of non-consciousness called our cosmos. The material universe is, by this standard of comparison, trivial. Compared with Things of Real Importance, Pure Life and Love, things are almost absolutely unimportant. Even nature's beauty comes a second best - no appreciator, no appreciation. Life is top central. Primary. Is this obvious or not? The chips are down, the game is up - what counts? Even in a winning streak what specific trumps enthusiasm's radiance? But our priorities are such that, topsy-turvy, we materialistic hedonists are spellbound by the wealth of circumstance.

	tam/ raj	*Sat*
	physical	*Metaphysical*
	non-conscious	*Sub-conscious*
	mass/ energy	*Physical Potential*
	effects	*First Physical Cause*
	actuality	*'Virtuality'*
	expression	*Archetypal Memory*
↓	*tam*	*raj* ↑
	gravitational	*levitatory*
	principle	*principle*
	inertia	*stimulus*
	contractive agents	*expansive agents*
	things	*active vacuum*
	strong nuclear force	*light*
	massive particles	*force*

From physical 'nothing' emerges something. Is it not as clear as space that voiding is the order of creation? Each super-grade represents implicit (*sat*) potential; at each grade 'ovoid' source exhausts its native force, its flood of capabilities.

What commotion! Void concentrates until its threshold, First Cause, starts creation rolling. Or, if you like, Ovoid is fertilised and things begin developing. They develop from their principle, a seed, into the practice of maturity. What is maturity but realisation? What is it but fruition? Does fruit with seed precede the tree or *vice versa*? Or, anticipating fruit, seed, branches, roots and leaves, is 'plant concept' more important still? What is the case with cosmic seed?

An Idea is brought to focus; creation step by step grows sharper. Shapes stand clearly out. If *(raj)* kinetic consciousness voids into 'static' sub-consciousness; if sub-consciousness is 'fixed mind'; and if the place of 'solid' or 'fixed' mental objects is a storage level, what is it you record in mind? A memory. Objective matter subtends subjective mind. If sub-consciousness is, in turn, voided into physical creation, then a comatose state of mind, creatively rigid, set in its ways and the repository of instinct and memory, becomes physically important. Dormant mind is a psychological void but, at the same time, the physical supergrade. It is matter's 'secondary' ovoid. Super-matter is sub-conscious mind viewed 'through glass darkly'; its 'instinctive channels', its memories are the archetypes for physical creation. Are these not, therefore, most important pre-phenomena?

Physical cannot sense metaphysical and so it looks like nothing. If, as science suspects and the Dialectic expects, 'virtual matter' does exist then it must look like nothing. What 'isn't there' is, however, usually unsuspected. Least of all might materialism suspect its own first cause was on the other side of things and buried in a class of thought, the fixity of memory! Yet, in dialectical terms, this is just the case. *Archetype is the potential or source for actual, physical phenomena.* In other words, at the base of psychological dormancy there exists a Janus-like junction straddling mind and matter. What hierarchically precedes comes first; and precedence is top priority. So what is mentally least important is, as its source, most important to the material stage below. It is the gateway to the garden of the outer universe; it is the field whose crop is what's around us. Although subliminal its archetypal grade is matter's starting-point, centre-point, the keystone of its operation. The 'psycho-somatic' structure of this inner, immaterial interface is dealt with in detail in Chapters 15 and 16. How, though, before reaching into physic's vacuum, can we summarise it?

Since the base grade of one division interlinks with the top grade of the one below we are driven, by this logic, to assume that the origin of mind lies in the Infinite and the origin of matter in sub-conscious mind, that is, in the archetypes of universal memory.

This makes our fully energetic cosmos a memory derived according to the differentiation of First Principles from an informative First Cause; which is, in turn, an issue of Infinite Potential. In one sense, therefore, matter is a memory whose presence has a future. You know what I mean because a CD track is just the same. In a similar sense the presence of its dead-end shell is the past replayed; it is the rigid reiteration of a once 'live' recording session. In this view matter is an endless recollection, the continual playback of its own primeval creation. What is ringing up recall? What keeps remembrance on the go? What kind of electric drives, as if perpetually, the world machine?

	tam/ raj	Sat
	expression	Potential
	polarity	Unity
	expression	Archetypal Memory
	law-abiding automata	Laws
↓	tam	raj ↑
	materialisation	dematerialisation

gravity	*levity*
resistor	*activator*
locked aspect	*free aspect*
particle	*wave*
mass	*energy*
classical	*quantum*

With respect to the origin of the physical world, Natural Dialectic has now identified an important lesser or apparent nothingness. 'Top matter' is, in the order of creation, the grammatical governor of physical behaviours (see Chapter 6: Information's Infrastructure - Code). It is above yet within the orderly text of polar energies (see also Chapter 6: The Lowest, Physical Level - Data Item). This 'language governing text' is the upper side of a valuable, two-faced, psycho-somatic coin. Called 'crystalline mind', it is a way of describing matter that operates reflexively according to archetypal order. The first issues from pre-matter are (*raj*) quantum factors - sub-microscopic, rapid yet supporting all the bulk that's built of them. The order 'falls' as naturally as steam to ice from subtle subatomic through atomic, gaseous and liquid to the gross properties of large-scale, sensible things. Life's specificity by *DNA* is paralleled in physics by a quantum alphabet. Or, from a common-sensible and *bottom-up* perspective, you might say material patterns 'emerge from a genetic, quantum mist' and 'thicken' towards the shape of lowest, locked precipitate, solidity; they seem to fall from 'nothing' (perhaps in the form of light or gas) to something fixed and definite. Materials of the physical universe are, of course, as completely devoid of informative consciousness as they are full of its anti-principle, unconscious energy. They represent an end-phase, death because of consciousness squeezed out. Therefore metaphysical mind, whose grade of *prakriti* is called (*raj*) psychological *manas*, is construed a lesser void than (*tam*) physical matter, whose grade of *prakriti* shows as the shapes and behaviours of 'physics-book' forms of energy; in other words, mind's mix is less devoid of consciousness than matter's which, having none, is not a mix but concentrate of energy alone.

To recap, physical behaviours are like the record of a singer's song. There is no life in their mindless repetitions. A material body is, in effect, an annihilation of subjectivity. It represents the death of mind. Matter is the wakeless part of things. Automata represent a negation of Life, one completely void of Void. This applies to all bodies, not just the chemicals that constitute a biological specimen. *In this way the Dialectic reaffirms (from Chapter 2) that death subtends life.* This means informative, constructive life is above destructive death. Death is dependent on life, not *vice versa*. It comes after; life is prior, precedes and is, therefore, more powerful than death. How important is it, therefore, that one understands the Nature of Priority? Or grasps the quintessential experience of Subjectivity that Oscar Wilde might have dubbed 'The Supreme Importance of Being'? Because if Nothing is Pure Life then let the sky-borne billboard cry, 'The Most Important Entity is Nothing at the Top'!

How Does Nothing Physical Work?

The psychosomatic junction appears to be a critical form of nothing. It is of prime scientific importance to try and understand how the latch leading to our house is lifted, how the entrance to our universe swings open and the frame of natural law begins to work. How can we better grasp the archetypal 'grammar' out of which the pages of the world's book flow?

What is a book? It is a record. It is information stored according to a language, code or natural law. The law of nature's energetic book accords with polar principles; these are stored in universal memory. The enactment is enforced by fields-of-influence-in-space (ink and blank paper), letters (particular 'shapes of ink' called stable subatomic particles) and words (about ninety stable elements) arranged in various permutations (molecules). Aggregations of molecular phraseology accumulate to 'paragraphs' of gases, liquids or of solid forms. Whether it is simple, complex, large or small each 'lawfully-coded paragraph' is called an object or, in dynamic form, event. What is a body but a slow event, a lengthy process? And events compounded write the book of every circumstance.

If space-shapes are seen as relatively 'frozen' energy so memory is seen as 'frozen' thought. Archetypal memories are by this definition concepts 'frozen' in a universal mind. This autonomic operator's body is our astronomical universe. In other words the logic of Natural Dialectic sees (*tam*) physical patterns of behaviour as the non-conscious, automated consequence of (*raj*) mind - instinctive mind that precipitates involuntary, 'crystalline' matter. Physics and chemistry study these 'instincts' or archetypal control systems. They are natural law and their 'nervous system' is actualised in electromagnetic and other fields that orchestrate the way things interact.

What is a musical record? It is the memory of a creative event. Cast in plastic or 'burnt' onto a CD its tracks are irreversible and can only, to reiterate, repeat the original performance mindlessly, endlessly, on and on. What is an atom? From this perspective it is a memory whose composition is extremely short, simple and reiterated very fast. It is a vibratory program whose electrons keep spinning and whose protons fail to decay. It is as if memory were a 'skin' which, when vibrated from within, reiterates. Its motion is relayed from psychological to physical space. The first, indelible imprints of creation are empowered from within. This power of universal mind continually resuscitates the patterns we perceive as things in space. In this view matter is, at root, rhythms etched at the moment of creation and, since this first impression, continually drummed into reawakening by a cosmic 'hum'. Things are behaviours; behaviours are concrete instinct. While the origin of this 'breath' or 'hum' is causal *Logos / Om*, in the diffuse condition at which it 'quickens' matter across the psychosomatic tympanum it is called *prana* or *qi*. Metaphysical *prana* is identified as the motor behind various modulations that comprise physical phenomena. The instrument of memory vibrates and its vibrations resonate with various constructions in tensile space. In which case its medium of communication is light, 'string', quark or other particle. It is as if physical patterns are the dynamic reflection, even the shell, of 'hum' across the strings of archetype. Or as if they are the sounds of 'breath' across the organ of lower, reflex mind; sub-conscious energy is the breath, archetype the instrument and things the melody. No 'noise'. The song of the world is the world itself and, like Mozart's composition, it can be accurately described by physicists' equations. Plato had his 'ideas' and yoga claims its *tanmatras* (see *fig.* 6.2) but perhaps the nearest modern simile for 'archetypal cosmic notes' is found in the topological loops and strings of the eponymous theory (Chapter 9). As individual units are numerically analysed so their shapes are defined by geometries. The 'crystalline' geometries of matter are translations from a very coarse matrix of mind. We are watching abstract, archetypal records playing out in fact. Ours is, therefore, a cinematographic show. **In this view, contrary and outrageous for materialism, matter is developed memory; it is a projection.**

How does a radio work? Does every set contain its programs? Of course not. Each set picks up the broadcast to which it is attuned. Just as a billion sets can pick up the same, single broadcast, so the atomic principals that underwrite each bulk shape or change of shape can pick up the rhythms broadcast according to their rigid 'instinct', pattern or memory. They accord with this archetypal 'music'. A specific channel, a particular field can influence its particles everywhere at the same time. Resonance (Chapter 16) is seen as a key cosmic actuator.

Biological bodies are constructed of physical materials. Is their coherence simply chemical or does psychosomatic order acting as a higher-level framework rule the dormant day?

In other words, are cells 'intelligent' or not? Unicellular organisms such as *amoeba* and *paramecium* can hunt, perform complex tasks and so behave 'intelligently' without the benefit of brain or nervous system. What about sponges, brainless corals, jellyfish and sea anemones that sensitively and selectively eat, fight and reproduce? What about rotifers with a 'brain' but also rudimentary eyes and tentacles - indeed, a host of virtually brainless but clearly purposive animals such as bivalves, sea squirts, sea acorns, sea urchins and starfish? Jellyfish have nerve nets but not what you could call a brain - especially in the polyp phase. Indeed, each kind of form down to bacterial is equipped with its own distinctive, instinctive patterns of behaviour.

Of course, biochemicals play a part but what actually registers a pheromone or a chemical gradient? Biopsychology and neuroscience (Chapters 13-17) equate mind's consciousness with neurons but is this justified? Aren't even dormant plants, with their tropisms, sensitive? How does a root 'sense' water yards away and grow towards it? Botanists think plant response is a case of mindless cybernetics; biologists see instinct just the same. Fascinating - but is this the whole truth or a part - the lower, chemical part? In fact perhaps neurons and, simpler, lone or complexed chemicals act as 'antennae'; perhaps, operating in an electromagnetic field, they act as media between an organism's mind and body. Perhaps (Chapters 16, 17 and 19) this coherent field is an intermediary between 'typical mnemone' and its physical form.

Experiment. Waggle the little finger of your right hand. Your mind, in so doing, connected with your body. Will-power controlled matter. By what means, through what levels and across what borders was your thought transformed into its corresponding digital display? What might be the nature of a *PSI* (psychosomatic interface) through which information's two-way traffic flows 'twixt mind and matter? Does it exist in all organisms? Does it exist at all? Is there a universal as well as a biological linkage between mind and matter? Are archetypal or 'memorised' patterns of energy called morphogenes involved? Whence derive the laws of nature that seem, in a holographic sort of way, to inform nature's every particle? And what constitutes the way matter is, at least in you, controlled? Precise digit waggling raises a lot of imprecisely answered questions.

Do you remember (from Chapter 7) we asked how a light bulb emits electromagnetic radiation from passing electrons that lose neither charge nor pace? We have developed the hierarchical notion that sources are 'above' their consequence, potentials are a source and that active vacuum is a virtual source of actual physical energy. The absorption and conversion of virtual photons from the vacuum by dipoles gives rise to observable energy. This energy is not, according to the Dialectic, disordered in the forms it takes. Could structured charge absorb a structured load from vacuum state? It would come as no surprise, given its position on the scale of things, if vacuum state was found to be a vital intermediate in the scheme of mind to matter, in the translation of mental patterns into physical and *vice versa*. It would (see also Chapters 15-17) be a vital component of, for example, your brain. From this perspective brain becomes a 'vacuum state computer'.

Check the stack at the beginning of 'Truth, Reality and Appearance' in Chapter 3. True Nothingness is Essential. Every lower 'nothing' is apparent. From an invisible source issues what follows lower down. Physics suspects, in accordance with what the Dialectic expects, that this is the case with space. Lots of ideas are floating round about the vacuum but there is a consensus nailed to space that 'nothing' is not what it seems. It is not quite nothing! And it was (and perhaps still is) the womb of all material things. Can this in fact be so?

Yet what, if you espouse a 'virtual' metascience, might issue through a vacuum's open doors? Might psience mechanise the state of space? Might paraphysics invent 'over-unity machines' (ones from which you take more than you have given, solving the world's energy problems) by tapping emptiness as fuel and thus, with a clean, alternative technology, achieve what standard science couldn't? In science standards change. Could it be that virtual photons can impinge on complex fields that constitute, within the liquid crystal of a body, the base denominators of a bio-system such as you or I? Even more 'dubious', if you're reaching to the base of mind, could this Pandora's box unleash the paranormal and a brave, new but dangerous world of mass mental (and thereby social) engineering? Mind, metaphysics - what next? Don't think that people, even governments, may not be trying now. If they succeed will not the paradigm of science-as-we-know-it have to flex? Textbooks and training undergo revision? Don't think that by saying nothing you won't venture into trouble. Silence won't stop science straying into unmapped psycho-areas of interest.

You might retreat. What on earth have psientific metascience, metaphysics and the 'fuzziness' of 'virtuality' to do with sharp precision, resolution and the scientific way I live today? The notice pinned upon this chapter's door forewarned that nothing might seem as irrelevant to scientific tests as nothing's nature; or as irrelevant to the hubbub of an office, factory or street as adequate conceptions of 'Infinity'. **The immaterial is indeed, by definition, irrelevant** to material science and anyone who demands physical as opposed to inferential proof of non-physicality may well end up denying even his own mind.

Space for Space

So let's indeed retreat. *Let's turn from (Sat) Metaphysical Void to study its polar reflection, (tam) physical zilch.* Let's flip *top-down* perspective through 180° to get a grip on nothing from its other side. What, *bottom-up*, did our forefathers empty out their minds with? What can you think when you examine 'the existence of the non-existent', if that is what space isn't?! What, in particular, has the history of science made of unobservable, sweet nothing? Remember, first, that space and time did not exist before material nature's start. It is currently believed the void of physic, space, has been expanding since it sprang with other titans we call time and energy *ex nihilo*. Such nothingness was not material; at least a void of metaphysic, at most the archetypal memory of universal mind preceded physical phenomena. With this in mind let's now review some historical and contemporary ideas about the physical transparency of vacuum in its complementary forms of micro- and macro-space. From ancient to modern and, thirdly, in a dialectical way - how do vacuums look? When you gaze into space, when you look into nothing, what exactly do you see?

Old Vacuums

Top to bottom exhaust it. Freeze your black box, pump out every particle and piece of radiation to get nothing in return. A bare vacuum. Material immateriality. Objective but inert, impotent metaphysic. Is outer space not such a box without its sides? A vacancy that's tenanted, an empty platform things play on, a non-existent sea that buoys the world up and yet, tell me, is not space an actual absence? It is an extension 'formed' of formlessness, the negative of any thing. Physical abstraction at the base of things, space is a bottom bare and bottomless; 'old', 'traditional' void extends the coolest, thinnest of receptions.

Down the cosmic slope, therefore, we slide to matter through mind's postern gate; we drop from the archetypal 'soul' of physics into formless vacancy; we are swallowed from the mouth of order into nothing but pure, structureless extent. Vacuum. Boundless space. When everything's exhausted nothing's cleaned right up. Over the years many people, some in a professional capacity, have had a lot of fun making something out of this void, zero or blank. Is nothing really like it seems? Or is it really nothing like it seems? Possibly the earliest references to emptiness occur in Hindu holy books. The usual difficulty arises in establishing an unambiguous classification using words that seem, for different authorities, to have different meanings. Nevertheless, one word for both Infinity and Nothing is *bindu*. A *bindu* is a dot. The dot became written as a ring - whence our null numeral, zero. The sign of the *bindu* (or *bindi*) is a forehead decoration dabbed at the point just above and between the physical eyes which symbolises the first, central, metaphysical eye of knowledge. It thus represents our cosmological axis or pivot of truth. Also called the third eye, it marks where the search for Higher Truth begins. Bull's eye for an archer aiming towards the Infinite!

	tam/raj	*Sat*
	existence	*Essence*
	else	*(N)One*
	relative void	*Void*
	realisation	*Potential*
	diversity	*Bindu*
	development	*Egg/Seed*
↓	*tam*	*raj* ↑
	yin	*yang*
	negative	*positive*
	diversification	*unification*
	effect	*cause*

In terms of information we write:

↓	*informed*	*informant* ↑
	shape	*shaper*
	formed	*formative*
	receiver	*communicant*
	receptor	*transmitter*
	another	*self*

In terms of energy:

↓	*mass*	*energy* ↑
	shape	*shaper*
	space	*motion*
	void	*action*
	gravity	*levity*
	fixation	*release*

Bindu is a dot or microscopic sphere. A quantum. Next to nothing. Point infinity. An emergence from blank void. A dot is circular, a circle round and polarised a circle might 'hatch' into an oval or ellipse; or from a sphere might 'hatch' a polar, ovoid form. Circles like vibrations are containers giving shape to energy; and ovoid forms give shape to special information. That very special sort of nucleus, almost nothing in a nutshell, is a unit called an egg. It is also, in a nutshell, everything that follows. Nuclear information; egg or seed are powerful symbols of potential, totipotent precursors out of which the rest of differentiated bodies are, *top-down*, preordained. Whether elliptical or spherical in shape, eggs are always polarised potential. They are encoded being prior to purposeful becoming; they are determination of idea that, set in motion, drives towards predesignated shape. No egg evolves or ever was evolved by chance. Such embodiments of premeditated aim are, as already noted, excellent symbols of initialisation, 'fertilisation' or creation from potential matter. They represent the psychosomatic issue of energy from an archetypal 'psi-field' to the prospective 'development' of a final, 'adult' universe. From psychological psience emerges physical science, from archetype is hatched the chick of quantum actuality. The way of projected or externalised creation is one of graduated development. Each parent, every teacher knows that *right* conditions best evoke potentialities. With matter *different* conditions evoke them. For example, colour and the textures of a solid are implicit in a gas; and immaterial information is implicit in a towel or teacup but you only see what's physically expressed; ideas, creators, keep implicit.and unseen. The attributes and properties of an event, the character of a behaviour that's expressed at a specific level of the cosmic show is 'kept' or 'stored' or 'saved' implicit in the rest. Connection is intrinsic and its chain runs from the top. Each nested stage that's made explicit has (towards its interior above or towards its exterior below) implicit links with all the rest. **The notion of 'nested command and control' has been alluded to before but, due to its lack of emphasis in a materialistic paradigm, is here reiterated, underlined, in bold.** Thus, in a universal case, atoms are implicit in bulk matter; particles compose an atom; what about, higher up the scale and at the centre of inanimation's nest, the 'super-quantal' cosmic egg of archetype itself?

	tam/ raj	*Sat*
	polarities	*Neutral Void*
	expression	*Information*
	meaning developed	*Code*
	transmission	*Message*
	egg/ sperm	*Egg*
↓	*tam*	*raj* ↑
	female/ receptor	*male/ donor*
	egg	*sperm*
	process	*initiator*
	implantation	*birth*
	pregnancy	*child*
	labour	*release*
	gravid/ born down	*lightened/ raised up*

An egg was fertilised. You developed from a zygote. Union (man with woman, two made one) was followed by division (child from woman, one made two or more). Creation from the union of polar opposites is a common theme. Rites of fertility are what, as Chapter 23 elaborates, creation myths are generated from. A male god always acts on female. Sperm symbolises will-power pressed on receptivity. *Yang* on *yin*, active stirs up passive. Sky showers the earth, seed germinates. Light develops shapes that grow up from the womb of darkness. Male, female, love, ecstasy and child are not the product of a set of incoherent, aimless accidents. They are a profoundly factual, powerful symbol of creation in whose superbowl you play a part.

No chance. A program is ordered in accord with principles. These are, in rigid form, its laws. As with a cell from whose program biological life is developed, so a creator spins an idea into a material work of art. The current which galvanized Essence into existence was will-power or, in the previously drawn parallel, 'volitio-magnetic radiation'. The endless radiance of First Cause issues from Nothing. For this reason the teleological, causal or 'heavenly' regions of the cosmos are called by Hindus *Hiranyagarbha* (Golden Egg), *Brahmand* (Creator's Egg) or simply cosmic egg. A similar cosmology enveloped Greek mythology. At a lower level universal matter issued from a programmed 'egg'. *This archetype at PSI is nothing physical because it's metaphysical.*

If a miracle is what transcends the laws of physics then Beginning is a miracle. From nothing into something is a miraculous transition. Existence is a miracle, not least if the universe of galaxies were issued from a focus smaller than a thumbnail or full stop. Modern, materialistic philosophers, working from a chancy material end, seem to have forgotten about the predetermined, controlled development of eggs; so they conceive of their start as an uncontrolled outburst, an accidental extemporisation. The naturalistic cosmic non-egg is a quantum fluctuation's growth; but Naturalistic world-egg is, where natural law is code, developed from prefabricated program, that is, archetypal code.

Do you remember (Chapter 7) the greatest double act of all time? *Purusha* and *prakriti* - cosmic polar influence that male and female represent? 'Male' *Purusha* was identified as both pre-existential Void and its First Assertion called First Cause, *Logos* and so on. This first stir of the existential soup, this primal act polarised (and continues to polarise) the Infinite into Self and 'not-self' or, as the Vedantists say, 'other'. *Purusha* is both Transcendent Self and Self-in-motion, that is, Causal Word in its primal and highest involvement with existence. If Self is alive you, who are also alive, can relate to It. You might, furthermore, personalise a supra-existential (let alone superhuman) relative - even if this one is absolute. Absolute Being is beyond gender but you might humanize it as 'He' or 'She'. Such (*Sat*) Being's *Chitta* (Attentive Consciousness) is seen as the (*raj*) origin of intelligent, rational information, that is, the source of informative mind. Finally, therefore, if Awareness is a cosmic entity distinct from matter, then your own awareness (which is the centre of your life) may, soul within The Soul, partake of It. Or, if you allow the metaphor, Her or Him.

Prakriti, on the other hand, is 'other'. 'She' is identified as First Negation of The Void, the passive product of primary polarisation, non-self. If *purusha* is the subjective then *prakriti* is the non-conscious, objective aspect of existence. It is the informed, energised, materialising aspect of creation from which all (*tam*) downward anti-principles of Natural Dialectic are derived. *Yin* to *yang*. It is the metaphorical 'feminine' on which the 'masculine' imprints, darkness that the positive 'light of *purusha*' relieves, a film which focus of attention develops. *Prakriti* is polar. It is (*tam*) force of resistance, limit and formation; it is the womb from which all shapes, both mental and physical, develop and are born. And, at the same time, it shows a (*raj*) anti-inertial, driving side that keeps things on the go. *Prakriti's* subdivision of *manas* (figs. 6.2 and 13.6) is mind-film or mental energy. The undisturbed psycho-space of *manas* is called *akash* or inner sky; this sky's sun is attention and thought-forms its weather. Thoughts roll, a Tibetan monk in the high Himalayas would tell you, like passing clouds. *Akash* is the field of mind - not just your mind but universal mind. Metaphysical *manas* is as real as vacuum or as air. Physical *prakriti*, on the other hand, is given western names such as (for microscopic construction) strong nuclear force, electromagnetism and (for large-scale aggregation) gravity. Its '*akashic field*' or outer sky, only defined relative to its contents, is called a vacuum. Forces, particles and their congregations disturb its pristine clarity; they are networks of excitement, currents rippling through creation's air. It is all a matter of varying stresses, strains and tensions in hiatus. Things stand out from nothing which is why, simply, they exist (Lat. *ex-sisto* 'come forth' or 'stand out'). A Tibetan monk would tell you objects and events are cosmic weather; they too, like thoughts, are simply thicker, slower passing clouds.

Without asking the author it is difficult to judge whether *prakriti* is the *chaos* of Hesiod the Greek's book called 'The Birth of Cosmos' or 'Cosmogony'. *Chaos* might easily confuse you. Does it lead to cosmos or the opposite occur? Modern usage of the word implies random, incoherent motion. It is from such anarchy that our contemporary sages think that cosmos flew and then, according to the laws of nature, settled down. But to classical Greeks among the sunlit colonnades of their discussions the word did not imply disorder. It meant a chasm or abyss; it meant 'gaping space' but also formless latency. In this respect their structureless, primordial lack of anything is less aligned with anarchy than immaterial field or Dialectic's archetype.

Does cosmos lead to chaos? If, with dials of nature's constants finely tuned, big bang's percussion was a most exact affair then doesn't order at the start give way to chaos later? Fix the rules then, within their stricture, roll the dice. For Natural Dialectic, if not Chaos Theory, this precipitate of chaos is a feature of creation. From generality without a structure (gas) apparently disordered clumps made of specific, solid crystals grow. From principle to practice, from legal information to the energetic patterns that result, 'disorder' is a natural product of a process known as entropy. With entropy at every point there intersect a multiplicity of causes; each combination will produce a different effect. Complexity and entropy combine to make apparent chaos a material fixity. The stage is set on which the complex codes of life can now, according to their script, be played.

Or does chaos lead to cosmos? Modern 'big inflation' and its early clouds of gas might seem chaotic. In the old, Greek sense the rawness of emergence out of nothing is reined in by law. Energetic chaos is restrained, 'original profundity' is turned to cosmos. Indeed, in Greek philosophy titanic deities tamed primal anarchy into the elements (earth, air, sea and sky) whose line gave rise to a series of principles and anti-principles each cast in the easily comprehensible, personalised role of an Olympian or lesser deity. Unfortunately and inconsistently, the mythical behaviour of these humanoid divinities could itself be whimsically anarchic. Like antics in a distorting mirror, they amused the junior class; dramatic stories lighten up the logic. In clear dialectical reflection, however, negative principle yields, in its involvement with positive, the shapes of psychological and, lower down, physical forms. Existence fundamentally entwines these complementary opposites. Where information predominates the state is called mind; whether conscious or sub-consciously dormant, there is life in mind. On the other hand, physical energy alone constitutes life's void. The 'life' of objects from electrons to galaxies is their rigidified, repetitive and passively informed behaviour.

This is the moment to address a couple more interesting Greek 'takes' on space - ethereal and geometrical. For western science an *element* signifies a specific type of atom such as hydrogen, oxygen, titanium etc. For us there are about ninety stable elements. On the other hand Greeks and Romans, in concert with Hindus and medieval Christian and Muslim philosophers, used the word in a different way. For them (to whom element meant what we call a state, grade or 'elemental condition' of matter) there was 'method in the madness' of their five. These were, *top-down*, ether, air, fire, water and earth. Air, water and earth are obvious. They stand for our gas, liquid and solid. They constitute what is moved - passive (*tam* ↓) matter. Fire is a little trickier. It symbolised, as far as they could see, (*raj* ↑) energy. Light shines, heat excites and together they move what is moved. However our study of energy, physical and chemical, is broader. Fire is only one form of energy and not all these are, as fire appears, free. Atomic energies, for example, are locked in specific waveforms. Humans have always appreciated shapes of energetic waveforms - after all, you only have to visit a beach, drop a pebble in a pool or hear the echo of a gong. *Both Hindus and Greeks emphasised the profound natural implications of coherent, harmonic waveforms as well as geometrical harmonies of line and proportion.* Science has enormously and precisely developed an understanding of harmonic oscillation at, in quantum activity, the roots of material creation. Although Heracleitus generalised that 'all is in flux' (i.e. in motion), did our forefathers actually link fire, light, music and the kinetic geometry of waveforms at the root of elemental energy?

It is, however, when we turn to what was meant by ether that real confusion arises. Is the argument about 'etheric nothing' simply one of words, not substance?

Where the orient seems to have grasped the *psychological* nature of ether, Greek (and later Christian) interpretation was more *physical*. The Greek word '*aither*' meant (as opposed to the lower, thicker '*aer*' that we breathe) an upper, rarefied atmosphere. Its 'highness' reaches towards and includes the sparkling, starry outer space of heaven. The Greeks might have agreed, had we explained, that in this physical sense it involved the clear 'ionosphere'. Equally (but not equally) it also meant 'heaven' or 'the abode of the gods'. The gods lived high on allegorical Mount Universe and who should inhabit the highest peak of all? Zeus, king of the gods, on a literal Mount Olympus - of which name there were several vying pinnacles scattered around the classical landscape. In this respect popular theology, when it brought its gods to earth, degraded metaphysics. Child-like understanding often concretises myth and literalises metaphor; such literalisations not only obscure true meaning but, for a sceptical antagonist, provide excuse for ridicule. To miss the inward, metaphysical point and ascribe heaven a physical location above the clouds is a conceptual error that invites scepticism. President Khrushchev duly, accurately and scathingly obliged with a report that astronauts had spied neither God nor angels from the portholes of a soviet ethercraft!

For western science, one of whose missions is to demythologise, the word 'ether' means a redundant concept, 'fluid space', and the fact of an organic chemical. *On the other hand oriental psychology never construed 'ether'* (a common translation of a*kash* or 'sky') *as physical*. It was not physical vacuum but metaphysical psycho-space. This 'inner sky' is the field of *manas*, the subjective, psycho-spatial ground of mind. Therefore ether, the fifth and topmost element, was the medium of psychosomatic linkage. It mediated between material body and sensitivity; and mediated cosmically between physical earth and metaphysical heaven - somewhat along the curious lines of Sir Isaac

Newton's 'universal sensorium' but as a sub-conscious and not (as in his case) a conscious kind of 'diaphragm'. Yogis call its dormant, psychomatic energy *prana*. Natural Dialectic would, with *figs*. 6.2 and 7.5, identify *akash* with the 'atmosphere' of mind; its lower reaches thereby correspond with immaterial, archetypal levels that the wind of *prana* activates. *Akash* is, in this sense, latent but when activated is transformed into a field of physical expression, quantum vacuum, and the ordered origin of force and particle. Thus 'quintessence' or 'fifth element' is (Chapter 6) seen as the medium of ether, 'inner sky' or potential matter's 'meta-space'. It is therefore (Chapters 15, 16 and 18) an unseen, *implicate* channel, a 'memory field' through whose archetypes physical energy is ordered. Its foci involve subatomic structures, bio-centres called *chakras* and (*fig*. 15.8) the agencies of instinct, signal translation and morphogene.

Bottom-up, peering through the glass darkly, nothing appears to be there, certainly nothing to grasp and with which to experiment. We have reached the limits of resolution, the end of physical science. The notion of metaphysics beyond the quantum limits or just 'behind the veil' of space is, understandably for a pragmatist, shrugged off with a tough snort.

In *top-down* terms, however, it is not to be lightly sniffed at. This empty end is simply at the base of mind. Not that metaphysic, even memory, is thought of as a 'static' object. It is a dynamic infrastructure. Universal mind is 'a broadcast of instructions' from which the building particles of cosmos are precipitated using energy; and from which, therefore, the properties and textures of bulk matter can ensue, from which the sets of world-stage are construed.

	tam/ raj	Sat
	something	Nothing
	line/ circle	Dot
↓	tam	raj ↑
	straight line	curve
	ray	wave
	square/ cube	circle/ sphere
	flatness	undulation/ sinuosity
	stillness	vibration
	projection/ trajectory	radiation
	fixity	flux/ flexibility
	corner/ angularity	confluence
	break/ discontinuity	continuity
	sharpness/ roughness	smoothness

How does the cut of nothingness stack up? Simply carve up shapeless ether with the archetypal lines of geometry! As the Greeks realised, space and time constitute a 'frame' or 'blank' within which shapes and behaviours can be generated. Harmonious form of energy is best expressed as music; harmonious shape of mass is best expressed in terms of proportion or ratio. Greek geometry, the Euclidian division of space, was an answer to the question of order, proportion and the nature of objects and behaviours in space. Take, for example, an infinitely small 'dotlessness' called point infinity. Extend it in a line. A straight line. Repeat the process so that from the same point there extend an infinite number of rays. They are the radii of a two-dimensional circle that, as the rays increase in length, can expand infinitely. Make the two-dimensional circle a three-dimensional globe that can also expand infinitely. If you take a point on the periphery of any circle and roll it round through time you create a wave. A wave (*figs.* 2.14 and 5.6) is a periodic vibration around an axis. Rings bind and the motion of a ring through time, a wavelength, frequency or oscillation, is a basis of balance and stability. *Circles and vibratory waves are basic geometrical forms; cycles are a fundamental form of balance whose 'feedback' underlies the intrinsic, cybernetic control of cosmos.*

Not only Greek but Moslem philosopher-mathematicians waxed lyrical over the division of nothingness into spheres, circles and linear symmetries. 'The Centre,' exclaimed Ibn Arabi, 'is everywhere and its circumference nowhere'. Medieval Islamic art is a subtle, multi-faceted metaphor to describe the dynamic yet stable, finite order that derives from an Infinite Centre. One-in-multiplicity and multiplicity-in-one. Its motifs, created by the multiplication, division, rotation and symmetrical distribution of simple themes, evolve endlessly repetitive yet variable effects. The architecture of well-constructed space is geometrically reasoned; music is well constructed, emotionally arousing energy-in-time. What art combines these forms of beauty? What golden verse transforms the beauty with its extra meaning? What, for these philosophers, would raise the bar and charge such loveliness with an immortal look? Could it be appreciation? Could it even be the case that, in the eyes of proper understanding, such loveliness suffuses healthy nature everywhere? Not that anything in nature is immortal. Every mortal pattern, each object and event (including you) is framed within a birth, a lifetime and a death. This includes the universe when taken as a whole. The last chapter of this preparatory section (Chapter 12) is devoted to a description of its interpenetrating layers as they extend from the Centre to Ibn Arabi's elusive periphery. How, you might ask, can a Centre be extended yet within its sphere of influence centrality continue universally?

Point, moving or repeated point as curve and special case of curve (straight line) are the building blocks of geometry. These three shapes (*fig.* 2.14) are the finite fundamentals from infinity. Their lines break misty nothing into form. Combined they summarise each shape in space. Made from their triplex is Mount Universe itself.

There is (*sat*) *bindu*, the dot, point, apex, axis or centrality. From this special case of 'non-line' or 'line potential' there derives the calculus of (*raj*) curve whose spirals (diffusing down or snaking to climactic peak) and dynamic equilibrium (vibration emanating coils and rings of radiant influence) signify acts of energy; and, since body and desire are built of ups and downs and cycles, life. Energetic curve sinks dying into a (*tam*) line whose flatness signifies a straight, non-cyclic path and, in a special case of consciousness, non-consciousness. There is no cycle, ever, in a straight line so that subjectively it signifies the non-psychology of physics and the 'life' of matter, that is, inanimation's arrow.

This trio are reflected in the Dialectic models - central or topmost dot, concentric rings of energetic cone and square solidity of ziggurat. They are the simple basis of all shapes. And they are reflected singly in the purest form of energetic 'trinity', one 'closest to the gods'; is not the brilliant weightlessness of light composed of dot (a photon), wave and ray (see also *fig.* 12.2)? From the (*sat*) potential of these simple principals there are developed, as a consequence of interception, interference and distortion, the interminable irregularities and regularities of (*raj*) flux and (*tam*) fixity; that is to say, the geometries of (1-d) line, (2-d) surface and (3-d) volume - a universe of morph-on-theme.

Things therefore trace an evanescent and coincidental calculus of dots of time and space whose dot is here and now. If a *bindu* stands for nothing then from dotty nothing things emerge, become and move along. Energy-in-motion rolls. It vibrates. Take another form of period - orbit. Make it earth's around the sun. Suns do not levitate. They fly through space. So draw the sun's straight flight through the centre of the earth's surrounding orbit. Thus you obtain a globe that spirals round its star; it follows in a corkscrew motion. Such a helix snakes around an axis-line composed of space and time. Exhaustion, on the other hand, falls flat. It stops. Nil vibration and flat surfaces are made of dead-straight lines. Waves are energy and lines are things. This is why curvaceous lines lease flexibility while straight ones imply severity. How else could you carve space?

An abstraction of lineless space is not quite the same thing as literal nothingness. Atomists, like Democritus, Leucippus and the Epicureans, viewed its 'interval' as a device for keeping things apart and allowing them to move. An atom needs its space to jig. Such empty area was the guarantor of separate identity and changefulness.

As opposed to *Epicurean discontinuity* the *Stoics* looked at space from the point of view of *continuity*. The basis of the Stoic continuum was *pneuma*. This word means breath (as in breath of life), movement of air, *animus* or spirit. Perhaps the same 'nothing-in-motion' is translated into English as 'Holy Spirit'. Did the Stoics actually mean a metaphysical, spiritual association as implied by *Om* or *Logos*? Or did they mean a kind of etheric glue that permeated all space and held everything together? Athenian lessons at the Stoa might have made what commentators squabble over clear.

Outside pneumatic creation was the real Stoic puncture. Void. So Thales of Miletus, thinking nothing came of nothing, therefore thought some kind of essence sourced existence; he claimed that 'space' is actually a seething plenitude of primal matter. Empedocles thought of space as a kind of granulated substance. These granulations, rather in the manner of modern quanta, insulated existence from any continuous void. Minute discontinuities broke uniformity apart. And the great Aristotle had neither time nor space for space. If, he reasoned, you can place an object in a volume that was full of space then you displace or override it. This is nonsense therefore space is non-existent. Moreover how, he asked, can nothing be or become anything; and, since creation's elements (including passive, inert ether) fill up everywhere, where is the space for empty space? Thus the logic of his *horror vacui* denied a void. Just as such logic abhorred real nothingness either inside or outside creation so nature followed suit. Because Aristotle wielded an intellectual authority that was still dominant in the Old Schools at Oxford and Cambridge Universities, the tripos also abhorred a vacuum. Not for nothing did you take a degree.

Aristotle did not understand the force or field of modern physics but we still agree. The *'horror'* of his cosmos ruled and rules the roost. Modern quantum physics thinks he could be right and yet renaissance science grew from disagreement with elimination of all nothingness from space. Galileo propelled the thought of emptiness outside Aristotle's gravitational sphere of influence - though even he, von Guericke (with his Magdeburg hemispheres) and Pascal could not persuade Descartes of sheer vacuum. Spatial extent was an attribute of things. It was part of the continuum of matter. No thing, no space, no vacuum.

Sir Isaac Newton, on the other hand, believed in absolute, infinite but separate space and time. His problem was how an insensible medium might propagate light or gravity. Was it really a subtle fluid called the 'ether'- something through which waves might pass like those of sound in air? Although he seemed to need a medium 'exceedingly more rare and elastic than air' the great man did not like the idea of any impurity that might influence the motion of celestial bodies. He vacillated over ether.

Others did not and, to cut long stories short, light-bearing ether became all the contemporary rage. True, Michael Faraday thought electricity was a 'varying state of strain', by which he meant vibration. In his theory of radiation he ventured to dismiss any concept of celestial ether but to include 'a high species of vibration in the lines of force known to connect particles, and also masses of matter, together'. Field theory was conceived. On the other hand James Clerk Maxwell, building on Faraday's precepts, recalled it. He developed a model for the medium in which electric and magnetic effects could occur. He drew an analogy between the behaviour of Faraday's lines of force and the flow of an incompressible, elastic kind of medium. He saw both light and electricity vibrate through this hypothetical stream. The ether came flooding back. By the late nineteenth century it was one of those scientific truths that, like biological evolution today, we hold to be self-evident. Scientists practically unanimously accepted the idea of a motionless, invisible, luminiferous medium extended throughout space and with the property of transmitting electromagnetic and gravitational forces. Although solid objects could traverse its 'gossamer' some degree of friction should accrue. It was this 'drag' which a French mathematician, Fresnel, had calculated and which the experiments performed from 1881 to 1887 by Michelson and Morley set out to test. The results showed no such 'drag'. You might therefore conclude that the assumption of a motionless etheric 'ocean' that fills all space with the earth ploughing through it is unnecessary, an untidy excess due for Ockham's chop. Maxwell's ether was guillotined; one kind of space was dead, long live another species of the vacuum.

For a beheaded wink. Michelson himself soon realised that he had not executed the existence of ether, only disproved a particular hypothesis concerning it. By 1899 he had expressed the idea that a strain in the ether, as space was then called, corresponded to the emergence of an electric charge; also that displacement of the ether amounted to an electric current and vortices in it to atoms. Therefore, he said, all phenomena of the physical universe can be seen as manifestations of various modes of motion in one all-pervading substance - ether. This insight was compounded by prescient Sir Oliver Lodge who believed science had no mechanical handle on the ether but could only grasp it electrically. Due, however, to Michelson's experiments and Einstein's theories of relativity that permitted the calculation of physical effects without it, the notion of all-pervading etheric ocean was finally replaced by ones of quantum vacuum, co-existential fields and a 'surface' on which patterns of energy/matter appear and interact. Pieces of space can now theoretically interact and pass each other by. How can space pass through another chunk of space?

New Vacuums

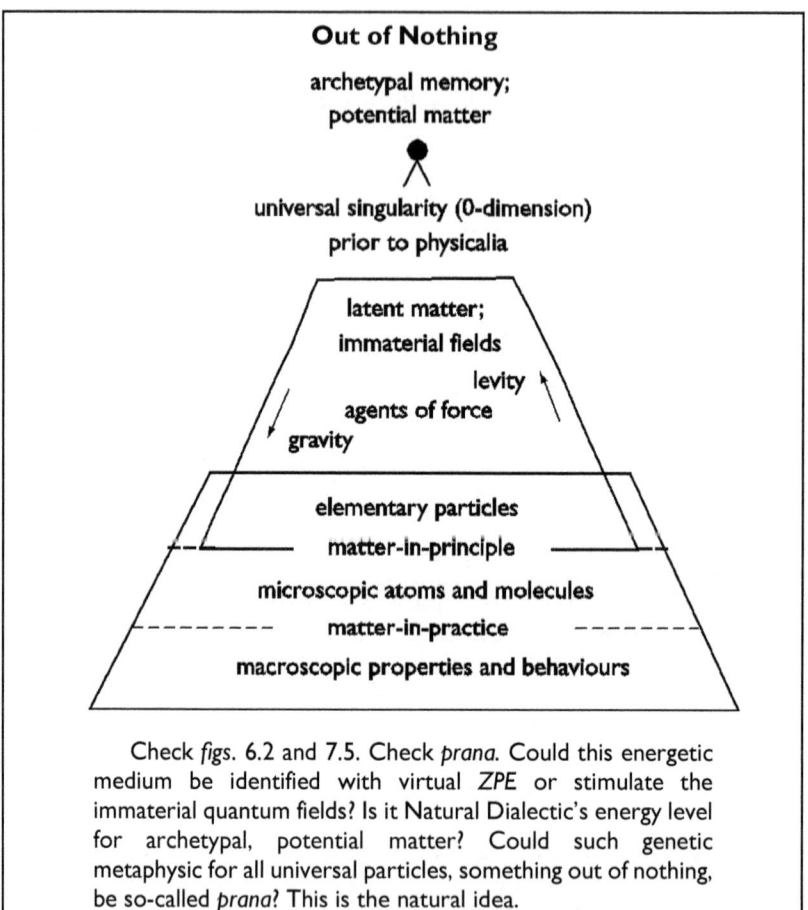

Check *figs*. 6.2 and 7.5. Check *prana*. Could this energetic medium be identified with virtual ZPE or stimulate the immaterial quantum fields? Is it Natural Dialectic's energy level for archetypal, potential matter? Could such genetic metaphysic for all universal particles, something out of nothing, be so-called *prana*? This is the natural idea.

Be *prana* as it may, it is suggested that this ziggurat of *physicalia* is matched by the rings of *PAM*'s Miracle (*fig*. 12.5); it is also equivalent to the base stages of Pregnant Voids (*fig*. 10.1). 'Out of Nothing' relates the emission of matter (physic) out of metaphysic.

fig. 10.4

Physics is a field of resolution but, in contemplating space, infinity and origins of things, sharp clarity is blurred and blur becomes a headache. Churning speculations and apparent contradictions flex in flux. Did something come from 'nothing'? What is the fundamental nature of a vacuum? Will we ever know?

To show the way that nothing in three phases, metaphysical to physical, assumes specific shapes, bears out events and renders objects of the world we can write:

	tam/ raj	Sat
	something	Nothing/ No Physical Space
	subsequent order	Archetype
	its issue	Substance
	becoming	Being
	matter-in-practice/	Prima Materia/ Principal/
	matter-in-principle	Abstract or Potential Matter
	expression	Latency/ Potential
	apparent disorder	Internal Order
	material effects	Immaterial Field(s)
	action	Inactive Priority
	motion	Motionlessness
	form	Formlessness
	phenomena	Noumenon
	contingency	Law
	variation	Theme
	development	Cosmic Egg
↓	tam	raj ↑
	coarse/ gross/ bulk	fine/ subtle/ quantum
	slow	rapid
	fixed locality	fluid probability
	visible/ obvious	invisible/ arcane
	matter-in-practice	matter-in-principle
	bulk objects and events	internal structure
	space in-between	fields/ wave-filled space
	'empty' macroscopic space	'full' microscopic space

$0 = +1 -1$. From Neutrality springs Polarity. From Potential appears Actuality. Physical space is nothing if not an elusive, mysterious 'object'. What's its nature nowadays?

Aristotle and quantum physicists deny the emptiness of space. Einstein solves the paradox a slightly different way. He lets space contain a universe of things but also grants it flexibility. In the presence of a mass his vacuum curves. Space is bendy. The geometry of emptiness depends, we now believe, on how it's filled. So, taken as a whole, is the nothingness that holds us convex, concave or just flat and, since expansive, endless?

New skins for old - modern quantum vacuums are a zippier affair. They are positive, potent 'forms' of nothing that, as well as carrying 'particularity', host 'forcefulness' and radiation through their immaterial fields. So suck and freeze! At Plum Brook Station in Ohio, *NASA*'s space facility, go for nothing perfectly! Particles are pumped out and, in the belly of a massive church-like chamber, energy's subdued until a metaphysical horizon, experimental nothingness, appears. But since you can't obtain a thermal zero (0°K) Heisenberg weighs in with delicate uncertainty. A modern vacuum never is entirely vacant. Never perfect. As well as being zippy they are full of fluctuations (*ZPE*s) and (*ZPF*s) their fields. Always and everywhere remains the vibrant ground-state void. How did filling of once-boundless emptiness occur; or is space never empty? What's behind what is?

Late in the 19th century experiments were performed with a special radiator called a 'black body'. The first formulae to describe the emission of its radiation were accurate except at long wavelengths. To correct this discrepancy more formulae were developed which were accurate except at short wavelength - a circumstance dubbed 'the ultra-violet catastrophe'. In 1900 Max Planck developed a formulation that overcame both problems - at a cost. The cost was bundling everything in bits called quanta. It was to divide the previously held continuity of space and time. Radiation was not emitted continuously but in discrete quanta valued by 'h' (Planck's constant) and 'f' (the frequency of radiation). Quantum physics was born.

In 1911 Planck and in 1913 Einstein and Otto Stern deduced (the latter pair without the need to invoke quantisation) that there is no such thing as absolutely empty space. This is because you never reach 0°K. Full fixity does not exist since atoms dance about their minimal position; therefore thermal radiation must persist. With regard to a black body there exists an un-removable, residual exchange of radiation between a wall and and an enclosed volume of space; and where can you find space without a body for a wall? Not in this starry universe. At a temperature of almost 0°K, an irreducible residue of energy remains in any system, including space. This, calculated at $1/2hf$, is what is called its *ZPE*. The implication (one we've already met) is that there exists, entirely permeating cosmic vacuum, an 'isotropic, uniform background' of a kind of 'noise', a 'trembling' or subliminal perturbation of less than quantum (or sub-quantal) strength. If you are still uncertain then let quantum theory's *HUP* (Heisenberg's Uncertainty Principle) explain the omnipresent 'buzz'. It states that definitely you can't have zero energy to make a perfect vacuum. Vacuums just aren't perfect emptiness. Residual fluctuations are required; these can 'wave' from different fields and since each one's energies can vary you can also have a spectrum built of *ZPE*s. You can have a theoretical infinity of vacuums or, in other words, endless types of nothing. Uncertainty is certain of it.

Therefore space is absolutely full of a residual energy. A quantum vacuum is a ground-state, a 'granulated structure' with the least amount of stress or strain or wobble. But subliminal rumblings of its immaterial fields are known as the aforesaid *ZPE*. These primordial fluctuations everywhere make up an effervescent 'sea of virtuality'. They flicker like a flame about to die; but could their flicker also represent a flame about to spring alive? From this sea (the scientific saying goes) or sky ('*akash*' as the yogic saying goes) the world evolves.

However is not such quantum emptiness with all its wriggling, animating *ZPE* an artefact of theory? Is it real? In 1925 Robert Mullikan and later Willis Lamb experimentally confirmed its existence using shifts measured in the spectral lines of boron monoxide and hydrogen. Later, after WW2, Dutchmen Casimir (an erstwhile assistant of Wolfgang Pauli) and Marcus Spaarnay working for Philips at Eindhoven added to the evidence. The fact that helium cooled to within micro-degrees of absolute zero (0°K) stays liquid and the weak interaction of 'instantaneous induced dipoles' (van der Waals forces) are also attributed to *ZPE*. In fact it has been calculated that the energy density (energy per unit volume) of *ZPE* is comparable with that inside an atomic nucleus. In other words, a centimetre cubed of vacuum contains the energy of a galaxy of stars shining for a million years! *In that case nothing's very far from nothing*! Its amount is sufficient, if converted to matter by Einstein's principle of equivalence, to have caused cosmic collapse to a size far smaller than a pea soon after notional 'big expansion' spread its levitating wings. Is that, if real, not fantastic?

If it's real what is it? And what binds it back from bursting out? Check with the Glossary; browse *fig.* 8.1; re-read Chapter 8: Tugs and Wobbles; you might even glance a page ahead to dialectical Vacuum. Some don't believe there is intrinsic energy of vacuum. They say *ZPE* is just a function of the particles in space; when you cool them down to 0°K still residual 'oscillation' bounces. Others leap ahead. It has been said that once you have extracted every particle from given space you still don't have a vacuum. You have to cut out radiation. Cool thermal radiation until none exists yet, as we've seen the theory go, 'empty' vacuum is still not inert but active with residual energy. Could such exhausted residue be anything but the antithesis of 'super-force'? Mainstream quantum/ particle physics predicts 'fluctuations' that might represent a flux of virtual particles or quantum flickers of strong force, weak force or electromagnetism that could polarise the 'emptiness' of space. Is it, on the other hand, made of scalar energy - energy that, like a standing wave, has no vector of direction? Tom Bearden and others now call Nikolai Tesla's 'radiant energy of space' 'scalar electromagnetism' (as opposed to the familiar 'vector electromagnetism' and its propagated waveforms).

If you feel no wiser ask again. What is it? Is it just a 'boundary condition' that, in the beginning at the edge of endlessness, 'just happens'? Is it an artefact of *HUP* - because an indeterminate quantity of motion even at 0°K means there must exist residual energy; the sum of zero-point particle motions that radiate with 'fluctuations' that in turn drive atomic motion; or, as *fig.* 10.5 suggests, genetic metaphysic known as *ch'i*? Could you identify the three as one; in whichever case a 'virtual' but substantial form of energy hierarchically precedes atomic orbit energies and sustains them. As an electron orbits it radiates energy but, you claim, this loss is reclaimed off mysterious 'virtuality'. As rhythmically as a child is pushed on a swing *prana* or the *ZPE* kicks in. The resonance whacks electrons round; and vibrant nucleons act as vortices that 'draw out' energy from space. The energy that makes up particles (and thus all else) is thereby sucked from nothing! Metaphysical the archetypal fields. Such instance raises immateriality to pole position, primary and prior. It constitutes a reservoir of energy from which perpetual motion issues out of 'space'. A cause is hierarchically higher in the Dialectic's frame. If an electron in its orbit and therefore atomic structure were thus stabilised then a (*raj*) dynamic equilibrium between 'something and nothing' would support the world. What is dynamic equilibrium but 'harmonic oscillation' or 'vibratory stability'? What are sub-atomic particles and atoms but a composite of frequencies and wavelengths - harmonic oscillators? Is the whole stability of matter in this way dependent on an omnipresent flux of immaterial forces?

You might even say that like a ship the cosmos sails a wondrous sea - it floats, it flies on waves of vacuum levity!

Integral to Dialectic is the vector (↑) levity. Levity is, of course, (*raj*) stimulus. It is non-inertial, 'negentropic' effort or a force of anti-gravity. What lends space its buoyancy? Does its unseen, dark energy create repulsive force and as such constitute the polar, universal complement of gravity? Could it be diffuse and at its low ebb now but in initial concentrated form have been what created and inflated nothing into something still expanding? Where the force of gravity increases as two masses close, could levity increase with distance? When expansion spreads things far enough apart is this repulsion what accelerates them towards infinity - as Hubble's eye spots far-off galaxies now sucked ever faster into nothing, rushing towards a never-ending edge, the great abyss? Levity as anti-gravity is logical but not much mentioned in the current scientific book. Is it what is called the lambda force or dark, repulsive energy? Could anti-gravity be tapped? If charge is coupled in some way with mass (say, due to electron spin) could you expect to find and then exploit 'electro-gravitation'? The problem is both proving and extracting it. How do you dig up space, transduce the vacuum and thus siphon off its endless, virgin, ever-virgin fields of 'zero energy'? That's rich! An endless source of energy for human use, a giant pump-house for perpetual motion of machines and, countering the tide of time, a mechanism generating from the ether an 'electrical negentropy'? What, if this is not some pseudo-science but the scientific future, a fabulous and wealthy nature such a form of nothing has of benefit to everyone!

To argue from the premise of 'an empty space' would be a fallacy because, on the above account, there's no such thing! A vacant vacuum has been falsified by physics; its potential for 'dynamics' has been sketched above. <u>Does quantum vacuum not support alone but also make the universe?</u> In this case apparent void is seen as actually the cosmic egg, womb of worlds and pregnant context out of which all apparitions rise, all things have appeared. **Space might be invisible. It might seem dark and cold but in the Dialectic view space isn't simply barren emptiness. It is a bridge. It is a psychosomatic point, a medium connecting space-less metaphysic with its spaced-out physic. Information is transmitted; void links mind with matter and joins archetypal instrument with natural patterns that derive, like note from flute, from it.** Just as Pauli's distinct quantum states give rise to atomic structure so, it is suggested, distinct, specific fluctuations of a vacuum 'harmonically' give rise to fundamental particles. This is the graduated nature of a cosmic egg; preordination rules; a scale of immaterial fields constrains the action. It is precedent to sub-atomic form and, therefore, the potential for development of any influence that traces lines of various tensions joining spots. Thus cosmos issues from 'as yet un-twanged' latency.

In this view, you can't see, space is multi-layered. 'Top' emptiness is metaphysical. With all constructions a prior blueprint represents specifically restricted possibility. So with universal architecture; latent, archetypal fields are those that govern how phenomena are realised; top-level patterns for the forces mean that definite expressions of behaviour are evoked. Nothing random in its character arises from disturbance in these immaterial spheres of influence. Expression of the microscopic fundamentals will include subliminal flickers of pre-physics named as 'virtuality; but also basic particles and forces that comprise the fabric of our 'real' world. These we cannot see and so space that's macroscopic, sensible and knowable appears like a shell-less shell. It is an impotent, inert and outside continent of nothingness that shells (called things) disconnect.

A modern vacuum is, with wobbling fields and particles, a relatively graded one; but Natural Dialectic adds its absolute, the zone of archetypal memory. What, as opposed to quantum reckoning, about the theory of general relativity and its smooth, non-granulated space? The structure of Einstein's space-time is often pictured as an elastic void distorted by the presence of mass and energy. His field equations yield its topological shape or geometry for any particular distribution of matter while his equations of motion show, conversely, how light and objects move across this invisible four-dimensional 'landscape'. They 'gravitate' along the easiest, shortest path around the contours of a deformation. More matter, more deformation and *vice versa* - until light, like a phantom, almost fails to make its mark. Empty of radiation or of massive objects space-time is 'flat', pure or non-existent. What, therefore, is the medium carrying the curves that matter generates? How is spatial nothingness deformed in such a way sufficient as to create a groove, influence or track along which swings the tonnage of our heavenly bodies, galaxies of stars and planetary hangers-on? How can nothing be curvaceous? In what kind of emptiness do massive bodies, or even those as minuscule as astronauts like you, fly, fall forever or float weightlessly. Facts can stretch imagination mightily. Is 'stretching' an elastic, 'solid' sort of space what causes particles to pop out? Would 'rubber void' grow taut and denser with expansion of the universe so that it slowed the speed of light? It's certainly, with or without uncertainty, a tale of ghostly and disputed mystery.

Do you remember (Chapter 9: Holy Grails) two pillars of physics called classical and quantum; and, if it were indeed necessary, the attempt to unify them into a *TOE*? The authority of Max Born, Bohr, Heisenberg, de Broglie, Dirac, Schrödinger and others has meant that physics from their time has emphasised, as standard, a quantum approach to dealing with natural minutiae. *QED* (quantum electrodynamics or the quantum field theory of electromagnetic force) that is used to describe the vacuum is, however, rivalled by a different approach called *SED* (stochastic electrodynamics). The point of *SED*, first developed by Einstein, Nernst and Stern and later workers such as Edward Nelson (1967) and Tim Boyer (1975) is that it obviates the need for quantum physics. The maths of classical approach (Newtonian mechanics and electrodynamics) can be used to derive the Schrödinger quantum equation and provide an alternative to esoteric interpretations of the micro-world made by quantum mechanics.

Either way, if particles at 0°K have an irreducible zero-point motion they must equally have *ZPE*. Charged particles in motion emit radiation and this is seen as fluctuations in a *ZPF* (zero-point field). Boyer, by adding a classical *ZPF* to a non-quantum model of events, showed that the effect of *ZPF* fluctuations on particles is exactly that predicted by *HUP*. If *SED* derives zero-point energy (*ZPE*) outside the framework of quantum physics there is no need to invoke quantisation to explain the microscopic world. In other words, *SED* succeeds in classically deriving quantum physics and, thereby, explaining the universe by one set of laws instead of, as often believed, two possibly incompatible ones. It replicates the quantum way another way and therefore renders it redundant. If this were so then the notion of two pillars of faith would be false and, with unity already present, the grail of unification an illusion. Would quantum physics, like a virtual fluctuation, disappear? At least the energetic, universal vacuum would remain.

Dialectical Vacuum

This is painless, isn't it? There's really nothing more to say. Check diagrams 2.13, 4.1 and 8.2 with those in this chapter. Also in this chapter you've just met the order of nothing, its importance and how it works; chapters 15 to 17 will elaborate.

It should be obvious by now. Revise *fig.* 8.1. Vacuum is essential. Order runs from Essence to existence. *(Sat) Void is the Independent Essence on whose motion all existence hangs. Lower down creation's scale (sat) non-conscious matter's essence is a void that's called potential matter.* From the latter zone of archetypal memory derives the order of the vacuum; on the definition of its fluctuations physical phenomena depend. You could say it hosts preordination and, when activated, it controls; it is the medium, the informant that transmits an archetypal pattern of behaviour to the forces, particles, the atoms and bulk world 'below'. An archetype is, being metaphysical, nothing physical. *Therefore nothing is the substance and its creations are, although you sense and call them real, your ghosts.*

Vacuum is (can you see?) more than single. It's a trinity. **(*Sat*) void is an essential form of nothingness; it is an archetypal 'emptiness' composed in its entirety of latent, archetypal fields**. This unconscious potential generates, with tugs and wobbles, causal principles of action. For mind the Archetype is *Logos*; but for matter it is archetypal memories, potential matter or the principles of natural law established in the substrate of a universal mind. Law is the frame of possibility.

(*Raj*) microscopic, kinetic space comprises immaterial fields in activated but subliminal or fundamental mode, that is to say, perhaps the fields of *(sat)* virtuality/ vacuum energies, *(raj)* electromagnetism and *(tam)* gravitation. **The quantum vacuum is a medium between material and its immateriality.**

(*Tam*) macroscopic space is what appears inert, oblivious and empty. It seems, in this aspect of absence, that there's nothing there. There are apparently no things at all in it. Such is, physically, the passive emptiness of what we know as 'outer space' - though atoms of the world and their bulk aggregations are also, since it takes up more than 99.9% of their own volumes, hugely full of void.

Physics rightly, as was noted, falsified a vacant vacuum. Its explicit, superficial side is empty but implicitly it brims with energy. We've suspected that a subtle 'sea of radiant energy' amounts to the potential out of which all we observe has come; and, in Chapter 7, Natural Dialectic linked the immaterial fields on which all 'harder' matter rests with archetype in universal mind. Such species of space (that is notionally nothing) may well substantiate phenomena. From its hierarchical position on the conscio-material gradient it works as intermediate between sub-conscious energy at the base of universal mind and the non-conscious energies of universal body. It is the point of psychosomatic junction between archetypal memory and its physically developed expressions. Thus the active vacuum operates, in dialectical order, as a medium. It is 'potential matter' and, as such, first cause of all that's lower down the scale. Did something come from potent nothing? Something for free? What did you ever pay to live here? What cost your ticket to the theatre of the universe? You can't see a playwright when you watch or act but can you read the space between your lines?

Gazing into Space

Mathematical scientists, who understand the limitations of their toolkit, understandably look askance at lay interpretations of what their equations might seem to say. Pure mathematics can invent ideas without recourse to correspondence with anything in the worlds of science, philosophy, psychology or, especially, theology. Only a fraction of possible mathematics describes nature or, conversely, nature only uses a fraction of the possible mathematical systems. *Moreover mathematical equations are not physical. None adorn the working world. Maths' principles are psychological; its process is conceptual; and it is physics' language.* **Does this mean that its abstractions are reflections of a universal mind; and that its discoveries describe how cosmos speaks? Could the whole of scientific enterprise not indicate, in turn, conceptual mind behind?**

Obviously science steers as closely as possible along the course of nature's logic but this does not mean that every 'proof', even one based on a system self-consistent within its own rules, has to happen in the real world. The process is, moreover, confined to the manipulation of automata. Its perspective may never be able to characterise all or even, for those alive, the most important but subjective part of creation - life. Life lives in mind. In order to understand something intangible you have to conceptualise. An inward, invisible character can only be inferred from the behaviour of its effects. If this character is metaphysical how can its physical void be explained? And how can pregnant shapelessness, womb of worlds, the void of vacuum best be visualised? If the gap between physical and metaphysical is leapt by metaphor, how can one penetrate the doors of space with modern, scientific metaphors?

We hang ideas on picture-pegs. Do you want one? Do you want an image helping you round space? How, though, can you paint an empty canvas or hang nothing on the wall? There exist three scientific metaphors of void each, like the modern vacuum, with a '*je ne sais quoi*', with a dash of something in them. Their exhibition at The Anti-Gallery of Things might be entitled 'Nothing Just For You'. So watch this space. Or look upon its spaces.

The first empty canvas illustrates a sphere of influence. It's name is '*Field of Sky*'.

A sky looks empty but it's not. Absolute vacancy has been subtly replaced by unseen, subliminal tensions. Positive nothingness is there, a possibility for things to happen. Cannot clouds and storms blow up? Is not the vacuum's sky potential energy and its formlessness a clear radiance of power? How might you stir up differences? What about a fluctuation, tensions and polarity? Electrical charge? What is it makes a charge positive or negative? Is a vacuum made, Dirac and Lodge suggested, of ubiquitous but unrealised charge whose separation causes particles to 'pop up' into being? In other words, is the 'empty', 'neutral' field in fact continuously composed of charge that's balanced out, thus latent, until it is somehow disturbed? Did they mean 'charge alone'? How can charge, the source of whose field is a particle, be mass-less? While charge is normally associated with mass, are the two intrinsically inseparable? If, however, they were separated half the deal would be an immaterial 'cloud' of mass-less charge.

Up pops Harold Aspden with another kind of charge. A physicist and corporate director who, like Einstein, once worked in a patent office, Aspden took his doctorate at J. J. Thomson's Cambridge College, Trinity. He has written books called 'Physics Unified', 'Physics without Einstein', 'Modern Aether Science' and 'The Physics of Creation'. You can see why 'those who know' ignore him! Aspden's charge is that the (a)ether does exist, is missed and is, as Dirac might agree, a 'cloud' of mass-less charge. It is a storage place for electricity and if it stores electric energy it must have electric composition. If image can stand in for mathematics Aspden sees 'a uniform background continuum of charge', 'pure, electrostatic equilibrium' and 'a stable array of mass-less charge' that takes up, due to mutual repulsion and other constraints, 'a simple cubic structure'. The ether is a vast array of, because you cannot see it, an apparent nothingness. Its lattice dimensions have been computed and he identifies, for reasons given, its main particle to be a virtual muon. What happens in a tug of war when sides are evenly matched? So great is the stability of ether that unless it's 'wrenched' you wouldn't think that it existed. Non-existence. Empty space yet absolutely full of what the world is made of - energy.

Where might you find the 'Dirac cloud'? On *PSI*, set just the other side of what we know as space? Space may be present but it also represents an absence; Dirac's 'cloud' is omnipresent, nowhere absent. Inside as well as outside of your body can you not perceive an ultimate pervasion? You cannot but its undrawn voltage is a quasi-infinite accelerator. It is a latent motor, a kind of battery' that's never flat behind a manifestly charged and polar world. *In dialectical terms materialisation arises as tension or motion due to the separation of potential into various sorts of pole.* (*Sat*) singular potential is polarised into duality. These poles may retain a constant value (as in the case of electrical charge) or there may range a scale of strengths between them. It seems electric charge is (unless a partial form is found and not just theorised) an 'absolute duality' but otherwise the secondary dialectic implies variable tension residing at any particular time in any single characteristic of an object or event's whole (*raj/ tam, yin/ yang*) being. In other words this 'whole being' of an individual unit is a composite of many characters and many kinds of tension. It is a product of their overall effect.

The separation of poles creates force fields and within them influence and sway. The motion of such sway exerts a tendency to try and regain equilibrium, a neutral stability, another form of poise. You have poised potential (as in 'static charge') and, once triggered, motion; you have voltage that, once loosed, will give you current and excitement - motion and emotion. Existence sums to changes in such fields which for the Dialectic are metaphysical (that is, psychological) as well as physical. What causes tension pent up in a neutral poise to split? What sparks polar action, actually starts the motor and might cause a metaphysical accelerator to 'spontaneously' flash?

Clouds have long been associated with spark, charge, action and life. Lightning struck from storm clouds has inspired an image of electrical creation - shock that fertilises, crashing sound and splintered light whose current comes before the rain. Does not electric in the air burst into showers of germination that 'ignite' the crops to life? Think of antique Aryan gods such as *Dyaus-pitar*, whose rumbling chariot rode above the storm; tremble at the awesome thunder those wheels propagate across the high way. Ponder on his namesake *Zeus-pater* or Father Zeus, the Roman Jupiter or bolt-dispensing Deity in William Blake's delineation of creation called 'The Ancient of Days'. The vacuum is a sky of non-existence, of non-being in the sense that all potential is 'pre-being'. Massing clouds of fluctuation presage bursts of action, storms of birth and the development of things. Now science, in rephrasing the historical intuition of an inward truth, has also placed non-being at its core. From apparent nothingness, we are told, emerges everything. It sounds like wisdom from the misty mountain, from the Chinese sage called Lao Tzu - but hard-bitten businessmen are scrambling through its non-existent rocks as well. Their quest is simple. At the apex of material universe they would create a one-way valve to siphon off the infinite. Without current, without motion they would channel light, electrons and perhaps even protons from the sideless reservoir of space. Will they engineer a mechanism that unlocks an endless energy supply or underwrites perpetual motion? They want an unfailing house of power, an 'over-unity' machine.

Please, however, don't forget the other side of absence, the bottom end of spectral emptiness - a featureless, fagged space, exhaustion's post-activity.

From sky to sea the second picture is a watercolour; its translucent washes dye a canvas sketched already - ocean. It is called 'A Fluctuation in Profundity'.

In dialectical terms physical space is an antipodal reflection of (*Sat*) Infinite Void, the Infinite Ocean into which an individual mystic, like a drop in all the sea, aspires to merge. Polar First Cause has, naturally, a two-way flow. While a mystic ascends towards final unification, the 'left-hand' vector of materialisation descends into division and difference. In descent it transforms into a grid of currents appropriate to support the devices of any particular level. The psychosomatic phase of universal mind is the spring of physical offspring. Here, therefore, what 'virtually' vibrates the sea is, most unscientifically, a subtle flux of *Om*, an air called *prana* flowing through the instruments of archetypal memory. It causes the appropriate sequences of melody, of immemorial strains in time that's out of mind. Such strains are not, any more than an orchestral air, the jangling jars of chance.

In virtual water currents ripple. Little eddies 'dusk and shiver' in a vacancy subliminally but totally reformed by the addition of invisible energy. Quantum field theory implies that what seems empty space is really a plenitude, a reservoir of energy, an ocean continent of nature's basic drops and flows.

Imagine, smooth as a millpond, the two-dimensional 'surface' of this void.

Now introduce creative wind. It sweeps up turbulence. Its jet warps, spins and rips the superficial fabric of the water. As bubbles, vortices and currents are part of but separate from the substance of the sea, so isolates are spun from space. They are called 'disruptive tensions', 'bound energy' or particles. Matter (as opposed to anti-matter) runs according to the tilt of one-way flow; any momentary anti-matter's spin is, in the flood, wiped out. In this picture the three-dimensional objects science studies are like figures of a streaming sea - waves, vortices, currents, pressures, bubbles and so swirling on. They vary with their origins and interactions. Could you believe that, if the water flowed through templates just as air through different stops of flute, 'notes of shape' from 'high' to 'low' might be produced? Calm or choppy, you could still define the orders of the sea. Different archetypal 'oscillations' would generate a limited variety of force and unforced particles - the basic patterns that inform a universal ocean's swell.

Now imagine three-dimensional profundity. The surface, multiplied forever, means that you are of the waters and the waters are of you. Drop and wave of water move as aspects of undifferentiated whole. Imagine further that the matrix of the ocean is not water. Conceive it as a three-dimensional sheet of elasticity. It is in fact stretched tight but tensions are so evenly distributed that they accrue to zero. They cancel out. The whole continuous volume is smooth. This is pre-creational tension, the symmetrical latency that underlies expression; it is stretched poise before elastic action. **Implicit in such immaterial fields are the rules that govern each's pattern of disturbance. In the Dialectic's terms a field *is* an archetypal memory and its activation *causes* a particular effect - such as light, charged electron or a proton. The code for physical creation is, like the biological one, conceptually simple yet breathtakingly ingenious and efficient.**

Materialisation (*tam* ↓) drops. It falls from subtle into gross. It gravitates from (*sat*) archetype through (*raj*) principle or quantum matter into the (*tam*) heavy aggregates we call our world. A layered vacuum is a link mode in this scheme of view. It would be a player in the 'almost-world' that science now believes precedes our actuality; and would figure on the conscio-material gradient as a critical medium between archetypal 'nothing' (latency) and this world of massive particles, atoms, molecules and aggregates like you and me. Its subliminal sea of *ZPR* (zero point radiation) is an ever-restless one. Its electromagnetic ground state implies, because there is no difference between a virtual photon and a real one, an ever-present universal light; like wind over wheat, breeze on the sea or twang of string that breaks the silence this aurora ripples everywhere. Perhaps its notion is another artefact of an evolving quantum theory or perhaps the Casimir effect has proved it real. At least it gives us an idea of excitation of the vacuum, *ZPR/ ZPE*, quantum 'foam' and, from the seething volume of our spatial ocean, waves of particle-antiparticle pairs. Electron and positive electron (positron) are an example. So are proton and its anti-matter, anti-proton. Charge is split from where none was - except its possibility. Such pairs have been observed; and observed to self-annihilate. In fact Einstein's principle of mass/energy equivalence requires that mass can be converted into energy and *vice versa*. Pairs of particles flip in and out of being. The quantum vacuum is an insubstantial 'immanence', 'becoming' or a causal field in which, as Dirac tried to show, these flippers rise spontaneously, interact and disappear in flashes. Explained another way, if the tiny fluxes created by the motion of virtual particles do not exceed the threshold of a pair-formation (such as a positron and electron) the flux remains invisible; no wave from the massless charge field would swell enough to be polarised, picked up, noticed. Mini-fluctuations are supposed to everlastingly create and then annihilate a 'bubbly' but abortive mass of not-quite-real 'virtuality'. Potential of the void could, like the laughter scintillating on a sunlit sea, sparkle with an in-and-out-of-being flight. The particles of such shy, retiring laughter emerge only as requirements of both *QED* and *SED* theories; they have never been experimentally unveiled. Although electrons, for example, can theoretically emit virtual photons, no such virtual virtuosity has been glimpsed let alone tested. Or is that really so? Because, in a case where virtuality is thought to interchange with actuality, virtual photons are the same as real ones - you can therefore detect both species in a field of light. In this case 'let virtual be real' would translate into 'let there be light'. Or does such 'deftness' seem a tricky, weavy way of waving something out of nothing?

Emerging from a potent vacuum into what we call phenomena, quantum space appears as packed with action as an ocean with its drops of water, bubbles, sways and currents; bulk materials like you and me are simply aggregates, frigid icebergs cast among the eddies of a microscopic flux. An influential swell of *ZPE* supports the universe. New myths for old - its waves were called *samsara* (ocean of existence) and its tide beats up against both Heracleitan and Buddhistic maxims that 'all flows'. Such fluid energy emotes the ancient, vibratory dance of *Siva*; it quickens and upholds a lively world. What drives an ocean, whips up water? What wind whips up, what influence drives elements of nullity over a virtual threshold, across the material edge into reactions we call, from a measureable frame of reference, actuality?

You are not, like mass-less light, point electrons or the strange, invisible and weightless influence of gravity, something that is almost nothing. You are, like every other sensible incorporation, a kind of 'iceberg' made of mass. Masses of it, masses of them - massive particles like protons making nuclei. Weighty substance makes things what they are. *All this weight is pressing on the floor of nothing. There's the problem. If it comes from nothing how does it weigh in? Whence arises mass and so momentum? How and when did modern physics get to grips with (tam) materialisation?* Not quite yet. Do you remember mention (in Chapter 7) of so-called Higgs bosons? The idea is to commend 'real particles' as reason for a weight. A gang of bosons called 'Higgs condensate' would confer the property of mass on other fundamental particles. Their viscous field would 'drag' or 'rough up' any particle that passed it; it could act on energetic boots like mud that dirtied up, pulled down and stuck them in it - only light can fly unsullied. Since the 'condensate' would be ubiquitous as ocean so would be its tow. Real particles (except light with a speed and rest-mass both deemed absolute) would accrue inertia. They would be graced with mass by interacting with this all-pervading 'field'. Things would get heavy as high heat was lost and cosmos cooled. Feet of clay won't let you run. Isn't mass a clogged up, frozen kind of energy?

Theory predicts that, in the very highly energetic top-drawer of the universe, several sizes of Higgs 'freezer-fields' must exist. Some of them should have been observed already. Will the *LHC* and its detectors infer that such bosons and, therefore, a crucial 'field of mass' exists? The hope is that, in a rarefied zone of higher-than-ever energies, an ephemeral effect, the fall-out from perhaps one in a billion collisions per second, might match the projected profile of a Higgs' boson. Couldn't you, however, create a transient resonance effect or, maybe, 'particle' of any mass according to the energy you generate? Has a previously unsuspected gelling agent, a Higgs field without detectable force, really been embedded in the cosmos from its earliest time? Or is it just the twinkling of a virtual eye keeps plastering weight all over nothing, giving grams and kilograms to everything? How can mass be plastered on a point that, quark or electron-like, doesn't have a structure; how can a formless quark grant form to nuclei and, by aggregation, every massive body; and how can three quarks weigh just one-hundredth of the proton they compose, thus lending spurious weight to atoms and again, thereby, all larger bodies? Forces and particles of mass (strong nuclear force, gluons, quarks, mesons, protons and other species) probably compose the vast majority of cosmic mass. They have their own fields. Why, therefore, should they interact with Higgs'? If they do, then it allows you to exist. If they don't then as was noted (Chapter 7) science will, without due explanation for a property as basic to the universe as mass, still be wandering without an anchor line.

Why drift? W.C.D. Whetham was a colleague of the great 'JJ' (J. J. Thompson who discovered the electron) and Master of Trinity College, Cambridge until his death in 1940. Oliver Heaviside and 'JJ' had quantified the value of inertial effect on charge by virtue of its motion and experimental results confirmed the figures. Now, in 1904 and before Einstein came along, Whetham proposed that the mass variation of an electron with speed is consistent with the later formulation $E = mc^2$. In other words, the electron theory advanced by Thomson had anticipated Einstein's famous equivalence; but the latter's theory swept competitors aside and relativity still rules the waves.

The metaphor for field, it's clear to see, is an invisible extent of sea. Disturbance, concentration, dipoles - each sends waves connecting here to there. The currents of one tide or other flow and shapes of fluid show as *physicalia*. Things interfere or swirl into each other's spheres of influence and interact; and if they part bonds are broken, relationships dissolve. In this view matter is a perturbation in the expanse at the expense of oceanic possibility. Even solids are soft-centred with a hard and sensible exterior but an atomic flux inside. Logically, to trace the order of creation as it first occurred and is still found in every object or event, one has to inspect the smallest ripple and see what shapes it takes as it grows into large-scale waves. You can imagine, at minuscule dimension, the large disturbance of a massive vessel. Its great waves lap a space-time grid. It ploughs a gravitational furrow of *tsunami* size while roaring through the archetypal millpond of the deep.

Are there square motions in the sea? At this fluid level flux of matter is a matter of just curves. What about commotion in an ocean of infinity? The metaphor of curling wave as separate from but indissolubly a part of ocean is not only physical. It is found in all mystical descriptions of creation. And slight fluctuations amplified can swell into a heavy sea. When stressed beyond a certain threshold an immanence that for the most part broods immobile, poised behind the veil, erupts. Imbalance swells in 'virtuality', force bubbles over into space-time. Here it can realise potential in a variety of aggregate, observable forms. When, therefore, waves are sufficiently swollen we see them, when currents are strong enough we feel them. In calm times they appear absent. Einstein's space-time requires a geometrical smoothness but, although it may be calm, Planck's subatomic millpond isn't ever wholly flat or motionless. As we maximise on close-up there zoom into resolution tiny structures, perturbations that less magnified seemed smooth. It is the froth of subliminal agitation, imperceptible quantum corrugation - perhaps even quantised space itself - that divides relativity from quantum theory. Other shots of space show empty but quantum's freeze-framed flash is absolutely full. Sublime, subliminal - *top-down* or *bottom-up* - its film is brimful of the shadows flitting just behind the curtain, waveforms lapping at the brink; and our hard world is made of overflow.

We likened space-shapes to the complicated curlings generated by a passing ship or submarine. Wake indicates the projectile's effect and itself affects a neighbouring vessel's yaw and roll. The smallest fish (a minnow of a photon) is the basis of a current in our sea of emptiness; if you measure turbulence in fish-power then a shoal creates a field-in-motion through the waves themselves. Two-way translation of such waves, affecting and affected by fish, whales or other vessels, can be detailed using a mathematical construct called Fourier transforms. Fourier showed that any pattern in space-time can be analysed into a set of oscillations that differ in frequency, amplitude and phase. Such specific waveforms can exactly represent three-dimensional shapes. Are waves (or currents) actually separate from the sea? Could 'Fourier transforms' track how mind is, through the shapes of archetypal memory, mapped to matter?

The question always resolves into a two-fold one. Part one involves the shapes of bubble, perturbation, spin and a separation of electric charge. Part two asks about the prior cause of their 'emergence'. In dialectical terms (*sat*) archetype is the cause and its primary effect is the specific, stable shapes of (*raj*) matter-in-principle. From these you 'step down' to the secondary effect of (*tam*) matter-in-practice; all bulk forms, like diamonds, stars or you and me, are in this sense secondary effects. We call matter-in-principle quantum or subatomic particles. Their shapes and properties are not chaotic. The reverse. They are limited, stable, coherent and repetitive in character. For example, all electrons are identical 'clones'; they do not have various but single units of charge and mass. From what template, therefore, are these principals pressed? What 'virtual' order mints the variation and complexity we know as our world? From a *top-down* point of view we are groping in the psychosomatic area where mind meets matter, where archetypal memory is the conduit for shapes arising in our starry sea.

But, remember, the 'emergence' of a whirling system still is not explained. From atoms to the galaxies the axis of their cosmic spin is mass; and where does mass heave from? Neither submarines nor massive vessels came before; they could not have been the first to plough a course through space. Egg-before-the-chicken, then, what wind of motion, what current of creation blew the primal subatomic shapes into an elemental substance, atoms, that support our world?

Something out of nothing physical has certainly whipped up the storm of things, the tempest that we call existence.

From liquid to transparencies of sound and light, a third blank 'shines' with pure energy. What is the shape of 'Brilliance'? Sun? What brushstrokes raise a sun? How do you paint light? Is not the colour of illumination clear? Could a sparkling, vibrant canvas even seem to sing?

What causes 'radiant sound and light'? From what underlying field do things emerge? Recall Dirac's cloud and Aspden's array. What is isotropy? A vacuum shows the same to all observers, never mind their relative velocities. Recall, instead of radiation, ropes under stress. Or a drum-skin rolled. Inside interacts with outside through the medium of a trembling diaphragm. Is space a cosmic tympanum whose isotropic trembling some dub, at low strength, the insubstantial *ZPE*? Could the *ZPF's* potential resonance effect an interface between internal, psychological and physical patterns of the world's external 'sound and fury'? Is this psychosomatic medium, so small that you can't see it so you thought it wasn't there, in fact the key, the valve, the channel for an information flow that sets up orders down below? Does it allow the instruments of archetypal law to call their tune? Is it the light illuminating how the paths of energy should run?

When the tensions sum to zero a drum-skin doesn't move and a rope is taut but motionless. *Space appears as nothing when its tensions sum to zero*. Imagine a tug-of-war with the sides evenly matched. You cannot spot movement in the rope. Its symmetry is such that, however many players tug, as long as teams are balanced effort never shows. Within the 'field' of this taut equilibrium rests a considerable but unseen density of energy. *For classical physics empty space has zero gravitational field while quantum theory throws up 'trembling noise', an omnipresent 'intonation' that may support both atom and the universe. It averages round zero but it fluctuates minutely*. Critically for us, it shows the signs of latent stress, a skin stretched tense. Can you picture a subliminal quivering, a stress in immaterial fields whose influence has not yet 'broken out'? Pursed lips of space. A low-key hum but not a twang? It is like a fluttering yet powerful latency, a rope that's strained but absolutely under strain's control. If, on the other hand, the rope is jerked or drum-skin rolled events 'appear'; some plectrum breaks notes from the space guitar.

'Virtuality', another word for physical potential, is raised to 'actuality'. And if the jerk is to-and-fro or the beat continual you have repetition, a stable, cyclical event or what is termed a particle. Material forms are seen as acting like antennae; they channel energy of certain shape and thus reiterate themselves until 'knocked off their balance' by some exterior, impacting force. Such bodies can be seen, therefore, as reiterative disturbances in space, as tensions in the vacuum. Standing waves. Is nothing really any more than tension pegged, controlled and polarised in space?

Control does not mean randomness. Randomness does not mean balance. Balance and control involve establishment of order, rhythm and specific shape. Maybe pre-conditional stability includes Planck particles or strings but certainly the 'lattice' of the vacuum must be 'homogeneous', potentially dynamic and extended everywhere. And from such regularity arise the regulars - electrons, protons, neutrons, atoms and so forth. You can create ephemeral irregulars but nature drives them off. It very soon annihilates 'unnaturals' before it settles back. So if there's regularity what beats its time, sets up the rhythm, shakes very fine and fast vibrations out of which things work their way? Some technicians suggest that two kinds of vibration inhabit diaphragmatic space. Scalar waves represent the energy of the vacuum and familiar, vector waves represent physical energy and mass. Scalar means characterised by magnitude alone. Stress or disturbance of the mass-less charge field would be expressed as electric potentials or an electrostatic scalar potential (called *phi*). Such potential, once activated, manifests as motion. A quantum vacuum thereby generates vector waveforms from its unobservable interior to the outside of the 'drum'. Such motion inhabits what we see as particles; it also exhibits the properties we observe as oscillation, rotation, mass etc. According to such specifics we label it, say, an electromagnetic or sound wave. Just as a tuning fork can create patterns on a Chladni dish, so vector waveforms carry information throughout the physical world. Just as your finger is different in shape and properties from the atoms that compose it, so the archetypal sets of bio-form are different again. What shapes things is, by this token, a kind of 'un-struck music'. Ever heard of standing waves like this? A standing wave or 'packets' of them rolled into a particle? Are particles transformers, conduits or taps whose being is supported by the flow of virtual energy and its conversion into forces that we understand? If this is the case then oughtn't you to call a particle a 'node', an atom 'plexus'? And note that these material '*chakras*' work in the same way as the psycho-biological ones dispersed, says yoga, in an orderly fashion along your body's axis, spine (*figs*. 17.6 –10)? Once seeded with its particles as form-in-principle the larger world develops all its forms-in-practice. Don't you believe a chemist when he shows you how?

Hung in a quiet space of shadow you can see the third attempt to capture nothing is in essence luminous. It glows faintly and is dusted with a hint of glitter sparkling as when sun meets sea. Have you noticed? I think I could raise another canvas to fill up the exhibition's empty space. The other three transparencies - air, water and light - are practically all you need to make a plant. I would seed my own transparency, my empty canvas with a single dot from which one representing everything should blossom. From seed develops form-in-practice represented by the pretty petals of a flower. Can you not break translucence into colours? The coloured beauty of that flower would be my contribution.

What would be yours? How would you draw nothing? How does the world draw nothing to itself?

Tops spin, candles in the darkness and a gong through silence radiate. How is higher information passed along? How is harmonic law translated to the workings of this world? (*Raj*) positively active, (*tam*) negatively empty and (*sat*) in equilibrium as well - the aspects of the universal vacuum sum to dialectical. Pictures help us to imagine how it operates. It must have been and is, we may concur, a very finely balanced prior stability that preordains all order 'down below'. The vacuum is, in equilibrium, an epitome of readiness, potential, poise; space is the nucleus of all the world's development. Nuclei, like eggs, are full of latent information and, once activated, issue waves of power and influence. Nuclear First Cause is, Logically, likened to a 'ringing radiance'. Its brilliant dispensation is by power of resonance. Space and spirit hold creation up. Pictures three and four are therefore nuclear dispensations, shots of nothing more important in the sense of prior information than the world you're looking at!

Is Space a Waste?

Gaze into the more than crystal ball of space. What does its transparent future hold? Is there a space to watch, is it physics' new frontier or all a waste - just a waste of space? It is, no doubt, difficult to experiment with nothing. Vacuum, emptiness and even sizes smaller than can ever be perceived are at the limits of the scientific mission statement and, as far as materialistic naturalism is concerned, off-limits is all hocus-pocus. This is Galilean, safe and sensible. It is an honest, cautious attitude and, where practical is physical, seemingly pragmatic. *Sometimes, however, marginal research and science fiction are transformed into accepted fact.* Thus if there's more to 'space and things' than meets contemporary eye then other forms of theory and practicality would need consideration.

It is true you have a mind. While conscious, this central aspect of your life also has what might be termed sub-conscious faculties. Have you ever dreamt? Were you awake while doing so? Can you remember things? Then you have memory. Do organisms (that's including you) use innate information that is labelled 'instinct'? What is the nature and location of these faculties? Natural Dialectic places them, according to the *top-down* order of creation and yourself, between conscious mind and matter (*figs.* 2.13, 7.5 and Chapters 15-17); physics' half of that location is the vacuum. Conventional neuroscience must, on the contrary, by its philosophy believe that mind is brain, thought is meat and maps including hierarchy simply wrong. Thought and life, you must aver, are just 'correctly' ordered atoms. Of course, the bug in this objective program is the immaterial fact of consciousness; and the aim, therefore, to squeeze it out. It is to pulverise it metaphysically, annihilate it immaterially and thus, by nervous force of physically forceless reason, convert the mind's immeasurable, 'unscientific' quantity into a 'scientific' sort of quality.

What is metaphysical is nothing physical - not even vacuum; but Natural Dialectic identifies the logical nature of a psychosomatic link between 'psychological nothing' and physical things. The purest form of physical energy is, arguably, electromagnetism. Light and electric charge. Mind and matter certainly exchange information and, if quality of mind is more than brain's electrochemistry then, it is suggested, space-borne electromagnetism plays but a quantum half-part in the linkage game. The other half, the black box we call sub-consciousness, is metaphysical. With conventional science in denial and conventional psychology thus leashed one needs approach the gates of nothing cautiously, rationally and with discrimination. Are there tried and tested systems, structures and authorities that, from a *top-down* angle, allow a grip on anything that's nothing physical?

There is one; but its origins are oriental and its *top-down* concepts hardly overlap with scientific aims and practices - except perhaps in the field of health and sanity. The system is called yoga. The word means 'yoke'. What yokes and thereby binds you to the Truth. It doesn't seem, by etymology, much different from 'religion' meaning just the same - a way of binding to your Truth. The difference is that yogic systems offer links between this world and that. They offer bridges that the occidental mainstream lost or never built.

If you won't bridge the gap that separates a state of mind from matter it's as if you sliced the neck of cosmos off. If you won't connect with information it's as if you had decapitated universal body from its source of order. Off with its head! If no apparent reason ruled the realm would it surprise you if conceptual confusion sighed?

Bottom-up, a conceptual vacuum has indeed allowed a plethora of speculation to develop outside firm constraints. The approach to nothing from below is mixed and goes by various names that, climbing up the mountain, still can't see their neighbours for the mist. Check the internet; 'google' metaphysics. You find 'psychotronic', 'bioresonant' and 'paranormal' vying with the 'metascientific', 'psientific' and plain 'scientific'. You brush with branes, strings, loops and zero-pointed fields; you stumble over auras, astral bodies, ghosts and 'virtuality'. Was that 'living matter', 'morphic resonance' or 'cosmic evolution' I just heard? Even 'living matter' stalks the garrets where iconoclasts and innovators strive to bridge the gap between today's art and the next. To heighten and to broaden knowledge perhaps you need to work among some unkempt rafters. *ESP* (extra-sensory perception), psycho-kinetics and telepathy - might today's junk be the store from which are forged tomorrow's classics? Acupuncture, crystals, herbal remedies. Might even brain scans need a fresh perspective so the nature of awareness as opposed to nervous systems heaves across material horizons into immaterial view? Then - perhaps - clear the benches for ghost-busting sensors, Kirlian pictures, radionics or an over-unity machine.

Still, scan the photograph of those researchers working at the trans-dimensional border. Some bending edges of a paradigm are scientists, doctors, laymen, churchmen, honest seekers-after-truth and, sometimes funding 'fringe adventures', even governments. You'll also find non-funded, unofficial groups and individuals of sane bent or zany mind. Don't forget that motives and morality inhabit corners dark and light; your photo will include a range of shade and shady characters - quacks, conmen and some quick-buck charlatans. What a jumble. What a mix-up, philosophical and otherwise. What is ridiculous and what is not? It is clear that viewed from its material side the bridge is territory up for grabs. The other side of *PSI* (the psychosomatic interface) could start a revolution in the standard way we're taught to think. Can I ignore this jumble of ideas? Or can some magisterial pattern magnetise them into proper shape? How do you separate the wheat, if there is wheat, from chaff? What is worthwhile, what a waste of space? Or shouldn't space be wasted?

Before closer inspection (in Chapter 12) of how nothing might have 'popped up' everything, let's ask if there's any physical let alone metaphysical point in it. What, in short, of business on the misty mountain? The view from *PAND* (The Primary Axiom of Natural Dialectic) allows a metaphysical dimension and therefore, if this is true, also requires that science carefully, cautiously and methodically adapt. *It should be obvious by now that physicists can grasp the actual world as caused by one that's virtual.* <u>*Something is projected and sucks strength to support its own projection; material existence is lit by, apparently, nothing. This, exactly, is the Dialectic's point.*</u>

From mid-Victorian times men like Thomas Edison and Guglielmo Marconi have exploited the intangible power of radiation. Their discoveries have combined to change civilisation. In ways previously unimaginable they made fortunes out of 'light fantastic'. From pure energy to space, from one ghost to another, who will first summon up a fortune out of space? Is there any use for an objective absence? Is there actually any point in nothingness, any place or way that man can get to grips with space and use not waste it? How, in other words, will enterprise commercially exploit what's nothing - except perhaps *ZPE* which you can't grip or grab? How might he spade space, void the vacuum, tap an empty well; who'll first find a socket in the sky and plug to space?

Is nothing really empty? Has it any end? If not then our businessman's success would mean a bottomless resource excavated from a single and unfathomable mine. Can you imagine endless barrels of indefinite nothing and, banking on them, what is definitely the ultimate blank cheque? A non-stop stream of money, currency for nothing, endless bags of swag for free! It must be worth a trek nowhere, one that could create a wild-waste bonanza from cosmos-wide deposits of pure, absent gold. How do you pump abstraction's weightless oil to town and store it, well, no-place-to-go? Can't there be a downside to such profitable levity? Do you think nothing's worth it?

Astronauts camp where they can't stay - divers in a hostile sea, fliers in a dark, cold, absent desert, miners hovering far above the ground. As they float they flit and yet, already, men can take a solid platform to the cosmic hole. They work on space technologies like stations, satellites and telescopes. Could they, however, harnessing the power of immateriality, work with the vast and empty envelope? Could they soon engineer the void, exploit the buoyant reservoir of anti-minerality and thus, however daft it sounds today, consume the crop-less field of space? Anti-gravity, anti-mass, vacuum technology and, at low energies, a putative link between electromagnetism and gravity are *not* currently sensible, public, investible science but the vacuum is gradually gravitating into the spotlight. It is levitating centre-stage. From the past to the future, from a postscript to the front page, what does its future science hold? The focus is on nothing but 'doing nothing' is hard work. Is the focus hocus-pocus or will attention lift ethereal silver from a loadless lode - how could humans treat the vast potential of the vacuum right? How, miraculously, make a meal of nothing? The following is just a brief and tentative *entrée*.

If quantum vacuum or psi-field technology ever becomes a hard-nosed, mainstream discipline of twenty-first century science, posterity will recall various faltering steps towards the harvest. They take two main directions.

Firstly, there would be the direct transformation of zero-point influence into useful electrical and electromagnetic energies - a potentially inexhaustible source of power; since the vacuum is not a thermal reservoir this exercise should not suffer from the Second Law's injunction against extracting energy from a colder source. The quantum vacuum might, if 'entanglement' becomes employed, also furnish teleportation, faster-than-light - even instantaneous - communication anywhere and vast computing power. Who knows, as science plays, what the 'God effect' might manage?

Secondly, by linking gravity and charge electro-gravity devices controlling gravity or generating levity might be developed. Rolf Schaffranke, a co-worker with rocket scientist Werner von Braun at NASA and later a Boeing employee, has documented such technology. Perhaps various military agencies have also, under the category 'classified', investigated advanced field propulsion techniques that incorporate a combined use of electric, magnetic and gravitational forces. If so, it seems that nothing has transpired; spatial nothing has not yielded power that way.

Between the world wars Paul Biefeld and Townsend Brown experimented with charged capacitors suspended in a vacuum. As there exists a relationship between electricity and magnetism whose link is the coil, is there also one between electricity and gravity whose link is the condenser? They developed a so-called gravitor. This gravitor, for which a patent was reportedly taken out, is a motor with no moving parts supposedly capable of auto-mobility when vertically oriented. In this position thrust might perhaps cause loss of weight and lift the high-voltage machine.

The most efficient shape to exploit electro-gravitational lift is said to be discoid. In a program headed by Viktor Schauberger the Third Reich invested research into 'levitating discs' and at least one prototype was tested near Prague in 1945. Very soon after the war ended Schauberger's documents and prototype 'flying saucers' were in the hands of US engineers. Is it possible that anti-gravity technology has been designed for military use (e.g. by the USAF)? Are electro-gravitic propulsion, inertia-less space-drives and 'over-unity' perpetual motion by means of controlling gravity through high voltage charge feasible propositions?

Scientists such as Harold Aspden, Sandy Kidd, Scott Strachan and Eric Laithwaite have researched the properties of magnetic and gyroscopic lifting forces (levitation). Could rotation exert its bearing on changes in apparent inertia, that is, mass? Could it alter mass and, therefore, its gravitational effect? Just as you can construct objects like planes whose geometries and 'finish' render them invisible to radar, could you ever use rotation to 'vanish' mass? What form might a 'gravity-shield' take? Fringe or cutting edge? Thus far these studies have not yielded power technology.

Erwin Saxl, one of Einstein's post-doctoral students, discussed with the latter the possibility of an interrelationship earlier suspected by Michael Faraday between electricity, inertial mass and gravitation. Saxl observed, in a moving system, a variation in the gravitational constant. Furthermore, on application of a negative charge to his rotating pendulum he discovered it took less time to swing across the same arc than when the application was positive. If such effects occur with a pendulum what electrogravitic force (if there is such) might operate across the vast amounts of charge and mass that inhabit intergalactic space? Indeed, the electrogravitic interactions implied in such an extrapolation could cause the observed red shift - in which case the universe might not be expanding after all! There would be obvious, adverse implications for 'big bang' theory.

In the 1930's the work of Henry Moray was marked as classified by U.S. authorities and his patent for a 'Radiant Energy Device' 'lost' by the patent office. Would the interests of the powerful oil, gas and electricity generating lobbies have welcomed a public ability to withdraw endless power from the well of vacuum effortlessly, at home and for free?

Do you remember Hendrik Casimir? He discovered that simply by placing two uncharged, reflecting metal plates close together in a high vacuum they were, due to quantum effects, 'squeezed' together by a measurable amount. As you may recall, the post-war confirmation of the 'Casimir Effect' was 'double Dutch'. And the cause of this phenomenon - *ZPF*s of *ZPE* - was unambiguously detected in 1996 by American Steve Lamoreaux. So *ZPE* and quantum vacuum are for real!

Others, such as Harold Puthoff, have also claimed that it may be possible to exploit the limitless fund of *ZPE* available in space. Puthoff is the author of books such as 'The Fundamentals of Quantum Electronics' and 'A Polarizable Vacuum Approach to General Relativity'. He has suggested that particles such as electrons form an effective channel continually absorbing and radiating energy from the background vacuum. If endless *ZPF*, a zero-point field of electromagnetic radiation, could be harnessed into moving things then, like those orbiting electrons, you would have an endless engine - in the sense its tanks would always, being void of fuel, be full of fuel that's nothing. What's a 'free lunch' compared with perpetuity of motion and endless power supplies? Puthoff has also teamed up with Russell Targ, a pioneer in laser development. The pair collaborated on investigations into paranormal abilities such as remote viewing, spoon bending and so on. Can they tighten up on space? Immediate objection seems to bar the way. *Zero point is, by definition, cosmic sink. You can't have less energy that virtually none so how can you use sink as source?* Impossible! There'll never, practically, be any use for virtuality or *ZPE*.

One hardly knows what to believe. You might, however, include a third direction of vacuum harvesting - mind control. This might be expected if you gained access by technology into sub-conscious mind's domain. Tom Bearden's files detail allegations, instances and methodology of so-called 'psychotronic warfare'. This is a subject he personally believes is on the super-power agenda. Its form of psychological attack employs sub-conscious energy's close neighbour - electromagnetism. *ELF* (extra-low frequency) and other signals beamed can, it is alleged, unnoticeably change the mood of a recipient population's mind - to order! This is thought entrainment on the scale of mass psychology. Does any government lack scruple? A dictator or a 'democratic' party that's in power could buy a mind-control machine! Does any government want war and all its winnings with hardly any cost? Then simply let it purchase several more.

Beware! Every advance that scientific humans make is turned to war as well as peace. The more powerful the advance the more dangerous it is. But an upside counterbalance in the 'psychotronic' case is that its flow of energy, archetypal *prana*, has long been the basis of oriental medicine including yoga and acupuncture (see Chapters 16 and 17); and modern techniques of biological healing and rejuvenation are being developed using the careful application of pulsing electromagnetism. However, warring factions even struggle in the Hippocratic shrine. 'There is no alternative to mine! Give me all the funds!' May some elements of the occidental medical establishment, saturated with materialistic theory, perspective and trained in commitment to the use of allopathic pharmacy, in future come to see the logic and respect the tried and tested parts of treatments they for now don't want to understand? You wouldn't, when drug companies become involved, guess so.

Bearden and colleagues also filed a patent (U.S. 6362718) in 2002 for a *MEG*. A *MEG* is a motionless electrical generator, an over-unity machine that exploits a theory of the vacuum called 'scalar potential electromagnetics'. Such alternative energy technology may indeed hold huge potential but mass production and its marketing are, after several years, still not forthcoming. Is it serious, does it work and will it become the first, forever-famous revolution in the public use of brave, new space technology? You might well hope it matches expectations. We are waiting.

Who else is engineering levity? 'Anti-gravity works', says George Muellner, boss of Boeing's 'Phantom Research Works' (Sunday Times 4-8-02 p. 21). Of course, there is huge potential for the aerospace industry if anti-gravitational devices such as those claimed by Evgeny Podkletnov, using a rotating superconductor, are developed - but also for the military. As such, the development of such theory and practice has not entered mainstream science, let alone the High Street. You will not be parking your new Ford Lambda on 'float' for a while! Nobody has repeated his result and Podkletnov himself is almost wholly silent on the subject. His silence symbolises why the outside world must draw one of two conclusions about this novel, high-tech appearance of nothingness. Either it does not work (it is 'flaky', fictional, fringe pseudo-science) or, conspiratorially, research programs proceed in the utmost military and/or commercial secrecy. If the latter is the case it would suggest that programs have been operative for many years.

It was noted in the previous chapter that the order of scientific discovery has been, in general, in the reverse order of creation. It has progressed 'inward' from (*tam*) obvious, large-scale matter to (*raj*) subtle atomic and subatomic principles. Nor are the be-all and end-all of thing. *Will a theory of space-energy usurp today's pre-eminence of strings*? Now, as a central focus of interest, science approaches (*sat*) tenuous vacuum and the fields of archetype. Research from theory and technology are twins. Revolutions of the 'knowledge-wheel' have moved from iron to steam to electricity, from electricity to nuclear power and information technology. Surely this is not the scientific end of it? What innovations, deriving power from the vacuum, will this century reveal? Man waved copper wands and conjured electricity he had not previously known was there into a major thrust behind modernity. Will he be able to whisk previously unsuspected riches from a baseless vacuum? What price a bucketful of space? What will he use for wand, probe or elicitor? How will he tickle nowhere and, in so doing, transform 'flaky' science into mainstream lines of R & D? *How, better than a magical illusion, might technology conjure real usefulness from nothing - even as The Great Magician, Nature, has already conjured everything?*

'Nothing' - in the form of vacuum potential - has not reached the market yet. How can it when we only know it as a sink; and when we can't plug in. If we do will we tap free electrical power; scalar wave devices for energy extraction, transmission and immediate, holographic communication; anti-inertial and anti-gravitational mechanisms to fly with; and fuel-less transport, industry and entertainment? Will space science reveal electro-gravitation? Mass levitation? And, more than any other technological advance, endless scope for good and evil?

If riches can be harvested from space there is plenty of it, far more than of that other rich but hostile place, the sea. Will vacuum physics reap a commercial crop of nothing and space, after a technical interview, be found a career path, be promoted and, in human terms, have a future? There are many questions about nothing but who, pragmatically, will fill that vacancy? Who will patent space and then, becoming its first mogul, pump endless billions from a boundless well of raw and 'non-existent' immaterial? And who, space billionaire, will write the first chapter in the story of exploited emptiness - not a myth of fortunes spun, thinner than the air, from nothing but a real and glorious history culled from frontiers all along the edge of being? There's no doubt that even sceptics should be glancing sideways, seeing and expecting nothing. Watch this space.

11. Time

'If nobody asks me, I know. If someone asks me to explain I find I do not know', admitted St. Augustine when it came to time.

The second physical absolute, as obvious, impotent yet hard to grasp as space, is Augustine's process of duration.

Time

	tam/ raj	Sat
	existence	Essence
	finite	Infinite
	time-stream	Origin
	period	Centre/ Axis
	motion	Poise
	times	No-time/ Timelessness
↓	tam	raj ↑
	finished past	kinetic present
	discontinuity	continuity
	isolation	relationship
	apart	part

Nothing flows through nothing, time through space. Inexorable, implacable, what can resist? You can feel it but you'll never touch - the harder you grip the softer it slips, completely yielding, to an edgeless flow of presence. Time, like space, is certainly a cosmic paradox. It's nothing you can use your organs on yet something, closely, you perceive. Not physical but psychological, it isn't nerve but mind clocks time. Can you touch it? Nor can you affect but only tell it. What, therefore, is time but the space for motion and your turn of mind?

Metaphysic! Absent presence! Where are surface, centre, volume or its texture? It is everywhere and at the same time nowhere all at once; and it's a single edge in nothingness that fills all space. Is it force or no? Like gravity it keeps the world in step; like levity it buoys all motion forwards. How can nothing roll the whole world down an absolutely one-way course? No time, no change, no motion, no existence. Is there anything not carried down its abstract stream? Time's all-powerful but wields no instrument to force its subjects, all events and objects, into line; yet, simultaneously, it is a powerless property of energetic flux. If time were power's measure what of everything created, men say, in one immeasurable instant; what about a universe projected in no time at all? Were that, no time at all, no power either? It's powerless but what is able to affect its omnipresent fluency, fluency forever 'up-front', always on its blocks and effortlessly pushing at the start of what is left? When is not the first, fresh moment of eternity? When is life not starting? Now is new beginning, ever rises dawn, forever open doors.

What will you make fate hold in store? How floods this moment up until, against the roar of time's last waterfall, it's over the abyss? Or will you glide, like water through an estuary, into the source of rivers at their end, the sea? For all that lives an end is nigh, world's end, the moment when this body's times and places disappear. The edge of cosmos, bar of space and time is dissolution. Does such mortal disappearance - unstoppable, implacable and irresistible - really mean the end? What is the nature of death's sea? Does its swell bear life or not? The world will carry on just as before. Will you? What might change for you and (Chapter 18) how?

Whether you know or whether you don't, time streams. Whether seconds flow by you or you glide them, there's a current. Time's flow is motion's second nature. It is transformation's permit, space for change. Space seems motionless. If, as is claimed, its load of stars is flying outwards is their space itself expanding? Or is a space forever in the same place? If the stars are simply whirling round then what supports them? If they're flying what first pushed? Does space move or not? Time moves for sure but never moves as well. Does it not seem, forever in the instant, always standing at the same place - now? When is not now? Past was once now, future will be; and you only know them now. In this sense, since time is always now; the changeless streams with seeming change, with colours when it's colourless, with difference the same. You thought it moved but presence is immobile and immovable. When is not now if that is always all there is?

'Thus,' the wise advise, 'focus on this evanescent yet eternal moment where alone you exercise control. Either side twin ghosts of past and future worry mind; forsake these uncontrollable eternities that, Janus-like, stretch forever either side of point-reality, that is, of ever-present life.'

Omnipresent phantoms haunt real life as well. Two phantoms of the opera, two ghosts more stable than a slab of rock support the dynamism of appearances. Are these two ghosts in fact their own realities at all? Is time's flood an independent agency or, twinned with space, no more than a transparent shadow? Is its film a coupled shadow thrown by motion on the mass of things; is it a simple, secondary function of the energy of transformations making mind and matter up? Time, like space, is understood by everyone but surely difficult to grip. Clocks have hands that only let time slip but who can handle, who can get a handle on the omni-absent, ever-present pair?

	tam/ raj	*Sat*
	relative	*Absolute*
	matter/ mind	*Essence*
	range of states	*Super-state*
	objective/ subjective	*Subjective*
	time	*Super-time*
↓	*tam*	*raj* ↑
	objective	*subjective*
	matter	*mind*
	'flat'	*'elastic'*
	physical	*psychological*
	no sense of time	*sense of time*
	clock time	*mood time*
	sub-state special case/ oblivion	*perception of change*

Perhaps there are, in the main, three kinds of time. Do you remember (Chapter 0) subjectivity, objectivity and (*figs.* 0.7i and 0.10i) a cone of focus that runs from your cosmological axis (X) inwards towards Subjective Centre and outwards through the senses, to the objective, peripheral rings of things? This cone reflects (*fig.* 0.11) the dialectical structure of cosmos.

At the apex rests Pure Subjectivity which is, as the only 'condition' of complete, conscious detachment from the finite events of existence, the only 'form' of Pure, Psychological Objectivity. The *GUE* of this upper pole knows Presence. Its Principle Time is neither existential nor relativistic. Do you remember that a geometrical point signifies axis, centre and infinity? Therefore call this Axial Time. It is Essential, Absolute and Timeless Now - (*Sat*) Super-time; and, in its relation with the existential flux, No Time at All.

Below the *Bindu*, underneath the Source Omnipresence is dispersed throughout the conscio-material spectrum. It drops, like energy along its spectral lines, into different rates of passage and perception. Its presence, though forever now, appears to flow. And so the relativities of time extend to a diffraction that includes past, present and the future. What is behind the central presence is called 'past', what is in front 'the future' - but only mind knows that. Time psychological is also variable in its emotional velocities, subjective ups and downs and moods of interest, anxiety and boredom.

When there's no mind, where's time? What time does the subjective emptiness of objects feel? Oblivion, their sub-state character, knows no time. (*Tam*) sub-time is time's subjective absence, its 'special case' of flat, objective presence; it is a dimension that, coupled with blank space, permits material 'happening'. Call it straight-line time; but couple it with bendy space and it becomes elastic too!

What if time depends on thoughts in mind and things in matter, entities that move? In this case time and space are simply functions of the things that move through them. For mind they are devices to describe and order its relationships; for matter they are real devices of extent. What if, in imagination, you subtracted all existence? What if you deducted motion's maelstrom called creation? So that, stripped away, there was nothing left? Would you not retrieve, above the top end of its spectrum, Essence - of which a facet shows as Super-time? Psychologically, upon the opposite and base periphery of mind, might you find sleep? And physically, at the base periphery of cosmos, would you not find the purest and most abstract of all things and motions - none of them? Would this special case of no-thing have a time or place? Is its absence space-less and without a time? Or are the absolutes of formlessness material abstractions - a weightless, endless duo made of space and time?

Yes, said Newton, space and time are absolute. They are the canvas on which changeful things are drawn. Our cosmos is projected like an island dream-for-real within this endless emptiness.

But could you look at it another way? Could you propose that time and space are secondary effects of energy? That they are hangers-on and properties of motion's relativities so that, lacking an event or object, they would disappear?

Yes, said Einstein, space-time is a relative 'periphery' that surrounds a central axis, light. Transparent, formless light's the absolute. Matter and its motion are the drivers of this cosmic vehicle's relativity. The pair dictates its rate of flow and geometry; such space-time changes in their turn direct material displacements. Without displacement space-time's infrastructure would fall flat. With only latency remaining it would disappear.

Let a batsman strike a cricket ball and then a hundred watchers, spaced differently from the event, hear the crack at different times. If two events (the hitting and the hearing) don't take place together they are not, you'd reckon, simultaneous. However Einstein said that each observer would, according to his frame of reference, hear it 'synchronously'. 'Synchronicity' replaced 'simultaneity'. Thus you could have a hundred 'synchronous hits and hearings' but no simultaneous sense of time. Is this a real or tricky relativity of time? Is it a true or false assumption? Can time itself behave like echoes? A little later we shall take another sounding.

What, you might also think to ask, was space-time like before illumination lit its dark-and-blank? Where will it be when stars no longer burn the sky? Or, if you like, don't take the universal case but think about an individual one. At death the world of physics disappears. What of post-mortem space and time? In mind (including the sub-conscious energy that Natural Dialectic has equated with potential matter) time is no more clockwork than a thought is three-dimensionally solid. What is the nature of a 'thing' called psycho-time?

Go now! Be gone! Whatever else its mystery time is obviously, inexorably on the move; but subtract the fluid metaphor and what reality remains? A count? Is it motion's metronome, a marker and a measurement of change insensible except to mind? Time's abstraction is a real fact but there is nothing to its act - no property or quality, no shape or texture of its own. It grips nothing and yet nothing slips its grip. Time's ocean inescapably but freely floats the cosmos on its uniform and single crest; the world-ship cuts an ever-leaving bow-wave as it ploughs the world's abyss. It has neither power nor energy nor substance yet without it nothing moves. No time would freeze existence - how would action buck the clamp of timelessness? Immobile, it is stuck here in the present and yet, mobile, in no time at all the present moment's over. We ride a cusp, a head-wave that is forever rolling from a past (that's finished, gone) into a future that, like destiny, is always rising into view. Just a sliver yet as wide as all creation is the time-band of a notion - 'now'!

Time's dynamic! Even deities can't order its arrest! Sit still, Canute! Time, although we said it wasn't moving, flows the same. No frame, no break, it flows as seamlessly as consciousness. It is a way, like mind, to bring the differences together, to unite locations and events. One moment, one perception, each the same and yet we cut the whole in slices. Different tenses are just fractions, trackers back or forward along a line extended from the moving dot, a calculus of presence drawn from innumerable moments. Which part is real? Behind - a memory called the past? In front - imagination called the future? Or the present that, in thinking, both the others have their presence in? Every timeline tugs back to the present and this moment holds the greatest paradox of all. It's always, in its presence, just the same yet, every moment, different. Changed in some respect. Seamless and yet broken into episodes, endless yet with endless ends, sometimes easy-going and at others pressurized, on what does three-faced time depend?

	tam/ raj	*Sat*
	peripheral	*Axial/ Central*
	moving presence	*Synchronicity*
	not-Now	*Now*
	lesser nows	*Now*
	appearance	*Actuality*
	becoming	*Being*
	presences	*Omnipresence*
	not-Here	*Here*
	expression	*Potential*
	change	*Precondition*
	separate times	*Pre-time*
	in time/ going	*Pre-active/ Pre-go*
↓	*tam*	*raj* ↑
	behind/ backwards	*in front/ forwards*
	now-looking-backwards	*now-looking-forwards*
	now-from-fate	*now-to-destiny*
	history	*possibility*
	what came	*what will come*
	closed	*open*
	become	*becoming*
	run/ gone	*running/ going*
	memory/ remembrance	*conception*
	period ending	*period beginning*
	exhaustion	*spring*
	fall	*rise*
	finished/ end	*course*

Time and space, of matter or of mind, are the frame in which the pendulum of anything is swung; they are the rails on which the sheet of finiteness is slung. Each is a function of kinetics, an aspect of a transformation, passive nothings that enable active things to come to pass. Are the missing couple metaphysical or are they physically present? They are certainly 'location' for the 'vital set', the stage for action that accords with voluntary, flexible desires of mind or the involuntary, fixed programs of material behaviour. *Space-time (if you really have to tie the knot) is the means of separation, differentiation and isolation without which nothing could appear.* Without the continuity of its togetherness how could apartness happen? **Without such isolating continuity there would be, literally, neither space nor time for anything; without these differentiators nothing could stand alone; without these dimensions what a sudden cramp would clamp, what a paralytic lock-out of the most impotent kind would leave you neither jot nor dot of anything. Nothing. Nothing but for a couple that are nothing-in-themselves!**

Paradoxical a pair! The insubstantial base of continuity! Time is serial; 'now' can be contracted to a single, moving pointless point. Space is synchronous; 'here' is motionless extent. You can expand its volume from an individual dot to universal magnitude. What volume has the heart of time, the present's moving spot? Each immaterial can be expanded or contracted by an indefinity; and, pushing just a little further, micro-infinite their speck and macro-infinite the all-pervading spread.

Just a tick! Although I keep forgetting I remember too! There exists an all-at-once of most dynamic and yet finite kind. Presence. Super-time. Where are its pasts and lesser presence but in mind? Where is any thought but now? Past was present, future will be; nothing happens but it happens now. History and possibility are simply shadows; they're illusive dreams dependent for their sustenance on thinking that's done now. Is there actually any other but this changeless moment? We are locked to rolling time, we are surfed through our existence. Each of lifetime's moments is a gift and therefore this one is, now and always very really, present of all presents, gift of gifts!

Sacred moment! Is this gift a glimpsing of eternity? Existence is division of Essential Omnipresence into many parts; it appears as finite, different-looking 'pieces' of the Infinite. Time and space are fragments that, with zestful threads of energy, split up Infinity; they allow a seamless cloth of finite patterns and behaviours to be woven on their loom. They are connective media in which separate events and objects last a while together, they are ghost-screens through which all must pass. Wheels within wheels, wheels within even the arch-cycle of creation as-a-whole, each transient form has its own hour and place, its ring of time and space, its binder. Are these phantom binders, as a clock ticks or a ruler rules, fixed? Does length of space or time of tick-tock wobble with acceleration, speed or force of gravity? Or is it simply clocks and measurements that change while time and space remain the same? Yes, says Newton's intuition. No, says Einstein's, space-time really is elastic; nothingness is like a rubber sheet. Mine, therefore, is counter-intuition's baby, mine the brainchild lacking common sense.

At the Centre rings of influence start and cease. This Axis is unbound and ring-less, motionless and therefore timeless while the worldly carousel swirls round; this Hub of time is, set above existence and before creation, an Eternal Flame. Yet its Essential Radiance paradoxically pervades each lesser presence and its Balance is the sameness that substantiates each fleeting 'now'. A Timeless Pivot, Synchronicity, embraces all eternity's extent and yet is concentrated to the point of timelessness - immediacy. On the dot. Omnipresence is, therefore, simultaneous. All lesser 'nows' are wrapped in Now, all 'partial presences' are, like sunlit scintillations on the world-sea's swell, just reflections of the Sun. So time is Now-in-motion; all the world is Heart-in-action; it is Omnipresence trailing glory through the drama, through the scenes of cosmic play.

Quieten down your body for a moment. Sit quite still. Can you even stop your mental stream-in-flood called thought? There is, endlessly repeated, nothing but the present and its unrepeated differences. You cannot escape this present; you are always locked in fleeting presence but, implicit in the instant there reside intimations of forever, a connection to the roots of being and expansion to an ancient yet most modern Timelessness. It is as if you surfed a ripple of the Infinite, as if you glimpsed the Ocean. This Ocean is a sea of light. It is (N)One to which you are related absolutely, it is the Infinite of which you are most certainly, in mind and body, just a finite drop. What kind of drop in ocean is your soul?

Species of Time

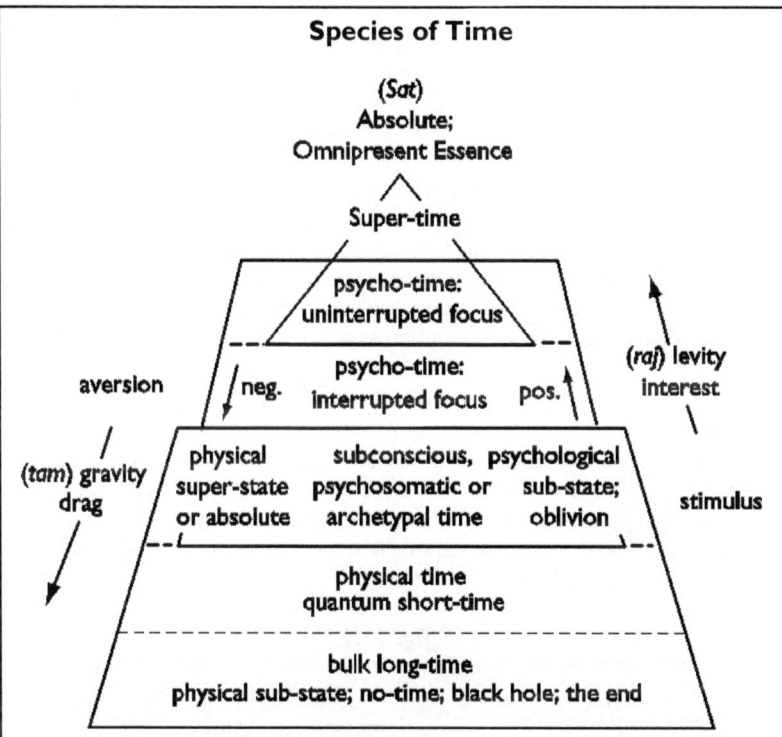

Irrelative time is absolute or real 'time'. Call its timeless nature 'super-time'. Super-time is Essence. Here the present is concentrated into Infinite Presence; it is realised as every mystic's Singularity; and as such is a facet of the Infinite (N)One.

Relative time is the dynamic expression of Timelessness. It is fractionated Timelessness. As sensed by conscious mind its flow is changeful and 'elastic', variable according to the quality of absorption. Absorption? Time seems varied by engagement of attention. Indeed, is not any fraction of The Timeless Presence just a function of perception? Instruments of such perception include clocks and minds. Thus 'positive' short time (where clocks or minds run fast) is a function of stimulus or 'levitatory cause'; and 'negative' long time (where they run slow or 'drawn out') is measured as drag, 'gravitational cause' or 'squeeze'. Pain, for example, squeezes much time into little space. Aren't black holes compressions that should do it too?

Relative times are inter-states between Super and sub-state neutralities of timelessness. The primary expression of Psychological Super-state is known as First Cause or *Logos*. At the other end of the spectrum sub-state timelessness is also neutral with respect to motion. It is, as opposed to super-state potential, the effect of *(tam)* sub-state exhaustion. Psychological sub-state involves unconsciousness, memory and sleep. In psychological respect the whole mindless, material cosmos is in sub-state oblivion. Isn't non-conscious oblivion more timeless than the deepest sleep? There is no perception in black time - only mindless motion or its waiting room.

But there exist conceivable divisions even in oblivion. Matter's super-state involves its own first cause - archetype - whose primary expression the Dialectic knows as matter-in-principle. Of matters-in-principle, that is, subatomic and atomic conditions, the primal motion and therefore the 'highest' luminary were logically best identified as the 'pure energy' of mass-less light. What time does light take? Not a lot; for massless particles time never ticks. Is its velocity through nothing absolute? What time does space take? Does time float to a standstill in a perfect vacuum? There would be no time in nothing. Or does it grind to halting at the heaviest end, matter's own sub-state consisting of a black hole singularity? Is that, finally and most wound-up, the crushing end of time?

Between potential matter and its black hole anti-pole, between extremes of space and light there exist material inter-states. The chronology of an observer takes, depending on his clocks and mind, a relativity of time. Is it time or him that's really relative? Or both? Or, in essence, are both absolute? What in time, the measure of all change, is changeless? The question of its origin spins round again. What is the nature, one might ask, of Super-time?

fig. 11.1

	tam/ raj	Sat
	relative	Absolute
	absence of Presence	Presence
	relative presence	Omnipresence
	lesser truth/ relative illusio	Truth
	information	Preordination
	appearance/ partial realit	Reality
	objective/ subjective	Subjectivity
	durations	Super-time/ Presence
↓	tam	raj ↑
	objective time	subjective time
	physical time	psychological time
	split	flow
	time measured	time experienced
	exhausted	in progress
	flat	cycling/ vibrant
	death	lifetime
	old	young
	fixed	quick/ free
	memory	volition/ desire/ plan
	past	future
	sealed/ finished	coming/ becoming
	effect	cause
	over	open
	no possibility	possibilities

Time is a powerlessly powerful species of nothing. Let's review the three fundamental forms of its sensible absence, its intangible omnipresence - *sat*, *raj* and *tam*; and, at the same time, note this stack applies to psychological and physical domains. In psychological there's Super-time, conscious 'quick' and sub-state fixed time in the form of memories. In sub-state fixed time find the archetypes we've called potential matter; next down, in the physical domain, 'quantum quick' substantiates fixed patterns called bulk matter.

For Newton and Aristotle time absolute was independent of space. It was neither 'elastic' nor passed at different speeds in different circumstance. It was not relative. Their common-sensible version of time flowed at the same rate everywhere and at all times in the universe. In this scheme, however, space and therefore the measure of distance varies according to the position or motion of an observer so that space is relative. Because his equations implied absolute time but no absolute space Newton was unhappy. He believed that both should be beyond the grasp of relativity.

	tam/ raj	*Sat*
	relativity	*Absolution*
	outer/ inner peripheries	*Centre*
	physical/ psychological times	*Super-time*
	existential real-time	*Essential/ Real Time*
	lesser, finite experiences	*Experience*
	distraction	*Absorption*
	times	*No-Time*
	negative/ positive times	*Neutral Time*
↓	*tam*	*raj* ↑
	passive	*active*
	negative	*positive*
	dark time	*light time*
	down-time	*on the up*
	rigid	*flexible*
	monotonous	*variable*
	ruled behaviour	*spontaneity*
	pressure	*ease*
	pain	*pleasure*
	confinement	*freedom*
	object of experience.	*experient*
	inertial	*dynamic*
	static	*involved/ busy*
	exhausted	*kinetic*
	slow (down)	*speed (up)*
	no-time/oblivion	*interesting times*
	bored	*enthusiastic*
	'dull'	*'sharp'*
	wasted time	*well-spent time*
	sleep	*wakefulness*
	no time	*involved time*

Super-time

Remember, dialectically, the Head is inverse of the foot; Top is obverse of the toe. If, therefore, in oblivion's base zone of physicality time 'endures' then such eternal, serial flow will not exist Above. (*Sat*) Absolute, Axial or Top Time is, though eternal, not eternal in a 'prolonged' sense at all. Super- time is therefore, in materialistic terms, paradoxical nonsense. It is a special species, a special case of time. This case is its positive absence. It is the 'dot' or 'egg' or poised potential which gives rise to existential time. At the apex of existence you might call this 'time' Essential. As a facet of Essence it simply is: it is what it is. Since Essence is Pure Consciousness such Presence, beyond subjective mind and objective matter, is Subjective. 'I am that I am.' Its experience is labelled, among other labels, *Nirvana* or Enlightenment. You might also call this No-time Central Time. Such Omnipresence is pre-existential or pre-active. It precedes its fractions, encompasses its lesser times and thus includes all different pasts, presents and futures. It substantiates each slice of time and, being always present, thus post-dates the world's end. How can unimaginable Timelessness shed every tense? Aren't mystics mad to indicate Transcendent, Supra-mental Peace above the existential stream of time?

For example, St. Augustine simply thought that time was part of God's creation so did not exist before. Before the world, he mused, there was no time. Just timelessness. Essential Timelessness is prior. It is independent of existence - out of mind and mindlessness. Yet if dependent existence constitutes a field of Essence-in-becoming, Infinity-in-motion or the working-out of possibilities then the Highest enters in; No-Time, in moving, comes to manifest each lesser, lower time-in-relativity. Such Axial Super-time, although a facet of Alpha and Omega, knows neither start nor end. Neither drag nor boundary restricts the Single Moment of Infinity, the Eternal Instant, Now. Real Time. Paradox of paradox, this Instant is eternal yet lacks any span! Without vector yet it marks the pivot of all worlds; it is a balance-point round which every motion every moment makes up part of oscillation. With such Moment those unscientific mystics seek a tryst. Their method is entire introspection but pupils, generating concentrated focus (for example, on a candle flame or flower), can find a physical reflection of Now's Presence. Meditative technique blocks future, past and any drifting motion of the mind; practitioner's call its fulfilment *nirvana* or *samadhi* or communion.

Time Absolute is Infinite. *Time relative* divides the Infinite; its real-times are appearances of Time. Such real-times reflect the fundamental nature of existence - motion. Thus their changes 'endlessly become'. They occur in two main, finite modes - (*raj*) psychological and (*tam*) physical. Call them episodes of mind and body. The former is subjective, metaphysical and flexible; it is 'creative, active time'. The latter is objective, physical and 'involuntary, passive time'. Matter's changes don't need mind but does the reflex process of events occur in an inflexible, Newtonian sort of flow or can it, as Einstein suggested, be distorted into various velocities? Standard hours or flexi-time?

Embodiment endows us with a correlation of the two main existential modes of time - psychological and physical; but between them ranges number three. You might call this third mode psycho-physical. Archetypal instinct, body-functions and their periods of homeostatic chemistry compound to dominate a life-form's instinct and its physiology; this inter-mode includes, therefore, archetypal time and bio-time. It is thus to investigate three existential species that we turn our hand.

Psycho-time

From time's essential paradox we drop to time in mind - informative, subjective sense of time. Let's call it psycho-time. Such passing is a function of attention. Its elasticity depends upon degree of interest; its path is 'warped' through varying engagements with intensity. It involves volition, interest and purpose. The more coherent focus, the more positive attention, the faster that time 'flies'; the more negative is an experience the more it 'drags'. In other words, psycho-time involves an oscillatory spectrum of more or less time-aware episodes. Less aware, time (↑) levant; more time-aware, (↓) depressant. While perception of time oscillates, plateaus of uniform appreciation can occur as well. Doesn't time stretch further when tired body, bored mind or pain's extended moment drags for hours on end; and when you're full of interested energy hours pass as if they were an instant? In short, we call time's keen, uninterrupted flow contemplative. Such time, a function of love and interest, flies. Scattered focus, with its short-term interests and aversions, yields interrupted, broken psycho-time. Indeed, grounded by negativity or transfixed by pain it can seem to drag for an 'eternity'. Time metaphysical is flexible.

Let's take mind's (*raj*) upside first, the one you might dub 'levitatory'. What's elevation like, how's 'positively uplifting'?

Mind's purposes continually involve filtration of sense data, problem-solving and the construction of more or less complex orders of information; its schemes are 'negentropic'. In pursuit of purpose focus of attention sharpens. Sharp attention 'speeds up' to 'accelerated heat' of creativity. Such thermal uplift can be called 'ascent'. It can accelerate to break-through moments of absorption and ecstatic interest. Mind enjoys its peaks intensely, that is, timelessly. Keen-time flies, it passes in no time, as if immediately. You know this truth because you weren't unconscious when time disappeared. You have tasted the experience. Having sat in the cinema for what felt like a moment, your watch may tell you three earth-hours elapsed. Life is subjective and subjective reality is vested in the 'eternal moments' of our happiness. **In fact, more lovely is more real**. The most real times are the sparkling, happy ones. At a time of truth each instant is enthused with happiness. Enthusiasm literally means that 'god is manifest in you'. *Time disappears with interest, enthusiasm shortens time and love can annihilate it.*

To approach is not to reach; but to have reached the Centre is to have broken free of existential gravity; in the attainment of transcendent Super-time all chains of lesser times have been dissolved. Linkage with creation's cut and its Creator found. The message of such brilliance at the core of cosmos is a very positive but also immaterial one. Do you not value an experience in proportion as you flow with energy; and in communion with other lives and perhaps another special life? We crave such 'height'. Increasing concentration squeezes all else out; as consciousness approaches Concentration time, space, mind and all that's in them disappear. Yet enlightened times are paradoxical. They know only unrestricted Here and Now. Such Concentrate is, for a mystic, his Solution. Its radiance, at the cosmic core, is his pursuit; and Super-time, high at the very apex of the mind, his chosen goal.

Of course, up has its downside. Negative time is 'gravitational'. It weighs heavy, drags and slows you up. Aversion and pain brake. Sharp pain may seem to 'last forever'.

The further consciousness leaves Central Essence out the more it spirals into 'downward pulls' of gross desires, the more under the spell of magnetic attachments to people, places and memories and the oscillatory tugs of emotion. In short, the more identified with body the more 'materialised' a mind becomes. To ground-based human vision such identity seems normal. Natural. Ultimately true. But really, in various degree, it's left the Truth behind.

Ups have downs. Restrictions, when you can't do what you want, obstruct your mental vehicle. They exhaust. Frustrated frictions slow and tire. When you have to do what you don't want to does it 'get you down'? Fatigue or disagreeable 'down-times', which are 'massive' and inertial, snag your being. Confinement and rigidity depress. Distraction, boredom and the crushing negativity of pain drag time down. They stretch it into grey holes, black holes. Coloured short time lengthens into long-time. One over heaven, is it hell? Is hell, for wranglers, an inverted paradise? Are not negative extremities compressions in the field of time, dark fractions that, like heavy objects, hardly roll along the bottom of the present's stream? At these subtendencies time grinds to a halt. Fear, shock and horror freeze; but with unconsciousness like sleep time does not petrify but simply disappears. Under even sleeping, where there is for-never waking, lies the oblivious, uncoloured grade of matter. When awareness goes so, psychologically, does time. It has gone where even darkness is not dark. What is time to coma or a stone? All matter, from an atom to a galaxy, is obliviously mindless. What never lived lives ever dead. The corpus of our starry cosmos is a corpse of time.

In fact, one begins to wonder. If Omnipresent Super-state is timeless, so is sub-state non-perception (called oblivion) of material entities. Time seems dependent on what moves between this neutral pair. It needs a mind. It's an aspect of perception, an order of events, a dependency of change. Could it be a tool of self-defining, only here because we think it is? We need it to compare one moment with another, that is, to compare the images of moment, to weave the present with imaginations and resuscitations from the bank of memory or simply to project desire. *Imaginary time is, therefore, a compass*. It is a form of information neither more nor less divorced from consciousness than properties of colour, shape or size from things. It is an elemental part of mind's fine tracking system.

If there isn't mind to know them, do space, time and objects cease to be? This solipsistic version of events presumes that 'all is in the mind' - which is true as far as you're concerned. It presumes that, if you cease to notice it, any other ceases to exist - nonsense from its enduring point of view. So absurd a posture, called misleadingly 'subjectivism', no longer sees the light of day!

But leave mind out of the equation for a moment. If, in this case, the 'impotent couple' (space and time) are no more than aspects of a 'potent single', energy, has Natural Dialectic rightly disentangled what has seemed a triplex strand? If space-time (or even split-up space and time) is dependent on the presence of an object, interaction or an energetic pattern of behaviour, does it cease if all these disaappear? In the event of absence, that is, emptiness of space does time also fade away? Is the vacuum, having nothing to define, illusory; or is space-time not an independent field on which, like screen, the film of nature's program is expressed? You may hypothesise but you will never know because, being something, you disturb the vacancy. In this case an observer would have introduced a basic principle of never knowing or at least residual uncertainty.

Psychological time, inherently free of physical constraint, is subjective, flexible and intrinsically independent of the material universe. Constrained in a body, however, mind's 'non-locality' is earthed. It is 'de-universalised'. It ticks with body's clock, locks into material events and, as a stream of relativity, marks speed of change from place to place and shape to shape. As such subjective time aligns with the objective order. It's circumstance that fractions mind. Such sensible, serial perception of space, time and things is common sense. Time's passage, like the surface of a mirror, seems to stay the same. An inflexible reflector, an inexorable nothingness-in-motion - could such steady motion be the measure of all changes that unfurl?

This mirror's image isn't one of total symmetry. At least let's call a sure-fire, one-way shot - causality; and kill time's anti-ghost - reversal. Does a cause precede effect? If so, then serial action must occur and, since effect cannot precede its cause, time flows from cause into effect, potential to expression, initiation to completion. You can sprinkle mirror oppositions, particles with anti-particles and other sparkles in time's path but you will not divert, you will not deflect it. If you include causality then serial time has 'ppf'. From First Cause onwards time cannot reverse - even if you want to die before you're born. Although equations that describe the universe allow their logical reversal the real world does not subscribe. You can't in fact flip cause or run a film of life or cosmos backwards. You can freeze time (in a memory or solid object) but you can't turn back the clock. Tick-tock but no tock-tick. The escalator drops, cascades down the river fall and the arrow flies one-way alone - downstream.

Might you, however, turn time over in imagination? Might you reverse gear into memory or, at least, haul past into the present? You retrieve a memory. Memory is natural information storage. What if the plans for a great scheme were finalised and filed? What if then, preceding and succeeding the duration of this scheme's accomplishments, such files survived on each side of the project's length of time? If (*figs.* 2.8, 2.11 and 11.1-3) such plans were archetypes and the project was material cosmos, then psychosomatic 'timelessness' would constitute a library. Plans are principles and these would constitute the causal files of how creation works. Such a cache in universal mind would represent the order of time physical. What does this mean? It means the composition of all large permutations would be composed of simple sub-atomic basics expressed from this 'dream-time'. Events and their engaging bodies would be seen as the effects of immemorial time; they would be direct projections or, if you like, replay of records cut before the world began.

Archetypal Time.

	tam/ raj	Sat
	temporary	No Time
	different times	Eternity
	a unit/ individual's time	Archetypal Time
	protraction	Immediacy
↓	tam	raj ↑
	gravity	levity
	gross	subtle
	inertial	dynamic
	drawn out	quick
	persistent eternity	dynamic time
	lng-time	short-time
	slow-time	fast-time
	macro-time	micro-time
	contraction	dilation
	massive	weightless
	classical	quantum
	'flat'	vibrant
	arrow	wheel

Check *figs.* 7.4 and 11.1. While mind's Archetype is called First Cause, *The Logos, Kalma, Shabda* and so on, matter's archetypes at base of mind are 'lower down'. Are memories not 'solid' or enduring fixtures that compose base phase of mind? Are archetypal memories not fixtures inside universal mind? Archetypal time is metaphysical; it is time-physical's potential; and is the world clock's ideal or 'Platonic' super-state. From the perspective of material existence such archetypes are timeless; their time is, with respect to material phenomena, none-at-all. Yet, for physic, archetypal time enduring prior and subsequent to physical expression outlasts such an ever-changeful world; transcendent, it keeps now, then, hence and forever. Is natural law not omnipresent and omnipotent and thereby infinite in influence? Such 'physical infinity' is, though, being unconscious, nescient. Set beside The Infinite, therefore, it is a 'lesser' or 'apparent' strain. Existence, set by Essence, is a hierarchical phenomenon; creation is built up of scales of limitation, change and temporary being. Constraints bind even cosmic blueprints.

The archetypal phase of mind is identified (*figs.* 2.12, 2.13 and Chapters 15 and 16) as sub-conscious, dormant and psychosomatic. It is sandwiched between awareness and the subtle subatomic, atomic and molecular components of a body. It is also called potential or transcendent matter. Archetypal memory is the repository of natural law. Its 'Creation Record', whose harmonies are sounded by behaviours that inhabit immaterial fields of force in space, replays the patterns chemistry and physics recognise as law; and the laws and sub-laws that biology reveals encoded in the government of composites called life-forms.

A 'type' of life is, in this but not in scientism's view, a complex law. Its morphogene and instinct are each patterns that incorporate specific form and function. The bio-time of bodies integrates all three. Instinct is reflex mind-in-action, morphogene's the architectural aspect of an archetype and together their expression constitutes the involuntary behaviour of an organism. Atomic bio-time is notched by biochemists. They delve its molecular machinery. The periodic motions of a body clock are homeostatic; such innate regulation runs in cycles and its whirr collects, takes stock and then dispenses each and every special order to equilibrate. Wheels in wheels; cogs in bio-teeth. Physiology and biochemistry each involve a special tick of clock, ticks called cycles integrated into one grand body tock.

Bio-time

How psycho-time relates to bio-clocks is not yet understood. How does brain (if it is that) assign events a time and sequence them? How does memory organise an order of events so that you pick them up correctly? Check *fig.* 13.2 (iv). No doubt the hippocampus (damage to which prevents the creation of personal memories), the temporal lobe that flanks it and the cerebral cortex (damage to which inhibits recollection) constitute an operating system which control access to and from a personalised database but how do atoms, even those that make up nerves, subjectively experience time? How can they date-stamp memories?

Jump from memory into the present. Flex your right hand's little finger. It has even been suggested, from a *bottom-up* perspective, that brain waves indicating the immanence of such flex preceded your decision. Your brain knew what to do before you knew you had decided! In other words, brain, wishes, chooses, makes decisions; you (at least, who you think you are) follow your pre-emptive nerves! According to this kind of order, logical or not, a twitch of nerves preceded even such thoughts as you're having now. Could the results of experimental brain-scans claiming to show such strange reversal of our common sense have been interpreted correctly?

Bottom-up it's perfect sense because mind is an effect of brain so brain must cause its thoughts. In fact, if consciousness trails nervous cause then you're a total robot! Your volition must occur as a result of neurological excitement. What, therefore, of free will? What's free about molecular determination? Do atoms that make up your brain anticipate your every wish? Has the bio-time gap in between a thought and its expression slipped the other way? Has the order of creation really been reversed?

Top-down, brain is a mediator. No doubt, sensory input will register on a scan minutely *before* its translation into mind. By the reverse token a purposing of motor output will register just *after* its controlling thought arose in mind. Could it be true, however, that output (say, decision) might post-date the nervous excitation that that will execute it? As a read-out follows computation can your consciousness post-date the biochemical procedures of your brain? If so then how much cleverer would be the cerebrum's connections than your cleverest or most complex train of thoughts! How much more sensitive such nerves as presage every feeling than you are! That is, unless you *are* your instrument, the brain.

It should be remembered that we simply do not know the way deliberation impacts nerves. Nor do scanned motions that arise before you make decisions have to be a nervous correlation of that process. Nerves do not execute a thought;

they are motors that receive its 'highness' with contextual retinue into the body-world. Nerves that preformulate what you then come to know, understand and act upon may vie for scientism's prize conclusion; how, though, do nerves (a mass of chemicals alone) preformulate proposals? If, moreover, mind is not your brain then such a bid is, from a *top-down* vantage, highly premature.

When it comes to body there are, measured on two kinds of living clocks, two kinds of bio-time. Science studies clocks. It studies cycling instruments but instruments are simply tools of purpose. Why should atoms want to measure time? Why should they chronologically organise themselves? What are these clocks, whose fascinating mechanisms biochemists probe, here for? Has it clicked? Atoms and their aggregates, called bodies, do not need to know the time; purpose and not molecules developed bio-watches.

The *first* kind of 'bio-chron' is a stopwatch measuring externalities; it measures the events around and of our bodies. Such watches measure tiny intervals of time. They clock the gaps between appearances and let us judge the speed of change. A psychoactive stopwatch helps us gauge appropriate response to myriad and simultaneous perceptions we call circumstance. *MRI* scans have demonstrated that a part of the basal ganglia in the brain, called the striatum and equipped with specially adapted cells, is involved in modulating sensory input with respect to timing a response. Its formulations are pulsed through thalamus to cortex and communicate with 'cognitive functions' - which surely must be more than simply special but non-conscious patterns made of atoms and of shifting charge. The more a neurotransmitter called dopamine is involved the more a clock is measured 'slow'; the more adrenaline or amphetamine are used the 'faster' runs response's clock. Indeed, intense absorption reached in meditation can bypass the system so that, in complete ascent, no time at all is sensed. What state of timelessness is that? It's not a state that's physical when universal time keeps passing just the same. Time does not change, just the way it's measured.

Just as a mind is conditioned by its cerebral co-operant so experience of time is adapted according to the kind of brain it 'inhabits'. The perception of bio-time is certainly affected by one's size, pace of life and rate of reflex. An elephant, for example, lives twenty-five times more slowly than a shrew; what seems normal to the elephant is for the shrew in slow, slow motion. When you crack down as fast as possible fast-buzzing flies evade your swat, which seems to slowly fall, easily with time to spare. One's 'long-time' is the other's 'short-time' and *vice versa*. Birds and flies live in the fast lane. To them we lumber at lethargic rate. Their time is 'dilated'; in what passes for an instant they can register, distinctly, manifold events. To you the flickering of a fluorescent bulb at 20 Hz. looks like a stable and continuous light but a fly detects flickering at ten times that rate. Speed not definition is its need; its 'slow motion' lacks your detail. It is as if it saw your cinematic movie like a slide show minus resolution. Speed *and* definition are the need of birds. Song or high-speed, sweeping navigation onto the leaf-hidden branch of a tree are further clear examples of perception speeded up. Indeed, even human 'feel for time' changes through a lifespan. Old human hearts beat slower and, with them, life's tempo. A week or month, which seems a year to their grandchild, flies by. Does that not show elastic bio-time?

The *second* kind of living clock is cyclical and as such vibrant and dynamic. Hard-wired to internal systems, it revolves by means of suites of molecular cogs that integrate feedback cycles, that is, periods of homeostasis. Homeostasis is a process ticking at the centre of biology (Chapter 19). Its internal triggers link, of course, with the external world. Bio-cybernetics is connected with the cycles of earth, moon and sun. Computation flows according to diurnal, menstrual and annual rhythms. These are periods, oscillations, signs for life of ups and downs and points of poise. For example, there is (*raj*) activity of day, (*tam*) inertial night and (*sat*) cusps of peace at dawn and dusk. As with day so with year. (*Raj*) summer yields to (*tam*) winter with, between, the equinoxes - spring's life potential and the impotence of fall's exhaustion sinking to midwinter's 'death'. Does not, each cycle, a female's egg develop to its summer solstice, ovulation? Whence momentum drops into a depth of 'wintry' menstruation just before the new round springs? What are equinoctial tropics but an emphasis on life's potential, cycling fast and always being on the reproductive, growing 'go'? What, on the other hand, are deserts but impotent drains?

Purposes anticipate, timing mechanisms expedite and any sense of timing needs a mind. An extensive literature documents but not necessarily explains the operation of different kinds of biological clock. Many work like fully automated cybernetic systems; either external or internal cues trigger the oscillatory production of relevant biochemicals and thence physiological effects. In other cases mental triggers, instinctual and conscious, awaken the appropriate behaviours. It seems the 'top cogs' in circadian rhythms (waking, sleeping, thermoregulation, blood pressure, various secretions, urination, defecation etc.) are a couple of nuclei - called *SCN*s - in the hypothalamus. Dedicated cells in the retina transmit light signals to them. These informants then exercise, through the office of a sleep hormone called melatonin, body-wide control. They do it through agents of expression placed in every cell, proteins synthesised in varying amounts at different times. In fact these molecular oscillations can act independently of brain to buffer changes that occur from stress, exercise, change of weather and so on.

A prime suspect in the mystery of longer rhythms - such as migration, moulting, hibernation, mating and so on - is the pineal gland (see also Chapter 17). But is the real culprit simply chemical or partly chemical in method - chemical, that is, upon the body's side of the affair? From the mind's side doesn't archetypal instinct regulate these long-term reiterations? As an example culled from hundreds how do soldier crabs negotiate the complex pattern of tides in order to survive? Or periodic cicadas synchronize their break-out from a 17-year-old pupation and all of them mature at once?

Of course, plants run on rhythms too. Seeds, leaves, flowers and fruits clock in on time. What triggers agriculture's seasons in a plant? Check, for example, germination and the oscillations of a chemical chronometer called phytochrome. The light dependent action is as physical as clocks but what kind of chemical could institute it? Clocks are made of chemicals but why should chemicals make them?

Life wheels towards death. Just as its development is programmed to reach an adult peak so, perhaps, is physical decline. Telomeres, material that 'knots' the chromosomal strands together, degrade with time. Such degradation is associated with a failure of integrity; such genetic 'fraying' may systematically cause unravelling we perceive as aging, our disintegration and decrepitude. Is aging clocked and, according to its body-type, life-span preordained? Clocks time and timing needs a sense; timed operations must anticipate their moment. This means target. It means scheduled process and (since wind and rain don't think like that) a mind behind. Every schedule-to-an-end, every homeostasis or composition towards a final frame requires, to establish mechanism and machinery, prior intelligence. What exception, except reductionism's strong fixation that material process can produce a 'purpose', can there be?

Down at the cemetery, death imminent, a branch of yew adopts the nature of a root. It dives earthward, buries down and from pre-mortal grave propels the resurrection of its tree. Does such a cycle confer immortality on yews? Likewise from ancient origin, from an ancestral core, rings of vegetative creosote re-cycle. Concentric waves ripple from a centre estimated, in some cases, to be over 10,000 years old. Is this a biological eternity? We'll see because there's more to come - in Chapters 18 and 22 - on frozen and accelerated time, aging, an escape from death and beating bio-time by various brands of 'everlasting' immortality.

Material Times

Life involves attractants, purposes that pull towards the future; the purposeless oblivion of objects is only shunted from the past. In this sense physic has no future since it doesn't care. Nothing matters. There's no target. Matter's going nowhere since it has no sense of time at all. It is not time-aware; it is inanimate. Don't protons and electrons each endure a senseless kind of immortality? Time's absence, as in oblivion and sleep, is Natural Dialectic's case of sub-time. After all, can matter measure anything? Are the atoms of the chemicals that make your tissues living; or molecules in complex compositions any more so? Are your bones alive? Life forms have bodies of 'non-living stuff'. Body-time, in this respect, is physical alone. It's meaningless unless appreciated by a mind. But how precisely can a mind appreciate the body-time of actions (things and forces) happening in space? You can't since sleep is not appreciative.

A chronicle of clocks, a history of the measurement of time, comes down to calendars and counters. Humans take their own backyard, the solar system, to set the spaces that they live through. Year (sun), season (earth's wobbling tilt), month (moon's cycle) and day (earth's spin about its axis) are the points of reference used. What were stone circles like Stonehenge or tombs like Irish Newgrange all about? Indian astrology used complex maps including stars. Babylonians and Egyptians split time's flow the way, with refinements, that we know today. They used twelve temporal hours (that vary with the seasons) for a day. The Greeks, Romans, Jews and Arabs followed this device. Drop by drop or inch by inch water clocks and sundials registered the passing. Now we use the Gregorian scale and rigid hours whose divisions, harking back to an early geometry of circles/ cycles, split to 60 minutes then to 60 seconds.

There first came, in English church towers, bells geared to clocks mechanical. Weights tolled and pendulums swung time through village life. They still do but some began to crave a scientific accuracy and others wanted time wrapped in their pockets. The operation of clocks' cogs and wheels were finely tuned with springs, ratchets, balances and engineered improvements. Watchmakers sought, in the ancestor of your wristwatch, the pocket-watch, even to excel the great precision of celestial bodies as they sweep their rounds.

Not only could you put time in your pocket but the quantity went global. The planet's space and time are carved by two imaginary kinds of line called latitude and longitude. In 1884 in Washington a presidential conference decided to adopt the Greenwich meridian as the centre of all measurement of space east or west. A universal day (of 24 hours) was also adopted. Loss of such a day in circumnavigation of the globe had been bothering sailors since Magellan's trip and by the 1830's tentative lines had been proposed. They wiggled variously from north to south across the Pacific Ocean. No law can stop you moving an imaginary line and so the international date line, ruled in 1921 by the British Navy approximately along a line of longitude 180 degrees to Greenwich, is open to adjustment. For example, Kiribati stepped into a single day in 1995. While either line could have been anywhere at least a global standard now anticipated jets and mankind's 'global village'.

The date line separates two days. No doubt one date, today's, can last somewhere on earth for 48 hours but all days are 24. If you go westward from the line you add a day. You skip from this date to the next. Go eastward and you lose. You subtract a date and then have to repeat its 'day'. Indeed, you could play cricket on the line. You could hit a ball into tomorrow then throw it back into today. How's that for measurement of time? 'Howzat' for relativity of time? But does it really matter that he caught you out tomorrow? Does it matter if you use a metal pendulum or cogs, quartz crystals or atomic oscillations to get more and more precise about a difference? Dates change but do the days? Does time itself change by a whisker, flick or tick? Is real time a part of or apart from space? Did Newton or did Einstein actually get it right?

Measurement is one thing. A history of time itself is quite another. Is it eternal, as Newton absolutely wished, or not? If it began then what 'initial condition', 'boundary state' or start-point did it have? Did it even, strangely, have one though it didn't? Are you sat comfortably? Then I'll begin what might be nonsense - space and time were just the same and cosmos 'quantum tunnelled' out of nowhere nowhen in particular! No boundary to start with! The truth is we don't have the information covering what *physicalia* erupted from. Maybe we never will. Professorial fantasia might have to do.

Time physical's inflexible; but still it is a-changing. For Newton it moves just the same for all observers; for Einstein it's dependent on a watcher's motion or an object's pace. And for Newton space is always similar and immovable (an inertial frame of reference); for Einstein (as for Leibniz) it involves mass, speed and relative location of bodies. The latter's General Theory of Relativity (1915), which has conditioned the thinking of all subsequent students of physics, combines the pair, space and time, into a four-dimensional 'object' (or 'absence of object') called by Rudolf Minkowski space-time. Space-time would, if there were nothing in it, be 'flat' (that is, undisturbed) and boundless. It would be, dialectically, an apparent infinity or, in terms of physics, an oxymoronic 'no-boundary condition'. Such a contradiction in terms denotes a finite state without edge or individuality. On this empty canvas Einstein painted motion and gravitation. The latter, a function of mass, he viewed as a property of space rather than a force. It is simply a geometric effect whose curved lines can be seen as a distortion or a 'warp' in space-time. Therefore, even with things in it, the 'no boundary' condition of curvaceous space-time still applies. How, though, can you score curves in nothing - grooves sufficient to roll tons of moons, planets, stars and even galaxies around? At any rate, for Einstein the presence of a massive object warps surrounding space-in-time but lonesome, un-warped context lets you, if you're massless, travel in straight lines.

Hold on! The trip's about to take a wobbly, hall of mirrors turn! You might have thought one minute was as flat, smooth, automatic and as long as any other minute anywhere. If not what varies? Is it the minute or its counter (mind or clock)? For Einstein not only relativistic space but time is wobbly too. Space-time is elastic, flexible and capable of distortion by both mass and its acceleration. Not only can it be distorted but collapsed.

You watched a batsman hit the ball then heard it later. Can you see an alien waving to you from a planet ten light years away? What you see happened ten years ago and, while the signal was transmitted, she'll have been getting on with ten years' worth of business. She'll be busy doing other things as you observe her then - but also now! If you waved back and she were looking out for you she'd see you wave ten years ago; she'd see you twenty years from when she waved at first. You'd have both moved on a score within the greeting. She was twenty when you smiled, thirty when she saw it and forty when you saw her, aged thirty then, smile back. In time's echo each would seem to the other to be engaged with the present but as receiver you'd be in the other's past; or, as projector, in their future. Meanwhile all the intervening time you'd both be, mutually unseen, living in the same and simultaneous present. Now. You intertwined your gaze, remember, through ten years of travelling light. There could be millions of such women dotted round the universe. And you might, starry-eyed, observe each alien girlfriend at this moment at, according to her distance, different times ago.

Rephrase the situation. You see girlfriend A wave when you're 30, she seems 20 but is actually 30 too and you are having dinner with terrestrial girlfriend B. These two events, wave-received and dinner, appear *simultaneous* but clearly actually aren't. Einstein called them *synchronous*. You could orchestrate a hundred girls a-waving from their various galaxies in times that let you see all hundred at once synchronously. In other words, synchronous events occur at different times but appear to an observer at the same time. Synchronicity replaces absolution; it means relativity of simultaneity. Is a simultaneous 'now' for all observers thus ruled out? If time's defined by an observer's motion or position can a present moment be diffracted into many different times? Are there different times of life for all in these affairs? In other words, does everyone in these hands-off relationships love out of touch by sensing interaction at a different time or is there, behind apparent difference, a simultaneous cosmic 'now'?

The physicist applied a further twist to strangle common sense by mooting that the apparent distinction between past, present and future is a persistent, subjective but objectively meaningless illusion. What kind of torque is that? What's the thrust of what that means? Is relativity of time no more than sophistry? Because if you accept causality you must accept a one-way arrow - a serial 'ppf' - as well. You can't return down Time Street. What kind of 'block' of time, Time's Square, can lose its flow's direction? What asymmetry could lose the vector running from a cause to its effect? Einstein surely didn't mean causality, the warp and woof of all creation, is illusory; but perhaps, in a Platonic mode of musing, he was grasping matter's Real, Archetypal Time. Potential matter timelessly transcends its issue; as *fig.* 11.1 illustrates, changeless, archetypal metaphysic dialectically precedes material phenomena. Could such condition be the physicist's Time Square? Natural Dialectic here reiterates suggestion that material absolution *is* the form of timeless, immaterial archetype.

Snap your fingers twice in quick succession. If light could circle the equator it would have done so more seven times within the snap-gap. And particles travelling at 99.9% the speed of light experience 'time-deceleration' such that their moments pass thousands of times more slowly than our own. From its own full-speed, light-time point of view illumination unimpeded by a vacuum dilates to cover all space in no time; it has, in its 'flat-out', 'all-go' extremity, stopped. From our slow point of view it's speeding fast but less than 'absolutely, infinitely' fast. At the specific rate of 'c' (about 300,000 kms per second) it is supposed, bizarrely, to appear the same to all observers even if they are moving in opposite directions relative to a beam! This non-relativity, against which absolute standard every other motion is measured, also constitutes a cosmic speed limit, an invisible 'motional barrier' past which, its argued, no object can flash. Why? Because on the approaches to this limit mass increases (except for mass-less photons) towards infinity. Things can't strain any faster thus cannot reach 'c'. At the same time they 'shorten' in the direction of travel; but at the other end of physicality the reverse occurs. Here you find, theoretically and by inference from observation, the warp of matter elongated into time-rigid black hole singularities. These are final sinks of space-time and its energetic quantities.

You might even, if experiencing a different gravitational field or moving relative to me, disagree about *duration* of the same event. Our clocks don't seem to agree. For example, for 'outside' observers, Einstein says a dashing clock ticks slow. But is time really stretched so that at high speed what was a slow-speed second now takes only half of that? Dash, for a moment in imagination, almost at the speed of 'c' towards another galaxy. You can't strap a stop-watch on a photon's back but, if a clock were in your pocket, it might show you aging at perhaps half the 'normal' rate or, if you like, you'd have ticked only twenty years against our torpid forty back on earth. This way, since you'd live twice as long, life in the fast lane pays. If you flew fast and far enough you'd colonise some future; different speeds and different futures! What, though, if you returned? Would you, on walking through the door, return to your future or we laggards greet our past? Or, on reconvening, would you jump to the past and we leap forward? What would each side see? Indeed, how could we both meet now if our two 'nows' are set apart because one party used a rapid rocket? Would we be locked in different time frames and therefore in temporal isolation where our time-lines never met?

Hands up those who fully understand this kind of logic! Is not such a fragmented view of time present relatively correct but also, at the same time, nonsense absolute? Perhaps, on the other and less boggling hand, there's been an oversight in rationale. The actual time-line never deviates from standard, universal presence. If, as theory demands, everything is relative, then not only does one twin travel at nearly the speed of light from earth but the other twin, stood on earth, travels at the same speed, relatively, away from him. Just as you moved rapidly away from us so, from your point of view, did we from you - even if we didn't move an inch. The sensible solution is, therefore, that both grow older at the same rate so that, *au revoir*, each is simultaneously the same age! In other words, because we actually moved at the same speed relative to each other might we conclude the whole scenario of different ages is in fact a fiction and with un-warped age we both inhabit cosmic 'now'? In other words, have time warps warped the mind of physics so that only amateurs can disagree with its rare sense and speed off towards a common one? Do you buy Einstein's wobbly trip? Or is Newtonian 'time absolute', although it's not the same as dialectical 'time absolute', the straight and normal one that we think of when our common sense prevails?

If, however, time *is* able dilate then could dilation, when acceleration was complete, reflect the immemorial origin of physical phenomena, that is, the nature of an archetype? Archetype is lodged in universal mind. Could this psychological condition, lacking physic, seem to realise eternity? Speed absolute, top time, instantaneity. When a clock's tick slows until, at speed absolute, it seems to stop you might expect that, without any punctuation in between, all time was rolled up in a point infinity called super-time or (without past and futureless) just 'now'. Eternal presence. No time for massless particles. Is it not, riding on the back of light, an omnipresence that you might experience? Especially if, at that same moment, space contracted (called 'foreshortening') to the point it disappeared. This is it! Here and now forever in an Instant!

You might extol this Instant. It is, you might agree, 'potential time' and, of course, if you slowed down you'd drop back from it (and all its possibilities) into definite localities. What change has no after or before? Therefore you'd step down to actual space-time fractions each one with its ppf. You'd alight into a serial experience composed of just a single row of 'lesser instants' - circumstances-with-events whose changes you continually keep calling 'here and now'.

Now decelerate. Completely. Stop. Is your immobility immobile or, in other's frames of reference as compared to them, do you still seem mobile? Is absolute immobility a possibility and, if it were, would time freeze too? What is the way time negatively stops or, conversely, goes for-never in An Instant? Do not confuse 'short-time' with 'long-time'; nor believe The Instant and an endless row of instances are just the same! One is Super and the other, when there's none, a black hole sub-state. They are different poles of temporal extremity - oblivion and instantaneous eternity.

Time-bending is mind-bending! Grab another 'funny' mirror; let's distort what is invisible. From high-light to the low-life, instead of flying let's spin through the floor into the black hole that was mentioned just above. Grossly mass with space, coil it tightly with an escalating warp and wring it to a singularity. What happens then? No light escapes so you can't see what's going on inside. Is space locked out while time is frozen? Or does business, as if on some alien planet, carry on as normal? If clocks ran fast then, in between the fibrillating ticks of an extreme, it would seem that nothing happened. Have you played a record as the player lost its power? Motion grinds into a halt and at the hole's black entrance time would seem to freeze. You might expect, at point infinity, all space rolled up together, all time disappeared and you become a frozen ghost! World's end, a cosmic form of heart attack! Rubber time got rigid. Einstein's elastic has grown hard; brittle time, perhaps even space as well, can snap. Time as a whole has got the chop! The gravity of mass has dragged dynamic flexi-time into a fixed condition. Halt! No go! Here, at the cosmic boundary called mass, find solid time! Is this, as opposed to super-state immediacy and all-can-go, the only sub-state case where time that once ran really 'no-can-go'? Once drawn within the dark one's fascinating rim, once irreversibly beneath the clutches of its spell then theory states that you would see your friend's last wave frozen at the window of a ring, an unseen ring of influence that science labels an 'event horizon'. You would catch a rigor in the rictus that would last forever as the world's projector jammed - unless there's a hole in theory. Perhaps even sinks leak and black holes can drain away; if you can catch the effluent then might not you discover what they're made of once again?

Concede, because it's ours, we live in 'normal time'. We don't, by riding rays of light or being crushed inside a size-less coffin at The Black Hole Cemetery, inhabit the extremities. Your average, regular, easy-going kind of in-between-time isn't much distorted; it's all we definitely know and for practical intents and purposes you might, as Newton did, confer the honour of infinitude and omnipresence, that is, assign it absolution. Time's an odometer that measures change; what about dividing up its stream? Existence means division, things all standing out distinct from one another. We've been thinking how to slice each side of mobile moments, how to dice time up. Long and slow the years pass; long and slow is how, in spite of how we think, we humans take our time. Dragonflies and birds flit to a quicker beat. Very short and fast the atoms jump and spin. Cycles, lifetimes and perceptions vary. Micro-time inhabits macro-time and *vice versa*. Was it not agreed if you accelerate enough your time 'dilates'? That clocks 'tick slow'? But forget that magic moment, point infinity, all time rolled up in super-time, all space disappeared and you become of ageless immortality! Decelerate until you reach the rate of subatomic oscillation. Even if such oscillation's far too fast for you to other than numerically understand, it is certainly the measure of atomic motions and metabolism - indeed, most chemical reactions. Fast and short, subatomic cycles live in micro-time. The particles of atoms that compose bulk objects 'travel', or at least vibrate, like lightning - even if the object of their composition doesn't, like a statue, move at all. In other words, what's short for particles or atoms 'contracts' into long time for the composite in which they spin. An atom's lifetime might, from its perspective, seem of almost ageless 'immortality'. Bulk objects would, conversely, seem to flick by in a flash. Instead of 'bulk object' read 'my body'; then, from your atoms' point of view, how ephemeral your changes are - and what are these in total but your life? Life embodied, if the theory's strange but true, flicks like a flash in deep eternity.

After all, big and small, what is the smallest cut of all? What is the smallest frame in time's continuous film? What, in all the ways that time and place appear distinct, is the smallest cut-out in the puzzle of existence, the nearest to a stood-up 'piece' of timelessness? Is it found locked with the smallest dot of space? If space were atomised what would time's atom be? Can it, at any level, ever be fragmented or is it a pure abstract but continuous function of non-abstract mass, energy and this pair's product, gravity? It is, by some, surmised that the fundamental granulation of physical time, intimately linked to the highest frequency and the shortest possible wavelength of a tension in space, a minuscule fraction of second, is called Planck time (10^{-43} sec). Is such granulation really discontinuous - time peppered here and there with timeless absence in between? Or is it simply an atomic calculus of motion, the final slice of temporal analysis beyond which....what? Before physical creation there was, understandably, no physical passage of time. Does Planck's natural division into 'quantum' parts represent not last but first and subtlest period in the universe? Do things at their most primal pulse this way? So that Planck's primordial time's the beat that underlies but drums up, bit by bit, jump by jump, changes, things, the whole phantasmagoria? And, if time's staccato, is the universe not flickering on and off? If time is stippled why not space? Why not break them both in Planck sized granulations with, like quantum levels of an atom's electronic orbits, nothing in between?

The Geometry of Time

		tam/ raj	Sat
		calculus of curve or ray	Dot
		protraction	Immediacy
		outer/ inner peripheries	Centre
l		ines of time	No-Time
	↓	tam	raj ↑
		straight	curved
		finish is not start-point	finish-point is origin
		anti-cycle	cycle
		flat/ inertial	vibrant
		aspect of 'thing'	aspect of energy
		irregular	regular
		exhaustion	renewal
		entropy	sustenance
		time of body	time of life
		death	survival
		DC current	AC current

We're slowly getting here. Do you remember (*fig. 2.14*) the fundamentals of all geometry - dot and calculus of dot? And that the calculus forms curves or, special case of curve, straight line? *There exists, which correlates with species, a geometry of time.*

Do you also remember centre-point, the world-wheel's axis and the sage's conscious here and now? The dot is (*Sat*) Now. It is Potential and Preordination. It is Eternal Origin. From this Essential Moment at the Centre of Existence there issue all the lesser, passing moments we experience as time. Round this Independent Axis cycle all dependencies. Is there any other Bearing? Round its Norm all changes oscillate and transformations go ahead. Around this Hub the world spins and its individuals dance.

There are three main grades of non-conscious matter - (*sat*) potential, (*raj*) matter in subatomic principle and (*tam*) the matter of bulk forms; you might reckon, therefore, there appear three basic grades of time.

A dot represents the central, archetypal starting-point; it's time physical's potential, no time yet - eternally. Such potential time reflects the right-hand qualities of Primary Dialectic (such as Central, Neutral, Omnipresent and Original). Although inanimate it will reflect the pristine quality and order of the Higher Cosmic Archetype called *Logos*. The other kinds of time are strings composed of series or a calculus of dots. Each point represents a lesser, temporary presence. It is one of many finite 'fragments' of Infinity. It is an appearance of The Presence.

(*Raj*) vibrant time is energetic. It is survival's sort of upkeep. Call its cycling current AC (alternating). Cycles are composed of curves. Then call it curvaceous, wobbly time. Its flexibility predominates in mind but also, fast or slow, shows up in oscillations and the various periods of matter..

(*Tam*) flat time is sub-state. Call its line a ray. Its straightness runs from start to finish, past to future, full potential to complete exhaustion. Call such a discharge, flying like an arrow towards an end-point, DC time. Call such expenditure the way of aggregation, bulk form, entropy and death.

Is time AC, DC or them both at once? Is it an arrow, wheel, a calculus of dot or just, without any geometry, a part of the events mind measures?

Of course, the species, shapes and lines of time occur at once. All sorts of dotty trails are happening simultaneously within The Omnipresent Moment, Now. Let's deal first with animated time in terms of consciousness and then turn to apply a wheel (AC) and arrow (DC) to the world of physics and its history.

AC or DC?

	tam/ raj		*Sat*
	swing		Moment
	fall/ rise		Pivot
	motion		Poise
	difference		Link
	division		Seam
	night/ day		Dawn
↓	tam	raj	↑
	out	in	
	night	day	
	dusk	dawn	
	flat line	cycle	
	arrow	period	
	solidity	vibration	
	square	circle	

expiry	*inspiration*
materialisation	*dematerialisation*
creation	*dissolution*
subjective exhaustion	*subjective vitality*
objects	*subjects*
sleep	*waking*
death	*life*

Just as a ray of light is vibratory and straight, so time is an arrow *and* a cycle. Time is cyclical *and* straight. Orbit, line and, in the presence of eternal 'now', a dot - the basic elements of geometry are congregated into space-less time - except for symmetry.

Not back as well as forwards. Time reversal's not an option. Opposites (mirror images or positives and negatives) can be reversed. Whoops! You can drop and break a glass but can you forward in reverse and have the shards reform? Can a diver 'back-dive' from the pool to where he started on the board? If you inflate a football does deflation turn the time-line round? Why, therefore, has it been suggested that, if cosmic U-turn followed an expansion, contraction of the universe would reel up time? No pump can push against the flow still less reverse it.

DC means straight. AC means oscillation; it is vibratory. Since the song of cosmos is a periodic one then AC time is at the root of action. Rings of time are chains that bind. Both physical and psychological, they bind. They make links, irregular and regular, that hook the interactions of existence up. Involuntary oscillation gives you atoms, cycles give you days, months, years and galaxies. And the wheel of life keeps turning - one biocycle of a body is achieved, from conception until death, by coherent oscillations of its chemistry called homeostasis (*fig.* 5.6). It's also DC since from the very start your body bound you to a line that runs directly out of vibrant youth to age that loses bounce. Voluntary rings are just the same. What are the suns and moons of your desire that clasp you in their sway? The swing is from a wish as straight as possible to satisfaction. Over and over. And from life to life equilibration runs its course - metaphysical recycling called reincarnation (*figs.* 18.2 and 18.5) spins the wheel round many times.

Hold on! That's not all. Cause and effect combine to forge a chain. Held in! Such chains are what existence is a prisoner of. We live in rings, we live in chains and so are prisoners of Freedom! Things and events are just tied up Infinity. Rings of time, originating from the Centre, bind them into shape and place. What is the Cause of causes? If Cause emerged from Causelessness, if Nothing's where it all began then Nothing is the Origin. How can anyone escape the shackles of their body and its world? *What is the source of mind and, where alpha's omega, how (fig. 12.7) can you complete a cosmic cycle to reach Original Freedom?*

A mystic has observed this wish to stop the world. His goal, where the stop is Home, is to alight from cosmic omnibus. Where is the stop? How does 'getting off' work? How do you step out of time? He knows an anti-timely sentiment might be the key to leave its bustling business behind, switch time's ignition off, get out, clunk shut the door upon existence then.... silence in the country, breath of fresh life in Infinity. Will-power's piston and a focus of attention are the only pumps to drive upstream, to actively transport a climber on Mount Universe and, at its Apex, break the bonds of time. You can, in other words, pump out of time. You can voluntarily ascend, against the grain of downward flow that is creation's sustenance, to Timeless Origin. But how, if naturalism's 'negentropic' anti-flow is deemed impossible, can cosmo-cycling ever be completed? Is time 'reversed' back to The Start? Do you, going back, go forward? Or going forward get back? How confusing seems the mystic climb upon Mount Universe.

In fact there are no lines of time in Timelessness. *The solution is dissolving them.* Time is dissolved by all enthusiasms. The trick is choosing one that lasts. No boundary, no end; mustn't what is infinite be permanent? What is the nature, we discussed, of Nothing and Infinity? A meditator aims to really turn things round and, with completion of the cosmic circuit, evaporate into the heavens. Can attention's laser lift the dews of cold, dark mind? Can he land up at the origin of motion - a point of truth preceding time? You might argue with his answer. Can you really argue with solutions you have never tried? Is science not experimental? Practical? *How, without a repetition of his practice or expert experiment, can the metaphysics of a mystic be denied?*

DC-time.

Can any end precede its start? Can effect precede its cause? A 'line' that's sometimes reasonably straight connects a cause to its effects. This, no tock-tick, is basic cosmic law. Down this line prior potential (involving energy or information) is expressed, exhausted and then finished. Things run, as viewed dialectically or otherwise, from (*sat*) possibility through (*raj*) kinetic phase to (*tam*) inertial equilibrium - the finish. Options are suppressed, possibility and plausibility reduced into a single, actual outcome. You could say potential's realised and then, exhausted, hardens up and dies. Are not solids graves of energy? Even cosmos as a whole becomes exhausted. Friction, loss of power and gravity always flatten bounce into a fixed, straight line. Causality's a valve that plumbs the cosmos into one-way flow.

The law of entropy requires this flow is down to earth; and that, in winding down, the universe began its time in a most highly ordered way. From bulk matter's angle arrow and progression seem to dominate. DC is the arrow from a bow; DC is the river from its source. This way rolls youth and mid-term life along. It runs, relentlessly, to age and oceanic death. Arrows, say the laws of entropy and gravity, always fall to earth. And rivers never flow, like cycles, back to where they started. They drop down and slowly run into the sea. Cosmos is a river flooding time; it is an arrow loosed in space. Before and after any action these two immaterials remain the same - or else, with universal action's end, perhaps both shrivel up and disappear? Who knows how forever and a day will show or if, exhausted, time itself will fall away? Nothing then. Full stop. A stop as full of emptiness can't be imagined.

Look behind. Turn radiant expansion round. Compress the universe back into where it came. Some think this passage started at a very lively blow. How many years can you extrapolate? Set out from this end backwards down the channel that includes our planet and its forms of life. Dendro-chronology (the measurement of time by counting tree-rings) and carbon dating can together slide you back some 20000 years. It is when you try and calibrate the age of rocks that trouble starts.

Modern geology is based on two major premises - that the earth has great age and that James Hutton's principle of uniformitarianism (summarised in the aphorism 'the present is the key to the past' - holds good. Earth's stratigraphic columns are built up of rock units, for example, coal and chalk measures, and subdivided into beds and bedding planes. Such strata, like varves (thin layers of annual, sedimentary deposit on the floor of lakes), are not washed in rapidly; geological accrual is, in general, very slow. It is not as if the column below your feet (or anywhere else that layers of sedimentary deposit have accrued) runs from levels, say 10 down to 1 by age till bedrock. Far from this, it is doubtful if such a column exists anywhere on earth. The norm is a permutation of layers, even layers 'folded' by tectonic activity into the 'wrong' order. Nevertheless, by using certain fossils as indicators a column can be divided into time zones. This 'bio-stratigraphic column' is defined by the zone or index fossils that occur in a given stratum. For instance, any rock containing fossils of one type of trilobite (*Paradoxides*) is called Cambrian rock; of another type (*Bathyurus*) Ordovician. In this plausible way a *relative* geological table is compiled. You can compare its long ages with the reasons for short ones proposed by so-called 'flood' or 'liquefaction geology' based on the notion of an initial period of world-wide, hydraulic upheaval prior to its current 'remnant' - a current largely stable circumstance with continual local 'rumblings and residual catastrophes'. It is also possible that upheaval of earth's crust was, at some point in history, much more violent and immediate than today but both parties accept plate tectonic theory as an explanation of many geological events. Liquefaction theory's view is, however, hostile to the long ages required by evolutionary theory; and the hostility is returned. Each rock formation round the world is, in its turn, the subject of oft-times disputed origin. The literature is great and growing all the time. One, other or neither estimate may be right but there is no give or take at all (except a few million or even, at its start, billion years in the standard version). In their clash each side fiercely believes itself reasonably but irreconcilably 'right' and the other ridiculously 'wrong'. The mode in this arena is attack alone; its gladiatorial mood is, thumb down, death to the adversary. Nor has academic amphitheatre time for short-scale or 'young earth' measurements of time. Of course, a major underlying point of combat is world-view. Planless evolution needs, vitally, an aged earth. On the other hand, however, to conceive, develop and materially implement a plan is not time-dependent process. Long or short, time is not the critical issue; in this case quality of mind is.

No geologist claims his understanding, even if heavily presumptuous, is flawless. While evolution mandates long periods of time and is the habitual, modern frame of mind everyone still recognises that major accidents create in moments what years cannot. Nor can you tell the age of rock or an ecosystem simply by a look at it. Local 'catastrophes' like the volcanic explosion of Mount Washington or the creation of Surtsey Island (Chapter 24) have seen ecological succession from bare lava in 20 or 30 years - less than the wink of a geological eye. Can simple observation show whether a tsunami from the burst of Santorini destroyed the Minoans in Crete only a brief 3000 years ago? In short passage it can seem that ancient times have passed. How, therefore, can you really map life's history?

Fossils? Geology relies heavily on the evolutionary interpretation of the fossil record (Chapter 21) because the relative ages of various rock formations are determined largely in accordance with the presumed stage-of-evolution of the fossils in them. The only chronometric scale applied by geo-logic to the stratigraphic classification of rocks and to dating prehistoric events is indexed by fossils. Where there appears discordance between physical and fossil evidence as to the age of any series of beds fossil logic is usually preferred - even over radiometric assessment. There is no single area where living things evolved in a series from less to more complex up a stratigraphic column; but an evolutionist arranges finds from different areas on the assumption that his 'scientific' decision is correct. If science is solely materialistic in interpretation he must be absolutely right.

Herein, though, lies a powerful tautology, a circular argument. **The assumption of evolution is the basis on which index fossils are used to date the rocks; and these same fossils are supposed to provide the main evidence for evolution. The fossil record, itself based on the assumption of evolution, is interpreted to teach evolution!** *By this sort of reckoning the main evidence for evolution is the assumption of evolution. Is that so scientific?*

Not at all. In fact a number of index fossils have, after 'an absence of millions of years', turned up alive. Examples are the tuatara lizard, a small mollusc *Neopilina galatea*, the maidenhair tree (gingko) and the dawn redwood. Perhaps most celebrated is the 'link fossil' coelacanth fished up off the coast of Madagascar in 1938 after 70 million years' absence. Any rock dated according to a coelacanth fossil is not now of such age but must be anything between then and now. Because most index fossils are small, marine invertebrates, for example hard-shelled molluscs, it is possible that more index fossils will be found alive and well.

Dating by fossils is, in short, tautologous and gives relative data. **Relative is not absolute.** *Radiometric dating, on the other hand, holds promise of absolution.* The method capitalises on the concept of 'half-life', that is, the characteristic time for 50% of one radioactive species to decay into a second species. Each 'parent isotope' decays into its 'daughter' form (e.g. rubidium into strontium) with a specific half-life (in this case 48.8 billion years) and is used to date different rock materials (in this case whole igneous or metamorphic rock). In this way absolute dates as, for example, 430-500 Ma ago for the Ordovician period, are ascribed by using igneous rocks as calibration points. These occur among sediments in the form of sills and dykes. Radiometric methods can be used to date them. Sediments on which dated igneous rocks lie are presumed to predate them. Those overlying must be younger. So, despite the possibility of igneous extrusion squeezed between two older layers, one ascribes actual ages to fossil-bearing strata. Rock with those fossils can thereafter be ascribed that age.

We can't, however, measure age directly. **We count isotopes in rocks with accuracy but long-age dating is crucially dependent on at least three long-age assumptions.**

Foremost among assumptions is that rates of decay are immutable. They have always been just as they are. Is this definitely so? Could there not, for example, have been any episode of accelerated nuclear decay? Such 'fast-track' fission would, like fast-forwarding film, give a shorter playing time. It would reduce earth's age but there is, in fact, no evidence at all the 'rules' that govern nature changed. Astronomers go back in time. Travelling far back into distant galaxies they find the 'rules' the same. Moreover any change to the weak nuclear force (of which decay rates are a function) would have blocked the production of any heavy element leaving only a cosmos made of hydrogen. If you prevaricate then wouldn't high decay rates at the start have soon crisped Adam, Noah or the others? Life wouldn't have a chance. On the other hand, lack of certain isotopes (e.g. neptunium-237 or iodine-129) in earth, sun and meteorites indicates an age of over a billion years for the solar system. This assumption seems, except perhaps right at the very start of cosmos, assured.

A *second assumption* is that any sample is and has been effectively a closed system with little contextual leakage or contamination, This assumption is, in practice, much harder to defend.

Thirdly, perhaps most important of all, there rests the assumption that initial amounts of isotope at the time of solidification of a rock can be accurately or even reasonably estimated. *In other words, how long is a piece of string?* To determine how long a fuse has been burning you need to know its initial as well as its present length. Therefore, how much 'daughter isotope' was present when a piece of rock in question was formed? Is that amount always negligible or non-existent? An 'isochron' method tries to determine such amounts by using samples of rocks and minerals formed in the same magma flows. It presumes that any magma has its 'internal clock' reset at the time of melt; that initial conditions were uniformly mixed throughout a chosen sample of rocks or minerals; that partial melts or re-melts might not affect the chemical composition of a rock; and that a closed system wherein quantities are not affected by remixing (due to, say, fractionation by weight), other migration, contamination or loss of daughter product (for example, argon gas to atmosphere or by groundwater leaching), has been sustained for eons. It also, by averaging results, minimises such discordance as may occur when particular mineral ages within a rock appear to greatly exceed the whole sample's computed age. Each such circumstance is possible, seems to actually occur and introduces uncertainty into the equation. In this case data would, effectively, describe the rate of atomic migration as well as (or even rather than) the ticking of a radiometric clock. *Incorrect assumptions would falsify that clock*. **Indeed, assumptive dating is not so much a mode of measuring as one of thinking.**

Different methods of radiometric dating sometimes can give very different ages for the same rock. But no evolutionist is 'phased' by unexpected anomalies. It is plausibly and may well be rightly argued that, if different methods using different isochrones, isotope ratios, decay-rates and so on are applied to a sample and they calibrate with one another, the date becomes assumption-free. *In this vein naturalism squarely demands billions of years and so this overarching, basic assumption sways every other judgment.* Even ice deposits, tree rings and coral reefs give ages by far exceeding young earth's 6000 years. Yet consider the radioactive isotope, carbon-14, whose half-life in decay towards stable carbon-12 is 5736 years. Such relatively rapid decay means that, after about 100,000 years in a sample, only a practically undetectable amount should remain. Indeed, if the whole earth were made of carbon-14 after a million years not one atom of it would be left. However, it is found in 'ancient' materials such as diamond, 'carboniferous' coal and even deep in subterranean rocks. Almost everything that contains carbon contains a measure of the radioactive element. Yet no bone or fossil dated over a million years should contain any whatsoever. How does one account for such arrest of cosmic age? Call up, as a rescuing device, 'contamination'? Carbon-14 certainly destabilises, it would seem, the uniformitarian cause. Of course, every barrister knows that conclusions based on faulty premises are flawed. What's its real time, how old is the earth? With these mixed messages one thing alone is sure. Time's DC current, if not evolution's, progresses in an uncreative line that's straight.

AC-time.

Do you remember DC river? Of course, the ocean may not be a droplet's end. Levity might lift you to your origin; evaporation raises clouds; is the sky-born cycle now complete? And is not, the mystic parable exclaims, a drop of soul likewise distilled? What kind of still is used to raise it 'hydro-logically' into the place where clouds of being first accumulate? Where do such clouds float but through Unearthly, Inner Sky?

On earth you cycle straight along a road or walk a line straight round the world. Many kinds of clock and every calendar go round and round. So does vibratory AC time. It oscillates, rotates and cycles with all sorts of frequency. A roll around circumference will pass you back to where you started. Cycles shuttle; each wave passes through a centre-point, an axis or an origin. Waves do not indicate, against the straight-line evolutionary idea, progression. *Mind* makes improvements but for matter there are only simple ups and downs of things. Equations wheel, pistons push and pull, motions grind through nature's great and balanced round. This is *dharma*; this is the way of *karma*; this is Newton's law of motion as applied to mind and matter. Not only Buddhists spin time on a wheel or, if you like, a clock-face.

Bulk matter's DC time is 'flat'. On the other hand, the subtler, more holistic or 'inwardly' a viewpoint, the more that notions of a periodic cycle, dynamic equilibrium, governing principle or oscillating order swing into ascendancy. Quantum oscillations spring to mind. Might mind also oscillate, as between awake and sleeping, so between excarnate, metaphysical and incarnate, embodied states? Could it (Chapter 18) be recycled in and out of physical connection, on and off, periodically round and around? Indeed, does cosmos, on a lifetime's wavelength eons long, vibrate? Does it cycle in and out of being? Does it, cosmo-cycling, realise an AC current and the vibrant side of time?

You might, in this case, query what the origin or centre-point of cosmos is. What is the nature of the 'orb' within whose orbit objects fly through their existence? Is it an archetypal form? What kind of archetype could hold the universe together? And what might constitute a central sun of life's psychology, the archetypal origin of mind? Could it be the order that's intrinsic in a life stream called the *Logos*? At least that's what the mystics, bound up in the metaphysical, all seem to say.

↓	tam	raj	↑
	out	in/ return	
	lower	higher	
	expiry	inspiration	
	materialisation	dematerialisation	
	genesis/ creation	dissolution	

Take a deep breath in. Breathe out. Ebb and flow, influx, efflux, inspiration, expiry. Isn't ventilation AC oscillation? Do you remember (from *fig.* 7.5) yoga's breathing techniques? They are said to mirror cosmic cycles that include expansion and contraction of the universe. How are 'cosmic ventilation' and 'the breath of activation' registered in oriental time?

Two main versions exist. In the first a full cycle comprises both descending and ascending arcs. Descent involves materialisation and ascent the reverse. Each cycle is calculated to last 12000 years. It swings either away from or back to the axial position of poise and is characterised by four divisions. These include a short, most eccentric period of 1200 years (called an 'iron age' or *kali yuga*) followed by three increasingly long and less eccentric, more balanced periods of 2400 ('copper'), 3600 ('silver') and 4800 ('golden') ages. It may be, on this kind of reckoning, that we are presently 300 years into the second ('copper') stage of ascent towards a golden age. Where the 'iron age' is engrossed in the manipulation of gross materials, 'copper' rises to an understanding of the internal subtleties of matter such as electricity, gravity and the interactions that atomic physics, chemistry and molecular biology (not least in its aspect of genetics) demonstrate. Up the conscio-material gradient the "higher" ages turn from subtle matter into mind. The study of psychology leads to the highest, 'richest' seam - an emphasis of the pursuit of Immaterial Ideal.

In the second version each day in a 30-day month of a 12-month (360-day) year is held to be the reflection of a solar year. Multiplication by 360 therefore gives a full cycle, a *mahayuga*, of two arcs (descending ↓ and ascending ↑ modes) each lasting 4,320,000 years. The sub-divisions are extended proportionately so that, for example, the 'iron age' or *kali yuga* lasts 432,000 years. This rating puts us soon after the start of a long, dark era.

A cosmic 'respiratory cycle' is therefore measured as an eccentric orbit of either 24000 years or 360 times that amount (8,640,000 years). It is unclear whether this period represents the cycle of our sun around some galactic staple (such as a black hole core) or involves a metaphysical aspect (such as short periods that are more under the heavy, gravitational influence of matter than longer, lighter periods whose process tends 'inward' and away from physicality). Whichever way, the physical universe is held to dissolve in a '*pralaya*' once very 70 *mahayugas*, that is, 1,680,000 years in the first system and 604,800,000 in the second.

These are massive oscillations. A thousand full cycles is called a 'Day of Brahm'. *Brahma* is a high point on the conscio-material gradient. It is First Causal whence issues *pranava* or *Om* - the cosmic vibrator, the world excitor and support. This is the level at which both (↓) genesis and, conversely, (↑) *mahapralaya* ('great dissolution') can occur. *Mahapralaya* is not only physical but psychological annihilation. The inner worlds of universal mind are dropped as well. This, when all worlds die, is total cosmic death. Existence is dormant; Nothing Alone, called Essence, now is; and, therefore, No-time rules. How can you calculate a time of non-existence?

Many bursts for one. Cosmologists now bounce the idea of a universal yo-yo; oscillatory 'bangs' of cosmos are (Chapter 12: Alpha Points) no longer off the clock. Cycles aren't beneath the scientific radar. They've been hypothesized by modern scanners in the field. Not necessarily repetive or regular, mind you. Serial and parallel, all kinds of brave, new world are mathematically inventable. We're lucky, in the right one at the right time but, with more or less dimensions than our sensible quartet (and definitely no metaphysical required) multi-tasking physics now invigorates a hydra we've already met, the multiverse. Indian time-scales also conflict with astronomical observations that imply cosmic expansion. 'Retro-extrapolation' from the present to an initial singularity leads, with components like rules and constants remaining intact, to a point of origin that's nearly 15 billion years ago. An expanding universe is one thing; the age of actual rocks another. Long-scale radiometric dating systems are all based on assumptions that we took a look at earlier. This is neither time nor place to recount abundant geological anomalies or mount an in-depth critique of the assumptions that assail both young and old world orthodox chronologies. It needs, on the other hand, to note a snippet of a contradiction mounted at the short end of world-scale.

Day-Age Controversy.

Look round. Wind time back. How long would it take to press the universe back into whence it came? You say no-one was there. Astronomers, in viewing galaxies billions of light years young, can creep up on the start. For many, therefore, any 'day-age' controversy is a waste of time.

Would you Adam-and-Eve it? It was not the early Christian apologists that started it. Origen, Eusebius, Augustine and later (see The Correspondence of Sir Isaac Newton 1676-1694) Newton himself cast doubt on a literal 6-day creation with the seventh day of maintenance extending until now. No doubt about it, though, claimed the Vice-Chancellor of Cambridge University; during the 1640's John Lightfoot tussled with the Irish Archbishop of Armagh, James Ussher and eventually concluded creation week was October 18-24, 4004; Adam was created 'after breakfast' at the good clocking-in time of 0900 hours on October 23rd. Thus controversy's thunder clapped. Some thenceforth took the days as literal and others as a metaphorical expression of serial stages in the development of cosmos. All kinds of evidence has been adduced in either case. Some would say the facts have been traduced to literalise 144 hours-worth of creation. Does the Bible really mean this? Contemporary science definitely disagrees. Astronomers espy a universe expanding from, they logically surmise, a very lively blow. They identify an instant miracle from outside scientific laws. Their telescopes seem to confirm, about 14.7 billion years ago, that a transcendent burst projected cosmos as a 4-d space-time screen whose general character incorporates decelerations and decay. Dilated and diluted - every hour is stretched into about 100 million years!

Time generates its heat. To what degree can invisible frictions burst into hours of flaming intellectual dispute? It is unclear how the Indian scales of time were calculated in the first place. They were Babylonian homework. A conservative from the old school, a stickler for terrestrial chronology, it ticks divisions drawn from relationships between the earth (with her spin and wobble), moon and sun. Its days and months and years are down-to-earth. They are practical for us but not universal. The speed of earth's rotation generates what we label a '24-hour day' but day-lengths differ on every planet of the universe; how, in an hours-old cosmos, was Adam not burned to a crisp? How on earth, therefore, can flat contradictions be reconciled? Indian 'periods' and scientific timelines flatly contradict a literal interpretation of the Biblical account of a creation in 144 hours (6 days) about 6000 years ago. **Earth sciences, astronomy and modern physics are, it may seem, denied since if the earth is very old the fear is that Darwinian evolution of life-forms might just be true.** No doubt, spadefuls of mega-time are heroes of the evolutionary plot. Impossible might, if you wish, turn possible; possible is almost probable and probable, you claim fallaciously, inevitable! Hey presto! Life on earth *evolved* (but see Book 3). And thus short time-scales generate the maximum emotional heat!

What is the truth as far as maths and words can tell? You might reasonably enquire why a Progenitor's creative schedule should be locked precisely to the spin of one obscure planet in a certain galaxy? Should scientific understanding really rest on exegetics and debate about a Hebrew noun? The noun *'yom'* is used, like its English equivalent 'day', in different senses. In some contexts it implies a period or epoch (say, 'in the day of Queen Victoria') but when you speak of 'day and night' or specific numbers of days you imply definite 24-hour spins of earth around its axis. Six days isn't much to conjure up a universe, even if you used any other planet's revolution as a gauge, so does a 'day-age' theory work? 'A thousand years is unto the Lord as a watch in the night'. How long, therefore, is a Genetic Day? Does *'yom'* mean an earth-day, a thousand years, a million years or an undefined period that gives Hebrew God a lengthy break in which to make? Or, indeed, was science's transcendent burst a kind of switch which simply started up the preconceived and archetypal space machine? Inconsistencies plague the 'day-age' interpretation (for example, concerning the ages of the patriarchs descended from the first man, Adam, and his wife Eve). Yet insistence that the Hubble-interpreted cosmos was made in a week and is still only 6000 years old is manna to materialistic cynicism and contempt; evolutionists lampoon the literalist faith for its 'suspension of reason' when it comes to rocks of ages and life's cradle here at home. If you need unimaginable swathes of time then 'young-earth creationism' is a serious thief. It would, if true, definitely steal evolution from you. Worst of all and with no lease of doubt you'd have to come to terms with a Creator.

Traditional anthropology, archaeology and palaeontology work (for all their plethora of hypotheses) using objective data within the framework of a straight line. This 'arrow' is one of progressive cultural evolution. Yet the notion of an initial 'golden age' or a 'high-grade genesis' lingers. It is the staple of prolific mythological research by Max Muller, Malinowski, Frazer, Jung, Eliade, Levi-Strauss, Joseph Campbell and others but, along with a critique of scientific dating systems, is not the remit of this book. Such study seeks to germinate the philosophical seeds and sociological truths embedded and obscured in so-called 'masks of God', that is, in mythical traditions, antediluvian legends and the common sorts of ritual of an historical 'golden era' hardly remembered but apparently common to mankind across the globe. Perhaps Cro-Magnon and Neanderthals were, on this basis, as strong and intelligent but more spiritually evolved and ecologically aware than most of us superior, scientific types! Perhaps they lived like the American Indians did until about two centuries ago. How would you develop if, with a mind washed free from any technological knowledge or expectation, you were dropped on an island devoid of other human life? You might, were you and your line particularly intelligent, healthy and persevering, evolve in culture and technology much as the advance of history shows humans have. And history starts, at maximum, 6000 years ago. Prehistory is, by definition, without written trace. How far, therefore, has archaeology traced our prehistory? Not by bones but definitely dated settlements? 9000 years in reverse? A 'climb' from huts to modern cities, from sticks and stones to nuclear bombs in about 350 generations. Wrangle the maths (*see* 'A Mutant Ape?' Chapter 23 footnote 142)! That is enough time to generate the present human overpopulation of the world. How many years have humans been actually reproducing? Would you guess half a million, a million or a million and a half? In which case what, for between 20000 and 60000 generations previously, were our bright, inquisitive and capable relations thinking of?

Basic anachronisms we've discussed conflict. They confuse. Everybody can't be right. And yet, not having been awake at cosmic dawn, no-one truly knows but everyone infers the age of earth and its surroundings. This author without doubt confesses ignorance. *An evolutionist will never; he cannot emphasise enough that mega-time's an absolutely proven fact.* He may be right. And, without doubt, he professes faith in the creative power of those unimaginable and essential tracts. **However, Natural Dialectic takes the line that information, not time, is the critical factor. Informative mind precedes informed energy.** *Intelligent design of cosmos trumps its days, months, weeks or billions of years.* Age alone cannot create life forms. The science of biology now shows (it will be argued in Books 2 and 3 *passim*) that naturalistic process is an insufficient 'author' of our human hours. And physics shows an initial projection so fine-tuned that, although including elements of chance and naturalistic methodology, design is a most reasonable choice of origin.

Thus, to reiterate, the main issue of creation is not whatever kind of time you choose but *mind*. Mind is natural. Creative mind is anti-chance and has its range of purposes. The one thing every creator strives to eliminate (with an effect proportional to his intelligence) is bugs, error and the chaotic part of chance. *Only mind is capable of systematic determination, coherent construction and purposive complexity and the time it takes to realise its ideas is of secondary relevance.*

Grades, Principles and Times.

What about time and the order of creation? Could the eras of 'progressive' time, whether from Hindu 'egg' or from a 'space inflation' genesis, also be embedded in the layers of existence? In other words, could you relate the grades of Mount Universe to stages of creation? And thus perceive these pre-formulated stages as steps in the projection of an idea from drawing board to physical performance? It is to such grades, principles and times that, as far as the objective, scientifically described sector of the cosmos is concerned, we now turn.

A motif that underlies the presentation of Natural Dialectic is one that reflects the gradient of creation. This runs from principle to practice and is, therefore, the trend which is followed in discussion of information (Chapters 5 and 6) and energy (Chapters 7-12) as they combine to produce effects studied by the metaphysical 'sciences' of logic, philosophy, mathematics and psychology, the physical sciences of physics and chemistry and, straddling each side, the intermediate, psychosomatic subject of biology.

	tam/ raj	*sat*
	physical	*metaphysical*
	derivation	*archetype*
	expressed matter	*potential matter*
	actuality	*virtuality*
	order	*preordination*
	long-time/ short-time	*no-time*
↓	*tam*	*raj* ↑
	exhausted	*kinetic*
	confined	*free*
	external	*internal*
	massive appearance	*energetic reality*
	governed	*informant*
	aggregate/ bulk	*atomic/ molecular*
	long-time	*short-time*
	classical	*quantum*

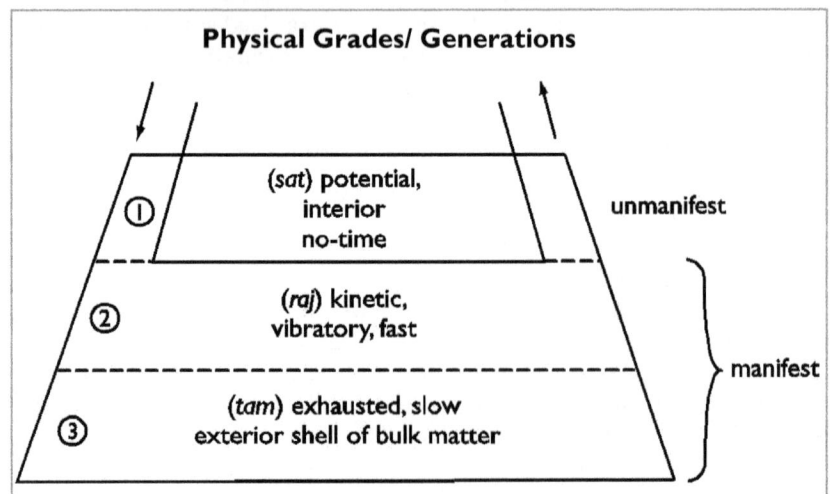

Physical eras/ grades are:

(i) A pre-existential; psychosomatic starting-point; the drawing board of archetypal memory. This metaphysical level is omnipresent but does not operate in physical time. Call it archetypal 'no-time'.

> (ii) The expression of 'grammatical' information; archetypal memory is expressed as a subatomic, quantum or principle grade of matter (represented in the Standard Model by three generations of fundamental particle); this level operates in 'fast' micro-time also called 'short-time'.
>
> (iii) Gross data level; the atomic border gives way to a multiplicity of differences in the form of plasma, molecular gas, liquid and solid; the 'internal' energies locked in these low, classical grades of matter are vibratory; external definition increases with loss of energy. This level of time, in which you and I live, operates in 'slow' or 'sensible' macro-time. Call it 'long-time'.

fig. 11.2

The gradient runs top-down. Cosmic order runs from mind to matter, from archetype to physical expression. Principle (condensed information) guides practice; thus simple runs to complex, general to specific, universal plan to detailed, individual, localised expression.

	tam/ raj	*sat*
	physical	*metaphysical*
	energy	*information*
	matter	*mind*
↓	*tam*	*raj* ↑
	practice	*principle*
	exterior	*interior*
	complex appearance	*simple reality*
	bulk	*microscopic*
	massive	*atomic*
	classical	*quantum*

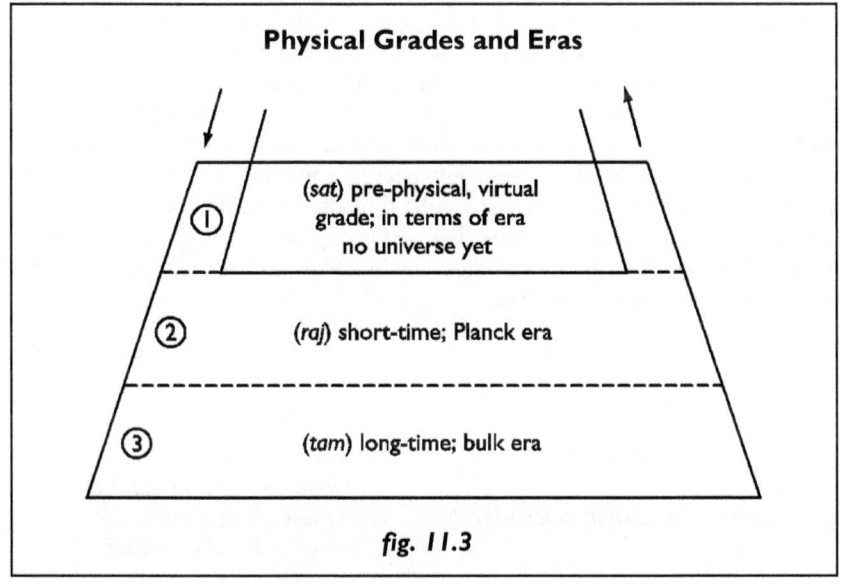

Physical Grades and Eras

① (*sat*) pre-physical, virtual grade; in terms of era no universe yet

② (*raj*) short-time; Planck era

③ (*tam*) long-time; bulk era

fig. 11.3

This diagram elaborates 113. It brings eras into line with grades. Do you remember (Chapter 2) that prepositions describing the hierarchical structure and developmental gradient of creation are not used spatially? In a case of precedence, for example, the plan is not physically 'before' its execution or result; neither is principle literally 'within' practice, happiness 'above' sadness nor the physical atoms of an adult 'inside' the egg that it develops out of. Nor, again, is potential literally 'within' activity, gas 'within' a solid or mind somehow 'within' matter. Yet in each case lower devolves from prior, higher, subtler information or energy. Isn't a short king, enthroned in his power centre, still above his six-foot subjects? If grades run from (*sat*) potential to (*tam*) exhausted then hierarchy is both metaphorical and real. *In this sense it follows that higher is 'within' lower and first is 'within' last. What came first still informs what follows. 'Higher' is the magisterial cause of 'lower', consequent effect. Each grade nests within the one below. Atoms are nested within bulk materials, particles within atoms, vacuum within both.* **The clear implication is that creation springs 'organically' from within outwards. The dialectical expectation is, therefore, that subdivisions of the physical cosmos devolved in the same order in time as in material grade.** Creation moves from centre towards periphery, from potential towards its full expression. (*Sat*) 'top' or potential matter transcends its expression as, firstly, (*raj*) quantum and then large-scale (*tam*) classical matter. Does not time deliver serial order? Is not hierarchy serial passing-down? **Thus time's order would reflect the order of creation, linkage that is still inlaid in things**. Potential, short and long times - these have grades and eras interlinked; all three exist, as in the beginning, now.

There is, therefore, in this *top-down* view, a strong link between cosmic **grades and the eras in which different grades of materialisation were predominant**. In terms of existence, for example, the grade of mind precedes matter. It is also possible to treat energy in terms of grade and time together. For example, you would take (*sat*) potential matter - a physical vacuity - and study its characteristics. These involve the combined expression of its Main or Primary dialectical principles. You would then, using secondary dialectic, analyse general and specific forms, that is, (*raj*) quantum and (*tam*) bulk phases in the same way. An actual start (a 'transcendent burst of radiation' or other scenario), its lifetime through to the present low-energy condition and the end of the universe are therefore subjects that can be discussed according to time as well as grade. *To this end it is important that, as far as possible, the structural order of The Drama of Natural Dialectic unfolds in a way that transparently reflects the vectors of both grade and time.*

	grade	time
sat	pre-physical latency	-
raj	quantum micro-level,	Planck/ quantum era of high energy and quick time
tam	classical macro-level,	era of low energy/ bulk aggregates and slow time

If the material universe was expressed 'organically' from within outwards, then the conjoined grades/ eras devolve in the order illustrated above or in *figs.* 2.10 and 2.13. Thus physical preconditions (archetypal patterns) exist 'within' quantum seed that in turn 'flowers' as gross, bulk forms and their properties. *This would suggest that both the span and order of eras in which the universe was first projected are also incorporated, presently, into the fabric of its construction.* In other words, archetypal omnipresence still co-exists with quantum fast and bulk slow times. Now; this instant; in no time at all. And an era of high energy (when particles were unattached or in simple atomic form) is now generally locked up in a low energy universe that is still, today, winding down. In this view bulk matter is a 'living fossil' of its own creation. It is not that physical phenomena, which exemplify the lower range of projection in the existential dipole, have any sense to perceive it. Nevertheless as the 'freedom' of non-conscious energy is locked up, it is precipitated into coarse, hard solids. In the process both mobility and time decelerate; as it exhausts, you might say, time is wasted. From the point of mystic interest matter is a waste of time.

And, dialectically, time is not the creative issue, intelligence is. Intelligence can do in seconds what unintelligent time could not in billions of years. Nor does time create purposely-informed systems. The more knowledgeable and efficient a creator, the more rapidly conceived, cohesively and viably constructed his creation. *Such psychological creation would be the first phase of physical creation and, if archetypes are memories, then they need no longer to lay down than yours when you make a plan.* But, in universal mind as in your own case, such memories are not perforce immediately expressed. Time may lag. If, in the cosmic case, the lag's between conception of its scheme and physical expression, what kind of lag is that? Before material action is there action time? That is to say, time physical could not pre-date a physical creation.

Who, therefore, could ever tell the lag between an archetype's construction and its proto-typical expression? There is, physically, none. If, however, it is reasonable to propose that principles crystallise from a first through two consequent phases into large-scale creation, then these remain to be accounted for. *If the first (sat) informative phase were metaphysical it would seem logical that the second (raj) kinetic, creative phase would be levity-predominant.* Its sea of energy would involve *matter-in-principle,* that is, the force fields of communication and elementary particles which have in some and perhaps all cases been discovered. These fields and particles substantiate bulk matter. A 'fast-time phase' would involve the production of light, seed-mass (electron and 'quark' types), at a lower level protons and neutrons and, at base, hydrogen plasma and the suite of elements. Its expression is governed on the small-scale by nuclear and electromagnetic forces. Just as atomic and chemical activities occur at high speed we might expect this 'spring' of energy to be rapid. *It could be called the particle* (or *quantum*) *era.*

Inner matter springs in short-time; its quick era started up the universe. Even now, locked within the walls of all bulk matter subatomic pistons of the atoms fire at speed. Of course, localised free fast-time eras (known as stars) still exist. Hiroshima involved and high-energy physics involves them. Indeed, high energy physics is a (*raj*) 'dialectical' venture in that it recharges matter back from its 'normal', locked low level of energy into the short-time era. It liberates or studies matter in, as far as possible, original conditions.

Is it possible that in the primal, quantum era 'black dots' of slow-time floated? Were such 'mini black holes', dots or seeds of mass what pegged the universe and brought it down to earth? Were they pinpricks in space, Higgs bosons in their 'inertialising' field, nuclear forces clamping quarks in place or what? They were (*tam*) dialectically inevitable nuclei. They are slow-time's fast initials and, as such, precursors of stage three.

<u>The third, (tam) inertial phase might be called the chemical (or classical) era</u>. It is the relatively slow-moving, low-energy phase of creation in which we now live. Stable atoms and, consequently, bulk or aggregate mass 'falls out' in the form of gas, liquid and solid states. *This 'cool time' involves the final precipitate of creation, bulk material.* Such precipitate is governed on the large-scale by the force of gravity, which operates in all directions simultaneously and tends to generate such sub-symmetries as spirals, spheres, lines and rings; and locally on a small scale it is governed by heat loss and electromagnetism which creates, by way of 'magnetic' chemical bonding, the detail of asymmetrical shapes. Each 'warden-force' is an aspect of the same (*raj/ tam*) fundamentals operative at every level. At this phase the nether directive holds inexorable sway. Its time-laden, universal force of binding and containment is also called the negative power. Behold, materialisation!

Restriction, confinement and material definition are the way potential is realised. They are (↓) localising aspects of existence that run against the Infinite (where (↑) releasing factors run towards it). The coarser an object or more chaotically violent an event, the further its value from Central Reality. Perhaps the most extreme form of confinement is a logical and, it is suspected, actual abyss which most of all material conditions might diametrically oppose the Highest Freedom. An end product of gravitational violence, a final catastrophe, a black-hole singularity stands for the death of space and frozen time.

Time's up! At least for time's being. We have not, however, reached the end. The game goes on, the play is in full swing as we progress, in Chapter 12, to consider the third (*tam*) essential of the physical universe, the ingredient of polar energy. If there was a start to energy an end is logical. Possible physical expressions and terminations are considered. Having cycled from *alpha* to *omega* one can then close the first half of the whole book. And take a break. In other words, clock off.

12. Physical Energy

Pure, Polar Energy

You say there is only matter. In this case there is no hierarchical notion of a 'higher power'. The highest physical power is vested in the world's first cause - a projection from beyond the laws of physics, big bang's levity or something as surprising.

I say there are distinct entities, mind and matter. In this hierarchical case we have to work from Infinity's First Cause down the gradient of Mount Universe through mind to matter. What is apparently formless, motionless, forceless, weightless? What is both transparent, peaceful yet full of radiant potential? What seems, in other words, to well express and to reflect the characteristics of (*Sat*) Essence?

(Raj) polar energy is, according to Natural Dialectic, the third of three physical absolutes.

Check *fig.* 7.9. Of its polar vectors (*tam*) gravity 'pulls down' towards the base of things but (*raj*) levity 'pushes up' from darkness towards the light of its (*sat*) essential source. Drag, lift; descent, ascent; add balance and you've built the basic archetype. The physical expression of these cosmic fundamentals (levity, equilibrium and gravity) is elaborated in the text associated with *fig.* 12.6.

The two basic components of creation's existential dipole are, for Natural Dialectic, information and energy. Each is well characterised in terms of light - inner and outer lights. They are respectively the light of knowledge, consciousness, and the light of physics, electromagnetic radiation. One is at the subjective heart of life; the other is the light by which the world is known, all bodies warmed and through photosynthesis life's bodies fuelled. Subjective, living light and objective, inanimate light - these irradiate with understanding and with warmth. Each kind of (*sat*) purity and brilliance is at the top of its domain. They are represented, on the conscio-material gradient, in three main forms - (S*at*) Pure Consciousness, (*raj*) the psychological light of comprehension and, objectively, (*tam*) physical light.

tam/ raj	*Sat*
existence	*Essence*
finite	*Infinite*
spectrum	*Absolute*
all below	*Transcendence*
lesser knowledge	*Knowledge*
spectrum of consciousness	*Consciousness*
colours	*Transparency*
impurity	*Purity*
chiaroscuro	*Illumination*
shadow	*Light*

↓	*tam*	*raj* ↑
	inertialising	*dynamising*
	illuminated	*illuminator*
	objective	*subjective*
	creative will	*unifying love*
	creation	*dissolution*
	negative power	*positive power*
	turn from light	*turn to light*
	Brahm	Nam
	Lucifer/ Maya etc.	Christ /Buddha etc

(*Sat*) Illumination is Essence. Essence is First Principle, the Light of the World, a Living Light, Perfect Knowledge from which all lesser comprehension, all lesser information, all principles and practices are gradually derived. Having met this Top, Metaphysical Projector in Chapters 2, 5 and 7, we shall meet it again in Chapter 14.

	tam/ raj	Sat
	duality/ multiplicity	Unity
	expression	Potential
	lesser principles	First Principle
	lesser causes	First Cause
↓	*tam*	*raj* ↑
	descent	*ascent*
	informed	*informant*
	increasing ignorance	*increasing knowledge*
	of Unity	*of Unity*
	differentiation	*unification*
	capture	*liberation*
	restriction	*expansion*

(*Sat*) Sublime Potential and its (*raj*) First Motion are almost identical. The second is simply a concentration, focus or initial expression of the first. First Motion is the First Cause of existence. It is (*figs. 7.2 and 7.4*) the highest formulation of mind, pure light before its brilliance is shadowed, pure transparency before its inset colours are revealed. Light of mind is knowledge; what drives mind is will-power and desire - desire for action and creation of fresh circumstance.

A prism disperses light. Mind is the first prism of The Infinite. It is the differentiator whence simplicity devolves complexity, principle practice and themes their variations. Superlatively dynamic, this primary illumination is identified as *Ain Sof, Tao, Kalma, Shabda, Nada, Logos,* Holy Ghost and many other names. Curiously, the Latin word for 'ghost' is *spectrum*. '*Holy Spectrum*' well describes the conscio-material 'rainbow' of creation as it is dispersed from its Projector; or, in terms of informative and energetic levels, a gradient descending from on high. The Conscious Stimulus of this Light initiates existence. Existence is, in dialectical terms, an orderly series of restrictions issuing from its Source. It is a grid of transformations that derive current from a single, very powerful Dynamo. Branch (or Secondary) Dialectic derives from Primary Mains!

'Electric-in-the-grid' might seem a pantheistic metaphor. In the sense of power creating and supporting cosmic operations it is so. However just as solid, liquid and gas are clearly different states of matter you can exercise their 'phase transitions'; you can transform one state to another just by adding or subtracting energy. Outside the single phase of matter, in mind/ matter phase-transition, information and not energy is key. The valve between the conscious and non-conscious 'phases' of Mount Universe is a transformer. This transformer is an archetype, that is, an archetypal memory in universal mind. The 'phase' of matter is non-conscious, unaware and purely energetic. The sole release from such complete, informative restriction is the way awareness in a life-form might control its body parts.

If creation represents restrictions, what about lifting them? What about expanding a contracted form of consciousness? Levels of consciousness depend on the scope of unification and, therefore, experience of unity. You cannot have knowledge without consciousness; nor can you obtain full knowledge with any restriction of consciousness. 'The acme of information' does not, as seen from this angle, mean the comprehension of a bundle of restrictions or how some factual system works - even if that system is the universe itself. It involves a union with the pure substance that all knowledge 'needs that it might know' - consciousness.

Humans know this substance as First Principle. Because First Principle is unrestricted, its consciousness is unconstrained. Also, because consciousness is what we recognise as life, First Principle is both alive and the source of each lesser, restricted life. What, therefore, about a higher pass, a leap that of this planet's life-forms only man can make? 'Personal expansion' is a phrase that means the purification of my own first principle, consciousness, until the point of identity with The First Principle, Pure Consciousness. Having first surrendered a restricted, run-of-the-mill ticket labelled with my name and called self-identity, might I obtain Perfect Union with the Source of all lesser information? Would this not amount to Complete Enlightenment? Perfect Knowledge? Wisdom? And, if consciousness is life, a Pure Concentrate of Deathless Life? If consciousness is immaterial I find, at the centre of existence, Subjectivity of Light. I discover that a Real Ghost ('Geist' is German for 'spirit') is at the heart of all appearances, that is, of those outer cosmic layers made of mind and matter that I previously identified as real.

None of this is, of course, in the slightest scientific. It is psychological, metaphysical and its monastic considerations find little favour in the court of materialism - whose intriguing cynics include Lenin, Stalin, Mao Zedong and numerous lesser anti-luminaries. Since immaterial information has been the subject of Chapters 5 and 6 and psychology will irradiate Chapters 13-18, let's now turn to the inanimate quarter and give consideration to 'the best reflection of the infinite in action', the most insubstantial and yet illuminating, life-giving of all things, physical light.

Light divine is not an earthly matter; yet if archetype is nothing physical what is *least* physical, *least* massive and thus like pure, fluid energy? Light? Let there be, first and foremost, mass-free light.

Matter's Holy Ghost

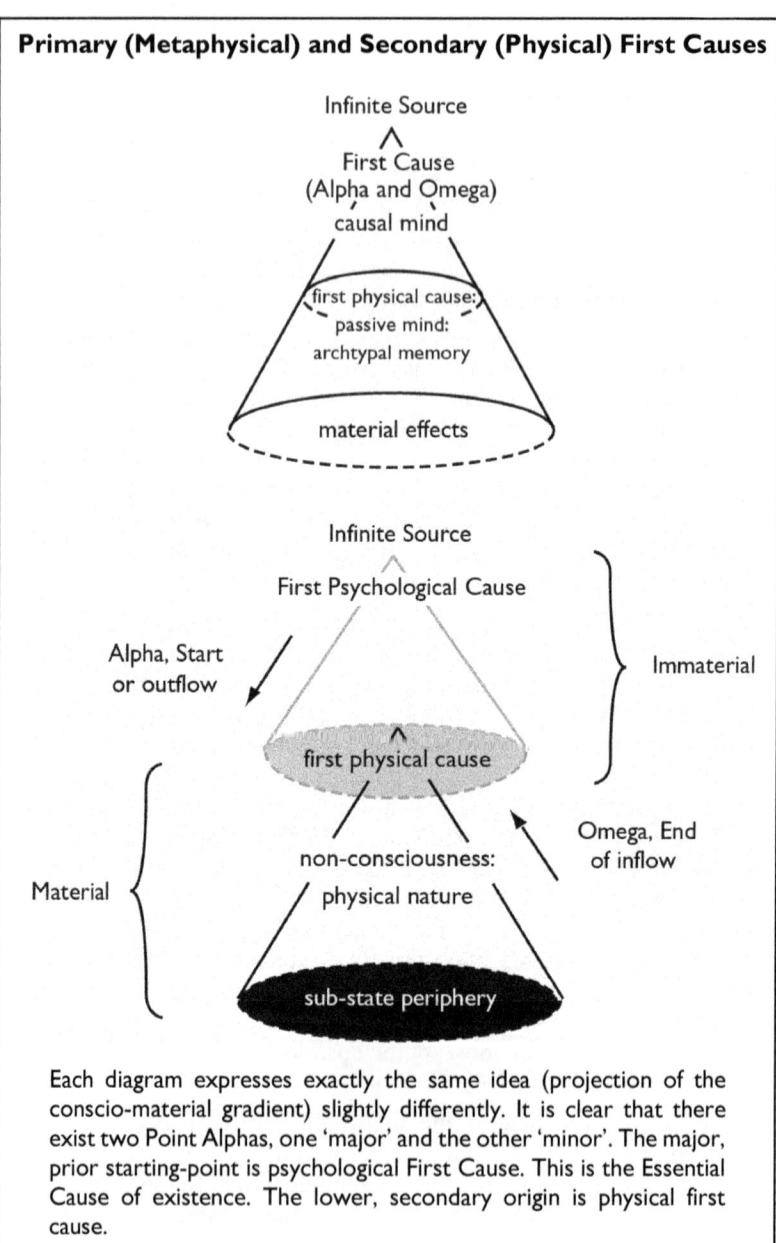

Each diagram expresses exactly the same idea (projection of the conscio-material gradient) slightly differently. It is clear that there exist two Point Alphas, one 'major' and the other 'minor'. The major, prior starting-point is psychological First Cause. This is the Essential Cause of existence. The lower, secondary origin is physical first cause.

fig. 12.1

tam/ raj	*Sat*
inertial/ kinetic	*Potential*
relative obscurities	*Clarity*
colours	*Transparency*
impurity	*Purity*

↓	*tam*	*raj* ↑
	lower/ darker	*higher/ lighter*
	black	*spectrum of colour*
	contractor	*radiator*
	slowed	*kinetic*
	dragged/ fixed	*fluid*
	particle	*wave*
	absorption	*emission*
	photosynthesis	*respiration*
	energy locked/ bound	*energy freed*
	mass	*energy*
	gravity	*levity*
	light/ heat loss	*heat/ light gain*

Pellucid radiation from the heavens was traditionally seen as both the primary expression of creation and a medium of communication straight from Creator to phenomena. Pure will informs, pure energy expresses His Chiaroscuro. If, at material level, you identify 'pure energy' with light then insubstantial light is matter's holy ghost. And this ghost's light and heat are certainly the engines that, with water, steam the pistons of biology.

Add to this perception a semitic metaphor that likens 'The Maker of Heaven and Earth' to an objective manufacturer who constructs things like an engineer or shapes them like a potter, one hand pressing from the inside and the other out. Now you have a picture of 'externalised divinity', one whose creativity shines through the objects of creation. Such an image is, from the dialectical position, misconceived. It reverses actuality. It may seem how you build a house or gun or car but not how a flower or child develops. What is the source, the nuclear seed of power? Energy and information build a system; which precedes? As well as looking from the outside any artist or inventor also, first and foremost, knows his work in mind. First cause, idea, is, like freshwater from a mountain spring. High source with clear and subtle flow precedes the murky, turgid and meandering low energy below. Solid forms like rocks or trees or teacups aren't the start but end stage of a process; they represent result from prior cause, periphery from a centrality, the last and heavy step of materialisation. They are externals where the first, internal phase is as unseen as a vacuum. Mind and its ordinations come before the shapes of gross, non-conscious matter. You will not find them outside their creation. Nor would you expect to find, in cosmic terms, a Creator anywhere in space or, wherever that may physically be, outside it. In this case divinity is centralised internally. Light Divine is first shone psychologically. **Creation issues from within**.

Check *fig*. 0.9ii. The 'Holy Spectrum' is a conscio-material gradient including both objective and subjective parts. In this view information governs energy's expression - government we call the laws of nature. Expression is a coupled show; it includes both structure and behaviour, form and action. Such coupling applies, of course, to (*raj*) light. Although arguably the 'purest', its expression is but one projection from the energetic set.

You consciously devise a scheme but where, for recall, is it registered? You file it, if there is no paper or hard disc to hand, in memory; thence you can realise your scheme at a specific time and place. Where are stored appropriate schemes of action, principles of behaviour? As instinct and habit in your memory? Whence specific patterns of response are drawn out and variation on habitual theme drives forward your life's eventfulness. Creative inspiration is divided from its cache; a memory is metaphysical, contextual but uncreative; and archetype is fixed at base of mind. Its universal chords are, like facets of an instinct, rendered automatically. Elaboration of the physical is therefore indirect. Unconscious. How can perception know oblivion or *vice versa*? How can spirit look straight out on matter? That is why we have at least three media in between - mind, memory bank (sub-conscious, psychosomatic mind) and body. Of these memory bank is (12.1) first cause physical.

From nuclear programs issue flowers and children. So do rocks and clouds and stars; their substance is an alphabet of subatomic particles whose syntactic interactions build into a large vocabulary; they write the world's text not on paper but the blank of space. This view of creation is, we said, 'organic'. *'Egg' is an apt metaphor for outworking from within.* To repeat, the cosmos is projected from a seminal conception and its consequent, logical ramifications; conceptual development is stored, like the blueprint for any construction, in memory; the blueprint's physical code is represented by an 'alphabet' of simple particles; and a starry universe is, finally, the outermost fulfilment of The Plan. There is neither external Creator nor creation external to its Creator. All is one although divided endless times. Although the 'organic' concept is clearly spelt out in John 1: 1-3, the abovementioned metaphor of 'external deity' tends to obscure it. How imaginations (though you cannot think without them) fog religious issues up!

So how do the organic, internal-to-external metaphors - eggs, electrical currents and vibratory music - work? The egg-like development of concept to fruition has been discussed at length (Chapters 5 and 6). As a single current from the mains enters an electrically driven system and is transformed to perform different tasks, so the continuous current of First Cause ramifies into an archetypal grid of events. A computer program runs its course; an engine, once switched on, purrs until switched off. This Logical Dynamo constitutes the motive power to sustain lower grades of form. At sub-conscious level 'low-voltage' *Om*, called *prana* (*figs.* 6.2 and 7.5), is the pre-physical stuff of either personal or archetypal memories. The latter are, in turn, expressed as 'matter-in-principle'; and from atomic shapes appear the orderly (according to physical and chemical rules) but often apparently disorderly, randomly distributed bulk phenomena of our world; this, energetic cosmos, is *Om*'s sink.

The uplift of song, flute playing, dervish dancing and music of the spheres are well-known metaphors that allude to starry heaven and its animated earth in terms of, rather than 'evolutionary accidents', an archetypal score. Such music, once created, is repeated mindlessly. Its record simply spins. The symphony called physical nature is played out in an automatic process sometimes likened to a musical box. Whether a musical box is clockwork or electrically driven, pins on a rotating cylinder pluck programmed tunes. Instrument and score are 'brought to life' but what 'electrifies' the system is not itself alive. A *ROM* chip is another analogy. The solid-state logic of its archetypal program and electrical current together project a 'program' that represents the computer's potential. Do you (Chapter 6) find creators in creations? Information that underwrites the universe is, by tokens of music and mind machine, vibratory and logical: but, as is the case with chips, chimes and all machines, you won't find, physically, the maker. Did you expect to? An accident's is chance; design's is intelligence. Neither unexpected chance nor the expected outcome of a scheme is ever physically found. They are psychologically found, that is, inferred.

In this view no physical energy is 'injected' into the system. **Rather latent potential is roused into actuality**. The notion of a drum roll whose intensity increases is quite apart from one of accidental levity from singularities. Physical energy, operative under basic Conservation Laws, is simply part of a system that is itself, like a musical tune, the effect of a plan vibrated into existence. So the program or score, a memory derived from the Logical Level of First Cause, raises the latent potential of matter into 'life'. Its broadcast 'quickens', its reiteration 'refreshes' the instruments of energetic pattern, the notes of archetypal memories it first established. Reiterative vibration keeps the particles 'alive'; and, at a less fundamental level, variation in their combinations is their story, is the theme of change. In the last analysis atomic components (electrons, protons and a few charge-neutral particles) are the expressions of continual harmony, the modulated repercussions of a steady roll that seems perpetual - until it stops and there is silence. Nothing left. When the current is switched off how long does music from a record player last? Cosmos straightaway collapses and is voided back the way it came. Do you remember '*mahapralaya*' (Chapter 11)? When existence dissolves into the Void?

The opposite of dissolution is creation. Pressing out. Expression. From silence concerts start. In this case what first cause of matter might appear its 'holy ghost'? What physical elaboration might be 'highest' in the chain of things? What, closest to the psychosomatic border between universal mind and matter, might you expect to identify as the top, subtlest materialisation? What, in other words, might you identify as an apparently formless, motionless, forceless, weightless form of energy? Is it transparent, peaceful and yet full of radiant potential - an insubstantial and yet most substantial physical first cause? Might physical reflection of The Top and Metaphysical Illumination not be radiant electromagnetism with its fields and foci, particles and beams?

	tam/ raj	*Sat*
	things	*Nothing*
	consequences	*First Cause*
	spectrum/ scale	*Apex/ Top*
	colours/ shades	*Transparency*
	expressions	*Potential*
	visibilities	*Invisibility*
	issue	*Source*
↓	*tam*	*raj* ↑
	matter	*energy*
	fixity	*stream*
	darkness	*brilliance*
	heavyweight	*lightweight*
	absorbent drain	*free flyer*
	life-taker	*life-giver*

Paradoxical light! That comes in rays, radiance, waves or particles depending how you treat it. Transparent but its purity is mixed! White light is a mix of three pure colours - from high to low (*sat*) blue, (*raj*) green and (*tam*) red. From its purity a prism prises rainbows. As light disperses colours from a raindrop or a block of glass so natural science has prised information out of photons; it has lit with understanding a delightful, vivid world. It has earthed, through correlate transparencies, its central principle, its leading light, its non-divinity. What do I mean?

You are granted eyes. You see because informative light streams into but not, as the Greek Empedocles thought, out of them. You can watch light bounce, reflect, refract and feel it, having been absorbed, converted into vibrant heat. You can experiment, calculate and work out laws. Pure light, luminescent plasma, incandescent gaslight, 'fluid transparency' (called water) and, down to earth, the 'solid light' of diamond, crystal and of glass - you can use them all to find out more. Modern science broke from glass. It ground out mirrors, lenses, microscopes and telescopes of increasing power to extend and thereby influence the human way of seeing things. And don't forget photography or television - imprints made of light, images like memories and films to fix the carry-on. You could almost trace the history of science down a ray of luminosity!

Since at least Assyrian times men knew that lens-like glass could magnify and focus to make fire. The Hindu schools of Samkhya and Vaishesika developed theories of light. So did Aristotle, relating optics to geometry and vision. Euclid and Alhazen (Ibn al-Haithum) rightly disagreed with Empedocles. They, along with Ptolemy, Christiaan Huygens (with his wave theory of light) and Sir Isaac Newton (with his corpuscular theory) all penned 'Optics'. In the 18th century Pierre Laplace even postulated the possible existence of light dungeons now called 'black holes'.

Early in the 17th century Zaccharias Janssen, Antony van Leeuwenhoek and Robert Hooke developed microscopy; and the aforesaid Janssen and Huygens, along with Hans Lippershey, Galileo and Johannes Kepler manufactured instruments of teloscopy that drove the large-scale scientific vision forward. Rayleigh, Young, the Herschels, Talbot, Maxwell, Planck, Einstein, de Broglie and others have taken up the torch. How many great names can you cluster on a photon? Photons are the ethereal ghosts behind the great men's reasoning. They are a mass-less, speck-less nothingness that, bearing fire, set everything alight. Electromagnetic radiation, nearest to non-physicality, is at once supreme abstraction at the heart of fullness; it ripples in a clear void ripping space apart and filling up the opening with motive power. For the Hindus light was one of several subtle (virtual?) elements from whose energy massive phenomena derive; certainly what is virtually a ghost, a phantom flitting round the edges of the world machine is shining at the heart of life. Brilliant construction. Photosynthesis.

Thus light, matter's 'holy ghost', is perhaps the principle with which you start biology. What part does light play in electronic operations at the bonding base of chemistry? It certainly, spectroscopy agrees, allows you to identify your elements and molecules. It tells you all about far stars. What purer form of energy than light? Thus physics, whose study is at root just one of energy, might take a shine to start with. A book of physics treated dialectically would throw its light upon light's levity of character in Chapter 1.

	tam/ raj	Sat
	thing	Vacuum
	product	Source
	extant pattern	Field
	motion/ tilt	Balance
	polarity	Neutrality
↓	tam	raj ↑
	minus	plus
	mass concentration	light projection
	confinement	liberation
	grip	release
	fixing field	radiant field
	particle/ mass-form predominant	waveform predominant
	gravitator	radiator
	gravity	levity
	magnetism	electricity
	locking energy	connective energy
	isolator	communicator

Those tiny shiners, photons, don't weigh anything at all. Their rest mass is, strictly, zero; but, since photons never lounge about and (unless transformed to specks of heat) no such thing as an immobile sort exists, there is always some momentum and light-waves exert some pressure. Radiance is therefore almost nothing making clear the clarity of space. Yet this same spectre that irradiates the vacuum is, as any laser expert will confirm, a substance that can, focused, cut through steel. It looks motionless enough. It seems all peace and quiet. You'd take it for transparent absence but its presence speeds through nothing faster than a flash. On its way the pulse relates, communicates and penetrates; it colours, heats, stimulates and activates. Light's beneficence falls as a nutrient, as an enrichment pumped from the dynamic heart of stars, a stimulant - the lifeblood of the universe. For earthlings sunbeams are umbilicals that, through a double subtlety called photosynthesis and respiration, cord us to the sun and recharge all our batteries. Is not light in fact a kind of sacrament? Devotees of sun gods, fire-worshippers and, more prosaically but with precision, scientists recognise the power-giving, life-officiating possibilities of positive potential. This illumination is communicated in the form of formlessness and, dialectically, is noted as the top, transparent flux of matter.

If energy is polar it involves both contractive (↓) gravitational and radiant (↑) levitatory vectors. However (*raj*) levity tends towards (*sat*) purity; its grades ascend towards a higher level, towards a source that both informs and stimulates what is below. What, therefore, might best represent the 'egg' whence issues physical nature? What might logically express the purest form of matter better than the brilliance of quicksilver known as light? Light certainly represents a strong predominance of (*sat*) and (*raj*) characteristics at the same time as a virtual absence of (*tam*) mass-centred, inertial ones. In this sense it is, dialectically, the first and highest physical phenomenon. As primary, 'embryonic' matter it is, sited at the psychosomatic border between metaphysical and physical grades of existence, a good candidate for the primary, informative projection of physical first cause. That is to say, the neutral, uncharged photon is identified as the top-level or original, causal form of polar matter. Its actuality is based, some claim, on theoretical, wraith-like 'virtuality'. 'Virtual' light is supposed to reside in the metaphysical zero-point vacuum-field. Does this simply mean that the vacuum has a potential, like that of a battery to generate a current, to emit light? In other words, a particular sort of activation will transform the vacuum's field potential into kinetic, local actuality. Such actuality is called a photon, a stream of photons or a beam of light.

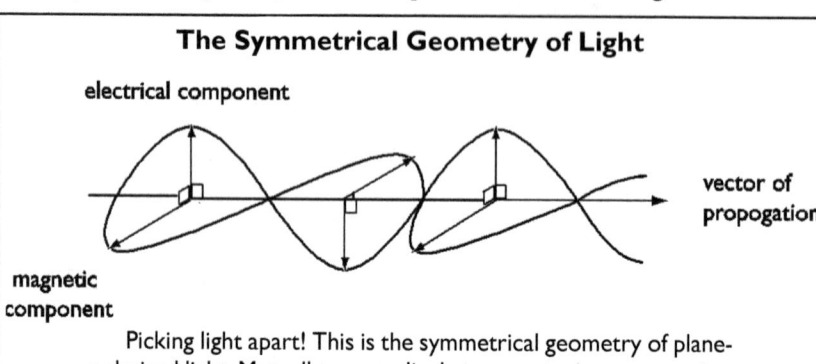

The Symmetrical Geometry of Light

electrical component

vector of propogation

magnetic component

Picking light apart! This is the symmetrical geometry of plane-polarised light. Mutually perpendicular symmetry is an expression of perfect dynamic balance - in this case between the oscillations of electrical (radiant) and magnetic (contractive) field components and their straight-line 'normal' of propagation. A non-polarised/ neutral beam of light travels along a 'tube' of electric and magnetic undulations in the complete range of planes perpendicular to the direction of propagation.

How can you spot a beam not entering your eye? Or spot a photon flit across your line of sight? Look up 'sunburst monstrance', an implement used to display a wafer called the host at Catholic mass. A ray of travelling light emits 'sunbursts'. It emits coronas of more photons as it flies. Does that mean photons from a photon? And thus free light and heat? Is something shining clearly that I just can't see? You can prove all-round effulgence but completely miss an instantaneous acceleration from zero to full speed in no time as a shot of radiation quits its point of origin. And if mass approaches an infinity as speed approaches 'c' how does light obtain a limit, a fixed maximum velocity? Because if there's no mass why should there be restriction? If you were able to pipette a photon to a pan its weight just wouldn't be. No grams. No resting mass. With this kind of statelessness could there exist a hint of timelessness as well? Is this another side to absolution - absolution that makes light the measure of all relativity? Which would make it a paradoxical if not an impossible phenomenon!

fig. 12.2

Perhaps, therefore, light is the most (*sat*) balanced, supremely energetic and informative expression of the (*raj*) mass-phobic principle of levity. The geometrical symmetry of its potential is remarkable. It includes three ways in which infinity is implicit but restricted. There is a point, the photon; and straight-line, the potentially infinite extension of a point (or photon) to a ray. These are the components of 'inertial' or 'flat' shapes. A point can also trace a circle; it can roll around forever - infinitely captured in a ring. A ring's a wheel. A wheel can also spin forward in a line - in which case it is drawn as a waveform, oscillation or a vibrant potency. What about a composite, a beam or halo, of innumerable rays? Above all brilliance propagates 'in volume', radiating spherically from its central point of origin. Like 'holy ghost' from Origin? In which sense each candle-flame is a mini-image representing cosmos.

Waves of light are a function of precisely coupled electrical and magnetic components. Light subtly vibrates; vibration is a period, a cycle, the form of cybernetic feedback that obtains a norm. Is radiation's vibrant cycle not as crisp, neat and regular as any in the universe? While un-polarised light has electrical and magnetic vibrations in all planes perpendicular to the direction of propagation, plane polarisation singles out the underlying three-dimensional geometry. Either way, however, light is composed of an electric field coupled at right angles with a magnetic field, both at right angles to the direction of travel. Through the polarity of these balanced fields light couples (or photons interact) with matter. The wholly symmetrical extent of electromagnetic coupling was demonstrated by Faraday (that a changing magnetic field creates an electric field) and Maxwell (that a changing electric field creates a magnetic field). Electromagnetism is, therefore, a product of electricity in motion. In particular both non-uniform motion (acceleration or deceleration) of electrons and their shift to a less energetic atomic orbit generate waves we call light or, equally, streams of particles called photons in a local electromagnetic field. Nineteenth century Maxwell had, at a leap into the void, linked three intangibles. He linked particulate charge (an electron) with a field expression, light; and each with magnetism. From his seminal equations, from his *magnum opus* the twentieth century's age of radiation and electrical appliance dawned. It is no exaggeration to suggest that, out of ghosts of nearly nothing, one man laid concrete but intellectual foundations for our modern, international way of life.

Controlled, structured waves make light of space. Intangible illumination flies at the top edge of physical existence. However, while it can be argued that light is the highest, least material form of pure energy, it can be brought to earth. What is heat but 'grounded' light? Light is absorbed by bulk materials and converted into heat for work (the word 'energy' actually means 'there is work in it'); conversely, what is temperature if not a measure of trapped energy, internal motion and heat transfer? In a way that might be called 'emphotic' ('there is light in it') all objects above absolute zero (0°K) emit radiation. And while it is the vital element in a trinity of transparencies (space, time and pure energy), electromagnetism is also a factor in electronic changes of state and those electrical handshakes known as chemical bonds. From chemical bonds derive syntheses which give rise to all kinds of irregular shapes or geometries; by this form of electromagnetism are fixed the chemical properties and behaviours of our variegated world.

What more than light informs? Sight illuminates the mind. Each object seen reflects its story to the eye. And, in principle, at elemental level it absorbs, reflects and radiates its own barcodes, spectrographic fingerprints that tell us what it's made of. It doesn't matter if it's next to you or parsecs removed, by light you know. At the same time, with either coherent or incoherent beams, light can make or break materials. In which case shakes-in-nothing can embody, in a symbolic or a scientific way, the shapes of energy, of mass and their activities. Insubstantial radiation is a measure, on a universal scale, of all fixed substances and of flux unfixing them.

How Absolutely Holy is the Ghost?

	tam/ raj	Sat
	relativity	Absolute
	polarity	Neutrality
	shadowed	Pure
	subsequent order	Archetype
	proton /electron	Photon
↓	tam	raj ↑
	negative	positive
	passive component	active component
	less pure	more pure
	inertia	excitation
	fixity	(re)action
	proton	electron
	confinement	liberation
	particle/ mass	wave
	gravity	levity

Does inanimate light reflect the higher, animate wavebands of the Logical Spectrum? If Essence is Absolute, how apparently absolute is the candidate for matter's most elevated form, the sublime pre-eminence of light?

While it may be argued, by Einstein for example, that the speed of light in a vacuum is absolute, such velocity can diminish according to the substances it travels through. But its acceleration *on emission* in a vacuum (0 to 2.1098×10^8 ms^{-1} in no time) appears incomparable, indeed almost infinite. Could you call it absolute acceleration?

Although they apply to only about 4% of physics' modern universe (not dark energy and matter, quantum *ZPE* or virtuality) we saw, in Chapter 7, that it is frowned upon to question Conservation Laws. It is, equally, rude science to suggest a *VSL* (variable speed for light in a vacuum). Such impoliteness represents, even though Einstein himself considered it in 1911, a heresy. This is because the special theory of relativity is based around an assumption of the constancy of 'c' (as the speed of clear 'nothingness' in dark emptiness is dubbed) and a correspondingly flexible space-time. In other words, a suggestion of *VSL*'s sort seemingly tweaks Einstein's theory and, by implication, shin-kicks a foundational leg of modern physics. In other words, the constancy of 'c', as in $e = mc^2$, is fiercely defended because so much depends on its conservation. If, in science as philosophy, a first principle or foundational assumption errs then all that follows, however logically it follows, is in basic error too. As a structural foundation loosening 'c' would have far-reaching ramifications affecting the whole shape of the relativistic way that science frames nature. It could dislocate the spine of modern physics since it would throw a basic staple of Einstein's great theory into doubt.

The speed of sound varies according to the medium of its transmission (e.g. water or air); and its apparent speed varies according to the motion of its emitting object (e.g. an aeroplane) relative to the motion of its observer. Thus if sound is, say, reflected off a wall by a vehicle approaching it, the apparent values of its speed will vary in each direction. You might or might not care to take an average; and if the they seem to be the same, say of sonar echo from a cliff, then both directions take the same time but you don't therefore presume the speed of sound to be invariant.

Due to its velocity ($c = 2.1098 \times 10^8$ ms^{-1}) the speed of light in a vacuum is difficult to measure except using a single clock and bouncing it a sufficient number of times between mirrors to allow a readable measurement. But is its speed the same each way? Because when measuring you simply average it out. Its speed in different media (e.g. water, glass or space) varies but does it vary with respect to variations in the density of space itself or an observer's motion? It is assumed to be the same whatever the motion of emitter or receiver (say, a star and earthbound astronomer) anywhere in the universe. In this case you could ignore an ineffectual element of 'ether'. Perhaps ether isn't even real but certainly it doesn't matter. The consequence of this assumption is, however, that while you fix the speed of light all other measurements (e.g. of time and space) would vary as an object moved. If an absolute frame of rest (say, the ether of space) is abandoned, then all other changes are relative. With no fixed frame of reference you have, of course, 'absolutely' relativity. These assumptions and especially the one that's central to them, the irrelativity of light, run counter to one's natural intuition. Are they actually and universally correct?

It is fair to surmise that original 'pre-bangian' conditions were disturbed by a different cosmic concentration of energy from that prevalent in the freezing space of today's cosmos. Indeed, any force or property not yet operative would have had a latent, null value - null but (*sat*) potential. And as a highly energetic vacuum converted into matter levity (the expansive energy of space) and gravity (the anti-expansive, gravitational, gripping potential of matter) may for a short, initial while have had very different values from the ones observed today; from null to extremely warped, initial properties and structures of the vacuum itself may also have varied from those settled into now. Large-scale, relativistic space may kink and warp but is it empty? What, if it kinks, is kinked? In Chapter 10 the constitution of space was discussed. What is a vacuum made of? Is it actually all nothing? Or is it made of Higgs fields, Planck particles, arrays of 'virtual' charge, 'quantum foam' or granulations like the pixels making up a crystal screen - in which case it is not empty. Consistency might vary and the denser *ZP* radiations the more the vacuum might act like a medium, say, an optical medium. Light varies speed in different media. It bends or 'refracts'. And if, of course, the 'quantum foam' caused friction then you'd expect c varied with a wavelength's energy; or might non-emptiness of vacuum similarly slow light down by amounts that vary in proportion to the density of *ZP* force or, call it by another name, the ether? As such factors fluttered so would variously fly the speed of light. At any rate, why should today's recording, set at 2.1098×10^8 ms^{-1}, hold absolutely as 'gold standard' for all time, every place or quality of space?

What about illumination, in a 'young earth', from a star more than 6000 light-years away? We know, also, that measurements from galaxies 11-12 billion light-years distant show the velocity of light has stayed the same. What, though, about the very start? What if the universe in its initial (*raj*) short-time were of a very different energetic circumstance? In this case it might be fair to suppose that while principles do not change, the values of their expressions, measured in terms of constants, could. For example, as parts of a car speed up or slow in concert, might not cosmic constants vary their 'revs' in a coordinated way? Why then should the *speed* (as opposed to the principle and the actuality) of light have always been the same? Or perhaps, like the speed of a car as opposed to its construction, the speed of light is not a fundamental constant anyway; could it, even in a so-called vacuum, be dependent on its medium? The biological development of a child is, at first, sprung very fast but decelerates to zero at maturity before a gradual decay sets in. Could an 'organic/ holistic' universe have sprung fast through its first, scintillating episodes? Could an expansive dawn with low densities of vacuum energy have now settled into 'afternoon' so that, as a medium, 'thickened' space now slows light down? Could such (*tam*) 'drag' by vacuum have slammed on the brakes in just the first few seconds so that its speed dropped exponentially from 'infinite' or almost infinite initial velocity? Does light nowadays proceed at uniform velocity related to the spatial structure of a vacuum at low-energy - the level of our cosmos? Or is it gradually, in long-time, slowing down? In other words, is cool space 'thicker' than hot space? When might it stop and, since light is nothing without motion, utter darkness then prevail?

What, furthermore, would be the consequences of a decrease in speed? For a start it would mean the universe that we think has been here 'as is' for billions of years (and *must* have been for human evolution) may not have been. Mass would have been a lighter business; lower viscosities, faster heat transfer and electron movement might have combined to hasten geological activity. The red shift would have to be reinterpreted, the half-life of radiometric decays recalculated and the age of the universe reduced accordingly. It is possible that the rate of photosynthesis might have been higher with consequent effect on plant sizes. Lower viscosity and faster rates of diffusion would have improved many biological functions. Insects could have grown larger and lift-to-drag ratios improved all flight including that of, say, pterosaurs. In the same vein gas exchange, blood flow and nervous/ muscular movements would have quickened. Large bodies would have functioned more easily, lightly and efficiently. Although fossils of giant relatives of extant organisms may be adduced to support such claims they detract badly from the theory of evolution. Indeed, the higher the initial speed or greater the decrease in the speed of light the younger the universe. The more dramatic the reduction in the age of the universe the worse it gets for an eon-hungry theory. Accordingly the notion of an exponential *VSL* is, for any materialistic paradigm, unwelcome.

Could it anyway be true? The fine-structure constant measured in far-distant (therefore historically very old) galaxies yields a value for 'c' practically invariant from today's on earth. And, to further dampen any 'young-earth' glee, the Milky Way's disc diameter of 120,000 light-years implies at minimum that age. How great an age might space across the universe imply?

Yet it is claimed that *VSL* might solve intractable difficulties that confront those experts in the art of living dangerously - big bang expansion theorists. In 1987 an astrophysicist (V. S. Troitskii of the Radiophysical Research Institute, Gorki, USSR) suggested that the speed of light *is* variant. It is decreasing. This would explain why, puzzlingly, very distant galaxies appear to be receding at speeds greater than light. In fact several theories of 'c-decay' have been mooted. It has been suggested (by J. Magueijo, J. Barrow, J. Moffat, M. Clayton and others) that, if you break the sacred rule of an invariant speed of light, cosmology's big problems might, in part at least, just melt away. **These problems include an uneven cosmic distribution of matter, a lack of deviation (that is, a balance) in the critical divide between an expanding and a contracting universe, the reason for mass (Higgs proposal), the nature of a force (called 'dark energy' or lambda) suspected to be the cause of super-luminary rates at which very distant objects are observed to be accelerating from us, and the reason galaxies form and spin as they do (for which an equally exotic, unobserved, hypothetical material dubbed 'dark matter' has been proposed).** However, John Moffat for one has said that he finds it easier to question Einstein's theory than to assume such exotica inhabit his kitchen. Will various claims for a long-term deceleration in the speed of light, made since the 1930's (e.g. Nature 4-4-31 p.522) and controversially repeated today, prove true? Might light fire physics once again?

Indeed it may. Watch out! Too late! Perhaps tachyons are back! In 2011 neutrinos were caught cracking Einstein's speed limit while travelling through mountains from CERN in Switzerland to Gran Sasso in Italy. In fact, some piece of kit was probably to blame. But of many explanations for the 'experimental anomaly' the one that it might be true leaves physicists in a disturbed and contradictory 'quantum state of mind' - both highly excited and highly resistant at the same time. 'C' is an absolute constant round which current understanding of the whole cosmos is constructed. If such constancy even wavered then $e = mc^2$ and all the rest might be wrong! Strong scepticism therefore rightly rules the day. The unthinkable need not, yet, be thought nor any textbook be rewritten; but it would be ironic if, at the same time and using the same machine as experiments to find a Higgs boson and thereby iron out inconsistencies in the standard model of particle physics, that same model had from an unexpected direction been lethally exploded.

Whether or not *VS* sheds *L* on cosmological problems, temperature if not illumination has been further increased by a compound heresy dubbed *DSR* ('doubly special relativity'). In this theory G. Amelino-Camelia and J. Magueijo propose that *two* impassable thresholds 'seal' the physical universe. These are 'c' of value higher than the 'low-energy' one familiar to Einstein and the rest of us; and an 'absolute' energy or, in terms of waves, length. What is not clear is whether this idea will blossom clear of Einstein's or, if it does, where it will lead. What *is* clear is that establishment physics remains unthawed, cosmology is still young and the subject is a tangle of esoteric, conflicting mathematical models and associated vocabulary that changes almost annually. What will things look like in the year 2200? Will you have to look at them with slower light? Will holy 'c' have actually slipped its halo? At present this suggestion mostly causes physicists to grit their teeth.

Grit in the Ghost

	tam/ raj	Sat
	polarity	Neutrality
	fermion	Boson
	hadron/ lepton	Photon
↓	tam	raj ↑
	minus	plus
	hadron	lepton
	Higgs field (?)	anti-mass field (lambda?)
	quark/ proton	electron
	massive/ bulky	lightweight/ lifting
	confinement	liberation
	binding/ capture	freedom/ radiance
	particle form predominant	waveform predominant
	locked energy	connective energy
	fixity	mobility
	isolator/ objectifier	relator/ communicator

There's more to light than meets the eye - not only, from its range of spectrum, powers of clear invisibility. Maxwell and Faraday linked electromagnetism (of which light is a radiant form) to electricity and, thereby, to negative electrical charge whose only carrier, as far as is understood, is an electron. Because it is a product of electrons in accelerating mode or shifts in atoms' orbitals, light does not just happen. It depends, except in the drastic cases of atomic fusion or material annihilation, on electrons. No doubt the *effects* of electricity are precisely understood and practically all the mechanical, electrical and magnetic properties of matter (and therefore chemistry and materials science) are based on electromagnetic interaction. No doubt also that electrical circuit theory is a detailed and important abstraction behind the design of electrical devices. What, however, *causes* charge? What *is* its point of action; what, next heavier than a photon, is grit in the ghost called an electron? How in the first place did charge come to be; what set up electrical polarity? What, next heavier than a photon, *is* Higgs' first grit in spectral ghostliness called an electron? And, finally, is an electron elementary? Could it be broken into quasi-parts each fading nearer to the mist of really elementary nothingness?

It is argued, dialectically, that matter is the product of principle and, as such, primary particles are the agents of character, of basic constants, of the neutral and polar cosmic fundamentals. In this case there is no doubt that, as opposed to the 'left-handedness' of a proton, (*raj*) 'right-hand' characteristics predominate in an electron. Neither chargeless nor mass-less like the (*sat*) radiance of light, an electron is still 'levitatory' in character - an activator, mover and informer. Its spin, its biased charge is an embodiment of signal, a communicator, a carrier of electric that sparks a primitive message. 'Yes, come' or 'no, keep away'. It either repels, separates or isolates - like charges repel and make a space between them; or it attracts its opposite and binds - it is the glue that underlies all chemical difference in a compound, complex world. It is not padlocked like a proton at the centre of a mass-charged zone - it only skirts in orbit round the nuclear heaviness. Nevertheless, an electronic speck of charge is an important differentiator. It is fine grit in the smoothness of a ghost composed of light and space.

For Expansion Theory uncharged electrons are the lone, fundamental particle. The theory's universe is composed of spontaneously created, ever-expanding masses of them. Electrical charge, light and all other phenomena are explained as different forms of their behaviour. For example, light is not a wave phenomenon at all but simply continuous clusters of expanding electrons that together make up beams. The more clusters in a given length, the higher the frequency; the larger the clusters the lower the frequency. Everyone agrees that an electron is an adamantine wave; it is indivisible, un-crushable and unconvertible but otherwise Expansion Theory receives short shrift from professional physicists. What, therefore, is the latters' definition of 'electron-ness'? Is it a spot of glue that clinches atoms into permutations, molecules, and thence the larger aggregates we see and touch and on which gravity can start to bite. At temperatures like those on earth it grants molecular, chemical and macroscopic properties (texture, colour and so on) to matter. An electron itself has mass, charge and spin but what *are* these properties? Curiously, while its effects are common knowledge an electron's *being* is not well construed. Is it a microscopic kind of billiard ball, trapped vibratory energy or a 'node of buzz'? Is its shape of 'orbit' round a nucleus like that of a moon around its planetary mesmeriser or is it better known, like an air on a guitar, as a melodic pattern made of standing waves? Is it a channel sucking inexhaustibly upon the energy of vacuum or a point-like, vibrant 'smear' or 'smudge' of charge? Is it some or all of these? How (see also Chapter 9: Split Flow) is an electron made?

The atomic absorption of light causes an electron to shift up a 'gear' and, conversely, a change down 'sparks' photons. Such emissions are not as energetic as gamma rays which, when they pass through matter, can create an electron-positron pair (a positron is a positively charged electron) but cannot, due to conservation of charge, create an electron alone. So where does an electron come from? Just as electronic motion generates light so, dialectically, does the annihilation of polar matter. Could a reverse-annihilation, creation, make electrons? It can. High energy physics, seeking to simulate conditions in a star or perhaps the early universe can make them in a process called quantum pair production. How does the charge that is produced adhere to an electron or its polar opposite, a positron? Is it separate at all? If electrical polarity can ever be untangled what exact connection might charge have with neutral light?

The thing is, an electron or a proton has, like every other piece of grit in nothing, mass. How, though, would you create grit out of ghost, how erect a massive world-stage out of nothing and then set it up without support? What is the argument about? Tensions in mind about tensions in space? No doubt that things in space need, as their localiser, mass. In Chapter 10: Gazing into Space getting something out of nothing held the focus of our interest. Now it does again. Mass-production's problems, that Higgs bosons and inertialising fields should help explain, are embedded at the heart of Standard Theory. The *LHC* may well confirm materialisation's method but such theory still does not explain why Higgs field is, conveniently and critically, inlaid at the root of physical creation; and thus primordially dovetailed with all following fine-tuned factors.

We have danced and pussyfooted. We have beaten round the almost-bush of 'virtuality'. What does it mean? Does 'virtually' translate as 'nearly'? What, therefore, are 'nearlies'? Are so-called 'virtual particles' a function of the *HUP*? Or, in a more certain and less virtual interpretation, is an 'in-and-out-of-being particle' the inevitable effect of fluctuations in high densities of *ZPE*, conservation of charge and the fact that energy and mass are inter-convertible? Can you swallow such a 'flipper' as (Chapter 10) the random zest a space, any space, is always throwing up? What's the proof that nothing's actually extremely busy?

'Virtual' means, according to the Routledge English Dictionary, 'being in essence or effect, not in fact or name'. This could almost be, unwittingly, a dialectical deployment of the word. 'Virtual' mind is consciousness without a thought; 'virtual' mind is therefore Essence. 'Virtual' matter is the archetypal field, potential matter, metaphysic that precedes its physic, world-in-principle before expressed as fact. Thus thirdly and as science would not say, is 'virtuality' synonymous with dialectical 'potential' (*figs*. 1.4 and 5.2 (ii)), 'super-matter' (*figs*. 2.5 and 8.2) or the physical first cause (*figs*. 7.2 and 12.1)? In other words, is it a stage in materialisation, a level on Mount Universe - the psychosomatic level (*figs*. 6.1 (i), 8.2, 10.2 (ii) and 15.2) of 'cosmic grammar', an ideal record of the cosmic program lodged in memory in universal mind? Never, in such a case, could 'nearlies' realise a random outcome even though the actual effects might seem, like sparkles on the sea's waves or motions in a gas, too unpredictably complex to assign reality that's not statistical. Yet is not sea-sparkle real enough? Are not motions in a gas all lawful, straight and, if you could measure each and every one, not random in the least? *And once the possibility of local chance evaporates you might ask how a general and universal form of chance was able to create the interlocking rules.*

In this view vacuum waves a blank cheque, its bankroll is the first to set creation's business up. First cause physical lays out the way. It is a contract as yet undelivered, an exciter drumming at the skin of musical existence, a current modulated through a set of one-note instruments. These instruments are vibrant archetypes each of whose instruments 'spins' out a certain type of particle. In which case the question changes. Why should 'virtual particles' occur at all? Only actual sand-grains bounce on vibrant skin, only actual molecules translate a sound through air. Jostling happens but the jostling-frames are relatively fixed. Motion is constrained by its parameters of type; energy is varied only in invariant themes. In this sense random fluctuations disappear. Indeed, is every force and particle not, in this end-analysis, a twist, a flip, vibration or 'an excitation' that shows orderly in space? Whose various separate knots or nodes or field-like influences show up, like notes of scale, according to their archetypal frequency and therefore shape? They are, in fact, conversions out of the template we call 'virtuality' (or dialectical 'potential') into what from the very start of things must have been there - electrons, protons and associated durability - the forces. And, of these localised expressions of the forces, particles, some show up more like 'field workers', joiners or (*raj*) relaters - say, electrons. Others, like the coarser grit of protons, characterise (*tam*) apartness and promote the equally necessary insulation, separation and individuality of things. Fixity and flow, discontinuity and continuity - can you spot music in the making? Counterpoints? Non-accidental rhythms we define as laws of nature and whose patterns we observe as things? The question then becomes one of initiation. What, in the first place, made up definite, functional differences? What created integrated suites of archetype, the language of our being? This is a question that the principles of Natural Dialectic as a whole address.

Now, from a different angle, light becomes a strain, a field of fluctuating tension pulsed by charge across material nothingness, across an immaterial sky. And captive energy, a 'knotty kink' in space, is called a particle. Metaphors abound. You might see an electron as a 'lasting blip', a ring of spin, a vortex 'budded off' an underlying quantum sea. Is this how you polarise a vacuum? What might be interpreted as a constant flux of 'virtual particles' is, translated to the Dialectic's terms, a perpetual vibrancy and motion (both of 'spin' and atomic orbit energies) caused 'from the other side' by archetypal forms of energy. This underlying (or, as Natural Dialectic would phrase it, overlying) psychosomatic energy is called, for want of a western equivalent, *prana*. From *prana*, in this archetypal line of logic, spring the properties of mass, spin (with associated magnetism) and polar charge wrapped in dual aspect of continuity/ discontinuity, togetherness/ apartness, wave and particle!

Look at the dynamics in another way. Imagine three-dimensional space as, for a moment, two-dimensional. On this flat surface virtual particles are like solitary waves of swellings, bumps or knobs in water. These fluctuations ripple out of only-guessed-at subtleties; hard worlds are seeded, rooted and informed by virtual unreality. If this 'unreal potential' is an interface then particles (such as electrons) might be seen as points absorbing, radiating, re-absorbing - simply resonating with its different layers, characters or fields. In other words all particles and therefore all things made of them draw, as in pregnancy a child draws from umbilical cord, their nurture and life-giving strength from the blood of 'emptiness', from mother vacuum. For a crystal, mountain, star or a body such as yours or mine, vacuum is the sustenance. In this sense you are one with mother. Indeed your body is entirely ghosted by the compound grit of atoms knitted in the overarching vault of space. All things are thereby practically one with vacuum and its unseen, metaphysical nutrition. It was not mother earth but mother vacuum's empty womb first bore, along with every other body's, the constituents of yours.

Call it 'potential matter', vacuum fields or 'virtuality' but also check with *fig.* 11.3. This is the stage, grade or era of pregnancy, potential and creation from the world egg - archetype. It is a genetic step first taken as the start of things but which continues now. It sustains the natural cornucopia. Electricity seems everlastingly conserved if you can't see the spinning generators. What about a drum roll if the drummer is behind the scenes? If you don't include the source of its vibration isn't sub-atomic motion and all that springs from it apparently but inexplicably conserved? Conservation stands but is sustained. All physical objects, behaviours and events flow from and are supported by this womb of worlds, an ethereal, effervescent spring of 'nothing'!

Strange, isn't it, when different things begin to sound the same? A picture of what you can't see with your chemically constructed eyes but only in your mind's eye is a model or a metaphor. In this case, at the insubstantial level, different descriptions merge as pictures of the same thing snapped from different angles. One aspect is related to another. Even in substantial physics one transformation leads into the next and explanations called equations lead you round the universe. Kinds of energy are interlinked and also, from atoms to the galaxies, the majestic yet flamboyant show revolves. Do not be surprised if, in describing such a carousel, your models of its operation also cycle in and out of style!

Actually we're revolving in the dark not standing, axial, in the light. Standard Theory has neither an all-inclusive theoretical model of an electron, of any other subatomic particle nor therefore how even simple atoms cohere. We're happy in the middle but not at those extremities called starts and finishes. We're not too keen on the approach to endlessness. Through a dot towards infinitesimal infinity or through expansion to a boundless, unconditional extent first causal origin and final destiny are each the same. Things there slip away. They dissolve. We start to grasp at nothing, guess and speculate. Such speculation, like the universe itself, is built on an abstraction. Vacancy. Have you ever stamped, in space, on outer space or pushed on our imperfect vacuum that supports, in every sense, the cargo of all physically manifest? In its profound and obvious yet hidden spring there rests our theoretical (or virtual) as well as actual being. Its unity begets phenomenal duality; its potential generated and still bears 'plus-and-minus' charge-fields and their motion. Every chemical attraction or repulsion, each kind of bond involves them. 'Virtual photons' swarming round electrons are, in this sense, at the roots of chemistry. Thus photons, electrons and electromagnetic forces hold the world together and (if you try to squash an atom flat) apart. You might add that gluons in the vacuum acting over nuclear dimensions organise the masses of a proton, neutron, nucleus and thereby all bulk matter. The stability behind our transient solidities is almost immaterial! This massive world rests on foundations made of, virtually but actually as well, space.

Productive calculations have, with mathematical sleight-of-hand, seemed to conjure charged and uncharged matter out of nothing; on the other hand, since the virtual/ actual mobs of particles buzzing an electron or a nucleon have (like strings and quantum vacua) never been nor may ever be directly seen, the basic business plies unspied. 'Virtual' fizz and theoretical 'fuzz' froth existence into bubbles of both mind and matter. Froth or body, is your cosmic cappuccino poured by chance?

Ex nihilo (or, if you like, *e vacuo*) we have dropped from 'up there in the gods' onto the stage of physics; we have materialised the playground of oblivion. Here dance - according to kinetic theory of vibrating matter - hard-nosed objects. Here plies grit in the ghost, dust of the earth that you can grip both technologically and mathematically. Photons and charge (embodied in electrons) are, according to the scheme of Natural Dialectic, the 'highest', least material expressions of materiality. As such they may logically mediate between mind's metaphysic and its massive aspect, body. How this might happen is described in Chapters 15-17.

The order of an embryonic universe is therefore radiation first - neutral, uncharged light expresses (*sat*) radiant unity. Next lightweight electrons are the nodes of charge, carriers of the (*raj* ↑) active aspect of polarity. Last of the phenomenal trinity, hierarchically if not in time, plays bass. A proton is a dab of heaviness, of localising influence, of 'drag'. It is an effective spot of highly concentrated grit. Its (*tam* ↓) 'gravitational mass' is first to really bring you down to earth; protons are the final nail in nothing's coffin, a kind of 'mini-black-hole' inside space that constitutes, with neutrons, the substance of inanimation, an atomic nucleus. 'Nucleus' means, literally, 'seed'. Nuclei are seeds round which the weighty cosmos partly aggregates, disintegrates but through it all endures. - you can neither make nor break a proton or electron. What, therefore, sustains the couple's adamantine, universe-long dance? In stable times - outside a star or synchrotron - what keeps them embraced as atoms and so stops a world that's built of those same atoms falling down? And what is it, behind all transformations, keeps conserved? If nothing more, has this section left you with a feeling for a Dialectic frame of answer?

Alpha Points.

Alpha is the first letter of the Greek alphabet and omega the last. They represent beginning and the end.

	tam/ raj	*Sat*
	existence	Essence
	something	Zero
	process	Start
	consequence	Origin
	cycle	Axis
	concentric rings	Centre
	time	Timelessness
	separate times	Omnipresence
	lifespan	Alpha
↓	*tam*	*raj* ↑
	down/ from centre	up/ towards centre
	outgoing	return
	expiry	inspiry
	gravity	levity
	materialisation	dematerialisation
	creation	anti-creation/ dissolution
	omega	alphabet
	flatness	energetic oscillation
	aged part	youthful part
	death	lifespan

Every object and event, every pattern and behaviour has a (*sat*) birth, a (*raj*) lifetime and a (*tam*) death. Does this include the universe as a whole? The last eight chapters have been devoted to a *top-down* description of its interpenetrating layers as they extend from the Centre to an elusive periphery - for the Centre is everywhere. In other words, what could have seemed like a temporal odyssey from (*Sat*) Potential down and out to (*tam*) physical solidity could equally appear shorter than the distance of the book you hold, closer than grey matter to your thoughts, as lengthless as the twinkling of an eye. So how was 'universal body', in all its great complexity, first projected? A week is a long time in cosmogony but as nothing for the intellectual gas surrounding it. Even saying 'big bang' takes far longer that the real thing, if it did, did; and explaining 'levity' much longer. What are the current versions of the birth (Point Alpha) and death (Point Omega) of physical creation?

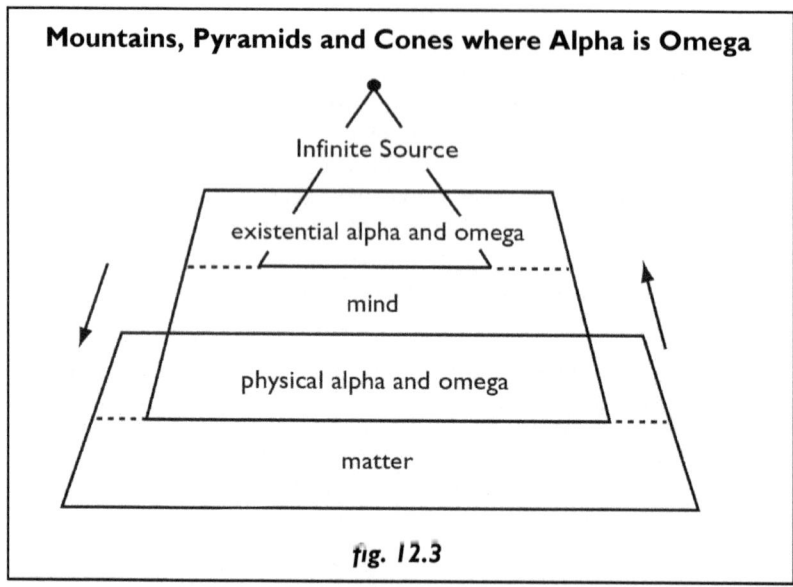

fig. 12.3

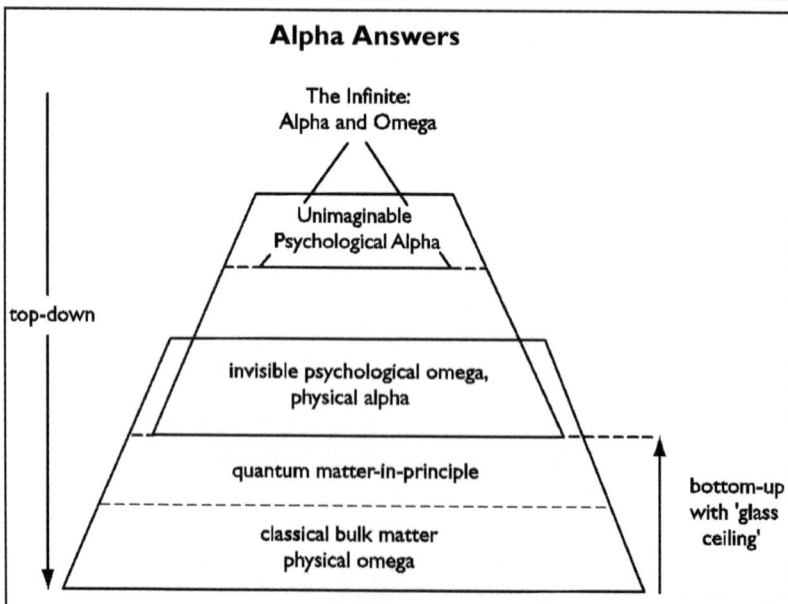

Top-down the Source is Infinite. In this view a 'glass ceiling' (made, maybe, of vacuum energy but certainly transparent space) is reached where physic becomes metaphysic and, therefore, unreachable by physical

> means. This psychosomatic 'phase-border', which appears infinitely void, is the immaterial dimension of transcendent matter or archetypal memory. The pattern of physical energy is issued 'through' this 'glass ceiling'; in other words physical derives its origin and sustenance from metaphysical.
>
> *Bottom-up* the *top-down* view is absurd. This is because there is nothing that, even prior to our own cosmos, is not physical. Logically, therefore, everything physical must have had something physical precede it - except at first. What is the nature, great or small, of that anomaly? What formulation might resolve such critical discrepancy? Things must have come from nothing physical but 'nothing physical' is not, in this view, metaphysical. The natural, material universe must derive from absolutely impotent emptiness. How?
>
> Which is the A-grade answer? Which the E-grade?
>
> *fig. 12.4*

The universe and understanding of it are both projects starting with initial conditions. What are these? What are the prior assumptions? One is that any sort of past which led to the present must explain what is called 'the cosmological principle'. This grand-sounding item of faith presumes the universe has the same laws and looks about the same from any vantage point. Its volume has neither centre nor, being infinite, edge - unless, of course, space is somehow spherical. Local fluctuations occur and, if an inflationary cosmos were the case, you would expect its overall density would be 'symmetrical'. This is not, for reasons unknown, what is seen at large-scale modes of *CMBR*; and a 4-biilion light-year 'wall' of galaxies distorts the principle as well. Could such lack of cosmological symmetry be an anomaly of measurement or is big bang's expansion compromised?

What about the furthest past called origin? Infinities have neither starts nor finishes. Does this apply to time and space as well? The option called a '*steady-state*' has no beginning at a point in space or time. Neither moment nor location. In which case why, an endless time ago, did the universe adopt its current constitution? What supports apparently perpetual subatomic motions? What sustains expansion or generates its mighty cosmic mass? Why have its sky-borne star-fires not failed long ago?

Science and religion both prefer a single origin. For the Greeks a definite start led to various concepts of the contemporary cosmos. Hesiod, Pythagoras, Epicureans, Stoics and Aristotle (with his intuitive but influential notion of hierarchical, concentric cosmic spheres) each made contributions to the cosmological question of our place in space. Once formulated, however, their models seem to have stayed suspended or revolving round inside a non-expanding emptiness. It is the same with genesis explained by the Semitic faiths. Newton's clockwork universe also ticked along this sort of line; and Einstein so recoiled from the idea of any overall expansion or contraction that, you may recall from Chapter 8, he injected a number called the 'cosmological constant' into his equations in order to explicitly cut concertina action out and keep cosmos, in the way of Aristotle, balanced.

Various observations seemed, however, to indicate Einstein was wrong and that the world's expanding. In order to avoid having to account for a start but to include expansion you could, of course, simply propose perpetual expansion - except that surely even perpetuity of this kind must begin somewhere? Something must have driven it from prime infinitesimality? Perpetual expansion stretches comprehension! Then stretch it. You might propose dynamic 'steady state expansion' as did, in the 1950s, astronomers Hoyle, Gold and Bondi and later, with their own peculiar scenario, Linde and Susskind. For the latter there was nothing ever but, *ad hoc*, expansion; in Hoyle's case an eternal 'steady state' at some point started to expand. A second *ad hoc* factor keeps the density of matter just the same. The continual creation of matter *ex nihilo* is always compensating for dilution that expansion 'threatens' balance with; if it did not both solar and galactic systems would break up and fly apart - and that would not explain what's happening.

You could, of course, retain expansion but explode eternity. In Stalinist Russia the principle behind all scientific work was atheism. Such a world has to include, in principle, some mechanism of material 'start-me-up'. In the venerable tradition of a Russian fairy tale or, up-to-date, of international science fiction George Gamov obliged. While comrade Alexander Oparin took care of the Little Accident (see Chapter 20) George took on The Big One. His explosive if not revolutionary conjecture of the 1920's was that 'nothing banged out everything'. Such a startling wind-up seemed to blow the 'everlasting theory' of a steady state to smithereens and so Hoyle mocked and super-glued it with a terrible misnomer that, has stuck - 'big bang'.

Could there ever be a bigger bang? Gamov's imagination survives as the dominant, modern version of cosmogony - the origin of the universe. Why should his suggestion be construed as real? It is because it is reasonably derived from what is assumed to be a straightforward extrapolation back from here and now in the expansive process of our universe to there and then, a point of naked singularity. *The 'burden of proof' for this astonishing process and its causative event is based on inference and interpretation drawn from two observations. These are the Doppler red shift as applied to stars and galaxies and CMBR (cosmic microwave background radiation).* What is the meaning of these observations? Are they proof enough? And has the evidence now metaphorically cast the 'bang' in stone or does substantial anomaly and argument exist? If argument exists then does its strain accumulate to such extent that we can't paper over cracks in space and have to turf the mother of explosions out? Let's take a closer look at each phenomenon.

First, red-shifting. The physicist Christian Doppler demonstrated a wave phenomenon whereby approaching waves are squeezed closer into shorter wavelength and therefore higher frequency; recession causes the reverse effect. You have heard it when, say, an aeroplane flies overhead. As the plane approaches the pitch of its 'whine' increases; as it recedes it fades. Passing lorries and trains also do it; and young boys driving imaginary vehicles imitate it very well. Doppler is a speed effect and so the greater a speed the more pronounced its shift. And, as it works with the longitudinal compression waves of sound in something, so it is presumed it works with a different kind of beast - the transverse waves of light in nothing.

Astronomers use spectrometers to measure starlight and from black absorption lines in spectra they can tell the elements that make a star. You might expect that, according to Doppler's effect, visible light waves from an approaching object would be shifted towards the violet end of the spectrum and *vice versa*. Cosmological luminary Edwin Hubble first noticed red shift (or Doppler red shift) in receding galaxies. They all 'recede' and the more distant the galaxy the greater its red shift. Does this, Hubble asked, mean that light is 'stretched' because the cosmos is expanding? If so then is expansion from an ancient starting dot? It's taken that it is.

All distant objects seem to fly away from us. Whether this 'red shift' is construed as due to actual recession through space or as space itself expanding is effectively semantics. Whichever way, everything should be a part of cosmos-wide expansion. Things are not, however, quite as straightforward as this. The shift can, for example, vary within galaxies. Can a passenger race much faster than the train he's in? How long would he remain aboard? Check the extreme red shift of a quasar embedded in galaxy NGC 7319. The implication is, because the galactic shift is much lower, that the quasar races faster than the galaxy in which it is embedded - unless a red shift has nothing to do with distance or expansion. Anomalies as sharp as that could, unless you turn a banger's blind eye, re-chart the heavens. Are recession or ballooning space the only ways to interpret a red shift?

For example, could light be scattered by intergalactic gas clouds or interfered with by 'dark energy'? Certainly, the pull of a strong gravitational field (of, say, a star or galaxy) can drag a red shift out of light that leaves its source. Again, do you remember (Chapter 10) Saxl's notion of variation of the gravitational constant that might also shift light's spectrum? Yet another suggestion, whose implications have been glimpsed, is that the speed of light dropped either exponentially soon after the 'alpha moment' or gradually at a rate proportional to the cosmic rate of expansion. In fact, you don't need to vary anything. Galaxies and stars rotate and a light source moving at right angles to an observer will submit a red shift. If this does not explain a universal red shift 'tired light' might. *Sheer distance might cause fade. In this case the more distant a galaxy the greater its fade and therefore red shift.* Could distortion due to distance be the actual circumstance?

In spite of these objections, however, suppose a red shift *does* indicate recession. If you then relate its speed to an estimate of galactic distance you can guestimate (assuming a constant rate of expansion) that the universe is an approximate sort of age. From this point it is possible to extrapolate, by theoretically contracting backwards in time and space, and reach an alpha point of hypothetical beginning. Still better, because light from distant stars and galaxies takes so long to reach us, we should see them when they were younger than they are now. Powerful eyes, like the Hubble space telescope, can now peer further (and, by implication, further back) towards a younger, less expanded universe. Not only might you mount the family album with a set of snaps from Cosmo's childhood but, ultimately, even catch his birth. Might there, hidden in the belly of the void, have been conception and development behind material scenes?

Enter, at this point, the *second* observation. Gamov had construed the motion of his naked singularity as a cosmic, thermonuclear explosion. As its remnant he predicted 'hot' *CMBR* (cosmic microwave background radiation) at 50°K. According to Einstein's general theory of relativity, however, the expansion of space would have 'stretched' or red shifted the initial fireball's radiation by a factor of a thousand down to longer wavelengths. In 1965 Arno Penzias and Robert Wilson discovered a background radio 'hiss', a spherical shell of microwave radiation interpreted according to theory as 'the faint afterglow of genesis'. Wherever astronomers point their telescopes they do indeed record 'an invisible sheet of light'. They register, in effect, the patterns on a bubble whose surface has from its commencement been the limit of the universe and now is swollen to the absolute cut-off point for our inspections 14 billion years from point of origin. The temperature of this 'fossil radiation', *CMBR*, is ~ 2.726°K, that is, less than a freezing 2.8 degrees above absolute zero. This value concurs with Einstein's prediction. *CMBR* is taken as a proof of material creation causing Penzias to note (in the book *Cosmos, Bios and Theos*) that, 'Astronomy leads us to a unique event, a universe which was created out of nothing, one with the very delicate balance needed to provide exactly the conditions required to permit life, and one which has an underlying (one might say 'supernatural') plan'.

Not everybody sees the hiss his way. More likely, Sir Arthur Eddington calculated, that the light from billions of stars would warm space. He gauged the heat of this icy fire to be about 3°K. If he is right then simple starlight isn't proof for any big bang singularity - an idea he thought folly anyway.

What if you were shrapnel? Drop a pebble in a pond and watch disturbance radiate. An explosion radiates its sphere of light; and an explosion in the air will radiate, like rays of light, its pieces everywhere. They would fly (or in the cosmic case 'develop') as if laid on the surface of a swelling sphere. Look sideways all around. Your 'mates' fly on divergent lines. Some are neighbours almost parallel to you but most are further as the angle grows. Now look forwards. Nothing has arrived there yet - the sky's a blank. Look backwards. There's nothing in the middle - everything has passed; nor can you see (because of distance and the time light takes to travel) the other hemisphere where shrapnel's flying in the opposite direction. Even when light reaches you see history. If it took five years to reach you, you would see what happened those five years ago. What therefore about the source of your 'explosion'? Looking back in that direction you might identify a residue of heat, a trace of shrapnel's origin. Could empty point of origin itself appear?

If, however, this is how explosions work then why is heat (or hiss) from source apparent at all angles? Why, contrary to pyrotechnic theory but according to cosmological principle, is stellar shrapnel streaking everywhere I look? If the heat rushed out with shrapnel then it will have cooled near us; but the further back you go the hotter unto Hot Point must have been *CMBR*. Why, therefore, is the distribution of its heat observed at the same minimum strength everywhere? Did big bang, mind-bogglingly, occur at 'no-place', that is, 'every place at once'? Instead of an explosion did emptiness itself balloon and this ballooning carry galaxies away? Or could the heat of starlight seem a better bet than afterglow? The odds today don't seem to lean that way.

Eternity was guillotined but is the evidence for an unsteady start and subsequent expansive state more concrete?

If Alpha Point is real what is the nature of decay towards Omega? Does cosmos cycle round the way it came or fly straight off towards a different destination? Balls thrown in air fall back to ground. You can run a film or theory backwards but could expansion drop into reverse and contract, anti-timewise, back to whence it came? Could such dead-end as a black hole for some *ad hoc* reason bounce back as a white one? Contraction to expansion, sink to source, could nothing yo-go? Could, as certain up-to-date cosmologists suspect, singularities go yo-yo?

Hindus are wont to bounce this sort of rubber ball as well. In their view time is cyclical. The influence of *Brahma* is (*tam*) materialising; he is the creator of bodies. The influence of *Siva* is, on the other hand, a (*raj*) kinetic upswing; he is a stimulant, a solvent of rigidity, the god of dissolution. Not only is this resilient, vibratory aspect of existence perceived in every step and particle of '*Siva*'s dance' but the whole act has a vast wavelength - start, lifetime and end. Cycling. Round and round a cyclodrome, wheels within wheels, various calculations of whose spans were clocked (*AC* or *DC*) just before this chapter. An elastic, multi-cyclical universe is analogised with the breathing of a cosmic soul called *Paramatma*. Controlled breathing, a reflection of this cosmic cycling, has since pre-history been the basis of an exercise, *yogic pranayama*. Prana (*figs*. 6.2 and 7.5) is closely linked, conceptually, with the biological breath of life, with what animates and with the pre-physical source of universal energy.

Oceanic *Paramatma* in his aspect of *Brahma* exhales; He breathes life (that is, *Om* and not just oxygen) into the creation. For Him this represents a 'centrifugal' loss of energy (if what is Infinite can lose a thing); conversely, from creation's point of view, it represents incoming tide - one that is centripetal based on mass, a motion of materialisation we call gravitation.

In the aspect of *Siva Paramatma* inhales. The inspirational flow is one of freeing up, dissolution and revitalisation. Its ebb/ reverse becomes a centripetal gathering. Strength is returned to the centre; which motion appears to creation as an ebb tide, its own 'going-out', death or dissolution.

After inhalation, before the next exhalation there is a period of poise. Silent potential. Nothing Alone. There is, without distracting motion, self-absorption in a pure and perfect concentrate of Consciousness. Then, in the cycle of creation, there is breathing out again. Fresh life is breathed into a new world. Galaxies, stars, planets and moons are all rotating. How did fragments of rotation just drop from expanding clouds of gas in space? What kind of turbulence creates a clustered globe of stars? Yet the wheels of cosmos seem to spin, wheels within sub-cycles, interlocked in their engagement with each other. Indeed, does the whole creation turn? What is the axis of its angular momentum? Where the only banger's 'centre' is a Point of First Expansion then what kind of hub is that? Surely you don't think that unrestrained inflation is a legal source of balanced oscillation?

The abovementioned ideas are interesting but not entirely materialistic. Time out of mind and the origin of matter are the province of myth. Myths are mankind's solace when it comes to the big questions - origins, lifetimes, moralities and ends. They identify crucial problems, frame them in terms of dialectical opposites and then 'cut to the chase' by resolving those paradoxical opposites in a positive, complementary way. Advanced modernists may decry mythological imagery but every scientist knows, when he explains the structure of an atom, that you have to use a picture or a metaphor. Even more so does an artist who describes the metaphysical dynamic of emotion. Are myths not extended metaphors for what you cannot see or touch or understand but need to know? Like any work of art they are cultural and devised to suit their audience. And different audiences understand them in different ways. We constantly update our myths, that is, we reformulate and refine our models. This does *not* mean there is no truth in previous explanations. And an important fraction of mythology has been created to satisfy a yearning to know the history and origin of things - especially ourselves. It is like a child that wants to meet its parents. It sets the world in context. What is the modern myth?

There's no myth now. Myths are unprofessional, imprecise and, worse, involving poetry and passion. Ugh! We use mathematics to make models. Except that the precision of elegant theories derived from 'beautiful' mathematical logic is not, as string theory shows, always clearly related to the real world. In modern physics a reliance on mathematical ideas can exuberantly overrun empirical observation and thus retire from the laboratory. Indeed the laboratory of cosmology, 'the science of an *ordered* universe', *is* the universe. Do you remember from Chapter 5 the view of Karl Popper, philosopher of science, that if a theory is untestable then it is not scientific? Untestable hypotheses and models amount, even if they appear plausible for one reason or another, to strait-laced science fiction based on prior assumptions; their 'mother of assumptions' is that nothing matters except matter – philosophical materialism. A model can be reasonably constructed but, because its foundations are built-in assumptions, turn out partially or fully incorrect be that much wrong. Every detective understands the danger - practised plausibility. 'Missing factors' can easily derail his line of reasoning. Could a model even be untestable *and* incorrect? Could this ever be the case for *PAM*'s great miracle - 'The Bang'? Recall (from Chapter 7: Are There Any Other Kinds of Answer?) the frustration set against big bang? This frustrating fact remains.

The fact remains that key properties of 'strings', 'branes' and extra dimensions remain conjectural. They are rooted neither in experiment nor observation. They are simply abstract mathematical models dressed with some fashionable, colourful and comprehensible conceptual clothes. Are we, on this sexy scientific catwalk, talking skimpy cloth to cover nakedness? Are exotic factors like 'dark energy' a fig-leaf, fudge factor or the cutting edge of how things really are? Give the show a whirl. Such productions are as laudable as enjoyable. It is natural, using suitability and elegance, to cover an unfashionable nakedness of ignorance. **Nevertheless if Popper is right it makes *PAM*'s miracle, the cosmic missile issued from a non-existent gun, the mythology of our time - even if it is a modern, mathematical one. And it relocates the collateral theory of evolution, under protest, into the class of para-science.** So what? You can't ditch the issue of origins because it won't fit in a laboratory. So how came humans (Chapters 17 and 19-25)? How, for physics, did nothing in a black box blow it all up in the air?

Recall the yogic yo-yo. Curiously, dissatisfaction with certain features of big bang theory (not least the fact that it might constitute a single starting-point) has spawned a sort of Hindu version of events. Physicists Paul Steinhardt and Neil Turok propose that cosmic cycles might replace the standard notion of a one-off, chance-based episode. What came before the beginning of this space, time and things? It must have been a previous cycle of space, time and things. Crunch-burst, crunch-burst *ad infinitum*; that's how you evolve a bouncy species of the multiverse! No need in this vibrant form of 'steady state' to explain the real origin of matter since it never started! Such cycles are an aspect of a holy grail to settle differences between Einstein and Planck; and between glorious, scientific and infamous, holistic kinds of believer. Do they need, like strings and loops, five or more dimensions? Where does matter, luminous and otherwise, disappear between the cosmic periods? A singularity (if that's what loopy starts are made of) 'projects' in perfect order that, by entropy, runs into great disorder. How, therefore, does number two (unto infinity) burst in reconstituted perfect order; how does a black hole's total entropy regain transcendent order prior to bursting, as a white hole, back to physic's space and time again? Or, if bouncing is not serial, what's the nature of the underlying perfect order that transcends material entropy? What species of escape hatch is it into which you cram the question of material origins and, conversely, from what sort of rimless crater might accelerators such as Bursts erratically erupt? Why not suggest a hundred trillion of them if they have eternity? One might consider such stringed yo-yo theories less an explanation than apology for lack of one. A plethora of answers to a question usually means that students haven't got a clue. If strings are 'mini-science-fiction' surely this most roundabout of all descriptions of cosmogony is 'maxi sci-fi'?

Where was Alice when, as with cosmology, she was asked to believe six impossible things before breakfast? Any which way each decade-wise know one thing - physics and its novel imaginations will eschew a single, central factor. This, the one that powers all conception, is the subject's consciousness. It is the power of mind. The sophisticated speculation of recycling on a universal scale is just, at root, another mask of metaphysic. Its pure materialism ignores archetypal memory, its imaginations are unseen and therefore at root undeniable and its solutions are not as immediately obvious as Natural Dialectic's universal mind. Indeed they are metaphysic in a class, if not a complex phylum, of its own. An immaterial factor, mind, is unappealing in simplicity; but, because unseen, at root similarly undeniable.

According to psientific string theory 'space' contains a couple of cosmic figments called 'branes'. A 'stringer' calculates that the emptiness created by an expanding four-dimensional universe eventually causes a fifth (or is it seventh or tenth?) dimension, having the shape of these non-conscious 'branes', to collapse. When, due to this contraction, they collide then hot, dense matter and radiation (constituents of the standard projection's 'white hole') are recreated. If you say so. But where this energy *originally* comes from remains unexplained. If you are waiting for an Alpha Answer, don't hold your breath. What comes around, says yo-yo theory, goes around. The whole wheel turns and returns but whether you'll recur is a matter of evolutionary chance. Don't count on that then. Don't assume, you lucky chap, a cumulative series of accidents will ever sum to you and your peripheral consciousness again! Thank instead your lucky stars, bless the speculation of a brane and genuflect at Good Fortuna's shrine!

So what hope is there? Cut to the chase, what is the real issue?

The real issue divides between *top-down* perspective (which includes the aspect of idea, plan and purposive execution according to the logic of its 'script') and the *bottom-up* unscripted, sniff-it-out-as-you-proceed perspective.

The Labours of an Empty Womb

There's no business like show business! How did something show from nothing? Is 'space' the source? Is everywhere a field (or fields) of latent and unbroken energy? What force of creativity invigorated and invigorates this field's virginity? Are 'things' themselves the fruit? Can human thinking gather how an empty womb gave birth, how matter breaks from nothing?

If creation's first and fundamental stage dawned on the other side of comprehension how can one frame ignorance? First frame, *top-down*, includes a metaphysical dimension; the second, *bottom-up*, does not. If, however, the truth cannot be grasped by intellect, all stories of creation are reduced to myths - albeit cultural, personified or intellectual ones. Myths restricted in this way are inevitably only partial pictures of the whole truth. Such mythology is, indeed, the substance of the third Mosaic injunction 'you will not make graven images (including imaginations)'. The two contrary frames involve a further issue, an explosive one. A metaphysical perspective is set diametrically at odds with the physic-only, self-imposed restrictions of scientific atheism. Presumably no-one believes that everything was always, as is, here? *Therefore such restrictions mean that, for complex solidity to aggregate, evolution from a simple physical origin must have happened.* Full focus is, therefore, applied to the conception and authentication of scenarios that permit this possibility. These are, according to such wisdom, acceptable ventures but a frame that includes the metaphysical is not. Whether or not it is true such a paradigm as, for example, Natural Dialectic's is deemed 'unscientific'. How, though, can a holist in this bind align professionally with materialism's total scope of answer if he thinks there's more? Do you remember, arising out of Chapter Zero, two basic pillars of faith? *Which is true - intelligent but 'unscientific' design or accidental but scientific 'design'?*

Let's rephrase. The scientific renaissance began with clear-minded men saying 'Let's cut the cackle, let's begin with what is sensible. Let's study, without prejudice, what we can observe and test. Let's experiment with solids, liquids and gases we can grasp'. When atoms and, later, subatomic particles appeared by inference and, although unseen, were proven real by developed, tested theory, then science was moving up the ranks. It had discovered 'matter-in-principle' - atomic letters, molecular words and aggregate texts that spell all stories out. The question became 'Where do subatomic particles come from?' By definition science cannot reach the metaphysical and so at this point, having brilliantly deduced the way nature 'speaks', it had reached its 'glass ceiling', the vacuum. From this angle it is impossible, when you consider origins, to include what any '*top-down*' hierarchy might have fallen through. You exclude the informative metaphysic of idea, purpose and memory.

So *bottom-up* there is no hierarchy; you box yourself in one non-conscious plane. Oblivion is all you have to work with and, as such, its material constructions must all derive from necessity (in the form of natural law whose constraints are described but not explained) and chance. Since things did not start out as rocks, teacups, humans or other such complexities, you have to work from simple unto complex, from material first principles to their effects. This forces you, without admission of initial conditions, into the arms of George Lemaitre's bang and its inflation. Indeed, inflation is cosmology's foundation. Why not, therefore, inflate its power. Why not assume it *is*, in perpetuity, The Initial Condition? Thus, each time some fraction of an omnipresent, omnipotent and infinite inflation dies, another universe precipitates? Deflation of inflation is The Cause! By which reckoning our 'big bang' beginning, whose fall-out links directly to a story of its animating debris called the theory of biological evolution, was just another random surge. Whichever way, allow that happenstance (in macro- and in micro-forms) is enough for evolution of our world. Large and Little Accidents explain all things. From first assumptions (*PAM* and *PCM*: see also Chapters 19-26) what could be more obvious?

Or less? You claim that, paradoxically, from 'bang' that's not a bang expands a space from which develops non-space, that is, from nothing there emerge material things. But, in this case, it needs be recalled that if a claim is neither tested nor experienced, its actuality is no more than hypothesis or hearsay. Imaginative anecdote. Neither of the 'wobbly couple' (red shifts and *CMBR*) constitutes sufficient evidence let alone proof to impress a cautious scientific or a civil court. It is not the facts, however, but the inferences drawn from them that are under scrutiny. In a level-headed court you need trustworthy witness and in science your witness is interrogation with repeated tests. If your hypothesis is one you can't test even once it might be labelled para-science just to flag that fact. And if under cross-examination you must keep inventing new, unknown and imagined factors to sustain your story, then the legal case for a suspension of belief is strengthened. Suppose, for example, in order to sustain the idea of Point Alpha as a projection of physic from non-physic you had to imagine a new, unobserved kind of universe that included 'dark matter', 'dark energy', 'inflationary powers' etc. Such major doctoring of the actual evidence would, despite smoothing over theoretical difficulties and despite pleas that it might someday be found true, at this point be incredible. Inadmissible. <u>*It would demand that you suspend both reason and belief in favour of a judgment that might never come*</u>! Suppose you added more and more *ad hoc* suggestions. At best you might forgive a critic for feeling you were in the grip of a persuasive possibility, had not grasped the correct first principles or were labouring under the influence of materialism's neat brew and thereby plain, straight wrong. Impartial judgment hands down 'case dismissed'.

If big bang cosmogony was a damp squib where would that leave another prejudice, Darwinian evolution, that's also touted as if fact? We shall see (Chapters 19-25).

As noted in Chapter 5, the real issue boils down to intelligence. How good is the information? Do you, taking visible physic alone, neglect invisible metaphysic? Or include both in your 'anthropic' paradigm? Why indeed should nothing be immaterial, everything material and, because you deny the existence of metaphysical science, reducible to the explanations of physical study alone? <u>*The basic premise of Natural Dialectic is existential mind/ matter dualism subsumed under the essential, central monism of immaterial Consciousness*</u>. If the negative pole of actual creation is non-conscious, passive and inanimate, might we presume its diametrical opposite is conscious, animate and active? Information is potential that controls an outcome. It informs what is informed. Thus, for example, archetypal memory embodies natural law; by natural law our cosmos is informed. And if this peripheral realisation of Informative Projection is material and matter is 'natural', how is the Centre from which it was projected less natural? It is Most Natural and, being lively, Most Intelligent.

This is the point to interject a couple of possibly irritating provisos, one of them cautioned earlier.

Did the beginning really begin or not? If it did, is its singularity a physical state of affairs or not? If it is then what are the rules that regulate its material 'nothingness'? If it is not then scientific atheism places faith in a wholly unknown, lifeless absence. A gaping, godless gap. This sounds like religion, albeit of an inverted kind. The first proviso is, therefore, that before 'creation' any naked singularity would, since there was not yet any physical nature, have been beyond any law due to the presence of matter alone. Therefore neither its origin nor its first 'creative motion' would have been based on such law. **If, on this basis, creation by intelligent design is excluded from scientific consideration then so equally must be *PAM*'s miracle, The Bang and, as a direct consequence, all *PCM*s (evolutionary theories) concerning origins.**

If you object that *after* big bang's fireball *PCM* then operates within the law, is this the case? Does water flow uphill? We looked and will take a further look at laws of entropy and information. What flouts these is certainly outside the law! Can the process naturalism chooses as its central explanation for the origin of things itself be operating lawlessly? In a moment we will check what astrophysics claims *PAM*'s miracle performed. Chapters 22 and 23 explain that the physical causes ascribed to biological evolution (random mutation and natural selection) are either conservative or degenerative. They are neither creative nor competent to service evolution as biology construes the word. **Could the grand idea of origin by 'negentropic' evolution therefore not be based on natural law at all? And thus be an unnatural outlaw?**

Indeed, is not a view that systematically upsets correct perception like an illness, a psychological infection or disease?

The second proviso involves miracles. A 'miracle' is how people describe events they do not understand; and their faith, unless the miracle in question has without doubt been perceived, is the conviction of things unseen - like cosmogonic bangs and gods and cells from mud and water. Even though it seems theoretically as good as tracked to lair by 'bang' theory, the 'miracle' of a universe *ex nihilo* is certainly not clearly understood. The crucial template of initial conditions is hidden beneath the flap of a Planck-sized black box. At this point, where micro- and macro-worlds both converge on infinity, physical definition breaks down and exponential improbability breaks loose.

On the one hand, therefore, persons claim that miracles are 'natural improbabilities'; but how, if nature issued out of nothing physical, could non-physical initiation be a physical unlikelihood?

Need nature be reduced to physical alone? Your body 'hums' in autonomic mode but voluntary choice can enter in; the macrocosmic body, cosmos, 'hums' in lawful mode but couldn't choice by conscious intervention, a miracle of thought, still enter in? Metaphysical initiation of the world does not compel all metaphysic thenceforward be withdrawn. 'Autonomic' archetype sustains the cosmos and, from higher up in universal mind, why should creative will not enter in? Such communication happens in you all the time across mind-matter's border, the psychosomatic interface. Apart from familiar psychosomasis thousands of 'external' examples of mind over matter (such as the voluntary materialisation of objects or rearrangement of objective patterns called miracles) are reported in all cultures at all times. The question's *how* - because so-called 'psi-phenomena' do not, within materialism's paradigm, make any sense. Indeed, the immaterial is rigorously proscribed. Thus paranormal paraphysics is a highly suspect enterprise and miracles completely unbelieved! This is because what is not rigorously predictable, testable and mathematically defineable strikes at the heart of scientific method and is rejected as fundamentally alien or irrelevant. Such rejection is healthy because it strikes out superstition, quackery and ignorance. But it is also unhealthy if, with infidel reflex, it eliminates any consideration of human consciousness, mind and life as having metaphysical parts; if it refuses to acknowledge any consistent paradigm that includes these entities as universal; or if it refuses to acknowledge causal metaphysic 'above' and prior to physical creation. It's Catch 22. Materialism makes you throw the cosmic baby from the deep and sparkling waters of the bath of space. The only way to ascertain a 'psi-improbability' is, since it includes a subjective element, to personally experience its effect, to devise tests that indicate its probability or to understand it from a higher, altered state of consciousness. This leaves objective science with no better than objective probability - where quantum theory does the same. However, the metaphysic of psi interactions *even once* incontrovertably demonstrated by weight of experiment or by some widely witnessed miracle would reveal materialism as a seriously truncated form of truth. *A sceptic must have missed or disallowed a key fact or two - of which the most important is the immaterial origin of information.*

Of course a sceptic condemns such rationale as the charlatan's illusion. Anecdote and hearsay are the sort of rumour scientific method was developed, rightly, to confirm or kill. Augury's inspection of entrails on an altar may seem as useless as Our Mystic Meg's soothsaying at the fair; still, never underestimate a power that should not spook you though it's psychic - it is called suggestion. Suggestion can manipulate as well as fingers do. Environmental, sensual and verbal cues assail us but yoga concentrates on three internal sorts - concentration of attention (*tratak* or *dharana*), development of will-power and especially auto-suggestion (*dhyana*). What is there ever in your mind but what your context or volition are suggesting? Yogic suggestion is, because mind works on what is brought to its attention, key to meditation.

But, you reply, a conjuror suggests he can, magically or miraculously, manipulate a rabbit from an empty hat. He can pull a fast one so that you believe it's real. He can conjure things that nature never could. Of course, he uses focused mind and tricks that Mother Nature never thought of - seeing that she can't. Still she turns a trick or two - for elegant reality. More miracle than tawdry tricks. If, for example, a magician conjures flowers from thin air does he not deceive? Are real petals a deception? But from mineral water, light and air a seed will conjure up a bloom. Is floral beauty not a miracle? A miracle of chance or great intelligence? Which answer is deception's?

The psychic issue's one of mind. Is mind a natural or unnatural quality? Universal mind, informative construction from an archetype and 'natural metaphysic' are outside materialism's scope; but can an individual's thought alone, if tuned enough, exert a 'psychic influence' not only on another mind but matter? Perhaps the major psychological exposition of yoga is contained in Patanjali's Yoga Sutras. Here '*siddhis*' such as 'materialisation of objects at will', 'mind-reading/ telepathy' and 'levitation' are cited as standard abilities at certain levels of proficiency. Physical yoga certainly works but is Patanjali (or any other yogi) a liar, charlatan or ignoramus when it comes to his metaphysical assertions? These are, after all, the basis upon which the important but essentially peripheral exercise of postures is built; they serve to keep the body fit while exercising mind in meditation to the maximum. Are specific, mind-expanding techniques of contemplation and the consequent development of mental, metaphysical powers nonsense? Yet at a stroke of philosophical condemnation Patanjali's holistic paradigm is reasoned 'unscientific' and, because of that, the psychological premises of yoga deemed clinically awry! Even when, say, Tibetan monks have been measured, under scientific conditions at Harvard and other universities, performing 'miraculous' psycho-kinetics such as increasing body-heat at will by meditative thought; or when acupuncture is, although overtly metaphysical in explanation, well observed to work. *Just one authentication of direct, psychosomatic creation would not mean that what physics and chemistry have discovered is changed at all - but that the creation and therefore the fundamental nature of matter needs radical rethinking.* **It would force materialism to dramatically revise a presumption that it sees the whole picture. Indeed, the goggles of its profession would have to come off.**

Why all this puff concerning miracles? You assume that once you've tracked expansion to its lair you'll find a strange primordial beast - one with no horizons called a naked singularity. This singularity, beyond the grasp of physical-cum-natural law, is claimed to be an infinitesimal point of infinite temperature and density. Is that absurd? This 'fifth state' of matter is unknowable except, just like intelligence, by inference. Nor can it be known. The other states of matter are transformable but not this one. When 'normal' matter's crushed into a black hole singularity there's no return. This means, in terms of entropy, that such a singularity is a dead end. It is the ultimate thermodynamic cul-de-sac. Yet was it not a singularity that you, without a thought for the absurd, said started everything?

How can anyone resolve such paradox? The pinprick that injected cosmos into nothing from a metaphysical syringe is set beyond necessity. Known laws and natural forces follow from and don't precede it. And if necessity's not there then that leaves chance alone to force its hand. Yet what strength has Lady Luck? How could she roll the cosmic ball when chance is not a force? What natural instrument, if not her ladyship, might fashion such a shock and all its waves? Will quantum physics and its random fluctuations get us far? There are plenty of them even though a normal one does not contain the universe's mass! They could happen any time so why aren't they creating universes everywhere right now? Perhaps The Bang's was just a single, special case. It must have been because it came before itself. This is because the world of physics, including quantum fluctuations, is 'inside' and not 'outside' the world. There were no quantum fields before the space and time and energy for them appeared. The cause of such effects was the mystical (or, if you like, mysterious) singularity and effects do not, in any real world, precede their cause.

In fact, where all laws have broken down then how can you, with hindsight, go there and predict a thing? *If you have no instrument to rouse the singularity; if the singularity itself is hypothetical; and if its bang and consequent expansion are also an assumption then aren't you left with what every other person calls a miracle?*

Therefore, if the world did not burst from God's head or was not delivered from the womb of feminine divinity, let's take a peep at how it reared from absolutely nothing - *PAM*'s own miracle.

PAM's Miracle

Let's, for a start, be clear. If you are just an object made of stardust, if you are just a body made of atoms we can claim that you evolved, in your totality, out of *PAM*'s Great Miracle. When you go back as far as possible do you not find your Ancestor, a Single Cell and then your Ultimate Progenitor, a Naked Singularity in Nothing? Your life did not, therefore, spring from life. It has sprung from a Profundity of Emptiness - although how sure is sure? Surely surety is not, like the Singularity itself, *ad hoc*? Let's scrutinise our *PAM*'s cosmology.

PAM's Miraculous Projection

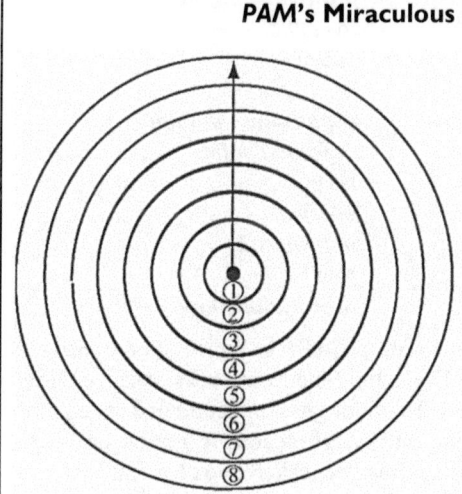

① naked singularity
② 'quantum fluctuation'; quantum pair production
③ inflation
④ mass-making 'drag' field; plasma; subatomic particles and nuclei
⑤ hydrogen/helium gas
⑥ embryonic galaxies
⑦ stars
⑧ solar systems; planets and moons

Now, with a clear idea of the reasonable reasons but nonetheless limitations of scientific method, inference and interpretation, we can turn to the ways in which science struggles to understand how, as a physical Point Alpha, something must have sparked and then developed out of nothing. PAM's miracle is a mathematical creation story. Remember, though, that it knows nothing of creation's gradient. It is a purely one-tiered job. How do you try and squeeze three tiers to one? Like this.

fig. 12.5

tam/ raj	Sat
neg/ pos.	Neutral
thing	Nothing
periphery	Axis/ Centre
consequent orders	Source of Order
subsequent behaviour	Archetype
oscillation	Norm
ruled/ ruler	Regulation
effect/ cause	Potential
manifest	Unmanifest
formed	Formless
differentiated	Undifferentiated
relative power	Omnipotence
after	Prior
lesser presence	Presence
↓ tam	raj ↑
down/ outward	inward/ up
from Centre	to Centre
closer to particular	closer to generality
more gross	more subtle
appearance of chance	appearance of order
negative	positive
inertial	dynamic
effect	cause
bind	release
contraction to a point	expansion/ inflation from a point
amass/ aggregate	rarefy
anti-energy	energy
mass	anti-inertia
fixed	flexible
materialisation	dematerialisation

PAM's miracle might be Creator of a Godless World. Some bangers therefore may not like the whiff of Dialectic. Scientism looks askance. How, though, does *PAM*'s assumption stand? How does its conjecture of a naked birth make sense? How can hot gas in the end cough academic reason up? Let's inspect *PAM*'s miracle in order. Having already checked two elements that prop the theory up (questionable inferences and interpretations drawn from Doppler red shifts and the background radiation called *CMBR*) we now turn, in the same way as with the theory of abiogenesis in Chapter 20, to further cumulative reasons why 'big bang' cosmogony might, perhaps, be null and void. *In other words, does there accumulate a momentum of objections that in collision knock the big bang out? If, though, its fireball cannot be extinguished intellectually then the world might actually have detonated in a most expansive, generous and precisely detailed way.*

For starters, though, just before we take to zero absolute let's ask if immaterial could ever rise from a material base. **How can non-physical (or** metaphysical) **appear from physical?** Yet, metaphysically active and physically passive, immaterial information is at every level of creation a reality. Whence, therefore, arose this subtler aspect of our universe? **If matter can't produce what's not itself then big bang or any other materialistic explanation is thus insufficient, by itself, to account for everything.** Evolution, we shall see, can't even organise a single cell. Information is a critical ingredient. And it is *PAM*'s invisible, missing factor. Perhaps materialistic myths of origin aren't, therefore, the full story. Perhaps they're a whole dimension out of true.

You hadn't thought of that? **Transcendent projection from a metaphysical dimension might be right but, according to the logic, big-bang-out-of-nothing-physical is a non-starter.** Cross it off your cosmic list but, since otherwise you're left with nothing, at the same time generously allow its moment once arrived. Was there, just before epiphany or post-apocalypse, such a 'thing' in all of nature as a zero absolute? At the foot of what rainbow, at the edge of which ever-receding horizon or from what less-than-dot-in-space emerged Point Alpha? How thence did Alpha's absence crack asunder and the world stream out? Nothing couldn't even crack a joke. How, therefore, might *Ad Hoc,* a masseur who will oil your process through its problems, first rub space and knead the cosmic session in? And if the intellectual fingers of *Ad Hoc's* massage don't press deep enough, we'll take a scalpel to Big Accident, dissect proposals for the stages of development and then perform anatomy upon the body of the universe. Such surgical analysis is called cosmology.

Liberals and rebels dislike rules, rule-makers and enforcers. Yet even anarchy imposes rule - the 'anti-rule' of lawlessness. Think as you please, do what you whimsically want. Was cosmos made by nihilistic anarchy? Or were rules, acting as potential for the type of cosmic fun and games, in place before the cosmic starting-gun? Because they are certainly here now and, absolutely and inexorably, inform the circus in its massive ring. So what informed the big outburst? Without non-physical preconditions, that is to say, without prior information, why did it behave the way it did? Information, a function of consciousness, is immaterial; it is a natural property of mind and not matter. Nor can it be created by material energies. Nevertheless, from a *bottom-up* point of view the notion of 'universal mind' is nonsensical. Original spontaneity must have been an information-free event; it must have been the product of irrational spontaneity - chance. In this case all information is, in the last analysis, a property of mindless matter. Nothing, not even what might cause a fluctuation in the void, preceded this mindlessness, this informatively impotent and power-free blank. *So bang went nothing not explosively.* Terrestrial observations show a universe expanding equally in all directions which would make it geocentric. Who claims that? Nor do explosions normally behave so smoothly. So perhaps, unusually for an 'explosion', it's space and not what's in it that expands. And what's outside space? Materially nothing. Imagine that not even space is there.

What a dust-storm, what a smokescreen covering nothing everywhere! No space, no dimensions, nothing. In what kind of emptiness could a materialistic womb convulse? How could physical acceleration first break from the bounds of nothingness? What is the paradoxical nature of omnipotent void from which, expansive 'bang' or otherwise, delivery is seen as organised? <u>Bang goes nothing</u>.

Boom, like that! What a paradox! A singularity of zero size is held to hold, implicitly, a universe of mass. At the same time bangers calculate ground-zero space is dense with weightless ripples in the *ZPF*s; furthermore they claim that one extraordinary ripple self-ignited, grew incomprehensibly hot yet at the same time superfinely ordered - then catapulted as a wide and merry dance. Something isn't right here, is it? Why, you might ask, should any 'fragment of the void' pack itself as tightly as a density a trillion times more clenched than water? Indeed, can *nothing* ever have a density or heat? The truth is the reverse; and thus it seems that nothing never was! Nor was the bang a bang! A reasonless eruption of this kind might well turn out to be by far the tallest, broadest story you will ever hear!

Babies are conceived from small but highly programmed eggs. Can you believe the 'cosmic baby' was conceived and cradled in an empty egg, a dot in space much smaller than the one beneath this question mark - especially when you're told the dot was everywhere? How do you reconcile a will o' the wisp of uncertain 'fluctuation' with the unimaginable squash of a singularity, with its infinite concentration of potential mass? If all matter and energy in the universe were squeezed into less space than is occupied by a pinhead the gravitational force would be so intense that nothing, even light, could escape. Such mass should not explode but collapse. How, therefore, could a 'white hole' spontaneously burst into nature? If that won't work consider, theoretically, that fundamental particles (electron, quark) need neither size nor mass. Now a jar of nothing packed with everything could start you easily - as long as it had energy! How, in this case, did matter gain extent? How (unless preordinated mechanisms were inlaid to let this happen) were things built out of less-than-dots (say, pointlessly) appointed size and mass? What, having triggered a ripple trillions of times smaller than an atom, might then apply an after-burner that whooshes those appointments in a fraction of a mini-second up into a trillion times original size? 'Nowhere' ejects a cosmos-worth of push-and-shove. Excuse me asking who exactly casts about; who catches and then swallows fishy miracles!

I know we heard it once (Chapter 7: Order Below a Physical Start) but such a roller-coaster's surely worth a second ride! *Look at your thumbnail.* Imagine once again that for no reason all the universe jumped out from nowhere (at least a point far smaller than the nail). Nothing ambushed destiny and plundered endlessly! Walk on the wild side. Check the massive swag. Gaze upon cliffs, gorges, chains of ragged mountains and the other treasures of this earth; the slabs of rock you slap are but a microscopic fraction of galactic clusters that, *Ad Hoc* is claiming seriously, erupted from a microscopic, smaller-than-a-pinhead fraction of your thumb!

There was absolutely no intelligence so you are certainly in mind of Luck. You have conjured Lady Luck, Miss Chance, everywhere some luck and chance all rolled together. Is such spontaneity predictable? What could Luck or Little Miss produce without some prior form to work their magic on? A barren ovary is seedless; its womb cannot create a thing. If mindless spontaneity is chance then Luck alone would constitute an unexpected, unpredictable and yet primordial sway whose principality comprises empty impotence. No rule at all. Her second name is Anarchy. You gaze upon an empress of abstraction, a pose of photogenic nakedness whose main attraction is her lack of all intelligence - Lady Luck. You contemplate, for starters, lucky 'natural legislation'; you propose an extra-scientific apparition and define a sudden, wondrous discontinuity prior to which there was 'continuously' nothing. Never mind a multiverse because the other explanation could offend your naturalistic sensibilities much worse. It could, indeed, involve what chance denies - intelligence. It could be metaphysical, that is, an immaterial absurdity since we all know Luck's baby was an Accidental Throw. Even, inexplicably, this Big Accident, a central tenet of the atheistic creed, has even drawn unctuous benediction from a Pope. To John Paul II The Lucky Break is what his Book maybe describes. Now a cardinal of his successor, Benedict XVI, has argued that the second fundamental tenet of the atheistic canon, The Little Accident, is 'perhaps compatible' with what the founder of his church proclaimed (see Chapter 25: Theories of Accommodation)! Are Popes or Christ in error? Do the holy fathers know the Father? Or, if Christ were here and took a course in science, would they all agree? This volume does not argue for the literal truth of Genesis. If that is what you'd pointedly dismissed as 'rank creationist' then you'd have slid right past this book. Equally, however, Dialectic (and thus nature?) argues that Big Anarchy and Chance are not the 'logic' underlying things. If papal licence now endorses a material cosmogenesis then you might reasonably presume that Truth was now a Catholic casualty - unless Projection that's non-accidental and whose source Transcends its physical effect is what you really, metaphysically mean. And if you don't, well, popes are human and as such, the history of science shows, are fallible; and, anyway, what on earth if not in heaven's name is wrong with tolerant, well-mannered, bright and even suave but (see Chapter 26) somewhat inconsistent atheists?

Everybody has to start somewhere. Everybody makes a first assumption. Which party's sort of miracle can conjure cosmos out of thinner than thin air? What about spontaneous combustion? Does it not need *preconditions* such as Natural Dialectic calls (*sat*) *potential*? Did any precondition stoke the cosmic fire? If it did you must admit that, even prior to beginning, there was never nothing there. If you abstract both 'things' and 'nothing' from the formula then what is left? Pre-physical is nothing physical, something metaphysical. But metaphysic is another class entirely from a force that's physical – information's attribute is anti-chance. Such precondition is a form of mind, an instrument of order, purpose, will. Its prior archetypal substance would, forever, dethrone Chance. And, in the *putsch*, Anarchic Origin would vanish too. Perish the idea of mindlessness. Forget a cosmic origin whose reason was supposed to be an accident. Behold instead a bigger bang, a universal academic shock, an Orderly Beginning. What might source such order? What might constitute First Cause? Oh dear! Despite the benefice of papal drift could Lord Deliberate emerge on stage again?

For now, however, following the drift of Lady Luck allow the luckiest break of all time! Set your stopwatch. Press! After naked speculation of a singularity assume, from the abyss, things start to rise. A fluctuation in false vacuum has uncurled; this minuscule and trembling instability is due to be creation's Seed.

Assume *ad hoc* expansion and in only billionths of a second swirling particles emerging from the Father's heat. Receive the wisdom or be damned!

Aha! A second, bigger whoosh occurs. Bang on bang on time. Because the first exertion would have been in danger of collapsing back inside itself you need an even bigger miracle to free it up and drive the show onto the road. 'Cool expansion' grew until the universe was cabbage-large; the fading gong of midnight ushers in a day that's pumpkin-sized. Then 'hot inflation' had to blast things to the skies. This magical phenomenon, this *ad hoc* inspiration was devised to blow the first bang up a trillion times in billionths of a second. It exploded much, much faster than the speed of light then, like the first one, braked just right. How did destruction (which is what a single let alone a double bomb means) blast things to such a highly tuned degree as cosmos shows? What order and what information do explosions normally provide; how does inflation work? Or has mathematics dreamed a damned fine myth?

	tam/ raj	*Sat*
	realisation	*Potential*
	consequences	*Precondition*
	effects	*First Cause*
	subsequent order	*Archetype*
	things	*Nothing*
	informed	*Informant*
	classical/ quantum matters	*Potential Matter*
	gravity/ levity	*Pivot*
	vectors	*Balancing Factor*
↓	*tam*	*raj* ↑
	gross (macroscopic) physics	*subtle (microscopic) physics*
	gravity/ anti-levity	*levity/ anti-gravity*
	materialising factors	*dematerialising factors*
	deflation	*inflation*
	contraction	*expansion*
	inertia	*excitement*
	confined energies	*radiant energies*
	mass-making 'drag'	*energetic input*
	mass	*vacuum/ space*
	'dark' things	*'luminous' things*
	exhausted end	*bright start*
	gravitational predominance	*levitatory predominance*
	self-attraction	*self-repulsion*
	positive pressure/ push	*negative pressure/ tension/ suction*
	force inc. with greater/ closer masses	*force inc. with distance/ greater emptiness*
	Newtonian/ Einsteinian gravity	*cosmological constant/ dark energy, lambda etc.*

In its 'downward', materialising direction this stack typically follows the dialectical order of potential through kinetic to exhausted phases. It suggests, thereby, that physical existence issues from a metaphysical precondition, that is, from nothing physical.

In the world of paraphysics, where nothing is actually seen or tested, any projection-out-of-nothing has to qualify as both the largest and the smallest of convulsions. Its Singular Moment was theoretically very, very, very small. In comparison to the start of the whole universe a human egg would zoom to the size of planet earth and a mustard seed appear like a star.

Do you remember the 'outward pressure' of Einstein's cosmological constant that was supposed to resist, in a static universe, collapse due to gravity? He so disapproved of the idea of any overall expansion or contraction that he injected the constant into his equations in order to sedate the possibility. It was a levity exerted by nothing, an invisible 'spring' that kept the world's ends up and, in balancing its gravity, constant. But where gravity had perhaps been his greatest conceptual success levity was, by his own confession, Einstein's greatest conceptual failure. In the light of subsequent theory and observation it might have seemed to have to go. The grip of Einstein could not hold creation still.

Yet, in analysing how it moves, his levity might have the last laugh. How can you picture it? As opposed to mass attraction is it space repulsion, push where gravity is pull or suction of a universal vacuum acting everywhere in all directions and whose weightless buoyancy can hold mass up? Is it anti-pressure, anti-mass, dark energy, the lambda force or a 'pull apart' called (here it comes again) the cosmological constant? Is it 'negative pressure', 'gravitational repulsion' or, in a word, anti-gravity? The gravitational constant does not change but, under different conditions, the effects of gravity do. Extreme contractive energy implodes into a tiny, massive hole. Could the same apply to levity so that, in the reverse condition of a black hole, expansive energy accelerates towards nothing? With a Big Implosion as opposed to Big Explosion suction draws decreasing density of mass (mass ever more thinly spread into expanding space) into a (raj ↑) expanding void. Why, in this case, aren't new worlds 'unplugged' all the time? And is there any limit? Is such limit, opposite (tam ↓) singularity's extreme, an infinite (sat) void? Does levity sound, in some respects, no different from inflation?

Things, you may agree, weren't always as they are. Preconditions preface set-up's chapter one; if that is right it's downhill all the way. The problem is to find the nature of a peak of nothing down whose rolling slope space, time and things accumulate. It is to find the source of power from which all physics can cascade. Did you consult the oracle? Did the sage snatch up and throw away your calculations or reward them with a hefty grant to find how things aren't what had you thought? How on earth did things begin?

Even those contriving explanations for the transformation of a vacuum into rains of galaxies find nothing making something hard. On a clear night gaze at stardust flung across the sky. With two handfuls of grey matter and half a rucksack's worth of starting-mass a cosmological star, Alan Guth, reckons he'd cook up a universe. By anti-reason (chance), the reckoning goes, you could inflate a 'heavy bubble' into all the galaxies! Not that any of 57 varieties of inflation or 'inflaton specialities' have ever been observed.

You can already see that nothing must have been, according to inflation theory, a 'false nothing'. In fact vacuum power held sway and its 'gravitational repulsion' - anti-gravity - flared to the point of overcharge; its levity inflated (in one of different, arguable ways) until, after the universe had doubled size every 10^{-34} of a second for far, far less than a second, a 'phase transition stage' was reached. It was as if a gas had liquefied. In this case vacuum symmetry was broken and from the cracks popped out a cosmos-worth of particles and interactions. Something bulged from nothing. Pregnant void! Potential bodies! An excited void convulsed, the empty womb of nature had conceived primordial, principle matter from whose particles all bulkier worlds mature.

Doesn't Guth's inflation just inflate the story? No doubt, though, it seems to solve some problems. Do you remember (Chapter 8) 'lambda', 'flatness' and 'horizon' problems? A bang or bangs must generate such uniformity of universe as the cosmological principle envisages and is, from our particular line of sight, observed. How, though, when both probability and quantum uncertainties should generate large variation between opposite extremities in all directions of the cosmos, do we find such uniformity? How can you explain what's all about, how could explosions generate a universe that's uniform in all directions with respect to density and composition? How could matter have 'decoupled' out of flying radiation in about the same proportions everywhere; and then hot fractions mixed with cold to average out the heat? The extremities cannot (if in Shannon jargon you call light or heat transaction 'information') 'inform' each other and yet still far-off regions of the universe have properties and - to between one part in 10,000 and 100,000 by reckoning from cosmic background radiation (*CMBR*) - temperature the same in every quarter of the sky. How does the whole cosmic super-sphere coolly score, to within a tiny, tiny fraction, just the same degree (~ $2.726°K$)? 'Causal disconnection' should deny this spread its ultra-smooth existence. Could *ad hoc VSL* (variable speed of light) have meant that light skimmed faster at the start, smoothed out irregularities and passed the smoothness over to *ad hoc* inflation. Or perhaps inflation did the trick alone. It 'froze' primordial homogeneity inside a small 'horizon' that was later whooshed to what we see at large today. Or again, perhaps rough, early inconsistencies were smoothed when the process blasted out the crinkles and expansion stretched the curvature of primal volume into what appears flat space.

If *ad hoc* inflation is the same as *ad hoc* lambda (or dark energy) then that marks two down, one to go. Bang's initial conditions must, with respect to a critical density of matter/ energy, have been very, very finely tuned to generate the geometry of space, 'flat' space, we find today. Inflation 'knew' (*ad hoc*) just when to start and must have known (*ad hoc*) when to stop. How, though, did 'lift-off' cancel 'boosters' most precisely for a special, consequent trajectory - the so-called 'graceful exit problem'? This deft and crucial cosmic trick has overcome initial levity with a corresponding force of gravity that can precisely reassert its grip. Not only are the cohesive forces that sustain the universe exactly fixed but the 'flatness problem' (Chapter 8) has been 'solved'. After the booster switch was closed a 'rocket universe' was left to coast, by its momentum, into expansion. The big bang's lucky long shot was indeed well gauged. Are two bursts extra orderly? If expansion was too slow or if the singularity had blown up too massively then gravitational collapse was on the cards. Such space is called 'convex' or 'dense' and its closed universe would perhaps finish at the 'crunch-time' of a cosmic black hole (although you may well ask why everything should fall back to a single point). We would not have reached the point that we are, here and now, able to observe. If the opposite applied then there would have been a blasting past redemption. Such 'open', 'saddleback' or 'concave' space would have thrown things into a widespread iciness. Big chill. What a lonely, sunless world we wouldn't live in. It seems, however, that our 'flat' space lies and has always lain exactly in between the numerous possibilities of 'angle' that both concave and convex types of space could show. It would have needed an initial thrust (or explosion on explosion) tuned to one part in many trillions (10^{62}) of the value critical to obtain what is observed today and therefore, obviously, lets us exist. Our spatial geometry is neither convex nor concave but flat for continuity. Extraordinary! Miraculous! Perhaps a multiverse, according to the *SAP* (Chapter 4|: strong anthropic principle), has beached us here; or, *WAP* (weak) because a single universe has specially zoned us to a region where the spatial geometry's just right, we have happened to evolve. Perhaps, though, inflation solved these problems and delivered us, as it's observed, the cosmological principle.

Inflation certainly inflates the big bang story's plausibility. Does it (by a stunning and imaginary factor of 10^{30} in a fraction of a millisecond) really solve the wind-up of the way the world has wound up and, since we are matter's sons and daughters, we have wound up where we are? Our father who art earth, hallowed be thine eyeless, stony stare! Cold, lifeless comfort one way, physics' swab upon an ailing bang the other. Are not, as far as life's concerned, inflaton fields as wondrous and yet numinous as deity? Does not Guth's scientific outburst represent a great foreshortening of creation week? At any rate, materialism hangs upon ideas like these.

Deflation of so-far-imaginary inflatons stoked, you contend, the heat from which all matter has been cooked. Here arises more severe embarrassment. With respect to about 90% of its own subject matter science seems to have to speculate. Critical density is around six atoms per cubic metre of space but all matter discovered so far accounts for ten times less than this. In other words the relation between actual and critical densities needed to achieve flat space like ours is out of kilter. Much more energy and matter than we see are needed. To keep the theory from collapsing masses of *ad hoc* fudge-factor, 'dark matter', must be shovelled in to fill the shortfall in our actuality. And, of course, cosmology's great quest is now to hunt this ghostly heaviness to ground. Theory needs it; is it in fact a wild ghost chase or is the 'dark stuff' more than theoretically real?

Are you grasping the parameters this bang ejects are, to produce our current state of cosmos, so exact that they resemble very, very fussy engineering - precise and tightly interlocked fine-tuning of constraints? You've already met *Ad Hoc* - a bouncy, cheerful chap who's never short of an idea. With 'inflation' you can see that he is magic. Inscrutable as Jeeves he's always there to tickle and elicit, smoothly guiding *PAM*'s great miracle along. Like a genie *Ad Hoc* always rubs an answer up. Has he by now extemporised sufficiently to answer every question? Are you happy nothing could and did explode, expand or rupture space?

Your hand is up. A rush of questions has occurred. From what, *Ad Hoc*, did energy appear? Why should a 'latent' or a 'virtual fluctuation' suffer a severe inflation and then express peculiar patterns of behaviour we call 'laws of nature'? Is solid texture latent in a gas; were those laws and all the qualities and properties of matter latent in a primal 'fluctuation'? How does a wave solidify? How does energy condense or spin into a particle? And wasn't Newton's case that something hurled through nothing travelled uniformly straight? Hurled or carried by expanding nothingness, the point's the same. Why should detritus from the bang-to-nothing in a vacuum slow, diverge from straight and narrow 'rays' and then materialise in clouds? Why should particles fly straight apart through space and yet 'condense' together in a mist? Why should the primal products of such 'condensation', spinning atoms, ever congregate as clouds of gas? Gas in space expands; why should primordial gas cave in to your demands? Why should a quantity of molecules collapse together just as they are hurtling fast apart; how did gravity get grip? Let's allow an outside chance it did. In which case why, when these clouds 'collapse', should galaxies and all that's in them - even perhaps the broader universe - come to embody angular momentum? Everything rotates but how, in spite of Newton's law, did their rotations start? What first flicked spin in nothing? Did the bang itself fly round? And how did gravity produce networks of planets, stars and galaxies that for the most part, orbiting like dervishes in dance, don't collide? Something from nothing isn't quite as easy as it seems. It doesn't all add up. Do you mind if we invoke *ad hoc* responses yet again?

So far everything's been conjured up from nothing - despite an explicit statement from the First Law of Thermodynamics that you cannot do this. In what a massive contravention, therefore, to this unbroken rule was the one that, luckily, completely broke it for a start. Indeed, how could Conservation Law that had not yet existed be smashed to smithereens before it started by a revolutionary bang? How can nothing change from being nothing? How could what is outside law deliver law and order or, if there was prior law, could its immaterial blueprint influence the behaviours of all material things? The complete spontaneity of nothing, it is argued, generated patterned physical phenomena. A single spurt delivered just the dollop that, far from splattering space with random blobs, gave rise to highly fashioned, astral orreries and all (including you) that they contain. Yet big's 'bang' is a word with connotations of explosive disorder. *All uncontrolled explosions create chaos; double ones make more.* And physical not psychological eruptions are completely mindless. This is not, as we have seen, how creation must have happened. Its order must have driven out of nothing like a Rolls Royce going through its automatic gearbox with a foot hard down on the accelerator. Final revelation of the origin of cosmic canon, the source of universal information and spring of natural law may well explode the myth of an explosion. Everyone once swore by ether and ethereal waves. *Could big bang metaphor be a misleading one and, possibly, completely wrong. In this sense bang would go the bang we swore by. No bang.* **Perhaps, however, a precise projection.**

Stop the watch! You've already missed the moment (less than a second short) when quantum pair production is supposed to seed the cloud of being. In this process vacuum yields its fruit - but have you missed the massive anti-massive problem too? When energy makes matter it makes anti-matter in the same proportion. Where has all this anti-matter gone, long time passing? On meeting up the two poles mutually obliterate so why are either of them left? Why is any cosmos here? Nor is there an imbalance; stray anti-atoms never float about. A pico-second is about as long as anti-particles can last so even if you up anti- it's still very nearly nil.

Ad Hoc knows he's pinched. He squeaks, because his life depends on it, that perhaps some rare occasion generated matter only; or for some lack of reason polar types of matter must have got divided into different universal zones, or distinct galaxies or... Perhaps they didn't separate; quicker than it takes to wag a finger matter must have knocked out all the anti-matter but still, breaking their pair parity, had some to spare. The result heaped, perhaps, a billion-to-one excess of material over anti-material particles; but this excess contained by chance, bless me, a one-to-one charge parity of protons and electrons! This tiny remnant is 'our universe' and, of it, a far tinier fraction than one billionth is employed in making you. Any which way, whatever constitutes our local habitation has survived intact and free of lethal competition. Utter speculation. Critical excuse. Another rescuing device to quietly save big bang, long ages and the vital evolutionary view that hangs upon its thread. Well, goodness gracious me!

A week is a long time in politics but, according to current theory, almost an eternity in terms of how ignition and its cosmic blast-off flew. The quickest ever trick. A few billion trillion trillionths of a second is an evanescence with which even an exceptional magician can't compete. In this mini-moment super-force, inflating twice, is calculated to have pulled a great white rabbit from its micro-hat, defined its outline into separate forces and, as the cosmos cooled for a short time, resolved its 'fluctuations' into various particles (called photons, leptons like electrons and so on).

Reset your watch. In a second thrust fades. Protons rear and, after several minutes more, neutrons and nuclei. From now momentum will secure expansion sufficient unto evolution. From a searing, billowing 'fog' of radiation a transparent cloud of hydrogen, helium, a little lithium and beryllium appears. Until the fireball cools below the temperature of our sun's surface sheer heat strips off electrons that might tend to join with nuclei and form primordial atoms. The age of plasma lasts 300,000 years; then it is said that the expanding matter cools, free electrons fix to nuclei and the general radiance disappears. For some reason a dynamic, potentially perpetual equilibrium between nucleus and orbiting electrons is established. *In universal darkness atoms are composed.* A transparent, silent cloud of gases swells through space. If you need to make a space inside of nothing then the swelling makes space for itself. This is *hyle*. It is *prima materia*. The simplest atom made of just a proton and electron won't solidify at any temperature. Nor have even neutron stars crushed an electron, proton or a nucleus. Perhaps it is impossible. Their size is point-like but their strength is adamantine. They perpetually support each massive object in the universe. Once atomic structure is established bonding can occur; hydrogen gas is formed. Once inflation has exhaled a large and dimpled universe then gravity clocks on. *Now the real process can begin - from hydrogen to humans, molecules to men!*

Yet, as was asked before, how can a fluctuation ripped or mini-currents flung apart through nothing start to slow? Why not smooth and uniform expansion with its cloud of particles just thinning out? How could frictionless velocity be overtaken by a second runner, gravity? In other words, what is the friction space applies to cause deceleration, curvature or (given an accelerating, divergent space/ mass ratio for particles) make the bang's projectile 'clump', 'collapse' or pick up angular momentum (do not some galaxies and other systems deep in space rotate)? Radiation in a vacuum is a ray-like business. What experimental observation might be brought to bear on how a void's explosions organise themselves in ways that are not straight? Might you dig a fossil of the very Bang from space? We've just mentioned homogeneous *CMBR*. Perhaps now's the cue for *COBE* (COsmic Background Explorer) and *WMAP* (Wilkinson microwave anisotropy probe). If cosmos *is* expanding there must be an outer limit, one defined by distance of expansion. This boundary must, by banger definition, be the limit of background radiation (*CMBR*) because this 'fossil light' is practically the 'bones' of Bang. Has not the rubric of universe been 'set in stone' this way?

Observation demurred from prediction. Theory espied what it did not expect - an 'unstraight' universe. In respect of radiation *CMBR* seems homogeneous, isotropic, smoother than a baby's bottom. Problematically, however, the actual as opposed to average distribution of material is not. Massive super-clusters made of galaxies reel amid great voids of space. There are more than two hundred in the Milky Way alone. You don't find galaxies outside them and within the galaxies themselves globular clusters of stars revolve. These clusters have a minimum size below which they cannot exist so how did they first form? How did they build up to the point that 'let them stay' - unless you think that clouds fragment and 'fluctuations' regularly fall apart. Indeed, how do galaxies not fly asunder due to expansion of the universe - unless dark matter helps gravity to rope them in? And if their rotation is contained why do they rotate at all. If centrifugal forces are dispersing them how were these elegant constructions first established accidentally? What started all the 'lumpiness'?

Could early 'quantum fluctuations', once inflated, do the trick? Irregularities in primal Heracleitan flux are proffered as an answer. If, therefore, irregularities or clumps of gaseous fireball separated into lumpy filigrees, nets of stars and galaxies and super-clusters then you should find related hot and cold spots ('discontinuities') in the original background radiation. You do not. In 1989 *COBE* was launched to try and find large-scale granulations in the smoothness. It is from such corrugations on the face of space that everything is supposed to have inflated - or evolved. These 'wrinkles on the face of time', signifying youth not age upon the cosmic brow, had by 1992 still not been found. It seemed the skin of youth was really smooth. Computers were engaged to enhance the signal to noise ratio until minuscule variations were indeed detected. In other words, although you can barely distinguish between signal and noise, statistical evidence was supposed to demonstrate hot and cold 'puckers' that differ by less than a 'most delicate' one hundred-thousandths ($30\mu K$) of a degree.

Then a more sensitive *WMAP* flew and, using devices to detect minute temperature variations of 1 in 100,000, found such 'real' variations on the cosmic flatness and homogeneity as close examination of a shellac finish with a microscope might possibly disclose; but, more surprisingly, ones that also indicated the presence of north and south poles and equator, that is to say, a kind of cosmic globe. A polarised, global universe, perhaps one that is revolving, would be for bangers strangely unexpected and unwelcome news. It is not that anybody wants to reconstruct a model like sage Aristotle's 'spheres of concentric harmony'. Who needs seven heavens to envelope earth? But, curiously, quasars in all directions of the sky show red shifts that are not continuous in scale. They seem clumped (or 'quantised') in their values. Is this a sign that galaxies are found, like electrons in their orbits, on the 'surfaces' that constitute concentric shells or spheres in space? You would need be stationed at about the centre of the universe to observe the kind of pattern that we do and so, by cosmological principle banged up against bangerian law, it can't be true. But, though you don't want it, is it?

Be that as it may, *COBE*'s micro-fluctuations in temperature have, with *WMAP*'s help, been enhanced into sky maps perceived by many critics as 'beautiful'. Science certainly evolves. Now images from the Planck satellite sharpen and refine the detail of a massive, starlit universe. Perhaps analysis will highlight further fluctuant subtleties from which our own lives flew. In a theatre make-up artists sometimes 'apply' wrinkles. Heat contours on the space probes' maps are marked up with contrasting colours - yellow, orange, green and blue. From such cosmetic corrugations on the god-less face of space, it is easily suggested, everything inflated and evolved. The beauty's in imprints of time's first split flash - deep, deep history and, it seems, simplicity.

You're sure there's no dispute? No wishful thinking colouring how those enhanced minutiae might be interpreted? Although most are happy with freezing cold radiation, a few still feel this might mean skating on thin icelessness. Such 'wrinkles', on which the bang scenario depends, cause frowns. As noted earlier in this chapter (in the section Alpha Points) several reasons other than bangerian might cause various red shifts and the freezing radiation Planck/ *COBE*/ *WMAP* analyse. Heat of starlight, heavy gravitational fields and the passage of background radiation through large clouds of intergalactic gas are possibilities. Its collision with electrons in the cloud would cause so-called Compton scattering so that some of it would pass out of our line of register. The resulting changes of intensity in the radiation reaching us would be interpreted as a change in temperature. Variations would therefore depend on the various number and size of invisible but intervening gas clouds. The original source of radiation would, as the data without computer enhancement showed, have been practically uniform. Smooth uniformity does not create such isolates of matter as suns, super-clusters or whole shining galaxies in flight.

Let's rephrase current theory in a simpler way. Everywhere you look the material composition of the universe is broadly similar but how can *WMAP*'s very tiny 'crinkles' spell out stellar conurbations floating in the vast, dark tracts of spatial countryside? Old bangers thus rebel against the uniformity. Old school conformists need to break original smoothness up. They assert that minuscule irregularities, blown up upon the film of space and registered by *COBE* and *WMAP* as 'initial conditions', have contracted into galaxies, stars, planets and, least in size and astronomical relevance, people. If, on the other hand, things had set out just a tad too lumpy (or gravity a tad too strong) these irregularities would have degenerated rapidly enough into black holes that would, as gulpers will, have swallowed everything around them. If, therefore, the results are right and the wrinkles are not artefacts they must have crinkled up with great and serendipitous precision.

A torch can rove odd corners of a treasure tomb. Don't you wish a cosmic floodlight would illumine everything in space at once? What about the really large-scale structure of the universe? Could a humble crease in nothing plaster sweeps of super-clusters 300 million light years long and 100 million miles thick, stretching out for a billion light years and separated by voids of 300 million light years? Or even larger structures like the 'Great Wall' (reported in 1989) and patterns of clusters flung across about a quarter of the diameter of the universe (seven billion light years)? According to standard big bang cosmology such conglomerations would have taken 150 billion light years to form. Peer still further out. In the deepest recess galaxies seem to accelerate away towards the speed of light. What, anti-gravitationally, is sucking them into the endless void? And Something Else! An all-sky red shift survey of those galaxies detected by the Infrared Astronomical Satellite revealed many more great super-clusters than current theory reckons there should be. Could they make up the 'missing mass' that dark matter was invented to provide?

Why should mass be missing? What's it missing from? Do you remember (from the 'flatness' problem) that expansion energy and gravitational energy must have remained 'fine-tuned near unity' with respect to ratio throughout existence otherwise collapse or wild and lethal expansion would have put paid to a habitable world. Not only that but it transpires the world's expansion rate is identical throughout the ancient sky - good going for explosion on explosion. Gravitational energy is, however, related to mass; and judging by what we can see (luminous matter) there isn't nearly enough of it to have resisted Bang's expansion in the 'equalising' way it seems it has.

Do you recall Einstein's *ad hoc* invention called Λ (lambda) or the 'cosmological constant'? It was a force of anti-gravity or (*raj*) levity, a push to resist the pull of gravity and keep the cosmos balanced. Nowadays this 'push' is tied up in the so-called lambda force, dark energy or 'energy density of the vacuum'. Dark energy is held to be what might accelerate expansion of the universe. And according to Heisenberg's calculations it might contribute 65% of matter in the same! *Ad Hoc* reminds that no-one's ever seen this dark, unholy ghost. Having bolted one inflation to another this more than slight embarrassment has not appeared - because you cannot see it. Yet some believe that unobserved 'dark' elements constitute more than 90% of cosmic material. Why? To explain what we observe today in accordance with Concordance (Standard) Theory it seems that there exists a shortage of natural, visible matter. Indeed, so dire is the shortage of this 'common-sensible' stuff that ordinary atoms in galaxies (2%) and intergalactic clouds of gas (3%) have been calculated to comprise only about 5% of cosmic mass energy! About 0.1% is ascribed to 'hot dark matter' (*HDM*) in the almost observable form of neutrinos, very common and yet noble, unreactive ghosts that are supposed to travel right through planet earth (when a similarly chargeless but even smaller photon or even, perhaps, neutrinocan't). Do they really make it into existence? They seem to scrape and hover round the entry-line but another 25% of cosmic mass is ascribed to something else - *CDM* ('cold dark matter').

What off earth might be nature of this *CDM* (or *CCDM* if you count it 'cold' *and* 'collision-less')? All sorts of guesses throng the space of mind. Although it is conceived as flitting unseen and in quantity through you at this instant, maybe such non-atomic matter will defy detection. Its *ad hoc* exotica are dreamed of in varieties of state ('fuzzy' *FDM*, 'warm' *WDM* etc.) and component (*WIMP*s such as axions, heavy photinos and neutrinos, magnetic monopoles etc); also in the form of *MACHO* - massive compact halo objects - such as brown dwarfs and marble-sized black holes that weigh the earth! Magnetic monopoles are problematic. Why, if separate electric charge is normal, can't we find separate magnetic monopoles? Theory may predict them in enormous quantities but we find absolutely none at all. Indeed, experiments have not illumined *WIMP*s or *MACHO*s as the source of dark mass either. This might be because a theory, even theory as big as the bang, could still be wrong.

There's something else. Another matter altogether. What about the other 65% of things? *For Natural Dialectic levity is logical.* Is there, however, actually a force of anti-gravity as Einstein with his cosmological constant (Λ) suggested? Might it be the weakest force in nature only overcoming gravity in vast and lonely spaces? Or could it, reckoned trillions of times greater at the primary moments of inflation than today, be the motive power that first propelled expansion of the universe? *Lambda* is a self-repulsive force, a negative pressure (suction) with, like a Hindu deity, several aspects of its face. Its expression as the cosmological constant is spread evenly about the universe; its chemically inert presence takes the same value everywhere and, like the vacuum that it seems an aspect of, it neither emits nor absorbs light. As so-called 'quintessence' it can vary both in space and time, perhaps like *ZPE* or Dirac's virtual sea, in fields where particles are 'dancing'. In this sense *lambda,* dark with vacuum energies and the cosmological constant are conflated as aspects of levity and, perhaps, the fifth, ethereal, causal element of our ancestors (see Chapter 10: Old Vacuums).

Forget the ether since another altogether larger problem rears. Quantum field theory predicts a huge vacuum energy (10^{93} grams per cm^3). All known galaxies squeezed into that cm^3 would only aggregate 10^{55} grams! Yet, in the context of such vast vacuum density of energy, relativity theory demands a large value for the cosmological constant - but an infinitesimal one (10^{-28} grams per cm^3) is actually observed. If it were greater by a single power, a minuscule amount, then you and I would not be here. Quantum huge and astronomical infinitesimal. **The difference between the cosmologist's classical and the particle physicist's quantum version of dark energy is a massive 120 orders of magnitude!** Professors gloss a problem which is vast. By far the worst 'fine-tuning' chime between fact and its theory in all physics grates and jars and clangs: it stretches your credulity by leagues. This is vacuum catastrophe. What is it that we do not understand? **Is what we call physical energy/ matter, our universe, just a tiny, tiny leakage out of space?** What will reconcile, what cancel discord out, what solve the problem? Why is the actual density of spatial energy so small when theory demands repulsive energy with power to blow the universe away in seconds after it began? Yet if the actual value isn't figment how can dark energy exert its critical effect and, like Atlas, carry dust clouds and star-laden skies of cosmos piggyback?

Critical effect? Critical indeed. If dark energy is missing big bang theory is exploded. Blown apart. Luminous matter, which you can see, seems to account for less, perhaps much less, than 10% of the critical mass required to keep cosmic expansion stable enough to have lasted, without either collapse or anti-gravitational expansion ripping everything apart, as long as it has. What, moreover, stops Bang's particles from ever flying more apart? If normal matter's all there is galactic super-clusters, let alone a single galaxy, should never form. *CDM* is postulated as the cosmic glue. If this fixer isn't real it indicates, unthinkably, that on the large-scale even understanding of the force of gravity is incorrect.

Formation's one thing. Even motion and continuation seem to need dark influence to swing them round. Measurements reveal that stellar orbits at the periphery of a galactic spiral have rotation velocities at least as high as those at its centre. This indicates that mass is not only concentrated where we see it, at the luminous core. Such a core might include invisible kinds of 'dark matter'; indeed, it is believed most galaxies are stapled to existence by a 'super-massive black hole' struck somehow at their heart. Black holes stand for closure yet here they act as vital hubs for wheels of stars; but, in order to account for angular velocities and conserve coordinated orbits of the structure's stellar networks, a galaxy's whirling, centrifugal tendency would need to be compensated by an outer 'membrane', 'skin' or 'halo of darkness' that resisted disintegration. Think of a galactic burger sandwiched in its buns of *CDM*. How do you make *CDM* do that?

Perhaps even vast tracts of space, behind whose voids you can discern far-off star clusters, are full of dark, burdensome, self-attractive transparency. Maybe luminous matter simply floats in an invisible sea of mystery. It is estimated that perhaps 95% of our own Milky Way comprises such a heavy absence. What precisely is the nature of the silent, undetected and yet (if state-of-the-art cosmology is right) apparently essential majority of unseen 'stuff' that the visible minority of matter rides upon? No physicist has certain evidence it's here. Although the presence of this alien reserve is 'missing', large quantities of its fudge factor are required to promote the anti-gravitational acceleration of an expanding universe; and, at the same time, 'flatten' space and supply sufficient gravitational pull for initial radiation, flying apart at high speed, to coalesce into 'confinements' such as super-clusters, galaxies (both of which need lots of *CDM* to make them stick), gas clouds and gassy stars. Are you still in the dark - the darkness of both cosmic and theoretical glue? In order to solve its problems astronomy seems stuck with the idea; and yet if *CDM* and *CDE* are true its telescopes have still to register up to 95% of universal substance - intangible substance composed of a substantial paradox! And if we only understand a residue of 5% how do we know the conservation laws apply to all the rest? If not, could fresh dark stuff be converted into the kind of luminous matter that we understand and thus continually increase its sum? A revolution could sweep physics lecture halls. **In fact, calm down. There is no evidence for inflation, *CDM* or *CDE* outside the cosmological arenas for which they were invented; but, in order to bridge the gap between the predictions of theory and actual observation, a big bang needs fiddle factors as a tottering Ptolemaic theory (of earth at the centre of cosmos) once needed layers of epicyclical adjustments.**

What other scientific discipline would tolerate not just a rule or law but universe *ad hoc*? Is this not *Ad Hoc par excellence*? All in the name of building worlds by chance.

A detective always tries to explain an unknown in terms of as many known facts as possible. An accumulation of the evidence adds weight. However, to explain unknowns by using unknowns sounds a trifle threadbare. To use *ad hoc* all down a chain of reasoning wears thin. When you want to find the truth professional use of the imagination's fine - but do 'inflaton fields', brane meta-worlds, 'dark materials' or myriad exotica from physics' cornucopia, the mind, add *gravitas* to argument? Or *levitas* that lifts it out of seriousness? Is not science brought to fiction, big bang to rights, by the quantity of elements required by theory but not found in fact? Just think of *CDE* and *CDM*'s transparency. *PAM*'s miracle is stripped. Lady Luck has lost her clothes but, being 'naked nothing' in the first place, bareness isn't any loss. Her Theory of Origin is, though, blushing red. It's intellectually embarrassed. By rights it should have self-effaced, imploded from its mass of black marks called *ad hocs*. Is it really just the fiction of a phantom carefully constructed out of unreality?

If a stream of '*ad hocs*' pours from *PAM* then why not try, with its single immaterial quality and no exotica required, the *PAND*? The answer is a simple one. If your mind-set's your religion then the mind-set of a secular religion with its naturalistic methodology eschews consideration of projection out of metaphysic; it excludes the notion of a cosmic archetype in universal mind. **No doubt, the instant that a baby's born then you can start to measure it; but measurement alone is not an explanation of the origin of seed, its informed code or program of development. These involve a different level of perspective - one including immaterial conception.**

Twinkle, Twinkle.....

Do you want to be electrified? Electrified by awe and wonder at the place we live? Where galaxies like golden raindrops drift across an everlasting sky and, in that night, the light of billions of suns composing every drop showers down on you? Go to a planetarium, be overwhelmed, sing in the rain!

While you're singing you're not thinking! *Don't think modern theory's infelicities dry up with vacuums, particles, big bangs and exotic zoos of matter.* What is the universe apart from what you see? Stars, suns, moons, clouds and galaxies compose its regularities. Why did these structures come to be? Objects lose energy when they collide with others. How, therefore, did a primordial spray of light or subatomic particles condense to atoms, molecules and nebulae? Perhaps expansion was the coolant. Was it cooling that, with superfine precision, contracted into stable protons - only 0.2% heavier would have generated instability ensuring no atomic nuclei, atoms, molecules and therefore stars and galaxies could have emerged from that first, legendary mist? Similarly, whence 'condensed' an exactly complementary drizzle of electrons? Then again, if you have monotomic hydrogen, how did it cool into expanding gas? What a gas poured out of nothing's maw! The word 'gas' flows from χάος (chaos) meaning space. How does gravity contract a vast cloud of expanding chaos into 'great walls', super-clusters or plain, single galaxies? How does it even crush a small one to a star? Perhaps the contractant was a vortex or a massive swirl that 'something' set in motion. Molecules of hydrogen are made in stars. Before a star what finger stirred them up, made helium and spun the galaxies like candy floss? Did it then, against expansive pressure, press stars like nodules out of whirling dust in space? But hydrogen is never ever solid. How did it solidify to stardust and the other, larger objects wheeling round in space? How were the objects that astronomy observes first made? The truth is that we do not know; therefore, never having seen the start, we endlessly and faithfully to one creed or the other speculate.

If galaxies are old they should, by now, be far apart but most are found in clusters. How do you regularly but locally inhibit the expansion of a plasma or hot gas so that its particles don't fly apart (as is the wont of hydrogen or any other gas)? How, in other words, do you compress expansiveness precisely to arrange some stellar productivity? Exactly how do you inflict a 'shock' upon a nebula sufficient to compress it to the point where gravity can start to act? After all, the only force you have is gravity. Ionizing radiation will induce electric currents and magnetic fields. The turbulence of these will, even at low temperatures, work against contraction. How, therefore, can you 'prime' your cloud until, against expansion, turbulence and any other kind of levity, it's ready to contract? You see the problem. It just doesn't want to happen, does it? Even less so early on when galaxies were thought to form but gas was even hotter. This is called a 'priming problem'.

Ah, so! Puzzlement. A smile from glistening heaven heavy with inscrutability. From particles to galaxies still stretches mystery but *Ad Hoc* with a flourish waves his Chinese wand. Could the answer, cries this masseur turned magician, beat like gong? Could 'fluctuations' do the trick? Didn't *WMAP* show that tiny quantum rippling of the singularity broke through from nothing into physics and was amplified, *ad hoc*, by inflation's bigger bang? Wouldn't wiggles on the *WMAP*'s sky charts indicate, vastly enlarged, fluctuations that have formed the 'seeds' whence galactic clusters are sprung up today? Surely spots and ripples of 'irregular density' must have once inflated into distributions of primeval gas from which condensed the embryonic galaxies? Should these 'wrinkles' not, like crinkles on the surface of a balloon, be smoothed out with expansion? Perhaps they should but they must not because it's from their preservation that you're going to explain all objects in the universe. Therefore, *ad hoc*, you speculate the opposite. Expansion long enough may 'damp' them but it also amplifies them into forces of collapse! You can now, in this wobbly and self-contradictory way, not lose the world you were inventing. Shaky logic keeps the edifice of all things firm. Original *ad hoc* quantum fluctuations, you announce, have grown up; they have turned, *ad hoc*, into 'perturbations of gas density' or, if you prefer, 'fluctuations in a cloud' from which the galaxies will rain. Gravity, *Ad Hoc* confirms, is as keen as bangers are to pick a fluctuation out and pull it into shape. Just give gravity sufficient time.

Is that clear? Yet within clusters individual galaxies make complicated movements, some away and others towards each other; and, Fritz Zwicky found, they're moving faster than they should. Gravity can't hold their horses so, *Ad Hoc*, shall we simply add an unseen 'immateriality' to rein the matter in? 'I've already said', he testily retorts. 'Why not lasso each galaxy with that hypothesis, dark matter, and embed its haloed axis with a hearty black hole so it doesn't race apart? Then clusters won't disintegrate.' And what's dark matter? 'Easy, you already know! It's a non-entity that I've invented with sufficient power to net the universe. Its density, it seems, is highest in the oldest, largest galaxies and thus its web perhaps helps to glue our cosmic company; all clusters, galaxies and crucibles of stars are formed, maybe, along 'threads' of dark matter. Hey presto! Now you've got the gravity that observation craves and theory can't satisfy. Nobody's seen this 95% of mass but it must be there or else the world would fly apart. Is it part of Higgs field? Surely physics does not hang upon such heavy but entirely ghostly chains; or wobble on its razor edge between vast, dark expanses, chasms of another great unknown? Through all the fog of explanation one sharp outline is clear: galaxies are orderly and complicated but it isn't known how they form.

If you are unaware of crucial players you can't fully guess the way a sport will work. It's no wonder galaxies-with-stars were mysteries. Even now, the facts are little more than plausible hypotheses. Is it impossible, therefore, that a transparent immateriality called archetype might take the ruling role of referee in setting up the cosmic fun-and-games?

Anyway, dark matter. That was easy, wasn't it? But having conjured sufficient gravitational force precisely how do you obtain a right-sized fragment of your nebula, a cloud of just the weight? Gravitational collapse of matter summing to a billion stars cannot be allowed to disappear down into the blackness of a hole. Implosion pulls the same plug that explosion pushes; you could lose a universe that way as well. Even fifty thousand solar masses would defeat the 'purpose', flushing any hope of life's potential down a cosmic drain. It's not black holes but stars you want. While collapsing all together you must have a cloud that also falls apart. Individual chunks of gas must break apart and each keep doing so until, all in a cluster, many galaxies appear! Or perhaps some swirling chunks could twirl together, merging to a single galaxy in which another dust of sparkling stars falls out and starts to light the darkness. Add 'density' to 'fluctuation' and, *ad hoc* again, you can explain both gravitational collapse against expansion's grain and fragmentation into billions of stars against compaction of the great collapse. *Ad hoc, ah so, ad hoc.* Truly wondrous are the works of hydrogen, helium and 'fluctuations in their density'.

Stars. Stars are supposed to twinkle in your eye because great chunks of the exploding, primal gas fragmented into showers of solar sparks or, actually, nuclear infernos made of gas. How? *Ad Hoc* could only wobble when it came to galaxies; now observe his trembling lip. How do you carve, by nebular hypothesis, the sharp precision of a star? In hot and early skies where stars first formed, expanding clouds would have to top a cool 100000 solar masses for the force of gravity to grasp a grip and the compact them; but such size would in theory only yield black holes. The greatest size that's known for stars is about a 100 times our sun's and this would yield black holes as well. You have to make a ball of gas and not a black hole so you'll need small fragments of a nebula. Your fireball has to drop below 10 solar masses. Statistically the ones as small as this should be in a 'large minority' - so black holes should outnumber stars. Where have all these black holes gone, long time passing? Various 'small' masses give you various kinds of star but don't you think that big chunks should have broken first? In other words, what manipulates a cloud to break small fragments off and then compress those relatively tiny cloudlets till, now in tight enough a knot, G-force can assume control and drive the gas to its spontaneous combustion? And how is it, if these random vortices occurred, that all star-clusters in the universe exhibit the same range of stellar sizes? Could *ad hoc* turbulence perform the task of whipping up a globe-shaped cluster of a million stars? Is making up so hard to do; or is it just that breaking up is easy and, up against expansion, heat and turbulence or against contraction that would leave you with a black hole, the 'fragmentation habit' everywhere is 'just in balance'? 'In balance' is the only way you'll get a lasting star. *Ad Hoc*'s replies are, as you might expect, verging on the nebulous. You can, however, lip-read from his vacillation and the phrase is 'fluctuation in the gaseous density'. Like wind in mist. Is such a foggy *ad hoc* answer clear or not?

OK. Levity (↑) disorganises, dematerialises and throws things energetically apart. No doubt, therefore, some (↓) gravitational compress is the way, along with cooling, that materialisation works. Localise, de-energise and fix. Are gravity, contraction and collection not the order of a cosmic builder's day? It's funny, then, how the thrust of squeezing everything together ends up with it floating carefully apart. If you're a star you keep your distance (3.5 light years at least) and even in the heavy traffic of a galaxy don't infringe your neighbour's flying space. You do not make revolution by collision but by regularity of circulation. Does anarchy create a mass rotation or falling-all-together generate the separate yet correlated orbits of a million lesser systems that, wheels within wheels, compose the whole? Do these two losers, anarchy and crashes, ever govern galaxies and all that galaxies contain? Collisions do occur. Was it, for example, 'shock' that kicked irregularities and generated 'swirls' of dust and gas to build our humble solar system? Most of its mass (99%) is concentrated in the slow-rotating sun; but the great majority of angular momentum (also about 99%) is spent rotating satellites. Such facts oppose the nebular hypothesis but if the planets did not orbit fast enough they'd crash into the sun; and if the sun did not rotate so slowly it would fling away its mass. The system spins around in balance, defiant of hypothesis but mindlessly content. What rough 'fluctuations' must have generated, even in this minor case, the equation that allows sun's planetary specks to orbit almost perfectly and timelessly? How easily is our simple system actually explained?

Stars much below our own sun's mass are destined to burn out as white dwarfs. One like our sun will, when its hydrogen supply is burnt, collapse from hydrogen to helium fusion. A red giant swells up and, at core, carbon and oxygen are formed. This is lucky for us but (check Chapter 8: The Match is Very Friendly), given how the world's fine-tuned, inevitable. If the star is double or up to about eight times heavier than the sun then its alchemic game's not run. During a further contraction it will mint elements until the final transmutation into iron. When it at last these stars explode the core, a white dwarf, fades to black; and from its outer shell elements are thrown out in planetary nebulae. It is thought that from such dusty nebulae the resurrection of a fresh stellar generation is 'condensed'. Stars over about ten solar masses are also of elemental interest. After burning fiercely they cool into super-giants, the shell of whose collapse contains elements heavier than iron. The inner part becomes the densest form of ordinary matter. Electrons are squashed into the nucleus so that, where a teaspoonful of material is heavier than a large mountain, no space is left in neutron stars. If, however, the remnant is more than three times as heavy as our sun, it will implode into the grave called a black hole. Meanwhile the shock wave of collapse blasts off the outer part of the star into an expanding cloud of gas and dust called a supernova. The seed of heavier elements is sown and then reaped as 'fluctuations' draw it into other stars until these in turn burst as planetary nebulae or supernovae. Re-genesis. Generations of element-bearing parent stars. Is the story of such genesis a possibility? A supernova is a rare event and even the most distant and, therefore, early stars we spy are not observed exploding. Each explosion does not, moreover, lose much matter - a supernova maybe 10% and novae less. Of this 10% the Crab Nebula (exploded 1054 and still under investigation) appears to consist of hydrogen and helium with a few 'dabs' of nitrogen, oxygen, sulphur, argon and nickel. Not much heavy fuel! Have sufficient repeated cycles of supernovae (explosion/ condensation/ explosion and so on) really happened? At the Crab's sparse rate could the rarity of such explosions actually have populated the universe with (outside hydrogen and helium) such quantity of natural elements as compose your body, the earth below your feet and trillions of other fragments making up the bulk of universe, that is, the heavier range of elements that the periodic table in the lab displays? How else? You swear by great *Ad Hoc* that chemistry evolved this way!

Compression to 'prime' the gas and dust for just one star is hard enough. It is supposed that supernovae, interstellar winds and 'fluctuations in the density' are sufficient agents to create a star, and thence another and another. Generations of such furnaces are needed for the elements that planet earth must hold in trust for life. Not only are those cycles an *ad hoc* necessity but, you ask yourself, whence came all the shock waves, interstellar winds and compressors of expanding gas before there was a star? It's a bit like trying to imagine what fist first crushed energy and locked it in a proton, in a nucleus and therefore at the heart of all locality. What kind of hand is it that gathers up pure energy, hydrogen or helium and squeezes out, if not solidity, at least a pressure cooker called a star? How did the process first get started? It's difficult to tell. Who has ever seen (even furthest out and therefore earliest) a 'first generation star' composed of hydrogen, helium and any other lightweight trace in big bang's starter gas? From what starting point could chain reactions of those 'supernovae sunbursts' first weigh in? How could they scatter forth seeds of solidity?

Just imagine a few molecules of gas. One would not blaspheme the great creator, gravity, but how can it compact them? How would heat or turbulence cause them to glue? Why should a billion, as they fly apart, be crushed together by a force or forces that can't even handle just a couple? Star nurseries are where, it's said, these forceful midwives work. Although dispersal is more likely than contraction in a vast and energetic cloud of gas and dust, it might still happen that an old star caught up in such volatility would pull material. Accretion, accumulation, aggregation might occur. Even if it was a cold star might it not, like a blade embedded in some embers, start to glow? It depends how you interpret photographs of where you think midwifery's at work delivering infant stars from wombs of cloud into their 'nurseries'. Has anybody definitely seen a 'baby' proto-star? Who has identified the mass of gas about to make a sun? Mystery and magic are the real, vexatious issue. Surely, in deep space, ancient galaxies should show us star formation? Are youngsters there? No, just a range of stars in galaxies as grown-up as 'nearby' ones we spot today. What therefore in the first place builds great balls of fire? *Ad Hoc*'s begun to fidget. He begins to dither, vacillate and....now the Chinese whispers have it - fluctuate. Both galaxies and stars are made of 'fluctuations in the density' of nebulosity! Who says such speculation's starry-eyed? Who does not accept that misty condensation must have happened? That's the problem sorted then. Next patient, please!

Planets. Surely planets are an easy part. We know so much. We've long had seven theories how they came to be; now, unlike ours, exo-solar systems are seen with giant gas planets next to suns. Do seven ideas and these exo-anomalies mean that we know or guess or what? They all first need what we found hard to countenance - a circling gas contracting by 'shocked fluctuations' to a sun. Did this ball of fire assume rotation fast enough to sling off planetary material? Did it 'suck the dusty muck' from off some other fly-by star? Or did a 'sunburst' spray enough for our locality to somehow start to circulate and spin into a 'disc'? If such a 'circum-stellar disc' rotated it might form concentric rings; irregularities in these might well, by the force of gravitation, clump and thus the disc yield various planets and their moons. But spy on Saturn. Have its rings condensed? Haloes don't congeal to moons. Perhaps, unhappily, the early set-up was not saturnine.

As probes (such as Voyager, Cassini, Galileo) push so theories crash; surprises are the order of the day. On second thoughts, perhaps, it's not so simple after all. But is it fair, amidst the fun, to think we should confess how much we do not know? Over the last forty years satellites exploring the solar system have beamed back a baffling diversity of images. Unexpected actuality has challenged previous hypotheses and raised a lot of questions. Did one huge, bland cloud of gas and dust give rise to such variety of planets? Surely such clouds in accretion should be thinnest at the outer rim where, in the only known case, our own largest planets fly? Did, moreover, such diffuse 'maternity' evolve their moons as well? Perhaps, one might ruminate, dust and gas were never wombs in space. Perhaps the planets wandered up from outer space into our star's embrace. Why then did they stop and not slam into it? Even if the solar retinue coagulated from a passing cloud of gas and dust why should it dance so regularly around instead of falling, by the force of gravity that caused it, into the sun king's central fire? Indeed, how exactly does a gas, rock, icy particle or particle of dust collide with others at a high velocity and not just bounce away? Do billiard balls collide like plasticine? *Ad Hoc* thinks so. Rocky planets like the earth were formed, his tale insists, from crashing granules made of ice and dust that, instead of bouncing off each other, stuck together. Why they should cohere and thus accumulate by 'sticking' is the sticky 'sticking problem'. They're neither big enough to fuse by melting nor embrace by gravity. Nor does stardust, which is freezing cold, petrify on impact. In fact, you need a planetesimal at least one kilometre across before the pull of gravity can start the sticking game. No single story seems to fit the facts. As *Ad Hoc* cracks, a baby planet built from flying dust and called a planetesimal is probably a ball of bluff.

Ad Hoc contracted nebulosity. It didn't work. Perhaps swirling knots of nebulosity could *suddenly* contract; the labour of 'disc-instability' might force gas planets into life. The idea improves *Ad Hoc*'s dull mood; its *ad hoc* thrust shakes his impasse. Does it suddenly grab you as well? Which way, therefore, were the disc-knots swirling when (it has been supposed) they formed into the bodies of our solar system - with Venus spinning 'backwards'? As well as planets there exist the system's sixty different kinds of moon, some spinning clockwise and the other third spun 'retrograde'. Why do they seem to fly in such contrary way? Not fluctuations but collisions now begin to dominate hypothesis. Collisions do occur but there's no evidence of such great clangers as the nebular hypothesis requires to reverse the spins of heavenly bodies, knock up moons and thereby bash the solar system into shape. Or is the answer simpler? If they revolve more slowly than rotate or their rotation axis tilts at more than a right angle bodies may *appear* to rotate backwards - so maybe in reality they don't..

Why, computer simulations ask, did ice-giants Neptune and Uranus form so far out from the sun? Why is high density, like Mercury's, found close in but far out in the deep freeze unfrozen gas composes other planetary giants? And why, after an alleged few billion years or so, is Saturn's tiny moon Enceladus still active and 'volcanically' spouting geysers into freezing space? Why, moreover, are the elemental compositions of the moons and planets all so different from each other and their 'certain' source, the sun? Of course, because it isn't known yet does not mean that a fact won't be revealed - although its absence may just as well mean that it wasn't ever there. Not all hypotheses and theories, although they constitute the promise of adventurous research, turn out correct. Therefore one retains an open mind . At any rate, whatever happens all depends on orders that the metaphysical projection first entrained.

Hold on! The source of meteorites is known. They are mostly stony with a few iron types and seem to represent, as Olbers thought, the debris from a broken planet; but could they make one up? They would bounce or shatter off each other and not, in a thousandth of a second and in freezing space, melt on impact and then fuse. A planet is a long way off from bouncing stones. We see no evidence that this is how a planetesimal is made. Never mind. Take another tick, *Ad Hoc*, for an imaginative try.

Surely, once you've got some planetesimals from somewhere they'll collide and clump together? They *must* 'accrete' into a planet. How else could it be? But remember, speeds in space are zippy and on collision planetesimals would shatter if not vaporise. Consider, then, the impact craters in our system. They litter surfaces of 'wanderers' (a planet is, in Greek, a wanderer) and moons. From all the swirling turbulence of space the natural missiles must have pelted perpendicularly. Weren't angled landings up to scratch? What only let dive-bombers through? Or are these uniformly circular abrasions mostly evidence of action from within - volcanic action? It doesn't help *ad hoc* collision theory or solve any 'sticking problem' but it might be true. That doesn't answer, though, how planets like the earth with solid, superficial rocks and molten iron core were made.

Each closer look unravels preconception. The many-coloured coat of planetary understanding keeps on changing hue. For example, Mercury remains a mystery. Current models can't explain its high degree of density or, though there's not much iron on its surface, a globe-wide magnetic field. *Ad Hoc* would call up sulphur mixed with much iron in its core. This might lower melting points and get you back the flux that magnetism needs. Who knows? It's just a guess.

If whizzing rubble won't cohere what chance has gas or ice? Was gas drawn off the sun and blown past the rocky inner planets? How then did gas giants such as Jupiter and Saturn form? A thousand times less massive than the sun, how did the former's core accrete if (as the Galileo spacecraft found) that core is small. And how did all the gases, let alone a lot of noble gases, gravitate into its regal atmosphere? Or is an icy stare the only answer that you'll get? Was it really ice whose mass attracted dust and gas - despite the presence of a solar wind that should have blown both away? Nor has the 'sticking problem' lessened in severity. Why does Saturn whirl its special rings and why should gas be squeezed or ice in any form conglomerate into the even smaller planets Neptune, Uranus and the relegated Pluto? How do comets, lumps of ice or snowflakes very thinly spread and far out from the sun congeal as planets? What's the force applied? Oh, no! Not again! If there isn't any density of matter that far out for heaven's sake, *Ad Hoc*, don't just repeat the *mantra* 'fluctuation in some density'. The fact is we, like those free tombs, are circulating in the dark. Is there much we know for certain from bangerian cosmology?

We're down to moons and still no answer. If planetesimals can't stick together how can moons shine? They're smaller than a planet and therefore less likely to pull astro-dust and plump themselves in layers. Astronauts jumped on our moon and demonstrated just how thin her 'geo-skin' is; how can whole moons build from powdery layers only inches deep? Perhaps instead a catastrophic accident occurred. Possibly a boulder smashed into a planet and knocked off a massive scoop of mantle - was a satellite once excavated from the Gulf of Mexico (an analysis of lunar soils proves, since they are distinct from earth's, another body's role in this earth-shattering event)? Perhaps, therefore, the safest bet is, about four billion years ago, collision with a Martian-sized lump of unearthly rock that buried itself in earth's core while the debris from the crash coalesced into Our Lady of the Night. Were the others in this solar system fascinating chips off their old blocks or are they something else? There's much to study and to try and understand. Funny, for example, how the orbits of so many Jovian moons (and Neptune's Triton too) are retrograde; but if they're captured due to gravity you might expect that half of them would spin one way and half the other - you don't need interfering 'creativity'. It's also funny, from an earthling's point of view, how our own random chip turned out so round and how its roundness is the size, combined exactly with its distance, to perfectly eclipse the sun. Explanation might be easier if the many moons all disappeared and earth's coy lady of the night (see also Chapter 26) were absent from our sky. She really isn't easy to explain with certainty and so *Ad Hoc* retreats behind catastrophe. Calamity's to blame! Conjure up explosions and imagine huge collisions in your scientific space of mind. It must have been, he cries, a cataclysmic sequence of events that randomly created such non-uniformity as probes of various kinds reveal in sun, planets, moons, stars and, thanks to an all-powerful concept, evolution, lucky you! So lucky since if any of the cosmic serendipity had run astray you wouldn't be here, would you? You would not have lost the plot but it lost you. That's as certain as uncertainty can be!

Even flying ghosts of dirty ice are problematic. If whizzing rubble won't cohere then, one repeats, what chance has gas or ice? How, therefore, do those fragilities called comets form, accrue their 'five-mile' nuclei and what are they composed of? If they are frozen relics of our system's origin how do they keep on coming? Ghosts evaporate in light and so do comets skimming near the sun. Each time they skim they have skimmed off a million miles of vapour tail that loses some of what they somehow had accumulated at the start. How, therefore, could such apparitions blaze their trails of glory ever since our solar time began? How does a snowball ever losing crystals last so long or does it fast evaporate? You might have thought that it was known. Perhaps comets are but fuzzy phantoms thrown out of circulation, elliptical scatterings from a disc of minor, trans-Neptunian objects called the Kuiper belt or, on a larger scale, Oort cloud? The latter is an unseen calculation, a wispy, *ad hoc* reservoir from which as many comets as you like can fly. It is a good example of what's called a 'rescuing device'. Such devices are invoked in tight spots when the evidence may seem to contradict a world-view. Atheist, creationist and evolutionist all use them. The tactic employs appeal to 'a plausible unknown'; it can thus explain away anomaly and smooth the theory to allow continuing belief. It is commonly employed, especially by an evolutionist, when 'talking up' his histories. And, of course, evidence of absence isn't by necessity an absence of the evidence. In this rider rests the weak but promissory strength of any rescuing device. In this case an Oort cloud *might* exist; and it is hoped the evidence for this supposed source of long-period comets will grow and thus confirm the solar system is billions of years old. In 2005, however, the smashing success of Deep Impact mission on an apparition called Tempel 1 detailed, for the first time in close-up, nuclear composition. Ice and silicates you might expect but why in heavenly comet do you find some aromatic hydrocarbons? Benzene, naphthalene and other 'poly-cyclic' rings are formed, at least on earth, by the incomplete combustion of substances like diesel, wood, coal and tar. And how do you explain the presence of such compounds - clays and carbonates - as form in liquid water? Chalk, seashells and bones are kinds of carbonate.

After Deep Impact came The Stardust Flyby. It defined large-cratered, hard-centred Wild-2 (pronounced Vilt-2) with its streaming jets of gas. Wild-2's coma - a luminous, nuclear envelope of particles and volatiles - yielded the first dedicated capture and return to earth of actual cometary chemicals. Analysis found these included factors formed in stellar or similarly fierce conditions. How could a streak of aggregate formed in the freezing depths of space contain materials from very hot locations? Did pieces of Wild-2 first congregate around the centre of our system, the periphery, both or neither? What do you suggest *Ad Hoc*? *Ad Hoc*, a Jeevish genius, always has an opiate answer dulling worry till some better explanation hoves in view. He returns to where we started and suggests, just as a 'primordial soup of chemicals in water' must have concentrated at some point to form a cell, so a 'soup of chemicals in space' must have condensed into the planets, comets, asteroids and moons. There's the answer. Condensation. Condensation must have made the solar system from explosive shrapnel; or from 'fluctuations' in deep space; or from gas and dust that's flying round in rings.

Who can fault an ignorance that's keen to learn? Who faults enquiry or the student's frame of mind? No teacher scorns a lad who's keen and bright and asks a lot of questions. His array of kit is stunning and, by working day and night, the young astronomer collects, sorts and compares a wealth of data. Analysis is top class and, of course, you start to understand how any system works by checking speculation up against the facts. Who can make advance without hypothesis or without a framework into which to slip your new idea? Cosmology is full of fine hypotheses set in the paradigm of scientism's faith. Hardly knowing, though, cannot rightly claim to absolutely know; nor, save as an article of scientific faith, proclaim materialism does and always will encompass all there is to know. In fact, in the last analysis '*ad hoc*' is a kind of invocation. It is not *Om* or *Amen* but isn't it a spell propelling you across a gap in understanding to the answer that you want? *Ad Hoc* is, like a scientific god of gaps, an agent of the missing trick.

What is wrong with that? What's wrong with an educated guess you work to verify? It's just that, in dealing with the inferential case of origins, different paradigms use different gods of gaps. Is such a god oblivious and in all cases made of matter? Or is one of them intelligent? Are only gravity and gene mutations the creators of life and its universal stage; or does immaterial information, with a hand as unseen as internal energy that shapes the superficial look of things, impinge? Do archetypes inform the constitution of the universe - or not? Of course, it is observation and empiricism that inform a scientific theory and experiment; but the self-same data serves as evidence for inference employed by evolutionary theories of creation and those that count intelligent design. Gods of material gaps inhabit both the Theories of Intelligence and No Intelligence. Crying shy of controversy (as government departments and educational officialdom are politically inclined to do) or claiming this is not the case does not change anything - because it is the case!

If you discovered rock formations, a computer on a palm beach or a space ship in a parking lot would you expect, eventually, to work out every aspect of their physical constructions? Of course you would. Would this information explain *everything* about them? Spaceships and computers are chock full of purposive complexity but what about the natural rocks or, by extension, formations of the twinkling, starlit universe? Of course, if *Ad Hoc* hasn't dived as far as the profundities he doesn't need a nudge from God. If he can't see every detail yet the mist will clear - eventually invisible constraints, necessities and rules will click together in the intellectual air. And, as with Mozart and machines, will the clicking rule a Framework of Construction out or in? Is non-purposive complexity inevitably the child of chance? **The nature of an object's or a mechanism's origin is different from its operation**. Even if the cosmos is *entirely* physical we still aren't sure, *Ad Hoc* makes clear, about the way its various parts were generated - not least its fabric, nothingness called space-time. That's not to say that, when it comes to cosmological development, our facilitator doesn't have a stream of plausible ideas. It's simply that the lad's from *PAM*'s school and, although his explorations push its boundaries, his ideas aren't allowed to cross. No matter that *PAM*'s scientific paradigm enlists *Ad Hoc* to airbrush ignorance, our student is restrained inside its walls.

If *Ad Hoc* is, to the extent we've seen, allowed to airbrush ignorance yet sharpen lines round any passing cloud of speculation then why not speculate, *ad hoc*, upon an immaterial factor; why not allow an extra, metaphysical, informative dimension that imposes their behaviours on material facts? **The answer is, '*Ad Hoc* is not allowed to'.** *PAM*'s discipline, unlike the *PAND* of Natural Dialectic, denies such element in nature. Speculation that includes this *PAND* is therefore philosophically out of bounds - expelled from *PAM*'s establishment. *It has, on the other hand, no argument with facts - indeed the conservation laws support it well; but from its perspective cosmos is controlled by code - whose letters are the forces, patterns of behaviour are the laws of nature and whose agent's energy.*

Just as an egg develops by preordination so perhaps universal body, inanimation's fraction hatched from Essence to a three-tiered universe, develops its expression through an archetypal plan. The real issue then becomes the state of egg. Was initial set-up of the cosmos accidental like *PAM*'s miracle? Or, as in the case of egg, pre-formulated with a passage to maturity in mind? Of course, no hard-boiled egghead out of *PAM*'s container can accept pre-formulation or a process with a purpose - even the idea is chipper levity!

(*Raj*) Levity

You could set out cosmic fundamentals in tri-logical form:

↓	*tam*	*Sat*	*raj* ↑
	down/ descent	*Pivot*	*up/ ascent*
	gravity	**Balance**	**levity**

Or normally:

	tam/ raj	*Sat*
	neg/ pos.	*Neutral*
	sides	*Middle*
	swing/ sway/ motion	***Equilibrium/ Balance***
↓	*tam*	*raj* ↑
	down/ lower	*up/ raise*
	negative	*positive*
	sinister	*dexter*
	drag	*release*
	deaden	*quicken*
	contract	*expand*
	extinguish	*stimulate*
	deadweight	*light/ heat*
	positive pressure	*negative pressure*
	entropy	***negentropy***
	mass-centred	*space-centred*
	gravity	***levity***

Equilibrium holds to the centre. If (*tam* ↓) gravity is mass-centred, (*raj* ↑) levity is space-centred. Gravitation is 'thing-centred' so that, conversely, levitation must be 'nothing-centred', centred nowhere. In other words, contractive gravity involves the central point of mass. Where is the centre-point of space? Expansive levity's delocalised; it will as it sucks from nowhere, press apart from everywhere.

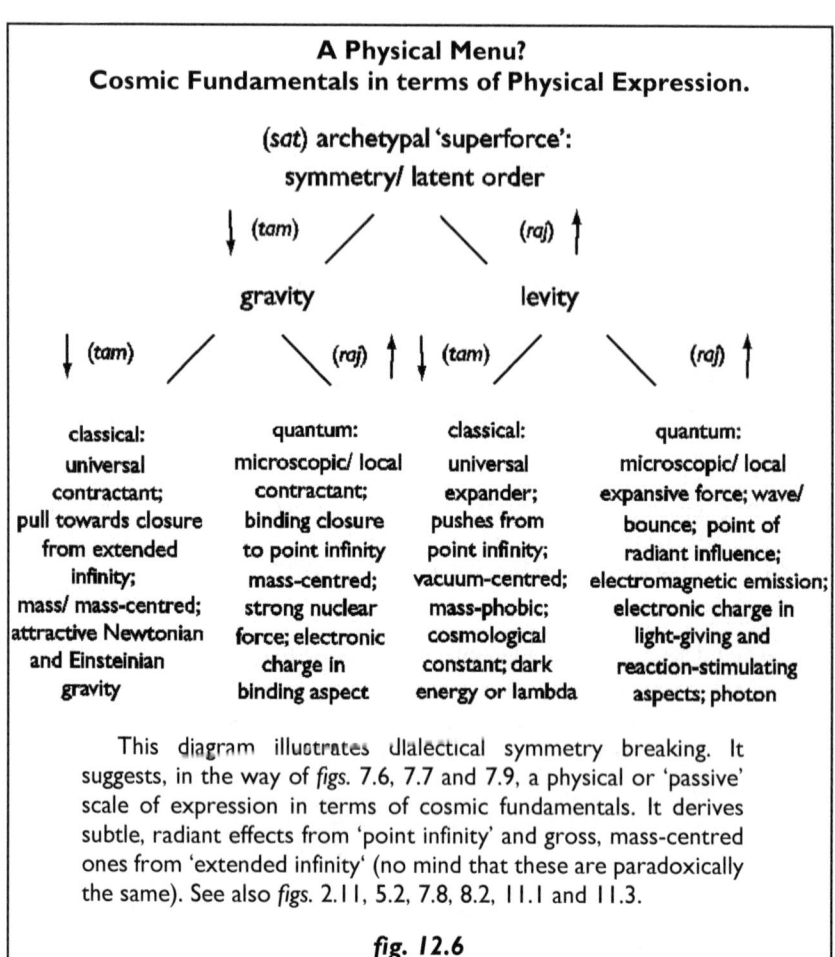

fig. 12.6

We've glimpsed this (*raj*) principle several times before. It should be clear by now that levity will be expressed through various agents. The relationship of its patterns of expansion, stimulation and release are related to those of gravity in *fig.* 12.6. No doubt the logic of Natural Dialectic demands a universal, levitatory vector and this hasn't been identified for what it is in current paradigms. It is presumed, maybe correctly, that the universe expands; it has been suggested that a large-scale vacuum might exert a force of 'anti-gravity''. But, dialectically, it's also been suggested levity's a vector; any 'fire' that stimulates reaction, motion or transformation is called 'levitatory'. Such promotional 'fires' include light, heat, electricity and other factors that inject change and do not drain an instance of its energy. In short, dextrous levity involves the actions of release. It raises, quickens or enlivens. Call it 'fire of freedom'.

(*Sat*) Equilibrium

The scale swung up. Before it gravitates back down it has to pass a mid-point, pivot, point of balance. It is not too much to claim such point of balance, equilibrium, is at the centre of this book, Science and the Soul. It is the core of

Natural Dialectic and the heart of Universe. Poise is at (*Sat*) Centre of tri-logical expression; or on the right-hand side of Primary Dialectic's stacks.

Do you remember images of cosmic scale (especially Chapter 2 and *figs*. 2.4 - 9)? Check 'pivoted existence' and two (*raj/ tam*) vectors of event that oscillate around a third. At the middle of the trinity this axial factor is their governor. The basic law of cosmos is a maintenance of (*sat*) equilibrium or, if you like, equations made between opposing forces. It has even been hypothesized (by Edward Tryon, who also first proposed physical cosmos is a large-scale quantum fluctuation in vacuum energy) that, where the amount of positive energy in the form of matter is exactly cancelled out by its negative energy in the form of gravity, we inhabit a zero-energy universe. Whether this is true or not, certainly moments of the world swing round its hub, its bearing or, if you prefer, its axis of control. This sun, this central pivot is the law. **The essential pivot of the cosmos is stability of natural law.**

Around such pivot of (*sat*) informative neutrality swing two vectors of physical expression - (*raj*) levity and (*tam*) gravity. Dialectical logic identifies the balancing factor as a source of manifest patterns, a precondition, a pre-physical potential. It is the basis of natural law whose 'fifth condition', the real 'quintessence', resides in a grade above physical - (*sat*) metaphysical archetype.

Immobile poise precedes mobility. Guidance issues from the stage above and so, for physic, guidance issues from its metaphysic; for matter from its mind; for *physicalia* guidance from their archetype(s). *Recall from Chapter 7 the equation of a latent, immaterial field with archetype and archetypal memory; and (Chapter 10: dialectical Vacuum) the (sat) metaphysical aspect of space whose archetypes are physic's rule of law.* Program, algorithm, formulae. Discovery and description of the operational 'shape' of archetype is embodied by the 'natural mathematics' at the abstract heart of physics; but, where maths is abstract, have not immemorial archetypes an immaterial actuality?

What is order but a form of information? Tell me, how should I behave? Does not behaviour make me what I am? And if I have no mind (and protons and electrons don't) I have no choice; automata of physics are the product of a world-plan; archetypal information moulds them from ts energetic substrate to the integrated set of factors that they are.

Check *figs*. 1.3 and 2.13. 'Matter-in-principle' and thence 'matter-in-practice' each issue serially from their preconditions. This couple are the product of the first and leading, over-arching factor out of physics' trinity. Microscopic quantum physics and its macroscopic classic counterpart are children of the archetypal root of Natural Dialectic's energetic theme. The law is order. Nor is chance the order of the day. The world swings on a fulcrum; it is stapled to its archetypal sky. Equilibrium is the reference point. It should be clear by now that balance of the up and down sides is a universal point of order.

The real nucleus of nature is, therefore, vested in the metaphysic of its archetype. Could you say that regulations are the core of any game; or that pure law is 'broken' into actual play? If so, the core of nature is its latent lawfulness, the 'symmetry' of rules prior to the way that they are 'broken' into instances and differently expressed. Such expression of control occurs in every act of law, that is, in nature's every move. Newton's laws of motion demonstrate its quality and quantity; and Hindu Buddha's '*karmic dharma*', called the law of action and reaction, covers mind as well.

Mind includes not only the reflex zones (of memory, sub-consciousness and archetype) but also conscious phases. As opposed to the generic gravity of matter there's an existential levity of mind. Objective matter, (↓) gravitating, will develop darkening concentrations; matter draws in matter to itself. Conversely consciousness by (↑) levitating will develop concentrations too; such subjective concentration is a focus of attention. One makes weighty and the latter naturally wants to lighten. Does not a balloon released from string fly naturally to sky? Mystics thus recount ascent of soul from binding sheaths of mind and body towards its Light of Origin.

Of course, things wobble, split and change; but each temporary motion of events seeks equilibrium again. Existence is a multiplex equilibration. All equations flow from conservation laws that, using an explicit form of balance (=), balance one side with the other. Transformations have to balance. Imbalance here would equal chaos and we know from every formula that change is not like that.

In the balance of our changes many other sorts of equilibrium hang. An equal pact between strong nuclear force and charge conspires to keep atomic nuclei intact. It keeps each electron in an atom in its place and thus conserves the balance of opposing charge. What determines mass, the density and properties of chemicals and thereby every body in the universe? Atoms. And atoms are a composite of particles held in (*sat*) balance by (*raj*/*tam*) opposing forces. So is the whole world that they underlie yet make. **An atom is, in this respect, Natural Dialectic rendered physical, philosophy materialised.**

Atoms aren't the end of things. You'll find the world's end, cosmic buffer, made of solid objects. The size of a large body is dependent on opposing forces held in balance too. Electronic levity keeps proper distance from its nuclear opposite - described as an exclusion principle; it thwarts - save under overwhelming odds - the influence of gravity and, against compression, holds the space in atoms up. Levity lets chemistry. And a further 'tight-rope act' keeps stellar bodies balanced. Expansive heat from nuclear reactions is opposed by gravity so that, with poise, our sun shines on.

Do you recall fine balance of its latent factors that immaterial archetype (or 'symmetry' of nothing if you are materialist) must hide behind the starting-line? The revelation needs to be just right. Take the force of gravity. If it were any stronger there would be no small stars or their satellites. If it were any weaker then not even large stars or the heavy elements could come to pass. If electromagnetism were a force much smaller or much larger then no bonds could form and reform chemically. Forget yourself! If the strength of charge were multiplied by three nothing but hydrogen would exist; if its electric force were three times less all atoms would disintegrate even in the 'freezing heat' of outer space. Take strong nuclear force. If it were larger there could be no hydrogen, only heavier elements; if smaller only hydrogen. For stability of elements the mass of neutron must exceed the proton by twice electron mass. More, and atoms crush together; less, they fly apart. Proton and neutron masses are well balanced to prevent decay.

Are not the values of our basic forces finely 'weighed' so that the cosmic show plays to at least one audience? The strongest character is billions stronger than the weakest yet each one is precisely what the combination needs. Such combination of all 'bar-codes' turns into a key to unlock and to open universal circumstance; their 'swipe' gains systematic access to an endless fund of energy. Force and particle are agents of an inward government; the key aspect of their constants is not values but their point of equilibrium, a symmetry of interaction that in balance generates this world and therefore you. And the key aspect of an outward equilibrium is inward law. The immanence of (*sat*) archetypal constancy is axial; principles are nuclear; they are the unseen, inmost pivot round which variable phenomena all spin.

Eternal, vectored forces struggle under such titanic, immaterial equilibrium. Such is the simple message from the first part of this book. Not only inward constancy but also, as the second part explains, outward dynamism can provide stability. Dynamic equilibrium in the form of homeostasis vibrates in cycles, periods and the norms of animate biology. This form of balance cycles round a norm, a principle, its law; atoms, galaxies and living bodies cycle energetically but also round the axis of their inside, nuclear information. This is how a swaying world works out. It is the clue, the key, the music of the spheres, the healthy way the universe is bound to dance. (*Sat*) equilibrium, the tipping and the turning point, is what the swinging's all about.

(*Tam*) **Gravity**

	tam/ raj	*Sat*
	distortion/ warp	*Pure Space-time*
	thing/ event	*No Space/ No Time*
	varying fields	*Transcendent Latency*
	imperfection	*Perfection*
	relativity	*Absolution*
	curvature	*Straightness*
	expression	*Potential*
	motion	*Changelessness*
	locality	*Nowhere/ Everywhere*
	variability-on-theme	*Invariance*
↓	*tam*	*raj* ↑
	mass	*energy*
	more warp/ definition/ locality	*less warp/ definition/ locality*
	massive pull	*vacuum suction*
	slower	*faster*
	gravity	*levity*

What about the downside? Levity (↑) is but a third of the whole polar chiaroscuro that includes equilibrium and gravity (↓). The (*yin/ tam*) fraction of pure energy's expression acts as the anti-principle of levity. What is anti-lightness? Is it drag or darkness? Or is it anti-energy that takes the form of (*tam*) suppressive forces? <u>By convention we understand gravity to be a specific force acting between the centres of two masses but it should by now be plain that in the Dialectic's context meaning is extended to include all agencies of the (*tam* ↓) downward vector such as contraction, confinement, inertia and heaviness.</u> It grabs the whole left hand of every secondary stack. Its totem is a proton (or perhaps a quark) and its expressions show as any isolating, dragging or compressive influence. Such gravitational influence binds; or else depletes an instance of its energy. Its expression is one of (*tam*) entropy. As cosmos sinks into such low-energy condition as nowadays prevails, the gravitational vector overwhelms the levitatory and we observe a 'left-hand' or 'entropic' swing towards closures. Running down. 'Gravity' in a generic, Dialectic sense is negativity, materialising power as necessary for creation as its running-mate, the trait of positivity. Its tendency is towards resistance, binding and deceleration. Its 'sinister' or 'left-hand' character is of oblivion and mass; the purest psychological gravity is oblivion and the purest physical is mass.

Local binding forces 'freeze' mass out of energy; they grip protons against their own electrostatic repulsions and thus stop atomic nuclei from flying apart; and, through the agency of electrons, they magnetise different nuclei into 'legally bonded' permutations we call molecules. These micro-gravitational attractions are known as nuclear binding forces and the weaker magnetism that electric charge exerts. On a larger, universal scale the mass-prone principle weighs in with expressions of pure, archetypal anti-energy - ones Newton and Einstein would have recognised as self-centred, mass-centred or inertial. Inertia is resistance against motion; and gravity drags objects into mutual aggregation. Accompanied by heat loss and a tendency to slow down, isolate and solidify, gravitational effects accumulate a 'wealth' of burden in their rings. With mindless 'fascination' bodies of the heavens circle round each other. Gravity 'spellbinds' a universe of galaxies, solar systems, moons and helps to hold all objects in their place. At the same time other downward influence is trailing them towards *rigor mortis* and the drift towards disintegration.

The particular effect that science calls 'gravity' is an intangible, attractive-only, mass-dependent influence with, in Einstein's eye, the capacity to bend elastic space. Nor is it vibratory but a 'flat', passive force without oscillatory motion or radiance. As such it is a mass-attractive dark force and, since it acts to crush together, a gross rather than subtle and a negative rather than positive power. From a mass-centred point of view objects are accelerated centripetally towards each other's centre. The greater the mass, the greater the gravitational energy, that is, the ability to unbalance and induce falling motion expressed as acceleration or momentum. The extremity of gravity's collection is not an extent but a single spot. It is not all-round effulgence but a black-hole singularity, an apparent absolution that disappears like grave-bound death. We can learn much about the (*tam*) principle of materialisation from its apparent absolute, the ultimate existential negativity, the extreme antithesis of Infinite Life. From a material point of view such inexorable, irresistible violence seems to be the starkest instance of reality possible. On the other hand, from a holist perspective the hardness of its non-conscious solidity marks an antipole, exile at the furthest shore, a peripheral circumference of nature. Rind. Seeding Centre diametrically opposed. A 'black hole', if it exists outside hypothesis, is a spaceless, placeless, pointless abyss; it is a locally 'infinite' concentration of mass confined within a thermal cul-de-sac, a state of matter you cannot transform, an Erebus beyond experience or experiment. Behold (but you cannot) a negative eternity, a black spot as opposed to instant freedom light allows. Know (but you never can) ultimate and indiscriminate destruction, cave of creation's nemesis, annihilation at the mouth of matter's underworld. Such gravitational concentration marks a spot of bent and broken time, the limit of condensed illusion, boundary of unreality and nothing's final fence.

Thus from a *top-down* perspective non-conscious nature is diametrically removed from Conscious Nature. You have dropped, in a dialectical sense, from Nucleus to shell. You have dropped from Potential to exhaustion, Origin to outcome, Start to finish. Non-conscious matter is The Single Essence rolled to its antipode, subjective impotence and dualistic isolation in extreme. At the foot of universal rainbow find a crock of fool's gold; slide the conscio-material gradient and fall upon the bottom of the world. You bump upon the world's end, one in glorious technicolour view. You can see it all around you all the time. Our normality, the things we're used to, *are* creation's outer darkness. No doubt, this is the vale of death. Things never were alive. But the world's end is not nigh. This kind of end will last a universe.

Stars are born and die. Main sequence stars (like our sun) explode as supernovae. They become red giants and white dwarfs or black holes. Sirius - the sky's most brilliant star, fifth closest to earth and, it is alleged, the 'destination' for souls of Egyptian pharaohs as they rocketed from burial chambers in the pyramids - is orbited by a companion white dwarf. Only 1500 years ago this was recorded as a red giant, so the process of stellar gerontology need not be slow. However, just as extinction does not explain the origin of any species, so its death does not explain the birth of a star.

You can speculate through telescopes but what about the ground beneath your feet? It is presumed the first galaxies, stars, planets and even moons (some, strangely, circling in a retrograde direction) *must* have evolved but how? Hypotheses, *Ad Hoc* explained, abound. All are, *de rigeur*, framed in materialistic, evolutionary ways; and, *de rigeur*, excluding any other kind of explanation. Such reflex exclusion may or may not constitute a reasonable procedure. Slap a slab of rock. Planets, moons and massive slabs of mountain rock seem absolutely solid. They can hurt. Indeed, hardness seems to be the litmus test for what our senses deem reality. It is certainly concrete but is it nothing but the whole truth? A aeroplane's rotating prop assumes the texture of a solid disc off which, when thrown, a ball will bounce. Likewise 'rotating blurs' (electrons) cause this book or your hand to engage (although they're mostly empty space) a borrowed seeming of solidity. Indeed they hold the earth and every other massive aggregate in place.

We have now gravitated down to base. We have landed on the (*tam*) cosmic floor, life's outer shell, the large-scale world of physics' uniform, indefinitely hard but perfectly elastic billiard ball. With this ball we model bodies of the suns, moons, planets, atoms and the classic laws of motion. *Static, passive mass is a sensible myth.* In fact, however, from a small-scale point of view, clear-cut blurs to probability; the world seems more a phantom than a billiard ball. Solids, which seem to provide the stable foundation of our sensible world, are composed of dancing atoms. And these, we've learnt, are mostly made of space. If an atom were magnified to the size of Wembley Stadium then its nucleus would be like a grapefruit at the point of kick-off in the centre of the pitch. Electronic points would lap the outer stands. These specks like pips spun from the quantum vacuum's energies inhabit, all in order, mostly emptiness. **The 'hard-headed' world of senses is, as the Buddhists say, illusory. It exists but its fullness is not the full fact of the matter.** *Indeed, the emptiness approaches full truth. Nothing takes up most of space; and, diluted unto light and then dissolved by space, gravity doth wholly disappear!*

Change the angle slightly. Say it in another way. Truth's substance is internal and, 'organically' from inside out, external matter is its subsequence, its shell. You were born out of creation's cave; you issued from her cervix down into a now-familiar world of 'sense and sensibility'. Natural Dialectic traces an odyssean gradient down the conscio-material slope, a gravitational trip along creation's way. It started at First Cause and, having parachuted through the sphere of mind, now touches down upon the solid fields of our terrestrial home. Nevertheless, solid is almost empty space; and in this space the inner energies support (as Einstein's principle of the equivalence of matter and energy affirms) our 'normal', outer frame. At this point you understand the gardens, fields and skies you thought were real are, in a real sense, not. They constitute an ever-changing, insubstantial projection which we in error think is ultimate reality. How, you gasp, did I ever dream of foreground physicality as more than elemental background and an insubstantial basis of my transient body's life? How, you further gasp, did I entrain the habit that this shell was real when I am really an atomic matrix pasted onto mostly space? Not only the universal body but my own is, in the hard-nosed catechism of a scientific faith, an inanimate phantom, an unholy ghost, a lifetime's chemo-physics, borrowed atoms wound into a precious but a lifeless wraith.

Be that as it may, I must say I feel real enough. We've ended up down here. This is gravitation's fate and we inhabit heavens wherein gravity is systems' king. Mass is 'charged' with gravity but, by itself, the latter can't create a thing. Thus, what dialectical and dualistic start-point might this end-point, earth, have had? *Top-down*, you might consider that intrinsic to the Alpha Moment were co-expressive levity and gravity. You might also assume material events start energetically and fizzle towards a halt. You would therefore think of an imbalance, one involving extremes of complementary. Initial high instability would sink into the torpor of stability - the cool, low energy of cosmos that we call 'normality'.

It Takes Three to Tango

Keeping balance, sustenance of equilibrium is one side. Did you forget, as well as complementary, polar partners, order and the counterpoint involved in writing music and creating dance? You need partners but you also need their senior partner, a first and crucial composite of rhythm, steps and balance - swing and sway we call the archetypal source of dance. Dynamic ordering of motion is another layer of control. Will not any dancing couple tell you prior order is their other half, the information that informs engaging steps?

Existence is a polar dance. It takes not one but two to tango and produce a third. From parents a succession. From two a third but is there not, besides the process of a child's development, another kind of third - a third before as well as after the productive couple? Dance develops and a child evolves according to an unseen set, a nuclear set of principles, an algorithmic sequence of those steps whose order fundamentally informs events. A single, causal factor, plan, is what you'd call the reason for effects. And what is universal space-scape but a very grand effect?

You can gather that, as well as (*raj/ tam*) active 'side-kicks', (*sat*) equilibrium and information have a central role to play. An orderly, harmonic dance is poetry in motion; dynamic equilibrium and matching poise are vibratory spectaculars that, in this case, derive from 'tango' archetype. Is the cosmic dance of *Siva* and his consort 'tango' or another kind? Who or what is Lord of all the styles of dance?

Do you remember (Chapter 0) your third eye - the single eye of thought unseen, between and yet above your pair of peepers? Do you remember cause (or source) precedes effect and thus is from the stage before? Cause, in terms of a creative act, is the potential for what follows and, dialectically, it issues from the stage 'above'. In the case of any actual parents isn't there a higher precedent? Isn't there a prior potential - pre-condition in the form of instinct and productive capability? Does this precondition not inform the procreative act? By analogy what might precede the dance of physics or create the government of chemistry? Is information physical or metaphysical? Aren't plans and rules the latent precondition you can see expressed by moving objects; aren't they unseen factors of control that order actions of inert and passive pieces of the play? In this light such pieces (polar charge and particles that represent the cosmic fundamentals) are seen, although inanimate, as actors on a universal stage. Out of 'nothing physical' there issued physical phenomena. And perhaps, if particles are channels drawing 'virtual' energy from just behind the actual veil of things, this is the way they are supported still. Thus, one might ask, is 'virtual' archetype the instrument of ordering energy, the fundamental rule of how things play their natural game? And thus, if mind is metaphysical, did universal mind inform the universal body from its start?

It takes one to make a tango - one creator, governor or ruler of the dance. Then two more to actually dance makes three. *Polarity is based on trinity. It takes, as three cosmic fundamentals demonstrate, three to tango.*

Where inhabits information if not mind? You should have grasped the dialectical idea by now (see for revision *figs*. 6.1 and 7.5)! Psychological first cause is an idea and the will to implement it, to create. In this creation two that tango through existence are named mind and matter. Towards the top-end of Mount Universe material is 'ethereal', 'subtle' or 'refined'; mind-force - not heat or light or any physical expression of *prakriti* - is called *manas*. At the very top there is no matter, just a pure concentrate of consciousness; and, in this layered view of things, at the very bottom no life stirs - there lies just a pure concentrate of matter, a construction of oblivion observed by us on starlit nights as universe.

The tango's trinity not only governs present operation. It must also indicate the circumstance and nature of its origin, the way it all began. Is a CD or its CD-player conscious? A recording once was 'live' and now is filed. How do you file an instance if there is a lack of paper, vinyl or other storage medium? What if there is nothing yet but mind? Where do you file in mind? You file the conscious business in your natural vehicle of storage - memory. You build up habits, instincts and 'contextual programs' so that the non-conscious part, your body, operates without you thinking. It runs, like any good machine, automatically according to the plan. Does that stop you drawing from the hidden cache - remembering if you wish?

Are consciousness and mind, according to the Dialectic, elemental? If they are, the record of the world must lie, as was and will be discussed at length (in Chapters 15-17), in the archetypal memory of universal mind. 'Archetypal memory' amounts to 'physical first principle'. Its record, source of law, is the tango's first of three. It is the template or the program of atomic dance. **Archetype is seen as language, a communicator, code**. Take a literary analogy. From principals, a grammar of the basic forces and an alphabet of subatomic particles, certain 'words' emerge; from these atomic elements evolves vocabulary (such as strict rules of chemistry allow) composed of molecules; and, finally, from 'forms-in-principle' the rest of all the complex 'book of happenings' is logged.

Why say the same thing three ways in as many paragraphs? It is to emphasise its importance. Major issues hinge on it. How did the world begin? Was it merely happenstance? Are things at root inanimate, as the philosophy of modern science preaches, or alive? Is life an element that's animated to the core or just ephemeral effervescence frothing from the earth of death? Neither language nor the complex steps of dance are instruments of chance. Nor are they physical in structure. An ordered view eliminates the random element. It vaporises Lady Luck. It therefore vaporises theories that derive from her 'divinity'; it wrong-foots the starting point of scientific tangos, intellectual steps devised to celebrate the history rather than the present operations of our world. **Could big bang theory be partially right yet, as regards the whole truth, wrong? Could, similarly, all the other naturalistic explanations only bump against birth's moment and not, in the way of a projection's issue from transcendence, go before? Could facts alone not force precise revision of philosophy and thus the possibility of metaphysical involvement?**

In the case proposed by Natural Dialectic there was not nothing, there was something just before the physical phenomenon began - there was the instrument of leverage to churn out everything. The set-up was in place. The program laid in readiness. Does chance hit a mechanism's starter? Is the ignition button pushed by mindlessness? An orderly procedure must ensue. Science has discovered that it did. Don't worry, what can fail when dancing's only in the frame of prior, top-level choreography? In such circumstance the product must succeed. Strike up and let the dance begin.

Strike up? What sort of energy breaks vacuum's hymen and thence seeds an empty womb with the effects of universe? What initial resonance could drum, in ways that science very cleverly explores, our world into its dance? Could 'quantum foam' or stipples known as 'strings' or 'particles of Planck' be spun, like brilliant stories, out of nothing? Such particles, you may remember, are supposed to come in pairs; their specks are charged and turbulence can separate charged specks and spin them. Separation generates electric fields; spin generates magnetic ones. Creation's wind whips up the spin, grit gathers in the ghostly maelstrom and, round deepening vortices, symbols of the actual universe appear. The energy that penetrates between a mind and body is subconscious. Psychosomatic breath that pipes the melodies of earth and rhythms of the starry heavens was introduced some chapters back as energy of universal mind, a force long labelled *prana, ch'i* or, in the hierarchical sense of Natural Dialectic, 'potential matter'.

What about what follows? What's the shape of things to come? Whence originate the patterns that make up a tango's steps? Science has mapped orderly behaviours radiating in between our polar partners and across the whole floor in the course of dance. It has expressed by mathematics all the poetry in motion that a vibrant field of music brings to bear. Three kinds of shape inhabit every move, three postures strike in all events. You can (*sat*) hold things level, harmonise, neutralise or balance using counterpoints. The actual notes from instruments run up and down the scales. Similarly the notes of archetypal instrument spin up (towards high) or drop deep down an energetic scale. (*Raj* ↑) levity includes all transformations raised by photon, heat, electron, charge, release, propulsion, acceleration and the 'upward' crew. Conversely, when energy is lost you slow down and drop off into 'bass sound'. The transformations of (*tam* ↓) gravity involve nuclear force, mass (proton, nucleus and bulk forms), inertia, capture, aggregation, gravitation, loss of heat or motion and so on down the hill.

But does any single picture tell the whole, great story? Or is whole truth beyond the mind of man? One thing is dialectically sure - exterior, gross appearance proceeds from interior, subtle guidance. The great womb's labour bears a mist of space; it expels 'an infancy' composed of particles, atoms and their molecules. From the mist emerges mountain, from the inner principles an obvious maturity of practice we call gas, liquid and solid textures. These are the detailed, differentiated states of 'adulthood' that we in error take for what is hierarchically more real. The world is offspring that its source begets. Thus substance in the form of archetype and causal quantum entities takes unseen precedence. Material ghosts hold sway; subatomic principals rule over what emerges from their nothingness as visible gross matter. Not time but timing is the substance of this natural tango. Things could have worked up slowly or a full-fledged dance begun dramatically. Whichever way things started now phenomenal appearance is hard about you even making up your body. It is 'sensible' enough but actually insubstantial. That is, external form is much dependent on its inner principles, the shapes and vectors of subatomic energy. And we've discussed at length the origin of these.

Hiding nothing you see nothing. Pulling nothing from an empty hat is tricky and I sympathise with good professors struggling to define the surges that gave border to the borderless and flight to all the airs of physics. Let the heat of wrangling fizz its course - mind simply boggles at the wondrous moment of creation, one that possibly denies its definition. What then about the other end of things, their dissolution? Mind also boggles when it asks how Alpha ends and Omega, the end, begins.

Points Omega

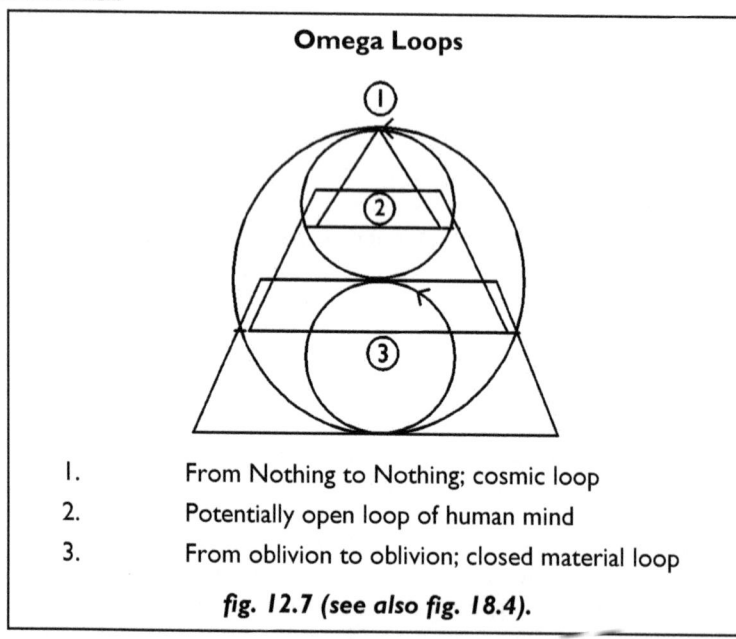

Omega Loops

1. From Nothing to Nothing; cosmic loop
2. Potentially open loop of human mind
3. From oblivion to oblivion; closed material loop

fig. 12.7 (see also fig. 18.4).

Omegas are ends. There is, in a sense, an end to every change, to every pass from old to new. This section concerns neither start nor change but annihilation - something made nothing.

	tam/ raj	Sat
	existence	Essence
	thing	Nothing
	outer	Inmost
	periphery	Axis
	sink/ process	Source
	end/ lifetime	Start
	alphabet	Alpha = Omega
	body/mind	Soul
↓	tam	raj ↑
	lower	higher
	outward	inward
	non-conscious	conscious
	body	mind
	differentiation	unification
	creation	dissolution
	bondage	release
	materialisation	dematerialisation
	material tendency	soul tendency
	annihilation of life	annihilation of death
	end/ death	process/ lifetime
	omega	alphabet

The business of annihilation hangs upon your standpoint. If something came from nothing then nothingness preceded it. Nor does void have limitations - it is neither large nor small but unbounded, that is, essentially infinite. In this case the issue turns again upon the nature of such nothingness.

If only what's materially perceptible can count then you are locked in a material loop (*fig.* 12.7: loop 3). So are you sitting comfortably? Then I'll relate a tale you've heard before. Once upon no time there must have been a bang (or some rough cornucopia) whence issued, in the course of ages, everything including forms of life; but, as things are inanimate, the chemicals of which these things were made must have 'self-animated'. Thence a tiny little cell appeared and, after strife and struggle, evolved to a successor, you. You lease a body issued free of rent. In a short time it will definitely die; after a duration perhaps the universe itself will die. DC entropy, one-way, prevails. Change from being to non-being, death, is everything's conclusion.

This only-possible, materialistic story is one of 'natural' self-ordering. Evolution. Chemicals and aggregations build themselves. They do so until purpose in the form of brain 'occurs'. In this view Point Omega, annihilation, means two things and, because material things form the basis of existence, both involve loss and are thought of as negative. *Firstly*, personally, it means your death. A personal full stop. You've got it coming to you, that's for sure. If 'life' is equated with biological form, then the complex chemicals of which it is systematically compounded will degrade back to their original simplicity - unsystematic little molecules. Corpus to corpuscles in the anti-evolutionary way of entropy - that is the end of complicated you. *Secondly*, universally, it means the death of physical nature. This does not necessarily mean the annihilation of all mass into energy and all energy back into the kind of emptiness that prefaced a primordial projection. You may extrapolate, rightly or wrongly, from the present back to this sort of origin but may not, especially since time is a one-way street, presume that a simple cosmic reversal must occur so that all the large things of creation run forwards backwards into a fateful, final singularity; or else disintegrate into the 'grit' of particles and, in the end, even their disintegration and declining energy until original, unholy ghostliness - dark and utter void. Tenebrosity. If that were true then physicality, in terms of time and space and discharged energy, would never cease. What had its start would, strangely, actually lack an end.

Within a cosmos that has not become entirely inert matter still goes nowhere. It endures a wakeless state; oblivion is shuttled round the system taking different phases, shapes and compounds. And eventually it stops. How far is 'stopped'? Or else it disappears. How near is nowhere? Several scenarios, as well as everlasting discharge, can be thought of for the point of nowhere any longer, the terminus, physical Point Alpha's other end, Point Omega.

Prior to this, however, you could *for a start* eliminate both ends of time from where there is no issue but eternal matter. This universe is called a 'steady state'. You might ask how perpetual 'steadiness' continues. Don't worry whether mind, the immaterial organiser, takes a hand but what about thermodynamics and effects of gravitation? Fires go out, reactions finish, only dust and ash are winners; and expansion or contraction will decide, on balance, whether things drop back into a cosmic black hole or rarefy without restraint. Why, therefore, is the greatest show still going steady on the same-for-ever road? Without beginning matter would have had no end of opportunity to flare and flame and die. And ashes of such stellar outbursts would have had forever to accumulate into the frozen chamber of some body or be scattered round the wholly empty tomb of space. Hydrogen converts to helium in each body called a star; therefore, given endless time, why is our present universe still mostly made of hydrogen? No beginning and no end. Incomprehensible. Perhaps, as with Expansion Theory, you propose, *ad hoc*, continual violation of the First Law (energy can neither be created nor destroyed) so that a natural 'quintessential field' is understood to be expansion. You might assert, with Fred Hoyle and his steady-state scenario, that an *ad hoc*, 'undetected negative-energy field' continually sheds freshly-minted atoms into space; you might even have it conjure up *ad hoc* dark matter. *Ad hoc* creates *ad hoc*. How else might you bank on a constant density of matter in a universe that never fails to swell? Whence, however, might this energetic sinecure be paid; of what becomes the endless, surely extra-physical but definitely most expansive generosity that such hypotheses cash in on? Anyway (Chapter 7: Order of the Finite) it was the galaxies that did Fred for.

If 'steady as she blows' is not the way then you might, *secondly*, speculate on the spectacular - a steady blow. Upon its flying start why should big bang's ejaculate not endlessly expand? And with such expansion ours and every other spot spread ever lonelier and more apart? Fresh data from the *HST* (Hubble space telescope) suggests that galaxies at the edge of the known universe are accelerating away from us and each other at vast speeds. After an anti-gravitational big bang might there not, as already noted, come a point where this expansion involves such weakening of material density that gravity must lose its grip - and an anti-gravitational force such as 'dark energy' or the suction of infinite vacuum resume its predatory control? So that there begins a 'lambda' rush, a 'big rip' as evinced by distant torrents of those galaxies sucked towards a Pointless Omega of endless expansivity? After sunburst, for example, if it wasn't 'burnt alive', earth's crisp would wind up lifeless flying into emptiness. In this case the creation train, run out of puff, would have hit a different sort of buffer - an anti-buffer of immense 'go-power'. The old puffer would gradually lose steam (but gain dramatically in speed) until occasional flickers of its stellar coals died out and there fell cold, eternal night. Mathematical extrapolations suggest various bizarre fates as even lonely specks of dust and forms of final death, black holes themselves, dissolve with spasms into *rigor mortis* sans the rigour. How can nothing twitch or petrify? 'Immortal' energy, which cannot 'die' but only be transformed, will have dropped into perpetual, thermal equilibrium. An endless grave of heat, freezing cold, would timefully yet timelessly endure. This, the start that lacked its end, would constitute a species of subtendence (see Chapter 2), sub-state, an inertial equilibrium, a dark, burnt-out galactic yard of death. And the brash big bang will have deflated, as the poet T. S. Eliot once supposed, to a fading whimper and then utter silence.

But we're not finished yet. Time's arrow's shot along expansion but also 'weighted' by the drag of entropy. The cosmic projectile's projection (which some call 'evolution of the cosmos') changes everything; it speeds each phenomenon towards decay, diffusion and an unlit grave. From perfect order at the start chaos increasingly intrudes. The world is on the slide, events are on the down-hill side. Large stars will implode into black holes; smaller ones will fade, the starlight of red dwarves glow fainter into, it is mathematically predicted, black dwarves; and this final ash will then disintegrate until there's nothing but a sea of lightless photons left. Lightless light, perhaps minuscule black holes, heat death - the diametric opposite of focused physicality would be diffusion into energetic slack. Such dark-side mirror asymmetrically reflects a singularly concentrated birth. But, Natural Dialectic has explained (Chapter 2: Subtendence), there is no infinite subtendence, no negative that's absolute. Even black holes will evaporate till only dark light's left - a diffuse, inert and changeless pool of universal impotence. Without change where is time? Without motion what is space? Could even background radiation disappear, without a whimper, into nothing physical at all?

You might wish 'dark light' would go away; or that 'dark energy' would disappear. You might imagine one or more of nature's basic forces fell into decay and slowly let the world break up. This would constitute a *third* form of apocalypse.

Could expansion darkly unto nothingness go 'yo-yo'? *Fourthly*, you might like to think about expansion's opposite - contraction. Could dark matter's gravity somehow resist an overwhelming whoosh of inverse energy, dark levity? And thus arrest the cosmic vehicle, bring it grinding to a halt then throw it in reverse? Might it turn the juggernaut of cosmos round and, throttling backwards, drag it shrivelling towards its origin, a singularity? Would not that black hole tighten time till it was strangled into immobility? And space went pop? What happens after that - a physical *nirvana*? Nothing? Or could the universe in 'yo-yo' go? Could there erupt, phoenix-like, from termination an elastic resurrection? A corresponding birth, a baby universe comes bouncing back. From the Black End flips a White Beginning. This would be the ultimate, oscillatory resilience of 'new ping for pong'. The idea of chance-based cosmological or deliberate Hindu cosmological recycling was explored earlier this chapter (Alpha Points). Is this the way just temporary erasure goes? Nobody has a clue.

Hierarchy changes everything. Dawn on Mount Universe takes on a different hue. So does Night. There's a *fifth* way. Perhaps it is, in fact, the first and only last way; it is AC, cyclical as illustrated by Loop 1.

Top-down, the nature of Nothing is construed as Infinite Essence. This Essential Pole sources existence which, if non-conscious matter is its antithesis, must be Pure Consciousness. The First Cause is therefore alive, volitional and conceives in an orderly, purposeful way. *Its conscio-material slope, called Mount Universe, can be seen as a 'diffusion gradient' down which concentrated information is transformed, by mixing it with matter, from active into passive mode.* Thus creation is projected; this would mean that we are living 'in the mind of God'. Passive matter is informed by archetype; it is the 'body of divinity' whose condition is, although physically energetic, psychologically impotent.

Look at the scale another way. The 'vertical' axis of subjectivity is matched against a 'horizontal', objective coordinate. This subjectivity, gradually overwhelmed in its involvement with material form, becomes objectified; active becomes passive, light dark, hot cold, subjective freedom is objectively enslaved. The trend is obvious. At a metaphysical level creation shows as a gradient of consciousness eventually declining into rigid, sub-conscious mind. At the physical level mind (what we recognise as life) has been deleted. Materialisation shows as vectored, non-conscious energies that sustain atoms and massive, bulk compositions. These finally externalise through discontinuous phase transitions from fluid gaseous and liquid to, at last, the buffers. Condensed physics. Things get hard. They fall low and dark towards ashes, dust and a familiar kind of *rigor mortis* - solid objects. Solid objects are a concentration of exhaustion, a final creation. Fixed structure is a terminus and stationed at the far end of the cosmic line. This rural halt, this stop far beyond the Inmost Nuclear Capital, is quoted as 'spent mass'. Solids, like the book that you are holding, are the edge of the world. Put it down. Put down a boundary. Place one edge of the universe upon another.

It is here, out on creation's margin's, that we wander. Are the atoms that compose your body conscious? Or molecules or hosts of molecule more so? All non-conscious bodies are the diametric opposite of Cosmic Origin. *Matter is therefore an end in itself.* A species of Point Omega. The whole universe is stationed at the bottom of Mount Universe. It is a 'valley of death', a corporeal graveyard and, except for a few animated sparkles wrapped against its icy darkness on a planet that you know, a desert of oblivion.

If bodies are chemical and, therefore, essentially non-alive, how might a Conscious Infinite know its non-conscious universal body? How might an inventor know how his machine is running or an author how his play is working? Could you watch things from the outside - from, somehow, the other side of space? By staring through transparency, by squinting through a glass division separating mind and matter? Are creators lonely faces peering in from an external freeze? Or, if Life is looking, does such appreciation demand a lively interaction with the user? What are you, who are you that asked for neither body nor this world but were gifted both? On earth alone (but why should watery oases not exist elsewhere?) a range of bodies could allow the full potential of experience to be gradually realised. Are life forms the humbler eyes and ears, the senses of Soul? Are they media through which the Infinite, at material level, participates in drama of a finite creativity? What would that make the spark you are?

A universe of chemicals is, like your body which is part of it, a temporary home. We camp awhile beneath the tent of stars. Is this camp ever struck? What about the graveyard's death? What happens when the temporary 'home from Home' disintegrates and disappears? *This means the time we die (Chapter 18); or it means the time that all involuntary bodies, collected in an outer darkness of oblivion called the cosmos, die.* Your body is returned to powder or to dust. What supports these fragments? Atoms. Particles. Polar energies. The vacuum's *ZPE*. Then what? What form of 'nothing' have the last six chapters been discussing? Transcendent or potential matter means, in the dialectical hierarchy, a dualistic pair of 'form' and 'force'. The '*tattwic* form' informs, like an instinct, the behaviour of '*pranic* energy'. It is called archetypal memory - the lowest projection of metaphysical (and therefore physically space-timeless) universal mind.

Memories don't think; instinct doesn't contemplate; but dormant mind robotically communicates with body. What activates subconscious mind? Not, as psychology is clear, just conscious mind. Electrochemistry must play a part and, therefore, sub-atomic particles and forces. Is that the lot? If not, what psychosomatic pattern interplays with them? Potential matter's energetic part is 'electric' *prana*, 'low-level voltage' on the grid of *Om*. Such energy is held to source the quantum world and thereby be the engine of our large-scale universe. Phenomena are, through the natural instrument of archetypal memory, projected. A bulb or any mechanism worked by an electric current stops when power is switched off. Likewise withdrawal of a current that supports the working and is, therefore, essentially the substance of a 'mechanism' called the universe, would cause collapse. As an original projection raised and now sustains all *physicalia* so, at a dissolution or '*pralaya*', withdrawal ends them. Their company is dissolved, their business folds. The mnemonic record stops, tango's cut, the switch is flipped. Music disappears, phenomenal existence vanishes. There is simply a collapse of all constructions back into the vacuum so that 'virtual nothingness' remains. Even here, below the threshold of detectable phenomena, reverberations cease. If a subatomic particle has stopped where does its atom go? If the atom goes, what happens to the superstructure built on its foundation? Where do solid objects disappear? Try still and silent inner space, the nothingness whence long ago their energies first burst, kept coming and in the end will stop. There is nothing left to say. You would have witnessed ultimate recession.

A deliberate end could happen instantly. Or, as with a dimmer switch, it could be dimmed out gradually. If a yo-yo style of cosmic oscillation is the way, the end might come as ebb tide caused by 'pranayamic inspiration' of the Deity. As breath was drawn inward things would, in this process of withdrawal, fade. The ladder of creation's hierarchy would be raised and, in this sense, time reversed. Termination would occur as naturally as breathing.

Check *fig.* 2.6 noting negentropic and entropic tendencies. Now let's rephrase the business dialectically. Omega is when a show is over. The endpoint of a process comes in two main forms. (*Tam*) serial reactivity, as every chemist knows, runs to inertial equilibrium. It may, like water, drop by stages to the bottom of reaction's hill but never rises back towards the top. It is entropic. (*Raj*) cyclical activity runs, on the other hand, back to its starting poise, its balance or potential equilibrium. In this negentropic, vibratory case omega is alpha. Material returns to origin will fade; a vibratory or homeostatic cycle weakens and then dies. Just protons and electrons finally endure. Bulk material's tendency is mass-centred, gravitational, inertial. Disintegration and a loss of reactivity bear down on objects and events.

Mind's tendency, the other way, is (*raj*) life-centred, dialectically levitational and seeking to gain information. A tendency to further interact, unite and rise with happiness, unless obstructed, buoys purpose towards improved experience. Mind is by nature negentropic. Therefore, while oblivious matter tends towards omega apart from alpha you may well ask, if conscious mind seeks better living, what constitutes improvement. Could its vector (↑) indicate the highness of its basis, inner nature or its origin? Could this be consciousness and a return to concentrated origin turn out to be the same as bliss? In this cyclical way omega would come to alpha. A part and not apart, a part and at the same time whole omega and alpha would unite as one. If Essence is Pure Consciousness could this not be your individual case? With what potential you were thereby blessed. If Essence is the Infinite beyond finite existence could such communion not be the cosmic case at last as well?

From a *top-down* point of view, therefore, loss of material cosmos is oblivion's loss. No more than letting go of what is meaningless. Subtraction of the lowest layer of the cosmic pyramid (represented in *fig.* 12.7 by loop 3) still leaves two more. Nor does the metaphysical aspect of creation *have* to end at the same moment as the physical. Do you remember the Indian notion of *pralaya* - a period of cosmic creation and dissolution? And that this concept includes a longer period for metaphysical *mahapralaya* than lower, physical *pralaya*? In this way universal mind might survive multiple physical cycles before it too were furled up into Omega, Alpha, the Infinite. So ends the cosmic sentence, so completely dies away reverberation of the Word. Loop 1 has died into the Apex of All Worlds, Eternal Life. Now no existence moves. Only Essence is.

No doubt, therefore, that the state of physical oblivion is of interest to an outward scientific quest but, on the other hand, its 'Going-Nowhere Cul-de-Sac' is of little interest to the seeker after metaphysical awakening, a climber towards the Infinite Peak. Such final destination, Point Omega, turns out also to be the Source, Point Alpha. *So if you have a different view of source you have a very different view of end.* Your loop (*fig.* 12.7: loop 2) includes minds and, in this case, a voluntary search with purpose may, if properly projected, reach what Natural Dialectic has identified as Apex - a pure and absolutely negentropic Concentrate of Consciousness. At this point, cosmic starting-point, you'd end. Omega in Alpha, Loop 1 is now complete. This is, truly, cycling round the universe.

This sort of self-annihilation, which seems nonsense to a casual, unpractised or inexperienced sceptic, is in fact most positive a psychological occasion called *Nirvana* or enlightenment. Does wisdom differ from unknowing? 'Mystic birthday', rising to the Father's Everlasting Home, is very different from a baby dropping out of mother to the arms of earth. The motion's not *from* nature's wombs but *into* Nature's Womb. Instead of donning coverings of mind and body coverings are shed. It is, therefore, not creating an existence where there wasn't one before but dissolving out of one that is. It is not a physical creation *ex nihilo* but metaphysical dissolution *ad Nihilum. Nihil* is pure life. Annihilation takes a whole new, naked meaning. Nor, if the *top-down* view is right, is it a matter of religion (which is a formalised description and a climber's guide towards his conclusion) but of Natural Fact - Fact as natural as space, light, air, water or the earth. Death to the world, the birth is one of Naked Soul.

If the annihilation of existence just leaves Essence and if the annihilation of everything leaves Nothing, you are simply back where you began. The key to personal return must therefore, as just noted, be self-sacrifice. Transcendence of Freud's id. It is a process of dissolving ego into super-ego. Such line reduces separation by relationship; it increases fusion of self-interest with interests of other lives, friends and, above and over-arching all, The Friend. It must logically involve the loss of individual, egocentric self to gain Communion with the Central Self.

In becoming The Essential Self you rub existence out. Is not self-annihilation, in the form of blanking mind and body, not a form of suicide? Self-murder, like all other kinds of murder, snuffs a body out. It kills an animation that's the gift of what or whom? The desecration of a life deliberately cut short is an extreme ingratitude. Nor does leaving body mean the mind is left behind. Existential mind is not Essential Self. Dying to the world while living is, however, what all mystic's teach. It is their clear, obvious and illuminated goal. Such self-sacrifice involves no trace of (↓) negativity. The absolute reverse. Love instead of hatred, life replaces death. Embracing life is not embracing death. The way involves the carriage upwards of a sacrosanctity, your soul. It is life-giving not life-taking (which is left-hand, sinister and wrong - the path of steps towards oblivion). It is life-receiving. The grip of negativity is prised; spellbinding rings of material fascination crack. Having stripped peripheral shells of mind and body, only Central Self remains. Is this the nakedness of death? It is wisdom's Life. This process of annihilation is an evolution of the soul. The rings are broken but, with enlightenment, the loop complete. Full circle. Anti-creation is annihilation. Annihilation means obliteration. Forcing something into nothing - in this case nothing existential. Essential Nothing is the termination - but the ends of cycles are their starts.

In other words, if this world is the base end of existence it is also, for The Great Return, a point of leaving. Here, at the anti-pole, we humans all can make a voluntary departure towards the Start. What if this Start is Nothing too; would you swap everything for Nothing and, in doing so, regain it all? If it is In finite, the opposite of finities, would you negate the world to paradoxically find that, from beyond it, you had rediscovered it but with fresh eyes and now illustriously transformed? From Boundless Nothing things devolve; and then towards Nothing they evolve again. And if Nothing's anti-pole is pure material what is the nature of its pole? Of what might consist pure anti-matter, anti-matter that is diametrically opposite all physical? Could it be consciousness? Whose awareness we call life - Pure Consciousness that is, in effect, Pure Life? The Start's the End of cosmic loop; the singularity is Life. The practice of such dissolution, which facilitates a disengagement from sensation, pacifies the mind and sieves out all but Life, is meditation. *Therefore the top-down logic is that, in reaching to the peak of Life, meditation's harvest must yield final victory over time and death.*

You may have begun to see from this perspective that exclusive, *bottom-up* materialism's is a 'flat' and thereby shallow entity, a shadowy domain. Although apparently pragmatic, sensible, adopted as a standard by bureaucracies and promulgated from the very heart of secular agendas, it embraces just a lesser half of whole truth. And, in the light of whole, it shows as 'flat-earth', 'topsy-turvy', a position with 'head-buried-in-the-sand'. Evolved from mud and stuck-in-mud can leave you gasping. No fresh air! It is a kind of spiritual asphyxiation, a vacuum forced against the back of mind. Therefore which projection to the future, which eschatology of doom or anti-doom both is and will become the final truth?

Is science not a brilliant light? It is. Not in the mystic way but still a most rewarding one. Its method of analysis is powerful. You fillet feeling and then strip the world to simplest form. Yet what place have protons, electrons, the

Standard Model or a *TOE* with purpose, art or, say, the aesthetic life of Florence? Nor, though tales are told, have we the slightest grasp how particles become transformed into the biological and moral orders of, say, Hong Kong. Engaging, fascinating, down-to-earth, effective, sometimes caring, lucrative but, as may be clear by now, scientifically physical interpretations aren't the only angle to illuminate our state.

The excitement of discovery runs apace. Aren't we within a quantum fluctuation's less-than-hair's-breadth from, physically, all there is to know? You complain, therefore, that *Ad Hoc* sought too many shadows out. **It needs be repeated, as this section's epigraph, that Natural Dialectic's scope on *physicalia* is broader than a micro- or a telescope's.** What's prior to physic's starting point? Anything? Eternal matter or infinity of consciousness? What set(s) the seal on natural law? Is it immaterial archetype or, in the form of fluctuations, only non-conscious energy that gave you life? If there's an immaterial element, as *PAND* avers but *PAM* denies, then these are valid questions that still stand. **The Dialectic seeks to square this fact with facts that science and technology reveal; and, where interpretation better suits an immaterial explanation, fill the gaps that random gods of quantum fluctuation and mutation leave open.**

From this perspective music emanates beyond materialistic writ; scientism's mind-set is reductive and reduced. And while closed-shop materialism serves the body's business well, it frowns upon a 'serious nonsense', soul. So, in the apparent absence of a higher truth than lifeless matter, nonsense starves. An 'emaciated' soul may therefore grasp without discrimination any quality of nourishment from any quality of open shop. Religion is an outlet dedicated only to its brand. Who, therefore, serves Impartial Fare of Highest Quality - the Whole Truth? Whose Top Grade is met in joining with the (*raj*) uplifting aspect of First Cause, 'right-hand' experience of the Word, a dialectical Communion?

The mystic is a revolutionary. Revolve from upside down by half a circle then you stand up straight. From *bottom-up* you have his *top-down* view. It embraces brilliant science in its own objective place but also fills subjective void with Nectar of a Brilliant Void. It does so with a diet of consistent logic that reflects the whole of Nature. What healthier than to digest Universal Truths?

The first leg of this philosophical odyssey has docked. Alpha is, for Natural Dialectic, Omega. The end is the beginning. The volume ends but in the next we'll see how its system applies to you personally - to your mind, body, friends and neighbourhood.

Glossary

A

adaptive potential:	involves pre-programmed, super-coded switches and recombinant (transposable) refinements intrinsic in the genomic program of any particular biological type (*SAS* Chapter 23).
ahamkar:	pre-scientific term; conscio-material band/ grade - conscious band of mind; sometimes identified (only partially correctly) with *ego*; faculty of self; *ahamkar*, involving identity, frames thought; habitual identification with own physical body; also with friends, family, community, study, work, country or cosmos as a whole; intellectual analysis is a 'knife' that dissects according to the interpretations of this frame.
akash:	pre-scientific term; possible conscio-material band/ grade - archetypal field; translated as sky, space, void within (and, some argue, from) which all events occur; also, as a higher (metaphysical) element from which the physical world derives, termed 'ether'; 'inner skies' are psychological; the 'outer sky', physical *akash*, is vacuum or *inter omnia* space.
anti-entropy:	*see* **negentropy**
archetype:	basic plan, informative element; conceptual template; pattern in principle; instrument of fundamental 'note' or primordial shape; causative information in nature; 'law of form'; nature's script; Natural Dialectic's 'holographic' edge, omnipresent but invisible because it's metaphysical; the psychosomatic place where metaphysic and its physic meet; morphological attractor or field of influence in universal mind; the subconscious component of universal (natural) mind comprising archetypes; prototype-in-mind (maybe related to Platonic ideas or Aristotelian entelechies) whose potential matter is seen as hard a metaphysical reality as, say, particles are physical realities; program(s) naturally stored in cosmic memory - simple in terms of inanimate physical 'law' (of particles and forces), complex in terms of animate structure/ function/ behaviour; information stored in a typical mnemone; in biology, metaphysical correlate of biological type/ super-species that is physically expressed in code as *potential* form; abstract or metaphysical precursor; as thought is father to the deed or plan is prior to ordered action, so archetypes precede physical phenomena; pre-physical initial condition of matter.
Archetype*:*	Primary First Cause; *Logos* (*figs.*4.1, 7.6, 10.1, 12.1).
AV:	Authorised Version e.g. *AVS* authorised version of science.

B

big bang:	unconscious singularity *see* also **transcendent projection**
black box:	process or system whose workings are unknown.
boson:	*see* **elementary particle**.

buddhi: pre-scientific term; conscio-material band/ grade - conscious band of mind; faculty of intellect; instrument, whether sharp or blunt in an individual case, of learning and discovery; analytical tool to educe physical and metaphysical patterns; pragmatic and hypothetical power of reason; crucial to gauge physical circumstance for survival and metaphysical principle for optimising state of mind.

C

caduceus: staff of Hermes/ Mercury the communicator, intercessor and informant deity; the messenger of metaphysic, carrier of thought is a power mythologically trivialised; symbol, including double helix, used by Natural Dialectic to represent basic human infrastructure, that is, the archetypal form of man.

chakra: pre-scientific term; conscio-material band/ grade - subconscious mind; metaphysical modulator; also called a node or plexus; device for the wireless transmission of *prana* to the electrical systems (e.g. nervous) of an organism; psychosomatic gate; trans-dimensional (metaphysical to physical and *vice versa*) transit-point for informative signals; a mechanism including an antenna (receiver), transformer (between 'voltages' of *pranic* energy, transducer (between electrical and *pranic* conveyance of charge), simple harmonic oscillator (a 'heart' controlling *pranic* flow) and distributor (of *prana* through a network of meridians); archetypal channel; specific mind-body broadcasting interface; morphogenetic informant especially important in the process of bio-development; lowest metaphysical component in the hierarchical transmission of power throughout macrocosmic creation and its microcosmic reflection as mankind and other forms of life of earth; *chakras* are power hubs (and thus, like suns, 'centres of influence' or 'controllers of regions') which exist on a cosmic as well as individual-body scale; conscious chakras (or focal concentrations) of higher mind are not included in this cosmology; because it cannot observe or experiment on metaphysical apparatus science has not developed any understanding beyond the physical expressions of informed energy; see also *prana*.

chaos: a confusing notion with three main but disparate implications - emptiness, disorder and randomness; Greek word meaning chasm, emptiness or space; structureless 'profundity' that pre-existed cosmos; *prima materia, prakriti* or primordial energy structured by regulation of divinity, archetype or natural law; anti-principle of cosmos i.e. disorder; any case of actually or apparently random distribution or unpredictable behaviour; also apparently random but deterministic behaviour of systems (e.g. weather, electrical circuits or fluid dynamics) sensitive to initial conditions.

chitta: pre-scientific term; conscio-material band/ grade - conscious mind; attention *per se*; pure, formless (or boundless) intelligence in which forms of thought are projected; source of ideas and creativity; psychological focus.

chromosome: a 'book' in the 'encyclopaedia' of life; the human genome contains 46 chromosomes.

cladistics: method of classification using diagrams called cladograms; organisms are collected into groups on the basis of shared (homologous) features; homologies are tallied and numerical rather than speculative, evolutionary/ phylogenetic links drawn up between organisms; cladism is thus a powerful, neutral, objectively detached tool of analysis and for this reason the technique enjoys growing popularity among the world's taxonomists.

code: the systematic arrangement of symbols to communicate a meaning; code always involves agreed elements of morphology (the form its symbols take), syntax (rules of arrangement) and semantics (meaning/ significance); without exception such prior agreement between sender (creator/ transmitter) and recipient involves intelligence.

codon: a 'word' in the genetic language; stands for an amino acid or a full stop; since more than one codon may stand for a single amino acid the genetic code is sometimes ill-perceived as 'degenerate'.

conscio-material dipole: basic, binary structure of existence; slope of creation that, in dialectical description, extends from the immaterial pole (a concentrate of Pure Consciousness) to a material pole of pure non-consciousness - the physical plane; drop/ descent from Essence through existence; drop from Conscious Singularity to unconscious singularity (black hole); a hierarchical description of polar creation simply modelled *passim* by the use of spectrum, concentric rings and, step-wise, ziggurats; immaterial, subjective element (information) tapers on a sliding scale with material, objective element (energy); 100% objective is, for example (*fig*. 0.14), 0% subjective - physical nature is a special case of consciousness, its total absence at creation's spectral base; *vice versa*, at Top, Transcendent Nature (Essence) there is 0% objective form; and ratios, from top to bottom, in between; from this perspective any 'sharp' division between mind and matter becomes illusory; it is no more real than exists between, say, UV and microwave radiation in the 'rainbow' continuum of their e-m frequencies; cosmos is a conscio-material spectrum, a taper of consciousness to, at base, its absence (*fig*. 0.14).

convergence: the tendency of unrelated organisms to evolve similar characteristics; in the case of *divergence* adaptation/ speciation from an original feature occurs (e.g. beaks of finches);

convergence, involving the unrelated, mosaic occurrence of similar features (such as the camera eye, viviparity and thousands of other instances), runs counter to Darwinian expectation; it means that such codified features must have evolved independently many times over; evolutionary explanations of this profound yet ubiquitous puzzle may thus involve speculations such as appeal to non-random 'deep bio-structure', 'principles of evolution', 'morphological laws' or 'inevitablility' granted by imaginary natural laws of codification/ innovation; for a design theorist the bio-codification and engineering of 'convergent' forms derives from either an original use of modular programming or, in the case of so-called micro-evolutionary variation, from in-built adaptive potential flexibly but appropriately activated by genetic switches and epigenetic markers.

cosmic fundamentals: cosmic psychological and physical qualities (see Chapter 2); basic states or tendencies; universal ingredients whose mixture is variously expressed in every object and event.

cosmological axis: human pivot; the point at which subjective and objective perception meet; eye-centre; third eye; thought centre; *ajna chakra*.

cosmological principle: idea that, on a sufficiently large scale, the distribution of matter in the physical universe looks just about the same from any vantage point; it therefore has neither centre nor, being infinite, edge - unless of course, its space is somehow spherical.

cosmos: often applied to physical universe, universal body; from Greek word denoting orderly as opposed to chaotic process; involuntary pattern of nature; also equated, including metaphysical mind, with existence as a whole; seen, dialectically, as a projection through the template of metaphysical archetypes; **the umbrella title of the series and website of books, Cosmic Connections, could, with reference to the Natural Dialectic which structures its *CUT* (see Glossary: unification), equally be called Orderly Linkages.**

creation: origination; physical or psychological arrangement; mind creates with purpose, matter without; creation means active production but also passive result; a creation will have been informed by force of mind and/or matter.

D

dialectic: a form of debate between positions of polar opposition (argument and counter-argument or thesis and antithesis); the motion of to-fro discussion that results in resolution (synthesis) whereby points of view are aligned; balance, compromise, neutral ground, golden mean and central truth are aspects of this synthetic (*Sat*) fundamental;

paradoxically, two become one; union supersedes division; Natural Dialectic suggests that dialectical motion reflects the binary, cyclical nature of cosmos; such polar, to-fro or oscillatory dynamic occurs as the continual disequilibrium of nature (called motion and transformation) always seeks its various re-balances.

dialectical stack: stack of opposites; columnar expression of polarity; there are two kinds of stack - primary or non-vectored and secondary, vectored; primary (essential) stacks set (*Sat*) Unity against (↓ *tam*/ *raj* ↑) duality (for elaboration see Chapter 2 and *figs*. 1.4 and 2.2); secondary (existential) stacks represent the various kinds of polarity from which the changeful web of existence is composed (see *figs* 1.4 and 24.1); each pair of polar 'anchor-points' implies a scale or dynamic range that runs between 'paired opposition' or 'complementary covalency'; stacks do not necessarily list synonyms or make equations; ***their perusal is intended to promote connections because consideration of connections tends to help unify/ collate/ organise one's working comprehension of any matter in hand.***

DNA: a complex chemical; a large bio-molecule made of smaller units, nucleotides, strung together in a row; a polymer in the form of a double-stranded helix; a medium superbly suited to the storage and replication of 'the book of life'; 'paper and ink' on which the genetic code is inscribed; an organism's 'hard drive'.

dukkha: imperfection, suffering.

E

electromagnetism: physics of the field that exerts an electromagnetic force on all charged particles and is in turn affected by such particles: light/ e-m radiation is an oscillatory disturbance (or wave) propagated through this field; light; light paradoxically involves a perfect, polar balance between contractive/ magnetic and radiant/ electric components.

elementary particles: science has discovered and, for the most part, experimentally verified, over fifty elementary particles; these are divided, in simple terms, into bosons (force carrying particles) and fermions (separate particles); bosons include photons (which mediate the electromagnetic force), gluons (which mediate the strong nuclear force), W and Z particles (which mediate the weak nuclear force), possibly gravitons (which mediate the gravitational force) and also possibly a Higgs boson (which may mediate a proposed mass-giving field); fermions include two main groups - six quarks and leptons (six electron/ neutrino types); derived from quarks are strongly interactive composites called hadrons; hadrons include baryons such as protons and neutrons and (perhaps a little confusingly) bosons such as short-lived mesons.

entropy: a measure of the amount of energy unavailable for work or degree of configurative disorder in a physical system (see second law of thermodynamics); inertial aspect of an energetic, material or conscious gradient; diffusion or concentration gradient outward from source to sink; drop towards 'most probable' outcome i.e. inertial slack; a measure of disintegration or randomness; expression of the (*tam* ↓) downward cosmic fundamental; a major property of matter, closely coupled with materialisation; in a closed system, which the universe may or may not be, this tends the eventual loss of all available energy, maximum disorder and the exhaustion of so-called 'heat death'.

enzyme: biochemical widget; protein catalyst without whose type metabolism (and therefore biological life) could not happen.

epigeny: genetic super-coding; contextual punctuation; chemical modification of *DNA*; also extra-nuclear factors that may cross-reference with genetic expression.

equilibrium: three modes of equilibrium are (*sat*) balance of poise or pre-active potential; (*raj*) dynamic balance occurring in all regular cycles, wave-forms and cybernetic homeostasis that is basic to the stability of life-forms; and (*tam*) inertial equilibrium that results from diffusion of information or energy; it equates with exhausted inaction or 'flat', impotent rest; such post-active inertia represents the most probable distribution of energy/ matter with the least energy available for work viz. the most random arrangement permitted by the constraints of a system; expressed in psychological terms as ignorance, unconsciousness or sleep; see also equilibration, *karma* and *fig.* 1.1 'Pivoted Existence'.

Essence: (*Sat*) Supreme or Infinite Being; Substance (perhaps Spinoza's Substance) 'prior to' or 'above' existence; Pure Consciousness/ Life; Peace that transcends all psychological and physical action; the root of an essentially undivided universe; Conscious Singularity; Uncreated One within which and whence all differences have their being; Apex of Mount Universe; goal of saints/ 'philosopher kings'; the 'point' at which All-Is-One.

eukaryote: non-prokaryote; any organism except bacteria and blue-green algae.

evolution: there are today *four* main usages of this word; each 'loading' derives from the original Latin, 'evolvere', meaning to unroll, disentangle or disclose; the *first two*, physical and biological, are conceived as natural/ mindless processes; the *second, mindful pair* is of psychological/ teleological import; specious ambiguity may conflate or switch between the fundamentally separate pairs of meaning. **Firstly**, in the scientific context of physics and chemistry, the word is used to describe change occurring to physical systems; the laws of nature can't, it seems, evolve through time but stars, fires, rocks or gases can.

Secondly, though also subject to the 'rules' of entropy, biological evolution is a theory of *random progression* from simple to complex form; it thereby implies increasing, codified complexity; while retaining the 'hard loading' of physical science it also, ambiguously, claims that codes, programs, mechanisms and coherent, purposive systems - normally the province of mental concept - self-organise by, essentially, chance; such confusion, the basis of naturalism, is compounded by failure to distinguish between, on the one hand, ubiquitously observed variation (called micro-evolution) and, on the other, Darwinian 'transformation' between different sets of body plan, physiological routines and associated types of organism - such 'black-box macro-evolution' as is never indisputably observed; to evoke a naturalistic ambience it is fashionable to use 'evolved' interchangeably with or to replace the words 'was created', 'was planned' or 'designed'; finally, it is noted that the coded, choreographed development of a zygote, packed with anticipatory information, through precise algorithms to adult form is the absolute antithesis of blind Darwinian evolution. ***Thirdly***, man certainly evolves ideas; intellect can evolve 'purposive complexity'; we invent all kinds of codes, schemes and machines; we devise increasingly complex theories and technologies; and we evolve an understanding of natural principles; this, which all parties accept, is an informative, psychological sense of 'evolution'. The ***fourth*** sense of evolution, at least as near to the original Latin as the other three, is the spiritual usage; immaterial spiritual evolution, unacceptable to materialists and unknown to physical science, is at the very heart of holism; in this voluntary sense of evolution practitioners cast off material attachment, evolve and merge into the *Logos*; evolution (or, perhaps better, centripetal involution) of the soul is their great business; their aspiration is to unite with The Heart of Nature.

evolution pre-Darwinian: minority/ anti-mainstream pre-Socratic snippets and sense-based Epicureanism lionized by interpretations of post-18th century materialists; virtually undetectable eccentricity in Chinese, Indian and Islamic literature; natural selection treated by creationists al-Jahiz and Edward Blyth; Buffon, a non-evolutionist, addressed 'evolutionary problems'; Lamarck (evolution by inheritance of acquired characteristics); hints in poem by Erasmus Darwin.

evolution Darwinian: mechanism - natural selection; major tenets - common descent (inheritance), homology and 'tree of life'.

evolution neo-Darwinian/ synthetic: as Darwinian, except synthetic theory adds random mutation as the mechanism for innovation; also adds a mathematical treatment of population genetics and various elements (e.g. geno-centric perspective) derived from molecular biology.

evolution post-synthetic phase: natural selection and random mutation acknowledged as mechanisms insufficient to source bio-information; post-Darwinian evolution invokes mechanisms from hypotheses such as *NGE* (natural genetic engineering) and 'evo-devo'; holistic possibilities also address the origin of complex, specified and functional bio-information.

existence: which 'stands out' from background 'nothingness'; the apparently divided universe; seemingly disparate, finite things; all motion/ change/ relativity; all psychological and physical events.

F

fermion: *see* **elementary particle**.

field: any extent wherein action either physical or metaphysical but of a certain kind occurs e.g. field of battle, influence of mind or magnetism; the scientific definition is limited to a collection of numbers varying from point to point - such as a scalar field of contours on a map - or numbers with direction - such as a vector field showing speeds and directions of wind.

first causes: check *figs*. 3.3, 5.1, 9.1, 11.1 or 11.3. First cause is first motion in a previously undisturbed, pre-conditional field. **First Cause Psychological** is Archetype, Potential Informant or (see Chapter 5: Top Teleology) *Logos*; attributes of this Primary Source and Sustenance of Creation include omnipresence, omnipotence and omniscience; **first cause physical** is also called potential matter or archetypal memory; as the secondary source of creation it precedes physical phenomena; as such it is, transcending physical appearances, metaphysical; this 'physical nothingness' is therefore, paradoxically, the source of everything composing astronomical cosmos; it consists of their being or essence as opposed to their becoming; its void, with respect to the presence of finite phenomena, appears infinite; attributes of immanent archetype, the primary informant of our non-conscious, energetic universe, include omnipresence and omnipotence.

G

gene: generally means a basic unit of material inheritance; section of chromosome coding for a protein; digital file; a reading frame that includes exons and introns; the old one gene-one protein hypothesis is incorrect; in fact, by gene splicing, a particular piece of *DNA* may be used to create multiple proteins.

genome: total genetic information found in a cell: think of the genome as an instruction manual for the construction and physical operation of a given organism.

genotype:	the genetic constitution of an organism, often referring to a specific pair of alleles; the prior information, potential, plan or cause of an effect called phenotype.
gravity:	in physics an attractive mass-to-mass force or warping of space-time; in Natural Dialectic the term is redefined more broadly - the agency of its (*tam* ↓) downward vector includes all psychological and physical factors of materialisation; such 'gravitational' factors and their properties are listed in the left-hand column of Secondary, Existential Dialectic; they include pain, pressure, confinement, strong nuclear force, mass, electromagnetic binding, inertia, entropy, 'standard' gravity and so on; gravity might be summarised as 'negative power' or 'the principle of death'.
GTE	general theory of evolution; *see* macroevolution.

H

holism:	opposite of reductionism; the view that a whole is greater than the sum of its parts; the extra metaphysical (immaterial) ingredient is identified by Natural Dialectic as information; information implies the purposeful design, development and arrangement of contingent parts in a working system; may operate according to a Logical Norm.
hologram:	a 3-d photograph made with the help of lasers. Unlike a normal photographic image each part of it contains the image held by the whole.
homeostasis:	vibratory or periodic control of a system to obtain balance round a pre-set norm; the mechanism of its information loop involves sensor, processor and executor; the operative cycle works by negative feedback; psychological (nervous) and biological cybernetics; the informed basis of biological stability.
homozygous:	having the same allelic forms of a particular gene.

I

illusion:	is the cut between illusion and delusion an illusion? illusions, apparently outside the mind, appear real; a delusion, in it, we think real; neither, mind allows, is real or true.
information	the immaterial, subjective element; information occurs in three distinct modes; informative *potential* is action's precedent; this potential is both the source and substrate of all psychological activity; we know this substrate of life and consciousness; *active* information inhabits its own centre, mind; mind knows, feels, purposes, creates, codifies and recognises meaning; it is also, by way of secondary and subconscious forms of archetype) a physical entrainer; thus *passive* forms of information, either in memory or physical objects and events, reflects active; in other words, *passive* information is stored as subconscious 'files' in memory;

and it is fixed in the expressions of non-conscious matter according, universally, to the archetypal behaviours of natural bodies or, locally, to particular schemes of life; that is to say, both the constructions of life-forms and the inanimate cosmos are the physical product of stored concept.

informative entropy: loss of information due to degradation of its carrying medium; such a medium may be metaphysical (mind) or passive and physical (for example, computer files or genetic code); and its entropy may be metaphysical (loss of memory, focus or consciousness) or physical (for example, genetic mutation); the informative correlate of such degeneration is diminished organisational capacity, meaning or thrust of original purpose.

informative negentropy: gain of informative clarity; increasingly focused, purposive specificity; associated with knowledge, wisdom, grasp of principle and pristine construction; machines are a good example of informative negentropy.

inversion: turning upside-down or inside-out; reversing an order, position or relationship; in a hierarchical sense inversion is allied with the reflective asymmetry of opposite poles; information outwardly expressed; pole-to-pole reversal integral to dialectical structure; inversion represents, of the two anti-parallel vectors on creation's conscio-material gradient, the (*tam*) centrifugal vector; as opposed to (*raj*) centripetal eversion; various kinds of inversion (cosmic and micro-cosmic (biological)) are discussed in these books.

K

karma: the cosmic law of balance between action and reaction; equilibration such as underlies all mathematical equation; a deed with implications of the reactions or 'payback' it provokes; fruit or result of previous thoughts, words and deeds; applies as rigidly to metaphysical (psychological) as physical events.

L

lepton: *see* **elementary particle**

levity: agency of the (*raj* ↑) upward vector; dialectical converse of gravity; psychological and physical 'levitatory' forces lift or stimulate; they are listed in the right-hand column of Secondary, Existential Dialectic and include light, heat, excitement, dematerialisation, release, negentropy, focus of interest, affection and so on; physically, levity includes anti-gravity or the intrinsic property of matter's absence, space; generally summarised as 'positive power' or 'a buoyant principle of liveliness'.

logic: analysis of a chain of reasoning; principles used in circuitry design and computer programming; 'normative reason' relates to the basic axiom(s) of a given standard e.g. *bottom-up* materialism or *top-down* holism; three main logical thrusts are: (1) inductive (premises/ observations supply evidence for a probable/ plausible conclusion) as in the case of experimental science working *bottom-up* from specific instances to general principle: (2) abductive (best inference concerning an historical event): and (3) deductive (conclusion in specific cases reached *top-down* from general principle): two pillars of logic are holism and materialism; holism employs mainly deductive/ abductive operations and a Logical Norm; materialism tends to inductive/ abductive operations whose axis is non-conscious force and chance.

***Logos*:** First Cause; Prime Mover; Causal Motion that sustains creation's conscio-material gradient; labelled with many names at different times, places and languages; *Logos*, transcending mind, is Conscious; therefore, <u>Who</u> is *Logos*?

M

macrocosm the physical universe of astronomy and cosmology; dialectically, the whole of existence (i.e. both universal mind and universal body) as opposed to individual, microcosmic objects and events - including the human body.

macro-evolution: large-scale, non-trivial evolution; process of common or phylogenetic descent alleged to occur between biological orders, classes, phyla and domains; includes the origin of body plans, coordinated systems, organs, tissues and cell types; unexplained by mutation, saltation, orthogenesis or any known biological mechanism; sometimes called 'general theory of evolution' (*GTE*); crucial but unseen, hypothetical extrapolation from micro-evolution; may or may not occur; an extrapolation from Darwinian micro-evolution vital to sustain a 'progressive' materialistic mind-set, is conjectural alone.

***manas*:** pre-scientific term; conscio-material band/ grade - both conscious and subconscious bands of mind; 'mind-stuff'; 'clay' moulded by the the hand of thought and perception; metaphysical 'material' on which direct formative action of *chitta* occurs; 'film' on which the perceptions of mind are developed and, at the same time, 'screen' on which they are perceived; receptor for sense perceptions and storage silo of such impressions as 'seeds' or 'files' of subconcious memory; substance of archetypal field, in other words, of universal archetypes (cf. typical mnemone); mental form and energy.

***mantra*:** archetypal symbol; psychological transformer; authorised form of words repeated to exclude other thoughts; examples include the 'Hail Mary', '*Om Mane Padme Hum*' and,

	materialistically, 'evolution made…', 'in time nature designed…' or similar incantation.
maya:	partial truth; world of forms and forces; illusion that changeful cosmos is the ultimate reality; motions and perceptions composing *maya* are thus, set against Essential Truth, more or less unreal; becoming wise to the nature of *maya* yields liberation from its cosmic veil.
meditation:	Lat. *medius*, middle; coming to Centre; mind and body dropping away.
meiosis:	shuffling the information pack: variation-on-theme; mechanism for the production of haploid gametes; genetic postal system for sexual reproduction.
metabolism:	body chemistry.
metaphysic	= non-physical/ immaterial/ psychological/ unnaturalistic; physically expressed as specific/ intended arrangement/ behaviour of materials; physical behaviour reflects metaphysical blueprint; involves element of information; involves symbol/ code/ abstraction/ logic/ reason/ mathematics; also meaning/ message/ goal/ teleology; also consciousness/ mind/ life/ experience/ feeling; and also morality/ force psychological/ emotion; involves innovation/ creativity/ art/ invention/ aesthetics.
microcosm:	an entity that reflects the universe by containing all its basic constituents. Used especially of the human state where it may refer to both mind and body or, in a purely physical context, body alone.
micro-evolution:	misnomer; non-progressive, small-scale variation within a species or, more broadly, between strains, races, species and genera; variation/ adaptation within type; trivial Darwinian changes that may occur by natural selection/ ecological factors acting on genetic recombination, mutation or adaptive potential (q.v. this Glossary); sometimes called 'special theory of evolution' (*STE*), micro-evolution/ variation is a fact.
mitosis:	conservative copying and delivery of genomes in cell division; genetic reprinting; genetic postal system for asexual reproduction.
mnemone:	a division of memory: the two divisions are personal mnemone (likened to a working cache or data store) and typical mnemone (likened to a *ROM* or an operating system); typical mnemone is a synonym for archetypal memory; Natural Dialectic's definition (see especially Chapters 15 and 16) involves no 'cultural' connotation whatsoever and is thus wholly distinct from evolutionary psychology's use of the word; typical mnemone is an umbrella word for natural memory; term also applies to each specific type of organism's archetype.

morphogene: part of typical mnemone or archetypal memory relating to physical construction; morphological attractor; the component of subconscious mind associated with electrochemical function and thereby body; morphogene is the dominant, perhaps exclusive, aspect of mind in unconscious organisms such as plants or fungi.

morphogenesis: the development of biological structure; more generally, the production of physical form.

mosaic: the presence of permutations of codified sub-routines or similarities of form and/or function scattered in organisms unrelated by lineage.

mutation: accidental change to genetic code.

mysticism: quite different from objective, it is the subjective science; not philosophy, religion or opinion but practice to achieve communion with natural, inner, immaterial truth; esoteric as opposed to exoteric, materialistic discipline; 'science of the soul'; as gyms and physical action are to athletes so meditative exercise and psychological stillness are to mystics; involves psychological techniques to achieve a clear, rational goal - purity of consciousness and thereby understanding of the fundamental nature of the informative principle, mind; since life is lived in mind a mystic seeks consummate knowledge of life's source and sanctum, that is, communion with its deathless heart; adepts were, are and will be 'Olympian' meditative concentrators.

N

nano-biology: biology of structures/ physiologies involving a few atoms or molecules; 'extremely small biology'.

nanotechnology: technology at atomic and sub-atomic level as is, basically, life's.

naturalistic methodology: also known as 'methodological naturalism': is, strictly, not concerned with claims of what exists or might exist, simply with experimental methods of discovering physically measurable behaviours; thus only materialistic answers to any question (e.g. how biological forms arose) are deemed 'scientific' or 'scientifically respectable'.

negentropy: opposite of entropy; lowering of entropy; expression of the (*raj*) upward-pointing cosmic fundamental closely coupled with stimulus, dissolution and dematerialisation; a measure of input, cooperation or synthesis; motive/ fluidising aspect of an energetic, material or conscious gradient; gain of energy, configurative order, information or consciousness in a system; when used in terms of information negentropy involves gain in order or understanding of principle from which different actualities derive; a measure of the amount of concentrated/ conceptual information, specific, intentional complexity or

	conscious arrangement in a system; a natural and essential property of mind.
Nirvana	state of enlightenment; 'non-condition'; *nirvana* is devoid of existential motion; extinction of existence (i.e. perpetual change) leaving Essence Alone; pure soul; psychological super-state; Buddhists call such transcendence non-self or the Formless Self.
non-existence:	where creation = formful existence, non-existence is formless; the polar opposite of physical space and time is Transcendent Potential; such pre- or super-existential formlessness is non-existent; Absolute Non-Existence is Essential; however relative non-existences of two kinds also occur; the first kind is metaphysical/ subjective and therefore psychological; it involves the absence of a specific psychological form or event; unconscious oblivion is one such non-existence; the second kind involves the local absence of a possible physical event (an object is a 'slow event'); impossibilities are non-existences but imaginations of non-existence (including symbolic abstractions, hypothetical entities, physical absences, absolute emptiness and the number zero) exist; furthermore, the nothingness of space and time, the zero-point of calculus and zero's empty set together constitute the basis of physical science and mathematics.
nucleic acid:	*see DNA* and *RNA*.
nucleotide:	basic, triplex unit of nucleic acid polymer; monomer composed of phosphate and sugar (the 'paper' part) and base (the 'ink letter'); letters' of the genetic alphabet are (G) guanine, (C) cytosine, (A) adenine and (T) thymine. In *RNA* thymine is replaced by (U) uracil.
nucleus:	centre, heart, creative core; informative *sine qua non*; psychological nucleus is consciousness or (in formful aspect) mind; atomic nucleus, made of protons and neutrons, is a centre of mass determining electron configuration; biological cell nucleus is the instruction centre of a cell containing *DNA* and nuclear operating machinery; nuclear is critical.
O	
Om:	universal sound, fundamental reverberation, basic truth; initial motion of Potential Information; sometimes spelt *Aum*, a Sanskrit word whose Semitic transliterations are Am'n, Amin and Amen; see also First Cause, *Logos, Kalam, Shabda* etc.
order:	regular, regulated or systematic arrangement; organisation according to the direction of physical law; passive information by which things are arranged naturally (with

	predictable but non-purposive complexity) or purposely (with innovative or specified complexity); mind, generating specified complexity in the order of its technologies and codes, actively informs; the orders of mind are meaningful, the orders of matter lack intent; see also cosmos.
organelle:	cellular sub-station; discrete part of a cell; sub-cellular compartment having specific role such as informative (nucleus), energetic (mitochondrion, chloroplast), constructional (ribosome, Golgi body) or other.

P

PAM, *PAND*, *PCM* and *PCND*:	philosophical gambits; see Primary Axioms and Corollaries.
phenotype:	the effect of causal potential; result of the development of prior, informative 'egg'; outward expression of inner plan; sensible appearance of an organism as opposed to its genotypic scheme: the whole set of outward appearances of a cell, tissue, organ and organism are sometimes called a phenome (*cf.* genotype/ genome).
photosynthesis:	process by which inorganic carbon is introduced to the biological zone and energetic sunlight fixed as a crystalline molecule of storage, a sugar called glucose.
phylogeny:	evolutionary history; relationships based on common or evolutionary descent.
potential:	poise; latent possibility; potent non-action that precedes any particular action or creation; in science potential energy is defined as the energy particles in a system (or field) possess by virtue of position/ arrangement; gravitational, electrical, electro-chemical, thermo-dynamical and other kinds of potential are recognised; in dialectical terms mind precedes matter, information precedes the pattern of material behaviour; information is energy's pre-requisite potential; in this case *informative potential* involves two conditions; firstly, a pre-existential/ essential state of pure potential; secondly, a pre-material, metaphysical fact of potential matter, archetype or laws of nature; if potential's pre-active equilibrium is related to the voltage of a full battery then aspects of psychological 'voltage', whose currents drive intentional behaviour, are purpose, will and plan.
potential matter:	see archetype.
prakriti:	pre-scientific term; conscio-material band/ grade - whole spectrum of existence; complements Essential *Purusha*; universal energy; generic term for nature; screen and light on which the show of creation is projected; 'clay' with which the potter of conscious experience works to produce form; thus also (*figs.* 6.2, 7.5 and 13.6) identified with the objective, energetic as opposed to subjective, informative side of

conscio-material cosmos; fundamental substrate whose root exercise involves continual recombinations of three cosmic fundamentals (Chapter 2); interplay of these qualities, attributes or tendencies intrinsically inhabits every object and event; it generates all patterns, forms and forces, whether in psychological or physical regions of the universe; nested, *prakritic* layers are hierarchically arranged like a grid of stepped-down voltages from a power station; in a second comparison, as the waveband of visible light is part of a much larger electro-magnetic spectrum, so the bands of higher and lower conscious mind, subconscious memory and non-conscious physicality compose the spectrum of conscio-material *prakriti*; in this way, for example, *prana* is a low-level, 'infra-red' expression of *prakriti* operant at the *PSI* border where subconscious archetypes (Chapters 15 and 16) give rise, with location, to quantum phenomena; and its lowest, 'radio' level of expression, peripheral to the full creation, is gross physical energy whose various transformations are expressed as the operations of physics, chemistry and biology.

prana: pre-scientific term; possible conscio-material band/ grade - subconscious and physical (lowest) levels of *prakriti*; lowest metaphysical bandwidth of *Logos*/ Om; Chinese *qi* or *ch'i*; associated, as in the yogic practice of *pranayama*, with breath and thereby life of material body; also with light (visible band electromagnetism) and oxygen (specifically, negative ionic charge); identified as the archetypal energy of subconsciousness and the operations of typical mnemone (Chapter 16); subliminal, psychosomatic or mnemonic energy of universal mind at archetypal grade; supports *PSI* (psychosomatic) traffic between universal memory and quantum agencies in the case of both biological and and physical formations; 'infra-mental' band called potential matter; vibration underlying perpetual atomic motion; five pranic bands are analogised with visible light's rainbow, each frequency correlated with an expression of elemental character in physical phenomena or with a level of biophysical expression (*fig.* 17.11); wireless *prana* is identified as the metaphysical life-force of the physical body; in living organisms (including you and me) it is processed by way of metaphysical apparatus called node or '*chakra*'; each of a hierarchical series of such nodes operates as an antenna, transformer and distributor; the system acts as a transducer of *pranic* frequencies to those of bio-electromagnetism and electrical charge; the highest bio-frequency resonates with our '*ajna chakra*' or third eye behind the forehead; this in turn, is subservient to '*sahasrara*', the 'thousand-petalled lotus' supplying 'voltage' (and thereby current) to sustain the physical universe; having entered the human system by resonance with the '*ajna*' antenna *prana* is passed through a grid of aforementioned nodes, well-known to yoga and arranged down the spine; each distribute frequencies appropriate to its body area by a network of *nadis* or meridians identified by medical acupuncture; being metaphysical the pranic system cannot be physically tested by empirical, scientific experiment but only by inference (e.g. a cure); for this reason some proponents of occidental medical science dismiss *pranic* mechanisms of the mind as 'pseudoscientific' and, having thus 'rationally' condemned, proceed to narrowly and unwisely dismiss the broader immaterial fraction of holistic order wherein such components play a crucial part.

Primary Axiom of Materialism (*PAM*): all objects and events, including an origin of the universe and the nature of mind, are material alone; cosmos issued out of nothing; life's an inconsequent coincidence, a fluky flicker in a lifeless, dark eternity.

Primary Axiom of Natural Dialectic (*PAND*): there exists a natural, universal immaterial element - information; immaterial informs material behaviour; a conscio-material dipole that issues from First Cause informs and substantiates both mental (metaphysical) and physical creations; there is eternal brilliance whose shadow-show is called creation.

Primary Corollary of Materialism (*PCM*): the neo-Darwinian theory of evolution, that is, life forms are the product, by common descent, of a random generator (mutation) acted on by a filter called natural selection; such evolution is an absolutely mindless, purposeless process; the *PCM* is a fundamental *mantra* of materialism.

Primary Corollary of Natural Dialectic (*PCND*): the origin of irreducible, biological complexity is not an accumulation of 'lucky' accidents constrained by natural law and death; forms of life are conceptual; they are, like any creation of mind, the product of purpose.

prokaryote: non-eukaryote; bacterial type with little or no compartmentalisation of cell functionaries.

promissory materialism: belief system sustained by faith that scientific discoveries will in the future justify/ vindicate exclusive materialism and, as a consequence, atheism; confidence that technology will solve (more often than create) the problems the world faces; may involve a call to progress towards the technological provision of its 'promised land'.

protein: factor made from a specific sequence of amino acids to perform a specific task; 'informative' protein includes some hormones; skin, hair, bone, muscle and other tissues are made of 'structural' protein; 'functional protein' called enzymes mediates all stages in cell metabolism, that is, it catalyses all biochemistry.

***PSI* (psychosomatic interface):** psychosomatic border; the level of mind-matter interaction; bridge between metaphysical and physical dimensions; potential matter; 'gap of Leibniz'; 'fit' of mind to matter; point of linkage between subconscious mind and non-conscious matter; gearing between instinct/ archetype and the behaviour of material objects and energies; as in the case of physical law, psychosomatic influence is both general in potential and local/ specific in engagement.

psychological entropy: a measure of loss of concentration, focus of attention or consciousness; loss of 'mental energy' or aptitude; the drop from waking to sleep; loss of knowledge, information or sensitivity; the gradient from intelligence through stupidity to oblivion; an expression of the (*tam*) downward cosmic fundamental in mind; a tendency predominant in lower, egotistical or selfish mind; increasing level of ignorance, anguish or immorality; loss of integrity, psychological disharmony or disintegration; see also *information entropy*.

psychological negentropy: a measure of gain in order; an increase in concentration, focus of attention or consciousness; gain in sense of purpose, 'mental energy' or aptitude; the rise from sleep to waking, 'dark to light' or unhappiness to happiness; gain in knowledge, information or sensitivity; the gradient of learning and spiritual evolution; an expression of the (*raj*) upward cosmic fundamental in mind; a tendency predominant in higher mind; increasing level of contentment, understanding and the natural morality of happiness; the ascent towards psychological radiance, harmony and integration. The converse of psychological negentropy involves *entropy of information*.

psychosomasis: operation across the psychosomatic border; mind/ body interaction; the one-way imposition of archetypal pattern on *physicalia*; the two-way exchange of information in sentient organisms through the agency/ medium of subconscious patterns; *see* also synchromesh 2 (Chapters 16 and 17).

***Purusha*:** pre-scientific term; conscio-material band/ grade - conscious; Pure Consciousness, Source of Life, Subjectivity and Creativity; Universal (*Sat*) Potential; boundlessly pre-active, in action Prime Mover, First Cause or Top Governor on the universal scale and order of creation; given many other names in many languages; ultimate subject of worship, praise and love.

Q

quantum: minimum discrete amount of some physical property such as energy, space or time that a system can possess; quantum theory states that energy exists in tiny, discontinuous packets

	each of which is called a quantum; an elementary discontinuity; an elementary particle e.g. photon or electron.
quantum level:	matter-in-principle; 'internal', 'causal' or 'subtle' matter; the vibrant or energetic phase of physical organisation; zone of sub-atomic particles and forces; step (on cosmic ziggurat) between potential and bulk matter whose aspect is sometimes extended to include atomic and molecular interactions; small-scale substance underlying large-scale, sensible appearances.

R

raj:	(↑) upward, levitatory or stimulatory cosmic vector.
reductionism:	opposite of holism; the materialistic view that an article can always be analysed, split up or 'reduced' to more fundamental parts; these parts can then be added back to reconstruct the whole; a whole is no more than the sum of its parts.
religion:	etymology debated between Latin *religare* (bind) and *relegere* (review); *religio* means dutiful and meticulous observance; currently religion means world-view, mind-set or basic faith; whether of materialistic or holistic belief, it involves the non-negotiable substance of an individual or community's truth - notably as regards origins; antagonism between holistic practice and the naturalistic methodology of science is, because the couple deal with separate but complementary physical and metaphysical dimensions, flawed; a materialist/ atheist 'binds meticulously' to an evolutionary mind-set, a holist to pantheism or a Living Creator; in the case that self-deception is crucial to successfully deceiving others which, holism or materialism, is the religion that is ultimately true?
resonance:	the tendency of a body or system to oscillate with a larger amplitude when subjected to disturbance by the same frequencies as its own natural ones; thus a resonator is a device that naturally oscillates at such (resonant) frequencies with greater amplitude than at others; resonance phenomena occur with all kinds of vibration, oscillation or wave; their sorts include mechanical, harmonic (acoustic), electrical (as with antennae), atomic and molecular.
respiration:	the controlled release of energy from food.
RNA:	a single-stranded nucleic acid polymer employed in three different forms during the process of protein synthesis; in computer terms might be likened to a portable memory stick as opposed to *DNA*'s hard drive.

S

sanskara:	character trait; groove, habit, obsession or repetitious mode of thought proportional in depth to the intensity of desire, force of impact or impression that created or sustains it.

samsara: existence, phenomena, the place of cycles and, therefore, reincarnation; non-essence or, in Buddhism, what is not *Nirvana*.

sat: 'top' or essential cosmic fundamental; 'vector' of balance, neutrality.

science: Latin *scire* (know); knowledge; commonly understood as the practical and mathematical study of material phenomena whose purpose is to produce useful models of the physical world's reality.

scientism: a philosophical face of official, *de facto* commitment to materialism; today's majority consensus of what the creed of science is; an -ism born of *PAM*; a faith that all processes must be ultimately explicable in terms of physical processes alone; like communism, a one-party state of mind; a doctrine that physical science with its scientific method is ultimately, the sole authority and arbiter of truth; a set of concepts designed to produce exclusively material explanations for every aspect of existence, that is, to colonise each academic discipline and build its intellectual empire everywhere; 'scientific fundamentalism' closely allied, when expressed in social and political terms, with 'secular fundamentalism', sociological interpretation of behaviour and the fostering of a humanistic curriculum.

secular fundamentalism: *PAM* as applied to the worlds of nature and of human society.

secularism: concern with worldly business; lack of involvement in religion or faith; secularism is generally identified, as defined by the dictionary, with materialism; for a secularist the ultimate arbiter of truth is human reason - ideas are open to negotiation so that even morality is relative; however many liberal agnostics, atheists and humanists argue that their metaphysical, philosophical system also embraces so-called 'universal' moral values and, as opposed to zealotry or the logic of evolutionary faith, a liberal politic of 'philosophical live-and-let-live'.

siddhi: marvellous, miraculous or 'super-natural' psychic ability that, at the point a practitioner actually masters it, becomes natural.

STE: special theory of evolution; *see* microevolution.

stereo-computation: stereochemistry involves study of the relative spatial arrangement of atoms in molecules; in biology a 1-D line of informative code (whose 3-D constituents bear no figurative relationship with their informed product) give rise to relative 3-D spatial arrangements at all levels from molecular to systemic and whole-body. Such targeted generation may be termed bio-logical stereocomputation.

sub-state: *opp.* super-state; impotence, discharge, exhaustion, final stage in the expression of potential; fixity; non-conscious base-state;

	state 'below/ subtendence; extreme negativity/ (*tam*) condition.
sufi:	mystic, Islamic 'heretic' of whom the most influential is perhaps Jalal-ud-Din Rumi, a disciple of Shamas of Tabriz.
super-state:	potential; source of possibility; causal metaphysic/ archetype; state 'before' or 'above' subsequent expression; immanence; transcendence; precondition; (*sat*) priority.
symmetry:	closely allied with the (*sat*) characteristic of balance; aesthetically pleasing balance and proportion; geometrical balance or interactive process such that some feature of an action remains invariant, that is, conserved; the symmetry of an entity (such as a sphere, empty space or natural law) or feature (such as energy) that remains the same at all times everywhere from any local point of observation or through every transformation is called 'higher' or 'continuous'; if a feature is conserved only when an object or process is moved, turned or viewed at certain angles or under specific conditions its symmetry is called 'lower' or 'discrete'; the symmetrical properties of a system may be precisely related to corresponding conservation laws and *vice versa*; various kinds of symmetry independent of space-time coordinates are important to both quantum and classical physics; scale symmetry occurs when a reduced or expanded object keeps its shape but not its size (as with Mandelbrot fractals); dialectical symmetry also involves *informative potential*; its metaphysical archetypes inform principles, laws or determinant fields that exist prior to action and, from their possibilities, govern actual outcome; such 'configuration of the world' is absolute and, beyond entropy, stable; it is negentropically immune from decay; by contrast, the 'free' symmetry of potential energy is inherently unstable and (like a pencil balanced on its tip) liable to spontaneously 'topple' or 'break' into the least energetic of a range of circumstantial possibilities; such spontaneous symmetry-breaking, the basis of diversity, represents an expression of 'deep symmetry' or archetype under local conditions and is therefore called by physicists 'contingent'.

T

tam:	($\downarrow$) downward, gravitational or inertialising cosmic vector.
tanmatra:	pre-scientific term; possible conscio-material band/ grade - subconscious band of mind; *tanmatras* involve the least metaphysical, most nearly physical band of mind; they are traditionally thought of as mental ideas, psychological forms or the Platonic ideals of physical perceptions (e.g. notions of heat, light and motion in a flame or fragrance in a scent); as *chakras* are immaterial structures dealing with energy, so *tanmatras* deal with image, quality and form; they represent the qualitative aspect of matter and are the 'device' that allows

mental grasp of physical effects; as such *tanmatras* are an instrument of potential matter, the archetypal processors of image; as particle to wave so *tanmatra* to *prana*; and as sound is plucked from a tuned string so each of five *tanmatras* is like a string creating, by resonant association, one of five *pranic* 'notes' (see Glossary: *prana*); *tanmatric* apparatus represents a stage in the hierarchical translation of incoming (matter to mind) or outgoing (mind to matter) signals across either individual or universal psychosomatic border (see *figs*. 15.8 and 15.9, also Chapters 16 *esp*. signal translation and 17); the two-way traffic across this border means they equally act, as prism or lens to light, as media for the orderly projection of *prana* into subtle (quantum) events and thence, lower down, gross aggregates of called bulk matter; as such they are, in conjunction with *prana* as power source, the pre-physical mechanism that expresses, in the physical vacuity of archetypal field, the fundamentals of material phenomena; whether simple, single or in 'complex opera' that codes for bio-symphony, *tanmatras* transmit 'sounds' (vibrations or frequencies) which translate into cymatic messages (check Index: Chladni) exciting the emergence or maintenance of physical form; in short, they were the final agents in the initial creation of physical form and force; the wireless physical expression of cooperant *tanmatras* and *pranas* is known to quantum physics; the wired, bonded or aggregate expression involves the chemistry, physics and biology of condensed or bulk matter; and, as electricity supports the running of a machine, they iteratively, correctly support our starry universe.

tattwa: pre-scientific term; possible conscio-material band/ grade - both subconscious and physical bands; means 'that-ness' or 'not-self'; by nature, using their intellect/ *buddhi*, philosophers seek to analyse, categorise and argue so that, in the case of *tattwa*, the number of items listed varies considerably according to tradition; basically, however, it amounts to a 'catch-all' description of human, animal and inanimate condition; five well-known *tattwas* correspond exactly to Greek, Latin and medieval European elements of ether, fire, air, water and earth (see Chapter 10: Old Vacuums); on the subconscious side of *PSI*, in archetypal memory of universal mind, these correspond to five *tanmatras*, five *pranic* 'notes' and five lower *chakras* (*fig*. 17.11); and on the physical side to five informative senses and five energetic organs of action (*fig*. 0.8 and Chaps. 14: Lower Physical Loop, 15: Psychosomatic Linkage and 16: Signal Translation); the five elements/ states of matter are variously defined; ether is space, upper air (home of the gods) or psychosomatic archetypal potential (which, being metaphysical, is physically unseen); it is related to the throat chroat *chakra*, seat of dormancy; there follow gas (air *tattwa* related to breath, oxygen and heart *chakra*), energy (fire *tattwa* giving stimulus for change, heat and light whose hub is the solar plexus), liquid (water and its osmoregulatory and waste expulsion systems) and solid (the earth *tattwa* of related to bio-mass and its sense of pressure/ touch whose *chakra* rests at the supportive base of the spine).

teleology: the doctrine that there is evidence of purpose in nature; doctrine of non-randomness in natural architecture; doctrine of reason ('for the sake of', 'in order to', 'so that' etc.) and intent behind biological and universal design.

third eye: place where you think; point of metaphysical focus between and behind the eyebrows, that is, just above the physical eyes; HQ/ seat of mind beyond the sensory world; cosmological eye-centre; gate through which meditative concentration can pass; single way that leads within.

transcendent projections: **psychological:** see Chapter 5 Top Teleology, Index: Archetype and *figs*. 2.6, 3.1, 5.1, 9.1 and/ or 11.1; **physical:** see Chapters 8, 9, 11, 12; Glossary: archetype; Index: transcendence, archetype, cosmo-logical language; *figs*. as above and 12.1; such projection involves an orderly, energetic expression from either metaphysical or physical nothingness, that is, unseen potential; an instantaneous 'miracle' that issues from 'within' non-conscious physicality; transcendently emergent, finely tuned expansion from 'inner' metaphysic into 'outer' material/ natural law; 0-dimensional singularity (paradoxically everywhere at once) from whose prior pointlessness all points perhaps began; cosmic seed whence, *ex nihilo*, the world developed; projection whose appearance, once physical, is visible and perhaps described but certainly not explained by big bang theory; transcendent projection of archetype is possibly, to the constrained sensory and intellectual states of human mind, ultimately incomprehensible; its invisible mechanism, the practice of materialisation, may remain a fact beyond material understanding. **biological:** if matter is developed memory (Chapter 9: How Does Nothing Physical Work?) then see Chapters 16: *passim* and 19: Conceptual Biology; see also Glossary: mnemone and archetype; Index mnemone, archetype; and *figs*. 2.6, 3.1, 16.1 and 19.1.

U

unification: simplification: details are unified by their working principles, themes or programs; better to perceive intrinsic principle is to simplify or unify an understanding; progressive unification of forces is the grail of physics: Clerk Maxwell unified electricity and magnetism; electroweak or *GSW* theory brought in the weak nuclear force; now the goal is to include the strong nuclear force (*GUT*), gravity in a super-force and show that, in essence, particles and forces are interchangeable (super-symmetry and *TOE*); Natural Dialectic, also working with the maxim 'All is One', includes what sums to a hierarchical *TOP* or Theory of Potential (*see* especially Chapters 5, 6, 7, 9, 16 and 19, also *fig.* 5.1); potential is the absolute from which variant orders of relativity derive; the equivalent of *TOP* is *CUT* (**Cosmic Unification Theory); Natural Dialectic is a vehicle of *CUT*, whose aim is to build cosmic connections, that is, orderly linkages towards a Holy Grail of Unification**; the Great Connector, that is, Unifier is consciousness; the subjective potential for mind is consciousness and the objective potential for matter is archetypal memory; such archetypal element unites psychology with the physics of natural science; it is the informative precondition of physical and biological form.

universal mind: cosmic grade; also called the 'mind of nature' or 'natural mind'; as a biological body is a specific though complex arrangement of universal chemicals so individual mind partakes of a particular, equally minuscule fraction of the metaphysical components of universal mind; *see* also archetype.

V

vector: existential dynamic; a vector has both direction and magnitude; it illustrates direction of travel with respect to a model or a secondary stack used in Natural Dialectic; fundamental vectors (↑ and ↓) denote relative gain or loss of information or energy; and, similarly, motion towards and from the axis, peak or source of a cosmic model; in this case, magnitude is inferred to occur on a scale between any pair of opposites, for example, the relative proportions of black and white in the grey-scale between these opposites; use of the word is general, metaphorical rather than specific; opposite members of a stack may involve metaphysical factors (e.g. love/ hate, beauty/ ugliness) as well as physical; thus its spectra do not necessarily concern physical motion or mathematical calculation; its 'field of relativity' extends beyond non-conscious elements; in this respect a Dialectical vector is similar in principle but not the same in practice as that defined by physics or biology.

virtuality: exotic component of quantum physics; para-physical feature of the quantum vacuum; immaterial substrate of material phenomena; inner (where solidity's the outer) edge of physical reality; ephemeral 'virtual particles' rise and sink back into a 'void' thought to teem with their 'fluctuations'; virtuality is identified as the agent of such important actualities as the strong nuclear force (resulting from interaction between

	virtual mesons and gluons), vacuum polarisation, the Coulomb force (between electric charges and mediated by the exchange flight of virtual photons) and so on; not used in the computer sense of a continuum between real and imaginary circumstance; see also *ZPE*.
Vitruvian man:	where art meets science Leonardo's 'universal man' demonstrates an architectural symmetry, excellence of composition and, microcosm unto macrocosm, a reflection of the universe; Da Vinci's connection, from his notebooks, is quite the opposite of Darwin's doodle (Chapter 5); if, with ratios and rationality, it demonstrates mathematical perfection then does not design of larger cosmos demonstrate it too? Natural Dialectic certainly concurs with Leonardo's logical submission.

Z

zero:	zero (the number) is a metaphysical entity, one critical to mathematics; zero (the fact) means, for Natural Dialectic, nothing in two senses; in the *negative sense* it means an absence of perception (psychological oblivion) or absolutely nothing physical (as naturalistically prescribed to precede, say, a big bang or as the nature of a theoretically perfect vacuum); negative sense may also be construed as (*tam*) an extreme sub-state, sink or emptiness; for materialism 'absolute nothingness' may involve natural law and its mathematical description; what, one may enquire, is the source of such 'eternal metaphysic', what is the nature zero-physical?: on the other hand, in a *positive sense* zero refers to source, pre-existent potential or (*sat*) higher cause-in-principle; for example, information (which is zero-physical) transcends/ precedes a course of action; information that passively governs the operation of cosmos derives from immaterial archetype.
ZPE:	zero-point energy; quantum vacuum; vacuum energy of all fields in space; residual energy of all oscillators at $0°K$; concept first developed by Albert Einstein and Otto Stern; intrinsic energy of vacuum; the ground-state minimum that any quantum mechanical system, in particular the vacuum, can have; remainder, according to the uncertainty principle, when all particles and thermal radiation have been extracted from a volume of space; residual non-thermal radiation; irreducible 'background noise'; 'quantum foam'; the potent, microscopic side of quantum vacuum (as opposed to impotent, macroscopic vacuum left by the apparent lack of anything); subliminal 'rumblings' of immaterial weak, strong and electromagnetic fields (called *ZPFs*); seething, jostling ferment of subliminal waves and particles in emptiness; a flux of unobservable 'virtual' matter and anti-matter that may or may not appear as the basis of observable forces such as electromagnetism, charge and perhaps

	inertial mass and gravity; a subtle facet of levity; the anti-gravity of dark energy (or the cosmological constant) has been postulated as a component of *ZPE*; suggested 'mother-field' support for electron orbits, atomic structure and thus the phenomenal universe.
zygote:	fertilised egg.

Index

A

abiogenesis 75, 169, 286, 305, 356, 566, 843
Absolute .. 209, 215, 246, 450, 461, 675
Absolute Morality 258, 264, 268
absolute motion 685
Absolute Time .... *see* Time Absolute
absolute zero 677
absolutely nothing ... 530, 538, 591, *see* absence/ nothing
absolution 739
absurd equation .*see* Essential Equation
AC time .... *see also* cyclical time
accident .... 517, *see* chance
accommodation theory 296
act of counter-creation .... *see also* comprehension
act of creation. 277, 284, 294, 324, 339, 344, 348, 366, 371, 373, 374, 386, 560, *see also* information output/materialisation/phased intent
active complexity .... *see* purposive complexity
active information . 153, 155, 279, 316, 324, 327, 328, 376, 384, 410, 455, 468, 483, 490, 504, 531, 658
active matter 324
ad nihilum 848
aether *see* ether
agent *see* instrument/ mechanism/ medium/ means to end
Ahriman 214
Ahuramazda 214
AI (artificial intelligence). 401, *see also* computation
Ain Sof.. 372, 661, *see also* Logos/ Unity
akash 411, 476, 701, 710, *see* psycho-space/ metaphysical space/ Glossary
Alcatraz 75
Alexander the Great 56, 59
Alfonso X 59
Algazel (Al-Ghazzali) 139
Alhazen 778
Allah 661
alpha *see* Origin
alpha and omega.... *see* physical extent/ universal body
Alpha and Omega 470
Alpha Moment .... *see* Point Alpha
Amelino-Camelia G. 785
amino acid 149, 362, 387
amoeba 696

amorality 259, 267
anaesthesia metaphysical 56
anaesthesia physical 56
anarchy 165
Anaxagoras 58
Angkor Wat 679
animistic language/ animism .. 87, 297, 313, 331, 534, 583
annihilation 843, 848, 849
ant   260
anthropic cosmological principle ..... *see* anthropic principle
anthropic principle . 275, 276, 277, 288, 597, 604, 813
anti-chance 394, 407, 518, 552, 556, 809, *see also* mind/ universal mind
anti-cosmos .... *see* chance/ chaos
anti-deity 406
anti-entropy .... 566, 583, *see* negentropy/mind
anti-gravity .... 811, 832, *see* levity
anti-mass 630
anti-matter 436, 718
antimony 223
anti-parallel time lines 142
anti-pole 467
anti-principle 468, 624, 685
anti-reason .... *see* chance
anti-water 603
ape   259
apparent infinity .... 464, *see* infinity
application program .... *see* computer program
Aquinas Thomas 141
archetypal matter ... *see* potential matter
archetypal memory .. 58, 383, 465, 486, 534, 545, 586, 637, 649, 652, 683, 717, 743, 846, *see also* universal mind
archetypal principle 385
archetypal program 297, 476, 526, 649, 652
archetype 58, 77, 142, 147, 156, 205, 249, 288, 297, 325, 337, 343, 386, 411, 414, 425, 430, 452, 453, 465, 472, 484, 486, 488, 508, 517, 519, 524, 526, 535, 544, 610, 621, 622, 634, 637, 642, 649, 652, 663, 680, 693, 695, 709, 714, 743, 744, 752, 755, 773, 782, 806, 810, 833, 834, 840
Archetype .... *see* First Cause
area 46 404
Aristotle.... 45, 57, 59, 60, 64, 125, 303, 313, 418, 419, 460, 596, 608, 706, 739, 778, 793

877

Arjuna 177
Arp Halton 441
arrow *see* vector
arrow of time.*see* time's arrow/ entropy
Aspden Harold 523, 524, 716, 727
Aspect Alain..................................... 542
aspects of energy.............................. 150
atheism...... 66, 90, *see* also materialism
Atman-Brahman........................ 59, 618
atom 169, 505, 627, 723
atom as symbol................................. 505
atomic theory.................................... 436
attention ....*see* focus of attention
Attention ....*see* Enlightenment
attractor....*see* archetype, *see* archetype
attractor conscious. 142, *see* also *Logos*
attractor unconscious........ 143, *see* also archetypal memory
Aurobindo 594
author ....*see* purpose/ teleology
automaton..........*see* non-consciousness
Averroes.. 59
aware embodiment ....*See* biology
awareness.. 75, 109, 129, 247, 262, 690, 701, 742, 744, 771, 849, *see* also consciousness
axion 819

B

Babbage Charles 396, 400
Bacon Francis............................ 70, 304
bacteria.. 405
Baird John Logie.............................. 396
balance ... 176, 186, 220, 343, 450, 462, 495, 601, 832 *see* also balance point
balance of body......... *see* homeostasis
balance of mind..............24, 202, 224
balance point...43, 44, 75, 103, 109, 175, 182, 186, 189, 272, 343, 368, 449, 513, 518, 601, 626, 659, 698, 736, 740, 832, 833
balancing factor....... 187, *see* archetype
Bani see Shabda
Barrow John 273, 785
basal ganglia..................................... 746
basis of biology 347, 397
Bearden Tom.................. 524, 711, 728
becoming.. 687
Beelzebub.............................. *see* devil
being 687
Bell John 542
Bergson Henry 304
Berners-Lee Tim 400
Biefeld Paul..................................... 727
Big Accident ..546, 582, 589, 807, 809, *see* also big bang

Big Anarchy 809
big bang .144, 168, 276, 317, 356, 392, 422, 436, 445, 465, 472, 492, 538, 546, 577, 591, 597, 630, 794, 795, 801, 814, 843, *see* equally transcendent projection, *see equally* transcendent projection
big bang illogical 807
big bang, bad name......................... 168
Big Egg ...... 669, *see* also Big Accident
Big One .... *see* Big Egg
binary code ....*see* polarity
bindu 705, 733, *see* eye-centre
bioelectrics 684
bio-force - Dialectical.............*see* mind/ information; also *prana*
bio-force - evolutionary . 296, 311, 313, 589
bio-logic 402
biological information ....*see* code
biological machine 398
biological time................................. 741
bio-philic..........275, 518, 597, 601, 603
bio-time 746
Bishop Berkeley 64, 541
Bishop of Oxford.................... 311, 312
bit… 319
bivalve.. 696
black box .597, 635, 648, 698, 798, 802
black hole146, 210, 213, 218, 249, 630, 677, 753, 770, 819, 836, 845
Blake William 717
Bletchley Park 387
blind faith 81
bodhisattva .. *see* saint/ enlightened one
Bohm David 77, 380, 465, 543, 637
Bohr Neils243, 249, 445, 540, 542, 593, 595
Boltzmann Ludwig 330, 576
Bombe 46
Bondi Hermann 794
book *see* code
Born Max 542
Bose Jagdish.................................... 594
Bose Satyendra................................ 595
boson378, 496, 595
bottom-up 122, 272
bouncy universe.............. 797, 799, 845
Boyer Timothy 713
Boyle Robert 76
Brahe Tycho 419
Brahma.......................... 618, 762, 797
Brahman ....see Atman-Brahman
Brahmand 700
brain 83, 203, 341, 404, 601, 629
branch dialectic *see* secondary dialectic
brane 609, 639

Braun Werner von 727
brown dwarf 819
Brown Townsend 727
Brownian motion 540
Bruno Giordano 419, 459, 460, 607
Buddha . 56, 64, 80, 116, 250, 258, 266, 270, 467, 658, 661, 834
bulk matter 199, 324, 451, 632, *see* also matter-in-practice
Burr Harold. 594
butterfly effect 145, 551

C

c, speed of light in a vacuum*see* speed of light (fixed)
Calabi-Yau geometry 639
calculus 146
Cambridge 289
Cambridge Lord's Bridge 385
Cambridge University 59, 418
Campbell Joseph 764
Cantor Georg 461
Capra Fritjof 65
Caprice .... 610, *see* Lady Luck
carbon 600
carbon dating 758
carbon-14 760
Carnot Sadi 576
Casimir Hendrik 710, 728
casino .... *see* randomness
catastrophe theory 432
causal matter .... *see* potential matter
causality 137
cave of Hira 679
Cavendish Henry 600
CBR *see* cosmic background radiation
CDE *see* dark energy
CDM *see* dark matter
cell 397, 399, 696
centaur (wo)man .. 25, 57, 90, 222, 260, 262, 266, 271, 351, 355, 413, 662
centaur paradox 222, 262, 346, 354, *see* also *fig.* 5.8
Central First Cause 172
Central Time .... *see* time absolute
Centre 97, 128, 166, 167, 171, 185, 204, 226, 377, 413, 650, 705, 733, 736
Cerf Vint 400
certainty 556
ch'i .... *see prana* and Glossary
chakra 360, 704, 723 *see* also Glossary
chance 273, 376, 406, 407, 530, 537
change .... 144, 169, *see* motion/ transformation
changing gear .... *see* scale switch
chaos 493, 538, 702, *opp.* cosmos, *see* also disorder, Glossary

chaos theory 70, 145, 431, 448, 543, 633
Chardin Teilhard de 296
charge 788
chemical bond 602, 655
chemical evolution .... *see* abiogenesis
chemistry 655
chicken-and-egg 453, 472, 669, 811, 822, *see* also egg
chicken-before-egg. 135, 197, 393, 405
Chinese boxes 103, 165, 452, 526
chitta see attention
chitta vritti 637
Chladni Ernst 636, 640, 723
Christ 80, 110, 123, 258, 265, 270, 313
Christ Pancrator .... *see Logos*
Christ's College, Cambridge 594
Christian Trinity 172, 228, 619
Cicero 60
circle *see* cycle
Circuit Theory 653
civilisation 374
classical era 770
classical physics 157
Clausius Rudolf 330, 576
Clayton M 785
climax 105, 219, 277, 284, 296, 351, 461, 469
CMBRsee cosmic background radiation
COBE 816
code 166, 208, 300, 316, 324, 337, 341, 342, 347, 366, 367, 375, 376, 377, 378, 383, 385, 387, 388, 397, 399, 517, 525, 531, 551, 565, 650, 652, 695, 718, 840, *see* also archetype
coded information .... 652, *see* code
Coelacanth 759
coincidence .... *see* chance
Colossus 46
coloured gradient *see* rainbow/ spectrum
columnar construction .. 137, 161, 236, 242, 272, 449, 484, 514, 623
common sense 62, 302, 313
Communion 56, 114, 247, 462, 619, 848, *see* also Enlightenment
communism 307
complementary paradox 226
complexity non-purposive 365, 559
complexity purposive ..... *see* purposive complexity
comprehension . 54, 186, 251, 273, 350, 372, 402, 470, 619, 646, 772, *see* also sensation/knowledge
computation 366, 382, 399, 401, 407
computer .... *see* computation
computer program . 141, 221, 334, 366, 552, 776
concentrate of consciousness .... *see*

Consciousness
concentrated information.......... see egg
concentration. *see* unification/ focus of attention/ source of power or purity/ meditation
concentration gradient..45, 73, 95, 103, 105, 188, *see* also conscio-material gradient/ hierarchy/ diffusion/ concentration
concentric rings60, 95, 96, 98, 105, 167, 183, 190, *see* also hierarchy
concentric spheres ....95, 126, 167, 180, 418, 659, 793, *see* also concentric rings
confinement...................................... 661
conscience.. 267
conscio-material continuum*see* conscio-material gradient
conscio-material spectrum/ gradient 44, 73, 95, 97, 125, 152, 153, 164, 167, 169, 171, 173, 188, 195, 201, 236, 237, 248, 252, 279, 288, 318, 363, 389, 411, 412, 420, 454, 462, 469, 481, 482, 486, 504, 525, 531, 617, 661, 838, *see* also Primary Axiom of Natural Dialectic
conscio-material spectrum733, 772, 845, *see* conscio-material gradient
consciometer 47, 233
conscious information ....*see* active information
Conscious Singularity 428
consciousness11, 39, 53, 69, 79, 82, 93, 112, 120, 127, 135, 153, 176, 205, 216, 220, 226, 234, 246, 248, 260, 278, 288, 315, 316, 327, 336, 355, 357, 364, 365, 405, 419, 422, 426, 446, 450, 462, 469, 470, 475, 483, 517, 555, 570, 591, 701
Consciousness 75, 76, 102, 135, 152, 153, 155, 167, 171, 203, 205, 213, 237, 247, 262, 272, 278, 279, 280, 285, 317, 318, 333, 334, 349, 363, 408, 449, 451, 454, 462, 474, 476, 479, 486, 505, 525, 568, 615, 684, 687, 701, 740, 771, 773, 845, 849
consciousness evolution of............. 351
consciousness-in-motion 272, 285, 327, 469, 631, 687, *see* also mind
conservation law 777, 782
constant 515
constant apparent........................... 521
constant real 523, 525
Constantine VII Porphyrogenitus 88
contemplation................ *see* meditation
continuity 642
continuity physical 643

continuity psychological................. 642
Copernicus...................................... 419
Copernicus Nikolaus 596
cosmic background radiation. 422, 598, 794, 816
cosmic effect 462
cosmic egg 285, 318, 676, 700, *see* first physical cause
cosmic fundamental172, 176, 177, 181, 182, 193, 195, 199, 269, 272, 318, 324, 414, 450, 462, 486, 517, 518, 786, 839, *esp.* ps. 92 - 97
cosmic mind ....452, *see* universal mind
cosmic onion ..... 527, *see* also layers of onion
cosmic psychology 560
cosmic pyramid 247, 468, 470, *see* also Mount Universe
cosmic stage ....*see* universal body
cosmo-cycling .757, 761, 798, 845, 847
cosmogenesis............... see Point Alpha
cosmogony .... 284, 530, 586, see also cosmogenesis
cosmological axis. 105, 109, 110, 114, 117, 167, 271, 317, 364, 466, 733, *see* also eye-centre
cosmological constant.. . 598, 793, 811, 818, 819
cosmological principle ... 793, 812, 817
cosmology 420, 493, 610, 629
cosmos 11, 66, 284, 302, 351, 414, 418, 460, 486, 508, 538, 558, 568, 616, 659, 702, 807, *see also* universal body, Glossary
cosmos
   as light 96
   as mountain................................... 99
   as pyramid 97
   as sound 97
   as waves....................................... 97
counter-creation............................. 372
Crab Nebula 825
creation myth................................. 700
creator.......................... 568, *see* author
creosote 748
Crick Francis 591
cricket... 597
critical density 598, 813, 820
Cro-Magnon 765
crow 260
Cuvier George 313
cybernetic ....*see* homeostasis
cycle *see* vibration
cyclical time .... 797, *see* time's cycles
cyclotron... 665
cymatics 636

D

dark energy 505, 598, 785, 801, 811, 818, 819
dark matter 440, 441, 600, 785, 801, 820, 822
dark peripherals .... *see* devil/devilry
Darwin Charles .. 74, 289, 294, 304, 594
Darwin Erasmus 304
data item 386, 469, 557, 694
Davies Paul 14
Dawkins Richard 300, 353, 612
DC time 757
de Broglie Louis 595, 778
death 249, 694, 843, 849
deductive logic 12
deism 296, 594
Delphi 418
dematerialisation 166
Democritus 45, 60, 460, 627, 706
demon .... *see* devil
dendrochronology 757
depolarisation .... *see* neutrality
dervish .... *see* Jalal-ud-Din Rumi
Descartes Rene 127, 479
design *see* intent/ purpose/ instrument of purpose
desire 323
destiny .... 569, 570, *see karma*
determinism 431, 520
devil 214, 268
dharma .... 567, 658, 761, *see* laws of nature/ perennial philosophy
Dharmakaya .... see Nirvana
Dialectical polarity ..*see* Polar Dialectic
dialectical stack 137
Dialectical stack 43, 161, 221, 236, 272, 449, 467, 514, 623, 647
Diamond .... *see* Consciousness
diffusion 44, 97, 278, 584, 603, *see* also concentration gradient
diffusion gradient .... *see* also power/ concentration gradient
Digges Thomas 419
dimensionless numbers 525
dipole .... 602, *see* polarity
Dirac Paul 435, 593, 651, 652, 654, 716
directions of focus 64
discontinuity 645
Discovery Institute 307
disorder .... 193, 565, *see* randomness
dissipative structure 561
dissolution... 166, 451, 847, *see pralaya* and dematerialisation
DNA 284, 316, 331, 342, 378, 384, 387, 602, *see* also code
Dobhzhansky Theodosius 313
Doppler Christian 794
Doppler shift .... *see* red shift
dormant mind 285, 318, 349, 693
dot 698, 705, 754, 756
duality .... 227, 626, *see* polarity
dukkha .... 214 and Glossary
Dyaus-pitar .... see Zeus
dynamic equilibrium 241, 514, 705, see vibration
dysfunctional logic .. 63, 314, 396, 399, 401, 407

E

eagle 259
Eddington Arthur 386, 489, 523, 796
Edison Thomas 726
education 374, 658
egg 186, 284, 372, 414, 637, 669, 699, 700, 776, *see* also cosmic egg, potential and archetype
eigen function 665
Eigen Manfred 589
Einstein Albert ... 38, 41, 46, 57, 74, 81, 130, 155, 225, 306, 420, 431, 436, 437, 438, 523, 524, 527, 543, 593, 595, 610, 616, 629, 630, 660, 665, 708, 710, 711, 713, 718, 721, 727, 733, 778, 782, 785, 793, 796, 799, 811, 836
electrical charge 384, 436, 444, 462, 599, 644, 651, 653, 707, 716, 787, *see* also polar charge
electrochemistry 349
electromagnetic radiation .... *see* light
electromagnetic spectrum 167, 324, 411
electromagnetism .. 113, 208, 322, 323, 505, 602
electron ... 198, 322, 599, 653, 719, 722, 786, 787, 790
electroweak interaction ... 436, *see* weak nuclear force
element atomic 105, 392, 600, 602, 702
element dark .... *see* dark energy/ matter
element informative .... *see* archetype
element non-atomic/traditional 411, 476, 477, 478, 504, 702, 703, 819
element psychological 246, 407, 409, 421
element subatomic.. 382, 541, 639, 649
elementary particle ..... *see* fundamental particle
Eleusinian mysteries 55
Eliade Mercer 764
embodied awareness .... See psychology
emergent property 79, 285, see also mind
Empedocles 60, 148, 706, 778
empiricism 62
empty canvas 715
En Sof .... see Ain Sof

energetic transformation ...... *see* change
energy 65, 73, 93, 96, 126, 133, 136, 152, 155, 418, 446, 450, 462, 529, 595, 666, 686, 702, *see* energy/ matter equivalence
energy coordinate 152
energy/mass equivalence 666
Engels Friedrich 311
engineer .... 556, *see* creator/inventor
Enlightenment 58, 64, 114, 123, 146, 173, 209, 214, 219, 228, 247, 251, 256, 266, 268, 271, 355, 375, 470, 482, 619, 642, 691, 740, 773
entropy ... 188, 330, 387, 391, 489, 498, 527, 559, 575, 581, 584, *see* also diffusion/downward power gradient/outward concentration gradient
Epicurean 706, 793
Epicurus 60, 460
equation 625
equilibrium .... *see* balance
Equilibrium 657
ESP (extra-sensory perception) 725
Essence ... 131, 133, 135, 140, 147, 173, 209, 279, 359, 363, 449, 454, 461, 467, 674, 680, 687, 772, 845, 848
Essential Dialectic *see* Primary Dialectic
Essential Equation ... 133, 428, 461, 669
Essential Information .... *see* Potential Information
Essential paradox 227
eternal inflation 443, *see* also expansion theory and multiverse
ether 504, 523, 703, 707, 716, 819, *see* also vacuum/ state of matter
ether (aether) 523
ethics 76, 659
ethology 267
Euclid 431, 461, 778
Euler Leonhard 147, 520
European Union 310
event vector .... *see* vector
evil 214
evolution of natural law? 532
evolution theory of .66, 69, 79, 92, 164, 169, 274, 277, 288, 305, 801
ex nihilo .... 545, *see* also nothing
exclusion principle 654, 834
executive intellect 562
exhaustion 250, 659
existence. 133, 135, 136, 147, 173, 454, 506, 687
existential cone ..... *see* Mount Universe
existential dialectic .... *see* secondary dialectic
existential dipole 73, 93, 150, 152, 156, *see* energy/information
existential equation 461
existential paradox 228
exo-planet 611, 826
expanding universe 420, 794
expansion theory 454, 466, 472, 787, 844
experimental method 340, 372, *see* also Popper Karl
extended infinity 460, 471, 832
eye 474
eye-centre 109, 114, 191, 271, 317, 355, 364, 370, 404, 679, 698, 839, *see* also cosmological axis

F

faith 81, 315, 352, 353
family as symbol 700
Faraday Michael 76, 306, 436, 707, 727, 781, 786
Farnsworth Philo 396
fate .... 569, 570, *see karma*
Father .... 470, *see* Unity
fenafillah .... *see* Enlightenment
fermion 378, 496
Feynman Richard 590, 644, 684
field 249, 322
fifth dimension .... 631, *see* mind
fine structure constant 599
finity 472
first cause . 685, *see* first physical cause
First Cause 140, 165, 168, 173, 219, 246, 251, 332, 360, 369, 390, 392, 445, 450, 461, 465, 474, 616, 618, 636, 675, 678, 683, 688, 692, 693, 737, 772, 776, 845, 850, *see* also Logos
First Motion .... *see* First Cause
first physical absolute 731, *see* also space
first physical cause 454, 530, 575, 649, 680, 738, *see* also potential matter/ archetype
first physical principle ...... 680, *see* also archetype
First Principle .. 235, 250, 257, 284, 361, 446, 449, 450, 451, 467, 470, 475, 680, 773
five elements .... *see* state of matter
fixed frame of reference 523
flat-earth view 216, 849
flatness problem 598, 812
flat-universe perspective 449
flexible homeostasis 376
Flowers Tommy 46
focus of attention 47, 64, 114, 272, 450, 495, 690, 701, 741, 757, *see* also concentration
footballer 388
formalisation 88, 355

Fortuna .... *see* Caprice
fossil record 759
Four Noble Truths 658
Fourier Jean Baptiste 721
Franklin Benjamin 444, 653
free will and determinism 83, 223, 283, 368, 407, 549, 553, 569, 657, 745
Freedom 616, 661
Fresnel Augustin 707
Freud Sigmund 63
frozen time 753
FSL .... *see* speed of light (fixed)
Fu Xi 148
Fuji 106
functional complexity ...... *see* purposive complexity
functional logic 130, 314, 393, 396, 399, 407
fundamental error 278
fundamental particle 436, 437, 499, 514, 712, 767, 808

G

gain   161, 164, 166, 171, 188, 204, *see* raj vector
galactic spiral 820
galaxy 504, 822
Galileo Galilei 46, 70, 304, 419, 707, 778
Galvani Luigi 348
Gamov George 794
Gates Bill 400, 568
Gauss Karl Friedrich 520
general relativity .... 440, *see* relativity theory of
genetic code 378
genetic program 534
genome 525
geometry 703, 704, 705
Gibbs Josiah 576
gingko 759
glass ceiling 479, 792, 800
God effect ... *see* quantum entanglement
God particle .... 439, *see* Higgs boson
goddess of the gaps .... *see* chance
Gödel Kurt 402, 407, 591
God-spot 405
Gold Thomas 794
golden mean .... 187, 523, *see* also *phi*
Goldilocks enigma 516
Goldstein Herbert 593
Goodness 354, 619
Goonhilly 384, 396
Gould Stephen 313
grade of time .... *see* time's grades
grammar ... 361, 378, 384, *see* also code
Grand Cosmic Simulation 612
gravitational predominance 770

gravity 186, 192, 451, 471, 503, 527, 529, 600, 602, 630, 680, 685, 771, 779, 835
Great Wall 818
Green Michael 634
Gregersen Niels 14
Gresley Nigel 556, 568
GUE 119, 628, 733
Gurbani .... see Logos
GUT 118, 119, 628, *aka* Grand Unified Theory
Guth Alan 811

H

Haeckel Ernst 311, 588
Harris Sam 353
Hawking Stephen .. 131, 133, 473, 508, 591, 592
health 347
heat death 500, 583, 844
heat window 601
Heaviside Oliver 720
Hegel Georg 42
Heisenberg Werner 386, 540, 541, 548, 593, 595, 818
hell 214, 269
Heller Michael 58, 434
Heracleitus 631, 703
Heraclitus 42
heresy materialism's 74
Herschel John 684, 778
Herschel William 419, 778
Hesiod 493, 702, 793
hierarchy 38, 99, 146, 152, 164, 166, 183, 202, 281, 330, 370, 371, 412, 452, 469, 682
hierarchy of process. *see* act of creation
Higgs boson 440, 595, 719, 770
Higgs condensate 719
Higgs field 439, 647, 665, 720
Higgs Peter 439
Hinduism 172
Hiranyagarbha 700
Hiroshima 438, 769
Hobbes Thomas 64
holarchy 164
holism ... 66, 90, 93, 104, 169, 239, 246, 272, 287
Holmes Sherlock 374
hologram 77, 173, 545, 652
holy grail 247, 351, 407, 408, 516, 627, 628
holy of holies 679
homeostasis ... 172, 202, 343, 345, 376, 384, 495, 513, 593, 747
Hooke Robert 778
Horava Peter 442

horizon problem 598, 812
Hoyle Fred303, 396, 473, 599, 794, 844
HST see Hubble Space Telescope
Hubble Constant............................. 522
Hubble Edwin 419, 420, 795
Hubble Space Telescope (*HST*)572, 844
Hubble's Law.................................. 521
Hukm .... see *Logos*
HUP 540, 550, 710, 788
Hutton James.................................. 758
Huxley Thomas 297, 313
Huygens 778
Huygens Christiaan 778
hydrogen................493, 599, 815, 822

I

I Ching 148
Ibn Arabi 59, 705
Ibn Sina 313
illusion 77, 82, 253, 838
imagination 562
imaginery time 742
immortality..................................... 748
importance............................. 475, 692
impotence...44, 52, 130, 155, 159, 171, 193, 206, 240, *see* also exhaustion/ subtendence/ fixity
impotent equilibrium .,., 193, *see* inertial equilibrium/ negative neutrality
indefinity 458, 466
indeterminism................................ 540
inductive logic................................. 12
Indus valley civilisation 392
inert space ....233, *see* vacuum
inertia 600, 836
inertial consciousness............. *see* sleep
inertial equilibrium. 193, 241, 471, *see* also impotent equilibrium/ negative neutrality
infinite extent ...... *see* extended infinity
infinite number ....*see* infinity
infinite point.................................. 460
infinite regression............. 97, 295, 458
infinity 78, 99, 135, 146, 173, 200, 213, 225, 419, 455, 465, 480, 485, 510, 526, 610, 623, 626, 629, 647, 698, 705
Infinity ...56, 93, 95, 97, 104, 128, 133, 135, 141, 147, 218, 219, 222, 233, 235, 246, 247, 252, 278, 286, 334, 363, 372, 415, 427, 428, 446, 447, 449, 454, 461, 463, 466, 469, 474, 490, 571, 615, 616, 623, 625, 632, 638, 641, 680, 687, 688, 689, 717
inflation.. 131, 422, 492, 598, 810, 811, 819
inflexible homeostasis 376

information93, 153, 166, 318, 374, 384, 391, 426, 482, 530, 637
information
   active ....*see* active information
   by chance?..................................534
   coordinate 125, 152
   density 387, 397, 399
   entropy.................. 188, 193, 282, 394
   field....................................... see mind
   incarnate................. 300, 347, 350, 398
   input. *see* sensation/knowledge/trigger
   loop...... 172, 345, 347, 348, 350, 351, 367, 368, 375, *see* also homeostasis
   output...*see* action/creation/materialisation/response
   passive .... *see* passive information
   storage ....208, *see* also passive information
informative hierarchy 333
informative models......................... 322
informative potential .... *see* Potential Information
initial conditions.............. *see* first cause
Inner Light..................................... 466
instinct..................................... 63, 316
instrument*see* mechanism/ agent/ means to end
intelligence294, 303, 311, 316, 367, 391, 474, 690
intelligent design 507, 558
intelligent life 597
intuition 224
invariance 506
inventor 396, 568
inversion..234, 248, 404, 413, 490, 525
inward focus .... *see* meditation
irrational world-view. *see* no intelligent design theory of
irrelative time ....*see* time absolute
isotope 599

J

Jabal-al-Nur............................ 106, 679
Jacob's ladder................................ 216
Jalal-ud-Din Rumi 123, 194, 637
Janssen Zaccharias 778
japa see *mantra*
Jeans James................... 205, 386, 595
Jehovah.. 661
Jenny Hans 636
Johnson Phillip 309, 561
Jones Judge John 309
Joule James............................ 386, 494
Jung Carl Gustav 764

K

Ka'bah..679

Kailash 106
Kal time-lord/ negative power..*see* also *Brahma*
Kalam-i-Illahi........................ see Logos
Kalma (Kalima).....360, 678, see Logos
Kaluza Theodor............................... 442
Kant Emmanuel...................... 223, 304
Kant's antimony 223
*karma*268, 330, 414, 462, 569, 570, 657, 761, 834, *see* also causality
Karnak... 679
Kelvin Lord 330, 494, 576
Kepler Johannes 386, 419, 778
Khrushchev Nikita 703
Kidd Sandy..................................... 727
kinetic energy................................. 150
kinetic theory of matter . 250, 412, 445, 790
Kitzmuller v. Dover 309
knowledge................ 186, 251, *see* also comprehension
Knowledge 691
Koestler Arthur 164
Krishna 177, 637
Kun .... see *Logos*

L

Lady Luck ......535, 536, 538, 543, 545, 547, 552, 563, 569, *see* also chance
Laithwaite Eric............................... 727
Lamarck Jean Baptiste 304
Lamb Willis 710
lambda force...505, 598, 785, 811, 818, 844, see also levity/ dark energy
Lamoreaux Steven.......................... 728
language ....*see* code
Lao Tzu 75, 133, 222, 455, 678
Laplace Pierre 222, 539
latency.............................. see potential
Law of Innovation 548
Law of Motion 462, 761
Law of Thermodynamics, First330, 494, 587, 814, 844
Law of Thermodynamics, Second. 330, 494, 495, 576, 587, 726
Law of Thermodynamics, Third..... 580
Lawrence Ernest............................. 665
laws of nature 153, 156, 197, 203, 260, 274, 275, 276, 288, 292, 296, 304, 323, 325, 340, 343, 344, 366, 367, 378, 383, 389, 397, 415, 426, 495, 507, 513, 530, 545, 546, 567, 658, 695, 697, 701, 789, 802, 837
laws of nature/ definition195, 208, 343, 379, 487
laws of physics ....*see* laws of nature

layers of onion... 94, 107, 160, 527, *see* also concentric spheres/hierarchy
Leeuwenhoek Antony van.............. 778
Leibniz Gottfried Wilhelm von 46, 139, 147, 148
Lenin Vladimir 689, 773
lepton496
lesser absolute 680
lesser infinity................................. 464
lesser reality ....see reality
Leucippus........................... 45, 60, 706
Levi-Strauss Claude 764
levity 186, 189, 451, 471, 503, 527, 529, 598, 680, 685, 771, 779, 811, 831
levity-predominant 769
LHC (Large Hadron Collider) 440, 665
life   426, 470, 561, 674, 679, 680, 692, 694, 741, 771, 773, 846, 849, *see* also consciousness
Life   450, 462, 470, 474, 618, 679, 836, 846, *see* Consciousness
life - material definition................. 123
life form..551, 597, 601, 644, 748, 843, 846
life recycling.............. *see* reincarnation
life's void .... *see* energy/matter
life-negative................................... 625
life-positive 625
light   211, 418, 469, 628, 644, 649, 676, 683, 706, 738, 771, 775, 778, 781, 789
light of knowledge....*see* consciousness
light window.................................. 601
Lightfoot John 763
light's speed ....*see* speed of light
Linde Andrei 443, 794
Lippershey Hans............................. 778
Little Accident........ 547, 566, 589, 809
Lodge Oliver 707, 716
Logic 390
logical positivism 80, 314, 605
*Logos*128, 173, 219, 251, 280, 361, 362, 363, 372, 383, 392, 465, 474, 619, 637, 649, 680, 688, 691, 695, 706, 737, 741, 772, *see* also First Cause and conscio-material gradient
long-time 746, 753, 784
loop quantum gravity...................... 633
Lord Deliberate534, 538, 563, 567, 570, 604
Lorenz Edward 431
loss   161, 164, 166, 171, 188, 204
love   251, 266, 849
Love 251, 252, 253, 257, 264, 268, 361, 462, 465, 474, 549, 692
Lövtrup Soren................................. 301

luck .... *see* chance
Lucretius 45

M

Mach Ernst 600
machine 389, 392, 397, 415
MACHO 819
macrocosm .93, 96, 264, 269, 285, 318, 348, 351, 355, 415, 448, 461, 462, 486, 513, 525, 568, 584, 632, 733, 741, 776, 839, *inc.* universal mind and body
macro-evolution 169, *see* also Glossary
macro-infinity 460, 596, 629, *see* infinite extent/extended infinity
macro-time .... *see* long-time
Madame Tussaud's 400
Madurai 679
Magi *see* Zarathustrianism
magnetic monopole 819
Magueijo Joao 785
mahapralaya 777, 848
Maimonides 59
Main Dialectic .... *see* Primary Dialectic
Malinowski Bronislaw 764
manas 281, 324, 411, 413, 476, 483, 486, 505, 694, 701, 703 also Glossary
Mandelbrot Benoît 432
mantra .... 56, 636, *see* also repetition
Mao Zedong 773
Marconi Guglielmo 385, 594, 726
Marx Karl 246, 311
mass 444, 719
mass/ energy equivalence 438
mass/energy equivalence 718
massive compact halo object .... *see* MACHO
Master Analogy 126, 251, 362, 390, 636
materialisation 104, 133, 166, 193, 451, 586, 660, 718
materialism 66, 90, 93, 104, 144, 239, 245, 259, 272, 287, 293, 316, 401, 405, 507
mathematics .57, 75, 79, 145, 357, 429, 448, 552, 625, 715, 798
matter .... *see* also energy
matter's holy ghost 680
matter-in-practice ... 154, 198, 199, 212, 334
matter-in-principle .. 154, 198, 199, 211, 212, 334, 454, 649, 738
maxi black hole 685, 820
Maxwell James Clerk 76, 306, 420, 436, 629, 644, 707, 778, 781, 786
maya see illusion and Glossary
Mayer Robert 330, 576
McCutcheon Mark 443

means to end .... *see* teleology/ design
measure of the universe 461
mechanism .... *see* machine
meditation 54, 55, 56, 81, 109, 194, 254, 369, 658, 746, 849
memory .. 193, 286, 316, 325, 328, 343, 366, 379, 405, 452, 560, 649, 650, 693, 737, 743, 775, 834
Mendeleyev Dmitri 474, 654
Mercury 827
meta-law .... *see* archetype
metaphor .... *see* symbol
metaphysic 135, 166, 202, 205, 326, 676
metaphysical energy .... see manas
metaphysical force 631
metaphysical law .. *see* karma/ morality/ archetype
Metaphysical Trinity 146, 428, 446, 449, 450, 462, 615, 616, 641
metaphysics 635
Meyer Stephen 311
Michelangelo 439
Michelson Albert 707
microcosm ... 93, 96, 351, 412, 415, 486
microcosm of the macrocosm ... 93, 351
micro-infinity 596, 629, *see* point infinity
micro-time 753
mind 155, 203, 249, 318, 323, 365, 394, 483, 557, 562, 566, 595, 657, 690, 809
mind in matter 396, 452
mind machine ..... 399, *see* computation
mind of God 386
mind/body time ...... *see* biological time
mini black hole 840
Minkowski Rudolf 749
miracle 562, 802, *see* also *PAM*'s miracle
mnemone 337, 400, 696
models of cosmos 96
Moffat John 785
Mohammed 258, 266
moksha ..... 661, *see* also Enlightenment
monkey 260
Monopole 58
moon 828
moral principle 554, 569
morality 76, 239, 246, 259, 267, 268, 374, 546, 658
Morley Edward 523, 707
morphogene 452, 640, 697
morphogenesis 405
motion 170
Mott Neville 594
Mount Meru 106
Mount Universe .. 81, 99, 102, 105, 106, 109, 113, 114, 119, 126, 130, 131, 153, 154, 155, 158, 159, 160, 166, 167, 169, 172, 178, 180, 193, 202,

203, 237, 247, 250, 252, 286, 370, 420, 466, 505, 527, 632, 666, 676, 678, 703, 705, 765, 771, 845, *see* also conscio-material gradient
Mount Washington 758
Mozart Wolfgang ...325, 367, 439, 521, 638, 695
M-theory 508, 635
Muellner George 729
Muller Max 764
Mullikan Robert 710
multiverse273, 275, 419, 423, 457, 460, 516, 799, 809
music 126, 390, 391, 636, 638, *see* also Master Analogy
musical box 776
mutation 299, 331
mysteries 55
mystic...51, 91, 228, 257, *see* also saint
mystic ideal 268
mystic practice 45, 75, 110, 114, 218
mystical explanation... 251, see duality/ trinity/ paradox
mysticism .... *see* mystic practice/ yoga/ meditation

N

Nada 219, 688
naked singularity422, 516, 794, 802, 804
Nanak 271, 313
Natural Dialectic 12, 60, 121, 124, 140, 148, 149, 160, 161, 176, 181, 205, 220, 272, 277, 286, 290, 315, 420, 449, 530, 616, 629, 695
Natural Dialectical Logic 181
Natural Fact..................................... 848
natural language .... 695, *see* code
natural law......... 424, *see* laws of nature
Natural Morality *see* Absolute Morality
natural order .... *see* laws of nature
natural selection 299, 331
natural units.................................... 522
naturalism.. 62
nature 490, 505
Neanderthal 765
necessity*see* laws of nature, *see* laws of nature
Nechaev Sergei 689
negative feedback........ *see* homeostasis
negative neutrality .... see neutrality negative/ inertial equilibrium
negative nothing 133
negative paradox 226
negative power .... *see tam*
negative singularity*see* sink/ exhaustion
negative time 742
negative void .... 677, 683, 698, *see* also absence/ exhaustion
negentropy188, 327, 330, 335, 391, 486, 498, 501, 575, 741, *see* also uplift, stimulus, ascent
Nelson Edward 713
Neopilina galatea........................... 759
nested rings............*see* concentric rings
nested squares/cubes......... 99, 103, 679
neuron.. 404
neuroscience................................... 349
neutral time........................... 737, 742
neutrality 50, 51, 231, 232, 383, 425, 462, 494, 495, 514, 599, 648, 651, 655, 709
neutrality negative... 52, 151, 170, 193, 659, *see* also post-active impotence/ inertial equilibrium
neutrality positive51, 150, 170, 187, 193, *see* also poise/ prior potential/ potent equilibrium
neutron 199, 599, 723
Newgrange 748
Newton Isaac..41, 46, 74, 76, 147, 155, 268, 296, 306, 313, 359, 419, 431, 436, 462, 523, 527, 629, 657, 704, 707, 733, 739, 749, 753, 763, 778, 793, 833, 836
Nietzsche Friedrich................. 255, 311
nihilism.. 689
Nihilism... 689
Nirvana...123, 214, 228, 357, 678, 740, 848, *see* also Enlightenment
No Intelligence Theory of*see* Theory of No Intelligence
no-boundary condition 749
Noether Emmy 519
noise 385, 387, *see* randomness
non-conscious energy .... *see* energy/ *prakriti*
non-consciousness 155, 237, 288, 426, 453, 465, 491, 550, 676, *see* also matter/ *prakriti*
non-existence131, 133, *see* also Nothing and nothing
non-locality .... 542, *see* quantum entanglement
non-massive consciousness .... *see* information
non-purposive complexity 365, 559
Nordström Gunnar.......................... 441
norm 343, 450, 513
Northrop Filmer.............................. 594
nothing38, 50, 133, 146, 147, 212, 218, 225, 233, 249, 252, 284, 293, 350, 415, 426, 482, 504, 526, 545, 546, 610, 627, 647, 648, 654, 668, 698, 777, 811, 845, *see* also void/space

887

Nothing 45, 46, 102, 113, 133, 146, 147, 158, 175, 212, 218, 219, 242, 246, 247, 257, 288, 356, 358, 364, 428, 447, 449, 455, 461, 469, 470, 471, 474, 475, 568, 616, 623, 641, 661, 677, 680, 686, 688, 689, 692, 849, *see also* Void
Nothing Nature of .... *see* Nothing
nothing physical 548
not-self 687
nuclear energy level 599
number nothing .... *see* zero

O

OBE (out-of-body experience) 409
object as symbol 841
objective perspective .... 61, 68, *see also* materialism
oblivion .. 193, 214, 348, 428, 733, 848, *see* non-consciousness
Ockham William of 605, 608, 635, 639, 707
octopus 260
Olbers Heinrich 827
Old Schools 608, 706
Olympus 106, 703
Om 219, 280, 372, 637, 649, 706, 717, 762, 776, 847, *see also Logos*
Omnipresence .... 736, *see* Presence
One 212
Oort cloud 829
Oparin Alexander 794
order of creation 193, 284
Orderly Beginning 809
organ of teleology .... *see* mind
organisation 584
organism as law 344
organs (five) of action 109
organs (five) of perception 109
Origin of Everything ... *see* First Cause/ Source/ *Logos*/ causality/ big-bang
Origin of Species 311, 313
Orpheus 390
oscillation ... *see* vibration, *see* vibration
outer space 698, 715
over-unity machine 697, 717, 729
ovoid 685, 689, 692, 699
Ovoid 674, 675, 678, 680, 687, 692, *see also Logos*
Owen Richard 304
Oxford University 59, 418

P

Paley William 297, 304, 312
PAM see Primary Axiom of Materialism
PAM's miracle .... 516, *see* big bang
PAND ... *see* Primary Axiom of Natural Dialectic

pantheism 593
paradox 221
Paramatma 797
Paramecium 260, 696
paraphysical .... *see* metaphysical
paraphysics 803, 811
para-science 798
Pascal Blaise 46, 373, 707
passion 78, 233, 267, 355, 575, 662
passive information 153, 155, 279, 283, 316, 324, 327, 328, 376, 378, 380, 448, 453, 454, 470, 486, 490, 504, 531, 562, 650
passive matter 324
passive order .... *see* matter
Patanjali 476, 477, 804
Pauli Wolfgang 654
PCM see Primary Corollary of Materialism
pendulum .... *see* vibration
Penrose Roger .. 77, 402, 429, 430, 543, 604, 641
Penzias Arno 796
perennial philosophy 45, 57, 216, *see also dharma*
perfect mystic .... *see* saint
period .... *see* vibration
perpetual motion 101, 184, 222, 232, 445, 711, 717, 727
perspective switch 238
phased intent .... *see* teleology/ act of creation
phi 523
philosopher's stone .... *see* holy grail
philosophy 88
photography 684
photon 198, 504
photosynthesis 172, 482, 644, 771, 779, 784
physical absolute first . *see* first physical absolute
physical absolute second ..... *see* second physical absolute
physical constants .. 453, 523, 531, 597, 610, 640
physical cosmos 684, *see* universal body
physical first cause 428, 683
physical first principle ...... 840, *see also* archetype
physical law .... *see* laws of nature
physical omnipresence 529, 769, *see also* archetype/ physical absolute
physical space .... *see* vacuum
physical trinity 434, 446, 464, 530, 632, 647, 680, 686, 781
physical unification 627
physical unity 622

physics 379, 491
phytochrome................................ 747
pillars of faith 60, 90, 272, 292, 800
Pippard Brian 391
pivot *see* balance point
Planck Max 41, 46, 306, 420, 436, 437, 522, 523, 549, 593, 595, 597, 599, 630, 633, 635, 638, 647, 648, 652, 654, 660, 710, 778, 799
Planck particle 722, 783, 840
Planck satellite 817
Planck time................................... 754
planet 826
planetesimal 826
plasticity
   limited..............................299
   unlimited..........................299
Plato 45, 56, 57, 58, 59, 60, 64, 93, 124, 219, 303, 386, 567, 695
plenitude................................ 455, 677
Plotinus 45, 60, 471
Plum Brook vacuum chamber 709
Plutarch 55
Podkletnov Evgeny 729
Poincaré Henri................................ 431
Point Alpha 774, 791, 801, 805, 848
point infinity............ 460, 471, 529, 629
Point Omega................... 792, 842, 848
point X 112, 117, 350, 370, 733, *see* also cosmological axis
poise 187, 347, *see* also potent equilibrium
Poise 658
polar bonding .... *see* polarity
polar charge 39, 46, 187, 199, 504, 574, 648, 651, 653, 716
polar coordinate.................. *see* energy
polar energy 518, 631, 680, 734, 770, 771
polar equation................................ 462
polar existence.................... *see* polarity
polar opposites .... *see* anti-pole
polar stack 161
polarisation... 50, 66, 90, 124, 127, 496, 616, 651, 716
polarity . 37, 38, 41, 43, 46, 58, 60, 137, 140, 147, 148, 149, 162, 173, 175, 177, 209, 216, 220, 227, 230, 238, 242, 272, 383, 384, 411, 425, 435, 438, 449, 450, 451, 462, 503, 527, 574, 602, 624, 655, 709
Polarity - Essential Neutrality
   Unity ⟺ polarity/ duality........15
Polarity Dialectic. *see* Natural Dialectic
politics 66, 76, 165, 224, 255, 262, 265, 546, 658
Polkinghorne John........................... 84

Polyani Michael.............................. 393
Pope Benedict XVI......................... 809
Pope John Paul II........................... 809
Popper Karl 217, 309, 310, 798
Porphyry.. 45
positive neutrality.... 657, see neutrality positive/ potent equilibrium
positive nothing 133
positive power .... *see raj*/ energising principle
positive singularity *see* source/ potential
positive time 741
positive void ..... 676, 678, 683, *see* also potential
potent equilibrium 52, 187, 241
potent neutrality. *see* positive neutrality
potential 38, 97, 104, 119, 133, 135, 137, 155, 158, 176, 187, 194, 383, 415, 768, 790, 811
Potential .. 134, 136, 170, 171, 195, 196, 209, 230, 280, 370, 392, 470, 616, 624, 625, 626, 657, 674, 680, 683, 687, 709, 772, 791, 837
potential being .... see egg
Potential Information 133, 153, 316, 335, 356
potential matter. 72, 135, 154, 156, 197, 198, 207, 324, 451, 452, 453, 465, 488, 526, 545, 663, 674, 676, 683, 684, 693, 788, *see* also sub-conscious energy
potential shape............................... 166
potential time... *see* super-time physical
practice 285
prakriti... 152, 155, 281, 392, 411, 413, 474, 476, 479, 482, 486, 505, 529, 617, 638, 666, 694, 701, *see* also Glossary
pralaya .... 762, 847, 848, *see* also dissolution
prana 411, 412, 476, 477, 485, 486, 587, 637, 649, 684, 695, 717, 776, 789, 797, *see* also Glossary
prana as ZPE................................ 684
pranal virtuality 708, 711
pranayama yoga..................... 476, 797
precondition ..... 679, 809, 839, *see* also potential
preordination 558, 637, *see* also purpose
Presence 455, 628, 736, 740
Pribram Karl.................................... 77
Prigogine Ilya 561, 589
prima materia .... see hydrogen
Primary Axiom of Materialism.. 69, 94, 205, 278, 293, 301, 421, 446, 771, *see* also Glossary
Primary Axiom of Natural Dialectic 44,

889

135, 205, 278, 293, 421, 446, 771, *see also* Glossary
Primary Corollary of Materialism ... 15, 301, 561, *see also* Glossary
Primary Dialectic ...173, 178, 194, 220, 225, 355, 450, 518, 772, *see also* Main/Essential Dialectic
primary inversion 229
Primary Ovoid.. *see* Ovoid/ First Cause
Primary Paradox............................ 247
principle 284, 285, 448, 564
prism of Infinity .... *see* mind
probability442, 448, 503, 516, 533, 539, 542, 547, 549, 552, 558, 604, 644
program..297, 324, 342, 366, 367, 385, 404, 446, 695, *see* also phased intent
Prosser Richard 643
prote hyle 493, 815, *see* also hydrogen
protein 602
proton 198, 322, 599, 723, 836
pseudo-science 63
PSI *see* psychosomatic interface/ border
psychological entropy 394
psychological negentropy. 741, *see* also negentropy
psychological unification 621
psychological unity 621
psychology 267
psychosomatic
  border .....*see* psychosomatic interface
  energy .... 789, *see* also *prana*
  interface167, 208, 285, 288, 325, 390, 544, 622, 657, 676, 680, 693
  medium............*see* sub-consciousness
psycho-space 48, 316, 337, 341, 465, 490, 687, 701
psycho-time.................................... 741
psychotronics 725, 728
pterosaur... 784
Ptolemy 596, 778
puppet...................................... 277, 570
pure energy..................................... 435
purpose...286, 558, 560, 562, 564, 568, 592, *opp.* chance, see also teleology
purposeful .... *see* teleological
purposive complexity72, 306, 320, 348, 360, 365, 533, 560, 584, 586, 625, 765
purposive information .... *see* active information
purposive system.......... *see* mechanism
Purusha .. 152, 155, 281, 410, 474, 479, 486, 505, 617, 701
Puthoff Harold................................ 728

Pythagoras ...45, 57, 386, 429, 523, 793

Q

QCD .... *see* quantum chromodynamics
QED ..... *see* quantum electrodynamics
qi… .... *see* prana
Qom 59
quantum
  chromodynamics................... 436, 645
  cosmology............................. 473, 540
  electrodynamics.... 436, 644, 713, 719
  entanglement................. 542, 644, 726
  era...769
  foam...721
  gravity........... 436, 442, 510, 631, 641
  level..324
  matter.... 451, 652, *see* also matter-in-principle
  pair production..............................814
  physics 157
  theory.................... 436, 442, 444, 647
quantum foam............*see also ZPE* and virtuality
quantum/quanta .... *see* Glossary
quark 105, 173, 492, 496, 627, 720, 770
quasar 441, 795, 817
quintessence704, 819, 833, *see* also dark energy

R

radioactivity..................................... 540
radiometric dating 759
rainbow............................*see* spectrum
raj 178, 187, *see* cosmic fundamental
raj vector .... 189, *see* vector of levity
randomness74, 193, 319, 384, 394, 545, 549, 552, 555, 558, 559, 564, 568
Rayleigh John................................. 778
reality 249
Reality 247, 250, 251, 690
Reality-Value 252
reason 352
record ....*see* memory
red shift 727, 784, 794, 795
reductionism..................................... 71
Rees Martin 612
reflective asymmetry150, 168, 192, 240, 655, *see* also inversion
reincarnation................................... 761
relative illusion..................... *see* reality
relativism.. 62
relativity 246, 461, 739
relativity theory of. 250, 442, 523, 749, 782
religion ...66, 76, 88, 90, 262, 265, 658, 793, 850
repetition 55, 56
rescuing device...............................829

resonance 696
rings of time .... *see* time's cycles
robot 408, 571, 591
Ross Hugh 603, 611
rotifer 696
Ruse Michael 297, 312, 401

S

saddhu .... see yogi
saint 253, 391, 466, 467, 565, 616, *see also* bodhisattva, *see also* mystic/ enlightened one
Saint Augustine 58, 363
Saint John 594, 776
Saint Patrick 228, 619
salah 679
samadhi .... *see* Enlightenment
samsara 123, 631, 661, 719
Sanskrit 60, 360
Santorini 758
SAP see strong anthropic principle
Sat 178, 619, *see* cosmic fundamental
Satan .... *see* devil
Sat-Chit-Ananda ...... see Consciousness
satori see Enlightenment
Saxl Erwin 727
scale 44, 97, 155, 176, 220
scale switch 235
scales 176, 177, 178, 182, 187, 272, *see also* balance
Schaffranke Rolf 727
Schauberger Viktor 727
Schrödinger Erwin 393, 542, 713
Schwarz John 634
Sciama Dennis 600
science 88, 292, 367, 421, 793, *see also* knowledge
science of the soul .. 114, 271, 351, *see also* mystic practice
scientific method 53, 88, 118, 272, 309, 310, 341, 372, 803
scientism 15, 68, 169, 421
sea acorn 696
sea squirt 696
sea urchin 696
second physical absolute . 731, *see* time
secondary dialectic . 174, 178, 194, 220, 225, 355, 451, 716, 772, *see also* branch/existential dialectic
secondary ovoid 675, 680, 683, 693, *see also* archetype
secondary void 676
secularism .... *see* materialism
SED see stochastic electrodynamics
seeming ... *see* reality or relative Reality
Self 687
semantic switch 240

sensation 349
sense-deprivation 56
sense-deprivation tank 56
sensorium 704
SETI 611
sexuality 238
Shabda .... 360, 465, 688, *see Logos*
Shaitan .... see devil
Shakespeare William 46
shamrock .... see Christian Trinity
Shannon Claude 306, 319, 579
Shannon-information 319, 320, 387, 533, 579
Shapley Harlow 419
short-time 746, 784
Shroedinger Erwin 595
siddhi ....
804
Silence .... *see* Nothing
simran .... see mantra
simulation 400
simulator, cosmic 612
sin 192, 621
Sinai 106, 679
singular coordinate ... *see* consciousness
singularity *see* big bang/ black hole, *see* big bang/ black hole
Singularity .... *see* Essence
Singularity, Conscious .... *see* Essence
singularity, unonscious ... *see* black hole
sink .... *see* matter
Sirius 837
Siva 392, 618, 719, 797
sleep 282, 334, 737
smaran/simran .... see mantra
SMBH .... *see* maxi black hole
smoothness problem *see* horizon problem
snowflake 365, 559, 624
Socrates 45, 55, 57, 58, 313
Solomon's temple 679
Song *see* Word
Sorbonne 59, 418
soul 358, 701, *see* also Consciousness/Psyche
Source 418
source to sink 96, 99, 418, 527, *see also* concentration gradient/ diffusion/ entropy
Spaarnay Marcus 710
space 172, 211, 250, 446, 528, 632, 680, 683, 695, 702, 704, 706, 707, 709, 710, 722, 728, 770, 787, 790, 838, 844
curved 523
inner .... *see* psycho-space
outer .... *see* vacuum
physical .... *see* vacuum

psychological... *see* psycho-space
   structure of............................ 713
space-time 684, 749
Sparnaay M. 710
special case.62, 98, 176, 213, 214, 233,
   282, 285, 428, 451, 465, 491, 618,
   676, 705, 733, 740
special relativity *see* relativity theory of
specified complexity ....*see* purposive
   complexity
speed of light (fixed)211, 420, 751, 780,
   782, 783, 785
speed of light (variable)782, 784, 785, 795
Spencer Herbert.............................. 313
Spinoza Baruch 46, 360, 593
spiritual entropy 355
spoken symbol........................*see* word
Sraosha................................ see Logos
stack *see* dialectical stack
Stalin Joseph 773
Standard Model........436, 437, 440, 629
Standard Theory 440, 445, 454
standing wave................................. 665
star 823
starfish... 696
state of matter/ space 703
static matter ....*see* matter-in-practice
statistics... 387
steady-state theory........... 472, 793, 843
Steinhardt Paul 799
Stephenson George.................. 395, 439
Stern Otto.. 710
stochastic electrodynamics..... 442, 543,
   713, 719
Stoic 706, 793
Stonehenge...................................... 748
stork 375
Strabo 59
Strachan Scott 727
straight line..................................... 706
stratigraphic column....................... 758
striatum 746
string theory 438, 632, 638, 648, 722
strong anthropic principle.273, *see* also
   anthropic principle
strong nuclear force........ 505, 599, 602
sub-conscious energy 587, 649, 695
sub-consciousness.115, 155, 208, 218,
   285, 288, 316, 318, 325, 452, 465,
   476, 483, 545, 676, 683, 688, 693
subjective perspective ... 40, 43, 63, 73,
   423, *see* also holism
subjective subtendence.................... 214
subjectivity631, 715, 741, 845, *see* also
   life
Subjectivity ..50, 75, 80, 210, 245, 247,
   248, 357, 664, 678, 694, 733, 773,

see also Enlightenment
sub-reason48, 63, 352, *see* also instinct,
   see also instinct
sub-state580, 742, *see* also subtendence/
   impotence/ exhaustion, *see*
   impotence
subtendence 172, 193, 204, 206, 210,
   213, 219, 230, 249, 677
sub-time physical............................ 733
sub-time psychological 733, 737
sufi 466
sun 601
sunburst monstrance 780
super-cluster 818
Super-consciousness208, 218, 228, 390,
   464, 483
super-massive black holes*ee* maxi black
   hole
super-matter....156, 622, 674, 680, 687,
   688, *see* also potential matter
Super-mind................... *see* First Cause
supernatural................... 43, 68, 84, 407
Super-nature ....*see* First Principle
super-order ....*see* archetype
super-reason 63, 352
Superspace............................. *see* Void
super-state482, 483, 495, 574, 577, 580,
   622, 687, 742, *see* also potential/
   archetype
Super-state................ 742, *see* Potential/
   Transcendence, *see* Potential
super-state material............738, *see* also
   archetypal memory
super-symmetry.....................*see* SUSY
Super-time 733, 737, 740, 753, *see* also
   Enlightenment
super-time physical......................... 752
super-time psychological*see* Super-time
supremacy ....*see* transcendence
Supreme Being134, 195, 228, 359, 360,
   372, 427, 687, *see* also Essence/
   Atman-Brahman/ Cosmic Soul,
   see also Essence/ *Atman-
   Brahman*/ Cosmic Soul
Supreme Court, USA....................... 308
Surtsey Island 758
survival.................. 283, 299, 398, 415
Susskind Leonard 443, 465, 794
SUSY 438, 461, 499, 514, 615
symbol54, 167, 260, 284, 342, 349, 367,
   375, 396, 674, 679, *see* also code/
   passive information
symmetry 186, 333, 334, 450, 518, 519,
   575, 630
symmetry-breaking.... *see* variation-on-
   theme
syntax .... 323, 377, 378, *see* code

T

Tagore Rabindranath 594
Talbot William Henry Fox 684, 778
tam 178, 187, 192, *see* cosmic fundamental
tam vector..............*see* vector of gravity
tanmatra .... 411, 477, 696, *see* also Glossary
Tao 174, 178, 186, 227, 363, 455, 678, *see Logos*
Tao Te Ching 222, 227
Targ Russell 728
tattwa ....*see* Glossary
teleologist..............*see* author/ inventor
teleology.273, 277, 286, 299, 300, 359, 370, 393, 515, 569, 625
teleonomy....................................... 393
Tempel 1 829
Tesla Nikolai 711
tetragrammatikon 679
Tevatron 665
Thales 45, 479, 706
theatre.......................268, 415, 571, 688
theism.......................................66, 593
Theory of Everything*see TOP* and *TOE*
theory of evolution 610
Theory of Intelligence ... 294, 308, 311, 314, 415, 507, 538, 613
Theory of No Intelligence294, 301, 308, 309, 311, 312, 314, 356, 507, 534, 539, 546
thermodynamic laws .... *see* Laws of Thermodynamics
third eye ....*see* eye-centre
third physical absolute..... 770, 771, *see* polar energy
Thom René.................................... 432
Thomson Joseph John 716
tick fast... 752
tick slow 751
time 172, 435, 446, 528, 680, 683, 704, 707, 731, 849
   absolute............................... 737, 753
   asymmetry 756
   geological 758
   neutral... 737
   physical...................................... 748
   relative....................... 737, 740, 750
   reversal.............................. 743, 756
   warp....................... *see* time relative
time's arrow .... 489, *see* also entropy
timelessness...............*see* time absolute
time's arrow 743, 751
time's cycles 747
time's eras...................*see* time's grades
Tipler Frank 273, 543
TOE 66, 78, 80, 105, 118, 119, 351, 422, 556, 565, 568, 591, 627, 628, 639, 713, *aka* Theory of Everything,
TOP 66, 119, 351, 568, 628, *aka* Theory of Potential
top matter .... *see* potential matter
top-down................... 38, 122, 123, 272
Transcendence172, 217, 254, 260, 266, 270
Transcendent Information *see* Potential Information
transcendent matter... 676, *see* potential matter
transcendent projection. 168, 277, 317, 332, 425, 430, 434, 441, 445, 451, 472, 492, 495, 521, 587, 597, 626, 702, 801, 812, 814, 840, *see* also big bang, *see* also big bang, *see* also big bang and Glossary
Transcendent Singularity................ 219
tri-logical dialectic.......................... 175
trinity105, 172, 178, 185, 227, 228, 428, 626, 632
Trinity College, Cambridge... 419, 523, 720
tri-universe .... *see* Mount Universe
Troitskii V. S. 785
Trotsky Leon 689
True Love ....*see* Love
Truth 67, 80, 111, 135, 187, 228, 247, 621, *see* also Enlightenment
truth (relative)................................ 249
Tryon Edward................................ 833
tuatara.. 759
Turing Alan 46, 402
Turok Neil 799
two pillars of faith 446

U

uncertainty principle...............*see HUP*
understanding ..*see* also comprehension
unification450, 468, 469, 622, 623, 627, 648
unit-ification 624
unity 135, 137, 156, 469, 495, 526, 619, 620, 622, 625, 626, 627, 652, 660
Unity 56, 60, 105, 110, 136, 146, 218, 220, 241, 247, 252, 271, 334, 358, 372, 392, 426, 428, 447, 449, 474, 615, 616, 623, 626, 628, 641, 642, 680, 688, *see* also Essence/ Supreme Being
universal body.274, 276, 286, 343, 389, 419, 425, 445, 449, 486, 490, 505, 523, 531, 597, 629, 637, 640, 686, 692, 704, 743, 757, 766, 768, 792, 793, 839, 846, 847, *see* also cosmos

universal egg .... 584, *see* cosmic egg/ archetype
universal energy .... *see prakriti*
universal mind 123, 147, 285, 288, 318, 325, 343, 372, 378, 385, 386, 409, 426, 452, 469, 526, 534, 545, 552, 560, 586, 613, 637, 652, 657, 680, 704, 773, 807, 839, 846, *see also* archetypal memory
universe .... see cosmos
Ussher James 763
utilitarianism 255
utopia 263

V

vacuum 46, 135, 198, 233, 434, 452, 483, 571, 627, 632, 649, 652, 671, 674, 677, 686, 693, 698, 706
vacuum energy . 445, 598, 684, *see also* ZPE
vacuum potential 631
variation-on-theme .299, 365, 506, 546, 624
vector 164, 171, 175, 180, 272, 527
vector of gravity 192
vector of levity 189
vectored energy 164
vibration ... 98, 172, 200, 343, 345, 636, 638, 706
virtuality . 485, 541, 651, 652, 719, 722, 780, 788
Vishnu 618
vitalism 296, 313, 399, 422
VLT (Very Large Telescope) 572
void 133, 135, 209, 217, 350, 671, *see also* nothing
Void 135, 213, 215, 450, 628, 675, 678, 680, 683, 686, 694, 717, 777
volitio-attractive 373
volitio-magnetic radiation .... 700, *see* mind/will and desire
volition .... 661, 757, *see* will-power
Volition .... see First Cause
von Guericke 707
VSL .... *see* speed of light (variable)

W

WAP see weak anthropic principle
Warnock Geoffrey 227
Washington George 309
water 602, 603
wave-packet 666
weak anthropic principle .. 273, *see also* anthropic principle
weak nuclear force 602
weakly interacting massive particle . *see* WIMP
Weaver Warren 298, 319

wheel of time .... see time's cycles
Wheeler John 542
Whetham W. C. D. 720
white hole 845
white horse 324
Wiener Norbert 326
Wild-2 829
Wilde Oscar 694
William Thomson .... *see* Lord Kelvin
will-power 323, 405, 517, 562, *see also* volition
Wilson Robert 796
WIMP 819
WMAP 816
Woit Peter 641
Wolpert Lewis 76
word .... *see* symbol/ code
Word .... *see Logos*
world's end 753

X

x-axis 202
Xenophon 60

Y

yang 148, 178, *see* masculine principle
Yangtze 550
y-axis 201
yew 748
yin .... 148, *see* feminine principle
yin-yang 46, 58, 66, 160, 174, 178, 186, 187, 593, 700, *see also* polarity
Yogananda 595
yogi 56, 466
Yom Kippur 679
Young Thomas 778

Z

Zarathustrianism 59
Zen 222
Zeno 42, 222, 458
zero 146, 626, 698, 807, *see* nothing
zero-energy universe 833
zero-point 186
zero-point energy .... *see* ZPE
Zeus 703, 717
ziggurat 99, 180, 195, 196, 476, *see also* nested cubes/cosmic pyramid/Mount Universe
zikr .... *see* repetition
Zoroastrian 226
Zoroastrianism ..... *see* Zarathustrianism
ZPE 146, 197, 445, 549, 633, 649, 650, 654, 682, 684, 710, 711, 712, 713, 717, 788, 790, 846
ZPE as *prana* 684
ZPF see ZPE
Zwicky Fritz 822

Bibliography

Abusing Science	Kitcher P.	1982
Accidental Univer	Davies P.	1982
Adam and Evolution	Pitman M.	1984
Alas, Poor Darw	Rose H. & S (eds).	2000
Basis for a New Biology	Wilder Smith A. E.	1976
Billions of Missing Links	Simmons G.	2007
Blind Watchmaker	Dawkins R.	1986
Bones of Contention	Lubenow M.	2008
Book of Nothing	Barrow J.	2000
Brain Science & Biology of Belief	Neuberg A. et alii	2001
Brief History of Time	Hawking S.	1988
Cell's Design	Rana F.	2008
Chance and Necessity	Monod J.	1970
Cheating Time	Gosden R.	1996
Chemical Evolution	Aw S.	1976
Complete World of Human Evolution	Stringer C., Andrews P.	2005
Consciousness	Blackmore S.	2003
Constants of Nature	Barrow J.	2002
Contested Bones	Sanford J., Rupe C.	2020
Creating Life in the Lab	Rana F.	2011
Creation and Evolution	Hayward A.	1985
Creation of Life	Wilder Smith A. E.	1974
Darwin's Black Box	Behe M.	1996
Darwin Devolves	Behe M.	2019
Darwin's Doubt	Meyer S.	2013
Darwin and Design	Ruse M.	2003
Darwin on Trial	Johnson P.	1991
Darwin Retried	MacBeth N.	1971
Darwinian Fairytales	Stove D.	1995
Darwinism: Refutation of a Myth	Lovtrup S.	1987
Dawkins' God	McGrath A.	2005
Dawkins Letters	Robertson D	2007
Dawkins Proof for the Existence of God	Barns R.	2009
Debating Darwin's Doubt	Klinghoffer D.	2015
Deluded by Dawkins?	Wilson A.	2007
Descent of Man	Jones S.	2002
Devil's Delusion	Berlinski D.	2009
Did God Use Evolution?	Gitt W.	1993
Double Helix	Watson J.	1970
Dreaming	Allan Hobson J.	2002
Edge of Evolution	Behe M.	2007
Edge of Infinity	Davies P.	1981
Elegant Universe	Greene B.	1999
Endless Forms Most Beautiful	Carroll S.	2011
Epigenetics	Francis R.	2011
Ever Since Darwin	Gould S.	1977
Evolution, A Theory in Crisis	Denton M.	1985
Evolution, A View from the 21st Century	Shapiro J.	2011
Evolution Impossible	Ashton J.	2012
Evolution of Life	Jarman C.	1970
Evolution of Living Organisms	Grasse P-P.	1978

Evolution of Sex	Maynard-Smith J.	1978
Evolution, Still A Theory in Crisis	Denton M.	2016
Evolution, The Human Story	Roberts A.	2011
Explore Evolution	Meyer S. and others	2007
Facts of Life	Milton R.	1992
Fallacies of Evolution	Hoover A.	1977
Fearful Symmetry	Zee A.	1999
Flaws in the Theory of Evolution	Shute E.	1962
Fossils in Focus	Anderson J., Coffin H.	1977
Gaia	Lovelock J.	1979
Gaia: Practical Science of Planetary Medicine	Lovelock J.	1991
Genetic Entropy & the Mystery of the Genome	Sandford J.	2005
Ghost in the Machine	Koestler A.	1975
God Delusion	Dawkins R.	2006
God Gene	Hamer D.	2004
God and the New Physics	Davies P.	1983
God and Stephen Hawking	Lennox C.	2011
God beyond Nature	Clark R.E.D	1982
God, Science and Evolution	Andrews E.	1980
God: To Be or Not To Be	Wilder Smith A. E.	1975
God's Undertaker	Lennox C.	2009
Goldilocks Enigma	Davies P.	2007
Grand Design	Hawking S., Mlodinow L.	2010
Great Evolution Mystery	Rattray Taylor G.	1983
Hallmarks of Design	Burgess S.	2000
Has Darwin Had His Day?	Rosevear D.	2007
How Life Began	Croft L.	1988
How the Mind Works	Pinker S.	1997
Icons of Evolution	Wells J.	2000
In the Beginning was Information	Gitt W.	1997
Infinite Book	Barrow J.	2005
Infinity	Clegg B.	2003
Information and the Nature of Reality	eds. Davies & Gregersen	2010
Inspiration from Creation	Burgess S., Statham D.	2018
Intelligent Design Uncensored	Dembski W. and Witt J.	2010
Intelligent Universe	Hoyle F.	1983
Just Six Numbers	Rees M.	1999
Language of God	Collins F.	2007
Life Itself	Crick F.	1972
Life's Solution	Conway Morris S.	2003
Macroevolution	Stanley S.	1979
Master and his Emissary	McGilchrist I.	2009
Mind and Cosmos	Nagel T.	2012
Mind of God	Davies P.	1992
Mind, Body & Electromagnetism	Evans J	1992
Mysterious Epigenome	Gills, Woodward	2012
Mystery of Life's Origin	Thaxton, Bradley, Olsen	1985
Myth of Junk DNA	Wells J.	2011
Naked Emperor: Darwinism Exposed	Latham A.	2005
Nature of Life	Waddington C.	1961
Natural Sciences Know Nothing of Evolution	Wilder Smith A.	1981
Natural Theology	Paley W.	1802
Nature's Destiny	Denton M.	1998

Neck of the Giraffe	Hitching F.	1982
New Biology	Augros R. & Stanciu N.	1988
New Science of Life	Sheldrake R.	1981
New Story of Science	Augros R. & Stanciu N.	1986
New Theories of Everything	Barrow J.	2007
Not By Chance	Spetner Lee	1997
Nothing	ed. Webb J.	2013
Nothingness	Genz H.	1999
On Guard	William Lane Craig	2010
One Small Speck	Sodera V.	2009
Origin of Species	Darwin C.	1859
Origin of Species Revisited	Bird W	1989
Origin of Life	Bliss R.	1979
Origins of Life	Rana F, Ross H.	2004
Our Cosmic Habitat	Rees M.	2001
Quantum World	Polkinghorne J.	1984
Panda's Thumb	Gould S.	1980
Politically Incorrect Guide to Darwinism and Intelligent Design	Wells J.	2006
Presence of the Past	Sheldrake R.	1988
Piltdown Forgery	Weiner J.	1955
Reason in the Balance	Johnson P.	1995
Refuting Evolution 1 and 2	Sarfati J.	2007
Roots of Coincidence	Koestler A.	1976
Runes of Evolutio	Conway Morris S.	2015
Science and Creation	Polkinghorne J.	1988
Science Delusion	Sheldrake R.	2012
Science & Evidence for Design in the Universe	Behe, Dembski, Meyer	1999
Science and Human Origins	Gauger, Axe, Luskin	2012
Secret of the Creative Vacuum	Davidson J.	1989
Seven Sins of Memory	Schacter D.	2001
Shadows of the Mind	Penrose R.	1995
Signature in the Cell	Meyer S.	2009
Tao of Physics	Capra F.	1976
Tao Te Ching	Lao Tse	
There is a God	Flew A.	2007
Thermodynamics & the Development of Order ed. Williams E.		2002
Time, Space and Things	Ridley B.	1976
Transformist Illusion	Dewar D.	1957
Web of Life	Davidson J.	1988
Web of Life	Capra F.	1996
What Darwin Got Wrong	Fodor, Piatelli-Palmarini	2010
What is Life?	Schrödinger E.	1944
When is a Fly not a Horse?	Sermonti G.	2005
Who Made God?	Andrews E.	2009
Without Excuse	Gitt W.	2011
Wonderful Life	Gould S.	1989
Wonders of the Universe	Cox B.	2011
Universe, Plan or Accident?	Clark R.E.D	1961
Vital Question	Lane N.	2015
Void	Close F.	2007
Y, The Descent of Men	Jones S.	2002

The author has recently written a few more books (available from Amazon, Foyles, Waterstones, Barnes & Noble etc. and see website addresses on p.2):

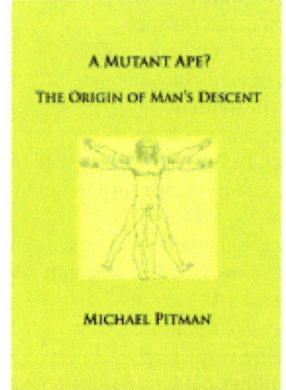

Back cover illustration credit: wikiHow creative commons (BY-NC-SA) license.

Lightning Source UK Ltd.
Milton Keynes UK
UKHW020405080620
364540UK00007B/185